T0317703

Flight Dynamics

FLIGHT DYNAMICS

Robert F. Stengel

Princeton University Press
Princeton and Oxford

Copyright © 2004 by Princeton University Press
Published by Princeton University Press,
41 William Street, Princeton, New Jersey 08540
In the United Kingdom: Princeton University Press,
3 Market Place, Woodstock, Oxfordshire
OX20 1SY
All Rights Reserved

Library of Congress Cataloging-in-Publication Data

Stengel, Robert F.
Flight dynamics / Robert F. Stengel.
p. cm.
Includes bibliographical references and index.
ISBN 0-691-11407-2 (cl : alk. paper)
1. Aerodynamics. 2. Flight. 3. Airplanes—
Performance. I. Title.

TL570.S73 2004 2003062310

British Library Cataloging-in-Publication Data is
available

This book has been composed in Sabon and Univers

www.pupress.princeton.edu

10 9 8 7 6 5 4 3 2 1

For Frank and Ruth

Contents

Chapter Four
Methods of Analysis and Design 273

Preface

Flight Dynamics takes a new approach to the science and mathematics of aircraft flight, unifying principles of aeronautics with contemporary systems analysis. It is a text and reference book for upper-level undergraduates and first-year graduate students as well as practicing engineers. While the book presents traditional material that is critical to understanding aircraft motions, it does so in the context of modern computational tools and multivariable methods. Particular attention is given to models and techniques that are appropriate for analysis, simulation, evaluation of flying qualities, and control system design. Bridges to classical analysis and results are established, and new territory is explored. The book is intended for readers with a broad background and interest in engineering and science. It begins with basic principles of trigonometry, calculus, linear algebra, and dynamics, and it continues with advanced topics, algorithms, and applications.

Dynamic analysis has changed dramatically during recent decades, with the introduction of powerful personal computers and scientific programming languages. Analysis programs have become so pervasive that it can be assumed that all students and practicing engineers working on aircraft flight dynamics have access to them. Therefore, this book presents the principles, derivations, and equations of flight dynamics with frequent reference to computed examples, and it includes MATLAB code for six-degree-of-freedom simulation and linear system analysis as on-line supplements.

Because it uses common notation and does not assume a strong background in aeronautics, *Flight Dynamics* is accessible to a wide variety of readers. Introductions to aerodynamics, propulsion, structures, and the atmospheric and gravitational environment precede the development of the aircraft's dynamic equations. *Système International* (SI) units are used for the majority of the text. These units take better advantage of decimal arithmetic than do U.S. Customary Units, the majority of world industry and academe uses them, and virtually all high-school and college students learn physics, chemistry, and mathematics with SI units.

Bibliographic reference lists are presented at the end of each chapter and in Appendix E. Not all the books, reports, and papers are explicitly cited in the accompanying chapters. The lists are intended as resources for aerodynamic data and analytical results that are related to the associated text. Of course, many of the references did contribute directly to the book, and I am grateful to all of the authors for the insights that they provided.

Flight Dynamics is organized in seven chapters and an epilogue. "Introduction" (Chapter 1) presents a summary of the important elements of the aircraft and its systems; examples of about a dozen aircraft ranging from an uninhabited air vehicle, through general aviation and transport aircraft, to combat aircraft, trans/supersonic aircraft, and the space shuttle; and an overview of the mechanics of flight. "Exploring the Flight Envelope" (Chapter 2) describes the flight environment and the essential ingredients for steady flight. "Dynamics of Aircraft Motion" (Chapter 3) provides comprehensive coverage of the nonlinear equations of motion, expressed in various coordinate systems and with alternative angular measures (direction cosines and quaternions). "Methods of Analysis and Design" (Chapter 4) addresses the fundamental equations used for both classical and modern analysis of aircraft stability and control, including stochastic methods, and it presents introductions to aeroelasticity, flying qualities criteria, and flight control system design. "Longitudinal Motions" (Chapter 5) presents motions in the vertical plane in a novel fashion, building insight with reduced-order linear models derived from the complete set. "Lateral-Directional Motions" (Chapter 6) parallels the previous chapter, describing rolling and yawing motion with increasingly complex models and subjecting each to a series of analyses. "Coupled Longitudinal and Lateral-Directional Motions" (Chapter 7) goes into considerable detail about the ways in which motions interact, using bifurcation analysis, linear analysis, and nonlinear simulation, and it introduces several approaches to nonlinear control system design that are essential for high-angle-of-attack flight. The "Epilogue" briefly discusses the implications of the book.

One feature of the book is the inclusion of a nonlinear, six-degree-of-freedom simulation model for a generic business jet aircraft, which can be found on line at http://www.pupress.princeton.edu/stengel. The model is used for most examples, beginning in Chapter 2. The inertial and aerodynamic properties of the model are representative of business jets from subsonic to high-transonic airspeed and for angles of attack up to 90 deg. Together with the MATLAB nonlinear simulation program FLIGHT, the model is used to demonstrate motions across the spectrum of flight, including effects of compressibility, stalling, spinning, and control nonlinearity (e.g., limit cycles and chaotic motion). FLIGHT can generate linear, time-invariant (LTI) dynamic models at arbitrary flight conditions. The LTI models can be subjected to various linear analyses in the MATLAB program SURVEY, including model order reduction through truncation or residualization, transient response, equilibrium response, controllability and observability, Bode, Nyquist, and Nichols plots, and root locus analysis. Longitudinal and lateral-directional motions can be analyzed separately or under fully coupled flight conditions. The FLIGHT and SURVEY programs are heavily commented, and they serve a tutorial role on their own.

While covering all the material in the book typically would require a two-semester course, a one-term undergraduate course can be based upon the book, leaving the more advanced topics for independent reading. Such a course would contain introductory material (Chapter 1), nonlinear equations of motion (Chapters 2 and 3), configuration aerodynamics (Chapters 2 and 3), aircraft performance (Chapter 2), linearized equations of motion and analysis (Chapter 4), longitudinal stability and control (Chapter 5), and lateral-directional stability and control (Chapter 6). Assignments may include the estimation of configuration aerodynamics for contemporary aircraft and various analyses of these aircraft conducted with FLIGHT, SURVEY, or student-generated computer code.

I am indebted to the many students, teachers, and colleagues with whom I have worked over the past years. It is (or should be) well known that teachers frequently learn at least as much from their students as the students learn from them. That is certainly the case for me, and I want to acknowledge the hundreds of graduate and undergraduate students who contributed to my education. Former teachers who introduced me to much of flight dynamics include Eugene Larrabee and Wallace VanderVelde at MIT, and Enoch Durbin, Dunstan Graham, David Hazen, Courtland Perkins, and Edward Seckel at Princeton. I also learned a great deal from many communications with Duane McRuer (Systems Technology), Albert Schy (NASA Langley), Malcolm Abzug (ACA Systems), and William Rodden (consultant).

Past and current associates provided a continuing flow of good ideas. These include Paul Berry and John Broussard at The Analytic Sciences Corporation, who had superb grasps of flight dynamics and control, as well as Princeton faculty and staff colleagues: Howard (Pat) Curtiss, Earl Dowell, David Ellis, Phillip Holmes, Jeremy Kasdin, Naomi Leonard, Michael Littman, Robert McKillip, Kenneth Mease, George Miller, and Barry Nixon.

Not to be forgotten are the research and consulting sponsors whose support was foundational. The NASA/FAA Joint University Program for

Air Transportation Research has provided vital support for over 25 years. Its technical monitors have included David Downing, William Howell, Wayne Bryant, Frederick Morrell, Richard Hueschen, and Leonard Tobias at NASA, and Albert Lupinettti, Donald Eldridge, John Fabry, James Remer, and Michael Paglione of the FAA. Additional support was overseen by Luat Nguyen (NASA), David Siegel (ONR), Ralph Guth (AFFDL), Phillipe Poisson-Quinton and Claudius LaBerthe (ONERA), Warren North (NASA), Thomas Meyers (STI), David Bischoff (NADC), Jagdish Chandra (ARO), Paula Ringenbach and Dy Le (FAA), and Thomas McLaughlin (Perkins Coie). Two grants from Princeton alumnus George E. Schultz played a significant role in advancing the Princeton flight research program.

Special thanks go to those who reviewed the book manuscript at various stages of its preparation. I am especially grateful to former student Jean-Paul Kremer, who read the manuscript critically and in remarkable detail, making countless discerning comments. Former students Dennis Linse, Axel Niehaus, and Muthathamby Sri Jayantha made a number of helpful suggestions, while Professor Haim Baruh (Rutgers) and Professor Eric Feron (MIT) perceptively reviewed the manuscript and provided valuable advice.

And finally, my warmest appreciation and love go to Pegi, Brooke, Christopher, and Brian, whose care, understanding, and support have been essential.

Chapter 1
Introduction

Flying has become so common that we tend to take many details of flight for granted. Nevertheless, flight is a complex process, involving the equilibrium, stability, and control of a machine that is both intricate and elegant in its design. All aircraft are governed by the same rules of physics, but the details of their motions can be quite different, depending not only on the shapes, weights, and propulsion of the craft but on their structures, control systems, speed, and atmospheric environment. This book presents the flight dynamics of aircraft, with particular attention given to mathematical models and techniques for analysis, simulation, evaluation of flying qualities, and control system design.

In this chapter, we introduce the basic components of configuration that are common to most aircraft (Section 1.1) and provide illustrative examples through descriptions of contemporary aircraft (Section 1.2). Notation that is used throughout the book is presented in Section 1.3, with an introductory example based on the flight of a paper airplane.

1.1 Elements of the Airplane

Aircraft configurations are designed to satisfy operational mission and functional requirements, with considerable consideration given to cost, manufacture, reliability, and safety. All aircraft have structures that generate aerodynamic lift, mechanisms for effecting control, and internal spaces for carrying payloads. Most aircraft also have propulsion systems and associated propellant tanks, undercarriage for takeoff and/or landing, and navigation systems. The layout and interactions of all these components have major influence on an aircraft's motions [e.g., B-1, C-1, G-2, H-1, H-2, K-2, N-1, R-1, S-1, S-2, S-3, T-1, and W-1].

Airframe Components

Consider the Cirrus SR20 aircraft shown in Fig. 1.1-1. A single-propeller-driven aircraft, the SR20 has all the elements of a conventional aircraft:

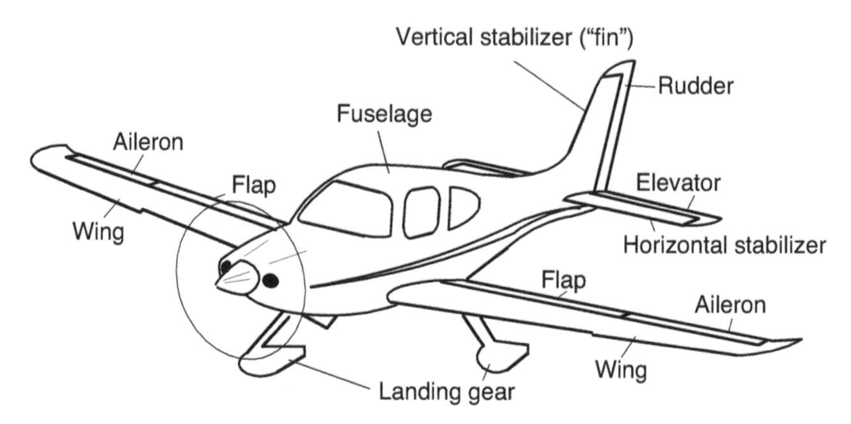

Fig. 1.1-1. Cirrus SR20 general aviation aircraft. (Courtesy of Cirrus Design Corp.)

wing, fuselage, horizontal tail, vertical tail, and control surfaces, as well as cockpit/cabin, engine, and landing gear. The *wing* provides the aircraft's largest aerodynamic force, used principally for supporting the vehicle's weight and changing the direction of flight. The wing's fore-aft location is near the vehicle's center of mass for good balance, and the wing itself is shaped to provide a large amount of lift with as little energy-absorbing drag as possible. The *fuselage* is the aircraft's principal structure for containing payload and systems. Its slender shape provides a usable volume for the payload and a mounting base for engine and tail components, with minimum drag and weight penalty.

Just like a weathervane, the *vertical tail* produces directional stability about an aircraft-relative vertical axis; it gives the aircraft a tendency to nose into the *relative wind* that results from forward motion. Similarly, the *horizontal tail* provides angular stability for rotations about an axis parallel to the wingspan. Aerodynamically, the vertical and horizontal tails are small wings whose lift acts through moment arms (between aerodynamic centers and the center of mass) to generate restoring torques that are proportional to angular perturbations. Collectively, the tail surfaces are called the *empennage*. Stability about the third (rolling) axis is a complex effect, dominated by the wing's vertical location on the fuselage and the upward inclination of the wing tips relative to the wing roots (the *dihedral angle*).

The SR20 has conventional control effectors: ailerons, elevator, rudder, flaps, and thrust setting. The *ailerons* are movable surfaces located near the wing tips that produce large rolling torques; the two surfaces are linked so that the trailing edge of one moves up when the trailing edge of the other moves down. The *elevator* is a movable surface that extends across the trailing edge of the horizontal tail for angular control. In addition to regulating the aircraft's angle of attack,[1] which, in turn, governs the amount of lift generated by the wing, the elevator setting controls the pitch angle[2] when the wings are level, an important function during takeoff and landing. The *rudder* has a similar effect for yawing motions, controlling the sideslip angle[3] for turn coordination and crosswind takeoffs and landings. It is a movable surface mounted on the trailing edge of the vertical tail.

The wing's trailing-edge *flaps* are movable surfaces mounted inboard of the ailerons for direct control of lift and drag during takeoff and landing. The left and right surfaces act in unison, deflecting downward from their cruising-flight (stowed) positions. Whereas the pilot exerts continuous control of the ailerons, elevator, and rudder, the flaps normally are adjusted to discrete settings that depend on the flight phase. Engine *thrust setting* is regulated only occasionally to achieve takeoff acceleration, climb, desired cruising conditions, descent, and landing sink rate.

A number of other subsystems have dynamic and control effects, including landing gear, trim tabs, leading-edge flaps, spoilers, speed brakes, thrust reversers, engine inlet shape and bleed air, jet exhaust deflection, and external stores. Functions may be combined or distributed, as for tailless aircraft or those possessing redundant control surfaces. These are configuration-dependent effects that are explored in later sections.

If each of the aircraft's elements performed only the functions described above and performed them perfectly, aircraft dynamics and control could be relatively simple topics; however, there are numerous complicating factors. Perhaps the greatest is that important physical phenomena are inherently *nonlinear*. The significance of nonlinearity is revealed in the remainder of the book, but the general notion is that doubling a cause does not always double the effect. For example, the wing's lift is linearly proportional to the angle of attack up to a point; then, the wing stalls and no greater lift can be achieved, even if the angle continues to increase.

A particular element (e.g., the wing) produces a primary effect (lift), but it may also produce secondary effects (drag and side forces, as well as pitching, yawing, and rolling moments). Because the components are in close proximity, the aerodynamic forces and moments that they

1. The *angle of attack* is the aircraft-relative vertical angle between the centerline and the relative wind. The *relative wind* is the velocity of the aircraft relative to the air mass through which it flies.
2. The *pitch angle* is the angle of the aircraft's centerline relative to the horizon.
3. The *sideslip angle* is the aircraft-relative horizontal angle between the centerline and the relative wind.

generate are interrelated. Hence, the airstream downwash and vorticity induced by the lifting wing have major effects on tail aerodynamics, thrust changes may upset pitch or yaw equilibrium, and so on. A good deal of useful analysis can be accomplished with single-input/single-output (scalar) models, but the actual dynamic system has multiple inputs and outputs, requiring a more comprehensive multi-input/multi-output (vector) approach.

There are challenging structural and inertial effects as well. The aircraft's structure must be lightweight for good overall performance, but it is very flexible as a result; consequently, the wing, fuselage, and tail deflect under air loads, and the changes in shape have both static and dynamic effects on equilibrium and motion. In modern aircraft, natural frequencies of significant vibrational modes may be low enough to be excited by and to interfere with aircraft maneuvers, leading to major problems to be solved in control system design. They also contribute to poor ride quality for the passengers and crew and to reduced fatigue life of the airframe. Although most aircraft possess mirror symmetry about a body-fixed vertical plane, inertial and aerodynamic coupling of motions can occur in asymmetric maneuvers. The loading of asymmetric wing-mounted stores on combat aircraft clearly presents a dynamic coupling effect in otherwise symmetric flight, and even a steady, banked turn couples longitudinal and lateral-directional motions.

Propulsion Systems

All operational aircraft other than unpowered gliders and rocket-powered "aerospace planes" produce thrust by moving air backward with power from air-breathing engines. Power is generated either by the combustion of fuel and air in *reciprocating engines*, which use crankshafts to transform the linear motion of pistons moving in hollow cylinders to rotary motion, or in *turbine engines*, which produce rotary motion by the action of combustion gasses flowing through a finned wheel (i.e., a turbine). Power is most efficiently transformed to thrust by moving large masses of air at low speed rather than small masses of air at high speed; therefore, large-diameter *propellers* spinning at low rotational rate are indicated. Propellers are driven by either reciprocating engines or turbines; the latter arrangement is called a *turboprop* propulsion system if the propeller is of conventional design or a *propfan* if the propeller is designed for high-speed flow.

Propellers absorb unacceptable amounts of power as airspeed increases, inducing tip speeds that approach or exceed the speed of sound. The *turbofan* engine overcomes these problems by enclosing a smaller-diameter fan within a streamlined duct, allowing higher aircraft speeds but reducing power conversion efficiency. The proportion of air that is merely accelerated by the fan compared to that which flows through the engine core is called the *bypass ratio*. The propfan mentioned earlier can be considered a turbofan with ultrahigh bypass ratio. With no bypass air accelerated by a fan, the turbine engine is called a *turbojet*—all the thrust-producing air mass passes through the combustion chambers and turbines. Turbojets were used in early supersonic[4] aircraft, where the drag losses of the fan outweighed the fan's conversion efficiency. Advances in fan design have led to supersonic aircraft with turbofan engines. *Afterburning* produces increased thrust by burning additional fuel in the oxygen-rich exhaust of a turbojet engine, at the expense of reduced combustion efficiency.

For flight at a Mach number $\mathcal{M}$ of two or more, the air can be satisfactorily compressed for combustion by a properly designed inlet, so the compressor blades and the turbine that turns them are no longer necessary. The *ramjet* is an engine consisting of an inlet (which compresses the air and slows it to subsonic speed), combustion chamber, and exhaust nozzle. It proves to be the most fuel-efficient powerplant for $\mathcal{M} = 3$–6. Compressing the air increases its temperature; at even higher speeds, this temperature rise is excessive, challenging the structural strength of engine materials. The *supersonic-combustion ramjet* (or *scramjet*) provides one possible solution to this problem, slowing the air enough for adequate compression, though yielding a supersonic flow through the combustion chamber.

The *rocket* engine is capable of producing thrust at any Mach number and in a complete vacuum. Moreover, its structural weight per unit of thrust is less than that of any of the previously mentioned devices. The principal penalty is that the oxidizer as well as the fuel must be carried onboard the aircraft, increasing the size and weight of the vehicle. Consequently, the rocket is useful for launch to orbit, but it has very restricted roles to play in atmospheric flight, such as in short-term thrust augmentation for takeoff, maneuvering, or braking.

Future aircraft that operate over a wide range of Mach numbers will almost certainly use *variable-* or *multiple-cycle engines* (referring to the thermal cycles of combustion). For example, a *turboramjet* would operate as a turbojet at low speeds and a ramjet at high speeds, with the percentage of inlet air ducted directly to the afterburner increasing as Mach number increases. A rocket could accelerate an aircraft to $\mathcal{M} = 3$, possibly driving a turbine to compress inlet air for combustion, with a ramjet taking over

4. The *Mach number* $\mathcal{M}$ is the ratio of airspeed V to the speed of sound a in the surrounding air. The flow is *supersonic* when $\mathcal{M}$ is greater than 1 and *subsonic* when $\mathcal{M}$ is less than 1. Air moves faster or slower than the *freestream airspeed* over different parts of the aircraft. The *transonic region* begins at the freestream Mach number for which the flow reaches sonic speed over some part of the aircraft and ends when the entire flow is supersonic, typically for $0.4 < \mathcal{M} < 1.4$.

at $\mathcal{M} = 3$, a scramjet coming on line at $\mathcal{M} = 6$, and the rocket resuming for orbital insertion. There are numerous alternatives; which ones are practical is a matter of technological development.

1.2 Representative Aircraft

Attributes of several aircraft types are presented in preparation for the more general study of flight dynamics and control. The principal objectives are to indicate the effect that flight conditions and missions have on aircraft configurations and, conversely, to suggest the latitude for motions and control afforded by existing designs. The following descriptions are approximate.

Light General Aviation Aircraft

The *Cirrus SR20* (Fig. 1.1-1) is a four-place aircraft with a 149 kW (200 hp) reciprocating engine. The wingspan is 10.8 m, the length is 7.9 m, and the wing area is 12.6 m^2. The empty and maximum takeoff masses are 885 and 1,315 kg; hence, the maximum wing loading[5] is 105 kg/m^2. The cruising airspeed[6] of this unpressurized aircraft is 296 km/hr (184 statute miles per hour [mph]) at 75% power and 1,980 m (6,500 ft) altitude, and its range is 1,480 km (920 mi). The maximum Mach number is 0.24. The SR20's straight wing has a moderate aspect ratio[7] (9.1) for efficient subsonic cruising, and it is mounted low on the fuselage with 5 deg dihedral angle. The horizontal tail is also mounted low, and the vertical tail is swept, with a full-length rudder whose throw is unrestricted by the horizontal tail location. The principal airframe material is a structural composite, and the structure uses semimonocoque construction (i.e., the skin carries a significant portion of the loads), with surfaces bonded to spars and ribs. A recovery parachute that can bring the entire aircraft safely to the ground is standard equipment.

Variable-Stability Research Aircraft

The *Princeton Variable-Response Research Aircraft* (VRA) is a highly modified single-engine North American Navion A airplane designed for research on flight dynamics, flying qualities, and control (Fig. 1.2-1). The

5. *Wing loading* is defined in SI units as aircraft mass (kg) divided by wing area (m^2). Using U.S. Customary Units, it is defined as weight (lb) divided by wing area (ft^2).
6. The *airspeed* is the magnitude of the aircraft's velocity relative to the surrounding atmosphere, i.e., of the *relative wind*. If the wind is blowing, airspeed and ground speed are not the same.
7. The wing *aspect ratio* is defined as the span squared divided by the wing area.

Fig. 1.2-1. Princeton variable-response research aircraft. (courtesy of Princeton University)

most distinguishing feature of the VRA is the pair of vertical side-force-generating surfaces mounted midway between wing roots and tips, but equally important elements of its design are the digital fly-by-wire (DFBW) control system, first installed in 1978, which parallels the standard Navion's mechanical control system, and the fast-acting wing flaps that produce negative as well as positive lift. In operation, a safety pilot/test conductor has direct control of the aircraft through the mechanical system, while the test subject controls the aircraft through the experimental electronic system.

Although limited to an airspeed of about 200 km/hr (125 mph), the VRA can simulate the perturbational motions of other aircraft types through independent, closed-loop control of all the forces and moments acting on the airplane. Feedback control of motion variables to the control surfaces allows the natural frequencies, damping ratios, and time constants of the Navion airframe to be shifted to values representative of other airplanes, while the direct-force surfaces (side force and lift) provide realistic accelerations in the cockpit.

Consider, for example, pure pitching and yawing motions of the 747 airliner. The aircraft rotates about its center of mass, which is considerably aft of the cockpit; consequently, substantial lateral and vertical motions and accelerations are felt in the cockpit when simple angular rotations occur. The VRA can simulate these lateral and vertical motions through

Fig. 1.2-2. Schweizer SGS 2-32 sailplane. (courtesy of Princeton University)

the actions of its direct-force control surfaces. Conversely, these surfaces make it possible to decouple or modify the VRA's natural modes of motion for experimentation. Thus, the VRA can execute a turn without banking or can bank without turning. Its control surfaces can also be used to simulate turbulence for test purposes.

Having been used for over twenty years of research at Princeton University, the VRA and its sister ship, the *Avionics Research Aircraft* (which is virtually identical to the VRA but does not have side-force panels) are currently owned and operated by the University of Tennessee Space Research Institute.

SAILPLANE

The *Schweizer SGS 2-32* sailplane (Fig. 1.2-2) is a three-place aircraft with a wingspan of 17.4 m, an empty mass of 380 kg, and a typical loaded mass of 600 kg. With an aspect ratio of 18.1 and a wing area of 16.7 m^2,

Fig. 1.2-3. Cessna Citation CJ2 executive jet transport. (courtesy of Cessna Aircraft Co.)

the aircraft's minimum sink rate is 0.6 m/s. The maximum speed is 255 km/hr, while its speed for maximum lift-drag ratio is about 80 km/hr. The aircraft held numerous speed and distance soaring records in the 1970s, and it was used for aerodynamic studies at Princeton's Flight Research Laboratory. Several 2-32s were purchased by the U.S. government and designated the X-26A. As the QT-2 or X-26B, Lockheed fitted a quiet engine to the aircraft for testing of stealthy reconnaissance concepts that led to the operational YO-3A. The mechanical flight control system is conventional, with all-moving horizontal tail, ailerons, rudder, flaps, and speed brakes (or "spoilers").

Business Jet Aircraft

The *Cessna Citation CJ2* is representative of many small executive jet transports (Fig. 1.2-3); it has a "T" tail, two 10.7 kN (2,400 lb) thrust turbofan engines in separate nacelles,[8] and an unswept wing. Mounting the engines on the fuselage aft of the rear pressure bulkhead has several advantages in this aircraft class, including good ground clearance, protection of the engines against ingesting foreign objects while on the ground, reduced cabin noise, low contribution to drag, and structural efficiency. The Model CJ2 carries one or two crew members and up to six passengers in its 14.3-m-long fuselage. The wingspan, area, and aspect ratio are 15.1 m, 24.5 m^2, and 9.3, while the minimum and maximum masses are 3,460 and 5,580 kg. The maximum wing loading is 228 kg/m^2. The economical cruise speed at 12,500 m (47,300 ft) altitude is 704 km/hr (437 mph), and the maximum Mach number is 0.72. With maximum fuel, the maximum range of the Model CJ2 is 3,110 km (1,935 mi).

Turboprop Commuter Aircraft

The *Bombardier Q200* (Fig. 1.2-4) exemplifies the many twin-turboprop designs used by regional airlines to move dozens of passengers on relatively short trips. The Q200 has a "T" tail and a high, straight wing with span, area, and aspect ratio of 25.9 m, 54.4 m^2, and 12.4, respectively. The craft is 22.3 m long, and it carries 37 passengers plus a crew of two; its empty and maximum masses are 10,380 and 16,465 kg, with a maximum wing loading of 305 kg/m^2. Typical cruise speed is 535 km/hr (335 mph), corresponding to $\mathcal{M} = 0.44$. The maximum operating altitude is at 7,620 m (25,000 ft), and the maximum range is 1,715 km (1,065 mi). The Q200's primary structure is aluminum, and its engines each generate 1,490 kW (2,000 hp). The aircraft has an active noise and vibration suppression system for quieting the passenger cabin.

Small Commercial Transport Aircraft

The *Airbus A318* (Fig. 1.2-5), a shortened derivative of the A320, is on the boundary in flight performance and accommodations between typical regional jets and trunkline aircraft. The A320 was the first subsonic commercial jet to use a DFBW control system and sidearm hand controllers in the cockpit, and it was the first to use composite materials for primary structures. Like its predecessor, the A318 has an advanced technology wing, winglets at its wing tips, and a high degree of redundancy in flight control. Hydraulically powered flight control surfaces include spoilers for roll control and speed braking, leading-edge slats for increased lift at higher angles of attack, and conventional surfaces for roll,

8. A *nacelle* is a streamlined fairing over an engine.

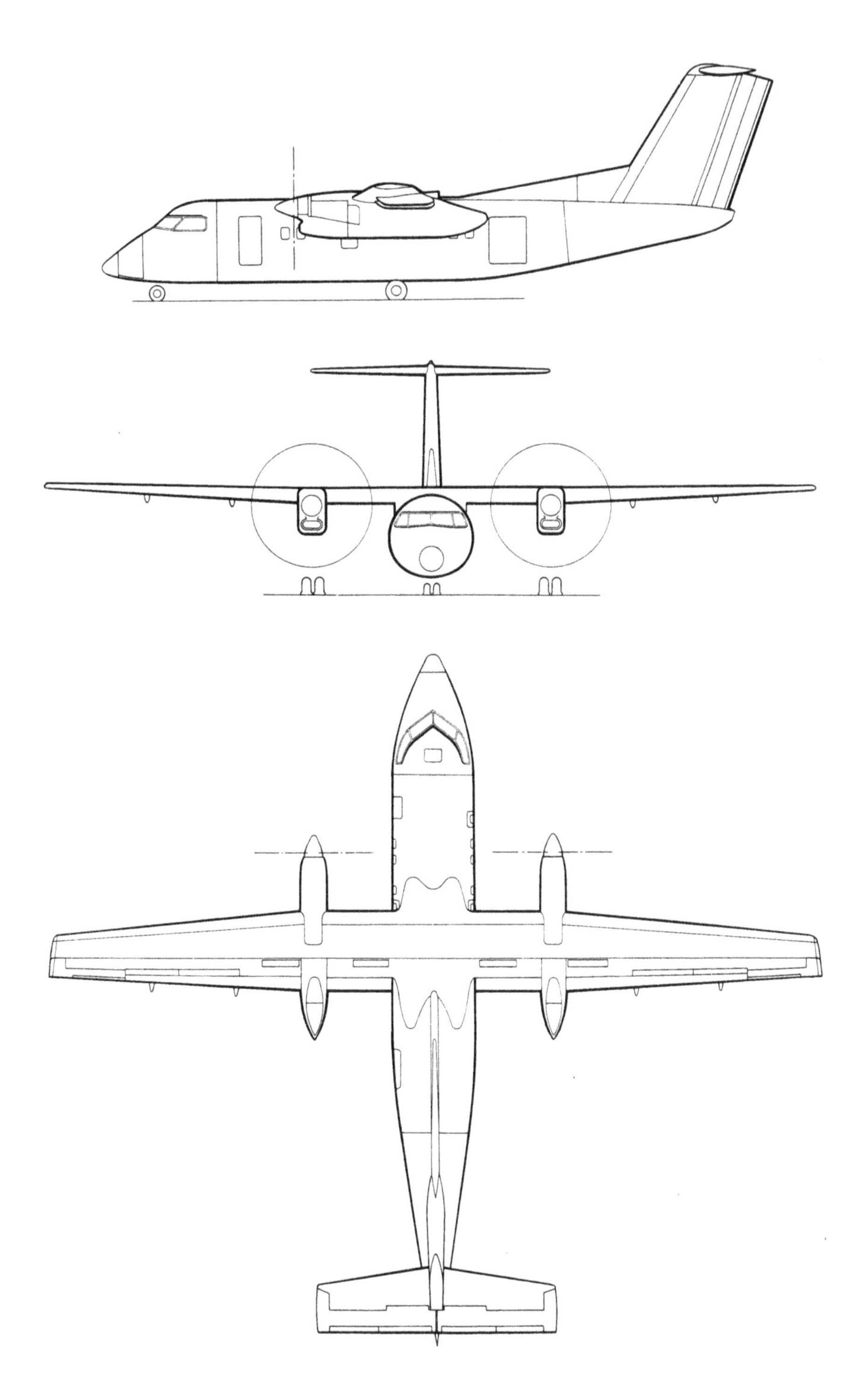

Fig. 1.2-4. Bombardier Q200 regional airliner. (courtesy of Bombardier Aerospace DeHavilland)

Fig. 1.2-5. Airbus A318 small transport aircraft. (courtesy of Airbus Industries)

pitch, and yaw control. There are mechanical backups for rudder control and pitch trim. The single-aisle A318 carries 107–129 passengers, depending on configuration, a flight crew of two, and a cabin crew of three. The empty mass is 85,900 kg, and normal maximum takeoff mass is 59,000 kg. The wingspan and fuselage length are 34.1 and 31.4 m, while the wing area, sweep angle, and aspect ratio are 122 m^2, 25 deg, and 9.5. Each of the A318's engines produces 98–107 kN (22,000–24,000 lb) of maximum thrust. The optimal cruising Mach number is 0.78. The maximum range with 107 passengers is over 2,750 km.

Large Commercial Transport Aircraft

At 73.9 m, the *Boeing 777-300* was the longest twin-engine aircraft built at its introduction. It has a maximum takeoff mass of over 299,000 kg and an empty mass of 160,530 kg (Fig. 1.2-6). It is powered by two 435 kN (98,000 lb) thrust turbofan engines mounted in nacelles under the wing. The aircraft has a maximum range of 6,715–15,370 km (3,625–8,300 nautical mi), assuming flight at the economic cruising Mach number of

Fig. 1.2-6. Boeing 777-300 commercial airliner. (courtesy of The Boeing Commercial Aircraft Co.)

0.84, with step climbs in altitude[9] as fuel is expended. The maximum certified altitude is 13,135 m (43,100 ft). The 777-300 has a wingspan of 60.9 m and its wing area is 427.8 m^2, giving an aspect ratio of 8.7 and a maximum wing loading of 700 kg/m^2. The span of the horizontal tail is about double the wingspan of the Cirrus SR20. The double-aisle 777-300 can be configured to carry 450 passengers in two-class seating or 550 passengers in single-class seating.

In spite of its size, the 777 is certified to be flown by a two-person crew. The wing, mounted low on the fuselage, is swept 31.5 deg. Like other large aircraft with long, flexible wings, the 777 has two sets of roll

9. Current air traffic regulations typically require aircraft to cruise at constant altitude, allowing discrete altitude changes between assigned flight levels. With improving navigation and air traffic control systems, continuous cruise-climb will become more common, as it can provide additional fuel savings.

Fig. 1.2-7. Aérospatiale/British Aerospace Concorde commercial airliner. (courtesy of NewsCast Limited)

controls: ailerons near the wing tips for low-speed roll control and mid-wing flaperons plus spoilers for high-speed roll control with reduced wing bending moments. The 777 is the first U.S. commercial transport to use DFBW controls, with a triply redundant system; each of these is internally triply redundant, with dissimilar microprocessors to minimize the impact of hardware or software failures. While control signaling is electronic, the control surface actuators are powered hydraulically.

Supersonic Transport Aircraft

The *Aérospatiale/British Aerospace Concorde* was the world's only operational supersonic transport aircraft (Fig. 1.2-7), flying from 1976 until 2003. It carried 128 passengers and a crew of three at a maximum Mach number of 2.2 and a typical cruising speed of 2,180 km/hr (1,350 mph). The cruising altitude was over 15,000 m (50,000 ft). Concorde also had an economical subsonic cruising point at $\mathcal{M} = 0.93$. The supersonic cruising range with a 10,100 kg payload and reserve fuel was 6,300 km (3,900 mi). Concorde was powered by four turbojet engines, each generating 169 kN (38,050 lb) of thrust with 17 percent afterburning for cruise; full afterburning was available for acceleration through the transonic region, where drag is especially high. The aircraft was 61.7 m long and had a wingspan of 25.6 m; its wing area and aspect ratio were 358.3 m^2 and 1.82, respectively. With empty and maximum-takeoff masses of 79,265 and 181,435 kg, its average wing loading was 365 kg/m^2.

Fig. 1.2-8. Lockheed-Martin F-35 joint strike fighter aircraft. (courtesy of Lockheed-Martin Corp.)

Concorde's delta wing had a sweep angle that varied from root to tip in an "ogee" curve. This shape is important both for supersonic cruising efficiency and for satisfactory placement of the strong vortices that form over the wing at high angle of attack during takeoff and landing. The need for good pilot visibility during landing and sharp-nosed streamlining in supersonic flight conflict; hence, the aircraft had a two-position fairing ("droop snoot") in front of the cockpit that accommodated both demands. As there was no separate fore or aft control surface, pitch and roll control were both supplied by the same set of "elevons" (deflected in sum for pitch and in difference for roll) at the wing's trailing edge. The control moment arms were relatively short, and without a separate surface for pitch control, there could be no flaps for landing. Fuel was transferred fore and aft during flight to maintain proper balance as the aerodynamic center shifted with Mach number. Concorde was the first commercial airliner to incorporate a dual-redundant, analog fly-by-wire (FBW) control system in parallel with a mechanical control system.

Fighter/Attack Aircraft

The *Lockheed-Martin Joint Strike Fighter* is a single-engine, single-seat, high-performance aircraft intended for ground attack and air-to-air combat (Fig. 1.2-8). It is designed for low observability ("stealth") as well as high performance. Three variants with 80–90 percent commonality are planned: a multirole conventional aircraft (F-35A), a carrier-based version (F-35B), and a short-takeoff and vertical-landing (STOVL) version (F-35C). The main turbofan engine for all variants produces an unaugmented maximum thrust of about 155 kN (35,000 lb), while the lift fan for the F-35C, driven by a shaft from the main engine, produces 80 kN (18,000 lb) of thrust for STOVL operations. The F-35A is 15.5 m long and it has a 10.7 m wingspan. The empty mass is 10,000 kg, while the maximum-takeoff mass is 22,700 kg. Given its wing area of 42.7 m^2, the aspect ratio is 2.7, and the maximum wing loading is 530 kg/m^2. The F-35's maximum Mach number is about 1.5, although its typical maneuvering speed would be in the transonic range below Mach 1. The F-35B has a larger wing that folds for storage on an aircraft carrier, as well as larger stabilizing and control surfaces, strengthened landing gear, and a tail hook for arrested landing. The dimensions of the F-35C are similar to those of the F-35A, although the lift-fan engine and attendant structure increase the empty weight. Distinguishing features include twin vertical tails, side-mounted air inlets, and trapezoidal wing and horizontal tail.

Bomber Aircraft

The *Northrop Grumman B-2 Spirit* (Fig. 1.2-9) is a high-subsonic flying wing configuration with an unrefueled range of about 11,500 km and a payload of 18,000 kg. The angular planform, with straight edges and sharp corners, and the lack of a vertical tail are beneficial characteristics for low observability, although the latter provides a unique flight control challenge: providing adequate yaw stability and control. An active system using deflection of outboard "drag rudders" achieves this goal. There are four control surfaces on each wing, with elevator, aileron, and rudder functions achieved by blending commands to these surfaces. The aircraft's wingspan, wing area, and length are 52.4 m, 465 m^2, and 21 m. The B-2 is powered by four 77 kN (17,300 lb) turbofan engines.

Space Shuttle

The *Rockwell Space Shuttle* (Fig. 1.2-10) is the second glider on our list, although it is powered into orbit by dual solid-fuel rockets and its three liquid-oxygen/hydrogen engines. The shuttle also has two orbital maneuvering rockets, as well as small thrusters for attitude control in space. During its return from orbit, the shuttle transitions from spacecraft to aircraft, reentering the atmosphere at $\mathcal{M} = 25$ and a high angle of attack to dissipate heat efficiently without damaging the structure, and performing a "dead-stick" horizontal landing at 335 km/hr (210 mph). The shuttle's planform and aerodynamic control surfaces are functionally similar to the Concorde's;

Fig. 1.2-9. Northrop Grumman B-2 Spirit bomber aircraft. (courtesy of Northrop Grumman Corp.)

however, the shuttle has a good deal more drag and thus glides at a much steeper flight path angle. The craft is controlled by a quintuply redundant DFBW system with no mechanical backup. The shuttle carries a flight crew of two plus up to five mission specialists. Its empty mass is 70,600 kg, and it can return to earth with a 14,500 kg payload. With a wing area

Fig. 1.2-10. Rockwell space shuttle. (courtesy of NASA)

of 250 m^2, the average wing loading during entry is 310 kg/m^2. The wingspan is 23.8 m, for an aspect ratio of 2.3, and the vehicle length is 37.2 m.

Uninhabited Air Vehicle

Many missions, including meteorological sampling, communications relay, surveillance, reconnaissance, target acquisition, and ground attack, may not require the presence of a human pilot on board the aircraft, may subject the pilot to high risk, or may require endurance or stress that exceeds normal human capabilities. Uninhabited air vehicles (UAV) are low-cost alternatives for performing these missions, and a number of designs have proved their worth to date. The *General Atomics RQ-1 Predator A* (Fig. 1.2-11) has physical and flight characteristics that are similar to those of a light general aviation aircraft, with a wingspan of 14.8 m, a length of 8.2 m, and maximum mass of 1,020 kg. Powered by a 75 kW (100 hp) reciprocating engine, the RQ-1 has a cruising speed of 130 km/hr, a top speed of 210 km/hr, and a maximum endurance of 40 hr. The Predator A carries a payload of 205 kg, and it has been outfitted to launch

Fig. 1.2-11. General Atomics RQ-1 Predator A uninhabited air vehicle. (courtesy of General Atomics International)

small missiles as well as conduct surveillance. The aircraft has a straight wing, inverted-V tail, and retractable tricycle landing gear. Its ceiling is 7,600 m.

1.3 The Mechanics of Flight

Aircraft flight is described by the principles of classical mechanics. *Mechanics* deals with the motions of objects that possess a substantive scalar inertial property called *mass*. Objects may be modeled as individual particles (also called *point masses*) or assemblages of particles called *bodies*. The translational motions of both point masses and bodies are of interest, as such objects occupy positions in space and may possess three linear velocity components relative to some frame of reference. The three scalar velocity components can be combined in a single three-dimensional column vector **v**, where

$$\mathbf{v} \triangleq \begin{bmatrix} v_1 \\ v_2 \\ v_3 \end{bmatrix} \tag{1.3-1}$$

The product of an object's mass and velocity is called *translational momentum*, also a three-dimensional vector. *Forces* may act on an object to change its translational momentum, which otherwise remains constant relative to an inertial (absolute) frame of reference.

Unlike point masses, bodies have three-dimensional shape and volume; the *position* of a body is defined by the coordinates of a particular reference point on or in the body, such as its *center of mass* (the "balance point"). The velocity of the body refers to the velocity of that reference point. The angular orientation and rotational motion of a body are also important descriptors of its physical state. *Angular momentum*, the rotational equivalent of translational momentum, remains unchanged unless a *torque* (a force that is displaced from the center of mass by a moment arm and that is perpendicular to the moment arm) acts on the body. A body can be characterized by six rotational inertial properties called *moments* and *products of inertia*; the former portray the direct relationship between angular rate and momentum about a rotational axis, while the latter establish coupling effects between axes.

Mechanics is further broken down into kinematics, statics, dynamics, and control. *Kinematics* is the general description of an object's motions without regard to the forces or torques that may induce change; thus, the geometry and coupling of position and velocity (both translational and rotational) are considered, while the means of effecting change are not. *Statics* addresses the balance of forces and torques with inertial effects to produce equilibrium. An aircraft can achieve static equilibrium when it is moving, as long as neither translational nor angular momentum is changing; for constant mass and rotational inertial characteristics, this implies unaccelerated flight. *Dynamics* deals with accelerated flight, when momentum is changing with time. While it is possible to achieve a steady, dynamic equilibrium (as in constant-speed turning flight), the more usual dynamics problem concerns continually varying motions in response to a variety of conditions, such as nonequilibrium initial conditions, disturbance inputs, or commanded forces and torques.

We refer to linear and angular positions and rates of the aircraft as its dynamic *state*; the corresponding twelve quantities are arrayed in the *state vector* described below. Motions occurring in the vertical plane are called *longitudinal motions*, while those occurring out of the plane are called *lateral-directional motions*. The variables of longitudinal motion, related to body axes, are axial velocity, normal velocity, pitching rate, and their inertial-axis integrals: range, altitude, and pitch angle. The lateral-directional variables are side velocity, roll rate, yaw rate, and their inertial-axis integrals, crossrange, roll angle, and yaw angle. *Stability* is an important dynamic characteristic that describes the tendency for the aircraft's state to return to an equilibrium condition or to diverge in response to inputs or initial conditions. *Control* is the critical area of mechanics that develops strategies and systems for achieving goals and assuring stability, given an aircraft's mission, likely disturbances, parametric uncertainties, and piloting tasks. The aircraft's natural stability can be augmented by *feedback control*, in which measured flight motions drive control effectors on a continuing basis.

For the most part, *kinematics* is well described by algebraic and trigonometric equations. Because many scalar variables often must be addressed simultaneously, matrix-vector notation is helpful for kinematic and other problems. For example, suppose $\mathbf{x}$ is a vector with three components (x_1, x_2, x_3) and $\mathbf{y}$ is a vector with two components (y_1, y_2) such that

$$y_1 = x_1 + \sin x_2 + \cos x_3 \tag{1.3-2}$$

$$y_2 = \cos x_1 + \sin x_3 \tag{1.3-3}$$

The two equations can be replaced symbolically by the single vector equation

$$\mathbf{y} = \mathbf{f}(\mathbf{x}) \tag{1.3-4}$$

where $\mathbf{f}(\mathbf{x})$ is a two-component vector composed of the right sides of eqs. 1.3-2 and 1.3-3. As a second example, let

$$y_1 = ax_1 + bx_2 + cx_3 \tag{1.3-5}$$

$$y_2 = dx_1 + ex_2 + fx_3 \tag{1.3-6}$$

The two scalar equations are represented by the matrix-vector equation

$$\mathbf{y} \triangleq \begin{bmatrix} y_1 \\ y_2 \end{bmatrix} = \mathbf{Hx} = \begin{bmatrix} a & b & c \\ d & e & f \end{bmatrix} \begin{bmatrix} x_1 \\ x_2 \\ x_3 \end{bmatrix} \tag{1.3-7}$$

where $\mathbf{H}$ is the (2×3) *matrix*

$$\mathbf{H} = \begin{bmatrix} a & b & c \\ d & e & f \end{bmatrix} \tag{1.3-8}$$

and the matrix-vector product is defined accordingly.[10]

The scalar velocity v is the time rate of change (or derivative with respect to time) of a scalar position r,

$$v = dr/dt \tag{1.3-9}$$

10. Generic scalar quantities are denoted by lower-case italic letters (a, b, c), while vectors are represented by lower-case bold letters ($\mathbf{a}$, $\mathbf{b}$, $\mathbf{c}$) and matrices by upper-case bold letters ($\mathbf{A}$, $\mathbf{B}$, $\mathbf{C}$). Italic letters (upper or lower case) and other fonts are used throughout to express commonly accepted symbols without rigid adherence to this convention.

while the position is the time integral of the velocity plus the appropriate constant of integration:

$$r(t) = r_0 + \int_0^t v(t)\,dt \tag{1.3-10}$$

The equations of *statics* can be combined in a single vector equation

$$0 = \mathbf{f}(\mathbf{x}, \mathbf{u}, \mathbf{w}) \tag{1.3-11}$$

which indicates that some combination of forces and inertial effects is simultaneously balanced along many coordinate directions. In eq. 1.3-11, $\mathbf{f}(\cdot)$ represents the vector of scalar momentum equations to be balanced, $\mathbf{x}$ is the solution variable (e.g., a six-component reduced state vector of translational and rotational velocities), $\mathbf{u}$ is a vector of constant control displacements, and $\mathbf{w}$ is a vector of constant disturbances.

In the *dynamic* case, the physical equations may vary in time t, and rather than equaling zero, they equal the rates of change of the motion variables, term by term:

$$\frac{d\mathbf{x}(t)}{dt} \triangleq \dot{\mathbf{x}}(t) = \mathbf{f}[\mathbf{x}(t), \mathbf{u}(t), \mathbf{w}(t), t] \tag{1.3-12}$$

This is an *ordinary differential equation* for the *state vector* $\mathbf{x}(t)$ as a function of the independent variable t. The definitions of $\mathbf{x}$ and $\mathbf{f}[\cdot]$ can be extended to include position variables and their kinematics. A first-order differential equation of this form is called a *state-space model* of system dynamics.

The object of control is to apply forces and moments that produce desired values of the state during the time interval of interest (t_0, t_f). The control vector $\mathbf{u}(t)$ may be a function of any measurable variables, of time, or of external inputs such as pilot commands $\mathbf{u}_c(t)$; hence, the dynamic equation becomes

$$\dot{\mathbf{x}}(t) = \mathbf{f}\{\mathbf{x}(t), \mathbf{u}[\mathbf{x}(t), \mathbf{w}(t), \mathbf{u}_c(t), t], \mathbf{w}(t), t\} \tag{1.3-13}$$

in the controlled case.

It is rarely adequate to know just the form of the aircraft's differential equation: we want to know the state as a function of time or some other variable. Therefore, equation 1.3-13 must be integrated subject to the initial condition of the state $\mathbf{x}(t_0)$:

$$\mathbf{x}(t) = \mathbf{x}(t_0) + \int_{t_0}^t \mathbf{f}\{\mathbf{x}(t), \mathbf{u}[\mathbf{x}(t), \mathbf{w}(t), \mathbf{u}_c(t), t], \mathbf{w}(t), t\}\,dt \tag{1.3-14}$$

Fig. 1.3-1. Classic paper airplane.

It is always possible to integrate eq. 1.3-13 numerically; however, simpler alternatives may provide greater insight into aircraft behavior. If, for example, $\mathrm{f}\{\cdot\}$ is a linear function of its arguments, the *initial condition response* can be written as the matrix-vector product

$$\mathbf{x}(t) = \mathbf{\Phi}(t, t_0)\,\mathbf{x}(t_0) \tag{1.3-15}$$

where $\mathbf{\Phi}(t, t_0)$ is a *state transition matrix* relating the state at time t_0 to the state at time t. It is also possible to write the steady sinusoidal response of a linear system to a steady sinusoidal control input $\mathbf{u}(t) = \mathbf{a}\sin\omega t$ with vector amplitude $\mathbf{a}$ at frequency ω as

$$\mathbf{x}(\omega) = \mathbf{A}(j\omega)\,\mathbf{u}(\omega) \tag{1.3-16}$$

where $\mathbf{A}(j\omega)$ is the system's complex-valued *frequency-response matrix* and $j \triangleq \sqrt{-1}$. Replacing $j\omega$ by the complex operator, $s \triangleq \sigma + j\omega$, $\mathbf{A}(s)$ is found to be the *transfer function matrix*, an important element of stability and control analysis.

Example 1.3-1. Longitudinal Motions of a Paper Airplane

The classic dart-shaped paper airplane (figure 1.3-1) provides an excellent introduction to flight mechanics, for both numerical analysis and experiment. Starting with a sheet of plain paper, which weighs[11] about 3 g,

11. Strictly speaking, its *mass* (not weight) is 3 g; "weighs" is used here and throughout in a colloquial sense.

the experimental vehicle has a wingspan of 12 cm and a length of 28 cm, yielding an aspect ratio AR and wing area S of 0.86 and 0.017 m^2, respectively. As shown in Chapter 3, the point-mass longitudinal motions of the paper airplane can be predicted by integrating four differential equations,

$$\dot{V} = -C_D\left(\frac{1}{2}\rho V^2\right)S/m - g\sin\gamma$$

$$\dot{\gamma} = \left[C_L\left(\frac{1}{2}\rho V^2\right)S/m - g\cos\gamma\right]/V$$

$$\dot{h} = V\sin\gamma$$

$$\dot{r} = V\cos\gamma$$

where V, γ, h, and r are the airspeed, flight path angle, height, and range, respectively. The parameters of this model are contained in the drag coefficient C_D, lift coefficient C_L, air density ρ (1.225 kg/m^3 at sea level), reference area S, mass m, and acceleration due to gravity, $g = 9.807$ m/s^2. The lift coefficient is modeled as a linear function of the angle of attack α,

$$C_L = C_{L_\alpha}\alpha$$

The lift-slope derivative is estimated as

$$C_{L_\alpha} = \frac{\pi AR}{1 + \sqrt{1 + (AR/2)^2}} \quad \text{(per radian)}$$

The drag coefficient is modeled as

$$C_D = C_{D_o} + \varepsilon C_L^2$$

with $C_{D_o} = 0.02$ and the *induced-drag factor* $\varepsilon = 1/\pi eAR$; e is known as the *Oswald efficiency factor*, estimated to be 0.9 here. In this example, attitude dynamics are neglected, so the angle of attack is taken as the control variable; its trimmed value is a function of the center-of-mass location and the amount of upward deflection of the wing's trailing edge.

This mathematical model can be put in the form of eq. 1.3-12 by making the following definitions:

$$\mathbf{x} \triangleq \begin{bmatrix} V \\ \gamma \\ h \\ r \end{bmatrix} \triangleq \begin{bmatrix} x_1 \\ x_2 \\ x_3 \\ x_4 \end{bmatrix}, \quad \mathbf{u} \triangleq \alpha \triangleq u$$

$$\mathbf{f}(\mathbf{x},\mathbf{u}) \triangleq \begin{bmatrix} -C_D(u)\left(\frac{1}{2}\rho x_1^2\right)S/m - g\sin x_2 \\ \left[C_L(u)\left(\frac{1}{2}\rho x_1^2\right)S/m - g\cos x_2\right]/x_1 \\ x_1 \sin x_2 \\ x_1 \cos x_2 \end{bmatrix}$$

$\mathbf{f}(\mathbf{x},\mathbf{u})$ is a nonlinear function of its arguments for several reasons: $C_D(u)$ is a quadratic function of u, and several nonlinear operations involve elements of the state, including the square of x_1, division by x_1, sines and cosines of x_2, and the product of x_1 with functions of x_2.

Three longitudinal paths are typical of paper airplane flight: *constant-angle descent*, *vertical oscillation*, and *loop*. Numerically integrating the equations of our mathematical model produces these paths (eq. 1.3-14), as shown in the accompanying figures. Four cases are simulated: (a) equilibrium glide at maximum lift-drag ratio (L/D), (b) oscillating glide due to zero initial flight path angle, (c) increased oscillation amplitude due to increased initial speed, and (d) loop due to a further increase in launch speed. In all cases, α is fixed at 9.3 deg, yielding the best L/D (5.2) for the configuration. This angle of attack produces the shallowest equilibrium glide slope (-11 deg) and a constant speed of 3.7 m/s. The equilibrium trajectory is represented by the straight descending path in Fig. 1.3-2a and the constant speed and flight path angle in Fig. 1.3-2b (case a).

Obtaining this smooth, linear path is dependent on starting out with exactly the right speed and angle; case b illustrates that having the right speed but the wrong initial angle (0 deg) perturbs the dynamic system, introducing a lightly damped oscillation about the equilibrium path. The period of the oscillation is 1.7 s, and the wave shape is very nearly sinusoidal.

The initial-condition perturbation is increased in case c, where the zero initial angle is retained, and the speed is increased. The vertical oscillation has greater height, speed, and angle amplitude, and the wave shape is perceptibly scalloped in the first two variables. Nevertheless, the rate of amplitude decay and the period are about the same as in the previous case. Further increasing the launch speed results in a loop, followed by sharply scalloped, decaying undulations with a 1.7 s period (case d). The loop provides an upper limit on the vertical oscillations that can be experienced by an airplane with fixed control settings.

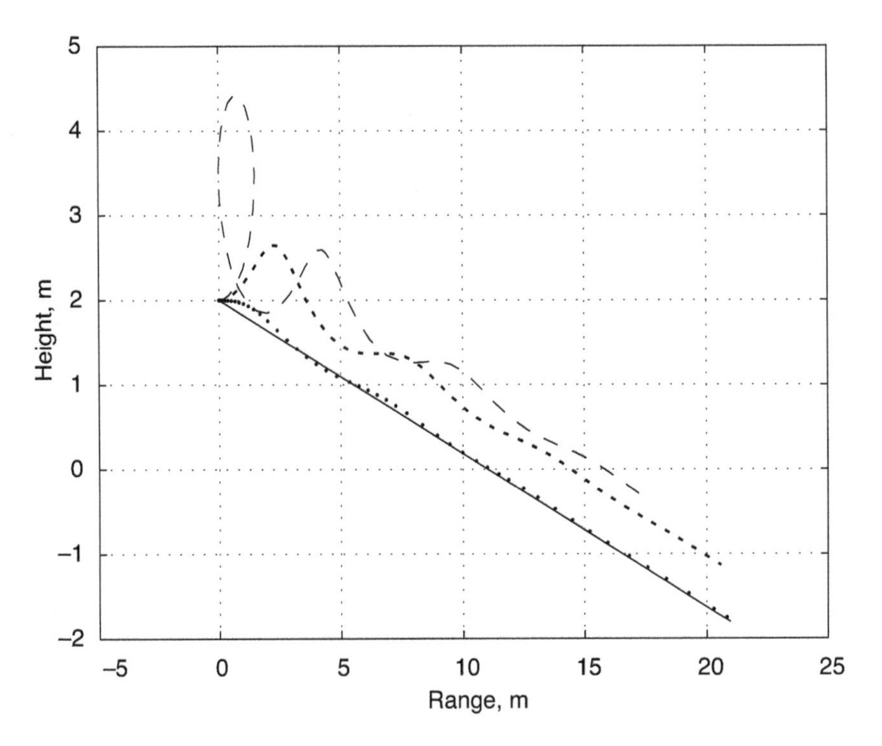

Fig. 1.3-2a. Flight of a paper airplane: Height versus range.

The MATLAB program used to generate these results can be found in Appendix D. You can replicate these flight paths with your own paper airplane, although you are not likely to get identical flight test results. The aerodynamic model used here is approximate, and the weight and dimensions of your airplane will be slightly different. It is difficult to trim the paper plane's α precisely, but it is relatively easy to sweep a range of values by bending the wing's trailing edge up or down. Similarly, launching by hand produces variable starting conditions, and it probably is not possible to launch a paper airplane at speeds much above 5 m/s without inducing significant aeroelastic deformations to the structure, changing its aerodynamics. Perhaps most important, attitude dynamics (not modeled above) play a role, particularly in large-amplitude maneuvers such as the loop. It is tricky to loop the classic dart-shaped paper airplane in practice, but other designs can be looped readily.

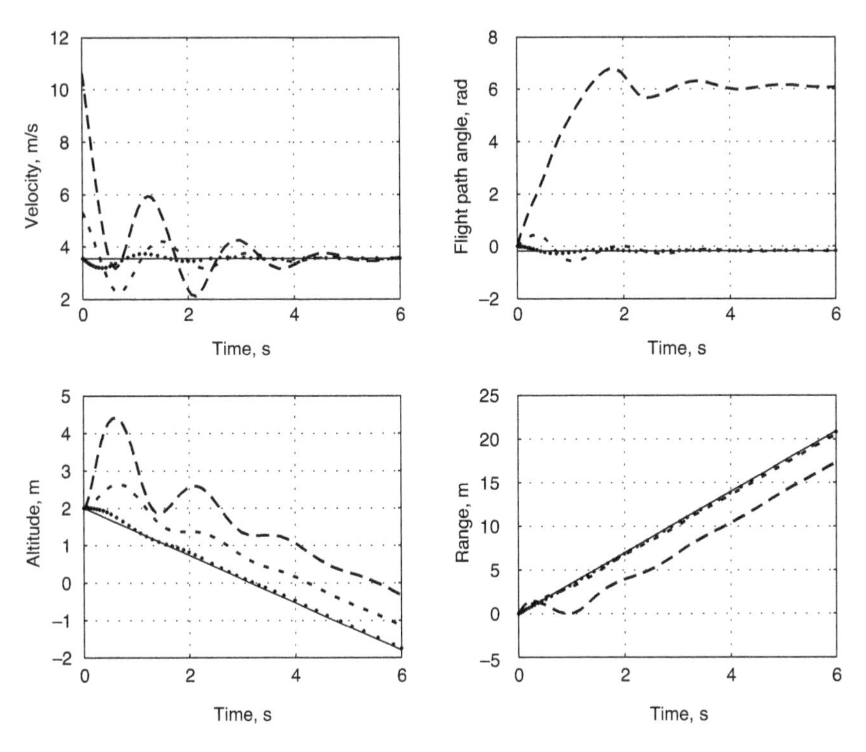

Fig. 1.3-2b. Flight of a paper airplane: Velocity, flight path angle, altitude, and range versus time.

References

B-1 Burns, B. R. A., "The Design and Development of a Military Combat Aircraft," *Interavia* Mar 1976, pp. 241–246; May 1976, pp. 448–450; June 1976, pp. 553–556; July 1976, pp. 643–646.

C-1 Corning, G., *Supersonic and Subsonic, CTOL and VTOL, Airplane Design*, published privately, College Park, Md., 1979.

G-1 Gatland, K., *The Illustrated Encyclopedia of Space Technology*, Crown Publishers, New York, 1981.

G-2 Gilbert, J., *The World's Worst Aircraft*, Coronet Books, London, 1975.

G-3 Green, W., and Swanborough, G., *The Illustrated Encyclopedia of the World's Commercial Aircraft*, Crown Publishers, New York, 1978.

H-1 Hale, F. J., *Introduction to Aircraft Performance, Selection, and Design*, J. Wiley & Sons, New York, 1984.

H-2 Huenecke, K., *Modern Combat Aircraft Design*, Naval Institute Press, Annapolis, Md., 1987.

J-1 Jackson, P., ed., *Jane's All the World's Aircraft, 2000–2001*, Jane's Information Group, Coulsdon, Surrey, U.K., 2000 (and earlier annual editions).
K-1 Kermode, A. C., *Mechanics of Flight*, Pitman Publishing, London, 1972.
K-2 Küchemann, D., *The Aerodynamic Design of Aircraft*, Pergamon Press, Oxford, 1978.
M-1 Mander, J., Dippel, G., and Gossage, H., *The Great International Paper Airplane Book*, Simon and Schuster, New York, 1967.
M-2 Miller, J., *The X-Planes, X-1 to X-29*, Specialty Press Publishers and Wholesalers, Marine on St. Croix, Minn., 1983.
M-3 Mondey, D., *The International Encyclopedia of Aviation*, Crown Publishers, New York, 1977.
N-1 Nicolai, L., *Fundamentals of Aircraft Design*, METS, San Jose, Calif., 1975.
P-1 Perkins, C. D., "Development of Airplane Stability and Control Technology," *Journal of Aircraft* 7, no. 4 (1970): 290–301.
R-1 Raymer, D., *Aircraft Design: A Conceptual Approach*, American Institute of Aeronautics and Astronautics, Washington, D.C., 1989.
S-1 Sacher, P. W., dir., *Special Course on Fundamentals of Fighter Aircraft Design*, AGARD-CP-R-740, Neuilly-sur-Seine, Oct. 1987.
S-2 Stinton, D., *The Anatomy of the Aeroplane*, American Elsevier Publishing, New York, 1966.
S-3 Stinton, D., *The Design of the Aeroplane*, Van Nostrand Reinhold, New York, 1983.
T-1 Torenbeek, E., *Synthesis of Subsonic Airplane Design*, Delft University Press, Delft, Netherlands, 1982.
W-1 Whitford, R., *Design for Air Combat*, Jane's Publishing, London, 1987.

Chapter 2

Exploring the Flight Envelope

The range of static and dynamic conditions within which an aircraft operates is called its *flight envelope*. To explore the flight envelope, we first must establish frames of reference and define basic elements of position, velocity, force, moment, and equilibrium. Our attention is directed to steady or quasisteady flight, and the aircraft is modeled as a rigid body flying over a flat earth. *Steady flight* is characterized by constant velocity and altitude, with unchanging aircraft mass, as in wings-level cruise, where there is a balance between lift, weight, thrust, and drag. Rigorously, steady flight is possible only if thrust is produced without changing the aircraft's mass. *Quasisteady flight* occurs when time variations in mass or flight condition are slow enough to produce negligible dynamic effects, such as small rate of weight loss due to fuel burn, shallow climbing and gliding, and turns with constant bank angle and yaw rate.

The *rigid-body assumption* allows the aircraft's translational position and angular attitude to be defined by the position of its center of mass and the orientation of a body-fixed coordinate system relative to an inertial frame of reference. It limits the degrees of freedom needed to describe aircraft motion and eliminates the dependence of aerodynamic and inertial effects on structural bending and torsion. Body flexibility affects structural loads and ride quality; it must be considered in flight control design, but it has little direct effect on the net position and velocity.

The *flat-earth assumption* neglects the variation of gravitational acceleration with distance from the center of the earth, and it disregards relatively subtle Coriolis and centrifugal effects. The description of position and velocity relative to the earth is simplified, a useful attribute for investigating the dynamics and control of flight within the atmosphere at suborbital speeds and for studying all modes of motion with periods or time constants less than several hundred seconds. *Round-earth effects* are important for analyzing navigation and guidance, especially for long-distance flights and for trajectories that enter the atmosphere from orbit.

Whatever the frame of reference, we assume in the remainder of the book that flight motions are described by *Newton's laws of motion*, which can be summarized as follows:

First law. If no force acts on a particle, it remains at rest or continues to move in a straight line at constant velocity, as measured in an inertial reference frame.

Second law. A particle of fixed mass acted upon by a force moves with an acceleration proportional to and in the direction of the force; the ratio of force to acceleration is the mass of the particle.

Third law. For every action, there is an equal and opposite reaction.

In this chapter, we focus on steady or quasisteady flight in various conditions, such as climbing, cruising, descending, and turning. We begin with a discussion of the atmosphere and its characteristics, noting the changes of air pressure, density, and temperature with altitude, as well as descriptions of the wind field (Section 2.1). We examine frames of reference and the relationships between them. Euler angles portray the orientation of one frame with respect to another; hence, transformations between reference frames must involve these angles (Section 2.2). The static force, moment, and inertial properties establish equilibrium conditions that are important for steady flight (Section 2.3). The corresponding aerodynamic effects and nondimensional coefficients commonly used to express aerodynamic properties and the generation of thrust by powerplants are presented (Sections 2.4 and 2.5). The chapter concludes with a review of the principles that govern aircraft performance (Section 2.6).

2.1 The Earth's Atmosphere

The actual values of the air's pressure, density, and temperature vary with the seasons and depend upon the weather and time of day, but standard atmospheric conditions can be used in many flight calculations. A brief summary of the U.S. Standard Atmosphere, 1976 [A-7] is presented in Fig. 2.1-1 and Table 2.1-1. A MATLAB function (`Atmos`) for interpolating air density, air pressure, temperature, and the speed of sound from the standard atmosphere table can be found in Appendix B. The systematic variability of air density from these average values is less than 1 percent below 10 km altitude and less than 20 percent below 50 km. Annual absolute temperature variations of 10 percent in this altitude region are typical. In the lower atmosphere, where aircraft normally fly, the mean free path of the air's molecules is short compared to aircraft dimensions.[1]

1. The *Knudsen number* equals the mean free path divided by some characteristic length of the vehicle. The mean free path of the air's constituent molecules is about 0.1 m at an altitude of 100 km.

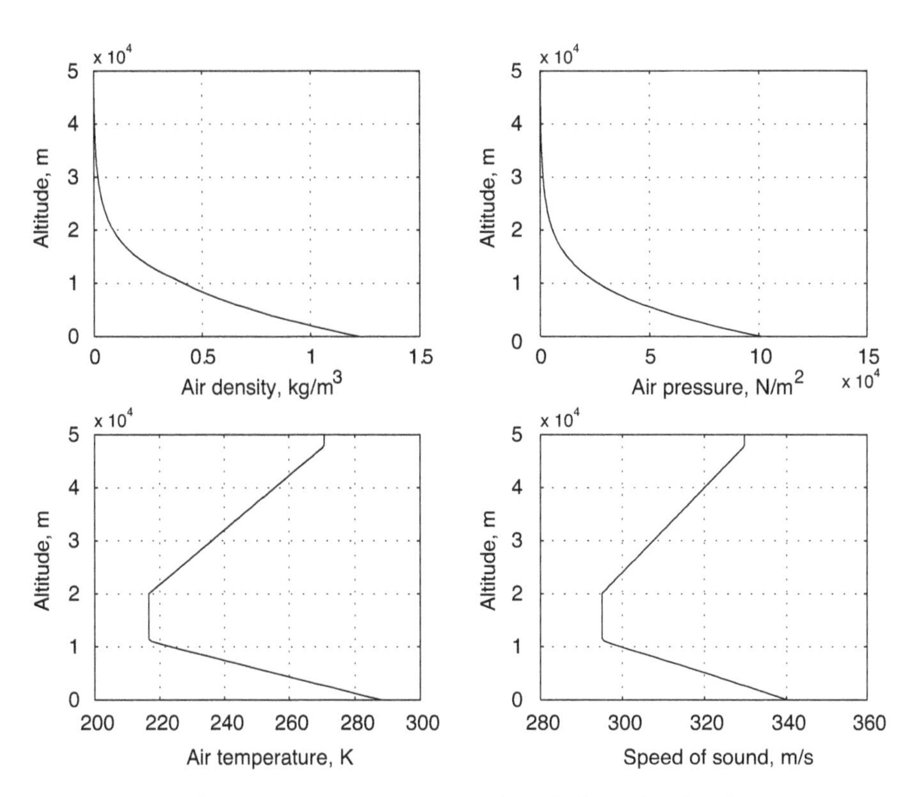

Fig. 2.1-1. Air density, pressure, temperature, and speed of sound at altitudes from 0 to 50 km.

TABLE 2.1-1 U.S. STANDARD ATMOSPHERE, 1976 [A-7]

Z, m	H, m	p/p_o*	ρ/ρ_o**	$\mathcal{T}$, K	a, m/s	ν/ν_o***
0	0	1.000	1.000	288.150	340.294	1.000
2,500	2,499	0.737	0.781	271.906	330.563	1.223
5,000	4,996	0.533	0.601	255.676	320.545	1.514
10,000	9,984	0.262	0.338	223.252	299.532	2.413
11,100	11,081	0.221	0.293	216.650	295.069	2.709
15,000	14,965	0.120	0.159	216.650	295.069	4.997
20,000	19,937	0.055	0.073	216.650	295.069	10.946
47,400	47,049	0.0011	0.0011	270.650	329.799	822.11
51,000	50,594	0.0007	0.0007	270.650	329.799	1,286.

*$p_o = 1.013 \times 10^5$ N/m^2 **$\rho_o = 1.225$ kg/m^3 ***$\nu_o = 1.461 \times 10^{-5}$ m^2/s

Therefore, it is generally acceptable to characterize the atmosphere as a gaseous continuum when studying aircraft flight.

Pressure, Density, and the Speed of Sound

Atmospheric conditions are related by an "equation of state" and are governed by hydrostatic forces arising from the earth's gravitational field. The *thermal equation of state* for a perfect gas is

$$p = \rho\left(\frac{R^*}{M}\right)\mathcal{T} = \rho R\mathcal{T} \tag{2.1-1}$$

where p is the *ambient pressure* (pascals, or N/m^2), R^* is the *universal gas constant* (8.314 J/mol-K),[2] M is the *mean molecular weight of air* (28.96 g/mol), R is the *gas constant of air* (287.05 N-m/kg-K), and $\mathcal{T}$ is the *absolute temperature* (K). The *hydrostatic equation* relates differential change in ambient pressure dp to a unit weight of air ρg_{eff} times the differential change in altitude dZ (positive up):

$$dp = -\rho g_{eff}\, dZ \tag{2.1-2}$$

The pressure at the earth's surface is the integral of eq. 2.1-2, that is, the weight per unit area of a column of air that extends from the surface to infinite height.

The effective gravity g_{eff} is a function of altitude. This is a consequence of the earth's mass attraction plus the centrifugal effect of the earth's rotation (assuming that the atmosphere rotates at the same rate), both of which can be described as potential functions (Section 3.1). Thus, a distinction is made between the *geometric altitude* Z, referenced to the earth's surface, and the *geopotential altitude* $H(Z)$, which is defined as

$$H(Z) \triangleq \int_0^Z \left(\frac{g(z)}{g_o}\,dz\right) = \frac{1}{g_o}\left[\mathcal{V}(Z) - \mathcal{V}(0)\right] = \frac{r_m Z}{r_m + Z} \tag{2.1-3}$$

g_o is the net gravitational acceleration at sea level, and $\mathcal{V}(\cdot)$ is the net *potential energy*, described in Chapter 3 and succeeding chapters. For a spherical earth with inverse-square gravitational field whose mean radius r_m is 6,367,474 m, the relationship between H and Z takes the simple form shown. The distinction between H and Z is not great in the lower atmosphere, as can be seen from Table 2.1-1.

The *U.S. Standard Atmosphere* characterizes the temperature by segments that are either constant with altitude or linear in $H(Z)$, as derived from experimental data. The speed of sound a (m/s) is related to the

2. J stands for *Joule*, a unit of energy that is identical to a Newton-meter. It is the work done by one Newton of force moving one meter (see Chapter 3).

absolute temperature by

$$a = \sqrt{\gamma R \mathcal{T}} \tag{2.1-4}$$

where γ, the *ratio of the specific heats* of air at constant pressure and volume (c_p/c_v), is 1.4. The *standard pressure* is modeled as a function of $H(Z)$. In *constant-temperature layers*, the function is

$$p = p_B e^{-g_o(H - H_B)/R\mathcal{T}} = p_B e^{-\beta(H - H_B)} \tag{2.1-5}$$

with $(\cdot)_B$ representing the value at the base of the layer. The quantity $\beta\,(= g_o/R\mathcal{T})$ is called the *inverse scale height* of the pressure function; it has units of m^{-1}. In a *linearly varying temperature layer*, the standard pressure is given by

$$p = p_B \left[\frac{\mathcal{T}_B}{\mathcal{T}_B + k(H - H_B)} \right]^{g_o/Rk} \tag{2.1-6}$$

where k is the slope of the temperature variation, $\partial\mathcal{T}/\partial H$. The density can then be obtained from the pressure using the equation of state (eq. 2.1-1).

The *dynamic pressure* $\bar{q}$ is a function of flight conditions; it can be expressed in terms of air density ρ and true airspeed V_A or of ambient atmospheric pressure p and Mach number $(\mathcal{M} = V_A/a)$. From the above equations,

$$\bar{q} \triangleq \frac{1}{2}\rho V_A^2 = \frac{\gamma}{2} p\, \mathcal{M}^2 \tag{2.1-7}$$

It is often useful to define an exponential model of air density for trajectory calculations within a given geometric altitude band (z_1, z_2), with positive altitude corresponding to negative z:

$$\rho(z) = \rho_1 e^{\beta(z - z_1)} \tag{2.1-8}$$

If the temperature is constant in the altitude band, β is defined by eq. 2.1-5. If the air temperature varies in (z_1, z_2), a least-squares curve fit to the tabulated data could be used to estimate the equivalent β. A simple expedient is to read ρ_1 and ρ_2 from the U.S. Standard Atmosphere and then compute

$$\beta = \frac{\ln(\rho_2/\rho_1)}{(z_2 - z_1)} \tag{2.1-9}$$

for the interval, as in the Atmos function of Appendix B.

Viscosity, Humidity, and Rain

Other atmospheric characteristics with flight-dynamic implications include kinematic viscosity, humidity, precipitation, icing, and wind [D-1].

The *kinematic viscosity* v, which is the ratio of dynamic viscosity to air density, characterizes the shear stress property of air (per unit mass). Large values encourage layered, sliding (*laminar*) flow between adjacent airstreams of differing velocity. At 100 percent relative *humidity* and 300 K, the water vapor concentration in the air can reach about 25 g/m^3. Given its lower molecular weight, the water vapor reduces the dry atmosphere's density by about 1 percent in this case.

Although relatively infrequent, precipitation rates of more than 1 m/hr have been measured at the earth's surface. Heavy *rain* can reduce aircraft performance by increasing effective drag, reducing maximum available lift, and adding weight to the aircraft [A-10, T-2, V-3]. In at least one case, rain is believed to have quenched a transport aircraft's jet engines. The effect is transient and presents an immediate hazard to most aircraft only when they are near the ground, operating at high angle of attack and power setting. Rain destroys the laminar flow on which many sailplanes depend for high performance, leading to pitch trim changes as well as lift and drag effects. *Icing*, which occurs when an aircraft flies into supercooled water droplets, is even more insidious, as the ice accumulates on the plane's skin, spoiling airflow, increasing weight, and possibly jamming control surfaces.

Wind Fields and Atmospheric Turbulence

Wind is the most prevalent and complex natural disturbance to aircraft flight. Changing wind fields have a number of meteorological origins. They can be characterized as gust fronts, sea-breeze fronts, air-mass fronts, gravity or solitary waves, terrain-induced wind shear, mountain waves (including wind rotors), low- and high-altitude jet streams, tornadoes, microbursts, and turbulence [T-4]. Flying aircraft leave a trail of intense, localized vorticity in their wakes that poses a hazard to other aircraft. Wind models produce force and moment inputs through the aircraft's aerodynamics; their vector-matrix descriptions are subject to the same matrix transformations used throughout this chapter.

The wind is a vector field that varies in space and time; its magnitude may approach the aircraft's normal airspeed during takeoff and landing. Furthermore, the change in the wind—the *wind shear*—can have persistent effects on aeroelastic modes, as well as dramatic and disastrous effects on the flight path when flying near the ground [T-4]. If the wind is modeled as

$$\mathbf{w}(\mathbf{r}, t) = \begin{bmatrix} w_x(\mathbf{r}, t) \\ w_y(\mathbf{r}, t) \\ w_z(\mathbf{r}, t) \end{bmatrix} \tag{2.1-10}$$

then the *spatial wind shear* must be represented as the *Jacobian matrix* of $\mathbf{w}$ with respect to the position $\mathbf{r}$,

$$\mathbf{w}_{\mathbf{r}} \triangleq \mathbf{W} \triangleq \begin{bmatrix} \dfrac{\partial w_x(\mathbf{r}, t)}{\partial x} & \dfrac{\partial w_x(\mathbf{r}, t)}{\partial y} & \dfrac{\partial w_x(\mathbf{r}, t)}{\partial z} \\ \dfrac{\partial w_y(\mathbf{r}, t)}{\partial x} & \dfrac{\partial w_y(\mathbf{r}, t)}{\partial y} & \dfrac{\partial w_y(\mathbf{r}, t)}{\partial z} \\ \dfrac{\partial w_z(\mathbf{r}, t)}{\partial x} & \dfrac{\partial w_z(\mathbf{r}, t)}{\partial y} & \dfrac{\partial w_z(\mathbf{r}, t)}{\partial z} \end{bmatrix} \tag{2.1-11}$$

while the *temporal wind shear* is the partial time derivative of $\mathbf{w}$:

$$\mathbf{w}_t \triangleq \begin{bmatrix} \dfrac{\partial w_x(\mathbf{r}, t)}{\partial t} \\ \dfrac{\partial w_y(\mathbf{r}, t)}{\partial t} \\ \dfrac{\partial w_z(\mathbf{r}, t)}{\partial t} \end{bmatrix} \tag{2.1-12}$$

The total time derivative of the wind at a point moving through the wind field with velocity $\mathbf{v}_\mathrm{I}$ is the sum of the two:

$$\dot{\mathbf{w}} = \mathbf{w}_t + \mathbf{W}\mathbf{v}_\mathrm{I} \tag{2.1-13}$$

The varying spatial wind distribution implied by eqs. 2.1-10 and 2.1-11 subjects different parts of an aircraft to different instantaneous winds, further complicating dynamic response. Consider a vertical wind field that has higher upward velocity at the left wing tip than at the right; the airplane is subjected to a positive rolling moment that cannot be computed from the wind velocity at a single point. If the wind distribution is linear across the span, then the rolling effect can be predicted from the stability derivatives presented in Section 3.4. If the distribution is more detailed, then the moment calculation involves a fine-scale integration of spanwise effects.

Isolated *atmospheric vortices*, including tornadoes and wind rotors [W-8], can be modeled as circular flows about a linear axis. The tangential velocity V_T of an inviscous, two-dimensional *ideal vortex* is expressed as

$$V_T = \frac{k}{r} \tag{2.1-14}$$

where k reflects the strength of the vortex and r is the radius from the vortex axis [S-1]. The viscosity limits the velocity as the radius becomes small, and real vortices are observed to have zero V_T at the axis. The *Lamb-Oseen vortex* imposes an exponential decay on eq. 2.1-14, producing

$$V_T = \frac{k[1 - e^{-(r/r_c)^2}]}{r} \tag{2.1-15}$$

where r_c is a reference radius. The tangential velocity is plotted in Fig. 2.1-2, with $k = r_c = 1$. The maximum value of V_T occurs when $dV_T/dr = 0$; the corresponding radius is $1.12r_c$, and the maximum tangential velocity is $0.64k/r_c$. The flow within the radius r_c approximates a *solid-body rotation* of the air.

A constant-strength, three-dimensional vortex is extended along a line that is perpendicular to the circular motion. Assuming that the axis is aligned with the x direction, the resulting cylindrical flow can be expressed as

$$\begin{bmatrix} v_x \\ v_y \\ v_z \end{bmatrix} = \begin{bmatrix} 0 \\ \dfrac{-k[1 - e^{-(r/r_c)^2}]z}{r^2} \\ \dfrac{k[1 - e^{-(r/r_c)^2}]y}{r^2} \end{bmatrix} \tag{2.1-16}$$

where y and z are the lateral and vertical distances from the axis. An aircraft leaves a *pair of vortices* of opposing rotational directions in its wake. Their intensity is proportional to the wing's lift and, consequently, to the aircraft's weight. The resulting wind disturbance can be modeled by adding two images of eq. 2.1-16, with r, y, and z referenced to the left and right rotational axes.

Atmospheric turbulence is a random variation that is superimposed on the slowly varying mean wind field:

$$\mathbf{w}(\mathbf{r}, t) = \mathbf{w}_o(\mathbf{r}, t) + \Delta\mathbf{w}(\mathbf{r}, t) \tag{2.1-17}$$

$\mathbf{w}_o(\mathbf{r}, t)$ characterizes a component that measurably persists over some time and region. While the statistical properties of $\Delta\mathbf{w}(\mathbf{r}, t)$ may be relatively constant, the actual wind-speed variations that occur are probabilistic and, therefore, not strictly predictable. $\Delta\mathbf{w}(\mathbf{r}, t)$ is a random process whose components are characterized by probability distributions, expected values, and power spectral densities, which are discussed in Section 4.5.

For the moment, it suffices to describe turbulence as a wavelike phenomenon whose component amplitudes vary with frequency. The von

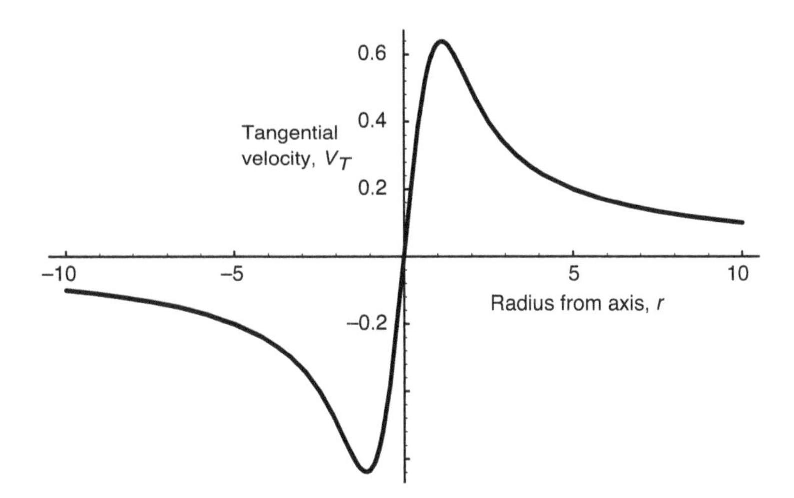

Fig. 2.1-2. Lamb-Oseen vortex.

Kármán turbulence spectra [V-3] are widely accepted representations of atmospheric turbulence in which the incremental wind "power" $\Phi(\Omega)$ (m^2/s^2) (i.e., velocity squared) is a function of wavelength or wave frequency Ω (rad/m). For example, a one-dimensional spectrum for disturbances that are transverse to the flight direction is

$$\Phi(\Omega) = \frac{\sigma^2 L}{2\pi} \frac{[1 + 8/3(aL\Omega)^2]}{[1 + (aL\Omega)^2]^{11/6}} \tag{2.1-18}$$

where σ, a (≈ 1.4), and L ($\approx 1{,}750\,\text{m}$) are constants of the random process. Note that the turbulence power decreases with increasing frequency (i.e., decreasing wavelength) as $\Omega^{-11/6}$, close to the $\Omega^{-5/3}$ decrease produced by the Kolmogorov law [L-4].

In later sections, we shall use rational transfer functions and their associated state-space models to describe aircraft response to inputs. The noninteger exponent of the von Kármán spectrum is not easily modeled in state-space format. The Dryden model, though a less accurate fit to experimental data, has no such problem [H-4, H-5]. For transverse turbulence, the Dryden spectrum takes the form

$$\Phi(\Omega) = \frac{\sigma^2 L}{\pi} \frac{[1 + 3(L\Omega)^2]}{[1 + (L\Omega)^2]^2} \tag{2.1-19}$$

while for axial turbulence (i.e., along the line of flight), it is

$$\Phi(\Omega) = \frac{2\sigma^2 L}{\pi} \frac{1}{[1 + (L\Omega)^2]^2} \tag{2.1-20}$$

with $L \approx 1{,}250\,\text{m}$.

When significant features of the wind field are large compared to aircraft size, it is satisfactory to consider eqs. 2.1-10 and 2.1-13 as *exogenous* (i.e., external) *disturbances* to the aircraft's dynamics, neglecting the converse effect of the aircraft's presence on the wind field itself. In this case, the aircraft is transparent to the wind; the wind's effects are assumed to be applied at the aircraft's center of mass and across the aircraft's surface. However, for intense small-scale features like the wing-tip vortices generated by another aircraft, the coupling between wind-field dynamics and aircraft dynamics cannot be ignored. Figure 2.1-3 illustrates the impingement of the tip vortex (marked by smoke) generated by a leading Lockheed-Martin C-130 aircraft on a trailing Boeing 737 aircraft. Note that the vortex core, which is essentially straight before reaching the 737, is deflected as it passes over the aircraft's surface.

2.2 Kinematic Equations

Kinematics deals with the geometry of translational and rotational motion. *Position* can be measured with respect to a *basic reference frame* that has three mutually perpendicular directions, customarily denoted by *unit vectors* $(\boldsymbol{i}, \boldsymbol{j}, \boldsymbol{k})$, and corresponding *lengths* (x, y, z) along the directions established by the unit vectors. Lengths are measured from an *origin*, where $(x, y, z) \triangleq (0, 0, 0)$ (Fig. 2.2-1a). The ordering of these axes follows the "right-hand rule": if the index finger represents $\boldsymbol{i}$, and the middle finger is bent 90 deg from the index finger to establish the $\boldsymbol{j}$ direction, then the upward pointing thumb represents the $\boldsymbol{k}$ direction. Scalar components of the *velocity* can be resolved *along* the $\boldsymbol{i}$, $\boldsymbol{j}$, and $\boldsymbol{k}$ directions, while right-handed rotations may occur *around* the $\boldsymbol{i}$, $\boldsymbol{j}$, and $\boldsymbol{k}$ axes.

If the origin is unaccelerated (but possibly moving with constant translational velocity) and the unit vectors are not rotating with respect to some absolute reference system of cosmic scale, then this basic frame is an *inertial reference frame*. Most "inertial" frames are not rigorously inertial, but the degree of error in making that assumption is, in some sense, acceptable. For calculating the trajectories of paper clips, or satellites, or interplanetary probes, the most appropriate "inertial" reference plane may be your desk top, or the earth's equator, or the solar system's ecliptic.

Having arrived at a basic reference frame, various *relative frames*, possibly accelerated and rotating, can be defined. These frames allow "local"

Fig. 2.1-3. Boeing 737 flying in the tip-vortex wake of a Lockheed-Martin C-130. (courtesy of NASA)

effects and variations to be described, while the motions of the relative frames (with respect to the basic frame) can be considered separately. Suppose the basic frame is an "inertial" frame whose origin is fixed at a point on the earth's surface, and an aircraft is flying near the basic frame origin. A relative frame can be fixed at the plane's center of mass (c.m.). A *body-fixed frame* is one such frame; its origin is at the plane's c.m., and its axes are aligned with the nose, the right wing tip, and the downward perpendicular to the two. Alternatively, a *velocity-fixed frame* is defined with the c.m. as origin but with one axis aligned to the direction of motion. The second axis may be confined to the horizontal plane, or it may be aligned with the right wing tip. Each of these relative frames has utility in characterizing and analyzing the aircraft's motions.

Translational Position and Velocity

The *translational (or rectilinear) position* of the center of mass is portrayed by the three components x, y, and z, representing northerly, easterly, and vertical displacement from the origin. By convention, z is positive *down* to preserve a "right-hand-rule" relationship with x and y (Fig. 2.2-1b). The aircraft's height h above the ground is a positive number; therefore, $h = -z$. The aircraft's position *vector* $\mathbf{r}$, measured from the origin,

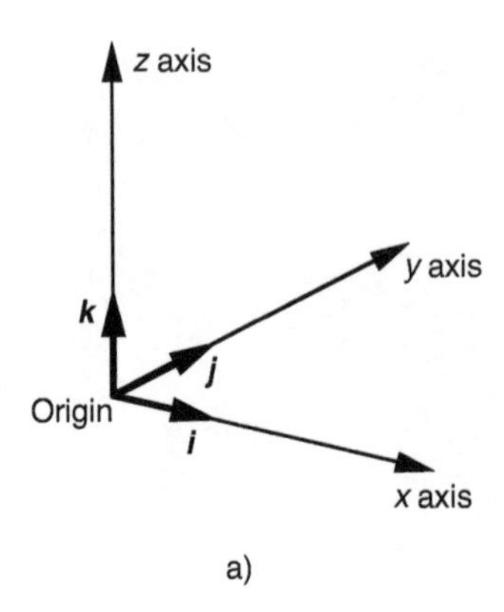

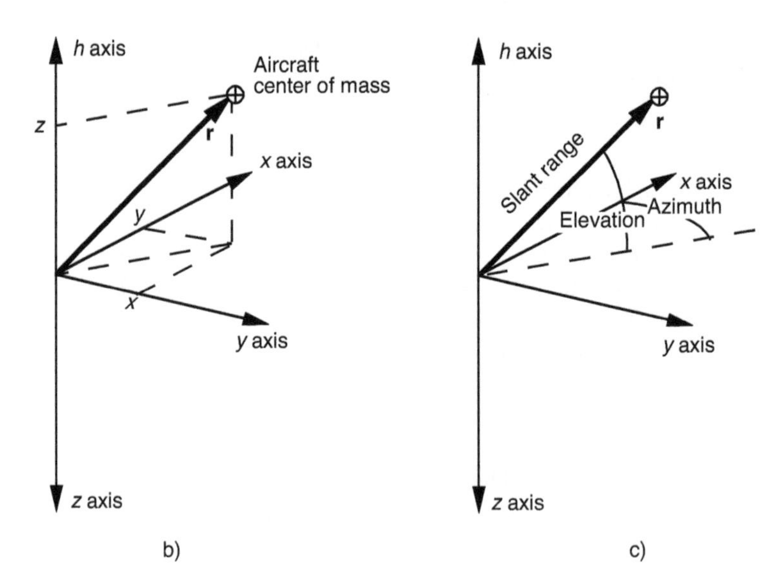

Fig. 2.2-1. Basic Cartesian reference frame. (a) Initial orientation. (b) Conventional orientation used in aircraft dynamics. (c) Corresponding polar coordinates.

is expressed as a column of three elements measured along the ***i***, ***j***, and ***k*** directions:

$$\mathbf{r} \triangleq \begin{bmatrix} r_1 \\ r_2 \\ r_3 \end{bmatrix} \triangleq \begin{bmatrix} x \\ y \\ z \end{bmatrix} \triangleq \begin{bmatrix} x \\ y \\ -h \end{bmatrix} \tag{2.2-1}$$

x, y, and z, are Cartesian components of $\mathbf{r}$, that is, their axes are orthogonal to each other. The aircraft's *distance from the origin*, r, is the magnitude (or absolute value) of the vector $\mathbf{r}$, calculated as the root sum square of its components:

$$r \triangleq |\mathbf{r}| = \sqrt{x^2 + y^2 + z^2} = \left[[x \quad y \quad z] \begin{bmatrix} x \\ y \\ z \end{bmatrix} \right]^{1/2} \triangleq [\mathbf{r}^{\mathrm{T}}\mathbf{r}]^{1/2} \tag{2.2-2}$$

The distance is constant if the scalar r is constant. For the position vector to be constant, all three components of $\mathbf{r}$ must not change. This is equivalent to saying that the position vector's magnitude and direction (from the origin) are constant, as $\mathbf{r}$ could be expressed in polar coordinates (Fig. 2.2-1c): *slant range* (= distance), *elevation angle*, and *azimuth angle*.

Equation 2.2-2 introduces the *vector transpose* operation $(\cdot)^{\mathrm{T}}$, which transforms the vector from a column to a row vector, and the *scalar* (or *dot* or *inner*) *product* operation $\mathbf{a}^{\mathrm{T}}\mathbf{b}$, in which the corresponding elements of two equal-dimension vectors are multiplied and summed. Even though $\mathbf{a}$ and $\mathbf{b}$ are vectors, the result is a scalar ($= a_1 b_1 + a_2 b_2 + \cdots + a_n b_n$, where n is the number of elements in each vector). The two vectors must be *conformable* for the product to exist, i.e., they must have the same number of elements.

Using MATLAB, eq. 2.2-2 would be expressed as `rmag = sqrt(r'* r)`, where the vector `r` would be defined as `[x; y; z]`. MATLAB does not use different notation for vectors and matrices (e.g., boldface or upper case for matrices). The transpose of a vector or matrix is denoted by the prime $(\cdot)'$, whereas this text uses the prime simply to denote a different version of a particular variable. Unless using the Symbolic Toolbox, all of MATLAB's computations are numerical; hence, the `x`, `y`, and `z` that define `r` normally would be numerical values.

The aircraft's inertial velocity vector $\mathbf{v}_{\mathrm{I}}$ is the time derivative of $\mathbf{r}$, $d\mathbf{r}/dt$, also symbolized as $\dot{\mathbf{r}}$:

$$\mathbf{v}_{\mathrm{I}} = \dot{\mathbf{r}} \triangleq \begin{bmatrix} v_1 \\ v_2 \\ v_3 \end{bmatrix} = \begin{bmatrix} dx/dt \\ dy/dt \\ dz/dt \end{bmatrix} = \begin{bmatrix} \dot{x} \\ \dot{y} \\ \dot{z} \end{bmatrix} \tag{2.2-3}$$

The time derivative of the vector is a vector of the time derivatives of the scalar Cartesian components. The subscript "I" indicates that the aircraft's velocity (measured relative to the inertial frame) is resolved into components that are parallel to the bases of the inertial frame of reference (Fig. 2.2-2a). The aircraft's earth-relative speed[3] is

$$V = v_I = |\mathbf{v}_I| = \sqrt{v_1^2 + v_2^2 + v_3^2} = \left[[v_1 v_2 v_3] \begin{bmatrix} v_1 \\ v_2 \\ v_3 \end{bmatrix} \right]^{1/2} = [\mathbf{v}_I^T \mathbf{v}_I]^{1/2} \qquad (2.2\text{-}4)$$

In steady, level flight, h, v_1, and v_2 are constant, and $v_3 = 0$. If the wind speed is zero, V is also the aircraft's *airspeed*, or speed relative to the surrounding air. If the wind is blowing, the simplest means of expressing the air-relative velocity vector is to add (or subtract, depending on sign conventions) the earth-relative wind velocity components to (from) v_1, v_2, and v_3.

The Cartesian description of the velocity vector can be converted to polar coordinates, with vector magnitude V, *heading angle* (from north) ξ, and *flight path angle* (from the horizontal) γ (Figure 2.2-2a):

$$\mathbf{v}_I \triangleq \begin{bmatrix} v_1 \\ v_2 \\ v_3 \end{bmatrix} = \begin{bmatrix} V\cos\gamma\cos\xi \\ V\cos\gamma\sin\xi \\ -V\sin\gamma \end{bmatrix} \qquad (2.2\text{-}5)$$

From trigonometry, the inverse relationships for the velocity vector are

$$V = \sqrt{v_1^2 + v_2^2 + v_3^2} \qquad (2.2\text{-}6)$$

$$\xi = \cos^{-1}[v_1/(v_1^2 + v_2^2)^{1/2}] = \sin^{-1}[v_2/(v_1^2 + v_2^2)^{1/2}] \qquad (2.2\text{-}7)$$

$$\gamma = \sin^{-1}(-v_3/V) \qquad (2.2\text{-}8)$$

Consequently, V is always a positive number, ξ does not depend on vertical velocity, and $\gamma = 0$ in steady, level flight. The angles γ and ξ can be depicted as arcs on a sphere surrounding the origin [K-1], as shown in Fig. 2.2-2b.

Angular Orientation and Rate

The angular orientation of the aircraft's body with respect to the (x', y', z') frame (the inertial frame translated to the aircraft's c.m.) is described by three *Euler angles* (Fig. 2.2-3), measured in degrees or radians: ψ (yaw angle), θ (pitch angle), and ϕ (roll angle). Euler angles represent an ordered set of sequential rotations from a reference frame to a frame fixed in the body. The ordering is arbitrary, but once chosen, it must be retained.

3. *Speed* is synonymous with velocity-vector *magnitude*.

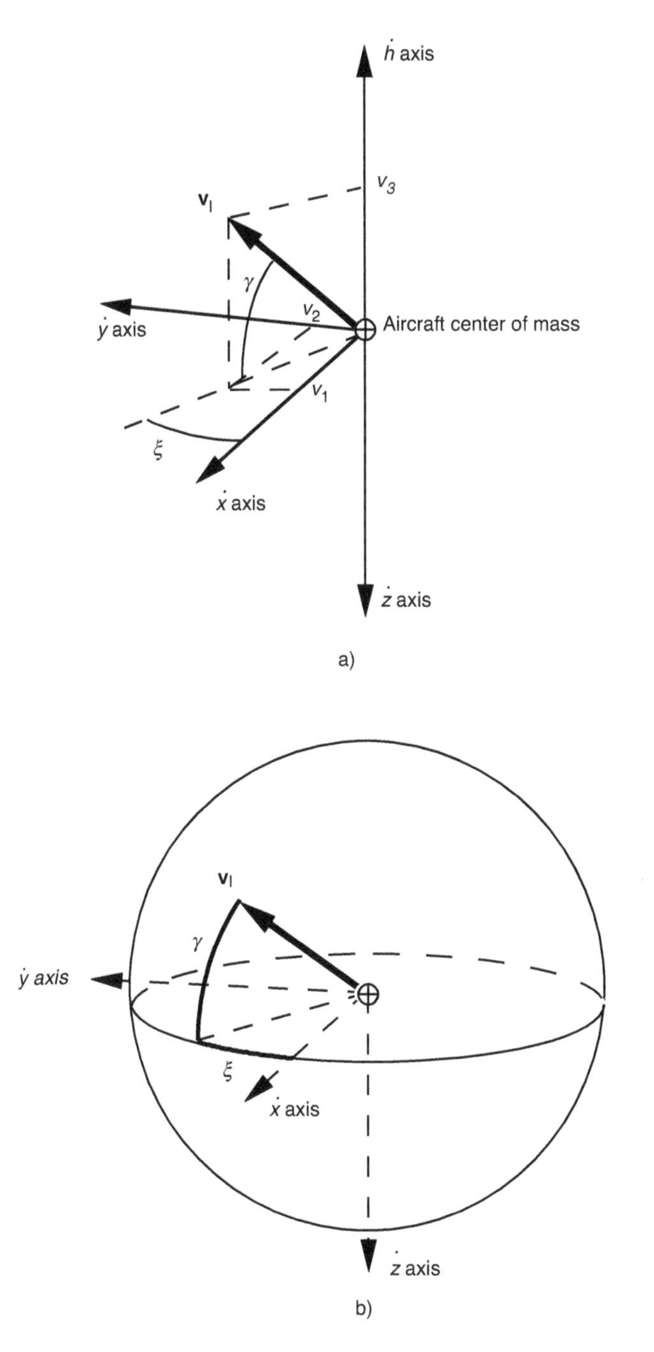

Fig. 2.2-2. Inertial velocity vector $\mathbf{v}_I$ and inertial-frame components (negative value of v_3 shown, corresponding to a positive altitude rate). (a) Definition of polar coordinates of $\mathbf{v}_I$. (b) Projection of γ and ξ on the unit sphere.

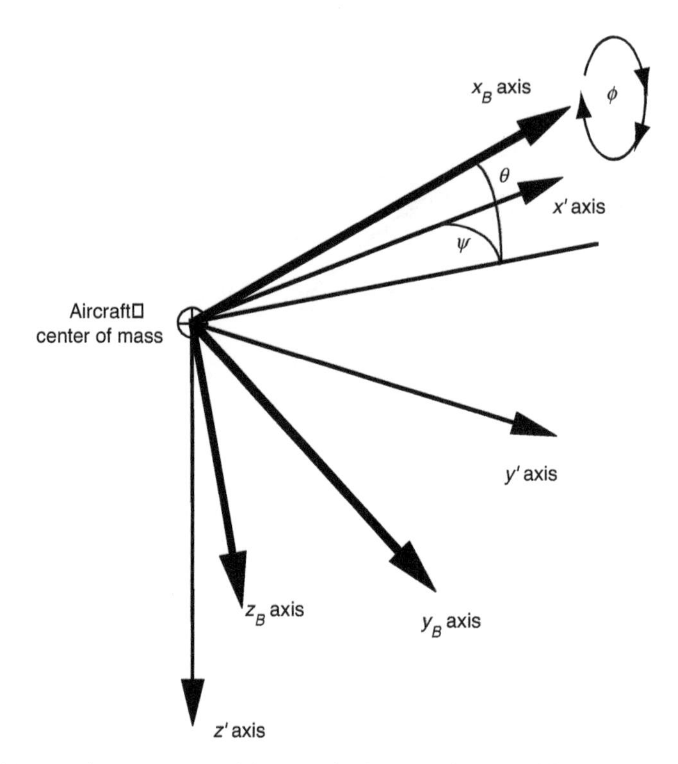

Fig. 2.2-3. Angular orientation of the aircraft relative to the inertial frame. (x_B, y_B, z_B) forms a Cartesian frame referenced to the body, with x_B running along the centerline, y_B pointing to the right, and z_B perpendicular (down) to the first two. The body-frame components of the aircraft's velocity vector (u, v, w) are parallel to (x_B, y_B, z_B).

Restricting ourselves to right-handed rotations, there are twelve possible rotational sequences. The *conventional sequence* from the inertial frame (Fig. 2.2-4a) to an aircraft's body frame begins with a right-handed *yaw rotation* ψ about the vertical (z') inertial axis, which is also the body (z_B) axis at the reference condition (Fig. 2.2-4b). This is followed by a right-handed *pitch rotation* θ about an inertial axis that is coincident with the intermediate spanwise (y_B) body axis (Fig. 2.2-4c); the y_B axis is still in the horizontal plane, but it is not the same as the y' axis unless $\psi = 0$. A right-handed *roll rotation* ϕ about an inertial axis that is coincident with the body's intermediate centerline (x_B axis) completes the sequence (Fig. 2.2-4d). The basic reference (x', y', z') directions and the body-axis (x_B, y_B, z_B) directions are aligned only when all Euler angles are zero.

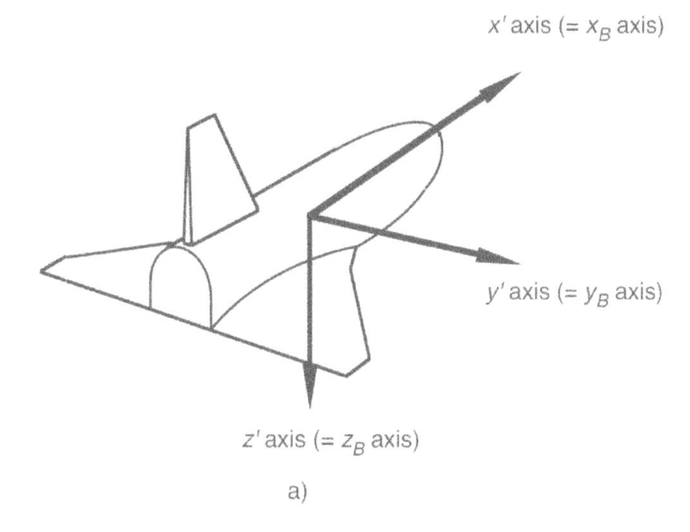

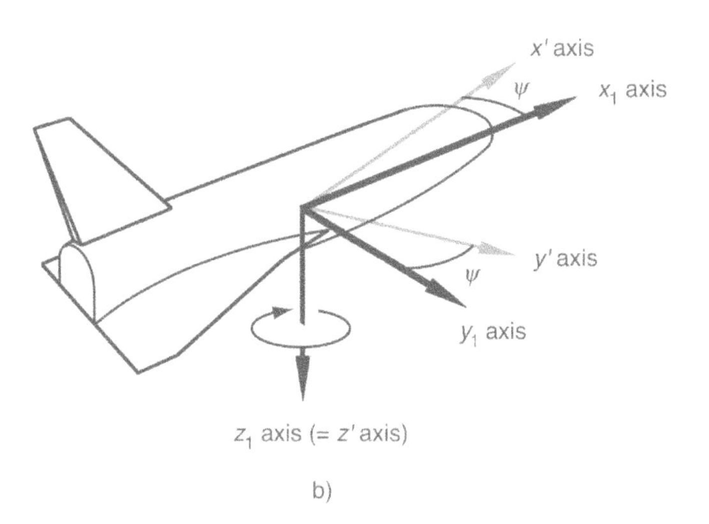

Fig. 2.2-4. Sequence of Euler-angle rotations and their effects on vector transformation. (a) Body frame aligned with inertial frame. (b) Yaw rotation from inertial frame to Intermediate Frame 1.

The three Euler angles are independent of each other, and they can be combined in the *attitude vector* **Θ**,

$$\mathbf{\Theta} \triangleq \begin{bmatrix} \phi \\ \theta \\ \psi \end{bmatrix} \tag{2.2-9}$$

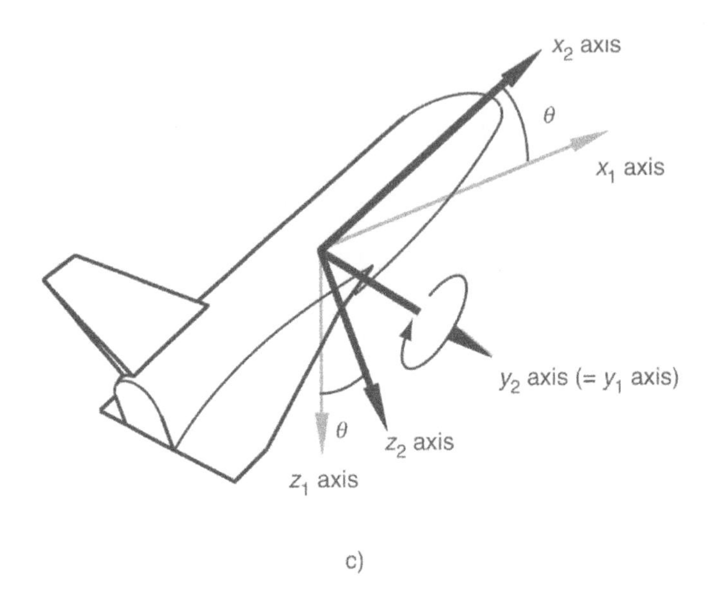

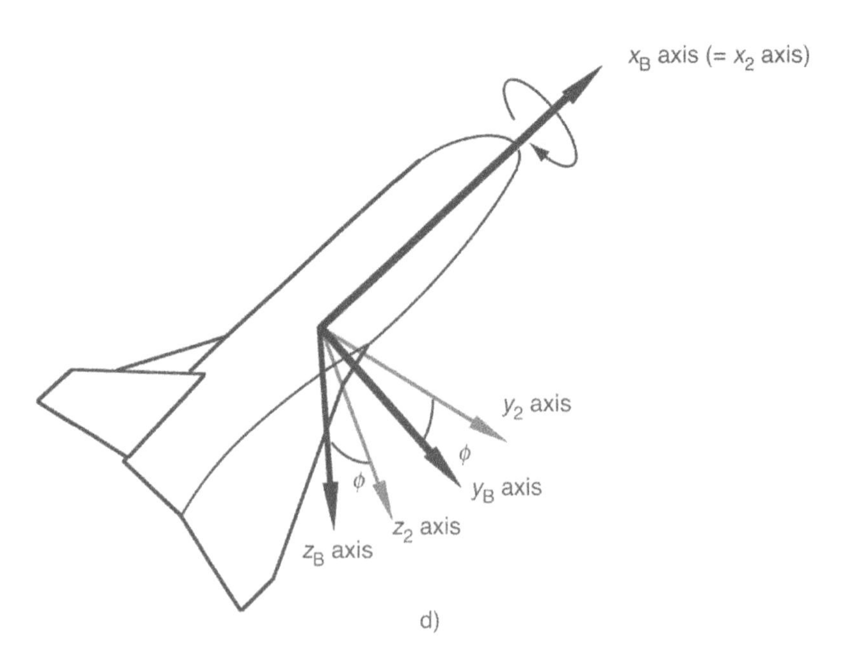

Fig. 2.2-4. *Continued.* Sequence of Euler-angle rotations and their effects on vector transformation. (c) Pitch rotation from Intermediate Frame 1 to Intermediate Frame 2. (d) Roll rotation from Intermediate Frame 2 to body axes.

where the $\Theta = 0$ reference point orients the aircraft with its x axis horizontal and pointing north and with the wings level. A sequence of 45-deg yaw-pitch-roll rotations is mapped on the unit sphere in Fig. 2.2-5; ψ and θ appear as arcs on the surface, while ϕ is represented by a shaded bar twisted away from the horizontal.

Inertial-frame vectors such as the velocity vector often must be resolved into body-fixed coordinates, and the Euler angles are central to making this transformation. $\mathbf{v}_B$ is the inertial velocity of the aircraft expressed in body-axis components:

$$\mathbf{v}_B \triangleq \begin{bmatrix} u \\ v \\ w \end{bmatrix} = \mathbf{H}_I^B \mathbf{v}_I \tag{2.2-10}$$

Here $\mathbf{v}_I$ is given by eq. 2.2-3 or 2.2-5, and $\mathbf{H}_I^B$ is a (3×3) matrix. (Note that $\mathbf{v}_B$ and $\mathbf{v}_I$ are the same vector observed in different axis systems.) More particularly, $\mathbf{H}_I^B$ is an *orthonormal transformation* from the inertial frame to the body frame. It transforms the components of a vector expressed in one orthogonal frame to the vector's components in another orthogonal frame while preserving the vector's magnitude:

$$\mathbf{H}_I^B = \begin{bmatrix} \cos\theta\cos\psi & \cos\theta\sin\psi & -\sin\theta \\ (-\cos\phi\sin\psi + \sin\phi\sin\theta\cos\psi) & (\cos\phi\cos\psi + \sin\phi\sin\theta\sin\psi) & \sin\phi\cos\theta \\ (\sin\phi\sin\psi + \cos\phi\sin\theta\cos\psi) & (-\sin\phi\cos\psi + \cos\phi\sin\theta\sin\psi) & \cos\phi\cos\theta \end{bmatrix} \tag{2.2-11}$$

The *matrix-vector product* in eq. 2.2-10 is a vector of the dot products of the matrix's rows with the column vector. Given an $(m \times n)$ matrix and an $(n \times 1)$ vector, the product is an $(m \times 1)$ vector. In this case, $m = n = 3$; therefore, u is given by

$$u = v_1 \cos\theta\cos\psi + v_2 \cos\theta\sin\psi - v_3 \sin\theta \tag{2.2-12}$$

and v and w are calculated accordingly.

The elements of $\mathbf{H}_I^B$ depend on the specified sequence of Euler angle rotations: yaw, pitch, roll. Yaw rotation from the Inertial Frame through the angle ψ to *Intermediate Frame 1* (Fig. 2.2-4b) produces the transformation matrix $\mathbf{H}_I^1$:

$$\mathbf{H}_I^1 \triangleq \begin{bmatrix} \cos\psi & \sin\psi & 0 \\ -\sin\psi & \cos\psi & 0 \\ 0 & 0 & 1 \end{bmatrix} \tag{2.2-13}$$

The first row of $\mathbf{H}_I^1$ projects the three elements of an inertial-frame vector onto the first element of a Frame 1 vector, and so on. The velocity vector resolved into Frame 1 coordinates is related to the inertial-frame expression as follows:

$$\mathbf{v}_1 \triangleq \begin{bmatrix} v_1 \\ v_2 \\ v_3 \end{bmatrix}_1 = \mathbf{H}_I^1\, \mathbf{v}_I = \mathbf{H}_I^1 \begin{bmatrix} v_1 \\ v_2 \\ v_3 \end{bmatrix}_I = \begin{bmatrix} v_1 \cos\psi + v_2 \sin\psi \\ -v_1 \sin\psi + v_2 \cos\psi \\ v_3 \end{bmatrix} \tag{2.2-14}$$

The pitch rotation from Intermediate Frame 1 through the angle θ to *Intermediate Frame 2* (Fig. 2.2-4c) produces the transformation matrix $\mathbf{H}_1^2$:

$$\mathbf{H}_1^2 \triangleq \begin{bmatrix} \cos\theta & 0 & -\sin\theta \\ 0 & 1 & 0 \\ \sin\theta & 0 & \cos\theta \end{bmatrix} \tag{2.2-15}$$

which operates on $\mathbf{v}_1$ to produce $\mathbf{v}_2$:

$$\begin{aligned} \mathbf{v}_2 &= \mathbf{H}_1^2\, \mathbf{v}_1 = \mathbf{H}_1^2 \mathbf{H}_I^1 \mathbf{v}_I \\ &= \begin{bmatrix} (v_1 \cos\psi + v_2 \sin\psi)\cos\theta - v_3 \sin\theta \\ (-v_1 \sin\psi + v_2 \cos\psi) \\ -(v_1 \cos\psi + v_2 \sin\psi)\sin\theta + v_3 \cos\theta \end{bmatrix} \end{aligned} \tag{2.2-16}$$

Alternatively, the same result is obtained through multiplying $\mathbf{v}_I$ by $\mathbf{H}_I^2$, defined as

$$\mathbf{H}_I^2 \triangleq \mathbf{H}_1^2 \mathbf{H}_I^1 = \begin{bmatrix} \cos\theta\cos\psi & \cos\theta\sin\psi & -\sin\theta \\ -\sin\psi & \cos\psi & 0 \\ -\sin\theta\cos\psi & \sin\theta\sin\psi & \cos\theta \end{bmatrix} \tag{2.2-17}$$

where the rows of $\mathbf{H}_1^2$ multiply the columns of $\mathbf{H}_I^1$ to form individual elements of the *matrix product* $\mathbf{H}_I^2$.

The rotational sequence is completed by the roll rotation from Frame 2 through the angle ϕ to the body frame (Fig. 2.2-4d). The transformation matrix $\mathbf{H}_2^B$ does the job, with

$$\mathbf{H}_2^B \triangleq \begin{bmatrix} 1 & 0 & 0 \\ 0 & \cos\phi & \sin\phi \\ 0 & -\sin\phi & \cos\phi \end{bmatrix} \tag{2.2-18}$$

$\mathbf{H}_{\mathrm{I}}^{\mathrm{B}}$ is then defined by the triple-matrix product

$$\mathbf{H}_{\mathrm{I}}^{\mathrm{B}}(\phi, \theta, \psi) = \mathbf{H}_{2}^{\mathrm{B}}(\phi)\mathbf{H}_{1}^{2}(\theta)\mathbf{H}_{\mathrm{I}}^{1}(\psi) \tag{2.2-19}$$

to yield eq. 2.2-11. A corresponding MATLAB equation involving matrix products is `HIB = H2B * H12 * HI1.`

If $\mathbf{v}_{\mathrm{B}}$ were known and it was necessary to know $\mathbf{v}_{\mathrm{I}}$, the inverse to eq. 2.2-10 could be invoked:

$$\mathbf{v}_{\mathrm{I}} = \mathbf{H}_{\mathrm{B}}^{\mathrm{I}}\mathbf{v}_{\mathrm{B}} \tag{2.2-20}$$

where $\mathbf{H}_{\mathrm{B}}^{\mathrm{I}}$ is the *matrix inverse* of $\mathbf{H}_{\mathrm{I}}^{\mathrm{B}}$, that is, $\mathbf{H}_{\mathrm{B}}^{\mathrm{I}} \triangleq (\mathbf{H}_{\mathrm{I}}^{\mathrm{B}})^{-1}$. Because $\mathbf{H}_{\mathrm{I}}^{\mathrm{B}}$ is a square, orthonormal matrix, its inverse is its *matrix transpose* $(\mathbf{H}_{\mathrm{I}}^{\mathrm{B}})^{T}$, formed by interchanging the rows and columns of the original matrix:

$$\mathbf{H}^{T} = \begin{bmatrix} h_{11} & h_{12} & h_{13} \\ h_{21} & h_{22} & h_{23} \\ h_{31} & h_{32} & h_{33} \end{bmatrix}^{T} \triangleq \begin{bmatrix} h_{11} & h_{21} & h_{31} \\ h_{12} & h_{22} & h_{32} \\ h_{13} & h_{23} & h_{33} \end{bmatrix} \tag{2.2-21}$$

The *Euler-angle rate vector* $\dot{\mathbf{\Theta}}$ is simply the term-by-term time derivative of the Euler-angle vector,

$$\dot{\mathbf{\Theta}} \triangleq \begin{bmatrix} \dot{\phi} \\ \dot{\theta} \\ \dot{\psi} \end{bmatrix} \tag{2.2-22}$$

Like the Euler-angle vector, the components of the Euler-angle rate vector are not measured along orthogonal axes. Unlike the translational velocity vector, transforming it into different axis systems is a complicated process. Nevertheless, it *is* possible to express the angular rate as an orthogonal vector; for example, the inertial angular rates measured about the body's (x_B, y_B, z_B) axes are defined by

$$\boldsymbol{\omega}_{\mathrm{B}} = \begin{bmatrix} p \\ q \\ r \end{bmatrix} \tag{2.2-23}$$

The inertial angular rate vector observed in the body-axis frame is related to the Euler angle rate vector by the nonorthogonal transformation $\boldsymbol{\omega}_{\mathrm{B}} = \mathbf{L}_{\mathrm{E}}^{\mathrm{B}}\dot{\mathbf{\Theta}}$, where

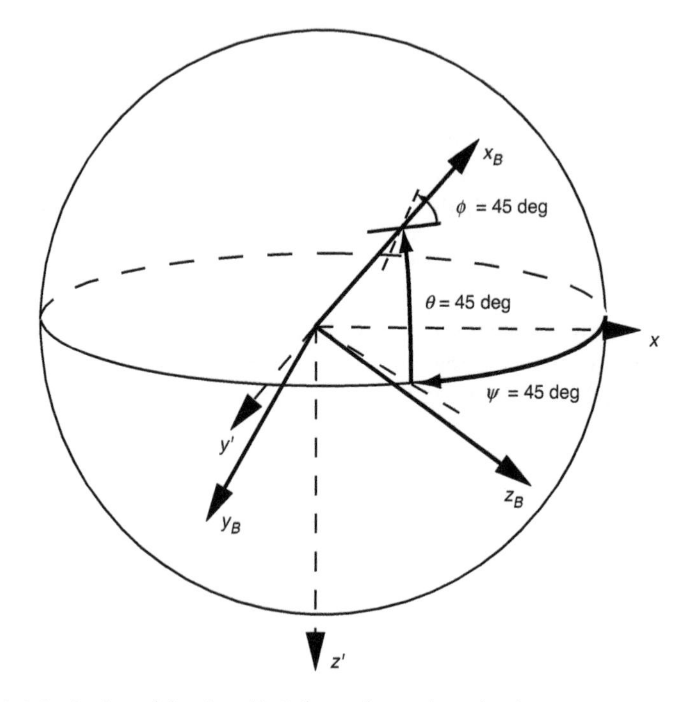

Fig. 2.2-5. Projection of the aircraft's Euler angles on the unit sphere.

$$\mathbf{L}_{\mathrm{E}}^{\mathrm{B}} \triangleq \begin{bmatrix} 1 & 0 & -\sin\theta \\ 0 & \cos\phi & \sin\phi\cos\theta \\ 0 & -\sin\phi & \cos\phi\cos\theta \end{bmatrix} \tag{2.2-24}$$

$\mathbf{L}_{\mathrm{E}}^{\mathrm{B}}$ represents separate transformations of the three Euler angle rate components by three different orthonormal matrices, the first of which is an identity matrix[4]:

$$\begin{bmatrix} p \\ q \\ r \end{bmatrix} = \mathbf{I}_n \begin{bmatrix} \dot{\phi} \\ 0 \\ 0 \end{bmatrix} + \mathbf{H}_2^{\mathrm{B}} \begin{bmatrix} 0 \\ \dot{\theta} \\ 0 \end{bmatrix} + \mathbf{H}_2^{\mathrm{B}}\mathbf{H}_1^2 \begin{bmatrix} 0 \\ 0 \\ \dot{\psi} \end{bmatrix} \tag{2.2-25}$$

4. The *identity matrix* $\mathbf{I}_n$ is an $(n \times n)$ matrix containing 1's on its principal diagonal (from upper left to lower right corners) and zeros elsewhere. Multiplying a vector by $\mathbf{I}_n$ does not change the vector in any way; hence, it provides an orthonormal transformation by default.

Only the first column of $\mathbf{I}_n$, the second column of $\mathbf{H}_2^{\mathrm{B}}$, and the third column of $\mathbf{H}_2^{\mathrm{B}}\mathbf{H}_1^2$ are used to define $\mathbf{L}_{\mathrm{E}}^{\mathrm{B}}$, as the remaining columns multiply zero vector elements in eq. 2.2-25.

The individual transformations are orthonormal, but the complete transformation is not; hence, the inverse of $\mathbf{L}_{\mathrm{E}}^{\mathrm{B}}$ is not simply its transpose. It is

$$\mathbf{L}_{\mathrm{B}}^{\mathrm{E}} \triangleq (\mathbf{L}_{\mathrm{E}}^{\mathrm{B}})^{-1} = \begin{bmatrix} 1 & \sin\phi\tan\theta & \cos\phi\tan\theta \\ 0 & \cos\phi & -\sin\phi \\ 0 & \sin\phi\sec\theta & \cos\phi\sec\theta \end{bmatrix} \tag{2.2-26}$$

where $\mathbf{A}^{-1} = \mathrm{Adj}(\mathbf{A})/\mathrm{Det}(\mathbf{A})$, as detailed below. The orthonormal transformations presented earlier do not have singularities, but $\mathbf{L}_{\mathrm{B}}^{\mathrm{E}}$ has four infinite elements when $\theta = \pm 90$ deg, that is, when the airplane's nose is pointed straight up or down.

The *inverse* of an $(n \times n)$ matrix is the $(n \times n)$ adjoint of the matrix divided by the scalar matrix determinant. The *determinant* is the sum of various permutations of products of the matrix elements. For $n = 3$,

$$\begin{aligned}\mathrm{Det}(\mathbf{A}) = |\mathbf{A}| &= a_{11}a_{22}a_{33} + a_{12}a_{23}a_{31} + a_{13}a_{21}a_{32} \\ &\quad - a_{11}a_{23}a_{32} - a_{12}a_{21}a_{33} - a_{13}a_{22}a_{31}\end{aligned} \tag{2.2-27}$$

For $n > 3$, Laplace expansion or pivotal condensation can be used to find the determinant (e.g., [S-17]). The *adjoint matrix* is the transpose of the matrix of cofactors of the original matrix,

$$\mathrm{Adj}(\mathbf{A}) = \mathbf{C}^T \tag{2.2-28}$$

where the *cofactors* are signed minors of the original matrix. The *minor* of element a_{ij} is the determinant of the matrix formed by the elimination of the ith row and jth column of $\mathbf{A}$; the cofactor's sign is positive if $(i + j)$ is even, negative if $(i + j)$ is odd. For $n = 3$,

$$\mathrm{Adj}(\mathbf{A}) = \begin{bmatrix} (a_{22}a_{33} - a_{23}a_{32}) & (a_{13}a_{32} - a_{12}a_{33}) & (a_{12}a_{23} - a_{13}a_{22}) \\ (a_{23}a_{31} - a_{21}a_{33}) & (a_{11}a_{33} - a_{13}a_{31}) & (a_{13}a_{21} - a_{11}a_{23}) \\ (a_{21}a_{32} - a_{22}a_{31}) & (a_{12}a_{31} - a_{11}a_{32}) & (a_{11}a_{22} - a_{12}a_{21}) \end{bmatrix} \tag{2.2-29}$$

This symbolic representation of the inverse is useful in later chapters; however, if only numerical values are required, MATLAB calculates the inverse and determinant as `inv(A)` and `det(A)`.

$\boldsymbol{\omega}_{\mathrm{B}}$ can be transformed to an orthogonal angular rate vector in the inertial frame, $\boldsymbol{\omega}_{\mathrm{I}} = \mathbf{H}_{\mathrm{B}}^{\mathrm{I}}\boldsymbol{\omega}_{\mathrm{B}}$, where the elements of $\boldsymbol{\omega}_{\mathrm{I}}$ are measured about the orthogonal inertial axes; therefore $\boldsymbol{\omega}_{\mathrm{I}}$ does not equal $\dot{\boldsymbol{\Theta}}$. The inertial angular rate vector $\boldsymbol{\omega}_{\mathrm{I}}$ has use in the description of steady, coordinated turns (Section 2.6).

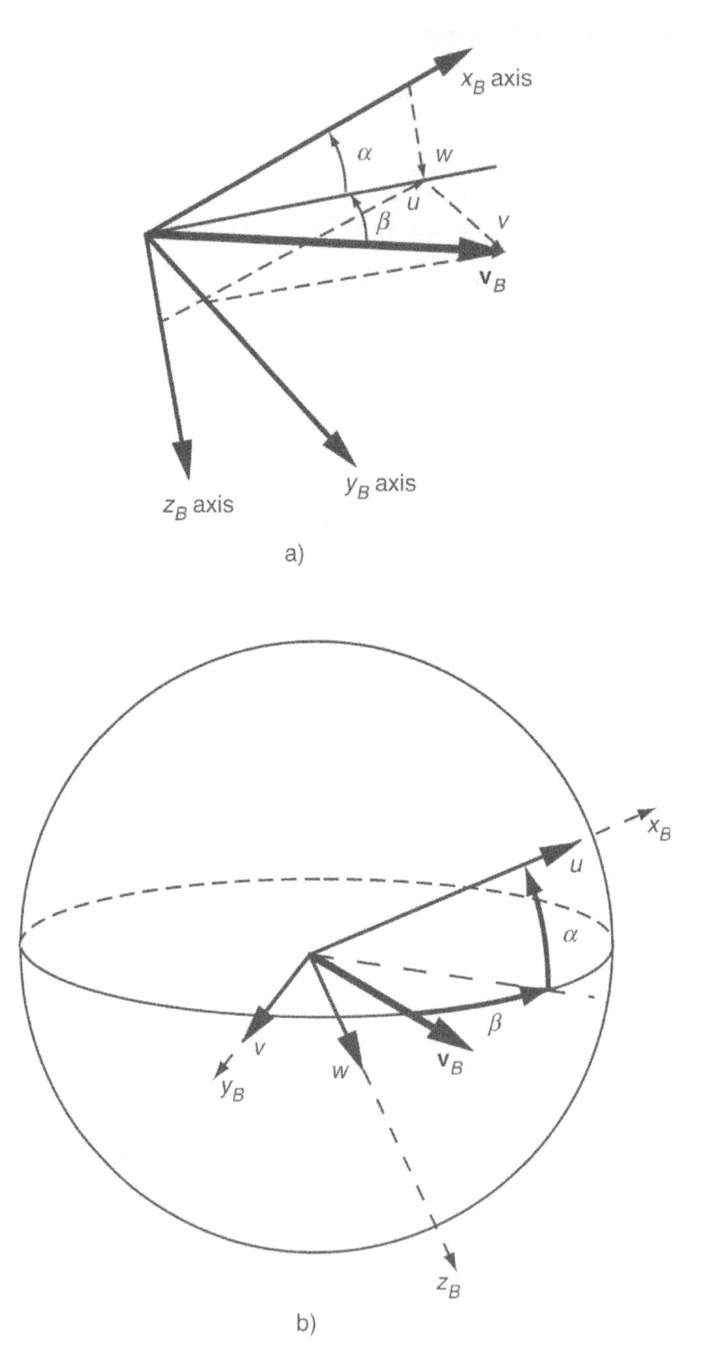

Fig. 2.2-6. Relationships between inertial velocity ($\mathbf{v}_B$), body-frame components (u, v, w), and airflow angles (α, β). The magnitude of $\mathbf{v}_B$ is V, assuming that the wind speed is zero. (a) Definition of polar coordinates of $\mathbf{v}_B$. (b) Projection of α and β on the unit sphere.

AIRFLOW ANGLES

The orientation of the velocity vector with respect to body axes plays an important role in determining the aircraft's aerodynamic forces and moments. Two angles express this orientation: the *sideslip angle* β and the *angle of attack* α (Fig. 2.2-6). They are defined such that

$$\begin{bmatrix} u \\ v \\ w \end{bmatrix} = \begin{bmatrix} V\cos\alpha\cos\beta \\ V\sin\beta \\ V\sin\alpha\cos\beta \end{bmatrix} \tag{2.2-30}$$

Therefore, a little trigonometry yields the inverse relationships:

$$\begin{bmatrix} V \\ \beta \\ \alpha \end{bmatrix} = \begin{bmatrix} \sqrt{u^2 + v^2 + w^2} \\ \sin^{-1}(v/V) \\ \tan^{-1}(w/u) \end{bmatrix} \tag{2.2-31}$$

Equations 2.2-6 and 2.2-31 give alternative definitions for airspeed V when the wind speed is zero. When the wind speed is not zero, it is important to distinguish between air-relative and earth-relative velocity vectors, as discussed in the next chapter.

Vectors expressed in wind axes can be represented in body axes, and vice versa; hence, another transformation matrix is needed. Initially aligning the aircraft with the velocity vector, the wind-relative attitude is defined by a *left-handed* rotation through the sideslip angle β, then a right-handed rotation through the angle of attack α. The transformation matrix $\mathbf{H}_W^B$ is thus defined as

$$\mathbf{H}_W^B \triangleq \mathbf{H}_3^B(\alpha)\mathbf{H}_W^3(-\beta) = \begin{bmatrix} \cos\alpha & 0 & -\sin\alpha \\ 0 & 1 & 0 \\ \sin\alpha & 0 & \cos\alpha \end{bmatrix} \begin{bmatrix} \cos\beta & -\sin\beta & 0 \\ \sin\beta & \cos\beta & 0 \\ 0 & 0 & 1 \end{bmatrix}$$

$$= \begin{bmatrix} \cos\alpha\cos\beta & -\cos\alpha\sin\beta & -\sin\alpha \\ \sin\beta & \cos\beta & 0 \\ \sin\alpha\cos\beta & -\sin\alpha\sin\beta & \cos\alpha \end{bmatrix} \tag{2.2-32}$$

SUMMARY OF AXIS SYSTEMS AND TRANSFORMATIONS

Four axis systems have value in the dynamic analysis of aircraft:

- Inertial axes
- Body axes

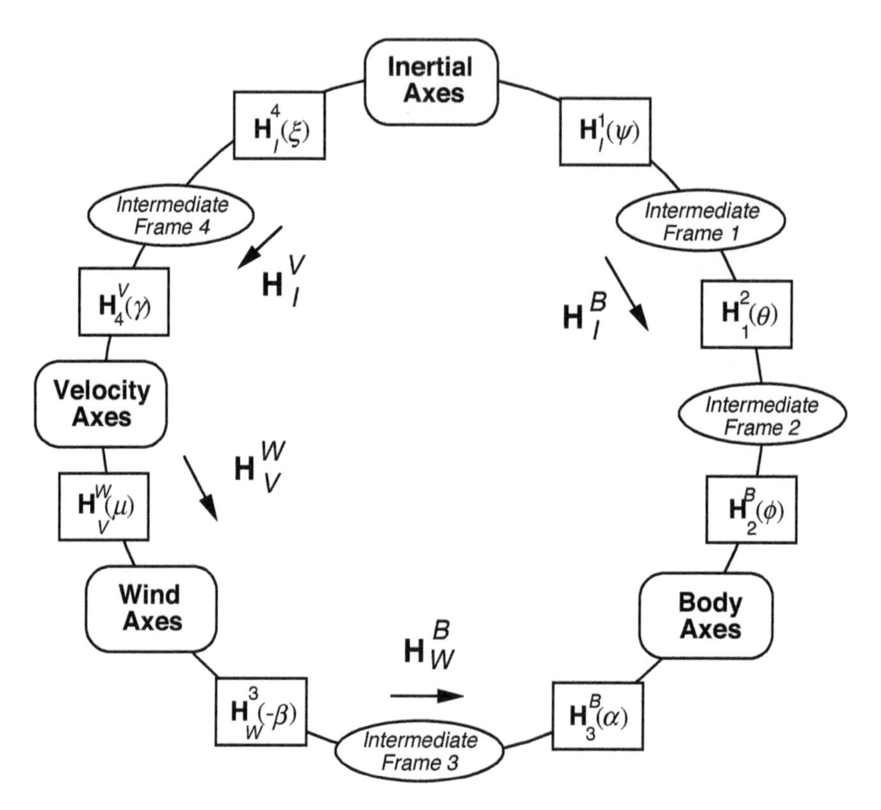

Fig. 2.2-7. Reference-axis transformations. Arrows indicate right-hand rotations. (adapted from [S-15]).

- Wind axes
- Velocity axes

Figure 2.2-7, adapted from [S-15], illustrates that each axis system is related to the others by a series of orthonormal transformations. Inertial-frame vectors are transformed to the body frame by $\mathbf{H}_I^B$ (eq. 2.2-11). Vectors expressed in the body frame are transformed to the wind-axis frame by $\mathbf{H}_B^W$, the inverse (and transpose) of eq. 2.2-32. Thus, a transformation from inertial to wind axes can be established through the body frame of reference as $\mathbf{H}_I^W \triangleq \mathbf{H}_B^W(-\beta, \alpha)\mathbf{H}_I^B(\phi, \theta, \psi)$.

Alternatively, $\mathbf{H}_I^W$ could be written as a function of the vertical and horizontal flight path angles γ and ξ as well as the *bank angle* μ. In typical flight conditions, the bank angle is very nearly the same as the roll angle; however, the former is measured about the velocity vector while the latter is measured about the body x axis. Using eq. 2.2-5 and Fig. 2.2-2, the transformation from inertial to velocity axes is

$$\mathbf{H}_I^V \triangleq \mathbf{H}_4^V(\gamma)\mathbf{H}_I^4(\xi) = \begin{bmatrix} \cos\gamma & 0 & -\sin\gamma \\ 0 & 1 & 0 \\ \sin\gamma & 0 & \cos\gamma \end{bmatrix} \begin{bmatrix} \cos\xi & \sin\xi & 0 \\ -\sin\xi & \cos\xi & 0 \\ 0 & 0 & 1 \end{bmatrix} \tag{2.2-33}$$

Transformation through the bank angle rotation completes the alternate route,

$$\mathbf{H}_I^W(\mu, \gamma, \xi) \triangleq \mathbf{H}_V^W(\mu)\mathbf{H}_I^V(\gamma, \xi) \tag{2.2-34}$$

where

$$\mathbf{H}_V^W \triangleq \begin{bmatrix} 1 & 0 & 0 \\ 0 & \cos\mu & \sin\mu \\ 0 & -\sin\mu & \cos\mu \end{bmatrix} \tag{2.2-35}$$

The various rotations are brought together by their projections on the unit sphere (Fig. 2.2-8), which relate the inertial x' direction to the vehicle's centerline direction and to the direction of travel. Note in particular the distinction between the roll angle ϕ (about the centerline) and the bank angle μ (about the velocity vector). Using spherical trigonometry, the bank angle can be derived from the remaining flight path and aerodynamic angles [K-1]:

$$\mu = \sin^{-1}\left\{\frac{[\cos\theta\sin\phi\cos\beta + (\cos\alpha\sin\theta - \sin\alpha\cos\theta\cos\phi)\sin\beta]}{\cos\gamma}\right\} \tag{2.2-36a}$$

or

$$\mu = \cos^{-1}\left[\frac{\cos\alpha\cos\theta\cos\phi + \sin\alpha\sin\theta}{\cos\gamma}\right] \tag{2.2-36b}$$

These equations can also be derived by equating the (3, 2) and (3, 3) elements of

$$\mathbf{H}_W^V = \mathbf{H}_I^V\mathbf{H}_B^I\mathbf{H}_W^B \tag{2.2-37}$$

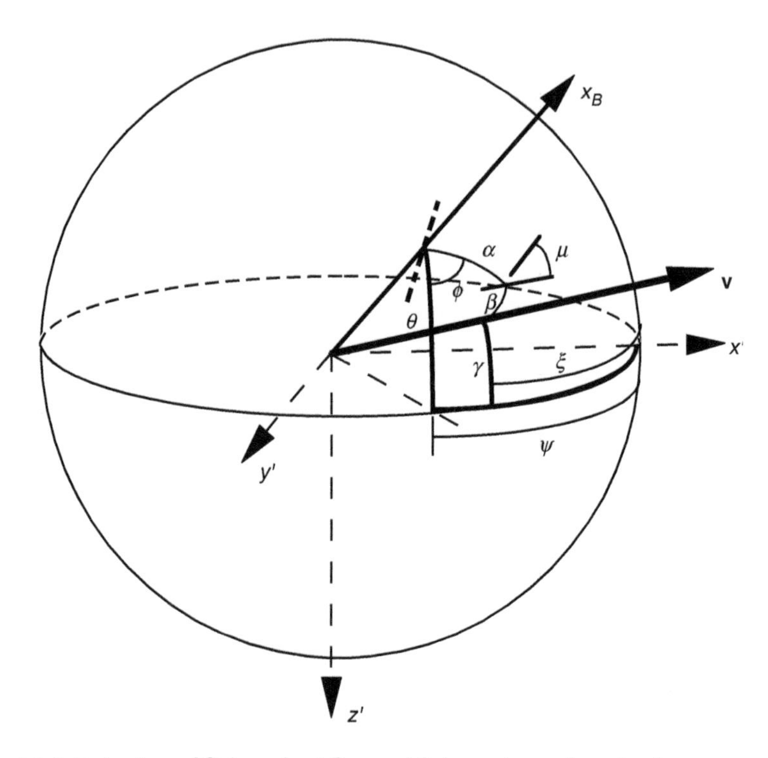

Fig. 2.2-8. Projections of flight path, airflow, and Euler angles on the unit sphere.

With the wing tips level, there are two simple relationships among the flight-path, aerodynamic, and body-attitude angles (Fig. 2.2-9). In the vertical plane, the flight-path angle is the difference between the pitch angle and the angle of attack:

$$\gamma = \theta - \alpha \tag{2.2-38}$$

In the horizontal plane and with horizontal velocity, the flight-path (headings) angle is the sum of the yaw angle and the sideslip angle:

$$\xi = \psi + \beta \tag{2.2-39}$$

Example 2.2-1. Flight Path and Aerodynamic Angles of a Jet Transport

Consider a jet transport that is maneuvering from level flight into a descending banked turn. It is flying in still air, with instantaneous inertial

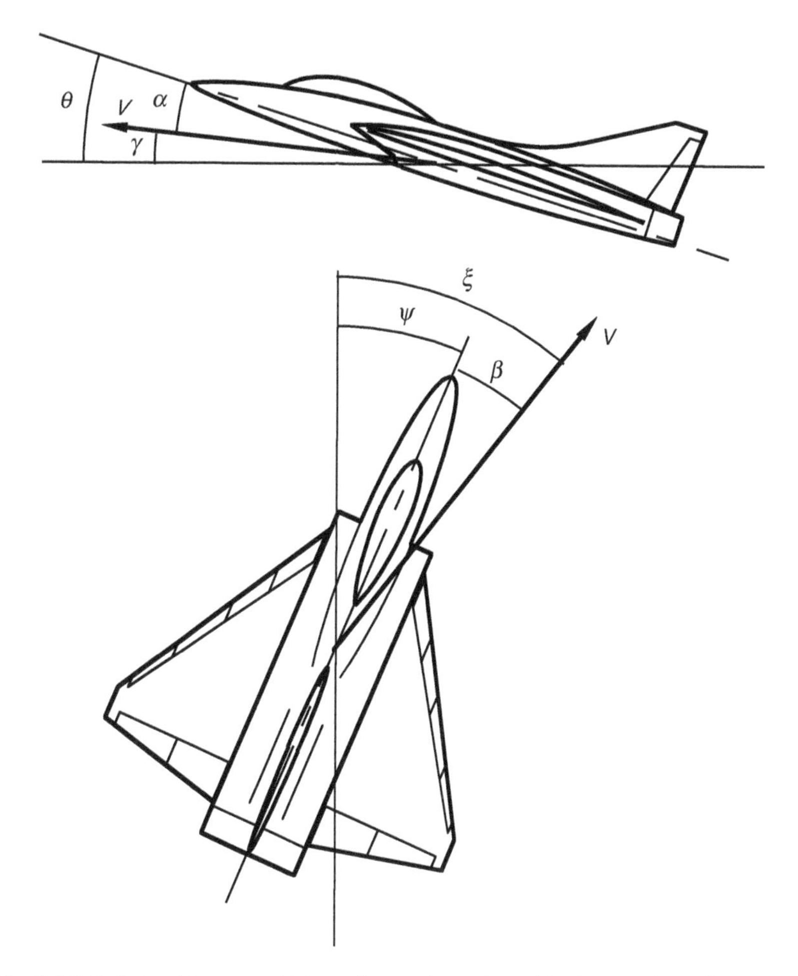

Fig. 2.2-9. Relationship among flight path, aerodynamic, and body-attitude angles with level wing tips.

velocity components and Euler angles

$$\mathbf{v}_I = \begin{bmatrix} 200 \\ 10 \\ 5 \end{bmatrix} \text{m/s}; \quad \Theta = \begin{bmatrix} 10 \\ 10 \\ 2 \end{bmatrix} \text{deg}$$

The flight path and aerodynamic angles at this point on the aircraft's trajectory are calculated by the following MATLAB code:

```
vi   = [200; 10; 5];         % Inertial velocity in inertial frame
dr   = pi / 180;             % Degrees-to-radian conversion
th   = [10 * dr;10 * dr;2 * dr]
                             % Euler angles entered as a column vector
V    = sqrt(vi' * vi)        % Airspeed (m/s)
gam  = asin(-vi(3) / V);     % Vertical flight path angle (rad)
gamd = gam / dr              % Vertical flight path angle (deg)
xi   = asin(vi(2) / sqrt(vi(1)^2 + vi(2)^2));
                             % Horizontal flight path angle (rad).
xid  = xi / dr               % Horizontal flight path angle (deg)
HI1  = [cos(th(3)) sin(th(3)) 0; -sin(th(3)) cos(th(3)) 0; 0 0 1]
                             % Yaw transformation matrix
H12  = [cos(th(2)) 0 -sin(th(2)); 0 1 0; sin(th(2)) 0 cos(th(2))];
                             % Pitch transformation matrix
H2B  = [1 0 0; 0 cos(th(1)) sin(th(1)); 0 -sin(th(1)) cos(th(1))];
                             % Roll transformation matrix
HIB  = H2B * H12 * HI1       % Inertial-to-body-axis transformation
vb   = HIB * vi              % Inertial velocity, body-axis frame
be   = asin(vb(2) / V);      % Sideslip angle (rad)
bed  = be / dr               % Sideslip angle (deg)
al   = atan(vb(3) / vb(1));  % Angle of attack (rad)
ald  = al / dr               % Angle of attack (deg)
mu   = asin(cos(th(2)) * sin(th(1)) * cos(be) + (cos(al) * sin(th(2)) ..
          - sin(al) * cos(th(2)) * cos(th(1))) * sin(be)/cos(gam));
                             % Bank angle (rad)
mud  = mu / dr               % Bank angle (deg)
```

The MATLAB m-file output is

```
th   =
          0.1745
          0.1745
          0.0349
V    =
          200.3123
gamd =
          -1.4303
xid  =
          2.864
HI1  =
          0.9994     0.0349     0
          -0.0349    0.9994     0
          0          0          1
HIB  =
          0.9842     0.0344     -0.1736
          -0.0042    0.9853     0.1710
          0.1770     -0.1676    0.9698
vb   =
          196.3170
          9.8609
          38.5667
bed  =
          2.8217
ald  =
          11.1143
mud  =
          9.7902
```

Example 2.2-2. Angular Rates of the Jet Transport

Suppose that the aircraft's body-axis angular rates in the previous example are

$$\omega_B = \begin{bmatrix} 10 \\ 4 \\ 2 \end{bmatrix} \text{deg/s}$$

What are the corresponding Euler-angle rates?

Equation 2.2-26 could be used to write L_E^B directly; then $\dot{\Theta} = L_E^B \omega_B$. However, for illustration, L_B^E is calculated using eq. 2.2-24, and the inverse is evaluated numerically by MATLAB:

m file

```
dr   =   pi / 180;
th   =   [10 * dr;10 * dr;2 * dr];
om   =   dr * [10; 4; 2]; % Body-axis rates converted to radians
LEB  =   [1 0 -sin(th(2)); 0 cos(th(1)) sin(th(1)) * cos(th(2)); ...
         0 -sin(th(1)) cos(th(1)) * cos(th(2))]
                                % Euler-rate-to-body-rate transformation
LBE  =   inv(LEB)               % Inverse of LEB matrix
eul  =   LBE * om / dr          % Euler-angle rates (deg)
```

Output

```
LEB  =
         1.0000     0          -0.1736
         0          0.9848     0.1710
         0          -0.1736    0.9698
LBE  =
         1.0000     0.0306     0.1736
         0          0.9848     -0.1736
         0          0.1763     1.0000
eul  =
         10.4698
         3.5919
         2.7053
```

2.3 Forces and Moments

A *force* is a physical influence that changes the translational motion of an unrestrained object, such as a flying aircraft. A *moment* (or *torque*) is the result of a force being applied at right angles to a *moment arm* that defines the distance from a rotational center to the force. The forces and moments acting on the aircraft are three-dimensional orthogonal vectors **f** and **m** that can be expressed in alternate frames according to their use. Thus, a force expressed in an inertial frame is

$$\mathbf{f}_I \triangleq \begin{bmatrix} X \\ Y \\ Z \end{bmatrix}_I \tag{2.3-1}$$

while the moment of that force expressed in the same frame is

$$\mathbf{m}_I \triangleq \mathbf{r}_I \times \mathbf{f}_I \triangleq \tilde{\mathbf{r}}_I \mathbf{f}_I = \begin{bmatrix} 0 & -z & y \\ z & 0 & -x \\ -y & x & 0 \end{bmatrix}_I \begin{bmatrix} X \\ Y \\ Z \end{bmatrix}_I = \begin{bmatrix} -zY + yZ \\ zX - xZ \\ -yX + xY \end{bmatrix}_I \tag{2.3-2}$$

$\mathbf{r}_I \triangleq [x\ y\ z]_I^T$ is the moment-arm vector between the rotational center and the point of force application. $\mathbf{r}_I \times \mathbf{f}_I$ is the *cross product* of the two vectors $\mathbf{r}_I$ and $\mathbf{f}_I$, and $\tilde{\mathbf{r}}_I$ is the *cross-product-equivalent matrix* of the moment-arm vector. The units of force are newtons,[5] so newton-meters are appropriate units for moments. The MATLAB function `cross` produces the numerical result: `mI=cross (rI, fI)`.

Alternative Axis Systems

Forces and moments expressed in noninertial frames can be referred to an inertial frame using the orthonormal transformations of Section 2.2. *Drag* is a force component that is parallel and opposed to the velocity vector, while *lift* is a force component perpendicular to the velocity vector and to the y_B axis; thus, these force components are wind-axis quantities. Yawing, pitching, and rolling moments are most directly associated with the body axes.

Consider the mapping of forces from wind to inertial axes in a two-step process. The drag (D), side force (SF), and lift (L) transform into body-axis axial, lateral, and vertical aerodynamic force components (X, Y, Z):

$$\mathbf{f}(\text{aero})_B = \begin{bmatrix} X \\ Y \\ Z \end{bmatrix}_B = \mathbf{H}_W^B \begin{bmatrix} -D \\ SF \\ -L \end{bmatrix} = \mathbf{H}_W^B \mathbf{f}(\text{aero})_W \tag{2.3-3}$$

5. One *newton* is the force required to accelerate one kilogram of mass at one meter per second per second.

The minus signs account for normal conventions (drag is positive aft, side force is positive to the right, and lift is positive up). Inertial-frame forces can then be written as

$$\mathbf{f}(\text{aero})_{\text{I}} = \begin{bmatrix} X \\ Y \\ Z \end{bmatrix}_{\text{I}} = \mathbf{H}_{\text{B}}^{\text{I}} \begin{bmatrix} X \\ Y \\ Z \end{bmatrix}_{\text{B}} = \mathbf{H}_{\text{B}}^{\text{I}} \mathbf{f}(\text{aero})_{\text{B}} = \mathbf{H}_{\text{W}}^{\text{I}} \begin{bmatrix} -D \\ SF \\ -L \end{bmatrix} \quad (2.3\text{-}4)$$

Body-axis roll, pitch, and yaw moments (L, M, N) can be transformed to the inertial frame, as needed for the examination of steady turns in Section 2.3:

$$\mathbf{m}(\text{aero})_{\text{I}} = \begin{bmatrix} L \\ M \\ N \end{bmatrix}_{\text{I}} = \mathbf{H}_{\text{B}}^{\text{I}} \begin{bmatrix} L \\ M \\ N \end{bmatrix}_{\text{B}} = \mathbf{H}_{\text{B}}^{\text{I}} \mathbf{m}(\text{aero})_{\text{B}} \quad (2.3\text{-}5)$$

The accepted symbols for lift force and roll moment are, unfortunately, the same; however, they can usually be differentiated by usage.

Aerodynamic Forces and Moments

The scalar force and moment components can be written as products of nondimensional coefficients and dimensionalizing quantities. Nondimensional coefficients facilitate the comparison of aerodynamic data for various aircraft configurations, and they allow subscale model wind tunnel data to be applied to full-scale aircraft. The dimensionalizing quantities for forces are the *dynamic pressure* $\overline{q}$ (n/m^2) and a *reference area* S(m^2). Thus, the drag, side force, and lift can be expressed as

$$D = C_D \overline{q} S \quad (2.3\text{-}6)$$

$$SF = C_{SF} \overline{q} S \quad (2.3\text{-}7)$$

$$L = C_L \overline{q} S \quad (2.3\text{-}8)$$

The reference area is normally the aircraft's wing area, while the dynamic pressure is

$$\overline{q} \triangleq \frac{\rho V_A^2}{2} \quad (2.3\text{-}9)$$

where ρ, the air density (kg/m^3), is a function of altitude. V_A is the *true airspeed* (or TAS), which equals the earth-relative speed V when the wind is zero.

Three additional definitions of airspeed can be found [K-2]. The *indicated airspeed* (IAS) is derived from the difference in stagnation and static pressure measurements made by a pitot tube and a static pressure port. (Static pressure also defines *pressure altitude*.) The *calibrated airspeed* (CAS) is the IAS corrected for instrumentation and position errors. The *equivalent airspeed* (EAS) is the CAS corrected for compressibility effects on the measurements. For a given dynamic pressure, TAS changes with altitude, while EAS is constant. Thus, EAS and TAS are the same at sea level, and $\text{EAS} = \text{TAS}\sqrt{\rho(z)\rho_{\text{SL}}}$ at altitude z. In common usage, the EAS and CAS are often called IAS.

Aerodynamic force coefficients are sensitive to the airplane's airflow angles. At low airflow angles and zero angular rates, the total lift coefficient due to the wing, fuselage, and horizontal tail is approximately linear in angle of attack:

$$C_L = C_{L_o} + C_{L_\alpha}\alpha \tag{2.3-10}$$

where the coefficients depend on the aircraft's configuration—notably the wing planform, chord section, and twist [A-6]—as well as the flight condition and control settings (Section 2.4). C_{L_o} is the lift coefficient at zero angle of attack, and the *lift slope* C_{L_α} is the partial derivative of the lift coefficient, evaluated at zero angle of attack:

$$C_{L_\alpha} \triangleq \left.\frac{\partial C_L}{\partial \alpha}\right|_{\alpha = 0^\circ} \tag{2.3-11}$$

At high α, the wing stalls, limiting the lift coefficient to a maximum value, and this linear lift characteristic breaks down (Fig. 2.3-1). The *angle of attack for zero lift, α_{zl}*, is

$$\alpha_{zl} = -C_{L_o}/C_{L\alpha} \tag{2.3-12}$$

C_{L_o} is usually positive, and α_{zl} is typically less than zero.

Similarly, the side-force coefficient can be written as a function of sideslip angle,

$$C_{SF} = C_{SF_o} + C_{SF_\beta}\,\beta \tag{2.3-13}$$

For most aircraft and intended flight conditions, side-force coefficients and derivatives are small.

The drag coefficient is very nearly proportional to the square of the lift coefficient at low airflow angles (Fig. 2.3-1):

$$C_D = C_{D_o} + \varepsilon C_L^2 \tag{2.3-14a}$$

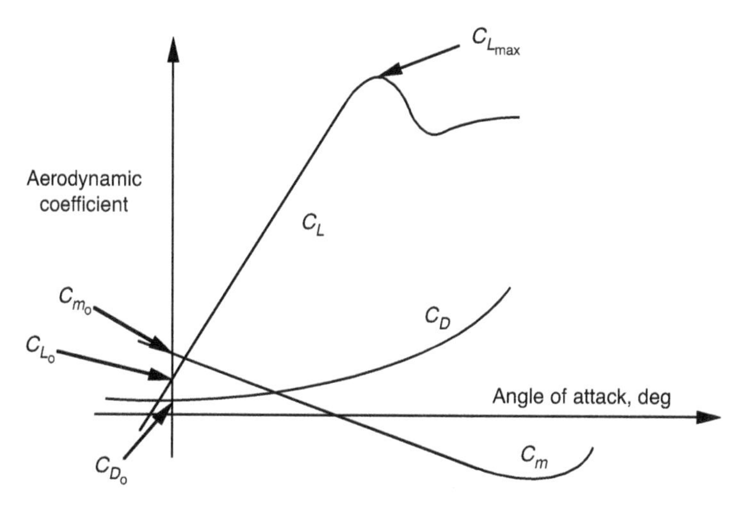

Fig. 2.3-1. Typical variations of lift coefficient C_L, drag coefficient C_D, pitching moment coefficient C_m, and lift-drag ratio C_L/C_D with angle of attack α.

where ε is an *induced drag factor* resulting from the efficiency with which lift is created [A-3]. From eq. 2.3-10, this leads to

$$C_D = (C_{D_o} + \varepsilon C_{L_o}^2) + C_{D_\alpha} + C_{D_{\alpha^2}}\alpha^2 \tag{2.3-14b}$$

where

$$C_{D_\alpha} \triangleq \left.\frac{\partial C_D}{\partial \alpha}\right|_{\alpha = 0°} = 2\varepsilon C_{L_o} C_{L_\alpha} \tag{2.3-15}$$

$$C_{D_{\alpha^2}} \triangleq \left.\frac{\partial C_D}{\partial(\alpha^2)}\right|_{\alpha = 0°} = \varepsilon C_{L_{\alpha^2}} \tag{2.3-16}$$

The *lift-drag ratio* C_L/C_D or *L/D* is a measure of aerodynamic efficiency that varies with α. Lift is required to support weight, drag is produced in the process, and thrust is required to oppose drag. It is normally desirable to maximize this ratio, thus minimizing the thrust and fuel used to offset drag (Section 2.6). The maximum *L/D* ranges from 2 or less for reentering space vehicles to over 50 for competition sailplanes.

Lift and drag coefficients are typically tabulated as functions of several motion and control variables, such as

$$C_L = C_L(\alpha, M, \delta F) + C_{L_{\hat{q}}}(\alpha, M, \delta F)\hat{q} + C_{L_{\delta E}}(\alpha, M, \delta F)\delta E \tag{2.3-17}$$

$$C_D = C_D(\alpha, M, \delta F) + C_{D_{\hat{q}}}(\alpha, M, \delta F)\hat{q} + C_{D_{\delta E}}(\alpha, M, \delta F)\delta E \tag{2.3-18}$$

where $\hat{q}$ is the normalized pitch rate (Section 3.4),

$$\hat{q} = q\bar{c}/2V \tag{2.3-19}$$

δF is the wing flap setting, and δE is the elevator angle.

The expressions for the aerodynamic rolling, pitching, and yawing moments are similar to the force equations, with the addition of a *reference length* to the dimensionalizing quantities:

$$L = C_l\bar{q}Sb \tag{2.3-20}$$

$$M = C_m\bar{q}S\bar{c} \tag{2.3-21}$$

$$N = C_n\bar{q}Sb \tag{2.3-22}$$

The reference length is the *wingspan* b for rolling and yawing moments and the *mean aerodynamic chord* $\bar{c}$ for pitching moments. For a wing whose chord length $c(y)$ varies with the spanwise location, y, the *mean aerodynamic chord* (or m.a.c.) of the wing is defined as

$$\bar{c} = \frac{2}{S}\int_0^{b/2} c^2(y)\, dy \tag{2.3-23}$$

Given a linearly tapered wing with the *taper ratio* λ defined as the tip chord c_T divided by the root chord c_R, the m.a.c. is

$$\bar{c} = \frac{2c_R}{3}\left(\frac{1+\lambda+\lambda^2}{1+\lambda}\right) \tag{2.3-24}$$

The aerodynamic coefficients are typically modeled as linear functions of the airflow angles, angular rates, and control settings,

$$C_l = C_{l_o} + C_{l_\beta}\beta + C_{l_{\hat{p}}}\hat{p} + C_{l_{\hat{r}}}\hat{r} + C_{l_{\delta A}}\delta A + C_{l_{\delta R}}\delta R \tag{2.3-25}$$

$$C_m = C_{m_o} + C_{m_\alpha}\alpha + C_{m_{\hat{q}}}\hat{q} + C_{m_{\delta E}}\delta E \tag{2.3-26}$$

$$C_n = C_{n_o} + C_{n_\beta}\beta + C_{n_{\hat{p}}}\hat{p} + C_{n_{\hat{r}}}\hat{r} + C_{n_{\delta A}}\delta A + C_{n_{\delta R}}\delta R \tag{2.3-27}$$

where the coefficients may be functions of flight condition and longitudinal control settings. The normalized roll and yaw rates (Section 3.4) are

$$\hat{p} \triangleq pb/2V \tag{2.3-28}$$

$$\hat{r} \triangleq rb/2V \tag{2.3-29}$$

The aileron, elevator, and rudder deflections for control are represented by δA, δE, and δR.

2.4 Static Aerodynamic Coefficients

The aerodynamic forces and moments of wings, tails, and bodies are central elements of aircraft motion. Here, we are less concerned with the phenomenology than with the gross effects of configuration, motion, and control settings in creating aerodynamic forces and moments. The flow about an entire aircraft is too complex to allow accurate estimates from simple formulas: precise wind tunnel tests or intricate computational fluid dynamics codes are needed for this purpose, and these topics deserve fuller treatment than space allows. Nevertheless, reasonable estimates can be made by adding up the aerodynamic effects of components and accounting for interference effects, using simple formulas to predict the qualitative effects of flight condition. Comprehensive presentations of this handbook approach to estimating aerodynamic coefficients can be found in [F-2, W-7].

Aerodynamic forces are modeled in a general form (exemplified by lift) such as $C_L S\rho V^2/2$, while moments (symbolized by pitch) take the form $C_m S\bar{c}\rho V^2/2$. The reference area and length S and $\bar{c}$ are airplane-specific constants, and it is clear that both forces and moments are expected to follow trends that are quadratic in airspeed V and linear in air density ρ. The application of data from subscale wind tunnel models to estimate full-scale characteristics is predicated on the use of nondimensional coefficients: the full-scale estimates are derived by replacing the model's reference area and length with the full-scale values, adjusting for differences in Reynolds number and other scaling parameters [W-9]. (The *Reynolds number* $\mathcal{R}e = Vl/v$, where v is the kinematic viscosity [Section 2.1], portrays the relative importance of inertial forces and viscous shear stress in the fluid flow, which govern the thickness and character of laminar [smoothly flowing] or turbulent *boundary layers* about the aircraft.) Consequently, *configuration aerodynamics* focuses on trends in the nondimensional coefficients (e.g., C_L and C_m) rather than the dimensional forces and moments.

Nondimensional aerodynamic coefficients are functions of

- Aircraft shape and proportion
- Aeroelasticity
- Mach and Reynolds numbers
- Airflow angles
- Angular rates
- Control settings

As noted in Section 1.2, the *aircraft configuration* varies greatly with mission and flight regime. Elements that have major impacts on the aerodynamic coefficients include aspect ratio, chord section, leading-edge sweep angle, incidence angle, and twist of the wing, comparable parameters of the horizontal and vertical tail, and fuselage fineness ratio.[6] Because it normally is generating a force equaling the aircraft's weight, the wing is a major factor in variations of all aerodynamic forces and moments. Air loads proportional to the *dynamic pressure* $\overline{q}$, which is defined as $\rho V^2/2$, bend and twist the airframe; these *aeroelastic distortions* alter the nondimensional coefficients.

The *Mach number* $\mathcal{M} = V/a$ gauges the importance of compressibility effects on airflow, including the formation of shock waves. A body moving through the air creates pressure disturbances that propagate at the speed of sound. As the speed increases, the body catches up with the disturbances propagated forward; hence, the air's compressibility has an important effect on the overall flow. *Shock waves*, which are discontinuities in the pressure field, form when the local flow about the body is accelerated to sonic velocity a or above, that is, when $\mathcal{M} \geq 1$. *Angles of attack* and *sideslip* affect the pressure differentials between windward and leeward surfaces and, therefore, the net forces and moments as well. *Angular rates* produce dynamic effects that are discussed in the next chapter. *Control deflections* reorient aerodynamic surfaces with respect

6. The *aspect ratio AR* is b^2/S, where b is the wingspan and S is the reference area. The *chord section* (or *airfoil*) is the cross-sectional shape of the wing. Principal descriptors are the ratio of maximum thickness to the distance between leading and trailing edges, relative location of the point of maximum thickness, *camber* (curvature) of the mid-thickness line, sharpness of leading and trailing edges, and general shape of the upper and lower wing surfaces. The wing's *sweep angle* Λ is the angle between the y_B axis and the leading edge of the wing. The wing's *incidence angle* i is the angle between the aircraft centerline and a line drawn from the leading to trailing edges of the wing at the root chord. Draw a straight line from the leading to the trailing edge. *Twist* is the net change in the incidence angle of this line (relative to the vehicle centerline) between the root chord and the tip chord. Negative twist (decreasing angle from root to tip) is referred to as *washout*. Assuming a circular cross section, the fuselage fineness ratio relates the length to the maximum diameter.

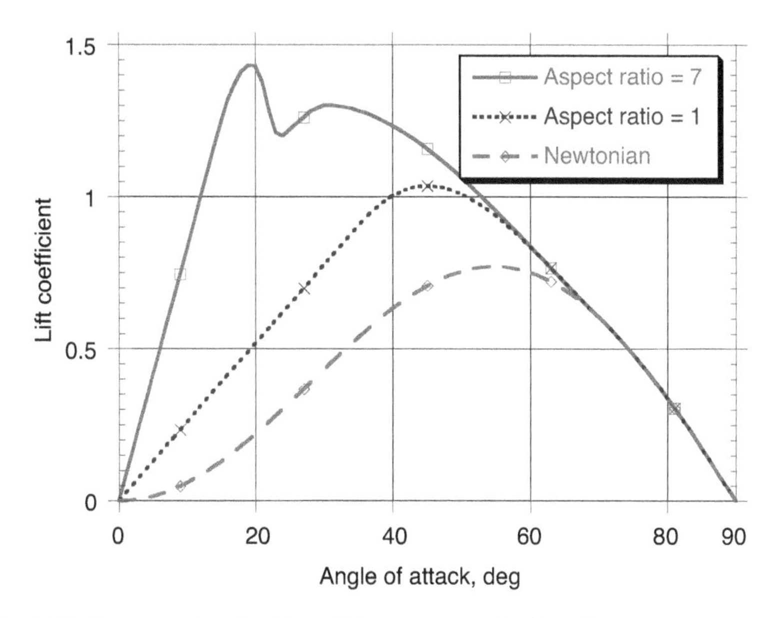

Fig. 2.4-1. Representative wing lift coefficients versus angle of attack.

to the flow, with results that are analogous to changes in the overall airflow angles. In the remainder of the section, some of the more significant effects of these factors are summarized, with particular attention given to angle of attack, sideslip angle, and Mach number variations.

Lift

Aerodynamic lift, the force perpendicular to the velocity vector in the aircraft's x_B-z_B plane, is generated primarily by the wing, with contributions from the fuselage, horizontal tail or canard, and engine nacelles. Typically, the wing lift coefficient varies with angle of attack α, as shown for three prototypical cases in Fig. 2.4-1. These curves are loosely representative of business jet and hypersonic transport wing planforms ($AR = 7$ and 1) plus the Newtonian approximation described below. At low α and Mach number, C_L is linear in α [A-3]. It may have a nonzero value (C_{L_o}) at $\alpha = 0$ deg due to the wing and tail incidence relative to the centerline, airfoil camber (curvature of the mean chord line between upper and lower surfaces), spanwise twist of the chord section, and fuselage shape. The lift coefficient necessarily reaches a maximum value $C_{L_{max}}$ at some α between 0 and 90 deg for reasons of flow geometry and separation.

Consider a simple geometrical example based on the *Newtonian flow* model [A-3]. An infinitesimally thin flat plate of area S is immersed in an airflow with density and speed of ρ and V. The *edge-on case* is represented by $\alpha = 0$ deg and the *broadside condition* by $\alpha = 90$ deg. Newton concluded (from his second law) that a flow uniformly brought to rest by a perpendicular surface ($\alpha = 90$ deg) exerts a normal force N on the surface of

$$\begin{aligned} N &= \text{(mass flow rate per unit area)(change in velocity)(area of surface)} \\ &= (\rho V)(V)(S) = \rho V^2 S \end{aligned} \tag{2.4-1}$$

Modeling this force as the product of a nondimensional coefficient, dynamic pressure, and reference area,

$$N \triangleq C_N \bar{q} S = C_N \left(\frac{\rho V^2}{2} \right) S \tag{2.4-2}$$

leads to the conclusion that $C_N = 2$. This judgment overlooks air viscosity and important details of the flow, and it ignores the leeward surface entirely, yet it produces a useful working hypothesis for some flow regimes. For angles less than 90 deg, the mass flow rate intercepted by the plate would be reduced to $\rho V \sin\alpha$, while the change in flow velocity would be $V \sin\alpha$; hence, the nondimensional coefficient expressing the normal force per unit area on the plate (i.e., the differential pressure) would be

$$C_N = C_{N_{max}} \sin^2\alpha = 2 \sin^2\alpha \tag{2.4-3}$$

The plate is at angle α to the flow, so the corresponding lift coefficient is

$$C_L = C_N \cos\alpha = 2 \sin^2\alpha \cos\alpha \tag{2.4-4a}$$

as plotted in Figure 2.4-1 (Newtonian). This function equals zero and has zero slope at zero angle of attack, and the lift coefficient's maximum value is 0.77 at $\alpha = 54.74$ deg. At higher α, C_N is larger, but the force component perpendicular to the flow (i.e., C_L) is smaller.

Equation 2.4-4a represents trends in the lift coefficient of all aircraft configurations with *plan area*[7] S_p in two circumstances:

- Large angle of attack (>60 deg) at all Mach numbers
- All angles of attack at hypersonic Mach numbers ($\mathcal{M} > 5$)

The fit to real data can often be improved by choosing $C_{N_{max}}$ other than 2, as well as by accounting for any differences between the reference and plan areas. Thus, eq. 2.4-4a would be modified to be

$$C_L = C'_{N_{max}} \left(\frac{S_p}{S} \right) \sin^2\alpha \cos\alpha \tag{2.4-4b}$$

7. In this usage, the *plan area* is the horizontal area of the entire aircraft projected onto the x_B, y_B plane, including wing, fuselage, and horizontal tail.

$C'_{N_{max}}$, is typically in the range (1.2, 2) except in the transonic regime, where it may exceed 2 [M-3]. For $AR = 7$ and 1 in Fig. 2.4-1, it is assumed that the high-angle lift coefficients approach the Newtonian model with $C'_{N_{max}} = 2$ and $S_p = S$.

For most other aircraft flight conditions, Newtonian theory greatly underpredicts the lift coefficient because the normal force depends on the negative differential pressure of the upper (leeward) surface as well as the positive contribution of the lower (windward) surface. It misses entirely the low-α linear proportionality of lift with angle of attack that characterizes typical wings in sub-, trans-, and supersonic flow. Neglecting shear-stress (skin-friction) effects, the normal-force coefficient of a wing is

$$C_N = \frac{1}{\bar{q}S}\int_{-b/2}^{b/2}\int_{x_{LE}(y)}^{x_{LE}(y)}[p_z(x, y)_{lower} - p_z(x, y)_{upper}]\,dx\,dy \qquad (2.4\text{-}5)$$

where p_z is the pressure normal to the lower and upper wing surfaces resolved along the body-axis z_B component. The inner integral extends from the trailing edge to the leading edge of the wing, and the outer integral extends from one wing tip to the other. In subsonic flow at low α, the pressure decrease on the upper surface of a high-aspect-ratio (AR) wing (relative to ambient pressure) is much greater than the lower-surface pressure increase.

The pressure differential occurs because the air traveling over the upper surface is faster than freestream speed, while the lower-surface air is slower. Looking at just the velocity differential between top and bottom, the flow about the lifting wing is circulating; it contains *vorticity*, a net rotational motion about some spanwise axis (or line) in the wing. This vorticity is bound to the wing, and it extends from tip to tip. At the tips of a straight, finite-span wing, the air's spiraling motion persists and is shed in counter-rotating helical vortices that trail downstream. Their formation absorbs energy from the flow, creating a complex field that may interact with aft portions of the aircraft. This vortical motion of the air is central to estimating the lift and drag of a wing using *lifting-line theory*, as in [A-3, M-3]. The wake vortex of a large aircraft presents a hazard to a small aircraft flying near the wake's core because its tangential velocity can exceed 100 m/s.

The favorable upper-surface effect of bound vorticity is dependent on the flow remaining attached, that is, flowing along the wing's surface contour. Airfoil camber and thickness distribution, as well as leading-edge slots and slats, have important effects on maintaining attached flow and inducing circulation. As the angle of attack increases, it becomes harder for the upper-surface flow to follow the surface, and a small vortex entirely above the wing forms, causing flow separation. When the upper-surface flow separates, pressure differential and lift are lost, and the wing ultimately "stalls." The stall can be relatively abrupt for moderate- to high-AR

wings [A-6], where localized "bubbles" of separation can first occur at leading or trailing edges or at root or tip chords (Fig. 2.4-1, Case a).

Details of the separation, including the spanwise location and spread of localized separation with increasing angle of attack [A-6, M-3, S-3, S-7, S-11, S-20], have great impact on the *maximum lift coefficient* $C_{L_{max}}$ of the wing and associated pitching, rolling, and yawing moments. The stalling properties of chord sections affect the overall properties of the wing. As the Reynolds number increases from 10^6 to 10^7, the maximum lift coefficient of typical chord sections and wings increases from ranges of 0.9–1.3 to 1.4–1.7 [J-1, S-3]. For thickness-to-chord-length ratios of 10–16 percent, increasing the leading-edge radius from 40 to 65 percent of the thickness increases the maximum section lift coefficient by 20 percent.

The wing aspect ratio and sweep play roles in defining stalling characteristics of the three-dimensional wing. At $\mathcal{R}e \approx 10^6$, rectangular and swept-back wings ($\Lambda = 45$ deg) with constant-length, symmetric chord sections have maximum lift coefficients within ± 10 percent of $[0.9 + 0.3e^{-(AR-1)}]$ for $(1 < AR < 6)$ [S-2]. The corresponding angles of attack for maximum lift are approximated by $17\,[1 + e^{-(AR-1)}]$ deg in the same range. Delta wings exhibit similar characteristics, although the greatest $C_{L_{max}}$ occurs for aspect ratios of 1.5–2, and the corresponding angle of attack is 35–40 deg [S-2]. We find somewhat higher values of $C_{L_{max}}$ and the corresponding α for tapered wings of $AR = 6$ and higher $\mathcal{R}e$ in [A-6], where typical values for swept and unswept wings (some with washout) are 1.4–1.6 and 20–25 deg.

Highly swept, low-AR planforms typically exhibit more gradual stall, as bound leading-edge vortices increase leading-edge "suction" and ameliorate the upper-surface flow (Fig. 2.4-1, Case b). The swept leading edge encourages a broader distribution of trailing vorticity that is generally associated with lower lift slope, $\partial C_L/\partial\alpha$ or C_{L_α}, and higher induced drag than that of straight wings. "Bursting" of trailing vortices at high α (the rapid transition from organized to disorganized flow) can have adverse (and unsteady) effects on moment characteristics if it occurs within the length of the aircraft, creating high dynamic air loads on the aircraft's aft structure as well. The swept wing's chord section and twist affect stalling properties but have little impact on the C_{L_α} at lower α, where aspect ratio, sweep, and Mach number are the dominant factors.

As shown in Fig. 2.4-1, the subsonic lift slope C_{L_α} of high-AR wings is higher than that of low-AR wings at low-to-moderate angles of attack, and both are greater than the Newtonian lift slope. $C_{L_{max}}$ also is higher in both cases and occurs at angles of attack below the Newtonian value of 55 deg. Classical thin-airfoil theory [S-2] predicts that an *unswept wing* of infinite aspect ratio would have a lift slope of 2π per radian in incompressible flow ($\mathcal{M} \approx 0$). If the wing has a symmetric chord section, its resultant lift is generated along a line that is 25 percent behind the leading edge, that is, along the *quarter-chord line*. The effective point of application of the wing's lift is called its *center of pressure* (c.p.). Measuring from the location of the root chord's two-dimensional (2-D) center of pressure, the *c.p.* of a linearly tapered wing with elliptical spanwise distribution is located at

$$\Delta x_{c.p.} = \frac{b}{6}\left(\frac{1+2\lambda}{1+\lambda}\right)\tan\Lambda \tag{2.4-6}$$

where b is the wing span, λ is the *taper ratio*,[8] and Λ is the sweep angle of the locus of local chord section centers of pressure [P-4, M-3]. For untwisted, symmetric chord sections, Λ is the sweep angle of the quarter-chord line, and the center of pressure is the same as the *aerodynamic center* (discussed below).

For wings with AR greater than 6–8 and elliptical spanwise lift distribution, lifting-line theory predicts a lift slope (per rad) of

$$C_{L_\alpha} = \frac{2\pi AR}{AR+2} \tag{2.4-7}$$

An *elliptical lift distribution* is maximum at the root chord and zero at the tip, varying as $\sqrt{1-(2y/b)^2}$ in between. The section lift coefficient, *downwash* (downward motion of the air associated with generating lift), and induced angle of attack are constant over the span in this case. For low AR (<1) and any Mach number, slender-wing theory predicts

$$C_{L_\alpha} = \frac{\pi AR}{2} \tag{2.4-8}$$

Both expressions are subsumed in the *Helmbold equation* [M-3, S-2, S-4]:

$$C_{L_\alpha} = \frac{\pi AR}{1+\sqrt{1+(AR/2)^2}} \tag{2.4-9}$$

This formula can be extended to wings of moderate sweep angle, including subsonic compressibility effects, as follows:

$$C_{L_\alpha} = \frac{\pi AR}{1+\sqrt{1+(AR/2\cos\Lambda_{1/4})^2(1-\mathcal{M}^2\cos\Lambda_{1/4})}} \tag{2.4-10}$$

$\Lambda_{1/4}$ is the sweep angle of the wing's quarter-chord line, measured positive aft of the y_B axis.

The *Prandtl-Glauert-Göthert-Ackeret similarity rules* of linear potential flow suggest that both incompressible and supersonic results can be

8. The *taper ratio* λ is the ratio of the tip chord length c_t to the root chord length c_r. A rectangular wing has a taper ratio of 1, while a triangular wing has $\lambda = 0$.

extended using the *Prandtl factor* $\sqrt{|1 - \mathcal{M}^2|}$. AR and cot Λ should be multiplied by the factor, while camber, thickness ratio, and angle of attack should be divided by it. As a consequence, the compressible lift slope $C_{L_{\alpha_C}}$ can be estimated from the incompressible slope $C_{L_{\alpha_I}}$ as [S-5]

$$C_{L_{\alpha_C}} = \frac{1.8 + AR}{1.8 + \sqrt{1 - \mathcal{M}^2}\, AR} C_{L_{\alpha_I}} \tag{2.4-11}$$

The similarity rules become singular and do not apply at $\mathcal{M} = 1$. All wings appear to be aerodynamically "slender" at $\mathcal{M} = 1$, and eq. 2.4-7 provides a usable approximation to C_{L_α} at sonic speed. At supersonic speed, an unswept rectangular wing's lift slope increases further before following the decreasing Prandtl trend. The center of pressure moves forward from the quarter-chord to the leading edge as the Mach number advances to 1, and then moves back toward 50 percent with further increase [W-7]. This large forward shift has adverse effect on transonic stability; large control changes to maintain trim (or pitching moment balance) are required in the transonic regime, although the reason is more complex, as discussed below. A slender delta (i.e., triangular) wing's lift is typically centered at the 60 percent root chord in subsonic flight, shifting smoothly to the 2/3 chord at supersonic speed.

Estimating C_{L_α} for finite-AR wings in transonic and supersonic flow is complicated by the dominant effects of shock-wave strength and position, as well as shockwave interactions with the boundary layers of real-gas flows. The *transonic regime* normally is defined as one in which there are local areas of both subsonic and supersonic flow over the aircraft. Typically, this corresponds to aircraft Mach numbers from 0.4 to 1.4, although especially blunt or complicated configurations could have a wider transonic region. In transonic flight, small $\mathcal{M}$ variations may lead to large variations in shock location. This Mach effect has particularly adverse effect on rectangular wings of moderate to high aspect ratio with thick airfoils. Thin swept and triangular wings generally adhere to the potential flow similarity rules.

In supersonic flight, the leading edge of a swept wing may be behind the conical shock wave produced by the fuselage nose (i.e., in locally subsonic flow) or it may be ahead of it (in locally supersonic flow). Different formulas apply for the two cases [M-3, S-4]. Measured aft of the y_B axis, the *Mach cone angle* is

$$\sigma = \cos^{-1} \frac{1}{\mathcal{M}} \tag{2.4-12}$$

This definition is the *complement* of the usual definition as the Mach cone's half angle μ, i.e., $\sigma = 90° - \mu$, in order to provide a common reference with the wing's leading-edge sweep angle Λ. In practice, the shock "cone" angle varies with distance from the nose as a consequence of aircraft cross-sectional area. For a delta wing with supersonic leading edge ($\sigma > \Lambda$),

$$C_{L_\alpha} = \frac{4}{\sqrt{\mathcal{M}^2 - 1}} \tag{2.4-13}$$

while for a subsonic leading edge ($0 < \sigma < \Lambda$)

$$C_{L_\alpha} = \frac{2\pi^2 \cot \Lambda}{\pi + a} \tag{2.4-14}$$

where $a \approx m(0.38 + 2.26m - 0.86m^2)$ and $m \triangleq \cot\Lambda / \cot\sigma$ [M-3, from B-2]; the polynomial is a least-squares fit to graphical data. Note that aspect ratio is not present in either of these estimates.

At hypersonic speeds, Newtonian flow provides a suitable model for aerodynamic lift. As noted in [A-3], the bow shock wave becomes very oblique, approaching the angle of the surface itself. The shock wave slows and deflects the flow so that it is nearly parallel to the wing's surface. The streamlines and flow speeds near the wing's surface are similar to the Newtonian assumptions, as are the resulting pressure distributions. A wedge-shaped wing airfoil with half angle i generates a surface force coefficient of

$$C_N = C_{N_{max}} \sin^2(\alpha + i) \tag{2.4-15}$$

on its windward sides for $\alpha > i$, and C_L is defined accordingly.

Movable flaps on the leading and trailing edges of a wing have different effects on the lift coefficient, as shown in Fig. 2.4-2. Neither type of flap has a significant effect on the slope of the lift curve in its linear region. Downward deflection of a leading-edge flap encourages the airflow over the wing to stay attached to a higher angle of attack; hence, the linear portion of the lift curve is extended, and the maximum lift coefficient is increased. Deflection of a trailing-edge flap tends to shift the entire lift curve up and down, with little change in the curve's shape. Beyond some angle, further deflection of the trailing-edge flap has little impact on C_L but continues to increase the drag coefficient C_D.

Flap configurations vary widely; in some cases, there is open space between the flap and the wing, forming a *slot*, and the flap itself has an airfoil shape. *High-lift flap assemblies* may have several movable surfaces, and slots between staggered surfaces may energize the leeward flow to keep it attached at higher angles. Flaps used to increase lift during slow-speed flight move slowly and in one direction only (relative to their stowed positions). Flaps used for angular maneuvers and aeroelastic control, including the wing's *ailerons*, the horizontal tail's *elevator*, and the

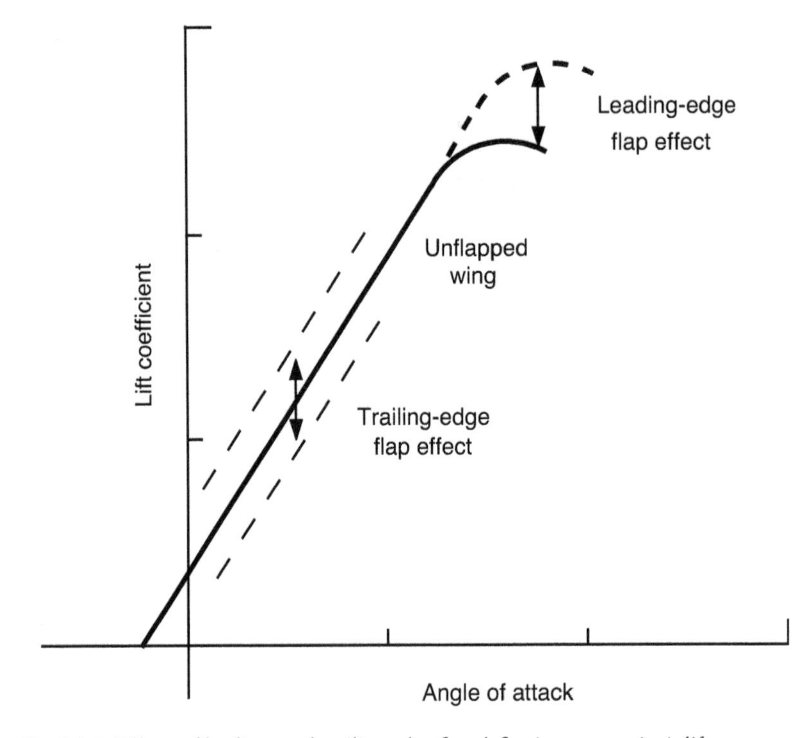

Fig. 2.4-2. Effects of leading- and trailing-edge flap deflections on a wing's lift coefficient.

vertical tail's *rudder* must be capable of moving quickly and in both directions. The lifting surface's area and planform are normally defined to include the undeflected flap area and shape. If there is a gap between the flap and the unmoving fixed surface, leakage may reduce the net lift slope, C_{L_α}. Flaps on the upper surface of the wing—called *spoilers*—move in an upward direction; their purpose is to *decrease* lift and *increase* drag for control purposes. Rolling moment is provided by deflecting the spoilers on one wing but not the other.

Horizontal tail and canard lift are governed by the same considerations as wing lift with one important modification: the smaller surface is immersed in a flow field that is dominated by the wing. At subsonic speed, there is an upwash ahead of a lifting wing and a downwash behind it, changing the airflow angle experienced by the smaller surface. The *downwash angle* ε of the flow is proportional to the amount of lift being produced by the wing and, therefore, to the angle of attack. For an elliptical lift distribution, this angle is predicted by [P-4, S-9]

$$\varepsilon(\text{rad}) = C_L/\pi AR \qquad (2.4\text{-}16)$$

This downwash reduces the angle of attack felt at the horizontal tail's center of pressure; thus,

$$\alpha_{tail} = \alpha_{wing} - \varepsilon \tag{2.4-17}$$

A perturbation in the wing's angle of attack, $\Delta\alpha_{wing}$, is felt at the tail as

$$\Delta\alpha_{tail} = (1 - \partial\varepsilon/\partial\alpha)\ \Delta\alpha_{wing} \tag{2.4-18}$$

where the *downwash rate of change* with respect to the angle of attack, $\partial\varepsilon/\partial\alpha$, is found by differentiating eq. 2.4-16. The upwash on a forward (canard) surface could be estimated in a similar manner, changing the sign of the effect in eqs. 2.4-17 and 2.4-18. Fuselage bending due to cruising-flight air loads tends to reduce the α at the tail, reducing the lift that aft tail surfaces generate. For both sub- and supersonic speeds, there are blockage (wing wake) and viscous (fuselage boundary layer or *scrubbing*) effects that decrease the airspeed felt by downwind surfaces, further modifying aerodynamic contributions. Conversely, the tail surfaces of propeller-driven aircraft may be in flows that are faster than the free stream, altering their sensitivities to α.

An aircraft's fuselage normally makes small contribution to lift at low α[J-1, S-7, S-20]. When its *fineness ratio* d_{max}/l (the ratio of maximum diameter to length) is much less than 1, it can be considered a slender body. If the fore and aft ends are pointed, *slender-body theory* predicts that the positive lift produced by the nose due to increasing cross-sectional area is canceled by negative lift associated with the tail cone's decreasing cross section. (This does, however, generate a destabilizing pitching moment.) In practice, many fuselages have blunt aft ends, and flow separation beyond the point of maximum diameter causes an effective bluntness, leading to a small but positive net contribution to C_{L_α}. For a fuselage with a circular-base cross section of area S_b, this contribution is estimated as

$$C_{L_{\alpha_{fus}}} = 2\frac{S_b}{S} \tag{2.4-19}$$

If the aft end is blunt and noncircular, S_b is typically calculated as $\pi d^2/4$, where d is the widest horizontal dimension of the base.

Fuselage-wing interference is dependent on the wing's vertical and horizontal position relative to the body; it is generally favorable to lift generation at low α; hence, the area of the wing within the fuselage can be considered to generate lift as if there were no fuselage at all [J-1]. At higher angles of attack, fuselage-wing interference may be more significant, with up to 25 percent reduction of maximum lift [J-1, S-7, S-20]. The loss of lift is most prominent for low- and mid-wing configurations and fuselages with circular cross section, as the fuselage creates a *burble* that interferes with the negative pressure differential on the wing's upper surface. This effect is reduced by *fillets* (smooth fairings) between the body and wing and by employing rectangular body cross sections.

The total lift coefficient and lift slope can be estimated as the sum of contributions from the components plus interference effects, all referenced to the aircraft's reference area, S:

$$C_L = C_{L_{wing}} + C_{L_{fus}} + C_{L_{tail}} + C_{L_{int}} \tag{2.4-20}$$

$$C_{L_\alpha} = C_{L_{\alpha_{wing}}} + C_{L_{\alpha_{fus}}} + C_{L_{\alpha_{tail}}} + C_{L_{\alpha_{int}}} \tag{2.4-21}$$

The component coefficients must be corrected for scrubbing, propeller-slipstream effects, downwash, aeroelasticity, and dissimilar reference areas. For example, the horizontal tail contribution to the lift slope would be

$$C_{L_{\alpha_{tail}}} = \left(\frac{\bar{q}_{tail}}{\bar{q}}\right)\left(1 - \frac{\partial \varepsilon}{\partial \alpha}\right)\eta_{elas}\left(\frac{S_{tail}}{S}\right)\left(C_{L_{\alpha_{tail}}}\right)_{tail} \tag{2.4-22}$$

where $\bar{q}_{tail}$ and S_{tail} and the tail's dynamic pressure and area, η_{elas} is an aeroelastic correction factor, and $(C_{L_{\alpha_{tail}}})_{tail}$ is the horizontal tail's lift slope referenced to tail area.

Drag

Aerodynamic drag falls into two principal categories: *induced drag*, which arises from the generation of lift, and *parasite drag*, a catch-all description for the remaining drag [H-4]. The wing is responsible for most of the induced drag, and its efficiency as a lifting surface is measured primarily by the amount of drag produced per unit of lift. The major contributors to parasite drag are *skin friction*, the result of aerodynamic shear stresses on the craft's *wetted area* (i.e., total surface area), and differential pressure that arises from blunt-base flow separation or supersonic shock waves.

A first approximation to total drag can be made by considering each major airframe component separately and then combining the results. Tail surfaces are treated as small wings and engine nacelles as small fuselages; nevertheless, *flow interference* between adjacent components has significant effect on both induced and parasite drag. Some portion of the parasite drag can be charged to the propulsion system and its aerodynamic interaction with the airframe, leading to a perennial accounting problem in the determination of thrust and drag: variations in vehicle drag and "installed" thrust, which depends on inlet and exhaust aerodynamics as well as the engine's thermal cycle, may be indistinguishable in flight testing data.

Three typical variations of induced drag coefficient with angle of attack in the range from 0 to 90 deg are shown in Fig. 2.4-3. Once again,

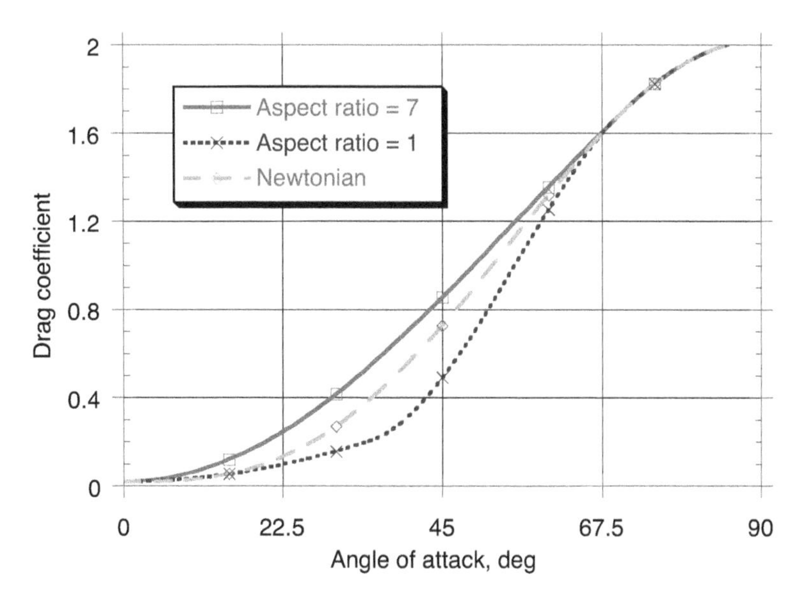

Fig. 2.4-3. Representative induced drag coefficients versus angle of attack.

Newtonian flow provides a reference point. For a flat plate with no parasite drag, eq. 2.4-3 leads to

$$C_{D_i} = C_N \sin\alpha = 2\sin^3\alpha \tag{2.4-23}$$

Hence, the drag due to lift (or induced drag) has its minimum and maximum values at the ends of the range. For $\alpha = 90$ deg, the flat plate produces no lift, $C_D = 2$, and the reference area S is the frontal area. In real flows, the C_D of finite flat plates is up to 40 percent less than this value, as pressure recovery occurs on the leeward side.

Lifting-line theory predicts that the drag produced by a finite wing with elliptical lift distribution in incompressible flow is

$$C_{D_i} = \frac{C_L^2}{\pi AR} \tag{2.4-24}$$

This comes about because the direction of the lift (always perpendicular to the flow) is rotated back from the vertical through an *induced angle of attack* α_i caused by the downwash, producing the drag component $C_L \sin\alpha_i \approx C_L\alpha_i$ (with α_i measured in rad). The downwash is itself proportional to the amount of lift generated. Hence, α_i is proportional to C_L through the factor $1/\pi AR$, leading to eq. 2.4-24. Clearly, the higher the aspect ratio is, the lower the induced drag. Sailplanes typically have aspect ratios of 15 or more to promote low induced drag.

Of course, not all wings have elliptical lift distributions; the net result is that the induced drag is increased, as expressed in various ways:

$$C_{D_i} \triangleq \frac{C_L^2}{\pi e AR} \triangleq \frac{C_L^2}{\pi AR}(1+\delta) \triangleq \varepsilon C_L^2 \tag{2.4-25}$$

A wing can have an essentially elliptical lift distribution without having a geometrically elliptical planform. Trapezoidal wings with taper ratios λ of 25–35 percent exhibit $\delta < 0.01$ for $AR < 10$, and decreasing AR has the overall effect of increasing the ellipticity of the distribution [M-3]. Fuselage interference tends to decrease the *Oswald* (or *span*) *efficiency factor* e, with values of 0.8 and 0.6 being representative of high-wing and low-wing general aviation configurations. A broad range of fuselage-wing interference effects on induced drag can be found in [J-1, S-7].

Lifting-line theory neglects the three-dimensional details of lifting-surface effects; it has been suggested that careful attention to planform shape, including the use of swept (or sheared) tips that begin to approximate the crescent shape of a swallow's wing, can yield equivalent wing-alone Oswald efficiency factors exceeding 1 [V-1]. Adding end plates or winglets to the wing tips increases the effective aspect ratio, thus lowering the induced drag; however, the two modifications are different in intent and result. A simple vertical *end plate* straightens the tip flow, creating the effect of more wing area beyond the tip. Unfortunately, the reduction in induced drag is more than offset by the increase in parasite drag. *Winglets*, which are small, vertical wings of moderate aspect ratio extending above (and sometimes below) the wing, oppose the tip vortices associated with induced drag. Their parasite drag penalty is small, and they afford significant induced drag reduction [W-4, H-2].

Figure 2.4-3 shows that the $AR = 1$ wing has lower induced drag than the $AR = 7$ wing at a given angle of attack, with the Newtonian model falling between. A more meaningful comparison is obtained from the *lift-drag polar*, where C_L is plotted against C_{D_i} (Fig. 2.4-4). For a fixed amount of lift, the high aspect ratio has the least drag, and the Newtonian model has the most. The reason for this apparent dichotomy is that the high-AR wing produces a given C_L at much lower α than the other wings.

The lift-drag ratio L/D should be large for efficient cruising flight. It equals the slope of a line drawn from the origin to a point on the lift-drag

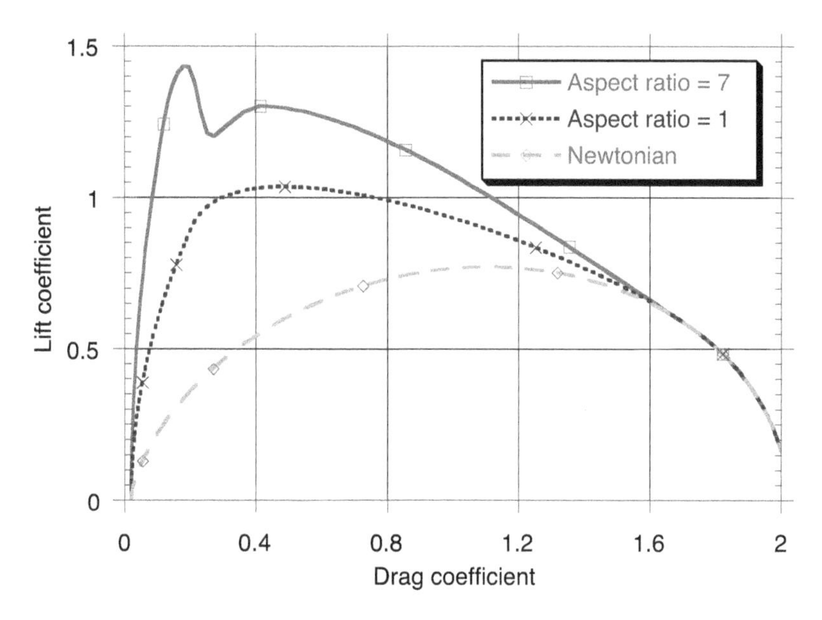

Fig. 2.4-4. Relationships between wing lift and drag coefficients.

polar. Figures 2.4-3 and 2.4-4 depict induced drag with a parasite drag coefficient of 0.02 added. Without parasite drag, the L/D for $AR = 7$ and 1 is modeled as

$$L/D = C_L/C_D = \frac{1}{\varepsilon C_L} \tag{2.4-26}$$

while for the Newtonian model, it is

$$L/D = \cot\alpha \tag{2.4-27}$$

For both equations, the slope is infinite at the origin (with $C_{L_o} = 0$). More realistically, all aircraft possess nonzero parasite drag, so the maximum L/D is finite, and the peak occurs at some positive angle of attack. Figure 2.4-5 presents a plot of L/D versus α, with $C_{D_o} = 0.02$ added to each drag model. It can be seen that the $AR = 7$ wing then has the largest L/D_{max} and that the peak occurs at the lowest angle of attack.

A natural laminar flow airfoil can maintain a low C_D that is essentially independent of α over a range of several degrees; C_D abruptly jumps to an εC_L^2 characteristic when the upper surface transitions from

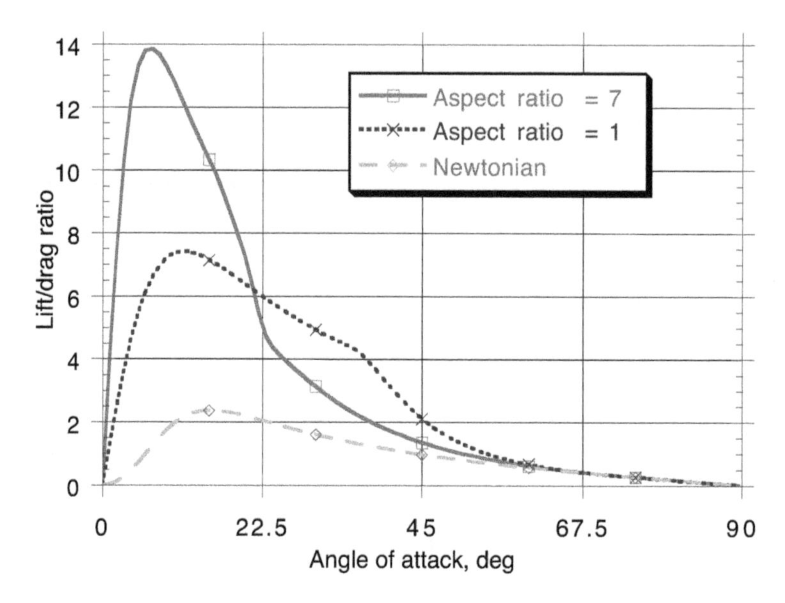

Fig. 2.4-5. Representative variation of wing lift/drag ratio with angle of attack.

laminar to turbulent flow. Plotting C_D versus C_L, as in Fig. 2.4-6, the laminar flow produces a *drag bucket* in the polar, that is, a region of very low drag. The challenge to wing designers is to assure that the C_L required for cruising flight remains within the drag bucket. Wing sweep produces pressure gradients that are adverse to laminar flow and drag buckets. Laminar flow and reduced drag can be encouraged by airfoil design and by using active suction to control the boundary layer on the leading portion of the wing.

Compressibility effects on the induced drag must be considered in transonic and supersonic flight, as reviewed in [M-3, S-4]. Linear potential flow theory allows the subsonic induced drag coefficient $C_{D_{i_C}}$ to be derived from its incompressible counterpart $C_{D_{i_I}}$ through similarity rules:

$$C_{D_{i_C}} = \frac{C_{D_{i_I}}}{(1 - \mathcal{M}^2)^{3/2}} \tag{2.4-28}$$

With eqs. 2.4-11 and 2.4-25, the compressible coefficient can be written as

$$C_{D_{i_C}} = \frac{\varepsilon C_{L_I}^2}{(1 - \mathcal{M}^2)^{3/2}} = \frac{\varepsilon C_{L_C}^2}{(1 - \mathcal{M}^2)^{1/2}} = \varepsilon' C_{L_C} \tag{2.4-29}$$

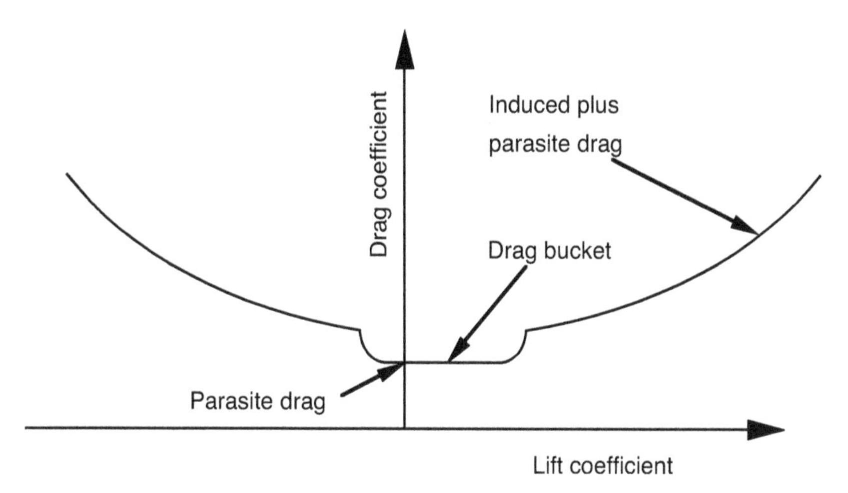

Fig. 2.4-6. Representative variation of drag coefficient with lift coefficient for a cambered, laminar-flow wing.

Hence, the induced drag's subsonic sensitivity to C_L^2 increases with Mach number. A slender delta wing's C_{L_α} and C_{D_i} are theoretically unaffected by $\mathcal{M}$.

The supersonic results depend on whether the leading edge is subsonic or supersonic, and only the delta-wing results are simple enough to present here. For a subsonic leading edge, the induced drag coefficient is

$$C_{D_i} = \left[2\left(1 + \frac{\lambda}{\pi}\right) - \sqrt{1 - \left(\frac{\cot\Lambda}{\cot\sigma}\right)^2} \right] \left(\frac{C_L^2}{\pi AR}\right) \triangleq \varepsilon' C_L^2 \tag{2.4-30}$$

where the symbols are defined as in eq. 2.4-14. For a supersonic leading edge,

$$C_{D_i} = C_L \tan\alpha \approx C_L \alpha = \frac{C_L^2}{C_{L_\alpha}} \triangleq \varepsilon' C_L^2 \tag{2.4-31}$$

Consequently, Mach effects for both transonic and supersonic flow can be expressed as modifications to the induced drag factor, and the quadratic dependence on C_L is retained at low to moderate α. For hypersonic flight, the Newtonian models of C_L and C_{D_i} provide reasonable estimates.

Even if no lift is produced, there is *parasite drag*, arising from the shear stress of air moving along the body's surface and from differential pressures between the windward and leeward sides. The mechanism of "skin friction" generation is complex; however, it is empirically associated with a *boundary layer* adjacent to the surface, where the body-relative flow velocity asymptotically transitions from zero at the surface (the *no-slip condition*) to the full velocity of a perfect flow [K-5]. As sketched in Fig. 2.4-7, the thickness of the boundary layer δ and the velocity profile $u(z)$ versus z within the layer depend on whether the flow is laminar or turbulent. The figure shows a sharp-edged, smooth flat plate that disturbs a fluid flow. Initially, the layers of fluid above the surface slide smoothly over one another, and the flow is laminar. Shear stress between the viscous layers gives rise to a tangential force at the surface, idealized as *skin friction*. Based on the wetted area S_{wet} of the surface, the *Blasius equation* yields the laminar friction drag coefficient of the flat plate [A-3]:

$$C_f = \frac{\text{friction drag}}{\bar{q} S_W} = 1.328 \mathcal{Re}^{-1/2} \tag{2.4-32}$$

$\mathcal{Re}$ is the Reynolds number Vl/v, based on the streamwise length l of the plate, and v is the kinematic viscosity of air.

As the distance from the leading edge (and $\mathcal{Re}$) increases, small flow disturbances are triggered by surface imperfections, and the flow becomes turbulent. In this case, the velocity profile shown in Fig. 2.4-7 represents the flow parallel to the surface, but there are small-scale velocity components perpendicular to the mean flow that increase the exchange of momentum between layers. The immediate effect is to increase the skin friction force along the surface. The friction drag coefficient for fully turbulent flow over a smooth flat plate is [V-7]

$$C_f = 0.455(\log_{10} \mathcal{Re})^{-2.58} \tag{2.4-33}$$

Surface roughness encourages the transition from laminar to turbulent flow, increasing the friction drag of a flat plate [H-4]. This effect is used in the wind-tunnel testing of subscale aircraft models. Strips of sandpaper or thin wires are glued to forward sections of the model at locations where the laminar-turbulent transition would normally be expected. These small protuberances "trip" the boundary layer, emulating flow at higher $\mathcal{Re}$.

For curved or angled surfaces, the boundary-layer transition can alter the mean flow direction and velocity, changing pressure profiles as well. The boundary-layer thickness δ is arbitrarily defined as the height at which $u = 0.99V$. The *displacement thickness* δ^* is $\delta/3$ for a laminar boundary layer above a flat plate and $\delta/8$ for a turbulent boundary layer [M-2]. The resulting pressure variations may be either detrimental or beneficial. For example, potential theory predicts that a streamlined body with a sharp trailing edge or point should have no drag at all because the

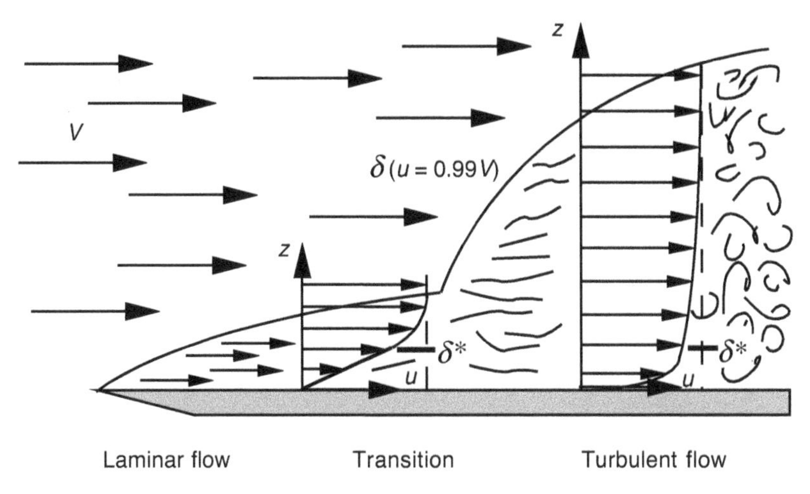

Fig. 2.4-7. Boundary layer structure above a sharp-edged flat plate.

forebody's pressure variation is precisely canceled by pressure variation on the afterbody. However, the δ^* of a real flow's boundary layer effectively fattens the afterbody and prevents full pressure recovery, producing pressure drag.

Just the opposite effect occurs on a blunt body, such as a sphere or a cylinder whose axis is transverse to the flow. At low Reynolds number, the laminar flow tends to separate near the points of maximum cross section, and the drag coefficient is high. At high Reynolds number, the flow is turbulent, and the boundary layer is thicker, but δ^* is lower. The flow near the surface retains higher velocity and remains attached to the body beyond the point of maximum thickness. This allows partial pressure recovery on the aft surface and decreases the cross-sectional area of the turbulent wake, thus reducing the drag coefficient.

Flow over a wing at high α has elements of blunt-body flow. Artificially tripping the boundary layer in the wind tunnel allows the flow over the wing's upper surface to stay attached to a higher angle of attack, resulting in subscale model measurements of C_{D_f} that more nearly capture the characteristics of the full-scale article.

An order-of-magnitude estimate of the friction drag C_{D_f} for a full aircraft can be expressed as

$$C_{D_f} = \left[C_f \frac{S_{wet}}{S}\right]_{wing} + \left[C_f \frac{S_{wet}}{S}\right]_{fus} + \left[C_f \frac{S_{wet}}{S}\right]_{tail} \qquad (2.4\text{-}34)$$

S_{wet} is the total surface area of each component, and C_f is the appropriate friction coefficient. As the flow over much of an aircraft's surface is likely to be turbulent, eq. 2.4-33 may provide reasonable values for C_f, although the flat-plate assumption is strained and interference effects are neglected.

Subsonic pressure drag on well-designed wings and bodies is almost entirely due to separated flow over the afterbody. Hoerner claims that an "ogive" nose section can have zero (even negative!) incremental drag, while flow past a blunt base has a "jet pump" effect that lowers the pressure in the wake [H-4]. Skin friction slows the relative flow speed, lessening the pressure drop. The subsonic base pressure drag for a blunt-based body is empirically determined to be

$$C_{D_p} = \frac{0.029}{\sqrt{C_f S_{wet}/S_b}} \frac{S_b}{S} \tag{2.4-35}$$

S_b is the equivalent base area, S_{wet} is the wetted area of the fuselage, and C_f is the skin friction coefficient of the fuselage. McCormick [M-3] shows that the base pressure drag must satisfy the inequality

$$C_{D_p} < \frac{2}{\gamma \mathcal{M}^2} \left(\frac{S_b}{S} \right) \tag{2.4-36}$$

as the base pressure can never drop below zero. An empirical formula for the parasite drag coefficient of wings with blunt trailing edges (or separated low-α wakes) is

$$C_{D_p} = \frac{0.135}{\sqrt[3]{C_f S_{wet}/S_b}} \frac{S_b}{S} \tag{2.4-37}$$

where S_b, S_{wet}, and C_f are defined appropriately.

The *Prandtl-Glauert-Göthert-Ackeret similarity rules* predict that local pressure coefficients vary with the *Prandtl factor* $1/\sqrt{|1 - \mathcal{M}^2|}$ (Fig. 2.4-8a), so it is not surprising that the pressure drag follows a similar trend. Note that the subsonic Prandtl factor would be applied to the incompressible C_{D_p}, while the supersonic Prandtl factor would be applied to C_{D_p} evaluated at $\mathcal{M} = \sqrt{2}$. The supersonic C_{D_p} typically is several times larger than the subsonic C_{D_p}, and the pressure drag curves are bounded (Fig. 2.4-8b).

Subsonic pressure drag usually stays relatively constant to higher Mach number than indicated by the Prandtl trend, although local flows about the aircraft are accelerated to higher speed than the free stream.

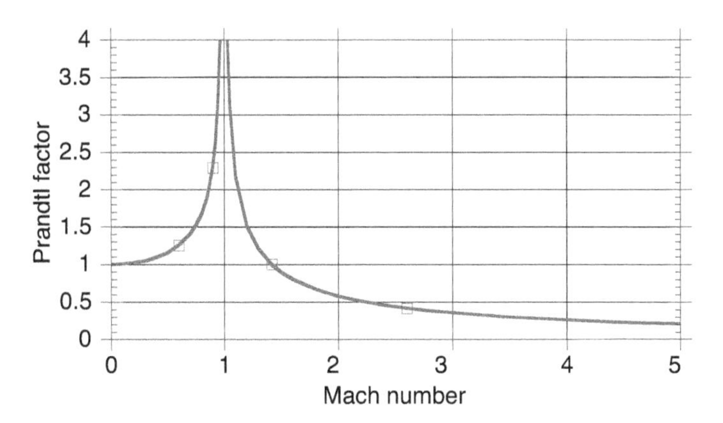

Fig. 2.4-8a. Compressibility effects on the drag coefficient: Prandtl factor versus Mach number.

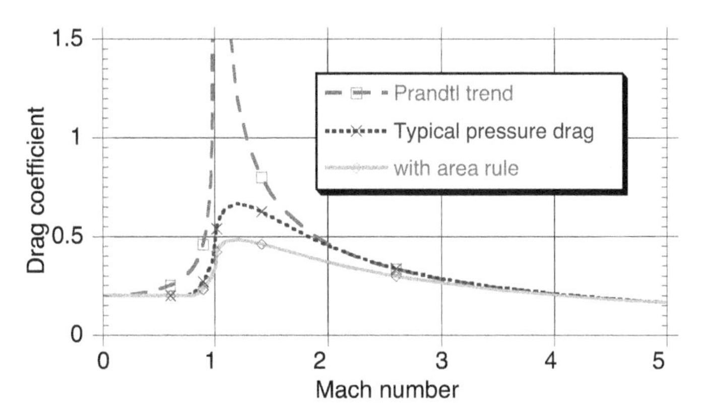

Fig. 2.4-8b. Compressibility effects on the drag coefficient: Parasite drag versus Mach number.

The freestream Mach number for which the local flow (usually on the wing's upper surface) first reaches sonic speed is called the *critical Mach number* $\mathcal{M}_c$. As speed increases further, a region of locally supersonic flow forms, and a *normal shock wave* appears. There is a dramatic increase in drag when this happens; hence, the corresponding freestream Mach number is called the *drag-divergence Mach number* $\mathcal{M}_{div}$. $\mathcal{M}_c$ is typically in the range from 0.4 to 0.7, while $\mathcal{M}_{div} = 0.7–0.85$ for current business aircraft and jet transports. The maximum cruising speed normally is determined by $\mathcal{M}$, as it is inefficient to fly faster. The critical and drag-divergence Mach numbers increase with decreasing wing thickness and increasing wing sweep. For sweep angle Λ,

$$\mathcal{M}_{c_{swept}} = \frac{\mathcal{M}_{c_{unswept}}}{\cos\Lambda} \qquad (2.4\text{-}38)$$

While the parasite drag at $\mathcal{M} = 1$ does not go to infinity as suggested by the similarity rules, it may go to very high values, creating what was called the "sound barrier" in the late 1940s. High parasite drag means that high thrust is required to fly at Mach 1 or faster. It was not until the 1950s that aircraft with low trans- and supersonic drag were designed and jet engines capable of providing sufficient sustained thrust were produced, thereby allowing sustained flight at supersonic speed.[9]

Following earlier theoretical work, Whitcomb conducted wind-tunnel experiments demonstrating that the drag rise is related to the axial distribution of an aircraft's cross-sectional area [W-2]. A straight wing superimposed on a cylindrical body produces an abrupt increase in the cross-sectional area distribution, and the resulting peak drag is quite large. The area distribution is somewhat better for a swept or triangular wing, but there are still discontinuities that lead to high drag. Peak drag can be reduced if the fuselage cross section is reduced near the wing so that a smooth areal distribution is maintained from nose to tail. This *area rule* reduces the thrust required for transonic and supersonic flight, and it is incorporated in most aircraft that operate in this flight regime.

Whitcomb made a second important contribution to efficient transonic flight by inventing the *supercritical airfoil* [W-3]. The basic idea is that $\mathcal{M}_{div}$ can be increased even though the local flow over parts of the wing has become supercritical ($\mathcal{M}_{local} > 1$). Pressure gradients on the upper surface must be tailored properly. The resulting airfoil has a leading edge with high upper-surface curvature, a relatively flat upper surface, and high downward camber on its aft portion. The lower-surface shape is less consequential, and the overall airfoil section can be designed to achieve ancillary structural and volumetric goals. The supercritical airfoil allows important design tradeoffs between cruising speed, wing thickness, sweep, and structural weight to be made.

At hypersonic speeds, the Newtonian model provides a good estimate of pressure drag. The local pressure coefficient is predicted by eq. 2.4-3 and is integrated over the nose and leading edges of the aircraft's surfaces. For a hemispherical nose, $C_{D_p} \approx S_o/S$, where S_c is the nose's cross-sectional area, while for a wing with cylindrical leading edge and sweep angle Λ, $C_{D_p} \approx (4S_c/3S)\cos^2\Lambda$, with S_c representing the leading edge's frontal area. Corrections can be made for a maximum pressure

9. The "sound barrier" was broken when *Chuck Yeager* flew the rocket-powered X-1 to Mach 1.02 in 1947.

coefficient smaller than 2, which is a consequence of centrifugal effects on the curved flow. A conical nose would have $C_{D_p} \approx 2\sin^2\theta$, where θ is the cone's half angle. The induced drag of a wing could be estimated as before, using eq. 2.4-15 to account for surface angle.

Pitching Moment

The aerodynamic forces generated by wing, fuselage, and tail can be idealized as acting at *centers of pressure* (c.p.) on their planforms. There is a moment arm between each of these forces and the aircraft's center of mass (c.m.); hence, the forces produce torques (or moments) relative to the plane's *rotational center*. Taking the c.m. as the origin, eqs. 2.3-1 and 2.3-2 indicate that the pitching moment (i.e., the torque's y component) due to a force $[X\ Y\ Z]_I^T$ generated at position $[x\ y\ z]_I^T$ is

$$M_{y_I} = z_I X_I - x_I Z_I \tag{2.4-39}$$

Equation 2.3-2 resolves forces and positions in an inertial frame, but they can be expressed in other frames, such as wind or body axes, as well. Figures 2.2-6–2.2-8 demonstrate that wind and body axes have the same y axis if the sideslip angle is zero; rotation through the angle of attack relates the two frames. The moment about the y axis due to the wing's force, shown in Fig. 2.4-9, can be expressed as either

$$M_{y_W} = z_W X_W - x_W Z_W \tag{2.4-40a}$$

or

$$M_{y_B} = z_B X_B - x_B Z_B \tag{2.4-41a}$$

From the definitions of eq. 2.3-3, X_W is the negative of drag, and Z_W is the negative of lift, so eq. 2.4-40a can be written

$$M_{y_W} = -z_W D + x_W L \tag{2.4-40b}$$

Similarly, the *axial force A* is defined as X_B and the *normal force N* is defined as $-Z_B$; therefore, eq. 2.4-41a becomes

$$M_{y_B} = z_B A + x_B N \tag{2.4-41b}$$

M_{y_B} is the same as M_B in eq. 2.3-5.

Although the force's c.p. is located by either (x_W, z_W) or (x_B, z_B), the latter is generally preferable because it is more nearly invariant with flight condition. For example, the c.p. of a slender triangular wing remains close to the 2/3-root-chord point for all α in (0, 90 deg); representing this c.p. in wind axes would involve the sine and cosine of α. Nevertheless, other planforms can experience large c.p. shifts with Mach number, as noted earlier in this section.

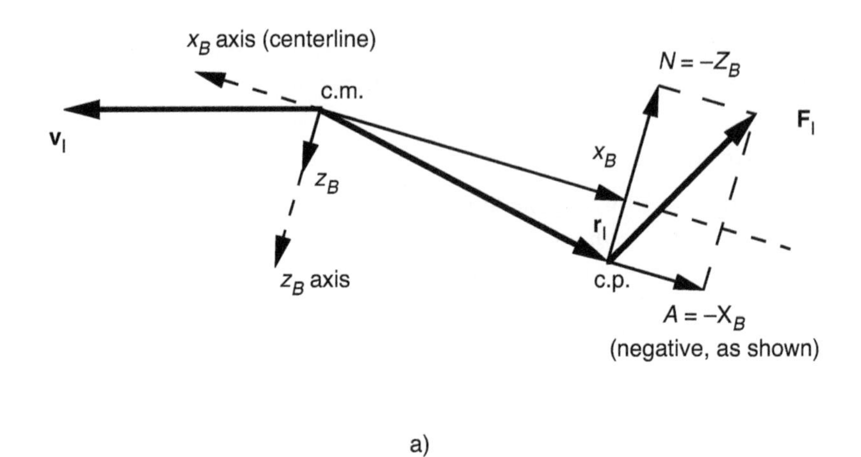

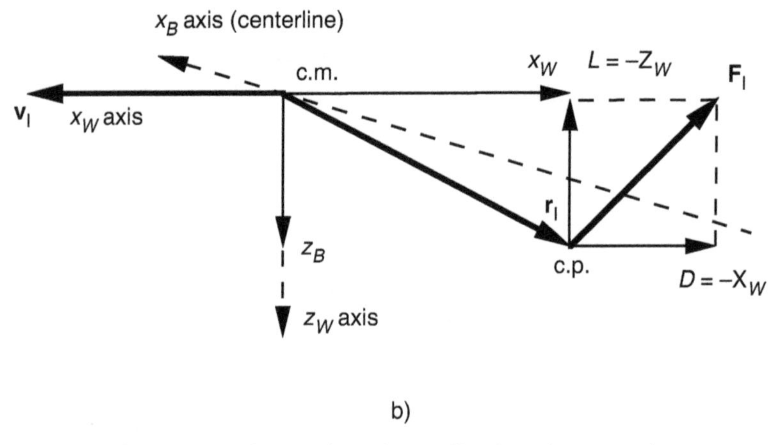

Fig. 2.4-9. Pitching moment due to a force that is offset from the center of mass. Resolution of force and movement arm in (a) body axes and (b) wind axes.

Because a lifting surface has chordwise length, it can provide a pure moment *couple* as well as lift and drag. Consider two small rockets thrusting with equal force T in opposite directions so that the net force is zero (Fig. 2.4-10a). If the rockets are separated by the distance Δx, the moment about the mid-point is $(T\Delta x/2 + T\Delta x/2) = T\Delta x$. However, the moment about either rocket is $(0 + T\Delta x) = (T\Delta x + 0) = T\Delta x$. For that matter, the moment about any point in the plane is $T\Delta x$; hence, the two rockets form a pure couple.

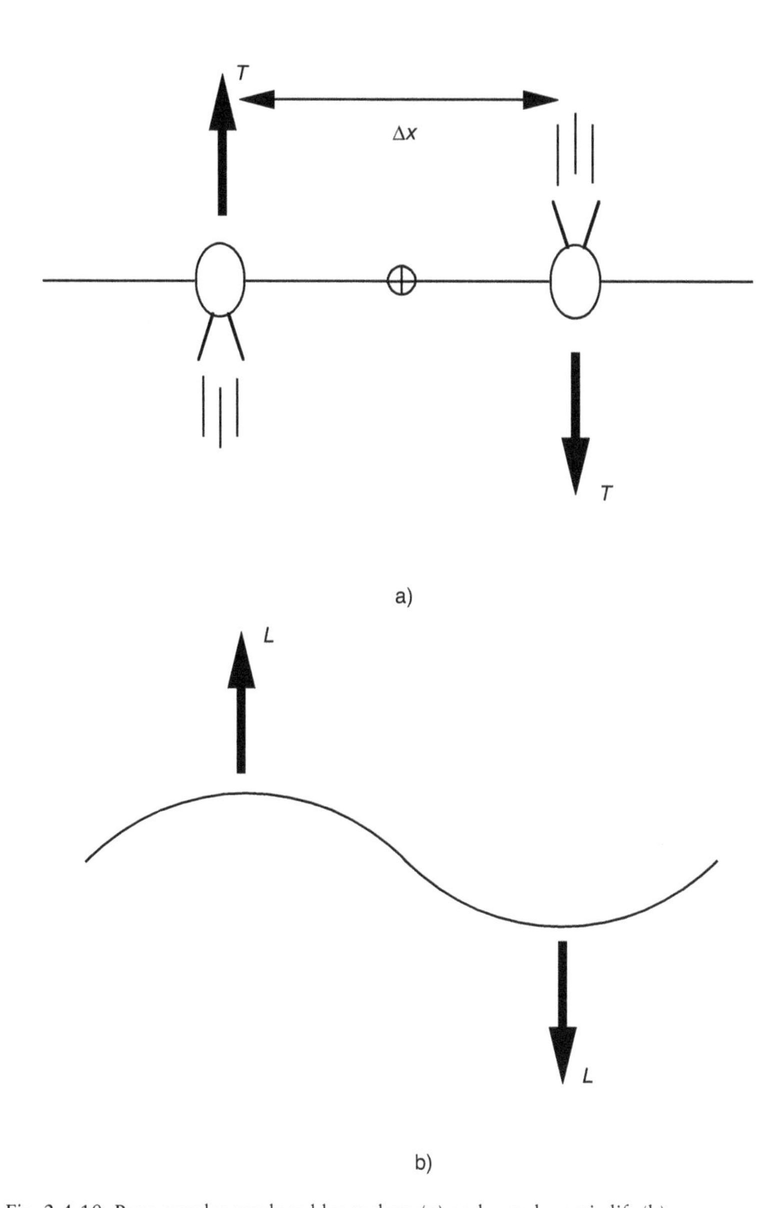

Fig. 2.4-10. Pure couples produced by rockets (a) and aerodynamic lift (b).

The aerodynamic equivalent is sketched in Fig. 2.4-10b. In this idealization, the positive lift produced by the forward half of the wing is opposed by the negative lift of the aft half, generating a pure couple M_c. Upward, or *reflex*, curvature of the aft chord section helps trim the pitching moment of a flying-wing aircraft. Wings of conventional aircraft rarely have the reflex curvature shown here; however, camber induces M_c through its effects on pressure and shear-stress distributions. Thus, the total pitching moment due to a wing can be represented by its pure-couple pitching moment plus its net axial and normal forces, each multiplied by the appropriate moment arm:

$$M_{y_B} = M_c + z_B A + x_B N \tag{2.4-42}$$

From equation 2.3-20, the corresponding contribution to the pitching moment coefficient is

$$C_m = \frac{M_{y_B}}{\bar{q}S\bar{c}} = C_{m_c} + \frac{1}{\bar{c}}(z_B C_A + x_B C_N) \tag{2.4-43}$$

where C_A and C_N are nondimensional axial- and normal-force coefficients. [C_A and $-C_N$ are related to C_D and C_L through $\mathbf{H}_W^B$ (eq. 2.2-33), as in eq. 2.3-3.]

The wing's pitching moment coefficient is the integrated result of pressure and shear-stress distributions over the entire wing surface. Neglecting the small contributions of axial force and shear stress, C_m can be estimated by

$$C_m = \frac{1}{\bar{q}S\bar{c}} \int_{-b/2}^{b/2} \int_{x_{TE}(y)}^{x_{LE}(y)} [p_z(x, y)_{lower} - p_z(x, y)_{upper}] x(y)\, dx\, dy \tag{2.4-44}$$

This equation captures the effects of the pure couple as well as the normal force. As the pressures vary with α, C_m also varies with α.

The total aerodynamic pitching moment is the sum of contributions from the wing, fuselage, horizontal tail, and mutual interference effects,

$$\begin{aligned} M_y &= M_{y_{wing}} + M_{y_{fus}} + M_{y_{tail}} + M_{y_{int}} \\ &= (M_{y_{wing}} + M_{y_{fus}} + M_{y_{tail}} + M_{y_{int}})_c + (zA + xN)_{wing} \\ &\quad + (zA + xN)_{fus} + (zA + xN)_{tail} + (zA + xN)_{int} \end{aligned} \tag{2.4-45}$$

where x and z describe the body-axis locations of the individual centers of pressure relative to the c.m. A and N are the forces produced by individual components with interference corrections. For many configurations, the zA contributions to the pitching moment are negligible: the vertical distances between the centers of mass and pressure are much less than the axial distances, and the axial forces are much smaller than the normal forces. In such case, the aircraft's pitching moment coefficient is approximately

$$C_m = \frac{M_y}{\bar{q}S\bar{c}} \approx C_{m_c} + \frac{1}{\bar{c}}[(xC_N)_{wing} + (xC_N)_{fus} + (xC_N)_{tail} + (xC_N)_{int}] \tag{2.4-46}$$

where the normal force coefficients are all referenced to S, as in eq. 2.4-21. Following eq. 2.4-20 and 2.4-21, the total normal force coefficient and its sensitivity to angle-of-attack variation are

$$C_N = C_{N_{wing}} + C_{N_{fus}} + C_{N_{tail}} + C_{N_{int}} \tag{2.4-47}$$

$$C_{N_\alpha} = C_{N_{\alpha_{wing}}} + C_{N_{\alpha_{fus}}} + C_{N_{\alpha_{tail}}} + C_{N_{\alpha_{int}}} \tag{2.4-48}$$

The *center of pressure* for the aircraft relative to its center of mass, $x_{c.p.}$, is the weighted average of component contributions

$$x_{cp} = \frac{[(xC_N)_{wing} + (xC_N)_{fus} + (xC_N)_{tail} + (xC_N)_{int}]}{C_N} \tag{2.4-49}$$

In the linear region near zero angle of attack, x_{cp} can be based upon the normal force sensitivities to α,

$$x_{cp} = \frac{[(xC_{N_\alpha})_{wing} + (xC_{N_\alpha})_{fus} + (xC_{N_\alpha})_{tail} + (xC_{N_\alpha})_{int}]}{C_{N_\alpha}} \tag{2.4-50}$$

The moment coefficient (eq. 2.4-46) can be expressed as

$$C_m = C_{m_c} + \left(\frac{x_{cp}}{\bar{c}}\right)C_N \tag{2.4-51}$$

The pitching moment sensitivity to the angle of attack is

$$C_{m_\alpha} = \frac{\partial C_m}{\partial \alpha} = \frac{\partial C_{m_c}}{\partial \alpha} + \left(\frac{C_N}{\bar{c}}\right)\frac{\partial x_{cp}}{\partial \alpha} + \left(\frac{x_{cp}}{\bar{c}}\right)C_{N_\alpha} \tag{2.4-52}$$

as C_{N_α} is approximately the lift slope, and both the pure-couple moment and the center of pressure could change with α. C_{m_α} is critical to the air craft's static stability in pitch, and it should be negative for stability. Thus, the sum of the terms in eq. 2.4-52 should be negative.

To this point, we have taken the center of mass as the origin for defining the axial length x, but the c.m. varies with aircraft loading. To examine the effects of center-of-mass variation, we choose a point fixed in the airframe, such as the nose tip or a particular fuselage bulkhead, as the origin (or *fiducial point*), replacing x in the prior equations by $(x - x_{cm})$. Distance *aft* of a forward fiducial point is *negative*, and $(x - x_{cm})$ is negative if x is behind x_{cm}. Equations 2.4-51 and 2.4-52 are rewritten as

$$C_m = C_{m_c} + \left(\frac{x_{cp} - x_{cm}}{\bar{c}}\right) C_N \tag{2.4-53}$$

$$C_{m_\alpha} = \frac{\partial C_{m_c}}{\partial \alpha} + \left(\frac{C_N}{\bar{c}}\right)\frac{\partial x_{cp}}{\partial \alpha} + \left(\frac{x_{cp} - x_{cm}}{\bar{c}}\right) C_{N\alpha} \tag{2.4-54}$$

The aircraft's *aerodynamic center* (or *neutral point*) is defined as the center-of-mass location x_{ac} that forces to zero. From eq. 2.4-54, the aerodynamic center (*a.c.*) is located at C_{m_α}

$$x_{ac} = \frac{\bar{c}}{C_{N\alpha}}\left[\frac{\partial C_{m_c}}{\partial \alpha} + \left(\frac{C_N}{\bar{c}}\right)\frac{\partial x_{cp}}{\partial \alpha} + \left(\frac{x_{cp}}{\bar{c}}\right) C_{N\alpha}\right] \tag{2.4-55}$$

Defining the nondimensional axial position $h = x / \bar{c}$, this can also be expressed as

$$\begin{aligned} h_{ac} &= \frac{\dfrac{\partial C_{m_c}}{\partial \alpha} + C_N \dfrac{\partial h_{cp}}{\partial \alpha} + h_{cp} C_{N\alpha}}{C_{N\alpha}} \\ &= h_{cp} + \frac{\dfrac{\partial C_{m_c}}{\partial \alpha} + C_N \dfrac{\partial h_{cp}}{\partial \alpha}}{C_{N\alpha}} \end{aligned} \tag{2.4-56}$$

If the pure couple and center of pressure do not vary with angle of attack, as is the case for simple wings with symmetric airfoils, $h_{ac} = h_{cp}$. Thus, for static stability ($C_{m_\alpha} < 0$), the actual center of mass should be in front of the aerodynamic center $[(h_{cm} - h_{ac}) > 0]$. The degree of static stability is indicated by the magnitude of the *static margin*, which is defined as $100(h_{cm} - h_{ac})$ (percent).

Representative moment coefficient variations over a 0–90 deg angle-of-attack range are sketched in Fig. 2.4-11, assuming that $C_{m_c} = 0$ and that rotational centers are adjusted to produce the same moment coefficients at $\alpha = 90$ deg. The aerodynamic center of a linearly tapered wing with elliptical lift distribution can be calculated by eq. 2.4-6, replacing section centers of pressure with section aerodynamic centers. For the example, the $AR = 7$ wing's a.c. is fixed at the quarter chord in the range of linear C_L, gradually shifting to the half chord by $\alpha = 70$ deg. The pitching moment becomes more negative with increasing angle of attack, except in the

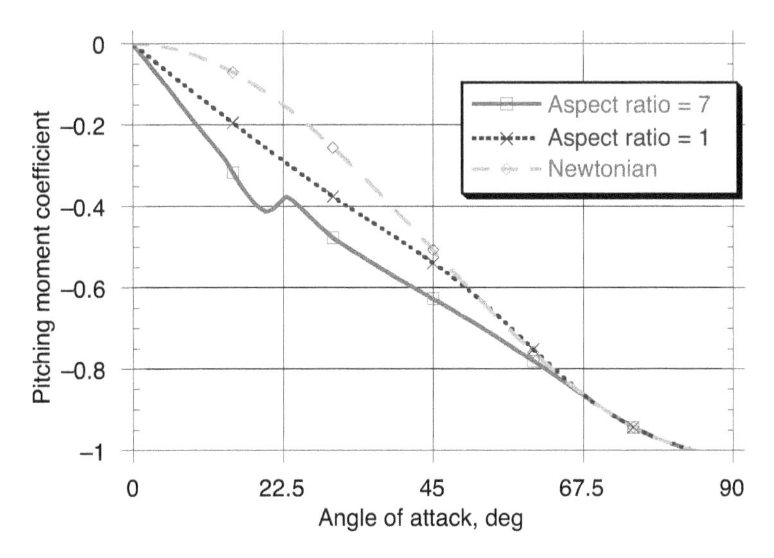

Fig. 2.4-11. Representative variations in pitching-moment coefficient with angle of attack.

region of stall, where decreasing lift causes a *pitch-up* characteristic in this simple model. The aerodynamic center for the other two cases is fixed at the half chord over the entire interval; hence, C_m is directly proportional to $-C_N$ and has a negative slope up to $\alpha = 90$ deg $AR = 1$. The Newtonian example has zero slope at $\alpha = 0$ and 90 deg, with negative slope in between.

In steady flight, pitching moments due to all sources, including aerodynamics, thrust, and control settings, must sum to zero, or the aircraft will not remain trimmed at the desired (generally positive) angle of attack α_{trim}. A negative moment slope (C_{m_α}) in the neighborhood of the trim point provides static stability in response to small angular perturbations. For a negative C_{m_α}, the C_m versus α curve can intercept the $C_m = 0$ axis at positive α only if the pitching moment at zero angle of attack, C_{m_0}, is greater than zero; hence, a positive zero-lift pitching moment must be supplied for trim.

C_{m_0} is affected by C_{m_c} and the net lift at zero α; typically, it is con trolled by adjustments in the elevator deflection or in the incidence of the entire horizontal tail. The ability to trim the pitching moment is essential, as variations in flight speed and aircraft weight change the desired angle of attack, while c.m. shifts due to fuel and payload variation alter C_{m_α}. Figure 2.4-12 illustrates typical effects of moment control and center-of-mass variations on C_m versus α. If the c.m. is behind the aircraft's a.c.,

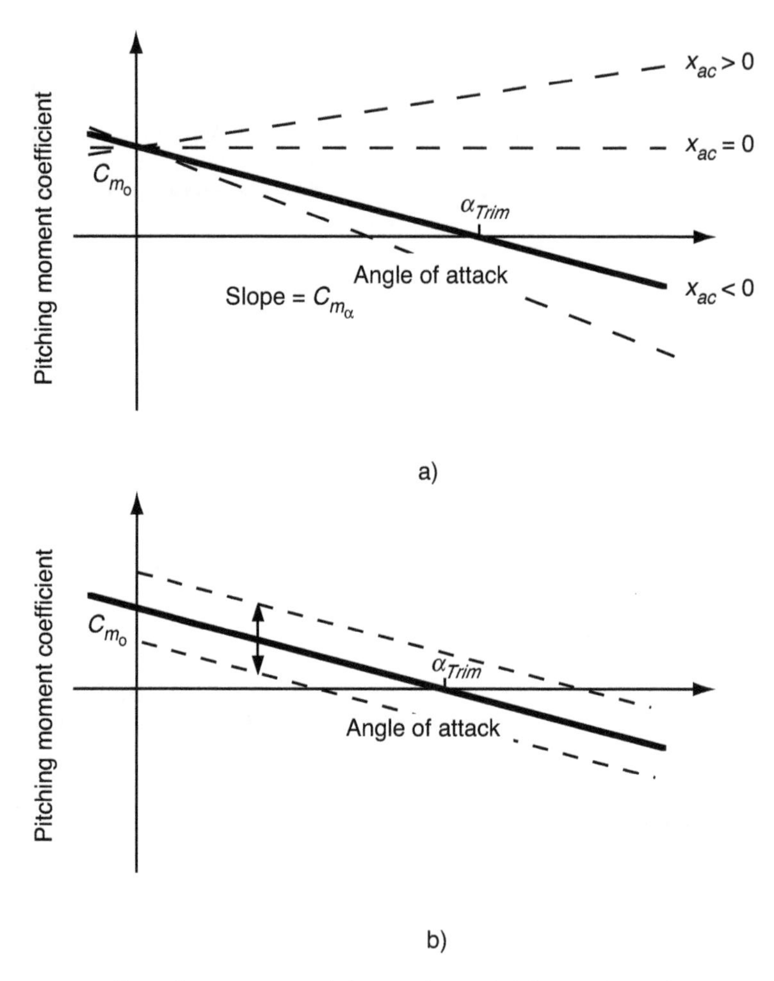

Fig. 2.4-12. Effects of center-of-mass shift (a) and control deflection (b) on the pitching-moment coefficient C_m.

C_{m_α} is positive, the configuration is statically unstable, and C_{m_o} must be negative for trim at positive α.

While canard surfaces are often associated with statically unstable aircraft (the surface itself is ahead of the c.m. and, therefore, destabilizing), stability is dictated by the relative positions of the center of mass and the vehicle's aerodynamic center. Put the c.m. far enough forward, and any configuration is statically stable. As detailed in later chapters, static stability is not necessary if the aircraft has a stability augmentation system, and an aft c.m. may be desired to reduce trim drag or to improve maneuvering performance. Most supersonic aircraft experience an aftward (stabilizing) aerodynamic-center shift in the transition from sub- to supersonic flight, requiring a pitch-trim control change. Fuel is sometimes pumped between forward and aft tanks to minimize the control change while maintaining proper balance.

The Mach effect on pitching moment for straight-wing aircraft is complex, often requiring major trim changes for relatively small changes in flight speed. Although the wing's aerodynamic center moves forward at high-subsonic (or low-transonic) speed, thereby decreasing static stability, the attendant reduction in lift slope reduces the wing's downwash variation with angle of attack, $\partial\varepsilon/\partial\alpha$, increasing the stabilizing effect of the horizontal tail. The net increase in pitching stability leads to a *tuck-under* or *Mach tuck* effect: as the nose pitches down, the airspeed increases, increasing stability, and so on. The large positive trim change required was greater than the pilots of World War II fighter aircraft could provide when the phenomenon was first experienced [A-1, S-8].

The *pitch-up characteristic* mentioned earlier is a characteristic common to many aircraft flying at moderate to high α. It can be the result of the wing's wake interacting with the horizontal tail on both straight-wing, low-tail and swept-wing, high-tail configurations [S-2, S-8]. At low α, configurations with long, slender noses and delta wings evidence a single, bound vortex pattern due to both nose and wing; separation of the nose and wing vortices with increasing α can cause a large local variation in C_m. The pitch-up feature introduces the dangerous possibility of *deep stall*, where the aircraft stabilizes at an angle of attack that is too high to sustain level flight.

The fuselage and nacelles normally produce destabilizing pitching moments. One reason is that a slender streamlined shape positioned at positive α generates a pure couple with positive C_{m_α}. From [M-7], the fuselage's contribution can be estimated as

$$C_{m_{\alpha_{fus}}} = 2K \frac{Vol_{fus}}{S\bar{c}} \tag{2.4-57}$$

where Vol_{fus} is the volume of the fuselage and K depends on the fuselage's fineness ratio:

$$K \approx 1 - (d_{max}/l)^{1.3} \tag{2.4-58}$$

Flow separation at a blunt base reduces the magnitude of the moment; the fuselage volume used in eq. 2.4-57 is reduced accordingly, accounting for the moment arm of the missing force. Nacelle moments can be estimated in the same way. The fuselage and nacelles are immersed in the curved flow field induced by the wing's lift. This creates an upflow on the forward section of the fuselage and a downwash on its aft section. Multhopp accounted for the wing's flow field by altering the computation of Vol_{fus} used above [M-7].

Side Force

Side force is the lateral equivalent of normal force, but it has much smaller magnitude and, therefore, less dynamic effect. It is desirable to keep the average value close to zero, and in symmetric flight (zero sideslip angle, roll rate, and yaw rate) at low angle of attack, the aerodynamic side force of a symmetric aircraft *is* zero unless control surfaces are deflected. Only those side forces arising from the sideslip angle β are addressed here. Because β is normally a small angle, it can be assumed that the side force is linear in β. This sensitivity is expressed as

$$Y_{B\beta} \triangleq Y_\beta \triangleq C_{Y_\beta}\bar{q}S \tag{2.4-59}$$

where C_{Y_β} is the partial derivative $\partial C_Y/\partial\beta$ evaluated at the trim condition. The derivative is negative with the usual sign conventions at low to moderate α. At moderate to high α, a symmetric aircraft may experience side force at zero β, as discussed in a later chapter.

The principal contributors to side force are the vertical tail (or *fin*), fuselage, and nacelles; their primary effects are analogous to the lift/normal force effects presented earlier. Thus, the fuselage's side-force sensitivity to β is, from eq. 2.4-19,

$$C_{Y_{\beta fus}} = -2\frac{S_b}{S} \tag{2.4-60}$$

$S_b = \pi d^2/4$, and d is defined by the widest vertical dimension of the base. Nacelle side force is predicted in the same way.

Following eq. 2.4-22, the vertical tail's contribution is

$$C_{Y_{\beta vt}} = \left(\frac{\bar{q}_{vt}}{\bar{q}}\right)\left(1+\frac{\partial\sigma}{\partial\beta}\right)\eta_{vt}\left(\frac{S_{vt}}{S}\right)\left(C_{Y_{\beta vt}}\right)_{vt} \tag{2.4-61}$$

$(C_{Y_{\beta vt}})_{vt}$ is the fin's side-force-coefficient sensitivity to β in an undisturbed flow, referenced to vertical tail area S_{vt}, and η_{vt} is an efficiency factor accounting for elasticity and end-plate effect. The horizontal tail may have an *end-plate effect* on the vertical tail if it is mounted either high ("T tail") or low relative to the fin's center of pressure; mounting in the middle of the vertical tail produces no such effect [S-4]. This effect increases the magnitude of $(C_{Y_{\beta vt}})_{vt}$ by straightening the tip or root flow, increasing the effective aspect ratio of the fin. As defined here, $(C_{Y_{\beta vt}})_{vt}$ is the negative of the lift slope of a typical wing section; otherwise, lift-prediction methods applied to wings can be applied to its estimation.

The fin's side force depends on the dynamic pressure ratio at the tail, $\overline{q}_{vt}/\overline{q}$, and the *sidewash angle* σ, which is due to wing-fuselage interference and propeller-slipstream effects. The sidewash sensitivity to the sideslip angle is expressed by $\partial\sigma/\partial\beta$. At high α, the vertical tail is immersed in the wing wake, reducing its effectiveness. A fighter aircraft's bubble canopy produces a wake that decreases the dynamic pressure at the vertical tail, while a propeller slipstream has the opposite effect. Swirl in a propeller slipstream changes the β experienced by the fin; the fin may be angled from the vehicle centerline to counter this effect at a single flight condition.

Propellers produce both longitudinal and lateral *fin effects* [P-4, S-4]. The longitudinal component is generally negligible compared to the wing's lift, while the lateral component may be significant compared to other side forces.

The wing produces little side force directly, but its flow field affects the side force generated by other parts of the aircraft. The wing's direct contribution is

$$C_{Y_{\beta_{wing}}} = -C_{D_{P_{wing}}} - k\Gamma^2 \tag{2.4-62}$$

where $C_{D_{P_{wing}}}$ is the wing's parasite drag coefficient and Γ is the dihedral angle (rad) [S-4]. For an unswept wing in subsonic flow, k is derived from eq. 2.4-9 with AR replaced by $AR/2$.

Yawing Moment

The moments about the aircraft's z_B axis are generated by the sum of axial and lateral forces, X_i and Y_i, that are displaced from the center of mass. Assuming body-axis coordinates centered at the *c.m.*, the moment is

$$M_z = -\sum_{i=1}^{I} y_i X_i + \sum_{i=1}^{I} x_i Y_i \tag{2.4-63}$$

for the I components that produce yawing moment (e.g., vertical tail, fuselage, and left and right wing surfaces). M_z is the same as N_B in eq. 2.3-5. The aft vertical fins are the primary sources of yawing moments on most aircraft, although differential axial forces on the left and right wing segments as well as fuselage lateral forces produce small yawing moments. In symmetric flight (zero sideslip angle, roll rate, and yaw rate), aerodynamic yawing moments are zero unless control surfaces are deflected. The sideslip angle is normally small, and N_B is approximately linear in β. Aircraft without vertical tails (e.g., Northrop-Grumman B-2 bomber) depend on yaw flaps (Section 3.5) and active feedback to generate yawing moments for stability and control.

The linear sensitivity of yawing moment to sideslip angle is expressed as

$$N_\beta = C_{n_\beta}\overline{q}Sb \tag{2.4-64}$$

The *directional stability effect* C_{n_β} is the partial derivative $\partial C_n/\partial\beta$ evaluated at the trim condition. Positive C_{n_β} is stabilizing because it opposes β perturbations. Its influence is analogous to the pitch-stabilizing function of negative C_{m_α}. Its accurate prediction is complicated by asymmetric flow effects. *Blanketing* of the flow over the tail by the fuselage and wing make C_{n_β} hard to predict at a high angle of attack without empirical data.

The vertical-tail effect is stated as

$$C_{n_{\beta_{vt}}} = \frac{x_{vt}}{b}C_{Y_{\beta_{vt}}} \tag{2.4-65}$$

The side-force coefficient is given by eq. 2.4-61, and x_{vt} is negative aft; hence, the vertical tail effect on the yawing moment is positive (stabilizing). All of the previously mentioned difficulties in estimating $C_{Y_{\beta_{vt}}}$ must be taken into account, and there is the added complexity of accurately estimating the aerodynamic moment arm x_{vt}.

The wing has a complex directional-stability effect that is not accurately estimated by simple theory. It acts through the differential induced drag of opposite wings, yet this term would be neglected in direct application of strip theory. Seckel [S-4] presents an equation of the form

$$C_{n_{\beta_{wing}}} = k_0 C_L \Gamma + k_1 C_L^2 \tag{2.4-66}$$

k_0 is about 0.075 for high-AR wings and k_1 ranges between 0.2 and 0.15 for wings of various aspect ratios and sweep angles in incompressible flow. This estimate is based on analytical studies reported in [P-3, C-1, A-2].

The fuselage effect on C_{n_β} is calculated in the same way as is C_{m_α} (eq. 2.4-57), with a change of sign to account for moment-angle relationships of the opposite sense. Nacelles are treated accordingly.

Rolling Moment

The moments about the aircraft's x_B axis are generated by the sum of lateral and normal forces, Y_i and Z_i, that are displaced from the center of mass. Assuming body-axis coordinates centered at the c.m., the moment is

$$M_x = -\sum_{i=1}^{I} z_i Y_i + \sum_{i=1}^{I} y_i Z_i \tag{2.4-67}$$

for the I components that produce rolling moment. M_x is the same as L_B in eq. 2.3-5. Differential normal forces on the left and right wing segments are the major sources of rolling moments. Vertical fins and horizontal tails also produce rolling moments, while the principal contribution of the fuselage is through interference effects. In symmetric flight (zero sideslip angle, roll rate, and yaw rate), the wing and tail do not produce rolling moments unless control surfaces are deflected. Only those rolling moments that arise from sideslip angle are addressed here, and L_B is characterized as a linear function of β.

The linear sensitivity of rolling moment to sideslip angle is

$$L_\beta = C_{l_\beta}\bar{q}Sb \tag{2.4-68}$$

where C_{l_β} is the partial derivative $\partial C_l/\partial\beta$ evaluated at the trim condition. Negative C_{l_β} is stabilizing, in that it makes the aircraft roll away from the relative wind, converting β to α and, therefore, reducing the magnitude of β. However, lateral-directional stability is complex: rolling and yawing motions are coupled, and effects that stabilize one mode of motion may destabilize another. It is shown in Chapter 6 that C_{l_β} can be too negative for overall stability.

C_{l_β} is called the *dihedral effect* because the geometric dihedral angle Γ of the wing (Fig. 2.4-13a) makes a significant contribution. Figure 2.4-13b presents an idealized flow at positive sideslip angle. Considering just the geometry of the situation, a streamline impinging on the wing's leading edge at Point (a) would leave the trailing edge at Point (b) in symmetric flight, but it leaves at Point (c) in the sideslipped condition. This produces an incremental increase in the local angle of attack α' that equals $\Gamma\beta$ for small angles. The windward wing surface produces more lift than the leeward surface, leading to a negative rolling moment.

Wing sweep also produces a dihedral effect, as suggested by Fig. 2.4-13d. Increasing the sweep angle reduces the lift slope of a wing (eq. 2.4-9). Sideslip effectively reduces the aft sweep of the windward wing while increasing the sweep on the leeward side. The result is a negative rolling moment for positive α and β. The sideslip effect is reversed for a forward-swept wing; hence, a positive rolling moment develops instead.

An estimate of C_{l_β} that accounts for the dihedral and sweep angles Γ and Λ, as well as the taper ratio λ and Mach number $\mathcal{M}$, can be derived using *strip theory*. Contributions of 2-D strips of the wing are integrated from one wing tip to the other to predict the dihedral effect. Assuming that the 2-D lift slope is constant across the wing,

$$C_{l_{\beta_{wing}}} = -\frac{1+2\lambda}{6(1+\lambda)}\left(\Gamma C_{L_\alpha} + \frac{C_L \tan\Lambda}{1-\mathcal{M}^2\cos^2\Lambda}\right), \quad \mathcal{M}^2\cos^2\Lambda \neq 1 \tag{2.4-69}$$

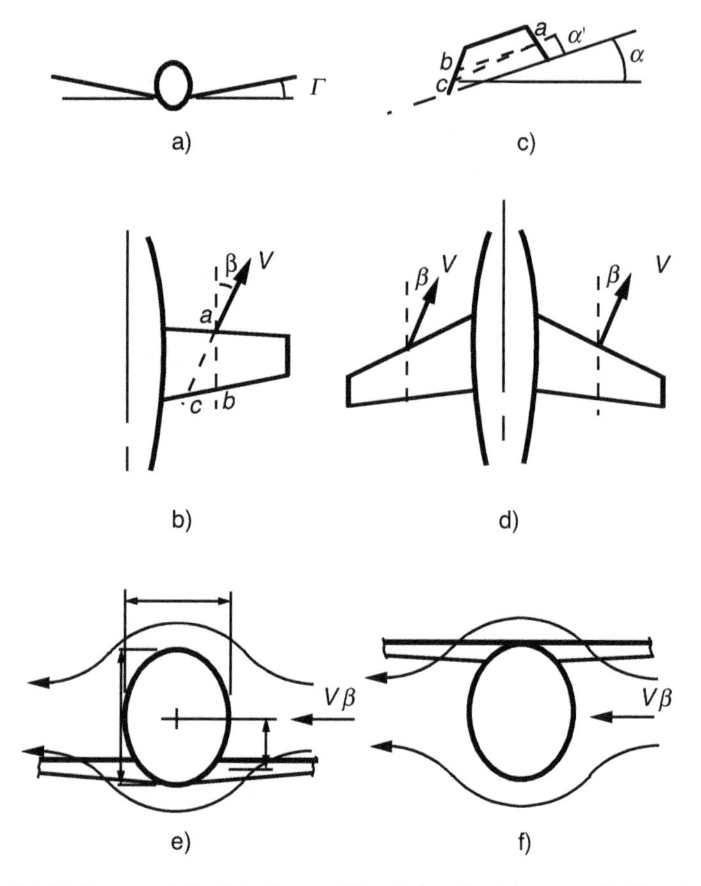

Fig. 2.4-13. Sources of dihedral effect. (a) Dihedral angle of the wing. (b) Plan view showing sideslip geometry. (c) Side view showing angle of attack geometry. (d) Plan view of sideslip angle for a swept wing. (e) Crossflow field for a low wing. (f) Crossflow field for a high wing.

where Λ is evaluated at the quarter chord for subsonic flight and at the leading edge for supersonic flight. C_L and C_{L_α} are the trimmed lift coefficient and lift slope of the wing.

The mounting of the wing relative to the fuselage centerline produces an interference dihedral effect. With airspeed V and small sideslip angle β, there is a freestream crossflow component of $V\beta$. The crossflow around the fuselage has upwash and downwash components, as shown in Figs. 2.4-13e and 2.4-13f, immersing the wing in a rotational flow whose sense depends on wing location. The crossflow over a wing mounted low on the fuselage encourages a positive rolling moment that can be estimated by

$$C_{l_{\beta_{wf}}} = 1.2\sqrt{AR}\,\frac{z_{wing}(h+w)}{b^2} \tag{2.4-70}$$

where h and w are the height and width of the fuselage, and z_{wing} is the vertical distance of the wing's quarter chord below the centerline. A high wing mounting produces negative interference moment. Small aircraft with low-mounted wings are typically designed with noticeable geometric dihedral to augment the dihedral effect, while high-wing configurations need little or no dihedral for satisfactory C_{l_β}. Aircraft with low-mounted aft-swept wings typically have no geometric dihedral on the wing at rest, although air loads bend the wing to slightly positive dihedral in flight. Some high-mounted swept-winged configurations have negative dihedral (or *anhedral*) to limit C_{l_β}. Horizontal-tail dihedral effects are similar to those of the wing, with consideration given to differences in size and location. A horizontal-tail dihedral or anhedral may be used, although the purpose may not be to enhance C_{l_β} but to improve the tail's pitching-moment effectiveness by removing it from the wing's wake at normal flight conditions.

The vertical tail contributes to C_{l_β} because it produces a side force when sideslipped, and its center of pressure is normally above the vehicle's centerline. This contribution is estimated as

$$C_{l_{\beta_{vt}}} = -\frac{z_{vt}}{b}C_{Y_{\beta_{vt}}} \tag{2.4-71}$$

where z_{vt} is the body-axis-vertical position of the tail's *a.c.* (positive down), and $C_{Y_{\beta_{vt}}}$ is its side-force derivative (eq. 2.4-61; negative with the y_B axis out the right wing). This expression provides little insight into the vertical tail's function, as it merely formalizes the nondimensional force-moment relationship and neglects possibly important fuselage interference effects.

Ground Effect

An aircraft flying very near the ground—at an altitude of one wingspan or less—experiences aerodynamic forces and moments that are quite different from those at greater height. The wing's downwash is constrained by the ground plane, so the pressure distributions on wing, tail, and fuselage are altered. While the ground effect on an isolated wing is often estimated using lifting-line theory, Küchemann warns that this is an oversimplification.

It ignores wing thickness effects, and it is inaccurate for swept wings, where the height of root and tip are quite different at even a modest angle of attack [K-5]. Furthermore, the ground effect is affected by the altitude rate [P-2]. Ground effects for a full aircraft configuration are reported in [C-3].

With these caveats in mind, the simplified approach predicts that ground proximity has the same effect on the flow as increasing the wing's aspect ratio. This, in turn, increases the lift slope and decreases the induced drag, as generally observed in practice. The effective aspect ratio AR_{eff} predicted in [F-3] can be modeled as

$$AR_{eff} = AR\frac{1 + 44(h/b)^2}{44(h/b)^2} \tag{2.4-72}$$

where h is the height of the wing quarter chord above the ground, and b is the wingspan. AR_{eff} is used to revise estimates of C_{L_α} and C_{D_i}. Küchemann presents an empirical correction to the normal force coefficient C_N of slender wings,

$$C_{N_{eff}} = C_N\left[1 + \frac{0.09}{\pi AR}C_{N_\alpha}\left(\frac{h}{b}\right)^{-1.4}\right] \tag{2.4-73}$$

where C_N and C_{N_α} are out-of-ground-effect values.

Pitching, rolling, and yawing moments are also subject to ground effects. For conventional wing/aft-tail configurations, the downwash on the tail is reduced by the ground effect, and the tail is in closer proximity to the ground than the wing. Both factors would normally produce a pitch-down moment, favorable in landing and adverse during takeoff. Slender delta wings also feel this effect, and they exhibit restoring rolling moments that tend to level the wings as the ground is approached. Differential-induced drag on the rolled wing would lead to yawing moments.

Following eq. 2.4-73, we can estimate the general trends in moments induced by ground effect. The pitching moment produced by an isolated horizontal tail would be

$$C_{m_{ht_{eff}}} = \frac{-l_{ht}C_{N_{ht}}}{\bar{c}}\left[1 + \frac{0.09}{\pi AR_{ht}}C_{N_{ht_\alpha}}\left(\frac{h_{ht}}{b}\right)^{-1.4}\right] \tag{2.4-74}$$

l_{ht} is the distance along the body axis between the horizontal tail's aerodynamic center and the aircraft's center of mass, $C_{N_{ht}}$ is the horizontal tails's normal-force coefficient, written for the aircraft's reference area, $C_{N_{ht_\alpha}}$ is the coefficient's sensitivity to angle of attack, AR_{ht} is the tail's aspect ratio, and h_{ht} is the height of the tail's aerodynamic center above the ground. The tail height can be expressed as

$$h_{ht} = -z_{ht} = -z_{cm} + (z_{cm} - z_{ht}) = -z_{cm} - l_{ht}\sin\theta \tag{2.4-75}$$

Hence, the moment is a function of both the aircraft's height and its pitch angle.

The rolling moment produced by the wing takes the form

$$C_{l_{w_{eff}}} = 0.045\frac{m_w C_{N_w}}{\pi AR_w b} C_{N_{w\alpha}}\left[\left(\frac{h_{w_L}}{b}\right)^{-1.4} - \left(\frac{h_{w_R}}{b}\right)^{-1.4}\right] \tag{2.4-76}$$

Here, m_w is the lateral distance of the left and right wing panels' aerodynamic centers from the centerline, C_{N_w} is the total wing's normal-force coefficient, $C_{N_{w\alpha}}$ is its sensitivity to angle of attack, AR_w is the wing's aspect ratio, and h_{w_L} and h_{w_R} are the heights of the left and right aerodynamic centers:

$$h_{w_L} = -z_{w_L} = -z_{cm} + (z_{cm} - z_{w_L}) = -z_{cm} + m_w\sin\phi \tag{2.4-77}$$

$$h_{w_R} = -z_{w_R} = -z_{cm} + (z_{cm} - z_{w_R}) = -z_{cm} - m_w\sin\phi \tag{2.4-78}$$

The rolling moment due to the ground effect is an odd function of roll angle, with zero magnitude and slope at zero roll angle. The yawing moment due to the ground effect would take a similar form.

2.5 Thrusting Characteristics of Powerplants

An aircraft requires *thrust* for takeoff, to climb to cruising altitude, and to sustain forward flight. Thrust is a reactive force, in which a *working fluid*, such as air or the products of a combustion process, is vigorously expelled from the aircraft. The powerplant that produces thrust actually performs two functions: *energy conversion* and *thrust generation.* In a combustion engine, the process begins with the conversion of chemical energy to heat, and that heat is used to accelerate the working fluid to produce thrust. Aircraft store fuel on board and use *air-breathing engines*, drawing oxygen for combustion and the bulk of the working fluid needed to produce the reactive force from the surrounding atmosphere. A *chemical rocket* stores all the energy-producing propellants and working fluid on board.

This section focuses on three air-breathing propulsion types: reciprocating engines driving propellers; turboprop, turbofan, and turbojet engines; and ramjet or scramjet engines. The mechanisms of energy conversion and thrust generation are separate in the *reciprocating engine.*

The fuel and air react in combustion chambers to produce linear motions of pistons; the linear motion is converted to rotary motion by a crankshaft, which is connected to a propeller that produces thrust. A *turbojet engine* combines the two functions. Air is compressed for combustion by rotating axial- or centrifugal-flow compressor blades; high-pressure air mixes and reacts with fuel in the combustion chamber; exhaust gas spins a turbine that turns the compressor; and the products of combustion are expelled at high speed. Driving the compressor removes some energy from the exhaust, but the flow is fast enough to produce a net thrust. A turbine extracts most of the energy from the exhaust of a *turboprop* or *turbofan engine* to drive the propeller or ducted fan that produces thrust.

The *ramjet* has neither compressor nor turbine; its inlet provides the necessary compression by forward motion through the atmosphere, so the engine cannot produce thrust at zero airspeed. A supersonic-combustion ramjet (*scramjet*) operates on the same principle, its primary distinction being the speed of the flow of air and fuel in the combustion chamber. The *pulsejet*, a variation on the ramjet, relies on a tuned flapper valve or detonation wave to periodically open and close the forward end of the combustion chamber. Air is sucked into the chamber between combustion pulses, and significant static thrust can be generated.

The effectiveness of a powerplant in converting propellant mass to thrust or power is conveyed by its specific impulse and specific fuel consumption. The *specific impulse* I_{sp}, measured in seconds, is defined as,

$$I_{sp} \triangleq \frac{T}{\dot{m}_{prop} g_o} = \frac{T}{\dot{W}_{prop}} \tag{2.5-1}$$

where T is the thrust (N), $\dot{m}_{prop}$ is the mass flow rate (kg/s) of the propellant carried *on the vehicle*, g_o is the sea-level gravitational constant (m/s^2), and $\dot{W}$ is the equivalent weight flow rate. For a rocket, $\dot{m}_{prop}$ includes oxidizer as well as fuel, whereas for an air-breathing engine it includes only the fuel. Consequently, the specific impulse of an air-breathing engine can be much higher than that of a chemical rocket. The *characteristic speed* of the exhaust flow, c, is related to the specific impulse as

$$c \triangleq g_o I_{sp} = \frac{T}{\dot{m}_{prop}} \tag{2.5-2}$$

Thrust-specific fuel consumption (*TSFC*) is effectively the inverse of eq. 2.5-1, although the units are slightly different: pounds of fuel burned per hour per pound of thrust, or the SI equivalent. The *brake-specific fuel consumption* (*BSFC*) measures the amount of fuel burned per hour per unit of power (e.g., horsepower), the "brake" referring to the manner in which power is measured. The TSFC of a turbojet engine and the BSFC of a reciprocating engine are approximately constant over a broad range of altitudes, although they increase at high altitude [A-5]

An aircraft's cruising range depends on aerodynamic efficiency (L/D), airspeed, the ratio of initial to final mass [$m_i/m_f = (m_f + m_{fuel})/m_f$], and the specific impulse, as reflected by the *Breguet range equation.* With fixed I_{sp}, V, and L/D, the cruising range is [K-4]

$$\text{Cruising Range} = I_{sp}V\left(\frac{L}{D}\right)\log\left(\frac{m_i}{m_f}\right) \tag{2.5-3}$$

(A slightly different form of the equation is presented in Section 2.6.) Representative values of I_{sp} are presented in Fig. 2.5-1 as functions of Mach number. Propeller-driven aircraft have the advantage at subsonic $\mathcal{M}$ (not shown), turbofan and turbojet engines hold the edge in transonic and low-supersonic flight, ramjets dominate in the region from $\mathcal{M} = 3$ to 6, and scramjets are best (theoretically) for hypersonic propulsion. Chemical rockets make a relatively poor showing by this measure.

Air-breathing engines are sensitive to many other factors, most notably air temperature, pressure, Mach number, and control settings. Powerplant performance is governed by three factors: mass flow of air, mass flow of fuel (typically 2–8 percent of the air mass flow), and powerplant geometry, which are subject to varying degrees of control that depend on engine type. For a fixed throttle setting, the thrust output of a given powerplant over some limited range of normal flight conditions is proportional to ρV^n, where n is less than zero for propeller-generated thrust, about zero for turbojets, and greater than zero for ramjets. At a fixed airspeed, the thrust of air-breathing engines decreases with altitude.

Thrust can be modeled in terms of a nondimensional *thrust coefficient* C_T similar to the aerodynamic coefficients of the previous section:

$$\text{Thrust} = C_T S \rho V^2/2 \tag{2.5-4}$$

In deriving C_T, S may be the propeller disk area or the engine exhaust area; however, in a specific aircraft installation, it may be preferable to define C_T in terms of the aircraft's reference area.

Unlike automobile engines, which must accelerate and perform well over a range of rapidly changing power settings, aircraft engines typically operate at just a few working points that are sustained for extended periods of time. The principal exception occurs in aerobatics or air combat, where rapid power changes must be carefully integrated with the control of flight motions. An increasing number of fast feedback control loops are needed for high performance, good transient response, and dynamic stability in modern engines, although external commands are

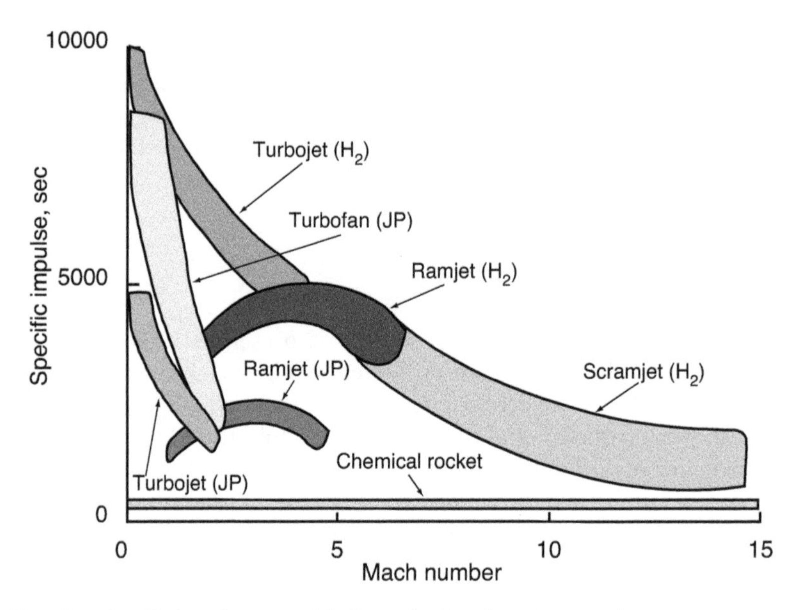

Fig. 2.5-1. Specific impulse versus Mach number for aircraft powerplants.

issued infrequently. These external commands may be expressed in terms of "pseudocontrols" [*intake manifold pressure* and *propeller rpm* in the case of reciprocating engines, and *high-pressure compressor rpm* (N_2) or *engine pressure ratio* (EPR) for turbine engines], which are readily measured output variables that have "tight" relationships with real control variables like fuel or airflow. In modern aircraft, the throttle lever may simultaneously change a number of real engine controls: airflow rate, mixture ratio, and ignition timing in a reciprocating engine, or fuel flow rate, exhaust nozzle area, guide vane and stator angles, bleed valve settings, and inlet ramp angle in a turbofan engine. Whether using real or pseudocontrols, the principal command objective is to change the thrust level while maintaining minimal fuel use.

Aircraft engines have both minimum and maximum throttle settings. The minimum (flight idle) setting is governed not only by the need to sustain the combustion process but by the ancillary requirement of powering accessories that provide heating, cooling, and electrical power. Thus, the allowable idle setting may be a significant percentage (up to 40) of the maximum setting. The upper throttle limit is set by high-temperature material properties or operating restrictions related to flight conditions.

Thrust or power levels corresponding to minimum and maximum throttle settings are functions of altitude and airspeed for a given installation.

Engine characteristics are traditionally portrayed in *performance maps*. These charts provide a basis for modeling the effects of control and flight conditions on engine performance. "Raw" performance data for uninstalled powerplants neglect the effects of nacelles, inlets, and exhaust nozzles. Performance characteristics must be modified to account for these factors, which are particularly significant in high-speed flight and at high angle of attack or sideslip.

For the greatest increase in mechanical energy, thrust is applied in the direction of flight; therefore, the *difference* between thrust and drag $(T - D)$ is an important quantity. The *thrust-to-weight ratio* T/W is an indicator of maximum acceleration, although an aircraft can always accelerate in level flight if $(T - D)$ is positive. Typical sea-level values of T/W are 0.3 for a transport airplane and more than 1 for high-performance aircraft.

Propellers

Actuator disk theory is a useful tool for studying the thrust characteristics of propellers [R-1, F-4, G-2]. The theory represents a spinning propeller as a uniform disk through which air flows, neglecting details of the actual flow over the propeller blades. It assumes that velocity is constant across the disk face, pressure is uniform over the disk, no rotation is given to the flow, the flow through the disk is a stream tube separate from the surrounding flow, and the flow is incompressible. Furthermore, the actuator disk is alone, with no fuselage or nacelle nearby. From considerations of flow continuity and momentum change [e.g., M-3], the thrust T of the propeller can be represented as

$$T = 2\rho A(V + \Delta V_i)\Delta V_i \tag{2.5-5a}$$

V is the freestream speed of the incoming flow, A is the disk area $\pi d^2/4$, d is the propeller diameter, and ΔV_i is the velocity increase across the actuator disk or *inflow velocity*. After passing through the disk, the stream-tube cross-sectional area decreases, and the final velocity increase of the flow is $\Delta V_f = 2\Delta V_i$. Therefore, eq. 2.5-5a indicates that the thrust equals the mass flow rate, $\dot{m} = \rho A(V + \Delta V_i)$, times the velocity increase $2\Delta V_i$, or

$$T = \dot{m}\Delta V_{f_i} \tag{2.5-5b}$$

The power $\mathcal{P}$ produced by the disk is

$$\mathcal{P} = T(V + \Delta V_i) \tag{2.5-6}$$

TV is *useful power* that can be applied to propelling the aircraft, and $T\Delta V_i$ is *induced power* that accelerates the flow downstream. The *ideal propulsive efficiency* η_I is then defined as

$$\eta_I = \frac{TV}{T(V + \Delta V_i)} = \frac{V}{V + \Delta V_i} = \frac{V}{V + \Delta V_f/2} \tag{2.5-7}$$

The lower is the inflow velocity, the higher the propulsive efficiency. For a fixed thrust level, inflow velocity is decreased by increasing the disk area. This definition of efficiency neglects the power lost to drag-induced torque on the propeller.

From eq. 2.5-5, the inflow velocity can be expressed as a function of the thrust,

$$\Delta V_i = -\frac{V}{2} + \sqrt{\left(\frac{V}{2}\right)^2 + \frac{T}{2\rho A}} \tag{2.5-8}$$

so the power can be written as a function of thrust, airspeed, and air density:

$$\mathcal{P} = T\left[\frac{V}{2} + \sqrt{\left(\frac{V}{2}\right)^2 + \frac{T}{2\rho A}}\right] \tag{2.5-9}$$

At zero airspeed,

$$\mathcal{P} = \frac{T^{3/2}}{\sqrt{2\rho A}} \tag{2.5-10}$$

and the static thrust can be written as

$$T = \mathcal{P}^{2/3}\sqrt[3]{2\rho A} \tag{2.5-11}$$

If $\mathcal{P}$ is the power applied to (and absorbed by) the actuator disk, eq. 2.5-11 computes the thrust that is generated. Equation 2.5-11 confirms that large-diameter propellers convert power to thrust more efficiently than small propellers. Finding the thrust corresponding to a given power input with nonzero airspeed requires that eq. 2.5-9 be rewritten as a cubic equation in T:

$$T^3 + 2\rho AV\mathcal{P}T - 2\rho A\mathcal{P}^2 = 0 \tag{2.5-12}$$

The positive real root of this equation is the propeller's thrust.

Real propellers are less efficient than actuator disks because, like wings, they generate drag. These drag effects are conveyed by the *propeller efficiency* η_P, which depends on chord section, thickness, taper, twist, and number of blades (typically 2–4). For a particular propeller, η_P is a function of blade angle (or *pitch angle* β, which should not be confused with either the aircraft's pitch or sideslip angle) and *advance ratio* $J = V/nd$, where n is the rotational rate of the propeller (revolutions per second). With fixed pitch, η_P is a strong function of J, taking very low values when J is both large and small, and reaching a maximum value that is typically above 80 or 85 percent at some intermediate value of J. A variable-pitch propeller keeps η_P high over a wide range of advance ratios by assuring that local angles of attack remain below stall, preventing buildup of induced drag. The actual thrust delivered to the aircraft can be written as

$$T = \eta_P(\beta, J)T_I(\mathcal{P}, \rho, V) = \eta_P\eta_I \frac{\mathcal{P}}{V} \tag{2.5-13}$$

Although η_P theoretically goes to zero when $J = 0$, a variable-pitch propeller's static thrust may be larger than its thrust in forward flight [M-3].

It has been assumed that the propeller is in one-dimensional (axial) flow and is located away from the influence of other bodies; however, propellers must generate thrust in real flows. The nonaxial-flow issue is complex, but it can be addressed using actuator disk theory, as in [S-18]. The effects of adjacent bodies are readily addressed by defining η_P as an *installed propeller efficiency*. Propeller efficiency should also account for slipstream swirl and Mach number, both of which reduce thrust. Propeller swirl effects can be reduced by downstream stator blades (with some drag penalty) or eliminated by counter-rotating propellers. The propeller-tip Mach number is

$$\mathcal{M}_{tip} = \frac{\sqrt{V^2 + (\pi nd)^2}}{a} \tag{2.5-14}$$

Hence, the tips approach sonic speed before the aircraft as a whole, decreasing propeller efficiency. Note that eq. 2.5-14 places limits on propeller rpm and diameter as well. The scimitar-shaped blades of the propfan engine have aft sweep relative to the flow, improving their transonic characteristics. Their diameters and aspect ratios are low relative to conventional propellers, and the number of blades is high (8–10). Blade aspect ratio is not a commonly used propeller parameter. The *solidity ratio* NcR/pR^2 reflects the product of the blade aspect ratio inverse, c/R, and the number of blades N, divided by p, where c is the average blade chord length and R is the blade radius.

Just as wing efficiency can be increased by end plates, thrust can be increased by surrounding the propeller with a duct; however, the increased thrust is offset by internal and external drag on the duct, so the net worth of the duct is installation dependent. Winglets on propeller tips may improve thrust, although high inertial loads limit their size.

Reciprocating Engines

Although reciprocating engines generating thousands of horsepower propelled large aircraft for decades, current attention focuses on smaller engines for general aviation. These engines are typically air cooled and have four to eight opposed cylinders. There is increasing interest in adapting automotive engines to small aircraft, and a stratified-charge Wankel rotary engine has been considered for aircraft application. Turbo-superchargers—centrifugal-flow compressors driven by engine exhaust—extend the power and operating altitude for this engine type. The BSFC is about 0.5 lb/hr per hp or 0.3 kg/hr per kW (± 20 percent) over a wide range of power levels [M-3], with large engines generally showing better efficiency than small engines.

The power output of a reciprocating engine is governed by intake manifold pressure (MAP) and crankshaft rpm. The sea-level power output traditionally was determined for a given MAP and rpm using a performance map. This power reading was then corrected for ambient air temperature $\mathcal{T}_a$ and carburetor heat ch (if turned on). For modern computation, engine power is stored as a table or polynomial of the form $\mathcal{P} = \mathcal{P}(\text{MAP, rpm}, h, \mathcal{T}_a, \text{ch})$.

Supercharging extends the range of available MAP, and the amount of pressure boost becomes an additional control variable. For a given throttle setting, propeller pitch has a controlling effect on engine rpm and, therefore, on power output; hence, it is possible to implement feedback control that regulates power (or percent of available power) directly.

Thrust is proportional to power (at fixed airspeed), inversely proportional to airspeed (at fixed power), and dependent on propeller and ideal propulsive efficiency factors. Alternatively, it can be expressed in terms of a thrust coefficient, as in eq. 2.5-4:

$$T = \eta_P \eta_I \frac{\mathcal{P}}{V} = C_T S \rho V^2 / 2 \tag{2.5-15}$$

The thrust coefficient for a reciprocating engine-propeller combination is expressed as a linear function of the power setting:

$$C_T = \frac{2\eta_P \eta_I \mathcal{P}}{S \rho V^3} \tag{2.5-16}$$

TURBOPROP, TURBOFAN, AND TURBOJET ENGINES

Turbine engines produce considerably more power per unit weight, burn a lower grade of fuel, and require less maintenance than reciprocating engines; therefore, they are the engines of choice in many classes of modern aircraft. Ducted fan or jet engines are the *only* choice (other than rockets) for flight above $\mathcal{M} = 0.9$ due to compressibility effects on propeller tips.

Actuator disk theory is applicable to turboprop engines, and the basic thrust equation (eq. 2.5-5b) applies to all three turbine-engine variants with a slight addition. Assuming that the pressure equilibrates in the prop or fan slipstream, the thrust can be written as

$$T = \dot{m}_{fan}\Delta V_{fan} + \dot{m}_{core}\Delta V_{core} + A_{core}\Delta p_{core} \tag{2.5-17}$$

where $\dot{m}_{fan}$ is the mass flow rate of air through the fan (or propeller), ΔV_{fan} is the ultimate speed differential produced by the fan, $\dot{m}_{core}$ and ΔV_{core} are similarly defined for the core engine, A_{core} is the core-engine exhaust area, and Δp_{core} is the difference between the pressure in the core-engine exhaust and the ambient air pressure. The *core engine* is the compressor-combustor-turbine unit; taken alone, it is a turbojet engine. The mass flow through the core contains fuel as well as air. The exhaust pressure is dominated by the expansion that occurs in the exhaust nozzle. Engine efficiency is maximized when Δp_{core} is zero, and this term is neglected in the remainder of the discussion.

The ideal thrust of a *turbojet engine*, assuming perfect compression and exhaust expansion, can be written as [K-4]

$$T = \dot{m}V\left\{\left[\left(\frac{\theta_o}{\theta_o - 1}\right)\left(\frac{\theta_t}{\theta_o\tau_c - 1}\right)(\tau_c - 1) + \frac{\theta_t}{\theta_o\tau_c}\right]^{1/2} - 1\right\} \tag{2.5-18}$$

where the mass flow rate is

$$\dot{m} \triangleq \dot{m}_{air} + \dot{m}_{fuel} = \dot{m}_{air}(1 + f) \tag{2.5-19}$$

and f is the fuel/air ratio. θ_o is the *ratio of freestream stagnation temperature to ambient temperature*,

$$\theta_o = \frac{\mathcal{T}_{stag}}{\mathcal{T}_{amb}} = 1 + \frac{1}{2}(\gamma - 1)\mathcal{M}_o^2 = \left(\frac{p_{stag}}{p_{amb}}\right)^{(\gamma - 1)/\gamma} \tag{2.5-20}$$

The temperature ratio is related to the freestream Mach number and to the stagnation/static pressure ratio through the *ratio of specific heats* γ as shown. θ_t is the *ratio of turbine inlet temperature to freestream ambient temperature*, and τ_c is the *compressor-outlet-to-inlet stagnation temperature ratio*. Consequently, the thrust is a function of Mach number, temperature rise across the compressor (which is directly related to the pressure increase imparted by the compressor), and temperature increase in the combustion chamber.

Comparing eq. 2.5-18 to eq. 2.5-17, it can be seen that ΔV_{core} has been expressed as $V\{\cdot\}$; thus the thrust is proportional to airspeed multiplied by a factor that reflects the effectiveness of the engine. The corresponding specific impulse is

$$I_{sp} = \frac{T}{\dot{m}_{fuel}\, g_o} = \frac{hT}{\dot{m} g_o \mathcal{T}_o(\theta_t - \theta_o \tau_c)} \tag{2.5-21}$$

where h is the *heat content of the fuel* (J/kg), c_p is the *constant-pressure specific heat of air* (1,004.7 J/kg-K), and $\mathcal{T}_o$ is the *freestream stagnation temperature* (K). The turbojet's static thrust is not zero as implied by eq. 2.5-18; it is given by

$$T = \dot{m}a\left[\left(\frac{2}{\gamma - 1}\right)\left(\frac{\theta_t}{\tau_c} - 1\right)(\tau_c - 1)\right]^{1/2} \tag{2.5-22}$$

where a is the freestream *speed of sound.*

Actual turbojet performance is degraded by pressure losses in the aircraft's inlet and the engine's diffuser, nonisentropic intake compression and exhaust expansion, incomplete burning, stagnation pressure losses due to viscosity and burning dynamics, variation in gas properties, and nozzle over- or underexpansion [K-4]. Bleed air is used to power accessories and could be applied for lift augmentation or drag reduction. Many of these effects are sensitive to flight conditions, particularly flight at high angles of attack or sideslip. Distortion of the flow at the face of the compressor can affect engine performance, and the engine may stall at very high angles of attack. Inlets designed for supersonic aircraft attempt to keep a plane or conical shock at an appropriate location just inside the inlet cowl. If there is a disturbance, the shock may be spit out, causing an engine *unstart*, a sharp loss of thrust that can lead to large perturbations in flight motions [S-13].

The principal means of controlling thrust is through the temperature (and pressure) increase across the compressor and the temperature increase due to burning fuel. The compressor's effect is proportional to its rotation rate N, while the burners' effect is directly related to fuel flow rate. The ratio of the turbine-outlet stagnation pressure to compressor-inlet stagnation pressure, or *engine pressure ratio EPR*, is affected by these quantities and is a primary measure of engine performance. The upper limit on turbojet thrust is set by the maximum allowable turbine-inlet temperature that can be sustained without damage. The limits are based on the duration of the temperature level and the urgency of the situation. These limits are stated as maximum continuous thrust, maximum cruise thrust, and various overthrusting conditions for emergency use only.

The *turbofan engine* differs from the turbojet in that a ducted axial-flow fan driven by an additional turbine downstream from the core generates a significant portion of the thrust. Fan "throughflow" is subsonic in current installations, although there is interest in supersonic-throughflow fans for high-speed applications. For the turbofan engine, the best propulsive efficiency is obtained when the fan and core exhaust speeds are the same. In this case, the ideal thrust is given by [K-4]

$$T = \dot{m}V(1+\beta)\left\{\left[\frac{\theta_t - (\theta_t/\theta_o\tau_c) - \theta_o(\tau_c - 1) + \beta(\theta_o - 1)}{(\theta_o - 1)(1+\beta)}\right]^{1/2} - 1\right\} \tag{2.5-23}$$

Here, β is the *bypass ratio*, the ratio of fan air to core air. The specific impulse can be estimated by eq. 2.5-20.

Accounting for the nonideal factors mentioned above, models of this sort can be used to generate performance maps, and these maps can be represented by tables or polynomials for flight path computations. For example, the normal rated (maximum cruise, uninstalled) thrust for the turbofan engine powering a four-engine military transport is represented by a third-degree polynomial in [S-14]

$$\begin{aligned} T_{max} = {} & 5{,}733 - 499\mathcal{T}' - 5{,}591h' - 124\mathcal{T}'^2 - 136\mathcal{T}'\mathcal{M}' + 78\mathcal{M}'^2 \\ & + 1{,}178\mathcal{T}'h' + 732\mathcal{M}'h' + 1{,}609h'^2 + 108\mathcal{T}'\mathcal{M}'h' - 875\mathcal{T}'h'^2 \\ & - 806\mathcal{M}'h'^2 + 386h'^3 \end{aligned} \tag{2.5-24}$$

where $\mathcal{T}'$ is the temperature deviation (°F) from the *U.S. Standard Atmosphere* [A-7] divided by 40, $h' = [h\ (\text{ft}) - 25{,}500]/29{,}500$, and $\mathcal{M}' = (\mathcal{M} - 0.55)/0.25$. The actual thrust is then $T = T_{max}\delta T$, where δT is the throttle setting ranging from 0 to 1 (subject to a minimum idle value). Alternatively, simpler models that reflect the effect of air density but neglect explicit dependence on airspeed or Mach number can be derived from empirical data. The business jet example of Appendix B uses the thrust model

$$T_{max}(h) = T_{max}(SL)(e^{-\beta h})^n\left[1 - e^{\alpha(h - h_z)}\right] \tag{2.5-25}$$

In the example, β is the air-density inverse scale height (1/9,042 m^{-1}), $n = 0.7$, h_z is the altitude at which the engine can no longer sustain adequate combustion to produce thrust (17,000 m), and α is a fitting constant (1/2,000 m^{-1}).

If all the oxygen in the air were consumed in reacting with fuel in the combustion chamber (i.e., if the combustion were *stoichiometric*), the exhaust temperature would be too high for the turbine materials. Turbine blades are made of high-temperature materials, including ceramics, and the blades subjected to the highest heat flux are often actively cooled by internal fluid flow. Therefore, a lean mixture is used in the core burners, and there is excess oxygen in the exhaust gas leaving the turbine. Extra fuel can be burnt with this oxygen to produce additional thrust in an *afterburner*. This is an inefficient way to produce thrust at subsonic speed, but it becomes more efficient with increasing Mach number. For $\mathcal{M} > 2$, the afterburner may be more efficient than the core engine in producing thrust [K-4]. The ideal thrust for an afterburning turbojet engine can be estimated from the temperature ratios,

$$T = \dot{m}V\left\{\left[\left(\frac{\theta_a}{\theta_o - 1}\right)\left(1 - \frac{\theta_t/\theta_o\tau_c}{\theta_t - \theta_o(\tau_c - 1)}\right)\right]^{1/2} - 1\right\} \tag{2.5-26}$$

where θ_a is the *ratio of the stagnation temperature at the afterburner outlet to the freestream stagnation temperature* and the other quantities are defined above. The corresponding specific impulse is

$$I_{sp} = \frac{hT}{\dot{m}g_o c_p \mathcal{T}_o(\theta_a - \theta_o)} \tag{2.5-27}$$

Afterburner fuel-flow rate becomes an additional control variable, and there is the option of ducting supplemental air to the afterburner to generate even greater thrust in the *turbine bypass engine*. With increasing Mach number, the afterburner becomes a ramjet, sustaining thrust as the turbojet core is phased out of operation. An engine with this capability would be considered a *hybrid* or *multi-thermal-cycle engine*.

An *afterburning turbofan engine* combines both modifications to the simple turbojet, yielding the ideal thrust

$$T = \dot{m}V(1 + \beta)\left\{\left[\left(\frac{\theta_a}{\theta_o\tau_f}\right)\left(\frac{\theta_o\tau_f - 1}{\theta_o - 1}\right)\right]^{1/2} - 1\right\} \tag{2.5-28}$$

with

$$\tau_f \triangleq \frac{\theta_t + \theta_o(1 + \beta - \tau_c)}{\theta_t/(\tau_c + \beta\theta_o)} \tag{2.5-29}$$

This engine type is installed in several operational supersonic aircraft, and it is likely to form the basis for higher-speed hybrid engines. The afterburning turbofan blends good subsonic and supersonic efficiency with high thrust at high speed. Its specific impulse is found from eq. 2.5-27.

Turboprop core-engine performance is usually expressed in terms of power rather than thrust, as this is consistent with the propeller thrust modeling discussed previously. If all the energy were extracted by the turbine leaving zero thrust in the exhaust, the core engine's ideal power output would be

$$\mathcal{P} = \dot{m} c_p \mathcal{T}_o (\theta_t - \theta_o \tau_c) \left(\frac{\theta_o \tau_c - 1}{\theta_o \tau_c} \right) \quad (2.5\text{-}30)$$

while the *brake-specific fuel consumption* would be

$$\text{BSFC} = \frac{\theta_o \tau_c}{\theta_o \tau_c - 1} \quad (2.5\text{-}31)$$

Although somewhat more fuel efficient than turbojet or turbofan engines at subsonic flight speed, the turboprop's BSFC is about 25 percent higher than that of a reciprocating engine. Nevertheless, the turboprop core burns kerosene (JP fuel) rather than gasoline, a considerable cost saving.

Ramjet and Scramjet Engines

Ramjets have powered several missiles, but they have not been used in operational manned aircraft because a separate engine would be needed to produce static and low-speed thrust. (The missiles are boosted to supersonic speed by solid-rocket boosters and have no need for extended subsonic operation.) Nevertheless, the ramjet thermal cycle is advantageous for high-speed flight and could be incorporated in future multicycle engines. Ramjet combustion can occur at high (stoichiometric) temperature because there is no turbine in the exhaust flow; therefore, the ramjet can operate at higher thermal efficiency. The ramjet's ideal thrust and specific impulse are as follows:

$$T = \dot{m} V \left(\sqrt{\theta_b / \theta_o} - 1 \right) \quad (2.5\text{-}32)$$

$$I_{sp} = \frac{V}{g_o f_s} \left(\sqrt{\theta_b / \theta_o} - 1 \right) \quad (2.5\text{-}33)$$

θ_b is the *ratio of the combustion chamber stagnation temperature to the freestream ambient temperature*, and f_s is the *stoichiometric fuel/air ratio*. For fixed mass flows and temperature ratios, T and I_{sp} typically reach maximum values when $\mathcal{M} = 2\text{–}4$ for hydrocarbon fuels and 2–6 for hydrogen [K-4]. Of course, the temperature ratio depends on the heating value of the fuel, h, as well as the details of the engine configuration:

$$\frac{\theta_b}{\theta_o} = \frac{f_s h}{c_p \mathcal{T}_o \theta_o} + 1 \tag{2.5-34}$$

Typical values of h are 42.8×10^6 J/kg for JP fuel, 49.1×10^6 J/kg for methane, and 120×10^6 J/kg for hydrogen. Thus, methane has a 15 percent heating advantage and hydrogen has a 180 percent advantage over JP. From an aircraft design viewpoint, the "downside" is that methane and hydrogen are gases at room temperature and require cryogenic (cooled) storage in the liquid state. Even then, liquefied methane and hydrogen are less dense, requiring 77 and 960 percent greater storage volume than JP. (Liquid hydrogen density can be increased by about 16 percent if it is stored as a supercooled "slush.") Increased storage volume and cryogenic systems add to aircraft size and weight.

With increasing Mach number, even the ramjet runs into heating problems. Recalling eq. 2.5-20, a flow that is decelerated adiabatically (without heat transfer) from freestream to zero speed is increased to stagnation temperature in the ratio

$$\frac{\mathcal{T}_{stag}}{\mathcal{T}_{amb}} = 1 + 0.2\mathcal{M}^2 \tag{2.5-35}$$

while a flow that is slowed to nonzero chamber Mach number $\mathcal{M}_c$ experiences a lower temperature increase:

$$\frac{\mathcal{T}_c}{\mathcal{T}_{amb}} = \frac{1 + 0.2\mathcal{M}^2}{1 + 0.2\mathcal{M}_c^2} \tag{2.5-36}$$

In a ramjet, the inlet diffuser slows the flow entering the combustion chamber to $\mathcal{M}_c \approx 0.2$–$0.3$; hence, the temperature is essentially the stagnation temperature. Freestream Mach numbers of 6 or more produce a chamber temperature $\mathcal{T}_c$ that is high enough to challenge typical combustor materials and that causes air to dissociate, precluding normal fuel burning.

Supersonic combustion presents an alternative to this heating problem, leading to the supersonic-combustion ramjet (scramjet). If, for example, $\mathcal{M} = 6$, $\mathcal{T}_{stag}/\mathcal{T}_{amb}$ is 8.2, but with $\mathcal{M}_c = 3$, $\mathcal{T}_c/\mathcal{T}_{amb}$ is only 2.9, about the same ratio experienced by a stoichiometric ramjet operating at its best Mach number.

The scramjet poses a considerable challenge for practical design [F-1, K-4, W-5]. Liquid hydrogen (LH_2) appears to be the best thermal choice for fuel, but methane may be a better choice from the perspective of vehicle design. Hydrogen's heating value is high, and it burns rapidly, an essential feature because the residence time in the combustor is brief. Heat flux from the engine is a major problem; for orbital launch, the duration of heating is short, and the hydrogen fuel itself could be used for cooling. Slush-liquid hydrogen (SLH_2) provides a heat sink advantage over conventional liquid hydrogen. Together with decreased fuel-tank pressure, weight, and size (in comparison to LH_2), SLH_2 is an attractive alternative for this application. Nevertheless, the infrastructure problem—availability of the fuel at the world's major airports—imposes a practical limitation on near-term use of hydrogen or methane as an aviation fuel.

The scramjet's combustor represents a small fraction of the engine's volume, with the inlet and exhaust nozzle occupying the majority of the structure. A scramjet might not have the cylindrical shape usually associated with jet engines: the drag associated with the nacelle would be high. It could have a rectangular or semicircular combustor located on the underside of the vehicle; the forward portion of the fuselage would be an integral part of the inlet, and the aft section would provide a "half-open" nozzle, the other half being the freestream flow. The inlet and nozzle shape would have to be carefully designed for near-isentropic compression (i.e., occurring at constant entropy and, therefore, adiabatic and reversible) and expansion over a wide range of Mach numbers, and there would be severe restrictions on the allowable angles of attack and sideslip. The scramjet's performance has an intimate relationship with the aircraft's flight dynamics and control, so it is not feasible to design the aircraft and powerplant separately.

General Thrust Models

From the previous sections, we see that the thrust of airbreathing engines can be modeled as (eq. 2.5-4)

$$\text{Thrust} = C_T(V, \delta T) S\rho^2/2 \tag{2.5-37}$$

The *thrust coefficient* C_T may be a function of the airspeed V and an *equivalent throttle setting* δT that varies between 0 and 1. We also note that $\dot{m}V$ is approximately proportional to ρV^2, as air makes up the majority of an engine's working fluid. This thrust equation takes the same form as those for lift and drag, facilitating analysis and discussion in following chapters.

Within the neighborhood of a nominal operating point, the thrust coefficient can be modeled as

$$C_T = (k_0 + k_1 V^{\eta})\ \delta T \tag{2.5-38}$$

Here, k_0 is the maximum-throttle thrust coefficient at zero velocity, k_1 weights the velocity dependence, and η is a function of the engine type. The partial derivatives with respect to velocity and throttle setting are used in linearized analysis of thrust effects; evaluated at nominal airspeed V_o and throttle setting δT_o, they are

$$\frac{\partial C_T}{\partial V} \triangleq C_{T_V} = \eta k_1 V_o^{\eta-1} \delta T_o = \frac{\eta C_T}{V_o}\left(1 - \frac{k_0 \delta T_o}{C_T}\right)$$

$$\approx \frac{\eta C_T}{V_o} \text{ (for small } k_0) \tag{2.5-39}$$

$$\frac{\partial C_T}{\partial \delta T} \triangleq C_{T_{\delta T}} = (k_0 + k_1 V_o^{\eta}) \tag{2.5-40}$$

where C_T is evaluated at the nominal condition.

Several idealized models can be identified; in each case, we assume that k_0 is zero. For *constant power* (i.e., unchanging with respect to airspeed variation), thrust times airspeed is constant, $\eta = -3$, and $C_{T_V} = -3C_T/V_o$. This condition is representative of propeller-driven aircraft and of high-bypass turbofan engines. For *constant thrust*, which is representative of turbojets and rockets, $\eta = -2$. With *constant mass flow* through the engine, $\eta = -1$, producing a linear increase in thrust with velocity. A *constant thrust coefficient* is equivalent to $\eta = 0$ and a thrust that increases as V^2, which is typical of ramjets up to a limiting Mach number [K-4].

2.6 Steady Flight Performance

Although aircraft flight adheres to Newton's laws of motion, there are complexities to be resolved in the interpretation and application of these simple principles. In *steady, rectilinear flight*, applied forces and moments would be balanced and constant, and the aircraft's velocity vector (magnitude and direction) would be constant. However, steady, rectilinear flight in the earth's atmosphere is not strictly possible because the earth is round, and its gravitational attraction is inversely proportional to the square of the distance from the earth's center. Flight at constant speed along an inertially straight line would involve a changing force vector, while flight at constant speed and altitude implies a curved path with rotating force and velocity vectors. Furthermore, the mass is changing as fuel is consumed, requiring force change for constant velocity. On a straight-line path with changing altitude, the air density varies; constant-velocity flight would call for changes in thrust and angle of attack.

We often neglect these effects, admitting the possibility of *quasisteady flight*. If the flight path is short, variations in the gravity vector are small, and rotation of the local vertical during the flight is negligible. If the speed is a small fraction of orbital speed, Coriolis and centrifugal effects are also negligible. In this case, we can use the *flat-earth assumption*, with a Cartesian reference frame and constant gravitational acceleration. If mass change and density variation during the flight are also small, straight-line, constant-speed motion can be treated as a steady flight condition.

In the remainder of the section, we consider takeoff and landing, as well as various quasisteady flight conditions (which, for conciseness, are called "steady"), including cruising, gliding, climbing, and turning flight. The concepts of steady and maneuvering flight envelopes are introduced, defining the regime within which normal flight occurs. We also examine energy methods for minimizing fuel or time during climb and descent.

Straight and Level Flight

Newton's second law tells us that the vector sum of forces $\Sigma \mathbf{f}_I$ acting on a body produces a rate of change of its *translational momentum* $m\mathbf{v}_I$, as observed in an inertial frame of reference:

$$\sum \mathbf{f}_I = \frac{d}{dt}(m\mathbf{v}_I) \tag{2.6-1}$$

Flight is at *static equilibrium* when the sums of forces along all inertial axes are zero,

$$\sum \mathbf{f}_I = \mathbf{f}(\text{aero})_I + \mathbf{f}(\text{thrust})_I + \mathbf{f}(\text{buoyancy})_I + m\mathbf{g}_I$$

$$\triangleq \mathbf{f}_I + \mathbf{t}_I + \mathbf{b}_I + m\mathbf{g}_I = 0 \tag{2.6-2}$$

where the aerodynamic, thrust, and buoyancy forces are identified explicitly, m is the vehicle's mass, and $\mathbf{g}_I$ is the gravity vector.

Similarly, the vector sum of moments $\Sigma \mathbf{m}_I$ acting on a body changes its *angular momentum* $\mathcal{I}_I \boldsymbol{\omega}_I$, as seen in the inertial frame:

$$\sum \mathbf{m}_I = \frac{d}{dt}(\mathcal{I}_I \boldsymbol{\omega}_I) \tag{2.6-3}$$

$\mathcal{I}_I$ is the vehicle's *inertia matrix*, expressed in the inertial frame as

$$\mathcal{I}_I = \begin{bmatrix} I_{xx} & -I_{xy} & -I_{xz} \\ -I_{xy} & I_{yy} & -I_{yz} \\ -I_{xz} & -I_{yz} & I_{zz} \end{bmatrix}_I \tag{2.6-4}$$

The rationale for this matrix representation of inertia is discussed in Chapter 3. At static equilibrium, the moments about all inertial axes are balanced,

$$\sum \mathbf{m}_I = \mathbf{m}(\text{aero})_I + \mathbf{m}(\text{thrust})_I + \mathbf{m}(\text{buoyancy})_I$$
$$\triangleq \mathbf{m}_I + \mathbf{n}_I + \mathbf{c}_I = 0 \tag{2.6-5}$$

Gravitational moments are negligible for atmospheric flight because the vehicle's center of mass is taken as the rotational reference, and moments due to the earth's inverse-square gravity field (i.e., arising from the gravity gradient) are several orders of magnitude smaller than the aerodynamic effects. With this supposition, moment equilibrium is unaffected by the inertia matrix.

Six scalar equations must be satisfied simultaneously for the force and moment equilibrium. With the flat-earth assumption, $\mathbf{g}_I$ is fixed and vertical ($= [0\ 0\ g]^T$, where g is the scalar acceleration due to gravity (m/s^2)), and the scalar equivalent of eq. 2.6-2 is

$$F_{x_I} + T_{x_I} + B_{x_I} = 0 \tag{2.6-6}$$

$$F_{y_I} + T_{y_I} + B_{y_I} = 0 \tag{2.6-7}$$

$$F_{z_I} + T_{z_I} + B_{z_I} + W = 0 \tag{2.6-8}$$

The aircraft is in "1g flight" (with *weight* $W = mg$) when the vertical force is equal and opposite to the weight; the *vertical load factor* is then 1. In this case, the three scalar moment equations derived from eq. 2.6-5 are also in equilibrium, providing zero angular rates and the angles necessary to maintain translational equilibrium.

Suppose that a symmetric, heavier-than-air craft[10] is in steady, level flight and that the wind speed is zero. The velocity vector is constant and horizontal, and the altitude is constant. Buoyancy effects are several orders of magnitude smaller than the other effects and can be neglected. Assume that the aircraft's velocity is directed along x (north) and that the nose also is pointing north[11]; the state-vector components then have the following values:

$$\begin{bmatrix} r_1 \\ r_2 \\ r_3 \end{bmatrix} = \begin{bmatrix} x \\ y \\ z \end{bmatrix} \tag{2.6-9}$$

10. As distinct from blimps, and hot-air balloons, i.e., lighter-than-air craft.
11. Alternatively, x can be defined to be any convenient horizontal direction, such as the initial (arbitrary) direction of flight; "north" is merely a tangible example.

$$\begin{bmatrix} v_1 \\ v_2 \\ v_3 \end{bmatrix} = \begin{bmatrix} V \\ 0 \\ 0 \end{bmatrix} \tag{2.6-10}$$

$$\begin{bmatrix} \phi \\ \theta \\ \psi \end{bmatrix} = \begin{bmatrix} 0 \\ \theta \\ 0 \end{bmatrix} \tag{2.6-11}$$

$$\begin{bmatrix} \dot\phi \\ \dot\theta \\ \dot\psi \end{bmatrix} = \begin{bmatrix} 0 \\ 0 \\ 0 \end{bmatrix} \tag{2.6-12}$$

x is continually changing but has no effect on the force and moment balance with the flat-earth model. A nonzero but constant pitch angle θ is allowed, and the angular rates must be zero to maintain a fixed attitude. Equations 2.2-7 and 2.2-8 show that γ and ξ are zero, while eq. 2.2-10 and 2.2-11 indicate that

$$\begin{bmatrix} u \\ v \\ w \end{bmatrix} = \mathbf{H}_\mathrm{I}^\mathrm{B}\,\mathbf{v}_\mathrm{I} = \begin{bmatrix} \cos\theta & 0 & -\sin\theta \\ 0 & 1 & 0 \\ \sin\theta & 0 & \cos\theta \end{bmatrix} \begin{bmatrix} V \\ 0 \\ 0 \end{bmatrix} = \begin{bmatrix} V\cos\theta \\ 0 \\ V\sin\theta \end{bmatrix} \tag{2.6-13}$$

From eq. 2.2-30 and with zero wind speed, $\alpha = \theta$, and $\beta = 0$. A positive angle of attack is consistent with the need to generate enough aerodynamic lift to support the aircraft's weight. Equilibrium requires that eq. 2.6-2 be satisfied, and the wind axes and inertial axes are aligned in this special case; therefore,

$$\begin{bmatrix} X \\ Y \\ Z \end{bmatrix}_I = \begin{bmatrix} -D + T_{x_I} \\ SF + T_{y_I} \\ -L + T_{z_I} + W \end{bmatrix} = \begin{bmatrix} 0 \\ 0 \\ 0 \end{bmatrix} \tag{2.6-14}$$

If the thrust vector $\mathbf{t}_\mathrm{B}$ is aligned with the aircraft's centerline (or x_B axis), it can be transformed to the inertial frame, with T representing the thrust magnitude:

$$\begin{bmatrix} T_x \\ T_y \\ T_z \end{bmatrix}_I = \mathbf{H}_\mathrm{B}^\mathrm{I} \begin{bmatrix} T \\ 0 \\ 0 \end{bmatrix} = \begin{bmatrix} T\cos\theta \\ 0 \\ -T\sin\theta \end{bmatrix} = \begin{bmatrix} T\cos\alpha \\ 0 \\ -T\sin\alpha \end{bmatrix} \tag{2.6-15}$$

Aircraft performance in the vertical plane is best understood using a simplified dynamic model. Attitude dynamics and wind effects are neglected.

We also ignore side force and motion, leaving just two components of velocity and two components of position to consider. It is convenient to express the velocity in polar coordinates V and γ, and to resolve forces along and perpendicular to the direction of motion rather than in the Cartesian inertial frame. The remaining four equations of motion are

$$\dot{V} = \frac{1}{m}(T\cos\alpha - D - mg\sin\gamma) \tag{2.6-16}$$

$$\dot{\gamma} = \frac{1}{mV}(T\sin\alpha + L - mg\cos\gamma) \tag{2.6-17}$$

$$\dot{h} = V\sin\gamma \tag{2.6-18}$$

$$\dot{r} = V\cos\gamma \tag{2.6-19}$$

where h and r are the height and range. In steady, level flight ($\gamma = 0$) and with small angle of attack,

$$D = C_D\bar{q}S = T\cos\alpha \approx T \tag{2.6-20}$$

$$L = C_L\bar{q}S = mg - T\sin\alpha \approx W \tag{2.6-21}$$

$$\dot{h} = 0 \tag{2.6-22}$$

$$\dot{r} = V \tag{2.6-23}$$

The lift and drag coefficients are modeled as

$$C_L = C_{L_o} + C_{L\alpha}\alpha \tag{2.6-24}$$

$$C_D = C_{D_o} + \varepsilon C_L^2 = C_{D_o} + \frac{C_L^2}{\pi e AR} \tag{2.6-25}$$

where the component coefficients may vary with Mach number. We consider two idealizations of thrusting effects that are appropriate for propeller- or turbojet-driven aircraft over their normal operating ranges: *constant power* or *constant thrust* (more about power in Section 3.1). In the first case, $TV = \text{constant}$, while in the second, $T = \text{constant}$. Maximum thrust or power is a function of altitude, and actual thrust or power is modulated by the throttle δT, which is limited as $0 \le \delta T \le 1$.

The lift coefficient necessary for equilibrium is, from eqs. 2.6-21 and 2.6-24,

$$C_{L_{trim}} = \frac{W/S}{\bar{q}} = \frac{2W}{\rho V^2 S} = \frac{2We^{\beta h}}{\rho_o V^2 S} \tag{2.6-26}$$

The trim lift coefficient depends on altitude as a consequence of the exponential air-density variation presented in Section 2.1. The required lift coefficient is proportional to wing loading, W/S, it increases with altitude, and it decreases with V^2. With fixed wing loading and dynamic pressure $\bar{q}$, the trimmed lift coefficient remains constant. Thus, $C_{L_{trim}}$ is constant along lines of equal $\bar{q}$, sketched in Fig. 2.6-1.

The corresponding *trimmed angle of attack* is

$$\alpha_{trim} = \frac{\dfrac{2W}{\rho V^2 S} - C_L}{C_{L_\alpha}} \tag{2.6-27}$$

The required angle of attack increases as the altitude increases (because the air density ρ decreases), and the angle decreases as the true airspeed increases.

The elevator establishes the equilibrium angle of attack by balancing the pitching moment (eq. 2.6-5); conversely, the α_{trim} required for steady level flight determines the necessary elevator deflection. With linear sensitivity to angle of attack and elevator deflection, the pitching moment coefficient is

$$C_m = C_{m_o} + C_{m_\alpha}\alpha + C_{m_{\delta E}}\delta E \tag{2.6-28}$$

The net pitching moment is zero when

$$\delta E_{trim} = -\frac{C_{m_o} + C_{m_\alpha}\alpha_{trim}}{C_{m_{\delta E}}} \tag{2.6-29}$$

Given constant values of C_{m_o}, C_{m_α}, and $C_{m_{\delta E}}$, the required elevator setting depends on altitude and airspeed only through their effects on the required angle of attack (eq. 2.6-27).

During takeoff, the weight of the aircraft is transferred from the undercarriage to the wings for airborne flight. Accelerating along the runway, the aircraft could, in principle, leave the ground once the required lift coefficient is below $C_{L_{max}}$, which occurs at the *stall speed* V_S. However, producing the needed lift at this indicated (or calibrated) airspeed would entail flying at a high angle of attack, where drag could exceed available thrust, tail clearance requirements restrict rotation to high pitch angle, control response could be poor, and there would be no margin for error. Consequently, the aircraft proceeds along the runway to higher speed through a sequence of formally defined values [M-1, M-3]: *minimum control speed* (V_{MC}), *critical engine failure speed* (V_1), *takeoff rotation speed* (V_R), *minimum unstick speed* (V_{MU}), and *lift-off speed* (V_{LOF}), and then gains altitude at *takeoff climb speed* (V_2). V_1 is also called the *decision speed*, where the aircraft is committed to flight. If an engine fails

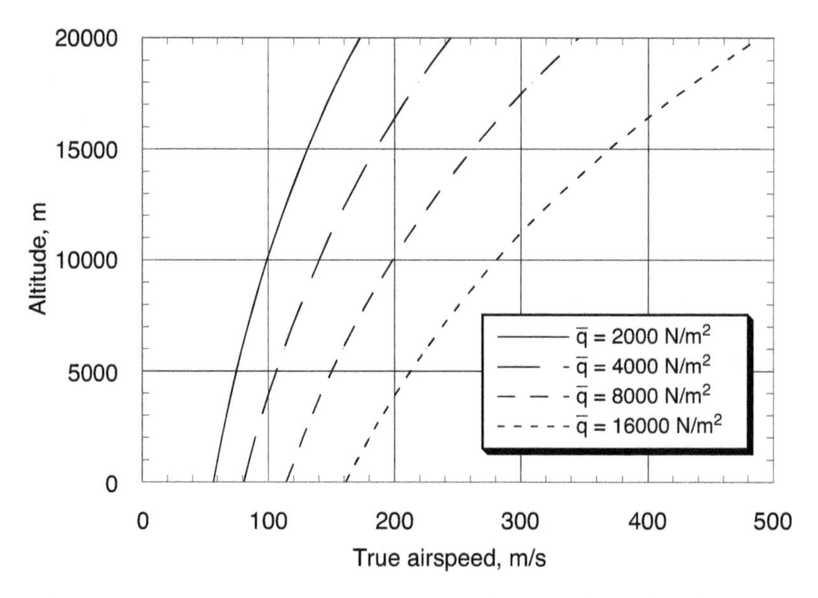

Fig. 2.6-1. Contours of constant dynamic pressure as functions of altitude and true airspeed.

below that speed, the aircraft should be brought to a halt on the runway. At V_R, the nose is pulled up to increase the angle of attack and establish lift on the wings for the *climbout*.

The *landing approach speed* is 30 percent higher than the stall speed ($1.3V_S$). The aircraft typically approaches the runway on a *glide slope* (i.e., flight path angle) of −3 deg, rounding out the flight path in a *flare* to touch down with negligible vertical velocity. During the flare, the angle of attack is increased to arrest the sink rate. Aircraft carrier landings are made without a flare to maximize the likelihood of being trapped by an arresting cable.

Steady Flight Envelope

The thrust required for steady, level flight depends on airspeed and altitude through their effects on drag (eqs. 2.6-20 and 2.6-25):

$$T_{trim} = C_{D_o} \frac{1}{2}\rho V^2 S + \frac{2\varepsilon W^2}{\rho V^2 S} \tag{2.6-30}$$

Increased airspeed normally requires increased thrust to offset the effect of increased dynamic pressure; however, below a certain airspeed, the induced drag $\varepsilon C_L^2 \rho V^2 S/2$ associated with flight at a higher angle of attack has a larger effect than the parasite drag, leading to increasing required thrust as airspeed decreases (Fig. 2.6-2).

The minimum thrust required for steady flight occurs at the airspeed V_{minT}, for which $\partial T_{trim}/\partial V = 0$ and $\partial^2 T_{trim}/\partial V^2 > 0$:

$$\frac{\partial T_{trim}}{\partial V} = 0 = C_{D_o} \rho V_{minT} S - \frac{4\varepsilon W^2}{\rho V_{minT}^3 S} \tag{2.6-31}$$

$$\frac{\partial^2 T_{trim}}{\partial V^2} = C_{D_o} \rho S + \frac{4\varepsilon W^2}{3\rho V_{minT}^4 S} > 0 \tag{2.6-32}$$

$$V_{minT} = \sqrt{\frac{2W}{\rho S}\sqrt{\frac{\varepsilon}{C_{D_o}}}} \tag{2.6-33}$$

Flight at higher or lower airspeed requires greater thrust; hence, $\partial T_{trim}/\partial V$ is positive above the velocity for minimum thrust, V_{minT}, and negative below. Navy fighter aircraft typically approach the aircraft carrier for landing at a flight speed just a few percent above the stall speed but below V_{minT}, in the region called the *back side of the thrust curve.* The pilot must increase thrust to decrease trimmed airspeed (or vice versa), a counterintuitive action, during this demanding task.

Excess thrust is the difference between thrust required and maximum thrust available. The excess thrust is zero at the two points of intersection of these curves (Fig. 2.6-2). Trimmed flight at any speed between these minimum and maximum airspeed values is possible by throttling the powerplant. The airspeed for minimum thrust increases with altitude, while the thrust available decreases. Consequently, the range of possible true airspeeds (V_{min}, V_{max}) decreases with altitude, producing the aircraft's *thrust-limited steady flight envelope.* The *thrust-limited altitude ceiling* occurs when the excess thrust goes to zero and $V_{min} = V_{max}$.

It is easy to see that the lift coefficient required for flight at an airspeed of V_{minT} depends only on the parasite drag coefficient and the induced-drag factor:

$$C_{L_{minT}} = \frac{2W}{\rho V_{minT}^2 S} = \sqrt{\frac{C_{D_o}}{\varepsilon}} \tag{2.6-34}$$

From equation 2.6-25, the corresponding drag coefficient is

$$C_{D_{minT}} = 2C_{D_o} \tag{2.6-35}$$

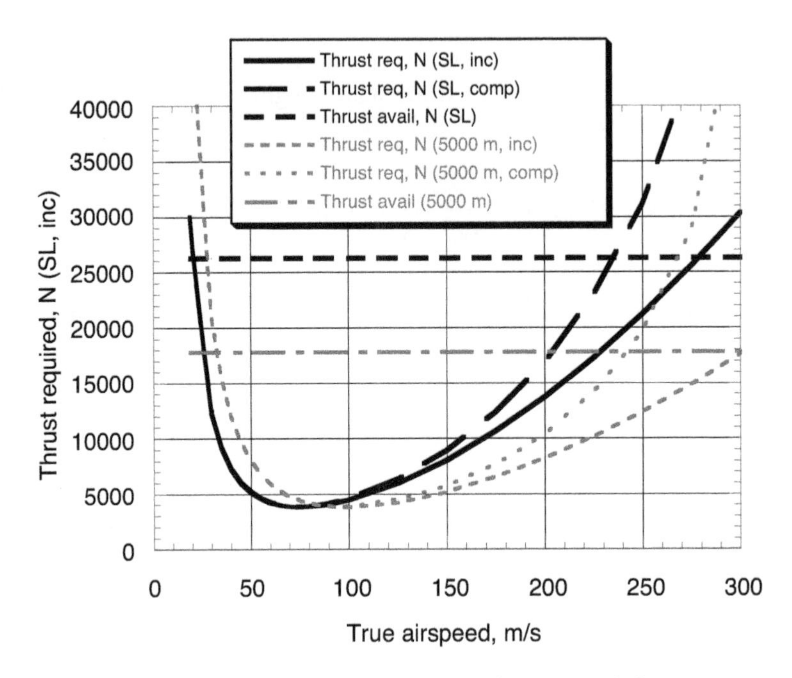

Fig. 2.6-2. Thrust required and available as functions of true airspeed (from Example 2.6-1).

For comparison, let us compute the lift coefficient that maximizes the *lift-to-drag ratio* C_L/C_D:

$$\frac{C_L}{C_D} = \frac{C_L}{C_{D_o} + \varepsilon C_L^2} \tag{2.6-36}$$

At the maximum, the change in the ratio with respect to change in the lift coefficient is zero:

$$\frac{\partial(C_L/C_D)}{\partial C_L} = 0 = \frac{1}{C_{D_o} + \varepsilon C_L^2} - \frac{2\varepsilon C_L^2}{(C_{D_o} + \varepsilon C_L^2)^2} \tag{2.6-37}$$

Solving for C_L, the result is identical to equation 2.6-34; therefore, steady flight at maximum lift-to-drag ratio,

$$(L/D)_{max} = \frac{1}{2\sqrt{\varepsilon C_{D_o}}} \tag{2.6-38}$$

minimizes the thrust required.

The *power required* for steady flight is given by

$$T_{trim}V = C_{D_o}\frac{1}{2}\rho V^3 S + \frac{2\varepsilon W^2}{\rho V S} \tag{2.6-39}$$

and it is minimum when

$$\frac{\partial(T_{trim}V)}{\partial V} = 0 = C_{D_o}\frac{3}{2}\rho V_{minP}^2 S - \frac{2\varepsilon W^2}{\rho V_{minP}^2 S} \tag{2.6-40}$$

It is evident that keeping weight to a minimum is important for power-limited aircraft, such as the human-powered Gossamer Condor and Daedalus and the high-altitude/long-endurance AeroVironment Pathfinder. The airspeed for minimum power, V_{minP}, is

$$V_{minP} = \sqrt{\frac{2W}{\rho S}\sqrt{\frac{\varepsilon}{3C_{D_o}}}} \tag{2.6-41}$$

The corresponding lift and drag coefficients are

$$C_{L_{minP}} = \frac{2W}{\rho V_{minP}^2 S} = \sqrt{\frac{3C_{D_o}}{\varepsilon}} \tag{2.6-42}$$

$$C_{D_{minP}} = 4C_{D_o} \tag{2.6-43}$$

and the associated lift-to-drag ratio is

$$(L/D)_{minP} = \frac{1}{4}\sqrt{\frac{3}{\varepsilon C_{D_o}}} = \frac{\sqrt{3}}{2}(L/D)_{max} \tag{2.6-44}$$

The excess power, back side of the power curve, power-limited steady flight envelope, and *power-limited ceiling* are analogous to the thrust-based definitions. Because thrust is largely independent of airspeed for jet engines, defining the flight envelope and ceiling on the basis of thrust is more convenient for jet-powered aircraft. The power-limited envelope and ceiling are more appropriate for propeller-driven aircraft, whose power is largely independent of airspeed (Section 2.5). Nevertheless, both power-plant types produce thrust *and* power, so the thrust- and power-based definitions of a particular aircraft's flight envelope are identical.

The actual steady flight envelope may be restricted by factors other than the thrust/power required and available (Fig. 2.6-3). The linear lift coefficient model (eq. 2.6-24) neglects aerodynamic stall, which limits the ability for lift to counter weight at low speed. Thus, $V_{\min}$ may be

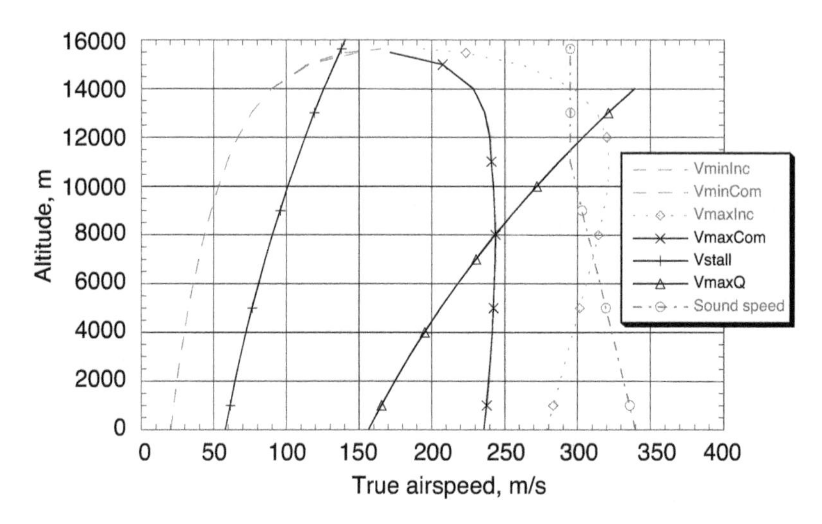

Fig. 2.6-3. Steady flight envelope as a function of altitude and true airspeed (from Example 2.6-1).

determined by the stall speed, with some margin above that speed to avoid *stall buffet* or actual loss of lift. $V_{\max}$ may be limited by several factors, including the maximum allowable dynamic pressure (to avoid structural overload) and Mach number (to avoid unsatisfactory aerodynamic effects such as excessive heating, increased drag, or loss of flight stability). The velocity limits also vary with aircraft configuration, such as flaps stowed/deployed or landing gear up/down. The operating ceiling may be limited by low ambient air density or pressure, which may degrade flight stability or exceed cabin pressurization capacity. The *absolute ceiling* is defined by the altitude at which the maximum attainable climb rate is zero. The *service* and *performance ceilings* are those altitudes for which the maximum climb rate is 100 (150) ft/min [0.51 (0.76) m/s].

Cruising Range

An aircraft's maximum range depends on its aerodynamic efficiency, fuel mass fraction, and specific fuel consumption (Section 2.5). The fuel mass flow rate for a jet engine can be expressed as

$$\dot{m}_{fuel} = -c_T T \tag{2.6-45}$$

where c_T is the thrust-specific fuel consumption (kg/N/hr) and T is thrust (N). The rate of change of range with fuel burn is

$$\frac{dr}{dm} = \frac{dr/dt}{dm/dt} = -\frac{V}{c_T T} = -\frac{V}{c_T D} = -\left(\frac{L}{D}\right)\frac{V}{c_T mg} \tag{2.6-46}$$

where velocity is expressed in km/hr. Therefore, the range traveled, R (km), with a fuel burn of $\Delta m = m_0 - m_f$, is

$$R = \int_0^R dr = -\int_{m_0}^{m_f}\left(\frac{L}{D}\right)\frac{V}{c_T g}\frac{dm}{m} \tag{2.6-47}$$

For constant L/D, V, and c_T, the range is

$$R = -\left(\frac{L}{D}\right)\frac{V}{c_T g}\ln m\Big|_{m_0}^{m_f} = \left(\frac{C_L}{C_D}\right)\frac{V}{c_T g}\ln\frac{m_0}{m_f} \tag{2.6-48}$$

which is the thrust-specific form of the *Breguet range equation*. The airspeed required for lift to equal weight is

$$V = \sqrt{\frac{2W}{C_L \rho S}} \tag{2.6-49}$$

The range is maximized with respect to the lift coefficient when $\partial R/\partial C_L = 0$ and $\partial^2 R/\partial C_L^2 < 0$, which, for fixed specific fuel consumption and mass ratio, is equivalent to maximizing VC_L/C_D. This occurs when the lift and drag coefficients are

$$C_{L_{maxP}} = \sqrt{\frac{C_{D_o}}{3\varepsilon}} \tag{2.6-50}$$

and

$$C_{D_{maxP}} = C_{D_o} + \varepsilon\left(\frac{C_{D_o}}{3\varepsilon}\right) = \frac{4}{3}C_{D_o} \tag{2.6-51}$$

Because the optimizing values of C_L and C_D are fixed (eqs. 2.6-50 and 2.6-51), the range is maximized by flying at the highest possible airspeed (eq. 2.6-48). As airspeed is proportional to the square root of W/ρ, for a given weight, the aircraft should fly at the highest possible altitude (eq. 2.6-49). Because airspeed is assumed to be constant in eq. 2.6-48, altitude should increase as fuel is burned in an *optimal cruise-climb*, keeping W/ρ constant.

Conversely, if altitude is held constant, the cruising velocity must decrease to maintain efficient flight (eq. 2.6-49). The range equation (eq. 2.6-47) can be written as

$$R = \int_0^R dr = -\int_{m_0}^{m_f} \left(\frac{C_L}{C_D}\right) \frac{1}{c_T g} \sqrt{\frac{2mg}{C_L \rho S}} \frac{dm}{m} = \frac{\sqrt{C_L}}{C_D} \left(\frac{2}{c_T g}\right) \sqrt{\frac{2g}{\rho S}} \left(m_0^{1/2} - m_f^{1/2}\right) \tag{2.6-52}$$

The range is maximized with respect to C_L when $\sqrt{C_L}/C_D$ takes its largest value, which occurs when eq. 2.6-50 is satisfied.

For a propeller-driven aircraft, the fuel mass flow rate is expressed in terms of power $\mathcal{P}$,

$$\dot{m}_{fuel} = -c_P \mathcal{P} = -c_P TV \tag{2.6-53}$$

where c_P is the power-specific fuel consumption, and the rate of change of cruising range with fuel burn is

$$\frac{dr}{dm} = \frac{dr/dt}{dm/dt} = -\frac{V}{c_P TV} = -\frac{V}{c_P DV} = -\frac{L/D}{c_P W} \tag{2.6-54}$$

The conditions for maximizing the cruising range of propeller-driven aircraft are left as an exercise for the reader.

Gliding Flight

An aircraft cannot maintain equilibrium flight at constant speed and altitude with zero thrust. It can hold altitude while losing airspeed, or it can fly on a descending path. With no thrust, zero wind, and constant airspeed and flight path angle, force balance (eq. 2.6-16 and 2.6-17) is achieved when

$$0 = \frac{1}{m}(D + mg \sin\gamma) \tag{2.6-55}$$

$$0 = \frac{1}{mV}(L - mg \cos\gamma) \tag{2.6-56}$$

These equations show that the gliding flight path angle is simply a function of the lift/drag ratio:

$$D = C_D \frac{1}{2} \rho V^2 S = -W \sin\gamma \tag{2.6-57}$$

$$L = C_L \frac{1}{2} \rho V^2 S = W \cos\gamma \tag{2.6-58}$$

$$\gamma_{\text{glide}} = -\tan^{-1}(D/L) = -\cot^{-1}(L/D) \tag{2.6-59}$$

Thus, the maximum (least-negative) flight path angle, or *minimum glide angle* $(-\gamma)$, occurs when the lift/drag ratio is a maximum (eq. 2.6-38). From the associated flight path geometry, this condition defines the *maximum-gliding-range* condition, as noted below.

Balancing the vertical force component

$$0 = D \sin\gamma - L \cos\gamma + W \tag{2.6-60}$$

the equilibrium-glide airspeed is found to be

$$V_{glide} = \sqrt{\frac{2W}{\rho S(C_L \cos\gamma_{glide} - C_D \sin\gamma_{glide})}} = \sqrt{\frac{2W}{\rho S\sqrt{C_L^2 + C_D^2}}} \tag{2.6-61}$$

Plotting the *lift-drag polar*, as in Fig. 2.6-4, the glide angle is portrayed by a straight line from the origin. In general, there are two (C_L, C_D) pairs that produce a given lift-drag ratio and glide angle. One solution occurs at higher airspeed than the other. The minimum glide angle occurs when the straight line is just tangent to the lift-drag polar, that is, at the maximum L/D.

From eqs. 2.6-18 and 2.6-19,

$$\frac{dh}{dr} = \tan\gamma \tag{2.6-62}$$

and

$$\int_{h_0}^{h_f} dh = (h_f - h_0) = \int_{r_0}^{r_f} \tan\gamma \, dr \tag{2.6-63}$$

hence, for fixed glide angle and an altitude loss of $-\Delta h$, the glide range Δr,

$$\Delta r = \frac{-\Delta h}{\tan\gamma} \tag{2.6-64}$$

is maximized when the glide angle is a minimum or the lift-drag ratio is a maximum.

Maximizing the glide range for a given loss in altitude does not maximize the gliding time. The *maximum endurance glide* is obtained

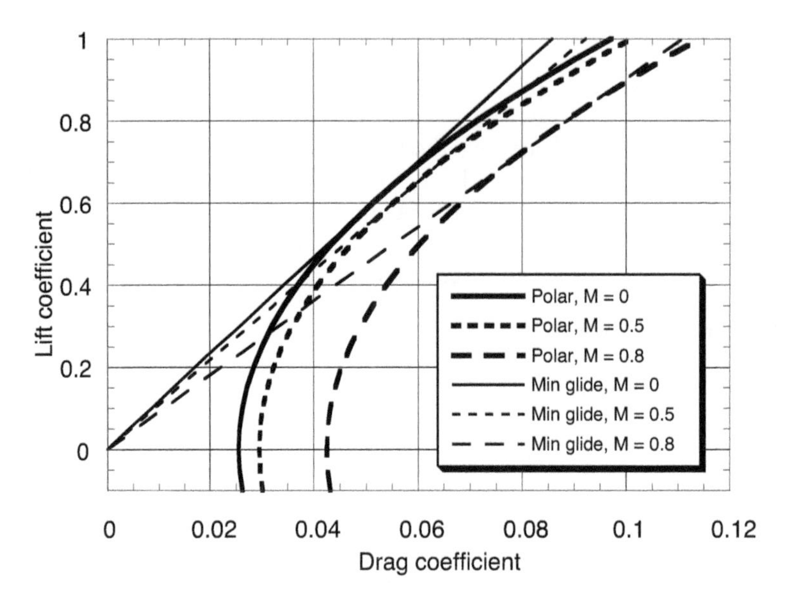

Fig. 2.6-4. Glide slope and the lift-drag polar (from Example 2.6-1).

by minimizing the sink rate $-dh/dt$. For fixed glide angle, the equilibrium altitude rate is

$$\dot{h} = V \sin\gamma = -\frac{DV}{W} \tag{2.6-65}$$

The airspeed and lift coefficient are directly related in an equilibrium glide (eq. 2.6-58); either one can be considered the controlling factor for minimizing the sink rate. Choosing the latter, the true airspeed is first expressed as a function of lift coefficient through the lift-weight balance (eq. 2.6-58):

$$V = \sqrt{\frac{2W\cos\gamma}{C_L \rho S}} \tag{2.6-66}$$

For $\cos\gamma \approx 1$, the minimum-sink-rate condition is defined by the minimum-power condition (eq. 2.6-40–2.6-43); however, a more precise calculation can be made. As $\cos\gamma$ and $\sin\gamma$ can be derived from eq. 2.6-57 and 2.6-58, the altitude rate can be written as

$$\dot{h} = -\sqrt{\frac{2W\cos\gamma}{C_L \rho S}}\left(\frac{C_D}{\sqrt{C_L^2 + C_D^2}}\right) = -\sqrt{\frac{2W/\rho S}{\sqrt{C_L^2 + C_D^2}}}\left(\frac{C_D}{\sqrt{C_L^2 + C_D^2}}\right) \tag{2.6-67}$$

Taking the derivative of the altitude rate with respect to lift coefficient, $d\dot{h}/dC_L$, and setting it equal to zero produces five solutions for C_L:

$$C_L = 0, \pm\frac{1}{2\varepsilon}\sqrt{(1 - 4\varepsilon C_{D_0}) \pm \sqrt{1 - 32\varepsilon C_{D_0}}} \tag{2.6-68}$$

The negative solutions are of no interest, and the two positive roots define a minimum and a maximum in vertical rate. (At a minimum, $\partial^2\dot{h}/\partial C_L^2 > 0$. At a maximum, $\partial^2\dot{h}/\partial C_L^2 < 0$.) The minimum sink rate occurs at the smaller of the two positive roots, and the corresponding airspeed can be found from eq. 2.6-66. From eq. 2.6-67, the minimum sink rate is proportional to the square root of wing loading, and it increases with altitude as a consequence of decreasing air density. What does the zero solution represent?

Climbing Flight

The force balance for unaccelerated flight with thrust aligned to the velocity vector is described by two equations:

$$0 = \frac{1}{m}(T - D - mg\sin\gamma) \tag{2.6-69}$$

$$0 = \frac{1}{mV}(L - mg\cos\gamma) \tag{2.6-70}$$

Consequently, the lift balance is unchanged from gliding flight, and the flight path angle is defined by eq. 2.6-69:

$$\gamma = \sin^{-1}\frac{T - D}{W} \tag{2.6-71}$$

The rate of climb is given by eq. 2.6-18; hence,

$$\dot{h} = V\sin\gamma = \frac{V(T - D)}{W} = \frac{\mathcal{P}_{thrust} - \mathcal{P}_{drag}}{W} = SEP \tag{2.6-72}$$

where SEP is the *specific excess power* or excess power per unit weight of the aircraft. Using eq. 2.6-58 and 2.6-25 to express the lift and drag coefficients, the altitude rate is

$$\dot{h} = V\left[T - (C_{D_0} + \varepsilon C_L^2)\frac{1}{2}\rho V^2 S\right]/W = \frac{TV}{W} - \frac{C_{D_0}\rho V^3 S}{2W} - \frac{2\varepsilon W}{\rho VS}\cos^2\gamma \tag{2.6-73}$$

The airspeed that maximizes climb rate can be found by evaluating $\partial\dot{h}/\partial V$ and setting it equal to zero; however, there are two complications. The first is that the optimum depends on the thrust sensitivity to velocity. For constant thrust, the sensitivity is zero; for constant power, the thrust is proportional to $1/V$. Thus, the equations for the best climb rate of jet- and propeller-driven aircraft are not the same. The second complexity is that the climb rate depends on the flight path angle, which itself is a function of the climb rate and airspeed.

A common approach to the flight path angle effect on climb rate is simply to ignore it, assuming that the climb rate is much smaller than the airspeed. Then $\cos^2\gamma \approx 1$, and the airspeed for best climb at constant power is derived from

$$\frac{\partial \dot{h}}{\partial V} = 0 = -\frac{3C_{D_o}\rho V^2}{2(W/S)} + \frac{2\varepsilon(W/S)}{\rho V^2} \tag{2.6-74}$$

There are four solutions to the equation, only one of which is real and positive:

$$V = \sqrt{\frac{2(W/S)}{\rho}\sqrt{\frac{\varepsilon}{3C_{D_o}}}} \tag{2.6-75}$$

The airspeed for best climb predicted by this equation is the same as that for minimum-power steady, level flight (eq. 2.6-41). The corresponding climb rate is found by substituting this solution in eq. 2.6-73.

For constant thrust, setting $\partial\dot{h}/\partial V$ to zero,

$$\frac{\partial \dot{h}}{\partial V} = 0 = \frac{T}{W} - \frac{3C_{D_o}\rho V^2}{2(W/S)} + \frac{2\varepsilon(W/S)}{\rho V^2} \tag{2.6-76}$$

leads to four possible airspeed solutions:

$$V = \pm\sqrt{\frac{T \pm \sqrt{T^2 + 12C_{D_o}\varepsilon W^2}}{3C_{D_o}\rho S}} \tag{2.6-77}$$

The positive, real solution defines the airspeed for best climb rate (with $T = T_{max}$), which is then determined by eq. 2.6-73.

There are two approaches to accounting for the flight path angle effect on steady climb rate. In both cases, we take note of the trigonometric identity $\sin^2\gamma + \cos^2\gamma = 1$, which, with eq. 2.6-18, leads to

$$\cos^2\gamma = 1 - \frac{\dot{h}^2}{V^2} \tag{2.6-78}$$

The first approach is iterative; ε is replaced by $\varepsilon \cos^2 \gamma$, and the airspeed and climb rate equations are solved recursively. As an example, for constant power, we choose $\gamma_0 = 0$, and iterate over $k = 0, 1, \ldots$ until satisfactory convergence is achieved:

$$V_k = \sqrt{\frac{2(W/S)}{\rho}\sqrt{\frac{\varepsilon \cos^2 \gamma_k}{3C_{D_o}}}} \tag{2.6-79}$$

$$\dot{h}_{k+1} = \frac{TV_k}{W} - \frac{C_{D_o}\rho V_k^3 S}{2W} - \frac{2\varepsilon W}{\rho V_k S}\cos^2 \gamma_k \tag{2.6-80}$$

$$\cos^2 \gamma_{k+1} = 1 - \frac{\dot{h}_{k+1}^2}{V_k^2} \tag{2.6-81}$$

The second approach is exact, although some testing of intermediate results is required. Substituting eq. 2.6-78 in eq. 2.6-73 produces an equation that is quadratic in altitude rate. For constant thrust, the two possible altitude rates are given by

$$\dot{h} = \frac{\rho S V^3}{4\varepsilon W}\left[1 \pm \sqrt{1 - 4\varepsilon\left(\frac{2T}{\rho S V^2} - C_{D_o} - \frac{4\varepsilon W^2}{(\rho S)^2 V^4}\right)}\right] \tag{2.6-82}$$

Setting $\partial\dot{h}/\partial V$ to zero produces four airspeed solutions for each root (five solutions in the constant-power case). From the eight (or ten) airspeed solutions, only those that are positive and real may correspond to the maximum climb rate. If more than one solution is positive and real, the sign of $\partial^2\dot{h}/\partial V^2$ must be checked to determine if the point is a local maximum or minimum, and results may be compared with an approximate or iterative solution.

Maneuvering Envelope

Limits on the maximum and minimum load factors that an aircraft may experience are expressed as functions of airspeed in the *maneuvering envelope*, or *V-n diagram*. These limits on instantaneous flight condition are derived from allowable structural loading and attainable lift, as shown in Fig. 2.6-5. The maximum and minimum instantaneous load factors, $n_{max/min_{aero}}$, that a specific configuration can generate aerodynamically are determined by its weight, altitude (through air density), true airspeed, and maximum/minimum lift coefficients:

$$n_{max/min_{aero}} = C_{L_{max/min}} \frac{1}{2}\rho V^2 S/W \tag{2.6-83}$$

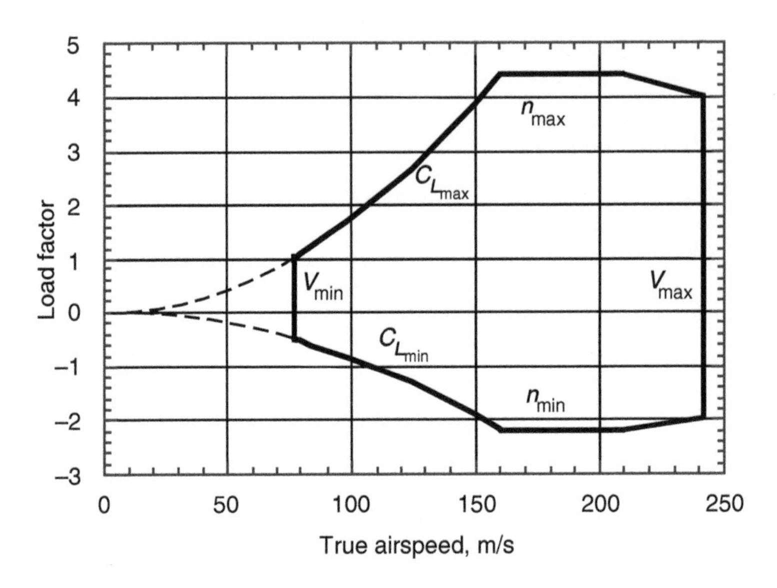

Fig. 2.6-5. Maneuvering envelope as a function of load factor and equivalent airspeed (from Example 2.6-1, altitude = 5,000 m).

The minimum airspeed V_{min} is that for which $n_{max_{aero}}$ is 1 (absolute stall speed), plus some margin of safety, as expressed in the steady flight envelope (Fig. 2.6-3). The maximum airspeed V_{max} is also derived from the steady flight envelope. It is conventional to use equivalent airspeed rather than true airspeed (Section 2.3) in the V-n diagram to subsume altitude and compressibility effects.

Structural design considerations determine the maximum and minimum allowable load factors $n_{max/min_{struct}}$, as summarized in [M-3]. Design specifications for U.S. civil aircraft require that there be no permanent damage to aircraft structure following exposure to maximum load factors $n_{max_{struct}}$ of 6.3 (aerobatic), 4.4 (utility), 3.8 (normal general aviation), or 2.5 (transport). Minimum load factors $n_{min_{struct}}$ are -1 for transports and from -1.5 to -3.2 for other aircraft types. Typically, there is a safety margin of 50 percent between the limit load factor and that which would actually fail the structure. Fighter aircraft are expected to sustain significantly higher load factors, such as $+9g$ and $-6g$; however, their limits when carrying external stores may be lower than those for a clean configuration.

The point at which $n_{max_{aero}}$ equals $n_{max_{struct}}$ is called the *corner velocity*. This is the slowest speed at which the maximum load factor can be achieved, which is of particular significance for turning flight, as described below. The high-speed corners of the *V-n* diagram are clipped to provide a margin against gusts overstressing the airframe during aggressive maneuvering.

Steady Turning Flight

One of the many discoveries made by the Wright brothers is that an aircraft is best turned by rolling (or "banking") the wings away from the horizontal, rather than trying to steer the craft like a boat with the rudder [A-1]. Rolling reorients the lift vector, producing a lateral acceleration that induces the turn. In a *coordinated turn*, the net side force in the aircraft's body frame of reference is zero. Not only is the coordinated turn more comfortable for the occupants; for a given turn rate, it induces less drag than the side force produced by sideslip and rudder. In some early general aviation and more recent human-powered ultralight aircraft, the dihedral effect was high enough so that the sideslip produced by rudder deflection induced significant rolling, allowing partially coordinated turns to be made.

If altitude is to be maintained during the turn, then the vertical component of lift must continue to equal the weight. For the bank angle μ,

$$L\cos\mu = W \tag{2.6-84}$$

and as shown in Fig. 2.6-6 the horizontal force available for turning the aircraft is $\sqrt{L^2 - W^2}$. Thrust-drag equilibrium is required to maintain airspeed; hence, in our simplified model, the thrust required can be expressed as

$$\begin{aligned} T_{req} &= (C_{D_0} + \varepsilon C_L^2)\frac{1}{2}\rho V^2 S = D_0 + \frac{2\varepsilon L^2}{\rho V^2 S} \\ &= D_0 + \frac{2\varepsilon}{\rho V^2 S}\left(\frac{W}{\cos\mu}\right)^2 = D_0 + \frac{2\varepsilon}{\rho V^2 S}(nW)^2 \end{aligned} \tag{2.6-85}$$

where D_0 is the parasite drag and n is the load factor L/W. From this equation, the steady-state bank angle can be expressed in various ways:

$$\mu = \cos^{-1}\left(\frac{W}{L}\right) = \cos^{-1}\left(\frac{1}{n}\right) = \cos^{-1}\left[W\sqrt{\frac{2\varepsilon}{(T - D_o)\rho V^2 S}}\right] \tag{2.6-86}$$

Consequently, the maximum steady bank angle is limited by L_{max}, T_{max}, or n_{max}, whichever produces the smallest value.

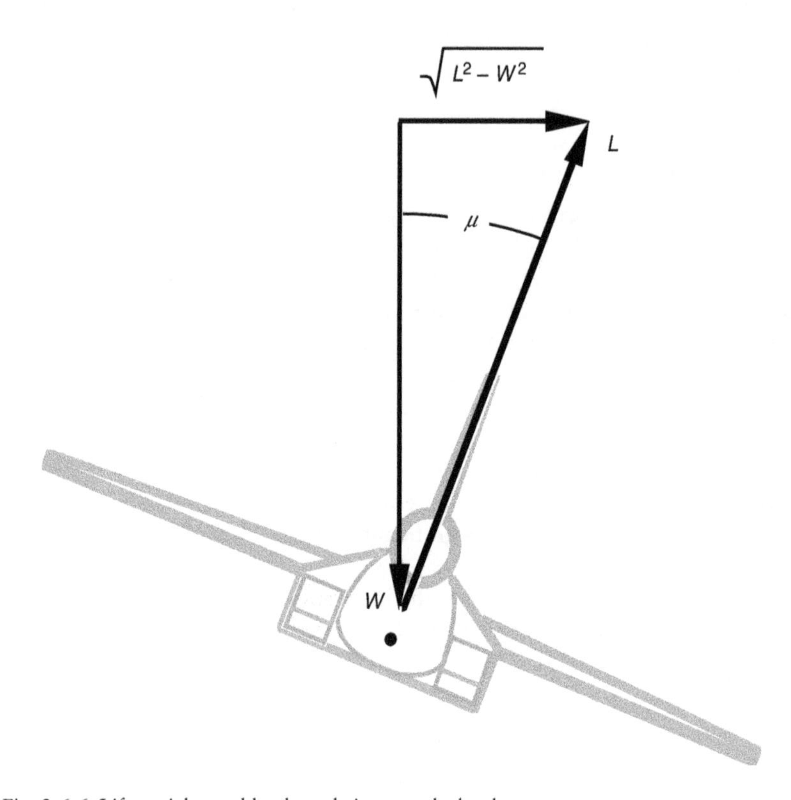

Fig. 2.6-6. Lift, weight, and bank angle in a steady, level turn.

The steady turn rate of the velocity vector $\dot{\xi}$ produced by the bank angle is then

$$\dot{\xi} = \frac{L\sin\mu}{mV} = \frac{g\tan\mu}{V} = \frac{\sqrt{L^2 - W^2}}{mV}$$

$$= \frac{g\sqrt{n^2 - 1}}{V} = \frac{\sqrt{((T - D_o)\rho V^2 S/2\varepsilon) - W^2}}{mV} \tag{2.6-87}$$

and it, too, is limited by L_{max}, T_{max}, or n_{max} (Fig. 2.6-7). Hence, we see that the *maximum attainable turn rate* is obtained by a *corner-velocity turn*, where $n_{max_{aero}}$ equals $n_{max_{struct}}$ and the ratio of load factor to velocity is as high as it can be. This condition also corresponds to the *minimum*

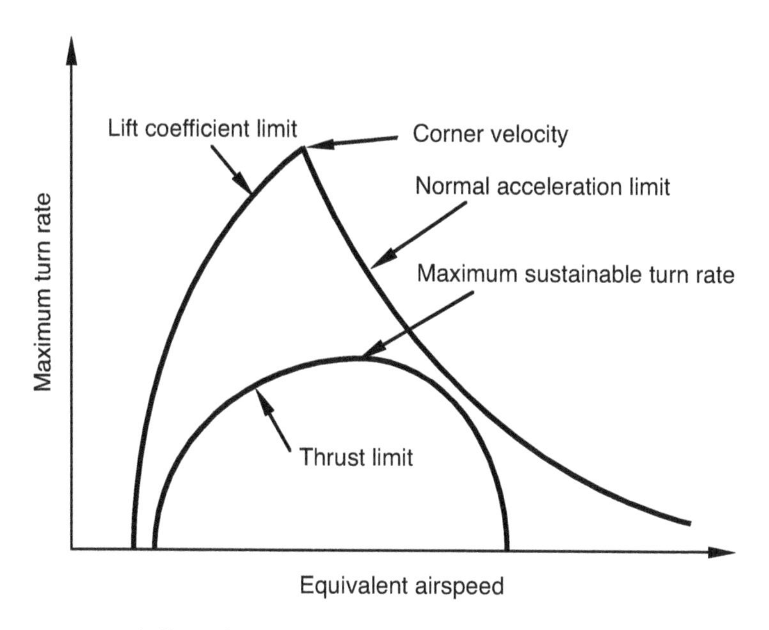

Fig. 2.6-7. Typical effects of maximum lift coefficient, normal acceleration, and thrust limits on maximum turning rate versus airspeed. (after [P-1])

turn radius R_{turn} because

$$R_{turn} = \frac{V}{\dot{\xi}} = \frac{V^2}{g\sqrt{n^2 - 1}} \tag{2.6-88}$$

Nevertheless, if thrust is not sufficient to oppose drag at a high angle of attack, the corner-velocity turn is not sustainable. The aircraft must either turn at a lower rate or lose altitude to maintain airspeed. The *maximum sustainable turn rate* occurs when thrust equals drag (eq. 2.6-85).

Example 2.6-1. Performance of a Business Jet Aircraft

Numerical examples of the performance equations presented in this section are calculated for the generic business jet aircraft model of Appendix B. The model is representative of an aircraft powered by two turbojet engines, with gross cruising weight of 10,000 lb (4,536 kg) and nominal cruising Mach number of 0.79. The model has a maximum lift-to-drag ratio of 11.7, with $C_L = 0.596$ and $C_D = 0.051$; the true airspeed required for level flight at this condition is 75.3 m/s at sea level (SL) and

130.9 m/s at an altitude of 10,000 m. Except as noted, this example uses the incompressible lift and drag coefficients of the model. Hence, the dynamic pressure, angle of attack, and required thrust are the same at both altitudes: 3,471 N/m^2, 4.92 deg, and 3,807 N. The maximum available thrust is 26,423 and 11,735 N at the two altitudes.

Minimum thrust is required for flight at L/D_{max}, but minimum power is achieved at an L/D of 10.1, with airspeeds of 57.2 m/s (SL) and 99.4 m/s (10,000 m). The required thrust increases to 4,396 N, but the required power decreases from 286,600 to 251,458 N-m/s at sea level and from 498,231 to 437,139 N-m/s at 10,000 m. Maximum constant-velocity cruising range is achieved at minimum power. With a starting mass of 4,536 kg, a 100 kg fuel burn allows the aircraft to fly 83.8 km (SL) and 145.7 km (10,000 m). With a fuel burn of 500 kg, these ranges increase to 439 and 763 km.

If the aircraft glides at L/D_{max}, its flight path angle is −4.9 deg, and its sink rate at the two altitudes is −6.4 and −11.1 m/s. The gliding range is 11,680 m for every 1,000 m of altitude lost. Minimum sink rate is achieved at $L/D = 10.1$ and $\alpha = 9.5$ deg, with values of −5.6 and −9.8 m/s at the two altitudes. The airspeed is reduced accordingly to 56.8 and 98.7 m/s. The flight path angle is −5.67 deg in both cases; hence, the gliding range is decreased.

The climb rate depends upon maximum available thrust as well as aerodynamic characteristics. After two iterations of eqs. 2.6-78–2.6-80, the best climb rate at sea level is found to be 56.8 m/s, while it is 28.9 m/s at 10,000 m. The corresponding airspeeds are 199 and 178 m/s, and the angles of attack are −0.25 deg (the necessary lift being produced by C_{L_o}) and 2.1 deg. The flight path angle is 16.6 deg in the first case and 9.3 deg in the second. The lift-to-drag ratios are 3.3 and 9.8, reflecting the increased importance of aerodynamic lift for climbing at higher altitude and lower thrust.

Assuming a maximum angle of attack of 9 deg, which corresponds to a lift coefficient of 1, and a maximum allowable load factor of 4.4, the corner velocities for this model are 121.7 m/s (SL) and 211.6 m/s (10,000 m). The thrust required to sustain a corner-velocity turn is 19,069 N. Consequently, this high-performance steady turn can be achieved at sea level but not at 10,000 m. At sea level, the bank angle, turn rate, and turn radius are 76.9 deg, 19.8 deg/s, and 353 m. At 10,000 m, the best steady values are 70.8 deg, 7.6 deg/s, and 1,587 m.

The service and performance ceilings calculated for the best climb rate of this model are 15,315 m and 15,275 m. At these altitudes, the amount of thrust available is 4,566 N and 4,648 N. If the compressibility effect on the parasite drag coefficient is considered, the service and performance ceilings reduce to 15,195 and 15,152 m.

Compressibility effects have a large impact on the high-speed portion of the flight envelope (Fig. 2.6-3). Without increasing drag coefficient, the maximum airspeed would exceed the speed of sound. However, the drag-coefficient rise with Mach number modeled by the Prandtl factor (Section 2.4) limits the maximum airspeed to 244 m/s (878 km/hr), or $\mathcal{M} = 0.79$. The absolute ceiling and low-speed flight-envelope boundaries are only slightly affected by the drag rise due to compressibility.

References

A-1 Abzug, M. J., and Larrabee, E. E., *Airplane Stability and Control*, Cambridge University Press, Cambridge, 2002.

A-2 Alksne, A., and Jones, A. L., "A Summary of Lateral-Stability Derivatives Calculated for Wing Plan Forms in Supersonic Flow," NACA Report 1052, Washington, D.C., 1951.

A-3 Anderson, J. D., Jr., *Fundamentals of Aerodynamics*, McGraw-Hill Book Co., New York, 1984.

A-4 Anderson, J. D., Jr., *A History of Aerodynamics and Its Impact on Flying Machines*, Cambridge University Press, Cambridge, 1997.

A-5 Anderson, J. D., Jr., *Aircraft Performance and Design*, McGraw-Hill Book Co., New York, 1999.

A-6 Anderson, R. F., *Determination of the Characteristics of Tapered Wings*, NACA Report 572, Washington, D.C., 1937.

A-7 Anon., *U.S. Standard Atmosphere, 1976*, Superintendent of Documents, Washington, D.C., Oct 1976.

A-8 Archer, R. D., and Saarlas, M., *An Introduction to Aerospace Propulsion*, Prentice-Hall, Upper Saddle River, N.J., 1996.

A-9 Ashley, H. *Engineering Analysis of Flight Vehicles*, Addison-Wesley Publishing Co., Reading, Mass., 1974.

A-10 Ashenden, R., Lindberg, W., Marwitz, J. D., and Hoxie, B., "Airfoil Performance Degradation by Supercooled Cloud, Drizzle, and Rain Drop Icing," *J. Aircraft* 33, no. 6 (1996): 1040–1046.

B-1 Bertin, J. J., and Smith, M. L., *Aerodynamics for Engineers*, Prentice-Hall, Englewood Cliffs, N.J., 1979.

B-2 Brown, C., "Theoretical Lift and Drag of Thin Triangular Wings at Supersonic Speeds," NACA Report 839, Washington, D.C., 1946.

C-1 Campbell, J. P., and McKinney, M. O., "Summary of Methods for Calculating Dynamic Stability and Response and for Estimating Lateral Stability Derivatives," NACA Report 1098, Washington, D.C., 1952.

C-2 Collinson, R. P. G., *Introduction to Avionics*, Chapman & Hall, London, 1996.

C-3 Condaminas, A., and Becle, J. P., "Determination de l'Effet de Sol sur les Caracteristiques de l'Avion A320," *Aerodynamics of Combat Aircraft Controls and of Ground Effects*, AGARD-CP-465, Neuilly-sur-Seine, April 1990, pp. 24-1–24-12.

C-4 Corning, G., *Supersonic and Subsonic, CTOL and VTOL, Airplane Design*, published privately, College Park, Md., 1979.

C-5 Covert, E. E., ed., *Thrust and Drag: Its Prediction and Verification*, American Institute of Aeronautics and Astronautics, New York, 1985.

C-6 Cumpsty, N., *Jet Propulsion*, Cambridge University Press, Cambridge, 1997.

D-1 Daniels, G., "Terrestrial Environment (Climatic) Criteria Guidelines for Use in Space Vehicle Development, 1971 Revision," NASA TM X-64589, Washington, D.C., May 1971.

D-2 Dole, C. E., and Lewis, J. E., *Flight Theory and Aerodynamics*, J. Wiley & Sons, New York, 2000.

E-1 Etkin, B., *Dynamics of Atmospheric Flight*, J. Wiley & Sons, New York, 1972.

E-2 Etkin, B., and Reid, L., *Dynamics of Flight—Stability and Control*, J. Wiley & Sons, New York, 1996.

F-1 Ferri, A., "Review of SCRAMJET Propulsion Technology," AIAA *Journal of Aircraft* 5, no. 1 (1968): 3–10.

F-2 Fink, R., and Hoak, D., "USAF Stability and Control DATCOM," U.S. Air Force Flight Dynamics Laboratory, Wright-Patterson AFB, Ohio, Apr 1978.

F-3 Fink, M. R., and Lastinger, J. L., "Aerodynamic Characteristics of Low-Aspect-Ratio Wings in Close Proximity to the Ground," NASA TN D-926, Washington, D.C., July 1961.

F-4 Froude, W., "On the Elementary Relation between Pitch, Slip, and Propulsive Efficiency, *Trans. Institute of Naval Architects* (1878): 47–57.

G-1 Gad-el-Hak, M., "Flow Control: The Future," *J. Aircraft* 38, no. 3 (2001): 402–418.

G-2 Glauert, H., *The Elements of Aerofoil and Airscrew Theory*, Cambridge University Press, Cambridge, 1926.

H-1 Hale, F. J., and Steiger, A. R., "Effects of Wind on Aircraft Cruise Performance," *J. Aircraft* 16, no. 6 (1979): 382–387.

H-2 Heyson, H. H., Riebe, G. D., and Fulton, C. L., "Theoretical Parametric Study of the Relative Advantages of Winglets and Wing-Tip Extensions," NASA TP-1020, Washington, Sept 1977.

H-3 Hill, P. G., and Peterson, C. R., *Mechanics and Thermodynamics of Propulsion*, Addison-Wesley Publishing Co., Reading, Mass., 1965.

H-4 Hoerner, S. R., *Fluid-Dynamic Drag*, privately published, Midland Park, N.J., 1965.

H-5 Houbolt, J. C., *Survey on Effect of Surface Winds on Aircraft Design and Operation and Recommendations for Needed Research*, NASA CR-2360, Washington, D.C., Dec. 1973.

H-6 Housh, C. S., Selberg, B. P., and Rokhsaz, K., "Aerodynamic Characteristics of Scissor-Wing Geometries," *J. Aircraft* 28, no. 4 (1991): 231–238.

J-1 Jacobs, E. N., and Ward, K. E., "Interference of Wing and Fuselage from Tests of 209 Combinations in the NACA Variable-Density Tunnel, NACA Report 540, Washington, D.C., 1936.

K-1 Kalviste, J., "Spherical Mapping and Analysis of Aircraft Angles for Maneuvering Flight," AIAA *Journal of Aircraft* 24, no. 8 (1987): 523–530.

K-2 Kayton, M., and Fried, W. R., *Avionics Navigation Systems*, J. Wiley & Sons, New York, 1997.

K-3 Kermode, A. C., *Mechanics of Flight*, Pitman Publishing, London, 1972.
K-4 Kerrebrock, J. L., *Aircraft Engines and Gas Turbines*, MIT Press, Cambridge, Mass., 1992.
K-5 Küchemann, D., *The Aerodynamic Design of Aircraft*, Pergamon Press, Oxford, 1978.
L-1 La Burthe, C., "Experimental Study of the Flight Envelope and Research of Safety Requirements for Hang Gliders," ONERA T.P. No. 1979-23, Chatillon, France, Mar 1979.
L-2 Lan, C-T. E., and Roskam, J., *Airplane Aerodynamics and Performance*, Roskam Aviation and Engineering, Ottawa, Kans., 1980.
L-3 Larrabee, E. E., "Aerodynamic Penetration and Radius as Unifying Concepts in Flight Mechanics," *J. Aircraft* 4, no. 1 (1967): 28–35.
L-4 Lecomte, P., *Mécanique du Vol*, Dunod, Paris, 1962.
L-5 Lowry, J. T., *Performance of Light Aircraft*, American Institute of Aeronautics and Astronautics, Reston, Va., 1999.
L-6 Lumley, J. L., and Panofsky, H. A., *The Structure of Atmospheric Turbulence*, J. Wiley & Sons, New York, 1964.
M-1 Mair, W. A., and Birdsall, D. L., *Aircraft Performance*, Cambridge University Press, Cambridge, 1992.
M-2 Mattingly, J. A., Heiser, W. H., and Daley, D. H., *Aircraft Engine Design*, American Institute of Aeronautics and Astronautics, New York, 1987.
M-3 McCormick, B. W., *Aerodynamics, Aeronautics, and Flight Mechanics*, J. Wiley & Sons, New York, 1995.
M-4 McLean, D., *Automatic Flight Control Systems*, Prentice-Hall, Englewood Cliffs, N.J., 1990.
M-5 Miele, A., *Flight Mechanics, Volume 1: Theory of Flight Paths*, Addison-Wesley, Reading, Mass., 1962.
M-6 Miranda, L., "Application of Computational Aerodynamics to Airplane Design," AIAA *Journal of Aircraft* 21, no. 6 (1984): 355–370.
M-7 Multhopp, H., "Aerodynamics of the Fuselage," NACA TM 1036, Washington, D.C., 1942.
M-8 Murthy, S. N. B., and Paynter, G. C., eds., *Numerical Methods for Engine-Airframe Integration*, American Institute of Aeronautics and Astronautics, New York, 1986.
M-9 Moler, C., Herskovitz, S., Little, J., and Bangert, S., *MATLAB for Macintosh Computers*, The MathWorks, Inc., Sherborn, Mass., 1987.
N-1 Nicolai, L., *Fundamentals of Aircraft Design*, METS, San Jose, Calif., 1975.
O-1 Oates, G. C., ed., *Aircraft Propulsion Systems Technology and Design*, AIAA, Washington, D.C., 1989.
P-1 Pamadi, B. N., *Performance, Stability, Dynamics, and Control of Airplanes*, AIAA Education Series, Reston, Va., 1998.
P-2 Paulson, J. W., Jr., Kemmerly, G. T., and Gilbert, W. P., "Dynamic Ground Effects," *Aerodynamics of Combat Aircraft Controls and of Ground Effects*, AGARD-CP-465, Neuilly-sur-Seine, April 1990, pp. 21-1–21-12.
P-3 Pearson, H. A., and Jones, R. T., "Theoretical Stability and Control Characteristics of Wings with Various Amounts of Taper and Twist," NACA Report 635, Washington, D.C., 1935.

P-4 Perkins, C. D., and Hage, R. E., *Airplane Performance, Stability and Control*, J. Wiley & Sons, New York, 1949.
P-5 Piland, R. O., *Summary of the Theoretical Lift, Damping-in-Roll, and Center-of-Pressure Characteristics of Various Wing Planforms at Supersonic Speeds*, NACA TN 1977, Washington, D.C., 1949.
P-6 Pittman, J. L., Bonhaus, D. L., Siclari, M. J., and Dollyhigh, S. M., "Euler Analysis of a High-Speed Civil Transport Concept at Mach 3," *J. Aircraft* 28, no. 4 (1991): 239–245.
P-7 Polhamus, E., "Applying Slender Wing Benefits to Military Aircraft," *J. Aircraft* 21, no. 8 (1984): 545–559.
R-1 Rankine, W. J. M., "On the Mechanical Principles of the Action of Propellers," *Trans. Institute of Naval Architects* (1865): 13–30.
R-2 Raymer, D. P., *Aircraft Design: A Conceptual Approach*, AIAA Education Series, Washington, D.C., 1989.
R-3 Roskam, J., *Airplane Design*, Roskam Aviation and Engineering Corp., Ottawa, Kans., 1990.
R-4 Rossow, V. J., and James, K. D., "Overview of Wake-Vortex Hazards During Cruise," *J. Aircraft* 37, no. 6 (2000): 960–975.
S-1 Saffman, P. G., *Vortex Dynamics*, Cambridge University Press, Cambridge, 1992.
S-2 Schlichting, H., and Truckenbrodt, E., *Aerodynamics of the Airplane*, McGraw-Hill International Book Co., New York, 1979.
S-3 Schrenk, O., "A Simple Approximation Method for Obtaining Spanwise Lift Distribution," NASA TM-948, Washington, D.C., Aug. 1940.
S-4 Seckel, E., *Stability and Control of Airplanes and Helicopters*, Academic Press, New York, 1964.
S-5 Shapiro, A. H., *The Dynamics and Thermodynamics of Compressible Flow, Vol. 1*, Ronald Press, New York, 1953.
S-6 Shen, J., Parks, E. K., and Bach, R. E., "Comprehensive Analysis of Two Downburst-Related Aircraft Accidents," *J. Aircraft* 33, no. 5 (1996): 924–930.
S-7 Sherman, A., "Interference of Wing and Fuselage from Tests of 28 Combinations in the N.A.C.A. Variable-Density Tunnel," NACA Report 575, Washington, D.C., 1937.
S-8 Shevell, R. S., *Fundamentals of Flight*, Prentice-Hall, Englewood Cliffs, N.J., 1983.
S-9 Silverstein, A., and Katzoff, S., "Design Charts for Predicting Downwash Angles and Wake Characteristics Behind Plain and Flapped Wings," NACA Report 648, Washington, D.C., 1940.
S-10 Smits, A., *A Physical Introduction to Fluid Mechanics*, J. Wiley & Sons, New York, 2000.
S-11 Soulé, H. A., and Anderson, R. F., "Design Charts Relating to the Stalling of Tapered Wings," NACA Report 703, Washington, D.C., 1940.
S-12 Spilman, D., and Stengel, R. F., "Jet Transport Response to a Horizontal Wind Vortex," *J. Aircraft* 32, no. 3 (1995): 480–485.
S-13 Stengel, R. F., "Altitude Stability in Supersonic Cruising Flight," *J. Aircraft* 7, no. 5 (1970): 464–473.

S-14 Stengel, R. F., and Marcus, F. J., *Energy Management Techniques for Fuel Conservation in Military Transport Aircraft*, AFFDL-TR-75-156, Wright-Patterson AFB, Ohio, Feb 1976.
S-15 Stengel, R. F., and Berry, P. W., "Stability and Control of Maneuvering High-Performance Aircraft", NASA CR-2788, Washington, D.C., Apr 1977.
S-16 Stengel, R. F., and Nixon, W. B., "Stalling Characteristics of a General Aviation Aircraft," AIAA *Journal of Aircraft* 19, no. 6 (1982): 425–434.
S-17 Stengel, R. F., *Optimal Control and Estimation*, Dover Publications, New York, 1994 (originally published as *Stochastic Optimal Control: Theory and Application*, J. Wiley & Sons, New York, 1986).
S-18 Stepniewski, W. Z., and Keys, C. N., *Rotary-Wing Aerodynamics*, Dover Publications, New York, 1984.
S-19 Stinton, D., *The Design of the Aeroplane*, Van Nostrand Publishing Co., New York, 1983.
S-20 Sweberg, H. H., and Dingeldein, R. C., "Summary of Measurements in Langley Full-Scale Tunnel of Maximum Lift Coefficients and Stalling Characteristics of Airplanes," NACA Report 829, Washington, D.C., 1945.
T-1 Teichmann, F. K., *Airplane Design Manual*, Pittman Publishing Co., New York, 1958.
T-2 Thompson, B. E., and Jang, J., "Aerodynamic Efficiency of Wings in Rain," *J. Aircraft* 33, no. 6 (1996): 1047–1053.
T-3 Thwaites, B., *Incompressible Aerodynamics*, Clarendon Press, Oxford, 1960.
T-4 Townsend, J., ed., *Low-Altitude Wind Shear and Its Hazard to Aviation*, National Academy Press, Washington, D.C., 1983.
V-1 van Dam, C., "Induced-Drag Characteristics of Crescent-Moon-Shaped Wings," AIAA *Journal of Aircraft* 24, no. 2 (1987): 115–119.
V-2 Van Dyke, M., *An Album of Fluid Motion*, Parabolic Press, Stanford, Calif., 1982.
V-3 Vicroy, D. D., "The Aerodynamic Effect of Heavy Rain on Airplane Performance," *Proc. AIAA Flight Simulation Technologies Conference*, Sept. 1990, pp. 78–85.
V-4 Vicroy, D. D., "Assessment of Microburst Models for Downdraft Estimation," *J. Aircraft* 29, no. 6 (1992): 1043–1048.
V-5 Vinh, N. X., *Flight Mechanics of High-Performance Aircraft*, Cambridge University Press, Cambridge, 1993.
V-6 Von Mises, R., *Theory of Flight*, McGraw-Hill Book Co., New York, 1945.
V-7 von Kármán, T., "Turbulence and Skin Friction," *J. Aeronautical Sciences* 1, no. 1 (1934).
W-1 Weissinger, J., *The Lift Distribution of Swept-Back Wings*, NACA TM 1120, Washington, D.C., 1947.
W-2 Whitcomb, R. T., "A Study of the Zero-Lift Drag-Rise Characteristics of Wing-Body Combinations Near the Speed of Sound," NACA Report 1273, Washington, D.C., 1956.
W-3 Whitcomb, R. T., and Clark, L. R., "An Airfoil Shape for Efficient Flight at Supercritical Mach Numbers," NASA TM X-1109, Washington, D.C., 1965.

W-4 Whitcomb, R. T., "A Design Approach and Selected Wind-Tunnel Results at High Subsonic Speeds for Wing-Tip Mounted Winglets," NASA TN D-8260, Washington, D.C., July 1976.

W-5 White, M. E., Drummond, J. P., and Kumar, A., "Evolution and Application of CFD Techniques for Scramjet Engine Analysis," *J. Propulsion and Power* 3, no. 5 (1987): 423–439.

W-6 Williams, J., *Prediction Methods for Aircraft Aerodynamic Characteristics*, AGARD-LS-67, Neuilly-sur-Seine, May 1974.

W-7 Williams, J., "The USAF Stability and Control Digital DATCOM," AFFDL-TR-76–45, Wright-Patterson AFB, Ohio, Nov 1976.

W-8 Wingrove, R. C., and Bach, R. E., Jr., "Severe Turbulence and Maneuvering from Airline Flight Records," *J. Aircraft* 31, no. 4 (1994): 753–760.

W-9 Wolowicz, C. H., and Brown, J. S., Jr., "Similitude Requirements and Scaling Relationships as Applied to Model Testing," NASA TP-1435, Washington, D.C., Aug. 1979.

Y-1 Yip, L. P., and Coy, P. F., "Wind-Tunnel Investigation of a Full-Scale Canard-Configured General Aviation Aircraft," ICAS-82-6.8.2, *Proc. of the International Council for the Aeronautical Sciences*, 1982, pp. 1470–1488.

Z-1 Zucrow, M. J., *Aircraft and Missile Propulsion, Vol. 1, Thermodynamics of Fluid Flow and Application to Propulsion Engines*, J. Wiley & Sons, New York, 1958.

Chapter 3

Dynamics of Aircraft Motion

Aircraft flight motions are described by solutions to ordinary differential equations. The dynamic equations present the rates of change of motion variables, as affected by external forces and moments such as gravity, wind disturbances, and control surfaces. These equations are nonlinear and must be integrated to portray the histories (or trajectories) of aircraft motion. The concept of *momentum* is central to the development of these equations through Newton's laws of motion, while *energy* and *power* relate to the range of possible motions (Section 3.1) and can be used directly to approximate certain maneuvering flight paths of aircraft.

We are interested in both linear and angular motion, as expressed in various frames of reference, including noninertial (or non-Newtonian) frames, and employing alternative descriptions of position, angular attitude, and rate. The equations of motion are presented for a flat-earth model, then recast taking the earth's oblate-spheroidal geometry, gravitational field, and rotation into account (Sections 3.2 and 3.3). Next the aerodynamic effects of angular motion, unsteady flows, and control-surface deflections are explained (Sections 3.4 and 3.5). The chapter concludes with a review of methods for solving nonlinear dynamic systems, including the numerical integration of ordinary differential equations, finding trimmed (equilibrium) solutions, and fitting smooth curves to tabulated data (Section 3.6).

3.1 Momentum and Energy

The most general statements that can be made about the dynamic condition of a moving body relate to its momentum and energy. *Momentum* extends the kinematic, vectoral concept of velocity to include the inertial properties of the body, while *energy* gives a scalar measure for the magnitude of motion and the potential for change.

We begin by considering a *point mass*, an idealized body of infinitesimal size whose angular motions are of no concern. The substantive property of mass governs the body's acceleration in response to an imposed force. Thus, a point mass may possess significant translational momentum and energy, but it has no rotational momentum and energy.

A *rigid body* can be viewed as an assemblage of point masses, internally constrained to maintain a given size and shape, free to rotate as a well-defined unit. The rotational momentum and energy of a rigid body are important, and they can be derived from the translational motions of the component point masses, taking into account the special constraints imposed by the body's rigid structure.

Translational Momentum, Work, Energy, and Power

A point mass m moving through space with inertial velocity $\mathbf{v}$ and inertial position $\mathbf{r}$ possesses the *translational momentum* $\mathbf{p} \triangleq m\mathbf{v} \triangleq md\mathbf{r}/dt$. Momentum is a three-dimensional vector that is proportional to the velocity vector through the *mass m*:

$$\begin{bmatrix} p_x \\ p_y \\ p_z \end{bmatrix} \triangleq m \begin{bmatrix} v_x \\ v_y \\ v_z \end{bmatrix} \triangleq m \begin{bmatrix} dx/dt \\ dy/dt \\ dz/dt \end{bmatrix} \tag{3.1-1}$$

By Newton's first law of motion, the linear momentum is constant when viewed in an inertial frame of reference as long as no forces act on the body; therefore, if the body has constant mass, its inertial velocity is also constant, and the body moves in a straight line.

Work $\mathcal{W}$ is done when an *external force* $\mathbf{f}$ is applied to a body that moves from one place to another. $\mathcal{W}_A^B$ is a scalar measure of the *change in energy* that occurs by traversing a path between two points A and B, properly defined by the inertial-frame vectors $\mathbf{r}_A = [x\ y\ z]_A{}^T$ and $\mathbf{r}_B = [x\ y\ z]_B{}^T$. The amount of work done by the force in moving from Point A to Point B is

$$\mathcal{W}_A^B = \int_A^B \mathbf{f}^T d\mathbf{r} = \int_A^B (Xdx + Ydy + Zdz) \tag{3.1-2}$$

The three-dimensional vector $\mathbf{f}$, whose components are measured in *newtons*, could be a function of $\mathbf{r}$, $\mathbf{v}$, and time t, so the work done may depend on the path taken, the speed of travel, and the time of the trip. From the integral's components, it can be seen that newton-meters (called *joules*) are suitable units for measuring work. If, for any choice of A and B and any two (different) paths between them,

$$\int_A^B \mathbf{f}^T d\mathbf{r} + \int_B^A \mathbf{f}^T d\mathbf{r} = 0 \tag{3.1-3}$$

then $\mathbf{f}$ is said to be a *conservative force*, and the work done in going from A to B is independent of the path taken, the speed, and the trip time. $\mathbf{f}$ is a conservative force if it is the negative transpose of the *gradient of a scalar potential*, $\mathcal{V}$:

$$\mathbf{f}^T = -\nabla_{\mathbf{r}}\mathcal{V} = -\mathcal{V}_{\mathbf{r}} = -\left[\frac{\partial\mathcal{V}}{\partial x}\ \frac{\partial\mathcal{V}}{\partial y}\ \frac{\partial\mathcal{V}}{\partial z}\right] \tag{3.1-4}$$

The *gradient of a scalar function of a vector variable* is a row vector containing the partial derivatives of the scalar with respect to the individual components of the vector.

Alternative notations for the gradient are shown in eq. 3.1-4. $\mathcal{V}$ is also called the *potential energy*, and $\mathcal{V}_{\mathbf{r}}$ is a *potential field*. $\mathcal{V}$ can be a function only of $\mathbf{r}$ (i.e., not of $\mathbf{v}$ and t); otherwise, eq. 3.1-3 would not hold for arbitrary paths.

If $\mathcal{V}_{\mathbf{r}}$ is the only force operating on the body, the work done in going from A to B is the difference between the potential energies at A and B:

$$\mathcal{W}_A^B = \int_A^B \mathbf{f}^T d\mathbf{r} = -\int_A^B d\mathcal{V} = \mathcal{V}_A - \mathcal{V}_B \tag{3.1-5}$$

This equation expresses only the *difference* in energies; hence, potential energy is normally defined with respect to the potential energy at some reference point such as A.

For a flat-earth model, the force of gravity on a body of mass m is $m\mathbf{g}$, where $\mathbf{g}$ is a constant vertical vector ($\mathbf{g} = [0\ 0\ g]^T$, and g is the scalar acceleration due to gravity [m/s^2]). Choosing A to be a point on the earth's surface (for which the height is defined as zero) and B to be a point at height $-z$ or h, the force applied in moving from A to B must oppose gravity ($\mathbf{f} = -m\mathbf{g}$), so $\mathcal{V}_A = 0$, and the work done is

$$\mathcal{W}_A^B = -\mathcal{V}_B = -\int_0^{-z} mg\,dz = mgz = -mgh \tag{3.1-6}$$

We conclude that the gravity force is a potential field, and $\mathcal{V}(z)$ is the *potential energy* of the body, referred to the earth's surface and measured in N-m or joules.

Newton's second law of motion equates the force acting on a point mass with the time rate of change of its translational momentum,

$$\mathbf{f} = \frac{d}{dt}\mathbf{p} = \frac{d}{dt}m\mathbf{v} = \frac{d}{dt}\left(m\frac{d\mathbf{r}}{dt}\right) \tag{3.1-7}$$

where $\mathbf{v}$ and $\mathbf{r}$ are referred to an inertial frame. This relationship is easily extended to assemblages of particles that form a rigid body. Rather than summing these effects for a collection of particles, the discrete model is replaced by a continuous distribution of mass elements whose effects are integrated over the body. Consider a differential element with density $\rho(x, y, z)$ and mass $dm = \rho(x, y, z)dx\, dy\, dz$. The mass of the entire body is found by integrating over the volume of the body i.e., between the minimum and maximum values of its (x, y, z) coordinates,

$$m = \int_{z_{min}}^{z_{max}} \int_{y_{min}}^{y_{max}} \int_{x_{min}}^{x_{max}} \rho(x, y, z)\, dx\, dy\, dz = \int_{body} dm \tag{3.1-8}$$

The body's center of mass c.m. is located at

$$\mathbf{r}_{cm} = \begin{bmatrix} x \\ y \\ z \end{bmatrix}_{cm} = \frac{1}{m} \int_{z_{min}}^{z_{max}} \int_{y_{min}}^{y_{max}} \int_{x_{min}}^{x_{max}} \rho(x, y, z) \begin{bmatrix} x \\ y \\ z \end{bmatrix} dx\, dy\, dz$$

$$= \frac{1}{m} \int_{body} \mathbf{r}\, dm \tag{3.1-9}$$

where the integrals extend over the body's volume. Inertial coordinates $\mathbf{r}' = [x'\ y'\ z']^T$ referred to the center of mass are defined as follows:

$$\mathbf{r} = \mathbf{r}_{cm} + \mathbf{r}' \tag{3.1-10}$$

Equation 3.1-9 then guarantees that

$$\int_{body} \mathbf{r}' dm = 0 \tag{3.1-11}$$

as every element to one side of the c.m. is precisely balanced by another on the opposite side.

Denoting the velocity of the elemental particle dm by $\mathbf{v}_m$, the differential force on a differential element can be written as (eq. 3.1-7)

$$d\mathbf{f} = \frac{d}{dt} d\mathbf{p} = \frac{d}{dt} d(m\mathbf{v}) = \frac{d}{dt}\left(dm \frac{d\mathbf{r}}{dt} \right) \tag{3.1-12}$$

Because eq. 3.1-11 holds,

$$\int_{body} \frac{d\mathbf{r}'}{dt} dm = 0 \tag{3.1-13}$$

and the integrated force on the body equates to the rate of change of the body's translational momentum:

$$\int_{body} d\mathbf{f} = \mathbf{f} = \frac{d}{dt}\left(\frac{d\mathbf{r}_{cm}}{dt} \int dm \right) = \frac{d}{dt}(m\mathbf{v}_{cm}) \tag{3.1-14}$$

Consequently, as long as the body is rigid, the net translational motion of the rigid body is totally described by the motion of its center of mass.

An equivalence between work done and the squares of the initial and final velocities can be drawn from Newton's second law and eq. 3.1-2 as $\mathbf{v}dt = d\mathbf{r}$. With constant mass and V representing the body's inertial velocity magnitude, the work done in going from A to B is

$$\mathcal{W}_A^B = m\int_A^B \left(\frac{d\mathbf{v}}{dt}\right)^T \mathbf{v}dt = \frac{m}{2}(\mathcal{V}_A^2 - \mathcal{V}_B^2) = \mathcal{T}_A - \mathcal{T}_B \tag{3.1-15}$$

$\mathcal{T}_A$ and $\mathcal{T}_B$ are defined as the *kinetic energies* of the equivalent point mass at points A and B. The integral does not depend on the functional form of $\mathbf{f}$, and the kinetic energy at a point, $mV^2/2$, has units of kg-m^2/s^2 or joules. For a conservative system, eqs. 3.1-5 and 3.1-15 are equal, so

$$\mathcal{T}_B + \mathcal{V}_B = \mathcal{T}_A + \mathcal{V}_A \tag{3.1-16}$$

The *total translational energy* of a body is the sum of its potential and kinetic translational energies,

$$\mathcal{E} = \frac{mV^2}{2} + mgh \tag{3.1-17}$$

The energy per unit weight (also called *specific energy, energy state*, or *energy height*) $\mathcal{E}'$ is dimensionally equivalent to altitude:

$$\mathcal{E}' = h + \frac{V^2}{2g} \tag{3.1-18}$$

An aircraft could trade altitude for speed without changing its total energy, and the energy height equals the actual height at zero airspeed. The relationship is nonlinear; at constant total energy, a linear altitude change would be offset by a quadratic velocity change.

The *power* absorbed or generated in doing useful work is the energy change per unit time. Dividing eq. 3.1-17 by the time interval and taking the limit as the interval between A and B vanishes,

$$\begin{aligned}\mathcal{P} &= \frac{d\mathcal{W}}{dt} = \frac{d\mathcal{E}}{dt} = \lim_{B\to A}\frac{\mathcal{W}_A^B}{t_B - t_A} = \frac{m}{2}\frac{d(V^2)}{dt} + mg\frac{dh}{dt} \\ &= m\left(\mathbf{v}^T\frac{d\mathbf{v}}{dt} + g\frac{dh}{dt}\right)\end{aligned} \tag{3.1-19a}$$

From eq. 3.1-7, this relation could also be written as a function of the applied force $\mathbf{f}_{applied}$:

$$\frac{d\mathcal{E}}{dt} = \mathcal{P} = m\left(\mathbf{v}^T\mathbf{f}_{applied} + g\frac{dh}{dt}\right) \tag{3.1-19b}$$

An airplane always experiences a nonconservative drag force D (a function of speed as well as altitude) in atmospheric flight; hence, energy is being absorbed and ultimately converted to heat that is either retained in the vehicle or rejected to the surroundings. Either energy must be added via thrust-generated power, or speed or altitude must be lost to compensate for energy consumed by drag. In steady, level flight, the power dissipated by drag equals the power produced by the thrust T; with greater thrust, the aircraft can climb.

If the thrust is parallel to the aircraft centerline, it is offset from the velocity vector by the angle of attack, and the *specific excess power*, or excess power per unit of weight, is

$$\mathcal{SEP} = \frac{V(T\cos\alpha - D)}{W} = \frac{(C_T\cos\alpha - C_D)(\frac{1}{2})\rho V^3 S}{W} \tag{3.1-20}$$

The $\mathcal{SEP}$ governs the aircraft's ability to gain specific energy through the application of thrust, in keeping with eq. 3.1-19. $\mathcal{SEP}$ is a function of altitude (which affects air density), airspeed, angle of attack, and throttle setting (through C_T). Substituting eqs. 2.6-16 and 2.6-17 in eq. 3.1-19 with $\cos\alpha \approx 1$ yields the total energy rate

$$\frac{d\mathcal{E}}{dt} = V(T - D - W\sin\gamma) + WV\sin\gamma = V(T - D) \tag{3.1-21}$$

Hence the rate of change of energy height (eq. 3.1-18) with respect to time is

$$\frac{d\mathcal{E}'}{dt} = \frac{V(T - D)}{W} = \mathcal{SEP} \tag{3.1-22}$$

Energy-Changing Maneuvers

Energy considerations give insight into the underlying mechanisms for climb, cruise, and descent. With zero thrust, the specific excess power $\mathcal{SEP}$ is negative, and it has a maximum (i.e., least negative) value at the V, C_L, and C_D for minimum power identified in Section 2.6 (eq. 2.6-40–2.6-43). This is the minimum-sink-rate gliding condition under the assumption that

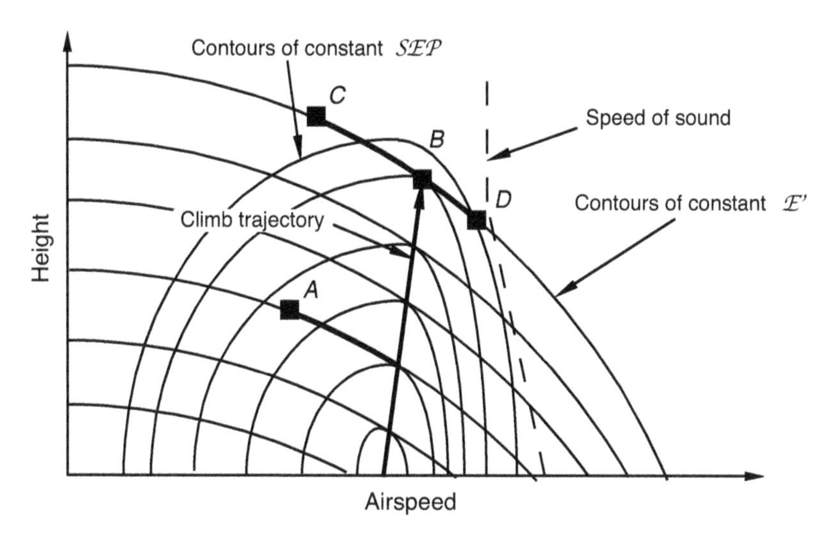

Fig. 3.1-1. Typical subsonic minimum-time climb path based on energy-state optimization. (after [B-7])

$\cos\gamma \approx 1$. Neglecting weight loss due to fuel burn, $\mathcal{SEP}$ is zero in cruising flight, as thrust equals drag; the steady flight condition could be chosen for maximum endurance or maximum range, as in Section 2.6.

The aircraft climbs and/or increases speed with positive values of $\mathcal{SEP}$, and the climb-speed profile can be chosen to minimize the time or fuel used to change the energy height [R-7, B-7]. From eq. 3.1-22, the time required to transition between two energy heights is

$$\int_{t_A}^{t_B} dt = t_B - t_A = \int_{\mathcal{E}'_A}^{\mathcal{E}'_B} \frac{d\mathcal{E}'}{\mathcal{SEP}} \tag{3.1-23}$$

Clearly, the time required to make the change is minimized by choosing the largest $\mathcal{SEP}$ possible. This leads to a climb strategy that is readily visualized as in Fig. 3.1-1, which idealizes the problem for a subsonic jet transport. Contours of constant $\mathcal{E}'$ are plotted as segments of parabolas whose maximum values define the zero-velocity heights, assuming negligible change in weight W. If there were no losses to drag, the aircraft could *zoom climb* from any point on a parabola to the maximum altitude without changing $\mathcal{E}'$, conversely, it could *zoom dive* to lower altitude and higher airspeed on the same parabola.

Contours of constant $\mathcal{SEP}$ with maximum available thrust are plotted by solving eq. 3.1-22 to find the (h, V) pairs that produce a given $\mathcal{SEP}$, with higher values toward the bottom of Fig. 3.1-1. $\mathcal{SEP}$ decreases with increased altitude due to typical thrust characteristics, at high airspeed due to drag increase, and at low airspeed due to the velocity multiplier in eq. 3.1-22. For a given $\mathscr{E}'$, the energy gain is highest where the $\mathscr{E}'$ locus is tangent to the largest possible $\mathscr{SEP}$; thus, the *h*-*V* profile for minimum-time climb is found by connecting these points of tangency. If the $\mathscr{SEP}$ values were plotted in a third dimension, the climb path would follow the ridge line of the surface.

Using the energy-state approximation, the minimum-time path from *A* to *B* in Fig. 3.1-1 contains two segments: a zoom dive to the ridge line, and a climb along the maximum-energy-rate ridge to *B*. Starting points near the ground would require a horizontal acceleration to the ridge line. If the final point is *C* or *D*, there is a final zoom climb/dive segment to match the desired end condition. The zoom times are neglected in formulating the strategy, and their values depend on particulars of the trajectory, such as the aggressiveness with which altitude is changed. As a practical matter, the sharp corners where segments intersect would be rounded. Exact optimal solutions exhibit asymptotic convergence to the energy-state approximations with smoothly varying *h*-*V* profiles [B-7], and optimization with singular perturbation (or multiple-time-scale) models provides simplified smooth approximations [K-2, C-1].

The initial time-minimizing climb profile is similar for a supersonic aircraft (Fig. 3.1-2), but the aircraft must penetrate the drag rise near $\mathcal{M} = 1$. Early jets had low thrust by contemporary standards, which caused difficulty in accelerating through the speed of sound. There was

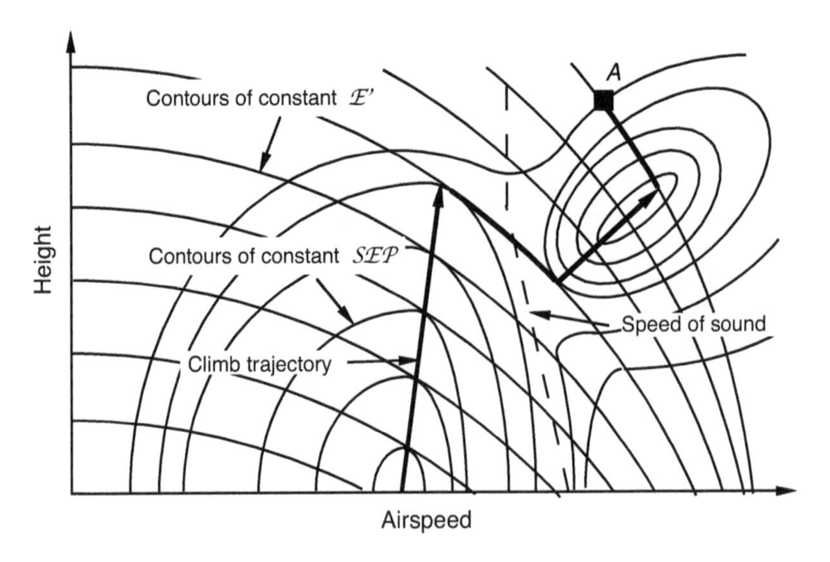

Fig. 3.1-2. Typical supersonic minimum-time climb path based on energy-state optimization. (after [B-7])

a relative minimum in $\mathcal{SEP}$ in that region, with increased $\mathcal{SEP}$ once the barrier had been overcome. The energy-state solution is to enlist gravity in a zoom dive, exchanging potential energy for kinetic energy when the $\mathcal{E}'$ contour is tangent to sub- and supersonic $\mathcal{SEP}$ contours of equal magnitude [B-7]. The sub- and supersonic equal-$\mathcal{SEP}$ contours of modern aircraft with "supercruise" capability or multicycle powerplants may be continuous, avoiding the need to perform the sonic zoom dive.

Energy climbs that minimize the amount of fuel used are based on

$$\frac{d\mathcal{E}'}{dm} = \frac{d\mathcal{E}'/dt}{dm/dt} = -\frac{V(T - D)}{Wc_T T} = -\frac{\mathcal{SEP}}{c_T T} \tag{3.1-24}$$

where c_T is the specific fuel consumption (Section 2.5). The procedure is the same as in the time-minimizing-climb case, with the contours of equal $\mathcal{SEP}$ replaced by contours of equal $\mathcal{SEP}/c_T T$. Commercial flight management systems combine minimum-fuel climb and descent with optimal cruise segments [E-1] for practical implementation. The optimality of cruise segments—as opposed to a sequence of optimal climbs and dives—has been questioned [S-2, S-7, S-3, S-8], although the demonstrated differences are small and dependent on assumptions, and the operational desirability of cruising flight is not in doubt. Energy-state approximations also have found application for optimization of three-dimensional maneuvers [K-1, H-3].

Angular Momentum and Energy

The angular momentum of a rigid body can be broken into two parts, one pertaining to the angular motion of the center of mass about some external origin and the other describing the angular motion of the body about its center of mass. The first part is useful for portraying orbital motion (as in a central gravitational force field), while the second part relates to changes in body attitude. Angular energy can be partitioned in the same way.

Choosing an arbitrary inertial-frame origin O for reference, the *angular momentum* (or *moment of momentum*) **h** of a point mass m about O is

$$\mathbf{h} = \mathbf{r} \times \mathbf{p} = \mathbf{r} \times m\mathbf{v} = \mathbf{r} \times m\frac{d\mathbf{r}}{dt} \tag{3.1-25}$$

The vector **r** describes the point mass's position relative to O, **v** is its velocity, and **p** is the particle's translational momentum. Newton's second law then yields

$$\mathbf{m} = \mathbf{r} \times \mathbf{f} = \frac{d}{dt}\mathbf{h} \tag{3.1-26}$$

These concepts can be extended to a continuum of point masses forming a rigid body. m is replaced by dm, and the position and velocity at a point are separated as follows:

$$\mathbf{r} = \mathbf{r}_{cm} + \mathbf{r}' \tag{3.1-27}$$

$$\mathbf{v} = \mathbf{v}_{cm} + \mathbf{v}' \tag{3.1-28}$$

The angular momentum of the point mass is

$$\begin{aligned} d\mathbf{h} &= (\mathbf{r}_{cm} + \mathbf{r}') \times d\mathbf{p} = (\mathbf{r}_{cm} + \mathbf{r}') \times dm\,(\mathbf{v}_{cm} + \mathbf{v}') \\ &= [(\mathbf{r}_{cm} \times \mathbf{v}_{cm}) + (\mathbf{r}_{cm} \times \mathbf{v}') + (\mathbf{r}' \times \mathbf{v}_{cm}) + (\mathbf{r}' \times \mathbf{v}')]\,dm \end{aligned} \tag{3.1-29}$$

and this equation can be integrated over the body to obtain the total angular momentum. Recall that $dm = \rho(x, y, z)dx\,dy\,dz$, where $\rho(x, y, z)$ is the density at the point (x, y, z), and "integration over the body" implies a triple integration over the x, y, and z directions. The position and velocity of the center of mass can be taken outside the integral, as they do not vary over the body; hence, the integral of eq. 3.1-29 is written as

$$\begin{aligned} \mathbf{h} &= (\mathbf{r}_{cm} \times \mathbf{v}_{cm})\int_{body} dm + \mathbf{r}_{cm} \times \int_{body} \mathbf{v}'\,dm \\ &\quad + \int_{body} \mathbf{r}'\,dm \times \mathbf{v}_{cm} + \int_{body} (\mathbf{r}' \times \mathbf{v}')\,dm \end{aligned} \tag{3.1-30a}$$

From eq. 3.1-11 and 3.1-13, the second and third terms are zero, leaving

$$\mathbf{h} = m\,(\mathbf{r}_{cm} \times \mathbf{v}_{cm}) + \int_{body} (\mathbf{r}' \times \mathbf{v}')\,dm \tag{3.1-30b}$$

Because the body is constrained to be rigid, the only relative motion about the center of mass that an elemental particle can have is pure, solid-body rotation. Thus, with a rigid-body angular rate $\boldsymbol{\omega}$ the translational velocity of a particle at $\mathbf{r}'$ is

$$\mathbf{v}' = \boldsymbol{\omega} \times \mathbf{r}' \tag{3.1-31}$$

so the angular momentum becomes

$$\begin{aligned} \mathbf{h} &= m\,(\mathbf{r}_{cm} \times \mathbf{v}_{cm}) + \int_{body} [\mathbf{r}' \times (\boldsymbol{\omega} \times \mathbf{r}')]\,dm \\ &= m\,(\mathbf{r}_{cm} \times \mathbf{v}_{cm}) - \int_{body} [\mathbf{r}' \times (\mathbf{r}' \times \boldsymbol{\omega})]\,dm \end{aligned} \tag{3.1-30c}$$

The first term of this equation represents the angular momentum due to the motion of the body's center of mass referred to an external origin, while the second portrays that due to the body's rotation about the center of mass. Choosing the center of mass as the origin, the first term vanishes. This is appropriate in the flat-earth frame, as no single origin (on the earth's surface, for example) is dynamically more significant than any other. Because $\boldsymbol{\omega}$ is constant across the rigid body, it can be taken outside the integral; interchanging $\boldsymbol{\omega}$ with the scalar dm, and using the cross-product-equivalent matrix (eq. 2.3-2),

$$\begin{aligned}
\mathbf{h}_{cm} &= -\int_{body} (\mathbf{r}' \times \mathbf{r}' \times dm)\,\mathrm{w} = -\left[\int_{body} \tilde{\mathbf{r}}'\tilde{\mathbf{r}}' dm\right]\boldsymbol{\omega} \\
&= -\left\{\int_{body} \begin{bmatrix} 0 & -z' & y' \\ z' & 0 & -x' \\ -y' & x' & 0 \end{bmatrix}\begin{bmatrix} 0 & -z' & y' \\ z' & 0 & -x' \\ -y' & x' & 0 \end{bmatrix} dm\right\}\begin{bmatrix}\omega_x \\ \omega_y \\ \omega_z\end{bmatrix} \\
&= -\left\{\int_{body} \begin{bmatrix} (y'^2+z'^2) & -x'y' & -x'z' \\ -x'y' & (x'^2+z'^2) & -y'z' \\ -x'z' & -y'z' & (x'^2+y'^2) \end{bmatrix} dm\right\}\begin{bmatrix}\omega_x \\ \omega_y \\ \omega_z\end{bmatrix} \\
&= \mathcal{I}_{\mathrm{I}}\boldsymbol{\omega}_{\mathrm{I}}
\end{aligned} \tag{3.1-32}$$

The integral defines the *inertia matrix* $\mathcal{I}_{\mathrm{I}}$ for an inertial frame, as (x', y', z') are measured parallel to the original inertial frame. For example, I_{xx} is $\int_{body}(y'^2 + z'^2)dm$, and I_{xz} is $\int_{body} x'z'dm$. The inertia matrix is *symmetric* (i.e., $\mathcal{I} = \mathcal{I}^T$), with the *moments of inertia* running along the diagonal and the *products of inertia* in the off-diagonal terms:

$$\mathcal{I} \triangleq \begin{bmatrix} I_{xx} & -I_{xy} & -I_{xz} \\ -I_{xy} & I_{yy} & -I_{yz} \\ -I_{xz} & -I_{yz} & I_{zz} \end{bmatrix} \tag{3.1-33}$$

The angular energy concepts follow from the translational discussion. Taking the center of mass as the origin, the work done in going between two angular orientations, A and B, must be defined. At each orientation, the elemental particles dm of the rigid body have locations $\mathbf{r}'_A = [x'\ y'\ z']_A{}^T$ and $\mathbf{r}'_B = [x'\ y'\ z']_B{}^T$, and they are acted on by elemental forces $d\mathbf{f} = [dX\ dY\ dZ]^T$; therefore, eq. 3.1-2 tells us that

$$\mathcal{W}_A^B = \int_A^B \int_{body} d\mathbf{f}^T d\mathbf{r}' = \int_A^B \int_{body} (dX dx' + dY dy' + dZ dz') \tag{3.1-34}$$

If the elemental force is the transpose of the gradient of a potential function (e.g., eq. 3.1-4), then

$$\mathcal{W}_A^B = \int_A^B \int_{body} d\mathbf{f}^T d\mathbf{r} = -\int_A^B \int_{body} d\mathcal{V}_\mathbf{r}\, d\mathbf{r} = \mathcal{V}_\mathcal{A} - \mathcal{V}_\mathcal{B} \tag{3.1-35}$$

thus defining the *angular potential energy*, although not in a particularly useful form. For a rigid body, internal forces are balanced, so $\mathcal{V}$ must arise from external forces whose net effect is a position-dependent torque. For example, pitching moments that depend on the angle of attack or yawing moments that depend on sideslip angle are sources of angular potential energy. In a round-earth model (Section 3.3), gravity-gradient torques also have a position-dependent trait.

As before (eq. 3.1-15), Newton's second law can be invoked to write the work done as

$$\begin{aligned}\mathcal{W}_A^B &= \int_A^B \int_{body} \left(\frac{d\mathbf{v}}{dt}\right)^T \mathbf{v}\, dt\, dm \\ &= \frac{1}{2}\int_{body} (V_B^2 - V_A^2)\, dm = \mathcal{T}_B - \mathcal{T}_A\end{aligned} \tag{3.1-36}$$

where $\mathcal{T}_A$ and $\mathcal{T}_B$ are the kinetic energies at the end points. If we restrict the allowable motions of elemental particles to solid-body rotation (eq. 3.1-31), the *angular kinetic energy* is

$$\begin{aligned}\mathcal{T} &= \frac{1}{2}\int_{body} V'^2\, dm = \frac{1}{2}\int_{body} \mathbf{v}^T \mathbf{v}'\, dm \\ &= \frac{1}{2}\int_{body} (\boldsymbol{\omega} \times \mathbf{r}')^T (\boldsymbol{\omega} \times \mathbf{r}')\, dm \\ &= \frac{1}{2}\int_{body} (\mathbf{r}' \times \boldsymbol{\omega})^T (\mathbf{r}' \times \boldsymbol{\omega})\, dm \\ &= \frac{1}{2}\int_{body} (\tilde{\mathbf{r}}'\boldsymbol{\omega})^T (\tilde{\mathbf{r}}'\boldsymbol{\omega})\, dm \\ &= \frac{1}{2}\int_{body} \boldsymbol{\omega}^T \tilde{\mathbf{r}}'^T \tilde{\mathbf{r}}'\boldsymbol{\omega}\, dm\end{aligned} \tag{3.1-37a}$$

Taking the angular rate vectors outside the integral and noting that $\tilde{\mathbf{r}}'^T = -\tilde{\mathbf{r}}'$, it can be seen that the angular kinetic energy is

$$\mathcal{T} = \frac{1}{2}\boldsymbol{\omega}^T \mathcal{I} \boldsymbol{\omega} \tag{3.1-37b}$$

where the inertia matrix $\mathcal{I}$ is defined by eqs. 3.1-32 and 3.1-33. The total angular energy then is described by $\mathcal{T} + \mathcal{V}$.

3.2 Dynamic Equations For a Flat Earth

Much of aircraft dynamics can be well understood using mathematical models that neglect round-earth effects. These models are developed for an inertial coordinate system and then are extended to various alternative axis systems.

Rigid-Body Dynamic Equations

Newton's second law describes the translational motions of an aircraft's center of mass through the differential equation

$$\frac{d\mathbf{p}_I}{dt} = \frac{d(m\mathbf{v}_I)}{dt} = \mathbf{f}_I \quad \text{(translational dynamics)} \tag{3.2-1}$$

and its rotational motion about the center of mass by

$$\frac{d\mathbf{h}_I}{dt} = \frac{d(\mathcal{I}_I \boldsymbol{\omega}_I)}{dt} = \mathbf{m}_I \quad \text{(rotational dynamics)} \tag{3.2-2}$$

The equations are expressed in an inertial frame of reference, the mass is m, and the remaining symbols are defined as follows:

$$\mathbf{p}_I = \begin{bmatrix} p_x \\ p_y \\ p_z \end{bmatrix}_I = m\mathbf{v}_I = \text{translational momentum vector} \tag{3.2-3}$$

$$\mathbf{v}_I = \begin{bmatrix} v_x \\ v_y \\ v_z \end{bmatrix}_I = \begin{bmatrix} dx/dt \\ dy/dt \\ dz/dt \end{bmatrix}_I = \text{translational velocity vector} \tag{3.2-4}$$

$$\mathbf{f}_I = \begin{bmatrix} X \\ Y \\ Z \end{bmatrix}_I = \text{external force vector} \tag{3.2-5}$$

$$\mathbf{h}_I = \begin{bmatrix} h_x \\ h_y \\ h_z \end{bmatrix}_I = \mathcal{I}_I \boldsymbol{\omega}_I = \text{angular momentum vector} \tag{3.2-6}$$

$$\mathcal{I}_I = \begin{bmatrix} I_{xx} & -I_{xy} & -I_{xz} \\ -I_{xy} & I_{yy} & -I_{yz} \\ -I_{xz} & -I_{yz} & I_{zz} \end{bmatrix}_I = \text{inertia matrix} \tag{3.2-7}$$

$$\boldsymbol{\omega}_I = \begin{bmatrix} \omega_x \\ \omega_y \\ \omega_z \end{bmatrix}_I = \text{angular velocity vector} \tag{3.2-8}$$

$$\mathbf{m}_I = \begin{bmatrix} L \\ M \\ N \end{bmatrix}_I = \text{external moment vector} \tag{3.2-9}$$

Equations 3.2-1 and 3.2-2 represent six scalar, first-order, differential equations whose integrals specify translational and rotational velocity. Another six scalar equations must be integrated to determine the translational and rotational position. These are expressed in vector form as

$$\frac{d\mathbf{r}_I}{dt} = \mathbf{v}_I \qquad \text{(translational kinematics)} \tag{3.2-10}$$

$$\frac{d\Theta}{dt} = \mathbf{L}_B^E \mathbf{H}_I^B \boldsymbol{\omega}_I \quad \text{(rotational kinematics)} \tag{3.2-11}$$

Here, the rectilinear and angular positions are denoted by

$$\mathbf{r}_I = \begin{bmatrix} x \\ y \\ z \end{bmatrix}_I = \text{translational position vector} \tag{3.2-12}$$

$$\boldsymbol{\Theta} = \begin{bmatrix} \phi \\ \theta \\ \psi \end{bmatrix} = \text{Euler angle vector} \tag{3.2-13}$$

$\mathbf{L}_B^E$ and $\mathbf{H}_I^B$, which contain trigonometric functions of the attitude Euler angles, are the body-to-Euler angle rate and inertial-to-body-axis transformation matrices defined earlier (eqs. 2.2-26 and 2.2-11). All of the quantities in eqs. 3.2.1–3.2-13 may be varying in time [e.g., $m = m(t)$, $\mathbf{f}_I = \mathbf{f}_I\,(t)$, etc.], although the explicit reference to time as an argument is suppressed for clarity. Together, the 12 scalar differential equations can be integrated to predict aircraft trajectories, but there are certain difficulties that make it desirable to modify the set.

The most pressing problem is that eqs. 3.2-1 and 3.2-2 express the time rates of change of momentum vectors, while the desired solution variables are the aircraft's translational and rotational velocity vectors. The derivatives can be expanded using the chain rule:

$$\frac{d(m\mathbf{v}_{\mathrm{I}})}{dt} = m\frac{d\mathbf{v}_{\mathrm{I}}}{dt} + \frac{dm}{dt}\mathbf{v}_{\mathrm{I}} \tag{3.2-14}$$

$$\frac{d(\mathcal{I}_{\mathrm{I}}\boldsymbol{\omega}_{\mathrm{I}})}{dt} = \mathcal{I}_{\mathrm{I}}\frac{d\boldsymbol{\omega}_{\mathrm{I}}}{dt} + \frac{d\mathcal{I}_{\mathrm{I}}}{dt}\boldsymbol{\omega}_{\mathrm{I}} \tag{3.2-15}$$

The translational momentum changes when either velocity or mass changes. If mass is gained or lost at some continuous rate dm/dt with no change in velocity, as might be idealized by simply "letting go" of detachable pieces of the structure, the momentum of the aircraft is reduced as $(dm/dt)\mathbf{v}_{\mathrm{I}}$, but the velocity of the remaining mass is unchanged. If the mass is expelled rearward along a positive-aft exhaust velocity vector $\mathbf{c}$ (measured with respect to the body), the momentum change is $(dm/dt)(\mathbf{v}_{\mathrm{I}} - \mathbf{c})$, and there is a net thrust on the aircraft. Whenever the mass is changing, additional equations must be integrated to determine $m(t)$ and $\mathcal{I}(t)$.

Consider a rocket-propelled vehicle that is unaffected by external forces. Newton's second law applies to the sum of masses contained in both the vehicle and its exhaust, so the total mass is partitioned into two parts: the vehicle mass m_1 and the exhaust mass m_2. We assume that the exhaust velocity is imparted to the exhaust mass instantaneously, without regard to the specific mechanisms of thrust generation. Equation 3.2-14 becomes

$$\sum_{i=1}^{2}\frac{d\mathbf{p}_i}{dt} = \sum_{i=1}^{2}\frac{d(m_i\mathbf{v}_i)}{dt} = \left(\frac{dm_1}{dt}\right)\mathbf{v}_1 + m_1\frac{d\mathbf{v}_1}{dt} + \left(\frac{dm_2}{dt}\right)\mathbf{v}_2 + m_2\frac{d\mathbf{v}_2}{dt}$$
$$= \mathbf{f}_1 + \mathbf{f}_2 = 0 \tag{3.2-16a}$$

where $\mathbf{v}_2 = \mathbf{v}_1 - \mathbf{c}$. By Newton's third law, the force that the exhaust produces on the vehicle $(\mathbf{f}_1)$ is equal and opposite to the force that the vehicle exerts on the exhaust $(\mathbf{f}_2)$. Vehicle mass decreases at the same rate that exhaust mass increases; therefore, $dm_1/dt = -dm_2/dt$; and

$$\sum_{i=1}^{2}\frac{d(m_i\mathbf{v}_i)}{dt} = \left(\frac{dm_1}{dt}\right)\mathbf{c} + m_1\frac{d\mathbf{v}_1}{dt} + m_2\frac{d\mathbf{v}_2}{dt} = 0 \tag{3.2-16b}$$

This equation is satisfied at all times in an interval beginning at the start time t_0, and ending at the final time t_f. Consider the balance at a particular instant of time t. At time $t-$(infinitesimally prior to t), the exhaust mass is subsumed in the vehicle mass, and m_2 is zero. At time $t+$ (infinitesimally past t), the exhaust mass has been expelled, and $d\mathbf{v}_2/dt$ is zero. Consequently, the momentum change of the exhaust that occurs at t is entirely described by $(dm_2/dt)\mathbf{v}_2$, and $m_2(d\mathbf{v}_2/dt)$ is zero. From eq. 3.2-16b,

$$m_1 \frac{d\mathbf{v}_1}{dt} = -\left(\frac{dm_1}{dt}\right)\mathbf{c} = \mathbf{f}_1 = \mathbf{f}_{thrust} \tag{3.2-16c}$$

where dm_1/dt is negative, leading to a positive force. The differential equation for change in the inertial velocity results from dividing eq. 3.2-1 by the time-varying mass $m(t)$. Including other external forces in the applied force, $\mathbf{f}_\mathrm{I} = \mathbf{f}_{\mathrm{I}_{aero}} + \mathbf{f}_{\mathrm{I}_{thrust}} + \mathbf{f}_{\mathrm{I}_{grav}}$

$$\frac{d\mathbf{v}_\mathrm{I}}{dt} = \frac{\mathbf{f}_\mathrm{I}}{m(t)} = \frac{\mathbf{f}_{\mathrm{I}_{aero}} + \mathbf{f}_{\mathrm{I}_{thrust}}}{m(t)} + \mathbf{g}_\mathrm{I} \tag{3.2-17}$$

where $\mathbf{f}_I$ contains the contact forces produced by aerodynamics and thrust, and $\mathbf{g}_\mathrm{I}$ is the acceleration due to gravity.

The mass variation also would be reflected in a changing inertia matrix $\mathcal{I}_\mathrm{I}(t)$, per eq. 3.1-28, and there is a corresponding torque due to mass expulsion. A more difficult problem is raised by the $(d\mathcal{I}_\mathrm{I}/dt)\boldsymbol{\omega}_\mathrm{I})$ term in eq. 3.2-15. The mass distribution of the aircraft is essentially fixed in a body frame of reference, so it is continually changing in an inertial frame. Therefore, $\mathcal{I}_\mathrm{I}$ is a function of $\boldsymbol{\Theta}$, found by integrating eq. 3.2-11, which depends on eq. 3.2-15.

The solution is to replace eq. 3.2-2 with an equivalent body-axis (non-inertial-frame) equation. The body-axis angular rate vector $\boldsymbol{\omega}_\mathrm{B}$ is related to the inertial-axis rate vector $\boldsymbol{\omega}_\mathrm{I}$ by

$$\boldsymbol{\omega}_\mathrm{B} = \begin{bmatrix} p \\ q \\ r \end{bmatrix} = \mathrm{H}_\mathrm{I}^\mathrm{B}\boldsymbol{\omega}_\mathrm{I} = \mathrm{H}_\mathrm{I}^\mathrm{B}\begin{bmatrix} \omega_x \\ \omega_y \\ \omega_z \end{bmatrix}_\mathrm{I} \tag{3.2-18}$$

where $\mathrm{H}_\mathrm{I}^\mathrm{B}$ is given by eq. 2.2-11; then $\boldsymbol{\omega}_\mathrm{I} = \mathrm{H}_\mathrm{B}^\mathrm{I}\boldsymbol{\omega}_\mathrm{B}$. The body's angular momentum with respect to the inertial frame, expressed in a body-axis frame of reference, is

$$\mathbf{h}_\mathrm{B} = \mathcal{I}_\mathrm{B}\,\boldsymbol{\omega}_\mathrm{B} \tag{3.2-19}$$

Note that the magnitudes of the vectors are unchanged by orthonormal transformation; therefore,

$$|\mathbf{h}_B| = |\mathbf{h}_I| \tag{3.2-20}$$

$$|\boldsymbol{\omega}_B| = |\boldsymbol{\omega}_I| \tag{3.2-21}$$

Furthermore, the (scalar) kinetic energy does not depend on the reference directions, so

$$\frac{1}{2}\boldsymbol{\omega}_B^T \mathcal{I}_B \boldsymbol{\omega}_B = \frac{1}{2}\boldsymbol{\omega}_I^T \mathcal{I}_I \boldsymbol{\omega}_I \tag{3.2-22}$$

The body-axis inertia matrix $\mathcal{I}_B$ is defined as

$$\begin{aligned}\mathcal{I}_B &\triangleq \int_{body} \begin{bmatrix} (y'^2 + z'^2) & -x'y' & -x'z' \\ -x'y' & (x'^2 + z'^2) & -y'z' \\ -x'z' & -y'z' & (x'^2 + y'^2) \end{bmatrix} dm \\ &\triangleq \begin{bmatrix} I_{xx} & -I_{xy} & -I_{xz} \\ -I_{xy} & I_{yy} & -I_{yz} \\ -I_{xz} & -I_{yz} & I_{zz} \end{bmatrix}_B \end{aligned} \tag{3.2-23}$$

with (x', y', z') referenced to the center of mass and aligned with the aircraft's body axes rather than the inertial axes. $d\mathcal{I}_B/dt$ is negligible, being principally due to fuel burn, and its incorporation in the moment due to thrust is analogous to the development of eq. 3.2-16c; hence, eq. 3.2-15 becomes

$$\frac{d\mathbf{h}_B}{dt} = \frac{d(\mathcal{I}_B\boldsymbol{\omega}_B)}{dt} = \mathcal{I}_B\frac{d\boldsymbol{\omega}_B}{dt} \tag{3.2-24}$$

The body-axis inertia matrix is related to the inertial-axis matrix by a *similarity transformation*. The transformation accounts for the fact that $\mathbf{h}_B$ and $\boldsymbol{\omega}_B$ are physically the same as $\mathbf{h}_I$ and $\boldsymbol{\omega}_I$, although expressed in a different coordinate system:

$$\mathbf{h}_B = \mathcal{I}_B\boldsymbol{\omega}_B = \mathbf{H}_I^B\mathbf{h}_I = \mathbf{H}_I^B\mathcal{I}_I\boldsymbol{\omega}_I = \mathbf{H}_I^B\mathcal{I}_I\mathbf{H}_B^I\boldsymbol{\omega}_B \tag{3.2-25}$$

Therefore, the similarity transform and its inverse are

$$\mathcal{I}_B = \mathbf{H}_I^B\mathcal{I}_I\mathbf{H}_B^I \tag{3.2-26}$$

and

$$\mathcal{I}_I = \mathbf{H}_B^I\mathcal{I}_B\mathbf{H}_I^B \tag{3.2-27}$$

The last equation illustrates the angle-dependent variability of $\mathcal{I}_I$ that occurs even when $\mathcal{I}_B$ is fixed, as H_I^B changes with the attitude Euler angles.

Similarity transformations also play a role in relating the time derivatives of the angular momentum vector that is observed in the two axis systems. Any vector can change in two ways: stretching (or shrinking) and rotating (a *Coriolis effect*). Given a body-axis reference frame that is rotating with respect to the inertial frame at angular rate $\boldsymbol{\omega}_I$, the time rate of change of the angular momentum can be expressed as

$$\frac{d\mathbf{h}_I}{dt} = \left[\frac{d\mathbf{h}_B}{dt}\right]_I + \boldsymbol{\omega}_I \times \mathbf{h}_I = \mathbf{H}_B^I \frac{d\mathbf{h}_B}{dt} + \tilde{\boldsymbol{\omega}}_I \mathbf{h}_I \tag{3.2-28a}$$

$[d\mathbf{h}_B/dt]_I$ connotes the change in $\mathbf{h}$ observed in the body frame but transformed to inertial coordinates, and $\tilde{\boldsymbol{\omega}}_I$ is the cross-product-equivalent matrix of $\boldsymbol{\omega}_I$:

$$\tilde{\boldsymbol{\omega}}_I \triangleq \begin{bmatrix} 0 & -\omega_z & \omega_y \\ \omega_z & 0 & -\omega_x \\ -\omega_y & \omega_x & 0 \end{bmatrix}_I \tag{3.2-29}$$

Applying the similarity transformation (eq. 3.2-27) to $\tilde{\boldsymbol{\omega}}_B$,

$$\tilde{\boldsymbol{\omega}}_I = \mathbf{H}_B^I \tilde{\boldsymbol{\omega}}_B \mathbf{H}_I^B \tag{3.2-30}$$

so the previous equation can be written as

$$\frac{d\mathbf{h}_I}{dt} = \mathbf{H}_B^I \frac{d\mathbf{h}_B}{dt} + \mathbf{H}_B^I \tilde{\boldsymbol{\omega}}_B \mathbf{H}_I^B \mathbf{h}_I = \mathbf{H}_B^I \frac{d\mathbf{h}_B}{dt} + \mathbf{H}_B^I \tilde{\boldsymbol{\omega}}_B \mathbf{h}_B \tag{3.2-28b}$$

Rearranging to solve for $d\mathbf{h}_B/dt$,

$$\frac{d\mathbf{h}_B}{dt} = \mathbf{H}_I^B \frac{d\mathbf{h}_I}{dt} - \tilde{\boldsymbol{\omega}}_B \mathbf{h}_B \tag{3.2-31}$$

Equation 3.2-2 indicates that $d\mathbf{h}_I/dt$ can be replaced by the sum of external torques, yielding

$$\frac{d\mathbf{h}_B}{dt} = \mathbf{H}_I^B \mathbf{m}_I - \tilde{\boldsymbol{\omega}}_B \mathbf{h}_B = \mathbf{m}_B - \tilde{\boldsymbol{\omega}}_B \mathbf{h}_B \tag{3.2-32a}$$

or

$$\mathcal{I}_B \frac{d\boldsymbol{\omega}_B}{dt} = \mathbf{m}_B - \tilde{\boldsymbol{\omega}}_B \mathcal{I}_B \boldsymbol{\omega}_B \tag{3.2-32b}$$

The external torques arising from aerodynamic and thrust effects have been transformed to body axes, a more convenient system for their specification; assuming gravity torques are negligible,

$$\mathbf{m}_{B} = \mathbf{m}_{B_{aero}} + \mathbf{m}_{B_{thrust}} + \mathbf{m}_{B_{grav}} = \mathbf{m}_{B_{aero}} + \mathbf{m}_{B_{thrust}} + 0 \qquad (3.2\text{-}33)$$

As our goal is a differential equation for the angular rate, both sides of eq. 3.2-32b are pre-multiplied by the inverse of the inertia matrix to yield

$$\frac{d\boldsymbol{\omega}_{B}}{dt} = \mathcal{I}_{B}^{-1}[\mathbf{m}_{B} - \tilde{\boldsymbol{\omega}}_{B}\mathcal{I}_{B}\boldsymbol{\omega}_{B}] \qquad (3.2\text{-}34)$$

The angular momentum of rotating machinery (e.g., propellers, engine compressors, or turbines) that is mounted on the airframe can be included in the calculation of the body-axis angular acceleration. Defining $\mathbf{h}_{R}$ as the net angular momentum of the rotating parts, referenced to the body frame, the term $\mathcal{I}_{B}\boldsymbol{\omega}_{B}$ is replaced by $(\mathcal{I}_{B}\boldsymbol{\omega}_{B} + \mathbf{h}_{R})$ in eq. 3.2-34.

Body axes also provide a more convenient frame for computing the translational velocity. The vector-derivative transformation applied to the angular momentum vector (eq. 3.2-31) can be applied to the linear momentum vector (eq. 3.2-1),

$$\frac{d\mathbf{p}_{B}}{dt} = \mathbf{H}_{I}^{B}\frac{d\mathbf{p}_{I}}{dt} - \tilde{\boldsymbol{\omega}}_{B}\mathbf{p}_{B} \qquad (3.2\text{-}35)$$

where $\mathbf{p}_{B} = \mathbf{H}_{I}^{B}\mathbf{p}_{I}$. Dividing by the mass, the linear acceleration expressed in body axes is

$$\frac{d\mathbf{v}_{B}}{dt} = \frac{\mathbf{f}_{B}}{m(t)} - \tilde{\boldsymbol{\omega}}_{B}\mathbf{v}_{B} \qquad (3.2\text{-}36)$$

where the external force contains aerodynamic, thrust, and gravitational forces that are measured in body axes:

$$\begin{aligned}\mathbf{f}_{B} &= \mathbf{f}_{B_{aero}} + \mathbf{f}_{B_{thrust}} + \mathbf{f}_{B_{grav}} = \mathbf{f}_{B_{aero}} + \mathbf{f}_{B_{thrust}} + m(t)\mathbf{H}_{I}^{B}\mathbf{g}_{I} \\ &= \begin{bmatrix} X \\ Y \\ Z \end{bmatrix}_{B_{aero}} + \begin{bmatrix} X \\ Y \\ Z \end{bmatrix}_{B_{thrust}} + m(t)\mathbf{H}_{I}^{B}\begin{bmatrix} 0 \\ 0 \\ g \end{bmatrix}\end{aligned} \qquad (3.2\text{-}37)$$

The four vector differential equations governing the aircraft's flight path are summarized in Table 3.2-1, allowing for the transformation of translational and rotational velocities to body axes in the kinematic equations.

The vector equations of Table 3.2-1 are deceptively simple by comparison to their 12 scalar equivalents. There are numerous matrix

TABLE 3.2-1 DIFFERENTIAL EQUATIONS FOR RIGID-AIRCRAFT FLIGHT

Equation	Description	No.
$\dfrac{d\mathbf{v}_B}{dt} = \dfrac{\mathbf{f}_{B_{aero}} + \mathbf{f}_{B_{thrust}}}{m(t)} + \mathbf{H}_I^B \mathbf{g}_I - \tilde{\boldsymbol{\omega}}_B \mathbf{v}_B$	(Translational Dynamics)	(3.2-36)
$\dfrac{d\boldsymbol{\omega}_B}{dt} = \mathcal{I}_B^{-1}(\mathbf{m}_{B_{aero}} + \mathbf{m}_{B_{thrust}} - \tilde{\boldsymbol{\omega}}_B \mathcal{I}_B \boldsymbol{\omega}_B)$	(Rotational Dynamics)	(3.2-34)
$\dfrac{d\mathbf{r}_I}{dt} = \mathbf{H}_B^I \mathbf{v}_B$	(Translational Kinematics)	(3.2-38)
$\dfrac{d\Theta}{dt} = \mathbf{L}_B^E \boldsymbol{\omega}_B$	(Rotational Kinematics)	(3.2-39)

multiplications to be carried out (particularly in eq. 3.2-34, which also contains a matrix inverse), and there are forces and moments to be specified in more detail. Nevertheless, these equations are the building blocks upon which much of the remainder of the book is based. This set of four vector equations can be further consolidated in a single nonlinear, ordinary differential equation of the form

$$\frac{d\mathbf{x}(t)}{dt} = \mathbf{f}[\mathbf{x}(t), \mathbf{u}(t), \mathbf{w}(t), t] \tag{3.2-40}$$

where $\mathbf{x}(t)$ is a 12-component *state vector* comprising the four three-component velocity and position vectors,

$$\begin{aligned}\mathbf{x}^T(t) &= [\mathbf{v}_B^T(t)\ \boldsymbol{\omega}_B^T(t)\ \mathbf{r}_I^T(t)\ \boldsymbol{\Theta}^T(t)] \\ &= [u(t)\ v(t)\ w(t)\ p(t)\ q(t)\ r(t)\ x_I(t)\ y_I(t)\ z_I(t)\ \phi(t)\ \theta(t)\ \psi(t)]\end{aligned} \tag{3.2-41}$$

$\mathbf{u}(t)$ is an m-component *control vector* (m depending on the particular aircraft, such as elevator, aileron, rudder, and throttle settings),

$$\mathbf{u}^T(t) = [\delta E(t)\ \delta A(t)\ \delta R(t)\ \delta T(t)] \tag{3.2-42}$$

$\mathbf{w}(t)$ is an s-component *disturbance vector* (where s is problem specific, such as the translational wind effect):

$$\mathbf{w}^T(t) = [u_w(t)\ v_w(t)\ w_w(t)] \tag{3.2-43}$$

Wind *shear* effects could be included by adding elements of the spatial wind shear matrix (eq. 2.1-11). The *dynamic function vector* f[·] (not to be confused with the force vector **f**) contains the right sides of the equations in Table 3.2-1.

Scalar Equations for a Symmetric Aircraft

The vector equations of Table 3.2-1 are expressed explicitly for an aircraft with mirror symmetry about its body-axis vertical plane. The inertia matrix may contain the xz product of inertia as well as the moments of inertia about all axes,

$$\mathcal{I}_B = \begin{bmatrix} I_{xx} & 0 & -I_{xz} \\ 0 & I_{yy} & 0 \\ -I_{xz} & 0 & I_{zz} \end{bmatrix}_B \tag{3.2-44}$$

and its inverse is

$$\mathcal{I}_B^{-1} = \begin{bmatrix} \dfrac{1}{I_{xx} - I_{xz}^2/I_{zz}} & 0 & \dfrac{I_{xz}}{I_{xx}I_{zz} - I_{xz}^2} \\ 0 & \dfrac{1}{I_{yy}} & 0 \\ \dfrac{I_{xz}}{I_{xx}I_{zz} - I_{xz}^2} & 0 & \dfrac{1}{I_{zz} - I_{xz}^2/I_{xx}} \end{bmatrix} \tag{3.2-45}$$

The cross-product-equivalent matrix for the body-axis angular rate is

$$\tilde{\omega}_B = \begin{bmatrix} 0 & -r & q \\ r & 0 & -p \\ -q & p & 0 \end{bmatrix} \tag{3.2-46}$$

From eq. 3.2-37, the three scalar equations for translational acceleration can be written as

$$\dot{u} = X/m + g_x + rv - qw \tag{3.2-47}$$

$$\dot{v} = Y/m + g_y - ru + pw \tag{3.2-48}$$

$$\dot{w} = Z/m + g_z + qu - pv \tag{3.2-49}$$

where X, Y, and Z include aerodynamic and thrust effects. There is inertial coupling of translational and rotational rates to translational accelerations associated with orthogonal axes: for example, $\dot{u}$ does not depend on the x-axis quantities u or p, but it is affected by cross-products of y and z variables. For the flat-earth case, gravity acts along the inertial z axis; hence,

$$\mathbf{g}_B \triangleq \begin{bmatrix} g_x \\ g_y \\ g_z \end{bmatrix}_B = \mathbf{H}_I^B \mathbf{g}_I = \mathbf{H}_I^B \begin{bmatrix} 0 \\ 0 \\ g \end{bmatrix} = \begin{bmatrix} -g\sin\theta \\ g\sin\phi\cos\theta \\ g\cos\phi\cos\theta \end{bmatrix} \tag{3.2-50}$$

The three scalar equations for angular acceleration are derived from eq. 3.2-34 using MATLAB's Symbolic Math Toolbox:

$$\dot{p} = (I_{zz}L + I_{xz}N - \{I_{xz}(I_{yy} - I_{xx} - I_{zz})\,p + [I_{xz}^2 + I_{zz}(I_{zz} - I_{yy})]r\}q) \div (I_{xx}I_{zz} - I_{xz}^2) \tag{3.2-51}$$

$$\dot{q} = [M - (I_{xx} - I_{zz})pr - I_{xz}(p^2 - r^2)] \div I_{yy} \tag{3.2-52}$$

$$\dot{r} = (I_{xz}L + I_{xx}N + \{I_{xz}(I_{yy} - I_{xx} - I_{zz})r + [I_{xz}^2 + I_{xx}(I_{xx} - I_{yy})]p\}q) \div (I_{xx}I_{zz} - I_{xz}^2) \tag{3.2-53}$$

L, M, and N contain all the external moments, expressed in body axes. The cross-coupling introduced by I_{xz} causes the roll and yaw accelerations $\dot{p}$ and $\dot{r}$ to respond to both rolling (L) and yawing (N) moments (eqs. 3.2-51 and 3.2-53). Furthermore, all of the inertial roll-yaw coupling sensitivity to angular rates is proportional to the pitch rate q. With zero pitch rate, $\dot{p}$ and $\dot{r}$ depend only on the external moments. The pitch acceleration $\dot{q}$ is insensitive to rolling and yawing moments, but it does respond to roll and yaw rates through I_{xz} and the difference between I_{xx} and I_{zz} (eq. 3.2-52).

Using the transpose of eq. 2.2-11 to define $\mathbf{H}_B^I$, the three translational kinematic equations obtained from eq. 3.2-36 are

$$\dot{x}_I = (\cos\theta\cos\psi)u + (-\cos\phi\sin\psi + \sin\phi\sin\theta\cos\psi)v + (\sin\phi\sin\psi + \cos\phi\sin\theta\cos\psi)w \tag{3.2-54}$$

$$\dot{y}_I = (\cos\theta\sin\psi)u + (\cos\phi\cos\psi + \sin\phi\sin\theta\sin\psi)v + (-\sin\phi\cos\psi + \cos\phi\sin\theta\sin\psi)w \tag{3.2-55}$$

$$\dot{z}_I = (-\sin\theta)u + (\sin\phi\cos\theta)v + (\cos\phi\cos\theta)w \tag{3.2-56}$$

Defining $\mathbf{L}_B^E$ with eq. 2.2-26, the three kinematic rotational equations that result from eq. 3.2-39 are

$$\dot{\phi} = p + (q \sin\phi + r \cos\phi) \tan\theta \tag{3.2-57}$$

$$\dot{\theta} = q \cos\phi - r \sin\phi \tag{3.2-58}$$

$$\dot{\psi} = (q \sin\phi + r \cos\phi) \sec\theta \tag{3.2-59}$$

In the flat-earth case, x_I and y_I do not feed back into the remaining dynamic equations, and z_I does so only through aerodynamic and thrust effects when air density and/or gravity variations with altitude are considered. However, the attitude Euler angles (ϕ, θ, ψ) figure in the transformation matrix $\mathbf{H}_B^I$ and its inverse, producing coupling with the translational position and velocity equations. The Euler angle rate equations contain $\tan\theta$ and $\sec\theta$; hence, they become singular when the pitch angle equals ± 90 deg.

Alternative Frames of Reference

The combination of body axes for velocity calculations and inertial axes for position calculations is generally satisfactory, but alternative frames of reference for one or more of the equation sets presented in Table 3.2-1 are sometimes useful. Such alternatives include

- Alternative inertial frames
- Alternative body frames
- Velocity axes
- Wind axes
- Air-mass-relative axes

The first two options merely change the directions of reference axes and the origins of axis systems without changing the forms of the equations found in the table. The last three options produce modifications in the dynamic equations as well.

Inertial Reference Frames

For flat-earth models, the most likely *alternative inertial frame* is one in which the reference x axis is rotated away from north by some azimuth angle ψ_o to a preferred direction, such as a nominal flight path heading. The z axis remains vertical down, and the y axis is perpendicular to the other two. The new yaw angle ψ_{new} equals $(\psi_{old} - \psi_o)$, while the other

Euler angles are unchanged. From eq. 2.2-13, the transformation matrix relating the old reference frame to the new is

$$\mathbf{H}_{\mathrm{Io}}^{\mathrm{In}} \triangleq \begin{bmatrix} \cos\psi_o & \sin\psi_o & 0 \\ -\sin\psi_o & \cos\psi_o & 0 \\ 0 & 0 & 1 \end{bmatrix} \tag{3.2-60}$$

Assuming that the body-to-inertial-axis transformation $\mathbf{H}_{\mathrm{B}}^{\mathrm{I}}$ relates the body axes to the new inertial-axis system, body-axis vectors (such as the translational velocity) can be transformed to the old inertial-axis system by the matrix product $\mathbf{H}_{\mathrm{In}}^{\mathrm{Io}}\mathbf{H}_{\mathrm{B}}^{\mathrm{I}}$. The transformation $\mathbf{L}_{\mathrm{E}}^{\mathrm{B}}$ relating Euler-angle and body-axis angular rate vectors does not contain ψ and is, therefore, unchanged.

The origin for translational position determination could also be shifted from the original reference point, while the origin for attitude determination normally remains at the vehicle's center of mass. The original position vector is then related to positions calculated in the new frame by

$$\mathbf{r}_{old} = \mathbf{H}_{\mathrm{In}}^{\mathrm{Io}}\mathbf{r}_{new} + \mathbf{r}_o \tag{3.2-61}$$

where the origin of the new frame $\mathbf{r}_o$ is expressed in the coordinates of the original frame.

Body-Axis Reference Frames

Alternative body frames of reference fall into three categories:

- General body axes
- Principal axes
- Stability axes

Body axes (Fig. 3.2-1) typically refer to a *centerline* (x axis) and two orthogonal axes that are convenient references for design, construction, and operation of the aircraft, one vertical down in a plane of mirror symmetry and another to the right (as viewed from the rear of the aircraft).

General Body Axes

It may be desirable to make an arbitrary change in the reference directions, either to adjust to engineering changes or to accommodate a subsystem (e.g., a sensor or camera) whose performance is to be investigated (Fig. 3.2-1). Assuming that the change in body axes is represented

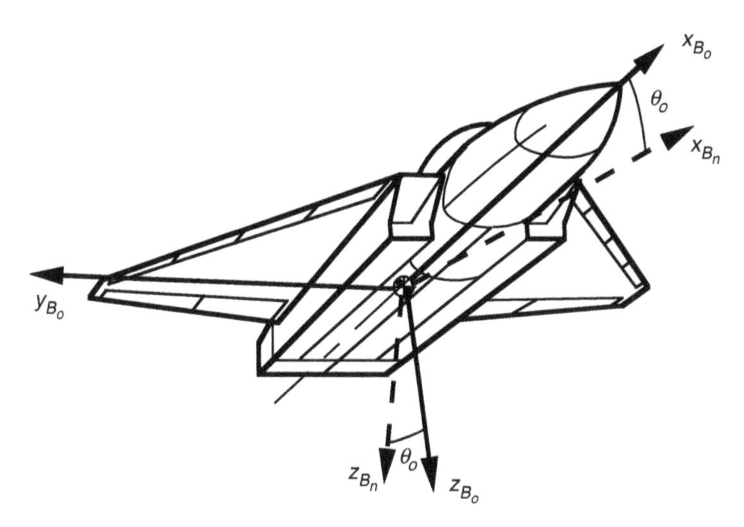

Fig. 3.2-1. Transformation from original to new body axes.

by three Euler angles (ϕ, θ, ψ) sequenced as before, the transformation matrix $\mathbf{H}_B^G$ that maps vectors from the original body axes to the *general body axes* is identical to the original definition of $\mathbf{H}_I^B$ (eq. 2.2-11). If the angle changes (ε_ϕ, ε_θ, ε_ψ) are small, we can approximate $\cos\varepsilon$ by 1 and $\sin\varepsilon$ by ε, and we can neglect products of ε, leading to

$$\mathbf{H}_B^G \approx \begin{bmatrix} 1 & \varepsilon_\psi & -\varepsilon_\theta \\ -\varepsilon_\psi & 1 & \varepsilon_\phi \\ \varepsilon_\theta & -\varepsilon_\phi & 1 \end{bmatrix} = \mathbf{I}_3 - \tilde{\boldsymbol{\varepsilon}} \tag{3.2-62}$$

where $\boldsymbol{\varepsilon}$ is the vector of angle perturbations and $\tilde{\boldsymbol{\varepsilon}}$ is its cross-product-equivalent matrix. The determinant of the exact $\mathbf{H}_B^G$ is precisely 1 because the matrix is an orthonormal transformation, which preserves vector length. The approximation produces the determinant $|\mathbf{I}_3 - \tilde{\boldsymbol{\varepsilon}}| = 1 + \varepsilon_\phi^2 + \varepsilon_\theta^2 + \varepsilon_\psi^2$. The error is of order (ε^2), which is consistent with the small-angle assumptions. (Of course, the error can be avoided by using the exact transformation matrix.) Because $\mathbf{H}_B^G$ is orthonormal, its transpose is its inverse. Then the two matrices

$$\mathbf{H}_I^G = \mathbf{H}_B^G \mathbf{H}_I^B \tag{3.2-63}$$

and

$$\mathbf{H}_G^I = \mathbf{H}_B^I \mathbf{H}_G^B \tag{3.2-64}$$

can replace $\mathbf{H}_I^B$ and $\mathbf{H}_B^I$ in the dynamic equations. Vectors are transformed in the process

$$\mathbf{v}_G = \mathbf{H}_B^G \mathbf{v}_B \tag{3.2-65}$$

$$\boldsymbol{\omega}_G = \mathbf{H}_B^G \boldsymbol{\omega}_B \tag{3.2-66}$$

$$\mathbf{L}_G^E = \mathbf{L}_B^E \mathbf{H}_G^B \tag{3.2-67}$$

$$\mathbf{f}_G' = \mathbf{H}_B^G \mathbf{f}_B' \tag{3.2-68}$$

$$\mathbf{m}_G = \mathbf{H}_B^G \mathbf{m}_B \tag{3.2-69}$$

and a similarity transformation is applied to the inertia matrix:

$$\mathcal{I}_G = \mathbf{H}_B^G \mathcal{I}_B \mathbf{H}_G^B \tag{3.2-70}$$

The translational position $\mathbf{r}_I$ and angular attitude $\boldsymbol{\Theta}_I$ remain referenced to an inertial frame.

Principal Axes

Shifting from the original body axes to principal axes is just a special case of the above. *Principal axes* are defined such that the inertia matrix is diagonal:

$$\mathcal{I}_P = \begin{bmatrix} I_{xx} & 0 & 0 \\ 0 & I_{yy} & 0 \\ 0 & 0 & I_{zz} \end{bmatrix}_P \tag{3.2-71}$$

Because the matrices $\mathcal{I}_B$ and $\mathcal{I}_P$ represent the same physical properties viewed in different Cartesian coordinate systems, they are related by a similarity transformation (eq. 3.2-70):

$$\mathcal{I}_P = \mathbf{H}_B^P \mathcal{I}_B \mathbf{H}_P^B \tag{3.2-72}$$

The trick is to find $\mathbf{H}_P^B$—or is it? Actually, it proves more direct to first compute $\mathcal{I}_P$ from $\mathcal{I}_B$ and then to calculate $\mathbf{H}_P^B$. The diagonal elements of $\mathcal{I}_P$ (i.e., the *principal-axis moments of inertia*) are the *eigenvalues* of $\mathcal{I}_B$, while $\mathbf{H}_B^P$ is derived from the *eigenvectors* of $\mathcal{I}_B$.

The eigenvalues of an $(n \times n)$ matrix $\mathbf{A}$ are solutions λ_i, $i = 1, n$, to the determinant equation,

$$|\lambda \mathbf{I}_n - \mathbf{A}| \triangleq \Delta(\lambda) = 0 \tag{3.2-73}$$

where the determinant of $(\lambda \mathbf{I}_n - \mathbf{A})$, the *characteristic matrix* of $\mathbf{A}$, produces the scalar *characteristic polynomial* $\Delta(\lambda)$,

$$\Delta(\lambda) = \lambda^n + c_{n-1}\lambda^{n-1} + \cdots + c_1\lambda + c_0 \tag{3.2-74}$$

$\Delta(\lambda)$ can be factored into n binomials, each of which contains a *root* (or eigenvalue) of the *characteristic equation* $\Delta(\lambda) = 0$:

$$\Delta(\lambda) = (\lambda - \lambda_1)(\lambda - \lambda_2)\cdots(\lambda - \lambda_n) = 0 \tag{3.2-75}$$

As there are n eigenvalues, they can be used to define an $(n \times n)$ diagonal matrix $\mathbf{\Lambda}$:

$$\mathbf{\Lambda} \triangleq \begin{bmatrix} \lambda_1 & 0 & .. & .. & 0 \\ 0 & \lambda_2 & .. & .. & 0 \\ .. & .. & .. & .. & .. \\ 0 & 0 & .. & .. & \lambda_n \end{bmatrix} \tag{3.2-76}$$

If $\mathbf{A}$ is a symmetric matrix with real-valued coefficients (i.e., not complex numbers, to be discussed in a later chapter), its eigenvalues and eigenvectors are also real. By inspection of eq. 3.2-23, the diagonal terms of an inertia matrix must be positive because the mass is positive and the coordinates of the integrands are squared. Hence, the eigenvalues of $\mathcal{I}_B$ and $\mathcal{I}_P$ are positive, and the inertia matrices are said to be *positive definite.*

Eigenvectors are n-vector solutions $\mathbf{e}_i$ to the equation

$$\mathbf{A}\,\mathbf{e}_i = \mathbf{e}_i\,\lambda_i = \lambda_i\,\mathbf{e}_i, \quad i = 1, n \tag{3.2-77}$$

With *distinct eigenvalues* (i.e., no two are the same), there are n eigenvectors. With repeated eigenvalues, there may be fewer eigenvectors [B-5]. A missile might have identical pitch and yaw moments of inertia, but it would be unusual for an aircraft to have precisely identical principal moments of inertia. Consequently, the eigenvalues of an aircraft's inertia matrix are normally distinct.

Eigenvectors are defined within an arbitrary scalar constant a because eq. 3.2-77 is satisfied by $a\mathbf{e}_i$ if it is satisfied by $\mathbf{e}_i$:

$$(\lambda_i \mathbf{I}_n - \mathbf{A})\,a\,\mathbf{e}_i = 0 \tag{3.2-78}$$

The inverse of the characteristic matrix is

$$(\lambda \mathbf{I}_n - \mathbf{A})^{-1} = \frac{\mathrm{Adj}(\lambda\,\mathbf{I}_n - \mathbf{A})}{|\,\lambda\,\mathbf{I}_n - \mathbf{A}\,|} \tag{3.2-79}$$

hence, using eq. 3.2-75, we set λ to λ_i, yielding

$$\begin{aligned}(\lambda_i\,\mathbf{I}_n - \mathbf{A})\,\mathrm{Adj}(\lambda_i \mathbf{I}_n - \mathbf{A}) &= (\lambda_i\,\mathbf{I}_n - \mathbf{A})\,(\lambda_i\,\mathbf{I}_n - \mathbf{A})^{-1}\,|\lambda_i\,\mathbf{I}_n - \mathbf{A}| \\ &= |\lambda_i\,\mathbf{I}_n - \mathbf{A}| = 0\end{aligned} \tag{3.2-80}$$

as $(\cdot)\,(\cdot)^{-1}$ equals the identity matrix **I**. From eq. 3.2-78 and 3.2-80, we conclude that, for a distinct eigenvalue λ_i, any nonzero column of the characteristic matrix adjoint (eq. 2.2-28)

$$\text{Adj}(\lambda_i \mathbf{I}_n - \mathbf{A}) = [a_1 \mathbf{e}_i \; a_2 \mathbf{e}_i \cdots a_n \mathbf{e}_i] \tag{3.2-81}$$

can serve as the corresponding eigenvector. Repeating this process for each eigenvalue, the n eigenvectors form the columns of the $(n \times n)$ *modal matrix* **E**:

$$\mathbf{E} = [\mathbf{e}_1 \; \mathbf{e}_2 \cdots \mathbf{e}_n] \tag{3.2-82}$$

From eqs. 3.2-74 and 3.2-75,

$$\mathbf{E}\Lambda = \mathbf{A}\mathbf{E} \tag{3.2-83}$$

Premultiplying both sides by $\mathbf{E}^{-1}$,

$$\Lambda = \mathbf{E}^{-1}\mathbf{A}\mathbf{E} \tag{3.2-84}$$

For **A** real and symmetric, normalizing the eigenvectors such that $|\mathbf{E}| = 1$ produces an orthonormal transformation, and $\mathbf{E}^{-1} = \mathbf{E}^T$. The similarity to eq. 3.2-72 is apparent, with $n = 3$, $\Lambda = \mathcal{I}_\mathrm{P}$, $\mathbf{A} = \mathcal{I}_\mathrm{B}$, and $\mathbf{E}^\mathrm{T} = \mathbf{H}_\mathrm{B}^\mathrm{P}$. Equations 3.2-65–3.2-70 then apply, replacing G with P in sub- and superscripts.

Example 3.2-1. Principal Inertia and Transformation Matrices for a Symmetric Aircraft

The principal moments of inertia and transformation matrix for the symmetric aircraft considered earlier can be calculated using MATLAB's Symbolic Math Toolbox in the following m-file:

```
Ixx = sym('Ixx'); Ixz = sym('Ixz'); Iyy = sym('Iyy'); Izz = sym('Izz');
IB  = sym('IB');
HPBint = sym('HPBint'); IPint = sym('IPint');
IB  = [Ixx 0 -Ixz; 0 Iyy 0; -Ixz 0 Izz];
[HPBint,IPint] = eig(IB)
```

The matrices and matrix elements are defined as symbolic variables, and $\mathcal{I}_B$ is entered as a symmetric matrix containing the moments of inertia and the single product of inertia I_{xz}. The transformation matrix `HPBint` and principal-axis inertia matrix `IPint` are calculated by the function `eig`$(\cdot)$, producing the following output:

```
HPBint =

[0,
   (1/2*Izz-1/2*Ixx-1/2*(Izz^2-2*Ixx*Izz+Ixx^2+4*Ixz^2)^(1/2))/Ixz,
   (1/2*Izz-1/2*Ixx+1/2*(Izz^2-2*Ixx*Izz+Ixx^2+4*Ixz^2)^(1/2))/Ixz]
[1, 0,                                                             0]
[0, 1,                                                             1]

IPint =

[Iyy,0,                                                            0]
[0,  1/2*Izz+1/2*Ixx+1/2*(Izz^2-2*Ixx*Izz+Ixx^2+4*Ixz^2)^(1/2),    0]
[0,  0,    1/2*Izz+1/2*Ixx-1/2*(Izz^2-2*Ixx*Izz+Ixx^2+4*Ixz^2)^(1/2)]
```

Each eigenvector list corresponds to the eigenvalue appearing in the same sequential position. The intermediate transformation and inertia matrices, `HPBint` and `IPint` must be reordered and normalized so that $\mathcal{I}_{\mathrm{P}}$ takes the form of eq. 3.2-71, while $\mathbf{H}_{\mathrm{B}}^{\mathrm{P}}$ is of the form $\begin{bmatrix} a & 0 & -b \\ 0 & c & 0 \\ b & 0 & a \end{bmatrix}$. The rotation from the original body axes to principal axes takes place about the y axis; from eq. 2.2-11, the angle of rotation θ can be determined, as $\cos\theta = a$ and $\sin\theta = b$. Numerical values of the inertia and transformation matrix coefficients are obtained by making the same function call with numerical rather than symbolic variables.

Stability Axes

Stability axes form yet another set of alternative body axes. The usual motivation is to align the aircraft's x axis to the nominal orientation of the air-relative velocity vector, as when examining perturbations from a steady flight condition. Steady flight normally implies symmetric flight, so a rotation back through the nominal angle of attack α_o is indicated. If the x axis is skewed from the heading angle, a reference-axis counter-rotation through the nominal sideslip angle β_o also must be performed.

The angles of attack and sideslip are defined from the velocity vector to the body x axis; therefore, the needed transformations derive from the inverse (or transpose) of $\mathbf{H}_{\mathrm{W}}^{\mathrm{B}}$ (eq. 2.2-32):

$$\mathbf{H}_B^S \triangleq \mathbf{H}_3^S(\beta_o)\mathbf{H}_B^3(-\alpha_o) = \begin{bmatrix} \cos\beta_o & \sin\beta_o & 0 \\ -\sin\beta_o & \cos\beta_o & 0 \\ 0 & 0 & 1 \end{bmatrix} \begin{bmatrix} \cos\alpha_o & 0 & \sin\alpha_o \\ 0 & 1 & 0 \\ -\sin\alpha_o & 0 & \cos\alpha_o \end{bmatrix}$$

$$= \begin{bmatrix} \cos\alpha_o \cos\beta_o & \sin\beta_o & \sin\alpha_o \cos\beta_o \\ -\cos\alpha_o \sin\beta_o & \cos\beta_o & -\sin\alpha_o \sin\beta_o \\ -\sin\alpha_o & 0 & \cos\alpha_o \end{bmatrix} \quad (3.2\text{-}85)$$

Once again, eq. 3.2-63–3.2-68 apply, with S replacing G in sub- and superscripts.

Velocity- and Wind-Axis Reference Frames

It may be preferable to solve for (V, ξ, γ) or (V, β, α) rather than $(u, v, w)_B$. The velocity and wind axes introduced in Section 2.2 always orient their x axes to the inertial velocity vector rather than to some body-fixed reference direction. Because the body's orientation within these frames may be continually changing, they are not appropriate for expressing the moment equations, which involve the inertia matrix. Therefore, their only practical application is to the force equations. Velocity axes are often used in point-mass trajectory calculations (where attitude dynamics are neglected), while wind axes provide direct solutions for the aerodynamic angles. Reference to the "wind" follows identification of air flowing over the aircraft as the "relative wind." Axis systems that account for natural wind fields are called "air-mass-relative" systems and are discussed below.

Because velocity and wind axes are rotating with respect to the inertial axes, they are noninertial frames, and the force equations must contain Coriolis terms. The velocity-axis vector differential equation takes the form

$$\frac{d\mathbf{v}_V}{dt} = \frac{\mathbf{f}_{V_{aero}} + \mathbf{f}_{V_{thrust}}}{m(t)} + \mathbf{H}_I^V \mathbf{g}_I - \tilde{\boldsymbol{\omega}}_V \mathbf{v}_V \quad (3.2\text{-}86a)$$

where $\mathbf{v}_V$ is the inertial velocity vector expressed in the velocity frame of reference, and $\mathbf{H}_I^V$ is given by equation 2.2-33. $\mathbf{f}_{V_{aero}} + \mathbf{f}_{V_{thrust}}$ expresses forces along the velocity vector, perpendicular to the vertical plane containing the velocity vector, and perpendicular to the velocity vector within the vertical plane (Figure 3.2-2). There is an apparent ambiguity: the x axis is aligned with the velocity vector,

$$\mathbf{v}_V \triangleq \begin{bmatrix} V \\ 0 \\ 0 \end{bmatrix} \quad (3.2\text{-}87)$$

so the y and z components of $\mathbf{v}_V$ are always zero! Equation 3.2-84a can be integrated to find the magnitude V, but there seem to be no

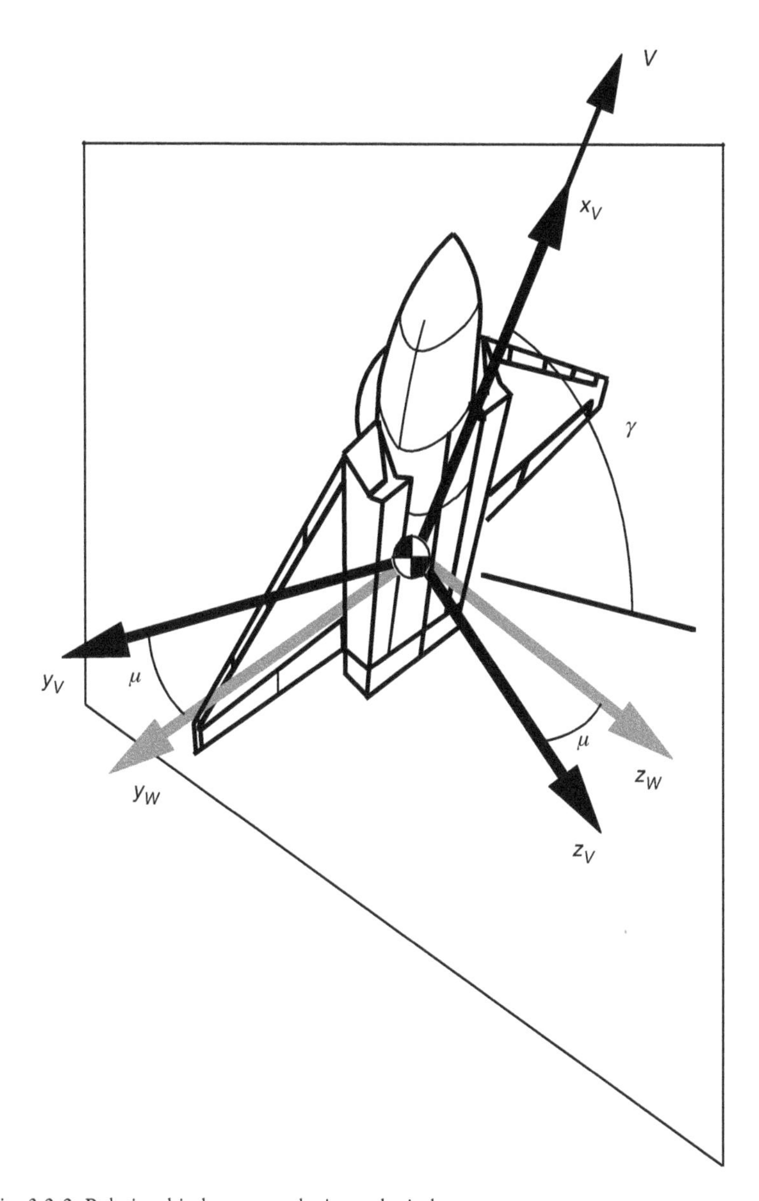

Fig. 3.2-2. Relationship between velocity and wind axes.

comparable equations for finding the horizontal and vertical flight path angles ξ and γ.

Actually the solution to this problem *is* contained in equation 3.2-86a, which can be rewritten as

$$\frac{d\mathbf{v}_V}{dt} + \tilde{\boldsymbol{\omega}}_V \mathbf{v}_V = \frac{\mathbf{f}_{V_{aero}} + \mathbf{f}_{V_{thrust}}}{m(t)} + \mathbf{H}_I^V \mathbf{g}_I \tag{3.2-86b}$$

The angular rate $\boldsymbol{\omega}_V$ of the velocity vector is related to the flight path angle rate vectors by a nonorthogonal transformation matrix that is constructed in the same fashion as $\mathbf{L}_E^B$ (equations 2.2-24, 2.2-25, and 2.2-33):

$$\begin{aligned} \begin{bmatrix} \omega_x \\ \omega_y \\ \omega_z \end{bmatrix}_V &= \mathbf{I}_n \begin{bmatrix} 0 \\ \dot{\gamma} \\ 0 \end{bmatrix} + \mathbf{H}_4^V \begin{bmatrix} 0 \\ 0 \\ \dot{\xi} \end{bmatrix} = \begin{bmatrix} 0 & 0 & -\sin\gamma \\ 0 & 1 & 0 \\ 0 & 0 & \cos\gamma \end{bmatrix} \begin{bmatrix} 0 \\ \dot{\gamma} \\ \dot{\xi} \end{bmatrix} \\ &= \begin{bmatrix} -\dot{\xi}\sin\gamma \\ \dot{\gamma} \\ \dot{\xi}\cos\gamma \end{bmatrix} \end{aligned} \tag{3.2-88}$$

The cross-product-equivalent matrix $\tilde{\boldsymbol{\omega}}_V$ is

$$\tilde{\boldsymbol{\omega}}_V = \begin{bmatrix} 0 & -\dot{\xi}\cos\gamma & \dot{\gamma} \\ \dot{\xi}\cos\gamma & 0 & \dot{\xi}\sin\gamma \\ -\dot{\gamma} & -\dot{\xi}\sin\gamma & 0 \end{bmatrix} \tag{3.2-89}$$

and

$$\tilde{\boldsymbol{\omega}}_V \mathbf{v}_V = \begin{bmatrix} 0 \\ V\dot{\xi}\cos\gamma \\ -V\dot{\gamma} \end{bmatrix} \tag{3.2-90}$$

Consequently,

$$\begin{aligned} \frac{d\mathbf{v}_V}{dt} + \tilde{\boldsymbol{\omega}}_V \mathbf{v}_V &= \begin{bmatrix} \dot{V} \\ V\dot{\xi}\cos\gamma \\ -V\dot{\gamma} \end{bmatrix} = \begin{bmatrix} 1 & 0 & 0 \\ 0 & V\cos\gamma & 0 \\ 0 & 0 & -V \end{bmatrix} \begin{bmatrix} \dot{V} \\ \dot{\xi} \\ \dot{\gamma} \end{bmatrix} \\ &\triangleq \mathbf{L}_{V'}^V \begin{bmatrix} \dot{V} \\ \dot{\xi} \\ \dot{\gamma} \end{bmatrix} \triangleq \mathbf{L}_{V'}^V \frac{d\mathbf{v}'_V}{dt} \end{aligned} \tag{3.2-91}$$

where the velocity-axis velocity vector (denoted by $\mathbf{v}'_{\mathrm{V}}$) is

$$\mathbf{v}'_{\mathrm{V}} = \begin{bmatrix} V \\ \xi \\ \gamma \end{bmatrix} \tag{3.2-92}$$

Equation 3.2-86b is equivalent to

$$\mathbf{L}^{\mathrm{V}}_{\mathrm{V}'} \frac{d\mathbf{v}_{\mathrm{V}}}{dt} = \frac{\mathbf{f}_{\mathrm{V}_{aero}} + \mathbf{f}_{\mathrm{V}_{thrust}}}{m(t)} + \mathbf{H}^{\mathrm{V}}_{\mathrm{I}} \mathbf{g}_{\mathrm{I}} \tag{3.2-93a}$$

leading to the *velocity-axis translational dynamic equation*,

$$\frac{d\mathbf{v}_{\mathrm{V}}}{dt} = \mathbf{L}^{\mathrm{V}'}_{\mathrm{V}} \left[\frac{\mathbf{f}_{\mathrm{V}_{aero}} + \mathbf{f}_{\mathrm{V}_{thrust}}}{m(t)} + \mathbf{H}^{\mathrm{V}}_{\mathrm{I}} \mathbf{g}_{\mathrm{I}} \right] \tag{3.2-93b}$$

with

$$\mathbf{L}^{\mathrm{V}'}_{\mathrm{V}} \triangleq \begin{bmatrix} 1 & 0 & 0 \\ 0 & 1/(V\cos\gamma) & 0 \\ 0 & 0 & -1/V \end{bmatrix} \tag{3.2-94}$$

$\mathbf{L}^{\mathrm{V}'}_{\mathrm{V}}$ rescales the forces from linear to angular units.

The derivation of the *wind-axis force equations* follows a similar path. As in the previous case, $\mathbf{v}^{T}_{\mathrm{W}} = [V\ 0\ 0]^{T}$, and we must define $\mathbf{v}'^{T}_{\mathrm{W}} = [V \quad \beta \quad \alpha]$. We require a dynamic equation of the form

$$\frac{d\mathbf{v}_{\mathrm{W}}}{dt} = \frac{\mathbf{f}_{\mathrm{W}_{aero}} + \mathbf{f}_{\mathrm{W}_{thrust}}}{m(t)} + \mathbf{H}^{\mathrm{W}}_{\mathrm{I}} \mathbf{g}_{\mathrm{I}} - \tilde{\boldsymbol{\omega}}_{\mathrm{W}} \mathbf{v}_{\mathrm{W}} \tag{3.2-95}$$

where the Cartesian components of the frame are rotated from the velocity-axis components by the angle μ (figure 3.2-2). Rather than beginning with the inertial-frame equation, it is more direct to begin with the body-axis dynamic equation (eq. 3.2-37), as α and β are the angles between the velocity vector and the body x axis. Relative to the body frame, the velocity vector has the angular rate

$$\begin{bmatrix} \omega_x \\ \omega_y \\ \omega_z \end{bmatrix}_A = \mathbf{I}_n \begin{bmatrix} 0 \\ -\dot{\alpha} \\ 0 \end{bmatrix} + \mathbf{H}^{\mathrm{B}}_{3} \begin{bmatrix} 0 \\ 0 \\ \dot{\beta} \end{bmatrix} = \begin{bmatrix} 0 & 0 & -\sin\alpha \\ 0 & 1 & 0 \\ 0 & 0 & \cos\alpha \end{bmatrix} \begin{bmatrix} 0 \\ -\dot{\alpha} \\ \dot{\beta} \end{bmatrix}$$

$$= \begin{bmatrix} -\dot{\beta}\sin\alpha \\ -\dot{\alpha} \\ \dot{\beta}\cos\alpha \end{bmatrix} \tag{3.2-96}$$

where $\mathbf{H}_3^{\mathrm{B}}$ is shown in eq. 2.2-32. Hence, the inertial angular rate of the velocity vector is $(\boldsymbol{\omega}_{\mathrm{A}} + \boldsymbol{\omega}_{\mathrm{B}})$ in the body frame, or

$$\boldsymbol{\omega}_{\mathrm{W}} = \mathbf{H}_{\mathrm{B}}^{\mathrm{W}}(\boldsymbol{\omega}_{\mathrm{A}} + \boldsymbol{\omega}_{\mathrm{B}}) \tag{3.2-97}$$

in the wind frame, where $\mathbf{H}_{\mathrm{B}}^{\mathrm{W}}$ is defined by eq. 2.2-32. The remainder of the derivation follows the velocity-axis development, noting that $\mathbf{H}_{\mathrm{I}}^{\mathrm{W}} = \mathbf{H}_{\mathrm{B}}^{\mathrm{W}}\mathbf{H}_{\mathrm{I}}^{\mathrm{B}}$.

A cautionary note: The α and β used here orient the body with respect to the *inertial* velocity vector, which is the same as the *air-relative* velocity vector in the absence of a natural wind field. If the aircraft is flying in a wind field, the two velocities, as well as the corresponding (α, β), are different, as noted in the section below. The aerodynamic effects discussed in Section 2.4 relate to the air-relative (α, β), not the inertial (α, β).

Air-Mass-Relative Reference Frame

Because aerodynamic forces and moments depend on the air-relative velocity, it may be desirable to solve for this velocity directly rather than compute the velocity in another coordinate frame and then account for the natural wind. The advantage of referring the velocity vector to the (possibly moving) air mass is that lift, drag, and moment calculations are simplified; the disadvantage is that the relationship of velocity to position is altered, and if the air-mass velocity is itself changing, the new reference frame is noninertial.

Consider how wind effects are treated in an inertial reference frame. Given an air-mass velocity $\mathbf{w}(\mathbf{x}, t)$ with respect to the inertial frame (eq. 2.1-10) and an inertial aircraft velocity $\mathbf{v}_{\mathrm{I}}(t)$ viewed in the inertial frame, the air-mass-relative (or simply air-relative) velocity vector of the aircraft is

$$\mathbf{v}_{\mathrm{A}} \triangleq \mathbf{v}_{\mathrm{I}} - \mathbf{w} = \begin{bmatrix} v_1 \\ v_2 \\ v_3 \end{bmatrix}_{\mathrm{I}} - \begin{bmatrix} w_1 \\ w_2 \\ w_3 \end{bmatrix} \tag{3.2-98}$$

The sign conventions are chosen so that a headwind increases the air-relative velocity of a forward-moving aircraft. In the body frame of reference, the air-relative velocity vector is

$$[\mathbf{v}_{\mathrm{A}}]_{\mathrm{B}} \triangleq \begin{bmatrix} u_A \\ v_A \\ w_A \end{bmatrix} = \mathbf{H}_{\mathrm{I}}^{\mathrm{B}}\mathbf{v}_{\mathrm{A}} = \mathbf{v}_{\mathrm{B}} - \mathbf{H}_{\mathrm{I}}^{\mathrm{B}}\mathbf{w} \tag{3.2-99}$$

with $\mathbf{v}_B$ representing the inertial velocity viewed in the body frame as before. Equation 2.2-31 can be applied to eq. 3.2-99 to obtain the air-relative velocity magnitude (i.e., the airspeed) and the angles of attack and sideslip:

$$\begin{bmatrix} V_A \\ \beta_A \\ \alpha_A \end{bmatrix} = \begin{bmatrix} \sqrt{u_A^2 + v_A^2 + w_A^2} \\ \sin^{-1}(v_A/V_A) \\ \tan^{-1}(w_A/u_A) \end{bmatrix} \quad (3.2\text{-}100)$$

This definition of V, α, and β is used to compute aerodynamic forces and moments, and the equations of motion presented previously are otherwise unchanged. (Of course, eq. 3.2-99 reverts to eq. 2.2-31 when the wind speed is zero.)

With constant wind speed,

$$\frac{d\mathbf{v}_A}{dt} = \frac{d\mathbf{v}_I}{dt} = \frac{\mathbf{f}_I}{m(t)} \quad (3.2\text{-}101)$$

Therefore, the translational dynamic equation applies to computing $\mathbf{v}_A$ as well as to computing $\mathbf{v}_I$, the only difference being in the specification of initial conditions. In other words, a reference frame translating at constant velocity with respect to an inertial frame is itself an inertial frame. Consequently, all the alternative frames for computing the translational velocity could be used, changing only the velocity vector definition.

The position equation (eq. 3.2-10) must be adjusted to accommodate the air-relative velocity; from eq. 3.2-98,

$$\frac{d\mathbf{r}_I}{dt} = \mathbf{v}_I = \mathbf{v}_A + \mathbf{w} \quad (3.2\text{-}102)$$

Because an aircraft cruising with a fixed power setting reaches equilibrium at a given airspeed rather than a given inertial speed, the air-mass-relative equations of motion may be preferred. However, care must be taken to differentiate between meaningful and meaningless quantities: just as the inertial (α, β) have little use when the wind is blowing, air-relative flight path angles (γ, ξ) are subject to misinterpretation.

If the wind field is changing in time either explicitly or implicitly (eq. 2.1-13), then this must be taken into account in the dynamic equation,

$$\frac{d\mathbf{v}_A}{dt} = \frac{d\mathbf{v}_I}{dt} - \frac{d\mathbf{w}}{dt} = \frac{\mathbf{f}_I}{m(t)} - \frac{d\mathbf{w}}{dt} \quad (3.2\text{-}103)$$

$d\mathbf{w}/dt$ produces an *apparent specific force* (i.e., force per unit mass) in the air-relative frame. The effect of microburst wind shear on an aircraft then can be compared to the aircraft's performance limits [M-1]; the *F factor* [H-5] expresses the microburst effect as a reduction in available specific excess power (Section 3.1). Nevertheless, the original equations of this section can also be used for that purpose, taking eq. 3.2-100 into account [P-7]. Furthermore, using air-relative equations in this case does not automatically account for unsteady aerodynamic effects that arise from wind shear, which are discussed in Section 3.4.

Direction Cosines and Quaternions

Euler angles are used as angular measures because they are the most compact means of expressing an aircraft's attitude in space. Two alternatives are presented in this section, not only to demonstrate different means of portraying attitude but to circumvent the singularities associated with the computation of Euler angles when $\theta = \pm 90$ deg.

Consider the vector $\mathbf{r}$ expressed in a Cartesian coordinate system (Fig. 3.2-3) with unit vectors $(\mathbf{i}, \mathbf{j}, \mathbf{k})$. The components (x, y, z) of the

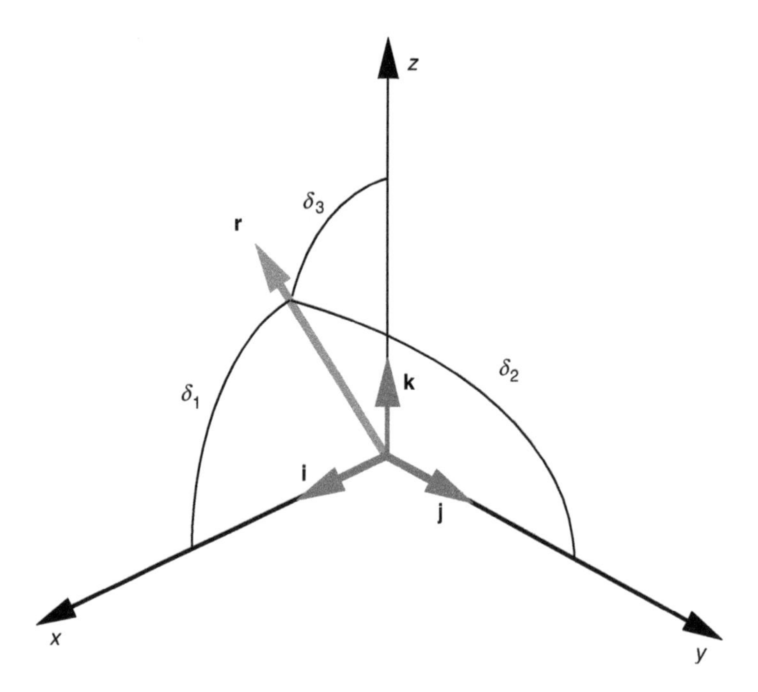

Fig. 3.2-3. Orientation of the vector $\mathbf{r}$ by its angles from the unit vectors. Projections of $\mathbf{r}$ on Cartesian axes.

vector measured along (**i**, **j**, **k**) can be viewed as *projections* of the vector on the axes, obtained from the dot product of **r** with each unit vector. For example,

$$\mathbf{i} \cdot \mathbf{r} \triangleq \mathbf{i}^T \mathbf{r} = r \cos \delta_1 \tag{3.2-104}$$

where δ_1 is the angle between **i** and **r**, and cos δ_1 is called a *direction cosine*. The three direction cosines that orient **r** can be identified as

$$\cos \delta_1 = \frac{\mathbf{i}^T \mathbf{r}}{r} \tag{3.2-105}$$

$$\cos \delta_2 = \frac{\mathbf{j}^T \mathbf{r}}{r} \tag{3.2-106}$$

$$\cos \delta_3 = \frac{\mathbf{k}^T \mathbf{r}}{r} \tag{3.2-107}$$

thus, the vector can be written as

$$\begin{bmatrix} x \\ y \\ z \end{bmatrix} = \begin{bmatrix} \cos \delta_1 \\ \cos \delta_2 \\ \cos \delta_3 \end{bmatrix} r \tag{3.2-108}$$

If there are three vectors (x, 0, 0), (0, y, 0), and (0, 0, z) orthogonal to each other in Frame A, which is different from the (**i**, **j**, **k**) frame (Frame B), the three vectors can be projected on Frame B by applying eq. 3.2-106 concurrently:

$$\begin{aligned} \begin{bmatrix} x \\ y \\ z \end{bmatrix}_B &= \begin{bmatrix} \cos \delta_1 \\ \cos \delta_2 \\ \cos \delta_3 \end{bmatrix} x + \begin{bmatrix} \cos \varepsilon_1 \\ \cos \varepsilon_2 \\ \cos \varepsilon_3 \end{bmatrix} y + \begin{bmatrix} \cos \eta_1 \\ \cos \eta_2 \\ \cos \eta_3 \end{bmatrix} z \\ &= \begin{bmatrix} \cos \delta_1 & \cos \varepsilon_1 & \cos \eta_1 \\ \cos \delta_2 & \cos \varepsilon_2 & \cos \eta_2 \\ \cos \delta_3 & \cos \varepsilon_3 & \cos \eta_3 \end{bmatrix} \begin{bmatrix} x \\ y \\ z \end{bmatrix}_A \end{aligned} \tag{3.2-109a}$$

There are nine angles and nine direction cosines in the *direction cosine matrix* $\mathbf{H}_A^B$ that relates the orthogonal Frames A and B. Equation 3.2-109a takes a familiar form:

$$\mathbf{r}_B = \mathbf{H}_A^B \mathbf{r}_A \tag{3.2-109b}$$

Replacing A by I, the direction cosine matrix is seen to be equivalent to the orthonormal transformation matrix synthesized from sines and cosines of the Euler angles (eq. 2.2-11). $\mathbf{H}_A^B$ (or $\mathbf{H}_I^B$) can be defined in terms of either a three-parameter set (Euler angles) or a nine-parameter set (direction cosines).

The difficulty with Euler angles is that singularities can be encountered in their propagation (eq. 3.2-11); there are no similar singularities involved in the propagation of $\mathbf{H}_I^B$. Equation 3.2-28 can be manipulated to yield

$$\frac{d\mathbf{h}_B}{dt} = \mathbf{H}_I^B \frac{d\mathbf{h}_I}{dt} - \tilde{\boldsymbol{\omega}}_B \mathbf{h}_B = \mathbf{H}_I^B \frac{d\mathbf{h}_I}{dt} - \tilde{\boldsymbol{\omega}}_B \mathbf{H}_I^B \mathbf{h}_I \tag{3.2-110a}$$

but this must be the same as

$$\frac{d\mathbf{h}_B}{dt} = \mathbf{H}_I^B \frac{d\mathbf{h}_I}{dt} + \dot{\mathbf{H}}_I^B \mathbf{h}_I \tag{3.2-110b}$$

Therefore,

$$\dot{\mathbf{H}}_I^B = -\tilde{\boldsymbol{\omega}}_B \mathbf{H}_I^B = -\begin{bmatrix} 0 & -r & q \\ r & 0 & -p \\ -q & p & 0 \end{bmatrix}_B \mathbf{H}_I^B \tag{3.2-111}$$

None of the terms of this equation go to infinity (with finite rates), so propagation of the direction cosine matrix is free of singularities.

It is relatively simple to propagate the direction cosine matrix over time. $\mathbf{H}_I^B$ can be initialized using eq. 2.2-11, and this equation can replace eq. 3.2-39 in Table 3.2-1. Consequently, *integration of the equations of motion does not require explicit knowledge of the Euler angles.* If the Euler angles must be known, they can be found by simultaneous solution of selected elements of $\mathbf{H}_I^B$ [e.g., the (1, 3), (1, 2), and (2, 3) elements, subject to numerical considerations]. Nevertheless, nine elements of $\mathbf{H}_I^B$ must be computed, and it is desirable to find a more efficient singularity-free set.

Quaternions form such a set. As shown in Fig. 3.2-4, a solid-body rotation from one attitude to another can be characterized as an *Euler rotation*, a single rotation about some axis in the reference frame. The rotation can be represented by four parameters: the three direction cosines of a unit vector $\mathbf{n}$ aligned with the rotational axis and the magnitude of the rotation angle χ. As noted in [A-5, S-9], the transformation matrix $\mathbf{H}_B^I$ can be expressed by

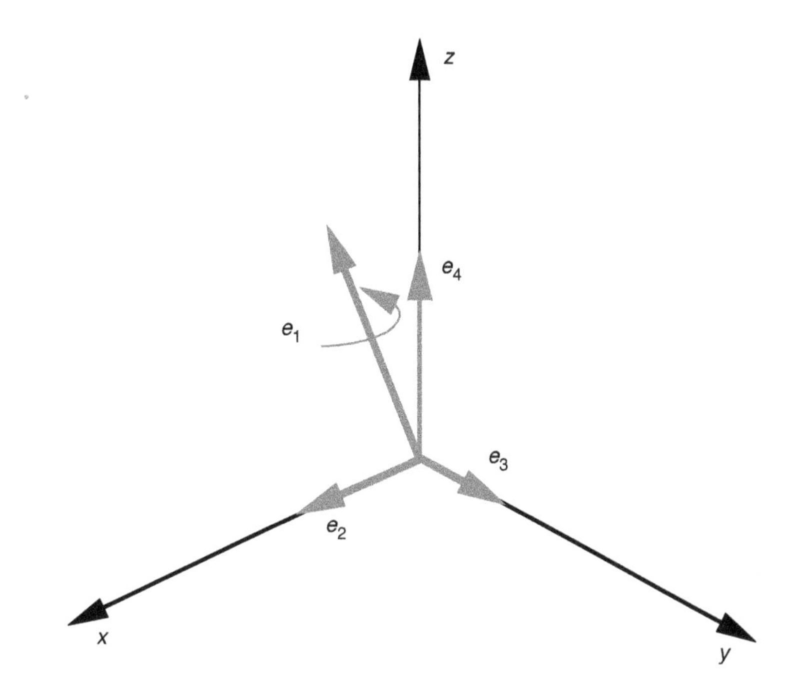

Fig. 3.2-4. Definition of the quaternion components.

$$\mathbf{H}_B^I = \exp(\chi\tilde{\mathbf{n}}) = \mathbf{I}_3 + (\chi\tilde{\mathbf{n}}) + \frac{1}{2}(\chi\tilde{\mathbf{n}})^2 + \cdots$$
$$= \mathbf{I}_3 \cos\chi + (1 - \cos\chi)\mathbf{n}\,\mathbf{n}^T + \sin\chi\,\tilde{\mathbf{n}} \tag{3.2-112a}$$

where $\tilde{\mathbf{n}}$ is the cross-product-equivalent matrix of $\mathbf{n}$ and $\mathbf{I}_3$ is a three-dimensional identity matrix. Using trigonometric identities, this transformation is

$$\mathbf{H}_B^I = \left(\cos^2\frac{\chi}{2} - \sin^2\frac{\chi}{2}\right)\mathbf{I}_3 + 2\sin^2\frac{\chi}{2}\,\mathbf{n}\,\mathbf{n}^T + 2\cos\frac{\chi}{2}\sin\frac{\chi}{2}\,\tilde{\mathbf{n}} \tag{3.2-112b}$$

suggesting that the four-element *quaternion vector* $\mathbf{e}$ be defined as

$$\mathbf{e} = \begin{bmatrix} \cos\frac{\chi}{2} \\ \sin\frac{\chi}{2}\mathbf{n} \end{bmatrix} \tag{3.2-113}$$

The inverse of eq. 3.2-110 then can be written as

$$\mathbf{H}_I^B = \begin{bmatrix} (e_1^2 - e_2^2 - e_3^2 + e_4^2) & 2(e_1e_2 + e_3e_4) & 2(e_2e_4 - e_1e_3) \\ 2(e_3e_4 - e_1e_2) & (e_1^2 - e_2^2 + e_3^2 - e_4^2) & 2(e_2e_3 + e_1e_4) \\ 2(e_1e_3 + e_2e_4) & 2(e_2e_3 - e_1e_4) & (e_1^2 + e_2^2 - e_3^2 - e_4^2) \end{bmatrix} \tag{3.2-114}$$

(Reference [J-1] notes that alternative relationships between the quaternion vector and the direction cosine matrix can be defined.)

Because four components are used to express a three-dimensional rotation, the quaternion elements must satisfy an equation of constraint. The *normality condition* constraint follows from eq. 3.2-113, which guarantees that the vector magnitude is 1:

$$e_1^2 + e_2^2 + e_3^2 + e_4^2 = 1 \tag{3.2-115}$$

The quaternion elements can be propagated by the following differential equation, which employs a (4×4) skew-symmetric matrix of the angular rates:

$$\begin{bmatrix} \dot{e}_1 \\ \dot{e}_2 \\ \dot{e}_3 \\ \dot{e}_4 \end{bmatrix} = \begin{bmatrix} 0 & -r & -q & -p \\ r & 0 & -p & q \\ q & p & 0 & -r \\ p & -q & r & 0 \end{bmatrix} \begin{bmatrix} e_1 \\ e_2 \\ e_3 \\ e_4 \end{bmatrix} \tag{3.2-116}$$

The quaternion propagation equation can replace eq. 3.2-39 in Table 3.2-1, but it is also necessary to compute eq. 3.2-114 and its inverse to support equations 3.2-37 and 3.2-38. As in the case of direction cosine matrix propagation, no singularities are encountered; however, there is an ambiguity in the rotational sense of χ, because e_1 is derived from the cosine of $\chi/2$. The quaternion's initial value can be specified in terms of Euler angles [G-4]:

$$e_1 = \cos\frac{\psi}{2}\cos\frac{\theta}{2}\cos\frac{\phi}{2} + \sin\frac{\psi}{2}\sin\frac{\theta}{2}\sin\frac{\phi}{2} \tag{3.2-117}$$

$$e_2 = \sin\frac{\psi}{2}\cos\frac{\theta}{2}\cos\frac{\phi}{2} - \cos\frac{\psi}{2}\sin\frac{\theta}{2}\sin\frac{\phi}{2} \tag{3.2-118}$$

$$e_3 = \cos\frac{\psi}{2}\sin\frac{\theta}{2}\cos\frac{\phi}{2} + \sin\frac{\psi}{2}\cos\frac{\theta}{2}\sin\frac{\phi}{2} \tag{3.2-119}$$

$$e_4 = \cos\frac{\psi}{2}\cos\frac{\theta}{2}\sin\frac{\phi}{2} - \sin\frac{\psi}{2}\sin\frac{\theta}{2}\cos\frac{\phi}{2} \tag{3.2-120}$$

Conversely, Euler angles can be recovered from quaternion propagation by relating appropriate elements of eq. 3.2-114 to the corresponding terms of eq. 2.2-11.

Acceleration Sensed at an Arbitrary Point

Acceleration is one of the easiest quantities to measure but one of the most difficult to interpret. An *accelerometer* is a simple instrument, containing a proof mass suspended by a spring. The deflection of the mass from its origin, or the rebalance force required to center the mass, is the measure of acceleration. Because the accelerometer is usually fixed to the body and is rarely mounted at the vehicle's center of mass, its output is a complex function of the state and control. The inertial acceleration of the center of mass is

$$\dot{\mathbf{v}}_{\mathrm{I}} = \mathbf{H}_{\mathrm{B}}^{\mathrm{I}}\,\mathbf{a}_{\mathrm{B}} + \mathbf{g}_{\mathrm{I}} \tag{3.2-121}$$

where $\mathbf{a}_{\mathrm{B}}$ is the body-axis output of a three-axis accelerometer. From eq. 3.2-35, $\mathbf{a}_{\mathrm{B}_{cm}}$ is seen to contain the forces produced by aerodynamics and thrust, divided by the mass. The output of an accelerometer mounted at the c.m. can be expressed in terms of the body-frame velocity, acceleration, Euler angles, and angular rate as

$$\mathbf{a}_{\mathrm{B}} = \mathbf{H}_{\mathrm{I}}^{\mathrm{B}}(\dot{\mathbf{v}}_{\mathrm{I}} - \mathbf{g}_{\mathrm{I}}) = \dot{\mathbf{v}}_{\mathrm{B}} + \tilde{\boldsymbol{\omega}}_{\mathrm{B}}\mathbf{v}_{\mathrm{B}} - \mathbf{H}_{\mathrm{I}}^{\mathrm{B}}\mathbf{g}_{\mathrm{I}} \tag{3.2-122}$$

If the accelerometer is mounted at $\mathbf{x}_a$ in the body frame of reference, the velocity, acceleration, and acceleration sensed at the point are

$$\mathbf{v}_{\mathrm{I}_a} = \mathbf{v}_{\mathrm{I}} + \mathbf{H}_{\mathrm{B}}^{\mathrm{I}}\tilde{\boldsymbol{\omega}}_{\mathrm{B}}\Delta\mathbf{x} \tag{3.2-123}$$

$$\dot{\mathbf{v}}_{\mathrm{I}_a} = \dot{\mathbf{v}}_{\mathrm{I}} + \mathbf{H}_{\mathrm{B}}^{\mathrm{I}}(\tilde{\boldsymbol{\omega}}_{\mathrm{B}}\tilde{\boldsymbol{\omega}}_{\mathrm{B}} + \dot{\tilde{\boldsymbol{\omega}}}_{\mathrm{B}})\Delta\mathbf{x} = \mathbf{H}_{\mathrm{B}}^{\mathrm{I}}\mathbf{a}_{\mathrm{B}_a} + \mathbf{g}_{\mathrm{I}} \tag{3.2-124}$$

$$\begin{aligned}
\mathbf{a}_{\mathrm{B}_a} &= \mathbf{H}_{\mathrm{I}}^{\mathrm{B}}(\dot{\mathbf{v}}_{\mathrm{I}} - \mathbf{g}_{\mathrm{I}}) + (\tilde{\boldsymbol{\omega}}_{\mathrm{B}}\tilde{\boldsymbol{\omega}}_{\mathrm{B}} + \dot{\tilde{\boldsymbol{\omega}}}_{\mathrm{B}})\Delta\mathbf{x} \\
&= \dot{\mathbf{v}}_{\mathrm{B}} + \tilde{\boldsymbol{\omega}}_{\mathrm{B}}\mathbf{v}_{\mathrm{B}} - \mathbf{H}_{\mathrm{I}}^{\mathrm{B}}\mathbf{g}_{\mathrm{I}} + (\tilde{\boldsymbol{\omega}}_{\mathrm{B}}\tilde{\boldsymbol{\omega}}_{\mathrm{B}} + \dot{\tilde{\boldsymbol{\omega}}}_{\mathrm{B}})\Delta\mathbf{x} \\
&= \mathbf{a}_{\mathrm{B}_{cm}} + (\tilde{\boldsymbol{\omega}}_{\mathrm{B}}\tilde{\boldsymbol{\omega}}_{\mathrm{B}} + \dot{\tilde{\boldsymbol{\omega}}}_{\mathrm{B}})\Delta\mathbf{x}
\end{aligned} \tag{3.2-125}$$

where $\Delta\mathbf{x} = \mathbf{x}_a - \mathbf{x}_{cm}$. Using eq. 3.2-47–3.2-49 and assuming that the gravitational acceleration $\mathbf{g}_{\mathrm{I}}$ is aligned with the z axis, the three body-axis components of the accelerometer measurement are

$$\begin{aligned}
a_{x_a} &= \dot{u} - rv + qw - (r^2 + q^2)\Delta x + (-\dot{r} + pq)\Delta y \\
&\quad + (\dot{q} + pr)\Delta z + g\sin\theta \\
&= a_{x_{cm}} - (r^2 + q^2)\Delta x + (-\dot{r} + pq)\Delta y + (\dot{q} + pr)\Delta z
\end{aligned} \tag{3.2-126}$$

$$a_{y_a} = \dot{v} + ru - pw + (\dot{r} + pq)\Delta x + (r^2 + p^2)\Delta y + (-\dot{p} + qr)\Delta z - g\sin\phi\cos\theta$$
$$= a_{y_{cm}} + (\dot{r} + pq)\Delta x + (r^2 + p^2)\Delta y + (-\dot{p} + qr)\Delta z \quad (3.2\text{-}127)$$

$$a_{z_a} = \dot{w} - qu + pv + (-\dot{q} + pr)\Delta x + (\dot{p} + qr)\Delta y - (p^2 + q^2)\Delta z - g\cos\phi\cos\theta$$
$$= a_{z_{cm}} + (-\dot{q} + pr)\Delta x + (\dot{p} + qr)\Delta y - (p^2 + q^2)\Delta z \quad (3.2\text{-}128)$$

Thus, the additional terms contain not only angular rates but angular accelerations, and they are proportional to the displacement of the accelerometer from the center of mass. An airplane's cockpit normally is located off the c.m., and these added effects may alter the flying qualities perceived by the pilot during rapid maneuvers.

3.3 Dynamic Equations for a Round, Rotating Earth

Accounting for the earth's roundness and rotation adds considerable complexity to the equations of motion, but such complexity is warranted when studying long-range navigation as well as when making precise calculations of flight to and from orbit. The dynamics are made more difficult by three factors: the shape of the earth, its rotation, and the variation of gravity with position. The planar reference surface of the flat-earth model must be replaced by a spherical or ellipsoidal surface. The earth's rotational rate must be added to the earth-relative velocity of the aircraft, producing a centrifugal effect not present in the flat-earth model. The inverse-square law for gravitation between point masses introduces varying gravitational force, while the nonspherical mass distribution of the earth adds angular dependence as well.

Geometry and Gravity Field of the Earth

If the earth is modeled as a perfect sphere of radius R, then positions on its surface can be identified by two earth-relative angles: *latitude* L_e and *longitude* λ_e (Fig. 3.3-1). The earth-fixed (rotating) coordinates of a point on the surface are

$$\mathbf{R}_E \triangleq \begin{bmatrix} x_o \\ y_o \\ z_o \end{bmatrix}_E = \begin{bmatrix} \cos L_e \cos\lambda_e \\ \cos L_e \sin\lambda_e \\ \sin L_e \end{bmatrix} R \quad (3.3\text{-}1)$$

where λ_e is zero at the Greenwich meridian and positive to the east, and L_e is zero at the equator and positive to the north; hence, z_o is positive in the direction of the north pole. In inertial (nonrotating) coordinates, the

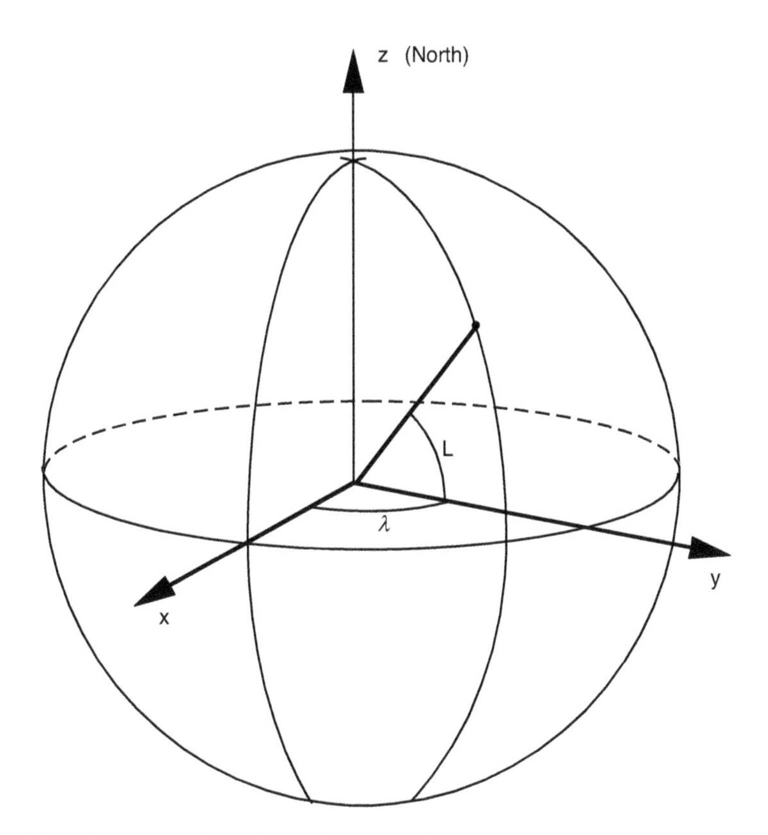

Fig. 3.3-1. Definition of latitude L and longitude λ.

equation is the same, but λ_e is replaced by λ_c, the *celestial longitude* (or *right ascension*), and it is referenced to the *First Point of Aries*, an astronomical reference. λ_e is related to λ_c by

$$\lambda_e = \lambda_c - \Omega(t - t_0) \quad \text{(modulo } 2\pi \text{ rad or } 360^\circ) \tag{3.3-2}$$

or

$$\lambda_c = \lambda_e + \Omega(t - t_0) \triangleq \lambda'_e + \Omega t \tag{3.3-3}$$

where Ω is the earth's rotational rate,[1] t is time, and t_0 is the *epoch*, a reference time for the Aries sighting. λ'_e subsumes the epoch in its definition. An inertial-frame position vector can be expressed in earth-fixed coordinates by the transformation H_I^E,

1. About 361 deg per day (0.004 deg/s) in a celestial frame 360 deg plus the daily change in sun angle, (which determines noon), about 1 deg per day. Given as $7{,}292{,}115 \times 10^{-11}$ rad/s in [A-7].

$$R_E = H_I^E R_I = \begin{bmatrix} \cos\Omega t & \sin\Omega t & 0 \\ -\sin\Omega t & \cos\Omega t & 0 \\ 0 & 0 & 1 \end{bmatrix} \begin{bmatrix} x_o \\ y_o \\ z_o \end{bmatrix}_I \quad (3.3\text{-}4)$$

assuming that $\lambda_e = \lambda_c$ at $t = 0$.

The earth-relative position of an aircraft moving above the spherical surface is derived by extending eq. 3.3-1. If the aircraft's altitude is h, then its earth-relative radius r is $R + h$, and its radius vector is

$$\mathbf{r}_E \triangleq \begin{bmatrix} x \\ y \\ z \end{bmatrix}_E = \begin{bmatrix} \cos L_e \cos\lambda_e \\ \cos L_e \sin\lambda_e \\ \sin L_e \end{bmatrix} (R + h) \quad (3.3\text{-}5)$$

L_e and λ_e are the latitude and longitude of the point directly beneath the aircraft, and the earth's *mean radius* is $R_m = 6{,}367{,}435$ m. The aircraft's inertial position vector is

$$\mathbf{r}_I = \mathbf{H}_E^I \mathbf{r}_E = \begin{bmatrix} \cos L_e \cos\lambda_c \\ \cos L_e \sin\lambda_c \\ \sin L_e \end{bmatrix} (R + h) \quad (3.3\text{-}6)$$

This is known as the *geocentric position vector.*

The earth is not a perfect sphere because the equatorial radius is greater than the polar radius. Thus, the earth is better modeled as an *oblate spheroid* or *ellipsoid.* Many reference ellipsoids have been defined for different purposes, such as characterization of the mean sea level of the earth or closest fit to the average surface level at a particular locale. The WGS-84 Ellipsoid [A-4, A-6, F-5] is the current international standard (Table 3.3-1).

From Fig. 3.3-2, the cross section is an ellipse that satisfies the equation

$$\frac{x_o^2}{R_e^2} + \frac{z_o^2}{R_p^2} = 1 \quad (3.3\text{-}7)$$

Here, R_e and R_p are the *equatorial* and *polar radii* given in Table 3.3-1. The x and z coordinates of a point on the earth's mean surface can then be expressed as

$$x_o^2 = R_o^2 \cos^2 L_o \quad (3.3\text{-}8)$$

TABLE 3.3-1 WORLD GEODETIC SYSTEM ELLIPSOID [A-7]

Equatorial Radius R_e, m	6,378,137
Polar Radius R_p, m	6,356,752
Mean Radius R_m, m	6,367,435
Ellipticity (or Flattening, f) e	1/298.257223563
Eccentricity k	0.08181919085

$$z_o^2 = R_o^2 \sin^2 L_o \tag{3.3-9}$$

where

$$\begin{aligned} R_o^2 &= \frac{R_p^2}{1-[1-(R_p/R_e)^2]\cos^2 L_o} \\ &= \frac{R_p^2}{1-k^2\cos^2 L_o} \end{aligned} \tag{3.3-10}$$

k is called the *eccentricity* of the ellipse, listed in Table 3.3-1. The earth's shape can also be characterized by its *ellipticity* e (or *flattening* f),

$$e = \frac{R_e - R_p}{R_e} \tag{3.3-11}$$

which is given in Table 3.3-1 and is related to the eccentricity by the formula

$$k^2 = 2e\left(1 - \frac{e}{2}\right) \tag{3.3-12}$$

From Fig. 3.3-2, it is clear that three definitions of latitude must be considered in connection with the ellipsoidal earth model and an object

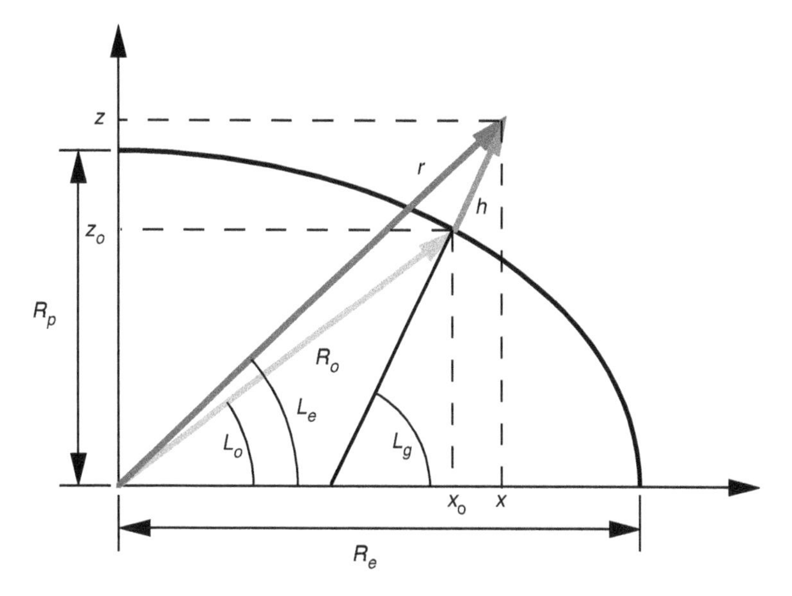

Fig. 3.3-2. Coordinates of the ellipsoidal reference frame.

not confined to its surface. Altitude is measured perpendicular to the local reference surface (along the *local vertical*). It is important to distinguish between altitude above mean sea level (MSL), altitude above the WGS 84 ellipsoid, and altitude above ground level (AGL) when flying near the ground, as there may be differences of several thousand meters in many areas of the world. Altitude is collinear with the radius vector to the center of the ellipsoid only at the equator and the poles. Consequently, there is a vector triangle connecting the center, the object, and its orthogonal projection to the surface (the *subpoint*). The *geographic* (or *geodetic*) *latitude* L_g, shared by the object and its subpoint, is the angle between the local vertical and the equatorial plane. The geocentric latitude of the object L_e is different from the geocentric latitude of the subpoint L_o, and both are different from the geographic latitude L_g. These differences are expressed by the *deviations of the normal* [B-4],

$$D = L_g - L_e \tag{3.3-13}$$

and

$$D_o = L_g - L_o \approx e \sin 2\, L_g \approx e \sin 2\, L_o \tag{3.3-14}$$

D_o has a maximum value of e rad, or about 0.2 deg, when $L_g = 45$ deg.

Because it is related to the local vertical, the geographic latitude of a point on the surface is readily measured from star sightings and is a useful quantity. It appears in the transformation matrix relating the inertial (nonrotating) frame to a local geographic frame (north, east, down):

$$\mathbf{H}_{\mathrm{I}}^{\mathrm{G}} = \begin{bmatrix} -\sin L_g \cos\lambda_c & -\sin L_g \sin\lambda_c & \cos L_g \\ -\sin\lambda_c & \cos\lambda_c & 0 \\ -\cos L_g \cos\lambda_c & -\cos L_g \sin\lambda_c & -\sin L_g \end{bmatrix} \tag{3.3-15}$$

Transformation to a local (rotating) geographic frame can also be defined using λ_e rather than λ_c. The x and y axes of the geographic frame establish the *local horizontal plane*. The reference ellipsoid's radius is expressed to within 1 m by

$$R_o \approx R_e\left[1 - \frac{e}{2}(1 - \cos 2L_g) + \frac{5}{16}e^2(1 - \cos 4L_g)\right] \tag{3.3-16}$$

and to within 30 m by

$$R_o = R_e(1 - e \sin^2 L_g) \tag{3.3-17}$$

The radii from the earth's center to the aircraft and its subpoint together with the altitude form a plane triangle (Fig. 3.3-2) for which

$$\begin{aligned} r &= \sqrt{R_o^2 + 2R_o h \cos D_o + h^2} \\ &= (R_o + h)\sqrt{1 - \frac{2R_o h(1 - \cos D_o)}{(R_o + h)^2}} \end{aligned} \tag{3.3-18}$$

With a "worst-case" D_o, the difference between r and $(R_o + h)$ remains below 1 m for all altitudes within the sensible atmosphere.

There is yet another definition of the "local vertical" and, therefore, of the latitude. A *plumb bob* (a weight suspended from a string connected to an earth-fixed point) measures the *true vertical*. It is deflected from the geographic local vertical by centrifugal force due to the earth's rotation, typically by less than 0.02 deg. The centrifugal effect adds to the gravity vector sensed in the rotating frame, producing

$$\mathbf{g}_E = \mathbf{g}_M + \Delta\mathbf{g}_E + \mathbf{\Omega} \times \mathbf{\Omega} \times \mathbf{r}_E = \mathbf{g}_I + \tilde{\mathbf{\Omega}}\,\tilde{\mathbf{\Omega}}\mathbf{r}_E \tag{3.3-19}$$

where $\mathbf{g}_M$ represents the gravitational acceleration due to mass attraction by the reference ellipsoid, and $\mathbf{H}_I^R$ contains anomalies arising from local mass concentrations (mountains, ore deposits, etc.). The earth's rotational vector and its cross-product-equivalent matrix are

$$\mathbf{\Omega} = \begin{bmatrix} 0 \\ 0 \\ \Omega \end{bmatrix} \tag{3.3-20}$$

$$\tilde{\mathbf{\Omega}} = \begin{bmatrix} 0 & -\Omega & 0 \\ \Omega & 0 & 0 \\ 0 & 0 & 0 \end{bmatrix} \tag{3.3-21}$$

Consequently, eq. 3.3-19 becomes

$$\begin{bmatrix} g_x \\ g_y \\ g_z \end{bmatrix}_E = \begin{bmatrix} g_x \\ g_y \\ g_z \end{bmatrix}_M + \begin{bmatrix} \Delta g_x \\ \Delta g_y \\ \Delta g_z \end{bmatrix}_E + \begin{bmatrix} -\Omega^2 x_o \\ -\Omega^2 y_o \\ 0 \end{bmatrix} \tag{3.3-22}$$

on the reference ellipsoid. This effective gravity vector resolves into a rotating geographic frame as

$$\begin{bmatrix} g_x \\ g_y \\ g_z \end{bmatrix}_G = \begin{bmatrix} \Delta g_x - \Omega^2 R_o \cos L_o \sin L_g \\ \Delta g_y \\ g + \Delta g_z - \Omega^2 R_o \cos L_o \cos L_g \end{bmatrix} \tag{3.3-23}$$

with anomalies redefined accordingly.

Newton's law of gravitation [D-1] indicates that the forces produced by two point masses on each other are equal and opposite in direction. In addition, the magnitudes of the forces are proportional to the product of their masses and the inverse square of the distance between them, giving

$$|\mathbf{f}_1^2| = |\mathbf{f}_2^1| = \frac{G\, m_1\, m_2}{r^2} \tag{3.3-24}$$

G is the *universal gravitational constant* $6.6695 \times 10^{-11}\,\mathrm{m^3/(kg\text{-}s^2)}$. The mass of the earth $m_\oplus$ is 5.977×10^{24} kg; therefore *earth's gravitational constant* $\mu = Gm_\oplus = 3.986 \times 10^{14}\,\mathrm{m/s^2}$. If the earth were a point mass located at (0, 0, 0), the gravitational force vector felt by a nearby point mass m at position $(x, y, z)^T$ would be

$$\mathbf{f}_I \triangleq m \begin{bmatrix} g_x \\ g_y \\ g_z \end{bmatrix}_I = -\frac{m\mu \mathbf{r}_I}{r_I^3} = -\frac{m\mu}{r_I^3} \begin{bmatrix} x \\ y \\ z \end{bmatrix}_I \tag{3.3-25}$$

$\mathbf{f}_I$ is the negative transpose of the gradient of a (scalar) potential field $\mathcal{V}(\mathbf{r}_I)$ (eq. 3.1-4), which is defined as

$$\mathcal{V}(\mathbf{r}_I) = \frac{m\mu}{\mathbf{r}_I} + \mathcal{V}_o = \frac{m\,\mu}{\sqrt{\mathbf{r}_I^T \mathbf{r}_I}} + \mathcal{V}_o \tag{3.3-26}$$

A body with spherically symmetric mass distribution exerts the same gravitational influence potential as a point mass does for particles outside the body's surface [B-2]; therefore, eq. 3.3-25 is a suitable formula for gravitational force with a spherical earth model.

Nonspherical, axisymmetric scalar functions of vector variables can be modeled using *Legendre polynomials* P_k. These polynomials are functions of the radius from the center to a point and the angle ϕ, the *colatitude* of the point $[\triangleq (90 - L_e)\,\mathrm{deg}]$, between the point and the north-south axis [B-4]. The gravity potential for a nonspherical earth can be expressed as

$$\mathcal{V}(\mathbf{r}_I) = \mathcal{V}(\mathbf{r}_I, \phi) = \frac{\mu}{r_I}\left[1 - \sum_{k=2}^{\infty}\left(\frac{R_o}{r_I}\right)^k J_k P_k(\cos\phi)\right] \tag{3.3-27}$$

The first three Legendre polynomials are

$$P_2 = \frac{1}{4}(3\cos 2\phi + 1) \tag{3.3-28}$$

$$P_3 = \frac{1}{8}(5\cos 3\phi + \cos 3\phi) \tag{3.3-29}$$

$$P_4 = \frac{1}{64}(35\cos 4\phi + 20\cos 2\phi + 9) \tag{3.3-30}$$

The J_k are empirically determined coefficients, currently estimated for the earth as [B-2]

$$J_2 = 1.0826 \times 10^{-3} \tag{3.3-31}$$

$$J_3 = -2.54 \times 10^{-6} \tag{3.3-32}$$

$$J_4 = -1.61 \times 10^{-6} \tag{3.3-33}$$

J_2 and J_4 are oblateness terms, resulting from the equatorial bulge in the earth's shape, while J_3 is the "pear-shaped" term, reflecting greater mass in the earth's southern hemisphere. Using eq. 3.1-4, the gravitational acceleration is modeled in geocentric inertial coordinates as [B-4]

$$g_r = -\frac{\mu}{r_I^2}\left[1 - \frac{3}{2}J_2\left(\frac{R_o}{r_I}\right)^2 (3\cos^2\phi - 1) - 2J_3\left(\frac{R_o}{r_I}\right)^3 \cos\phi(5\cos^2\phi - 3) - \frac{5}{8}J_4\left(\frac{R_o}{r_I}\right)^4 (35\cos^4\phi - 30\cos^2\phi + 3)\right] \tag{3.3-34}$$

$$g_\phi = -3\frac{\mu}{r_I^2}\left(\frac{R_o}{r_I}\right)^2 \sin\phi\cos\phi\left[J_2 + \frac{1}{2}J_3\left(\frac{R_o}{r_I}\right)\sec\phi(5\cos^2\phi - 3)\right] + \frac{5}{6}J_4\left[\left(\frac{R_o}{r_I}\right)^2 (7\cos^2\phi - 3)\right] \tag{3.3-35}$$

and in Cartesian inertial coordinates (to the J_2 term only) as

$$\begin{bmatrix} g_x \\ g_y \end{bmatrix} = -\frac{\mu}{r_I^2}\left\{1 + \frac{3}{2}J_2\left(\frac{R_o}{r_I}\right)^2\left[1 - 5\left(\frac{z}{r_I}\right)^2\right]\right\}\begin{bmatrix} \frac{x}{r_I} \\ \frac{y}{r_I} \end{bmatrix} \tag{3.3-36}$$

$$g_z = -\frac{\mu}{r_I^2}\left\{1 + \frac{3}{2}J_2\left(\frac{R_o}{r_I}\right)^2\left[3 - 5\left(\frac{z}{r_I}\right)\right]\right\}\left(\frac{z}{r_I}\right) \tag{3.3-37}$$

RIGID-BODY DYNAMIC EQUATIONS

The equations of motion for an aircraft flying near a round, rotating earth derive from the inertial-frame equations presented before, taking account of the geometry and gravitational field models introduced above. Three vector equations are integrated to obtain the translational velocity $\mathbf{v}_I$, translational position $\mathbf{r}_I$, and angular momentum $\mathbf{h}_I$:

$$\frac{d\mathbf{v}_I}{dt} = \frac{\mathbf{f}_I}{m(t)} + \mathbf{g}_I \tag{3.3-38}$$

$$\frac{d\mathbf{r}_I}{dt} = \mathbf{v}_I \tag{3.3-39}$$

$$\frac{d\mathbf{h}_I}{dt} = \mathbf{m}_I \tag{3.3-40}$$

These equations can be transferred to a rotating reference frame R using $\mathbf{H}_I^R$, $\boldsymbol{\Omega}$, and $\tilde{\boldsymbol{\Omega}}$ from eqs. 3.3-4, 3.3-20, and 3.3-21, noting from eq. 3.2-109 that

$$\dot{\mathbf{H}}_I^R = -\tilde{\boldsymbol{\Omega}}\mathbf{H}_I^R \tag{3.3-41}$$

The translational equations must recognize the significant displacement of the aircraft's position from the earth's rotational axis. Given in inertial (nonrotating) coordinates as $\mathbf{r}_I$, the position vector is expressed in the rotating frame as

$$\mathbf{r}_R = \mathbf{H}_I^R\mathbf{r}_I \tag{3.3-42}$$

Using the chain rule, the velocity observed in the rotating frame is

$$\begin{aligned} \dot{\mathbf{r}}_R &= \mathbf{H}_I^R\dot{\mathbf{r}}_I + \dot{\mathbf{H}}_I^R\mathbf{r}_I \\ &= \mathbf{H}_I^R\dot{\mathbf{r}}_I - \tilde{\boldsymbol{\Omega}}\mathbf{H}_I^R\mathbf{r}_I \\ &= \mathbf{H}_I^R\dot{\mathbf{r}}_I - \tilde{\boldsymbol{\Omega}}\mathbf{r}_R \triangleq \mathbf{v}_R \end{aligned} \tag{3.3-43}$$

and this equation replaces eq. 3.3-39. The translational acceleration is the time derivative of $\mathbf{v}_R$:

$$\begin{aligned}\ddot{\mathbf{r}}_R &= \mathbf{H}_I^R\ddot{\mathbf{r}}_I + \dot{\mathbf{H}}_I^R\dot{\mathbf{r}}_I - \dot{\tilde{\boldsymbol{\Omega}}}\mathbf{H}_I^R\mathbf{r}_I - \tilde{\boldsymbol{\Omega}}\mathbf{H}_I^R\dot{\mathbf{r}}_I - \tilde{\boldsymbol{\Omega}}\dot{\mathbf{H}}_I^R\mathbf{r}_I \\ &= \mathbf{H}_I^R\ddot{\mathbf{r}}_I - 2\tilde{\boldsymbol{\Omega}}\mathbf{H}_I^R\dot{\mathbf{r}}_I + \tilde{\boldsymbol{\Omega}}\,\tilde{\boldsymbol{\Omega}}\mathbf{H}_I^R\mathbf{r}_I \\ &= \mathbf{H}_I^R\ddot{\mathbf{r}}_I - 2\tilde{\boldsymbol{\Omega}}(\mathbf{v}_R + \tilde{\boldsymbol{\Omega}}\mathbf{r}_R) + \tilde{\boldsymbol{\Omega}}\,\tilde{\boldsymbol{\Omega}}\mathbf{r}_R \\ &= \mathbf{H}_I^R\ddot{\mathbf{r}}_I - 2\tilde{\boldsymbol{\Omega}}\mathbf{v}_R - \tilde{\boldsymbol{\Omega}}\,\tilde{\boldsymbol{\Omega}}\mathbf{r}_R \triangleq \dot{\mathbf{v}}_R \end{aligned} \tag{3.3-44}$$

Consequently, the translational dynamic equation is

$$\begin{aligned}\ddot{\mathbf{r}}_R = \dot{\mathbf{v}}_R &= \mathbf{H}_I^R\left(\frac{\mathbf{f}_I}{m(t)} + \mathbf{g}_I\right) - 2\tilde{\boldsymbol{\Omega}}\mathbf{v}_R - \tilde{\boldsymbol{\Omega}}\,\tilde{\boldsymbol{\Omega}}\mathbf{r}_R \\ &= \frac{\mathbf{f}_R}{m(t)} + \mathbf{g}_R - 2\tilde{\boldsymbol{\Omega}}\mathbf{v}_R - \tilde{\boldsymbol{\Omega}}\,\tilde{\boldsymbol{\Omega}}\mathbf{r}_R \end{aligned} \tag{3.3-45}$$

As in the transformation from inertial to body axes, the angular momentum's rate of change, referenced to the aircraft's center of mass, becomes

$$\dot{\mathbf{h}}_R = \mathbf{H}_I^R\mathbf{m}_I - \tilde{\boldsymbol{\Omega}}\mathbf{h}_R \tag{3.3-46}$$

The rate of change of total angular momentum, referenced to the earth's center of mass, also includes terms due to the motion of the aircraft's c.m., following eq. 3.1-26. These are significant terms when studying orbital motion, as the angular momentum of the c.m. is preserved; however, they have no effect on the angular attitude of the aircraft.

The next step is to transform the rotating-earth equations to a rotating geographic frame (Frame EG) using eq. 3.3-15 with the earth-fixed longitude λ_e:

$$\mathbf{H}_R^{EG} = \begin{bmatrix} -\sin L_g \cos\lambda_e & -\sin L_g \sin\lambda_e & \cos L_g \\ -\sin\lambda_e & \cos\lambda_e & 0 \\ -\cos L_g \cos\lambda_e & -\cos L_g \sin\lambda_e & -\sin L_g \end{bmatrix} \tag{3.3-47}$$

$\dot{\mathbf{H}}_R^{EG} = 0$, so the transformed translational velocity equation becomes

$$\begin{aligned}\dot{\mathbf{v}}_{EG} = \begin{bmatrix} \dot{v}_n \\ \dot{v}_e \\ \dot{v}_d \end{bmatrix} &= \mathbf{H}_R^{EG}\left[\frac{\mathbf{f}_R}{m(t)} + \mathbf{g}_R - 2\tilde{\boldsymbol{\Omega}}\,\mathbf{v}_R - \tilde{\boldsymbol{\Omega}}\,\tilde{\boldsymbol{\Omega}}\mathbf{r}_R\right] \\ &= \frac{\mathbf{f}_{EG}}{m(t)} + \mathbf{g}_{EG} - 2\mathbf{H}_R^{EG}\,\tilde{\boldsymbol{\Omega}}\mathbf{H}_{EG}^R\mathbf{v}_{EG} - \mathbf{H}_R^{EG}\tilde{\boldsymbol{\Omega}}\,\tilde{\boldsymbol{\Omega}}\mathbf{r}_R \end{aligned} \tag{3.3-48}$$

where the EG components are directed north, east, and down, and $\mathbf{r}_R$ is related to the geocentric radius ($\approx R_o + h$), latitude ($= L_g - D \approx L_g - D_o$), and longitude by eq. 3.3-5.

The earth-relative translational position is found by integrating the velocity vector. In Cartesian coordinates, the rate of change is

$$\dot{\mathbf{r}}_R = \mathbf{H}_{EG}^{R}\mathbf{v}_{EG} \tag{3.3-49}$$

The geocentric coordinate rates are written as

$$\begin{bmatrix} \dot{L}_e \\ \dot{\lambda}_e \\ \dot{r} \end{bmatrix} = \mathbf{L}_R^C\, \mathbf{H}_{EG}^R\, \mathbf{v}_{EG} \tag{3.3-50}$$

where $\mathbf{L}_R^C$ can be found by differentiating and manipulating eq. 3.3-5 term by term:

$$\mathbf{L}_R^C = \begin{bmatrix} -\left(\frac{1}{r}\right)\sin L_e \cos\lambda_e & -\left(\frac{1}{r}\right)\sin L_e \sin\lambda_e & \left(\frac{1}{r}\right)\cos L_e \\ -\left(\frac{1}{r}\right)\frac{\sin\lambda_e}{\cos L_e} & \left(\frac{1}{r}\right)\frac{\cos\lambda_e}{\cos L_e} & 0 \\ \cos L_e \cos\lambda_e & \cos L_e \sin\lambda_e & \sin L_e \end{bmatrix} \tag{3.3-51}$$

The geographic frame coordinates can then be expressed as

$$L_g = L_e + D \approx L_e + D_o \tag{3.3-52}$$

where D_o is given by eq. 3.3-14, and the altitude is approximated by

$$h = r - R_o \tag{3.3-53}$$

with R_o computed from eq. 3.3-16 or 3.3-17.

The angular momentum rate of change is rotated to the earth-fixed geographic frame as

$$\begin{aligned} \dot{\mathbf{h}}_{EG} &= \mathbf{H}_R^{EG}[\mathbf{H}_I^R\mathbf{m}_I - \tilde{\boldsymbol{\Omega}}\mathbf{h}_R] \\ &= \mathbf{m}_{EG} - \mathbf{H}_R^{EG}\tilde{\boldsymbol{\Omega}}\,\mathbf{H}_{EG}^R\mathbf{h}_{EG} \end{aligned} \tag{3.3-54}$$

Finally, the translational and rotational dynamic equations are transformed from the geographic frame to body axes. Body angular rates must be taken into account as in the flat-earth development; hence,

$$\dot{\mathbf{v}}_B = \mathbf{H}_{EG}^B\dot{\mathbf{v}}_{EG} - \tilde{\boldsymbol{\omega}}_B\mathbf{v}_B \tag{3.3-55}$$

$$\dot{\mathbf{h}}_B = \mathbf{H}_{EG}^B\dot{\mathbf{h}}_{EG} - \tilde{\boldsymbol{\omega}}_B\mathbf{h}_B \tag{3.3-56}$$

where $\dot{\mathbf{v}}_{EG}$ and $\dot{\mathbf{h}}_{EG}$ are obtained from eqs. 3.3-48 and 3.3-54, and $\boldsymbol{\omega}_B$ is measured with respect to the Frame EG. The dynamic equations can then be written as

$$\dot{\mathbf{v}}_B = \frac{\mathbf{f}_B}{m(t)} + \mathbf{H}_{EG}^{B}[\mathbf{g}_{EG} - 2\mathbf{H}_R^{EG}\tilde{\boldsymbol{\Omega}}\mathbf{H}_B^R\mathbf{v}_B - \mathbf{H}_R^{EG}\tilde{\boldsymbol{\Omega}}\tilde{\boldsymbol{\Omega}}\mathbf{r}_R] - \tilde{\boldsymbol{\omega}}_B\mathbf{v}_B \qquad (3.3\text{-}57)$$

$$\dot{\mathbf{h}}_B = \mathbf{m}_B - \mathbf{H}_R^B\,\tilde{\boldsymbol{\Omega}}\mathbf{H}_B^R\mathbf{h}_B - \tilde{\boldsymbol{\omega}}_B\mathbf{h}_B \qquad (3.3\text{-}58)$$

The angular momentum equation is replaced by the angular rate equation

$$\dot{\boldsymbol{\omega}}_B = \mathcal{I}_B^{-1}[\mathbf{m}_B - \mathbf{H}_R^B\tilde{\boldsymbol{\Omega}}\mathbf{H}_B^R\mathcal{I}_B\boldsymbol{\omega}_B - \tilde{\boldsymbol{\omega}}_B\mathcal{I}_B\boldsymbol{\omega}_B] \qquad (3.3\text{-}59)$$

where

$$\mathbf{H}_R^B = \mathbf{H}_{EG}^B\,\mathbf{H}_R^{EG} \qquad (3.3\text{-}60)$$

$\mathbf{H}_{EG}^B$ is defined in terms of earth-fixed geographic-frame Euler angles $\boldsymbol{\Theta}_{EG}$ using eq. 2.2-11. Alternatively, it can be propagated directly or via quaternions as in the previous section.

The differential equations for position variables can now be expressed in terms of the body-axis rates. From eq. 3.3-50, the rates of change of the geocentric coordinates are

$$\begin{bmatrix} \dot{L}_e \\ \dot{\lambda}_e \\ \dot{r} \end{bmatrix} = \mathbf{L}_R^C\mathbf{H}_B^R\mathbf{v}_B \qquad (3.3\text{-}61)$$

while from eq. 3.2-11, the Euler angles can be propagated as

$$\frac{d\boldsymbol{\Theta}_{EG}}{dt} = \mathbf{L}_B^{EG}\boldsymbol{\omega}_B \qquad (3.3\text{-}62)$$

with $\mathbf{L}_B^{EG}$ defined as in eq. 2.2-26.

3.4 Aerodynamic Effects of Rotational and Unsteady Motion

Steady translational flight produces steady aerodynamic forces and moments (Section 2.4), but maneuvering flight produces additional forces and moments arising from rotational and unsteady motions. These aerodynamic effects have two distinct causes: spatial and temporal variations in the flow pattern. In the first instance, a continuing *rotational velocity* causes a different flow pattern from that experienced in steady flight; if it were possible to sustain constant angular motion, the flow would adjust to a new equilibrium shape, producing constant (but different) forces and moments. In the second case, the aircraft's *translational or rotational acceleration* perturbs the flow; equilibrium is disturbed, and the flow produces transitory forces and moments that depend not only on the instantaneous acceleration but on its state history. *Wind shear* produces aerodynamic effects that are closely related to these rotational and unsteady effects.

It is commonly assumed that rotational and unsteady effects can be modeled as linear perturbations to steady aerodynamic forces and moments [F-3, W-3]. For contrast, consider a general nonlinear model of the pitching moment coefficient's dependence on the motion variables and time. At some time t, C_m would depend not only on $\mathcal{M}(t)$, $\alpha(t)$, $\beta(t)$, $\dot{\mathcal{M}}(t)$, $\dot{\alpha}(t)$, $\dot{\beta}(t)$, $p(t)$, $q(t)$, $r(t)$, $\dot{p}(t)$, $\dot{q}(t)$, and $\dot{r}(t)$ but on the specific histories of these variables. If $C_m(t)$ were characterized by instantaneous motion variables alone, its evaluation would require a 12-dimensional table lookup; if time histories were important, a 12-dimensional "functional" (i.e., a "function of functions") would have to be integrated over a period of time.

Fortunately, experimental results suggest that great simplification is possible for most aircraft. On a typical flight, variations in Mach number $\mathcal{M}$ and angle of attack α may be large enough to require that their nonlinear effects be considered, but variations in the remaining variables may be qualitatively small. For a symmetric aircraft in symmetric flight, small perturbations in sideslip angle, roll rate, yaw rate, and their time derivatives (the lateral-directional variables) have no effect on the pitching moment. The functional of temporal effects can be represented by a series of the motion variables' time derivatives; hence, the instantaneous dependence of C_m on the longitudinal variables is well modeled as

$$C_m(\mathcal{M}, \alpha, \dot{\mathcal{M}}, \dot{\alpha}, q, \dot{q}) \approx C_m(\mathcal{M}, \alpha) + \frac{\partial C_m}{\partial \dot{\mathcal{M}}}\dot{\mathcal{M}} + \frac{\partial C_m}{\partial \dot{\alpha}}\dot{\alpha} + \frac{\partial C_m}{\partial q}q + \frac{\partial C_m}{\partial \dot{q}}q \tag{3.4-1}$$

where the numerical values of the partial derivatives depend upon the current values of $\mathcal{M}$ and α. In practice, $\partial C_m/\partial \dot{\mathcal{M}}$ and $\partial C_m/\partial \dot{q}$ are usually negligible, leading to further simplification.

The remainder of this section is devoted to aerodynamic lift and moments that arise from rotation and unsteady motion (drag and side force effects are small and can be ignored). The assumption of a rigid body is retained, but it should be cautioned that aeroelasticity can have major impact on aerodynamic derivatives. Related wind-shear effects are then derived from these results.

Pitch-Rate Effects

Consider an aircraft with airspeed V that is pitching at a rate of q rad/s about a spanwise axis through the center of mass (Fig. 3.4-1). The pitch rate induces a local normal velocity (positive down) on the aircraft that is proportional to the distance Δx (positive forward) from the rotational center:

$$\Delta w = -q\Delta x \tag{3.4-2}$$

This, in turn, produces a local angle of attack

$$\Delta\alpha \approx \frac{\Delta w}{V} = -\frac{q\Delta x}{V} \tag{3.4-3}$$

that modifies the aircraft's lift and pitching moment. As the rotational center is moved fore and aft, the $\Delta\alpha$ distribution changes, affecting both the lift and pitching moment sensitivity to pitch rate.

The angle-of-attack perturbation at the tail's center of pressure is

$$\Delta\alpha \approx \frac{ql_t}{V} \tag{3.4-4}$$

where l_t is the horizontal tail's moment arm (positive aft), that is, the distance from the aircraft's center of mass to the tail's center of pressure. Neglecting the $\Delta\alpha$ variation across the horizontal tail's chord, the differential lift produced by the tail is

$$\Delta L_t = \Delta C_{L_t}\frac{1}{2}\rho V^2 S \tag{3.4-5}$$

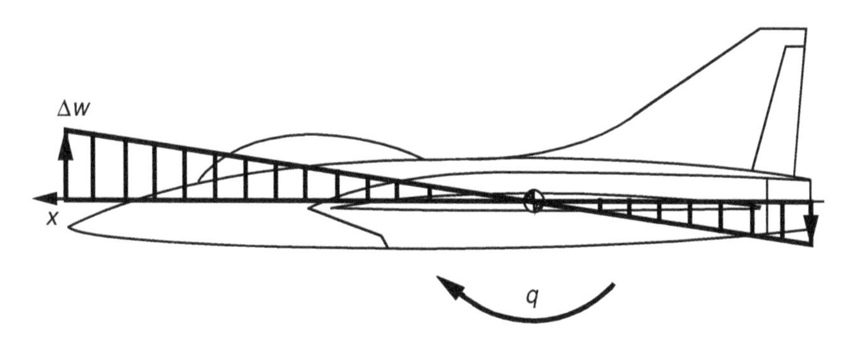

Fig. 3.4-1. Normal-velocity distribution due to pitch rate.

where

$$\Delta C_{L_t} = C_{L_{\alpha_t}}\left(\frac{ql_t}{V}\right) \tag{3.4-6}$$

and the tail's lift slope $C_{L_{\alpha_t}}$ is referenced to the wing area S. If $C_{L_{\alpha_t}}$ is referenced to the horizontal tail area S_t, it should be multiplied by S_t/S in the following equations. Therefore, the lift coefficient's sensitivity to pitch rate can be expressed as

$$C_{L_{q_t}} = \frac{\partial \Delta C_{Lt}}{\partial q} = C_{L_{\alpha_t}}\left(\frac{l_t}{V}\right) = (C_{L\alpha_t})_{tail}\left(\frac{S_t l_t}{SV}\right) \tag{3.4-7}$$

The differential pitching moment due to pitch rate is

$$\Delta M_t = \Delta C_{m_t}\frac{1}{2}\rho V^2 S\bar{c} \tag{3.4-8}$$

where $\bar{c}$ is the aircraft's reference length, and

$$\Delta C_{m_t} = -\left(\frac{l_t}{\bar{c}}\right)C_{L_{\alpha_t}}\left(\frac{ql_t}{V}\right) = -\left(\frac{l_t}{\bar{c}}\right)^2 C_{L_{\alpha_t}}\left(\frac{q\bar{c}}{V}\right) \tag{3.4-9}$$

The pitching-moment perturbation is linear in pitch rate, and its sensitivity to horizontal tail lift is expressed as a function of either the dimensional pitch rate q (rad/s) or its dimensionless equivalent $\hat{q} = q\bar{c}/2V$:

$$C_{m_{q_t}} \triangleq \frac{\partial \Delta C_{m_t}}{\partial q} = -\left(\frac{l_t}{\bar{c}}\right)^2 C_{L_{\alpha_t}}\left(\frac{\bar{c}}{V}\right) \tag{3.4-10}$$

$$C_{m_{\hat{q}_t}} \triangleq \frac{\partial \Delta C_{m_t}}{\partial \hat{q}} \triangleq \frac{\partial \Delta C_{m_t}}{\partial\left(\dfrac{q\bar{c}}{2V}\right)} = -2\left(\frac{l_t}{\bar{c}}\right)^2 C_{L_{\alpha_t}} \tag{3.4-11}$$

Equation 3.4-10 shows the *pitch-rate damping derivative* directly, while eq. 3.4-11 is independent of airspeed. Although the effect of pitch rate on lift changes sign if the aft tail is moved forward to become a canard, the pitching-moment effect is always negative.

The wing's lift and pitching-moment responses to pitch rate also depend on the rotational center, with the $\Delta\alpha$ distribution modifying the flow over the wing's chord. The pitch-rate effects are significant only if the fore-aft dimension of the wing is long, as in the case of a delta or swept wing. Etkin presents an analysis that likens the angle-of-attack distribution induced by pitch rate to the warping of a flat airfoil in a steady, linear flow [E-2]. His analysis of a thin, two-dimensional (infinite aspect ratio) wing predicts

$$C_{L_{\hat{q}}} = -2C_{L_\alpha}(h_{cm} - 0.75) \tag{3.4-12}$$

$$C_{m_{\hat{q}}} = -2C_{L_\alpha}(h_{cm} - 0.5)^2 \tag{3.4-13}$$

for subsonic flow, where C_{L_α} is the wing's lift slope, $h_{cm} = x_{cm}/\bar{c}$, and x_{cm} is the axial location of the aircraft's center of mass (positive aft from the leading edge). For supersonic flow,

$$C_{L_{\hat{q}}} = -2C_{L_\alpha}(h_{cm} - 0.5) \tag{3.4-14}$$

$$C_{m_{\hat{q}}} = -\frac{2}{3\sqrt{\mathcal{M}^2 - 1}} - 2C_{L_\alpha}(h_{cm} - 0.5)^2 \tag{3.4-15}$$

An alternative approach based on potential flow theory produces estimates for slender triangular wings. The method of Bryson [N-2, after B-6] predicts

$$C_{Z_{\hat{q}}} = -\frac{2\pi}{3} \tag{3.4-16}$$

$$C_{m_{\hat{q}}} = -\frac{\pi}{3AR} \tag{3.4-17}$$

for a rotational center at the centroid of the wing, located aft of the apex at the 2/3-root-chord point.

The aircraft's net moment due to pitch rate can be expressed as

$$\begin{aligned}\Delta M(q) &= C_{m_q}\frac{1}{2}\rho V^2 S\bar{c}q = C_{m_{\hat{q}}}\frac{1}{2}\rho V^2 S\bar{c}\left(\frac{q\bar{c}}{2V}\right)\\ &= C_{m_{\hat{q}}}\frac{1}{4}\rho V S\bar{c}^2 q\end{aligned} \tag{3.4-18}$$

where $C_{m_{\hat{q}}}$ (or C_{m_q}) incorporates the effects of the wing and fuselage as well. Following this convention, the lift increment due to pitch rate is

$$\begin{aligned}\Delta L(q) &= C_{L_q}\frac{1}{2}\rho V^2 S q = C_{L_{\hat{q}}}\frac{1}{2}\rho V^2 S\left(\frac{q\bar{c}}{2V}\right)\\ &= C_{L_{\hat{q}}}\frac{1}{4}\rho V S\bar{c}q\end{aligned} \tag{3.4-19}$$

Angle-of-Attack-Rate Effects

Pitch rate and angle-of-attack rate are easily confused, as both involve changing longitudinal angles; however, for flight in the vertical plane, the flight path angle is the difference between pitch angle and angle of attack:

$$\gamma = \theta - \alpha = \theta - \tan^{-1}\frac{w}{u} \tag{3.4-20}$$

Therefore, their time rates of change are similarly related:

$$\dot{\gamma} = q - \dot{\alpha} \approx q - \frac{\dot{w}}{u} \tag{3.4-21}$$

Pitch rate represents changing vehicle orientation relative to the inertial frame; angle-of-attack rate represents changing vehicle orientation relative to the velocity vector. The two rates would be equal if the center of mass were constrained to follow a straight line ($\dot{\gamma} = 0$). A heaving (vertically moving) aircraft could experience nonzero $\dot{\alpha}$ with zero q; a pitching and hearing aircraft could experience nonzero q with $\dot{\alpha} = 0$.

If a lifting wing undergoes a step change in vertical velocity (and, therefore, in angle of attack) without changing its pitch angle, there is a transient response in the wing's lift, and the lift asymptotically approaches a new steady-state value. Tobak developed an elegant way of handling the relationship between angle-of-attack variation and unsteady lift [T-4], assuming a linear dynamic process and applying the superposition (or convolution) integral to indicial functions; however, we revert to two simpler explanations of $\dot{\alpha}$ effects: lag of the downwash and apparent mass (Fig. 3.4-2).

The first of these concepts refers to the effect that changing angle of attack has on a wing's downwash field in the vicinity of the horizontal tail. As noted in Section 2.5, the angle of the flow over an aft tail is modified from the freestream flow by the wing's downwash, and the downwash angle ε itself varies with α through the sensitivity factor $\partial\varepsilon/\partial\alpha$. The theory is that pressure disturbances due to wing-lift variation (Δp_w) convect downstream, producing disturbances Δp_t that alter the downwash at the tail $\Delta t = l_t/V$ s after the wing lift variation. The *lag of the downwash* adjustment produces a negative incremental downwash angle that is proportional to angle-of-attack *rate*:

$$\Delta\varepsilon = -\frac{\partial\varepsilon}{\partial\alpha}\Delta t\,\dot{\alpha} = -\frac{\partial\varepsilon}{\partial\alpha}\frac{l_t}{V}\dot{\alpha} \tag{3.4-22}$$

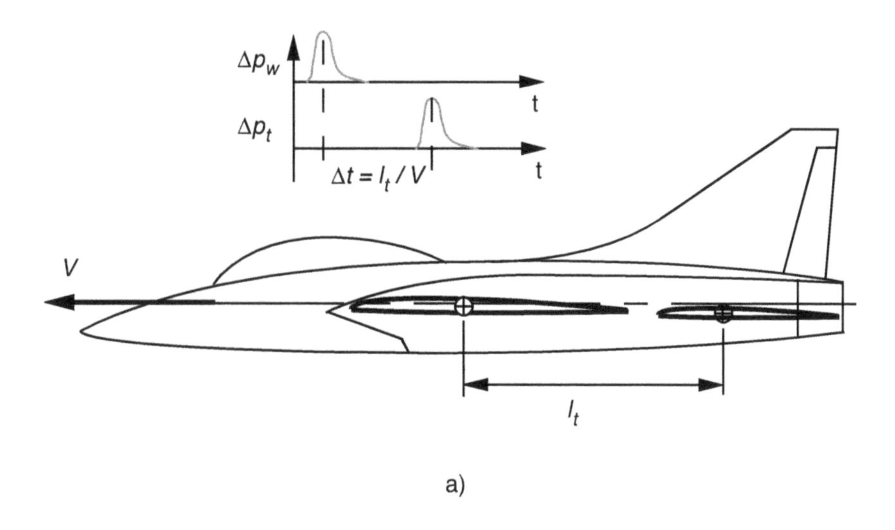

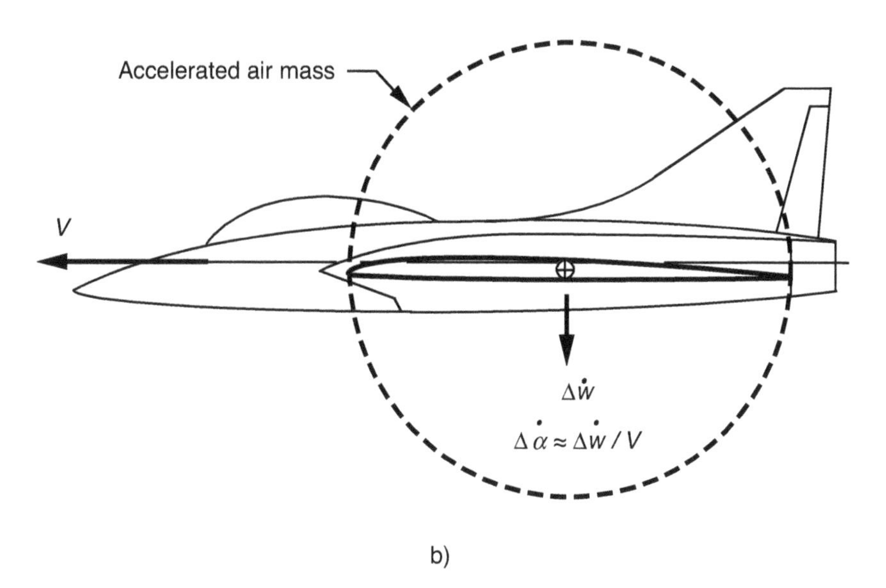

Fig. 3.4-2. Physical sources of angle-of-attack-rate effects. (a) Lag of the downwash effect. (b) Apparent mass effect.

Because the angle of attack felt by the horizontal tail is $\alpha - \varepsilon$, the tail lift is higher than it would have been without $\dot{\alpha}$. Consequently,

$$\Delta C_{L_t} = C_{L_{\alpha_t}} \left(\frac{\partial \varepsilon}{\partial \alpha} \frac{l_t}{V} \dot{\alpha} \right) \tag{3.4-23}$$

and the lift sensitivity to angle-of-attack rate is written as

$$C_{L_{\hat{\dot{\alpha}}_t}} \triangleq \frac{\partial \Delta C_{L_t}}{\partial \hat{\dot{\alpha}}} \triangleq \frac{\partial \Delta C_{L_t}}{\partial(\dot{\alpha}\bar{c}\,/\,2V)} = 2C_{L_{\alpha_t}}\left(\frac{\partial \varepsilon}{\partial \alpha}\frac{l_t}{\bar{c}}\right) \tag{3.4-24}$$

where $\hat{\dot{\alpha}} = \dot{\alpha}\bar{c}/2V$, following the pitch-rate convention. The tail's pitching-moment sensitivity to angle-of-attack rate follows directly as

$$C_{m_{\hat{\dot{\alpha}}_t}} \triangleq \frac{\partial \Delta C_{m_t}}{\partial \hat{\dot{\alpha}}} \triangleq \frac{\partial \Delta C_{m_t}}{\partial(\dot{\alpha}\bar{c}/2V)} = -2C_{L_{\alpha_t}}\left(\frac{\partial \varepsilon}{\partial \alpha}\right)\left(\frac{l_t}{\bar{c}}\right)^2 \tag{3.4-25}$$

The angle-of-attack-rate effect on the wing can be visualized as the result of *apparent* (or *virtual*) *mass*. If the aircraft accelerates vertically at $\dot{w}$ m/s, it forces a mass of air above and below the wing to accelerate as well; therefore, more force is required for the acceleration than would have been required to accelerate the aircraft alone. The mass of air involved is roughly contained in a cylinder whose diameter equals the wing's chord length; thus, the relative effect is quite small unless the aircraft itself is light. Apparent-mass effects are significant for a lighter-than-air craft and very small airplanes like micro uninhabited air vehicles, where the vehicle's mass is low relative to the air it displaces. For a slender triangular wing, apparent-mass considerations lead to the following estimates [N-2]:

$$C_{Z_{\hat{\dot{\alpha}}}} = -\frac{2\pi}{3} \tag{3.4-26}$$

$$C_{m_{\hat{\dot{\alpha}}}} = -\frac{\pi}{9AR} \tag{3.4-27}$$

Apparent-mass theory also predicts pitch acceleration sensitivity for the slender wing, with $\hat{\dot{q}} \triangleq \dot{q}\bar{c}^2/2V^2$:

$$C_{Z_{\hat{\dot{q}}}} = -\frac{\pi}{9AR} \tag{3.4-28}$$

$$C_{m_{\hat{\dot{q}}}} = -\frac{16\pi}{135AR^2} \tag{3.4-29}$$

Yaw-Rate Effects

The principal aerodynamic effects of yaw rate r are yawing and rolling moments produced by the vertical tail and the wing. Because the sideslip-angle rate $\dot{\beta}$ is small on most flight paths and its physical effect is small, its aerodynamic derivatives are either combined with yaw-rate derivatives or neglected. The vertical tail's yawing moment due to yaw rate is analogous to the horizontal tail's pitch-rate effect; hence,

$$\Delta C_{n_{vt}} = -C_{n\beta_t}\left(\frac{rl_{vt}}{V}\right) = -\left(\frac{l_{vt}}{b}\right)C_{n\beta_t}\left(\frac{rb}{V}\right) \quad (3.4\text{-}30)$$

Here, l_{vt} is the distance between the center of mass and the vertical tail's center of pressure, and the reference length is the wingspan b. Taking the derivative with respect to $\hat{r} \triangleq rb/2V$, the vertical tail's contribution to the yaw-rate damping derivative is estimated as

$$C_{n_{\hat{r}_{vt}}} \triangleq \frac{\partial \Delta C_{n_{vt}}}{\partial \hat{r}} = -2\left(\frac{l_{vt}}{b}\right)C_{n\beta_{vt}} \quad (3.4\text{-}31)$$

The wing's yaw-rate effect has no simple analogy to the longitudinal case. The yaw rate induces different amounts of drag on the advancing and receding sides of the wing, leading to yaw damping derivatives of the form

$$C_{n_{\hat{r}}} = k_0 C_L^2 + k_1 C_{D_{0_{wing}}} \quad (3.4\text{-}32)$$

The first term arises from differential induced drag, the second from differential profile drag $C_{D_{0_{wing}}}$ Values of k_0 range from -0.1 to 0.1, depending on the aspect ratio and sweep angle, while k_1 is in the range of -0.3 to -1 [S-2]. Negative values of $C_{n_{\hat{r}}}$ are stabilizing, but high sweep angles can make the wing's contribution positive, thereby reducing directional stability. The net effect of yaw rate on yawing moment is

$$\Delta N(r) = C_{n_r}\frac{1}{2}\rho V^2 Sbr = C_{n_{\hat{r}}}\frac{1}{2}\rho V^2 Sb\left(\frac{rb}{2V}\right)$$
$$= C_{n_{\hat{r}}}\frac{1}{4}\rho V S b^2 r \quad (3.4\text{-}33)$$

where $C_{n_{\hat{r}}}$ contains the wing and tail components.

Rolling moments are brought about by yaw rate in complex ways. In addition to producing differential drag, the yawing wing produces differential lift. Strip theory estimates the wing's contribution to this coupling derivative to be [S-4]

$$C_{l_{\hat{r}_{wing}}} = \frac{\partial \Delta C_{l_{wing}}}{\partial \hat{r}} = \frac{C_L}{12}\left(\frac{1+3\lambda}{1+\lambda}\right)\left(\frac{\mathcal{M}^2\cos^2\Lambda - 2}{\mathcal{M}^2\cos^2\Lambda - 1}\right) \quad (3.4\text{-}34)$$

λ is the wing's taper ratio, and Λ is the sweep angle of the quarter chord for subsonic application and of the leading edge in supersonic flow. The method of Bryson [N-2] predicts

$$C_{l_{\hat{r}_{wing}}} = \frac{\pi\alpha_o}{9\,AR} \tag{3.4-35}$$

for the slender triangular wing, where α_o is the nominal angle of attack.

The horizontal tail has a similar effect, complicated by the wing's wake, and its dihedral effect produces a term like equation 3.4-31 (replacing n by l and "vt" by "ht"). To the extent that the vertical tail's aerodynamic center is off the aircraft's x axis, it produces a rolling moment proportional to $C_{n_{\hat{r}_{vt}}}$:

$$C_{l_{\hat{r}_{vt}}} = -\frac{z_{vt}}{l_{vt}} C_{n_{\hat{r}_{vt}}} \tag{3.4-36}$$

with z_{vt} positive down and l_{vt} positive aft.

Roll-Rate Effects

The wing is the dominant factor in generating rolling and yawing moments from roll rate, although the tail surfaces also contribute. The rolling moment counters the roll rate, as the down-moving half of the wing has higher local angles of attack, producing more lift than the up-moving half and tending to dampen the motion. The roll rate p at wing station y adds a span-dependent increment $\Delta\alpha = py/V$ to the local angle of attack; hence, the wing tips trace a helix with a slope of $pb/2V$ rad. Strip theory predicts a roll-rate damping derivative of

$$C_{l_{\hat{p}_{wing}}} \triangleq \frac{\partial \Delta C_{l_{wing}}}{\partial \hat{p}} = -\frac{C_{L\alpha}}{12}\left(\frac{1+3\lambda}{1+\lambda}\right) \tag{3.4-37}$$

in sub- or supersonic flow [S-4], with $\hat{p} \triangleq pb/2V$. For a thin triangular wing, this derivative can be estimated as

$$C_{l_{\hat{p}_{wing}}} = -\frac{\pi AR}{32} \tag{3.4-38}$$

The rolling moment due to roll rate can be written as

$$\begin{aligned}\Delta L(p) &= C_{l_p}\frac{1}{2}\rho V^2 Sbp = C_{l_{\hat{p}}}\frac{1}{2}\rho V^2 Sb\left(\frac{pb}{2V}\right)\\ &= C_{l_{\hat{p}}}\frac{1}{4}\rho V S b^2 p\end{aligned} \tag{3.4-39}$$

These formulas presume a positive section lift slope across the wingspan. At constant roll rate, the angle of attack of the down-moving wing half is higher than that of the up-moving half (Fig. 3.4-3). If the mean angle of attack is high, the down-moving wing half may stall toward the wing tip, while the up-moving half remains unstalled. In addition to the net decrease in lift, this asymmetry reduces roll-rate damping and may even reverse the sign of the normally negative $C_{l_{\hat{p}}}$. The reduced damping effect can produce *autorotation*, and it may *destabilize the roll mode* of motion, which is introduced in Section 4.3.

Yawing moment due to roll rate is a coupling derivative that tends to have opposite signs in subsonic and supersonic flight. Differential lift again leads to differential induced drag; Seckel gives the formula

$$C_{n_{\hat{p}_{wing}}} \triangleq \frac{\partial \Delta C_{n_{wing}}}{\partial \hat{p}} = \frac{1}{12}\left(\frac{1+3\lambda}{1+\lambda}\right)\left(\frac{\partial C_{D_{o_{wing}}}}{\partial \alpha} \pm C_L\right) \tag{3.4-40}$$

where the minus sign applies to subsonic flight and the plus sign to supersonic flight. $\partial C_{D_{o_{wing}}}/\partial \alpha$ reflects the sensitivity of the wing parasite drag to the angle of attack. The thin triangular wing estimate is

$$C_{n_{\hat{p}_{wing}}} = -\frac{\pi \alpha_o}{9AR} \tag{3.4-41}$$

The suggested vertical tail contribution is [C-2, S-4]

$$C_{n_{\hat{p}_{vt}}} = -2\alpha_o\left(\frac{l_{vt}}{b}\right)C_{n_{\beta_{vt}}} \tag{3.4-42}$$

Effects of Wind Shear and Wake Vortices

The accommodations made for air-mass motion in Section 3.2 account for the difference between air-relative and inertial velocity of an aircraft's center of mass, but they do not address the aerodynamic effects of a *sheared wind field* varying over the surface of the aircraft. Sheared wind fields can be generated naturally, as by microbursts, tornadoes, and horizontal wind rotors, or by other aircraft, whose flight leaves a trail of concentrated vortical flow that may take several minutes to dissipate.

Wind shear induces time-varying aerodynamic angles and airspeed that change forces and moments through the static aerodynamic coefficients presented in Section 2.5. It also changes pressure distributions over the wing, fuselage, and tail, with particular impact on the aerodynamic moments. Wind-shear-induced moments can challenge the structural and

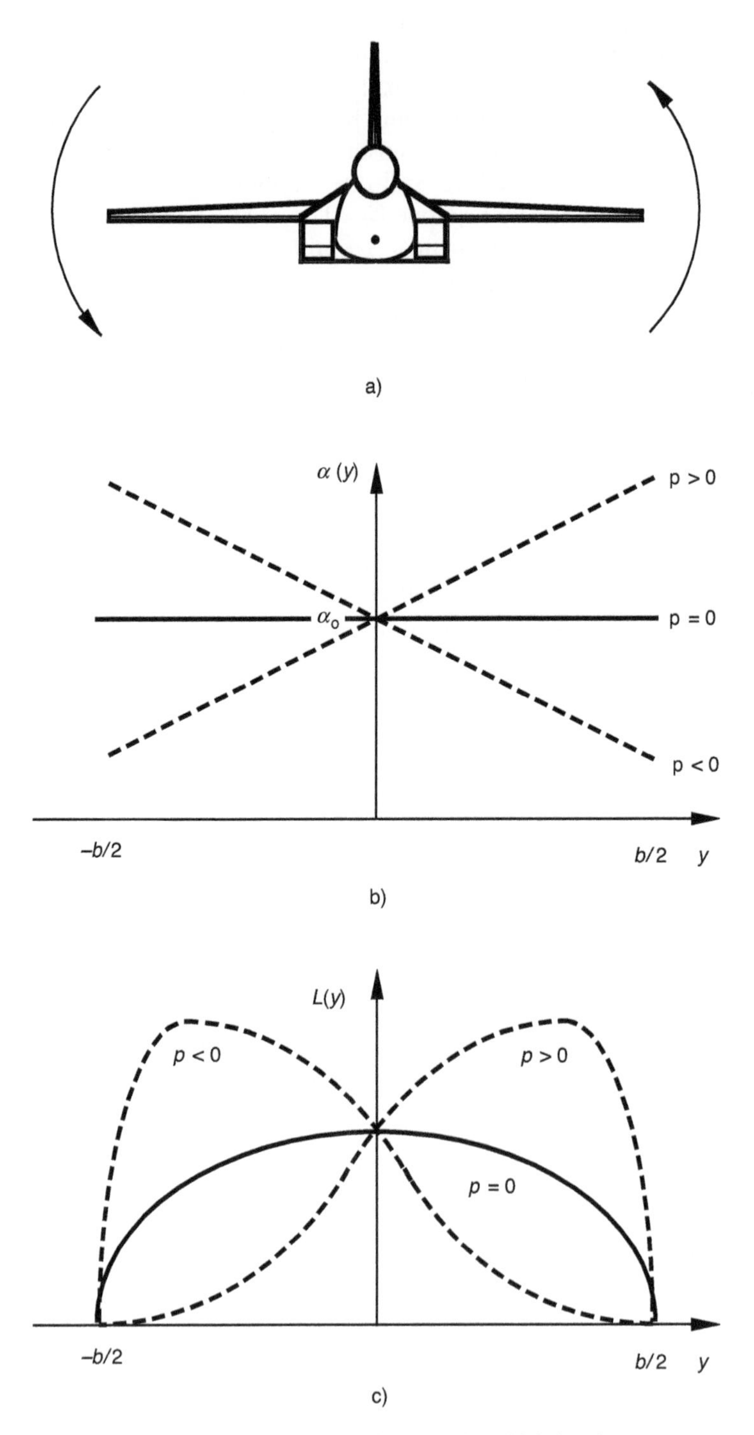

Fig. 3.4-3. Effect of roll rate on spanwise angle of attack and lift distribution. (a) Positive aircraft roll rate. (b) Geometric spanwise angle of attack distribution. (c) Spanwise lift distribution.

control limits of the aircraft for short periods of time during encounter, and they may produce large attitude excursions from which even experienced pilots may not recover.

Wind-shear effects can be measured in wind tunnels [R-5], and they can be estimated computationally using strip theory [J-2, M-1,V-1] or vortex-lattice methods [V-1]. For large-scale wind shear, the similarities between wind shear and the flow fields of rotation and unsteady motion make an approximate analysis of wind-shear aerodynamics possible.

The *spatial wind shear* at a point **r** can be expressed in inertial (actually earth-relative) coordinates as a (3×3) Jacobian matrix (eq. 2.1-11):

$$\mathbf{W}_{\mathrm{I}} \triangleq \begin{bmatrix} \dfrac{\partial w_x(\mathbf{r},t)}{\partial x} & \dfrac{\partial w_x(\mathbf{r},t)}{\partial y} & \dfrac{\partial w_x(\mathbf{r},t)}{\partial z} \\ \dfrac{\partial w_y(\mathbf{r},t)}{\partial x} & \dfrac{\partial w_y(\mathbf{r},t)}{\partial y} & \dfrac{\partial w_y(\mathbf{r},t)}{\partial z} \\ \dfrac{\partial w_z(\mathbf{r},t)}{\partial x} & \dfrac{\partial w_z(\mathbf{r},t)}{\partial y} & \dfrac{\partial w_z(\mathbf{r},t)}{\partial z} \end{bmatrix} \tag{3.4-43}$$

w_x, w_y, and w_z are components of the wind resolved along earth-relative axes, and the partial derivatives reflect the rate of change of each wind component along each direction. The spatial-wind-shear matrix can be resolved in body axes using the similarity transformation

$$\mathbf{W}_{\mathrm{B}} = \mathbf{H}_{\mathrm{I}}^{\mathrm{B}} \mathbf{W}_{\mathrm{I}} \mathbf{H}_{\mathrm{B}}^{\mathrm{I}} \triangleq \begin{bmatrix} u_{w_x} & u_{w_y} & u_{w_z} \\ v_{w_x} & v_{w_y} & v_{w_z} \\ w_{w_x} & w_{w_y} & w_{w_z} \end{bmatrix} \tag{3.4-44}$$

where, for example, u_w is the wind component along the x body axis and u_{u_z} is the rate of change of u_w in the z direction. The nine components portray the change in each of the three components of the wind along each of the three body axes. Given a linear shear field, the change in the body-relative vertical component of the wind between the wing tips would be modeled as $w_{w_y} b$, where b is the wing span. The fact that the aircraft's presence changes the wind field locally is neglected in this model.

If the scale of the wind shear is large compared to the dimensions of the aircraft, an equivalence between angular rates and shear flows allows wind-shear aerodynamic effects to be estimated from components of the rotary derivatives. In this case, we neglect the solidity and shape effects of the aircraft on its surrounding wind field, assuming that the wind's velocity vector and shear matrix take effect at the aircraft's center of mass (i.e., that the aircraft is "transparent"). This assumption is appropriate for calculating the most significant aspects of encounters with microbursts or wind rotors, with length scales of hundreds to thousands of meters.

Figure 3.4-4 illustrates that body-axis angular rates produce linearly increasing local translational flows perpendicular to the body axes.

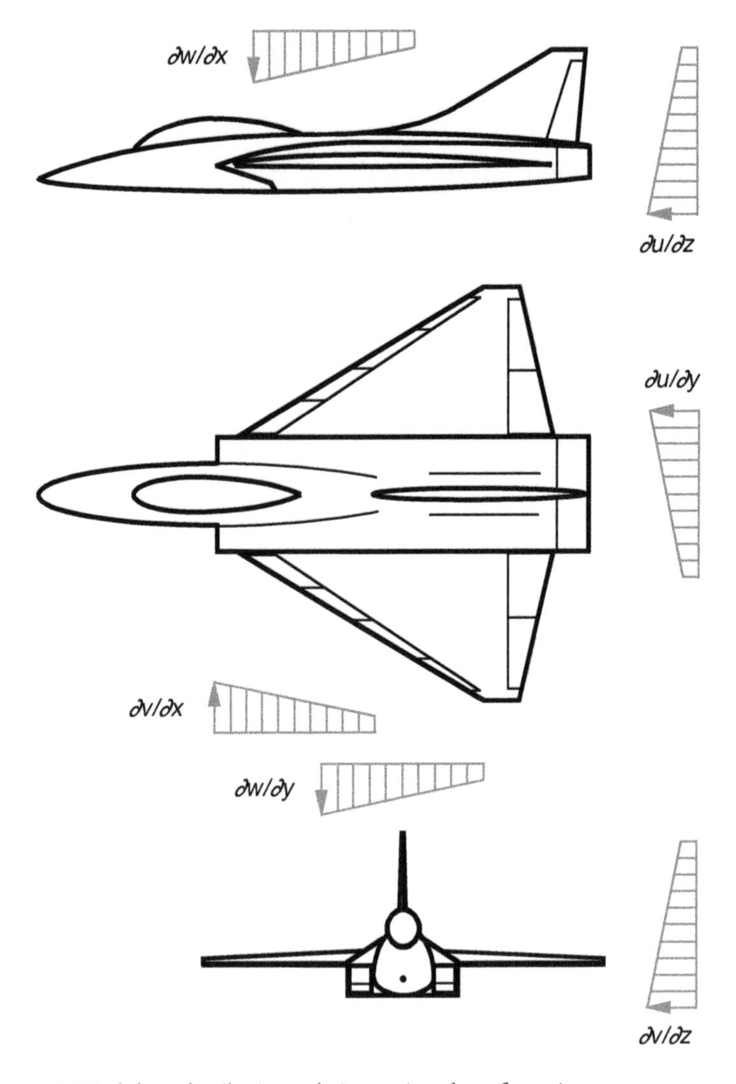

Fig. 3.4-4. Wind shear distributions relative to aircraft configuration.

Referring to the roll-rate effect, $w = py_B$ and $v = -pz_B$; because the rate of change is constant,

$$\frac{\partial w}{\partial y} \triangleq w_y = p \tag{3.4-45}$$

$$\frac{\partial v}{\partial z} \triangleq v_z = -p \tag{3.4-46}$$

Along the other axes,

$$w_x = -q \tag{3.4-47}$$

$$u_z = q \tag{3.4-48}$$

$$v_x = r \tag{3.4-49}$$

$$u_y = -r \tag{3.4-50}$$

Assume that the moments from wing and tail surfaces can be separated. Using the roll rate for an example, w_{w_y} is equivalent to p along the y_B axis but not along the z_B axis, where v_{w_z} is equivalent to p. Therefore, w_{w_y} leads to the same wing and horizontal-tail rolling and yawing moments that roll rate produces, but it does not cause the vertical tail to generate moments. Conversely, v_{w_z} acts on the vertical tail in the same way that p does, but not on the wing and horizontal stabilizer. By this reasoning, and noting that $\mathbf{v}_A = \mathbf{v}_I - \mathbf{w}$ (eq. 3.2-96), the moment perturbations produced by spatial wind shear can be estimated as follows:

$$\Delta L = -(\Delta L_{p_{wing}} + \Delta L_{p_{ht}}) w_{w_y} + \Delta L_{p_{vt}} v_{w_z} - \Delta L_{r_{vt}} v_{w_x} + \Delta L_{r_{wing}} u_{w_y} \tag{3.4-51}$$

$$\Delta M = (\Delta M_{q_{wing}} + \Delta M_{q_{ht}})(w_{w_x} - u_{w_z}) \tag{3.4-52}$$

$$\Delta N = -(\Delta N_{p_{wing}} + \Delta N_{p_{ht}}) w_{w_y} + \Delta N_{p_{vt}} v_{w_z} - \Delta N_{r_{vt}} v_{w_x} + \Delta N_{r_{wing}} u_{w_y} \tag{3.4-53}$$

The moment subscripts w, ht, and vt indicate wing, horizontal tail, and vertical tail contributions, respectively. Here, $\Delta L_{p_{wing}}$ is defined as in equation 3.4-39, and the other derivatives are defined accordingly.

Temporal wind shear effects derive from equation 2.1-13, which becomes

$$\dot{\mathbf{w}}_B = \mathbf{w}_{B_t} + \mathbf{H}_I^B \mathbf{W}_I \mathbf{H}_B^I \mathbf{v}_B \tag{3.4-54}$$

in body axes. As only $\dot{\alpha}$ has a significant unsteady effect, and $\dot{\alpha} \approx \dot{w}/V$, just the third row of eq. 3.4-54 is important. Time variation of the vertical component produces the incremental pitching moment

$$\Delta M = -\Delta M_{\dot{\alpha}}(w_{w_t} + w_{w_x}u + w_{w_y}v + w_{w_z}w)/V \tag{3.4-55}$$

where u, v, and w are body-axis components of the aircraft's inertial velocity.

Naturally occurring wind rotors may have diameters that are large compared to the wingspan of the aircraft (Section 2.1). The maximum tangential velocities of vortices produced in the wake of other aircraft have much smaller scale, on the order of the generating aircraft's wing-tip chord length times the sine of the angle of attack. Wake vortices occur in closely spaced counter-rotating pairs, with an area of intense downdraft between. Stability derivatives can be used to estimate the disturbance effects for wind rotors, but they have limited value for wake vortices, where fine-scale estimates of forces and moments based on strip theory or the vortex-lattice method are required. A further complication is that the receiving aircraft itself can no longer be considered transparent: its own flow field distorts the location and velocity field of the impinging vortices, coupling the response to the disturbance (Fig. 2.1-3).

3.5 Aerodynamic Effects of Control

Controlled forces and moments are needed to trim the aircraft for equilibrium flight, to respond to pilot commands, to reject external disturbances, and to augment natural stability. In addition to thrust control (Section 2.6), the principal means of controlling the aircraft is through aerodynamic forces generated by *control surfaces*, which are movable flaps located on the fuselage, wing, and tail [M-5, R-4, S-1, T-2, T-5, Y-1]. The primary purpose of certain control surfaces (e.g., elevator, rudder, and ailerons) is to create control *moments*; hence, their resultant forces act at some distance from the aircraft's center of mass and are largely perpendicular to their moment arms. Other surfaces, like wing flaps and speed brakes, are meant to produce control *forces* with as little moment as possible; their resultant forces pass close to the center of mass, usually as a result of two or more symmetrically positioned surfaces being deflected equally.

In most cases, it is desirable for a control surface deflection to produce a *single* control force or moment that is *linearly related* to the deflection; however, as foreshadowed by Sections 2.4 and 3.4, many factors limit the range within which uncoupled, linear response can be obtained [R-4, Y-1]. Control moments are rarely generated without residual control forces, and a varying center of mass assures that a command force is accompanied by a residual moment in some conditions. The net moment created by a single surface typically is felt about more than one rotational axis and may vary with flight conditions. Ancillary control effects are not always adverse; beneficial coupling can also occur.

More generally, the constants of proportionality between the control deflection δ and the force change with dynamic pressure, Mach number, and angle of attack. They are affected by flow interference and airframe elasticity. Forward-mounted control surfaces (e.g., horizontal canards) deflect the flow over the wing, creating large indirect effects, while aft-mounted controls are subject to downwash, sidewash, reduced dynamic pressure, lagged flow adjustment, and flow "blanketing" from the wing and fuselage.

Forces and moments generated by large control deflections are subject to nonlinear effects arising from induced drag, stall, and flow separation. Trailing-edge control surfaces operate in the wakes of main surfaces, while spoilers are immersed in the wing's boundary layer. In both instances, small-angle nonlinearities can result. Unsteady aerodynamic effects have been associated with high control deflection rates; therefore, transient control forces and moments may be functions of $\partial\delta/\partial t$ as well as of δ.

After brief summaries of control surface characteristics, their principles of operation are discussed in the following two sections.

Elevators, Stabilators, Elevons, and Canards

An *elevator* is a trailing-edge flap installed on the horizontal tail. Its principal purpose is to produce pitching moments for longitudinal rotational control. A pitch-up moment is generated by negative lift on the (aft) tail; the wing lift increase that results from increased angle of attack is opposed initially by the elevator's download. This leads to a "non-minimum-phase" normal-acceleration response, in which the immediate download produced by the deflected tail surface opposes the upload eventually produced by the wing as the angle of attack increases. *Stabilators* are all-moving horizontal tails that produce a similar effect. They have found widespread use as primary pitch control surfaces on supersonic aircraft, and they are used for pitch-trim control on a wide range of aircraft types. Like *elevons*, which are trailing-edge flaps mounted on delta wings for both pitch and roll control, stabilators may be commanded differentially for roll control.

Canards are pitch-control surfaces mounted forward of the aircraft's c.m. A pitch-up moment is accompanied by positive lift; hence, there is no "non-minimum-phase" control effect. The canard has the advantage of operating more nearly in the freestream flow rather than in the wake of fuselage and wing. However, canard downwash can adversely affect flow over the wing and tail, particularly at large deflection angles, and fixed surfaces mounted forward of the center of mass are inherently destabilizing. Asymmetric canard deflection generates small rolling moments, but, in some configurations, the differential pressures on the nose lead to large controllable yawing moments. Some aircraft configurations (e.g., the Piaggio P.180) have included both elevator (or elevon) and canard for "three-surface control," affording independent control of lift and pitching moment by coordinated deflection of the two surfaces. Canard surfaces are well positioned to minimize normal forces due to gust inputs. Consequently, they are good control effectors for improving ride quality and for controlling aeroelastic modes of the fuselage.

Rudders

Rudders are mounted on an aircraft's vertical fin (or fins) to exert a controlled yawing moment, although they produce roll moment and side force in the process. Unlike control surfaces on the wing and horizontal tail, which normally operate about some nonzero angle of attack, the rudder usually operates near zero sideslip angle, so it is susceptible to small-angle nonlinear effects (for example, small motions could be masked by a thick boundary layer; at high-supersonic speeds and above, the Newtonian nature of the flow results in low rudder effectiveness at low sideslip angle). The problem of nonlinearity can be reduced and directional stability can be increased by employing a single wedge-shaped fin and rudder (as on the North American X-15 rocket research aircraft) or a pair of toed-in fins and rudders (as on the Northrop YF-17A fighter prototype), at the expense of increased drag. All-moving rudders, as employed on the Lockheed-Martin SR-71, have increased effectiveness; however, there is a large weight penalty associated with this design [W-1], and trailing-edge rudders are preferred for most aircraft.

In addition to the geometric effects revealed by Newtonian analysis, rudder effectiveness may be reduced at high angles of attack as a consequence of blanketing by the fuselage, wing, and horizontal tail. Rudder effectiveness can be retained by mounting the horizontal tail at the top of the fin ("T tail") and by extending the rudder to the bottom of the fuselage, as on many single-engine general aviation aircraft. Conversely, the rudder effect may be too high in some flight conditions, producing large structural loads on the fin and excessive yaw moment at maximum deflection. Rudder deflections may be limited, either by mechanical stops that engage automatically when the dynamic pressure exceeds a certain value or by *blowdown* (or aerodynamic) hinge moments that overpower mechanical actuation.

The rudder may produce strong *roll* control at moderate-to-high angle of attack through a combination of direct effect and the sideslip-induced dihedral effect of the wing. With the large flap settings and low flight speed characteristic of landing approach, the rudder's maximum rolling moment may exceed that of the ailerons. The rudder can generate a controllable side force when used in conjunction with tip-mounted *yaw flaps* (or *drag petals*) to negate yawing moment. Yaw flaps mounted at the wing tips take the place of rudders on tailless aircraft [G-1], such as the Northrop-Grumman B-2 bomber.

Ailerons

Differential deflection of ailerons produces rolling moments and may lead to significant yawing moments as well. One method of reducing the yaw effect is to gear the opposing surfaces to deflect differently in up and down travel: for example, a 10 deg trailing-edge up deflection of the right aileron might be accompanied by a 5 deg down deflection of the left aileron. In general, there is a net lift as a result. Aileron differential can also reduce the control hinge moment required for a given rolling moment [A-2].

Ailerons traditionally have been located near the wing tips to gain maximum rolling moment with minimum surface deflection; however, this location can be problematical. At high speed, aileron deflection produces aeroelastic twisting of a straight or swept-back wing, decreasing control effect and possibly causing a net reversal of rolling moment. Swept-back wings tend to stall first near the wing tips with increasing α; consequently, outboard ailerons lose effectiveness. Mid-span ailerons reduce both of these adverse effects (many transport aircraft have both mid-span and outboard ailerons for high- and low-speed roll control), although they can produce strong interference effects on tail surfaces. All-moving wing tips (or *tip vanes*) overcame the adverse effects of conventional ailerons in the "aero-isoclinic wing" of the experimental Short Brothers Sherpa (G-5) and the Boeing Bomarc missile but have not been used in operational aircraft. Forward-swept wings retain tip aileron effectiveness to higher angles of attack, but the wings must be elastically tailored to prevent torsional divergence, as on the Grumman X-29. Aileron effectiveness can be improved by simultaneous deflection of leading-edge flaps (or "slats"), which improve lifting performance.

Spoilers and Wing Flaps

Spoilers present an alternative to ailerons for roll control when deflected individually, and they provide lift control when deflected symmetrically.

A spoiler obstructs the flow above the wing, reducing the lift. It is inherently nonlinear in its operation, having little or no effect until the surface emerges from the boundary layer. A net $\pm$ linear effect can be achieved by partially deflecting the spoilers to a nominal position, although there is an associated drag penalty; then, retraction increases lift and extension decreases it. Spoilers can be placed on the elastic axis of a slender wing, eliminating the torsional effects of outboard ailerons. Stealth aircraft designed with little or no vertical tail may use spoilers that are positioned to produce significant combinations of pitching, rolling, and yawing moments without increasing the aircraft's radar observability. As spoilers mounted on the upper surface of the wing tend to lose effect with increasing angle of attack, lower-surface spoilers could be considered, although deflecting them could increase the radar cross section.

Trailing-edge *wing flaps* augment lift (per Fig. 2.4-2), and they produce a large amount of drag at high deflection angles [P-3]. Transport aircraft normally use the flaps only during takeoff and landing, and a slow actuation rate is permissible. The actuation rate is increased for the maneuvering flaps of high-performance aircraft, which find value in rapid decreases as well as increases in lift. The former high-lift devices are often complex, many-surfaced affairs, while the latter are akin to ailerons in shape and design. (The two are integrated in *flaperons*.)

Wing flaps have been used in conjunction with powerplants to produce powered lift. The flaps deflect engine exhaust vertically to increase lift, either by direct immersion in the flow of an underwing engine or by using the *Coanda effect* on an overwing exhaust stream, where the flow follows the upper surface of the downward-deflected flap.

Other Control Devices

A number of other surfaces and fluid dynamic mechanisms have usable aerodynamic control effects. Side force can be generated either by the addition of movable vertical surfaces near the c.m. that produce the side force directly (e.g., those used on Princeton University's Variable-Response Research Aircraft or Veridian's Total In-Flight Simulator) or by devices that produce a yaw moment that is offset by rudder deflection, in turn generating side force (as employed by Veridian's variable-stability T-33 aircraft).

Speed brakes increase the aircraft's frontal area, thereby increasing its drag. They normally extend from the fuselage, although the space shuttle uses a split rudder that opens to a wedge shape during landing. Deployed *landing gear* also have a measurable speed-brake effect.

Suction and *jet blowing* can provide boundary layer and vortex control to achieve a variety of purposes. The intent of suction is to remove low-energy boundary layer air so that drag can be reduced. Transverse jets blowing across the wing's upper surface or from the aircraft's nose have important effects on lift and moment generation. Controlling the vortices generated by the aircraft's nose can have a strong effect on rolling and yawing moments at high angles of attack [W-5]. Both concepts require either a source or a sink for the controlled air, bleeding power from the aircraft's engine(s).

Thrust has a potential role to play in attitude control, in addition to providing power for sustained flight. Attitude control can be effected by jets ("puffers") located at the plane's extremities (as on the Harrier vertical-takeoff-and-landing aircraft), but the power required is large in comparison to aerodynamic surfaces for conventional flight. A number of high-performance aircraft (e.g., SR-71 reconnaissance aircraft and F-15 fighter aircraft) have large, movable engine inlet ramps whose pitching and yawing moments cannot be ignored in flight control design. *Thrust vector control*, afforded by petals (or paddles) surrounding the exhaust stream or by deflecting nozzles can provide powerful attitude control in all flight conditions. A single nozzle can be deflected for pitching and yawing control, while rolling control requires differential deflection of two nozzles or differential vanes within a single nozzle. This potential is especially important at high angles of attack, where conventional control surfaces lose their effectiveness. *Thrust reversal*, commonly used for postlanding deceleration, has been suggested as a control for air combat maneuvering [P-2].

Isolated Control Surfaces

Individual movable surfaces that are not attached to the leading or trailing edges of larger surfaces would generate lift, drag, and moments as small wings do. The principal distinction is that a control surface's net angle of attack depends on a variable deflection angle relative to the body in addition to the flow direction measured with respect to the body (Fig. 3.5-1). Keeping in mind the caveats of Section 2.4, Newtonian aerodynamic models provide an introduction to the effects of vehicle angle of attack α and control deflection angle δ on control forces and moments, but in many flight conditions they do not produce accurate estimates.

Attention is restricted to small flat-plate surfaces that are isolated from any other bodies or lifting surfaces and whose centers of pressure (c.p.) are displaced from the aircraft's center of mass (c.m.). The first case (Fig. 3.5-1a) represents an idealized all-moving elevator or horizontal canard surface, located on the vehicle centerline at a distance l_h (positive) aft of the c.m.; its deflection angle δ is measured about an axis parallel to the vehicle's y axis, trailing-edge down being positive. Within this framework, wing-flap effects are analogous to elevator effects. The

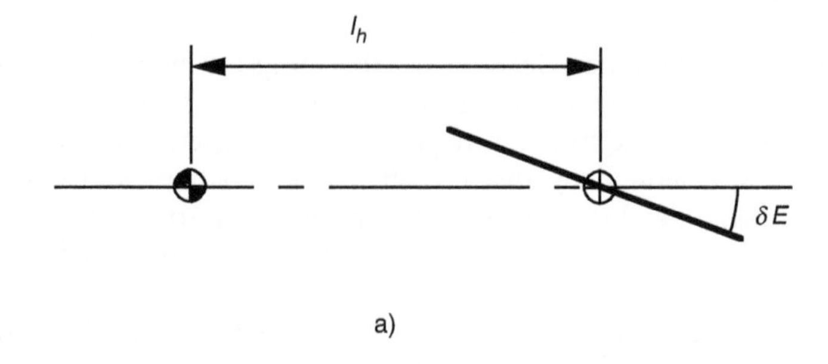

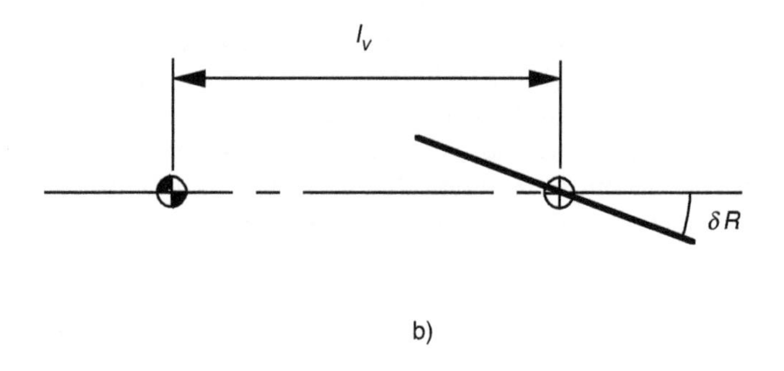

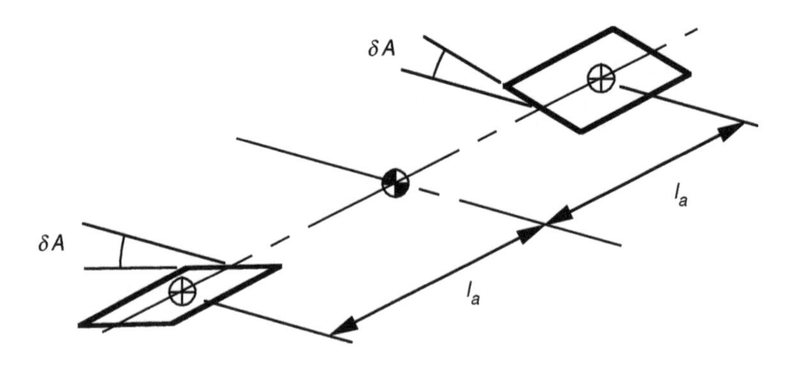

Fig. 3.5-1. Illustrations of idealized control surfaces: elevator (a), rudder (b), and ailerons (c).

second case (Fig. 3.5-1b) represents an idealized rudder or vertical canard surface on the centerline, with the c.p. displaced by l_v from the c.m.; the deflection δ is about an axis parallel to the aircraft's z axis, right-hand rotation positive. The third case represents an idealized pair of ailerons with aerodynamic centers symmetrically disposed about the c.m. at distances of $\pm l_a$ (Fig. 3.5-1c). Surface rotation occurs about the y axis, with positive deflection δ of the right aileron (trailing-edge down) accompanied by negative deflection $-\delta$ of the left aileron.

In all cases, Newtonian theory predicts the normal force coefficient for a flat plate as

$$C_N = 2\,\mathrm{sgn}\theta \sin^2\theta \tag{3.5-1}$$

where θ is the direction angle at which the flow impinges on the surface ($=0$ when the flow is tangent to the surface). C_N is referenced to the flat-plate area S_s. The *signum function sgn*$(\cdot)$ (± 1 depending on the sign of the argument) is introduced to give the proper direction of the normal force for both positive and negative impingement angles. C_N is resolved into body-axis normal and axial force coefficients and referenced to the aircraft's reference area S as follows:

$$C_{Z_B} = -C_N\left(\frac{S_s}{S}\right)\cos\delta = -2\left(\frac{S_s}{S}\right)\mathrm{sgn}\theta\,\sin^2\theta\,\cos\delta \tag{3.5-2}$$

$$C_{X_B} = -C_N\left(\frac{S_s}{S}\right)\sin\delta = -2\left(\frac{S_s}{S}\right)\mathrm{sgn}\theta\,\sin^2\theta\,\sin\delta \tag{3.5-3}$$

The pitching-moment coefficient produced by the idealized elevator (Fig. 3.5-1a) is

$$C_m = \frac{l_h}{\bar{c}}C_{Z_B} = -2\frac{l_h}{\bar{c}}\left(\frac{S_s}{S}\right)\mathrm{sgn}\theta\,\sin^2\theta\,\cos\delta \tag{3.5-4a}$$

Because the impingement angle is

$$\theta = \alpha + \delta \tag{3.5-5}$$

the moment coefficient becomes

$$C_m = -2\frac{l_h}{\bar{c}}\left(\frac{S_s}{S}\right)\mathrm{sgn}(\alpha+\delta)\sin^2(\alpha+\delta)\cos\delta \tag{3.5-4b}$$

where $\bar{c}$ is the reference length (mean aerodynamic chord) introduced earlier. The control effect is expressed as the incremental pitching-moment coefficient due to elevator deflection, that is, the difference between eq. 3.5-4b and its value at $\delta = 0$, for a given α:

$$\Delta C_m = \frac{l_h}{\bar{c}} \Delta C_{Z_B}$$
$$= -2 \frac{l_h}{\bar{c}} \left(\frac{S_s}{S} \right) [\text{sgn}(\alpha + \delta) \sin^2(\alpha + \delta) \cos\delta - \text{sgn}\,\alpha \sin^2\alpha] \tag{3.5-6}$$

Several plots of $-\Delta C_{Z_B}/2$ referred to S_s are shown in Fig. 3.5-2 for α in $(0, \pi/2$ rad) and δ in $(-\pi/2, \pi/2$ rad). Figure 3.5-2a is an oblique view of the $-\Delta C_{Z_B}$ surface showing the transition from the antisymmetric control effect at $\alpha = 0$ to the symmetric effect at $\alpha = \pi/2$ rad. A contour plot of $-\Delta C_{Z_B}$ reveals its maximum effect in the vicinity of $\alpha = 20$ deg and $\delta = 50$ deg (figure 3.5-2b). Conventional two-dimensional plots of $-\Delta C_{Z_B}$ versus δ for $\alpha = 0$, 20, 40, 60, and 90 deg illustrate the asymmetric elevator effect, as well as reversal of the slope of the control effect that occurs at decreasing deflection angles with increasing α.

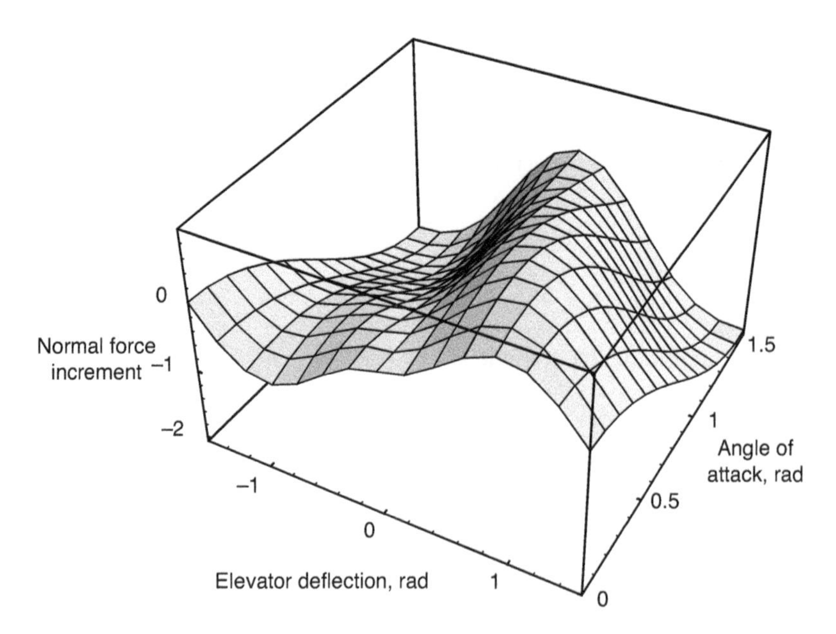

Fig. 3.5-2a. Newtonian estimate of isolated elevator normal force as a function of angle of attack and deflection angle: Surface plot.

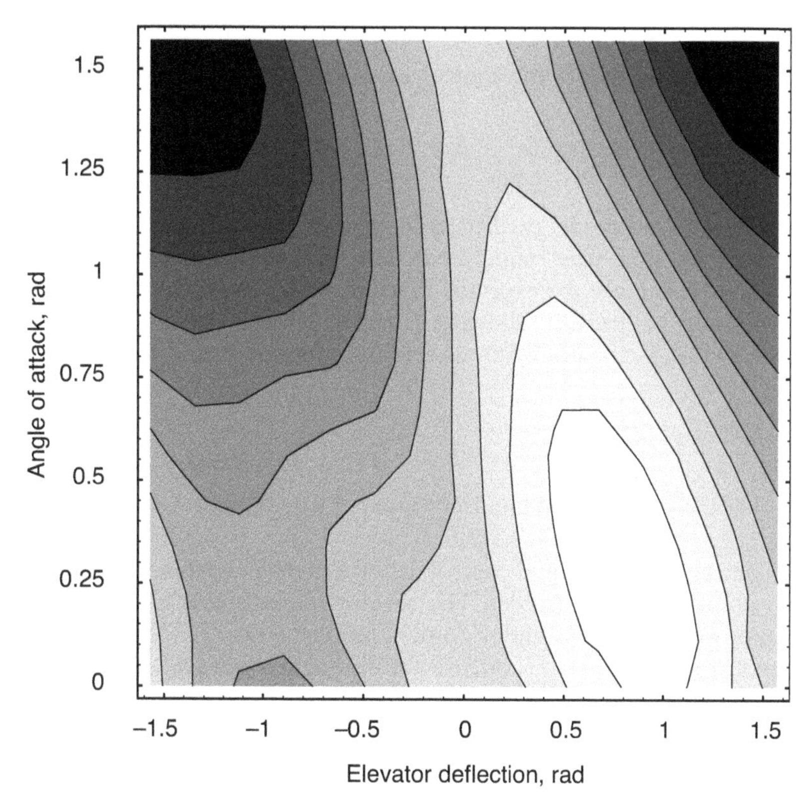

Fig. 3.5-2b. Newtonian estimate of isolated elevator normal force as a function of angle of attack and deflection angle: Contour plot.

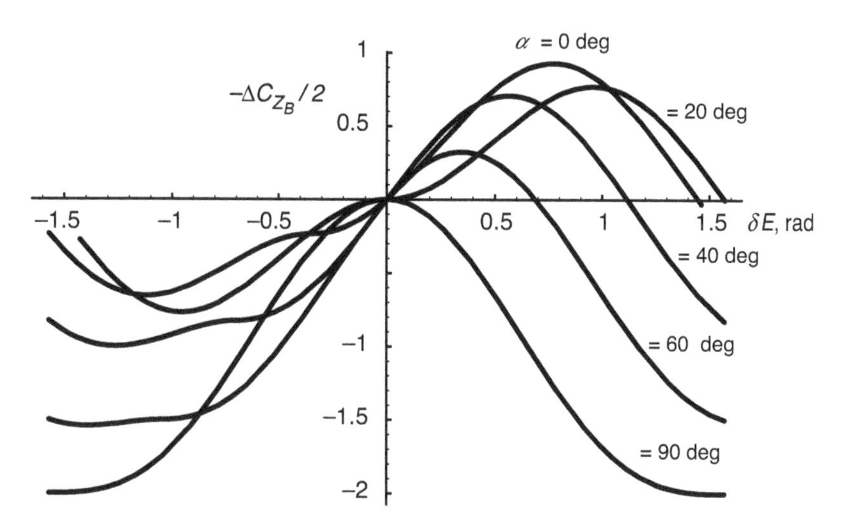

Fig. 3.5-2c. Newtonian estimate of isolated elevator normal force as a function of angle of attack and deflection angle: $-\Delta C_{Z_B}/2$ versus δE for five angles of attack.

For given α_o and δ_o, the control sensitivity is

$$\frac{\partial C_m}{\partial \delta} = -4\frac{l_h}{\bar{c}}\left(\frac{S_s}{S}\right)[\operatorname{sgn}(\alpha_o + \delta_o)\sin(\alpha_o + \delta_o)\cos(\alpha_o + \delta_o)] \qquad (3.5\text{-}7)$$

This Newtonian model predicts zero slope at zero α, which, as noted in Section 2.4, is a gross underprediction for less-than-hypersonic Mach numbers. It can also be expected that maximum control effects in a real flow would be greater, following Fig. 2.4-1. Note that the Newtonian control effect is not strictly linear for any values of α_o and δ_o, although

$$\Delta C_m(\alpha_o, \delta) \approx \left[\frac{\partial C_m(\alpha_o, \delta)}{\partial \delta}(\delta - \delta_o) + \Delta C_m(\alpha_o, \delta)\right] \qquad (3.5\text{-}8)$$

would be a good approximation for small values of $(\delta - \delta_o)$ throughout the region plotted in Fig. 3.5-2.

The angle of attack and control deflection angle combine in a different way for the isolated rudder. The surface impingement angle θ, rather than being formed by simple addition, as in equation 3.5-5, is the projection of δ on a plane that is oriented at the angle α from the aircraft's horizontal plane; thus,

$$\tan\theta = \tan\delta\cos\alpha \qquad (3.5\text{-}9)$$

Using a trigonometric identity, the surface's normal force coefficient (eq. 3.5-1) becomes

$$C_N = 2\operatorname{sgn}\theta\left[\frac{\tan\theta}{\sqrt{1+\tan^2\theta}}\right]^2 \qquad (3.5\text{-}10)$$

The body-axis side-force coefficient can then be written

$$C_{Y_B} = C_N\left(\frac{S_s}{S}\right)\cos\delta = 2\left(\frac{S_s}{S}\right)\operatorname{sgn}\theta\left[\frac{\tan\theta}{\sqrt{1+\tan^2\theta}}\right]^2\cos\delta$$

$$= 2\left(\frac{S_s}{S}\right)\frac{\operatorname{sgn}\delta\tan^2\delta\cos^2\alpha\cos\delta}{1+\tan^2\delta\cos^2\alpha} \qquad (3.5\text{-}11)$$

The moment coefficient and the incremental coefficient due to rudder deflection are the same ($C_n = 0$ when $\delta = 0$), leading to

$$C_n = \Delta C_n = 2\frac{l_v}{b}\left(\frac{S_s}{S}\right)\left[\frac{\operatorname{sgn}\delta\tan^2\delta\cos^2\alpha\cos\delta}{1+\tan^2\delta\cos^2\alpha}\right] \qquad (3.5\text{-}12)$$

where the wingspan b is taken as the reference length.

Three views of $C_{Y_B}/2$ (= $\Delta C_{Y_B}/2$) referred to S_s are shown in Fig. 3.5-3 for α in (0, $\pi/2$ rad) and δ in (0, $\pi/2$ rad); negative δ produces antisymmetric results. The overall effect of the Newtonian isolated rudder decreases to zero at $\alpha = 90$ deg, with the maximum effect occurring on a ridge line (Fig. 3.5-3a, b) that begins at $\delta = 54.7$ deg for $\alpha = 0$ deg, following increasing δ with increasing α. The Newtonian model predicts zero control effect and zero control slope for $\delta = 0$ deg; the latter is a pessimistic result for all but hypersonic aircraft.

The isolated ailerons produce rolling and yawing moments as a consequence of differential normal and axial forces (eqs. 3.5-2 and 3.5-3). With antisymmetric aileron deflection, the flow impingement angle θ is $(\alpha + \delta)$ for the right aileron and $(\alpha - \delta)$ for the left aileron. *Spoilers* may be commanded asymmetrically to augment roll control. Upper-surface

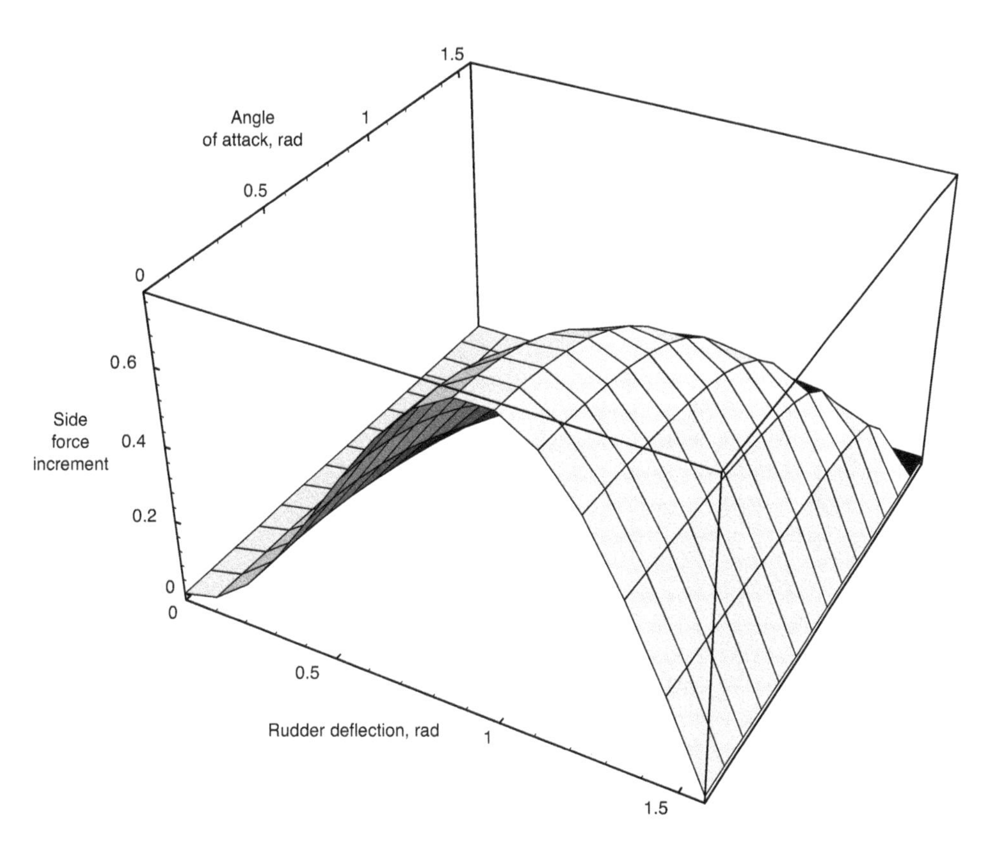

Fig. 3.5-3a. Newtonian estimate of isolated rudder side force as a function of angle of attack and deflection angle: Surface plot.

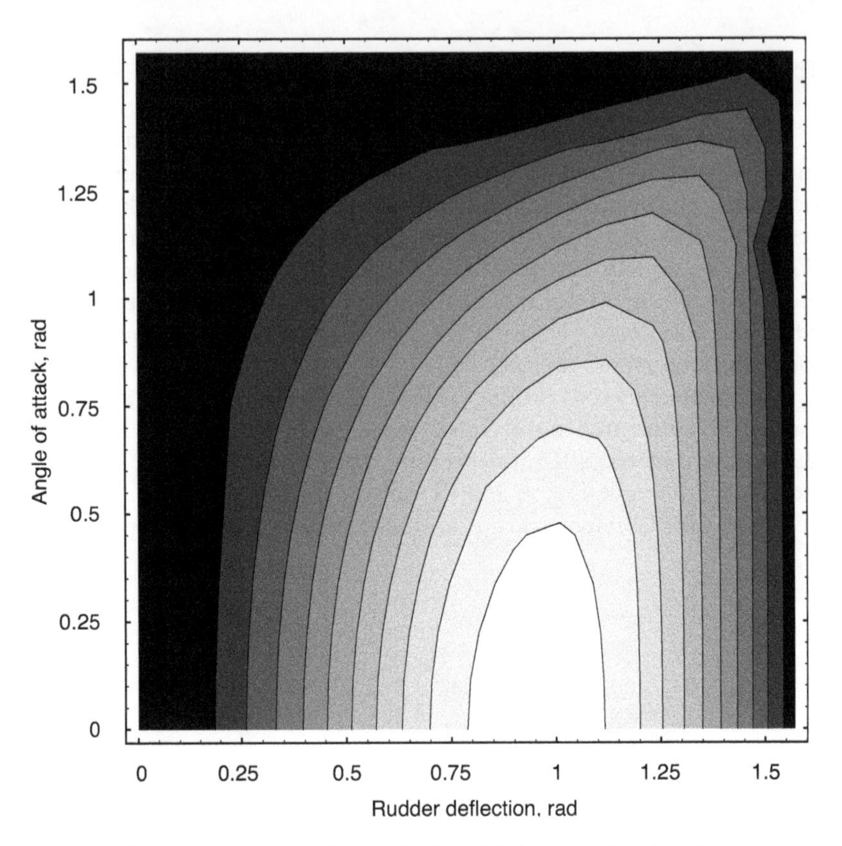

Fig. 3.5-3b. Newtonian estimate of isolated rudder side force as a function of angle of attack and deflection angle: Contour plot.

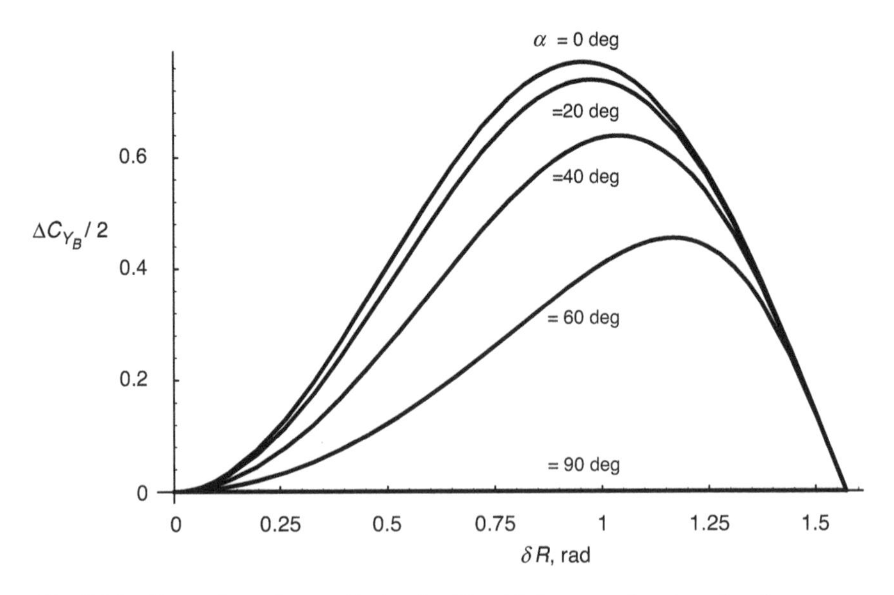

Fig. 3.5-3c. Newtonian estimate of isolated rudder side force as a function of angle of attack and deflection angle: $-\Delta C_{Y_B}/2$ versus δR for five angles of attack.

spoilers have effect only at negative impingement angle, producing a moment of the same sense as the trailing-edge-up aileron. Assuming a constant spanwise distribution of surface area, the net (and incremental) rolling moment produced by the two surfaces (each of area S_s) is

$$\begin{aligned} C_l = \Delta C_l &= \frac{l_a}{b}(C_{Z_{B_R}} - C_{Z_{B_L}}) \\ &= -2\frac{l_a}{b}\left(\frac{S_s}{S}\right)\cos\delta\,[\mathrm{sgn}(\alpha+\delta)\sin^2(\alpha+\delta) - \mathrm{sgn}(\alpha-\delta)\sin^2(\alpha-\delta)] \end{aligned} \tag{3.5-13}$$

Plots of $(C_{Z_{B_R}} - C_{Z_{B_L}})/2$ referred to S_s are shown in Fig. 3.5-4 for α in (0, $\pi/2$ rad) and δ in (0, $\pi/2$ rad). For $\alpha = 0$ deg, the pair of deflected ailerons behaves like a windmill in an axial flow. The zero-α plot is proportional to the zero-α Newtonian elevator effect, failing to indicate the linear control sensitivity that normally occurs near $\delta = 0$ deg. The rolling moment is a complex function of α and δ, with two distinct maxima in

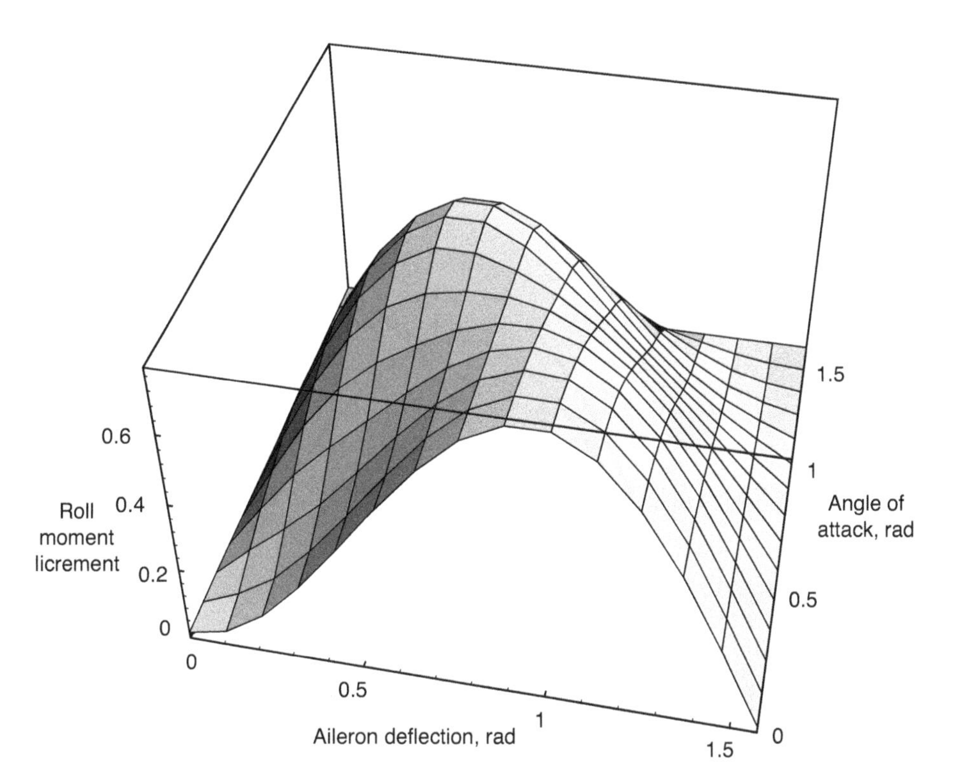

Fig. 3.5-4a. Newtonian estimate of isolated aileron-pair roll moment as a function of angle of attack and deflection angle: Surface plot.

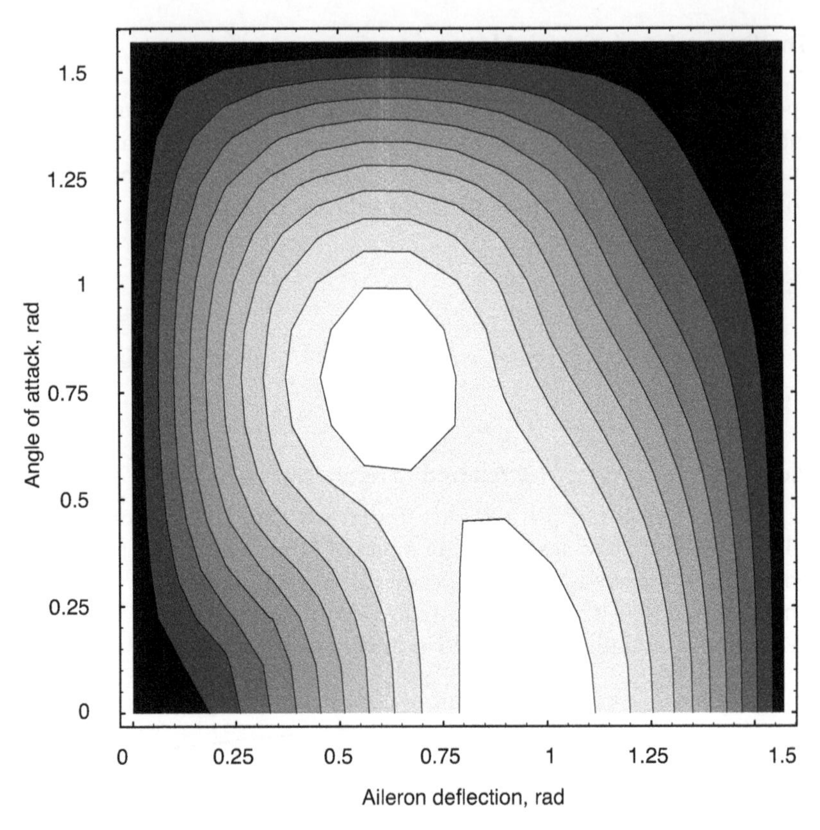

Fig. 3.5-4b. Newtonian estimate of isolated aileron-pair roll moment as a function of angle of attack and deflection angle: Contour plot.

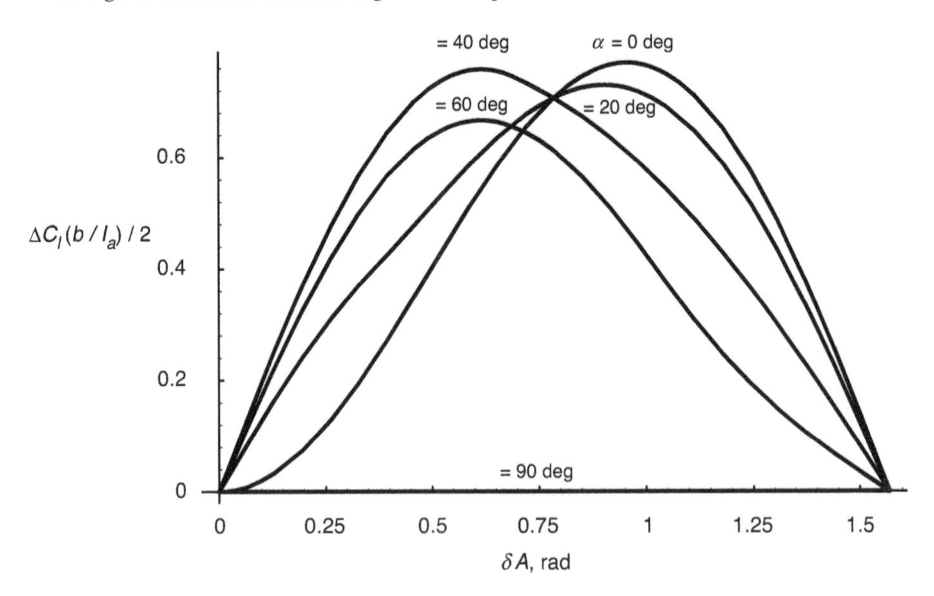

Fig. 3.5-4c. Newtonian estimate of isolated aileron-pair roll moment as a function of angle of attack and deflection angle: $C_l(b/l_a)/2$ versus δA for five angles of attack.

the plotted range that are due to asymmetric effects on the right and left surfaces.

Ailerons are not fitted to aircraft to provide yawing moments, but they can produce significant ancillary yawing moments, particularly at high angles of attack. Following the previous development, the yawing moment generated by isolated Newtonian ailerons is

$$\begin{aligned} C_n = \Delta C_n &= -\frac{l_a}{b}(C_{X_{B_R}} - C_{X_{B_L}}) \\ &= \frac{2l_a}{b}\left(\frac{S_s}{S}\right)\sin\delta[\text{sgn}(\alpha+\delta)\sin^2(\alpha+\delta) + \text{sgn}(\alpha-\delta)\sin^2(\alpha-\delta)] \end{aligned} \tag{3.5-14}$$

as $\sin(-\delta) = -\sin\delta$. At zero α, the yawing effects of the two ailerons cancel for all deflections (Fig. 3.5-5); however, the peak yawing moments reached at higher angles of attack are comparable to the peak rolling moments. The peak values are, in fact, identical; eqs. 3.5-13 and 3.5-14 are duals of each other, with (α, δ) of one corresponding to $(\pi/2 - \alpha, \pi/2 - \delta)$

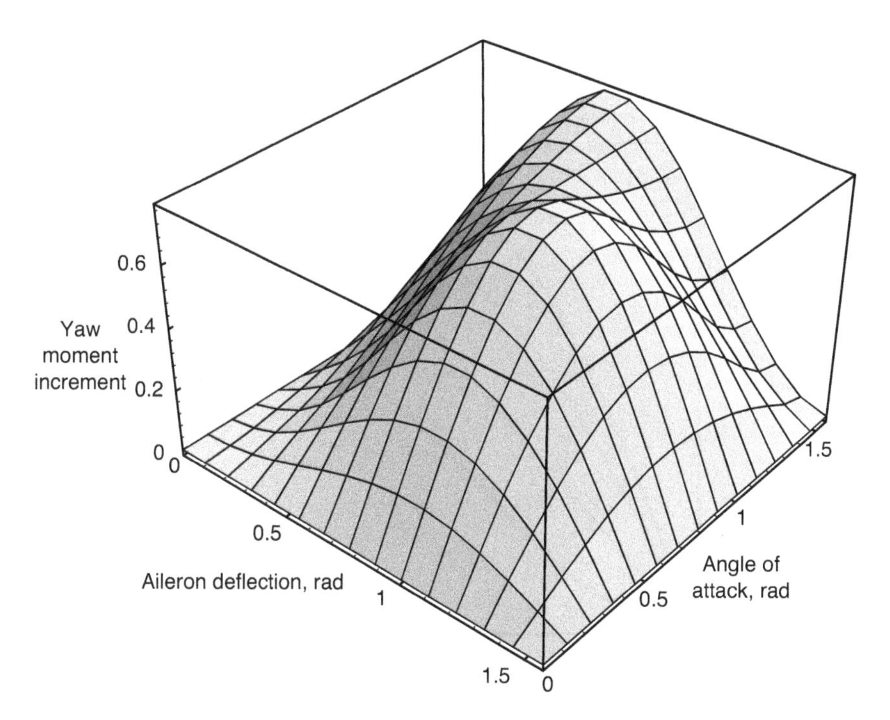

Fig. 3.5-5a. Newtonian estimate of isolated aileron-pair yaw moment as a function of angle of attack and deflection angle: Surface plot.

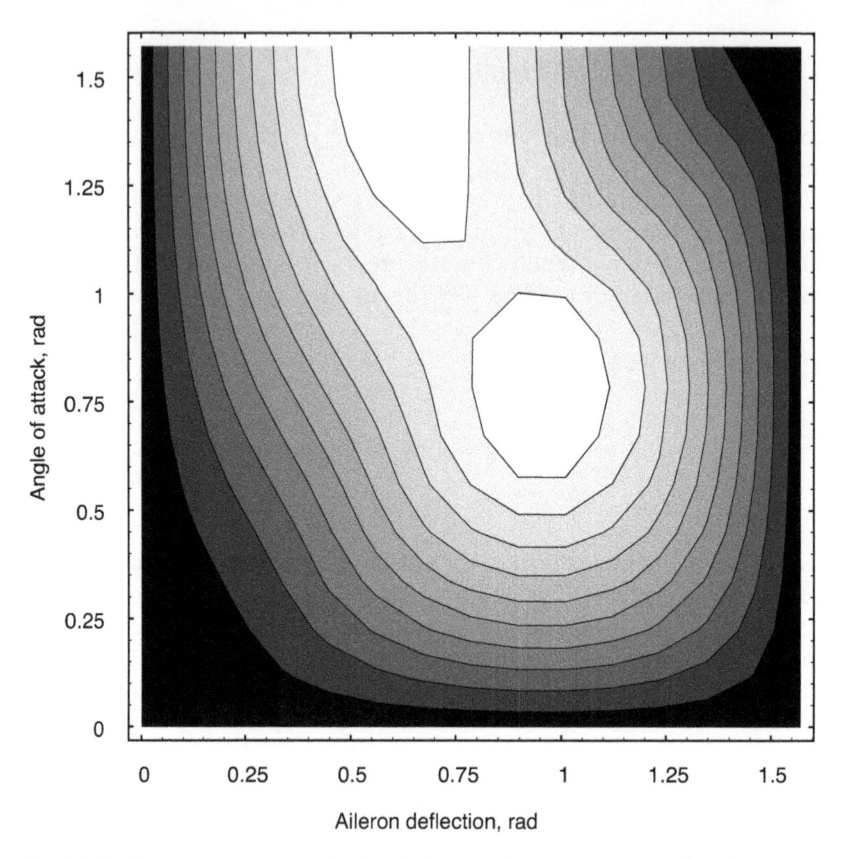

Fig. 3.5-5b. Newtonian estimate of isolated aileron-pair yaw moment as a function of angle of attack and deflection angle: Contour plot.

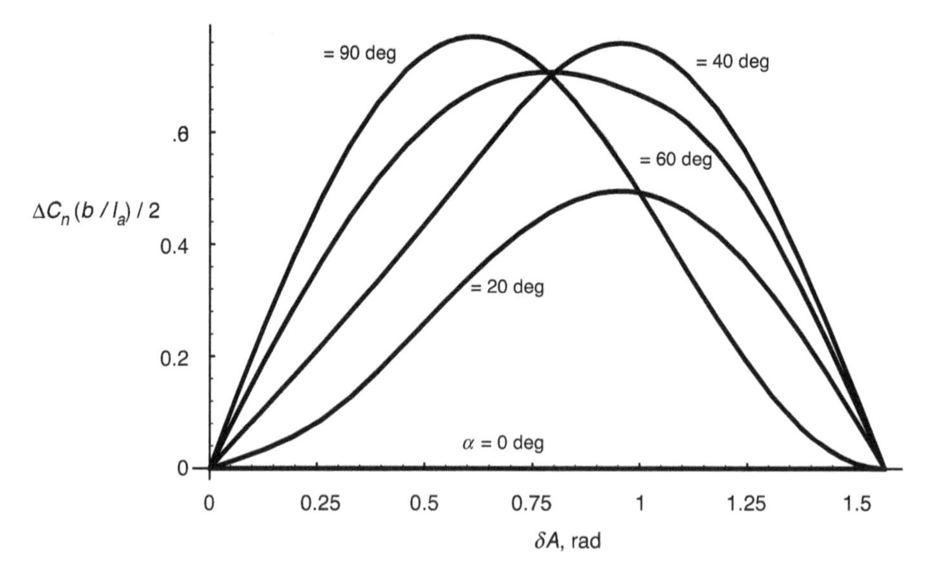

Fig. 3.5-5c. Newtonian estimate of isolated aileron-pair yaw moment as a function of angle of attack and deflection angle: $C_n(b/l_a)/2$ versus δA for five angles of attack.

of the other. (Rotate the contour plot Fig. 3.5-4b by 180 deg and compare with Fig. 3.5-5b.) For $\alpha = 90$ deg, the deflected aileron pair behaves like a windmill in an axial flow.

Aircraft without vertical tails rely on *yaw flaps* mounted near the wing tips to provide yaw moments for control and stability augmentation. The flaps are split surfaces whose trailing edges spread out to produce asymmetric drag with negligible lift. Deflecting the flaps on the left wing produces a negative yawing moment.

The net force produced by symmetric isolated wing flaps or spoilers is analogous to the elevator effect presented above. The principal distinction is that $l_h = 0$ in the ideal case, and we are interested in the net lift and drag rather than the pitching moment resulting from surface deflection. From eqs. 3.5-1 and 3.5-5, the lift and drag coefficients are

$$\begin{aligned} C_L &= C_N \cos(\alpha + \delta) \\ &= 2\left(\frac{S_s}{S}\right) \operatorname{sgn}(\alpha + \delta) \sin^2(\alpha + \delta) \cos(\alpha + \delta) \end{aligned} \quad (3.5\text{-}15)$$

$$\begin{aligned} C_D &= C_N \sin(\alpha + \delta) \\ &= 2\left(\frac{S_s}{S}\right) \operatorname{sgn}(\alpha + \delta) \sin^2(\alpha + \delta) \sin(\alpha + \delta) \end{aligned} \quad (3.5\text{-}16)$$

Substituting $(\alpha + \delta)$ for α in Fig. 2.5-1 and 2.5-3, it can be seen that the wing flaps and spoilers become significant drag control devices for deflections on the order of $|20 \text{ deg}|$ or more.

More generally, isolated control surface deflections create higher normal force, lift, and drag than Newtonian theory predicts as a consequence of induced circulation. The formulas of Section 2.4 apply to these surfaces, replacing α by θ in the calculation of control forces and subtracting the $\delta = 0$ deg solution to obtain the incremental effect.

Trailing-Edge Flaps

Control surfaces are usually appended to the trailing edges of larger surfaces (Fig. 3.5-6a), not only for reasons of structure, weight, and actuation but for aerodynamic efficiency [S-5, A-3]. For small deflections, the trailing-edge flap can be viewed as increasing or decreasing the camber of the larger airfoil section, thereby altering its lift. In subsonic flow, the deflected flap creates a pressure field that carries over to the primary surface, increasing the normal force per unit of flap deflection. Hence, it is reasonable to express the control effect as some proportion of the main surface's lift slope. For example, the lift and pitch effects of the elevator can be expressed as [M-1]

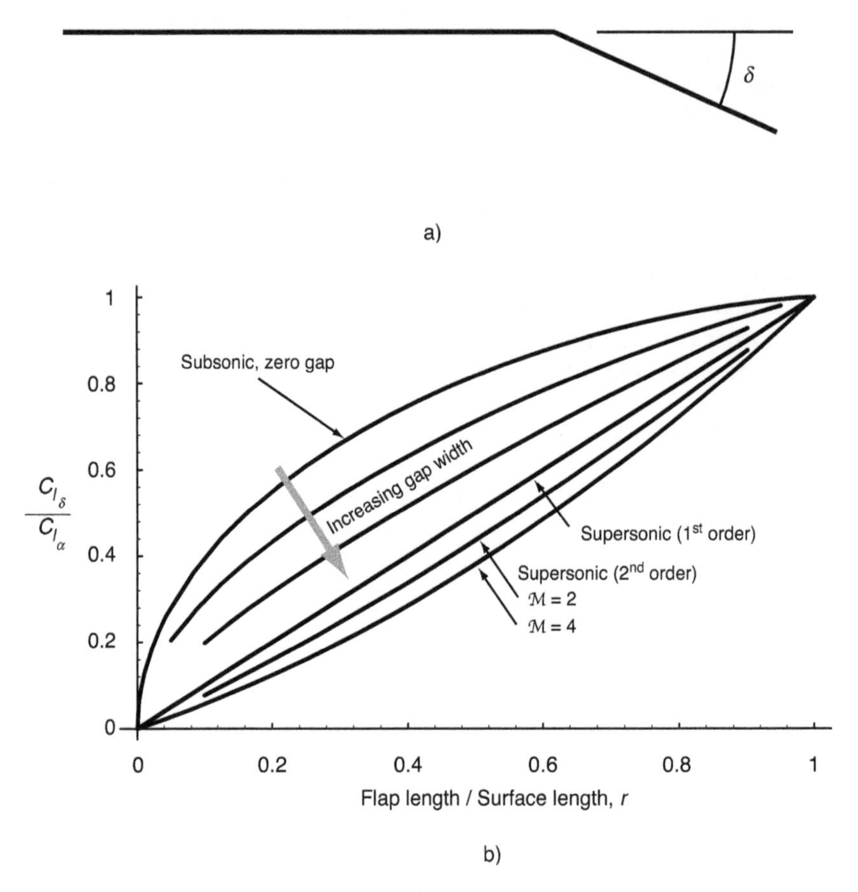

Fig. 3.5-6. Carryover effect of trailing edge flap in sub- and supersonic flow. (a) Trailing-edge flap. (b) Ratio of flap effectiveness to section lift slope.

$$C_{L_{\delta E}} = \tau_{ht} C_{L_{\alpha_{ht}}} = \tau_{ht} \eta_{ht} (C_{L_{\alpha_{ht}}})_{ht} \frac{S_{ht}}{S} \tag{3.5-17}$$

$$\begin{aligned} C_{m_{\delta E}} &= \tau_{ht} C_{L_{\alpha_{ht}}} = -\,\tau_{ht} \eta_{ht} (C_{L_{\alpha_{ht}}})_{ht} \frac{S_{ht} l_{ht}}{S\bar{c}} \\ &= -\,\tau_{ht} \eta_{ht} (C_{L_{\alpha_{ht}}})_{ht} V_{ht} \end{aligned} \tag{3.5-18}$$

where τ_{ht} is the carryover effect described below, η_{ht} is a tail efficiency factor, as in Section 2.4, S_{ht} is the horizontal tail area, and V_{ht} is the *horizontal tail volume.* Rudder effects can be related to the vertical tail's side force and yaw moment in the same way, defining equivalent parameters in terms of the vertical tail lift slope, area, distance from the center of mass, efficiency, and carryover factors, with b replacing $\bar{c}$ as a reference length.

First analyzed by Glauert [G-2], this *carryover effect* can be expressed by the ratio of the two-dimensional flap lift coefficient C_{l_δ} to the section lift slope coefficient C_{l_α} (Fig. 3.5-7b). (Here, the lower-case subscript "l" indicates a two-dimensional lift coefficient, not a rolling moment coefficient.) The subsonic *flap effectiveness* $C_{l_\delta}/C_{l_\alpha}$ is always larger than the ratio of flap chord to airfoil chord $r = c_f/c_a$, being given by [S-1]

$$\frac{C_{l_\delta}}{C_{l_\alpha}} = \frac{2}{\pi}[\sqrt{r(1-r)} + \sin^{-1}\sqrt{r}] \tag{3.5-19}$$

The carryover effect is reduced if there is a gap between the trailing-edge flap and the main surface [E-2]. It becomes negligible or even negative in supersonic flight, where thin-airfoil potential theory predicts no upstream pressure disturbances for flap deflection [M-6]. Airfoil thickness and boundary layer effects contribute to the negative supersonic carryover, and "shock stall" can reduce small-deflection effectiveness to zero [K-4]. A second-order theory due to Busemann accounts for thickness effects. Applied to a parabolic-arc airfoil, it produces [D-4, after B-7 and M-4]

$$\frac{C_{l_\delta}}{C_{l_\alpha}} = r\left[1 - 4\frac{c_2}{c_1}\frac{t}{c}(1-r)\right] \tag{3.5-20}$$

where

$$c_1 = \frac{2}{\sqrt{\mathcal{M}^2 - 1}} \tag{3.5-21}$$

$$c_2 = \frac{\gamma\mathcal{M}^4 + (\gamma\mathcal{M}^2 - 1)^2}{2(\gamma\mathcal{M}^2 - 1)^2} \tag{3.5-22}$$

γ is the ratio of specific heats (1.4), and t/c is the airfoil thickness ratio.

Strip theory can be used to extrapolate two-dimensional (2-D) flap effects to 3-D aileron effects. As an example, Seckel presents the following equation for the subsonic effectiveness $C_{l_{\delta A}}$ of ailerons extending from lateral stations $\pm y$ to the wing tips [S-4]:

$$C_{l_{\delta A}} = \left(\frac{C_{L\delta}}{C_{L\alpha}}\right)_{2-\mathrm{D}} \frac{C_{L\alpha}}{1+\lambda}\left\{\frac{1-k^2}{2} - \frac{1-k^3}{3}(1-\lambda)\right\} \tag{3.5-23}$$

where $k = y/(b/2)$ and λ is the wing's taper ratio.

Trailing-edge flaps are sometimes installed *on* flaps for one of two purposes. A small flap called a *tab* is deflected to produce a hinge moment on the control flap—in effect, controlling the control surface. This is a means of exerting aerodynamic leverage, as the force required to deflect the tab is typically smaller than the force required to deflect the control flap directly. For this application, the tab and control flap move in opposite directions: a trailing-edge-up tab forces the trailing edge of the control flap down. Conversely, a "tab" extending across the full span of the control flap could be deflected in the same direction, producing "variable camber," encouraging the flow to remain attached to higher deflection angles, but increasing the flap hinge moment. In this case, both tab and flap are driven directly.

The forces on control flaps produce hinge moments that affect surface positioning. The surface position depends not only on these hinge moments but on those produced by the control system, as discussed in Sections 5.5 and 6.5.

3.6 Solution of Nonlinear Differential Equations

Deriving state histories from nonlinear equations of motion requires the numerical solution of ordinary differential equations. In special cases, analytical solutions can be found, but in the general case, only numerical solutions are tractable. There are two functions to be performed: estimating forces and moments from tabulations of inertial, aerodynamic, thrust, and atmospheric data, and approximating integrals of differential equations [A-1]. In addition, it is important to have the capability of finding equilibrium ("trimmed") solutions to the differential equations without propagating them over time, so that simulations can be initialized at desired working points and further analyses of cruising flight can be conducted.

Numerical Algorithms for Integration

The equations of motion presented in Sections 3.2 and 3.3 can be written in the general vector form (eqs. 1.3-12 and 3.2-38)

$$\frac{d\mathbf{x}(t)}{dt} \triangleq \dot{\mathbf{x}}(t) = \mathbf{f}[\mathbf{x}(t), \mathbf{u}(t), \mathbf{w}(t), t] \tag{3.6-1}$$

where $\mathbf{f}[\cdot]$ is the n-dimensional vector function that expresses the effects of the motion variables $\mathbf{x}$, controls $\mathbf{u}$, disturbances $\mathbf{w}$, and time t on the state vector's rate of change $d\mathbf{x}/dt$. Our immediate goal is to use eq. 3.6-1 to find $\mathbf{x}(t)$ through some process of numerical integration; hence, we want to identify algorithms (or numerical procedures) that are suitable for evaluating the integral of eq. 3.6-1 in the interval between initial and final times (t_0, t_f). At some interior point t_k, the state is expressed as

$$\mathbf{x}(t_k) = \mathbf{x}(t_o) + \int_{t_0}^{t_k} \mathbf{f}[\mathbf{x}(t), \mathbf{u}(t), \mathbf{w}(t), t]dt \tag{3.6-2}$$

where $\mathbf{x}(t_0)$ is the initial value of the state, and $\mathbf{u}(t)$ and $\mathbf{w}(t)$ are specified in (t_0, t_k).

The state at a single interior point is rarely required; more often, a sequence of states $\{\mathbf{x}(t_k), k=1, k_f\}$ is sought, with the sampling index k_f corresponding to the final time t_f. This sequence approaches the continuous distribution $\mathbf{x}(t)$ in $[t_0, t_f]$ as the number of samples approaches infinity and the interval between successive points approaches zero. Consequently, it is useful to generate $\mathbf{x}(t_k)$ recursively, beginning at t_0 and stepping forward until $\mathbf{x}(t_k)$ is reached:

$$\mathbf{x}(t_k) = \mathbf{x}(t_{k-1}) + \int_{t_{k-1}}^{t_k} \mathbf{f}[\mathbf{x}(t), \mathbf{u}(t), \mathbf{w}(t), t]dt \tag{3.6-3}$$

Unless $\mathbf{f}[\cdot]$ is a special function, it is not possible to evaluate the integral exactly, so the integration algorithm provides an approximate solution. Shortening the interval over which the integral is evaluated from (t_0, t_k) to $\Delta t = (t_{k-1}, t_k)$ assures that good numerical accuracy can be obtained with relatively simple computations.

The simplest approximation to eq. 3.6-3 is *rectangular integration* or *Euler's method* [R-1]:

$$\begin{aligned}\mathbf{x}(t_k) &= \mathbf{x}(t_{k-1}) + \mathbf{f}[\mathbf{x}(t_{k-1}), \mathbf{u}(t_{k-1}), \mathbf{w}(t_{k-1}), t_{k-1}](t_k - t_{k-1}) \\ &= \mathbf{x}(t_{k-1}) + \mathbf{f}[\mathbf{x}(t_{k-1}), \mathbf{u}(t_{k-1}), \mathbf{w}(t_{k-1}), t_{k-1}]\Delta t \\ &\triangleq \mathbf{x}(t_{k-1}) + \Delta\mathbf{x}(t_{k-1}, t_k)\end{aligned} \tag{3.6-4}$$

The function $\mathbf{f}[\cdot]$ is evaluated at time t_{k-1} and multiplied by the time increment Δt to estimate the change in the state $\Delta\mathbf{x}(t_{k-1}, t_k)$ that occurs during the interval. In practice, the integration accuracy provided by Euler integration is low, although it may be good enough for preliminary studies or for "real-time" computation (as in cockpit simulation or integration of dynamic equations that form part of an aircraft's feedback control system), where computational speed is placed at a premium.

During the time interval (t_{k-1}, t_k), $\mathbf{f}[\cdot]$ may be changing; this change is taken into account using *trapezoidal integration* (also called *modified Euler integration*). The idea is to make a first estimate of $\mathbf{x}(t_k)$ using eq. 3.6-4, evaluate $\mathbf{f}[\cdot]$ again at this point, and average the two results:

$$\mathbf{x}_1(t_k) = \mathbf{x}(t_{k-1}) + \Delta\mathbf{x}_1(t_{k-1}, t_k) \tag{3.6-5}$$

$$\Delta\mathbf{x}_2\,(t_{k-1}, t_k) = \mathbf{f}[\mathbf{x}_1(t_k), \mathbf{u}(t_k), \mathbf{w}(t_k), t_k]\;\Delta t \tag{3.6-6}$$

$$\mathbf{x}_2(t_k) \triangleq \mathbf{x}(t_k) = \mathbf{x}(t_{k-1}) + \frac{1}{2}[\Delta\mathbf{x}_1(t_{k-1}, t_k) + \Delta\mathbf{x}_2(t_{k-1}, t_k)] \tag{3.6-7}$$

Two function evaluations are required, effectively doubling the amount of computation needed to go from t_{k-1} to t_k. MATLAB's `ode23` integration function is based on this algorithm, with variable-time-step logic for error correction [A-6]. Equation 3.6-6 assumes that **u** and **w** are known at t_k; in real-time simulations, these control and disturbance inputs might not be available in time for the computation, making it necessary to use "stale" values, such as $\mathbf{u}(t_{k-1})$ and $\mathbf{w}(t_{k-1})$.

We may wish to improve the speed or accuracy of the numerical integration. One approach is to use a *predictor-corrector algorithm*, which predicts the state at the next point by extrapolating past estimates, then computes a single function evaluation to correct the result [H-8]. As an example, consider the following variation on the modified Euler method, assuming equal time intervals (t_{k-2}, t_{k-1}) and (t_{k-1}, t_k):

Predictor

$$\begin{aligned}\mathbf{x}_1(t_k) &= \mathbf{x}(t_{k-1}) + \Delta\mathbf{x}_1(t_{k-1}, t_k) = \mathbf{x}(t_{k-1}) + \frac{d\mathbf{x}(t_{k-1})}{dt}\Delta t \\ &\approx \mathbf{x}(t_{k-1}) + [\mathbf{x}(t_{k-1}) - \mathbf{x}(t_{k-2})] = 2\,\mathbf{x}(t_{k-1}) - \mathbf{x}(t_{k-2})\end{aligned} \tag{3.6-8}$$

Corrector

$$\Delta\mathbf{x}_2(t_{k-1}, t_k) = \mathbf{f}[\mathbf{x}_1(t_k), \mathbf{u}(t_k), \mathbf{w}(t_k), t_k]\;\Delta t \tag{3.6-9}$$

$$\mathbf{x}_2(t_k) \triangleq \mathbf{x}(t_k) = \mathbf{x}(t_{k-1}) + \frac{1}{2}[\Delta\mathbf{x}_1(t_{k-1}, t_k) + \Delta\mathbf{x}_2(t_{k-1}, t_k)] \tag{3.6-10a}$$

$$= \mathbf{x}_1(t_k) + \frac{1}{2}\left\{\mathbf{f}[\mathbf{x}_1(t_k), \mathbf{u}(t_k), \mathbf{w}(t_k), t_k]\Delta t - \Delta\mathbf{x}_1(t_{k-1}, t_k)\right\} \tag{3.6-10b}$$

If the state change is relatively slow and smooth, this algorithm takes advantage of past computations to project the state ahead, averaging that essentially geometric result with the dynamic calculation at the projected future time. Equation 3.6-10b expresses the corrective nature of the algorithm, adding the *difference* between the dynamic and projected state increments to the predicted value.

Note that the predictor-corrector algorithm is not "self-starting" because it requires the value $\mathbf{x}(t_{k-2})$ in eq. 3.6-8 (which is not available when $k=1$). It is necessary to take at least one beginning integration step using a "memoryless" routine like Euler or modified Euler integration to get started. Higher-degree extrapolations can be made, involving even earlier values of $\mathbf{x}$; however, the polynomial prediction that results has limited utility for flight simulation if high-frequency (i.e., rapidly changing) command or disturbance inputs have significant effect.

The *Adams-Bashforth algorithm* is an efficient approach to numerical integration that requires just one new function evaluation at each step. This method derives from a series expansion for the integral [D-2]; like the previous method, it requires prior estimates of $\mathbf{x}$, so the method is not self-starting. To first order, it can be expressed as

$$\mathbf{x}(t_k) = \mathbf{x}(t_{k-1}) + \{1.5\ \mathbf{f}[\mathbf{x}(t_{k-1}), \mathbf{u}(t_{k-1}), \mathbf{w}(t_{k-1}), t_{k-1}] \\ - 0.5\ \mathbf{f}[\mathbf{x}(t_{k-2}), \mathbf{u}(t_{k-2}), \mathbf{w}(t_{k-2}), t_{k-2}]\}\ \Delta t \qquad (3.6\text{-}11)$$

MATLAB's `ode13` is a modified Adams-Bashforth algorithm [A-6].

Runge-Kutta (R-K) algorithms are self-starting, and they offer improved accuracy at the expense of additional function evaluations. pth-order R-K integrators require p function evaluations, although more evaluations may be required in connection with adaptive step-size control. The principal idea is that $\mathbf{f}[\cdot]$ is evaluated not only at the extremes of the time interval but within the interval, providing an improved curve fit to $\mathbf{f}[\cdot]$ during that time. The fourth-order R-K algorithm is widely used in scientific analysis, as it makes a good compromise between accuracy and speed, and it is robust. The algorithm is

$$\Delta\mathbf{x}_1 = \mathbf{f}[\mathbf{x}(t_{k-1}), \mathbf{u}(t_{k-1}), \mathbf{w}(t_{k-1}), t_{k-1}]\ \Delta t \qquad (3.6\text{-}12)$$

$$\Delta\mathbf{x}_2 = \mathbf{f}\{[\mathbf{x}(t_{k-1}) + \Delta\mathbf{x}_1/2], \mathbf{u}(t_{k-1/2}), \mathbf{w}(t_{k-1/2}), t_{k-1/2}]\ \Delta t \qquad (3.6\text{-}13)$$

$$\Delta\mathbf{x}_3 = \mathbf{f}\{[\mathbf{x}(t_{k-1}) + \Delta\mathbf{x}_2/2], \mathbf{u}(t_{k-1/2}), \mathbf{w}(t_{k-1/2}), t_{k-1/2}]\ \Delta t \qquad (3.6\text{-}14)$$

$$\Delta\mathbf{x}_4 = \mathbf{f}\{[\mathbf{x}(t_{k-1}) + \Delta\mathbf{x}_3], \mathbf{u}(t_k), \mathbf{w}(t_k), t_k]\ \Delta t \qquad (3.6\text{-}15)$$

$$\mathbf{x}(t_k) = \mathbf{x}(t_{k-1}) + \frac{1}{6}[\Delta\mathbf{x}_1 + 2\Delta\mathbf{x}_2 + 2\Delta\mathbf{x}_3 + \Delta\mathbf{x}_4] \qquad (3.6\text{-}16)$$

where $t_{k-1/2}$ represents the mid-point of the time interval. MATLAB's `ode45` is a fourth-order R-K algorithm that achieves fifth-order accuracy by variable time stepping (see below). Rectangular and trapezoidal integrators are examples of first- and second-order R-K algorithms. Higher-order Runge-Kutta algorithms are described in [B-2, P-6, R-1].

Computational efficiency can be improved by varying the integration time interval. During quiescent periods, longer time steps can be taken without losing accuracy, while in the vicinity of abrupt changes, shorter steps are required. Step-adjustment procedures typically evaluate a test function (e.g., a weighted sum of the absolute values of state increments), doubling or halving the time interval to keep state increments small:

$$\Delta t = \begin{cases} \Delta t/2, & X > X_{max} \\ \Delta t, & X_{min} < X < X_{max} \\ 2\Delta t, & X < X_{min} \end{cases} \tag{3.6-17}$$

where

$$X \triangleq \sum_{i=1}^{n} a_i \, |\Delta x_i| \tag{3.6-18}$$

X_{min}, X_{max}, and the coefficients a_i are determined empirically.

The *state history* $\mathbf{x}(t_k)$, $0 \leq k \leq k_f$, may drive cockpit simulator displays [B-1], or it may be presented in various analytical formats. The most direct format is a tabulation or plot of $\mathbf{x}(t_k)$ using time as the independent variable. Alternatively, an element of the state, $x_i(t_k)$, or some combination of the elements, may be chosen as the independent variable for the tabulation $\mathbf{x}[x_i(t_k)]$, $0 \leq k \leq k_f$. For example, range along the runway centerline or path length in the approach pattern could be useful for examining aircraft landing. The dynamic interactions of two elements of the state can be portrayed in *phase-plane plots* $x_j[x_i(t_k)]$, while interactions of three elements can be shown in three-dimensional (oblique view) *state-space plots* of $\{x_a[x_i(t_k)], x_b[x_i(t_k)]\}$.

Example 3.6-1. Numerical Integration of Simple Functions

Two simple examples demonstrate the sorts of errors that can occur in numerical integration. We know that the integral of cos t is sin t. Figure 3.6-1 compares the exact integral over one period with the results of rectangular and trapezoidal integration with time steps of one-twentieth of the period. The rectangular integration overpredicts during the first half of the cycle and underpredicts during the second, with the errors canceling by the end of the period. The trapezoidal integration matches the exact result closely. In the second case, the integrand is 1 for t in (0, 10), −10 for t in (10, 11), and zero thereafter. The results are shown in Fig. 3.6-2. The exact integral is a sawtooth. Both numerical estimates are exact during the buildup but overpredict during the rapid falloff, and they retain bias

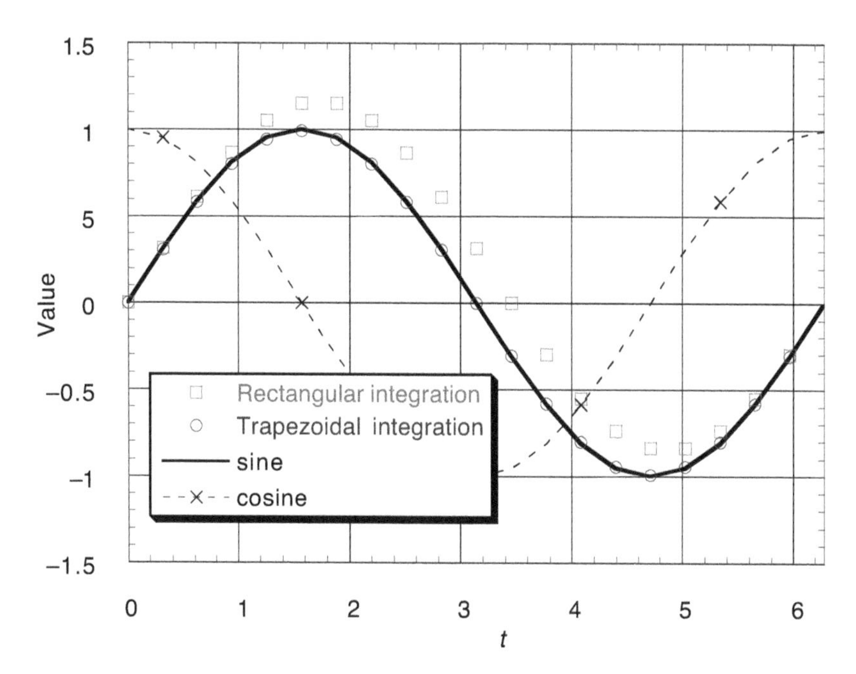

Fig. 3.6-1. Exact integral of cos t and numerical approximations.

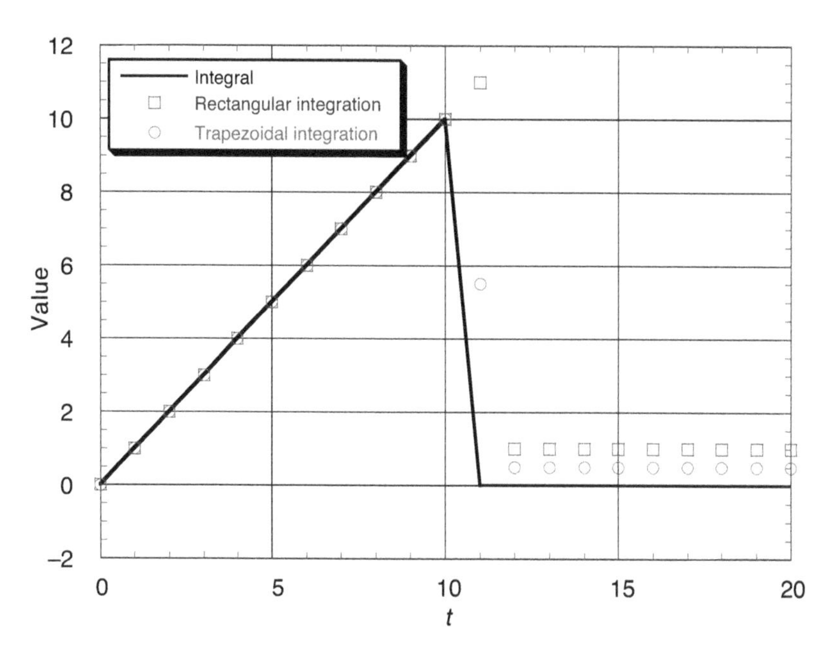

Fig. 3.6-2. Exact integral of piecewise-constant integrand and numerical approximations.

errors when the input is zero. Overall, the trapezoidal integrator is more accurate than the rectangular integrator.

Equations of Motion

The vector notation of the previous section tends to overstate the simplicity of numerical integration, as function evaluation can be an involved process. For the flat-earth, symmetric-aircraft example given in Section 3.2, the vector function of eq. 3.6-1 represents the following 12 scalar equations of motion:

$$\dot{u} = X/m + g_x + rv - qw \tag{3.6-19}$$

$$\dot{v} = Y/m + g_y - ru + pw \tag{3.6-20}$$

$$\dot{w} = Z/m + g_z + qu - pv \tag{3.6-21}$$

$$\dot{p} = (I_{zz}L + I_{xz}N - \{I_{xz}(I_{yy} - I_{xx} - I_{zz})p + [I_{xz}^2 + I_{zz}(I_{zz} - I_{yy})]r\}q) \div (I_{xx}I_{zz} - I_{xz}^2) \tag{3.6-22}$$

$$\dot{q} = [M - (I_{xx} - I_{zz})pr - I_{xz}(p^2 - r^2)] \div I_{yy} \tag{3.6-23}$$

$$\dot{r} = (I_{xz}L + I_{xx}N + \{I_{xz}(I_{yy} - I_{xx} - I_{zz})r + [I_{xz}^2 + I_{xx}(I_{xx} - I_{yy})]p\}q) \div (I_{xx}I_{zz} - I_{xz}^2) \tag{3.6-24}$$

$$\dot{x}_I = (\cos\theta\cos\psi)u + (-\cos\phi\sin\psi + \sin\phi\sin\theta\cos\psi)v + (\sin\phi\sin\psi + \cos\phi\sin\theta\cos\psi)w \tag{3.6-25}$$

$$\dot{y}_I = (\cos\theta\sin\psi)u + (\cos\phi\cos\psi + \sin\phi\sin\theta\sin\psi)v + (-\sin\phi\cos\psi + \cos\phi\sin\theta\sin\psi)w \tag{3.6-26}$$

$$\dot{z}_I = (-\sin\theta)u + (\sin\phi\cos\theta)v + (\cos\phi\cos\theta)w \tag{3.6-27}$$

$$\dot{\phi} = p + (q\sin\phi + r\cos\phi)\tan\theta \tag{3.6-28}$$

$$\dot{\theta} = q\cos\phi - r\sin\phi \tag{3.6-29}$$

$$\dot{\psi} = (q\sin\phi + r\cos\phi)\sec\theta \tag{3.6-30}$$

where

$$\begin{bmatrix} g_x \\ g_y \\ g_z \end{bmatrix} = \begin{bmatrix} -g\sin\theta \\ g\sin\phi\cos\theta \\ g\cos\phi\cos\theta \end{bmatrix} \tag{3.6-31}$$

Applied forces (X, Y, Z) and moments (L, M, N) are functions of airspeed and aerodynamic angles,

$$\begin{bmatrix} V_A \\ \beta_A \\ \alpha_A \end{bmatrix} = \begin{bmatrix} \sqrt{u_A^2 + v_A^2 + w_A^2} \\ \sin^{-1}(v_A/V_A) \\ \tan^{-1}(w_A/u_A) \end{bmatrix} \tag{3.6-32}$$

and the air-relative body-axis velocity components must account for the wind field,

$$\begin{bmatrix} u_A \\ v_A \\ w_A \end{bmatrix} = \begin{bmatrix} u \\ v \\ w \end{bmatrix} - \mathbf{H}_I^B \mathbf{w} \tag{3.6-33}$$

with $\mathbf{H}_I^B$ and $\mathbf{w}$ as defined previously. The body-axis forces and moments are expressed as sums of aerodynamic and thrusting effects:

$$X = C_X \bar{q} S + T_x \tag{3.6-34}$$

$$Y = C_Y \bar{q} S + T_y \tag{3.6-35}$$

$$Z = C_Z \bar{q} S + T_z \tag{3.6-36}$$

$$L = C_l \bar{q} S b + M_{T_x} \tag{3.6-37}$$

$$M = C_m \bar{q} S \bar{c} + M_{T_y} \tag{3.6-38}$$

$$N = C_n \bar{q} S b + M_{T_z} \tag{3.6-39}$$

The dynamic pressure is

$$\bar{q} \triangleq \frac{\rho(z) V_A^2}{2} \tag{3.6-40}$$

and the air density $\rho(z)$ is a function of altitude. The Mach number, needed below, is defined as

$$\mathcal{M} = \frac{V_A}{a(z)} \tag{3.6-41}$$

where the speed of sound $a(z)$ is a function of altitude.

Equations 3.6-19–3.6-41 apply to all aircraft; the distinctions between mathematical models of different aircraft come in the expressions for inertial, aerodynamic, and thrust effects. While some of these effects may be modeled by closed-form analytic expressions, high-fidelity data for modern aircraft are usually described by multivariate tables, and it may be necessary to transform data from one coordinate frame to another (e.g., wind-axis force coefficients to body-axis coefficients). The complexity of these operations depends on the level of detail required in the simulation, as well as the flight envelope within which the simulation takes place.

Representation of Data

An aerodynamic coefficient is normally modeled as the sum of several effects, each of which may vary with flight conditions. For example, suppose the pitching-moment coefficient C_m of eq. 3.6-38 is a function of center-of-mass location, Mach number, angle of attack, pitch rate, angle-of-attack rate, and elevator deflection. It could be expressed as

$$\begin{aligned} C_m &= C_m(x_{cm}, \mathcal{M}, \alpha, q, \dot{\alpha}, \delta E) \\ &= C_m(\mathcal{M}, \alpha) + C_z(\mathcal{M}, \alpha, q, \dot{\alpha}, \delta E)\frac{x_{cm} - x_{cm_O}}{\bar{c}} \\ &\quad + C_{m_q}(\mathcal{M}, \alpha)\frac{q\bar{c}}{2V_A} + C_{m_{\dot{\alpha}}}(\mathcal{M}, \alpha)\frac{\dot{\alpha}\bar{c}}{2V_A} + C_{m_{\delta E}}(\mathcal{M}, \alpha)\delta E \end{aligned} \quad (3.6\text{-}42)$$

$C_m(\mathcal{M}, \alpha)$ is the underlying static moment coefficient for the nominal center-of-mass location x_{cm_0}, the second term conveys the effect of the normal force when the center of mass is displaced from its nominal position, and the remaining three terms express linear sensitivities to angular rates and control. All four components of C_m are assumed to be two-dimensional functions of Mach number and angle of attack, while C_Z is computed by an expression similar to eq. 3.6-42. The static moment and sensitivity coefficients are derived from experiment or computational analysis, and the form of the aerodynamic model must conform to the available data. Other significant effects, like moments due to flap deflection and aeroelastic deformations, can be added to the model as required.

Given a table such as $\{C_m(\mathcal{M}_i, \alpha_j), i = 1–I, j = 1, J\}$, values of $C_m(\mathcal{M}, \alpha)$ between the tabulated points must be generated during simulation. This can be done either by fitting the data with a function or by interpolating between the tabulated points. The former approach has the advantage of modeling the entire data set by a single expression (e.g., one polynomial or computational neural network), while the latter may be more expeditious, particularly for multidimensional data sets.

Consider the generation of a power series for a function of a single variable $z = f(x)$, given the table $\{z_i = f(x_i), i = 1, I\}$. In principle, z can be reproduced precisely at the I points by an $(I - 1)$th-degree polynomial

$$\hat{z} = c_1 + c_2x + c_3x^2 + \cdots + c_Ix^{I-1} \quad (3.6\text{-}43)$$

where the series coefficients are contained in a vector **c**. Defining the vector $\hat{\mathbf{z}}$ by its I elements $\hat{\mathbf{z}}_i$ and the vector **c** by the I polynomial coefficients c_i, eq. 3.6-43 leads to

$$\hat{\mathbf{z}} = \mathbf{A}\mathbf{c} \tag{3.6-44}$$

where **A** contains powers of the table's x_i values:

$$\mathbf{A} \triangleq \begin{bmatrix} 1 & x_1 & x_1^2 & \ldots & x_1^{I-1} \\ 1 & x_2 & x_2^2 & \ldots & x_2^{I-1} \\ \ldots & \ldots & \ldots & \ldots & \ldots \\ 1 & x_I & x_I^2 & \ldots & x_I^{I-1} \end{bmatrix} \tag{3.6-45}$$

The polynomial coefficients can be computed from the inverse,

$$\mathbf{c} = \mathbf{A}^{-1}\,\mathbf{z} \tag{3.6-46}$$

where **z** contains the tabulated points.

The same approach can be applied to functions of two (or more) variables, $z = f(x, y)$ by fitting the polynomial

$$\begin{aligned} \hat{z} &= c_{11} + c_{12}x + c_{13}x^2 + \cdots + c_{1I}x^{I-1} + c_{21}y + c_{22}xy + c_{23}x^2y \\ &\quad + \cdots + c_{2I}x^{I-1}y + c_{31}y^2 + \cdots + c_{IJ}x^{I-1}y^{J-1} \end{aligned} \tag{3.6-47}$$

Equations 3.6-43–3.6-45 then apply with the IJ-dimensional vector **z** defined by the data points, **c** defined as a vector of the IJ coefficients, and **A** defined accordingly.

Series representation of the tabulated points assures continuity throughout the range of x, and derivatives of $\hat{z}$ with respect to x, (e.g., $\partial\hat{z}/\partial x$ and $\partial^2\hat{z}/\partial x^2$) are easily defined by symbolic manipulation of eq. 3.6-43; however, there are a number of problems with this approach. All the terms of eq. 3.6-43 must be computed for each function evaluation, and the number of polynomial terms grows with the number of tabulated points; for example, a (10×10) table requires 100 coefficients and terms. Numerical accuracy places a practical limit on the number of terms in the series: every evaluation is affected by the precision with which all the coefficients have been computed, as well as by the precision of the polynomial multiplications during evaluation. All the coefficients depend on all the tabulated data; therefore, the "support" for individual function evaluations is distributed across all points. There is no guarantee that the resulting polynomial is well behaved between tabulated points; computed values of $\hat{z}$ may take wild excursions from reasonable curve fairings, as a consequence of small variations in the data.

One solution to these problems is to define a lower-degree least-squares approximation to the data set. This approach is explained using the table $\{z_i = f(x_i),\ i = 1, I\}$. Given I tabulated points, a polynomial of degree $K-1$ $(<I-1)$ is defined:

$$\hat{z} = c_1 + c_2 x + c_3 x^2 + \cdots + c_K x^{K-1} \tag{3.6-48}$$

It is still possible to define an equation of the form of eq. 3.6-44, but $\mathbf{c}$ has lower dimension than $\hat{\mathbf{z}}$; hence $\mathbf{A}$ must be nonsquare; therefore, it is not directly invertible. A minimum-square-error (or "least-squares") fit to the data results from choosing $\mathbf{c}$ to minimize the quadratic fit error,

$$J \triangleq (\mathbf{z} - \hat{\mathbf{z}})^T (\mathbf{z} - \hat{\mathbf{z}}) = (\mathbf{z} - \mathbf{A}\mathbf{c})^T (\mathbf{z} - \mathbf{A}\mathbf{c}) \tag{3.6-49}$$

The minimizing solution is specified by $\partial J/\partial \mathbf{c} = 0$, which occurs when

$$\mathbf{c} = (\mathbf{A}^T \mathbf{A})^{-1} \mathbf{A}^T \mathbf{z} \tag{3.6-50}$$

The polynomial fit (eq. 3.6-48) passes close to but not through the original data points. If a closer fit near certain points is required, a *weighted-least-squares curve fit* can be used instead. Greater weight is assigned to the fit error at selected points using the diagonal weighting matrix $\mathbf{R}$. The cost function is expressed as

$$J \triangleq (\mathbf{z} - \mathbf{A}\mathbf{c})^T \mathbf{R} (\mathbf{z} - \mathbf{A}\mathbf{c}) \tag{3.6-51}$$

and its minimizing solution is

$$\mathbf{c} = (\mathbf{A}^T \mathbf{R}\, \mathbf{A})^{-1} \mathbf{A}^T \mathbf{R}\, \mathbf{z} \tag{3.6-52}$$

$(\mathbf{A}^T \mathbf{A})^{-1} \mathbf{A}^T$ is the *left pseudoinverse* of $\mathbf{A}$, while $(\mathbf{A}^T \mathbf{R}\, \mathbf{A})^{-1} \mathbf{A}^T \mathbf{R}$ is its *weighted left pseudoinverse.*

Table lookup provides an alternative estimate that is guaranteed to pass through all the tabulated data and that employs fewer coefficients in function evaluation. Given the table $\{z_i = f(x_i),\ i = 1, I\}$, the value of z at a point x in the interval (x_i, x_{i+1}) can be estimated using *linear interpolation*,

$$\begin{aligned} \hat{z}(x) &= f(x_i) + \frac{f(x_{i+1}) - f(x_i)}{x_{i+1} - x_i}(x - x_i) \\ &\triangleq c_1 + c_2 \Delta x \end{aligned} \tag{3.6-53}$$

The coefficient c_2 is called a *first-order divided difference* of $f(x)$. Linear interpolation provides function evaluation with *minimum support*, as it depends only on the bracketing tabulated points. This interpolation can be recognized as eq. 3.6-43, with $\Delta x = (x - x_i)$ replacing x and $I = 2$. The coefficients for each of the $(I-1)$ intervals are obtained directly from the table or by evaluating eq. 3.6-46, with $z^T \triangleq [f(x_i) f(x_{i+1})]$ and $\Delta x = (x - x_i)$ replacing x in eq. 3.6-45. The interpolation proceeds in two steps: the first is a search to determine the interval (x_i, x_{i+1}) containing x, and the second is calculation of eq. 3.6-53 with the appropriate values of c_1 and c_2. Equation 3.6-53 can be evaluated using I stored (x_i, z_i) doublets or from $I-1$ (x_i, c_1, c_2) triplets, the latter allowing faster computation.

Linear interpolation can be extended to functions of two or more variables. Given the two-dimensional table $\{z_{jk} = f(x_j, y_k), j = 1, J, k = 1, K\}$, the interpolation takes the bilinear form

$$\hat{z}(x) = c_1 + c_2 \Delta x + c_3 \Delta y + c_4 \Delta x \Delta y \tag{3.6-54}$$

where $\Delta x = (x - x_j)$, $\Delta y = (y - y_k)$, and the c_i can be obtained from eq. 3.6-46, with $I = 4$ and $z^T \triangleq [f(x_j, y_k)\ f(x_{j+1}, y_k)\ f(x_j, y_{k+1})\ f(x_{j+1}, y_{k+1})]$. Alternatively, the recurrence relations of *Newton's interpolation method* [P-5] can be used to define the coefficients; with constant increments between points in both x and y directions, the coefficients are

$$c_1 = f(x_j, y_k) \triangleq z_{jk} \tag{3.6-55}$$

$$c_2 = \frac{(z_{j+1,k} - z_{jk})}{(x_{j+1,k} - x_{jk})} \tag{3.6-56}$$

$$c_3 = \frac{(z_{j,k+1} - z_{jk})}{(y_{j,k+1} - y_{jk})} \tag{3.6-57}$$

$$c_4 = \frac{(z_{j+1,k+1} - z_{j,k+1}) - (z_{j+1,k} - z_{jk})}{(x_{j+1,k} - x_{jk})(y_{j,k+1} - x_{jk})} \tag{3.6-58}$$

A two-dimensional search is required to find the proper x and y intervals.

One-dimensional *cubic interpolation* for x in $(x_i, x_i + 1)$ is defined in the same way as Newton interpolation, choosing $I = 4$ and $z^T \triangleq [f(x_{i-1})\ f(x_i)\ f(x_{i+1})\ f(x_{i+2})]$. Then,

$$\hat{z}(x) = c_1 + c_2 \Delta x + c_3 \Delta x^2 + c_4 \Delta x^3 \tag{3.6-59}$$

where the c_i are specific to the interval $(i, i+1)$. The cubic formula provides *compact support* for the interpolation: function evaluation depends only on tabulated points in the neighborhood of x rather than on the entire data set. Nevertheless, the support region extends beyond the bracketing region. When $i = 1$ or $I-1$, the coefficients for $i = 2$ or $I-2$ are used, and Δx is defined accordingly. $I-3$ sets of (x_i, c_{1-4}) must be stored for evaluation. Cubic interpolation can be extended to *bicubic interpolation* for functions of two variables, following eq. 3.6-47.

Linear interpolation is simple to compute and provides a conservative estimate of the function between tabulated points; however, the slope of

$\hat{z}(x)$ is discontinuous across tabulated points. The effect may not be noticeable in a flight-path simulation, but it can cause great difficulty in trajectory optimization or stability analysis, where smooth, accurate gradients of nonlinear functions must be generated. Cubic interpolation reduces (but does not eliminate) the discontinuity across tabulated points, as external points are considered in the fairing. It provides a smoothly varying function with continuous second derivatives in the interval $(i, i+1)$, although wide swings can occur in the interpolation if the progression in $f(x_i)$ is not relatively smooth or the x_i are not evenly distributed.

Interpolation discontinuities are eliminated if *B-splines* (or "basis" splines) are used for data representation. B-splines are truncated polynomials defined over intervals of the data set; the function is represented by a weighted linear combination of B-splines in the neighborhood of x. Following [C-4], we consider interpolation of the table $\{z_i = f(x_i), i = 1, I\}$, where

$$x_{min} \leq x_1 < x_2 < \cdots < x_I \leq x_{max} \tag{3.6-60}$$

The interpolation is based on nth-order splines that are expressed with reference to N *interior knots* λ_i on the same interval,

$$x_{min} < \lambda_1 \leq \lambda_2 \leq \cdots \leq \lambda_N < x_{max} \tag{3.6-61}$$

where $N \triangleq I - n$. The *degree* of a spline is one less than the *order* of the spline; thus a cubic B-spline has order $n = 4$, and it has continuous derivatives up to order $n - 2 = 2$. Choosing two or more knots to be coincident (e.g., $\lambda_2 = \lambda_3$) purposely introduces discontinuity in the function or its derivatives. The x_i and λ_i are different in number and need not be coincident; however, for $n = 2k$ (i.e., if n is an even number), coefficient solutions are aided by choosing

$$\lambda_j = x_j + k, \quad j = 1, \cdots, N \tag{3.6-62}$$

B-splines are defined by a recurrence relation that expresses splines of order n in terms of $(n-1)$th-order splines [C-4]. The recursion begins with the first-order spline $N_{1,j}(x)$:

$$N_{1,j}(x) = \begin{cases} 1, & x \text{ in } (\lambda_{j-1}, \lambda_j) \\ 0, & x \text{ not in } (\lambda_{j-1}, \lambda_j) \end{cases} \tag{3.6-63}$$

The higher-order splines are then defined as

$$N_{n,j}(x) = \left(\frac{x - \lambda_{j-n}}{\lambda_{j-1} - \lambda_{j-n}} \right) N_{n-1,j-1}(x) + \left(\frac{\lambda_j - x}{\lambda_j - \lambda_{j-n+1}} \right) N_{n-1,j}(x) \tag{3.6-64}$$

To implement a cubic B-spline, the recursion continues until $n = 4$:

$$
\begin{array}{llll}
N_{1,j}(x) \rightarrow N_{2,j}(x) & \rightarrow N_{3,j}(x) & \rightarrow N_{4,j}(x) \\
& N_{2,j+1}(x) \rightarrow N_{3,j+1}(x) & \rightarrow N_{4,j+1}(x) \\
& & N_{3,j+2}(x) \rightarrow N_{4,j+2}(x) \\
& & & N_{4,j+3}(x)
\end{array}
$$

As an example, suppose that x lies between λ_3 and λ_4. Among the first-order splines, only $N_{1,4}(x)$ is nonzero, and it equals 1. There are two nonzero second-order splines:

$$N_{2,4}(x) = \frac{\lambda_4 - x}{\lambda_4 - \lambda_3} \tag{3.6-65}$$

$$N_{2,5}(x) = \frac{x - \lambda_3}{\lambda_4 - \lambda_3} \tag{3.6-66}$$

The three nonzero third-order splines are calculated from eq. 3.6-64, and the four nonzero fourth-order splines are

$$N_{4,4}(x) = \left(\frac{\lambda_4 - x}{\lambda_4 - \lambda_1}\right)\left(\frac{\lambda_4 - x}{\lambda_4 - \lambda_2}\right)\left(\frac{\lambda_4 - x}{\lambda_4 - \lambda_3}\right) \tag{3.6-67}$$

$$\begin{aligned} N_{4,5}(x) = &\left(\frac{x - \lambda_1}{\lambda_4 - \lambda_1}\right)\left(\frac{\lambda_4 - x}{\lambda_4 - \lambda_2}\right)\left(\frac{\lambda_4 - x}{\lambda_4 - \lambda_3}\right) \\ &+ \left(\frac{\lambda_5 - x}{\lambda_5 - \lambda_2}\right)\left[\left(\frac{x - \lambda_2}{\lambda_4 - \lambda_2}\right)\left(\frac{\lambda_4 - x}{\lambda_4 - \lambda_3}\right) + \left(\frac{\lambda_5 - x}{\lambda_5 - \lambda_3}\right)\left(\frac{x - \lambda_3}{\lambda_4 - \lambda_3}\right)\right] \end{aligned} \tag{3.6-68}$$

$$\begin{aligned} N_{4,6}(x) = &\left(\frac{x - \lambda_2}{\lambda_5 - \lambda_2}\right)\left[\left(\frac{x - \lambda_2}{\lambda_4 - \lambda_2}\right)\left(\frac{\lambda_4 - x}{\lambda_4 - \lambda_3}\right) + \left(\frac{\lambda_5 - x}{\lambda_5 - \lambda_3}\right)\left(\frac{x - \lambda_3}{\lambda_4 - \lambda_3}\right)\right] \\ &+ \left(\frac{\lambda_6 - x}{\lambda_6 - \lambda_3}\right)\left(\frac{x - \lambda_3}{\lambda_5 - \lambda_3}\right)\left(\frac{x - \lambda_3}{\lambda_4 - \lambda_3}\right) \end{aligned} \tag{3.6-69}$$

$$N_{4,7}(x) = \left(\frac{x - \lambda_3}{\lambda_6 - \lambda_3}\right)\left(\frac{x - \lambda_3}{\lambda_5 - \lambda_3}\right)\left(\frac{x - \lambda_3}{\lambda_4 - \lambda_3}\right) \tag{3.6-70}$$

The cubic B-spline interpolation equation is

$$\hat{z} = \sum_{j=1}^{I} c_j N_{4,j}(x) \tag{3.6-71}$$

where the coefficients c_j remain to be determined; however, only four of the terms in the summation are nonzero for any given value of x. For the example above,

$$\hat{z} = c_4 N_{4,4}(x) + c_5 N_{4,5}(x) + c_6 N_{4,6}(x) + c_7 N_{4,7}(x) \tag{3.6-72}$$

Near the ends of the range (x_{min}, x_{max}), interpolation may call for splines with knots below 1 or above N; therefore, it is necessary to define *exterior knots* such that

$$\lambda_j = \begin{cases} x_{min}, & j \leq 0 \\ x_{max}, & j \geq N+1 \end{cases} \tag{3.6-73}$$

When x lies in (x_{min}, λ_1), the interpolation is supported by the first four splines; when x lies in (λ_N, x_{max}), $\hat{z}$ is supported by the last four.

Cubic B-spline coefficient solutions follow the inversion procedure used in series expansion. Equation 3.6-71 can be applied for each tabulated point; hence,

$$\hat{z}_i = \sum_{j=1}^{I} c_j N_{4,j}(x_i); \quad i = 1, I \tag{3.6-74}$$

which can be expressed as

$$\hat{\mathbf{z}} = \mathbf{Ac} \tag{3.6-75}$$

where $\mathbf{A}$ takes the form of a square *band matrix*. For $I = 8$,

$$\mathbf{A} \triangleq \begin{bmatrix} x & x & x & x & 0 & 0 & 0 & 0 \\ x & x & x & x & 0 & 0 & 0 & 0 \\ 0 & x & x & x & x & 0 & 0 & 0 \\ 0 & 0 & x & x & x & x & 0 & 0 \\ 0 & 0 & 0 & x & x & x & x & 0 \\ 0 & 0 & 0 & 0 & x & x & x & x \\ 0 & 0 & 0 & 0 & x & x & x & x \\ 0 & 0 & 0 & 0 & x & x & x & x \end{bmatrix} \tag{3.6-76}$$

with "x" representing a nonzero term. The elements of $\mathbf{A}$ are $a_{ij} = N_{4,j}(x_i)$, $\mathbf{c} = [c_1\ c_2\ \cdots\ c_I]^T$, and $\hat{\mathbf{z}} = [\hat{z}_1\ \hat{z}_2\ \cdots\ \hat{z}_I]^T$; hence, equation 3.6-75 will fit the tabulated points exactly if $\mathbf{c}$ is chosen as

$$\mathbf{c} = \mathbf{A}^{-1}\mathbf{z} \tag{3.6-77}$$

where $\mathbf{z}$ contains the tabulated points. Matrix inversion is not needed to compute $\mathbf{c}$; because $\mathbf{A}$ is a band matrix, $\mathbf{c}$ can be found by Gaussian elimination [C-4, D-3]. Following the polynomial series development, *weighted-least-squares B-spline interpolation* uses the pseudoinverse of $\mathbf{A}$, allowing $\mathbf{c}$ to have smaller dimension than $\mathbf{z}$. Multivariate B-spline interpolation is described in [C-5].

Once the cubic B-spline coefficients have been computed, derivatives of $\hat{z}$ are obtained easily [C-4]. The first derivative is

$$\frac{\partial \hat{z}}{\partial x} = \sum_{j=1}^{I-1} 3\left(\frac{c_{j+1} - c_j}{\lambda_j - \lambda_{j-n+1}}\right) N_{3,j}(x) \triangleq \sum_{j=1}^{I-1} c_j^{(1)} N_{3,j}(x) \tag{3.6-78}$$

while the second derivative is

$$\frac{\partial^2 \hat{z}}{\partial x^2} = \sum_{j=1}^{I-2} 2\left(\frac{c_{j+1}^{(1)} - c_j^{(1)}}{\lambda_j - \lambda_{j-n+2}}\right) N_{2,j}(x) \tag{3.6-79}$$

Although the process of calculating the spline coefficients is involved, once the coefficients have been calculated, they allow smooth, efficient representation of functions in simulations.

Computational neural networks provide an alternate method of modeling multivariate nonlinear functions. They consist of *nodes* that simulate the *neurons* and *weighting factors* that simulate the *synapses* of a living nervous system. The nodes are nonlinear basis functions, and the weights contain knowledge of the system. Neural networks approximate multivariate functions of the form

$$\mathbf{y} = \mathbf{h}(\mathbf{x}) \tag{3.6-80}$$

where $\mathbf{x}$ and $\mathbf{y}$ are input and output vectors and $\mathbf{h}(\cdot)$ is the relationship between them. Neural networks can be considered *generalized spline functions* that identify efficient input-output mappings from observations [P-8].

An N-layer *feedforward neural network* represents the function by a sequence of operations

$$\mathbf{r}^{(k)} = \mathbf{s}^{(k)}[\mathbf{W}^{(k-1)}\mathbf{r}^{(k-1)}] \triangleq \mathbf{s}^{(k)}[\eta^{(k)}], \quad k = 1, N \tag{3.6-81}$$

where $\mathbf{y} = \mathbf{r}^{(N)}$ and $\mathbf{x} = \mathbf{r}^{(0)}$. $\mathbf{W}^{(k-1)}$ is a matrix of weighting factors determined by the learning process, and $\mathbf{s}^{(k)}[\cdot]$ is an *activation-function vector* whose elements normally are identical, scalar, nonlinear functions $\sigma_i(\eta_i)$ appearing at each node:

$$\mathbf{s}^{(k)}[\eta^{(k)}] = [\sigma_1(\eta_1^{(k)}) \cdots \sigma_n(\eta_n^{(k)})]^T \tag{3.6-82}$$

One of the inputs to each layer may be a unity threshold element that adjusts the bias of the layer's output. Networks consisting solely of linear activation functions are of little interest, as they merely perform a linear transformation $\mathbf{H}$, thus limiting eq. 3.6-80 to the form $\mathbf{y} = \mathbf{Hx}$.

Figure 3.6-3 represents two simple feedforward neural networks. Each circle represents an arbitrary, scalar, nonlinear function $\sigma_i(\cdot)$ operating on the sum of its inputs, and each arrow transmits a signal from the previous node, multiplied by a weighting factor. A scalar network with a single hidden layer of four nodes and a unit threshold element (Fig. 3.6-3a) is clearly parallel, yet its output can be written as the series

$$y = a_0\sigma_0(b_0x + c_0) + a_1\sigma_1(b_1x + c_1) + a_2\sigma_2(b_2x + c_2) + a_3\sigma_3(b_3x + c_3) \quad (3.6\text{-}83)$$

illustrating that parallel and serial processing may be equivalent.

Consider a simple example. Various nodal activation functions σ_i have been used, and there is no need for each node to be identical.

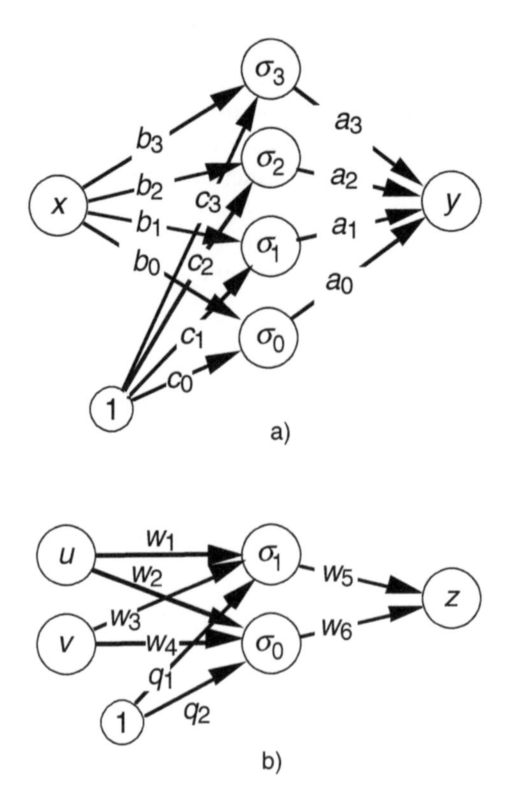

Fig. 3.6-3. Static neural networks. (a) Single-input/single-output network. (b) Double-input/single-output network.

Choosing $\sigma_0(\cdot) = (\cdot)$, $\sigma_1 = (\cdot)^2$, $\sigma_2 = (\cdot)^3$, $\sigma_3 = (\cdot)^4$, eq. 3.6-80 is represented by the truncated power series

$$y = a_0(b_0x + c_0) + a_1(b_1x + c_1)^2 + a_2(b_2x + c_2)^3 + a_3(b_3x + c_3)^4 \tag{3.6-84}$$

It is clear that network weights are redundant [i.e., that the (a, b, c) weighting factors are not independent]. Consequently, more than one set of weights could produce the same functional relationship between $\mathbf{x}$ and $\mathbf{y}$. Training sessions starting at different points could produce different sets of weights that yield identical outputs. This example also indicates that the unstructured feedforward network may not have compact support (i.e., its weights may have global effects) if its basis functions do not vanish for large magnitudes of their arguments.

The *sigmoid* is commonly used as the artificial neuron. It is a saturating function defined variously as

$$\sigma(\eta) = \frac{1}{1 + e^{-\eta}} \tag{3.6.-85}$$

for output in (0, 1) or as

$$\sigma(\eta) = \frac{(1 - e^{-2\eta})}{(1 + e^{-2\eta})} = \tanh\eta \tag{3.6-86}$$

for output in (−1, 1). Any continuous mapping can be approximated arbitrarily closely with sigmoidal networks containing a single hidden layer (N = 2) [C-6, F-6]. Symmetric functions like the *Gaussian radial basis function*

$$\sigma(\eta) = e^{-\eta^2} \tag{3.6-87}$$

have better convergence properties for many functions and have more compact support as a consequence of their near-orthogonality [P-8, H-7].

The scalar output of a two-input, single sigmoid with unit weights is shown in Fig. 3.6-4a. More complex surfaces can be generated by increasing the number of sigmoids. For comparison, a two-input, radial-basis-function network produces the output shown in Fig. 3.6-4b. Whereas the sigmoid has a preferred input axis and simple curvature, the radial basis function admits more complex curvature of the output surface, and its effect is more localized. The most efficient nodal activation function depends on the general shape of the surface to be approximated.

Neural networks can be trained from either tabulated data or flight measurements. Feedforward neural networks are furnished with typical

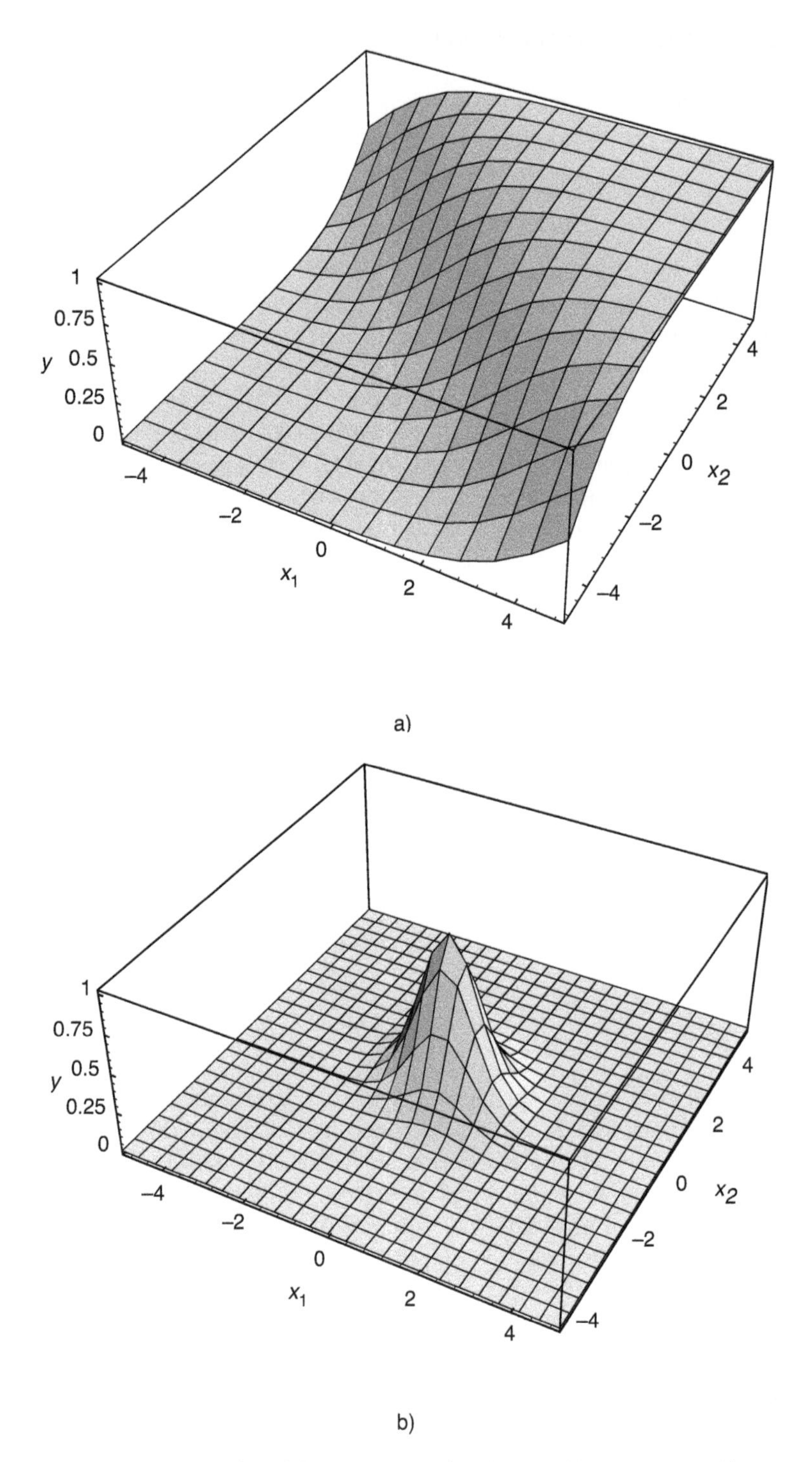

Fig. 3.6-4. Neural network nodal (or activation) functions. (a) Two-input sigmoid. (b) Two-input radial basis function.

inputs and outputs, and the training algorithm computes the weights that minimize fit error. *Back-propagation* learning algorithms for the elements of $\mathbf{W}^{(k)}$ typically involve a *gradient search* [R-6, W-2] that minimizes the mean-square output error

$$\mathrm{E} = [\mathbf{r}_d - \mathbf{r}^{(N)}]^T [\mathbf{r}_d - \mathbf{r}^{(N)}] \tag{3.6-88}$$

where $\mathbf{r}_d$ is the desired output. The error gradient with respect to the weight matrix is calculated for each input-output example presented to the network, and the weights are updated by

$$\mathbf{W}^{(k)}_{new} = \mathbf{W}^{(k)}_{old} + \beta\, \mathbf{r}^{(k-1)}[\mathbf{d}^{(k)}]^T \tag{3.6-89}$$

β is the learning rate, and $\mathbf{d}$ is a function of the error between desired and actual outputs. For the output layer, the error term is

$$\mathbf{d}^{(N)} = \mathbf{S}'\, [\mathbf{W}^{(N-1)}\mathbf{r}^{(N-1)}]\, (\mathbf{r}_d - \mathbf{r}^{(N)}) \tag{3.6-90}$$

where the prime indicates differentiation with respect to $\mathbf{r}$. For interior layers, the error from the output layer is propagated from the output error using

$$\mathbf{d}^{(k)} = \mathbf{S}'\, [\mathbf{W}^{(k-1)}\mathbf{r}^{(k-1)}]\, [\mathbf{W}^{(k-1)}]^T\mathbf{d}^{(k-1)} \tag{3.6-91}$$

The search rate can be modified by adding momentum or conjugate-gradient terms to eq. 3.6-89.

The learning speed can be further improved using an *extended Kalman filter* [S-6, L-4]. The dynamic and observation models for the filter are

$$\mathbf{w}_k = \mathbf{w}_{k-1} + \mathbf{q}_{k-1} \tag{3.6-92}$$

$$\mathbf{z}_k = \mathbf{h}(\mathbf{w}_k, \mathbf{r}_k) + \mathbf{n}_k \tag{3.6-93}$$

where $\mathbf{w}_k$ is a vector of the matrix $\mathbf{W}_k$'s elements, $\mathbf{h}(\cdot)$ is an observation function, and $\mathbf{q}_k$ and $\mathbf{n}_k$ are noise processes. If the network has a scalar output, then $\mathbf{z}_k$ is scalar, and the extended Kalman filter minimizes the fit error between the training hypersurface and that produced by the network (eq. 3.6-88). The fit error can be dramatically reduced by considering the *gradients* of the surfaces as well [L-4]. The observation vector becomes

$$\mathbf{z}_k = \begin{bmatrix} \mathbf{h}(\mathbf{w}_k, \mathbf{r}_k) \\ \dfrac{\partial \mathbf{h}}{\partial \mathbf{r}}(\mathbf{w}_k, \mathbf{r}_k) \end{bmatrix} + \mathbf{n}_k \tag{3.6-94}$$

with concomitant increase in the complexity of the filter. The relative significance given to the function and derivative errors during training can be adjusted through the measurement-error covariance matrix used in filter design.

It may speed the computation of weights to use a *random search*, at least until convergent regions are identified. Such methods as *simulated annealing* or *genetic algorithms* can be considered [D-5]. The first of these is motivated by statistical mechanics and the effects that controlled cooling has on the ground states of atoms (which are analogous to the network weights). The second models the reproduction, crossover, and mutation of biological strings (e.g., chromosomes, again analogous to the weights), in which only the fittest combinations survive. Conversely, the search can be eliminated entirely using an algebraic approach to neural network training, fitting well-defined numerical data precisely or with an arbitrary level of smoothing [F-1, F-2].

Problems that may be encountered in neural network training include proper choice of the input vector, local versus global training, speed of learning and forgetting, generalization over untrained regions, and trajectory-dependent correlations in the training sets. An aerodynamic model based on neural networks could span the entire flight envelope of an aircraft, including post-stall and spinning regions. The model would contain six neural networks with multiple inputs and scalar outputs, three for force coefficients and three for moment coefficients. If input variables are not restricted to those having plausible aerodynamic effects, false correlations may be created in the network; hence, attitude Euler angles and horizontal position should be neglected in training, while physically meaningful terms like elevator deflection, angle of attack, pitch rate, Mach number, and dynamic pressure should be included [L-4]. Elements of the input vector may be strongly correlated with each other through the aircraft's equations of motion; hence, networks may not be able to distinguish between highly correlated variables (e.g., pitch rate and normal acceleration). Training sets should provide inputs that are rich in frequency content, that span the state and control spaces, and that are as uncorrelated as possible. Generalization between training points may provide smoothness, but it does not guarantee accuracy.

Example 3.6-2. Data Representation for Simple Functions

Suppose that the sine wave and sawtooth function used in the previous example are stored with a small number of samples. How well can these wave forms be reconstructed using interpolating procedures contained in MATLAB? In the first case, we represent the sine wave at six equally spaced points in the cycle (including the end points). Figure 3.6-5 compares the exact sine wave with the four interpolation options made available in the function `interp1`. Linear interpolation (`linear`) and

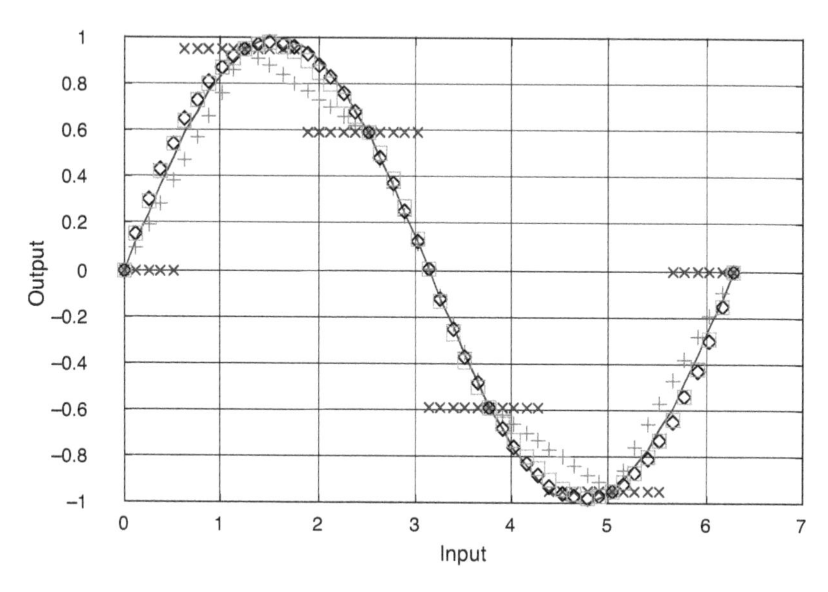

Fig. 3.6-5. Exact sine wave and approximations by interpolation between six points.

nearest-neighbor assessment (nearest), shown by "+" and "×," produce the most error, and their slopes are discontinuous across the tabulated points. The cubic spline (spline) and cubic interpolation (cubic), represented by diamonds and squares in the figure, provide good approximations to the sine wave.

In the second case, the six data points have been stored at unequal intervals (0, 5, 10, 11, 15, 20), providing a challenge for the more sophisticated interpolators (Fig. 3.6-6). The linear interpolation provides excellent results, and those of the cubic interpolation are very good. The neighboring solution lies close to the input function, but its slopes are not usable (except where the function is zero). The cubic-spline interpolation produces excessive error because it is ill suited to handling sharp discontinuities and unequally spaced data points. Interpolation with all the algorithms would be improved if more points were stored.

Trimmed Solution of the Equations of Motion

The natural motions of most aircraft are oscillatory. Given arbitrary initial conditions and fixed control settings, the body attitude and flight path fluctuate about some equilibrium cruise condition. If the modes of

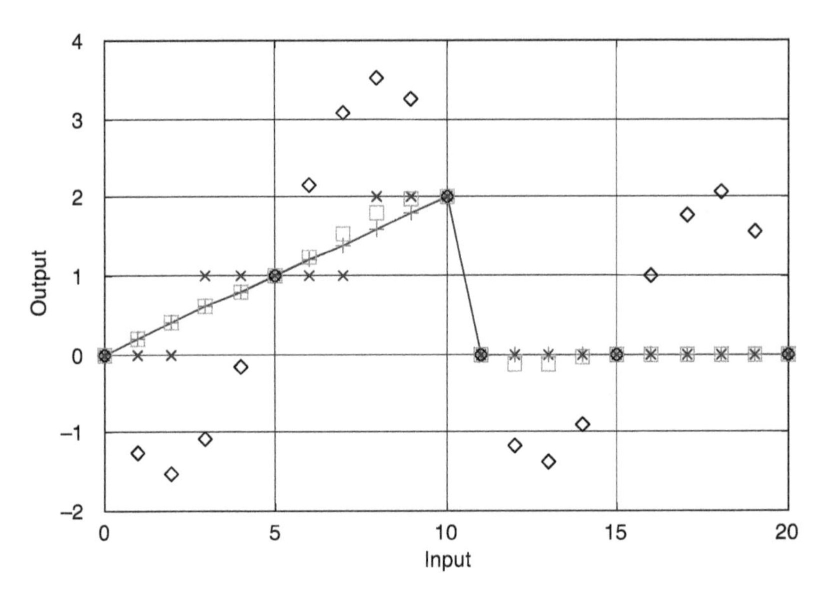

Fig. 3.6-6. Ramp function and approximations by interpolation between six points.

motion are unstable, the velocity and angles diverge from the equilibrium, no matter what the initial conditions and control settings. If the modes of motion are asymptotically stable, the transients ultimately die out, and the aircraft flies in the equilibrium condition; however, it may take a significant period of time before the oscillations are negligibly small. In either case, it is desirable to be able to compute equilibrium conditions and control settings without generating a state history.

An aircraft is in trimmed, equilibrium flight when its velocity is fixed and its pitch and roll angles are unchanging. The trim problem is to find the control settings that yield a steady flight condition. With six dynamic equations (eqs. 3.6-19–3.6-24), there are six force and moment components to be specified. The number of control settings may be larger or smaller than six, depending on the particular aircraft, and the effects of specific control effectors may be coupled by aerodynamics. If the number is larger, control redundancy admits a range of dynamically equivalent solutions; if it is smaller, the number of state components that can be independently specified is constrained by the available controls.

The object is to set the vector equation

$$\dot{\mathbf{x}}_d = \mathbf{f}_d(\mathbf{x}_d, \mathbf{u}, \mathbf{w}) \tag{3.6-95}$$

to zero by the proper choice of control **u**, subject to constant values of the state $\mathbf{x}_d$ and disturbance **w**. Equation 3.6-95 represents the first six (dynamic) equations of motion (eqs. 3.6-19–3.6-24), and $\mathbf{x}_d$ is defined accordingly:

$$\mathbf{x}_d = [u\ v\ w\ p\ q\ r]^T \tag{3.6-96}$$

The remaining (kinematic) state components

$$\mathbf{x}_k = [x\ y\ z\ \phi\ \theta\ \psi]^T \tag{3.6-97}$$

may be treated as fixed or free parameters, as described below. For a symmetric aircraft flying in a vertical plane, *longitudinal trim* is specified by just three equations (eqs. 3.6-19, 3.6-21, and 3.6-23), and $\mathbf{x}_d$ is $[u\ w\ q]^T$. The *lateral-directional trim* calculation is coupled to the longitudinal trim condition. Simple models such as those discussed in Section 2.6 admit closed-form analytical trim solutions, but trimming more complex models requires a numerical approach.

One numerical method is to find the control settings that minimize a *scalar cost function* $J(\mathbf{u})$ derived from the dynamic equations. A good choice is the quadratic function

$$J(\mathbf{u}) = \mathbf{f}_d^T(\mathbf{u})\mathbf{R}\,\mathbf{f}_d(\mathbf{u}) \tag{3.6-98}$$

where the parameters are absorbed in the definition of $\mathbf{f}_d(\mathbf{u})$. **R** is a symmetric, conformable, positive-definite[2] square matrix that weights the elements of $\mathbf{f}_d(\mathbf{u})$ to account for differences in units of measure and to provide control of the minimization. The minimized value of $J(\mathbf{u})$ is denoted by J^* or $J(\mathbf{u}^*)$, where $\mathbf{u}^*$ is the minimizing value of the control. J^* is zero at the trimmed condition.

If $\mathbf{f}_d(\mathbf{u})$ is linear in the m-vector **u**, then $J(\mathbf{u})$ is quadratic in **u**, and it can be expanded about any control setting $\mathbf{u}_o$ using a Taylor series containing three terms:

$$J(\mathbf{u}) = J(\mathbf{u}_o) + \frac{\partial J}{\partial \mathbf{u}}(\mathbf{u} - \mathbf{u}_o) + \frac{1}{2}(\mathbf{u} - \mathbf{u}_o)^T \frac{\partial^2 J}{\partial \mathbf{u}^2}(\mathbf{u} - \mathbf{u}_o) \tag{3.6-99}$$

Here, $\partial J/\partial \mathbf{u}$ is the *gradient* of J with respect to **u**, a row vector with m components,

$$\frac{\partial J}{\partial \mathbf{u}} \triangleq J_{\mathbf{u}} = \left[\frac{\partial J}{\partial u_1}\frac{\partial J}{\partial u_2}\cdots\frac{\partial J}{\partial u_m}\right] \tag{3.6-100}$$

2. A *positive-definite matrix* **R** assures that the product $\mathbf{x}^T\mathbf{R}\mathbf{x}$ is definitely positive for all real values of **x** except zero. The symmetric, real matrix **R** is positive definite if all of it eigenvalues (Section 3.2) are positive.

and $\partial^2 J/\partial \mathbf{u}^2$ is the symmetric $(m \times m)$ *Hessian matrix* of J with respect to $\mathbf{u}$:

$$\frac{\partial^2 J}{\partial \mathbf{u}^2} \triangleq J_{\mathbf{uu}} = \begin{bmatrix} \frac{\partial^2 J}{\partial u_1^2} & \frac{\partial^2 J}{\partial u_1\, du_2} & \cdots & \frac{\partial^2 J}{\partial u_1\, du_m} \\ \frac{\partial^2 J}{\partial u_1\, du_2} & \frac{\partial^2 J}{\partial u_2^2} & \cdots & \frac{\partial^2 J}{\partial u_2\, du_m} \\ \cdots & \cdots & \cdots & \cdots \\ \frac{\partial^2 J}{\partial u_1\, du_m} & \frac{\partial^2 J}{\partial u_2\, du_m} & \cdots & \frac{\partial^2 J}{\partial u_m^2} \end{bmatrix} \tag{3.6-101}$$

The scalar components of $J_{\mathbf{u}}$ and $J_{\mathbf{uu}}$ are derived from eq. 3.6-101 and are evaluated with $\mathbf{u}_o$ and the given values of fixed parameters inserted in the equations.

The first derivative of $J(\mathbf{u})$ with respect to $\mathbf{u}$ is

$$\frac{\partial J(\mathbf{u})}{\partial \mathbf{u}} = J_{\mathbf{u}}\big|_{\mathbf{u}=\mathbf{u}_o} + (\mathbf{u} - \mathbf{u}_o)^T J_{\mathbf{uu}}\big|_{\mathbf{u}=\mathbf{u}_o} \tag{3.6-102}$$

Because the slope equals zero at the minimized value of J, $\mathbf{u}^*$ is defined by

$$\mathbf{u}^* = \mathbf{u}_o - \left[J_{\mathbf{uu}}\big|_{\mathbf{u}=\mathbf{u}_o}\right]^{-1}\left[J_{\mathbf{u}}\big|_{\mathbf{u}=\mathbf{u}_o}\right]^T \tag{3.6-103}$$

Thus, the trim control settings can be obtained in a single step if J is quadratic in $\mathbf{u}$.

If J is not quadratic in $\mathbf{u}$, the first three terms of the Taylor series nevertheless should be a good approximation to $J(\mathbf{u})$ in the vicinity of the minimum, suggesting that equation 3.6-103 forms the basis of an iterative solution for the trim control:

$$\mathbf{u}_k = \mathbf{u}_{k-1} - \left[J_{\mathbf{uu}}\big|_{\mathbf{u}=\mathbf{u}_{k-1}}\right]^{-1}\left[J_{\mathbf{u}}\big|_{\mathbf{u}=\mathbf{u}_{k-1}}\right]^T \tag{3.6-104}$$

This algorithm is referred to as a *Newton-Raphson iteration*. It depends on $J_{\mathbf{uu}}$ being a positive-definite matrix (such that $\mathbf{u}^T J_{\mathbf{uu}} \mathbf{u} > 0$ for any nonzero choice of $\mathbf{u}$), which may not be the case when $\mathbf{u}$ is far from $\mathbf{u}^*$. If the diagonal elements of $J_{\mathbf{uu}}$ are positive, the off-diagonal elements can be suppressed (i.e., either set to zero or attenuated) in applying eq. 3.6-104.

Another solution is to begin any iterative process using a *steepest-descent algorithm*,

$$\mathbf{u}_k = \mathbf{u}_{k-1} - \varepsilon\left[J_{\mathbf{u}}\big|_{\mathbf{u}=\mathbf{u}_{k-1}}\right]^T \tag{3.6-105}$$

where $\boldsymbol{\varepsilon}$ is a parameter large enough to provide rapid convergence but small enough to prevent oscillations or divergence. When the solution is close to converging, a switch is made to the Newton-Raphson algorithm. Using either algorithm, an upper bound on $|\mathbf{u}_k - \mathbf{u}_{k-1}|$ prevents physically unrealistic excursions in control solutions.

Analytical evaluation of $J_{\mathbf{u}}$ and $J_{\mathbf{uu}}$ is desirable but often impractical, so repeated numerical evaluation of J with small variations in the elements of $\mathbf{u}$ can be employed. For example, the first element of $J_{\mathbf{u}}$ can be approximated as

$$\frac{\partial J}{\partial u_1} \triangleq J_{u_1} \approx \frac{J(u_1 + \Delta u_1, u_2, \ldots, u_m) - J(u_1 - \Delta u_1, u_2, \ldots, u_m)}{2\,\Delta u_1} \quad (3.6\text{-}106)$$

while the (1, 2) element of $J_{\mathbf{uu}}$ is approximately

$$\frac{\partial^2 J}{\partial u_1 \partial u_2} \approx \frac{J_{u_1}(u_2 + \Delta u_2) - J_{u_1}(u_2 - \Delta u_2)}{2\,\Delta u_2} \quad (3.6\text{-}107)$$

MATLAB's `fmins` operation uses a gradient-free algorithm to minimize functions like J, and more advanced routines can be found in the Optimization Toolbox.

The desired trim condition can be specified by various combinations of velocity and angle components. With the conventional four controls (elevator δE, ailerons δA, rudder δR, and throttle δT), a typical specification would be airspeed V, flight path angle γ (or climb rate $\dot{h}$), sideslip angle β, and turn rate $\dot{\psi}$. The pitch and bank angles $\boldsymbol{\theta}$ and μ would be free to take necessary values. The trim variables and attitude angles specify the corresponding u, v, w, p, q, and r.

Although they are free variables, $\boldsymbol{\theta}$ and μ do not change unless forced to do so. In other words, an adjustment rule is required. That rule is provided by incorporating the free variables in the control vector; hence, $\mathbf{u}$ is defined as

$$\mathbf{u} \triangleq [\delta E \;\delta T \;\theta \;\delta A \;\delta R \;\mu]^T \quad (3.6\text{-}108)$$

for three-axis trim, or

$$\mathbf{u} \triangleq [\delta E \;\delta T \;\theta]^T \quad (3.6\text{-}109)$$

for longitudinal trim.

Example 3.6-3. Trimmed Solution of the Longitudinal Equations of Motion

The generic business jet model of Appendix B provides two examples of the longitudinal trimming calculation. The first flight condition is steady, level flight at an altitude of 3,050 m and a Mach number of 0.3 (true airspeed = 102 m/s). The program FLIGHT employs MATLAB's `fmins` function for the minimization. With starting values close to the final values, it achieves acceptable convergence after 113 iterations.

The convergence history for throttle (δT, percent/100), pitch angle (θ, rad), and stabilator angle (δS, rad), shown in Fig. 3.6-7 in descending order, reveals that pitch angle is the first variable to converge. The throttle setting jumps around until the last 25 iterations, and the stabilator angle increases for the first 60 iterations. Trim parameter variations are not readily discernible after 80 or 90 iterations; however, the plot of trim-error cost (eq. 3.6-64), which represents an unweighted sum of the squares of axial, normal, and pitch acceleration, exhibits continued logarithmic decrease to the stopping point. The trim-error cost is reduced by about seven orders of magnitude during the calculation (Fig. 3.6-8). At the trimmed condition, $\delta S = -1.948$ deg, $\delta T = 19.18$ percent and $\theta = 3.6268$ deg.

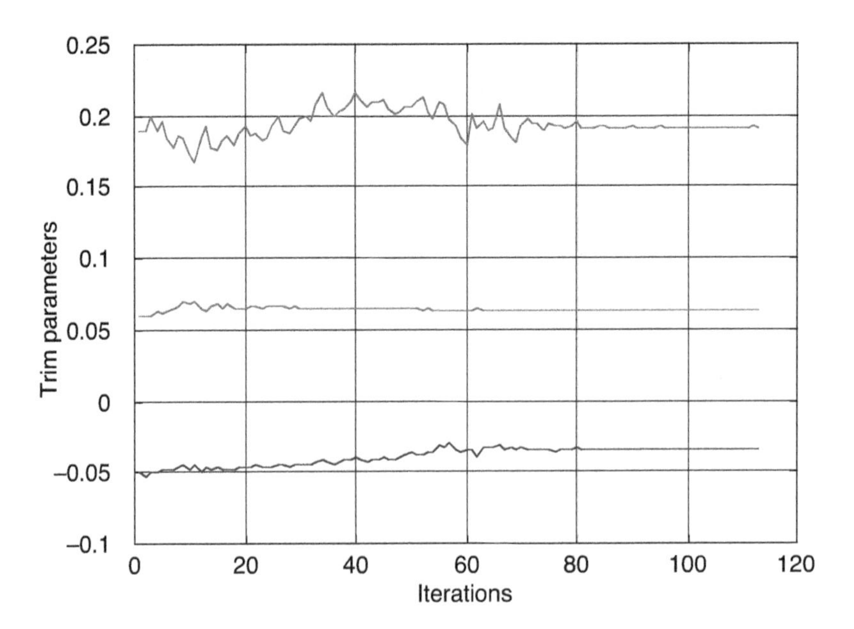

Fig. 3.6-7. Convergence of trim parameters for the low-subsonic case.

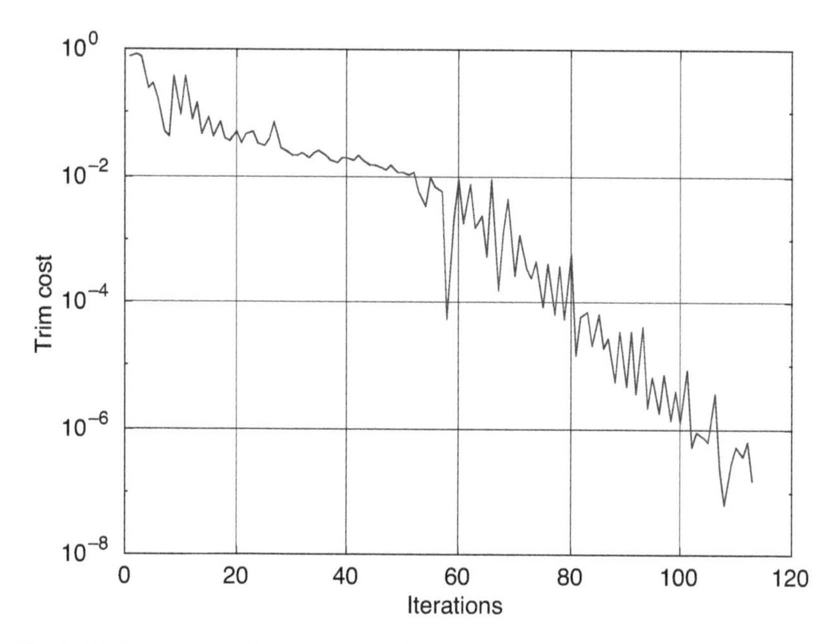

Fig. 3.6-8. Convergence of trim-error cost function for the low-subsonic case.

Starting from the same initial conditions, the trim solution for cruising at $\mathcal{M} = 0.81$ and an altitude of 9,150 m is sought. Satisfactory convergence takes 257 iterations, during which there are long periods of apparent inaction (Fig. 3.6-9). Once the throttle setting is in the right region, the solution proceeds quickly. Note that the algorithm tries throttle values greater than 1 before it settles at 98 percent.

Example 3.6-4. Results of Flight Path Simulation

Having trimmed the business jet model at a steady flight condition, we examine the effects of initial condition perturbations on the aircraft's motion. Thirty-second flight paths are simulated by FLIGHT with initial pitch, roll, and yaw rates of 0.1 rad/s (5.73 deg/s). The computations, described in Appendix B.1, use the nonlinear, flat-earth equations of motion presented in this chapter. In the next chapter, we see that classical *modes of motion* can be derived from the linearized dynamic equations. These modes are readily discernible in the nonlinear responses shown here. The state histories also reveal the level of cross-coupling between longitudinal and lateral-directional motions, which normally is neglected in linearized analysis.

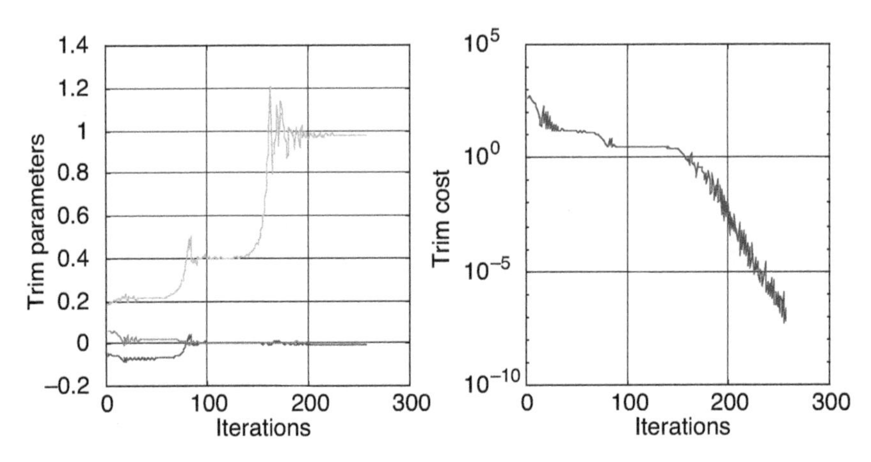

Fig. 3.6-9. Convergence of trim parameters and trim-error cost function for the transonic case.

Because the trimmed condition is symmetric and pitch rate introduces no asymmetry, the aircraft's response to pitch rate is restricted to the vertical plane Fig. 3.6-10. The lateral-directional response is precisely zero and is not shown. The initial pitch rate (third panel) is rapidly damped, but it triggers a long-period response that is best seen in the histories of altitude (second panel) and pitch angle (fourth panel). The normal velocity w (first panel) experiences a significant short-period transient from its trim value w_o, while the long-period mode has a more significant effect on the axial velocity perturbation $u - u_o$. These fast and slow motions are dominated by the *short-period* and *phugoid modes*.

An initial roll rate of 0.1 rad/s is rapidly damped out (third panel), but it also triggers a lightly damped oscillation that is seen in several of the traces (Fig. 3.6-11). The oscillation involves roll and yaw rates of comparable magnitude and is known as the *Dutch roll mode*. The decaying roll-rate response, governed by the *roll mode*, produces a bias in the roll angle, while the steady offset in yaw rate produces a ramp in yaw angle and a parabolic crossrange response. Not evident in this 30s history is the *spiral mode*, which affects the long-term growth or decay of these responses.

The corresponding longitudinal response to initial roll rate is small but not zero, as the four panels of Fig. 3.6-12 demonstrate. Most notable are the altitude drop and pitch-angle oscillation, which occur on the time scale of the phugoid mode. The pitch rate experiences a small short-period transient, also seen in the pitch-angle trace. The normal-velocity

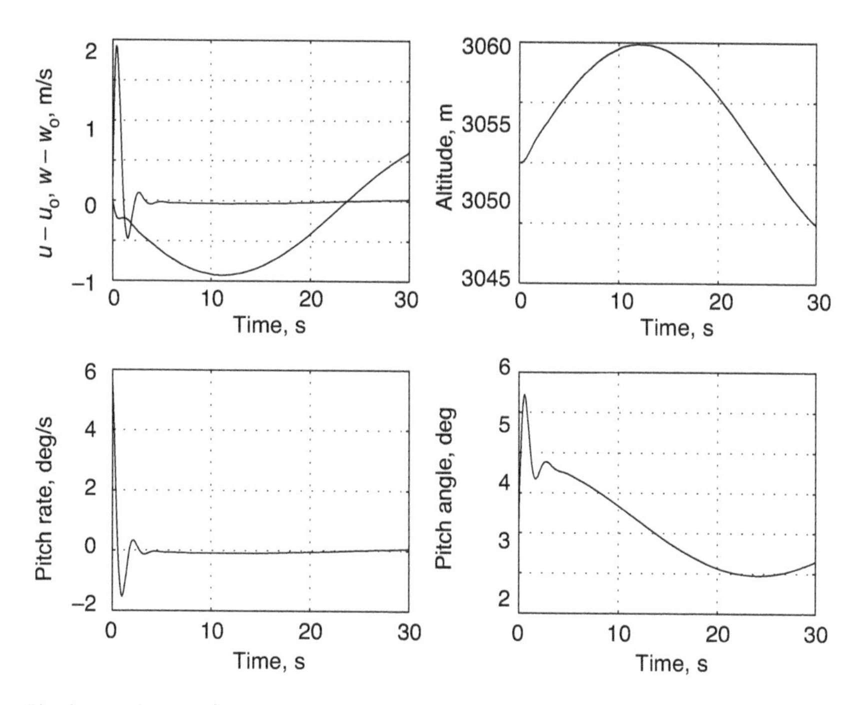

Fig. 3.6-10. Longitudinal transient response to initial pitch rate at $\mathcal{M} = 0.3$, $h = 3{,}050$ m.

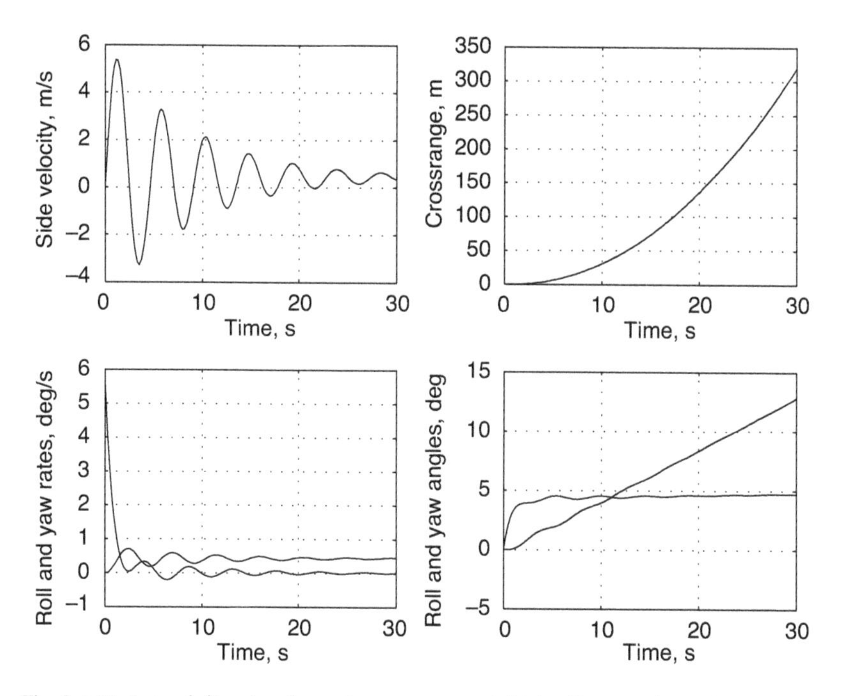

Fig. 3.6-11. Lateral-directional transient response to initial roll rate at $\mathcal{M} = 0.3$, $h = 3{,}050$ m.

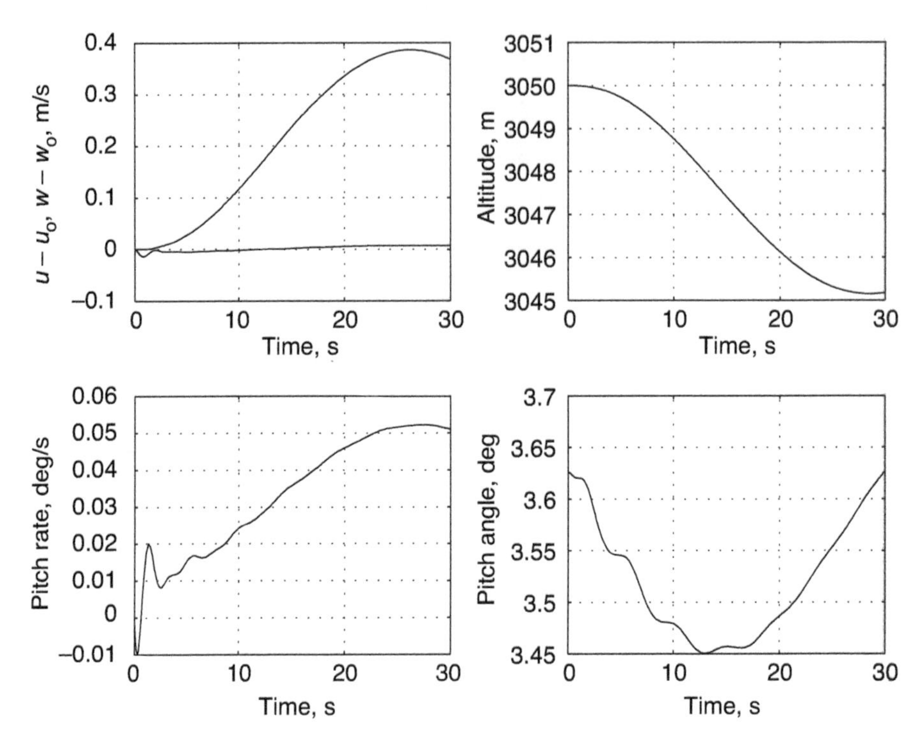

Fig. 3.6-12. Longitudinal transient response to initial roll rate at $\mathcal{M} = 0.3$, $h = 3{,}050$ m.

response is negligible, while the axial velocity responds on the long-period scale. Thus, the lateral initial condition disturbs the longitudinal equilibrium through a pitching-moment imbalance that excites the phugoid mode.

An initial yaw rate of 0.1 rad/s produces equivalent Dutch roll transients in both yaw and roll rate (Fig. 3.6-13). The amplitudes of both are much higher than in the previous case, and the side velocity is about ten times higher than before. There is no apparent involvement of the roll mode, but the damped step response of the roll angle indicates the presence of the spiral mode.

The associated longitudinal response reveals *period doubling*, a nonlinear phenomenon that was not apparent in the prior example (Fig. 3.6-14). The pitch-rate-oscillation frequency is about twice that of the Dutch roll mode. The frequency is similar to the longitudinal short-period frequency, but the slow decay—comparable to that of the Dutch roll—indicates that the motion is being driven by lateral-directional motion. The scalloped

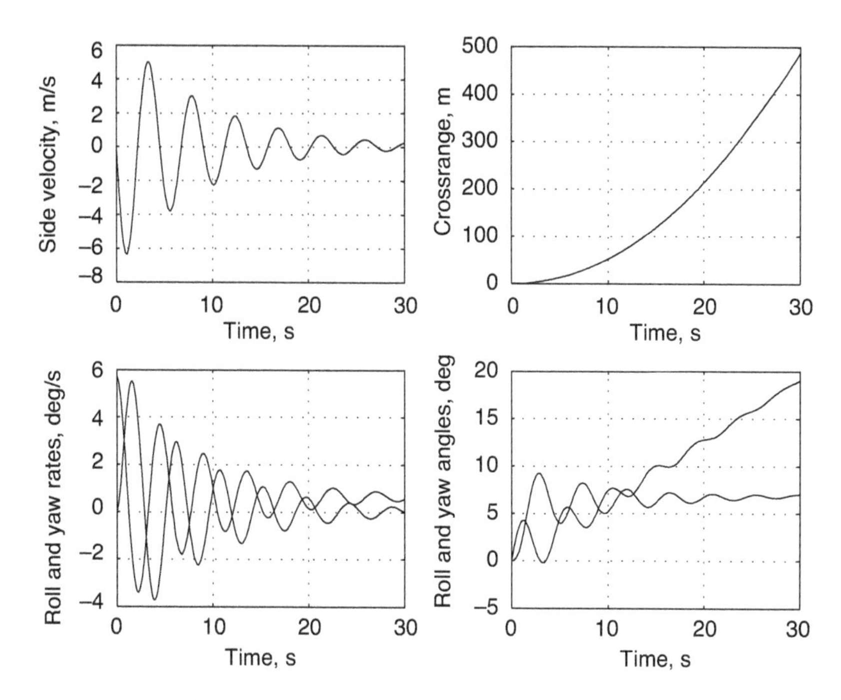

Fig. 3.6-13. Lateral-directional transient response to initial yaw rate at $\mathcal{M} = 0.3$, $h = 3{,}050$ m.

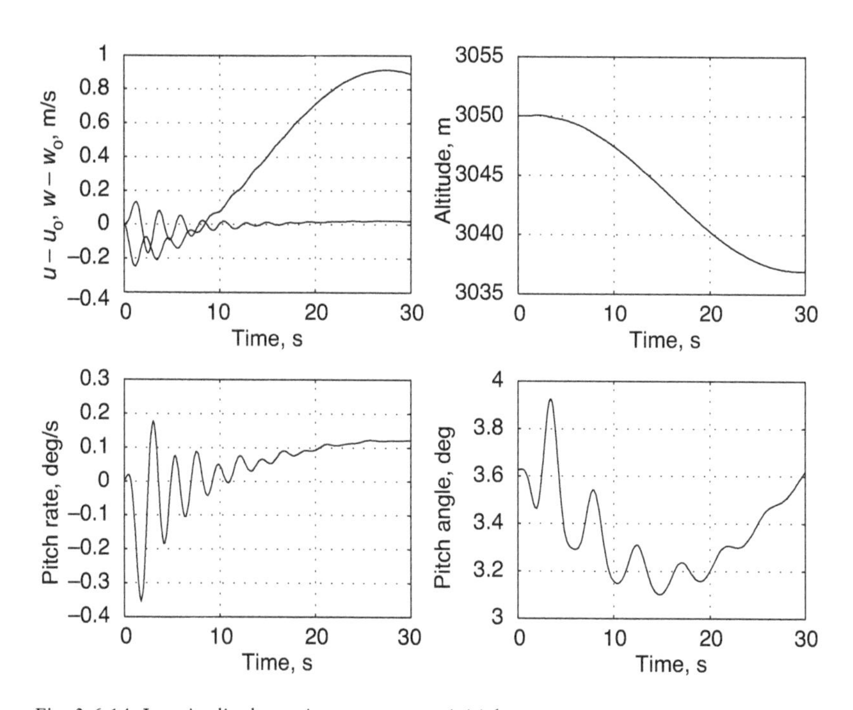

Fig. 3.6-14. Longitudinal transient response to initial yaw rate at $\mathcal{M} = 0.3$, $h = 3{,}050$ m.

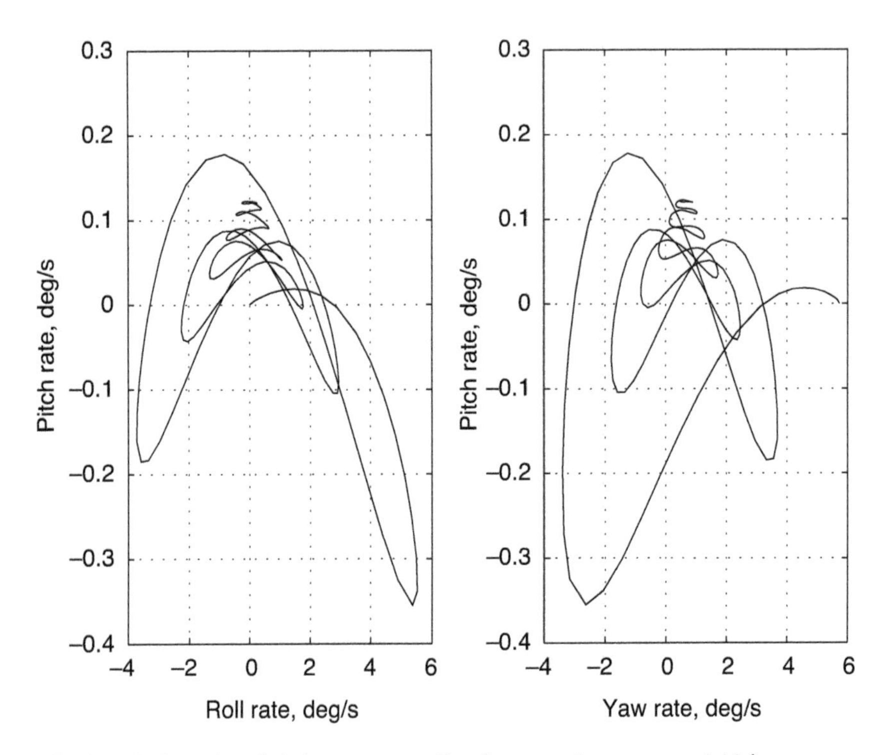

Fig. 3.6-15. Crossplot of pitch rate versus roll and yaw rate in response to initial yaw rate at $\mathcal{M} = 0.3$, $h = 3{,}050\,\text{m}$.

oscillation evident in the pitch angle (rather than a more sinusoidal shape) is a further indication of nonlinear response.

Period doubling suggests that the lateral-directional input is rectified: yawing and/or rolling to the left *or* the right produces a pitch-down moment. Crossplots of pitch rate against roll and yaw rates confirm this conjecture (Fig. 3.6-15). Referring to eq. 3.6-23, we see the reason: squares and products of p and r cause the effect. (Recall that $\sin^2(\omega t) = [1 - \cos(2\omega t)]/2$.)

References

A-1 Abzug, M. J., *Computational Flight Dynamics*, American Institute of Aeronautics and Astronautics, Reston, Va., 2002.

A-2 Abzug, M. J., and Larrabee, E. E., *Airplane Stability and Control*, Cambridge University Press, Cambridge, 2002.

A-3 Ames, M. B., Jr., and Sears, R. I., "Determination of Control Surface Characteristics from NACA Plain-Flap and Tab Data," NACA Report 721, Washington, D.C., 1945.

A-4 Ananda, M., "The Global Positioning System (GPS) Accuracy, System Error Budget, Space and Control Segment Overview," in *The NAVSTAR GPS System*, E. Lassiter, dir., AGARD-LS-16, Neuilly-sur-Seine, Sept. 1988.

A-5 Angeles, J. *Spatial Kinematic Chains*, Springer-Verlag, Berlin, 1982.

A-6 Anon., *MATLAB, The Language of Technical Computing: Using MATLAB, Version 5*, The MathWorks, Inc., Natick, Mass., 1997.

A-7 Anon., *World Geodetic System (WGS) 84 Ellipsoid*, DOD WGS 198-1, Defense Mapping Agency TR 8350.2, Washington, D.C., 1987.

B-1 Baarspul, M., "A Review of Flight Simulation Techniques," *Progress in the Aerospace Sciences* 27, Jan. (1990): 1–120.

B-2 Battin, R. H., *An Introduction to the Mathematics and Methods of Astrodynamics*, American Institute of Aeronautics and Astronautics, New York, 1987.

B-3 Boyden, R. P., Dress, D. A., Fox, C. H., Jr., Huffman, J. K., and Cruz, C. I., "Subsonic Static and Dynamic Stability Characteristics of a NASP Configuration," *J. Aircraft* 31, no. 4 (1994): 879–885.

B-4 Britting, K. R., *Inertial Navigation Systems Analysis*, J. Wiley & Sons, New York, 1971.

B-5 Brogan, W. L., *Modern Control Theory*, Prentice-Hall, Englewood Cliffs, N.J., 1985.

B-6 Bryson, A. E., Jr., "Stability Derivatives for a Slender Missile with Application to a Wing-Body-Vertical Tail Configuration," *Journal of the Aeronautical Sciences* 20, no. 5 (1953): 297–308.

B-7 Bryson, A. E., Jr., Desai, M. N., and Hoffman, W. L., "The Energy State Approximation in Performance Optimization of Supersonic Aircraft," *J. Aircraft* 6, no. 6 (1969): 481–488.

B-8 Busemann, A., "Aerodynamic Lift and Supersonic Speeds," British Aeronautical Research Council Report 2844, Feb 1937 (translation from *Luftfahrtforschung 12*, 1935).

C-1 Calise, A. J., "Extended Energy Management Methods for Flight Path Optimization," *AIAA Journal* 15, no. 3 (1977): 314–321.

C-2 Campbell, J. P., and McKinney, M. O., "Summary of Methods for Calculating Dynamic Stability and Response and for Estimating Lateral Stability Derivatives," NACA Report 1098, Washington, D.C., 1952.

C-3 Cook, M. V., *Flight Dynamics Principles*, J. Wiley & Sons, New York, 1997.

C-4 Cox, M. G., "Practical Spline Approximation," in *Topics in Numerical Analysis*, P. R. Turner, ed., Springer-Verlag, Berlin, 1982, pp. 79–111.

C-5 Cox, M. G., "Data Approximation by Splines in One and Two Independent Variables," in *The State of the Art in Numerical Analysis*, A. Iserles and M.J.D. Powell, eds., Clarendon Press, Oxford, 1987, pp. 111–138.

C-6 Cybenko, G., "Approximation by Superposition of a Sigmoidal Function," *Math. Cont., Sig., Sys.* 2, no. 4 (1989): 303–314.

D-1 Danby, J. M. A., *Fundamentals of Celestial Mechanics*, Macmillan, New York, 1962.

D-2 Davis, H. T., *Introduction to Nonlinear Differential and Integral Equations*, Dover Publications, New York, 1962.
D-3 Dongarra, J. J., Bunch, J. R., Moler, C. B., and Stewart, *Linpack User's Guide*, Society of Industrial and Applied Mathematics, Philadelphia, 1982.
D-4 Donovan, A. F., Lawrence, H. R., Goddard, F., and Gilruth, R. R., *High Speed Problems of Aircraft and Experimental Methods*, Vol. VIII of *High Speed Aerodynamics and Jet Propulsion*, Princeton University Press, Princeton, N.J., 1961.
D-5 Davis, L., ed., *Genetic Algorithms and Simulated Annealing*, Morgan Kaufmann, Palo Alto, Calif., 1987.
D-6 Dress, D. A., Boyden, R. P., and Cruz, C. I., "Measured and Theoretical Supersonic Dynamic Stability Characteristics of a National Aero-Space Plane Configuration," *J. Aircraft* 31, no. 3 (1994): 597–602.
E-1 Erzberger, H., McLean, J. D., and Barman, J. F., "Fixed-Range Optimum Trajectories for Short-Haul Aircraft," NASA TN D-8115, Washington, D.C, Dec. 1975.
E-2 Etkin, B., *Dynamics of Atmospheric Flight*, J. Wiley & Sons, New York, 1972.
F-1 Ferrari, S., and Stengel, R. F., "Algebraic Training of a Neural Network, *Proc. 2001 American Control Conference*, Arlington, Va., June 2001, pp. 1605–1610.
F-2 Ferrari, S., and Stengel, R. F., "Classical/Neural Synthesis of Nonlinear Control Systems, *J. Guidance, Control, and Dynamics* 25, no. 3 (2002): 442–448.
F-3 Fink, R., and Hoak, D., "USAF Stability and Control DATCOM," U.S. Air Force Flight Dynamics Laboratory, Wright-Patterson AFB, Ohio, Apr. 1978.
F-4 Fischel, J., Rodger, L. N., Hagerman, J. R., and O'Hare, W. M., "Effect of Aspect Ratio on the Low-Speed Lateral Control Characteristics of Untapered Low-Aspect-Ratio Wings Equipped with Flap and with Retractable Ailerons," NACA Report No. 1091, Washington, D.C., 1952.
F-5 Forsell, B., *Radionavigation Systems*, Prentice-Hall, Englewood Cliffs, N.J., 1991.
F-6 Funahashi, K.-I., "On the Approximate Realization of Continuous Mappings by Neural Networks," *Neural Networks* 2 (1989): 183–192.
G-1 Gillard, W. J., and Dorsett, K. M., "Directional Control for Tailless Aircraft Using All Moving Wing Tips," *Proc. 1997 Atmospheric Flight Mechanics Conference*, AIAA-97-3487, Washington, D.C., Aug 1997, pp. 51–58
G-2 Glauert, H., "Theoretical Relationships for an Airfoil with Hinged Flap," British A.R.C. R&M 1095, London, Apr. 1927.
G-3 Goldstein, H., *Classical Mechanics*, Addison-Wesley, Reading, Mass., 1950.
G-4 Grubin, C., "Derivation of Quaternion Scheme via the Euler Axis and Angle," AIAA *Journal of Spacecraft and Rockets* 7, no. 1 (1970): 1261–1263.
G-5 Gunston, B., "Short's Experimental Sherpa," *Aeroplane Monthly* 5, no. 10 (1977): 508–514.
H-1 Halfman, R. L., *Dynamics*, Vol. 1: *Particles, Rigid Bodies, and Systems*, Addison-Wesley, Reading, Mass., 1962.
H-2 Hedrick, J. K., and Bryson, A. E., Jr., "Minimum Time Turns for a Supersonic Airplane at Constant Altitude, *J. Aircraft* 8, no. 3 (1971): 182–187.

H-3 Hedrick, J. K., and Bryson, A. E., Jr., "Three-Dimensional, Minimum-Time Turns for a Supersonic Aircraft," *J. Aircraft* 9, no. 2 (1972): 115–121.

H-4 Heffley, R. K., and Jewell, W. F., *Aircraft Handling Qualities Data*, NASA CR-2144, Washington, D.C., Dec. 1972.

H-5 Hinton, D. A., "Flight-Management Strategies for Escape From Microburst Encounters," NASA TM-4057, Washington, D.C., Aug. 1988.

H-6 Holcomb, T., and Morari, M., "Local Training for Radial Basis Function Networks: Towards Solving the Hidden Unit Problem," *Proc. Amer. Cont. Conf.*, June 1991, pp. 2331–2336.

H-7 Holland, A. S. B., and Sahney, B. N., *The General Problem of Approximation and Spline Functions*, R. E. Krieger, Huntington, N.Y., 1979.

H-8 Howe, R. M., "An Improved Numerical Integration Method for Flight Simulation," AIAA Paper No. 89-3306-CP, Boston, Aug. 1989.

J-1 Jiang, Y. F., and Lin, Y. P., "Error Analysis of Quaternion Transformations," *IEEE Transactions on Aerospace and Electronic Systems* 27, no. 4 (1991): 634–638.

J-2 Johnson, W. A., Teper, G. L., and Rediess, H. A., "Study of Control System Effectiveness in Alleviating Vortex Wake Upsets," *J. Aircraft* 11, no. 3 (1974): 148–154.

K-1 Kelly, H. J., and Edelbaum, T. N., "Energy Climbs, Energy Turns, and Asymptotic Expansions," *J. Aircraft* 7, no. 1 (1970): 93–95.

K-2 Kelly, H. J., "Flight Path Optimization with Multiple Time Scales," *J. Aircraft* 8, no. 4 (1971): 238–240.

K-3 Kolk, W. R., *Modern Flight Dynamics*, Prentice-Hall, Englewood Cliffs, N.J., 1961.

K-4 Körner, H., "Theoretical Aerodynamic Methods for Active Control Devices," *Aerodynamic Characteristics of Controls*, A. D. Young, chair, AGARD-CP-262, Neuilly-sur-Seine, France, 1979, pp. 1-1–1-28.

L-1 Lagerstrom, P. A., and Graham, M. E., "Linearized Theory of Supersonic Control Surfaces," *Journal of the Aeronautical Sciences* 16, no. 1 (1949): 31–34.

L-2 Lane, S. H., "Theory and Development of Adaptive Flight Control Systems Using Nonlinear Inverse Dynamics," Ph.D. thesis, Princeton University, MAE-1806T, Princeton, N.J., June 1988.

L-3 Libove, C., "Coordinated Turn Relations: A Graphical Representation," *J. Aircraft* 23, no. 9 (1986): 725–726.

L-4 Linse, D., and Stengel, R. F., "Identification of Aerodynamic Coefficients Using Computational Neural Networks," *J. Guidance, Control, and Dynamics* 16, no. 6 (1993): 1018–1025.

M-1 McCormick, B. W., *Aerodynamics, Aeronautics, and Flight Mechanics*, J. Wiley & Sons, New York, 1995.

M-2 McCormick, B. W., "Wake Turbulence," *Flight in an Adverse Environment*, B. W., McCormick, dir., AGARD-LS-197, Neuilly-sur-Seine, 1994, pp. 1-1–1-15.

M-3 Miele, A., *Flight Mechanics, Volume 1: Theory of Flight Paths*, Addison-Wesley, Reading, Mass., 1962.

M-4 Miele, A., Wang, T., and Melvin, W. W., "Optimization and Acceleration Guidance of Flight Trajectories in a Windshear," AIAA *Journal of Guidance, Control, and Dynamics* 10, no. 4 (1987): 368–377.
M-5 Morgan, M. B., and Thomas, H. H. B. M., "Control Surface Design in Theory and Practice," *The Journal of the Royal Aeronautical Society* 49, no. 416 (1945): 431–510.
M-6 Morrissette, R. R., and Oborny, L. F., "Theoretical Aspects of Two-Dimensional Supersonic Control Surfaces," NACA TN 2486, Washington, D.C., 1951.
N-1 Nichols, M. A., and Hedrick, J. K., "Minimum Time and Minimum Fuel Flight Path Sensitivity," *J. Aircraft* 11, no. 6 (1974): 320–325.
N-2 Nielsen, J. N., *Missile Aerodynamics*, McGraw-Hill Book Co., New York, 1960.
P-1 Pamadi, B. N., *Performance, Stability, Dynamics, and Control of Airplanes*, American Institute of Aeronautics and Astronautics, Reston, Va., 1998.
P-2 Peckham, D. H., and Leynaert, J., chairs, *Aerodynamics of Combat Aircraft Controls and of Ground Effects*, AGARD-CP-465, Neuilly-sur-Seine, Apr. 1990.
P-3 Pearson, H. A., and Anderson, R. F., "Calculation of the Aerodynamic Characteristics of Tapered Wings with Partial-Span Flaps," NACA Report No. 665, Washington, D.C., 1939.
P-4 Phillips, W. F., Hailey, C. E., and Gebert, G. A., "Review of Attitude Representations Used for Aircraft Kinematics," *J. Aircraft* 38, no. 4 (2001): 718–737.
P-5 Powell, M. J. D., *Approximation Theory and Methods*, Cambridge University Press, Cambridge, 1981.
P-6 Press, W. H., Teukolsky, S. A., Vetterling, W. T., and Flannery, B. P., *Numerical Recipes in C*, Cambridge University Press, Cambridge, 1992.
P-7 Psiaki, M. L., and Stengel, R. F., "Optimal Flight Paths Through Microburst Wind Profiles," AIAA *Journal of Aircraft* 23, no. 8 (1986): 629–635.
P-8 Poggio, T., and Girosi, F., "Regularization Algorithms for Learning that Are Equivalent to Multilayer Networks," *Science* 247, no. 4945 (1990): 978–982.
R-1 Ralston, A., and Wilf, H. S., *Mathematical Methods for Digital Computers*, J. Wiley & Sons, New York, 1965.
R-2 Renier, O., "Outils pour la Caractérization Aérodynamique et l'Evaluation des Performances à Haute Incidence," *Technologies for Highly Manoeuverable Aircraft*, L. M. B. da Costa Campos, chair, AGARD-CP-548, Neuilly-sur-Seine, Mar. 1994, pp. 9-1–9-13.
R-3 Rolfe, J. M., and Staples, K. J., *Flight Simulation*, Cambridge University Press, Cambridge, 1986.
R-4 Ross, A. J., and Thomas, H. H. B. M., "A Survey of Experimental Data on the Aerodynamics of Controls, In the Light of Future Needs," *Aerodynamic Characteristics of Controls*, AGARD-CP-262, Neuilly-sur-Seine, France, 1979, pp. 2-1–2-48.
R-5 Rossow, V. J., "Prospects for Alleviation of Hazard Posed by Lift-Generated Wakes," *Proc. Aircraft Wake Vortex Conf.*, Washington, D.C., Oct. 1991, pp. 221–22-40.

R-6 Rumelhart, D., Hinton, G., and Williams, R., "Learning Internal Representations by Error Propagation," *Parallel Distributed Processing: Explorations in the Microstructure of Cognitions, Vol. 1: Foundations*, D. Rumelhart and J. McClelland, ed., MIT Press, Cambridge, Mass., 1986.
R-7 Rutowski, E., "Energy Approach to the General Aircraft Performance Problem," *J. Aeronautical Sciences* 21, no. 3 (1954): 187–195.
S-1 Schlichting, H., and Truckenbrodt, E., *Aerodynamics of the Airplane*, McGraw-Hill, New York, 1979.
S-2 Schultz, R. L., and Zagalsky, N. R., "Aircraft Performance Optimization," *J. Aircraft* 9, no. 2 (1972): 108–114.
S-3 Schultz, R. L., "Fuel Optimality of Cruise," *J. Aircraft* 11, no. 9 (1974): 586–587.
S-4 Seckel, E., *Stability and Control of Airplanes and Helicopters*, Academic Press, New York, 1964.
S-5 Silverstein, A., and Katzoff, S., "Aerodynamic Characteristics of Horizontal Control Surfaces," NACA Report 688, Washington, D.C., 1940.
S-6 Singhal, S., Wu, L., "Training Feed-Forward Networks with the Extended Kalman Algorithm," *Proc. Int'l. Conf. Acous., Speech, Sig. Proc.*, Glasgow, May 1989, pp. 1187–1190.
S-7 Speyer, J. L., "On the Fuel Optimality of Cruise," *J. Aircraft* 10, no. 12 (1973): 763–765.
S-8 Speyer, J. L., "Nonoptimality of the Steady-State Cruise for Aircraft," *AIAA Journal* 14, no. 11 (1976): 1604–1610.
S-9 Spring, K., "Euler Parameters and the Use of Quaternion Algebra in the Manipulation of Finite Rotations: A Review," *Mechanism and Machine Theory* 21, no. 5 (1986): 365–373.
S-10 Stengel, R. F. and Marcus, F. J., *Energy Management Techniques for Fuel Conservation in Military Transport Aircraft*, AFFDL-TR-75-156, Wright-Patterson AFB, Ohio, Feb. 1976.
S-11 Stengel, R. F., and Berry, P. W., *Stability and Control of Maneuvering High-Performance Aircraft*, NASA CR-2788, Washington, D.C., Apr. 1977.
S-12 Stengel, R. F., *Optimal Control and Estimation*, Dover Publications, New York, 1994 (originally published as *Stochastic Optimal Control: Theory and Application*, J. Wiley & Sons, New York, 1986).
T-1 Teper, G. L., *Aircraft Stability and Control Data*, Technical Report 176-1, Systems Technology, Inc., Hawthorne, Calif., Apr. 1969.
T-2 Thomas, H. H. B. M., "The Aerodynamics of Aircraft Control," *Special Course on Aerodynamic Characteristics of Controls*, A. D. Young, dir., AGARD-R-711, Neuilly-sur-Seine, July, 1983, pp. 2-1–2-40.
T-3 Thomasson, P. G., "Equations of Motion of a Vehicle in a Moving Fluid," *J. Aircraft* 37, no. 4 (2000): 630–639.
T-4 Tobak, M., "On the Use of the Indicial Function Concept in the Analysis of Unsteady Motions of Wings and Wing-Tail Combinations," NACA Report 1188, Washington, D.C., 1954.
T-5 Toll, T. A., compiler, "Summary of Lateral-Control Research," NACA Report 868, Washington, D.C., 1947.
V-1 Vicroy, D. D., and Bowles, R. L., "Effect of Spatial Gradients on Airplane Dynamics," AIAA *Journal of Aircraft* 26, no. 6 (1989): 523–530.

W-1 Walker, D. J., "Fin Design with A.C.T. in the Presence of Strakes," *Aerodynamic Characteristics of Controls*, A. D. Young, chair, AGARD-CP-262, Neuilly-sur-Seine, France, 1979, pp. 12-1–12-6.
W-2 Werbos, P. J., "Backpropagation Through Time: What It Does and How to Do It," *Proc. IEEE* 78, no. 10 (1990): 1550–1560.
W-3 Williams, J., "The USAF Stability and Control Digital DATCOM," AFFDL-TR-76-45, Wright-Patterson AFB, Ohio, Nov. 1976.
W-4 Wolfram, S., *The Mathematica® Book*, Wolfram Research, Champaign, Ill, 2003.
W-5 Wood, N. J., and Crowther, W. J., "Yaw Control by Tangential Forebody Blowing," *Technologies for Highly Manoeuverable Aircraft*, L. M. B. da Costa Campos, chair, AGARD-CP-548, Neuilly-sur-Seine, Mar. 1994, pp. 10-1–10-10.
Y-1 Young, A. D., dir., *Special Course on Aerodynamic Characteristics of Controls*, AGARD-CP-262, Neuilly-sur-Seine, France, 1979.
Z-1 Zagalsky, N. R., Irons, R. P., Jr., and Schultz, R. L., "Energy State Approximation and Minimum-Fuel Fixed-Range Trajectories, *J. Aircraft* 8, no. 6 (1971): 488–490.
Z-2 Zagalsky, N. R., "Aircraft Energy Management," AIAA Paper No. 73-228, New York, Jan. 1973.
Z-3 Zipfel, P. H., *Modeling and Simulation of Aerospace Vehicle Dynamics*, American Institute of Aeronautics and Astronautics, Reston, Va., 2001.

Chapter 4

Methods of Analysis and Design

With the ready availability of digital computation, it is increasingly easy to analyze the dynamics of aircraft by integrating their nonlinear equations of motion. The numerical results are as good as the data that are used for computation, and even subtle differences in response that arise from changing parameters or conditions can be stored, printed, and plotted for interpretation. What is missing in this direct approach is an underlying framework for simplifying, unifying, and generalizing the results, for learning the underlying lessons that the numbers only imply. In a tradition that predates digital computation by several decades, we can gain considerable insight by referring to linearized models that are derived from the nonlinear equations and by investigating the probabilistic effects of system uncertainty.

Small perturbations from a nominal flight path can be simulated and evaluated using the linearized equations of motion. Linear models provide particular advantages for analysis because the solutions are additive and underlying modes of motion can be identified. These perturbations may arise from small variations in initial conditions, control inputs, or disturbance effects. The nominal path may represent a changing flight condition, but more often it is a steady or quasisteady path. In the first case, linear, time-varying (LTV) models are appropriate, while in the second, even simpler linear, time-invariant (LTI) models are satisfactory. LTI models receive emphasis as the chapter evolves, as they form the basis for classical stability-and-control analysis.

Methods for deriving and analyzing linear models of aircraft dynamics are presented here. *Local linearization* of the previous chapter's nonlinear models produces linear ordinary differential equations that apply in the neighborhood of a nominal path, and it allows either time-varying or time-invariant *stability-and-control derivatives* to be defined (Section 4.1). These derivatives are discussed in some detail, and partitioned and reduced-order linear models suitable for analysis are identified. Analytical methods are described, including integration of the equations, concepts of stability and control, transformation from the time domain of ordinary differential equations to the frequency domain of Fourier and Laplace transforms, and transfer functions (Sections 4.2–4.4).

Most mathematical models contain some degree of imprecision, and Section 4.5 introduces ways of addressing uncertainty in analyses. Elements of probability theory are reviewed, and the propagation of uncertain inputs through linear dynamic systems is presented. A *system survey* based upon deterministic and stochastic components is described, and the use of Monte Carlo evaluation for generating probabilistic information about system response characteristics is explained.

Aeroelasticity plays a critical dynamical role in modern aircraft, as structures are lightweight and flexible. Interactions between elastic distortions of the airframe and its otherwise rigid-body motions may be significant for flying qualities and flight control system design. After briefly reviewing basic static properties of material stress and strain, linear aeroelastic models that incorporate rigid-body motions, structural vibration, and coupling aerodynamics are derived (Section 4.6). These equations fall into the general category of LTI state-space models presented earlier in the chapter; hence, they are amenable to frequency-domain descriptions as well.

Section 4.7 provides an introduction to aircraft flying qualities and flight control systems, focusing largely on LTI dynamic models. The human pilot possesses a versatility for aircraft guidance, navigation, and control that is hard to match in fully automated systems. Nevertheless, to maximize the effectiveness of piloting actions in achieving mission goals, the dynamic characteristics of the aircraft must lie within proscribed boundaries. These boundaries and the reasons for choosing them are reviewed in this section. The section concludes with a systematic approach to designing flight control systems that satisfy quantitative flying qualities criteria, in the process, assuring that closed-loop stability is maintained for the assumed dynamic models.

4.1 Local Linearization of Differential Equations

The nonlinear equations of motion derived earlier take the vector form

$$\dot{\mathbf{x}}(t) = \mathbf{f}[\mathbf{x}(t), \mathbf{u}(t), \mathbf{w}(t), t] \tag{4.1-1}$$

where $\mathbf{x}(t)$ is the n-dimensional state vector, $\mathbf{u}(t)$ is the m-dimensional control vector, and $\mathbf{w}(t)$ is the s-dimensional disturbance vector. The vector equation is solved subject to the initial condition $\mathbf{x}(t_0) = \mathbf{x}_0$. The r-dimensional *output vector* $\mathbf{y}(t)$,

$$\mathbf{y}(t) = \mathbf{h}[\mathbf{x}(t), \mathbf{u}(t), \mathbf{w}(t), t] \tag{4.1-2}$$

is an algebraic or transcendental function of the state, control, and disturbance that is chosen by the analyst. It may represent the set of measurements made by a tracking or control system, a set of variables commanded by the pilot or autopilot, or a set of variables that have special meaning in control system design. As detailed in later sections, the output vector could simply be the state ($\mathbf{y} = \mathbf{x}$, $r = n$), a subset of the state ($\mathbf{y} = [\mathbf{I}_r\ 0]\,\mathbf{x}$, $r < n$), or a more involved set.

All of the above variables can be expressed as the sums of nominal values $(\cdot)_o$ plus perturbations $\Delta(\cdot)$:

$$\mathbf{x}(t) \triangleq \mathbf{x}_o(t) + \Delta\mathbf{x}(t) \tag{4.1-3}$$

$$\mathbf{u}(t) \triangleq \mathbf{u}_o(t) + \Delta\mathbf{u}(t) \tag{4.1-4}$$

$$\mathbf{w}(t) \triangleq \mathbf{w}_o(t) + \Delta\mathbf{w}(t) \tag{4.1-5}$$

$$\mathbf{y}(t) \triangleq \mathbf{y}_o(t) + \Delta\mathbf{y}(t) \tag{4.1-6}$$

hence, equations 4.1-1 and 4.1-2 can be rewritten as

$$\dot{\mathbf{x}}_o(t) + \Delta\dot{\mathbf{x}}(t) = \mathbf{f}[\mathbf{x}_o(t) + \Delta\mathbf{x}(t), \mathbf{u}_o(t) + \Delta\mathbf{u}(t), \mathbf{w}_o(t) + \Delta\mathbf{w}(t), t] \tag{4.1-7}$$

$$\mathbf{y}_o(t) + \Delta\mathbf{y}(t) = \mathbf{h}[\mathbf{x}_o(t) + \Delta\mathbf{x}(t), \mathbf{u}_o(t) + \Delta\mathbf{u}(t), \mathbf{w}_o(t) + \Delta\mathbf{w}(t), t] \tag{4.1-8}$$

The right sides of these equations can be expanded in Taylor series; retaining just the first-order terms, the approximate equations are

$$\begin{aligned}\dot{\mathbf{x}}_o(t) + \Delta\dot{\mathbf{x}}(t) \approx\ & \mathbf{f}[\mathbf{x}_o(t), \mathbf{u}_o(t), \mathbf{w}_o(t), t] \\ & + \frac{\partial \mathbf{f}}{\partial \mathbf{x}}\Delta\mathbf{x}(t) + \frac{\partial \mathbf{f}}{\partial \mathbf{u}}\Delta\mathbf{u}(t) + \frac{\partial \mathbf{f}}{\partial \mathbf{w}}\Delta\mathbf{w}(t)\end{aligned} \tag{4.1-9}$$

$$\begin{aligned}\mathbf{y}_o(t) + \Delta\mathbf{y}(t) \approx\ & \mathbf{h}[\mathbf{x}_o(t), \mathbf{u}_o(t), \mathbf{w}_o(t), t] \\ & + \frac{\partial \mathbf{h}}{\partial \mathbf{x}}\Delta\mathbf{x}(t) + \frac{\partial \mathbf{h}}{\partial \mathbf{u}}\Delta\mathbf{u}(t) + \frac{\partial \mathbf{h}}{\partial \mathbf{w}}\Delta\mathbf{w}(t)\end{aligned} \tag{4.1-10}$$

The six partial-derivative (or *Jacobian*) matrices express the linear sensitivities of $\mathbf{f}$ and $\mathbf{h}$ to perturbations in $\mathbf{x}$, $\mathbf{u}$, and $\mathbf{w}$, and they are evaluated along the nominal trajectory. For example, the Jacobian of $\mathbf{f}$ with respect to $\mathbf{x}$ is an $(n \times n)$ matrix, also denoted by $\mathbf{F}$:

$$\begin{aligned}\frac{\partial \mathbf{f}}{\partial \mathbf{x}} &= \left.\frac{\partial \mathbf{f}}{\partial \mathbf{x}}\right|_{\substack{\mathbf{x} = \mathbf{x}_o(t) \\ \mathbf{u} = \mathbf{u}_o(t) \\ \mathbf{w} = \mathbf{w}_o(t) \\ t}} \triangleq \mathbf{F}[\mathbf{x}_o(t), \mathbf{u}_o(t), \mathbf{w}_o(t), t] \\ &= \left.\begin{bmatrix} \partial f_1/\partial x_1 & \partial f_1/\partial x_2 & \cdots & \cdots & \partial f_1/\partial x_n \\ \partial f_2/\partial x_1 & \partial f_2/\partial x_2 & \cdots & \cdots & \partial f_2/\partial x_n \\ \cdots & \cdots & \cdots & \cdots & \cdots \\ \partial f_n/\partial x_1 & \partial f_n/\partial x_2 & \cdots & \cdots & \partial f_n/\partial x_n \end{bmatrix}\right|_{\substack{\mathbf{x} = \mathbf{x}_o(t) \\ \mathbf{u} = \mathbf{u}_o(t) \\ \mathbf{w} = \mathbf{w}_o(t) \\ t}}\end{aligned} \tag{4.1-11}$$

Because $\mathbf{x}_o(t)$, $\mathbf{u}_o(t)$, and $\mathbf{w}_o(t)$ are functions of time and f itself may be an explicit function of time, $\mathbf{F}$ is, in general, a time-varying matrix. If $\mathbf{x}_o$, $\mathbf{u}_o$, and $\mathbf{w}_o$ are constant and f is not a function of time (as in the case of cruising flight with negligible fuel-burn rate), then $\mathbf{F}$ is approximately constant. The remaining Jacobian matrices are defined as follows:

$$\mathbf{G} = \frac{\partial \mathbf{f}}{\partial \mathbf{u}} \quad (n \times m) \tag{4.1-12}$$

$$\mathbf{L} = \frac{\partial \mathbf{f}}{\partial \mathbf{w}} \quad (n \times s) \tag{4.1-13}$$

$$\mathbf{H} = \frac{\partial \mathbf{h}}{\partial \mathbf{x}} \quad (r \times n) \tag{4.1-14}$$

$$\mathbf{J} = \frac{\partial \mathbf{h}}{\partial \mathbf{u}} \quad (r \times m) \tag{4.1-15}$$

$$\mathbf{K} = \frac{\partial \mathbf{h}}{\partial \mathbf{w}} \quad (r \times s) \tag{4.1-16}$$

Each Jacobian is evaluated at time t with $\mathbf{x} = \mathbf{x}_o(t)$, $\mathbf{u} = \mathbf{u}_o(t)$, $\mathbf{w} = \mathbf{w}_o(t)$.

The linearization proceeds by equating like parts of eqs. 4.1-9 and 4.1-10; thus the *nominal dynamic and output equations* are

$$\dot{\mathbf{x}}_o(t) = \mathbf{f}[\mathbf{x}_o(t), \mathbf{u}_o(t), \mathbf{w}_o(t), t] \tag{4.1-17}$$

$$\mathbf{y}_o(t) = \mathbf{h}[\mathbf{x}_o(t), \mathbf{u}_o(t), \mathbf{w}_o(t), t] \tag{4.1-18}$$

with initial conditions $\mathbf{x}_o(t_0)$, while the linear, time-varying *perturbation equations* are

$$\Delta\dot{\mathbf{x}}(t) = \mathbf{F}(t)\Delta\mathbf{x}(t) + \mathbf{G}(t)\Delta\mathbf{u}(t) + \mathbf{L}(t)\Delta\mathbf{w}(t) \tag{4.1-19}$$

$$\Delta\mathbf{y}(t) = \mathbf{H}(t)\Delta\mathbf{x}(t) + \mathbf{J}(t)\Delta\mathbf{u}(t) + \mathbf{K}(t)\Delta\mathbf{w}(t) \tag{4.1-20}$$

with $\Delta\mathbf{x}(t_o) = \Delta\mathbf{x}_0$. The nominal and perturbation equations can be solved separately; however, the nominal equations must be solved first, so that the sensitivity matrices of the perturbation equations can be assigned their proper values. The perturbation equations then convey the effects of small variations in $\mathbf{x}_o(t_0)$, $\mathbf{u}(t)$, and $\mathbf{w}(t)$. The sum of $\mathbf{x}_o(t)$ and $\Delta\mathbf{x}(t)$ renders a close approximation to $\mathbf{x}(t)$.

STABILITY AND CONTROL DERIVATIVES

Referring to Section 3.2, the aircraft's rigid-body equations can be written as

$$\frac{d\mathbf{v}_{\mathrm{B}}}{dt} = \frac{\mathbf{f}_{\mathrm{B}\,aero} + \mathbf{f}_{\mathrm{B}\,thrust}}{m(t)} + \mathbf{H}_{\mathrm{I}}^{\mathrm{B}}\mathbf{g}_{\mathrm{I}} - \tilde{\boldsymbol{\omega}}_{\mathrm{B}}\mathbf{v}_{\mathrm{B}} \tag{4.1-21}$$

$$\frac{d\boldsymbol{\omega}_{\mathrm{B}}}{dt} = \mathcal{I}_{\mathrm{B}}^{-1}(\mathbf{f}_{\mathrm{B}\,aero} + \mathbf{f}_{\mathrm{B}\,thrust} - \tilde{\boldsymbol{\omega}}_{\mathrm{B}}\,\mathcal{I}_{\mathrm{B}}\boldsymbol{\omega}_{\mathrm{B}}) \tag{4.1-22}$$

$$\frac{d\mathbf{r}_{\mathrm{I}}}{dt} = \mathbf{H}_{\mathrm{B}}^{\mathrm{I}}\mathbf{v}_{\mathrm{B}} \tag{4.1-23}$$

$$\frac{d\boldsymbol{\Theta}}{dt} = \mathbf{L}_{\mathrm{B}}^{\mathrm{E}}\boldsymbol{\omega}_{\mathrm{B}} \tag{4.1-24}$$

Note that $\mathbf{L}_{\mathrm{B}}^{\mathrm{E}}$ is the body-to-Euler-rate transformation matrix, while $\mathbf{L}(t)$ is the disturbance-effect matrix. These equations take the form of eq. 4.1-1, with state, control, and wind-disturbance vectors defined as follows:

$$\mathbf{x}(t) = [\mathbf{v}_{\mathrm{B}}^T(t)\ \boldsymbol{\omega}_{\mathrm{B}}^T(t)\ \mathbf{r}_{\mathrm{I}}^T(t)\ \boldsymbol{\Theta}^T(t)]^T = [\mathbf{x}_1^T\ \mathbf{x}_2^T\ \mathbf{x}_3^T\ \mathbf{x}_4^T]^T$$

$$= [u\ v\ w\ p\ q\ r\ x_{\mathrm{I}}\ y_{\mathrm{I}}\ z_{\mathrm{I}}\ \phi\ \theta\ \psi]^T \tag{4.1-25}$$

$$\mathbf{u}(t) = [\delta E\ \delta A\ \delta R\ \delta T]^T \tag{4.1-26}$$

$$\mathbf{w}(t) = [u_{\mathrm{W}}\ v_{\mathrm{W}}\ w_{\mathrm{W}}]^T \tag{4.1-27}$$

Consequently, the (12×12) *stability matrix* F can be written in terms of (3×3) blocks:

$$\mathrm{F} = \begin{bmatrix} \mathrm{F}_{11} & \mathrm{F}_{12} & \mathrm{F}_{13} & \mathrm{F}_{14} \\ \mathrm{F}_{21} & \mathrm{F}_{22} & \mathrm{F}_{23} & \mathrm{F}_{24} \\ \mathrm{F}_{31} & 0 & 0 & \mathrm{F}_{34} \\ 0 & \mathrm{F}_{42} & 0 & \mathrm{F}_{44} \end{bmatrix} \tag{4.1-28}$$

By the form of eqs. 4.1-21–4.1-24, several of the (3×3) blocks are zero for the flat-earth model. The *control-effect* and *disturbance-effect matrices* G and L are similarly defined, with the number of columns dependent on the number of control and disturbance inputs:

$$\mathrm{G} = \begin{bmatrix} \mathrm{G}_1 \\ \mathrm{G}_2 \\ 0 \\ 0 \end{bmatrix} \tag{4.1-29}$$

$$L = \begin{bmatrix} L_1 \\ L_2 \\ 0 \\ 0 \end{bmatrix} \tag{4.1-30}$$

The zero elements of **G** and **L** arise because forces and moments have no direct effect on translational and rotational position, $\mathbf{r}_I$ and $\boldsymbol{\Theta}$.

Given a nonlinear model of aircraft aerodynamic, thrust, and inertial properties, the Jacobian matrices of the linear system equations can be evaluated numerically. The operating point for linearization is defined by a feasible nominal flight condition $\mathbf{x}_o$, $\mathbf{u}_o$, and $\mathbf{w}_o$. If the nominal condition represents a trim point, the Jacobian matrices are constant, and eqs. 4.1-19 and 4.1-20 are *linear, time-invariant* (LTI) *equations*. If the flight condition represents an arbitrary point along a changing flight path, the matrices also vary from one instant to the next, and the dynamic equations are *linear time-varying* (LTV). The *principle of superposition* applies for both LTI and LTV models, allowing solutions of the differential equations with different initial conditions and inputs to be added; however, only LTI models have fixed *modes of motion* that can be analyzed by the eigenvalue-based and "frequency-domain" methods of the following sections.

We assume that aircraft data representation provides smooth, continuous, first partial derivatives of **f** and **h** with respect to **x**, **u**, and **w**. Linear interpolation does not provide continuous derivatives across tabulated points (Section 3.6). A higher-degree interpolation or representation by polynomials or neural networks is needed to produce smooth variation. Equation 4.1-11 provides an example for approximation of the stability matrix **F** by *first central differences*. The ith element of the vector **f** is f_i, the jth element of **x** is x_j, and the ijth element of **F** is F_{ij}, where each index runs from 1 to n. The derivative F_{ij} is approximated by the slope of points that bracket the function at the operating point,

$$\begin{aligned} F_{ij} &= \frac{\partial f_i}{\partial x_j} \approx \frac{\Delta f_i}{\Delta x_j} \\ &= \frac{f_i(\mathbf{x}_o + \Delta x_j, \mathbf{u}_o, \mathbf{w}_o) - f_i(\mathbf{x}_o - \Delta x_j, \mathbf{u}_o, \mathbf{w}_o)}{2\Delta x_j} \end{aligned} \tag{4.1-31}$$

where Δx_j produces a small variation in f_i. There is an inevitable tradeoff in defining "small." Δx_j must be small enough to assure that the curvature between points does not introduce significant error in the approximation, but it must be large enough to preserve accuracy with finite-precision data. The notation $(\mathbf{x}_o \pm \Delta x_j)$ indicates that only the jth term of $\mathbf{x}_o$ is perturbed to become $(x_{o_j} \pm \Delta x_j)$: all other elements of $\mathbf{x}_o$ keep their nominal values. Equation 4.1-31 is evaluated by function calls to f over all i and j, as through nested "FOR loops" in C or MATLAB. The remaining Jacobian matrices are generated in the same way, replacing **f** by **h** for **H, J,** and **K** and Δx_j by Δu_j and Δw_j for **G, L, J,** and **K.** The procedure automatically reveals any coupling between longitudinal and lateral-directional motions because all first-order sensitivities are evaluated.

The elements of **F, G,** and **L** can also be evaluated by symbolic manipulation of eqs. 4.1-21–4.1-24 or their scalar equivalents (e.g., eqs. 3.6-19–3.6-41). Using the partitioning established above, the block elements are as follows [S-9]:

$$\mathbf{F}_{11} = \frac{1}{m}\frac{\partial \mathbf{f}_B}{\partial \mathbf{v}_B} - \tilde{\boldsymbol{\omega}}_{B_o} \tag{4.1-32}$$

$$\mathbf{F}_{12} = \frac{1}{m}\frac{\partial \mathbf{f}_B}{\partial \boldsymbol{\omega}_B} + \tilde{\mathbf{v}}_{B_o} \tag{4.1-33}$$

$$\mathbf{F}_{13} = \frac{1}{m}\frac{\partial \mathbf{f}_B}{\partial \mathbf{r}_I} + \mathbf{H}^B_{I_o}\frac{\partial \mathbf{g}_I}{\partial \mathbf{r}_I} \tag{4.1-34}$$

$$\mathbf{F}_{14} = \mathbf{H}^B_{I_o}\tilde{\mathbf{g}}\mathbf{I}_o\,\mathbf{H}^I_{B_o}\mathbf{L}^B_{E_o} \tag{4.1-35}$$

$$\mathbf{F}_{21} = \mathcal{I}_B^{-1}\frac{\partial \mathbf{m}_B}{\partial \mathbf{v}_B} \tag{4.1-36}$$

$$\mathbf{F}_{22} = \mathcal{I}_B^{-1}\left[\frac{\partial \mathbf{m}_B}{\partial \boldsymbol{\omega}_B} + \tilde{\mathbf{a}} - \tilde{\boldsymbol{\omega}}_{B_o}\mathcal{I}_B\right], \quad (\mathbf{a} \triangleq \mathcal{I}_B\boldsymbol{\omega}_{B_o}) \tag{4.1-37}$$

$$\mathbf{F}_{23} = \mathcal{I}_B^{-1}\frac{\partial \mathbf{m}_B}{\partial \mathbf{r}_I} \tag{4.1-38}$$

$$\mathbf{F}_{24} = \mathcal{I}_B^{-1}\frac{\partial \mathbf{m}_B}{\partial \boldsymbol{\Theta}} \tag{4.1-39}$$

$$\mathbf{F}_{31} = \mathbf{H}^I_{B_o} \tag{4.1-40}$$

$$\mathbf{F}_{34} = -\tilde{\mathbf{b}}\mathbf{H}^I_{B_o}\mathbf{L}^B_{E_o}, \quad (\mathbf{b} \triangleq \mathbf{H}^I_{B_o}\mathbf{v}_{B_o}) \tag{4.1-41}$$

$$\mathbf{F}_{42} = \mathbf{L}^E_{B_o} \tag{4.1-42}$$

$$\mathbf{F}_{44} = \begin{bmatrix} \dot{\theta}_o \tan\theta_o & \dot{\psi}_o \sec\theta_o & 0 \\ -\dot{\psi}_o \cos\theta_o & 0 & 0 \\ \dot{\theta}_o \sec\theta_o & \dot{\psi}_o \tan\theta_o & 0 \end{bmatrix} \tag{4.1-43}$$

$$\mathbf{G}_1 = \frac{1}{m}\frac{\partial \mathbf{f}_B}{\partial \mathbf{u}} \tag{4.1-44}$$

$$\mathbf{G}_2 = \mathcal{I}_B^{-1}\frac{\partial \mathbf{m}_B}{\partial \mathbf{u}} \tag{4.1-45}$$

$$\mathbf{L}_1 = \frac{1}{m}\frac{\partial \mathbf{f}_B}{\partial \mathbf{w}} \tag{4.1-46}$$

$$\mathbf{L}_2 = \mathcal{I}_B^{-1}\frac{\partial \mathbf{m}_B}{\partial \mathbf{w}} \tag{4.1-47}$$

The elements of **F** contain *dimensional stability derivatives*, which can be expressed in terms of the nondimensional force and moment derivatives described in Chapters 2 and 3. The external force vector $\mathbf{f}_B$ is defined as

$$\mathbf{f}_B \triangleq \frac{1}{2}\rho V_{A_o}^2 S \begin{bmatrix} C_X \\ C_Y \\ C_Z \end{bmatrix} + \begin{bmatrix} T_X \\ T_Y \\ T_Z \end{bmatrix} \tag{4.1-48}$$

where

$$V_{A_o}^2 = (u_o - u_{W_o})^2 + (v_o - v_{W_o})^2 + (w_o - w_{W_o})^2 \tag{4.1-49}$$

and the air density ρ is a function of altitude. To simplify the remainder of the discussion, thrust effects are incorporated in the nondimensional derivatives (as in equation 2.5-37; thus, for the thrust aligned with the x axis, $C_X = C_{X_{aero}} + C_{X_{thrust}} = C_{X_{aero}} + C_T$). It is assumed that the nominal wind field is zero, so the nominal airspeed V_{A_o} is V_o. Nominal values are denoted by the subscript "o," and partial derivatives evaluated at the nominal condition are denoted by the subscripts "u," "v," or "w"(e.g., $\partial C_X/\partial u \triangleq C_{X_u}$). The matrix of translational acceleration sensitivities to the body-axis velocity is, with zero mean wind velocity,

$$\frac{1}{m}\frac{\partial \mathbf{f}_B}{\partial \mathbf{v}_B} = \frac{\rho_o V_o^2 S}{2m}\begin{bmatrix} \left(\dfrac{2u_o}{V_o^2}C_{X_o} + C_{X_u}\right) & \left(\dfrac{2v_o}{V_o^2}C_{X_o} + C_{X_v}\right) & \left(\dfrac{2w_o}{V_o^2}C_{X_o} + C_{X_w}\right) \\ \left(\dfrac{2u_o}{V_o^2}C_{Y_o} + C_{Y_u}\right) & \left(\dfrac{2v_o}{V_o^2}C_{Y_o} + C_{Y_v}\right) & \left(\dfrac{2w_o}{V_o^2}C_{Y_o} + C_{Y_w}\right) \\ \left(\dfrac{2u_o}{V_o^2}C_{Z_o} + C_{Z_u}\right) & \left(\dfrac{2v_o}{V_o^2}C_{Z_o} + C_{Z_v}\right) & \left(\dfrac{2w_o}{V_o^2}C_{Z_o} + C_{Z_w}\right) \end{bmatrix}$$

$$\triangleq \begin{bmatrix} X_u & X_v & X_w \\ Y_u & Y_v & Y_w \\ Z_u & Z_v & Z_w \end{bmatrix} \tag{4.1-50}$$

The dimensional stability derivatives in eq. 4.1-50 (e.g., X_u and Z_w) represent the sensitivity of the *specific force* (i.e., force per unit of mass) to changes in the state, evaluated at the nominal condition. Two effects lead to this senitivity; using X_u as an example, the term $[\rho_o(2u_o)S/2m]C_{X_o}$ arises from change in the dynamic pressure, while the second term $(\rho_o V_o^2 S/2m)C_{X_u}$ results from change in the coefficient C_X due to Mach number, Reynolds number, and aeroelastic effects. Note the departure in notation: elsewhere, (X, Y, Z) represents force (N), while here it represents force per unit mass (N/kg), or the *acceleration* induced by the force.

Specific force sensitivities to angular rate, translational position, and control are given by

$$\frac{1}{m}\frac{\partial \mathbf{f}_B}{\partial \boldsymbol{\omega}_B} = \frac{\rho_o V_o S}{4m}\begin{bmatrix} bC_{X_{\hat{p}}} & \bar{c}C_{X_{\hat{q}}} & bC_{X_{\hat{r}}} \\ bC_{Y_{\hat{p}}} & \bar{c}C_{Y_{\hat{q}}} & bC_{Y_{\hat{r}}} \\ bC_{Z_{\hat{p}}} & \bar{c}C_{Z_{\hat{q}}} & bC_{Z_{\hat{r}}} \end{bmatrix}$$

$$\triangleq \begin{bmatrix} X_p & X_q & X_r \\ Y_p & Y_q & Y_r \\ X_p & Z_q & Z_r \end{bmatrix} \tag{4.1-51}$$

$$\frac{1}{m}\frac{\partial \mathbf{f}_B}{\partial \mathbf{r}_I} = \frac{\rho_o V_o^2 S}{2m}\begin{bmatrix} \left(\frac{\rho_x}{\rho}C_{X_o} + C_{X_x}\right) & \left(\frac{\rho_y}{\rho}C_{X_o} + C_{X_y}\right) & \left(\frac{\rho_z}{\rho}C_{X_o} + C_{X_z}\right) \\ \left(\frac{\rho_x}{\rho}C_{Y_o} + C_{Y_x}\right) & \left(\frac{\rho_y}{\rho}C_{Y_o} + C_{Y_y}\right) & \left(\frac{\rho_z}{\rho}C_{Y_o} + C_{Y_z}\right) \\ \left(\frac{\rho_x}{\rho}C_{Z_o} + C_{Z_x}\right) & \left(\frac{\rho_y}{\rho}C_{Z_o} + C_{Z_y}\right) & \left(\frac{\rho_z}{\rho}C_{Z_o} + C_{Z_z}\right) \end{bmatrix}$$

$$\triangleq \begin{bmatrix} X_x & X_y & X_z \\ Y_x & Y_y & Y_z \\ Z_x & Z_y & Z_z \end{bmatrix} \tag{4.1-52}$$

$$\frac{1}{m}\frac{\partial \mathbf{f}_B}{\partial \mathbf{u}} = \frac{\rho_o V_o^2 S}{2m}\begin{bmatrix} C_{X_{\delta E}} & C_{X_{\delta A}} & C_{X_{\delta R}} & C_{X_{\delta T}} \\ C_{Y_{\delta E}} & C_{Y_{\delta A}} & C_{Y_{\delta R}} & C_{Y_{\delta T}} \\ C_{Z_{\delta E}} & C_{Z_{\delta A}} & C_{Z_{\delta R}} & C_{Z_{\delta T}} \end{bmatrix}$$

$$\triangleq \begin{bmatrix} X_{\delta E} & X_{\delta A} & X_{\delta R} & X_{\delta T} \\ Y_{\delta E} & Y_{\delta A} & Y_{\delta R} & Y_{\delta T} \\ Z_{\delta E} & Z_{\delta A} & Z_{\delta R} & Z_{\delta T} \end{bmatrix} \quad (4.1\text{-}53)$$

From Section 3.4, recall that the careted values of p, q, and r are defined as $\hat{p} \triangleq pb/2V$, $\hat{q} \triangleq q\bar{c}/2V$, and $\hat{r} \triangleq rb/2V$, causing the reference lengths b and $\bar{c}$ to appear in eq. 4.1-51. The origins of the rotational rate effects are found in Section 3.4, while control effects appear in Section 3.5. Translational position effects are usually quite small, arising principally from ground effects and from altitude gradients in air density and sound speed. Gravity-gradient effects are also modeled in eq. 4.1-34.

The dimensional stability derivatives for rotational accelerations (or *specific moments*, i.e., moments per unit of inertia) are defined in the same way, beginning with the aerodynamic and thrust moments of Sections 2.4 and 2.5:

$$\mathbf{m}_B \triangleq \frac{1}{2}\rho V_A^2 S \begin{bmatrix} b\,C_l \\ \bar{c}\,C_m \\ b\,C_n \end{bmatrix} + \begin{bmatrix} N_X \\ N_Y \\ N_Z \end{bmatrix} \quad (4.1\text{-}54)$$

The specific moment sensitivities to translational velocity, rotational rate, translational position, and control are expressed as

$$\mathcal{I}_B^{-1}\frac{\partial \mathbf{m}_B}{\partial \mathbf{v}_B} = \mathcal{I}_B^{-1}\frac{\rho_o V_o^2 S}{2}\begin{bmatrix} b\left(\frac{2u_o}{V_o^2}C_{l_o} + C_{l_u}\right) & b\left(\frac{2v_o}{V_o^2}C_{l_o} + C_{l_v}\right) & b\left(\frac{2w_o}{V_o^2}C_{l_o} + C_{l_w}\right) \\ \bar{c}\left(\frac{2u_o}{V_o^2}C_{m_o} + C_{m_u}\right) & \bar{c}\left(\frac{2v_o}{V_o^2}C_{m_o} + C_{m_v}\right) & \bar{c}\left(\frac{2w_o}{V_o^2}C_{m_o} + C_{m_w}\right) \\ b\left(\frac{2u_o}{V_o^2}C_{n_o} + C_{n_u}\right) & b\left(\frac{2v_o}{V_o^2}C_{n_o} + C_{n_v}\right) & b\left(\frac{2w_o}{V_o^2}C_{n_o} + C_{n_w}\right) \end{bmatrix}$$

$$\triangleq \begin{bmatrix} L_u & L_v & L_w \\ M_u & M_v & M_w \\ N_u & N_v & N_w \end{bmatrix} \quad (4.1\text{-}55)$$

$$\mathcal{I}_B^{-1}\frac{\partial \mathbf{m}_B}{\partial \boldsymbol{\omega}_B} = \mathcal{I}_B^{-1}\frac{\rho_o V_o S}{4}\begin{bmatrix} b^2 C_{l_{\hat{p}}} & b\bar{c}C_{l_{\hat{q}}} & b^2 C_{l_{\hat{r}}} \\ b\bar{c}C_{m_{\hat{p}}} & \bar{c}^2 C_{m_{\hat{q}}} & b\bar{c}C_{m_{\hat{r}}} \\ b^2 C_{n_{\hat{p}}} & b\bar{c}C_{n_{\hat{q}}} & b^2 C_{n_{\hat{r}}} \end{bmatrix}$$

$$\triangleq \begin{bmatrix} L_p & L_q & L_r \\ M_p & M_q & M_r \\ N_p & N_q & N_r \end{bmatrix} \tag{4.1-56}$$

$$\mathcal{I}_B^{-1}\frac{\partial \mathbf{m}_B}{\partial \mathbf{r}_I} = \mathcal{I}_B^{-1}\frac{\rho_o V_o^2 S}{2}\begin{bmatrix} b\left(\frac{\rho_x}{\rho}C_{l_o} + C_{l_x}\right) & b\left(\frac{\rho_y}{\rho}C_{l_o} + C_{l_y}\right) & b\left(\frac{\rho_z}{\rho}C_{l_o} + C_{l_z}\right) \\ \bar{c}\left(\frac{\rho_x}{\rho}C_{m_o} + C_{m_x}\right) & \bar{c}\left(\frac{\rho_y}{\rho}C_{m_o} + C_{m_y}\right) & \bar{c}\left(\frac{\rho_z}{\rho}C_{m_o} + C_{m_z}\right) \\ b\left(\frac{\rho_x}{\rho}C_{n_o} + C_{n_x}\right) & b\left(\frac{\rho_y}{\rho}C_{n_o} + C_{n_y}\right) & b\left(\frac{\rho_z}{\rho}C_{n_o} + C_{n_z}\right) \end{bmatrix}$$

$$\triangleq \begin{bmatrix} L_x & L_y & L_z \\ M_x & M_y & M_z \\ N_x & N_y & N_z \end{bmatrix} \tag{4.1-57}$$

$$\mathcal{I}_B^{-1}\frac{\partial \mathbf{m}_B}{\partial \Theta} = \mathcal{I}_B^{-1}\frac{\rho_o V_o^2 S}{2}\begin{bmatrix} bC_{l_\phi} & bC_{l_\theta} & 0 \\ \bar{c}C_{m_\phi} & \bar{c}C_{m_\theta} & 0 \\ bC_{n_\phi} & bC_{n_\theta} & 0 \end{bmatrix}$$

$$\triangleq \begin{bmatrix} L_\phi & L_\theta & 0 \\ M_\phi & M_\theta & 0 \\ N_\phi & N_\theta & 0 \end{bmatrix} \tag{4.1-58}$$

$$\mathcal{I}_B^{-1}\frac{\partial \mathbf{m}_B}{\partial \mathbf{u}} = \mathcal{I}_B^{-1}\frac{\rho_o V_o^2 S}{2}\begin{bmatrix} bC_{l_{\delta E}} & bC_{l_{\delta A}} & bC_{l_{\delta R}} & bC_{l_{\delta T}} \\ \bar{c}C_{m_{\delta E}} & \bar{c}C_{m_{\delta A}} & \bar{c}C_{m_{\delta R}} & \bar{c}C_{m_{\delta T}} \\ bC_{n_{\delta E}} & bC_{n_{\delta A}} & bC_{n_{\delta R}} & bC_{n_{\delta T}} \end{bmatrix}$$

$$\triangleq \begin{bmatrix} L_{\delta E} & L_{\delta A} & L_{\delta R} & L_{\delta T} \\ M_{\delta E} & M_{\delta A} & M_{\delta R} & M_{\delta T} \\ N_{\delta E} & N_{\delta A} & N_{\delta R} & N_{\delta T} \end{bmatrix} \tag{4.1-59}$$

Elsewhere, (L, M, N) represent torques (N-m); here, they represent the accelerations caused by those moments. Moment sensitivity to Euler-angle perturbations (eq. 4.1-59) occurs only in ground effects (Section 2.5) and from gravity-gradient torques, which are not significant in atmospheric flight.

Wind perturbations $\Delta \mathbf{w}^T(t) = [\Delta u_W \; \Delta v_W \; \Delta w_W]^T$ have the same aerodynamic effects as airspeed variations (with opposite sign, per eq. 3.6-33); hence, $(1/m)\partial \mathbf{f}_B/\partial \mathbf{w} = -(1/m)\partial \mathbf{f}_B/\partial \mathbf{v}_B$, and $\mathcal{I}_B^{-1}(\partial \mathbf{m}_B/\partial \mathbf{w}) = -\mathcal{I}_B^{-1}(\partial \mathbf{m}_B/\partial \mathbf{v}_B)$. Note that $\mathbf{F}_{11}$ and $\mathbf{F}_{12}$ contain Coriolis (nonaerodynamic) components, while $\mathbf{L}_1$ and $\mathbf{L}_2$ do not. Wind *shear* perturbations (e.g., $\partial \Delta w_W/\partial y$), have separate disturbance effects involving spatial distribution (Section 3.4) as well as unsteady flow.

Incorporating Unsteady Aerodynamic Effects

Translational and rotational *accelerations* of the aircraft may produce measurable aerodynamic effects, as noted in Section 3.4. They are not modeled in the above equations, but the effects can be added. Time variation of the wind field also has an effect that is neglected here. The specific force and moment sensitivities to accelerations are expressed by four (3×3) matrices:

$$\mathbf{K}_{11} = \frac{1}{m}\frac{\partial \mathbf{f}_B}{\partial \dot{\mathbf{v}}_B} \tag{4.1-60}$$

$$\mathbf{K}_{12} = \frac{1}{m}\frac{\partial \mathbf{f}_B}{\partial \dot{\boldsymbol{\omega}}_B} \tag{4.1-61}$$

$$\mathbf{K}_{21} = \mathcal{I}_B^{-1}\frac{\partial \mathbf{m}_B}{\partial \dot{\mathbf{v}}_B} \tag{4.1-62}$$

$$\mathbf{K}_{22} = \mathcal{I}_B^{-1}\frac{\partial \mathbf{m}_B}{\partial \dot{\boldsymbol{\omega}}_B} \tag{4.1-63}$$

Adding these effects to the velocity and rotational-rate perturbation equations produces

$$\begin{aligned}\Delta \dot{\mathbf{v}}_B = {} & \mathbf{F}_{11}\Delta \mathbf{v}_B + \mathbf{F}_{12}\Delta \boldsymbol{\omega}_B + \mathbf{F}_{13}\Delta \mathbf{r}_I + \mathbf{F}_{14}\Delta \boldsymbol{\Theta} + \mathbf{G}_1\Delta \mathbf{u} + \mathbf{L}_1\Delta \mathbf{w} \\ & + \mathbf{K}_{11}\Delta \dot{\mathbf{v}}_B + \mathbf{K}_{12}\Delta \dot{\boldsymbol{\omega}}_B\end{aligned} \tag{4.1-64}$$

$$\begin{aligned}\Delta \dot{\boldsymbol{\omega}}_B = {} & \mathbf{F}_{21}\Delta \mathbf{v}_B + \mathbf{F}_{22}\Delta \boldsymbol{\omega}_B + \mathbf{F}_{23}\Delta \mathbf{r}_I + \mathbf{F}_{24}\Delta \boldsymbol{\Theta} + \mathbf{G}_2\Delta \mathbf{u} + \mathbf{L}_2\Delta \mathbf{w} \\ & + \mathbf{K}_{21}\Delta \dot{\mathbf{v}}_B + \mathbf{K}_{22}\Delta \dot{\boldsymbol{\omega}}_B\end{aligned} \tag{4.1-65}$$

The two equations can be combined into a single six-dimensional equation

$$\Delta \dot{\mathbf{x}}_d = \mathbf{F}_d\Delta \mathbf{x}_d + \mathbf{F}_k\Delta \mathbf{x}_k + \mathbf{G}_d\Delta \mathbf{u} + \mathbf{L}_d\Delta \mathbf{w} + \mathbf{K}\Delta \dot{\mathbf{x}}_d \tag{4.1-66a}$$

where $\Delta\mathbf{x}_d$ and $\Delta\mathbf{x}_k$ are defined as in eqs. 3.6-44 and 3.6-45, and the matrix definitions are apparent. This dynamic equation can be rearranged as follows:

$$(\mathbf{I}-\mathbf{K})\Delta\dot{\mathbf{x}}_d = \mathbf{F}_d\Delta\mathbf{x}_d + \mathbf{F}_k\Delta\mathbf{x}_k + \mathbf{G}_d\Delta\mathbf{u} + \mathbf{L}_d\Delta\mathbf{w} \tag{4.1-66b}$$

$$\Delta\dot{\mathbf{x}}_d = (\mathbf{I}-\mathbf{K})^{-1}\left[\mathbf{F}_d\Delta\mathbf{x}_d + \mathbf{F}_k\Delta\mathbf{x}_k + \mathbf{G}_d\Delta\mathbf{u} + \mathbf{L}_d\Delta\mathbf{w}\right]$$

$$\triangleq \mathbf{F}_{d_U}\Delta\mathbf{x}_d + \mathbf{F}_{k_U}\Delta\mathbf{x}_k + \mathbf{G}_{d_U}\Delta\mathbf{u} + \mathbf{L}_{d_U}\Delta\mathbf{w} \tag{4.1-66c}$$

The new system matrices reveal unsteady aerodynamics effects on the control and disturbance response as well as on the state perturbation sensitivity.

Symmetric Aircraft in Wings-Level Flight

The stability and control matrices are greatly simplified for a symmetric aircraft in wings-level flight. The principal reasons are that small longitudinal perturbations are decoupled from lateral-directional perturbations and that numerous nominal values are zero, including angular rates, roll and yaw angles, side velocity, and all acceleration components. Perturbations from steady climb and descent paths are also described by the equations.

The nominal state, control, and disturbance vectors are defined as follows:

$$\begin{aligned}\mathbf{x}_o(t) &= [u\ v\ w\ p\ q\ r\ x_I\ y_I\ z_I\ \phi\ \theta\ \psi]_o^T \\ &= [V_o\cos\alpha_o\ 0\ V_o\sin\alpha_o\ 0\ 0\ 0\ x_{I_o}\ y_{I_o}\ z_{I_o}\ 0\ \theta_o\ 0]^T\end{aligned} \tag{4.1-67}$$

$$\mathbf{u}_o(t) = [\delta E_o\ \delta A_o\ \delta R_o\ \delta T_o]^T \tag{4.1-68}$$

$$\mathbf{w}_o(t) = [u_{W_o}\ v_{W_o}\ w_{W_o}]^T = [0\ 0\ 0]^T \tag{4.1-69}$$

For symmetric flight and inertial symmetry about the body-axis vertical plane (eq. 3.2-42), the submatrices of $\mathbf{F}$ identified in eq. 4.1-28 are

$$\mathbf{F}_{11} = \frac{\rho_o V_o^2 S}{2m}\begin{bmatrix}\left(\dfrac{2u_o}{V_o^2}C_{X_o}+C_{X_u}\right) & 0 & \left(\dfrac{2w_o}{V_o^2}C_{X_o}+C_{X_w}\right)\\ 0 & C_{Y_v} & 0\\ \left(\dfrac{2u_o}{V_o^2}C_{Z_o}+C_{Z_u}\right) & 0 & \left(\dfrac{2w_o}{V_o^2}C_{Z_o}+C_{Z_w}\right)\end{bmatrix}$$

$$\triangleq \begin{bmatrix}X_u & 0 & X_w\\ 0 & Y_v & 0\\ Z_u & 0 & Z_w\end{bmatrix} \tag{4.1-70}$$

$$\mathbf{F}_{12} = \frac{\rho_o V_o S}{4m}\begin{bmatrix} 0 & \bar{c}C_{X_{\hat{q}}} & 0 \\ bC_{Y_{\hat{p}}} & 0 & bC_{Y_{\hat{r}}} \\ 0 & \bar{c}C_{Z_{\hat{q}}} & 0 \end{bmatrix} + \begin{bmatrix} 0 & -w_o & 0 \\ w_o & 0 & -u_o \\ 0 & u_o & 0 \end{bmatrix}$$

$$\triangleq \begin{bmatrix} 0 & (X_q - w_o) & 0 \\ (Y_p + w_o) & 0 & (Y_r - u_o) \\ 0 & (Z_q + u_o) & 0 \end{bmatrix} \tag{4.1-71}$$

$$\mathbf{F}_{13} = \begin{bmatrix} 0 & 0 & X_z \\ 0 & 0 & 0 \\ 0 & 0 & Z_z \end{bmatrix} \tag{4.1-72}$$

$$\mathbf{F}_{14} = \begin{bmatrix} 0 & -g_o \cos\theta_o & 0 \\ g_o \cos\theta_o & 0 & 0 \\ 0 & -g_o \sin\theta_o & 0 \end{bmatrix} \tag{4.1-73}$$

$$\mathbf{F}_{21} = \mathcal{I}_B^{-1}\frac{\rho_o V_o^2 S}{2}\begin{bmatrix} 0 & bC_{l_v} & 0 \\ \bar{c}C_{m_u} & 0 & \bar{c}C_{m_w} \\ 0 & bC_{n_v} & 0 \end{bmatrix}$$

$$\triangleq \begin{bmatrix} 0 & L_v & 0 \\ M_u & 0 & M_w \\ 0 & N_v & 0 \end{bmatrix} \tag{4.1-74}$$

$$\mathbf{F}_{22} = \mathcal{I}_B^{-1}\frac{\rho_o V_o S}{4}\begin{bmatrix} b^2C_{l_{\hat{p}}} & 0 & b^2C_{l_{\hat{r}}} \\ 0 & \bar{c}^2C_{m_{\hat{q}}} & 0 \\ b^2C_{n_{\hat{p}}} & 0 & b^2C_{n_{\hat{r}}} \end{bmatrix}$$

$$\triangleq \begin{bmatrix} L_p & 0 & L_r \\ 0 & M_q & 0 \\ N_p & 0 & N_r \end{bmatrix} \tag{4.1-75}$$

$$\mathbf{F}_{23} = \begin{bmatrix} 0 & 0 & L_z \\ 0 & 0 & M_z \\ 0 & 0 & 0 \end{bmatrix} \tag{4.1-76}$$

$$\mathbf{F}_{24} = \begin{bmatrix} L_\phi & L_\theta & 0 \\ M_\phi & M_\theta & 0 \\ N_\phi & N_\theta & 0 \end{bmatrix} \tag{4.1-77}$$

$$\mathbf{F}_{31} = \begin{bmatrix} \cos\theta_o & 0 & \sin\theta_o \\ 0 & 1 & 0 \\ -\sin\theta_o & 0 & \cos\theta_o \end{bmatrix} \tag{4.1-78}$$

$$\mathbf{F}_{34} = \begin{bmatrix} 0 & (u_o \sin\theta_o - w_o \cos\theta_o) & 0 \\ -w_o & 0 & -(u_o \cos\theta_o + w_o \sin\theta_o) \\ 0 & (u_o \cos\theta_o + w_o \sin\theta_o) & 0 \end{bmatrix} \tag{4.1-79}$$

$$\mathbf{F}_{42} = \begin{bmatrix} 1 & 0 & \tan\theta_o \\ 0 & 1 & 0 \\ 0 & 0 & \sec\theta_o \end{bmatrix} \tag{4.1-80}$$

$$\mathbf{F}_{44} = 0 \tag{4.1-81}$$

The control- and disturbance-effect matrices reduce to the following:

$$\mathbf{G}_1 = \frac{\rho_o V_o^2 S}{2m} \begin{bmatrix} C_{X_{\delta E}} & 0 & 0 & C_{X_{\delta T}} \\ 0 & C_{Y_{\delta A}} & C_{Y_{\delta R}} & 0 \\ C_{Z_{\delta E}} & 0 & 0 & C_{Z_{\delta T}} \end{bmatrix} \triangleq \begin{bmatrix} X_{\delta E} & 0 & 0 & X_{\delta T} \\ 0 & Y_{\delta A} & Y_{\delta R} & 0 \\ Z_{\delta E} & 0 & 0 & Z_{\delta T} \end{bmatrix} \tag{4.1-82}$$

$$\mathbf{G}_2 = \mathcal{I}_B^{-1} \frac{\rho_o V_o^2 S}{2} \begin{bmatrix} 0 & bC_{l_{\delta A}} & bC_{l_{\delta R}} & 0 \\ \bar{c}C_{m_{\delta E}} & 0 & 0 & \bar{c}C_{m_{\delta T}} \\ 0 & bC_{n_{\delta A}} & bC_{n_{\delta R}} & 0 \end{bmatrix} \triangleq \begin{bmatrix} 0 & L_{\delta A} & L_{\delta R} & 0 \\ M_{\delta E} & 0 & 0 & M_{\delta T} \\ 0 & N_{\delta A} & N_{\delta R} & 0 \end{bmatrix} \tag{4.1-83}$$

$$\mathbf{L}_1 = \begin{bmatrix} -X_u & 0 & -X_w \\ 0 & -Y_v & 0 \\ -Z_u & 0 & -Z_w \end{bmatrix} \tag{4.1-84}$$

$$\mathbf{L}_2 = \begin{bmatrix} 0 & -L_v & 0 \\ -M_u & 0 & -M_w \\ 0 & -N_v & 0 \end{bmatrix} \tag{4.1-85}$$

As noted in the previous section, these matrices can be modified to account for unsteady aerodynamic effects. Typically, only the effects of $\dot{\alpha}$ (or $\dot{w}$)on the normal force and pitching moment and of $\dot{\beta}$ (or $\dot{v}$) on the yawing moment are significant; hence, $\mathbf{K}_{12}$ and $\mathbf{K}_{22}$ are negligible, and

$$\mathbf{K}_{11} = \begin{bmatrix} 0 & 0 & 0 \\ 0 & 0 & 0 \\ 0 & 0 & Z_{\dot{w}} \end{bmatrix} \tag{4.1-86}$$

$$\mathbf{K}_{21} = \begin{bmatrix} 0 & 0 & 0 \\ 0 & 0 & M_{\dot{w}} \\ 0 & N_{\dot{v}} & 0 \end{bmatrix} \tag{4.1-87}$$

Consequently, $(\mathbf{I}-\mathbf{K})$ of eq. 4.1-66b is

$$(\mathbf{I}-\mathbf{K}) = \begin{bmatrix} 1 & 0 & 0 & 0 & 0 & 0 \\ 0 & 1 & 0 & 0 & 0 & 0 \\ 0 & 0 & (1-Z_{\dot{w}}) & 0 & 0 & 0 \\ 0 & 0 & 0 & 1 & 0 & 0 \\ 0 & 0 & -M_{\dot{w}} & 0 & 1 & 0 \\ 0 & -N_{\dot{v}} & 0 & 0 & 0 & 1 \end{bmatrix} \tag{4.1-88}$$

and its inverse is

$$(\mathbf{I}-\mathbf{K})^{-1} = \begin{bmatrix} 1 & 0 & 0 & 0 & 0 & 0 \\ 0 & 1 & 0 & 0 & 0 & 0 \\ 0 & 0 & \dfrac{1}{(1-Z_{\dot{w}})} & 0 & 0 & 0 \\ 0 & 0 & 0 & 1 & 0 & 0 \\ 0 & 0 & \dfrac{M_{\dot{w}}}{(1-Z_{\dot{w}})} & 0 & 1 & 0 \\ 0 & N_{\dot{v}} & 0 & 0 & 0 & 1 \end{bmatrix} \tag{4.1-89}$$

The sensitivity matrices of eq. 4.1-66c are

$$[\mathbf{F}_{d_U} \;\; \mathbf{F}_{k_U}] = (\mathbf{I}-\mathbf{K})^{-1} \begin{bmatrix} \mathbf{F}_{11} & \mathbf{F}_{12} & \mathbf{F}_{13} & \mathbf{F}_{14} \\ \mathbf{F}_{21} & \mathbf{F}_{22} & \mathbf{F}_{23} & \mathbf{F}_{24} \end{bmatrix} \quad (6 \times 12) \tag{4.1-90}$$

$$\mathbf{G}_{d_U} = (\mathbf{I} - \mathbf{K})^{-1}\begin{bmatrix} \mathbf{G}_1 \\ \mathbf{G}_2 \end{bmatrix} \quad (6 \times 4) \tag{4.1-91}$$

$$\mathbf{L}_{d_U} = (\mathbf{I} - \mathbf{K})^{-1}\begin{bmatrix} \mathbf{L}_1 \\ \mathbf{L}_2 \end{bmatrix} \quad (6 \times 3) \tag{4.1-92}$$

while the full model matrices are

$$\mathbf{F} = \begin{bmatrix} & \mathbf{F}_{d_U} & \mathbf{F}_{k_U} & \\ \mathbf{F}_{31} & 0 & 0 & \mathbf{F}_{34} \\ 0 & \mathbf{F}_{42} & 0 & \mathbf{F}_{44} \end{bmatrix} \quad (12 \times 12) \tag{4.1-93}$$

$$\mathbf{G} = \begin{bmatrix} \mathbf{G}_{d_U} \\ 0 \\ 0 \end{bmatrix} \quad (12 \times 4) \tag{4.1-94}$$

$$\mathbf{L} = \begin{bmatrix} \mathbf{L}_{d_U} \\ 0 \\ 0 \end{bmatrix} \quad (12 \times 3) \tag{4.1-95}$$

Longitudinal Equations of Motion

It is convenient to separate the linearized equations into *longitudinal equations* and *lateral-directional equations*. This is possible because the two sets of perturbation motions are uncoupled for a symmetric aircraft in steady, level flight. The longitudinal state, control, and disturbance vectors are

$$\Delta\mathbf{x}_{\mathrm{Lo}} = [\Delta u \ \Delta w \ \Delta q \ \Delta x_I \ \Delta z_I \ \Delta\theta]^T \tag{4.1-96}$$

$$\Delta\mathbf{u}_{\mathrm{Lo}} = [\Delta\delta E \ \Delta\delta T]^T \tag{4.1-97}$$

$$\Delta\mathbf{w}_{\mathrm{Lo}} = [\Delta u_W \ \Delta w_W]^T \tag{4.1-98}$$

The linearized longitudinal equations of motion take the general form

$$\Delta\dot{\mathbf{x}}_{\mathrm{Lo}} = \mathbf{F}_{\mathrm{Lo}}\Delta\mathbf{x}_{\mathrm{Lo}} + \mathbf{G}_{\mathrm{Lo}}\Delta\mathbf{u}_{\mathrm{Lo}} + \mathbf{L}_{\mathrm{Lo}}\Delta\mathbf{w}_{\mathrm{Lo}} \tag{4.1-99}$$

Unsteady aerodynamic effects are included in the system matrices, which are defined as follows:

$$
\mathbf{F}_{\mathrm{Lo}} = \begin{bmatrix}
X_u & X_w & (X_q - w_o) & 0 & X_z & -g_o\cos\theta_o \\
\dfrac{Z_u}{(1-Z_{\dot{w}})} & \dfrac{Z_w}{(1-Z_{\dot{w}})} & \dfrac{(Z_q + u_o)}{(1-Z_{\dot{w}})} & 0 & \dfrac{Z_z}{(1-Z_{\dot{w}})} & \dfrac{-g_o\sin\theta_o}{(1-Z_{\dot{w}})} \\
[M_u + aZ_u] & [M_w + aZ_w] & [M_q + a(Z_q + u_o)] & 0 & 0 & M_\theta - ag_o\sin\theta_o \\
\cos\theta_o & \sin\theta_o & 0 & 0 & 0 & b \\
-\sin\theta_o & \cos\theta_o & 0 & 0 & 0 & c \\
0 & 0 & 1 & 0 & 0 & 0
\end{bmatrix} \quad (4.1\text{-}100)
$$

$$
a \triangleq \frac{M_{\dot{w}}}{(1-Z_{\dot{w}})}
$$

$$
b \triangleq -u_o\sin\theta_o + w_o\cos\theta_o
$$

$$
c \triangleq -u_o\cos\theta_o - w_o\sin\theta_o
$$

$$
\mathbf{G}_{\mathrm{Lo}} = \begin{bmatrix}
X_{\delta E} & X_{\delta T} \\
\dfrac{Z_{\delta E}}{(1-Z_{\dot{w}})} & \dfrac{Z_{\delta T}}{(1-Z_{\dot{w}})} \\
[M_{\delta E} + aZ_{\delta E}] & [M_{\delta T} + aZ_{\delta T}] \\
0 & 0 \\
0 & 0 \\
0 & 0
\end{bmatrix} \quad (4.1\text{-}101)
$$

$$
\mathbf{L}_{\mathrm{Lo}} = \begin{bmatrix}
-X_u & -X_w \\
\dfrac{-Z_u}{(1-Z_{\dot{w}})} & \dfrac{-Z_w}{(1-Z_{\dot{w}})} \\
-[M_u + aZ_u] & -[M_w + aZ_w] \\
0 & 0 \\
0 & 0 \\
0 & 0
\end{bmatrix} \quad (4.1\text{-}102)
$$

Lateral-Directional Equations of Motion

The remaining variables belong to the lateral-directional set, representing out-of-plane effects. The state, control, and disturbance vectors are,

$$
\Delta\mathbf{x}_{\mathrm{LD}} = [\Delta v \; \Delta p \; \Delta r \; \Delta y_I \; \Delta\phi \; \Delta\psi]^T \quad (4.1\text{-}103)
$$

$$\Delta \mathbf{u}_{LD} = [\Delta\delta A\ \Delta\delta R]^T \tag{4.1-104}$$

$$\Delta \mathbf{w}_{LD} = [\Delta v_{W_o}]^T \tag{4.1-105}$$

The corresponding linear equation of motion is

$$\Delta \dot{\mathbf{x}}_{LD} = \mathbf{F}_{LD}\Delta \mathbf{x}_{LD} + \mathbf{G}_{LD}\Delta \mathbf{u}_{LD} + \mathbf{L}_{LD}\Delta \mathbf{w}_{LD} \tag{4.1-106}$$

where the system matrices (including unsteady effects) are

$$\mathbf{F}_{LD} = \begin{bmatrix} Y_v & (Y_p + w_o) & (Y_r - u_o) & 0 & g_o \cos\theta_o & 0 \\ L_v & L_p & L_r & 0 & L_\phi & 0 \\ [N_v + Y_v N_{\dot{v}}] & [N_p + (Y_p + w_o)N_{\dot{v}}] & [N_r + (Y_r - u_o)N_{\dot{v}}] & 0 & N_\phi & 0 \\ 1 & 0 & 0 & 0 & w_o & a \\ 0 & 1 & -\sin\theta_o & 0 & 0 & 0 \\ 0 & 0 & \cos\theta_o & 0 & 0 & 0 \end{bmatrix} \tag{4.1-107}$$

$$a \triangleq u_o \cos\theta_o + w_o \sin\theta_o$$

$$\mathbf{G}_{LD} = \begin{bmatrix} Y_{\delta A} & Y_{\delta R} \\ L_{\delta A} & L_{\delta R} \\ [N_{\delta A} + Y_{\delta A} N_{\dot{v}}] & [N_{\delta R} + Y_{\delta R} N_{\dot{v}}] \\ 0 & 0 \\ 0 & 0 \\ 0 & 0 \end{bmatrix} \tag{4.1-108}$$

$$\mathbf{L}_{LD} = \begin{bmatrix} -Y_v \\ -L_v \\ -[N_v + Y_v N_{\dot{v}}] \\ 0 \\ 0 \\ 0 \end{bmatrix} \tag{4.1-109}$$

Stability-Axis Equations of Motion

The linearized equations of motion can be simplified further by choosing a set of body axes that are aligned with the nominal velocity vector (Section 3.2). The body-axis velocity and angular-rate vectors $\mathbf{v}_B$ and $\boldsymbol{\omega}_B$ are multiplied by $\mathbf{H}_B^S$ (eq. 3.2-85) to arrive at their stability-axis representations. In the symmetric case, $\beta_o = 0$, and the transformation is achieved by a pitch rotation through the nominal angle of attack:

$$\mathbf{H}_B^S(-\alpha_o) = \begin{bmatrix} \cos\alpha_o & 0 & \sin\alpha_o \\ 0 & 1 & 0 \\ -\sin\alpha_o & 0 & \cos\alpha_o \end{bmatrix} \tag{4.1-110}$$

Having realigned the axes, it is convenient to replace the Cartesian velocity components (Δu, Δv, Δw) with their polar-coordinate counterparts (ΔV, $\Delta\alpha$, $\Delta\beta$) (eqs. 2.2-30 and 2.2-31). These are related by

$$\begin{bmatrix} \Delta V \\ \Delta\beta \\ \Delta\alpha \end{bmatrix} = \begin{bmatrix} \dfrac{u_o\Delta u + v_o\Delta v + w_o\Delta w}{V_o} \\ \dfrac{V_o\Delta v - v_o\Delta V}{\sqrt{V_o^2 - v_o^2}} \\ \dfrac{u_o\Delta w - w_o\Delta u}{u_o^2 + w_o^2} \end{bmatrix}$$

$$= \begin{bmatrix} u_o/V_o & v_o/V_o & w_o/V_o \\ \dfrac{-u_o v_o}{V_o^2\sqrt{V_o^2 - v_o^2}} & \dfrac{1 - v_o^2/V_o^2}{\sqrt{V_o^2 - v_o^2}} & \dfrac{-v_o w_o}{V_o^2\sqrt{V_o^2 - v_o^2}} \\ \dfrac{-w_o}{u_o^2 + w_o^2} & 0 & \dfrac{u_o}{u_o^2 + w_o^2} \end{bmatrix} \begin{bmatrix} \Delta u \\ \Delta v \\ \Delta w \end{bmatrix} \tag{4.1-111a}$$

which, for small angles, is

$$\begin{bmatrix} \Delta V \\ \Delta\beta \\ \Delta\alpha \end{bmatrix} \approx \begin{bmatrix} \Delta u \\ \Delta v/V_o \\ \Delta w/V_o \end{bmatrix} \tag{4.1-111b}$$

Furthermore, the coordinate change makes it appropriate to express specific forces in terms of lift and drag rather than axial and normal force (with $D = -X$ and $L = -Z$). In the following discussion, thrust effects are contained in D and L. Thus, D is the specific force (force per unit of mass) due to both drag and thrust along and opposed to the velocity vector,

$$D = \frac{1}{m}\left(C_D \frac{1}{2}\rho V^2 S - T_x \cos\alpha - T_z \sin\alpha\right) \tag{4.1-112}$$

while the specific force normal to the velocity vector is

$$L = \frac{1}{m}\left(C_L \frac{1}{2}\rho V^2 S + T_x \sin\alpha - T_z \cos\alpha\right) \tag{4.1-113}$$

where

$$\begin{bmatrix} T_x \\ T_z \end{bmatrix} = \text{Thrust}\begin{bmatrix} \cos i \\ \sin i \end{bmatrix} \tag{4.1-114}$$

and i is the angle between the thrust line and the body x axis. Stability derivatives are then defined as in the following example:

$$D_V = \frac{1}{m}\left[\frac{\partial[C_D(1/2)\rho V^2 S]}{\partial V} - \frac{\partial T_x \cos\alpha}{\partial V} - \frac{\partial T_z \sin\alpha}{\partial V}\right] \tag{4.1-115}$$

The linearized longitudinal equations of motion take the form of equation 4.1-99, with

$$\Delta\mathbf{x}_{\mathrm{Lo}} = [\Delta V\ \Delta\alpha\ \Delta q\ \Delta x_{\mathrm{I}}\ \Delta z_{\mathrm{I}}\ \Delta\theta]^T \tag{4.1-116}$$

$$\Delta\mathbf{u}_{\mathrm{Lo}} = [\Delta\delta E\ \Delta\delta T]^T \tag{4.1-117}$$

$$\Delta\mathbf{w}_{\mathrm{Lo}} = [\Delta V_{\mathrm{W}}\ \Delta\alpha_{\mathrm{W}}]^T \tag{4.1-118}$$

Because $u_o = V_o$, $w_o = 0$, and $\theta_o = \gamma_o$ in the stability-axis description, the system matrices can be expressed as

$$\mathbf{F}_{\mathrm{Lo}} = \begin{bmatrix} -D_V & -D_\alpha & -D_q & 0 & -D_z & -g_o\cos\gamma_o \\ \dfrac{-L_V}{(V_o + L_{\dot\alpha})} & \dfrac{-L_\alpha}{(V_o + L_{\dot\alpha})} & \dfrac{(V_o - L_q)}{(V_o + L_{\dot\alpha})} & 0 & \dfrac{-L_z}{(V_o + L_{\dot\alpha})} & \dfrac{-g_o\sin\gamma_o}{(V_o + L_{\dot\alpha})} \\ [M_V - aL_V] & [M_\alpha - aL_\alpha] & [M_q - a(L_q - V_o)] & 0 & 0 & M_\theta - ag_o\sin\gamma_o \\ \cos\gamma_o & V_o\sin\gamma_o & 0 & 0 & 0 & b \\ -\sin\gamma_o & V_o\cos\gamma_o & 0 & 0 & 0 & c \\ 0 & 0 & 1 & 0 & 0 & 0 \end{bmatrix} \tag{4.1-119}$$

$$a \triangleq \frac{M_{\dot\alpha}}{(V_o + L_{\dot\alpha})}$$

$$b \triangleq -V_o\sin\gamma_o$$

$c \triangleq -V_o \cos\gamma_o$

$$\mathbf{G}_{\mathrm{Lo}} = \begin{bmatrix} -D_{\delta E} & T_{\delta T} \\ \dfrac{-L_{\delta E}}{(V_o + L_{\dot{\alpha}})} & \dfrac{-L_{\delta T}}{(V_o + L_{\dot{\alpha}})} \\ [M_{\delta E} - a L_{\delta E}] & [M_{\delta T} - a L_{\delta T}] \\ 0 & 0 \\ 0 & 0 \\ 0 & 0 \end{bmatrix} \tag{4.1-120}$$

$$\mathbf{L}_{\mathrm{Lo}} = \begin{bmatrix} D_V & D_\alpha \\ \dfrac{L_V}{(V_o + L_{\dot{\alpha}})} & \dfrac{L_\alpha}{(V_o + L_{\dot{\alpha}})} \\ -[M_V - a L_V] & -[M_\alpha - a L_\alpha] \\ 0 & 0 \\ 0 & 0 \\ 0 & 0 \end{bmatrix} \tag{4.1-121}$$

The linearized lateral-directional equations of motion take the form of eq. 4.1-106, with

$$\Delta\mathbf{x}_{\mathrm{LD}} = [\Delta\beta\ \Delta p\ \Delta r\ \Delta y_I\ \Delta\phi\ \Delta\psi]^T \tag{4.1-122}$$

$$\Delta\mathbf{u}_{\mathrm{LD}} = [\Delta\delta A\ \Delta\delta R]^T \tag{4.1-123}$$

$$\Delta\mathbf{w}_{\mathrm{LD}} = [\Delta\beta_W\ \Delta p_W]^T \tag{4.1-124}$$

$$\mathbf{F}_{\mathrm{LD}} = \begin{bmatrix} \dfrac{Y_\beta}{V_o} & \dfrac{Y_p}{V_o} & \dfrac{(Y_r - V_o)}{V_o} & 0 & \dfrac{g_o \cos\gamma_o}{V_o} & 0 \\ L_\beta & L_p & L_r & 0 & L_\phi & 0 \\ \left[N_\beta + \dfrac{Y_\beta N_{\dot{\beta}}}{V_o}\right] & \left[N_p + \dfrac{Y_p N_{\dot{\beta}}}{V_o}\right] & \left[N_r + \dfrac{(Y_r - V_o) N_{\dot{\beta}}}{V_o}\right] & 0 & N_\phi & 0 \\ 1 & 0 & 0 & 0 & 0 & a \\ 0 & 1 & -\sin\gamma_o & 0 & 0 & 0 \\ 0 & 0 & \cos\gamma_o & 0 & 0 & 0 \end{bmatrix} \tag{4.1-125}$$

$a \triangleq V_o \cos\gamma_o$

$$\mathbf{G}_{\mathrm{LD}} = \begin{bmatrix} \dfrac{Y_{\delta A}}{V_o} & \dfrac{Y_{\delta R}}{V_o} \\ L_{\delta A} & L_{\delta R} \\ \left[N_{\delta A} + \dfrac{Y_{\delta A} N_{\dot\beta}}{V_o}\right] & \left[N_{\delta R} + \dfrac{Y_{\delta R} N_{\dot\beta}}{V_o}\right] \\ 0 & 0 \\ 0 & 0 \\ 0 & 0 \end{bmatrix} \tag{4.1-126}$$

$$\mathbf{L}_{\mathrm{LD}} = \begin{bmatrix} \dfrac{-Y_{\beta}}{V_o} & \dfrac{-Y_{p}}{V_o} \\ -L_{\beta} & -L_{p} \\ -\left[N_{\beta} + \dfrac{Y_{\beta} N_{\dot\beta}}{V_o}\right] & -\left[N_{p} + \dfrac{Y_{p} N_{\dot\beta}}{V_o}\right] \\ 0 & 0 \\ 0 & 0 \\ 0 & 0 \end{bmatrix} \tag{4.1-127}$$

Note that the stability-axis matrices **F**, **G**, and **L** could be obtained from the equivalent body-axis matrices using similarity transformations (Section 3.2). Given state, control, and disturbance vectors ($\Delta\mathbf{x}_1$, $\Delta\mathbf{u}_1$, $\Delta\mathbf{w}_1$) and ($\Delta\mathbf{x}_2$, $\Delta\mathbf{u}_2$, $\Delta\mathbf{w}_2$) of equal dimension and related by constant transformation matrices

$$\Delta\mathbf{x}_2 = \mathbf{T}_{\mathbf{x}}\, \Delta\mathbf{x}_1 \tag{4.1-128}$$

$$\Delta\mathbf{u}_2 = \mathbf{T}_{\mathbf{u}}\, \Delta\mathbf{u}_1 \tag{4.1-129}$$

$$\Delta\mathbf{w}_2 = \mathbf{T}_{\mathbf{w}}\, \Delta\mathbf{w}_1 \tag{4.1-130}$$

we want to derive the dynamic equation

$$\Delta\dot{\mathbf{x}}_2 = \mathbf{F}_2\Delta\mathbf{x}_2 + \mathbf{G}_2\Delta\mathbf{u}_2 + \mathbf{L}_2\Delta\mathbf{w}_2 \tag{4.1-131}$$

from the known relation

$$\Delta\dot{\mathbf{x}}_1 = \mathbf{F}_1\Delta\mathbf{x}_1 + \mathbf{G}_1\Delta\mathbf{u}_1 + \mathbf{L}_1\Delta\mathbf{w}_1 \tag{4.1-132}$$

Because

$$\Delta\dot{\mathbf{x}}_2 = \mathbf{T}_{\mathbf{x}}\Delta\dot{\mathbf{x}}_1 \tag{4.1-133}$$

the desired equation is

$$\Delta\dot{\mathbf{x}}_2 = \mathbf{T}_{\mathbf{x}}\mathbf{F}_1\mathbf{T}_{\mathbf{x}}^{-1}\Delta\mathbf{x}_2 + \mathbf{T}_{\mathbf{x}}\mathbf{G}_1\mathbf{T}_{\mathbf{u}}^{-1}\Delta\mathbf{u}_2 + \mathbf{T}_{\mathbf{x}}\mathbf{L}_1\mathbf{T}_{\mathbf{w}}^{-1}\Delta\mathbf{w}_2 \tag{4.1-134}$$

and $\mathbf{F}_2$, $\mathbf{G}_2$, and $\mathbf{L}_2$ are defined accordingly.

Example 4.1-1. Stability and Control Matrices for a Business Jet Aircraft

The function `LinModel` of the program `FLIGHT` (Appendix B) calculates the Jacobian matrices for **F** and **G** (eqs. 4.1-28 and 4.1-29) and stores the results in the `mat` files `Fmodel` and `Gmodel`. Rather than using the literal equations presented in this section, the matrices are generated numerically using the MATLAB function `numjac`, as in equation. 4.1-31.

For the $\mathcal{M} = 0.3$, $h = 3{,}050$ m flight condition used in earlier examples, the business jet model has the following stability and control matrices:

$$\mathbf{F} = \begin{bmatrix} -0.12 & 0 & 0.096 & 0 & 0 & 0 & 0 & -6.45 & 0 & 0 & -9.79 & 0 \\ 0 & -0.16 & 0 & 0 & 0 & 0 & 6.45 & 0 & -101.8 & 9.79 & 0 & 0 \\ -0.12 & 0 & -1.28 & 0 & 0 & -0.001 & 0 & 100.8 & 0 & 0 & -0.62 & 0 \\ 1.0 & 0 & 0.06 & 0 & 0 & 0 & 0 & 0 & 0 & 0 & 0 & 0 \\ 0 & 1.0 & 0 & 0 & 0 & 0 & 0 & 0 & 0 & -6.45 & 0 & 102 \\ -0.063 & 0 & 1.0 & 0 & 0 & 0 & 0 & 0 & 0 & 0 & -102 & 0 \\ 0 & -0.025 & 0 & 0 & 0 & 0 & -1.18 & 0 & 0.18 & 0 & 0 & 0 \\ 0.005 & 0 & -0.078 & 0 & 0 & 0 & 0 & -1.279 & 0 & 0 & 0 & 0 \\ 0 & 0.017 & 0 & 0 & 0 & 0 & -0.011 & 0 & -0.093 & 0 & 0 & 0 \\ 0 & 0 & 0 & 0 & 0 & 0 & 1 & 0 & 0 & 0 & 0 & 0 \\ 0 & 0 & 0 & 0 & 0 & 0 & 0 & 1 & 0 & 0 & 0 & 0 \\ 0 & 0 & 0 & 0 & 0 & 0 & 0 & 0 & 1.002 & 0 & 0 & 0 \end{bmatrix}$$

$$\mathbf{G} = \begin{bmatrix} 0.0065 & 0 & 0 & 4.67 & 0 & 0 & 0.012 \\ 0 & -0.16 & 3.51 & 0 & 0.60 & 0 & 0 \\ -13.17 & 0 & 0 & 0 & 0 & 0 & -24.63 \\ 0 & 0 & 0 & 0 & 0 & 0 & 0 \\ 0 & 0 & 0 & 0 & 0 & 0 & 0 \\ 0 & 0 & 0 & 0 & 0 & 0 & 0 \\ 0 & 2.31 & 0.25 & 0 & -0.46 & 0 & 0 \\ -9.07 & 0 & 0 & 0 & 0 & 0 & -14.86 \\ 0 & 0.12 & -1.11 & 0 & -0.16 & 0 & 0 \\ 0 & 0 & 0 & 0 & 0 & 0 & 0 \\ 0 & 0 & 0 & 0 & 0 & 0 & 0 \\ 0 & 0 & 0 & 0 & 0 & 0 & 0 \end{bmatrix}$$

The corresponding state vector is defined by eq. 4.1-25, and the control vector is defined by eq. 4.1-26. The extra three columns of the control-effect matrix account for asymmetric spoilers, flaps, and stabilator. The business jet model simulates flap settings of 0 or 38° by discrete changes in aerodynamic coefficients; hence, flap effects are not included in the control vector, and the sixth column of **G** is zero.

The SURVEY program (Appendix C) accepts `Fmodel` and `Gmodel` as inputs and performs a variety of analyses on them. It first provides the option of reducing these matrices into smaller models using the function `LonLatDir`. The function produces a single model at each call for either longitudinal or lateral-directional motions, in body or stability axes, and with four or six elements in the state vector. The remaining survey functions operate on either the original 12-dimensional model or the reduced model produced by `LonLatDir`.

For the model shown above, eight possible subsets can be produced. The sixth-order stability-axis models derived from the original model are

$$\mathbf{F}_{\mathrm{Lo}} = \begin{bmatrix} -0.0185 & 3.9375 & -0.0560 & -9.8067 & 0 & 0 \\ -0.0008 & -1.2709 & 0.4139 & 0 & 0 & 0 \\ 0 & 19.1029 & -1.2794 & 0 & 0 & 0 \\ 0 & 0 & 1 & 0 & 0 & 0 \\ 1 & -0.0181 & 0 & 0 & 0 & 0 \\ 0.0001 & 244 & 0 & -102 & 0 & 0 \end{bmatrix}$$

$$\mathbf{G}_{\mathrm{Lo}} = \begin{bmatrix} -0.8271 & 4.6645 & 0 & -1.547 \\ -0.0538 & -0.0012 & 0 & -0.1007 \\ -9.069 & 0 & 0 & -14.862 \\ 0 & 0 & 0 & 0 \\ 0 & 0 & 0 & 0 \\ 0 & 0 & 0 & 0 \end{bmatrix}$$

$$\mathbf{F}_{\mathrm{LD}} = \begin{bmatrix} -0.1567 & -0.0076 & -102. & 0 & 9.7674 & -0.6196 \\ -0.0236 & -1.1616 & 0.2501 & 0 & 0 & 0 \\ 0.0186 & 0.0566 & -0.1079 & 0 & 0 & 0 \\ 1. & 0 & 0 & 0 & 0.0205 & 102.2038 \\ 0 & 1. & 0.0001 & 0 & 0 & 0 \\ 0 & 0.0001 & 1.002 & 0 & 0 & 0 \end{bmatrix}$$

$$\mathbf{G}_{\mathrm{LD}} = \begin{bmatrix} -0.1591 & 3.512 & 0.6008 \\ 2.3106 & 0.1782 & -0.4721 \\ -0.0288 & -1.1196 & -0.1348 \\ 0 & 0 & 0 \\ 0 & 0 & 0 \\ 0 & 0 & 0 \end{bmatrix}$$

4.2 Solution of Linear Differential Equations

Linear differential equations are special cases of nonlinear differential equations, so the methods of Section 3.6 for finding state histories and equilibrium solutions are applicable. However, there are simpler, more analytical methods for solving the linear equations. The *principle of superposition* applies to linear equations: doubling the input causes the output to double, and responses arising from initial conditions and forced inputs are additive. Solution techniques can make use of these attributes, and global characterizations of system behavior can be derived.

Numerical Integration and State Transition

Given the linear, time-varying ordinary differential equation

$$\Delta\dot{\mathbf{x}}(t) = \mathbf{F}(t)\,\Delta\mathbf{x}(t) + \mathbf{G}(t)\,\Delta\mathbf{u}(t) + \mathbf{L}(t)\,\Delta\mathbf{w}(t) \tag{4.2-1}$$

eq. 3.6-2 computes the state at time t_k:

$$\Delta\mathbf{x}(t_k) = \Delta\mathbf{x}(t_0) + \int_{t_0}^{t_k} [\mathbf{F}(t)\,\Delta\mathbf{x}(t) + \mathbf{G}(t)\,\Delta\mathbf{u}(t) + \mathbf{L}(t)\,\Delta\mathbf{w}(t)]\,dt \tag{4.2-2}$$

The state history $\{\mathbf{x}(t) \text{ in } (t_0, t_k)\}$ is the sum of the linear system's initial-condition (*homogeneous*) and forced (*nonhomogeneous*) responses. The latter is derived from the fundamental solutions of the former; therefore, we begin by considering the case for which $\Delta\mathbf{u}(t) = \Delta\mathbf{w}(t) = 0$. Equation 4.2-2 reduces to

$$\Delta\mathbf{x}(t_k) = \Delta\mathbf{x}(t_0) + \int_{t_0}^{t_k} \mathbf{F}(t)\Delta\mathbf{x}(t)\,dt \tag{4.2-3}$$

whose solution has the general form [B-10, D-1, S-14]

$$\Delta\mathbf{x}(t_k) = \boldsymbol{\Phi}\,(t_k, t_0)\Delta\mathbf{x}(t_0) \tag{4.2-4}$$

$\boldsymbol{\Phi}(t_k, t_0)$ is an $(n \times n)$ *state transition matrix* that predicts the system response at time t_k from the value of the state at time t_0. If the system matrix $\mathbf{F}$ is time varying, $\boldsymbol{\Phi}$ depends not only on the difference between t_k and t_0 but on t_0 itself, and its computation is complicated [S-14]. If, however, $\mathbf{F}$ is a constant matrix, then $\boldsymbol{\Phi}$ depends only on $\Delta t \triangleq t_k - t_0$, and it can be calculated as

$$\boldsymbol{\Phi}(t_k - t_0) = \boldsymbol{\Phi}(\Delta t) = e^{\mathbf{F}(\Delta t)} = \mathbf{I}_n + \mathbf{F}(\Delta t) + \frac{1}{2!}\mathbf{F}^2(\Delta t)^2 + \frac{1}{3!}\mathbf{F}^3(\Delta t)^3 + \cdots \tag{4.2-5}$$

e is the *Naperian base* ($=2.71828\ldots$), $e^{\mathbf{F}(\Delta t)}$ is a *matrix exponential function* with the power series representation shown, and $\cdot!$ is the *factorial* operation ($n! = 1 \cdot 2 \cdot 3 \cdot \cdots n$). Clearly, $\Phi(\Delta t)$ is represented accurately with fewer terms when Δt is small. Given a fixed time interval Δt, $\Phi(\Delta t)$ need be computed just once, and samples of the (exact) homogeneous response can be generated sequentially, beginning with the known initial condition $\Delta\mathbf{x}(t_0)$:

$$\Delta\mathbf{x}(t_k) = \Phi(\Delta t)\Delta\mathbf{x}(t_{k-1}), \quad \Delta t \triangleq t_k - t_{k-1}, \quad \Delta\mathbf{x}(t_0) \text{ given} \tag{4.2-6}$$

The sum of the linear system's unforced and forced solutions (eq. 4.2-2) can be obtained by evaluating

$$\Delta\mathbf{x}(t_k) = \Phi(\Delta t)\Delta\mathbf{x}(t_{k-1}) + \int_{t_{k-1}}^{t_k} \Phi(t_k, t)[\mathbf{G}(t)\,\Delta\mathbf{u}(t) + \mathbf{L}(t)\,\Delta\mathbf{w}(t)]\,dt,$$
$$\Delta\mathbf{x}(t_0) \text{ given} \tag{4.2-7}$$

If $\mathbf{F}$, $\mathbf{G}$, and $\mathbf{L}$ are constant, and if the inputs can be considered fixed during the interval $(t_k - t_{k-1})$, then the forcing terms can be taken outside the integral, leading to

$$\Delta\mathbf{x}(t_k) = \Phi(\Delta t)\,\Delta\mathbf{x}(t_{k-1}) + \int_{t_{k-1}}^{t_k} e^{\mathbf{F}(t_k - t)}\,dt\,[\mathbf{G}\,\Delta\mathbf{u}(t_{k-1}) + \mathbf{L}\,\Delta\mathbf{w}(t_{k-1})]$$

$$\Delta\mathbf{u}(t) = \Delta\mathbf{u}(t_{k-1}) \text{ in } (t_{k-1}, t_k)$$

$$\Delta\mathbf{w}(t) = \Delta\mathbf{w}(t_{k-1}) \text{ in } (t_{k-1}, t_k) \tag{4.2-8a}$$

or

$$\Delta\mathbf{x}(t_k) = \Phi(\Delta t)\,\Delta\mathbf{x}(t_{k-1}) + \Phi(\Delta t)\int_0^{\Delta t} e^{-\mathbf{F}t}dt\,[\mathbf{G}\,\Delta\mathbf{u}(t_{k-1}) + \mathbf{L}\,\Delta\mathbf{w}(t_{k-1})]$$
$$\triangleq \Phi(\Delta t)\,\Delta\mathbf{x}(t_{k-1}) + \Gamma(\Delta t)\,\Delta\mathbf{u}(t_{k-1}) + \Lambda(\Delta t)\,\Delta\mathbf{w}(t_{k-1}) \tag{4.2-8b}$$

This state propagation equation is equivalent to the integral equation (eq. 4.2-2), yet it is also the *discrete-time* equivalent of the *continuous-time* differential equation for the system (eq. 4.2.-1). It is a *difference equation* whose *control-effect* and *disturbance-effect matrices* Γ and Λ are defined as

$$\Gamma(\Delta t) = \Phi(\Delta t)\int_0^{\Delta t} e^{-\mathbf{F}t}dt\,\mathbf{G} = [\Phi(\Delta t) = \mathbf{I}_n]\mathbf{F}^{-1}\mathbf{G} \tag{4.2-9a}$$

$$\Lambda(\Delta t) = \Phi(\Delta t)\int_0^{\Delta t} e^{-\mathbf{F}t}dt\,\mathbf{L} = [\Phi(\Delta t) = \mathbf{I}_n]\mathbf{F}^{-1}\mathbf{L} \tag{4.2-10a}$$

Using the series expansion for $\Phi(\Delta t)$, the input matrices become

$$\Gamma(\Delta t) = \left[\mathbf{I}_n + \frac{1}{2}\mathbf{F}\Delta t + \frac{1}{6}\mathbf{F}^2\Delta t^2 + \frac{1}{24}\mathbf{F}^3\Delta t^3 + \cdots\right]\mathbf{G}\,\Delta t \tag{4.2-9b}$$

$$\Lambda(\Delta t) = \left[\mathbf{I}_n + \frac{1}{2}\mathbf{F}\Delta t + \frac{1}{6}\mathbf{F}^2\Delta t^2 + \frac{1}{24}\mathbf{F}^3\Delta t^3 + \cdots \right] \mathbf{L}\,\Delta t \qquad (4.2\text{-}10b)$$

For very small Δt, $\Phi(\Delta t) \approx \mathbf{I}_n + \mathbf{F}\Delta t$, $\Gamma(\Delta t) \approx \mathbf{G}\Delta t$, and $\Lambda(\Delta t) \approx \mathbf{L}\Delta t$. Inserting these approximations in eq. 4.2-8 produces a formula for rectangular integration (Section 3.6). The state transition matrix is calculated by `expm` in MATLAB. Discrete-time control-effect and disturbance matrices can then be found from eq. 4.2-9 and 4.2-10. If $\mathbf{F}$ is not invertible, Γ and Λ can be calculated using eq. 4.2-9b and 4.2-10b.

The discrete-time state equation plays a number of useful roles in linear system analysis, and it is easily solved with a digital computer. It portrays the initial condition response precisely; hence, it provides fundamental information about stability and transient behavior. Equation 4.2-8 also computes the step response correctly, assuming that the step input occurs at a sampling instant t_k. If the sampling interval Δt is short compared to the time scale of inputs, it provides a good approximation for continuous-system response to arbitrary inputs. Of course, if the inputs occur in stepwise fashion, as they would if the aircraft were controlled by a digital computer, the predicted response is precise.

Example 4.2-1. Initial Condition and Input Response of a Linear Dynamic Model

The SURVEY program (Appendix C) uses the function Trans to calculate the transient response to an arbitrary array of initial conditions and step control inputs. Trans calls the MATLAB function `lsim` to perform the simulation. Typical results are generated for the business jet $\mathcal{M} = 0.3$, $h = 3{,}050$ m flight condition used in earlier examples.

The responses to 0.1 rad/s initial conditions on pitch, roll, and yaw rate are presented in Figs. 4.2-1–4.2-3. Comparing these results to those for the fully nonlinear model (Figs. 3.6-10, 3.6-11, and 3.6-13), it can be seen that the differences are quite small.

The longitudinal response to elevator and throttle step inputs is shown in Figs. 4.2-4 and 4.2-5. There is no linear response to flap setting in this model (a flap setting of 38 deg makes discrete changes to the coefficients contained in the nonlinear function AeroModel used by FLIGHT), and response to the stabilator is similar to elevator response. The 1 deg elevator step produces a nose-down pitching moment, and it excites both the short-period and phugoid modes (Fig. 4.2-4). There is a large increase in axial velocity and decrease in altitude and pitch angle as a consequence of the resulting dive, all occurring at the long time scale. The pitch-rate transient is moderately damped.

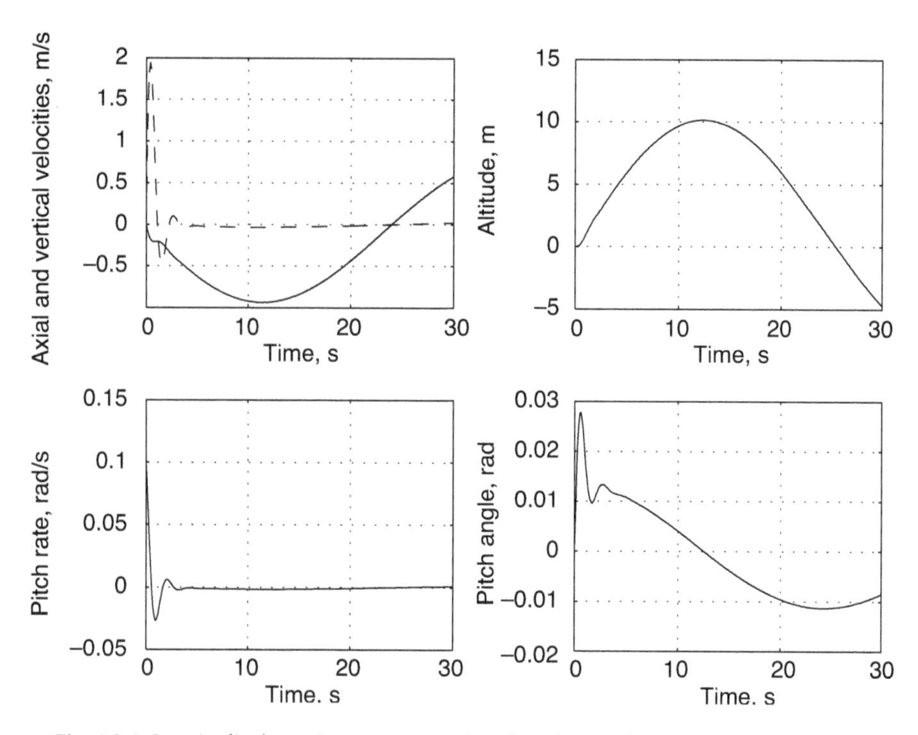

Fig. 4.2-1. Longitudinal transient response to initial pitch rate of 0.1 rad/s.

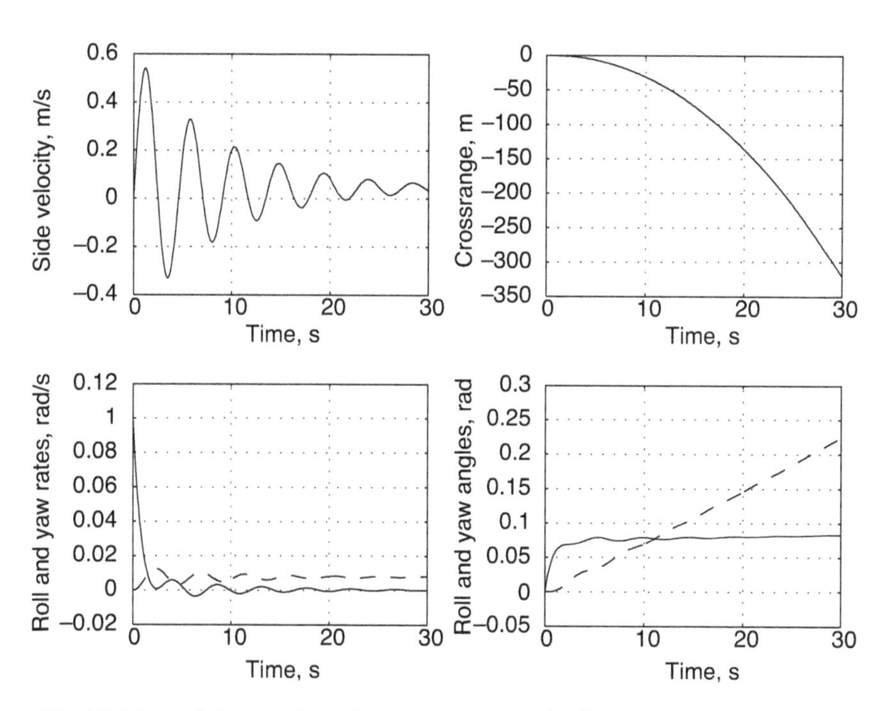

Fig. 4.2-2. Lateral-directional transient response to initial roll rate of 0.1 rad/s.

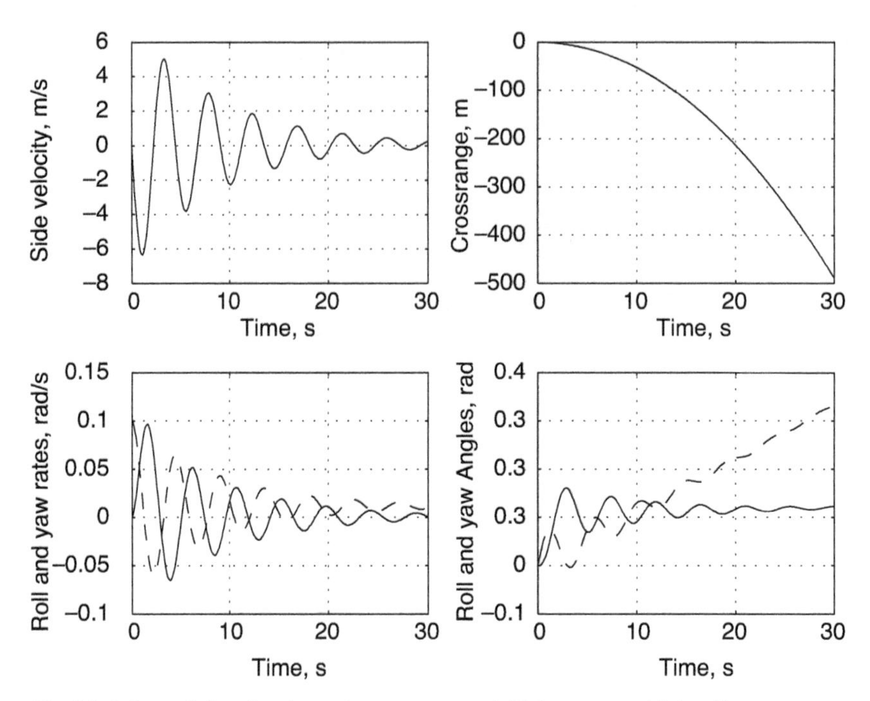

Fig. 4.2-3. Lateral-directional transient response to initial yaw rate of 0.1 rad/s.

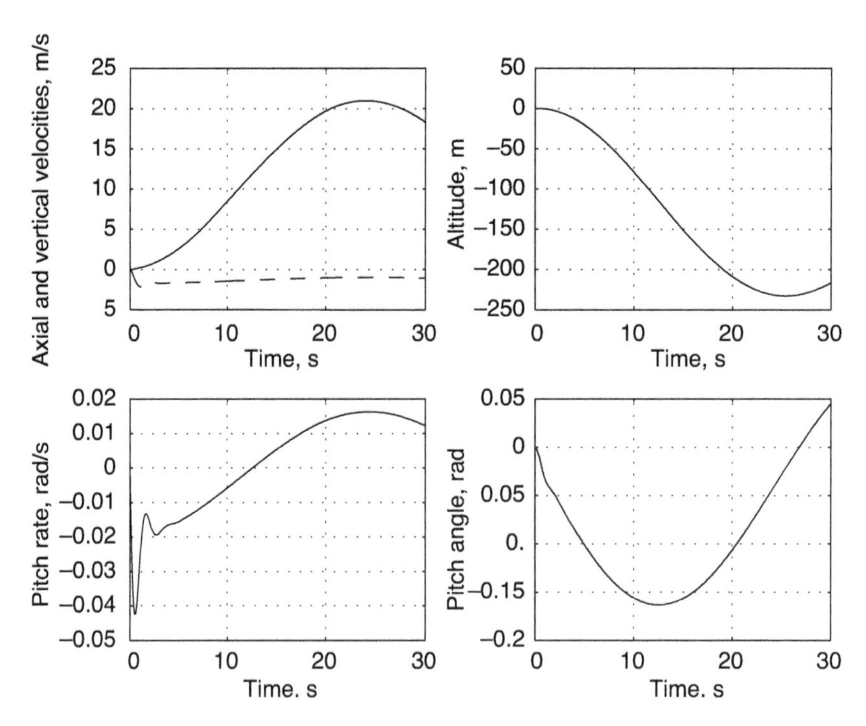

Fig. 4.2-4. Longitudinal transient response to 1 deg elevator step input.

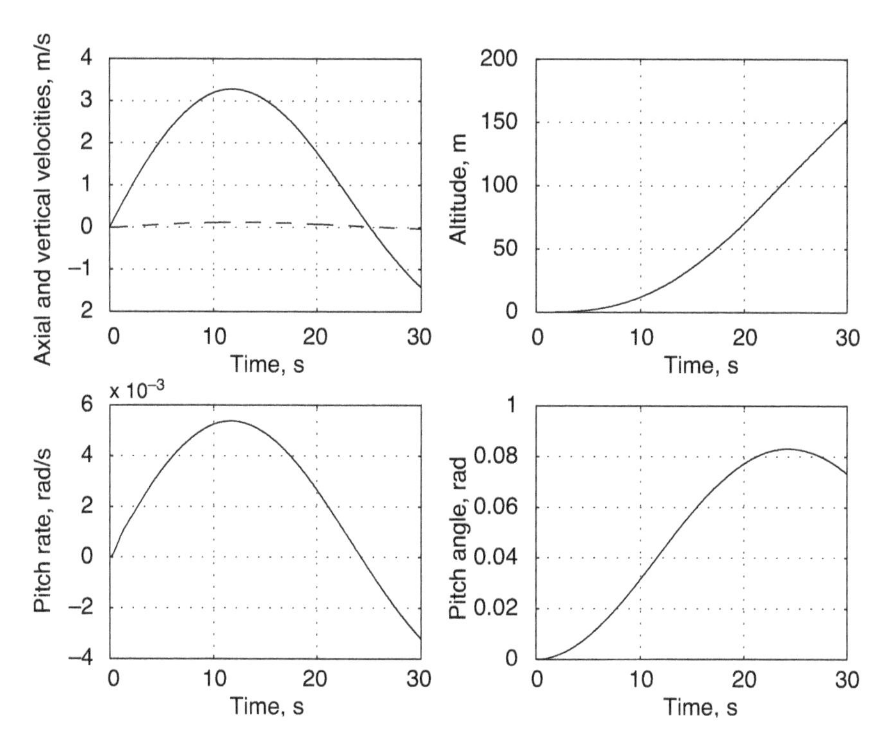

Fig. 4.2-5. Longitudinal transient response to 10 percent throttle step input.

A 10 percent throttle increase produces no discernible attitude response (Fig. 4.2-5). The phugoid mode is energized, and the aircraft begins to climb as a consequence of increasing airspeed; however, there are indications that the airspeed perturbation is oscillatory, as it begins to decrease at the end of the period shown.

A 0.5 deg aileron deflection introduces a steady roll rate that is accompanied by an oscillatory side-velocity transient and a steadily increasing yaw rate (Fig. 4.2-6). As a consequence, roll angle increases linearly and yaw angle increases quadratically during the period shown. Keep in mind, however, that the linearized model is valid only for small angular perturbations, so the actual response will be somewhat different as the predicted roll angle becomes large.

A 1 deg rudder deflection also produces a coupled response (Fig. 4.2-7). The lightly damped Dutch roll oscillation is more apparent than in the previous case. The side velocity and yaw rate are approaching steady offsets after 30 s, producing a linearly increasing yaw angle. The roll rate is

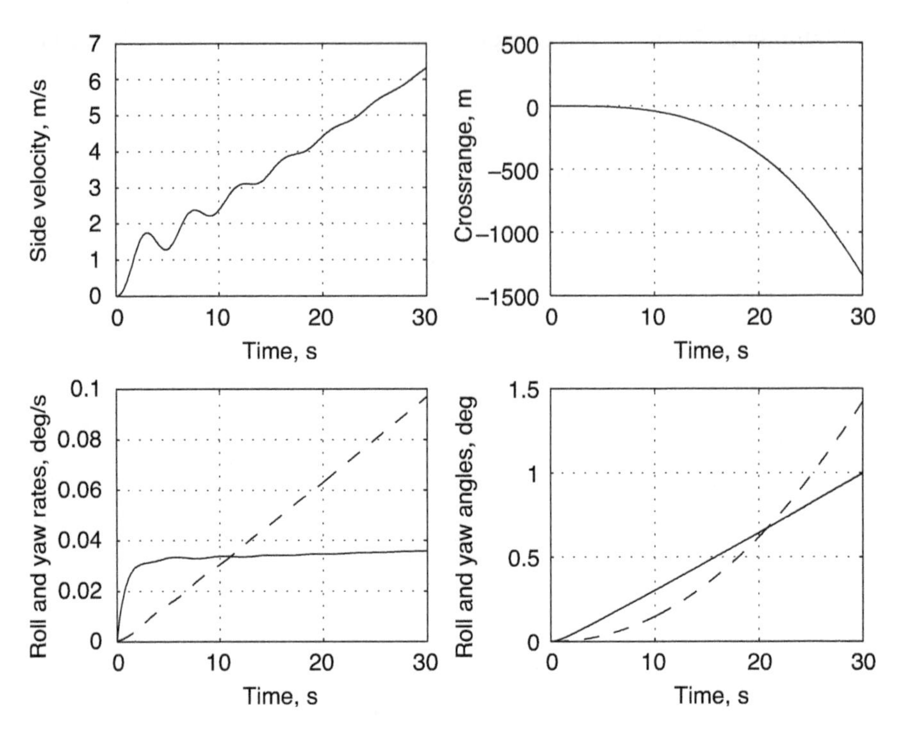

Fig. 4.2-6. Lateral-directional transient response to 0.5 deg aileron step input.

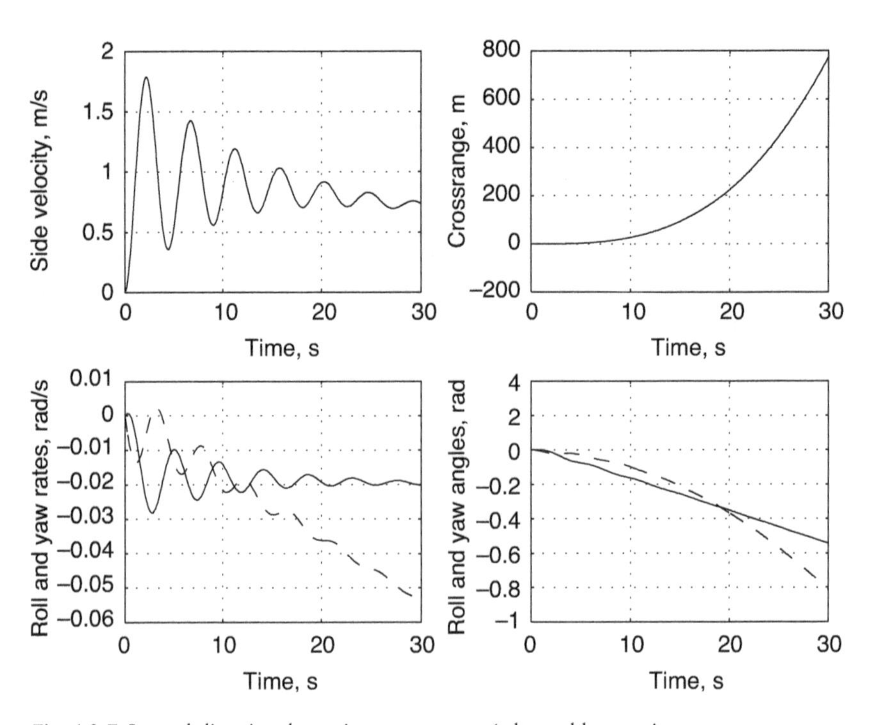

Fig. 4.2-7. Lateral-directional transient response to 1 deg rudder step input.

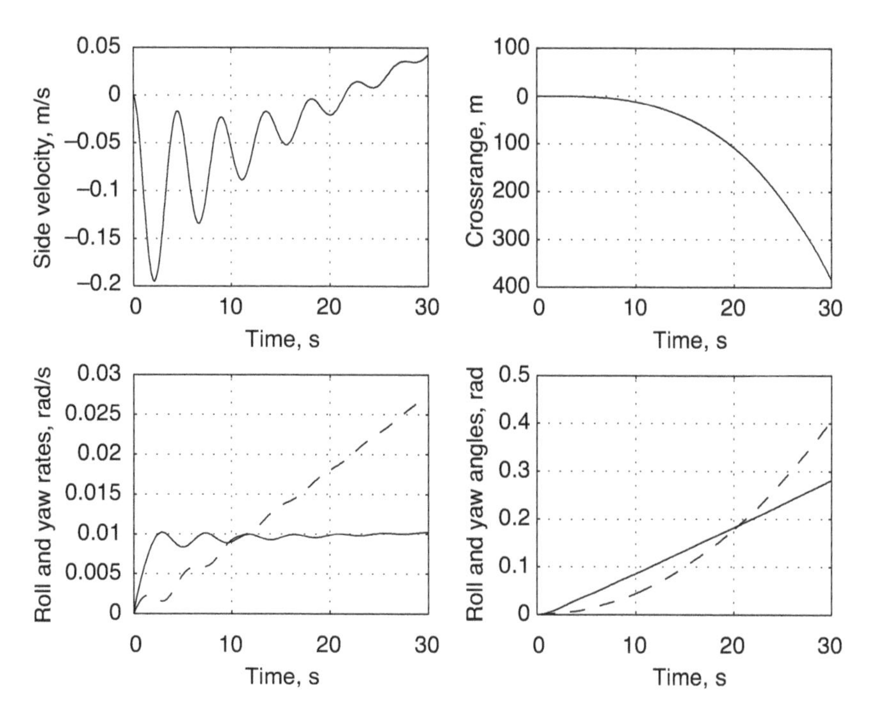

Fig. 4.2-8. Lateral-directional transient response to − 1 deg asymmetric-spoiler step input.

gradually increasing, as is the roll angle. The roll response is due primarily to the dihedral effect of the aircraft's wing; hence, it is less direct than the response due to the aileron, but the rudder-induced roll rate becomes larger than the yaw rate after 12 s.

The linearized asymmetric spoiler effect is similar to the aileron effect, but there is a higher degree of Dutch roll oscillation in the response (Fig. 4.2-8). The effect on side velocity is smaller and has opposite sign.

Static and Quasistatic Equilibrium Response to Inputs

When a linear dynamic system is at equilibrium, the rate of change of the state is zero; hence, its differential equation (eq. 4.2-1) becomes an algebraic one:

$$0 = \mathbf{F}\,\Delta\mathbf{x} + \mathbf{G}\,\Delta\mathbf{u} + \mathbf{L}\,\Delta\mathbf{w} \tag{4.2-11}$$

Here, we discuss characteristics of the equilibrium solutions of linear differential equations without regard to their stability, which is addressed in Section 4.3.

Assuming that control and disturbance inputs take constant values $\Delta\mathbf{u}^*$ and $\Delta\mathbf{w}^*$, the equilibrium state perturbation is a function of the two,

$$\Delta\mathbf{x}^* = -\mathbf{F}^{-1}(\mathbf{G}\,\Delta\mathbf{u}^* + \mathbf{L}\,\Delta\mathbf{w}^*) \tag{4.2-12}$$

provided that the inverse $\mathbf{F}^{-1}$ exists. Unfortunately, $\mathbf{F}^{-1}$ may not exist; in fact, we see such examples in $\mathbf{F}_{\text{Lo}}$ and $\mathbf{F}_{\text{LD}}$ (eqs. 4.1-100 and 4.1-107), which have columns of zeros that prevent inversion. A physical reason for the singularity in $\mathbf{F}_{\text{Lo}}$ is fairly obvious: in steady, cruising flight, the horizontal range Δx_I is continually increasing, while in climb or descent, Δz_I is changing as well. Therefore, there are no constant values of Δx_I (and possibly Δz_I) corresponding to the constant inputs.

Rearranging and partitioning the state vector as

$$\begin{aligned}\Delta\mathbf{x}_{\text{Lo}}^T &= [\Delta u\ \Delta w\ \Delta q\ \Delta\theta \mid \Delta x_I\ \Delta z_I]^T \\ &\triangleq [\Delta\mathbf{x}_1^T\ \Delta\mathbf{x}_2^T]\end{aligned} \tag{4.2-13}$$

the corresponding $\mathbf{F}_{\text{Lo}}$ can be partitioned as

$$\mathbf{F}_{\text{Lo}} = \begin{bmatrix}\mathbf{F}_{11} & 0 \\ \mathbf{F}_{21} & 0\end{bmatrix} \tag{4.2-14}$$

$\mathbf{F}_{11}$ is a (4×4) matrix, $\mathbf{F}_{21}$ is a (2×4) matrix, and the 0 matrices together represent two zero columns. Rearranging and partitioning $\mathbf{G}_{\text{Lo}}$ and $\mathbf{L}_{\text{Lo}}$ in the same way, static equilibrium values for the first four state elements are given by

$$\Delta\mathbf{x}_1^* = -\mathbf{F}_{11}^{-1}(\mathbf{G}_1\Delta\mathbf{u}^* + \mathbf{L}_1\Delta\mathbf{w}^*) \tag{4.2-15}$$

while the steadily increasing (quasistatic) response of $\Delta\mathbf{x}_2^*(t)$ is simply

$$\begin{aligned}\Delta\mathbf{x}_2^*(t) &= \Delta\mathbf{x}_2(t_0) + \int_{t_0}^{t}\mathbf{F}_{21}\Delta\mathbf{x}_1^* d\tau \\ &= \Delta\mathbf{x}_2(t_0) - \mathbf{F}_{21}\mathbf{F}_{11}^{-1}(\mathbf{G}_1\Delta\mathbf{u}^* + \mathbf{L}_1\Delta\mathbf{w}^*)(t - t_0)\end{aligned} \tag{4.2-16}$$

It may be desirable to control the aircraft's motions to produce a steady value of an r-dimensional *command response vector* expressed as the output vector,

$$\Delta\mathbf{y}^* = \mathbf{H}\,\Delta\mathbf{x}^* + \mathbf{J}\,\Delta\mathbf{u}^* \tag{4.2-17}$$

This equation must be solved simultaneously with eq. 4.2-12. The two are combined in the equation

$$\begin{bmatrix}-\mathbf{L}\Delta\mathbf{w}^* \\ \Delta\mathbf{y}^*\end{bmatrix} = \begin{bmatrix}\mathbf{F} & \mathbf{G} \\ \mathbf{H} & \mathbf{J}\end{bmatrix}\begin{bmatrix}\Delta\mathbf{x}^* \\ \Delta\mathbf{u}^*\end{bmatrix} \tag{4.2-18}$$

Equilibrium values of $\Delta\mathbf{x}^*$ and $\Delta\mathbf{u}^*$ can be found through inversion,

$$\begin{bmatrix}\Delta\mathbf{x}^* \\ \Delta\mathbf{u}^*\end{bmatrix} = \begin{bmatrix}\mathbf{F} & \mathbf{G} \\ \mathbf{H} & \mathbf{J}\end{bmatrix}^{-1} \begin{bmatrix}-\mathbf{L}\Delta\mathbf{w}^* \\ \Delta\mathbf{y}^*\end{bmatrix} \triangleq \begin{bmatrix}\mathbf{B}_{11} & \mathbf{B}_{12} \\ \mathbf{B}_{21} & \mathbf{B}_{22}\end{bmatrix} \begin{bmatrix}-\mathbf{L}\Delta\mathbf{w}^* \\ \Delta\mathbf{y}^*\end{bmatrix} \tag{4.2-19}$$

and elements of the inverse matrix can be specified by elimination:

$$\mathbf{B}_{11} = \mathbf{F}^{-1}(-\mathbf{G}\,\mathbf{B}_{21} + \mathbf{I}_n) \quad (n \times n) \tag{4.2-20}$$

$$\mathbf{B}_{12} = -\mathbf{F}^{-1}\,\mathbf{G}\,\mathbf{B}_{22} \quad (n \times r) \tag{4.2-21}$$

$$\mathbf{B}_{21} = -\mathbf{B}_{22}\,\mathbf{H}\,\mathbf{F}^{-1} \quad (m \times n) \tag{4.2-22}$$

$$\mathbf{B}_{22} = (-\mathbf{H}\,\mathbf{F}^{-1}\,\mathbf{G} + \mathbf{J})^{-1} \quad (m \times r) \tag{4.2-23}$$

Equilibrium values of the state and control corresponding to the command vector and the constant disturbance can then be calculated as

$$\Delta\mathbf{x}^* = \mathbf{B}_{11}\,(-\mathbf{L}\,\Delta\mathbf{w}^*) + \mathbf{B}_{12}\,\Delta\mathbf{y}^* \tag{4.2-24}$$

$$\Delta\mathbf{u}^* = \mathbf{B}_{21}\,(-\mathbf{L}\,\Delta\mathbf{w}^*) + \mathbf{B}_{22}\,\Delta\mathbf{y}^* \tag{4.2-25}$$

In an open-loop control law, $\Delta\mathbf{u}^*$ is the command sent to the controls to produce $\Delta\mathbf{y}^*$. In a closed-loop control law, $\Delta\mathbf{x}^*$ specifies the nominal value against which feedback data should be compared. For example, with linear feedback through the $(m \times n)$ gain matrix $\mathbf{C}$, the following is a suitable control law:

$$\Delta\mathbf{u}(t) = \Delta\mathbf{u}^* - \mathbf{C}\,[\Delta\mathbf{x}(t) - \Delta\mathbf{x}^*] \tag{4.2-26}$$

Three conditions must be satisfied for the commanded equilibrium solutions to exist:

1. $\mathbf{F}$ must be invertible, else the basic relationship between $\Delta\mathbf{x}^*$ and $\Delta\mathbf{u}^*$ is not well defined.

2. The command vector dimension must equal the control vector $(r = m)$, or the matrix to be inverted is not square.

3. The matrix $(-\mathbf{H}\,\mathbf{F}^{-1}\,\mathbf{G} + \mathbf{J})$ must be nonsingular; otherwise, the important term $\mathbf{B}_{22}$ cannot be calculated.

The third condition can be deceptive, as the following example shows. Choose $\mathbf{F}$ and $\mathbf{G}$ as the fourth-order longitudinal model ($\mathbf{F}_{11}$ and $\mathbf{G}_1$) defined above; then $n = 4$ and $m = 2$. In the first instance, let the command vector specify desired values of axial velocity and pitch angle; then

$$\mathbf{H} = \begin{bmatrix}1 & 0 & 0 & 0 \\ 0 & 0 & 0 & 1\end{bmatrix} \tag{4.2-27}$$

$$\mathbf{J} = \begin{bmatrix} 0 & 0 \\ 0 & 0 \end{bmatrix} \tag{4.2-28}$$

Defining $\mathbf{F}_{11}$ and $\mathbf{G}_1$ from the elements of eqs. 4.1-100 and 4.1-101, $(-\mathbf{H}\,\mathbf{F}^{-1}\mathbf{G}+\mathbf{J})$ is nonsingular for typical values of the stability and control derivatives, and $\mathbf{B}_{22}$ can be calculated. In the second case, choose axial velocity and pitch rate as the command variables; then

$$\mathbf{H} = \begin{bmatrix} 1 & 0 & 0 & 0 \\ 0 & 0 & 1 & 0 \end{bmatrix} \tag{4.2-29}$$

$$\mathbf{J} = \begin{bmatrix} 0 & 0 \\ 0 & 0 \end{bmatrix} \tag{4.2-30}$$

With the same assumptions, $(-\mathbf{H}\,\mathbf{F}^{-1}\mathbf{G}+\mathbf{J})$ is always singular, and $\mathbf{B}_{22}$ is undefined.

The second case presents another example of quasistatic equilibrium: a constant pitch rate Δq^* implies steadily changing pitch angle $\Delta\theta^*$ [S-15]. The solution is to partition $\Delta\mathbf{x}$ as

$$\Delta\mathbf{x}_1 = [\Delta u \; \Delta w \; \Delta q]^T \tag{4.2-31}$$

$$\Delta\mathbf{x}_2 = \Delta\theta \tag{4.2-32}$$

and to redefine the state and output equations as follows:

$$\begin{bmatrix} \Delta\dot{\mathbf{x}}_1 \\ \Delta\dot{\mathbf{x}}_2 \end{bmatrix} = \begin{bmatrix} \mathbf{F}_{11} & \mathbf{F}_{12} \\ \mathbf{F}_{21} & 0 \end{bmatrix} \begin{bmatrix} \Delta\mathbf{x}_1 \\ \Delta\mathbf{x}_2 \end{bmatrix} + \begin{bmatrix} \mathbf{G}_1 \\ 0 \end{bmatrix} \Delta\mathbf{u} \tag{4.2-33}$$

$$\Delta\mathbf{y} = \mathbf{H}_1\,\Delta\mathbf{x}_1 + \mathbf{J}\,\Delta\mathbf{u} \tag{4.2-34}$$

If $\mathbf{F}_{12}$ were zero, the static equilibrium of $\Delta\mathbf{x}_1$ could be calculated for zero disturbance as

$$\Delta\mathbf{x}_1^* = \mathbf{F}_{11}^{-1}\,\mathbf{G}_1\,(\mathbf{H}_1\,\mathbf{F}_{11}^{-1}\,\mathbf{G}_1)^{-1}\Delta\mathbf{y}^* \tag{4.2-35}$$

$$\Delta\mathbf{u}^* = (-\mathbf{H}_1\,\mathbf{F}_{11}^{-1}\,\mathbf{G}_1)^{-1}\Delta\mathbf{y}^* \tag{4.2-36}$$

and the increasing pitch angle would be given by

$$\Delta\mathbf{x}_2^*(t) = \Delta\mathbf{x}_2(t_0) + \int_{t_0}^{t} \mathbf{F}_{21}\Delta\mathbf{x}_1^* d\tau \tag{4.2-37}$$

Unfortunately, at least one term of $\mathbf{F}_{12}$ is always nonzero in this example, so the following equation must be satisfied:

$$\begin{bmatrix} \Delta\mathbf{x}_1^* \\ \Delta\mathbf{u}^* \end{bmatrix} = \begin{bmatrix} \mathbf{F}_{11} & \mathbf{G}_1 \\ \mathbf{H}_1 & 0 \end{bmatrix}^{-1} \begin{bmatrix} -\mathbf{F}_{12}\Delta\mathbf{x}_2^*(t) \\ \Delta\mathbf{y}^* \end{bmatrix} \triangleq \begin{bmatrix} \mathbf{B}_{11} & \mathbf{B}_{12} \\ \mathbf{B}_{21} & \mathbf{B}_{22} \end{bmatrix} \begin{bmatrix} -\mathbf{F}_{12}\Delta\mathbf{x}_2^*(t) \\ \Delta\mathbf{y}^* \end{bmatrix} \tag{4.2-38}$$

together with eq. 4.2-37. This result implies that a continually changing control is required to maintain constant $\Delta\mathbf{y}^*$. Further discussion is contained in [S-14].

Example 4.2-2. Static Response of the Linear Dynamic Model

The program SURVEY (Appendix C) computes the static control and command response for fourth-order systems with no disturbances and two control inputs (elevator and throttle in the longitudinal case, aileron and rudder for the lateral-directional model). Static control equilibrium does not exist for the higher-order models presented here because **F** is singular.

The fourth-order, body-axis, longitudinal model of the business jet at $\mathcal{M} = 0.3$ and $h = 3{,}050$ m is

$$\mathbf{F} = \begin{bmatrix} -0.0121 & 0.096 & -6.45 & -9.787 \\ -0.116 & -1.277 & 100.8 & -0.6202 \\ 0.005 & -0.0781 & -1.279 & 0 \\ 0 & 0 & 1 & 0 \end{bmatrix}$$

$$\mathbf{G} = \begin{bmatrix} 0.0065 & 4.674 \\ -13.17 & 0 \\ -9.069 & 0 \\ 0 & 0 \end{bmatrix}$$

with state and control vectors $\Delta\mathbf{x}^T = [\Delta u \;\; \Delta w \;\; \Delta q \;\; \Delta\theta]$ and $\Delta\mathbf{u}^T = [\Delta\delta E \;\; \Delta\delta T]$. The sensitivity of the static response to control is expressed by $-\mathbf{F}^{-1}\mathbf{G}$, which is

$$-\mathbf{F}^{-1}\mathbf{G} = \begin{bmatrix} 690.7 & -1.507 \\ -72.3 & -0.0955 \\ 0 & 0 \\ -1.565 & 0.4785 \end{bmatrix}$$

Thus, the static axial-velocity perturbation for constant elevator and throttle settings $\Delta\delta E^*$ and $\Delta\delta T^*$ is

$$\Delta u^* = 690.7\,\Delta\delta E^* - 1.507\,\Delta\delta T^*$$

and so on.

Choosing the command as $\Delta\mathbf{y}^{*T} = [\Delta u^* \;\; \Delta w^*]$, the output matrix is

$$\mathbf{H}_{\text{out}} = \begin{bmatrix} 1 & 0 & 0 & 0 \\ 0 & 1 & 0 & 0 \end{bmatrix}$$

The sensitivity of the static state and control response to command is expressed by $\mathbf{B}_{12}$ and $\mathbf{B}_{22}$, which are

$$\mathbf{B}_{12} = \begin{bmatrix} 1 & 0 \\ 0 & 1 \\ 0 & 0 \\ -0.1987 & -1.877 \end{bmatrix}$$

$$\mathbf{B}_{22} = \begin{bmatrix} 0.0005 & -0.0086 \\ -0.4135 & -3.95 \end{bmatrix}$$

For $\Delta\mathbf{y}^{*T} = [0 \;\; 1]$, the corresponding constant state and control are $\Delta\mathbf{x}^{*T} = [0 \; 1 \; 0 \; -1.877]$ and $\Delta\mathbf{u}^{*T} = [-0.0086 \; -3.95]$.

Initial Response to Control Inputs

It is easy to overlook one of the simplest response characteristics of dynamic systems, but it can have important consequences for piloted or automatic control. With zero initial conditions and disturbance input, the initial control response is (eq. 4.2-1)

$$\Delta\dot{\mathbf{x}}(t_0) = \mathbf{G}(t_0)\, \Delta\mathbf{u}(t_0) \tag{4.2-39}$$

and the corresponding command response (eq. 4.2-17) is

$$\Delta\dot{\mathbf{y}}(t_0) = \mathbf{H}(t_0)\, \mathbf{G}(t_0)\, \Delta\mathbf{u}(t_0) \tag{4.2-40}$$

$$\Delta\mathbf{y}(t_0) = \mathbf{J}(t_0)\, \Delta\mathbf{u}(t_0) \tag{4.2-41}$$

Some elements of the state rate $\Delta\dot{\mathbf{x}}(t)$ have immediate response to the control input, while others must wait for the system to integrate the control effect. For example, an elevator input evokes an immediate pitch acceleration, but the initial pitch rate is zero. Given a step input, it is desirable for the initial and final responses to have the same sign (this is known as a "minimum-phase" response). Otherwise, the initial response provides a deceptive ("non-minimum-phase") cue to the eventual motion, confusing the pilot and limiting the gains for stable feedback control.

Controllability and Observability of Motions

Two additional characteristics of the linear dynamic system represented by eqs. 4.1-19 and 4.1-20 portray the ability of the control vector $\Delta\mathbf{u}(t)$ to affect all the components of the state $\Delta\mathbf{x}(t)$ and the ability of an output vector $\Delta\mathbf{y}(t)$ to observe all the elements of the state. *Controllability* at some time t_0 implies that there is a control history $\{\Delta\mathbf{u}(t)$ in $[t_0, t_f]\}$ capable of forcing every element of an arbitrary initial condition $\Delta\mathbf{x}(t_0)$ to zero at time $t_f < \infty$. *Observability* at t_0 implies that the output history $\{\Delta\mathbf{y}(t)$ in $[t_0, t_f]\}$ is sufficient to reconstruct $\Delta\mathbf{x}(t_0)$.

Neglecting disturbance inputs and direct observation of control effects, the linear, time-invariant model of aircraft motions and their measurement consists of the two equations

$$\Delta\dot{\mathbf{x}}(t) = \mathbf{F}\,\Delta\mathbf{x}(t) + \mathbf{G}\,\Delta\mathbf{u}(t) \tag{4.2-42}$$

$$\Delta\mathbf{y}(t) = \mathbf{H}\,\Delta\mathbf{x}(t) \tag{4.2-43}$$

The dimensions of $\Delta\mathbf{x}$, $\Delta\mathbf{u}$, and $\Delta\mathbf{y}$ are n, m, and r. (Note that $\Delta\mathbf{y}$ need not be the command vector considered earlier in the section.)

If $n = m$ and $\mathbf{G} = \mathbf{I}_n$, the control vector has a direct effect on every element of $\Delta\dot{\mathbf{x}}(t)$, and the LTI system is completely controllable. For example, choosing the feedback control law $\Delta\mathbf{u} = (\mathbf{F}_m - \mathbf{F})\,\Delta\mathbf{x}$ would allow us to model any system dynamics $\mathbf{F}_m$ of the same dimension as $\mathbf{F}$. Similarly, if $n = r$ and $\mathbf{H} = \mathbf{I}_n$, every element of $\Delta\mathbf{x}$ is directly observed in $\Delta\mathbf{y}$. More generally, if $\mathbf{G}$ and $\mathbf{H}$ have rank n,[1] the system is completely controllable and observable. The challenge arises when m or r is smaller than n, for then there cannot be direct, independent control or observation of every state component. However, there may be indirect paths that produce controllability and observability.

From eqs. 4.2-7 and 4.2-8 and letting $t_0 = 0$, the state's response to control at t_f can be expressed as

$$\Delta\mathbf{x}(t_f) = e^{\mathbf{F}t_f}\left[\Delta\mathbf{x}(0) + \int_0^{t_f} [e^{-\mathbf{F}t}\mathbf{G}\,\Delta\mathbf{u}(t)]\,dt\right] \tag{4.2-44}$$

If the system is completely controllable, $\Delta\mathbf{x}(t_f)$ could be forced to zero by control actions, leaving

$$-\Delta\mathbf{x}(0) = \int_0^{t_f} [e^{-\mathbf{F}t}\mathbf{G}\,\Delta\mathbf{u}(t)]\,dt \tag{4.2-45}$$

The conditions that allow this to occur must be established. $e^{-\mathbf{F}t}$ can be expanded in a power series (eq. 4.2-5); as a consequence of the Cayley-Hamilton theorem [B-10], this series can be expressed in the remainder form:

$$e^{-\mathbf{F}t} = r_0(t)\mathbf{I}_n + r_1(t)\mathbf{F} + r_2(t)\mathbf{F}^2 + \cdots + r_{n-1}(t)\,\mathbf{F}^{n-1} \tag{4.2-46}$$

1. The *rank* of a matrix is the dimension of the largest nonzero determinant of any submatrix of the matrix. This is equivalent to the number of linearly independent rows or columns of the matrix.

Substituting in eq. 4.2-45, the right-side integral can be written as a matrix-vector product

$$-\Delta\mathbf{x}(0) = \mathbf{G}\int_0^{t_f} [r_o(t)\,\Delta\mathbf{u}(t)]\,dt + \mathbf{F}\mathbf{G}\int_0^{t_f} [r_1(t)\,\Delta\mathbf{u}(t)]\,dt$$

$$+ \mathbf{F}^2\mathbf{G}\int_0^{t_f} [r_2(t)\,\Delta\mathbf{u}(t)]\,dt + \cdots + \mathbf{F}^{n-1}\mathbf{G}\int_0^{t_f} [r_{n-1}(t)\,\Delta\mathbf{u}(t)]\,dt$$

$$= [\mathbf{G}\ \mathbf{F}\mathbf{G}\ \mathbf{F}^2\mathbf{G} \cdots \mathbf{F}^{n-1}\mathbf{G}]\int_0^{t_f} \begin{bmatrix} \mathbf{v}_0 \\ \mathbf{v}_1 \\ \mathbf{v}_2 \\ \cdots \\ \mathbf{v}_{n-1} \end{bmatrix} dt \tag{4.2-47}$$

where the $\mathbf{v}_i$ define the control settings needed to match every element of $-\Delta\mathbf{x}(0)$. The actual control values need not be found to demonstrate controllability. However, these control values can equal $-\Delta\mathbf{x}(0)$ only if the matrix $[\mathbf{G}\ \mathbf{F}\mathbf{G}\ \mathbf{F}^2\mathbf{G} \cdots \mathbf{F}^{n-1}\mathbf{G}]$ has n linearly independent columns, that is, if it has rank n. We call this $(n \times nm)$ matrix the *controllability matrix* $\mathscr{C}$:

$$\mathscr{C} \triangleq [\mathbf{G}\ \mathbf{F}\mathbf{G}\ \mathbf{F}^2\mathbf{G} \cdots \mathbf{F}^{n-1}\mathbf{G}] \tag{4.2-48}$$

As shown in [S-14] and elsewhere, observation (or measurement) of the dynamic system is the *mathematical dual* of its control. Given an output vector defined by eq. 4.2-43, there is a corresponding $(nr \times n)$ *observability matrix* $\mathscr{O}$:

$$\mathscr{O} = \begin{bmatrix} \mathbf{H} \\ \mathbf{H}\mathbf{F} \\ \cdots \\ \mathbf{H}\mathbf{F}^{n-1} \end{bmatrix} \tag{4.2-49}$$

$\mathscr{O}$ must have rank n for complete observability, which allows the n elements of $\Delta\mathbf{x}(0)$ to be derived from the r measurements in $\Delta\mathbf{y}(t)$. Using MATLAB's Control System Toolbox, the controllability matrix is calculated as `ctrb(F, G)`, and the observability matrix is calculated as `obsv(F,H)`.

It is necessary to have complete controllability *and* observability to implement closed-loop control of any system or subsystem. Of course, complete closed-loop control is not always required; uncontrolled elements that have stable dynamics and are not subject to excessive excitation may be acceptable. A system whose uncontrolled elements are stable is called *stabilizable*, while one whose unobserved elements are stable is called *detectable*.

For the most part, the longitudinal and lateral-directional models of aircraft dynamics are controllable by these standards, as **F** and **G** are dense. The extent of controllability can be understood by selectively zeroing elements of these matrices and evaluating the rank of $\mathscr{C}$. For example, the elevator alone, which produces a pitching moment, imparts controllability to longitudinal motions. It is only when a massive number of terms in $\mathbf{F}_{\mathrm{Lo}}$ are zeroed that controllability is lost.

Nevertheless, this test for controllability says nothing about the *quality* of control. It takes the LTI assumption quite literally, ignoring the fact that real control surfaces have limits on deflection and rate of movement. It does not account for errors that may occur in the transmission of control commands to control surfaces. Even when controllability is adequate for small perturbations, the $\mathbf{v}_i$ needed to satisfy eq. 4.2-47 may not be available if $\Delta\mathbf{x}(0)$ is sufficiently large. An element of **F** or **G** could be nonzero but quite small, providing a path for control action that has little practical significance. Furthermore, as a maneuver evolves, **F** and **G** may change, modifying the control response.

Similar comments can be made about observability. In principle, the complete state can be estimated from relatively sparse measurements, given typical definitions of **F** and **H** for aircraft rigid-body motions. However, sensors have finite limits on measurement range, and they are imperfect, possibly introducing errors due to miscalibration, noise, and unmodeled dynamics (e.g., higher-order elastic effects). Controllability and observability are necessary but not sufficient conditions for satisfactory closed-loop control.

Example 4.2-3. Controllability and Observability of a Linear Dynamic Model

The rank tests are applied to the sixth-order, body-axis version of the business jet model considered in prior examples using the SURVEY program (Appendix C). The longitudinal and lateral-directional states are $\Delta\mathbf{x}_{\mathrm{lon}}{}^{\mathrm{T}} = [\Delta u\ \Delta w\ \Delta q\ \Delta\theta\ \Delta x\ \Delta z]^{\mathrm{T}}$ and $\Delta\mathbf{x}_{\mathrm{LD}}{}^{T} = [\Delta v\ \Delta p\ \Delta r\ \Delta\phi\ \Delta\psi\ \Delta y]$. The two models are found to be completely controllable with any single control contained in each model.

Selectively observing a single state element at a time, the longitudinal case has a rank deficiency of 1 except when Δx is observed; thus the full state initial condition could be reconstructed from observing that single element alone but not from the others. For the lateral-directional case, there is no rank deficiency when Δy is observed. The deficiency is 1 when $\Delta\psi$ is observed and 2 when each of the remaining variables is observed. By comparison, the fourth-order models are completely observable with any single observation. What can you conclude about the unobservable variables?

TRUNCATION AND RESIDUALIZATION

Reduced-order dynamic models play a large role in analysis and design, simplifying calculations either by lowering the dimension of the equations to be solved or by partitioning a single large problem into several smaller ones. We may, for example, want to identify the critical factors contributing to short-period aircraft motions without addressing slow variations in flight condition. If rigid-body motions are our principal concern, we may choose either to neglect aeroelastic effects or to account for them in an approximate fashion. The dynamics of control surface motion may be studied without regard to gross motions of the aircraft. In each case, we focus on a piece of the problem rather than the entire solution.

It should be clear that simplifying a problem implies approximation, that allowable assumptions are problem specific, and that simplification is not always appropriate. The principal basis for partitioning a dynamic model is *spatial* or *time-scale separation of response characteristics*. Two subsystems might operate at the same time scale but be physically distant from each other, allowing each to be analyzed without regard to the other. Vertical acceleration felt in the cockpit can readily be separated into fast and slow components due to fuselage bending and aircraft maneuvering. Considering the latter two effects separately may be permissible for evaluating fatigue loads on the aircraft but not for designing a high-gain feedback control system.

If the state vector is partitioned into two parts, the linear system (e.g., eq. 4.2-42) is separated into four components:

$$\begin{bmatrix} \Delta\dot{\mathbf{x}}_1 \\ \Delta\dot{\mathbf{x}}_2 \end{bmatrix} = \begin{bmatrix} \mathbf{F}_{11} & \mathbf{F}_{12} \\ \mathbf{F}_{21} & \mathbf{F}_{22} \end{bmatrix} \begin{bmatrix} \Delta\mathbf{x}_1 \\ \Delta\mathbf{x}_2 \end{bmatrix} + \begin{bmatrix} \mathbf{G}_{11} & \mathbf{G}_{12} \\ \mathbf{G}_{21} & \mathbf{G}_{22} \end{bmatrix} \begin{bmatrix} \Delta\mathbf{u}_1 \\ \Delta\mathbf{u}_2 \end{bmatrix} \tag{4.2-50}$$

$\mathbf{F}_{11}$ and $\mathbf{F}_{22}$ represent the direct effects that each segment of the state has on its own evolution, and $\mathbf{F}_{12}$ and $\mathbf{F}_{21}$ represent the coupling that one dynamic subsystem has on the other. We also may associate elements of the control more strongly with one subsystem or the other via $\mathbf{G}_{11}$ and $\mathbf{G}_{22}$, accounting for coupling through $\mathbf{G}_{12}$ and $\mathbf{G}_{21}$. The system is *truncated* if the coupling elements are neglected,

$$\begin{bmatrix} \Delta\dot{\mathbf{x}}_1 \\ \Delta\dot{\mathbf{x}}_2 \end{bmatrix} = \begin{bmatrix} \mathbf{F}_{11} & 0 \\ 0 & \mathbf{F}_{22} \end{bmatrix} \begin{bmatrix} \Delta\mathbf{x}_1 \\ \Delta\mathbf{x}_2 \end{bmatrix} + \begin{bmatrix} \mathbf{G}_{11} & 0 \\ 0 & \mathbf{G}_{22} \end{bmatrix} \begin{bmatrix} \Delta\mathbf{u}_1 \\ \Delta\mathbf{u}_2 \end{bmatrix} \tag{4.2-51}$$

leaving two independent systems:

$$\Delta\dot{\mathbf{x}}_1 = \mathbf{F}_{11}\Delta\mathbf{x}_1 + \mathbf{G}_{11}\Delta\mathbf{u}_1 \tag{4.2-52}$$

$$\Delta\dot{\mathbf{x}}_2 = \mathbf{F}_{22}\Delta\mathbf{x}_2 + \mathbf{G}_{22}\Delta\mathbf{u}_2 \tag{4.2-53}$$

The only requirement for satisfactory approximation is that the coupling matrix terms are negligible. In such case, dynamics and control of the two subsystems can be considered separately.

Systems with *block-triangular coupling* are not strictly independent, although their dynamic characteristics are uncoupled. If $\mathbf{F}_{12}$ is not zero but $\mathbf{F}_{21}$ is, then $\Delta\mathbf{x}_2$ affects $\Delta\dot{\mathbf{x}}_1$, but $\Delta\mathbf{x}_1$ does not affect $\Delta\dot{\mathbf{x}}_2$. The evolution of $\Delta\mathbf{x}_2$ is totally uncoupled from $\Delta\mathbf{x}_1$ (eq. 4.2-53). The effect of $\Delta\mathbf{x}_2$ on $\Delta\dot{\mathbf{x}}_1$ is that of an external input, and the dynamic modes of $\Delta\mathbf{x}_1$ (defined in the next section) are unchanged.

If coupling is not negligible but the time scales of motion are widely separated, we can apply *residualization*. Let us assume that the response of Subsystem 2 is much faster than that of Subsystem 1. On 2's time scale, the slower motions of 1 are insignificant, and the evolution of $\Delta\mathbf{x}_2$ is governed by eq. 4.2-53. On 1's time scale, Subsystem 2 responds so quickly that it always appears to be at its equilibrium state. Therefore, $\Delta\mathbf{x}_1$ evolves according to the following equation:

$$\begin{bmatrix}\Delta\dot{\mathbf{x}}_1\\ 0\end{bmatrix}=\begin{bmatrix}\mathbf{F}_{11} & \mathbf{F}_{12}\\ \mathbf{F}_{21} & \mathbf{F}_{22}\end{bmatrix}\begin{bmatrix}\Delta\mathbf{x}_1\\ \Delta\mathbf{x}_2{}^*\end{bmatrix}+\begin{bmatrix}\mathbf{G}_{11} & \mathbf{G}_{12}\\ \mathbf{G}_{21} & \mathbf{G}_{22}\end{bmatrix}\begin{bmatrix}\Delta\mathbf{u}_1\\ \Delta\mathbf{u}_2\end{bmatrix} \tag{4.2-54}$$

The equation is separated into two parts, a differential equation and an algebraic equation:

$$\Delta\dot{\mathbf{x}}_1=\mathbf{F}_{11}\Delta\mathbf{x}_1+\mathbf{F}_{12}\Delta\mathbf{x}_2{}^*+\mathbf{G}_{11}\Delta\mathbf{u}_1+\mathbf{G}_{12}\Delta\mathbf{u}_2 \tag{4.2-55}$$

$$\Delta\mathbf{x}_2{}^*=-\mathbf{F}_{22}^{-1}(\mathbf{F}_{21}\Delta\mathbf{x}_1+\mathbf{G}_{21}\Delta\mathbf{u}_1+\mathbf{G}_{22}\Delta\mathbf{u}_2) \tag{4.2-56}$$

Substituting eq. 4.2-56 in eq. 4.2-55 produces a revised differential equation for $\Delta\mathbf{x}_1$:

$$\begin{aligned}\Delta\dot{\mathbf{x}}_1&=(\mathbf{F}_{11}-\mathbf{F}_{12}\mathbf{F}_{22}^{-1}\mathbf{F}_{21})\Delta\mathbf{x}_1+(\mathbf{G}_{11}-\mathbf{F}_{12}\mathbf{F}_{22}^{-1}\mathbf{G}_{21})\Delta\mathbf{u}_1\\&\quad+(\mathbf{G}_{12}-\mathbf{F}_{12}\mathbf{F}_{22}^{-1}\mathbf{G}_{22})\Delta\mathbf{u}_2\end{aligned} \tag{4.2-57}$$

There are two requirements for residualization to be useful: $\mathbf{F}_{22}$ must possess an inverse, and Subsystem 2 must be stable. The first requirement follows from the need to establish an equilibrium state, and the second results from the need to stay there. We shall discuss more about stability in the next section.

4.3 Stability and Modes of Motion

The traditional area of *aircraft stability and control* is largely based on linear, time-invariant (LTI) models of aircraft motion and "frequency-domain" descriptions of aircraft dynamics. LTI models allow broad statements to be made about the stability of an aircraft's response to initial conditions, disturbances, and control. In addition to supporting calculations of quantitative effects, these models facilitate a qualitative understanding of flight behavior. In the abstract, all aircraft possess the same modes of motion, but the specifics of these modes vary widely from one aircraft type to the next. For example, a mode that is stable in one aircraft may be unstable in another, speeds of response may vary with flight condition, and the coupling of response variables within a particular mode may change with aircraft configuration. It is important, therefore, to develop a good understanding of the definitions of stability, the means for transforming differential equations to equivalent frequency-domain models, and the ways in which the basic modes are reflected in aircraft motion.

Stability of Transient Response

Suppose that an aircraft is flying at steady cruise and a small gust of wind perturbs the equilibrium. If the aircraft's motion tends to return to the original condition without active control, then that condition is *stable*. If the perturbation produces a motion that diverges from equilibrium with time, the condition is *unstable*. The stability characteristics of these perturbed motions can be examined using linear models. If the perturbed motions are unstable, the linear equations approximate actual motions during their early phase, but the resulting divergence produces large perturbations that violate the assumptions required for linearization.

A scalar metric is needed to quantify the "size" of motions so that we can determine whether the perturbed motions are shrinking or growing with time. Individual components of $\Delta\mathbf{x}(t)$ can be examined, but it is plausible that some components will decay while others diverge. A *quadratic* (or *Euclidean*) *norm* can be defined for this purpose:

$$\|\Delta\mathbf{x}(t)\| \triangleq [\Delta\mathbf{x}^T(t)\Delta\mathbf{x}(t)]^{1/2} = [x_1^2 + x_2^2 + \cdots + x_n^2]^{1/2} \tag{4.3-1}$$

If *any* component of $\Delta\mathbf{x}(t)$ diverges, then $\|\Delta\mathbf{x}(t)\|$ diverges, indicating instability (Fig. 4.3-1).

There are many definitions of stability [N-1]; for linear, time-invariant systems, *asymptotic stability* is guaranteed with any initial condition $\Delta\mathbf{x}(t_0)$ if

$$\lim_{t\to\infty} \|\Delta\mathbf{x}(t)\| = 0 \tag{4.3-2}$$

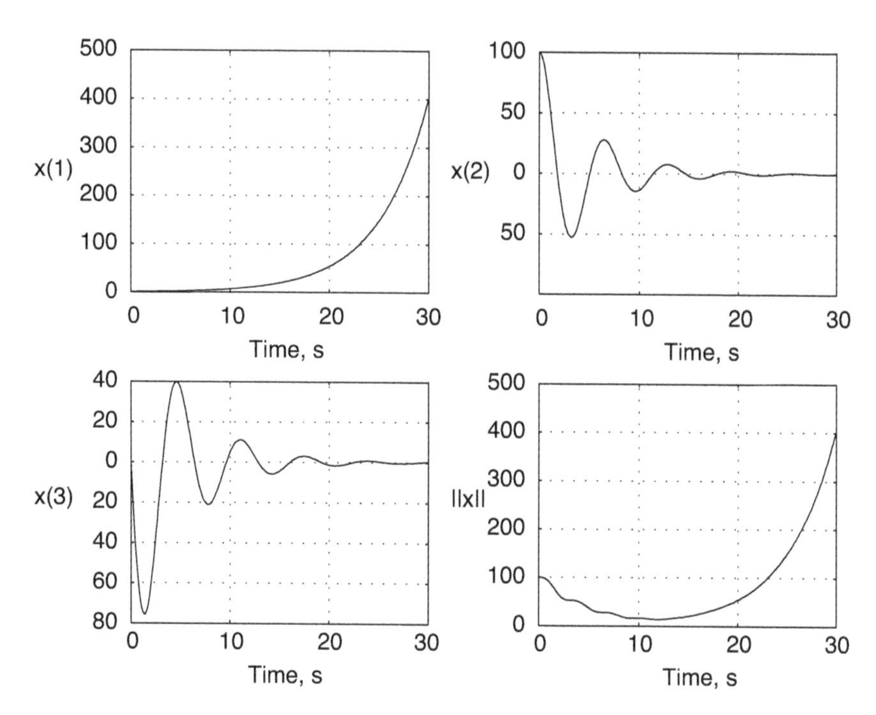

Fig. 4.3-1. Example of quadratic norm divergence.

From eqs. 4.2-4 and 4.2-5, the state transition matrix governs the evolution of $\Delta\mathbf{x}(t)$ for an unforced system,

$$\Delta\mathbf{x}(t) = \mathbf{\Phi}(t - t_0)\Delta\mathbf{x}(t_0) = e^{\mathbf{F}(t - t_0)}\Delta\mathbf{x}(t_0) \tag{4.3-3}$$

so the norm of $\Delta\mathbf{x}(t)$ is

$$\|\Delta\mathbf{x}(t)\| = \|e^{\mathbf{F}(t - t_0)}\Delta\mathbf{x}(t_0)\| \tag{4.3-4}$$

As $\Delta\mathbf{x}(t_0)$ is a constant, $e^{\mathbf{F}(t - t_0)}$ must be a decaying function of time for the linear system to be asymptotically stable.

Exponential asymptotic stability is guaranteed if

$$\|\Delta\mathbf{x}(t)\| \le k e^{-\alpha(t - t_0)}\ \|\Delta\mathbf{x}(t_0)\|,\quad k, \alpha > 0 \tag{4.3-5}$$

The norm of $\Delta\mathbf{x}(t)$ then must lie within the decaying envelope specified by $ke^{-\alpha(t - t_0)}$ (Fig. 4.3-2). The relationship to eq. 4.3-4 is apparent: if the

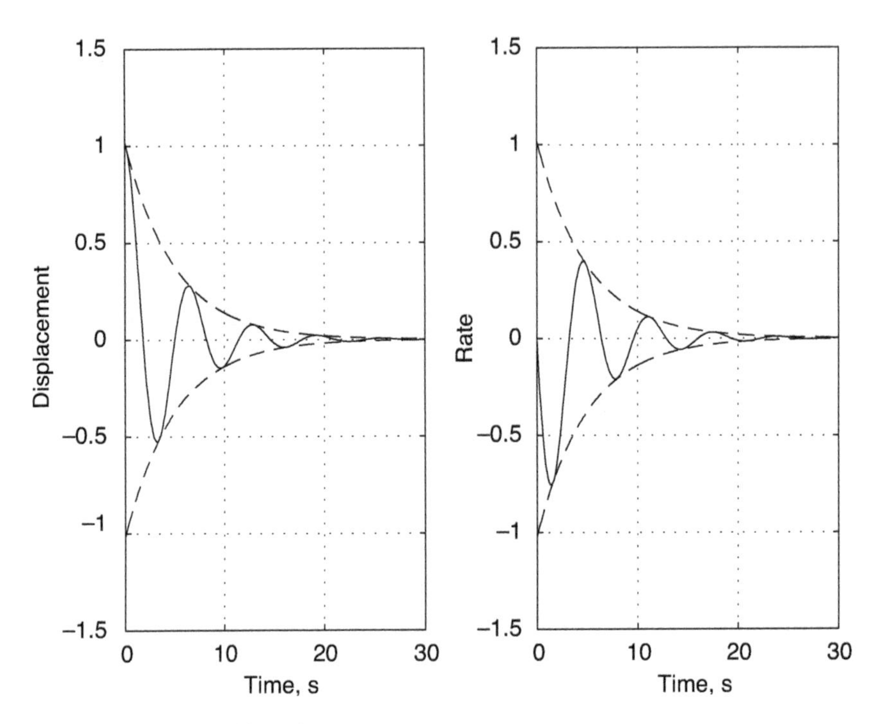

Fig. 4.3-2. Exponential bounds on response.

LTI system is stable, it possesses exponential asymptotic stability. *Neutral stability* is defined as the special case in which $\alpha = 0$; perturbations are guaranteed not to diverge, but they may not converge to zero.

How can we determine whether or not $e^{\mathbf{F}(t-t_0)}$ is a decaying function of time without generating the state history? One approach is to invoke *Lyapunov's second stability theorem.*[2] Define a scalar function $\mathcal{V}[\Delta\mathbf{x}(t)]$ (called a *Lyapunov function*) that is positive definite over the region of interest. For the unforced LTI system, the time derivative of $\mathcal{V}[\Delta\mathbf{x}(t)]$ is

$$\frac{d\mathcal{V}[\Delta\mathbf{x}(t)]}{dt} = \left\{\frac{d\mathcal{V}[\Delta\mathbf{x}(t)]}{d\Delta\mathbf{x}}\right\}\frac{d\Delta\mathbf{x}(t)}{dt} = \left\{\frac{d\mathcal{V}[\Delta\mathbf{x}(t)]}{d\Delta\mathbf{x}}\right\}\mathbf{F}\Delta\mathbf{x}(t) \qquad (4.3\text{-}6)$$

2. *Lyapunov's first stability theorem* states that a nonlinear system is asymptotically stable at its origin if the linearized model of the system is asymptotically stable at the origin.

If the time derivative is always negative, then $\mathcal{V}[\Delta\mathbf{x}(t)]$ is decreasing, and the system is stable. Choosing a quadratic norm of $\Delta\mathbf{x}(t)$ as the Lyapunov function,

$$\mathcal{V}[\Delta\mathbf{x}(t)] \triangleq \Delta\mathbf{x}^T(t)\mathbf{P}\,\Delta\mathbf{x}(t) \tag{4.3-7}$$

where $\mathbf{P}$ is a symmetric, positive-definite *defining matrix*, and the gradient is

$$\frac{d\mathcal{V}[\Delta\mathbf{x}(t)]}{d\Delta\mathbf{x}} = 2\Delta\mathbf{x}^T(t)\mathbf{P} \tag{4.3-8}$$

The time derivative of this Lyapunov function is

$$\frac{d\mathcal{V}[\Delta\mathbf{x}(t)]}{dt} = 2\Delta\mathbf{x}^T(t)\mathbf{P}\Delta\dot{\mathbf{x}}(t) = 2\Delta\mathbf{x}^T(t)\mathbf{P}\mathbf{F}\Delta\mathbf{x}(t) \tag{4.3-9a}$$

Because the result is a scalar, 2 $\mathbf{PF}$ can be expressed in an equivalent symmetric form:

$$\frac{d\mathcal{V}[\Delta\mathbf{x}(t)]}{dt} = \Delta\mathbf{x}^T(t)(\mathbf{PF} + \mathbf{F}^T\mathbf{P})\Delta\mathbf{x}(t) \tag{4.3-9b}$$

For asymptotic stability, $d\mathcal{V}[\Delta\mathbf{x}(t)]/dt$ must be less than zero, so eq. 4.3-9 must take the form

$$\frac{d\mathcal{V}[\Delta\mathbf{x}(t)]}{dt} = -\Delta\mathbf{x}^T(t)\mathbf{Q}\Delta\mathbf{x}(t) \tag{4.3-10}$$

where $\mathbf{Q}$ is a positive-definite matrix. Therefore, $\mathbf{F}$ represents a stable system if positive-definite $\mathbf{P}$ and $\mathbf{Q}$ can be found such that the *Lyapunov equation*

$$\mathbf{P}\,\mathbf{F} + \mathbf{F}^T\mathbf{P} = -\mathbf{Q} \tag{4.3-11}$$

has a solution. Given $\mathbf{F}$ and positive-definite $\mathbf{P}$, the solution for $\mathbf{Q}$ is straightforward; all the eigenvalues of $\mathbf{Q}$ must be positive to demonstrate stability of $\mathbf{F}$. Given $\mathbf{F}$ and positive-definite $\mathbf{Q}$, the MATLAB Control System Toolbox computes $\mathbf{P}$ by `P=lyap(F,Q)`. Then $\mathbf{P}$ must be tested for positive definiteness.

While Lyapunov's second theorem is a powerful way to define system stability, extending to nonlinear as well as linear systems, it provides little insight regarding the degree of stability or the modal properties of the system. For linear, time-invariant systems, that information is provided by the *eigenvalues* of **F** (Section 3.2). Forming the determinant equation (eqs. 3.2-71 and 3.2-72),

$$|\lambda \mathbf{I}_n - \mathbf{F}| \triangleq \Delta(\lambda) = \lambda^n + c_{n-1}\lambda^{n-1} + \cdots + c_1\lambda + c_0 = 0 \tag{4.3-12}$$

$\Delta(\lambda)$ is the *characteristic polynomial* of **F**, and $\Delta(\lambda) = 0$ is its *characteristic equation.* The characteristic equation can be factored (eq. 3.2.-73) to obtain n roots (or zeros) of the polynomial

$$\Delta(\lambda) = (\lambda - \lambda_1)(\lambda - \lambda_2)\cdots(\lambda - \lambda_n) = 0 \tag{4.3-13}$$

The roots of this equation, λ_1–λ_n, are the eigenvalues of **F**, and whenever $\lambda = \lambda_i$, $\Delta(\lambda) = 0$. Because **F** need not be a symmetric matrix, the eigenvalues may be either real or complex numbers. A *complex number*, $c = a + jb$, has two elements; a and b are real numbers and $j \triangleq \sqrt{-1}$. Real numbers multiplied by j are called *imaginary numbers.* A complex number can also be written in polar-coordinate form as $Ae^{j\phi}$, where the amplitude A is $\sqrt{a^2 + b^2}$, and the phase angle ϕ is $\tan^{-1}(b/a)$. **F** is a real-valued matrix, and any complex root ($\lambda_i = \sigma_i + j\omega_i$) must be accompanied by its *complex conjugate* ($\lambda_{i+1} = \sigma_i - j\omega_i$) in the factored polynomial.

As we shall see later in this section, the rates of growth or decay of the aircraft's individual modes of motion are governed by their eigenvalues. If all the eigenvalues of **F** have negative real parts, the system is asymptotically stable. Eigenvalues with zero real parts (i.e., roots that are imaginary numbers) represent modes with neutral stability. If any eigenvalue has a positive real part, the system is unstable.

Example 4.3-1. Stability of Small-Amplitude Motions

The business jet flight condition considered before ($\mathcal{M} = 0.3$, $h = 3{,}050$ m) is characterized by twelve eigenvalues, which are calculated using SURVEY (Appendix C). The six longitudinal eigenvalues are

$$\lambda_1 = 0$$

$$\lambda_2 = -0.0007$$

$$\lambda_{3,4} = -0.007 \pm 0.1287j$$

$$\lambda_{5,6} = -1.277 \pm 2.811j$$

The zero root signifies a neutrally stable first-order mode, and the remaining roots have negative real parts, indicating asymptotic stability. The second root also is real, and the remaining roots are arrayed in complex pairs. Each complex pair represents a second-order mode of motion. The eigenvalues are associated with the range, height, phugoid, and short-period modes.

The six lateral-directional eigenvalues are

$$\lambda_1 = 0.0035$$

$$\lambda_2 = 0$$

$$\lambda_3 = 0$$

$$\lambda_{4,5} = -0.113 \pm 1.39j$$

$$\lambda_6 = -1.204$$

Here, there is just one complex mode, and the remaining modes have real roots. λ_1 is small but positive, indicating instability. The next two roots are neutral, and the remaining roots are stable. The eigenvalues are associated with the spiral, crossrange, yaw, Dutch roll, and roll modes. We shall see the significance of the twelve eigenvalues in the next example.

Fourier and Laplace Transforms

Many of an aircraft's motions are oscillatory, so it is natural to associate these motions with waves that persist, grow, or decay. If a scalar variable $x(t)$ is periodic in a finite time interval $[-T/2, T/2]$, it can be characterized by a *Fourier series*, an infinite series of sine and cosine waves:

$$x(t) = \frac{a(0)}{2} + \sum_{n=1}^{\infty} [a(n\omega_o)\cos n\omega_o t + b(n\omega_o)\sin n\omega_o t] \tag{4.3-14}$$

ω_o is the *fundamental frequency* of the interval ($= 2\pi/T$), and $a(n\omega_o)$ and $b(n\omega_o)$ are coefficients chosen to fit $x(t)$:

$$a(n\omega_o) = \frac{2}{T}\int_{-T/2}^{T/2} x(t)\cos n\omega_o t\, dt \tag{4.3-15}$$

$$b(n\omega_o) = \frac{2}{T}\int_{-T/2}^{T/2} x(t)\sin n\omega_o t\, dt \tag{4.3-16}$$

As an alternative, the $n\omega_o$ components can be specified by the *amplitude* $c(n\omega_o)$ and a *phase angle* $\xi(n\omega_o)$ of a cosine wave,

$$x(t) = \frac{a(0)}{2} + \sum_{n=1}^{\infty} \{c(n\omega_o)\cos[n\omega_o t + \xi(n\omega_o)]\} \tag{4.3-17}$$

where

$$c(n\omega_o) = \sqrt{a^2(n\omega_o) + b^2(n\omega_o)} \tag{4.3-18}$$

$$\xi(n\omega_o) = \tan^{-1}\left[\frac{-b(n\omega_o)}{a(n\omega_o)}\right] \tag{4.3-19}$$

The Fourier series can also be written in *complex-variable form* as

$$x(t) = \frac{1}{T}\sum_{n=-\infty}^{\infty}\{g(n\omega_o)\, e^{jn\omega_o t}\} \tag{4.3-20}$$

where $g(n\omega_o)$ is a complex number defined as

$$g(n\omega_o) = \begin{cases} T[a(n\omega_o) - jb(n\omega_o)]/2, & n \geq 0 \\ T[a(-n\omega_o) + jb(-n\omega_o)]/2, & n < 0 \end{cases} \tag{4.3-21}$$

with $b(0) = 0$. The coefficient can be calculated directly as

$$g(n\omega_o) = \int_{-T/2}^{T/2} x(t)\, e^{-jn\omega_o t}\, dt \tag{4.3-22}$$

If $x(t)$ is aperiodic, the Fourier series is inappropriate; however, a *Fourier integral* that performs a similar function can be defined. The integral is derived by stretching the time interval to infinity and taking the limit as the fundamental frequency ω_o diminishes to zero. The shrinking value of ω_o is denoted by $\Delta\omega$, and eq. 4.3-20 can be written as

$$x(t) = \frac{1}{2\pi}\sum_{n=-\infty}^{\infty}\{g(n\Delta\omega)\, e^{jn\Delta\omega t}\,\Delta\omega\} \tag{4.3-23}$$

In the limit as T goes to infinity, this summation becomes the Fourier integral,

$$x(t) = \frac{1}{2\pi}\int_{-\infty}^{\infty} x(\omega)\, e^{j\omega t}\, d\omega \tag{4.3-24}$$

where $x(\omega)$ is a complex variable called the *Fourier transform* of $x(t)$:

$$x(\omega) = \int_{-\infty}^{\infty} x(t)\, e^{-j\omega t}\, dt \tag{4.3-25}$$

The "frequency content" of $x(t)$ is portrayed by $x(\omega)$, which is a continuous function of ω in $(-\infty, \infty)$. The Fourier integral is seen to be the *inverse Fourier transform*.

The Fourier transform could represent $x(t)$ in the frequency domain; however, the Laplace transform presents a more flexible alternative that includes the Fourier transform as a special case. The imaginary variable $j\omega$ is replaced by the complex *Laplace operator* $s \triangleq \sigma + j\omega$, where σ is a convergence factor that is helpful in evaluating certain integrals. Assuming that $x(t)$ begins at $t = 0$, the one-sided *Laplace transform* (extending from 0 to $+\infty$ rather than $-\infty$ to $+\infty$) and its inverse are defined as follows:

$$\mathscr{L}[x(t)] \triangleq x(s) = \int_0^{\infty} x(t)\, e^{-st} dt \tag{4.3-26}$$

$$\mathscr{L}^{-1}[x(s)] = x(t) = \frac{1}{2\pi j} \int_{\sigma - j\infty}^{\sigma + j\infty} x(s)\, e^{st} ds \tag{4.3-27}$$

A few more facts about the Laplace transform are needed. Laplace transformation is a *linear operation*; therefore, the transform of a sum of time-based functions is the sum of their transforms, and multiplication by a constant can be taken outside the transform:

$$\mathscr{L}[x_1(t) + x_2(t)] = \mathscr{L}[x_1(t)] + \mathscr{L}[x_2(t)] = x_1(s) + x_2(s) \tag{4.3-28}$$

$$\mathscr{L}[a\, x(t)] = a\, \mathscr{L}[x(t)] = a\, x(s) \tag{4.3-29}$$

The Laplace transform can be applied to each term of a time-based vector or matrix; therefore,

$$\mathscr{L}[\mathbf{x}(t)] = \mathbf{x}(s) = \begin{bmatrix} x_1(s) \\ x_2(s) \\ \cdots \\ x_n(s) \end{bmatrix} \tag{4.3-30}$$

$$\mathscr{L}[\mathbf{A}(t)] = \mathbf{A}(s) = \begin{bmatrix} a_{11}(s) & a_{12}(s) & \cdots & \cdots & a_{1r}(s) \\ a_{21}(s) & a_{22}(s) & \cdots & \cdots & a_{2r}(s) \\ \cdots & \cdots & \cdots & \cdots & \cdots \\ a_{n1}(s) & a_{n2}(s) & \cdots & \cdots & a_{nr}(s) \end{bmatrix} \tag{4.3-31}$$

The Laplace transform of the time derivative of $\mathbf{x}(t)$ is

$$\mathscr{L}[\dot{\mathbf{x}}(t)] \triangleq \dot{\mathbf{x}}(s) = s\mathbf{x}(s) - \mathbf{x}(0) \tag{4.3-32}$$

Frequency-domain analysis begins by finding the Laplace transforms of the LTI equations governing system perturbations and output:

$$\Delta\dot{\mathbf{x}}(t) = \mathbf{F}\,\Delta\mathbf{x}(t) + \mathbf{G}\,\Delta\mathbf{u}(t) + \mathbf{L}\,\Delta\mathbf{w}(t) \tag{4.3-33}$$

$$\Delta\mathbf{y}(t) = \mathbf{H}\,\Delta\mathbf{x}(t) + \mathbf{J}\,\Delta\mathbf{u}(t) + \mathbf{K}\,\Delta\mathbf{w}(t) \tag{4.3-34}$$

The matrices are constant, so the Laplace transforms of these equations are

$$s\,\Delta\mathbf{x}(s) - \Delta\mathbf{x}(0) = \mathbf{F}\,\Delta\mathbf{x}(s) + \mathbf{G}\,\Delta\mathbf{u}(s) + \mathbf{L}\,\Delta\mathbf{w}(s) \tag{4.3-35}$$

$$\Delta\mathbf{y}(s) = \mathbf{H}\,\Delta\mathbf{x}(s) + \mathbf{J}\,\Delta\mathbf{u}(s) + \mathbf{K}\,\Delta\mathbf{w}(s) \tag{4.3-36}$$

Equation 4.3-35 can be rearranged as

$$(s\mathbf{I}_n - \mathbf{F})\,\Delta\mathbf{x}(s) = \Delta\mathbf{x}(0) + \mathbf{G}\,\Delta\mathbf{u}(s) + \mathbf{L}\,\Delta\mathbf{w}(s) \tag{4.3-37}$$

leading to expressions for the state and output transforms:

$$\Delta\mathbf{x}(s) = (s\mathbf{I}_n - \mathbf{F})^{-1}[\Delta\mathbf{x}(0) + \mathbf{G}\,\Delta\mathbf{u}(s) + \mathbf{L}\,\Delta\mathbf{w}(s)] \tag{4.3-38}$$

and

$$\begin{aligned}\Delta\mathbf{y}(s) = {} & \mathbf{H}\,(s\mathbf{I}_n - \mathbf{F})^{-1}[\Delta\mathbf{x}(0) + \mathbf{G}\,\Delta\mathbf{u}(s) + \mathbf{L}\,\Delta\mathbf{w}(s)] \\ & + \mathbf{J}\,\Delta\mathbf{u}(s) + \mathbf{K}\,\Delta\mathbf{w}(s)\end{aligned} \tag{4.3-39}$$

The *characteristic matrix* of $\mathbf{F}$, $(s\mathbf{I}_n - \mathbf{F})$, plays a key role in portraying initial-condition and input response in the frequency domain. Its inverse (eq. 3.2-77)

$$\mathbf{A}(s) \triangleq (s\mathbf{I}_n - \mathbf{F})^{-1} = \frac{\mathrm{Adj}(s\mathbf{I}_n - \mathbf{F})}{|s\mathbf{I}_n - \mathbf{F}|} \tag{4.3-40}$$

is an $(n \times n)$ matrix of fractions (real or complex valued) whose numerators are determined by the adjoint matrix. All elements have the same denominator (prior to canceling any identical factored terms in the numerator and denominator): the *characteristic polynomial* of $\mathbf{F}$, $s^n + c_{n-1}s^{n-1} + \cdots + c_1 s + c_0$, symbolized as $\Delta(s)$. The *trace* of $\mathbf{F}$, $\mathrm{Tr}(\mathbf{F})$, is the sum of the diagonal elements of $\mathbf{F}$, and it is the sum of the eigenvalues of $\mathbf{F}$. Sometimes called the *total damping* of the system, $\mathrm{Tr}(\mathbf{F})$ also equals $-c_{n-1}$.

We now begin to see the significance of the eigenvalues of $\mathbf{F}$: they appear in $|s\mathbf{I}_n - \mathbf{F}|$, the denominator of $\mathbf{A}(s)$. From eq. 4.3-38, the Laplace transform of the initial-condition response is

$$\Delta\mathbf{x}(s) = (s\mathbf{I}_n - \mathbf{F})^{-1}\,\Delta\mathbf{x}(0) = \mathbf{A}(s)\,\Delta\mathbf{x}(0) \tag{4.3-41}$$

Because the initial-condition response depends only on the state transition matrix (eq. 4.2-4),

$$\Delta\mathbf{x}(t) = \mathbf{\Phi}(t, 0)\,\Delta\mathbf{x}(0) \tag{4.3-42}$$

it can be concluded that $\mathbf{A}(s)$ is the Laplace transform of $\mathbf{\Phi}(t, 0)$. Conversely, the elements of $\mathbf{\Phi}(t, 0)$ are the inverse Laplace transforms of the scalar elements of $\mathbf{A}(s)$:

$$\mathbf{\Phi}(t, 0) = e^{\mathbf{F}t} = \mathcal{L}^{-1}[(s\mathbf{I}_n - \mathbf{F})^{-1}] = \mathcal{L}^{-1}[\mathbf{A}(s)] \tag{4.3-43}$$

An analytical alternative to the generation of $\mathbf{\Phi}(t, 0)$ by power series (Section 4.2) is to express the elements of $\mathbf{A}(s)$, then find their inverse transforms in a suitable table [G-1, F-1].

A typical element of $\mathbf{A}(s)$, $a_{ij}(s)$, is a ratio of polynomials, each of which can be factored to yield

$$a_{ij}(s) = \frac{k_{ij}(s^{n-1} + b_{n-2}s^{n-2} + \cdots + b_1 s + b_0)}{s^n + c_{n-1}s^{n-1} + \cdots + c_1 s + c_0}$$

$$= \frac{k_{ij}(s - \beta_1)(s - \beta_2)\cdots(s - \beta_{n-1})}{(s - \lambda_1)(s - \lambda_2)\cdots(s - \lambda_n)} \tag{4.3-44}$$

The coefficients b_k are specific to the a_{ij}th element, as are the roots (or *zeros*) of the *numerator polynomial* β_k. There are, at most, $(n - 1)$ zeros in the numerator of a_{ij}. The zeros may be real or complex; if the latter, the zeros appear in complex-conjugate pairs.

Initial conditions have the following effect on the ith element of $\Delta\mathbf{x}(s)$:

$$\Delta x_i(s) = a_{i1}\Delta x_1(0) + a_{i2}\Delta x_2(0) + \cdots + a_{in}\Delta x_n(0)$$

$$\triangleq \frac{p_i(s)}{\Delta(s)} \tag{4.3-45}$$

$p_i(s)$ is a linear combination of the polynomial numerators in the ith row of $\mathbf{A}(s)$, each weighted by the corresponding element of the initial condition, and $\Delta(s)$ is their common denominator. If the roots are *distinct* (i.e., no two are the same), $p_i(s)/\Delta(s)$ can be expanded in *partial fractions* as

$$\Delta x_i(s) = \frac{p_i(s)}{\Delta(s)} = \frac{d_1}{(s - \lambda_1)} + \frac{d_2}{(s - \lambda_2)} + \cdots + \frac{d_n}{(s - \lambda_n)} \tag{4.3-46}$$

where

$$d_i = (s - \lambda_i)\frac{p_i(s)}{\Delta(s)}\bigg|_{s = \lambda_i} \tag{4.3-47}$$

The individual terms of eq. 4.3-46 describe *modes of response* to the specified initial condition $\Delta\mathbf{x}(0)$, and the process is repeated for every element of $\Delta\mathbf{x}(s)$.

The corresponding time response can be determined using the inverse Laplace transform. The inverse Laplace transform of each partial fraction is

$$\mathcal{L}^{-1}\left[\frac{d_i}{(s-\lambda_i)}\right] = d_i e^{\lambda_i t} \tag{4.3-48}$$

Consequently, each eigenvalue of $\mathbf{F}$ specifies an exponential mode of growth or decay in $\Delta\mathbf{x}(t)$. If λ_i is real, then d_i is real; the time response of $\Delta x_i(t)$ in the mode defined by λ_i is an exponential signal that decays if $\lambda_i < 0$ and grows if $\lambda_i > 0$. The mode is *exponentially asymptotically stable* in the first case and *unstable* in the second (Fig. 4.3-3). An alternative expression for the exponential mode is $d_i e^{-t/\tau_i}$, where τ_i is the *time constant* of the mode (positive, if stable).

$\Delta\mathbf{x}(t)$ is a real variable, and the net contribution of all partial fractions must be real. If λ_i is complex, it must be accompanied by its complex conjugate λ_{i+1}; then d_i and d_{i+1} are also complex conjugates of each other. Denoting the complex conjugate by $(\cdot)^\dagger$,

$$\lambda_i \triangleq \mu + j\nu \tag{4.3-49}$$

$$\lambda_{i+1} \triangleq \lambda_i^\dagger = \mu - j\nu \tag{4.3-50}$$

$$d_i \triangleq a + jb \tag{4.3-51}$$

$$d_{i+1} \triangleq d_i^\dagger = a - jb \tag{4.3-52}$$

the complex pair has the following second-order equivalent:

$$\begin{aligned}\frac{d_i}{(s-\lambda_i)} + \frac{d_{i+1}}{(s-\lambda_{i+1})} &= \frac{(s-\lambda_i^\dagger)\,d_i + (s-\lambda_i)\,d_i^\dagger}{[s^2 - (\lambda_i + \lambda_i^\dagger)\,s + \lambda_i\lambda_i^\dagger]} \\ &= \frac{2a\,[s - (\mu + \nu b/a)]}{s^2 - 2\,\mu s + (\mu^2 + \nu^2)}\end{aligned} \tag{4.3-53}$$

The partial-fraction pair has the inverse Laplace transform,

$$\mathcal{L}^{-1}[\cdot] = 2(a^2 + b^2)^{1/2}\, e^{\mu t}\cos(\nu t + \psi) \tag{4.3-54}$$

with the phase angle

$$\psi = \tan^{-1}\left(\frac{b}{a}\right) \tag{4.3-55}$$

Equation 4.3-54 describes an oscillatory mode of response that is exponentially asymptotically stable if $\mu < 0$ and unstable if $\mu > 0$ (Fig. 4.3-4).

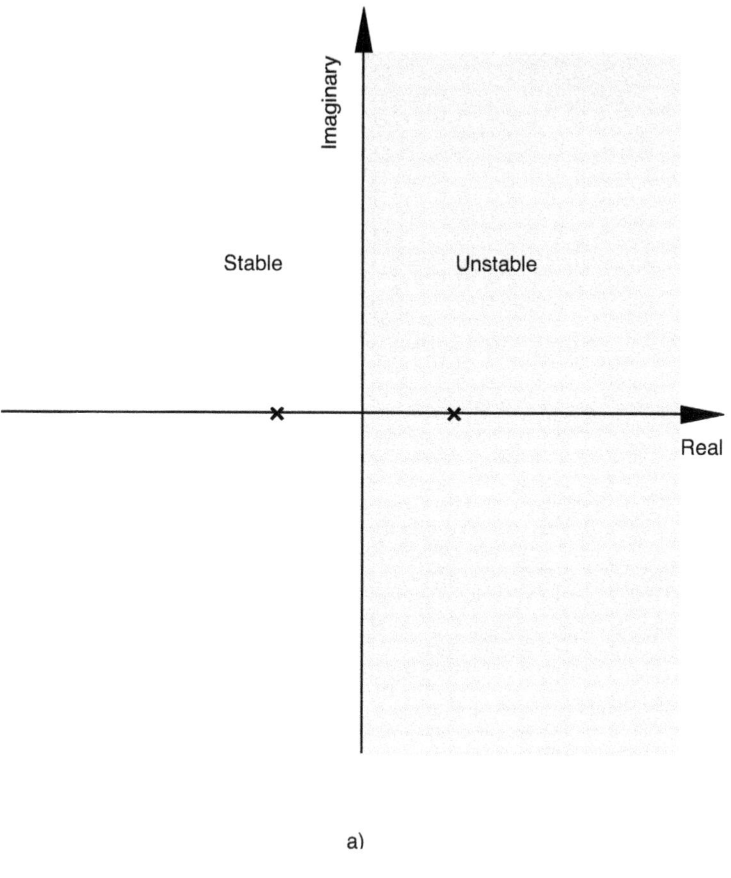

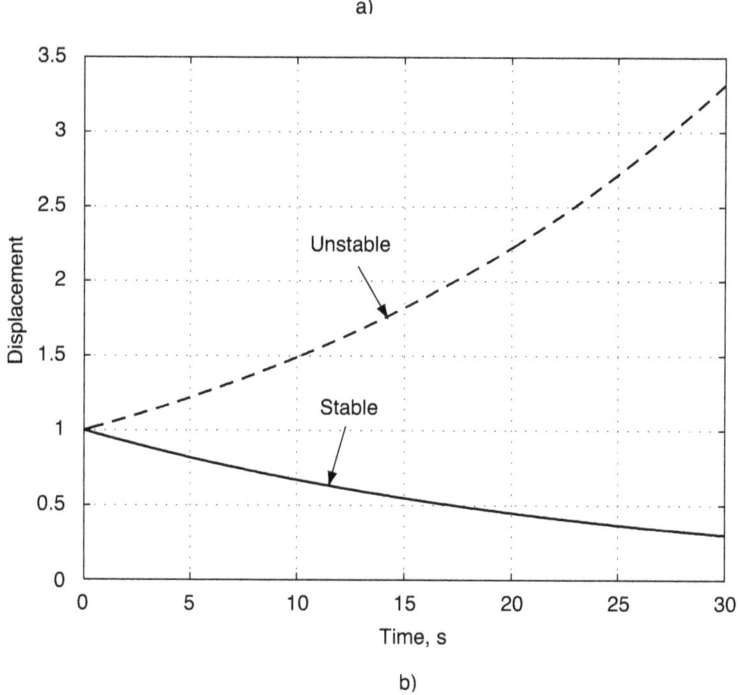

Fig. 4.3-3. Descriptions of first-order dynamics. (a) Stable and unstable locations of individual roots in the s plane. (b) Stable and unstable time responses to an initial condition.

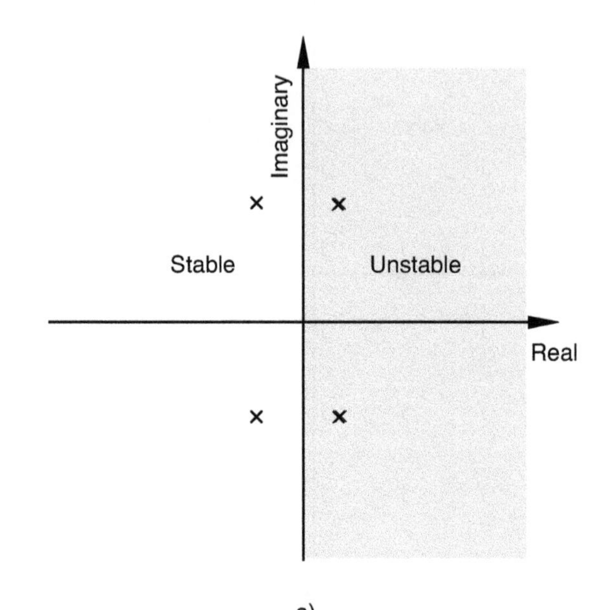

a)

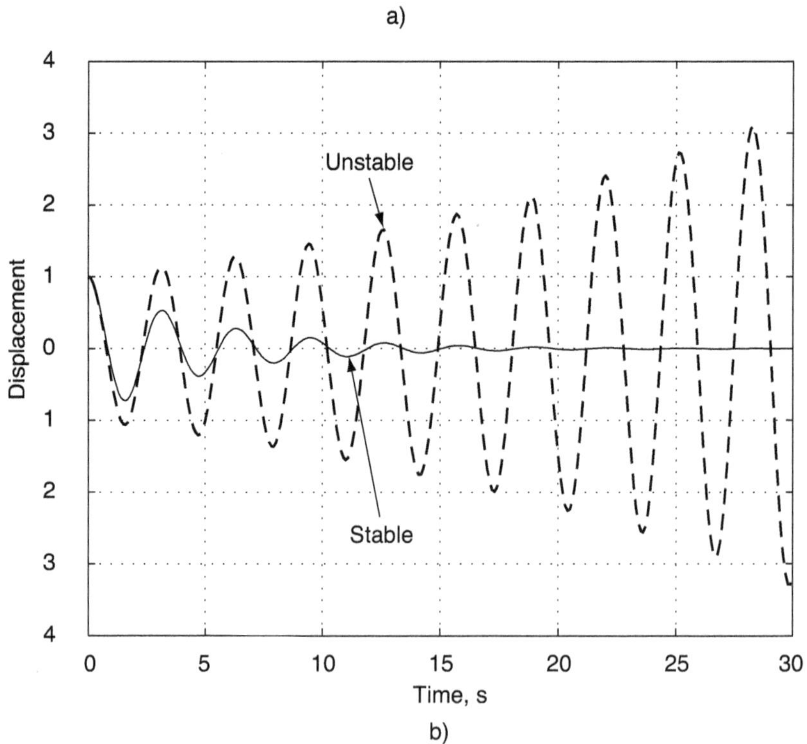

b)

Fig. 4.3-4. Descriptions of second-order dynamics. (a) Stable and unstable locations of complex root pairs in the s plane. (b) Stable and unstable time responses to an initial condition.

An alternative expression for the oscillatory mode is $ke^{-\zeta\omega_n t} \cos[(\omega_n\sqrt{1-\zeta^2})t+\psi]$, where ω_n is the *natural frequency* of the mode, ζ is its *damping ratio*, and $\omega_n\sqrt{1-\zeta^2}$ is its *damped natural frequency*. The characteristic polynomial for this second-order mode can be expressed as

$$s^2 - 2\mu s + (\mu^2 + v^2) = s^2 + 2\zeta\omega_n s + \omega_n^2 \tag{4.3-56}$$

and the natural frequency and damping ratio are

$$\omega_n = \sqrt{\mu^2 + v^2} \text{ rad/s} \tag{4.3-57}$$

$$\zeta = -\frac{\mu}{\sqrt{\mu^2 + v^2}} \tag{4.3-58}$$

The *period* associated with ω_n is $T = 2\pi/\omega_n$ s. A positive damping ratio indicates stability ($\mu < 0$). If $|\zeta| > 1$, the complex pair of roots (eqs. 4.3-49 and 4.3-50) actually consists of two real roots.

Combining these results, it can be seen that the time response of $\Delta\mathbf{x}(t)$ to initial conditions is the sum of individual responses that have exponential or sinusoidal form. For an *n*th-order system, there are at least *n*/2 modes of motion (if all the modes are oscillatory) and at most *n* modes of motion (if all the modes are real). The modes can be individually stable or unstable, but if *any* mode is unstable, then the system must be considered unstable. While it would appear that every element of $\Delta\mathbf{x}(t)$ responds in every mode, pole-zero cancellations in selected a_{ij} may restrict the response modes of individual elements.

Modes of Aircraft Motion

For symmetric flight, small-amplitude longitudinal and lateral-directional motions are uncoupled, and each set normally is characterized by well-established modes of motion. The *longitudinal modes* appear in the characteristic equation,

$$\begin{aligned} |\mathbf{A}_{\text{Lo}}(s)| &= |s\,\mathbf{I}_n - \mathbf{F}_{\text{Lo}}| \triangleq \Delta_{\text{Lo}}(s) \\ &= s^6 + c_5 s^5 + c_4 s^4 + c_3 s^3 + c_2 s^2 + c_1 s + c_0 = 0 \\ &= s^2(s - \lambda_{Ph})(s - \lambda_{Ph}^{\dagger})(s - \lambda_{SP})(s - \lambda_{SP}^{\dagger}) = 0 \\ &= s^2[s^2 + 2\,\zeta\,\omega_n s + \omega_n^2]_{\text{Ph}}[s^2 + 2\,\zeta\,\omega_n s + \omega_n^2]_{\text{SP}} = 0 \end{aligned} \tag{4.3-59}$$

with slower modes conventionally written to the left of faster modes. The s^2 term represents the separate pure integrations that produce Δx_I and

Δz_I response (neglecting atmospheric gradient effects). The first pair of complex roots λ_{Ph} and $\lambda_{Ph}^{\dagger}$ combines to form an oscillatory mode called the *phugoid mode.* It is a slow, lightly damped mode that reflects the interchange of kinetic and potential energy through velocity-altitude oscillations, and it is the linearized equivalent of the paper airplane's scalloped flight path (Example 1.3-1).

The second pair of complex roots λ_{SP} and $\lambda_{SP}^{\dagger}$ produces the *short-period mode*, which describes rotations about the aircraft's center of mass that involve angle of attack, pitch rate, and pitch angle. Modern military aircraft are often balanced so that the short-period mode degenerates into two real modes, one of which is unstable. Compensation of the instability must be provided by closed-loop control.

The *lateral-directional modes* of motion appear in the characteristic equation

$$
\begin{aligned}
|\mathbf{A}_{\mathrm{LD}}(s)| = |s\mathbf{I}_n - \mathbf{F}_{\mathrm{LD}}| &\triangleq \Delta_{LD}(s) \\
&= s^6 + c_5 s^5 + c_4 s^4 + c_3 s^3 + c_2 s^2 + c_1 s + c_0 = 0 \\
&= s^2(s - \lambda_S)(s - \lambda_{DR})(s - \lambda_{DR}{}^{\dagger})(s - \lambda_R) = 0 \\
&= s^2(s - \lambda_S)[s^2 + 2\zeta\omega_n s + \omega_n^2]_{\mathrm{DR}}(s - \lambda_R) = 0
\end{aligned} \tag{4.3-60}
$$

There are two pure integration modes corresponding to lateral-position and yaw-angle perturbations, Δy_I and $\Delta\psi$, which do not couple into the other lateral-directional components. The *spiral mode*, represented by λ_S, is a very slow, often slightly unstable, mode that governs the aircraft's ability to maintain wings level without control inputs. The complex roots, λ_{DR} and $\lambda_{DR}^{\dagger}$, form the *Dutch roll mode*, which arises from the "weather-cocking" tendency produced by sideslip on the vertical tail (Section 2.4). It usually is a lightly damped mode of moderate time scale. Because yawing motion induces differential lift on opposing wings, there often is significant rolling as well as yawing motion. The *roll mode*, represented by λ_R, portrays damping effects that normally oppose roll rate. For most aircraft at most flight conditions, it is a relatively fast mode. In some cases, the roll and spiral modes coalesce in an oscillatory *roll-spiral oscillation.*

How can we generalize about the degree to which each state element responds in each mode? By examining the *eigenvectors* of $\mathbf{F}$. From Section 3.2, we recall that n representations of each mode's eigenvector are defined by the adjunct of the characteristic matrix,

$$
\mathrm{Adj}(\lambda_i \mathbf{I}_n - \mathbf{F}) = [a_1\mathbf{e}_i \; a_2\mathbf{e}_i \cdots a_n\mathbf{e}_i], \quad i = 1 \text{ to } n \tag{4.3-61}
$$

where the a_i are arbitrary real or complex multipliers. The *modal matrix* contains all of the system's eigenvectors,

$$\mathrm{E} = [\mathrm{e}_1\ \mathrm{e}_2 \cdots \mathrm{e}_n] \tag{4.3-62}$$

which can be normalized such that $|\mathbf{E}| = 1$. From eq. 3.2-82, the system matrix can be expressed as the product of a diagonal matrix of eigenvalues, $\mathbf{\Lambda}$ (eq. 3.2-74), and the modal matrix, $\mathbf{E}$:

$$\mathbf{F} = \mathbf{E}\,\mathbf{\Lambda}\,\mathbf{E}^{-1} \tag{4.3-63}$$

Because the elements of $\mathbf{F}$ are real-valued, real modes are accompanied by real-valued eigenvectors and complex modes by complex-valued eigenvectors. Complex eigenvectors occur in complex-conjugate pairs, so the modal response represented by their sum always is a real number.

Let's examine the nature of the eigenvectors for two fourth-order longitudinal models, neglecting the integrations for altitude and range. We derived body-axis and stability-axis linear models in Section 4.1. The corresponding state vectors are

$$\Delta\mathbf{x}_{\mathrm{B}}^{T} = [\Delta u\ \Delta w\ \Delta q\ \Delta\theta]^{T} \tag{4.3-64}$$

$$\Delta\mathbf{x}_{\mathrm{S}}^{T} = [\Delta V\ \Delta\alpha\ \Delta q\ \Delta\theta]^{T} \tag{4.3-65}$$

which are related by the transformation

$$\mathbf{H}_{\mathrm{B}}^{\mathrm{S}} = \begin{bmatrix} 1 & 0 & 0 & 0 \\ 0 & 1/V_o & 0 & 0 \\ 0 & 0 & 1 & 0 \\ 0 & 0 & 0 & 1 \end{bmatrix} \begin{bmatrix} \cos\alpha_o & \sin\alpha_o & 0 & 0 \\ -\sin\alpha_o & \cos\alpha_o & 0 & 0 \\ 0 & 0 & 1 & 0 \\ 0 & 0 & 0 & 1 \end{bmatrix}$$

$$= \begin{bmatrix} \cos\alpha_o & \sin\alpha_o & 0 & 0 \\ -(\sin\alpha_o)/V_o & (\cos\alpha_o)/V_o & 0 & 0 \\ 0 & 0 & 1 & 0 \\ 0 & 0 & 0 & 1 \end{bmatrix} \tag{4.3-66}$$

The stability matrices are related by the similarity transformation

$$\mathbf{F}_{\mathrm{S}} = \mathbf{H}_{\mathrm{B}}^{\mathrm{S}}\ \mathbf{F}_{\mathrm{B}}\ \mathbf{H}_{\mathrm{S}}^{\mathrm{B}} \tag{4.3-67}$$

and the eigenvalues for these two representations of the same system are identical:

$$|s\mathbf{I}_n - \mathbf{F}_S| = |s\mathbf{I}_n - \mathbf{F}_B| = \Delta(s)$$
$$= (s - \lambda_{Ph})(s - \lambda_{Ph}^{\dagger})(s - \lambda_{SP})(s - \lambda_{SP}^{\dagger}) \tag{4.3-68}$$

The stability-axis modal matrix $\mathbf{E}_S$ could then be written as $\mathbf{H}_B^S\mathbf{E}_B$, where $\mathbf{E}_B$ is the body-axis modal matrix.

The modal matrices for body and stability axes are similar to each other in form, although the numerical values depend on the state-vector definition. For the body-axis set,

$$\mathbf{E} = [\mathbf{e}_{Ph} \quad \mathbf{e}_{Ph}^{\dagger} \quad \mathbf{e}_{SP} \quad \mathbf{e}_{SP}^{\dagger}]$$
$$= \begin{bmatrix} (a+jb)_u & (a-jb)_u & (c+jd)_u & (c-jd)_u \\ (a+jb)_w & (a-jb)_w & (c+jd)_w & (c-jd)_w \\ (a+jb)_q & (a-jb)_q & (c+jd)_q & (c-jd)_q \\ (a+jb)_\theta & (a-jb)_\theta & (c+jd)_\theta & (c-jd)_\theta \end{bmatrix} \tag{4.3-69}$$

where a and b are the phugoid eigenvector components and c and d are the short-period eigenvector components. Complex numbers can be expressed in polar form; hence, the modal matrix can also be written as

$$\mathbf{E} = \begin{bmatrix} A_u\, e^{j\phi_u} & A_u\, e^{-j\phi_u} & B_u\, e^{j\psi_u} & B_u\, e^{-j\psi_u} \\ A_w\, e^{j\phi_w} & A_w\, e^{-j\phi_w} & B_w\, e^{j\psi_w} & B_w\, e^{-j\psi_w} \\ A_q\, e^{j\phi_q} & A_q\, e^{-j\phi_q} & B_q\, e^{j\psi_q} & B_q\, e^{-j\psi_q} \\ A_\theta\, e^{j\phi_\theta} & A_\theta\, e^{-j\phi_\theta} & B_\theta\, e^{j\psi_\theta} & B_\theta\, e^{-j\psi_\theta} \end{bmatrix} \tag{4.3-70}$$

with *amplitudes* $A = (a^2 + b^2)^{1/2}$ and $B = (c^2 + d^2)^{1/2}$ and *phase angles* $\phi = \tan^{-1}(b/a)$ and $\psi = \tan^{-1}(d/c)$.

The degree of response is determined by the relative magnitudes of the components of each eigenvector. For the phugoid mode, the series [A_u (m/s): A_w (m/s): A_q (rad/s): A_θ (rad)] provides this information, while for the short period, the comparison is made in [B_u (m/s): B_w (m/s): B_q (rad/s): B_θ (rad)]. Note that the components are dimensioned; this must be taken into account when assessing "large" and "small" elements in each mode. Because eigenvectors can be multiplied by an arbitrary constant, their amplitudes are not absolute, and no direct comparison of magnitudes of different modes (e.g., phugoid and short period) can be made.

For a given initial condition $\Delta\mathbf{x}(0)$, the evolution of $\Delta\mathbf{x}(t)$ is governed by the eigenvectors as well as the eigenvalues. The unforced dynamic equation

$$\Delta\dot{\mathbf{x}}(t) = \mathbf{F}\,\Delta\mathbf{x}(t) \tag{4.3-71}$$

is accompanied by the *normal* (diagonalized) *form*

$$\Delta\dot{\mathbf{q}}(t) = \mathbf{\Lambda}\,\Delta\mathbf{q}(t) \tag{4.3-72}$$

where $\Delta\mathbf{q}(t)$ is the *normal-mode-coordinate vector*, and $\mathbf{\Lambda}$ is related to $\mathbf{F}$ by equation 4.3-63. The initial condition on $\Delta\mathbf{q}(0)$ is

$$\Delta\mathbf{q}(0) = \mathbf{E}^{-1}\,\Delta\mathbf{x}(0) \tag{4.3-73}$$

and the elements of $\Delta\mathbf{q}(t)$ take the form of equation 4.3-48. Then, $\Delta\mathbf{x}(t)$ can be represented as

$$\Delta\mathbf{x}(t) = \mathbf{E}\,\Delta\mathbf{q}(t) \tag{4.3-74}$$

Thus, a typical element of the state, the pitch angle $\Delta\theta(t)$, can be expressed as follows:

$$\begin{aligned}\Delta\theta(t) &= A_\theta e^{j\phi_\theta}\,\Delta q_1(t) + A_\theta e^{-j\phi_\theta}\,\Delta q_2(t) \\ &\quad + B_\theta e^{j\psi_\theta}\,\Delta q_3(t) + B_\theta e^{-j\psi_\theta}\,\Delta q_4(t) \\ &= A_\theta e^{j\phi_\theta}\,\Delta q_1(0)\,e^{\lambda_{Ph}t} + A_\theta e^{-j\phi_\theta}\,\Delta q_2(0)\,e^{\lambda^\dagger_{Ph}t} \\ &\quad + B_\theta e^{j\psi_\theta}\,\Delta q_3(0)\,e^{\lambda_{SP}t} + B_\theta e^{-j\psi_\theta}\,\Delta q_4(0)\,e^{\lambda^\dagger_{SP}t}\end{aligned} \tag{4.3-75}$$

Each state element has components rotating in the complex plane with a circular frequency determined by the damped natural frequency ν (eq. 4.3-49) of the mode and amplitude growth or decay governed by the damping constant μ. The first and third components rotate in a positive direction, and their complex conjugates rotate in a negative direction such that the sum of elements is always a real number, as illustrated for pitch rate and angle in Fig. 4.3-5.

Phase-angle comparisons can be made within an eigenvector but not between eigenvectors because the damped natural frequencies of modes are different; hence, the phase difference is continually changing. The phase angles tell whether one element of the state "leads" or "lags" another element in the initial-condition response. In the present case, pitch rate leads pitch angle because it is the derivative of pitch angle.

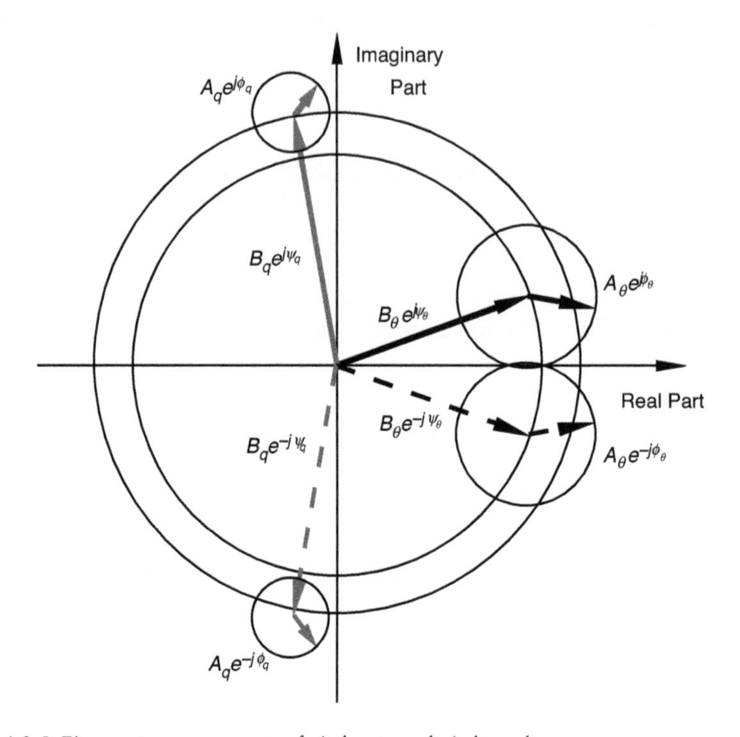

Fig. 4.3-5. Eigenvector components of pitch-rate and pitch-angle response.

Example 4.3-2. Modes and Mode Ratios of Small-Amplitude Motion

The modes of motion at the $\mathcal{M} = 0.3$, $h = 3{,}050\,\text{m}$ business jet flight condition are calculated using SURVEY (Appendix C). We begin by noting the time constants, natural frequencies, periods, and damping ratios of the modes identified in Example 4.3-1. For the sixth-order longitudinal model, the roots of the two real and two oscillatory modes are expressed as

$$\tau_{range} = \infty\,\text{s}$$

$$\tau_{height} = 142.9\,\text{s}$$

$$\omega_{n_{phugoid}} = 0.1289\,\text{rad/s} = 0.021\,\text{Hz}$$

$$T_{phugoid} = 48.74\,\text{s}$$

$$\zeta_{phugoid} = 0.0543$$

$$\omega_{n_{short\ period}} = 3.087 \text{ rad/s} = 0.49 \text{ Hz}$$

$$T_{short\ period} = 2.04 \text{ s}$$

$$\zeta_{short\ period} = 0.4137$$

The phugoid mode is a long-period oscillation that is lightly damped, and the short-period mode is a faster, moderately damped oscillation. The range mode time constant is ∞ because λ_1 is zero, denoting pure integration. The atmospheric gradient is considered in the linearization; hence, the height mode (λ_2) is slightly negative (rather than zero, as assumed in the discussion of eq. 4.3-59), leading to a large, positive time constant.

MATLAB's concurrent calculation of eigenvalues and eigenvectors produces the diagonal matrix of eigenvalues **Λ**

$$\mathbf{\Lambda} = \begin{bmatrix} 0 & 0 & 0 & 0 & 0 & 0 \\ 0 & -1.28+2.81j & 0 & 0 & 0 & 0 \\ 0 & 0 & -1.28-2.81j & 0 & 0 & 0 \\ 0 & 0 & 0 & -0.007+0.129j & 0 & 0 \\ 0 & 0 & 0 & 0 & -0.007-0.129j & 0 \\ 0 & 0 & 0 & 0 & 0 & -0.0007 \end{bmatrix}$$

and the corresponding modal matrix **E**

$$\mathbf{E} = [\mathbf{e}_1\ \mathbf{e}_2\ \mathbf{e}_3\ \mathbf{e}_4\ \mathbf{e}_5\ \mathbf{e}_6]$$

$$= \begin{bmatrix} 0 & -0.046+0.003j & -0.046-0.003j & 0.008+0.003j & 0.008-0.003j & -0.0007 \\ 0 & 0.961+0.236j & 0.961-0.236j & 0.0002-0.003j & 0.0002+0.003j & 0 \\ 0 & -0.007+0.027j & -0.007-0.027j & 0-0.0001j & 0+0.0001j & 0 \\ 0 & 0.009-0.002j & 0.009+0.002j & -0.001-0j & -0.001+0j & 0 \\ 1 & 0.003-0.007j & 0.003+007j & -0.595-0.031j & -0.595+0.031j & 0.992 \\ 0 & 0.11-0.074j & 0.11+0.074j & -0.018-0.799j & -0.018+0.799j & 0.13 \end{bmatrix}$$

The eigenvectors, each normalized to unit magnitude, are columns of the modal matrix. The row order of elements follows the state vector (Δu, Δw, Δq, $\Delta\theta$, Δx, Δz), and the column order follows the eigenvalue ordering in **Λ**. The magnitude of each element reveals its degree of initial-condition response in the associated mode. The phase angles of the responses are derived from the complex variables, as described in the text. The range perturbation (Δx) dominates the response of the two real modes, an effect of measurement units, as described below.

The relative importance of motion variables in the phugoid and short-period modes is best seen in the eigenvectors of the corresponding fourth-order model (which neglects Δx and Δz). Denoting the vectors of amplitudes as $\mathbf{a}_{\mathrm{Ph}}$ and $\mathbf{a}_{\mathrm{SP}}$,

$$\mathbf{a}_{\mathrm{Ph}} = \begin{bmatrix} |\Delta u_{Ph}| \\ |\Delta w_{Ph}| \\ |\Delta q_{Ph}| \\ |\Delta \theta_{Ph}| \end{bmatrix} = \begin{bmatrix} 0.999 \\ 0.037 \\ 0.002 \\ 0.012 \end{bmatrix}$$

$$\mathbf{a}_{\mathrm{SP}} = \begin{bmatrix} |\Delta u_{SP}| \\ |\Delta w_{SP}| \\ |\Delta q_{SP}| \\ |\Delta \theta_{SP}| \end{bmatrix} = \begin{bmatrix} 0.047 \\ 0.999 \\ 0.028 \\ 0.009 \end{bmatrix}$$

These amplitude vectors carry the dimensions of the state vector elements (m/s, m/s, rad/s, rad); dividing the first two terms by airspeed V_o normalizes the vectors and allows the components to be expressed in consistent units:

$$\mathbf{a}'_{\mathrm{Ph}} = \begin{bmatrix} |\Delta u'_{Ph}| \\ |\Delta w'_{Ph}| \\ |\Delta q'_{Ph}| \\ |\Delta \theta'_{Ph}| \end{bmatrix} = \begin{bmatrix} 0.603 \\ 0.022 \\ 0.099 \\ 0.791 \end{bmatrix}$$

$$\mathbf{a}'_{\mathrm{SP}} = \begin{bmatrix} |\Delta u'_{SP}| \\ |\Delta w'_{SP}| \\ |\Delta q'_{SP}| \\ |\Delta \theta'_{SP}| \end{bmatrix} = \begin{bmatrix} 0.015 \\ 0.317 \\ 0.902 \\ 0.292 \end{bmatrix}$$

We see that the phugoid mode is primarily an oscillation in $\Delta u/V_o$ and $\Delta\theta$, while the short-period mode is expressed largely by oscillations in $\Delta w/V_o$, Δq, and $\Delta\theta$. The modal period has an effect on relative amplitudes; the pitch rate is the derivative of the pitch angle in both cases, yet the ratio of amplitudes is one-eighth for the phugoid mode and three for the short period mode.

For the sixth-order lateral-directional model, there are four real roots and one oscillatory mode whose time constants, natural frequency, period, and damping ratio are

$$\tau_{spiral} = -\,285.7\ \mathrm{s}$$

$$\tau_{crossrange} = \infty\ \mathrm{s}$$

$$\tau_{yaw} = \infty\,\text{s}$$

$$\tau_{roll} = 0.8309\,\text{s}$$

$$\omega_{n_{Dutch\ roll}} = 1.394\ \text{rad/s} = 0.222\,\text{Hz}$$

$$T_{Dutch\ roll} = 4.507\,\text{s}$$

$$\zeta_{Dutch\ roll} = 0.0811$$

The spiral mode is unstable, but its negative time constant is so large (over 4 minutes) that it is easily controlled and can be considered essentially neutral. The crossrange and yaw modes are neutral, roll response is slow, and the Dutch roll oscillation is slow and lightly damped.

The lateral-directional modal matrix is

$$\mathbf{E} = [\mathbf{e}_1\ \mathbf{e}_2\ \mathbf{e}_3\ \mathbf{e}_4\ \mathbf{e}_5\ \mathbf{e}_6] = \begin{bmatrix} 0 & 0 & 0 & 0.17 & 0.74 - 0.67j & 0.74 + 0.67j \\ 0 & 0 & 0 & 0.17 & 0 + 0.01j & 0 - 0.01j \\ 0 & 0 & 0 & 0 & -0.01 - 0.01j & -0.01 + 0.01j \\ 0 & 0 & 0 & -0.14 & 0.01 - 0j & 0.01 + 0j \\ 0 & 0 & 0 & 0 & 0 + 0.007j & 0 - 0.007j \\ 1 & 1 & 1 & -0.96 & 0 - 0.05j & 0 + 0.05j \end{bmatrix}$$

where the row order of elements follows the state vector (Δv, Δp, Δr, $\Delta\phi$, $\Delta\psi$, Δy) with dimensions (m/s, rad/s, rad/s, rad, rad, m). The crossrange dominates the first four eigenvectors, a consequence of the disparate units of the elements.

We drop $\Delta\psi$ and Δy in order to gain insights about the three most significant modes. The relative amplitudes of the spiral, roll, and Dutch roll (Δv, Δp, Δr, $\Delta\phi$) eigenvector elements are

$$\mathbf{a}_S = \begin{bmatrix} |\Delta v_S| \\ |\Delta p_S| \\ |\Delta r_S| \\ |\Delta\phi_S| \end{bmatrix} = \begin{bmatrix} 0.473 \\ 0.003 \\ 0.084 \\ 0.877 \end{bmatrix}$$

$$\mathbf{a}_R = \begin{bmatrix} |\Delta v_R| \\ |\Delta p_R| \\ |\Delta r_R| \\ |\Delta\phi_R| \end{bmatrix} = \begin{bmatrix} 0.624 \\ 0.601 \\ 0.004 \\ 0.5 \end{bmatrix}$$

$$\mathbf{a}_{DR} = \begin{bmatrix} |\Delta v_{DR}| \\ |\Delta p_{DR}| \\ |\Delta r_{DR}| \\ |\Delta \phi_{DR}| \end{bmatrix} = \begin{bmatrix} 1 \\ 0.014 \\ 0.012 \\ 0.01 \end{bmatrix}$$

Normalizing Δv by V_o, the eigenvector amplitudes are

$$\mathbf{a}'_S = \begin{bmatrix} |\Delta v'_S| \\ |\Delta p'_S| \\ |\Delta r'_S| \\ |\Delta \phi'_S| \end{bmatrix} = \begin{bmatrix} 0.005 \\ 0.003 \\ 0.095 \\ 0.995 \end{bmatrix}$$

$$\mathbf{a}'_R = \begin{bmatrix} |\Delta v'_R| \\ |\Delta p'_R| \\ |\Delta r'_R| \\ |\Delta \phi'_R| \end{bmatrix} = \begin{bmatrix} 0.008 \\ 0.769 \\ 0.005 \\ 0.639 \end{bmatrix}$$

$$\mathbf{a}'_{DR} = \begin{bmatrix} |\Delta v'_{DR}| \\ |\Delta p'_{DR}| \\ |\Delta r'_{DR}| \\ |\Delta \phi'_{DR}| \end{bmatrix} = \begin{bmatrix} 0.416 \\ 0.603 \\ 0.525 \\ 0.433 \end{bmatrix}$$

The roll angle $\Delta\phi$ is the most significant component for the spiral mode, while the roll rate Δp and roll angle dominate the roll mode. All four of the elements are significantly involved in the Dutch roll response to initial conditions.

Phase Plane

An alternative way of presenting the time response of a second-order LTI system is to cross-plot one state element $\Delta x_1(t)$ against the other $\Delta x_2(t)$. The resulting *phase-plane plot*[3] hides the explicit dependence on time [which can be displayed by adding tick marks to the $\Delta x_1(t) - \Delta x_2(t)$ contour], but its shape reveals organized behavior that can be used to classify types of motion according to the eigenvalues. The second-order

3. The system *state* was once called the system *phase*, and the early name continues to be used for this plot.

model can be an isolated system, or it can represent one complex mode or two real modes of a higher-order system.

Consider the unforced second-order system

$$\Delta\dot{\mathbf{x}}(t) = \mathbf{F}\Delta\mathbf{x}(t), \quad \Delta\mathbf{x}(0) = \Delta\mathbf{x}_o \tag{4.3-76}$$

with eigenvalues (λ_1, λ_2). For uncoupled real roots, we define the system matrix as

$$\mathbf{F} = \begin{bmatrix} \lambda_1 & 0 \\ 0 & \lambda_2 \end{bmatrix} \tag{4.3-77}$$

For complex roots, the response is oscillatory, and we define $\mathbf{F}$ so that Δx_2 is the derivative of Δx_1:

$$\mathbf{F} = \begin{bmatrix} 0 & 1 \\ -\omega_n^2 & -2\zeta\omega_n \end{bmatrix} \tag{4.3-78}$$

Examples of the corresponding phase-plane plots are shown in Fig. 4.3-6. For all response classes, the origin is a *singular point*, that is, a point that formally satisfies eq. 4.3-76, although it may not lie on a state trajectory. A second-order LTI system has one singular point. If the singular point lies on a trajectory, it is an *equilibrium point* of the system. If the system is completely stable, the equilibrium point represents the *steady state* (or *final value*) of the system. If one or both eigenvalues are unstable, the singular point is a possible equilibrium point (e.g., for zero initial condition), but the slightest perturbation causes the state to diverge from that point. If one or both eigenvalues are zero, the corresponding initial condition(s) must be zero for the trajectory to pass through the origin.

For nonzero real roots,

$$\Delta x_1(t) = \Delta x_1(0)e^{\lambda_1 t} \tag{4.3-79}$$

$$\Delta x_2(t) = \Delta x_2(0)e^{\lambda_2 t} \tag{4.3-80}$$

Writing $\Delta x_2(t)$ as a function of $\Delta x_1(t)$ by eliminating t,

$$\Delta x_2(\Delta x_1) = \Delta x_2(0)\left[\frac{\Delta x_1}{\Delta x_1(0)}\right]^{\lambda_2/\lambda_1} \tag{4.3-81}$$

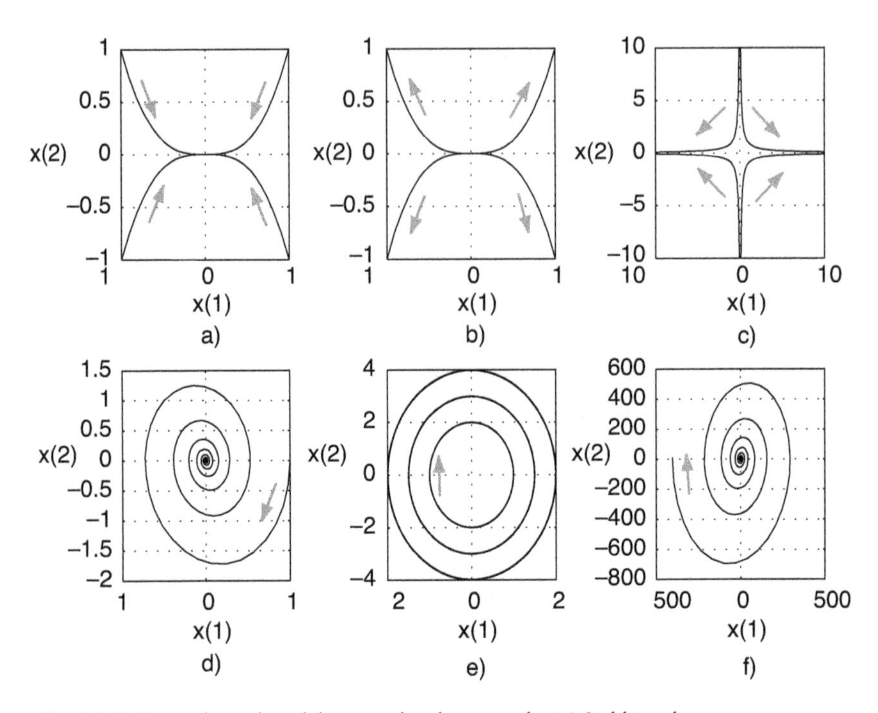

Fig. 4.3-6. Phase-plane plot of the second-order example. (a) Stable node. (b) Unstable node. (c) Saddle point. (d) Stable focus. (e) Unstable focus.

There are three *singular-point classes*. If both eigenvalues are stable, the singularity is a *stable node* (Fig. 4.3-6a); if both are unstable, it is an *unstable node* (Fig. 4.3-6b). If λ_1 and λ_2 have the same magnitudes but opposite signs, the two figures have the same shape, but the arrows denoting the trajectory direction are reversed. If one root is stable and the other is not, the singularity is a *saddle point*. Figure 4.3-6c illustrates the plot for stable λ_2 and unstable λ_1. For arbitrary initial conditions, $\Delta x_2(t)$ converges on zero, while $\Delta x_1(t)$ diverges; hence, typical trajectories veer away from the origin.

With oscillatory roots, the initial-condition response takes the general form

$$\Delta x_1(t) = ke^{-\zeta\omega_n t}\sin\left(\omega_n\sqrt{1-\zeta^2}\right)t \tag{4.3-82}$$

$$\Delta x_2(t) = \Delta\dot{x}_1(t) = ke^{-\zeta\omega_n t}\left(\omega_n\sqrt{1-\zeta^2}\right)\cos\left(\omega_n\sqrt{1-\zeta^2}\right)t \tag{4.3-83}$$

The norm of $\Delta\mathbf{x}(t)$ is

$$\|\Delta\mathbf{x}(t)\| = [\Delta x_1^2(t) + \Delta x_2^2(t)]^{1/2}$$
$$= ke^{-at}[\sin^2 bt + \omega_n^2 \cos^2(bt + \psi)] \quad (4.3\text{-}84)$$

where $a = \zeta\omega_n$, $b = \omega_n\sqrt{1-\zeta^2}$, and $\psi = \tan^{-1}(\zeta/\sqrt{1-\zeta^2})$, leading to the second three classifications. In all cases, the state response is periodic. For the diagram, assume that $\omega_n = 1$. If the damping ratio (ζ) is positive, the norm decreases over time, the trajectory spirals into the equilibrium point, and the singularity is a *stable focus* (Fig. 4.3-6d). If there is no damping ($\zeta = 0$), the norm is constant, and the trajectory circles the singularity, now called a *center* (Fig. 4.3-6e). With negative damping, the trajectory spirals away from the origin (Fig. 4.3-6f), and the singularity is an *unstable focus*.

Phase-plane plots can be drawn for *nonlinear* as well as linear systems. Although the trajectories deviate from those described by eqs. 4.3-79–4.3-85, their behavior in the neighborhood of singular points falls into one of the six classifications. If the system is time invariant and second order, trajectories can cross only at singularities. *Projections* of the trajectories of higher-order systems (e.g., a crossplot of two elements of the state or an oblique view of three elements of the state—a "3-D plot") can be displayed in two dimensions. Trajectories of time-invariant systems do not intersect in the higher-dimensional space; however, their projections may appear to cross on a 2-D plot.

Nonlinear systems may have more than one singular point, and they can experience *limit cycles*, which are closed, periodic trajectories (or *limit sets*) in the state space. A limit cycle is *stable* if trajectories converge to it; hence, it may be viewed as a state of *dynamic equilibrium* whose amplitude is independent of the amplitude of a neighboring initial condition. Unstable limit cycles can be defined, too. A small perturbation from an unstable closed trajectory diverges to another stable equilibrium point or limit set.

A *piecewise-linear system* contains one or more discrete nonlinearities (e.g., saturation or step functions) that separate the state space into regions with different linear models. References [H-2], [H-3], and [S-6] illustrate the multiple singular points that arise when the feedback control of a statically unstable aircraft encounters saturation limits on control-surface displacement and rate. The longitudinal case produces a stable focus within the unsaturated control region and saddle points in the saturated regions. Section 6.5 presents the effect of friction, a steplike nonlinearity, on the rudder-free dynamics of the Dutch roll mode. In this case, singularities occur outside the regions of their effect; hence, they cannot be equilibrium points. The system exhibits limit cycles for certain combinations of its parameters.

4.4 Frequency-Domain Analysis

Frequency-domain descriptions of response to control and disturbance inputs are useful adjuncts to the time-domain solutions described in Section 4.2. The stability and modal properties presented in the previous section are evident in the expression of transfer functions between inputs and outputs. The transfer characteristics of dynamic subsystems in an input-output chain are represented by the products of transfer functions of the individual subsystems. These compound transfer functions facilitate the analysis of feedback control effects on stability and response, and they allow the effects of random inputs such as atmospheric turbulence to be evaluated.

Transfer Functions and Frequency Response

Forced response of the linear, time-invariant system produces the output transform given by eq. 4.3-39. Neglecting initial conditions and disturbance effects, and assuming that the control $\Delta\mathbf{u}$ is not observed directly in $\Delta\mathbf{y}$, the output can be expressed as

$$\Delta\mathbf{y}(s) = \mathbf{H}(s\mathbf{I}_n - \mathbf{F})^{-1}\mathbf{G}\Delta\mathbf{u}(s) \triangleq \mathbf{H}\,\mathbf{A}(s)\mathbf{G}\;\Delta\mathbf{u}(s) \triangleq \mathbf{Y}(s)\Delta\mathbf{u}(s) \tag{4.4-1}$$

The dimensions of $\Delta\mathbf{y}$ and $\Delta\mathbf{u}$ are r and m, and $\mathbf{Y}(s)$ is the $(r \times m)$ *transfer function matrix* that relates outputs to inputs. Because each element of $\mathbf{H}$ and $\mathbf{G}$ is a real constant and $\mathbf{A}(s)$ is an $(n \times n)$ polynomial matrix, each element of $\mathbf{Y}(s)$ takes the form of a *scalar transfer function* relating Δu_j to Δy_i:

$$\frac{\Delta y_i(s)}{\Delta u_j(s)} \triangleq \mathrm{Y}_{ij}(s) = [h_{i1}\; h_{i2} \cdots h_{in}]\,\mathbf{A}(s)\begin{bmatrix} g_{1j} \\ g_{2j} \\ \cdots \\ g_{nj} \end{bmatrix}$$

$$= \frac{k_{ij}(s^{p-1} + b_{p-2}s^{p-2} + \cdots + b_1 s + b_0)}{s^n + c_{n-1}s^{n-1} + \cdots + c_1 s + c_0}$$

$$= \frac{k_{ij}(s - z_1)(s - z_2)\cdots(s - z_{p-1})}{(s - \lambda_1)(s - \lambda_2)\cdots(s - \lambda_n)} \tag{4.4-2}$$

The z_k ($k = 1, p, p \leq n$) are the *transfer function zeros*, k_{ij} is the *transfer function gain*, and the λ_l ($l = 1, n$) are the *system eigenvalues*, which characterize the system's *modes of motion*. If s equals any z_i, the transfer function is zero. If any z_k equals any λ_l, there is *pole-zero cancellation*; the associated mode does not appear in the response represented by $\Delta y_i(s)/\Delta u_j(s)$, but the response mode itself still exists.

The scalar transfer function denominator is the characteristic polynomial of $\mathbf{F}$, while the numerator polynomial is derived from the numerators of $\mathbf{A}(s)$, weighted by the ith row of $\mathbf{H}$ and the jth column of $\mathbf{G}$. In the simplest case, $r = n$, and $\mathbf{H}$ is an identity matrix of dimension n; a single component of the state Δx_k is measured, and the effect of a single control on a single state element is represented by g_{ij}; then,

$$\frac{\Delta y_i(s)}{\Delta u_j(s)} = \frac{\Delta x_i(s)}{\Delta u_j(s)} = a_{ij}(s) g_{ij} \tag{4.4-3}$$

where $a_{ij}(s)$ is defined by eq. 4.3-44.

As shown in Fig. 4.4-1, the eigenvalues (or *poles*) and zeros of a scalar transfer function can be plotted in the complex *s plane*. For the example, the transfer function is

$$Y(s) = \frac{K(s - z)}{[s^2 + 2\zeta\omega_n s + \omega_n^2]} = \frac{1.25(s + 40)}{[s^2 + 2(0.3)(7)s + (7)^2]} \tag{4.4-4}$$

To aid interpretation, ζ and ω_n are identified explicitly in the second-order denominator term. In the figure, the poles are denoted by "×" and the zeros by "○." Real singularities appear on the σ axis, and imaginary singularities fall on the $j\omega$ axis. Complex singularities $\mu \pm j\nu$ are plotted following eqs. 4.3-49 and 4.3-50, with units of rad/s for both components. The radius from the origin to a complex pole equals ω_n, the *natural frequency* of an oscillatory mode of motion, while the angle δ that the radius forms with the negative real axis depends only on the *damping ratio* ζ ($\delta = \cos^{-1}\zeta$). The σ intercept is the damping constant $\mu = -\zeta\omega_n$, and the $j\omega$ intercept is the *damped natural frequency* ($= \omega_n\sqrt{1 - \zeta^2}$). For notational convenience, we also refer to "damping ratios" and "natural frequencies" of complex transfer function zeros, although zeros do not represent modes of motion.

Stable poles fall in the left half of the s plane, neutral poles occur at the origin or on the $j\omega$ axis, and unstable poles plot in the right half of the plane. Zeros that fall in the left half plane are called *minimum-phase zeros*, while those in the right half plane are called *non-minimum-phase zeros*.

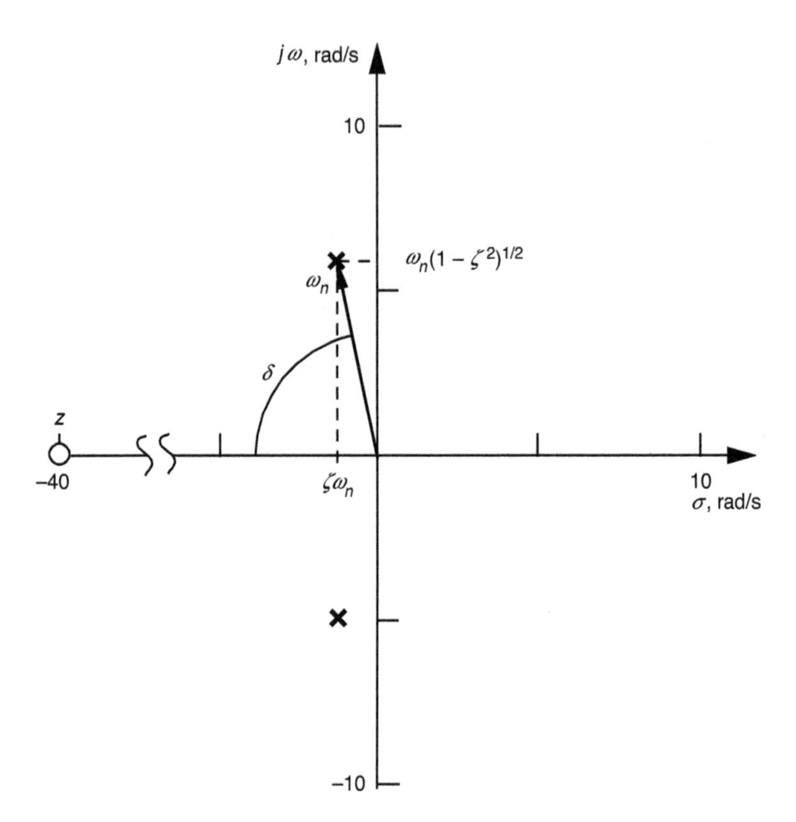

Fig. 4.4-1. Poles and zeros of a scalar transfer function.

The *frequency response* for an input-output pair is the steady-state response of the output to a sinusoidal input, and it is realizable only if the system is stable. For a LTI system, the input and output oscillations have the same frequency but differ in amplitude and phase. Consider the scalar sinusoidal input of unit amplitude,

$$\Delta u_j(t) = \sin \omega t \tag{4.4-5}$$

which has the following Laplace transform:

$$\mathcal{L}[\Delta u_j(t)] = \Delta u_j(s) = \frac{\omega}{s^2 + \omega^2} = \frac{\omega}{(s + j\omega)(s - j\omega)} \tag{4.4-6}$$

The partial-fraction expansion for the ith output can be expressed as

$$\Delta y_i(s) = Y_{ij}(s)\Delta u_j(s) = \frac{d_a}{(s + j\omega)} + \frac{d_b}{(s - j\omega)} + \frac{d_1}{(s - \lambda_1)} + \frac{d_2}{(s - \lambda_2)} + \cdots \tag{4.4-7}$$

and the inverse Laplace transform is

$$\Delta y_i(t) = \mathcal{L}^{-1}\left[\frac{d_a}{(s + j\omega)} + \frac{d_b}{(s - j\omega)} + \frac{d_1}{(s - \lambda_1)} + \frac{d_2}{(s - \lambda_2)} + \cdots\right] \tag{4.4-8}$$

Under the assumption of stability, all of the transient responses associated with the system's roots λ_i decay over time, and the output approaches

$$\Delta y_i(t) \rightarrow \mathcal{L}^{-1}\left[\frac{d_a}{(s + j\omega)} + \frac{d_b}{(s - j\omega)}\right] \quad \text{as} \quad t \rightarrow \infty$$

$$= d_a e^{-j\omega t} + d_b e^{j\omega t} \tag{4.4-9}$$

where

$$d_a = \left[\frac{\omega}{s - j\omega} Y_{ij}(s)\right]_{s = -j\omega} = \frac{j}{2} Y_{ij}(-j\omega) \tag{4.4-10}$$

$$d_b = \left[\frac{\omega}{s + j\omega} Y_{ij}(s)\right]_{s = +j\omega} = \frac{-j}{2} Y_{ij}(j\omega) \tag{4.4-11}$$

Substituting $s = j\omega$ in eq. 4.4-2 and noting that $Y_{ij}(j\omega)$ is a complex variable of the form

$$Y_{ij}(j\omega) = \frac{a + jb}{c + jd} = \frac{(ac + bd) + j(bc - ad)}{c^2 + d^2}$$

$$= \text{Re}(\omega) + j\,\text{Im}(\omega) = |Y_{ij}(\omega)|\, e^{j\phi(\omega)} = A(\omega) e^{j\phi(\omega)} \tag{4.4-12}$$

eq. 4.4-9 becomes

$$\Delta y_i(t) = \frac{1}{2j}[A(\omega) e^{j\phi(\omega)} e^{j\omega t} - A(\omega) e^{-j\phi(\omega)} e^{-j\omega t}]$$

$$= \frac{A(\omega)}{2j}[e^{j[\omega t + \phi(\omega)]} - e^{-j[\omega t + \phi(\omega)]}]$$

$$= A(\omega)\sin[\omega t + \phi(\omega)] \tag{4.4-13}$$

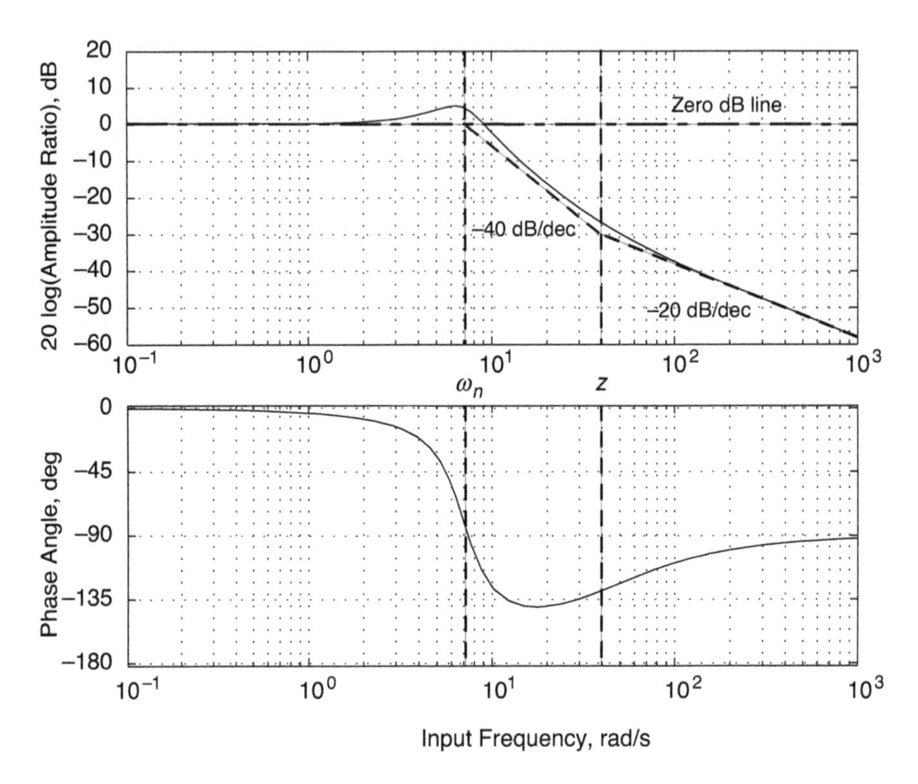

Fig. 4.4-2. Bode plot (frequency response) of a scalar transfer function.

Therefore, a plot of the function must portray both amplitude and phase angle as functions of frequency.

The *Bode plot* presents 20 $\log A(\omega)$ (measured in decibels, dB) and $\phi(\omega)$ (deg) plotted against $\log \omega$ (ω in rad/s) [B-9]. Logarithmic units allow large order-of-magnitude variations to be plotted on a single chart, as shown by the transfer function of eq. 4.4-4 (Fig. 4.4-2), and they facilitate sketching for informal analysis.

To understand the rationale for the Bode plot's logarithmic scales, consider the first-order transfer function

$$Y(s) = \frac{K}{s - \lambda} \tag{4.4-14}$$

whose frequency-response amplitude ratio is

$$A(\omega) = \frac{K}{|j\omega - \lambda|} = \frac{K}{\sqrt{\omega^2 + \lambda^2}} \tag{4.4-15}$$

Assuming $\omega > 0$, the limits of $A(\omega)$ for ω much larger or smaller than $|\lambda|$ are

$$A(\omega) \rightarrow \begin{cases} \dfrac{K}{|\lambda|}, & \omega \ll |\lambda| \\ \dfrac{K}{\omega}, & \omega \gg |\lambda| \end{cases} \tag{4.4-16}$$

and $A(\omega) = K/(\sqrt{2}|\lambda|)$ when $\omega = |\lambda|$. The limits of the frequency-response phase angle are

$$\phi(\omega) \rightarrow \begin{cases} 0\,\text{deg}, & \omega \ll |\lambda| \\ 90\,\text{sgn}(\lambda)\,\text{deg}, & \omega \gg |\lambda| \end{cases} \tag{4.4-17}$$

and $\phi(\omega) = 45\text{sgn}(\lambda)$ deg when $\omega = |\lambda|$.

The natural and base-10 logarithms of $Y_{ij}(\omega)$ are

$$\ln Y_{ij}(\omega) = \ln A(\omega) + j\phi(\omega) \tag{4.4-18}$$

$$\log Y_{ij}(\omega) = \log A(\omega) + 0.434 j\phi(\omega) \tag{4.4-19}$$

and we know that

$$\log \frac{ab}{cd} = \log a + \log b - \log c - \log d \tag{4.4-20}$$

Therefore,

$$20 \log A(\omega) \approx \begin{cases} 20(\log K - \log|\lambda|), & \omega \ll |\lambda| \\ 20(\log K - \log|\lambda|) - 3.01, & \omega = |\lambda| \\ 20(\log K - \log \omega), & \omega \gg |\lambda| \end{cases} \tag{4.4-21}$$

where $3.01 \approx 20 \log \sqrt{2}$.

The underlying trends in the frequency-response amplitude ratio are revealed by straight-line *asymptotes* on the log-log plot, as shown in Fig. 4.4-3. For $\omega < |\lambda|$, the asymptote is a horizontal line located $20(\log K - \log|\lambda|)$ dB above the 0 dB line. For $\omega > |\lambda|$, it is a straight line with a slope of -20 dB/decade. A *decade, dec*, is a factor of ten in frequency. As ω varies from 1 to 10 to 100, *dec* ($\triangleq \log \omega$) equals 0, 1, and 2. A plot that is logarithmic in ω is linear in decades of ω.

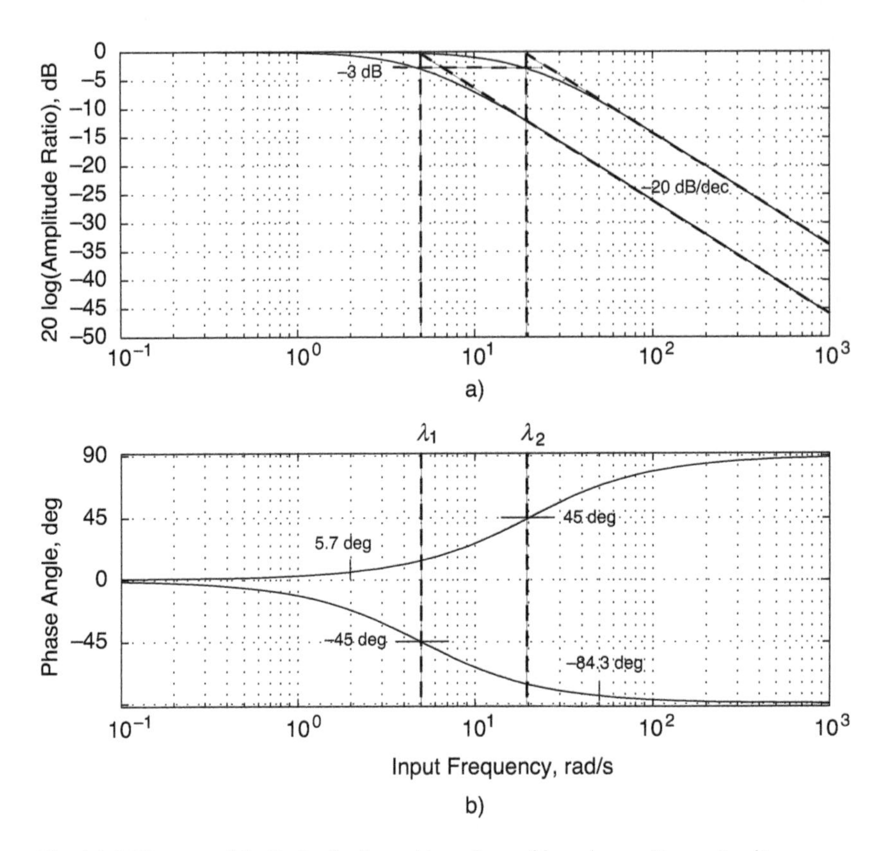

Fig. 4.4-3. Elements of the Bode plot for stable and unstable real roots ($\lambda_1 = -5$ rad/s, $\lambda_2 = +20$ rad/s). (a) First-order singularity. (b) Second-order singularity.

From eq. 4.4-21, the frequency-response asymptotes intersect when $\omega = |\lambda|$. Hence, $A(\omega)$ can be sketched quickly by plotting the asymptotes, locating a point about 3 dB below the intersection, and fairing a gradual curve through this point to the two asymptotes. The corresponding phase-angle response is sketched by locating the -45 deg point (stable root) or $+45$ deg (unstable root) at $\omega = |\lambda|$, then fairing a curve to points 5.7 deg above or below the 0 deg and ± 90 deg limits (as appropriate) a decade above and below. MATLAB easily plots the amplitude ratio and phase angle in precise fashion, but being able to sketch Bode plots by hand is useful for building insight.

The Bode plot construction procedure is similar for a second-order transfer function,

$$Y(s) = \frac{K}{s^2 + 2\zeta\omega_n s + \omega_n^2} \tag{4.4-22}$$

whose frequency-response amplitude ratio is

$$A(\omega) = \frac{K}{|-\omega^2 + 2\zeta\omega_n j\omega + \omega_n^2|} = \frac{K}{\sqrt{(\omega_n^2 - \omega^2)^2 + (2\zeta\omega_n\omega)^2}} \tag{4.4-23}$$

At the limits,

$$A(\omega) \to \begin{cases} \dfrac{K}{\omega_n^2}, & \omega \ll |\omega_n| \\ \dfrac{K}{\omega^2}, & \omega \gg |\omega_n| \end{cases} \tag{4.4-24}$$

$$\phi(\omega) \to \begin{cases} 0\,\text{deg}, & \omega \ll |\omega_n| \\ -180\,\text{sgn}(\zeta)\,\text{deg}, & \omega \gg |\omega_n| \end{cases} \tag{4.4-25}$$

When $\omega = \omega_n$, $A(\omega) = K/2\,\zeta\,\omega_n^2$ and $\phi(\omega) = -90\ \text{sgn}\ (\zeta)$ deg. Then

$$20\log A(\omega) \approx \begin{cases} 20(\log K - \log\omega_n^2), & \omega \ll \omega_n \\ 20(\log K - \log\zeta\omega_n^2), & \omega = \omega_n \\ 20(\log K - \log\omega^2), & \omega \gg \omega_n \end{cases} \tag{4.4-26}$$

The second-order asymptotes consist of a horizontal straight line at $20(\log K - \log\omega_n^2)$ dB for $\omega < \omega_n$ and a line with a slope of -40 dB/dec for $\omega > \omega_n$ (Fig. 4.4-4). The asymptotes intersect at $\omega = \omega_n$. The departure of the actual amplitude ratio from the asymptotes depends on the damping ratio. With $|\zeta| < \sqrt{2}/2$, there is a resonant peak of

$$20\log A(\omega) \approx (-20\log 2\,|\zeta\sqrt{1-\zeta^2}\,|) + 20(\log K - \log\omega_n^2) \tag{4.4-27}$$

at $\omega = \omega_n\sqrt{1-2\zeta^2}$. The actual *amplitude-ratio departures* from the asymptotes are elegantly portrayed in [M-7]. Similarly, the phase-angle variation with frequency is dependent on damping, with more gradual change as damping ratio increases. For $\zeta > 1$, the second-order pair is best described by two real roots. The sign of the phase variation is reversed for $\zeta < 0$, representing an unstable mode. However, the steady-state frequency response of an unstable system is overshadowed by the transient response of unstable roots and is not observable, as indicated by the partial-fraction expansion (eq. 4.4-8) of the output.

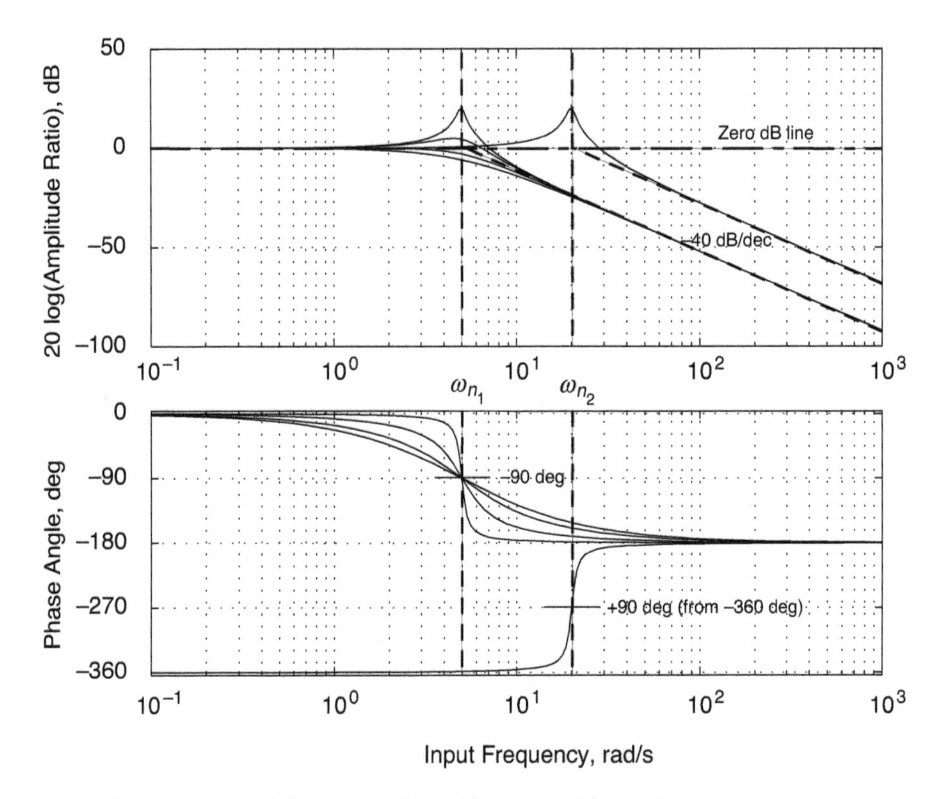

Fig. 4.4-4. Elements of the Bode plot for stable and unstable complex roots ($\omega_{n_1} = 5$ rad/s, $\zeta = 0.05, 0.3, 0.70711, 1$; $\omega_{n_2} = 20$ rad/s, $\zeta = -0.3$).

The frequency response of transfer functions containing many poles and zeros can be synthesized from individual contributions to the Bode plot because logarithms of products and divisions are additive (eq. 4.4-20). From eq. 4.4-18, amplitude and phase-angle effects can be considered separately.

In comparison to the shaping of frequency response by transfer function poles, transfer function zeros have inverse effect. The amplitude-ratio asymptote *increases* slope by 20 dB/dec at a first-order zero and +40 dB/dec at second-order zeros. The phase angle is *increased* by a minimum-phase (negative) zero and *decreased* by a non-minimum-phase (positive) zero.

The *stability margins* of a transfer function are significant in classical control system design and analysis. Consider a single-input/single-output system with negative unit feedback (i.e., feedback with a gain of positive one to a negative summing junction) around the transfer function $Y(j\omega)$

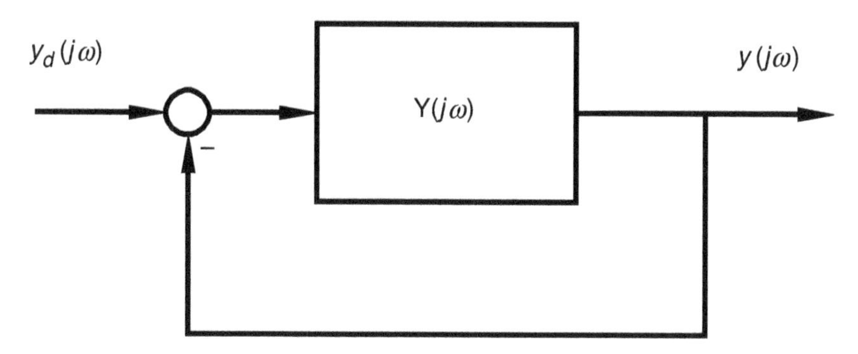

Fig. 4.4-5. Unit-feedback closed-loop system.

(Fig. 4.4-5). The effects of loop gains other than 1 and of forward compensation can be represented by defining $Y(j\omega) = kY_c(j\omega)\ Y_s(j\omega)$, where k is the gain, $Y_c(j\omega)$ is the compensation filter, and $Y_s(j\omega)$ is the system transfer function. $Y_c(j\omega)$ may express a *human pilot transfer function* as well as the effect of electronic control.

The closed-loop transfer function for this system relating the desired output $y_d(j\omega)$ to the actual output $y(j\omega)$ is

$$\frac{y(j\omega)}{y_d(j\omega)} = \frac{Y(j\omega)}{1 + Y(j\omega)} \tag{4.4-28}$$

Clearly, if $Y(j\omega) = -1$ (or the phase angle is ± 180 deg), the closed-loop relationship goes to infinity, indicating that the closed-loop system is unstable.

Of course, $Y(j\omega)$ changes with frequency; its amplitude may be 1 at one frequency, and its phase angle may be ± 180 deg at another. The *gain margin* is defined as the difference between $|Y(j\omega)|$ and 1 when the phase angle is -180 deg. A positive value [for $|Y(j\omega)|$ less than 1], either in the original units or in dB, indicates stability. The *phase margin* is defined as the difference between -180 deg and the phase angle when $|Y(j\omega)|$ equals one. The phase margin is positive for a phase angle more positive than -180 deg, indicating stability. The *crossover frequency* ω_{cross} is that for which $|Y(j\omega)|$ equals 1.

Example 4.4-1. Bode Plots for a Business Jet Aircraft

The program SURVEY (Appendix C) calculates single-input/single-output transfer functions and Bode plots using MATLAB functions, first defining `zpkSys = zpk(Sys)`, then applying `bode(zpkSys)`. The $\mathcal{M} = 0.3$, $h = 3{,}050$ m business jet flight condition can be characterized by the $(n \times m)$ transfer functions between the control and state elements. We review a few of the more significant relationships here. Using the fourth-order, body-axis model, the transfer function from elevator to normal velocity is calculated is

$$\frac{\Delta w(s)}{\Delta \delta E(s)} = \frac{-13.1653[s^2 + 2(0.063)(0.108)s + (0.108)^2](s + 70.7)}{[s^2 + 2(0.059)(0.126)s + (0.126)^2][s^2 + 2(0.414)(3.09)s + (3.09)^2]}$$

producing the Bode plot shown in Fig. 4.4-6. The low-frequency amplitude-ratio asymptote has zero slope, and the high-frequency asymptote decreases at −20 dB/dec. In between, there is a low-frequency complex pair of zeros that almost cancels the phugoid mode; however, the two are offset and lightly damped, producing a significant notch and peak at 0.108 and 0.126 rad/s. This is accompanied by a large spike in phase at intermediate frequencies. Resonant response at the short-period frequency is limited by the moderate damping of that mode, and a high-frequency zero reduces the rate of amplitude-ratio rolloff as frequency becomes large. For frequencies below $\omega_{n_{sp}}$, the elevator effectively commands normal velocity (or angle of attack). The phase angle begins at 180 deg because of the sign convention for elevator deflection (negative angle when the trailing edge is up).

A markedly different pattern occurs for the elevator-to-pitch-rate Bode plot (Fig. 4.4-7), whose transfer function is

$$\frac{\Delta q(s)}{\Delta \delta E(s)} = \frac{-9.069\,s(s + 1.153)(s + 0.02249)}{[s^2 + 2(0.059)(0.126)s + (0.126)^2][s^2 + 2(0.414)(3.09)s + (3.09)^2]}$$

The free s in the numerator assures that the transfer function behaves as a differentiator at low frequency, leading to a +20 dB/dec amplitude-ratio slope. The two real zeros, called $1/T_{\theta_1}$ and $1/T_{\theta_2}$ in [M-7], are, respectively, below the phugoid and short-period natural frequencies. There are resonant peaks at both modal frequencies, with a relatively flat frequency response between. Between the phugoid and short-period frequencies, the elevator is a pitch-rate control, with amplification of the effect at frequencies above $1/T_{\theta_2}$. Hence, angle-of-attack and pitch-rate command response are closely related in this frequency band.

The aileron effect on roll rate is similar to the pitch-rate response to the elevator; however, the region of flat amplitude-ratio response is bounded by the spiral and roll modes (Fig. 4.4-8). In this broad frequency band, the

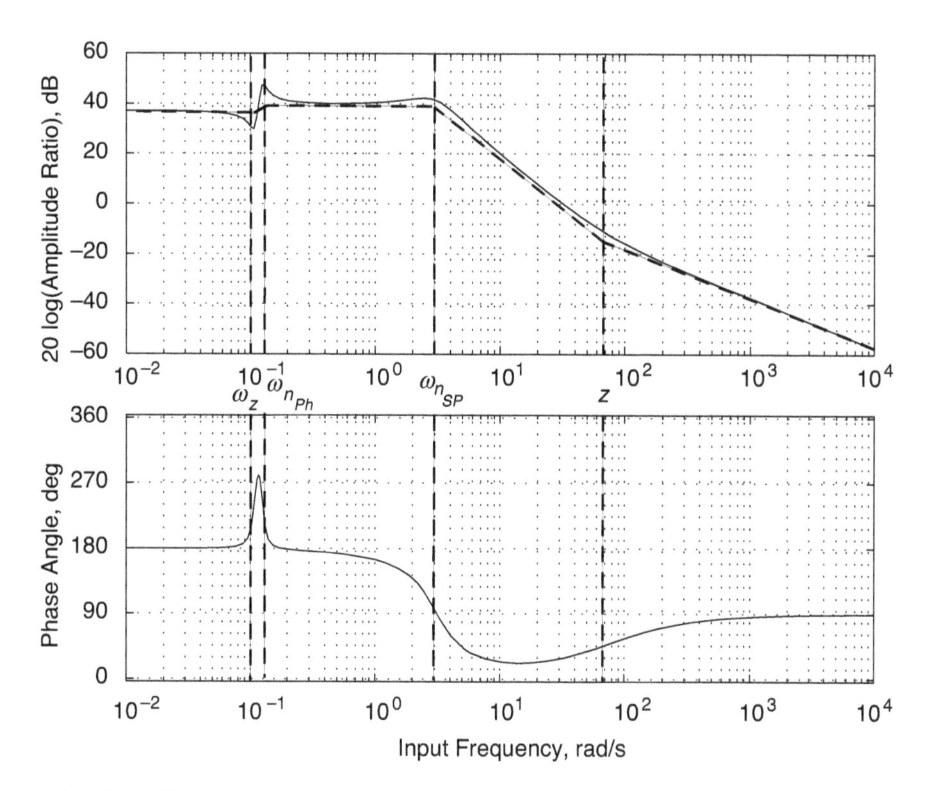

Fig. 4.4-6. Elevator-to-normal-velocity Bode plot.

ailerons produce roll-rate control. The complex zeros found in the transfer function,

$$\frac{\Delta p(s)}{\Delta \delta A(s)} = \frac{2.3078\, s[s^2 + 2(0.095)(1.37)s + (1.37)^2]}{(s - 0.003453)(s + 1.203)[s^2 + 2(0.081)(1.39)s + (1.39)^2]}$$

$$\frac{\Delta p(s)}{\Delta \delta A(s)} = \frac{2.3078\, s[s^2 + 2(0.095)(1.37)s + (1.37)^2]}{(s - 0.003453)(s + 1.203)[s^2 + 2(0.081)(1.39)s + (1.39)^2]}$$

effectively cancel the Dutch roll effect; their locations are plotted by a single dashed line. The roll-rate response is "washed out" below the spiral frequency, and it rolls off at −20 dB/dec at high frequency.

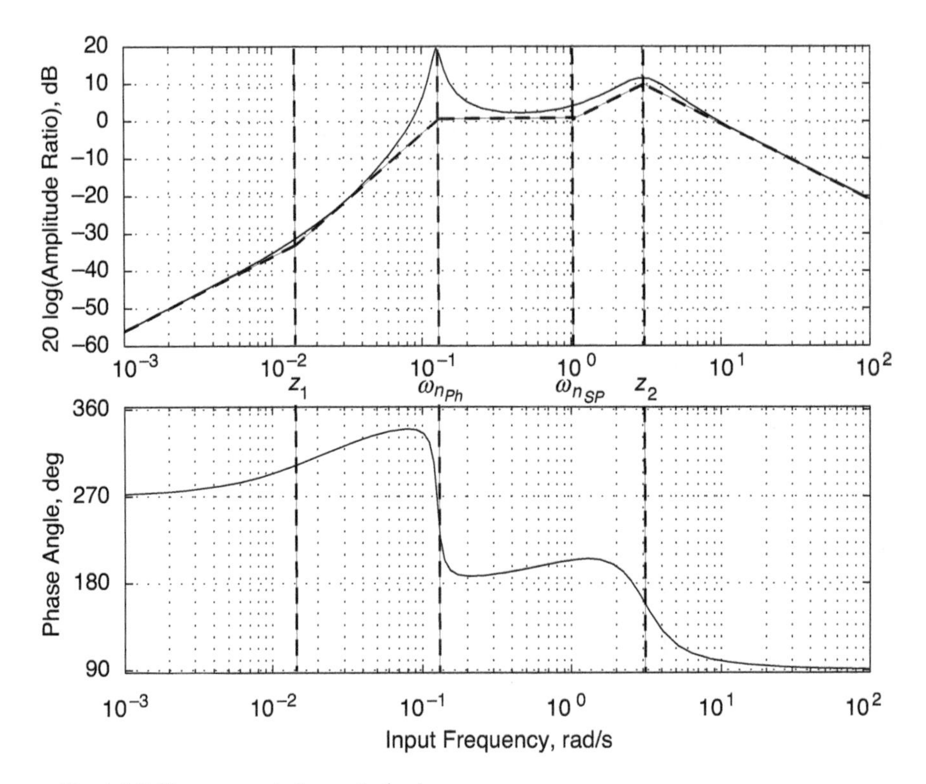

Fig. 4.4-7. Elevator-to-pitch-rate Bode plot.

The rudder-to-side-velocity transfer function is

$$\frac{\Delta v(s)}{\Delta \delta R(s)} = \frac{3.512(s - 0.01286)(s + 1.187)(s + 32.61)}{(s - 0.003453)(s + 1.203)[s^2 + 2(0.081)(1.39)s + (1.39)^2]}$$

The rudder commands side velocity (or sideslip angle) over a broad frequency band (Fig. 4.4-9). The low-frequency amplitude-ratio response is relatively flat, with the decreasing asymptote caused by the spiral mode λ_S eventually countered by z_1. The roll mode λ_R is effectively canceled by z_2, and the Dutch roll mode produces a resonant peak near $\omega_{n_{DR}}$ and a -40 dB/dec asymptote. z_3 increases the high-frequency asymptote to -20 dB/dec.

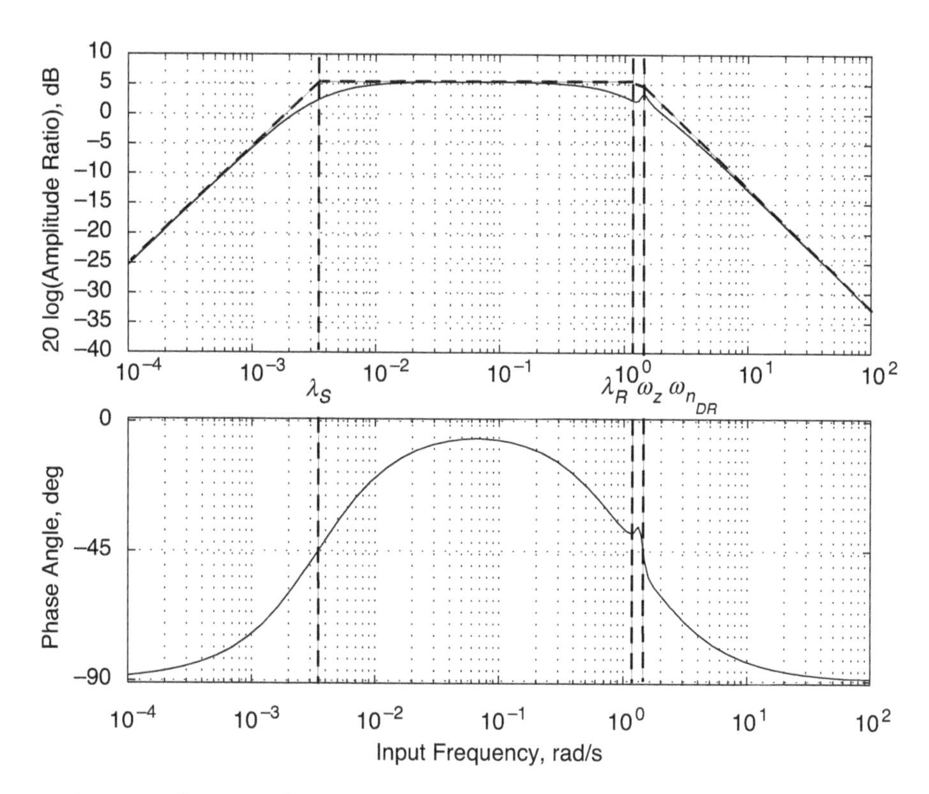

Fig. 4.4-8. Aileron-to-roll-rate Bode plot.

Nyquist Plot and Nichols Chart

Another way of presenting a system's transfer function is to crossplot the frequency-response amplitude ratio against the phase angle. The explicit dependence on frequency no longer appears in the single figure, although tick marks representing frequency can be added to the amplitude-phase angle contour. Such graphs portray the stability margins of a transfer function directly and are useful for classical closed-loop control design.

The *Nyquist plot* presents the amplitude and phase of the scalar transfer function in a polar plot or *Argand diagram*, as depicted for eq. 4.4-4 in Fig. 4.4-10a. The real and imaginary components of $Y(j\omega)$ are plotted on the Cartesian axes for frequencies in $(-\infty, \infty)$ in the $Y(j\omega)$ *plane* (as distinct from the s-plane graph of roots described in the next section). The amplitude and phase angle appear as the radius and angle from the

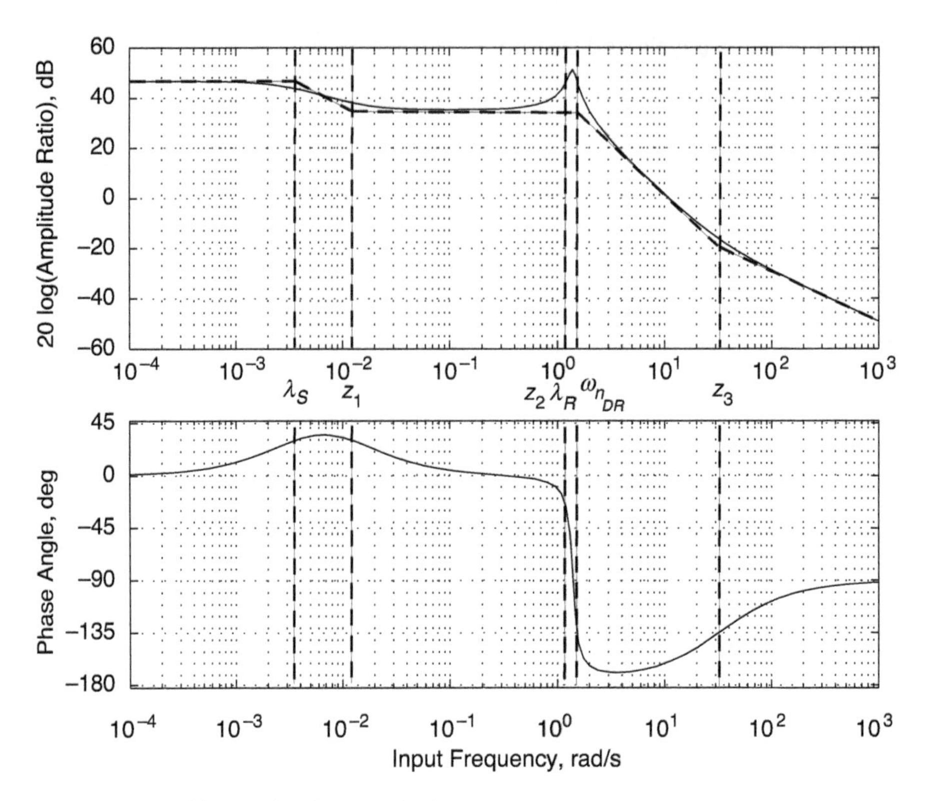

Fig. 4.4-9. Rudder-to-side-velocity Bode plot.

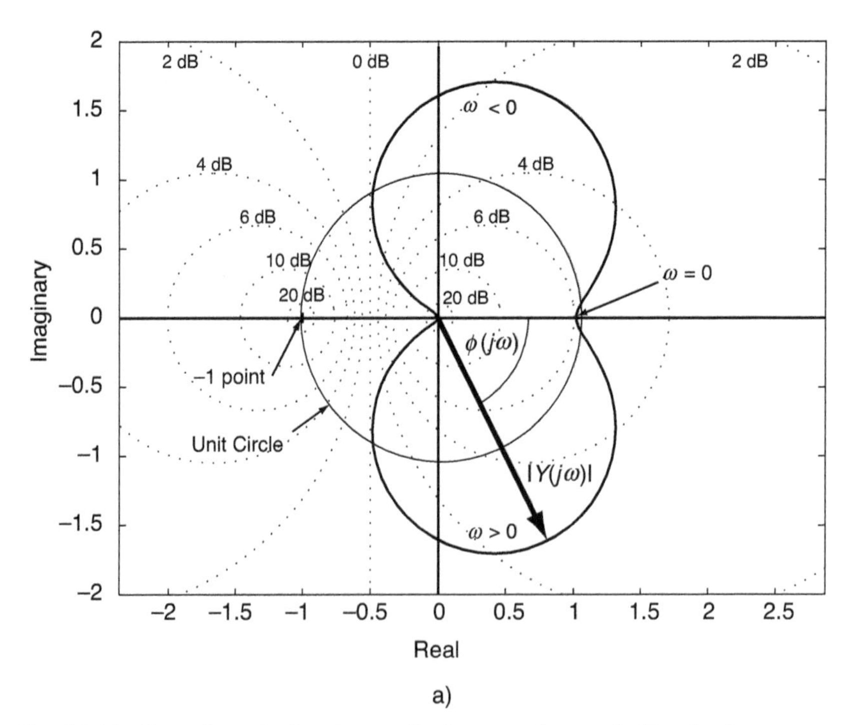

Fig. 4.4-10a. Plots of transfer function amplitude versus phase angle: Nyquist plot.

origin to the contour, and the plot is symmetric about the real axis. The "-1 point" noted above appears at $(-1, 0)$; hence, the gain margin would be determined by the magnitude of $Y(jw)$ at the crossing of the negative real axis. The transfer function shown in the figure has infinite gain margin, as $Y(jw)$ never crosses the negative real axis. The phase margin is the difference between the angle at which $Y(j\omega)$ crosses the unit circle, for $\omega > 0$ and -180 deg.

The *Nichols chart* presents the transfer function amplitude ratio, in dB, along the vertical axis and the phase angle (deg) along the horizontal axis for $\omega > 0$ (Fig. 4.4-10b). The origin is taken as (0 dB, -180 deg); hence, it corresponds to the -1 point. The gain and phase margins are read by the intersections of the $Y(j\omega)$ locus with the vertical and horizontal axes. Assuming unity feedback, *contours of constant closed-loop gain and phase angle* can be plotted as a curved grid on the Nichols chart, allowing graphical assessment of the effect of control-loop closure. The Nichols chart finds widespread use in the analysis of aircraft flying qualities.

Example 4.4-2. Nyquist and Nichols Charts for a Longitudinal Transfer Function

The Nyquist plot and Nichols chart for the elevator-to-normal velocity transfer function of the previous example, produced by the SURVEY program, are shown in Fig. 4.4-11 and 4.4-12. The sign of the transfer function is reversed to produce positive "D.C." (zero-frequency) gain. The phase angle never exceeds |160 deg|; hence, positive stability margin is maintained at all frequencies.

Root Locus

Variations in system parameters affect the roots of the characteristic equation; the plot of these variations in the s plane is called a *root locus*. The most common use of the root locus is to assess the effects of changing feedback control gain on system stability. However, an equally important application in aircraft dynamics is to illustrate the effects that changing stability derivatives have on system stability.

Root loci are easily calculated and plotted in MATLAB by repeated calculations of the eigenvalues as the elements of $\mathbf{F}$ are varied. Given the feedback law $\Delta \mathbf{u} = -\mathbf{C}\,\Delta \mathbf{x}$, where $\mathbf{C}$ is a *feedback control gain matrix*, the eigenvalues of $(\mathbf{F} - \mathbf{G}\,\mathbf{C})$ can be calculated for changing values in the control effect matrix $\mathbf{G}$ or in the feedback gain matrix $\mathbf{C}$. It then remains to plot the *loci of roots* (i.e., the curves traced by the roots as parameters vary) in the *s plane* for graphical interpretation.

If just a single parameter varies and the characteristic equation is linear in that parameter, *Evans's rules* for root locus construction can be

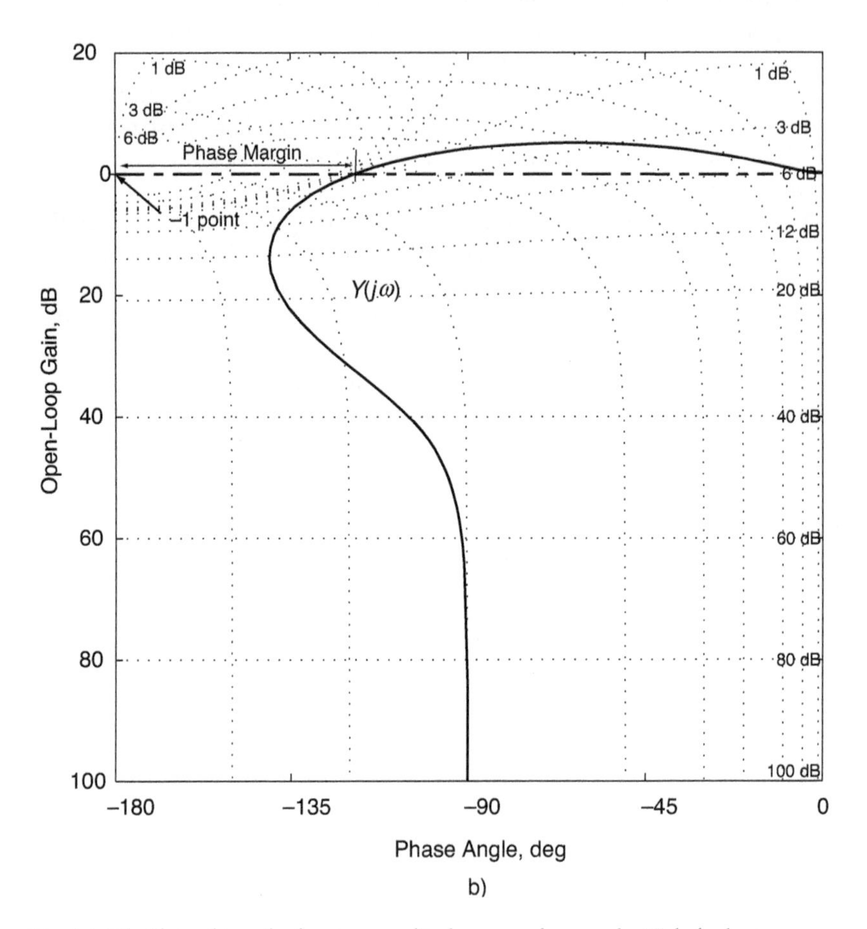

Fig. 4.4-10b. Plots of transfer function amplitude versus phase angle: Nichols chart.

applied [E-3], as in MATLAB's `rlocus` command. In the current context, the principal value in reviewing these rules is for conceptual understanding, not precise computation. The skilled analyst should be able to look at a pole-zero configuration and sketch the effect of parameter variation without falling back on the computer.

The principal goal is to find solutions to the scalar equation

$$1 + Y(s) = 1 + \frac{kn(s)}{d(s)} = 0 \tag{4.4-29}$$

or

$$D(s) = d(s) + k\,n(s) = 0 \tag{4.4-30}$$

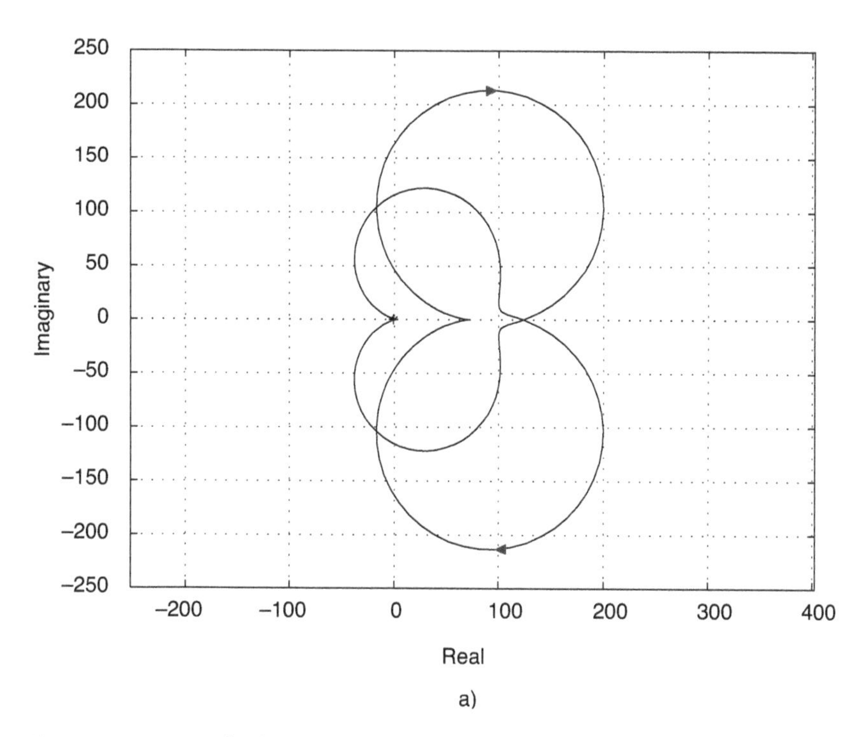

Fig. 4.4-11. Nyquist plot for elevator-to-normal-velocity transfer function.

as k varies between $(0, \infty)$ or $(-\infty, \infty)$. We assume that there are n poles of $d(s)$ and q zeros of $n(s)$, where $q \leq n$. Because there are n poles of $d(s)$, there are n roots of $D(s)$, and there are n branches of the root locus. For $k = 0$, the roots of the equation are the poles of $d(s)$ (e.g., those shown in Fig. 4.4-1), as evident from eq. 4.4-30. As k approaches $\pm\infty$, q roots approach the zeros of $n(s)$, and the magnitudes of the remaining $(n - q)$ roots go to ∞. The surplus roots radiate along linear asymptotes that originate at the root locus's *center of gravity*.

The *center of gravity* (c.g.) of the roots lies on the real axis because all complex roots are accompanied by complex conjugates. Identifying poles as $(a + jb)$ and zeros as $(u + jv)$, the c.g. location depends on the real parts of the poles, a, and zeros, c, as follows:

$$\text{c.g.} = \frac{\sum_{i=1}^{n} a_i - \sum_{j=1}^{q} u_j}{n - q} \tag{4.4-31}$$

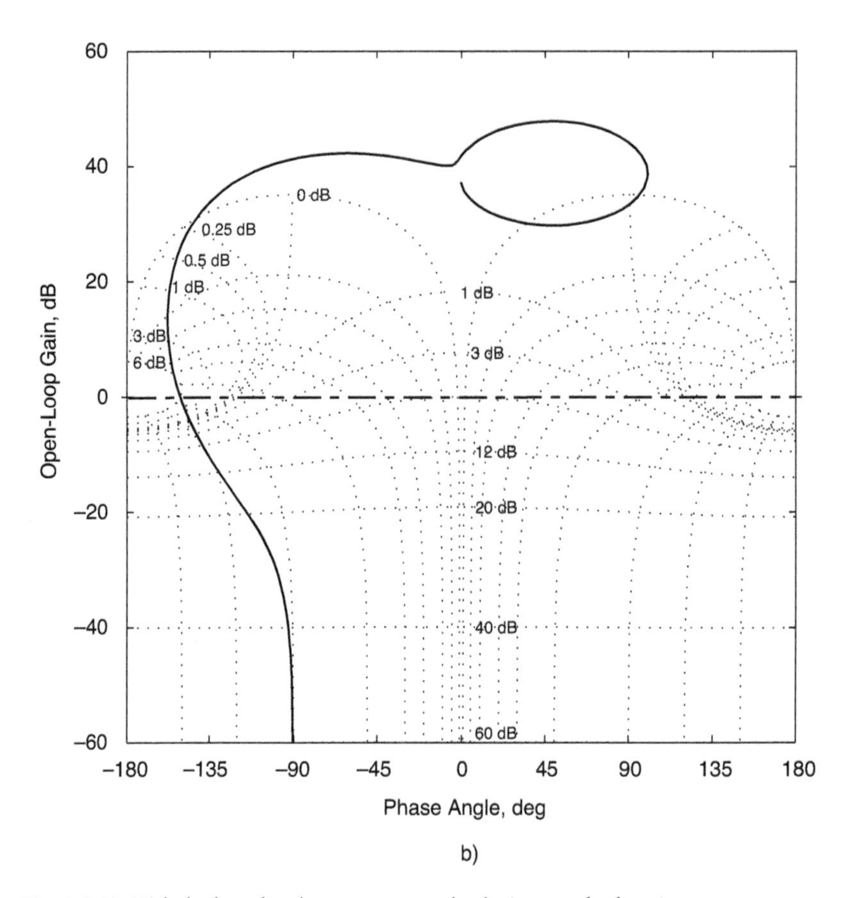

Fig. 4.4-12. Nichols chart for elevator-to-normal-velocity transfer function.

The angles of the asymptotes derive from a phase-angle criterion imposed by eq. 4.4-29. For positive gain, the net phase angle must be 180 deg, as $kn(s)/d(s) = -1$, while for negative gain, the phase angle must be 0 deg. The poles and zeros are symmetric about the real axis, so the $(n - q)$ roots going to ∞ must also be symmetric. Expressing the roots in polar form, $re^{j\phi}$, they must approach

$$\phi = \frac{(2m + 1)180}{n - q} \text{deg}, \quad m = 0, 1, 2, \ldots, (n - q - 1) \tag{4.4-32}$$

for positive gain and

$$\phi = \frac{2m\,180}{n - q} \text{deg}, \quad m = 0, 1, 2, \ldots, (n - q - 1) \tag{4.4-33}$$

for negative gain. These angles obtain only if the real and imaginary axes are scaled identically. MATLAB automatically scales the root-locus axes, often with different values; to preserve the angles, identical scaling must be selected for real and imaginary axes.

The phase-angle criterion also identifies branches of the root locus that must lie on the real axis. For the 180 deg (positive gain) criterion, there is a root locus branch on any real-axis segment that has an odd number of poles or zeros to the right. Segments that do not contain the locus for the positive-gain criterion do contain the locus when using the negative criterion, and vice versa. Branches can begin only at poles and end only at zeros (or $\pm$ infinity). If a branch lies between two poles, as the gain increases, the locus must split and break away from the real axis. Thus, two real roots become a complex pair. Conversely, a branch between zeros must receive a complex pair as the gain increases, turning them into two real roots. Breakaway branches for a single pair always leave or arrive at the real axis at ± 90 deg. Angles of departure from complex poles or arrival at complex zeros also are governed by the phase-angle criterion.

The root locus for negative feedback around the transfer function of eq. 4.4-4 is shown in Fig. 4.4-13. The closed-loop roots begin at the open-loop poles, initially increasing natural frequency while the damping ratio decreases. As the feedback gain increases, ω_n continues to increase, and ζ increases as well. Eventually, the two roots coalesce and reform as two real roots. One root goes toward the transfer function zero, and the other goes to $-\infty$.

4.5 Dealing with Uncertainty

We generally assume that the constants and variables used in the aircraft's dynamic equations are known without error; however, no quantities are known precisely. Our models are, at best, good approximations to imperfectly known physical phenomena. There are limits to our ability to describe and to predict. We must understand imprecision, when to analyze it, how to cope with it, and when to ignore it.

This section presents a brief introduction to the treatment of uncertain parameters in the dynamic models and of uncertain disturbances to an aircraft's motion; more details can be found in [S-14]. Most of the models in this book are *deterministic*, that is, they assume perfect knowledge of model structures, parameters, inputs, and outputs. It is easy enough to compute the bounds of behavior, given bounds on parameters and inputs, using these deterministic models. For example, for a lift-slope coefficient that may be 10 percent above or below the nominal value, the methods of this chapter can be reapplied to evaluate the ranges of equilibrium responses, eigenvalues, and so on.

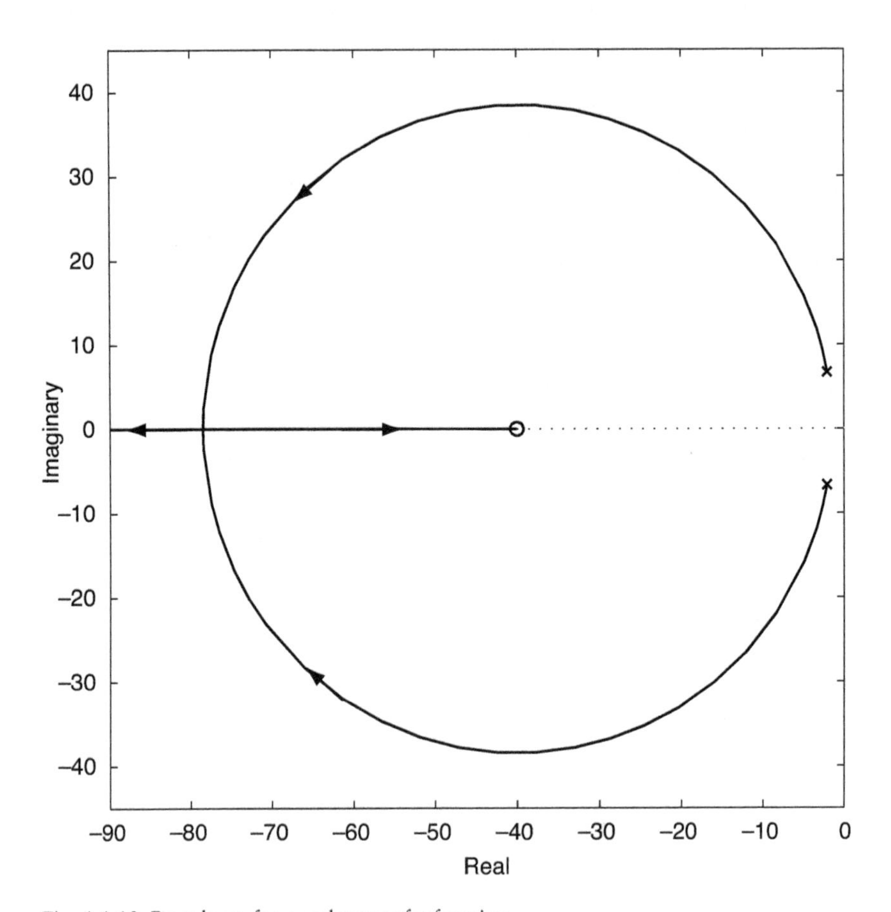

Fig. 4.4-13. Root locus for a scalar transfer function.

Uncertainty implies randomness, and the deterministic models can be extended to *stochastic models*, which predict not only the most likely response characteristics of the system but an expected range of response characteristics. Deterministic models predict specific values, while stochastic models estimate probability distributions of similar quantities. Numerical estimates of the effects of probabilistic parameter variations can be made using Monte Carlo evaluation, which uses conventional analysis tools repetitively while modeling uncertainty via random number generating routines.

Random Variables and Processes

A *random variable* x takes measurable but possibly different values each time it is sampled. For example, the color of a traffic light may change with a one-minute cycle, dwelling for 24 s on green, 6 s on amber, and 30 s on red. The cycle may be rigidly fixed, but we may look at random to see the light's color. Each time that the light is observed, the chances of seeing green, amber, or red are 40 percent, 10 percent, and 50 percent. Looking at random 100 times, we might expect to see 40 greens, 10 ambers, and 50 reds, although the actual numbers would vary due to the randomness of the sampling. As the number of samples, N, grows large, the observation percentages approach the theoretical probabilities. In the limit ($N = \infty$), the two sets are the same.

Assigning numbers to the colors ($x_1 = 1$ for green, $x_2 = 2$ for amber, and $x_3 = 3$ for red), the *probability mass functions* for the example are $\Pr(1) = 0.4$, $\Pr(2) = 0.1$, and $\Pr(3) = 0.5$. Then

$$\sum_{i=1}^{3} \Pr(x_i) = 1 \qquad (4.5\text{-}1)$$

More generally, for a random variable with I distinct values,

$$\sum_{i=1}^{I} \Pr(x_i) = 1 \qquad (4.5\text{-}2)$$

The sum of the probability mass functions is always 1, as x is assumed to take no value outside the set of I values. $\Pr(x_i)$ could be plotted as a column chart, or *histogram*, which could be compared with the *sampled estimates* of the probability mass functions, N_i/N (or *frequencies of occurrence*), derived from an experiment:

$$\Pr_{est}(x_i) = \frac{N_i}{N} \qquad (4.5\text{-}3)$$

Here N_i is the number of observations of the ith value, and N is the total number of samples of x (Fig. 4.5-1).

Suppose that x is a real number that can take any value in (x_{min}, x_{max}) and that the interval is broken into I segments of width Δx identified by their mid-points x_i. Each segment has a probability mass function $\Pr(x_i)$; we define the corresponding *probability density function* $\mathrm{pr}(x_i)$ such that

$$\mathrm{pr}(x_i)\Delta x = \Pr(x_i) \qquad (4.5\text{-}4)$$

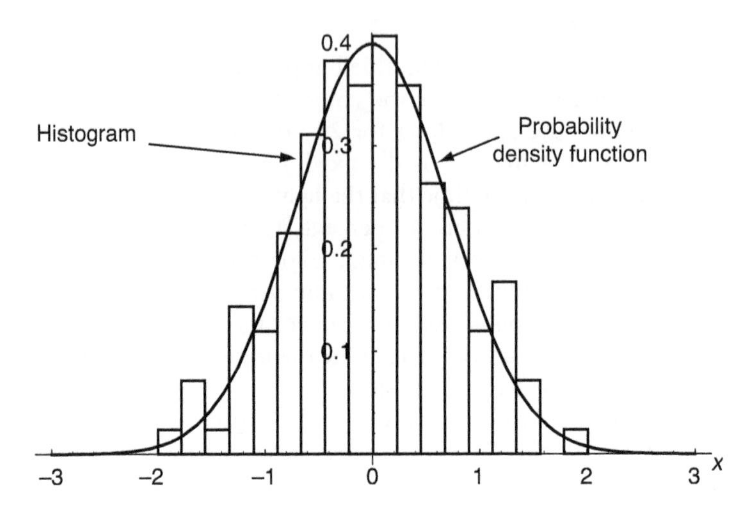

Fig. 4.5-1. Comparison of histograms and probability distributions.

This definition can be applied in eq. 4.5-2, and the number of segments in (x_{min}, x_{max}) can be increased (with a corresponding reduction in Δx). In the limit, the summation approaches the integral of the probability density function over all x:

$$\lim_{I\to\infty}\sum_{i=1}^{I}\Pr(x_i) = \int_{x_{min}}^{x_{max}} \mathrm{pr}(x)\,dx = \int_{-\infty}^{\infty} \mathrm{pr}(x)\,dx = 1 \tag{4.5-5}$$

The limits of integration can be expanded to $(-\infty, \infty)$ because there are no values outside (x_{min}, x_{max}). Alternatively, we can simply set the limits at $(-\infty, \infty)$ if there are no limits or the limits are unknown. This equation tells us that the probability that x lies in $(-\infty, \infty)$ is 1. All probability density functions must satisfy eq. 4.5-5.

The probability that x equals X or less, $\Pr(x \le X)$ is called the *cumulative probability distribution*, and it is

$$\Pr(x \le X) = \int_{-\infty}^{X} \mathrm{pr}(x)\,dx \tag{4.5-6}$$

The *median* of x, X_{med}, is that value for which $\Pr(x \le X_{med}) = 0.5$.

The histogram can be viewed as a quantized version of the probability density function plot (Fig. 4.5-1). A *Gaussian* (or *normal*) *probability density function*, with *mean value* $\bar{x}$ and *standard deviation* σ,

$$\mathrm{pr}(x) = \frac{1}{\sqrt{2\pi}\,|\sigma|}\,e^{-[(x-\bar{x})/2\sigma]^2} \tag{4.5-7}$$

is compared with a corresponding histogram in Fig. 4.5-1. Note that the Gaussian distribution is unbounded; the density is finite for all finite values of $|x - \bar{x}|$, approaching zero only as $|x - \bar{x}|$ approaches infinity.

We can use the difference between cumulative probability distributions to determine the probability that a Gaussian random variable x is within one, two, or more standard deviations of the mean. Referring to a tabulation of values [F-1],

$$\Pr[(\bar{x} - \sigma) \leq x \leq (\bar{x} + \sigma)] = \int_{-\infty}^{(\bar{x} + \sigma)} \mathrm{pr}(x)\,dx - \int_{-\infty}^{(\bar{x} - \sigma)} \mathrm{pr}(x)\,dx \approx 0.6827 \tag{4.5-8}$$

$$\Pr[(\bar{x} - 2\sigma) \leq x \leq (\bar{x} + 2\sigma)] = \int_{-\infty}^{(\bar{x} + 2\sigma)} \mathrm{pr}(x)\,dx - \int_{-\infty}^{(\bar{x} - 2\sigma)} \mathrm{pr}(x)\,dx \approx 0.9545 \tag{4.5-9}$$

and the probability of being within $\pm 3\sigma$ of the mean is about 0.9973. The probabilities represent the area under the Gaussian probability density function curve, bounded by $\pm 1\sigma$, $\pm 2\sigma$, and $\pm 3\sigma$ variations from the mean (Fig. 4.5-2).

The *expected value* of x, $E(x)$, is the integral of x over $(-\infty, \infty)$, weighted by its probability density function,

$$E(x) = \int_{-\infty}^{\infty} x\,\mathrm{pr}(x)\,dx = \bar{x} \tag{4.5-10}$$

and the result is called the *mean value* or *first moment* of x. Higher moments are integrals of powers of x weighted by the same probability density function. For example, the *second central moment* of x, defined for variations from the mean, equals the *variance* (or standard deviation squared) of the variable:

$$E[(x - \bar{x})^2] = \int_{-\infty}^{\infty} (x - \bar{x})^2\,\mathrm{pr}(x)\,dx = \sigma^2 \tag{4.5-11}$$

Note that the Gaussian distribution is fully defined by its first and second central moments. The *mode* of a random variable is the value of x that has the highest probability density function. The mean, median, and mode of the Gaussian distribution, which is symmetric and *unimodal* (i.e., has a single peak), are the same.

To this point, we have not specified that x varies over time. Although the traffic light color changes from one instant to the next, we could have considered an *ensemble* of identical but unsynchronized lights, each sampled just once, and have come to the same conclusions. A random variable could be an *uncertain constant*, such as the weight of a car coming off an assembly line: the value is fixed, but the actual weight is unknown. If the random variable x varies explicitly in time, denoted by $x(t)$, we call

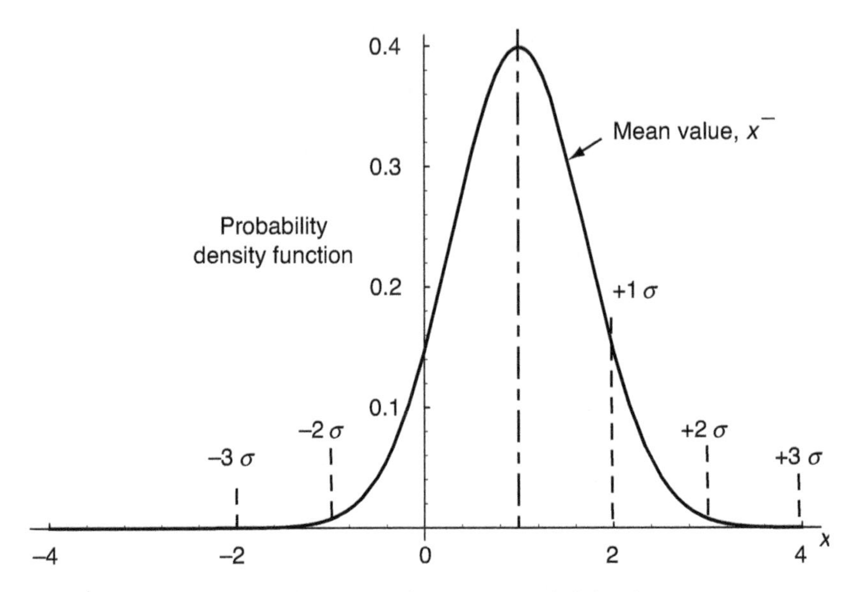

Fig. 4.5-2. Mean and standard deviation of a Gaussian probability density function.

it (the output of) a *random process*. The wind speed at the top of a flag pole is an element of the ensemble of random processes formed by the wind field throughout the adjacent space.

The probability density function of a random process, pr[$x(t)$], may vary over time, or it may be constant. In the first instance, the random process is *nonstationary*: an ensemble of unsynchronized traffic lights whose cycles change over the course of the day provides an example. In the second, cycles of the ensemble of lights are fixed (though individual lights may have different cycles) throughout the day, and the random process is *stationary*. If all the lights have identical (but unsynchronized) cycles, the process also is *ergodic*: identical probabilities would be estimated (in the limit) by sampling the entire ensemble or by sampling a single traffic light over time.

As a practical matter, it is relatively easy to define and operate on nonstationary random statistics (e.g., a Gaussian distribution with time-varying mean and standard deviation), but it is difficult to measure the statistical properties of physical data without assuming ergodicity. The reasons are that replications of the same random process are rarely available and that time durations of records are finite. Of course, a time-varying trend could be assumed if an underlying model of the process is at hand, and *ad hoc* measurements of slowly varying statistics can be estimated from short "windows" of data. For example, an ergodic disturbance may force a dynamic system with time-varying coefficients; the output is a time-varying random process.

In addition to their probability density functions, random processes are described by autocovariance and power spectral density functions. The *autocovariance function* $\phi_{xx}(t)$ of a stationary process $x(\tau)$ is defined as

$$\phi_{xx}(\tau) = E[x(t)\, x(t+\tau)] \tag{4.5-12}$$

and, for an ergodic process,

$$\phi_{xx}(\tau) = \lim_{T\to\infty} \frac{1}{T}\int_0^T x(t)\,x(t+\tau)\,dt \tag{4.5-13}$$

where τ is the *lag* from the current value of t. The maximum value occurs for $\tau = 0$; this is the *mean-square value* $\phi_{xx}(0)$, which is always positive and equals the *variance* σ^2 of $x(t)$. The autocovariance function is symmetric about $\tau = 0$, and it represents the relative similarity of the variable from one instant to the next (Fig. 4.5-3). $\phi_{xx}(\tau)$ may be negative, although its magnitude is never greater than $\phi_{xx}(0)$. Normalizing eq. 4.5-12 by $\phi_{xx}(0)$ produces the *autocorrelation function* of $x(t)$.

The *power spectral density function* $\Phi_{xx}(\omega)$ of a stationary process is the Fourier transform of the autocovariance function:

$$\Phi_{xx}(\omega) = \int_{-\infty}^{\infty} \phi_{xx}(\tau)e^{-j\omega\tau}d\tau \tag{4.5-14}$$

$\Phi_{xx}(\omega)$ represents the frequency distribution of the mean-square value of the process. Physical processes typically have high "power" at low frequency, and they "roll off" at high frequency. The logarithm of the power spectral density is plotted against the logarithm of the frequency, much like the transfer function amplitude ratios of Section 4.4. (Also see the spectral models of atmospheric turbulence presented in Section 2.1.)

Dynamic Response to Random Inputs and Initial Conditions

Given a spectral description of the stationary random input to a physical system, the spectrum of a stable, linear, time-invariant system's output is easily calculated. From Section 4.4, the Laplace transform of a system's output Δy given an input Δu is

$$\Delta y(s) = Y(s)\, \Delta u(s) \tag{4.5-15}$$

where $Y(s)$ is the scalar transfer function between the two. Letting $s = j\omega$ produces the equivalent Fourier transform relationship; multiplying each side by its complex conjugate

$$\Delta y(j\omega)\, \Delta y^*(j\omega) = Y(j\omega)\, Y^\dagger(j\omega)\, \Delta u(j\omega)\, \Delta u^\dagger(j\omega) \tag{4.5-16}$$

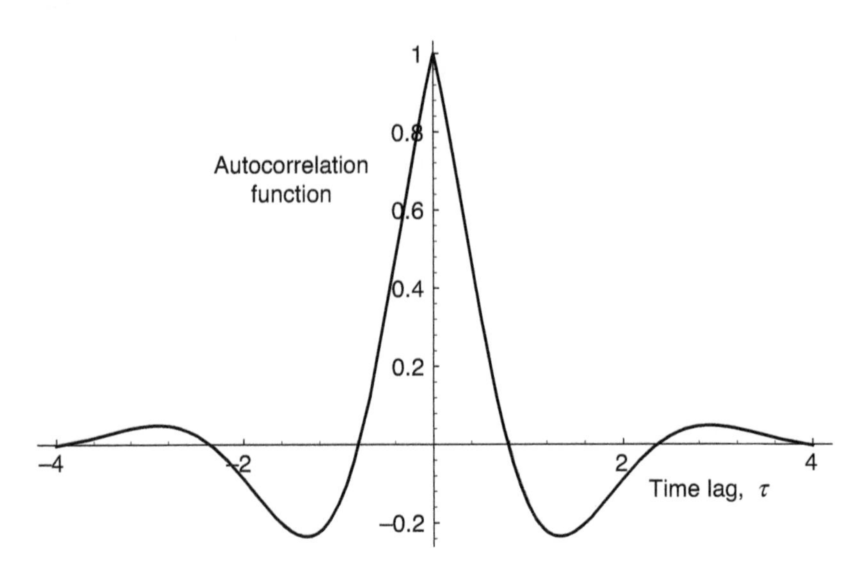

Fig. 4.5-3. Autocovariance function of a state history.

leads to

$$\Phi_{yy}(\omega) = Y(j\omega)\ Y^{\dagger}(j\omega)\ \Phi_{uu}(\omega) = |Y(j\omega)|^2\ \Phi_{uu}(\omega) \tag{4.5-17}$$

where $\Phi_{uu}(\omega)$ is the power spectrum of the input, and $\Phi_{yy}(\omega)$ is the power spectrum of the output [B-3]. This is a stationary result, assumed to hold at all times; starting transient effects are not considered, and bounded inputs are presumed to produce bounded outputs (hence, the system must be stable). There is no phase information in eq. 4.5-17; the spectra are real functions of the input frequency.

This transfer function effect can be used to create "colored"—or correlated—spectral disturbances from completely uncorrelated processes, as well as to compute aircraft spectral response to random inputs. Consider the Dryden spectrum for axial turbulence presented in Section 2.1; the spatial frequency Ω is replaced by the time-based frequency $\omega = V_o\Omega$ to account for the aircraft's forward motion through the turbulent wind field. If we let the input spectrum $\Phi_{uu}(\omega)$ be σ^2, the input is a *white noise process*, which is independent of frequency. Choosing $Y(j\omega)$ as

$$Y(j\omega) = \frac{(2\,L/\pi)^{1/2}}{[1 + (L/V_o)\,j\omega]^2} \tag{4.5-18}$$

the $\Phi_{yy}(\omega)$ produced by eq. 4.5-17 is the turbulence spectrum of eq. 2.1-20. $\Phi_{yy}(\omega)$ can then be multiplied by the appropriate aircraft transfer function and its complex conjugate to represent the spectral response of the aircraft to Dryden turbulence (e.g., the axial velocity response to axial wind input). A representative time trace of axial turbulence can be simulated by driving the state-space equivalent of eq. 4.5-18 with a Gaussian random number generator set to zero mean and standard deviation σ.

The scalar concepts of probability distributions and expected values can be carried over to vectors as well. The $(n \times 1)$ random vector $\mathbf{x}$ has a *joint* or *multivariate probability density function* $\mathrm{pr}(\mathbf{x})$, defined such that

$$\int_{-\infty}^{\infty} \mathrm{pr}(\mathbf{x})\,d\mathbf{x} \triangleq \int_{-\infty}^{\infty} \cdots \int_{-\infty}^{\infty} \mathrm{pr}(x_1, x_2, \ldots, x_n)\,dx_1\,dx_2 \ldots dx_n = 1 \qquad (4.5\text{-}19)$$

where $\mathrm{pr}(\mathbf{x})$ is a scalar function of a vector variable. The integration must be taken independently across all the dimensions of $\mathbf{x}$. Accordingly, the mean value of $\mathbf{x}$ is defined by its first moment,

$$E(\mathbf{x}) \triangleq \bar{\mathbf{x}} = \int_{-\infty}^{\infty} \mathbf{x}\,\mathrm{pr}(\mathbf{x})\,d\mathbf{x}$$

$$\triangleq \begin{bmatrix} \int_{-\infty}^{\infty} \cdots \int_{-\infty}^{\infty} x_1\,\mathrm{pr}(x_1, x_2, \ldots, x_n)\,dx_1\,dx_2 \cdots dx_n \\ \int_{-\infty}^{\infty} \cdots \int_{-\infty}^{\infty} x_2\,\mathrm{pr}(x_1, x_2, \ldots, x_n)\,dx_1\,dx_2 \cdots dx_n \\ \int_{-\infty}^{\infty} \cdots \int_{-\infty}^{\infty} x_n\,\mathrm{pr}(x_1, x_2, \ldots, x_n)\,dx_1\,dx_2 \cdots dx_n \end{bmatrix} \qquad (4.5\text{-}20)$$

The second central moment defines the $(n \times n)$ *covariance matrix* of $\mathbf{x}$, denoted by $\mathbf{P}$:

$$E[(\mathbf{x} - \bar{\mathbf{x}})(\mathbf{x} - \bar{\mathbf{x}})^T] \triangleq E[\tilde{\mathbf{x}}\tilde{\mathbf{x}}^T] \triangleq \mathbf{P} = \int_{-\infty}^{\infty} \tilde{\mathbf{x}}\tilde{\mathbf{x}}^T\,\mathrm{pr}(\mathbf{x})\,d\mathbf{x}$$

$$= \begin{bmatrix} p_{11} & p_{12} & \cdots & p_{n} \\ p_{21} & p_{22} & \cdots & p_{2n} \\ \cdots & \cdots & \cdots & \cdots \\ p_{n1} & p_{n2} & \cdots & p_{nn} \end{bmatrix} \qquad (4.5\text{-}21)$$

where the ijth element of $\mathbf{P}$ is defined as $\int_{-\infty}^{\infty} \tilde{x}_i \tilde{x}_j\,\mathrm{pr}(x_1, x_2, \ldots, x_n)\,dx_1\,dx_2 \cdots dx_n$. The diagonal of $\mathbf{P}$ contains the variances of the individual elements of $\mathbf{x}$, and the off-diagonal elements contain the covariances of the elements of $\mathbf{x}$, representing, for example, the correlation between x_i and x_j. The *multivariate Gaussian probability density function*

$$\mathrm{pr}(\mathbf{x}) = \frac{e^{-(1/2)[(\mathbf{x} - \bar{\mathbf{x}})^T \mathbf{P}^{-1} (\mathbf{x} - \bar{\mathbf{x}})]}}{(2\pi)^{n/2}\,|\mathbf{P}|^{1/2}} \qquad (4.5\text{-}22)$$

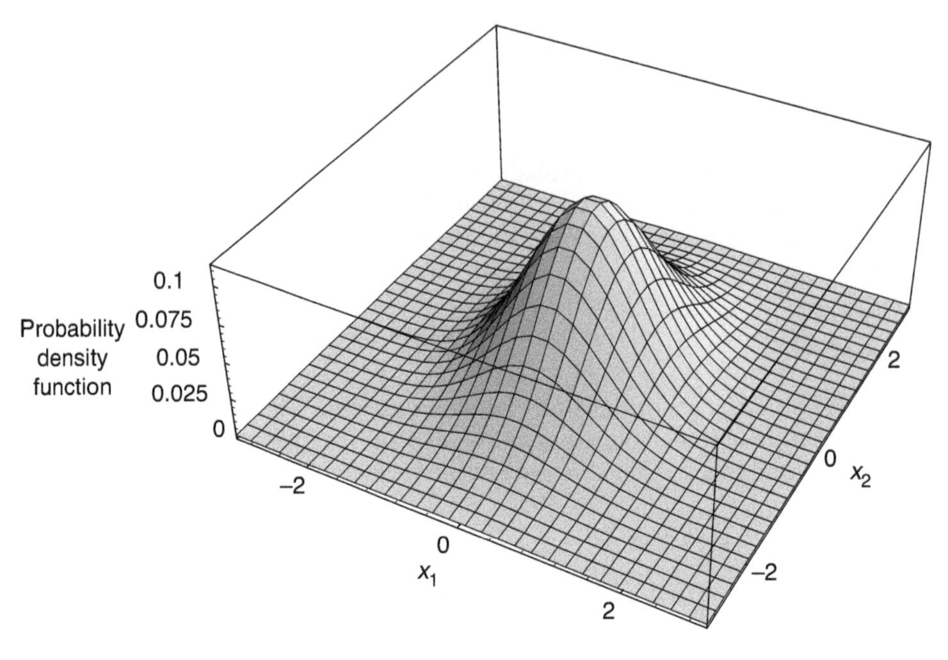

Fig. 4.5-4. Multivariate probability density function ($n = 2$).

satisfies eq. 4.5-19. It is fully described by its mean vector $\bar{\mathbf{x}}$ and covariance matrix $\mathbf{P}$ (Fig. 4.5-4, plotted for a two-dimensional distribution with zero mean and standard deviations of 1 and $\sqrt{2}$).

The aircraft's time response to Gaussian random disturbances and initial conditions is readily computed from the definitions of first and second central moments and the linear dynamic model:

$$\Delta\dot{\mathbf{x}}(t) = \mathbf{F}(t)\Delta\mathbf{x}(t) + \mathbf{G}(t)\Delta\mathbf{u}(t) + \mathbf{L}(t)\Delta\mathbf{w}(t), \quad \Delta\mathbf{x}(0) = \Delta\mathbf{x}_0 \tag{4.5-23}$$

We begin by taking the expected value of the system equation:

$$\begin{aligned} E[\Delta\dot{\mathbf{x}}(t)] &\triangleq \Delta\dot{\mathbf{m}}(t) \\ &= E[\mathbf{F}(t)\,\Delta\mathbf{x}(t) + \mathbf{G}(t)\,\Delta\mathbf{u}(t) + \mathbf{L}(t)\,\Delta\mathbf{w}(t)] \\ &= \mathbf{F}(t)\,E[\Delta\mathbf{x}(t)] + \mathbf{G}(t)\,E[\Delta\mathbf{u}(t)] + \mathbf{L}(t)\,E[\Delta\mathbf{w}(t)] \\ &= \mathbf{F}(t)\,\Delta\mathbf{m}(t) + \mathbf{G}(t)\Delta\bar{\mathbf{u}}(t) + \mathbf{L}(t)\Delta\bar{\mathbf{w}}(t), \quad \Delta\mathbf{m}(0) = \Delta\mathbf{m}_0 \end{aligned} \tag{4.5-24}$$

The expected value of $\Delta\mathbf{x}(t)$ is denoted by $\Delta\mathbf{m}(t)$ and the expected values of the control and disturbance are $\Delta\bar{\mathbf{u}}(t)$ and $\Delta\bar{\mathbf{w}}(t)$. Because the expectation operation is linear and the system matrices are assumed to be known without error, $\mathbf{F}(t)$, $\mathbf{G}(t)$, and $\mathbf{L}(t)$ can be taken outside the expectation operations. Equation 4.5-24 propagates an estimate of the *mean value of the state* using the original system equation.

Given a linear system with Gaussian inputs and initial condition, the distribution of $\Delta\mathbf{x}(t)$ is Gaussian. Consequently, the mean value estimate plus an estimate of the covariance of $\Delta\mathbf{x}(t)$ specifies the probability distribution completely. This is a nonstationary estimate that includes the effects of starting transients, that admits time variations in the coefficients of the system itself, and that does not require stability of the system (though estimates for an unstable system ultimately diverge).

The estimate of the covariance in $\Delta\mathbf{x}(t)$ due to uncertain inputs and initial conditions derives from the second central moment. If $\Delta\mathbf{u}(t)$ and $\Delta\mathbf{w}(t)$ are white-noise processes with *spectral density matrices* $\mathbf{R}(t)$ and $\mathbf{Q}(t)$, they are uncorrelated with $\Delta\mathbf{x}(t)$ and each other; if the control is known without error, $\Delta\bar{\mathbf{u}}(\tau) = \Delta\mathbf{u}(\tau)$, and $\mathbf{R}(t) = 0$. The differential equation for the covariance of $\Delta\mathbf{x}(t)$ about its mean is [S-14]

$$\dot{\mathbf{P}}(t) = \mathbf{F}(t)\mathbf{P}(t) + \mathbf{P}(t)\mathbf{F}^T(t) + \mathbf{G}(t)\mathbf{R}(t)\mathbf{G}^T(t) + \mathbf{L}(t)\mathbf{Q}(t)\mathbf{L}^T(t),$$
$$\mathbf{P}(0) = E[(\Delta\mathbf{x}(0) - \Delta\mathbf{m}_0)(\Delta\mathbf{x}(0) - \Delta\mathbf{m}_0)^T] = \mathbf{P}_0 \qquad (4.5\text{-}25)$$

This is a symmetric, linear equation in the symmetric matrix $\mathbf{P}(t)$; hence, its solution has the properties discussed earlier in the chapter. The equation allows the effects of uncertain initial conditions, time-varying inputs, and time-varying coefficients to be evaluated, expressing the uncertainty in path following during a dynamic maneuver. For an unstable system, the covariance estimate ultimately diverges; hence, the estimate of the mean becomes entirely uncertain.

Given a time-invariant system, a stationary solution for $\mathbf{P}(t)$, is found from eq. 4.5-25 by setting $\dot{\mathbf{P}}(t) = 0$. The equation becomes the *Lyapunov equation* of Section 4.3 (eq. 4.3-11). Positive-definite input matrices produce bounded, positive-definite state covariance only if $\mathbf{F}$ is stable. Kinematic elements of the aircraft state are not forced directly by inputs; hence, $[\mathbf{G}(t)\ \mathbf{R}(t)\ \mathbf{G}^T(t) + \mathbf{L}(t)\ \mathbf{Q}(t)\ \mathbf{L}^T(t)]$ can be positive semidefinite but not positive definite when these terms are included in the dynamic equations.

The spectral density matrices contain the variances of scalar inputs along the diagonal and the covariances in off-diagonal elements. As explained in [S-14], the spectral density matrix is related to the second central moment of the disturbance by

$$E\{[\Delta\mathbf{u}(t) - \Delta\bar{\mathbf{u}}(t)][\Delta\mathbf{u}(\tau) - \Delta\bar{\mathbf{u}}(\tau)]^T\} = \mathbf{R}(t)\delta(t - \tau) \qquad (4.5\text{-}26)$$

$$E\{[\Delta\mathbf{w}(t) - \Delta\bar{\mathbf{w}}(t)][\Delta\mathbf{w}(\tau) - \Delta\bar{\mathbf{w}}(\tau)]^T\} = \mathbf{Q}(t)\delta(t - \tau) \qquad (4.5\text{-}27)$$

where the *Dirac delta function* $\delta(t - \tau)$ equals ∞ when $t = \tau$ and zero elsewhere.

Given a correlated disturbance like the Dryden spectrum, the state covariance is propagated by augmenting the system model to include the dynamics that transform white noise to the input spectrum (e.g., the state-space equivalent of eq. 4.5-18). Denoting the aircraft model by the subscript "A" and the disturbance model by the subscript "D," the combined equation has upper-block-triangular structure,

$$\begin{bmatrix} \Delta\dot{\mathbf{x}}_A(t) \\ \Delta\dot{\mathbf{x}}_D(t) \end{bmatrix} = \begin{bmatrix} \mathbf{F}_A & \mathbf{L}_A \\ 0 & \mathbf{F}_D \end{bmatrix} \begin{bmatrix} \Delta\mathbf{x}_A(t) \\ \Delta\mathbf{x}_D(t) \end{bmatrix} + \begin{bmatrix} \mathbf{G}_A & 0 \\ 0 & \mathbf{L}_D \end{bmatrix} \begin{bmatrix} \Delta\mathbf{u}(t) \\ \Delta\mathbf{w}(t) \end{bmatrix} \tag{4.5-28}$$

indicating that the spectral content of the disturbance appears in the aircraft's response, while aircraft motion does not reshape the disturbance. The disturbance dynamics may be of higher order; hence, the dimension of $\Delta\mathbf{x}_D \geq s$, while the dimension of $\Delta\mathbf{w} = s$. The aircraft's disturbance effect matrix $\mathbf{L}_A$ must be expanded accordingly.

Effects of System Parameter Variations

Uncertain inputs and initial conditions have no effect on the stability and transfer characteristics of the linear system, but uncertain elements of the system matrices (**F**, **G**, **L**) do. This is a particular concern in the design of *robust* flight control systems, which must perform satisfactorily in the presence of system uncertainty.

We begin by considering linear systems with deterministic variations in parameters. The *original system* model is described by

$$\Delta\dot{\mathbf{x}}(t) = \mathbf{F}\,\Delta\mathbf{x}(t) + \mathbf{G}\,\Delta\mathbf{u}(t) + \mathbf{L}\,\Delta\mathbf{w}(t), \quad \Delta\mathbf{x}(0) = \Delta\mathbf{x}_0 \tag{4.5-29}$$

whereas the *variant system*, driven by the same control and disturbance, is

$$\begin{aligned} \Delta\dot{\mathbf{x}}_V(t) &= \mathbf{F}_V\,\Delta\mathbf{x}_V(t) + \mathbf{G}_V\,\Delta\mathbf{u}(t) + \mathbf{L}_V\,\Delta\mathbf{w}(t) \\ &= (\mathbf{F} + \Delta\mathbf{F})\,\Delta\mathbf{x}_V(t) + (\mathbf{G} + \Delta\mathbf{G})\,\Delta\mathbf{u}(t) + (\mathbf{L} + \Delta\mathbf{L})\,\Delta\mathbf{w}(t), \\ \Delta\mathbf{x}_P(0) &= \Delta\mathbf{x}_0 \end{aligned} \tag{4.5-30}$$

We define the error between the variant state and the original state as

$$\boldsymbol{\varepsilon}(t) = \Delta\mathbf{x}_V(t) - \Delta\mathbf{x}(t) \tag{4.5-31}$$

$$\dot{\boldsymbol{\varepsilon}}(t) = \Delta\dot{\mathbf{x}}_V(t) - \Delta\dot{\mathbf{x}}(t) \tag{4.5-32}$$

Subtracting eq. 4.5-29 from eq. 4.5-30, the differential equation describing the propagation of the error is

$$\begin{aligned} \dot{\boldsymbol{\varepsilon}}(t) &= \mathbf{F}_V\,\Delta\mathbf{x}_V(t) + \mathbf{G}_V\,\Delta\mathbf{u}(t) + \mathbf{L}_V\,\Delta\mathbf{w}(t) - [\mathbf{F}\,\Delta\mathbf{x}(t) + \mathbf{G}\,\Delta\mathbf{u}(t) + \mathbf{L}\,\Delta\mathbf{w}(t)] \\ &= \mathbf{F}_V\,\boldsymbol{\varepsilon}(t) + \Delta\mathbf{G}\,\Delta\mathbf{u}(t) + \Delta\mathbf{L}\,\Delta\mathbf{w}(t) + \Delta\mathbf{F}\,\Delta\mathbf{x}(t), \\ \boldsymbol{\varepsilon}(0) &= \Delta\mathbf{x}_V(0) - \Delta\mathbf{x}(0) = 0 \end{aligned} \tag{4.5-33}$$

Assuming the same starting point, there is no initial-condition effect on the error, and the inputs force the error dynamics only through the variations in the control- and disturbance-effect matrices. The dynamic modes of response are determined by $\mathbf{F}_V$, although the original modes appear in the error response through the forcing produced by $\Delta\mathbf{F}\Delta\mathbf{x}(t)$. The dynamics of the original system are, of course, unaffected by the error model.

A single equation combines the state and error models, revealing a lower-block-triangular structure:

$$\begin{bmatrix} \Delta\dot{\mathbf{x}}(t) \\ \dot{\boldsymbol{\varepsilon}}(t) \end{bmatrix} = \begin{bmatrix} \mathbf{F} & \mathbf{0} \\ \Delta\mathbf{F} & \mathbf{F}_V \end{bmatrix} \begin{bmatrix} \Delta\mathbf{x}(t) \\ \boldsymbol{\varepsilon}(t) \end{bmatrix} + \begin{bmatrix} \mathbf{G} & \mathbf{L} \\ \Delta\mathbf{G} & \Delta\mathbf{L} \end{bmatrix} \begin{bmatrix} \Delta\mathbf{u}(t) \\ \Delta\mathbf{w}(t) \end{bmatrix}, \quad \begin{bmatrix} \Delta\mathbf{x}(0) \\ \boldsymbol{\varepsilon}(0) \end{bmatrix} = \begin{bmatrix} \Delta\mathbf{x}_0 \\ \mathbf{0} \end{bmatrix} \tag{4.5-34}$$

All of the linear-system analysis tools presented in this chapter, including the calculation of eigenvalues and eigenvectors, transfer functions, discrete- time models, output spectra, and covariance equations, can be applied to this augmented system. For example, an error transfer function between Δu_i and ε_j could be formed for given values of $\Delta\mathbf{F}$, $\Delta\mathbf{G}$, and $\Delta\mathbf{L}$. For all combinations of variations $\{\pm\Delta\mathbf{F}, \pm\Delta\mathbf{G}, \pm\Delta\mathbf{L}\}$, a *worst-case error envelope* about the nominal frequency response $\Delta x_j(j\omega)/\Delta u_i(j\omega)$ could be derived. For the same case, a step response envelope could be computed as well.

System Survey

More generally, it may be desirable to analyze the effects of some number of parameter sets. The nonzero elements of $\{\Delta\mathbf{F}, \Delta\mathbf{G}, \Delta\mathbf{L}\}$, expressed as $\mathbf{p}_F$, $\mathbf{p}_G$, and $\mathbf{p}_L$, are arrayed in a single l-dimensional parameter vector:

$$\mathbf{p} \triangleq \begin{bmatrix} \mathbf{p}_F \\ \mathbf{p}_G \\ \mathbf{p}_L \end{bmatrix} \tag{4.5-35}$$

Alternatively, we could choose other scalar parameters, such as dynamic pressure, Mach number, altitude, lift slope, static margin, mass, or angle of attack, as elements of $\mathbf{p}$, and calculate the elements of $\{\Delta\mathbf{F}, \Delta\mathbf{G}, \Delta\mathbf{L}\}$ accordingly. If we are content to examine all possible combinations of the maximum and minimum values of $\mathbf{p}$'s elements, there are $N = 2^l$ parameter sets $\mathbf{p}_n$, $n = 1, N$. Each definition of the parameter vector defines a *vertex* (or *corner*) of a *hypercube*[4] in the space of $\mathbf{p}$. This leads

4. A *hypercube* is an l-dimensional structure in l-space comprised of intersecting planes or *hyperplanes* (i.e., "flat" surfaces in spaces of dimension greater than three) that enclose a volume. In our three-dimensional world, a cube is a hypercube.

to a *worst-on-worst-case analysis* of the system: combinations of extreme values are assumed to produce the largest variation from nominal conditions.

While this compounding of effects may overestimate likely variations in performance, extreme values of a parameter set may not represent the largest dynamic variations to be expected. For example, suppose that the parameter vector includes the Mach number. If the Mach number ranges from subsonic to supersonic values, a performance metric may achieve its largest value at an interior point (e.g., in the vicinity of Mach 1) rather than at either extreme.

If we wish to sample points within the interior of the hypercube—say q additional points between each maximum and minimum—there are $N = (2 + q)^l$ sets to be investigated. With more than a few parameters or if many points per parameter are to be evaluated, an exhaustive sampling of parametric effects requires much calculation. Monte Carlo evaluation, described in the next section, can provide a satisfactory alternative.

At this point, we must choose the set of evaluations to be conducted, the attributes of concern, and the criteria for assessing results. As an example, we choose the following *evaluation set*:

- Eigenvalue calculation
- Simulation of response to a step control input
- Calculation of response to a sinusoidal control input
- Propagation of state covariance induced by turbulence

Real-valued *performance metrics* based on the computed quantities must be specified, such as

- Natural frequency, damping ratio, time constant of each mode
- Overshoot, settling time, rise time, initial slope of each important input-output pair
- Crossover frequency, phase margin, gain margin, peak amplitude of each important input-output pair
- Variance of important state elements, correlation between elements, rise time, settling time

The evaluations produce *cardinal values* and *statistics of performance metrics*, including

- Maximum/minimum/average value
- Histogram (i.e., distribution) of values

- Probability of achieving target value
- Probability of exceeding acceptable limits

The cardinal values and statistics can be associated with different *levels of acceptability* (as in flying qualities criteria) that are appropriate for normal and emergency operations.

Each evaluation is a mapping from the parameter-vector set to the performance-metric set. In most cases, the mapping is nonlinear: doubling a parameter rarely doubles the performance metric precisely, and the effects of two scalar parameters are not necessarily additive. As a consequence, numerical experiment, such as the evaluation of cardinal values and statistics for N parameter-vector sets, provides the only general alternative for a system survey.

The evaluations serve different analytical purposes. *Stability* is determined by the first evaluation (eigenvalue calculation), and it is a necessary adjunct of the remaining evaluation types. Without a sufficient degree of stability, the other evaluations may be irrelevant. Each extreme value of a performance metric can be associated with a single $\mathbf{p}_n$, identifying parameter combinations that should be favored or avoided. Average values, histograms, and probability estimates of performance metrics are associated with the entire parameter-vector set, and they have meaning only to the extent that the set is a valid representation of all possible values of the parameters.

Monte Carlo Evaluation

Exhaustive computation of the effects of N parameter sets is an option only if the parameters take a small number of distinct values and the number of scalar parameters ($= l$) is small. For example, if six parameters each take 10 values, there are a million combinations. If the parameters are not distinct but real valued, $N \rightarrow \infty$, even for a single parameter. Evaluation of the effects of all parameter values is a *computationally intractable* (or *complex*) *problem* because the number of computations grows by the power of l [P-1] Without analytical relationships between the parameter-vector set and the cardinal values and statistics, the most practical means of characterizing the system is to use *probabilistic approximation*.

Monte Carlo evaluation is a numerical method for making probabilistic approximations to performance metrics that is ideally suited to digital computation. Statistics of performance metrics are estimated from a sequence of evaluation *trials*, under the assumption that $\mathbf{p}$ is a random variable. Each element of $\mathbf{p}$ is modeled as a scalar random variable with an assumed or known probability distribution. For each trial, a sample p is formed from the outputs of l random number generators, and the evaluation (e.g., eigenvalue calculation or step response simulation) is conducted. Averages, histograms, and probabilities of discrete results are calculated from the ensemble of trial results.

Statistical models of uncertain parameters often are not available, but common distributions provide plausible starting points. A tacit assumption of worst-case analysis is that there is a *uniform distribution* between maximum and minimum points, with a rectangular probability density function:

$$\mathrm{pr}(x) = \begin{cases} 0, & x < x_{min} \\ \dfrac{1}{x_{max} - x_{min}}, & x_{min} \leq x \leq x_{max} \\ 0, & x > x_{max} \end{cases} \tag{4.5-36}$$

The *Gaussian distribution* (eq. 4.5-7) is motivated by a wide range of physical phenomena and by the simplicity of its expression (using just first and second moments). By the *central limit theorem* [P-2], it is the outcome when several random processes of arbitrary distribution are summed to produce a single random variable; hence, it occurs often. A *Gaussian-but-bounded distribution* is formed when an otherwise Gaussian distribution is limited, as by quality control in a manufacturing process. For example, if all parameter values outside $\pm 2\sigma$ are excluded, the probability density function is

$$\mathrm{pr}(x) = \begin{cases} \dfrac{1.048}{\sqrt{2\pi}\,|\sigma|} e^{-[(x-\bar{x})/2\sigma]^2}, & |x - \bar{x}| \leq 2\sigma \\ 0, & |x - \bar{x}| > 2\sigma \end{cases} \tag{4.5-37}$$

Uniform and Gaussian random number generators are included in MATLAB as `rand` and `randn`. Figure 4.5-5 compares a 2-D Gaussian distribution with $\sigma = 1$, a 2-D uniform distribution in (± 1, ± 1), and (± 1, ± 1) worst-case boundaries, based on 5,000 trials for each distribution. The second distribution is contained within the box, while the first is not.

One particularly powerful application of Monte Carlo evaluation is in determining the probability of a *binary outcome*, such as whether or not a system is stable or a time response is inside a satisfactory envelope. Without regard to the quantitative details, we ask, "What is the probability of instability?" or "What is the probability of unsatisfactory performance?" In both cases, the Monte Carlo evaluation is called a *Bernoulli*

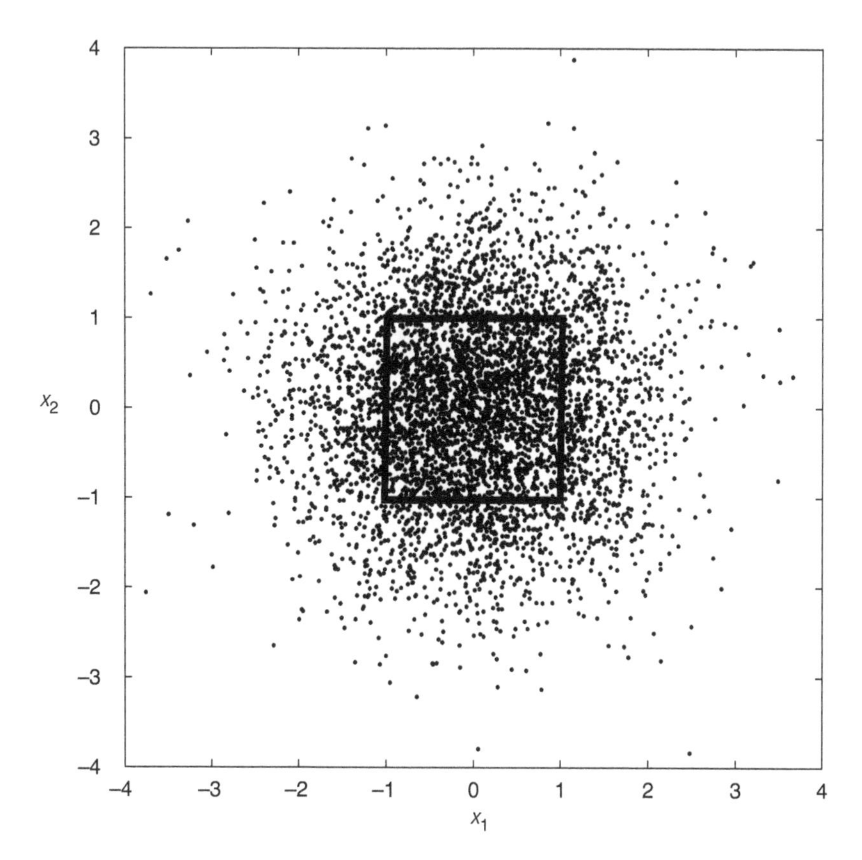

Fig. 4.5-5a. Comparison of worst-case bounds and random sampling; Gaussian distribution; 5,000 trials.

trial [P-2], and the probability estimate that results follows a *binomial distribution*. The significance is that the error in the estimated probability is governed by a well-defined *confidence interval* [R-2] that depends on just two things: the actual probability and the number of trial evaluations. Thus it is possible to say, "If the actual probability is x, then the estimate of the probability, given N trials, is within y percent of the actual value, with a confidence level of z percent." A graph of this relationship is shown in Fig. 4.5-6. When the actual probability is very near zero or 1, a large number of trials is needed to assure an accurate probability estimate. For intermediate values or for greater allowable estimation error, the number of necessary trials is less, with the minimum occurring when the probability is 0.5.

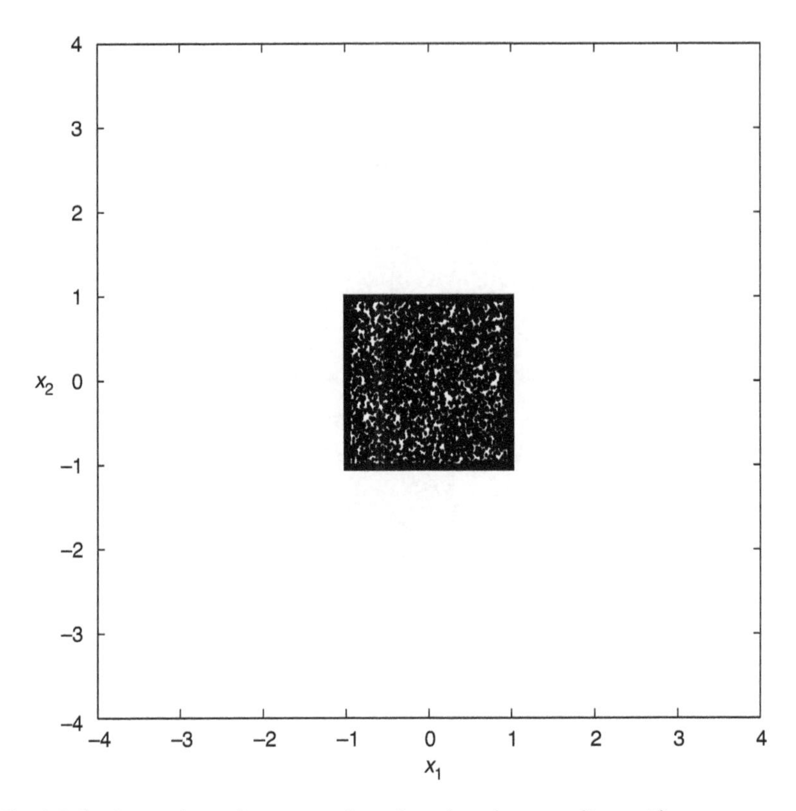

Fig. 4.5-5b. Comparison of worst-case bounds and random sampling; uniform distribution; 5,000 trials.

One important corollary is this: *The confidence in the performance-metric probability estimate is independent of the number of uncertain parameters or their individual probability distributions* [R-2]. The error in the estimate derives from the form of the performance-metric probability, not from the form of the parameter probabilities. Useful results do not require the assumption of Gaussian distributions. A mix of parametric probability distributions (e.g., some Gaussian and some uniform) can be used without affecting the nature of the performance-metric estimate error.

Furthermore, the solution complexity does not grow with the lth power. As the number of parameters increases in a given problem, the only added computation is through function calls to the random number generator.

Stochastic Root Locus

Plots derived from Monte Carlo evaluation reveal relationships between parameter variations and evaluation criteria that are useful in system

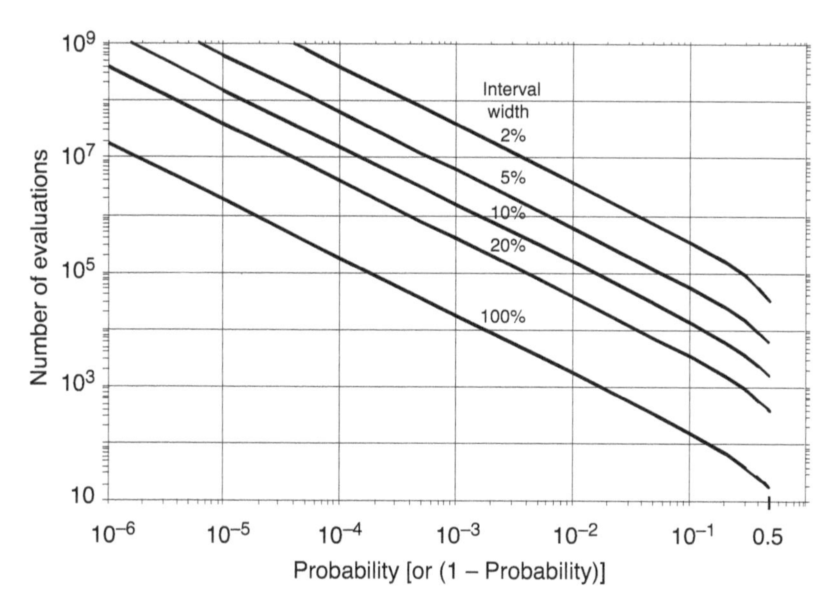

Fig. 4.5-6. Relationship between underlying probability (Pr), desired confidence level of the probability estimate, and number of required trials [R-2]. Horizontal axis represents probabilities between 0 and 0.5 or (1 − probability) if the probability is between 0.5 and 1.

design and analysis. The stochastic root locus for linear, time-invariant systems provides a good example of the insights that can be gained from the numerical results.

Consider a second-order system whose characteristic polynomial is

$$\Delta(s) = s^2 + 2\zeta\omega_n s + \omega_n^2 \tag{4.5-38}$$

and whose eigenvalues are

$$\lambda_{1,2} = -\zeta\,\omega_n \pm \sqrt{(\zeta\,\omega_n)^2 - \omega_n^2} \tag{4.5-39}$$

The roots of the equation are readily calculated and plotted for arbitrary values of ζ and ω_n. Because $\Delta(s)$ is linear in ζ, Evans's rules of construction could generate the ζ root locus. The ω_n root locus requires direct evaluation, as both ω_n and its square appear in eq. 4.5-38. With fixed ω_n, the ζ root locus is a circle about the origin with radius ω_n for $|\zeta| \le 1$, and a straight line along the real axis for $|\zeta| > 1$. For fixed $|\zeta| \le 1$, the ω_n root locus is a pair of straight lines running through the origin at angles of $\pm\cos^{-1}|\zeta|$ from the real axis. For $|\zeta| > 1$, the ω_n root locus is confined to the real axis.

If either ζ or ω_n is a random variable, the shape of the root locus is unchanged, but the probabilities of achieving eigenvalues along the locus are determined by the probability distribution of the uncertain parameter (Fig. 4.5-7). The *stochastic root locus* for single-parameter variation is thus defined by the probability density functions of the n eigenvalues, $\mathrm{pr}(\lambda_i)$, $i = 1, n$ (where $n = 2$ in the example), whose projection on the s plane follows the deterministic root locus. It is clear that there is a finite probability of instability if either parameter is Gaussian, as the tail of the distribution eventually drives one or both eigenvalues into the right half plane. With a bounded parameter distribution, root variation is limited and may not extend across the $j\omega$ axis.

The corresponding cumulative probabilities could, in principle, be evaluated by integrating eq. 4.5-6 along the deterministic root locus. However, the analysis is complicated, not only because λ_i is a complex value but because eigenvalues do not retain their identities when they coalesce and branch. Fortunately, we need not track individual eigenvalues; we merely want to estimate numerically the likelihood that eigenvalues will occupy segments of each root locus branch. Because

$$\int_{-\infty}^{\infty} \mathrm{pr}(p)\,dp = 1 \tag{4.5-40}$$

and each λ_i is a function of the scalar parameter p (either ζ or ω_n in the example)

$$\mathrm{pr}(\lambda_i) = \mathrm{pr}[\lambda_i(p)] \tag{4.5-41}$$

the sum of the eigenvalue probability density function integrals must be n:

$$\sum_{i=1}^{n} \int_{-\infty}^{\infty} \mathrm{pr}[\lambda_i(p)]\,dp = n \tag{4.5-42}$$

Regardless of their identity, there are always n eigenvalues and n branches of the root locus. Given N Monte Carlo trials, nN eigenvalues are calculated. The average probability density function within the jth segment of the ith root-locus branch is estimated as

$$\mathrm{pr}(\lambda_i)_j \approx \frac{N_{ij}}{N\Delta a}, \quad j = 1, J \tag{4.5-43}$$

where N_{ij} is the number of eigenvalues in the ijth root-locus branch segment, and J is the number of segments in each branch. The segment's *arc length* is

$$\Delta a \approx \sqrt{\Delta\sigma^2 + \Delta\omega^2} \tag{4.5-44}$$

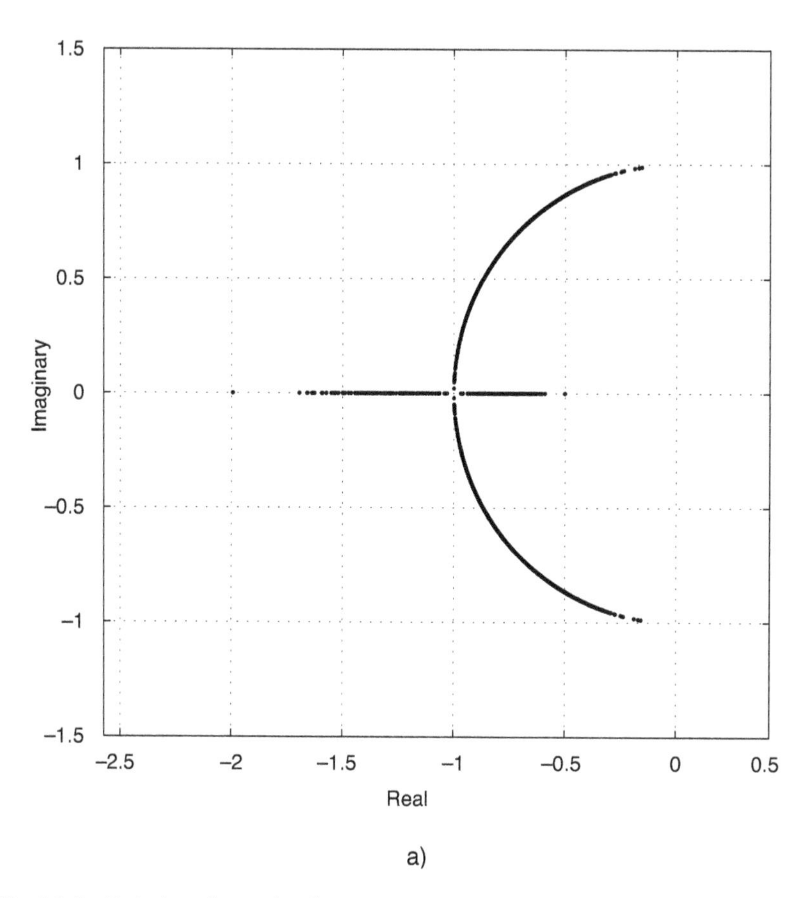

Fig. 4.5-7a. Variation of second-order roots as Gaussian parameters change; 5000 trials. Effect of random damping ratio. $\zeta_{mean} = 0.707$. $\sigma = 20$ percent of ζ_{mean}.

where $\Delta\sigma$ and $\Delta\omega$ are the real and imaginary components of the segment. A *scatter plot* of the nN eigenvalues calculated in N trials could be generated, although the density of roots would be hard to discern. As an alternative, the probability densities could be portrayed in a *"3-D" plot*, with the densities represented by height above the horizontal s plane.

If ζ and ω_n are both random variables, the locus of roots is not a line but a surface over the s plane. For two or more random parameters, eigenvalue samples form "clouds" of points that no longer fall into distinct lineal branches; the density of points is related to eigenvalue probability distributions, although roots associated with specific modes are indistinguishable. A scatter plot for Gaussian distributions in both damping ratio

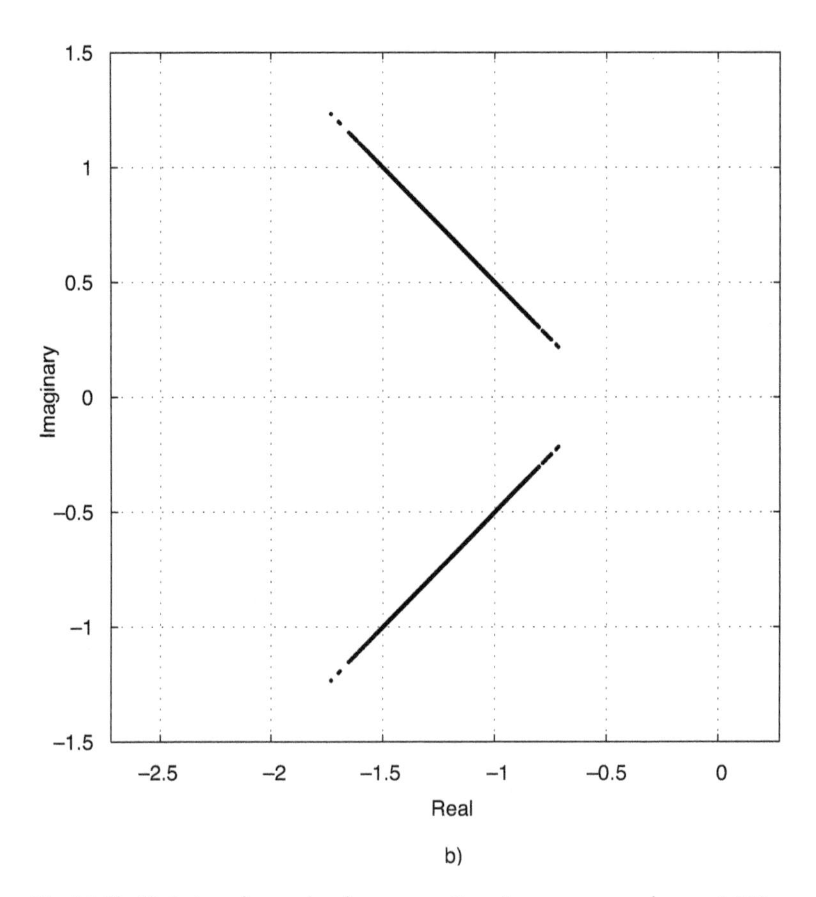

Fig. 4.5-7b. Variation of second-order roots as Gaussian parameters change; 5,000 trials. Effect of natural frequency. $\omega_{n_{mean}} = 1$. $\sigma = 20$ percent of $\omega_{n_{mean}}$.

and natural frequency is shown in Fig. 4.5-8a. The scatter plot illustrates the extent of significant root variation over the s plane, regions of clustering, and the frequencies at which instability is most likely to occur.

Dividing the s plane into $(I \times J)$ subspaces, the average eigenvalue probability density function is

$$\mathrm{pr}(\lambda)_{ij} \approx \frac{N_{ij}}{N\,\Delta\sigma\,\Delta\omega}, \quad i = 1, I,\ j = 1, J \tag{4.5-45}$$

where $\Delta\sigma\Delta\omega$ represents the *area* of the *ij*th subspace. A contour of the upper-half-plane root density identifies the general shape of the cluster (Fig. 4.5-8b). Several roots lie on the real axis; the oblique-view, three-dimensional plot of the root density reveals these roots (Fig. 4.5-8c).

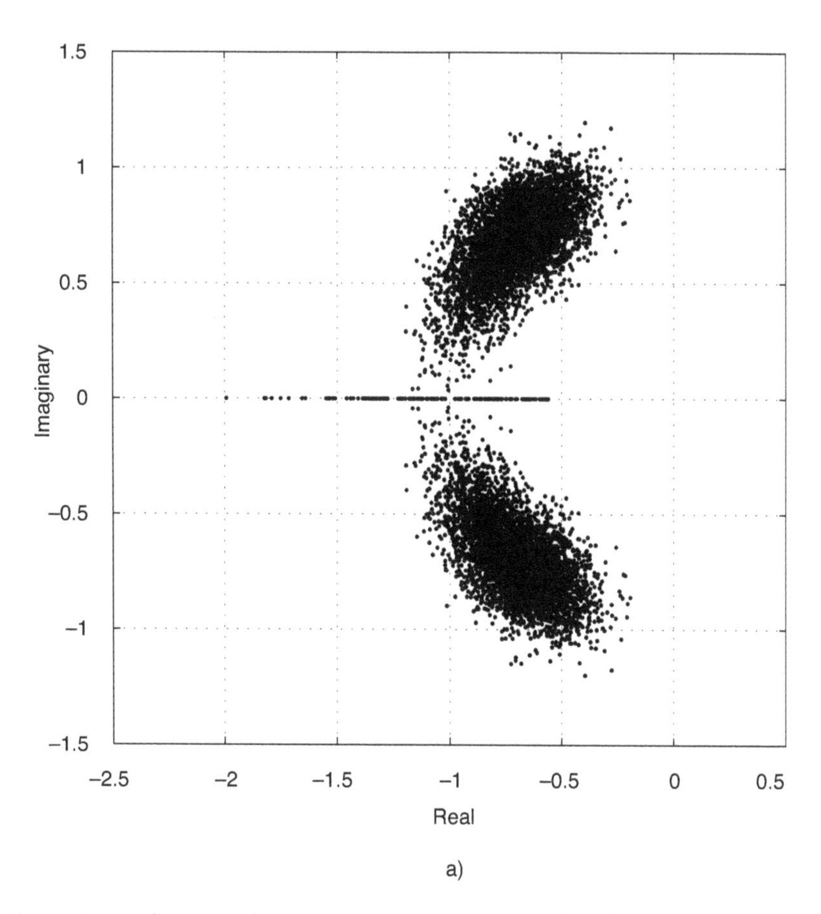

Fig. 4.5-8a. Stochastic root locus plot for combined Gaussian damping ratio and natural frequency variations. 5000 trials. $\sigma_\zeta = 0.2$ (0.707). $\sigma_\omega = 0.1$. Scatter plot of eigenvalues.

The stochastic root locus is recalculated and plotted for uniform distributions of damping ratio and natural frequency (Fig. 4.5-9). The roots fill circular arc segments bounded by ± 20 percent variations in ζ and ± 10 percent variations in ω_n. The contours and oblique view show that the eigenvalue distribution is not uniform even though the parameter uncertainties are. Furthermore, the centroids of the clusters need not correspond to the mean parameter values.

Tabulations made during Monte Carlo evaluation can help identify the sources of instability or poor performance. For example, the parameter values associated with instability could be recorded, then plotted as histograms to indicate the involvement of specific parameters in the

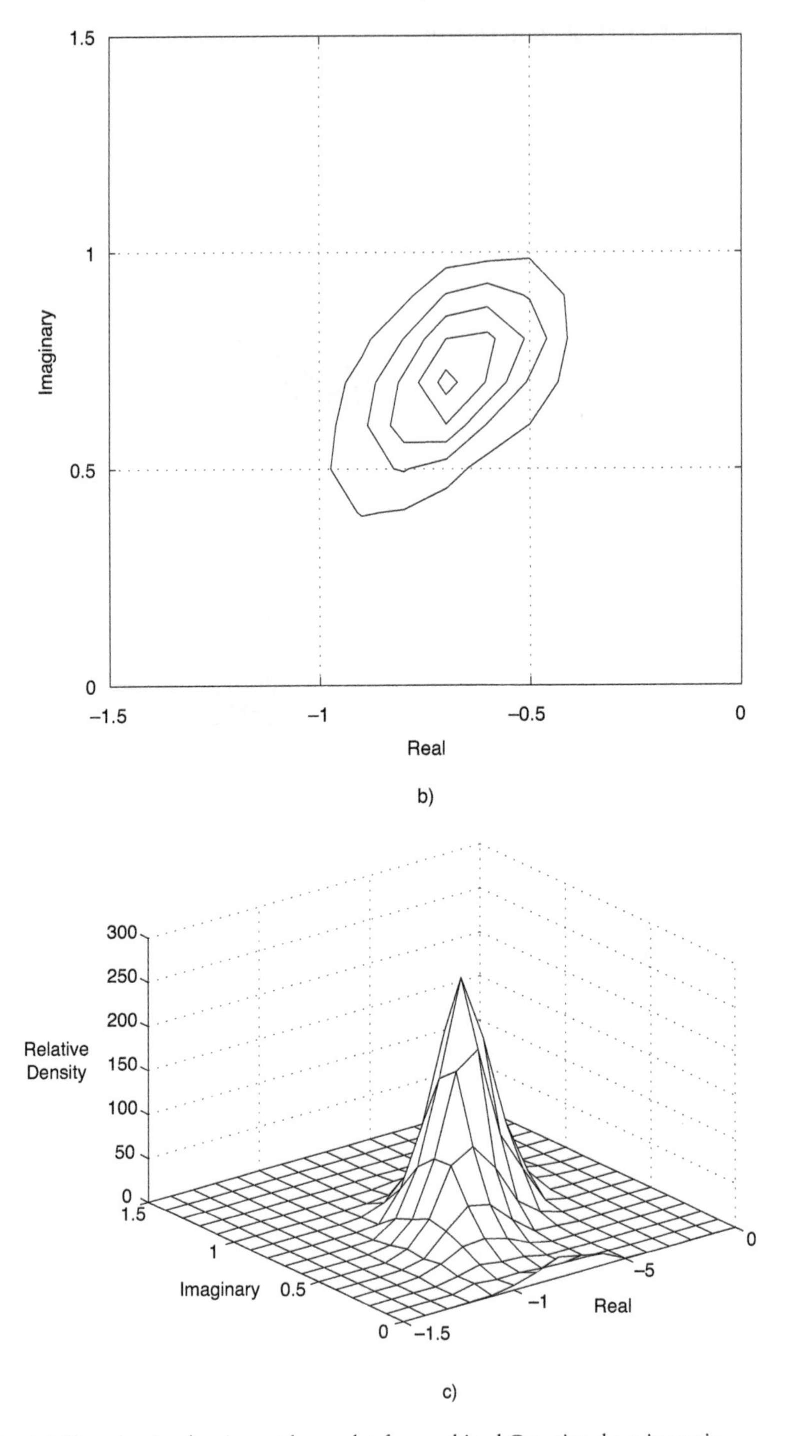

Fig. 4.5-8b and c. Stochastic root locus plot for combined Gaussian damping ratio and natural frequency variations. 5000 trials. $\sigma_\zeta = 0.2$ (0.707). $\sigma_\omega = 0.1$. (b) Contour plot of eigenvalue distribution; upper half plane. (c) "Three-dimensional" plot of eigenvalue distribution; upper half plane.

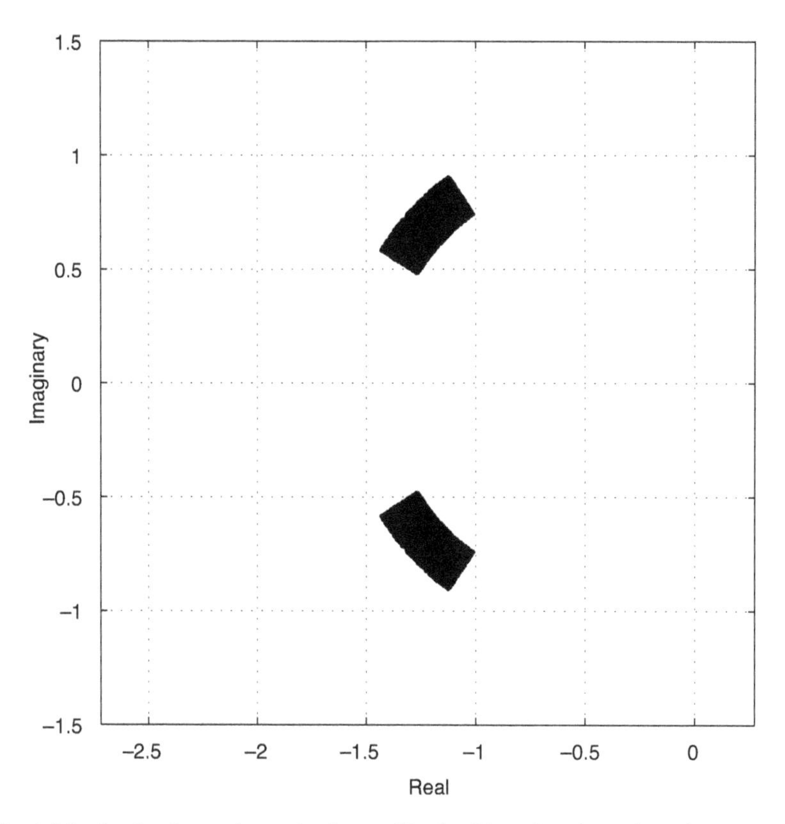

Fig. 4.5-9a. Stochastic root locus plot for combined uniform damping ratio and natural frequency variations. 5000 trials. $\Delta\zeta = 0.2$ (0.707). $\omega = 0.1$. Scatter plot of eigenvalues.

problem. *Crossplots* of one parameter against another uncover correlated effects that lead to instability. Thus, while the amount of computation required for Monte Carlo evaluation is large, the results are valuable for design and analysis.

4.6 Linear Aeroelasticity

The aircraft is comprised of slender bodies, plates, and beams that may be lightly damped and that deform elastically under aerodynamic load. Under normal operating conditions, aerodynamic and inertial properties vary as the shape changes and the structure vibrates about its equilibrium position. This combination of aerodynamic loading, inertial properties, and elasticity of the airframe—*aeroelasticity*—has been a central characteristic and design problem of aircraft since the beginning of aviation [F-3, B-6, A-2].

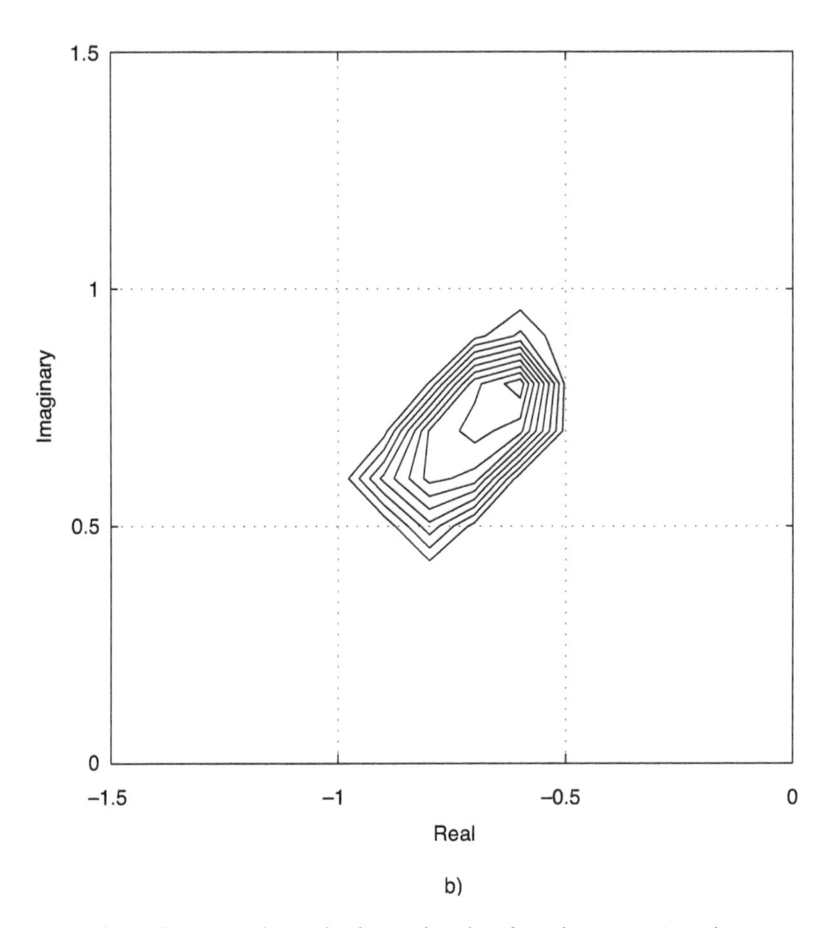

Fig. 4.5-9b. Stochastic root locus plot for combined uniform damping ratio and natural frequency variations. 5000 trials. $\Delta\zeta = 0.2$ (0.707). $\omega = 0.1$. Contour plot of eigenvalue distribution; upper half plane.

If structural stiffness is high and vibrational modes are well damped, aeroelasticity can be neglected when analyzing flight performance, stability, and control. Quasisteady deflections are small, and structural natural frequencies are well above those of the phugoid and short-period modes. However, modern aircraft are quite flexible, and their vibrational frequencies may be low enough to couple with the otherwise rigid-body modes of motion.

The amplitude and frequency of low-order structural motion is readily observed in the deflection of a jet transport's wing and wing-mounted engines from takeoff through landing. During steady flight, the vertical position of the wing tip may be a meter or two higher than it is when the aircraft is parked at the gate, and relative motions are apparent when

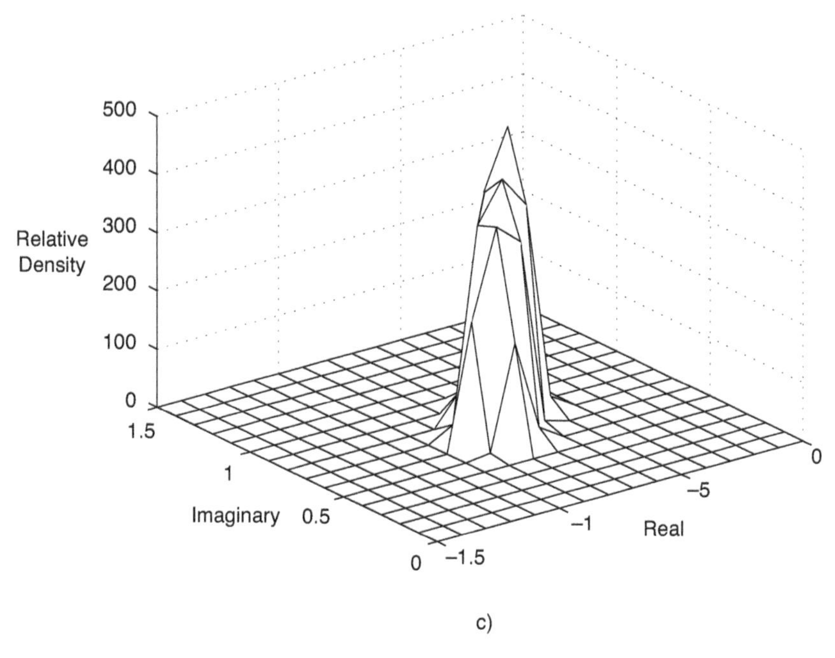

Fig. 4.5-9c. Stochastic root locus plot for combined uniform damping ratio and natural frequency variations. 5000 trials. $\Delta\zeta = 0.2$ (0.707). $\omega = 0.1$. "Three-dimensional" plot of eigenvalue distribution; upper half plane.

flying through turbulence. While higher-order structural motion is not so easily observed, torsion and bending of the wing, tail, and fuselage may have significant effects on the stability and control properties of the aircraft. Structural deflection and the resulting change in local angle of attack may alter the damping of rigid-body modes and reduce the effects of control surfaces. Conversely, rigid-body motions may force the structural modes, closing the dynamic loop to produce aeroelastic coupling.

This section addresses aeroelasticity from the perspective of its effect on flight motions. We focus on the static and dynamic interactions of structural and rigid-body motion, as distinct from the stresses within the structure, plasticity associated with strain beyond the elastic limit, details of distributed aerodynamics, and structural design. We briefly review structural aspects of the aircraft and its materials, introduce descriptions of static and dynamic structural deflections under load, and discuss distributed air loads, all of which may influence control system design and aircraft flying qualities. We motivate the form of the dynamic equations by considering bodies of simple shape, recognizing that actual aircraft structures are considerably more complex. Important details of aeroelastic theory and computation can be found in [B-6, B-7, S-1, F-5, D-6, D-7], as well as other publications listed at the end of the chapter.

Stress, Strain, and Material Properties

At very small scale, we can imagine a structural object as an assemblage of atoms (i.e., as crystals of atoms for most structural materials) fastened together by springs. When opposing forces pull on the assemblage, the atoms stretch apart: a *stress* produces a *strain.* If the forces are relaxed, the atoms spring back into shape, and the deformation is said to be *elastic.* Pull a little harder and the atoms shift with respect to each other, producing an *inelastic* or *plastic* deformation that does not return to the original form when the force is released. Pull even harder, and the atoms break apart in a *fracture.* Stress, strain, and fracture relationships vary from one material to another, and they depend upon the way in which stress is applied.

Pulling produces *tension*; pushing produces *compression.* The force per unit area in either case can be expressed as a *pressure* or *stress* $\sigma(\mathrm{N/m^2})$. We can imagine a tensile test specimen (Fig. 4.6-1) with thick ends for support, a test section of length l (m) and uniform cross-sectional area a (m^2) across the test section. Applying the force f (N) produces an average stress of f/a on the test section and stretches the specimen, producing the elongation δl (m). The *longitudinal strain* is defined to be the elongation per unit length, ε, which equals $\delta l/l$ (m/m). Pushing rather than pulling produces compression and contraction (negative elongation). Given constant volume and density, an elastic length increase produced by tension is accompanied by a cross-sectional area decrease, or *lateral strain.* Conversely, compression produces a thickened cross section. *Poisson's ratio* v describes the ratio of lateral to longitudinal strain for the cylindrical test specimen shown in Fig. 4.6-1.

Tensile tests of a structural material typically reveal a linear relationship between stress and strain at low loading, $\sigma = E\varepsilon$; the slope of that relationship, E, is called the *modulus of elasticity*, or *Young's modulus*, with units of $\mathrm{N/m^2}$. With increasing loading, the *proportional limit* is the point at which the stress-strain relationship is no longer linear. The relationship is elastic below that point, and it may be elastic (but nonlinear) with increased loading; hence, the *elastic limit* may be greater than the proportional limit. Aerospace vehicles are designed to keep normal stresses within the elastic limit to maintain proper shape of the vehicle and to provide safety margins against structural failure. At higher loading, the relationship becomes plastic, and the specimen has a permanent set when loading is released. We assume that structural deflections remain below the proportional limit in this section.

Pressure stresses (or normal stresses) like tension and compression act perpendicular to a surface, while shear stresses τ_{xy} act along (i.e., parallel

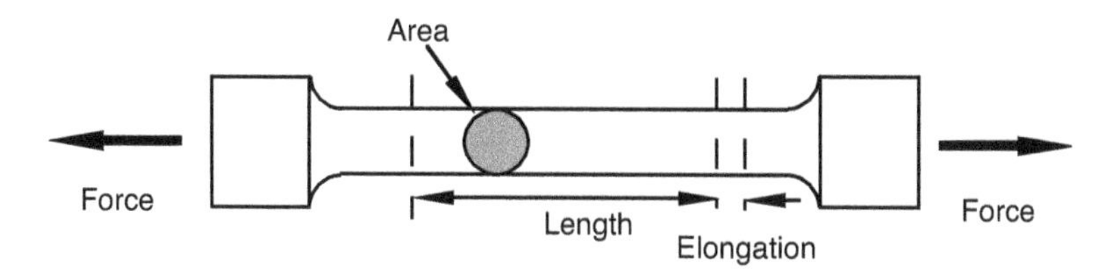

Fig. 4.6-1. Test specimen for determining stress and strain properties of a material.

to) the surface. Shear stress would be applied through friction or across a bonded or welded contact to another element. To achieve a pure force with no moment, the shear stresses on a square element must be oriented as in Fig. 4.6-2a. Shear stresses produce shear strains that deform a square into a parallelogram and that are represented by the angles $\gamma_{xy}/2$. In the linear region, the shear stress is proportional to the strain through the *modulus of rigidity* G, which is related to Young's modulus by

$$G = \frac{E}{2(1+\nu)} \tag{4.6-1}$$

If the shear stresses are equal in magnitude but oppose each other, as shown in Fig. 4.6-2b, the forces cancel, a pure moment results, and there is no deformation of the element's shape. Take away either the horizontal or vertical pair, and the remaining shear stress pair deforms the element and creates a net moment. A moment either is opposed by adjacent elements, or it produces a rotational acceleration. More generally, a combination of normal and shear stresses acts on every structural element. An object like a beam or thin wall is made up of a distribution of differential elements. The net forces, moments, deflections, and accelerations are found by integrating over the entire object.

Monocoque and Semimonocoque Structures

In modern aircraft, the thin walls of the fuselage, wing, and tail are load-bearing structures. A *monocoque structure* is supported by its skin (or shell) alone. Most aircraft have a *semimonocoque structure*, which augments the load-carrying abilities of the shell with stiffening members.

Thin shells readily carry tensile loads, but they buckle under compressive loads and require added rigidity. It is necessary to introduce lightweight bracing without eliminating the internal spaces required for passengers and crew, tankage, and subsystems, to employ wall geometry (especially curvature) that is inherently stiff, and to design internal structure into the walls themselves to increase stiffness.

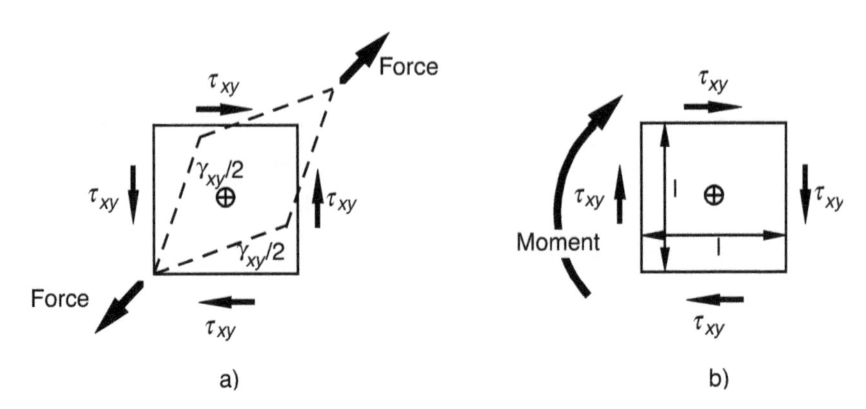

Fig. 4.6-2. Shear stress and strain on a square element. (a) Force produced by shear stress. (b) Moment produced by shear stress.

Stiffening beams are rarely solid bars; their cross-sectional geometries are "relieved" to provide high stiffness with relatively low weight. Examples include compound (or prismatic) shapes, like I-beams, rectangular tubes ("box beams"), or U- ("hat"), J-, or S-shaped stringers (Fig. 4.6-3). The *caps* of the I-beam carry virtually all of the bending moments, while the *web* carries the shear stress; the same principle is followed for the other cross sections.

The geometric stiffening property of a section is quantified by its *area moment of inertia*, *I*. For bending about an *x* axis, the area moment of inertia is computed as

$$I_x = \int_{z_{min}}^{z_{max}} \int_{x_{min}}^{x_{max}} x(z) z^2 \, dx \, dz \tag{4.6-2}$$

where x runs from left to right in the figure, and z is the vertical axis. $x(z)$ defines the beam's cross-sectional contour; as Fig. 4.6-3 shows, $x(z)$ may be multivalued, as in the case of a "hat" cross section or, more generally, a semimonocoque structure. For a rectangular cross section of width w and height h, $I_x = wh^3/12$, while the moment of inertia for a solid circular cross section of radius r is $\pi r^4/4$. A circular tube with inner and outer radii of r_o and r_i has a moment of inertia of $\pi(r_o^4 - r_i^4)/4$. The area moment of inertia about the z axis, I_z, is defined similarly.

Corrugated metal, honeycomb, and plastic foam cores can add stiffness to a structure with small increase in weight. For the latter two, thin skins (or face sheets) "sandwich" the lightweight core, serving as caps to take bending loads, while the core itself acts as the web. The result is a

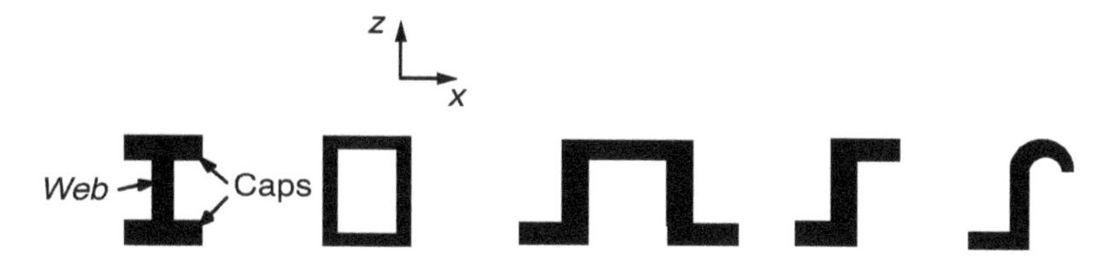

Fig. 4.6-3. Cross sections of stiffening structural elements.

composite wall that has the structural characteristics of a thick wall but the weight of a thin wall.

When we refer to composite materials, we usually mean materials that consist of long, load-bearing fibers embedded in a stabilizing, bonding matrix. Here, a fabric or tape of fiberglass, graphite, boron, or Kevlar fiber is held in place by thermosetting or thermoplastic resin. Metal-matrix composites, in which fibers of boron, silicon carbide, or Kevlar are held by a matrix of aluminum or titanium, have potential application for high-temperature applications. The structural characteristics of composites can be tailored to the application by laying up multiple plies of fabric or tape in different directions. As a consequence, the stress and strain properties may be *anisotropic*: the modulus of elasticity and ultimate strength may vary with the orientation of the stress.

Additional components of the aircraft structure affect its strength, inertial properties, and stiffness. *Bulkheads* are internal walls, open rings, or frames that transfer concentrated loads and maintain the cross-sectional form of the fuselage. *Ribs* are chordwise bulkheads in the wings. Stiffening beams attached to the skin that are approximately perpendicular to bulkheads and ribs are called *stringers*. *Spars* are large beams that provide attachment points for bulkheads and ribs, carrying major loads that are not readily distributed into or from the shell. Consequently, the actual structures are substantially more complicated than the simple models that we address here.

Force and Moments on a Simple Beam

We examine a uniform cantilever beam as an idealization of a straight (i.e., unswept) aircraft half wing. A force applied perpendicular to the free end of a rigid cantilever beam produces shear and bending stress distributions in the beam that must be reacted at the fixed-end support. The *shear force* at any cross section (or station) along the beam is constant (Fig. 4.6-4a). Because the *bending moment* is the product of the force and the moment arm, it increases from zero at the free end to its maximum value at the base (Fig. 4.6-4b). The shear force produced by the base support is equal and opposite to the applied force, and the reacting moment is simply the force times the beam length. In order to exert the bending moment, the base produces local tensile and compressive stresses in the length direction of the beam. If the force is applied off the centerline of the beam, it produces a *torsional moment* (the product of the force and its moment arm about the lengthwise axis) that is constant along the beam's length and that must be reacted at the base. The x-offset force shown in Fig. 4.6-4c has a complex effect, producing shear force and bending moment, as well as torsional moment. Adding an equal but opposite force with $-x$ offset would produce a pure torsional moment (or *couple*), with no shearing or bending effects.

More generally, the applied force could be distributed across the length and width of the beam (e.g., wing weight and the differential air pressure loads between the upper and lower surfaces of an airplane wing) as a stress field resolved into the z direction, denoted by $p(x, y)$. Here, we measure x in the chordwise (width) direction into the figure, y in the spanwise (length) direction to the left and right, and z down. Neglecting forces out of the y-z plane (e.g., shear stress), the net force in the z direction, $f_z(y_s)$, and moments about x and y axes, $m_x(y_s)$ and $m_y(y_s)$, due to the pressure field outboard of any station along the wing (including the base), y_s, are integrals of the stress distribution. For an idealized winglike beam fixed at the aircraft's centerline, the force and moments at the spanwise station measured from the centerline, y_s, are

$$f_z(y_s) = \int_{y_s}^{b/2} \int_{x_f}^{x_a} p(x, y)\, dx\, dy = \int_{y_s}^{b/2} q(y)\, dy \tag{4.6-3}$$

$q(y)$ is the normal force (or load) per unit length (i.e., df_z/dy) in the spanwise direction. The wingspan is b (so the beam length is $b/2$), and the fore and aft locations of the wing's leading and trailing edges are x_f and x_a. The corresponding bending moment about an x axis located at y_s is

$$M_x(y_s) = \int_{y_s}^{b/2} \int_{x_f}^{x_a} p(x, y)\, y\, dx\, dy = \int_{y_s}^{b/2} q(y)\, y\, dy = \int_{y_s}^{b/2} \mu(y)\, dy \tag{4.6-4}$$

The torsional moment about a y axis (which we call a pitching torque in reference to the corresponding rigid-body moment) at the station y_s is

$$M_y(y_s) = \int_{y_s}^{b/2} \int_{x_f}^{x_a} p(x, y)\, x\, dx\, dy = \int_{y_s}^{b/2} \tau(y)\, dy \tag{4.6-5}$$

$\tau(y)$ is the pitching torque (or moment) per unit length (i.e., dM_y/dy) in the spanwise direction. Whereas M_x tends to bend the wing, M_y tends to twist it.

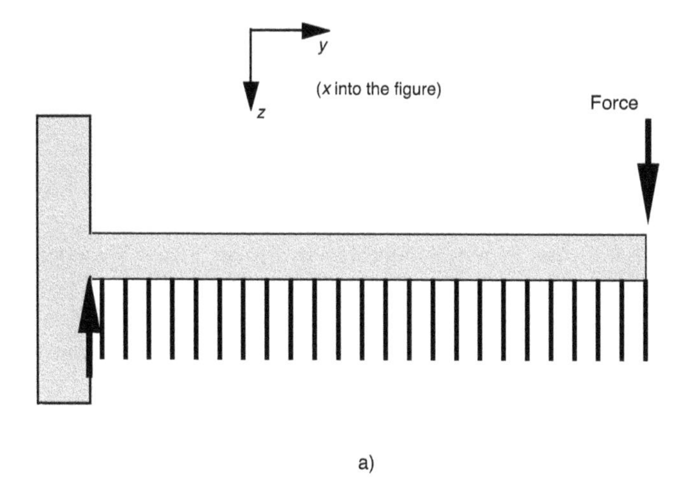

Force

b)

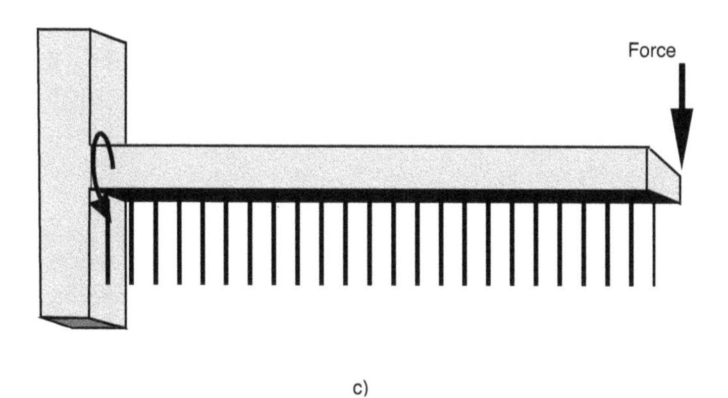

Fig. 4.6-4. Force and moment distributions due to a point force on a slender beam. (a) Shear distribution. (b) Bending moment distribution. (c) Torsion moment distribution.

Static Deflection of a Simple Beam under Load

An aircraft's shape in cruising flight is different from its shape while sitting on the landing gear, which is different from the *jig shape* that the aircraft possesses while in the factory's assembly forms or "jigs." We continue to examine a slender beam as an idealization of the fuselage or wing. The bending curvature or torsional twist at each station of a slender cantilever beam is proportional to the moment at that station, and deflection due to shear stress is negligible. The degree of deflection depends upon the material and the cross-sectional shape and area of the beam (expressed as its local area moment of inertia, I), as well as the end constraints.

We consider z-ϕ deflections of a *fixed-free beam*, whose bending angle (ϕ) and displacement (principally z, as changes in any element's y position due to flexing are negligible) are constrained at the base and are free at the tip. $z(y_s)$ represents the vertical displacement of the cross-sectional center of mass at station y_s; hence, y in $[0, b/2]$ describes the *center-of-mass axis* of the undeflected uniform beam. Separately, we examine twisting-angle deflections (θ) for the same beam. We assume that bending and twisting deflections allow small-angle assumptions to be applied.

The cross section of the beam has a significant effect on the bending and torsional strains that result from transverse loading. In upward bending of a uniform horizontal beam, the lower portion of the beam lengthens, and the upper portion contracts; there is a *neutral axis* in between that neither stretches nor shrinks. Within the spanwise beam's proportional limit, the local *radius of curvature* of the neutral axis, r, about an axis aligned with the aircraft's centerline (i.e., parallel to the x direction) can be expressed as

$$r = \frac{EI_x}{M_x} \tag{4.6-6}$$

where I_x is the *area moment of inertia in bending* of the beam about the x axis, and EI_x represents the *bending stiffness* of the cross section. Accordingly, the second derivative of the neutral axis's transverse linear deflection and the first derivative of bending angle are proportional to the bending moment:

$$\frac{d^2 z}{dy^2} = \frac{d\phi}{dy} = \frac{M_x}{EI_x} \tag{4.6-7}$$

Rearranging eq. 4.6-7, replacing M_x by eq. 4.6-4, and differentiating twice, we obtain the classical differential equation describing the effect of load distribution on the beam's bending deflection:

$$\left.\frac{d^2}{dy^2}\left[EI_x \frac{d^2 z}{dy^2}\right]\right|_{y = y_s} = q(y_s) \tag{4.6-8a}$$

For static (i.e., time-invariant) loading due to gravity and aerodynamics, eq. 4.6-8a is an *ordinary differential equation*, which can be expressed in first-order form as

$$\begin{bmatrix} z_1' \\ z_2' \\ z_3' \\ z_4' \end{bmatrix} = \begin{bmatrix} 0 & 1 & 0 & 0 \\ 0 & 0 & 1/(EI_x) & 0 \\ 0 & 0 & 0 & 1 \\ 0 & 0 & 0 & 0 \end{bmatrix} \begin{bmatrix} z_1 \\ z_2 \\ z_3 \\ z_4 \end{bmatrix} + \begin{bmatrix} 0 \\ 0 \\ 0 \\ q(y_s) \end{bmatrix} \tag{4.6-8b}$$

$z_1 = z$, $z_2 = z' = \phi$, $z_3 = z''$, $z_4 = z'''$, and $(\cdot)'$ denotes differentiation with respect to y. Equation 4.6-8 can be integrated to determine the linear z deflection, subject to boundary conditions on z and its first three derivatives:

$$z(y_s) = \int_0^{y_s} \int_0^{y_s} \left[\frac{1}{EI_x} \int_0^{y_s} \int_0^{y_s} q(y_s)\, dy^4 \right] \tag{4.6-9}$$

Equation 4.6-8 is linear in z and its derivatives, and it can be integrated from root to tip (0 to $b/2$) to determine static bending deflection. Applying this to a straight wing in trimmed, level flight, the load distribution is

$$q(y_s) \approx m'(y_s) g - C_l(y_s) \bar{q} c(y_s) \tag{4.6-10}$$

where m' is the wing mass per unit length (i.e., dm/dy), and g is the acceleration due to gravity. Invoking aerodynamic strip theory, C_l is the two-dimensional section lift coefficient of the wing's chord section, $\bar{q}$ is the dynamic pressure, and c is the chord length at station y_s.

If EI_x does not change with y, the term can be brought outside the integrals. EI_x does change if the beam's cross section depends on y or if different structural materials are used in different portions of the beam. For example, wing chord section, taper ratio, ribs, stringers, internal fuel tanks, and control surfaces are likely to vary along the span. An integrally mounted engine would change both EI_x and m' locally; fuel in a wing tank would increase m' but would have no effect on EI_x. The materials of a composite structure may have different moduli of elasticity; however, from eq. 4.6-2, an average bending rigidity EI_x, at y_s could be calculated as

$$(EI_x)_{\text{av}} = \int_{z_{min}}^{z_{max}} \int_{x_{min}}^{x_{max}} E(x,z)\, x(z)\, z^2 dx\, dz \tag{4.6-11}$$

where $E(x, z)$ describes the distribution of the modulus of elasticity across the section.

Torsional deflection of the beam is described in a similar fashion. The *area moment of inertia in torsion* (or *polar moment of inertia*) J is simply the sum of I_x and I_y for any cross section. The derivative of the beam's twist angle θ about the y axis is

$$\frac{d\theta}{dy} = \frac{M_y}{GJ} \tag{4.6-12}$$

and GJ represents the *torsional rigidity* of the cross section. Hence, from eq. 4.6-5, the twisting deflection is related to the torsional load $\tau(y_s)$ as

$$\left.\frac{d}{dy}\left[GJ\frac{d\theta}{dy}\right]\right|_{y=y_s} = \tau(y_s) \tag{4.6-13a}$$

or

$$\begin{bmatrix}\theta' \\ \theta''\end{bmatrix} = \begin{bmatrix}0 & 1/(GJ) \\ 0 & 0\end{bmatrix}\begin{bmatrix}\theta \\ \theta'\end{bmatrix} + \begin{bmatrix}0 \\ \tau(y_s)\end{bmatrix} \tag{4.6-13b}$$

Subject to boundary conditions on θ and θ', the integral of eq. 4.6-13 determines the beam twist θ at any station y_s:

$$\theta(y_s) = \int_0^{y_s}\left[\frac{1}{GJ}\int_0^{y_s}\tau(y_s)\,dy^2\right] \tag{4.6-14}$$

The torsional rigidity GJ of a wing is likely to change along the span for the same reasons that EI_x changes.

Applied to an aircraft wing in steady, level flight, the strip-theory estimate of aerodynamic torque distribution is

$$\tau(y_s) \approx C_m(y_s)\bar{q}c^2(y_s) \tag{4.6-15}$$

where C_m is the wing's 2-D sectional pitching moment coefficient at station y_s, referenced to the section's shear center. The *shear center* of any cross section is the point at which an applied force would produce no twisting torque to the section. A lengthwise line connecting local shear centers is called the *elastic axis*. For small displacements of a simple beam, twisting rotation occurs about the elastic axis, while bending can be viewed as deflection of the center-of-mass axis.

Vibrations of a Simple Beam

The nature of the beam equations changes dramatically when vibrations are considered. The inertial effect of acceleration adds derivatives of the displacement with respect to time, so z is a function of both y and t, and bending and twisting calculations involve *partial differential equations*. To motivate inclusion of elastic effects in flight dynamic calculations, we examine bending and twisting vibrational modes of a uniform beam *in vacuo*, that is, unforced by aerodynamics or gravitation and with no internal damping.

Bending Vibrations of a Uniform Beam

With constant bending stiffness EI_x, eq. 4.6-8a becomes

$$EI_x \frac{d^4 z}{dy^4}\bigg|_{y=y_S} = q(y_s) = -m' \frac{d^2 z}{dt^2}\bigg|_{y=y_S} \tag{4.6-16a}$$

or

$$EI_x z'''' = -m'\ddot{z} \tag{4.6-16b}$$

The mass variation with length, m', is also assumed to be constant. Because the system is linear, the beam displacement exhibits *synchronous motion*, with well-defined spatial mode shapes for vibrations at particular modal natural frequencies; consequently, $z(y, t)$ is the product of separable spatial and temporal functions $u(y)$ and $v(t)$:

$$z(y, t) = u(y)\, v(t) \tag{4.6-17}$$

Recognizing that $u(y)$ does not depend on time and $v(t)$ does not depend on position, the scalar dynamic equation becomes

$$EI_x u'''' v = -m' u \ddot{v} \tag{4.6-18}$$

which can be expressed as

$$\frac{EI_x u''''}{u} = -\frac{m'\ddot{v}}{v} \tag{4.6-19a}$$

The two sides of this equation are equal, but the left side does not vary in time and the right side does not vary in position; therefore, the equation holds only when both sides equal a *separation constant* k:

$$\frac{EI_x u''''}{u} = k = -\frac{m'\ddot{v}}{v} \tag{4.6-19b}$$

The equation can be satisfied with more than one separation constant; in fact, there are an infinity of such constants k_i, $i = 1, \infty$, and one goal of analysis is to find satisfactory values of k_i. Thus, each side of the equation can be equated separately to k_i, although the actual value of each k_i is determined by the spatial equation, as described below. The two new equations are linear, ordinary differential equations:

$$u_i'''' - \left(\frac{k_i}{EI_x}\right) u_i = 0, \quad i = 0, \infty \tag{4.6-20}$$

$$\ddot{v}_i + \left(\frac{k_i}{m'}\right) v_i = 0, \qquad i = 0, \infty \tag{4.6-21}$$

The eigenvalues of the spatial equation (eq. 4.6-20) are

$$(\lambda_u)_{1,2,3,4} = \pm \sqrt[4]{\frac{k_i}{EI_x}}, \pm j \sqrt[4]{\frac{k_i}{EI_x}} \triangleq \pm \Omega_i, \pm j\Omega_I \tag{4.6-22}$$

while those for the temporal equation (eq. 4.6-21) are

$$(\lambda_v)_{1,2} = \pm j \sqrt{\frac{k_i}{m'}} \triangleq \pm j\omega_{n_i} \tag{4.6-23}$$

From the temporal equation, which describes an undamped, linear oscillator, we see that k_i acts as a spring constant, and $\sqrt{k_i/m'}$ is the corresponding *natural frequency* ω_{n_i}. Thus, solutions of eq. 4.6-21 take the form

$$v_i(t) = A \sin \sqrt{\frac{k_i}{m'}} t + B \cos \sqrt{\frac{k_i}{m'}} t \tag{4.6-24}$$

where A and B depend on the initial conditions $v(0)$ and $\dot{v}(0)$. Solutions to eq. 4.6-20 could be written in terms of exponential functions of the *spatial frequency* Ω_i, but it is more convenient to use the corresponding trigonometric and hyperbolic trigonometric functions:

$$\begin{aligned} u_i(y) = C \sinh \sqrt[4]{\frac{k_i}{EI_x}} y + D \cosh \sqrt[4]{\frac{k_i}{EI_x}} y \\ + F \sin \sqrt[4]{\frac{k_i}{EI_x}} y + H \cos \sqrt[4]{\frac{k_i}{EI_x}} y \end{aligned} \tag{4.6-25}$$

Thus, from eq. 4.6-17, the general solution for vibrational bending deflection is

$$z(y,t) = \sum_{i=1}^{\infty} \left[A \sin \sqrt{\frac{k_i}{m'}} t + B \cos \sqrt{\frac{k_i}{m'}} t \right] \cdot \left[C \sinh \sqrt[4]{\frac{k_i}{EI_x}} y + D \cosh \sqrt[4]{\frac{k_i}{EI_x}} y + F \sin \sqrt[4]{\frac{k_i}{EI_x}} y + H \cos \sqrt[4]{\frac{k_i}{EI_x}} y \right]$$

$$\triangleq \sum_{i=1}^{\infty} [A \sin \omega_{n_i} t + B \cos \omega_{n_i} t] \cdot [C \sinh \Omega_i y + D \cosh \Omega_i y + F \sin \Omega_i y + H \cos \Omega_i y] \quad (4.6\text{-}26)$$

and it is expressed in terms of the temporal and spatial frequencies ω_{n_i} and Ω_i.

The coefficients of eq. 4.6-25 and the family of corresponding separation constants are found by evaluating the equation at the boundary conditions of eq. 4.6-20. C, D, F, and H, which determine the mode shape, depend on the boundary conditions at the ends of the beam. For the cantilever beam of length $b/2$, the boundary conditions at the fixed-end support are $u(0) = u'(0) = 0$; at the free end, they are $u''(b/2) = u'''(b/2) = 0$. Satisfying the first two conditions leads to the finding that $C = -F$ and $D = -H$. With these substitutions, the remaining two spatial boundary conditions can be expressed as

$$\begin{bmatrix} (\sin \Omega_i b/2 + \sinh \Omega_i b/2) & (\cos \Omega_i b/2 + \cosh \Omega_i b/2) \\ (\cos \Omega_i b/2 + \cosh \Omega_i b/2) & (\sinh \Omega_i b/2 - \sin \Omega_i b/2) \end{bmatrix} \begin{bmatrix} C \\ D \end{bmatrix} = \begin{bmatrix} 0 \\ 0 \end{bmatrix} \quad (4.6\text{-}27)$$

C is a unique function of D if the determinant of the matrix is zero, leading to the condition

$$\cos \Omega_i b/2 \cosh \Omega_i b/2 + 1 = \cos \sqrt[4]{\frac{k_i}{EI_x}}\, b/2 \cosh \sqrt[4]{\frac{k_i}{EI_x}}\, b/2 + 1 = 0, \quad i = 1 \text{ to } \infty \quad (4.6\text{-}28)$$

This equation provides the means for determining all k_i. Equation 4.6-28 can be evaluated numerically, by a zero-crossing algorithm, or graphically, by plotting $\cos(\Omega_i b/2)$ and $-1/\cosh(\Omega_i b/2)$ on the same chart and noting the intercepts. Beyond the first few intersections, $1/\cosh(\Omega_i b/2)$ is approximately zero, and the *harmonic progression* of solutions is given by $\cos(\Omega_i b/2) \approx 0$. Having determined the separation constant, the corresponding shape coefficients of the mode can be found (within an arbitrary constant) from either row of eq. 4.6-27, as the two rows are linearly dependent. The mode shape (or eigenfunction) is then given by substitution in eq. 4.6-25. The first four values of $\Omega_i b/2$ are 1.875, 4.694, 7.855, and 10.996 ($i = 1, 4$). Given the bending stiffness and mass variation, the corresponding separation constants, oscillatory frequencies, and modal coefficients are readily calculated.

While the cantilever beam is representative of an aircraft's half wing, the *free-free beam* approximates the full wing in spanwise bending or the fuselage in lengthwise bending. It is unconstrained at both ends and free to translate at the mid-point. The corresponding boundary conditions for a beam of length b are $u''(0) = u''(b) = u'''(0) = u'''(b) = 0$. The displacement and slope at both ends are unspecified, but the shear stresses and moments, which are proportional to the second and third derivatives of u, must be zero. From the boundary conditions at $y = 0$, we find that $C = F$ and $D = H$, while the boundary conditions at $y = b$, expressed as a matrix equation, yield

$$\begin{bmatrix} (-\sin\Omega_i b + \sinh\Omega_i b) & (-\cos\Omega_i b + \cosh\Omega_i b) \\ (-\cos\Omega_i b + \cosh\Omega_i b) & (\sinh\Omega_i b + \sin\Omega_i b) \end{bmatrix} \begin{bmatrix} C \\ D \end{bmatrix} = \begin{bmatrix} 0 \\ 0 \end{bmatrix} \qquad (4.6\text{-}29)$$

Substituting for the coefficients in eq. 4.6-25, the mode shape is

$$u_i(y) = \left[\frac{(\cos\Omega_i b \cosh\Omega_i b)}{(\sin\Omega_i b + \sinh\Omega_i b)} (\sinh\Omega_i y + \sin\Omega_i y) + (\cosh\Omega_i y + \cos\Omega_i y) \right] D \qquad (4.6\text{-}30)$$

where D is an arbitrary constant. The determinant of the matrix in eq. 4.6-29 is

$$\cos\Omega_i b \cosh\Omega_i b - 1 = \cos\sqrt[4]{\frac{k_i}{EI_x}}\, b \cosh\sqrt[4]{\frac{k_i}{EI_x}}\, b - 1 = 0, \quad i = 1, \infty \qquad (4.6\text{-}31)$$

The first five roots of eq. 4.6-31 occur when $\Omega_i b$ equals 0, 4.73, 7.853, 10.996, and 14.137, $i = 0, 4$. The corresponding mode shapes ($i = 1, 4$) are plotted in Fig. 4.6-5 for $b = D = 1$. The shapes alternate between even and odd functions of y as the mode index increases. Following the

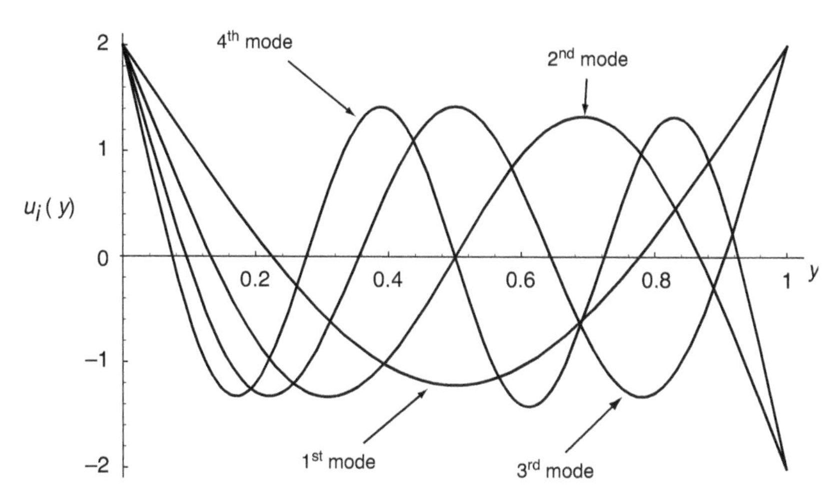

Fig. 4.6-5. First four bending mode shapes of a uniform free-free beam. $u_i(y)$ versus y in $[0, b\ (= 1)]$, $i = 1, 4$.

cantilever beam example, the solutions to eq. 4.6-31 are given by the points at which $\cos\Omega_i b = 1/\cosh\Omega_i b$, which is approximately zero for $i > 3$. This occurs when $\Omega_i b \approx (2i - 1)\ \pi/2$, corresponding to oscillation frequencies of $\omega_{n_i} \approx [(2i-1)\pi/2]^2 \sqrt{EI_x/m}$.

If the beam represents a high-aspect-ratio wing, the even deflections could affect longitudinal aerodynamics of the aircraft by changing lift and induced drag symmetrically, while the odd deflections could affect lateral-directional aerodynamics through asymmetric lift and drag. Because only z and not θ is affected by bending, the local angle of attack does not vary, and aerodynamic effects would be due to heaving (or plunging) velocity $\dot{z}$. With varying stiffness or mass along the span, including the addition of lumped masses representing the fuselage or wing-mounted engines, the bending mode shapes and natural frequencies are quantitatively different [B-7] although qualitatively similar to those shown in Fig. 4.6-5.

Reference [M-11] notes that there are two possible mode shapes when $\Omega_i b = 0$. The free-free boundary conditions specify values of u'' and u''', leaving u and u' open. For $k_i = 0$, eq. 4.6-20 is then a pure integral whose solution takes the form $u(y) = a_0 + a_1 y$. The solution admits a rigid-body heaving motion ($a_1 = 0$), in which the transverse displacements at 0 and b are identical, and a solid-body rotation ($a_1 = -2\ a_0/b$, e.g., roll for a wing or pitch for a fuselage), in which the displacements at 0 and b are equal but of opposite sign.

If external damping $d\dot{v}$ is added, with d a constant, the oscillation of the uniform beam is damped without changing the structural mode shapes. Equation 4.6-19b becomes

$$\frac{EI_x u''''}{u} = k = -\frac{m'\ddot{v}}{v} - d\dot{v} \tag{4.6-32}$$

and the separation of spatial and temporal effects is unaffected. The oscillation is then described by

$$\ddot{v}_i + \left(\frac{d}{m'}\right)\dot{v}_i + \left(\frac{k_i}{m'}\right)v_i = 0, \quad i = 1, \infty \tag{4.6-33}$$

which is the equation for a second-order damped oscillator. Consequently, we see that the bending vibration of the uniform beam can be expressed by an *infinite number of second-order modal equations*, each of which takes the form

$$u_i(y)\left[\ddot{v}_i(t) + \left(\frac{d}{m'}\right)\dot{v}_i(t) + \left(\frac{k_i}{m'}\right)v_i(t)\right] = 0, \quad i = 1, \infty \tag{4.6-34a}$$

or

$$\ddot{z}_i(y, t) + \left(\frac{d}{m'}\right)\dot{z}_i(y, t) + \left(\frac{k_i}{m'}\right)z_i(y, t) = 0; \quad i = 1, \infty \tag{4.6-34b}$$

Torsional Vibrations of a Uniform Beam

The general scheme for investigating torsional vibration is similar to that for bending vibration, but the structural component of the dynamic equation is second order, and the sign convention for the inertial effect is reversed:

$$\left.\frac{d}{dy}\left[GJ\frac{d\theta}{dy}\right]\right|_{y=y_s} = \tau(y_s) = I'_{yy}\left.\frac{d^2\theta}{dt^2}\right|_{y=y_s} \tag{4.6-35}$$

or

$$GJ\theta'' = I'_{yy}\ddot{\theta} \tag{4.6-36}$$

I'_{yy} is the rate of change of the torsional mass moment of inertia with y. The twist angle $\theta(y, t)$ can be separated into spatial and temporal components

$$\theta(y, t) = r(y)\, s(t) \tag{4.6-37}$$

Equation 4.6-36 can then be written as

$$GJr''\, s = I'_{yy}\, r\, \ddot{s} \tag{4.6-38a}$$

Dividing both sides by rs, the equation is separated, and each side equals a separation constant $-c$ $(c > 0)$:

$$\frac{GJr''}{r} = -c = \frac{I'_{yy}\, \ddot{s}}{s} \tag{4.6-38b}$$

An infinite number of separation constants c_i satisfy the equation; the corresponding linear differential equations for $r(y)$ and $s(t)$ are

$$r_i'' + \left(\frac{c_i}{GJ}\right) r_i = 0, \quad i = 1, \infty \tag{4.6-39}$$

$$\ddot{s}_i + \left(\frac{c_i}{I'_{yy}}\right) s_i = 0, \quad i = 1, \infty \tag{4.6-40}$$

The eigenvalues of the spatial equation (eq. 4.6-39) are

$$(\lambda_r)_{1,2} = \pm j \sqrt[4]{\frac{c_i}{GJ}} \triangleq \pm j\, \Xi_i \tag{4.6-41}$$

while those for the temporal equation (eq. 4.6-40) are

$$(\lambda_s)_{1,2} = \pm j \sqrt{\frac{c_i}{I'_{yy}}} \triangleq \pm j\, \xi_{ni} \tag{4.6-42}$$

The solutions for $r(y)$ and $s(t)$ take the forms

$$r_i(y) = C \sin \sqrt{\frac{c_i}{GJ}}\, y + D \cos \sqrt{\frac{c_i}{GJ}}\, y \tag{4.6-43}$$

$$s_i(t) = A \sin \sqrt{\frac{c_i}{I'_{yy}}}\, t + B \cos \sqrt{\frac{c_i}{I'_{yy}}}\, t \tag{4.6-44}$$

where A and B are determined by the initial conditions $s(0)$ and $\dot{s}(0)$, and C and D are set by the boundary conditions on the beam. The general form for the solution of the twist angle $\theta(y, t)$ is

$$\theta(y,t) = \sum_{i=1}^{\infty}\left[A\sin\sqrt{\frac{c_i}{I'_{yy}}}t + B\cos\sqrt{\frac{c_i}{I'_{yy}}}t\right]\cdot\left[C\sin\sqrt{\frac{c_i}{GJ}}y + D\cos\sqrt{\frac{c_i}{GJ}}y\right]$$

$$= \sum_{i=1}^{\infty}[A\sin\xi_{n_i}t + B\cos\xi_{n_i}t]\cdot[C\sin\Xi_i y + D\cos\Xi_i y] \qquad (4.6\text{-}45)$$

For a *fixed-free beam* of length $b/2$, the boundary conditions are $r(0) = r'(b/2) = 0$, indicating that $D = 0$ and $\cos(\Xi_i b/2) = 0$. Therefore, omitting the zero-deflection case, the admissible mode shapes are sine waves with spatial frequency

$$\Xi_i b/2 = \sqrt{\frac{c_i}{GJ}}b/2 = \frac{(2i-1)\pi}{2}, \quad i = 1, 2, \ldots, \infty \qquad (4.6\text{-}46)$$

Rearranging this equation, the separation constants are

$$c_i = GJ\left[\frac{(2i-1)\pi}{b}\right]^2, \quad i = 1, 2, \ldots, \infty \qquad (4.6\text{-}47)$$

and the corresponding oscillation frequencies are

$$\xi_{n_i} = \sqrt{\frac{c_i}{I'_{yy}}} = \sqrt{\frac{GJ}{I'_{yy}}}\left[\frac{(2i-1)\pi}{b}\right], \quad i = 1, 2, \ldots, \infty \qquad (4.6\text{-}48)$$

For a *free-free beam* of length b, $r(0)$ and $r(b)$ are open, and $r'(0) = r'(b) = 0$, indicating that $C = 0$ and $\sin \Xi_i b = 0$. The mode shapes are cosine waves

$$r_i(y) = D_i\cos\Xi_i y = D_i\cos\sqrt{\frac{c_i}{GJ}}y \qquad (4.6\text{-}49)$$

with spatial frequency

$$\Xi_i b = \sqrt{\frac{c_i}{GJ}}b = i\pi, \quad i = 1, 2, \ldots, \infty \qquad (4.6\text{-}50)$$

Therefore, the separation constants are

$$c_i = GJ\left(\frac{i\pi}{b}\right)^2, \quad i = 1, 2, \ldots, \infty \tag{4.6-51}$$

and the oscillation frequencies are

$$\xi_{n_i} = \sqrt{\frac{c_i}{I'_{yy}}} = \sqrt{\frac{GJ}{I'_{yy}}}\left(\frac{i\pi}{b}\right), \quad i = 1, 2, \ldots, \infty \tag{4.6-52}$$

The first four torsional mode shapes (Fig. 4.6-6) illustrate the interchange between odd and even functions as the mode index increases. The twistings of wing tips for odd modes oppose each other, and the resulting angle-of-attack variations would generate aerodynamic rolling moments. Conversely, the angle-of-attack variations of even modes would produce symmetric lift changes, forcing longitudinal motion.

External torsional damping $e\dot{s}$ can be added, as in the bending case. The temporal equation becomes

$$\ddot{s}_i + \left(\frac{e}{I'_{yy}}\right)\dot{s}_i + \left(\frac{c_i}{I'_{yy}}\right)v_i = 0, \quad i = 1, \infty \tag{4.6-53}$$

and the differential equations for twist-angle modes are written as

$$r_i(y)\left[\ddot{s}_i(t) + \left(\frac{e}{I'_{yy}}\right)\dot{s}_i(t) + \left(\frac{c_i}{I'_{yy}}\right)s_i(t)\right] = 0, \quad i = 1, \infty \tag{4.6-54a}$$

or

$$\ddot{\theta}_i(y, t) + \left(\frac{e}{I'_{yy}}\right)\dot{\theta}_i(y, t) + \left(\frac{c_i}{I'_{yy}}\right)\theta_i(y, t) = 0, \quad i = 1, \infty \tag{4.6-54b}$$

Coupled Vibrations of an Elastically Restrained Rigid Airfoil

If the elastic axis and the center-of-mass axis are parallel but not aligned, there is coupling between a uniform beam's bending and twisting motion. With an offset distance of x_m, linear inertia affects torsional motion and torsional inertia affects linear motion [F-5]:

$$EI_x z'''' = -m'\ddot{z} - x_m m'\ddot{\theta} \tag{4.6-55}$$

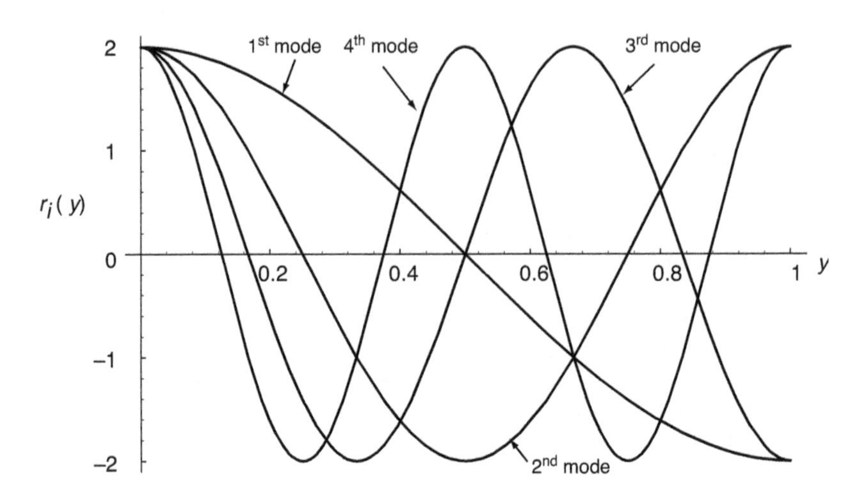

Fig. 4.6-6. First four torsional mode shapes of a uniform free-free beam. $r_i(y)$ versus y in $[0, b\ (= 1)]$, $i = 1, 4$.

$$GJ\theta'' = I'_{yy}\ddot{\theta} + x_m m'\ddot{z} \tag{4.6-56}$$

The difficulty is that the spatial and temporal components of z and θ are no longer separable, as substitution of eqs. 4.6-17 and 4.6-37 in eqs. 4.6-55 and 4.6-56 reveals. The bending and twisting mode shapes are not the same; consequently, both sides of the equations retain dependence on space and time after any manipulations. The method of successive approximations can be used to derive a coupled model for selected bending and torsional pairs [F-5], although coupling between more than two modes is neglected.

Nevertheless, a simplified model supporting bending and torsional coupling is useful. A rigid, two-dimensional airfoil that is supported by linear and torsional springs and is forced by eccentric lift can be viewed as a surrogate for the more complex wing (Fig. 4.6-7) [B-6, D-7, F-5, S-1]. It normally is assumed to represent the motions of a cantilevered half-wing at a point about 3/4 of the distance from the aircraft centerline to the wing tip, although the spanwise tilt of the wing is neglected.

Here, we build a rigid-airfoil model based on approximations for the free-free uniform beam at the mid-point (Fig. 4.6-8). The beam's symmetric bending and twisting mode shapes are approximately constant at that point; therefore, eqs. 4.6-55 and 4.6-56 are separable in the region, and the spatial effects can be represented by the separation constants,

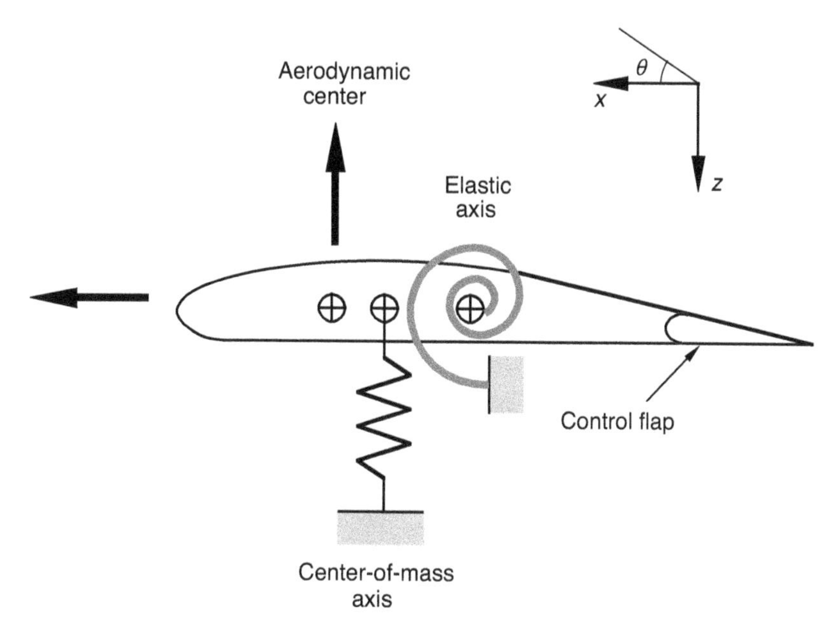

Fig. 4.6-7. Elastically restrained rigid airfoil model.

which appear as spring constants in the temporal equations. For a coupled model consisting of the first bending and second torsional modes (Fig. 4.6-8a), the unforced, coupled equations describe motions that are like those of the uniform beam's mid-point. Adding the 2-D lift force due to angle of attack and the 2-D pitching moments due to angle of attack and pitching rate, the elastically restrained rigid airfoil's dynamic motions, referenced to its equilibrium position, are solutions to the following equations:

$$k_1 z = -m'\ddot{z} - x_m m'\ddot{\theta} - [C_{l_\alpha}(\dot{z}/V + \theta) + C_{l_\delta}\delta]\overline{q}\overline{c} \tag{4.6-57}$$

$$-c_2\theta = I'_{yy}\ddot{\theta} + x_m m'\ddot{z} - [C_{m_\alpha}(\dot{z}/V + \theta) + C_{m_q}\dot{\theta} + C_{m_\delta}\delta]\overline{q}\overline{c}^2 \tag{4.6-58}$$

We view the airfoil as moving along the x axis with airspeed V, although the assemblage could be mounted in a wind tunnel with the flow impinging on the wing. The total angle of attack α is $(\dot{z}/V + \theta)$, with increments due to the vertical velocity $\dot{z}$ and the pitch angle θ. The dynamic pressure $\overline{q}$ equals $\rho V^2/2$, where ρ is the air density. The sensitivities of 2-D lift and moment coefficients C_l and C_m to α and pitch rate $q(=\dot{\theta})$ are

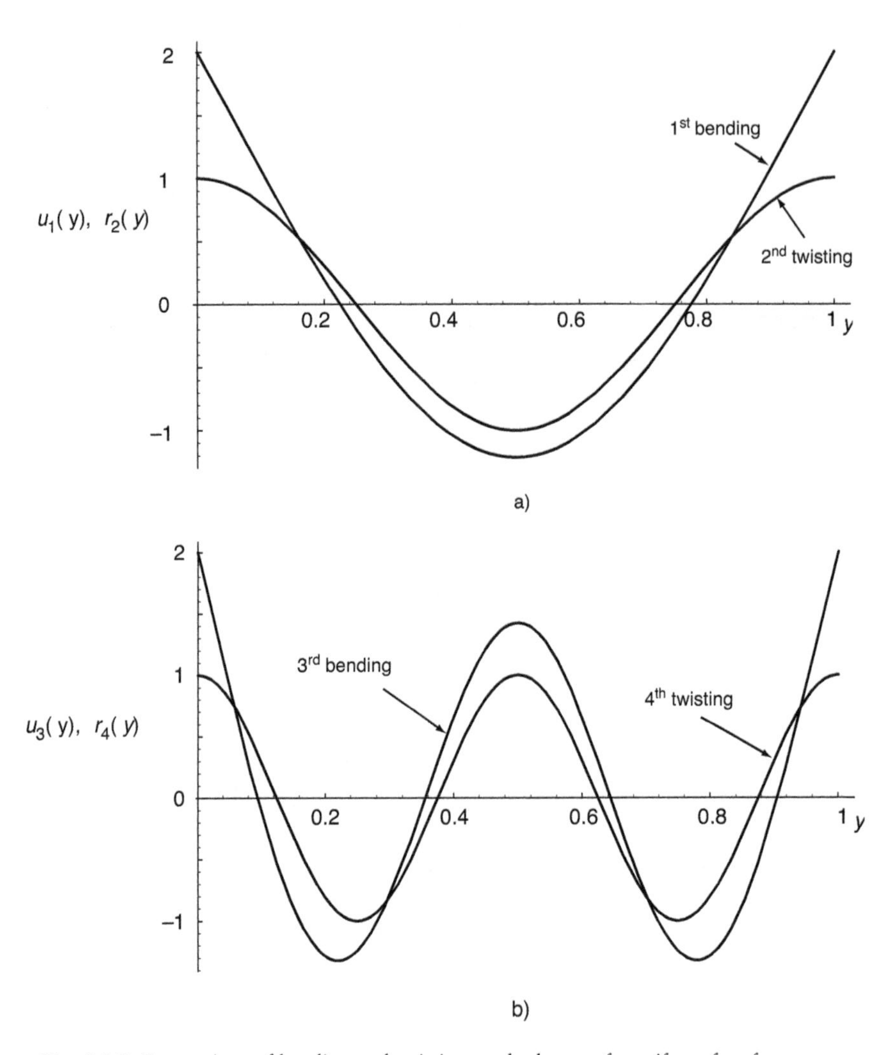

Fig. 4.6-8. Comparison of bending and twisting mode shapes of a uniform free-free beam. (a) First bending and second twisting mode shapes. $u_1(y)$ and $r_2(y)$ versus y in $[0, b\ (=1)]$. (b) Third bending and fourth twisting mode shapes. $u_3(y)$ and $r_4(y)$ versus y in $[0, b\ (=1)]$.

C_{l_α}, C_{m_α}, and C_{m_q}. They are dimensionalized by the chord length $\bar{c}$ and its square. $C_{m_\alpha} = C_{l_\alpha} x_a$, where x_a is the distance of the aerodynamic center ahead of the rotational (elastic) axis. The control flap deflection is δ (positive, trailing edge down), and it produces both lift and pitching moment, with proportionality given by $C_{l_\delta} \delta \bar{q} \bar{c}$ and $C_{m_\delta} \delta \bar{q} \bar{c}^2$. The heaving and torsional stiffness are provided by k_1 and c_2, while the inertial properties are represented by m' and I'_{yy}.

Equations 4.6-57 and 4.6-56 can be written as a matrix equation

$$\begin{bmatrix} -m' & -x_m m \\ x_m m' & I'_{yy} \end{bmatrix}\begin{bmatrix} \ddot{z} \\ \ddot{\theta} \end{bmatrix} = \begin{bmatrix} k_1 z + [C_{l_\alpha}(\dot{z}/V + \theta) + C_{l_\delta}\delta]\overline{q}\overline{c} \\ -c_2\theta + [C_{m_\alpha}(\dot{z}/V + \theta) + C_{m_q}\dot{\theta} + C_{m_\delta}\delta]\overline{q}\overline{c}^2 \end{bmatrix} \tag{4.6-59}$$

which can be inverted to find $\ddot{z}$ and $\ddot{\theta}$:

$$\begin{bmatrix} \ddot{z} \\ \ddot{\theta} \end{bmatrix} = \frac{\begin{bmatrix} I'_{yy} & x_m m' \\ -x_m m' & -m' \end{bmatrix}}{-m'I'_{yy} + (x_m m')^2}\begin{bmatrix} k_1 z + [C_{l_\alpha}(\dot{z}/V + \theta) + C_{l_\delta}\delta]\overline{q}\overline{c} \\ -c_2\theta + [C_{m_\alpha}(\dot{z}/V + \theta) + C_{m_q}\dot{\theta} + C_{m_\delta}\delta]\overline{q}\overline{c}^2 \end{bmatrix} \tag{4.6-60a}$$

This equation has the general form

$$\begin{bmatrix} \ddot{z} \\ \ddot{\theta} \end{bmatrix} \triangleq \begin{bmatrix} a & b \\ c & d \end{bmatrix}\left\{\begin{bmatrix} e & f & g & 0 \\ 0 & h & i & j \end{bmatrix}\begin{bmatrix} z \\ \dot{z} \\ \theta \\ \dot{\theta} \end{bmatrix} + \begin{bmatrix} k \\ l \end{bmatrix}\delta\right\} \tag{4.6-60b}$$

Defining the state vector, $\mathbf{x} = [z, \dot{z}, \theta, \dot{\theta}]^T$, the dynamic equation is written as $\dot{\mathbf{x}} = \mathbf{F}\mathbf{x} + \mathbf{G}\delta$, or

$$\begin{bmatrix} \dot{x}_1 \\ \dot{x}_2 \\ \dot{x}_3 \\ \dot{x}_4 \end{bmatrix} = \begin{bmatrix} 0 & 1 & 0 & 0 \\ f_{21} & f_{22} & f_{23} & f_{24} \\ 0 & 0 & 0 & 1 \\ f_{41} & f_{42} & f_{43} & f_{44} \end{bmatrix}\begin{bmatrix} x_1 \\ x_2 \\ x_3 \\ x_4 \end{bmatrix} + \begin{bmatrix} 0 \\ g_2 \\ 0 \\ g_4 \end{bmatrix}\delta \tag{4.6-61}$$

where the coefficients of $\mathbf{F}$ are determined from eq. 4.6-60:

$$f_{21} = ae \tag{4.6-62}$$

$$f_{22} = af + bh \tag{4.6-63}$$

$$f_{23} = ag + bi \tag{4.6-64}$$

$$f_{24} = bj \tag{4.6-65}$$

$$f_{41} = ce \tag{4.6-66}$$

$$f_{42} = cf + dh \tag{4.6-67}$$

$$f_{43} = cg + di \tag{4.6-68}$$

$$f_{44} = dj \tag{4.6-69}$$

$$g_2 = ak + bl \tag{4.6-70}$$

$$g_4 = ck + dl \tag{4.6-71}$$

The eigenvalues of the system are solutions to the fourth-order characteristic equation $|\lambda\mathbf{I} - \mathbf{F}| = 0$ (eq. 3.2-71):

$$\Delta(\lambda) = |\lambda\mathbf{I} - \mathbf{F}| = \lambda^4 + a_3\lambda^3 + a_2\lambda^2 + a_1\lambda + a_0$$
$$= (\lambda - \lambda_1)(\lambda - \lambda_2)(\lambda - \lambda_3)(\lambda - \lambda_4) = 0 \tag{4.6-72}$$

If $x_m = 0$, then $b = c = 0$, the off-diagonal blocks of $\mathbf{F}$ are zero, and the bending and rotating motions are uncoupled. The squares of their uncoupled natural frequencies are

$$(\omega_n^2)_{\text{bending}} = -f_{21} = -ae = k_1/m' \tag{4.6-73}$$

$$\xi_n^2 \triangleq (\omega_n^2)_{\text{torsion}} = -f_{43} = -di = (c_2 - C_{m_\alpha}\overline{q}\overline{c}^2)/I'_{yy} \tag{4.6-74}$$

The characteristic polynomial factors as

$$\Delta(\lambda) = (\lambda + 2\zeta\omega_n\lambda + \omega_n^2)_{\text{bending}}(\lambda + 2\zeta\omega_n\lambda + \omega_n^2)_{\text{torsion}} \tag{4.6-75}$$

with damping ratios ζ_{bending} and ζ_{torsion} determined by C_{l_α} and C_{m_q}.

Although the uncoupled bending natural frequency ω_n is not affected by aerodynamics, the torsional natural frequency ξ_n is. If the aerodynamic center is in the vicinity of the quarter chord and the elastic axis is near the half chord, C_{m_α} is positive, and ξ_n becomes smaller with increasing dynamic pressure. For $C_{m_\alpha}\overline{q}\overline{c}^2 > c_2$, the mode is described by two real roots; one of these roots is in the right half plane, and the torsional mode is statically unstable. The airspeed for which the two terms are equal is called the *torsional divergence speed.*

The bending and torsional modes are coupled by the off-diagonal blocks of $\mathbf{F}$ when $x_m \neq 0$. The total damping of the two modes is expressed by a_3 in the characteristic polynomial (eq. 4.6-70), where a_3 is the sum of the elements on the diagonal of $\mathbf{F}$ $(= f_{22} + f_{44})$. a_3 is approximately constant, and the effect of the offset is to redistribute damping between the two modes. Thus, with increasing x_m, both modes involve all elements of $\mathbf{x}$; one mode's damping increases, and the other's decreases. The system becomes susceptible to *flutter*, which occurs when one mode loses enough damping to become neutrally stable. At low airspeed, the inertial offset may reduce but not destroy system stability; however, as airspeed increases, aerodynamic effects become larger, leading to instability at the *flutter speed*. While torsional divergence arises from twisting effects alone, flutter is induced by coupling of the bending and twisting motions.

Torsional elasticity reduces the roll-control effects of ailerons mounted near an aircraft's wing tips. For this reason, high-speed aircraft with long, flexible wings use outboard ailerons only in low-speed flight, using inboard ailerons and/or spoilers for roll control in high-speed flight. Control-induced twisting increases with increasing flight speed, as the aerodynamic effects become larger while the structural rigidity remains fixed. At a particular speed, the aileron's direct rolling moment is counterbalanced by the twisted wing's rolling moment. This is the speed for *aileron reversal*, one of the critical problems of high-speed flight. "Reversal" refers to the rolling-moment effect of the aileron, not the sense of aileron deflection itself. This phenomenon can be understood using the elastically restrained rigid airfoil to model the twisting deflection of the outboard wing in response to aileron control. The lift produced by the airfoil times the distance of its aerodynamic center from the aircraft's centerline is the corresponding rolling moment.

The steady-state response to constant control is readily obtained from eq. 4.2-12:

$$\mathbf{x}^* = -\mathbf{F}^{-1}\mathbf{G}\,\delta^* \tag{4.6-76}$$

Defining **F** and **G** by eq. 4.6-59, the perturbed equilibrium response is

$$\begin{bmatrix} x_1{}^* \\ x_2{}^* \\ x_3{}^* \\ x_4{}^* \end{bmatrix} = \begin{bmatrix} z^* \\ \dot{z}^* \\ \theta^* \\ \dot{\theta}^* \end{bmatrix} = \frac{1}{f_{21}f_{43} - f_{23}f_{41}} \begin{bmatrix} -f_{43}g_2 + f_{23}g_4 \\ 0 \\ f_{41}g_2 - f_{21}g_4 \\ 0 \end{bmatrix} \delta^* \tag{4.6-77}$$

The perturbed rates are zero, while the perturbed linear and angular positions are proportional to the control-flap deflection δ^*.

The steady-state lift produced by control deflection includes not only the direct effect of the control surface but its effect on the wing's angle of attack, which in this case is its pitch angle:

$$\begin{aligned} \text{Lift}^* &= [C_{l_\alpha}\theta^* + C_{l_\delta}\delta^*]\overline{q}\overline{c} \\ &= \left[C_{l_\alpha}\frac{f_{41}g_2 - f_{21}g_4}{f_{21}f_{43} - f_{23}f_{41}} + C_{l_\delta}\right]\overline{q}\overline{c}\delta^* \end{aligned} \tag{4.6-78}$$

Substituting the previous definitions for the f_{ij}, the control response is

$$\text{Lift}^* = \left[C_{l_\alpha}\frac{C_{m_\delta}\overline{q}\overline{c}}{c_2 - C_{m_\alpha}\overline{q}\overline{c}^2} + C_{l_\delta}\right]\overline{q}\overline{c}\delta^* \tag{4.6-79}$$

In this equation, only C_{m_δ} is negative. Below the torsional divergence speed, the general effect of wing torsion is to reduce net lift (and above the divergence speed, where the aircraft cannot operate, the effect is unimportant). As airspeed and dynamic pressure grow, the torsional control effect increases. The onset of control reversal occurs when the term in brackets vanishes, driving the net lift to zero, that is, when

$$\frac{C_{l_\alpha}\overline{q}\overline{c}}{c_2 - C_{m_\alpha}\overline{q}\overline{c}^2} = -\frac{C_{l_\delta}}{C_{m_\delta}} \tag{4.6-80}$$

Vibrations of a Complex Structure

Simple beams and elastically restrained rigid airfoils serve to introduce the basic concepts of vibrational motion, but they are not adequate to describe the coupled motions of more elaborate shapes. From these models, we expect the aircraft structure to bend and twist with distinct symmetric and asymmetric mode shapes, each mode vibrating at a prescribed natural frequency. We have also seen that aerodynamic effects brought about by deflections couple into the vibrations, changing not only the frequency but the damping of these oscillations. Nevertheless, aircraft structures are substantially more complex. Elastic stiffness and torsional rigidity vary, engines, payload, and fuel produce mass concentrations at distinct locations, and the aircraft shape is far from uniform. Consequently, bending and twisting deflections frequently are coupled, and even the concept of separate bending and torsion may not adequately represent the multidimensional structural deformation.

Here, we establish a general framework for understanding the effects of an aircraft's aeroelastic deflections on flight stability and control. We restrict our attention to deflections out of the aircraft's reference horizontal plane, although elastic deformations from the vertical plane could be handled in a similar fashion. The structure is represented as an *elastic plate* cut to the aircraft's planform (Fig. 4.6-9), with variable mass distribution, elastic properties, and vertical forces. Although these properties vary continuously, with occasional discontinuities at cutouts or mass concentrations, we divide the plate into contiguous segments that are small in comparison to the overall planform. This allows the construction of a *lumped-parameter model* of the elastic plate that can be characterized by a finite number of second-order ordinary differential equations, each representing the vertical accelerations of a segment's center of mass [B-6]. Ultimately, we apply Newton's second law to each mass point, spacing the grid points close enough to give good approximations and

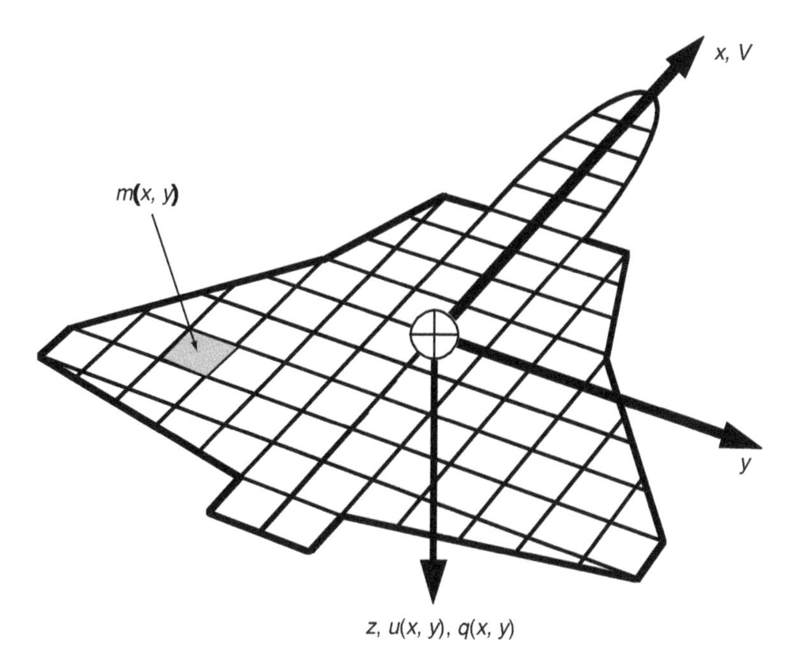

Fig. 4.6-9. Elastic plate representation of the aircraft structure.

accounting for the elastic and aerodynamic coupling that takes place between segments. The modes and mode shapes of this linear model can be evaluated using the methods presented earlier in this chapter. From these high-order models, truncation or residualization (Section 4.2) can be used to form lower-order models for analyzing the most significant aeroelastic effects on flight stability and control.

We begin by defining I segments within the airframe planform and their corresponding center-of-mass locations in the x-y plane. The vector of vertical deflections from the equilibrium position is

$$\mathbf{z}(x, y) = [z(x_1, y_1)\; z(x_2, y_2) \cdots z(x_I, y_I)]^T \tag{4.6-81}$$

Generalizing our earlier results, each deflection $z_i(x, y)$ is related linearly to forces $f_j(x, y)$ applied at all points in the center-of-mass grid. Writing the coupled elastic sensitivities (or *flexibility influence coefficients*) in the matrix $\mathbf{C}_\mathrm{f}^\mathrm{z}$, the deflection vector is

$$\mathbf{z} = \mathbf{C}_\mathrm{f}^\mathrm{z}\mathbf{f} \tag{4.6-82}$$

where f is the vector of forces applied to each of J segments ($J = I$), resolved to the center of mass locations using a curve-fitting method such as "splining" (Section 3.6):

$$\mathbf{f}(x, y) = [f(x_1, y_1)\ f(x_2, y_2) \cdots f(x_J, y_J)]^T \tag{4.6-83}$$

These forces include inertial and aerodynamic components, and eq. 4.6-82 becomes

$$\begin{aligned}\mathbf{z} &= \mathbf{C}_\mathrm{f}^\mathrm{z}(\mathbf{f}_\mathrm{inertial} + \mathbf{f}_\mathrm{aero}) \\ &= \mathbf{C}_\mathrm{f}^\mathrm{z}(-\mathbf{M}\ddot{\mathbf{z}} - \mathbf{D}\dot{\mathbf{z}} - \mathbf{K}\mathbf{z} - \mathbf{A}\mathbf{x}_\mathrm{rb} - \mathbf{B}\boldsymbol{\delta} - \mathbf{L}\mathbf{w})\end{aligned} \tag{4.6-84}$$

M is an $(I \times I)$ diagonal matrix containing the mass of each segment,

$$\mathbf{M} = \begin{bmatrix} m_1 & 0 & \dots & 0 \\ 0 & m_2 & \dots & 0 \\ \dots & \dots & \dots & \dots \\ 0 & 0 & \dots & m_I \end{bmatrix} \tag{4.6-85}$$

while **D** and **K** are $(I \times I)$ matrices that reflect distributed aerodynamic damping and spring effects. **A** is a matrix of forcing sensitivities that portray the effects of rigid-body aerodynamic state on structural deflections. **B** is the matrix of direct control effects on the structure, and $\boldsymbol{\delta}$ is a vector of control-surface deflections. An external disturbance **w** that is scaled by **L** is included in the model. The calculation of **A**, **B**, $\mathbf{C}_\mathrm{f}^\mathrm{z}$, **D**, **K**, and **L** is a principal topic of [B-6, B-7] that is beyond the present scope.

D and **K** are not necessarily diagonal; in fact, we may expect off-diagonal terms in **K** to represent changes in the local angle of attack, which, in turn, affects the pressure distribution. For example, suppose that the $(i-1)$th, ith, and $(i+1)$th centers of mass are aligned with the x axis. The differential angle of attack at point i is approximately $\Delta\alpha \approx (z_{i-1} - z_{i+1})/(x_{i+1} - x_{i-1})$, and the lift on segment i, $k\Delta\alpha$, is expressed in the ith row of **K** as

$$[\cdots k_{i(j-1)}\ k_{ij}\ k_{i(j+1)} \cdots] = \left[\cdots \frac{k}{(x_{i+1} - x_{i-1})}\ 0\ \frac{-k}{(x_{i+1} - x_{i-1})} \cdots\right] \tag{4.6-86}$$

$\mathbf{C}_\mathrm{f}^\mathrm{z}$ is inverted to form the *stiffness influence coefficient matrix* $\mathbf{C}_\mathrm{z}^\mathrm{f}$ and eq. 4.6-82 can be expressed as either

$$\mathbf{M}\ddot{\mathbf{z}} + \mathbf{D}\dot{\mathbf{z}} + (\mathbf{K} + \mathbf{C}_\mathrm{z}^\mathrm{f})\mathbf{z} - \mathbf{A}\mathbf{x}_\mathrm{rb} - \mathbf{B}\boldsymbol{\delta} - \mathbf{L}\mathbf{w} = 0 \tag{4.6-87a}$$

or

$$\mathbf{M}\ddot{\mathbf{z}} = -\mathbf{D}\dot{\mathbf{z}} - (\mathbf{K} + \mathbf{C}_\mathrm{z}^\mathrm{f})\mathbf{z} + \mathbf{A}\mathbf{x}_\mathrm{rb} + \mathbf{B}\boldsymbol{\delta} + \mathbf{L}\mathbf{w} \tag{4.6-87b}$$

Defining the state vector $\mathbf{x} = [\mathbf{z}^T \dot{\mathbf{z}}^T]^T \triangleq [\mathbf{x}_1\ \mathbf{x}_2^T]^T$, eq. 4.6-85b leads to the $2I$-dimensional state vector equation

$$\begin{bmatrix} \dot{\mathbf{x}}_1 \\ \dot{\mathbf{x}}_2 \end{bmatrix} = \begin{bmatrix} 0 & \mathbf{I} \\ -\mathbf{M}^{-1}(\mathbf{K} + \mathbf{C}_z^f) & -\mathbf{M}^{-1}\mathbf{D} \end{bmatrix} \begin{bmatrix} \mathbf{x}_1 \\ \mathbf{x}_2 \end{bmatrix} + \begin{bmatrix} 0 \\ \mathbf{M}^{-1}\mathbf{A} \end{bmatrix} \mathbf{x}_{rb} + \begin{bmatrix} 0 \\ \mathbf{M}^{-1}\mathbf{B} \end{bmatrix} \boldsymbol{\delta} + \begin{bmatrix} 0 \\ \mathbf{M}^{-1}\mathbf{L} \end{bmatrix} \mathbf{w} \tag{4.6-88}$$

which takes the standard form $\dot{\mathbf{x}} = \mathbf{F}\mathbf{x} + \mathbf{G}\mathbf{u} + \mathbf{L}\mathbf{w}$, with the control $\mathbf{u} = \boldsymbol{\delta}$ and $\dim(\boldsymbol{\delta}) = (r \times 1)$. The rigid-body input effectively acts as an additional disturbance. Alternatively, the dynamic equation could be expressed segment by segment, interchanging each $[z(x_i, y_i), \dot{z}(x_i, y_i)]$ pair, as in eq. 4.6-61.

We find the *eigenvalues* and *eigenvectors* of this aeroelasticity equation in the usual way (Sections 3.2 and 4.3), and the system then can be expressed using normal-mode coordinates. The $2I$ eigenvalues are solutions to the equation (eq. 3.2-71),

$$|\lambda\mathbf{I} - \mathbf{F}| = \Delta(\lambda) = 0 \tag{4.6-89}$$

and the I natural frequencies and damping ratios corresponding to each complex pair are calculated by eqs. 4.3-57 and 4.3-58. Ordering the eigenvalues from lowest to highest values, the associated eigenvectors (eq. 3.2-79) are used to form the frequency-ordered *modal matrix* $\mathbf{E}$ (eq. 3.2-80). The modal matrix allows the *normal-mode coordinate vector* $\mathbf{q}(t)$ to be written as a linear function of the original state (eq. 4.3-73),

$$\mathbf{q}(t) = \begin{bmatrix} \mathbf{q}_1 \\ \mathbf{q}_2 \end{bmatrix} = \mathbf{E}^{-1}\mathbf{x}(t) = \begin{bmatrix} \mathbf{E}_{11} & \mathbf{E}_{12} \\ \mathbf{E}_{21} & \mathbf{E}_{22} \end{bmatrix}^{-1} \begin{bmatrix} \mathbf{x}_1 \\ \mathbf{x}_2 \end{bmatrix} \tag{4.6-90}$$

Neglecting the rigid-body motion input, the diagonalized dynamic equation is obtained by similarity transformation of eq. 4.6-88,

$$\dot{\mathbf{q}} = \mathbf{\Lambda}\mathbf{q} + \mathbf{\Gamma}\boldsymbol{\delta} \tag{4.6-91}$$

where

$$\mathbf{\Lambda} = \mathbf{E}^{-1}\,\mathbf{F}\,\mathbf{E} \tag{4.6-92}$$

$$\mathbf{\Gamma} = \mathbf{E}^{-1}\,\mathbf{G} \tag{4.6-93}$$

$\mathbf{\Lambda}$ is a diagonal matrix of the eigenvalues of $\mathbf{F}$, λ_i (eq. 3.2-74). Assuming that all of the structural modes are oscillatory, the $2I$ eigenvalues occur in complex pairs, $\lambda_{i,i+1} = \mu_i \pm j\nu_i$, and as shown in Section 4.3, the normal-mode coordinates are complex. There are I modes, and $i = 1, 2, \ldots, 2I$; we assign the *mode number* n_m as $(i+1)/2$, with a maximum value of I.

Two transformations of eqs. 4.6-90 and 4.6-91 bring the normal-mode equations into more useful form. Given a complex (2×2) diagonal block of $\mathbf{\Lambda}$,

$$\mathbf{\Lambda}_i \triangleq \begin{bmatrix} (\mu_i + jv_i) & 0 \\ 0 & (\mu_i - jv_i) \end{bmatrix}, \quad i = 1, 3, \ldots, 2I - 1 \tag{4.6-94}$$

an equivalent real-valued block possessing the same eigenvalues results from the similarity transformation

$$\mathbf{\Phi}_i = \mathbf{V}_i \mathbf{\Lambda}_i \mathbf{V}_i^{-1}, \quad i = 1, 3, \ldots, 2I - 1 \tag{4.6-95a}$$

$$\mathbf{\Phi}_i = \begin{bmatrix} \mu_i & v_i \\ -v_i & \mu_i \end{bmatrix} = \begin{bmatrix} (1-j) & (1+j) \\ (1+j) & (1-j) \end{bmatrix} \begin{bmatrix} (\mu_i + jv_i) & 0 \\ 0 & (\mu_i - jv_i) \end{bmatrix} \times \begin{bmatrix} (1-j) & (1+j) \\ (1+j) & (1-j) \end{bmatrix}^{-1} \tag{4.6-95b}$$

The corresponding real-valued state vector is

$$\begin{bmatrix} v_i \\ v_{i+1} \end{bmatrix} = \mathbf{V}_i \begin{bmatrix} q_i \\ q_{i+1} \end{bmatrix}, \quad i = 1, 3, \ldots, 2I - 1 \tag{4.6-96}$$

The second similarity transformation

$$\mathbf{\Psi}_i = \mathbf{X}_i \mathbf{\Phi}_i \mathbf{X}_i^{-1}, \quad i = 1, 3, \ldots, 2I - 1 \tag{4.6-97a}$$

takes $\mathbf{\Phi}_i$ to the canonical form for a second-order oscillator,

$$\mathbf{\Psi}_i = \begin{bmatrix} 0 & 1 \\ -\omega_{n_i}^2 & -2\zeta_i \omega_{n_i} \end{bmatrix} = \begin{bmatrix} 1 & 1 \\ (\mu_i - v_i) & (\mu_i + v_i) \end{bmatrix} \begin{bmatrix} \mu_i & v_i \\ -v_i & \mu_i \end{bmatrix} \begin{bmatrix} 1 & 1 \\ (\mu_i - v_i) & (\mu_i + v_i) \end{bmatrix}^{-1} \tag{4.6-97b}$$

which displays the natural frequency ω_n and damping ratio ζ explicitly. The two components of the associated state vector,

$$\begin{bmatrix} x_i \\ x_{i+1} \end{bmatrix} = \mathbf{X}_i \begin{bmatrix} v_i \\ v_{i+1} \end{bmatrix}, \quad i = 1, 3, \ldots, 2I - 1 \tag{4.6-98}$$

are the n_mth mode's displacement and rate.

These block transformations are applied to the entire structural model (eqs. 4.6-90 and 4.6-91), forming a bank of second-order oscillators that represent the combined bending and torsional modes of the system. The

$(2I \times 2I)$ transformation matrices $\mathbf{V}$ and $\mathbf{X}$ contain $\mathbf{V}_i$ and $\mathbf{X}_i$ in the (2×2) diagonal blocks, and the frequency-ordered, real-valued structural model can be written as

$$\dot{\mathbf{x}}_s = \mathbf{F}_s \mathbf{x}_s + \mathbf{G}_s \boldsymbol{\delta} \tag{4.6-99}$$

where

$$\mathbf{F}_s = \mathbf{X}\,\mathbf{V}\,\boldsymbol{\Lambda}\,\mathbf{V}^{-1}\,\mathbf{X}^{-1} \qquad (2I \times 2I) \tag{4.6-100}$$

$$\mathbf{G}_s = \mathbf{X}\,\mathbf{V}\,\boldsymbol{\Gamma} \qquad (2I \times r) \tag{4.6-101}$$

$$\mathbf{x}_s = \mathbf{X}\,\mathbf{V}\mathbf{q} = \mathbf{X}\,\mathbf{V}\begin{bmatrix} \mathbf{q}_1 \\ \mathbf{q}_2 \end{bmatrix} \qquad (2I \times 1) \tag{4.6-102}$$

The net aerodynamic force and moments due to structural vibration, which couple to the aircraft's flight motion, remain to be determined. From eqs. 4.6-87b, 4.6-88, and 4.6-90, the vector of vertical forces at each grid point is

$$\begin{aligned} \mathbf{f}_{\text{aero}} &= -\mathbf{D}\dot{\mathbf{z}} - \mathbf{K}\mathbf{z} = -\mathbf{D}\mathbf{x}_2 - \mathbf{K}\mathbf{x}_1 \\ &= -\mathbf{D}(\mathbf{E}_{21}\mathbf{q}_1 + \mathbf{E}_{22}\mathbf{q}_2) - \mathbf{K}(\mathbf{E}_{11}\mathbf{q}_1 + \mathbf{E}_{12}\mathbf{q}_2) \end{aligned} \tag{4.6-103}$$

The aerodynamic force is related to the state of the structural model by eq. 4.6-102

$$\begin{bmatrix} \mathbf{q}_1 \\ \mathbf{q}_2 \end{bmatrix} = (\mathbf{X}\,\mathbf{V})^{-1}\mathbf{x}_s \triangleq \begin{bmatrix} \mathbf{T}_1 \\ \mathbf{T}_2 \end{bmatrix}\mathbf{x}_s \tag{4.6-104}$$

which provides the values of $\mathbf{q}_1$ and $\mathbf{q}_2$ for eq. 4.6-103:

$$\mathbf{f}_{\text{aero}}(\mathbf{x}_s) = -\mathbf{D}(\mathbf{E}_{21}\mathbf{T}_1\mathbf{x}_s + \mathbf{E}_{22}\mathbf{T}_2\mathbf{x}_s) - \mathbf{K}(\mathbf{E}_{11}\mathbf{T}_1\mathbf{x}_s + \mathbf{E}_{12}\mathbf{T}_2\mathbf{x}_s) \tag{4.6-105}$$

If the structural model is truncated or residualized to the first n_t modes (see Section 4.2), then the last $2(I - n_t)$ elements of $\mathbf{x}_s$ are replaced by zeros. The aerodynamic force due to a single structural mode n_m, is computed by setting all but the $(2n_m - 1)$th and $2n_m$th elements of $\mathbf{x}_s$ (i.e., the displacement and rate of that mode) to zero. In the latter case, we find $\mathbf{f}_{\text{aero}}$ as separate linear functions, $\mathbf{f}_{\text{aero}}(x_i)$ and $\mathbf{f}_{\text{aero}}(x_{i+1})$, where $i = 2n_m - 1$.

Once the force vector has been determined for all modes, a truncated set, or just one mode, the net force and moments corresponding to $\mathbf{f}_{\text{aero}}$ are found by summation over the I grid points. The normal force due to structural deflection Z_s is calculated as

$$Z_s = \sum_{i=1}^{I} f_i \triangleq C_{Z_s}\bar{q}S \tag{4.6-106}$$

where f_i is the ith element of $\mathbf{f}_{aero}$. The result can be expressed in the conventional form, $C_{Z_s}\bar{q}S$, where $\bar{q}$ is the dynamic pressure, S is the aircraft's reference area, and C_{Z_s} is the nondimensional z-force coefficient. Defining the I-dimensional vectors **x** and **y** that encode the center-of-mass locations, (x_i, y_i), referenced to the aircraft's rotational center, the pitching and rolling moments due to structural deflection M_s and L_s, are

$$M_s = \mathbf{f}_{aero}^T x \triangleq C_{m_s}\bar{q}S\bar{c} \tag{4.6-107}$$

$$L_s = \mathbf{f}_{aero}^T y \triangleq C_{l_s}\bar{q}Sb \tag{4.6-108}$$

Taken one mode at a time, this procedure generates the forcing terms of the stability derivatives (Section 4.1) that close the loop of structural coupling to rigid-body modes. From eqs. 4.6-106–4.6-108, the sensitivities of force and moments to modal displacement and rate can be expressed (e.g., $\partial Z_s/\partial x_i$, $\partial Z_s/\partial x_{i+1}$, etc.). The force sensitivities are divided by vehicle mass and the moment sensitivities are multiplied by the inverse of the vehicle inertia matrix to complete the calculations.

The Four-Block Model

Having derived linear equations of structural dynamics that are driven by rigid-body motions and that produce net forces and moments on the aircraft, the stage is set for expressing a fully coupled dynamic model. Throughout the book, linear coupling of related motions is portrayed by a *four-block structure*, with primary dynamic parameters in the diagonal blocks and coupling parameters in the off-diagonal blocks of a square stability matrix **F**. In the present application, the primary blocks represent the rigid-body and elastic models, with rigid-body-to-elastic and elastic-to-rigid-body effects in the secondary blocks (Table 4.6-1).

The four-block model is hierarchical; each primary block can be subdivided into groups of smaller blocks until the smallest primary blocks contain scalar parameters. For example, the rigid-body block can contain phugoid and short-period dynamics, while the elastic block contains bending and twisting modes. We build on this basic structure in Sections 5.6 and 6.6, where the coupling of symmetric and asymmetric structural vibrations with longitudinal and lateral-directional motions is addressed.

Fuel Slosh

Although not strictly an aeroelastic problem, sloshing of fuel in partially filled wing, wing-tip, or fuselage tanks produces a dynamic mass variation that can couple with the aircraft's rigid-body motions. As noted in

[A-2], fuel slosh contributed to unstable and sometimes uncontrollable oscillations during flight testing of the Douglas A4D, Lockheed P-80, Boeing KC-135, and Cessna T-37 aircraft, while aftward shift of fuel during takeoff acceleration led to static pitch instability during flight testing of a North American YF-100 equipped with partially filled external tanks.

Because the fuel center of mass in a tank is related to the orientation and shape of the free surface between fuel and air, there are an infinite number of slosh modes [L-2]. There is no dynamic problem when tanks are nearly full or nearly empty. Characterizing the first mode of fuel motion by an equivalent pendulous mass [G-4, L-4, A-2], the pendulum is short, the oscillation frequency is high, and the relative displacement of the fuel center of mass is small for a nearly full tank. For a nearly empty tank, the center-of-mass travel and equivalent pendulum length may be high, but the oscillation frequency and fuel mass are low. Thus, there is little likelihood of significant coupling in either case. If the lowest and highest slosh natural frequencies bracket the aircraft's short-period natural frequency, there will be coupling at some point as the fuel is consumed. As a practical matter, the problem can be solved by adding internal baffles to the fuel tanks, which increases the natural frequency of sloshing; however, baffles add weight and cut down on the volume available for fuel storage.

Abzug derives equations describing fuel slosh for the long, thin fuel tanks contained in swept wings [A-1], giving conditions in which lateral-directional coupling can produce instability. He expresses the slosh dynamics in the form of eqs. 4.6-87 and 4.6-88, and then concatenates them to the rigid-body equations as in the four-block structure of Table 4.6-1. Consequently, fuel slosh can be investigated using techniques that are similar to those employed in aeroelastic analysis.

4.7 Introduction to Flying Qualities and Flight Control Systems

Providing satisfactory flying (or handling) qualities for pilots and automated flight management systems may require that command inputs be processed and that motions be sensed and blended with the command inputs to the control actuators. The overall purpose is to modify the natural transient and steady-state response of the aircraft and to provide adequate

TABLE 4.6-1 FOUR-BLOCK COUPLING STRUCTURE OF THE STABILITY MATRIX FOR AN ELASTIC-BODY DYNAMIC MODEL

Rigid-Body Parameters	Elastic-to-Rigid-Body Coupling
Rigid-Body-to-Elastic Coupling	Elastic Parameters

stability. These functions are called *command* and *stability augmentation*, and they have been the subject of much analysis and design. It is commonly thought that control systems properly designed to satisfy well-described flying qualities specifications have similar characteristics no matter what the design approach, and a recent paper [T-3] appears to confirm that opinion. Hence, the choice of design method can be based on ease of use and the flexibility with which design objectives can be specified and realized. References for flight control design include [B-8, B-11, C-2, M-6, M-7, P-4, S-17, T-2].

Factors that contribute to flying qualities criteria for piloted operation are reviewed, beginning with an overview of biological and cognitive characteristics of the human operator. This is followed by a discussion of subjective and objective assessment of flying qualities, including the dependence on aircraft type and flight condition. Flying qualities for uninhabited air vehicles (UAV) are of comparatively recent interest and are less well developed than for piloted flight. While some of the factors that are of importance to human pilots are insignificant for fully automated systems, such as visual and vestibular cues that are consistent with human experience, we may expect desirable flying qualities for both types to be comparable.

A flight control design approach based on optimization and state estimation is presented as an introduction to the field. We focus on designing the flight control logic for small-perturbation motions at a trimmed flight condition, assuming that the aircraft's flight characteristics are known without error and that control actions are continuous over time. Thus, methods for adapting to changing flight conditions or aircraft system failures, for providing robust performance in the face of uncertainty, for accommodating significant nonlinearities, and for implementing control logic in digital computers and physical hardware are left to further reading. Motivation and development of design equations can be found in [S-14, B-11] as well as many other chapter references.

The control design goal is regulation of the aircraft's state about a *nonzero set point*, that is, a desired state that is specified by a steady command input, keeping state excursions and control actions as small as is feasible. In practice, the set point is defined by pilot or flight management system commands, and it is free to vary. The closed-loop system's transient and steady-state response should be close to that of an ideal system, perhaps specified by flying qualities requirements. The control system design process must be applicable with an arbitrary number of command inputs, control actuators, and feedback sensors. It must guarantee closed-loop stability, even if the unaugmented airframe is unstable, and it must minimize the effect of feedback measurement errors. A family of flight control law alternatives is developed, including controllers for stability augmentation, command augmentation, ideal model-following, and integral compensation.

Cognitive/Biological Models and Piloting Actions

A human pilot can interact with the aircraft at several levels, performing distinct functions of *sensing, regulation,* and *decision-making* (Fig. 4.7-1). These tasks exercise different human characteristics: the ability to see and feel, the ability to identify and correct errors between desired and actual states, and the ability to decide what needs to be done next. The first of these depends on the body's sensors and the neural networks that connect them to the brain. The second relies on motor functions enabled by the neuromuscular system to execute learned associations between stimuli and desirable actions. The third requires more formal, introspective thought about the reasons for taking action, drawing on the brain's deep memory to recall important procedures or data. Sensing and regulation are high-bandwidth tasks that allow little time for deep thinking. Decision-making is a low-bandwidth task that requires concentration. Each of these tasks exacts a workload toll on the pilot.

Pilot workload has become a critical issue as the complexity of systems has grown, and furnishing ideal flying qualities throughout the flight envelope is an imperative. It is desirable to reduce the need to perform high-bandwidth, automatic functions, giving the pilot time to cope with unanticipated or unlikely events. At the same time, the pilot must maintain proficiency in flying the aircraft, so some continuing attention to manual tasks during normal operations is warranted.

Piloting actions can be identified according to a cognitive/biological hierarchy of declarative, procedural, and reflexive functions. *Declarative functions* are performed in *outer control loops*, and *reflexive functions* are performed in *inner control loops. Procedural functions* have well-defined input-output characteristics of intermediate structure. Traditional design principles suggest that the outer-loop functions should be dedicated to low-bandwidth, large-amplitude control commands, while inner-loop functions should have high bandwidths and relatively lower-amplitude actions. There is a logical progression from the sweeping, flexible alternatives associated with satisfying mission goals to more local concerns for stability and regulation about a desired path or equilibrium condition.

Research on the flying qualities of aircraft has identified ways to make the pilot's job easier and more effective. The first flying qualities specification simply stated, "(the aircraft) must be steered in all directions without difficulty and all time (be) under perfect control and equilibrium" [A-6, M-15]. Further evolution of flying qualities criteria based on

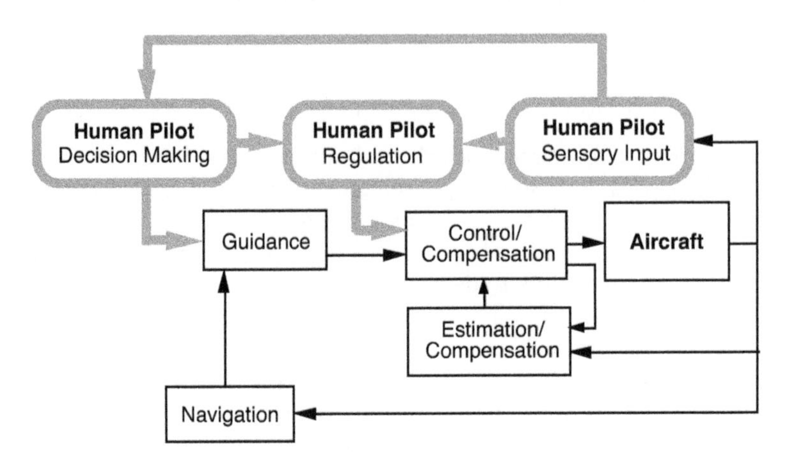

Fig. 4.7-1. Pilot interactions with guidance, navigation, and control of the aircraft.

dynamic modeling and control theory resulted in development of the widely used U.S. military specification [A-7], perhaps the most influential document in the field. Handling qualities ratings (HQR) assigned by skilled test pilots provide the basis for such specifications; *Cooper-Harper ratings* (CHR) are standardized HQR metrics for assessing the handling qualities of all types of aircraft [H-4].

The development of *control-theoretic models of piloting behavior* proceeded in parallel with flying qualities research. Most of these models dealt with reflexive, compensatory tracking tasks using time lags and transfer functions [M-8, M-9, T-5]. In the simplest scalar model, piloting actions are modeled by a gain K_p between the airplane output being tracked and the pilot's control variable (or input to the airplane), and the effects of different gain values can be evaluated using root locus or Bode plots. Delay between perception of aircraft motion and piloting action can be modeled as a first-order lag

$$P(s) = K_P \frac{1/T_P}{s + 1/T_P} \tag{4.7-1}$$

where T_P is the time constant for pilot response, or a *pure time delay* τ_P,

$$P(s) = K_P e^{-\tau_P s} \approx K_P \frac{-s + 2/\tau_P}{s - 2/\tau_P} \tag{4.7-2}$$

where the first-order *Padé approximation* to the pure delay can be employed for root-locus analysis (see Section 6.5 for more details). Minimum values of T_P and τ_P are on the order of a few tenths of a second, while the pilot could increase these values procedurally to adjust controlled response.

Some of the pilot modeling approaches go into considerable detail about neuromuscular system dynamics [M-8, M-9, S-7, S-8, V-1]. These models often show good correlation with experimental results, not only in compensatory tracking but in more intricate tasks. Procedural (or discretionary) *lead-lag compensation*, as well as a more detailed model of neuromuscular dynamics, can be included in the model [M-9]:

$$P(s) = K_P \frac{T_L T_{K_2} \omega_n^2}{T_I T_{K_1} T_N} e^{-\tau_P s} \left(\frac{s + 1/T_L}{s + 1/T_I} \right)_{pro} \times \left[\frac{s + 1/T_{K_2}}{(s + 1/T_{K_1})(s + 1/T_N)(s^2 + 2\zeta\omega_n s + \omega_n^2)} \right]_{NM} \quad (4.7\text{-}3)$$

Piloting actions have been associated with linear-quadratic-Gaussian (LQG) optimal controllers, modified to account for neuromuscular effects [K-4, S-10]. Optimal-control models facilitate the analysis of multi-input/multi-output strategies as the complexity of the task increases, and they are well suited for studying tradeoffs between display and controller alternatives. Pilot models have been used to predict test-pilot opinion ratings, as in the "Paper Pilot" computer program [A-5, S-19], and they have been incorporated in flying qualities criteria [H-6].

At the highest level of the declarative/procedural/reflexive hierarchy, four types of thought can be defined (Fig. 4.7-2, from [S-16]). *Conscious thought* occupies the pilot's attention, requiring focus, awareness, reflection, and rehearsal. *Unconscious thought* "describes those products of the perceptual system that go unattended or unrehearsed, and those memories that are lost from primary memory through display or displacement" [K-3]. Within the unconscious, there are two important compartments. *Subconscious thought* is procedural knowledge that is below the pilot's level of awareness but central to the implementation of intelligent behavior. It facilitates communication with the outside world and with other parts of the body, providing the principal home for learned skills such as piloting. Perceptions may take a subliminal (subconscious) path to memory. *Preconscious thought* is preattentive declarative processing that helps choose the objects of our conscious thought, operating on larger chunks of information or at a more symbolic level. It forms a channel to long-term and implicit memory, and it may play a role in judgment and intuition.

The central nervous system supports a hierarchy of intelligent and automatic functions. Declarative thinking occurs in the brain's cerebral

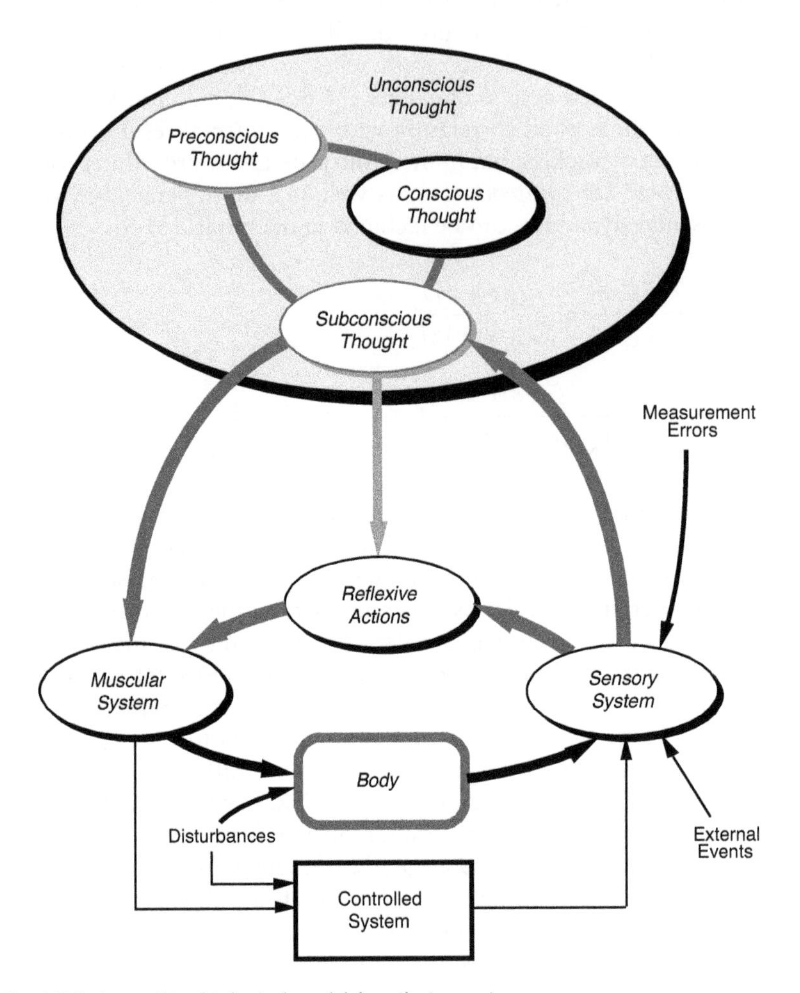

Fig. 4.7-2. A cognitive/biological model for piloting actions.

cortex, which accesses the interior limbic system for memory [A-4, H-5]. Together, they provide the processing unit for conscious thought. Regions of the cerebral cortex are associated with different intellectual and physical functions; the distinction between conscious and preconscious function may depend on the activation level and duration in regions of the cerebral cortex.

The working memory of conscious thought has access to the spinal cord through other brain parts that are capable of taking procedural action (e.g., the brain stem for autonomic functions, the occipital lobes for vision, and the cerebellum for movement). Procedural action can be associated with subconscious thought, which supports voluntary automatic processes like movement and sensing. These voluntary signals are sent over the somatic nervous system, transmitting to muscles through the motor neural system and from receptors through the sensory neural system.

The spinal cord itself "closes the control loop" for reflexive actions long before signals could be processed by the brain. Nevertheless, these signals are available to the brain for procedural and declarative processing. We are all aware of performing some automatic task (e.g., riding a bicycle) without effort, only to waver when we focus on what we are doing. Involuntary regulation of the body's organs is provided by the autonomic nervous system, which is subject to unconscious processing by the brain stem. *Bio-feedback* can be learned, allowing a modest degree of higher-level control over some autonomic functions.

A number of physical factors contribute to the complexity of actions that a pilot may take. First, there is a *richness of sensory information* that is hard to fathom, with millions of sensors providing information to the pilot. This results in high signal-to-noise ratio in some cases, and it allows symbolic/image processing in others. Those signals requiring high-bandwidth, high-resolution channel capacity (vision, sound, and balance) have *short, dedicated, parallel runs* from the sensors to the brain. This enhances the security of the channels and protects the signals from noise contamination. *Dissimilar but related sensory inputs facilitate interpretation of data.* A single motion can be sensed by the eyes, by the inner ear, and by the "seat-of-the-pants" (i.e., by sensing forces on the body itself), the signals corroborating each other and suggesting appropriate actions. When some of the signals are absent, as in fixed-base simulation of rapidly maneuvering flight, a pilot may experience confusion and disorientation.

There are *hierarchical and redundant structures* throughout the body. The nervous system is a prime example, bringing inputs from myriad sensors (both similar and dissimilar) to the brain, and performing low-level reasoning as an adjunct. Many sensing organs occur in pairs (e.g., eyes, ears, inner ears), and their internal structures are highly parallel. *Pairing allows graceful degradation in the event that use of an organ is lost.* Stereo vision vanishes with the loss of an eye, but the remaining eye can provide both foveal and peripheral vision, as well as a degree of depth perception through object size and stadiametric processing. Control effectors (arms, hands, legs, feet) also occur in pairs, providing alternative pathways for piloting actions.

Aircraft Flying Qualities

We distinguish three types of flying (or handling) qualities criteria in general use: pilot opinion, quantitative objectives for aircraft and flight control system design, and requirements for procuring aircraft. The opinions of test pilots are subjective and susceptible to personal preferences but are, perhaps, the most important of all three. Design criteria are objective and technical, though they are intended to elicit good pilot opinions in simulation and flight test. Design requirements are design criteria that provide a formal basis for accepting an aircraft type as satisfying contractual terms.

The *Cooper-Harper rating* used in cockpit simulation and flight testing quantifies pilot opinions of handling qualities on a best-to-worst scale from 1 to 10 [H-4]. The ratings are intended to be mission-specific and to minimize the role of personal preference. A rating of 1 signifies highly desirable characteristics, while a rating of 10 indicates that "control will be lost during some portion of required operation." Nine scores are grouped in three levels:

- Satisfactory without improvement (1–3)
- Adequate performance with a tolerable pilot workload (4–6)
- Performance not adequate with a tolerable workload (7–9)

In evaluating an aircraft configuration for a particular mission segment, the test pilot is encouraged to proceed through a *decision tree* that begins with the question, "Is the aircraft controllable?" With an affirmative answer, the pilot must decide if performance is adequate with a tolerable workload, and if so, if it is satisfactory without improvement. Thus, the pilot first chooses the level in a "bottom-up" procedure, then decides where in the chosen level the configuration should be rated. For example, an assessment in the second level may be further identified as

- Minor but annoying deficiencies (4); desired performance requires moderate pilot compensation
- Moderately objectionable deficiencies (5); adequate performance requires considerable pilot compensation
- Very objectionable but tolerable deficiencies (6); adequate performance requires extensive pilot compensation

Referring to the previous section, the pilot must make a declarative judgment about the aircraft's ability to respond to his procedural and reflexive actions. While the CHR scale is subjective, the meanings of the ratings are precise, and skilled test pilots evaluating the same configuration for the same task typically give the same ratings with a standard deviation of one or less. The CHR has been so successful as a realization of the decision-making process that it has been used as the model for other evaluation processes both in and outside the aeronautical community [V-3]. Nevertheless, the ratings ultimately reflect opinions, which may be biased by prior experience (e.g., large aircraft versus small), inadequate briefing about the evaluation task, lack of appreciation for closed-loop control modes that are unlike natural response modes, and insufficient training in the assignment of CHRs.

For many years, the "Military Specification MIL-F-8785C, Flying Qualities of Piloted Airplanes" [A-7], served the dual role of establishing design criteria and requirements. It is the most comprehensive and widely used source of criteria, presenting response characteristics that are suitable over a wide range of conditions. Civil aviation authorities also publish regulations, such as the European Joint Aviation Requirements, governing the flying qualities of aircraft. In the United States, the Federal Aviation Administration has established Federal Aviation Regulations Part 23 for general aviation aircraft and Part 25 for transport aircraft.

MIL-F-8785C no longer plays a direct role in the procurement of U.S. military aircraft, having been superseded by the program-specific Military Standard (MIL-STD-1797A) [A-8]. The "Mil Spec" tacitly assumes that closed-loop response should be like that of an equivalent open-loop system. In principle, the "Mil Standard" is more performance oriented, not strictly tied to key parameters of an open-loop dynamic model. It allows flight control systems to tailor response without regard to equivalency, possibly creating response modes that have no parallel in rigid-body dynamics. MIL-STD-1797A is augmented by "Military Specification MIL-F-9490D, Flight Control Systems—General Specification for Design, Installation, and Test of Piloted Aircraft" [A-9].

The requirements of MIL-F-8785C depend upon aircraft type and flight conditions [A-7]. There are four *aircraft classifications*:

I. Small, light airplanes, e.g., utility aircraft and primary trainers

II. Medium-weight, low-to-medium maneuverability airplanes, e.g., small transports or tactical bombers

III. Large, heavy, low-to-medium maneuverability airplanes, e.g., heavy transports, tankers, or bombers

IV. Highly maneuverable aircraft, e.g., fighter and attack airplanes

There are three *flight phases*:

A. Nonterminal flight requiring rapid maneuvering, precise tracking, or precise flight path control, e.g., air-to-air combat, ground attack, in-flight refueling (receiver), close reconnaissance, terrain following, and close formation flying

B. Nonterminal flight requiring gradual maneuvering, including climb, cruise, in-flight refueling (tanker), and descent

C. Terminal flight, including takeoff (normal and catapult), approach, wave-off/go-around, and landing

Three *levels of flying qualities* are specified:

1. Flying qualities clearly adequate for the mission flight phase
2. Flying qualities adequate to accomplish the mission flight phase, with some increase in pilot workload or degradation of mission effectiveness
3. Flying qualities such that the aircraft can be controlled safely, but pilot workload is excessive or mission effectiveness is inadequate

In addition, *operational, service*, and *permissible flight envelopes* are defined according to limits on airspeed, altitude, and load factor. Level 1 flying qualities are required within an aircraft's operational envelope when the aircraft is in its normal state, that is, when there are no system failures [A-7]. With failed systems, Levels 2 and 3 are acceptable within the operational envelope once per 100 flights and once per 10,000 flights, respectively.

The principal MIL-F-8785C metrics for longitudinal flying qualities are static speed stability, phugoid stability, flight path stability, short-period frequency and its relationship to command acceleration sensitivity, short-period damping, and control-force gradients. For lateral-directional flying qualities, the major concerns are natural frequency and damping of the Dutch roll mode, time constants of the roll and spiral modes, rolling response to commands and Dutch roll oscillation, sideslip excursions, maximum stick and pedal forces, and turn coordination. In both cases, there should be no tendency for *pilot-induced oscillations* (PIO), where the uncommanded aircraft is stable but piloting actions couple with aircraft dynamics to produce instability. Additional flying qualities metrics are presented in [A-3, A-8, G-2, H-6, H-7, P-4].

Linear-Quadratic Regulator

Turning to the design of control systems that provide satisfactory flying qualities, the aircraft's linearized equations of motion can be expressed in vector form as,

$$\Delta\dot{\mathbf{x}}(t) = \mathbf{F}\Delta\mathbf{x}(t) + \mathbf{G}\Delta\mathbf{u}(t) \tag{4.7-4}$$

where $\Delta\mathbf{x}$ is the $(n \times 1)$ state vector, $\Delta\mathbf{u}$ is the $(m \times 1)$ control vector, and $\mathbf{F}$ and $\mathbf{G}$ are the constant stability and control matrices that apply for a given trimmed condition, and they may include the aeroelastic effects presented in Section 4.6. The control system is designed to minimize a scalar, *quadratic cost function J*,

$$J = \phi[\Delta\mathbf{x}(t_f)] + \int_{t_0}^{t_f} L[\Delta\mathbf{x}(t), \Delta\mathbf{u}(t)]\,dt = \frac{1}{2}\{\Delta\mathbf{x}^T(t_f)\mathbf{S}_f\Delta\mathbf{x}(t_f) + \int_{t_0}^{t_f} [\Delta\mathbf{x}^T(t)\mathbf{Q}\Delta\mathbf{x}(t) + 2\Delta\mathbf{x}^T\mathbf{M}\Delta\mathbf{u} + \Delta\mathbf{u}^T(t)\mathbf{R}\Delta\mathbf{u}(t)]\,dt\} \tag{4.7-5}$$

with respect to control actions over the period from t_0 to t_f. The cost function contains a *terminal cost* ϕ that penalizes final error and an *integral cost* that trades off perturbations in the state against required control actions. $L[\Delta\mathbf{x}(t), \Delta\mathbf{u}(t), t]$ is called the *Lagrangian* of the cost function, and the matrices $\mathbf{S}_f$, $\mathbf{Q}$, $\mathbf{M}$, and $\mathbf{R}$ establish the relative weighting for state and control contributions to the scalar cost. $\mathbf{S}_f$, $\mathbf{Q}$, and $\mathbf{R}$ are square, positive definite, symmetric matrices of dimension $(n \times n)$, $(n \times n)$, and $(m \times m)$, while $\mathbf{M}$ is an $(n \times m)$ matrix. A straightforward means for specifying these weighting matrices is given later in the section.

The state equation (eq. 4.7-4) serves as a dynamic constraint on the minimization of the cost with respect to the control history $\mathbf{u}(t)$ in $[t_0, t_f]$. This constraint is "adjoined" to the Lagrangian by the $(n \times 1)$ adjoint vector $\Delta\boldsymbol{\lambda}$; the augmented cost function allows the aircraft's dynamics to be considered in the optimization:

$$\begin{aligned} J &= \phi[\Delta\mathbf{x}(t_f)] + \int_{t_0}^{t_f} \{L[\Delta\mathbf{x}(t), \Delta\mathbf{u}(t)] + \Delta\boldsymbol{\lambda}^T(t)[\mathbf{F}\Delta\mathbf{x}(t) + \mathbf{G}\Delta\mathbf{u}(t) - \dot{\mathbf{x}}(t)]\}\,dt \\ &= \phi[\Delta\mathbf{x}(t_f)] + \int_{t_0}^{t_f} \{H[\Delta\mathbf{x}(t), \Delta\mathbf{u}(t)] - \Delta\boldsymbol{\lambda}^T(t)[\dot{\mathbf{x}}(t)]\}\,dt \end{aligned} \tag{4.7-6}$$

where H is the corresponding *Hamiltonian* function:

$$\begin{aligned} H[\Delta\mathbf{x}(t), \Delta\mathbf{u}(t), \Delta\boldsymbol{\lambda}(t), t] &= L[\Delta\mathbf{x}(t), \Delta\mathbf{u}(t), t] + \Delta\boldsymbol{\lambda}^T(t)[\mathbf{F}\Delta\mathbf{x}(t) + \mathbf{G}\Delta\mathbf{u}(t)] \\ &= \frac{1}{2}[\Delta\mathbf{x}^T(t)\mathbf{Q}\Delta\mathbf{x}(t) + 2\Delta\mathbf{x}^T\mathbf{M}\Delta\mathbf{u} + \Delta\mathbf{u}^T(t)\mathbf{R}\Delta\mathbf{u}(t)] \\ &\quad + \Delta\boldsymbol{\lambda}^T(t)[\mathbf{F}\Delta\mathbf{x}(t) + \mathbf{G}\Delta\mathbf{u}(t)] \end{aligned} \tag{4.7-7}$$

Three *necessary conditions*, or *Euler-Lagrange equations*, must be satisfied to minimize the cost function [S-14]:

$$\Delta\boldsymbol{\lambda}^T(t_f) = \left.\frac{\partial\phi}{\partial\Delta\mathbf{x}}\right|_{t=t_f} = \Delta\mathbf{x}^T(t_f)\mathbf{S}_f \tag{4.7-8}$$

$$\frac{\partial \Delta\boldsymbol{\lambda}(t)}{\partial t} = -\left[\frac{\partial H}{\partial \Delta\mathbf{x}}\right]^T = -\mathbf{Q}\Delta\mathbf{x}(t) - \mathbf{M}^T\Delta\mathbf{u}(t) - \mathbf{F}^T\Delta\boldsymbol{\lambda}(t) \tag{4.7-9}$$

$$\frac{\partial H}{\partial \Delta\mathbf{u}} = \Delta\mathbf{u}^T(t)\mathbf{R} + \Delta\mathbf{x}^T(t)\mathbf{M} + \Delta\boldsymbol{\lambda}^T(t)\mathbf{G} = 0 \tag{4.7-10}$$

The first Euler-Lagrange equation (eqs. 4.7-8) specifies a terminal boundary condition on the value of $\Delta\boldsymbol{\lambda}$, while the second (eq. 4.7-9) provides a linear, ordinary differential equation that could be integrated back from $\Delta\boldsymbol{\lambda}(t_f)$ to find $\Delta\boldsymbol{\lambda}(t)$ over the entire time interval. The final equation (eq. 4.7-10) defines the optimizing control as a function of $\Delta\mathbf{x}(t)$ and $\Delta\boldsymbol{\lambda}(t)$:

$$\Delta\mathbf{u}(t) = -\mathbf{R}^{-1}[\mathbf{G}^T\Delta\boldsymbol{\lambda}(t) + \mathbf{M}^T\Delta\mathbf{x}(t)] \tag{4.7-11}$$

Assuming that the adjoint relationship between $\Delta\mathbf{x}(t)$ and $\Delta\boldsymbol{\lambda}(t)$ applies over the entire time interval, eq. 4.7-11 leads to an *optimal feedback control law*,

$$\Delta\mathbf{u}(t) = -\mathbf{R}^{-1}[\mathbf{G}^T\mathbf{S}(t) + \mathbf{M}^T]\Delta\mathbf{x}(t) = -\mathbf{C}(t)\Delta\mathbf{x}(t) \tag{4.7-12}$$

where $\mathbf{C}(t)$ is the $(m \times n)$ *optimal feedback gain matrix.*

Rather than solve for $\Delta\boldsymbol{\lambda}(t)$, we seek a solution for the time-varying adjoining matrix, $\mathbf{S}(t)$, which allows $\mathbf{C}(t)$ to be computed directly. $\mathbf{S}(t)$ is derived from the simultaneous solution of eqs. 4.7-4 and 4.7-9 subject to the boundary condition provided by eq. 4.7-8. Differentiating the adjoint relationship $\Delta\boldsymbol{\lambda}(t) = \mathbf{S}(t)\,\Delta\mathbf{x}(t)$ with respect to time and rearranging,

$$\dot{\mathbf{S}}(t)\Delta\mathbf{x}(t) = \Delta\dot{\boldsymbol{\lambda}}(t) - \mathbf{S}\Delta\dot{\mathbf{x}}(t) \tag{4.7-13}$$

Substituting the equations for $\Delta\mathbf{x}(t)$ and $\Delta\boldsymbol{\lambda}(t)$ and the adjoint relationship in eq. 4.7-13, all terms are found to be postmultiplied by $\Delta\mathbf{x}(t)$. Because the solution should hold for arbitrary $\Delta\mathbf{x}(t)$, this term can be canceled throughout, leading to the *matrix Riccati equation* for $\mathbf{S}(t)$,

$$\dot{\mathbf{S}}(t) = -[\mathbf{F} - \mathbf{G}\mathbf{R}^{-1}\mathbf{M}^T]^T\mathbf{S}(t) - \mathbf{S}(t)[\mathbf{F} - \mathbf{G}\mathbf{R}^{-1}\mathbf{M}^T] + \mathbf{S}(t)\mathbf{G}\mathbf{R}^{-1}\mathbf{G}^T\mathbf{S}(t) - \mathbf{Q} + \mathbf{M}\mathbf{R}^{-1}\mathbf{M}^T \tag{4.7-14}$$

with the terminal boundary condition given by $\mathbf{S}(t_f)$.

Suppose that t_f is fixed and t_0 is allowed to become more and more negative. As eq. 4.7-14 is integrated backward in time with t_0 going toward $-\infty$, $\mathbf{S}(t)$ approaches a constant, positive-definite value $\mathbf{S}_0$. Now, let a control response begin at t_0. As time runs forward from t_0, $\mathbf{S}(t)$ remains essentially equal to $\mathbf{S}_0$ until the final moments; hence, in the infinite-final-time problem, we can choose $\mathbf{S}(t) = \mathbf{S}_0$ to yield a constant-gain optimal control law or *linear-quadratic* (*LQ*) *regulator*. In the steady state, $\dot{\mathbf{S}}(t) = 0$, and the differential Riccati equation can be replaced by the *algebraic Riccati equation* to find $\mathbf{S}_0$:

$$0 = -[\mathbf{F} - \mathbf{G}\mathbf{R}^{-1}\mathbf{M}^T]^T\mathbf{S}(t) - \mathbf{S}(t)[\mathbf{F} - \mathbf{G}\mathbf{R}^{-1}\mathbf{M}^T] + \mathbf{S}(t)\mathbf{G}\mathbf{R}^{-1}\mathbf{G}^T\mathbf{S}(t) - \mathbf{Q} + \mathbf{M}\mathbf{R}^{-1}\mathbf{M}^T \quad (4.7\text{-}15)$$

There are many possible solutions to this algebraic equation, but only one is positive definite. Consequently, the terminal condition $\mathbf{S}(t_f)$ plays no part in determining $\mathbf{S}_0$, and it can be assumed to be zero. In the remainder, we drop the subscript on $\mathbf{S}_0$, and the *constant-gain LQ regulator* can be expressed as

$$\Delta\mathbf{u}(t) = -\mathbf{R}^{-1}[\mathbf{G}^T\mathbf{S} + \mathbf{M}^T]\Delta\mathbf{x}(t) = -\mathbf{C}\Delta\mathbf{x}(t) \quad (4.7\text{-}16)$$

The LQ regulator is a *proportional controller* that requires full-state feedback: every element of $\Delta\mathbf{x}(t)$ must be available to compute the control. This is advantageous from the perspective of achieving desired stability and performance, and it can be shown that full-state feedback provides the equivalent of *proportional-derivative compensation*. If it is not practical to measure every element of the state, $\Delta\mathbf{x}(t)$ can be estimated from fewer measurements, as noted below. Alternatively, it may be sufficient to develop the regulator for a reduced-order model. Dominant characteristics of the short-period mode are described by a second-order model (Section 5.3), which could be used in place of a fourth- or higher-order model during design. In such case, it is especially important to evaluate the control law with full-order dynamics.

The dimension of the control vector is not fixed by eq. 4.7-16; it could be a scalar (e.g., elevator deflection) or a vector (e.g., elevator, throttle, flaps, and canards). The constant LQ gain matrix is computed by the MATLAB function `lqr`, making the practical application of optimal feedback control easy.

The LQ regulator could be incorporated in a *stability augmentation system*, with the pilot's command $\mathbf{u}_c$ simply added to eq. 4.7-16. Thus, $\mathbf{u}_c$ serves as the desired value of control surface deflection, and $\Delta\mathbf{u}$ is a dynamic correction to that command. Whereas the stability of the open-loop system is governed by the eigenvalues of $\mathbf{F}$, the stability of the closed-loop system depends on the eigenvalues of $(\mathbf{F} - \mathbf{G}\mathbf{C})$, as deduced by substituting eq. 4.7-16 in eq. 4.7-4. With precise knowledge of $\mathbf{F}$ and $\mathbf{G}$, closed-loop stability is guaranteed when $\mathbf{R}$ and $(\mathbf{Q} - \mathbf{M}\mathbf{R}^{-1}\mathbf{M}^T)$ are positive-definite matrices, even if the open-loop system is unstable. Less restrictive conditions for guaranteed stability are given in [S-14]. As a starting point, one might define $\mathbf{Q}$ and $\mathbf{R}$ to be diagonal matrices and $\mathbf{M}$ to be zero. Choosing high values for the magnitude of $\mathbf{Q}$ (with respect to $\mathbf{R}$)

leads to rapid response and high control gains, while low values do just the opposite. Large individual elements of **Q**'s diagonal generally tighten the response of corresponding state elements, while increasing individual elements of **R**'s diagonal decreases the use of the corresponding control effector. If **F** is stable, the elements of **C** approach zero as **R** approaches infinity; however, if **F** is unstable, **C** remains large enough to stabilize the unstable aircraft.

Steady-State Response to Command Input

The LQ controller augments stability, regulating state perturbations about a trimmed or maneuvering condition specified by the pilot. Commands that specify a desired state, providing command augmentation, can be added, where the command gain matrix is an algebraic function of the system model and the feedback gain matrix. In equilibrium, $\Delta\dot{\mathbf{x}}(t) = 0$, and the state equation (eq. 4.7-4) is

$$0 = \mathbf{F}\Delta\mathbf{x}(t) + \mathbf{G}\Delta\mathbf{u}(t) \tag{4.7-17}$$

This equation can be rearranged to express the equilibrium state perturbation $\Delta\mathbf{x}^*$ that corresponds to a constant control input $\Delta\mathbf{u}^*$:

$$\Delta\mathbf{x}^* = -\mathbf{F}^{-1}\mathbf{G}\Delta\mathbf{u}^*(t) \tag{4.7-18}$$

A command input from the pilot or autopilot can be interpreted as a linear combination of state and control perturbations,

$$\Delta\mathbf{y}_c(t) = \mathbf{H_x}\Delta\mathbf{x}(t) + \mathbf{H_u}\Delta\mathbf{u}(t) \tag{4.7-19}$$

where $\Delta\mathbf{y}_c(t)$ is the $(r \times 1)$ command-input vector, and $\mathbf{H_x}$ and $\mathbf{H_u}$ define the relationships between the command, state, and control [S-14]. At equilibrium, not only must eq. 4.7-18 be satisfied, but $\Delta\mathbf{x}^*$ and $\Delta\mathbf{u}^*$ must take values that satisfy eq. 4.7-19. Defining a constant command input $\Delta\mathbf{y}^*$, the two equations can be expressed and solved simultaneously as

$$\begin{bmatrix} 0 \\ \Delta\mathbf{y}^* \end{bmatrix} = \begin{bmatrix} \mathbf{F} & \mathbf{G} \\ \mathbf{H_x} & \mathbf{H_u} \end{bmatrix} \begin{bmatrix} \Delta\mathbf{x}^* \\ \Delta\mathbf{u}^* \end{bmatrix} \tag{4.7-20}$$

and

$$\begin{bmatrix} \Delta\mathbf{x}^* \\ \Delta\mathbf{u}^* \end{bmatrix} = \begin{bmatrix} \mathbf{F} & \mathbf{G} \\ \mathbf{H_x} & \mathbf{H_u} \end{bmatrix}^{-1} \begin{bmatrix} 0 \\ \Delta\mathbf{y}^* \end{bmatrix} = \begin{bmatrix} \mathbf{B}_{11} & \mathbf{B}_{12} \\ \mathbf{B}_{21} & \mathbf{B}_{22} \end{bmatrix} \begin{bmatrix} 0 \\ \Delta\mathbf{y}^* \end{bmatrix} \tag{4.7-21}$$

where

$$\mathbf{B}_{11} = \mathbf{F}^{-1}(-\mathbf{G}\mathbf{B}_{21} + \mathbf{I}_n) \tag{4.7-22}$$

$$\mathbf{B}_{12} = -\mathbf{F}^{-1}\mathbf{G}\mathbf{B}_{22} \tag{4.7-23}$$

$$\mathbf{B}_{21} = -\mathbf{B}_{22}\mathbf{H}_{\mathbf{x}}\mathbf{F}^{-1} \tag{4.7-24}$$

$$\mathbf{B}_{22} = (-\mathbf{H}_{\mathbf{x}}\mathbf{F}^{-1}\mathbf{G} + \mathbf{H}_{\mathbf{u}})^{-1} \tag{4.7-25}$$

Consequently, the equilibrium values of the state and control are

$$\Delta\mathbf{x}^* = \mathbf{B}_{12}\Delta\mathbf{y}^* \tag{4.7-26}$$

$$\Delta\mathbf{u}^* = \mathbf{B}_{22}\Delta\mathbf{y}^* \tag{4.7-27}$$

Clearly, $\mathbf{F}$ and $(-\mathbf{H}_{\mathbf{x}}\mathbf{F}^{-1}\mathbf{G} + \mathbf{H}_{\mathbf{u}})$ must be invertible for the desired equilibrium to exist. $\mathbf{F}$ is not invertible if some elements of the state are pure integrals of others and have no dynamic effect on the motion. For example, horizontal position is the integral of horizontal velocity, and it does not effect the aircraft's dynamic characteristics in the flat-earth model. Consequently, a design model containing Δx or Δy cannot have a satisfactory definition of $\Delta\mathbf{x}^*$.

If the command vector is a linear combination of the state elements alone, then $\mathbf{H}_{\mathbf{u}} = 0$, and eq. 4.7-25 is invertible only if $\mathbf{H}_{\mathbf{x}}\mathbf{F}^{-1}\mathbf{G}$ is nonsingular. The matrix is square if $r = m$, that is, if the number of command variables equals the number of control variables. (If $r < m$, the *pseudoinverse* or *weighted pseudoinverse matrix* can be employed to specify $\mathbf{B}_{22}$.) The matrix must be of full rank (m), and because the maximum rank of the matrix is n (it contains $\mathbf{F}$, whose maximum rank is n), $m \leq n$ for the inverse to exist. Even then, care must be exercised in selecting the command vector. It is impossible for both pitch rate and pitch angle to have nonzero steady-state values with wings level; hence, both could not be members of the same command vector. If either of the conditions for a proper equilibrium are violated, the control-design problem must be redefined to conform to physics. As noted earlier, control design could be based on a reduced-order dynamic model (as by eliminating Δx or Δy), or the command vector could be modified, or additional logic could be added to account for the quasisteady state, as in [S-14].

To regulate the aircraft about its constant commanded value, we define the perturbations,

$$\Delta\tilde{\mathbf{x}}(t) = \Delta\mathbf{x}(t) - \Delta\mathbf{x}^* \tag{4.7-28}$$

$$\Delta\tilde{\mathbf{u}}(t) = \Delta\mathbf{u}(t) - \Delta\mathbf{u}^* \tag{4.7-29}$$

and write the feedback control law in terms of these variables:

$$\Delta\tilde{\mathbf{u}}(t) = -\,\mathbf{C}\Delta\tilde{\mathbf{x}}(t) \tag{4.7-30}$$

Substituting the four prior equations in eq. 4.7-30,

$$\begin{aligned}\Delta\mathbf{u}(t) &= \mathbf{B}_{22}\Delta\mathbf{y}^* - \mathbf{C}[\Delta\mathbf{x}(t) - \mathbf{B}_{12}\Delta\mathbf{y}^*] \\ &= (\mathbf{B}_{22} + \mathbf{C}\mathbf{B}_{12})\Delta\mathbf{y}^* - \mathbf{C}\Delta\mathbf{x}(t) \\ &= \mathbf{C}_F\Delta\mathbf{y}^* - \mathbf{C}_B\Delta\mathbf{x}(t) \qquad (4.7\text{-}31)\end{aligned}$$

where $\mathbf{C}_F$ is the *forward gain matrix* and $\mathbf{C}_B$ is the *feedback gain matrix*. $\mathbf{C}_F$ depends only indirectly on the optimization and could be used with any alternative definition of the feedback matrix. A $\Delta\mathbf{y}^*$ command to eq. 4.7-28 produces a step response in the closed-loop system whose transient response is governed by $\mathbf{C}_B$ and whose steady-state response yields $\Delta\mathbf{y}^*$ through eqs. 4.7-19, 4.7-26, and 4.7-27. In application, $\Delta\mathbf{y}^*$ is replaced by $\Delta\mathbf{y}_c(t)$ to allow continuous command inputs.

Substituting eq. 4.7-31 in eq. 4.7-4, with $\Delta\mathbf{y}^*$ replaced by $\Delta\mathbf{y}_c(t)$,

$$\begin{aligned}\Delta\dot{\mathbf{x}}(t) &= \mathbf{F}\Delta\mathbf{x}(t) + \mathbf{G}[\mathbf{C}_F\Delta\mathbf{y}_c(t) - \mathbf{C}_B\Delta\mathbf{x}(t)] \\ &= (\mathbf{F} - \mathbf{G}\mathbf{C}_B)\Delta\mathbf{x}(t) + \mathbf{G}\mathbf{C}_F\Delta\mathbf{y}_c(t)\end{aligned} \qquad (4.7\text{-}32)$$

The LQ controller provides the structure for a *command augmentation system*, in which the pilot's input represents not a control surface deflection but a desired value of the state (e.g., commanded pitch rate rather than elevator angle). Taking the Laplace transform, as in eq. 4.3-35, and setting the initial condition to zero, the transfer function matrix relating the closed-loop response of the state, $\Delta\mathbf{x}(s)$, to the command input, $\Delta\mathbf{y}_c(s)$, is

$$\Delta\mathbf{x}(s) = [s\mathbf{I} - (\mathbf{F} - \mathbf{G}\mathbf{C}_B)]^{-1}\,\mathbf{G}\mathbf{C}_F\Delta\mathbf{y}_c(s) \qquad (4.7\text{-}33)$$

Consequently, the frequency-domain techniques of Section 4.4 can readily be applied to the LQ-regulated system, with stability and control-effect matrices modified as indicated.

Implicit Model-Following and Integral Compensation

The LQ regulator provides a good platform for addressing practical flight control problems in the context of command augmentation for assuring satisfactory flying qualities. One reasonable goal for closed-loop control is to cause the aircraft to respond like an ideal aircraft, as defined by flying qualities criteria. The quadratic cost function then minimizes the error between ideal and actual closed-loop dynamics. If there is error in the knowledge of the aircraft's dynamic model or if there are disturbances, the resulting variations in steady-state response can be reduced using integral compensation. Alternatively, integration can be used to attenuate high-frequency control commands, producing smaller excitation of aeroelastic modes.

As its name implies, model-following control is intended to make the closed-loop system behave like an ideal model. In *explicit model-following*, a mathematical model of the ideal system is included in the control logic, and a high-gain feedback loop forces the aircraft to follow that model [T-6, S-14]. This structure adds a *command prefilter* to the controller logic. In *implicit model-following*, the model is used only to choose values of the weighting matrices in the previously defined cost function (eq. 4.7-5), and the control law takes the same form as before (eq. 4.7-31).

An *n*th-order homogeneous model of ideal system dynamics is specified as

$$\Delta\dot{\mathbf{x}}_M(t) = \mathbf{F}_M \Delta\mathbf{x}_M(t) \tag{4.7-34}$$

while the dynamics of the actual system are given by eq. 4.7-4. The cost function is designed to minimize the squared error between the *rates of change* of the two system's states. Minimizing the error between the rates implies that error between the states is minimized as well. Assuming that $\Delta\mathbf{x}_M(t) \approx \Delta\mathbf{x}(t)$, the integrand of the cost function can be written as

$$\begin{aligned}&\frac{1}{2}\{[\Delta\dot{\mathbf{x}}(t) - \Delta\dot{\mathbf{x}}_M(t)]^T \mathbf{Q}_M[\Delta\dot{\mathbf{x}}(t) - \Delta\dot{\mathbf{x}}_M(t)] + \Delta\mathbf{u}^T(t)\mathbf{R}_o\Delta\mathbf{u}(t)\}\\ &= \frac{1}{2}\left\{[\Delta\mathbf{x}^T(t)\Delta\mathbf{u}^T(t)]\begin{bmatrix}(\mathbf{F}-\mathbf{F}_M)^T\\ \mathbf{G}^T\end{bmatrix}\mathbf{Q}_M[(\mathbf{F}-\mathbf{F}_M)\mathbf{G}^T]\begin{bmatrix}\Delta\mathbf{x}(t)\\ \Delta\mathbf{u}(t)\end{bmatrix} + \Delta\mathbf{u}^T(t)\mathbf{R}_o\Delta\mathbf{u}(t)\right\}\end{aligned} \tag{4.7-35}$$

Expanding and setting like terms equal in eq. 4.7-5, the weighting matrices for the standard quadratic cost function are seen to be

$$\mathbf{Q} = (\mathbf{F} - \mathbf{F}_M)^T \mathbf{Q}_M (\mathbf{F} - \mathbf{F}_M) \tag{4.7-36}$$

$$\mathbf{M} = (\mathbf{F} - \mathbf{F}_M)^T \mathbf{Q}_M \mathbf{G} \tag{4.7-37}$$

$$\mathbf{R} = \mathbf{G}^T \mathbf{Q}_M \mathbf{G} + \mathbf{R}_o \tag{4.7-38}$$

$\mathbf{Q}_M$ could be chosen as a diagonal matrix, such as the identity matrix or one that emphasizes important elements of the state. $\mathbf{R}_o$ may be added to assure positive-definiteness of $\mathbf{R}$ and to adjust the overall magnitude of control gains.

The implicit-model-following LQ regulator provides *proportional compensation*, as feedback control is proportional to the state and to the command input (Fig. 4.7-3a). *Proportional-integral (PI) compensation* reduces error in the long-term response to commands by integrating the error between the commanded and actual response and adjusting the control accordingly (Fig. 4.7-3b). This error is defined as

$$\Delta\tilde{\mathbf{y}}(t) = \mathbf{H}_x \Delta\mathbf{x}(t) + \mathbf{H}_u \Delta\mathbf{u}(t) - \Delta\mathbf{y}_c(t) \tag{4.7-39}$$

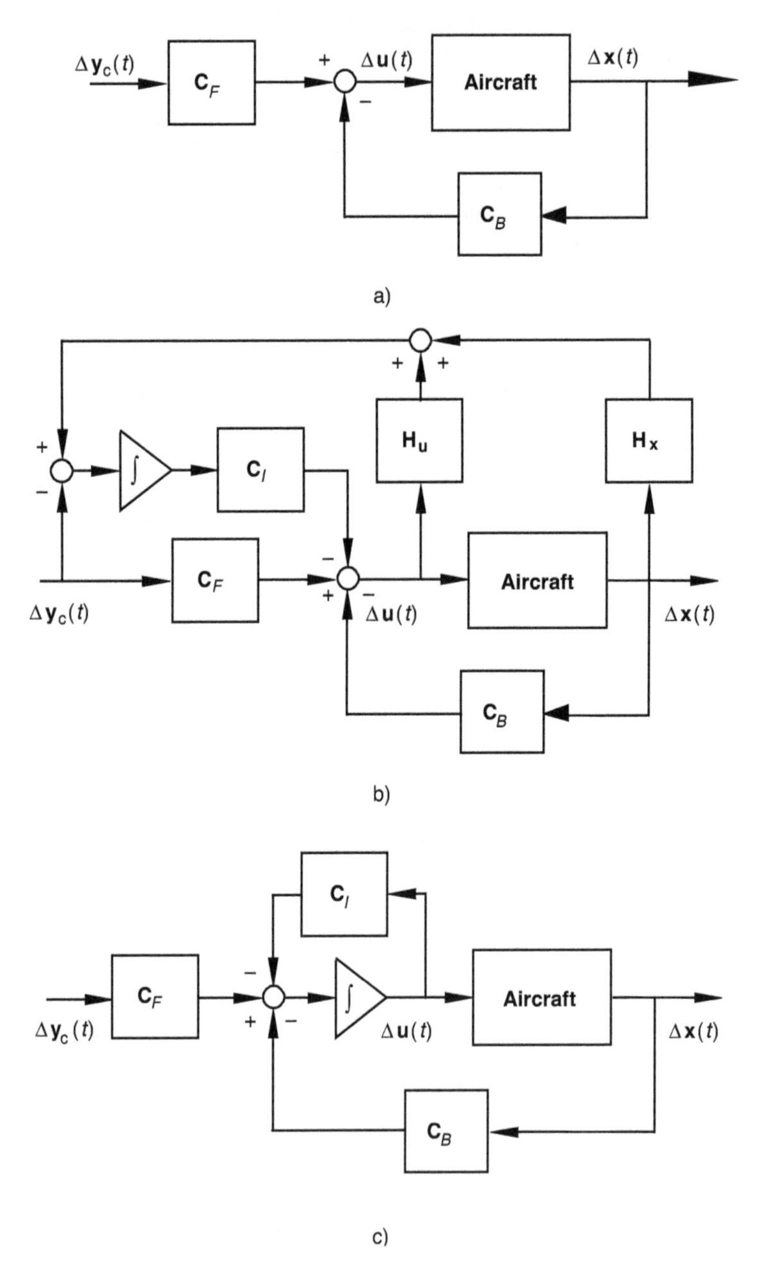

Fig. 4.7-3. Linear-quadratic models for flight control. (a) Proportional LQ regulator. (b) Proportional-integral LQ regulator. (c) Proportional-filter LQ regulator.

and its integral is

$$\Delta\xi(t) = \Delta\xi(t_0) + \int_{t_0}^{t} \Delta\tilde{y}(\tau)\,d\tau \tag{4.7-40}$$

The state equation used for control system design must represent not only the aircraft's dynamics but the dynamics of the integrator as well. Therefore, the system model is augmented to include both:

$$\begin{bmatrix} \Delta\dot{\tilde{x}}(t) \\ \Delta\dot{\xi}(t) \end{bmatrix} = \begin{bmatrix} \mathbf{F} & 0 \\ \mathbf{H_x} & 0 \end{bmatrix} \begin{bmatrix} \Delta\tilde{\mathbf{x}}(t) \\ \Delta\xi(t) \end{bmatrix} + \begin{bmatrix} \mathbf{G} \\ \mathbf{H_u} \end{bmatrix} \Delta\tilde{\mathbf{u}}(t) \tag{4.7-41}$$

We define an *augmented state variable*

$$\Delta\chi(t) = \begin{bmatrix} \Delta\tilde{\mathbf{x}}(t) \\ \Delta\xi(t) \end{bmatrix} \tag{4.7-42}$$

and apply the LQ regulator equations to the modified system,

$$\Delta\dot{\chi}(t) = \mathbf{F}'\Delta\chi(t) + \mathbf{G}'\Delta\tilde{\mathbf{u}}(t) \tag{4.7-43}$$

with $\mathbf{F}'$ and $\mathbf{G}'$ given by eq. 4.7-41. The cost functions $\mathbf{Q}$ and $\mathbf{M}$ are expanded to include weighting on $\Delta\xi(t)$, and the minimizing control law takes the form

$$\begin{aligned} \Delta\mathbf{u}(t) &= \Delta\mathbf{u}^* - \mathbf{C}_1[\Delta\mathbf{x}(t) - \Delta\mathbf{x}^*] - \mathbf{C}_2\left\{\int_{t_0}^{t} [\mathbf{H_x}\Delta\mathbf{x}(\tau) + \mathbf{H_u}\Delta\mathbf{u}(\tau) - \Delta\mathbf{y}^*]\,d\tau\right\} \\ &= \mathbf{C}_F\Delta\mathbf{y}^* - \mathbf{C}_B\Delta\mathbf{x}(t) - \mathbf{C}_I\left\{\int_{t_0}^{t} [\mathbf{H_x}\Delta\mathbf{x}(\tau) + \mathbf{H_u}\Delta\mathbf{u}(\tau) - \Delta\mathbf{y}^*]\,d\tau\right\} \end{aligned} \tag{4.7-44}$$

Over time, the control command builds up to counter persistent command response error, eventually forcing the error to zero when the input is constant, as the augmented system is guaranteed to be stable. Adding integration to a *rate-command* augmentation system converts it to a *rate-command/attitude-hold* system. Proportional-integral compensation is combined with implicit model-following in [H-9, M-5].

Integral compensation is introduced for a different purpose in the *proportional-filter (PF) controller* (Fig. 4.7-3c). Here, the immediate objective is to minimize a quadratic cost function using the rate of change of control rather than the control itself. In the process, a *low-pass filter* is introduced to the control command. The cost function is written as

$$J = \lim_{T\to\infty} \frac{1}{2T} \int_0^T \left\{ [\Delta\tilde{\mathbf{x}}^T(t)\ \Delta\tilde{\mathbf{u}}^T(t)] \begin{bmatrix} \mathbf{Q} & \mathbf{M} \\ \mathbf{M}^T & \mathbf{R}_1 \end{bmatrix} \begin{bmatrix} \Delta\tilde{\mathbf{x}}^T(t) \\ \Delta\tilde{\mathbf{u}}^T(t) \end{bmatrix} + \Delta\dot{\tilde{\mathbf{u}}}^T(t)\mathbf{R}_2\Delta\dot{\tilde{\mathbf{u}}}^T(t) \right\} dt \tag{4.7-45}$$

The augmented state variable is

$$\Delta\chi(t) = \begin{bmatrix} \Delta\tilde{\mathbf{x}}(t) \\ \Delta\tilde{\mathbf{u}}(t) \end{bmatrix} \tag{4.7-46}$$

the new control variable is $\Delta\mathbf{v}(t) = \Delta\dot{\tilde{\mathbf{u}}}(t)$, and the system equation is

$$\begin{bmatrix} \Delta\dot{\tilde{\mathbf{x}}}(t) \\ \Delta\dot{\tilde{\mathbf{u}}}(t) \end{bmatrix} = \begin{bmatrix} \mathbf{F} & \mathbf{G} \\ 0 & 0 \end{bmatrix} \begin{bmatrix} \Delta\tilde{\mathbf{x}}(t) \\ \Delta\tilde{\mathbf{u}}(t) \end{bmatrix} + \begin{bmatrix} 0 \\ \mathbf{I}_m \end{bmatrix} \Delta\mathbf{v}(t) \tag{4.7-47}$$

The system equation takes the standard form (eq. 4.7-43); if the dynamics of control actuation and linkage are of importance (Sec. 5.5 and 6.5), their models can be entered in place of the zeros in the lower rows of eq. 4.7-47. The corresponding LQ regulator is

$$\Delta\dot{\mathbf{u}}(t) = -\mathbf{C}_1[\Delta\mathbf{x}(t) - \Delta\mathbf{x}^*] - \mathbf{C}_2[\Delta\mathbf{u}(t) - \Delta\mathbf{u}^*] \tag{4.7-48}$$

The smoothing effect of this control law is revealed by its transfer function. Neglecting initial conditions, the control law's Laplace transform is

$$\begin{aligned} s\Delta\mathbf{u}(s) &= -\mathbf{C}_1[\Delta\mathbf{x}(s) - \Delta\mathbf{x}^*] - \mathbf{C}_2[\Delta\mathbf{u}(s) - \Delta\mathbf{u}^*] \\ &= \mathbf{C}_F\Delta\mathbf{y}^* - \mathbf{C}_B\Delta\mathbf{x}(s) - \mathbf{C}_C\Delta\mathbf{u}(s) \end{aligned} \tag{4.7-49}$$

or

$$\Delta\mathbf{u}(s) = (s\mathbf{I}_m + \mathbf{C}_C)^{-1}[\mathbf{C}_F\Delta\mathbf{y}^* - \mathbf{C}_B\Delta\mathbf{x}(s) - \mathbf{C}_C\Delta\mathbf{u}(s)] \tag{4.7-50}$$

which is seen to contain a matrix *low-pass filter* in $(s\mathbf{I}_m + \mathbf{C}_C)^{-1}$. This $(m \times m)$ matrix attenuates high-frequency signals while passing low-frequency signals with negligible change. Closed-loop stability meets the same guarantees as does the basic LQ regulator.

Both forms of integral compensation can be combined in the *proportional-integral-filter (PIF) controller* [S-14]. The PIF controller provides both good steady-state characteristics and smooth control commands. A gain-scheduled, digital PIF controller was successfully flight tested in a tandem-rotor helicopter; it is described in [D-8, S-11, S-12].

Optimal State Estimation

Often, only a subset of the state is measured, the state is transformed in the measurements, or the measurements contain significant error. The state must be estimated from these measurements so that it can be used for feedback control. The *Kalman-Bucy filter* [K-1, S-14] is a three-equation algorithm for making this estimate. It minimizes the expected error between the state and its estimate when the dynamic system is linear and random initial conditions, disturbances, and measurement errors are Gaussian. Derivations of the filter equations are presented in the references, and only the main results are presented here.

The $(s \times 1)$ perturbation measurement vector $\Delta\mathbf{z}(t)$ contains a linear combination of the aircraft state plus measurement error (or "noise") $\Delta\mathbf{n}(t)$:

$$\Delta\mathbf{z}(t) = \mathbf{H}\Delta\mathbf{x}(t) + \Delta\mathbf{n}(t) \tag{4.7-51}$$

The effect of external disturbances such as turbulence $\Delta\mathbf{w}(t)$ is added to the state equation (eq. 4.7-4):

$$\Delta\dot{\mathbf{x}}(t) = \mathbf{F}\Delta\mathbf{x}(t) + \mathbf{G}\Delta\mathbf{u}(t) + \mathbf{L}\Delta\mathbf{w}(t) \tag{4.7-52}$$

The expected values of the initial conditions, disturbances, and measurement errors are expressed as the following means, covariance, and spectral density matrices:

$$E[\Delta\mathbf{x}(0)] = \Delta\hat{\mathbf{x}}_0 \tag{4.7-53}$$

$$E\{[\Delta\mathbf{x}(0) - \Delta\hat{\mathbf{x}}_0][\Delta\mathbf{x}(0) - \Delta\hat{\mathbf{x}}_0]^T\} = \mathbf{P}_0 \tag{4.7-54}$$

$$E[\Delta\mathbf{n}(t)] = 0 \tag{4.7-55}$$

$$E\{[\Delta\mathbf{n}(t)\Delta\mathbf{n}^T(\tau)\} = \mathbf{N}\delta(t-\tau) \tag{4.7-56}$$

$$E[\Delta\mathbf{w}(t)] = 0 \tag{4.7-57}$$

$$E\{[\Delta\mathbf{w}(t)\Delta\mathbf{w}^T(\tau)\} = \mathbf{W}\delta(t-\tau) \tag{4.7-58}$$

where $\delta(t-\tau)$ is the Dirac delta function. We further assume that the control is known without error.

The state estimate of the Kalman-Bucy filter (actually an estimate of the state's expected or mean value) is given by

$$\Delta\dot{\hat{\mathbf{x}}}(t) = \mathbf{F}\Delta\hat{\mathbf{x}}(t) + \mathbf{G}\Delta\mathbf{u}(t) + \mathbf{K}(t)[\Delta\mathbf{z}(t) - \mathbf{H}\Delta\hat{\mathbf{x}}(t)], \quad \Delta\hat{\mathbf{x}}(0) = \Delta\mathbf{x}_0 \tag{4.7-59}$$

where the filter gain matrix $\mathbf{K}(t)$ is

$$\mathbf{K}(t) = \mathbf{P}(t)\mathbf{H}^T\mathbf{N}^{-1} \tag{4.7-60}$$

The estimate error covariance $\mathbf{P}(t)$ is found by solving a *matrix Riccati equation* that is the *mathematical dual* of eq. 4.7-14, with no cross-coupling terms (i.e., $\mathbf{M} = 0$) [S-14]:

$$\dot{\mathbf{P}}(t) = \mathbf{F}\mathbf{P}(t) + \mathbf{P}(t)\mathbf{F}^T + \mathbf{L}\mathbf{W}\mathbf{L}^T - \mathbf{P}(t)\mathbf{H}^T\mathbf{N}^{-1}\mathbf{H}\mathbf{P}(t), \quad \mathbf{P}(0) = \mathbf{P}_0 \tag{4.7-61}$$

Like the matrix Riccati equation for control, the solution of eq. 4.7-61 eventually reaches a positive-definite steady state; however, the integration runs forward in time, and the limit is approached as $t \to +\infty$. In the limit, **P**, and therefore **K**, reaches a constant value; the solution can be found from an algebraic Riccati equation (i.e., with $\dot{\mathbf{P}}(t) = \mathbf{0}$) and from `kalman` in MATLAB. Hence, there is some period of time between the estimator's starting transient and the controller's final transient where essentially constant values of **C** and **K** are optimal. In this region, the constant-gain, *linear-quadratic-Gaussian (LQG) regulator* described below is optimal.

The dimension of $\Delta\mathbf{z}$ is not directly tied to the dimension of $\Delta\mathbf{x}$; however, $\Delta\mathbf{x}$ must be *completely observable* from $\Delta\mathbf{z}$ for the estimate to be of value for full state feedback. As noted in [S-14] and elsewhere, this condition is satisfied if the *observability matrix* (eq. 4.2-29)

$$\mathcal{O} = \begin{bmatrix} \mathbf{H} \\ \mathbf{HF} \\ \cdots \\ \mathbf{HF}^{n-1} \end{bmatrix} \tag{4.7-62}$$

has rank n. Complete observability guarantees that the full state at some time t_0 can be estimated from the history of measurements made between t_0 and some later time t_1. However, it makes no statement about the quality of the estimate, which depends on measurement uncertainty, disturbance uncertainty, and the estimation algorithm.

The estimation error for a constant-gain Kalman-Bucy filter is guaranteed to be stable when all matrices are constant and known precisely [S-14], so it eventually goes to zero. Defining the estimation error as

$$\Delta\boldsymbol{\varepsilon}(t)] = \Delta\mathbf{x}(t) - \Delta\hat{\mathbf{x}}(t) \tag{4.7-63}$$

its dynamics are described by the difference between eq. 4.7-52 and 4.7-59, using eq. 4.7-51:

$$\begin{aligned} \Delta\dot{\boldsymbol{\varepsilon}}(t) &= \mathbf{F}\Delta\mathbf{x}(t) + \mathbf{G}\Delta\mathbf{u}(t) + \mathbf{L}\Delta\mathbf{w}(t) - \mathbf{F}\Delta\hat{\mathbf{x}}(t) \\ &\quad - \mathbf{G}\Delta\mathbf{u}(t) - \mathbf{K}[\Delta\mathbf{z}(t) - \mathbf{H}\Delta\hat{\mathbf{x}}(t)] \\ &= (\mathbf{F} - \mathbf{KH})\Delta\boldsymbol{\varepsilon}(t) + \mathbf{L}\Delta\mathbf{w}(t) - \mathbf{K}\Delta\mathbf{n}(t) \end{aligned} \tag{4.7-64}$$

Consequently, stability is governed by the eigenvalues of $(\mathbf{F} - \mathbf{KH})$, all of which have negative real parts when **K** is given by eq. 4.7-60 and when $\mathbf{LWL}^T$ and **N** are positive-definite matrices. Alternative requirements for guaranteed stability are given in [S-14].

Linear-Quadratic-Gaussian Regulator

If the measurement set is too small or noisy to allow an LQ regulator to be implemented directly, the optimal control system must incorporate a Kalman-Bucy filter for state estimation. With constant gains, **C** and **K** can be evaluated off-line, and the basic *LQG regulator* reduces to three on-line equations:

$$\Delta\mathbf{u}(t)] = \mathbf{C}_F\Delta\mathbf{y}(t) - \mathbf{C}_B\Delta\hat{\mathbf{x}}(t) \tag{4.7-65}$$

$$\Delta\dot{\hat{\mathbf{x}}}(t) = \mathbf{F}\Delta\hat{\mathbf{x}}(t)\mathbf{G}\Delta\mathbf{u}(t) + \mathbf{K}[\Delta\mathbf{z}(t) - \mathbf{H}\Delta\hat{\mathbf{x}}(t)] \tag{4.7-66}$$

plus eq. 4.7-51. A corresponding set of equations incorporating integral compensation could be formed. Whereas the LQ regulator uses strictly proportional feedback of the state, the LQG regulator contains nth-order *dynamic feedback compensation* in the form of the estimator (Fig. 4.7-4); hence, the combined dynamics of the controlled system and the feedback loop have $2n$ eigenvalues.

Writing the homogeneous system in terms of $\Delta\mathbf{x}(t)$ and $\Delta\boldsymbol{\varepsilon}(t)$ rather than $\Delta\hat{\mathbf{x}}(t)$ and using eq. 4.7-63,

$$\begin{bmatrix}\Delta\dot{\mathbf{x}}(t)\\ \Delta\dot{\boldsymbol{\varepsilon}}(t)\end{bmatrix} = \begin{bmatrix}(\mathbf{F}-\mathbf{GC}) & \mathbf{GC}\\ 0 & (\mathbf{F}-\mathbf{KH})\end{bmatrix}\begin{bmatrix}\Delta\mathbf{x}(t)\\ \Delta\boldsymbol{\varepsilon}(t)\end{bmatrix} \tag{4.7-67}$$

The eigenvalues of the system are seen to be solutions of the following determinant equation:

$$\begin{vmatrix}[s\mathbf{I}_n - (\mathbf{F}-\mathbf{GC})] & -\mathbf{GC}\\ 0 & [s\mathbf{I}_n - (\mathbf{F}-\mathbf{KH})]\end{vmatrix} = 0 \tag{4.7-68}$$

Because the combined system's matrix is upper block triangular, its $2n$ eigenvalues consist of the n eigenvalues of the upper left block and the n eigenvalues of the lower right block. The LQG-regulated system's eigenvalues are those of the LQ-regulated system plus those of the Kalman-Bucy filter. As both are independently stable when the system model is known perfectly, the LQG-regulated system is stable as well. By way of example, a pitch-rate-command/attitude-hold controller could be implemented as an LQG controller commanding elevator deflection with a reduced set of inputs (e.g., pitch rate and normal load factor) using either a reduced-order or full-state model of aircraft dynamics.

Design for Stochastic Robustness

LQG regulation is a powerful starting point for flight control system design, but *control system robustness* in the presence of reasonable system uncertainties must be assured. With the design model ($\mathbf{F}$, $\mathbf{G}$, $\mathbf{H}$) and the actual model ($\mathbf{F}_A$, $\mathbf{G}_A$, $\mathbf{H}_A$), eq. 4.7-68 becomes

$$\begin{vmatrix}[s\mathbf{I}_n - (\mathbf{F}_A - \mathbf{G}_A\mathbf{C})] & -\mathbf{G}_A\mathbf{C}\\ [(\mathbf{F}-\mathbf{F}_A) - (\mathbf{G}-\mathbf{G}_A)\mathbf{C} - \mathbf{K}(\mathbf{H}-\mathbf{H}_A)] & [s\mathbf{I}_n - [\mathbf{F} + (\mathbf{G}_A - \mathbf{G})\mathbf{C} - \mathbf{KH}]]\end{vmatrix} = 0 \tag{4.7-69}$$

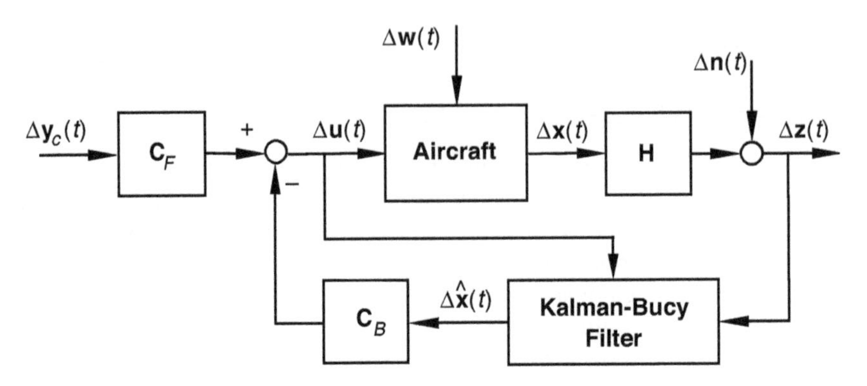

Fig. 4.7-4. Linear-quadratic-Gaussian regulator.

If the assumed system and the actual system are the same, cancellations assure that the lower left block of eq. 4.7-69 is zero, preventing modal coupling of the state and the estimation error in the LQG regulator. The LQ controller alone could be unstable if actual system dynamics stray too far from the values that are used for control system design, and coupling of the LQG controller and estimator when there is mismatch could be destabilizing. Inability to guarantee stability under system mismatch is not a unique failing of LQ/LQG design; *any* nonadaptive design method that uses an incorrect system model is subject to error and instability. Small differences normally would not cause a problem, but large differences could degrade stability and performance. For this reason, the robustness of the flight control system within the limits of expected dynamic variations should be investigated during the design process.

Gain and phase margins have long been used to portray the ability of closed-loop single-input/single-output systems to remain stable when there are changes in loop gain or phase of the feedback signal. These concepts have been extended to multi-input/multi-output systems through singular-value analysis ([S-14] and elsewhere). While these metrics may be useful in specifying desired performance, they are not good indicators of stability robustness for arbitrary variations in system parameters [M-1, M-3].

With uncertain parameter variations, control system robustness is a stochastic issue, and a probabilistic definition of robustness is appropriate. Stochastic analysis can be used as the basis for designing control systems that minimize the probability of instability and inadequate performance (or, conversely, of satisfying flying qualities criteria), given the statistics of parameter uncertainty. In all but the simplest cases, these probability estimates must be numerical because the relationship between parameter variations and the probability of unsatisfactory behavior is inherently nonlinear. Computation time does not present a substantial problem given the speed of contemporary computers, and it will be even less important in the future. A stochastic design approach has the further advantage of using the software already developed for simulation and design as the core of an iterative process. Thus, the procedure for stochastic design is wrapped around the existing deterministic design code, and the process of design is merged with the process of evaluation.

Stochastic control system design adds a numerical search algorithm, such as a line search, genetic algorithm, or simulated annealing, to the Monte Carlo evaluation described in Section 4.5 [M-2, M-4, M-5, M-16]. The purpose of the search is to find a control design vector **d** that minimizes a stability and performance metric, such as

$$J = a_1 P_i^2 + a_2 P_p^2 + a_3 P_c^2 \tag{4.7-70}$$

P_i, P_p, and P_c are the probabilities of instability, unsatisfactory performance, and excess control usage, and a_1, a_2, and a_3 weight the relative importance of these factors in the design. Here, the probabilities are squared to reduce the significance of small values and to increase the consequences of large values. P_i, P_p, and P_c are estimated numerically, using repeated simulation of the controlled system, with the uncertain parameter vector **p** specified on each trial by calls to a random number generator.

The stochastic design approach is applied to a flight control problem in [M-5]. The application is a proportional-integral, linear-quadratic controller for a hypersonic cruising aircraft. There are 28 uncertain aerodynamic, inertial, and atmospheric parameters in **p** for the nonlinear model of longitudinal dynamics. The stability and performance metric J contains 39 criteria, including probability of instability, 16 velocity response metrics, 16 altitude response metrics, and four control usage metrics. A genetic algorithm is used to find minimizing values of ten design parameters in **d**: the diagonal elements of **Q** and **R** and a normal acceleration weight used to design the LQ regulator, plus loop gain of the feedback controller. Hence, the principal result is to specify design parameters that allow the PI-LQ controller to minimize the probability of unsatisfactory stability and performance. A robust controller based on nonlinear-inverse-dynamics (Section 7.4) for the same problem is presented in [W-1].

References

A-1 Abzug, M. J., "Fuel Slosh in Skewed Tanks," *J. Guidance, Control, and Dynamics* 19, no. 5 (1996): 1172–1177.

A-2 Abzug, M. J., and Larrabee, E. E., *Airplane Stability and Control*, Cambridge University Press, Cambridge, 2002.

A-3 A'Harrah, R. C., dir., *Flying Qualities*, AGARD-CP-508, Neuilly-sur-Seine, France, Feb. 1991.

A-4 Anderson, J. R., *Language, Memory, and Thought*, Erlbaum, Hillsdale, N.J., 1976.

A-5 Anderson, R. O., "A New Approach to the Specification and Evaluation of Flying Qualities," AFFDL-TR-69-120, Wright-Patterson AFB, Ohio, 1970.

A-6 Anderson, R. O., "Flying Qualities Yesterday Today and Tomorrow," *AIAA Atmospheric Flight Mechanics Conf.*, Danvers, Mass., Aug. 1980.

A-7 Anon., "Military Specification, Flying Qualities of Piloted Airplanes," MIL-F-8785C, Wright-Patterson Air Force Base, Ohio, Nov. 1980.

A-8 Anon., "Department of Defense Interface Standard, Flying Qualities of Piloted Aircraft," MIL-STD-1797A Update, Wright-Patterson Air Force Base, Ohio, June 1995.

A-9 Anon., "Flight Control Systems - General Specification for Design, Installation, and Test of Piloted Aircraft," MIL-F-9490D, Wright-Patterson Air Force Base, Ohio, 1975.

A-10 Ashkenas, I. L., dir., *Advances in Flying Qualities*, AGARD-LS-157, Neuilly-sur-Seine, France, May 1988.

A-11 Atzhorn, D., and Stengel, R. F., "Design and Flight Test of a Lateral-Directional Command Augmentation System," *J. Guidance, Control, and Dynamics* 7, no. 3 (1984): 361–368.

B-1 Babister, A. W., *Aircraft Stability and Control*, Pergamon Press, Oxford University Press, Oxford, 1961.

B-2 Barber, M. R., Haise, F. W., Jones, C. K., and Sisk, T. R., "An Evaluation of the Handling Qualities of Seven General-Aviation Aircraft," NASA TN-D-3726, Washington, D.C., Nov. 1966.

B-3 Bendat, J. S., and Piersol, A. G., *Measurement and Analysis of Random Data*, J. Wiley & Sons, New York, 1966.

B-4 Benepe, D. B., Sr., "Aircraft Maneuvers and Dynamic Phenomena Resulting in Rapid Changes of Load Distribution or/and Fluctuating Separation," *Prediction of Aerodynamic Loading*, AGARD-CP-204, Neuilly-sur-Seine, France, Feb. 1977.

B-5 Binnie, W. B., and Stengel, R. F., "Flight Investigation and Theory of Direct Side-Force Control," *J. Guidance and Control* 2, no. 6 (1979): 471–478.

B-6 Bisplinghoff, R. L., and Ashley, H., *Principles of Aeroelasticity*, Dover Publications, New York, 1975.

B-7 Bisplinghoff, R. L., Ashley, H., and Halfman, R. L., *Aeroelasticity*, Addison-Wesley, Reading, Mass., 1955.

B-8 Blakelock, J. H., *Automatic Control of Aircraft and Missiles*, J. Wiley & Sons, New York, 1991.

B-9 Bode, H. W., *Network Analysis and Feedback Amplifier Design*, Van Nostrand, New York, 1945.

B-10 Brogan, W. L., *Modern Control Theory*, Prentice-Hall, Englewood Cliffs, N.J., 1985.

B-11 Bryson, A. E., Jr., *Control of Spacecraft and Aircraft*, Princeton University Press, Princeton, N.J., 1994.
B-12 Buttrill, C. S., Zeiler, T. A., and Arbuckle, P. D., "Nonlinear Simulation of a Flexible Aircraft in Maneuvering Flight," AIAA Paper No. 87-2501-CP, Washington, D.C., Aug. 1987.
C-1 Chalk, C. R., "Excessive Roll Damping Can Cause Roll Ratchet," *J. Guidance, Control, and Dynamics* 6, no. 3 (1983): 218–219.
C-2 Cook, M. V., and Rycroft, M. J., *Aerospace Vehicle Dynamics and Control*, Oxford University Press, Oxford, 1994.
C-3 Coupry, G., chair, *The Flight of Flexible Aircraft in Turbulence—State-of-the-Art in the Description and Modelling of Atmospheric Turbulence*, AGARD-R-734, Neuilly-sur-Seine, France, Dec. 1987.
C-4 Cox, T. H., and Jackson, D., "Supersonic Flying Qualities Experience Using the SR-71," NASA TM 4800, Washington, D.C., Aug. 1997.
C-5 Crimaldi, J. P., Britt, R. T., and Rodden, W. P., "Response of B-2 Aircraft to Nonuniform Spanwise Turbulence," *J. Aircraft* 30, no. 5 (1993): 652–659.
C-6 Crouse, G. L., Jr., and Leishman, J. G., "Transonic Aeroelasticity Analysis Using State-Space Unsteady Aerodynamic Modeling," *J. Aircraft* 29, no. 1 (1992): 153–160.
C-7 Curtis, H. D., *Fundamentals of Aircraft Structural Analysis*, Richard D. Irwin, Chicago, 1987.
D-1 DeRusso, P. M., Roy, R. J., and Close, C. M., *State Variables for Engineers*, J. Wiley & Sons, New York, 1965.
D-2 Dillow, J. D., "The 'Paper Pilot'—A Digital Computer Program to Predict Pilot Rating for the Hover Task," AFFDL-TR-70-40, Wright-Patterson AFB, Ohio, 1971.
D-3 Donaldson, B. K., *Analysis of Aircraft Structures: An Introduction*, McGraw-Hill, New York, 1993.
D-4 Donlan, C. J., compiler, "An Interim Report on the Stability and Control of Tailless Airplanes," NACA Report 796, Washington, D.C., Aug. 1944.
D-5 Donovan, A. F., Lawrence, H. R., Goddard, F., and Gilruth, R. R., *High Speed Problems of Aircraft and Experimental Methods*, Vol. VIII of *High Speed Aerodynamics and Jet Propulsion*, Princeton University Press, Princeton, N.J., 1961.
D-6 Dornfeld, G. M., Dusto, A. R., Rubbert, P. E., Pilet, S. C., Brune, G. W., Smutny, P., Webber, J. A., Mercer, J. E., and Tinoco, E. N., "A Method for Predicting the Stability Characteristics of an Elastic Airplane, Vol. 1, FLEXSTAB 3.01.00 Theoretical Description," AFFDL-TR-77-55, Vol. 1, Wright-Patterson AFB, Ohio, June 1977.
D-7 Dowell, E. H., Curtiss, H. C., Jr., Scanlan, R. H., and Sisto, F., *A Modern Course in Aeroelasticity*, Sitjhoff & Noordhoff, Alphen aan den Rijn, The Netherlands, 1978.
D-8 Downing, D. R., and Bryant, W. H., "Flight Test of a Digital Controller Used in a Helicopter Autoland System," *Automatica* 23, no. 5 (1987): 295–300.
D-9 Duncan, W. J., *Control and Stability of Aircraft*, Cambridge University Press, Cambridge, 1952.

D-10 Dykman, J. R., and Rodden, W. R., "Structural Dynamics and Quasistatic Aeroelastic Equations of Motion," *J. Aircraft* 37, no. 3 (2000): 538–542.
E-1 Etkin, B., *Dynamics of Atmospheric Flight*, J. Wiley & Sons, New York, 1972.
E-2 Etkin, B., *Dynamics of Flight—Stability and Control*, J. Wiley & Sons, New York, 1982.
E-3 Evans, W. R., "Graphical Analysis of Control Systems," *Trans. American Institute of Electrical Engineers* 67 (1948): 547–551.
F-1 Fogiel, M., ed., *Handbook of Mathematical, Scientific, and Engineering Formulas, Tables, Functions, Graphs, Transforms*, Research and Education Assoc., Piscataway, N.J., 1988.
F-2 Försching, H., chair, *Static Aeroelastic Effects on High Performance Aircraft*, AGARD-CP-403, Neuilly-sur-Seine, France, July 1987.
F-3 Fraser, C. D., ed., "Special Section: Flying Qualities," *J. Guidance, Control, and Dynamics* 9, no. 5 (1986): 514–584.
F-3 Frazer, R. A., and Duncan, W. J., "The Flutter of Aeroplane Wings," British Aeronautical Research Council R&M 1155, London, 1928.
F-4 Frazer, R. A., and Duncan, W. J., "Wing Flutter as Influenced by the Mobility of the Fuselage," British Aeronautical Research Council R&M 1207, London, 1929.
F-5 Fung, Y. C., *An Introduction to the Theory of Aeroelasticity*, Dover Publications, New York, 1969.
G-1 Gardner, M. F., and Barnes, J. L., *Transients in Linear Systems*, J. Wiley & Sons, New York, 1942.
G-2 Gibson, J. C., *The Definition, Understanding and Design of Aircraft Handling Qualities*, Delft University Press, Delft, The Netherlands, 1997.
G-3 Gilruth, R. R., "Requirements for Satisfactory Flying Qualities of Airplanes," NACA Report 755, Washington, D.C., Mar. 1941.
G-4 Graham, E. W., *The Forces Produced by Fuel Oscillation in a Rectangular Tank*, Douglas Aircraft Report No. ES 29441, Long Beach, Calif., 1950.
G-5 Grunwald, S., and Stengel, R. F., "Design and Flight Testing of Digital Direct Side-Force Control Laws," *J. Guidance, Control, and Dynamics* 8, no. 2 (1985): 188–193.
H-1 Hacker, T., *Flight Stability and Control*, American Elsevier, New York, 1970.
H-2 Hanson, G., and Stengel, R. F., "Effects of Displacement and Rate Saturation on the Control of Statically Unstable Aircraft," *J. Guidance, Control, and Dynamics* 7, no. 2 (1984): 197–205.
H-3 Hanson, G., and Stengel, R. F., "Effects of Control Saturation on the Command Response of Statically Unstable Aircraft," AIAA Paper No. 83-0065, Washington, D.C., Jan 1983.
H-4 Harper, R. P., Jr., and Cooper, G. E., "Handling Qualities and Pilot Evaluation," *J. Guidance, Control, and Dynamics* 9, no. 5 (1986): 515–529.
H-5 Hinto, D. E., and Anderson, J. A., *Parallel Models of Associative Memory*, Erlbaum, Hillsdale, N.J., 1981.
H-6 Hodgkinson, J., *Aircraft Handling Qualities*, American Institute of Aeronautics and Astronautics, Reston, Va., 1999.

H-7 Hoh, R. H., and Mitchell, D. G., "Handling-Qualities Specification—A Functional Requirement for the Flight Control System," in *Advances in Aircraft Flight Control*, M. B. Tischler, ed., Taylor & Francis, London, 1996, pp. 3–34.

H-8 Houbolt, J. C., Steiner, R., and Pratt, K. G., *Dynamic Response of Aircraft to Atmospheric Turbulence Including Flight Data on Input and Response*, NASA TR R-199, Washington, D.C., June 1964.

H-9 Huang, C., and Stengel, R. F., "Restructurable Control Using Proportional-Integral Implicit Model Following," *J. Guidance, Control, and Dynamics* 13, no. 2 (1990): 303–309.

K-1 Kalman, R. E., and Bucy, R. S., "New Results in Linear Filtering and Prediction," *ASME Trans., J. Basic Engineering* 83D, Mar. (1961): 95–108.

K-2 Kayten, G. G., "Analysis of Wind-Tunnel Stability and Control Tests in Terms of Flying Qualities of Full-Scale Airplanes, NACA Report 825, Washington, D.C., Apr. 1943.

K-3 Kihlstrom, J. F., "The Cognitive Unconscious," *Science*, 237 (1987): 1445–1452.

K-4 Kleinman, D. L., Baron, S., and Levison, W., "An Optimal Control Model of Human Response," *Automatica* 9, no. 3 (1970): 357–383.

L-1 Lamar, W. E., ed., "Handling Qualities Criteria," AGARD-CP-106, London, June 1972.

L-2 Lamb, H., *Hydrodynamics*, Dover Publications, New York, 1945.

L-3 Lane, S., and Stengel, R. F., "Flight Control Design Using Non-linear Inverse Dynamics," *Automatica* 24, no. 4 (1988): 471–483.

L-4 Luskin, H., and Lapin, E., "An Analytical Approach to the Fuel Sloshing and Buffeting Problems of Aircraft," *J. Aeronautical Sciences* 8, no. 4 (1952): 217–228.

M-1 Marrison, C., and Stengel, R. F., "Robustness of Solutions to a Benchmark Control Problem," *J. Guidance, Control, and Dynamics* 15, no. 5 (1992): 1060–1067.

M-2 Marrison, C., and Stengel, R.F., "Stochastic Robustness Synthesis Applied to a Benchmark Control Problem," *Int'l. J. Robust and Nonlinear Control* 5, no. 1 (1995): 13–31.

M-3 Marrison, C., and Stengel, R. F., "Reply by Authors to J. J. Gribble," *J. Guidance, Control, and Dynamics* 18, no. 6 (1995): 1468–1470.

M-4 Marrison, C., and Stengel, R. F., "Robust Control System Design Using Random Search and Genetic Algorithms," *IEEE Trans. Automatic Control* 42, no. 6 (1997): 835–839.

M-5 Marrison, C., and Stengel, R. F., "Design of Robust Control Systems for a Hypersonic Aircraft," *J. Guidance, Control, and Dynamics* 21, no. 1 (1998): 58–63.

M-6 McLean, D., *Automatic Flight Control Systems*, Prentice-Hall, Englewood Cliffs, N.J., 1990.

M-7 McRuer, D., Ashkenas, I., and Graham, D., *Aircraft Dynamics and Automatic Control*, Princeton University Press, Princeton, N.J., 1973.

M-8 McRuer, D. T., Graham, D., Krendel, E., and Reisiner, W., Jr., "Human Pilot Dynamics in Compensatory Systems," AFFDL-TR-65-15, Wright-Patterson Air Force Base, Ohio, 1965.

M-9 McRuer, D. T., "Development of Pilot-In-The-Loop Analysis," *J. Aircraft* 10, no. 9 (1973): 515–524.

M-10 McRuer, D. T., chair, *Aviation Safety and Pilot Control: Understanding and Preventing Unfavorable Pilot-Vehicle Interactions*, National Academy Press, Washington, D.C., 1997.

M-11 Meirovitch, L., *Principles and Techniques of Vibrations*, Prentice-Hall, Englewood Cliffs, N.J., 1997.

M-12 Meyer, G., and Cicolani, L., "Application of Nonlinear System Inverse to Automatic Flight Control System Design—System Concepts and Flight Evaluations," *Theory and Application of Optimal Control in Aerospace Systems*, AGARD-AG-251, Neuilly-sur-Seine, France, 1981, pp. 10-1–10-29.

M-13 Milne, R. D., "Dynamics of the Deformable Airplane," British Aeronautical Research Council R&M No. 3345, London, 1962.

M-14 Miranda, L. R., Elliott, R. D., and Baker, W. M., "A Generalized Vortex Lattice Method for Subsonic and Supersonic Flow Applications," NASA CR-2865, Washington, D.C., Dec. 1977.

M-15 Moorhouse, D. J., "The History and Future of U.S. Military Flying Qualities Specification," AIAA Paper No. 79-0402, New Orleans, Jan. 1979.

M-16 Motoda, T., Stengel, R. F., and Miyazawa, Y., "Robust Control System Design Using Simulated Annealing," *J. Guidance, Control, and Dynamics* 25, no. 2 (2002): 267–274.

M-17 Myers, T. T., Johnston, D. E., and McRuer, D. T., "Space Shuttle Flying Qualities and Criteria Assessment," NASA CR-4049, Washington, D.C., Mar. 1987.

N-1 Narendra, K. S., and Taylor, J. H., *Frequency Domain Criteria for Absolute Stability*, Academic Press, New York, 1973.

O-1 Olsen, J. J., *Aeroelastic Considerations in the Preliminary Design of Aircraft*, AGARD-CP-354, Neuilly-sur-Seine, France, Sept. 1983.

P-1 Papadimitriou, C. H., and Steiglitz, K., *Combinatorial Optimization: Algorithms and Complexity*, Dover Publications, New York, 1998.

P-2 Papoulis, A., *Probability and Statistics*, Prentice-Hall, Englewood Cliffs, N.J., 1990.

P-3 Perry, D. J., and Azar, J. J., *Aircraft Structures*, McGraw-Hill, New York, 1982.

P-4 Pratt, R., ed., *Flight Control Systems*, American Institute of Aeronautics and Astronautics, Reston, Va., 2000.

R-1 Raveh, D. E., Levy, Y., and Karpel, M., "Efficient Aeroelastic Analysis Using Computational Unsteady Aerodynamics," *J. Aircraft* 38, no. 3 (2001): 547–556.

R-2 Ray, L. R., and Stengel, R. F., "Stochastic Robustness of Linear-Time-Invariant Control Systems," *IEEE Trans. Automatic Control* 36, no. 1 (1991): 82–87.

R-3 Ray, L. R., and Stengel, R. F., "Application of Stochastic Robustness to Aircraft Control Systems," *J. Guidance, Control, and Dynamics* 14, no. 6 (1991): 1251–1259.

R-4 Ray, L. R., and Stengel, R. F., "Stochastic Measures of Performance Robustness in Aircraft Control Systems," *J. Guidance, Control, and Dynamics* 15, no. 6 (1992): 1381–1387.

R-5 Rea Company, Inc., J. B., "Aeroelasticity in Stability and Control," USAF WADC TR-55-173, Wright-Patterson AFB, Ohio, Mar. 1957.
R-6 Roberts, L., *Unsteady Aerodynamics—Fundamentals and Applications to Aircraft Dynamics*, AGARD-CP-386, Neuilly-sur-Seine, Nov. 1985.
R-7 Rodden, W. P., and Love, J. R., "Equations of Motion of a Quasisteady Flight Vehicle Utilizing Restrained Static Aeroelastic Characteristics," *J. Aircraft* 22, no. 9 (1985): 802–809.
R-8 Rodden, W. P., and Johnson, E. H., *MSC/NASTRAN Aeroelastic Analysis User's Guide*, Version 68, The MacNeal-Schwendler Corp., Los Angeles, 1994.
S-1 Scanlan, R. H., and Rosenbaum, R., *Introduction to the Study of Aircraft Vibration and Flutter*, Dover Publications, New York, 1968.
S-2 Schmidt, D. K., and Raney, D. L., "Modeling and Simulation of Flexible Flight Vehicles," *J. Guidance, Control, and Dynamics* 24, no. 3 (2001): 539– 546.
S-3 Schwanz, R. C., "Formulations of the Equations of Motion of an Aeroelastic Aircraft for Stability and Control and Flight Control Applications," AFFDL/FGC-72-14, Wright-Patterson AFB, Ohio, Aug. 1972.
S-4 Schy, A. A., "A Theoretical Analysis of the Effects of Fuel Motion on Airplane Dynamics," NACA Report 1080, Washington, D.C., 1952.
S-5 Seckel, E., *Stability and Control of Airplanes and Helicopters*, Academic Press, New York, 1964.
S-6 Shrivastava, P., and Stengel, R. F., "Stability Boundaries for Aircraft with Unstable Lateral-Directional Dynamics and Control Saturation," *J. Guidance, Control, and Dynamics* 12, no. 1 (1989): 62–70.
S-7 Smith, J. W., and Montgomery, T., "Biomechanically Induced and Controller Coupled Oscillations Experienced on the F-16XL Aircraft During Rolling Maneuvers," NASA TM 4752, Washington, D.C., July 1996.
S-8 Stark, L., *Neurological Control Systems*, Plenum Press, New York, 1968.
S-9 Stengel, R. F., and Berry, P. W., "Stability and Control of Maneuvering High-Performance Aircraft," NASA CR-2788, Washington, D.C., Apr. 1977.
S-10 Stengel, R. F., and Broussard, J. R., "Prediction of Pilot-Aircraft Stability Boundaries and Performance Contours," *IEEE Trans. Systems, Man, and Cybernetics* SMC-8, no. 5 (1978): 349–356.
S-11 Stengel, R. F., Broussard, J., and Berry, P., "Digital Controllers for VTOL Aircraft," *IEEE Trans. Aerospace and Electronic Systems* AES-14, no. 1 (1978): 54–63.
S-12 Stengel, R. F., Broussard, J., and Berry, P., "Digital Flight Control Design for a Tandem-Rotor Helicopter," *Automatica* 14, no. 4 (1978): 301–311.
S-13 Stengel, R. F., "In-Flight Simulation with Pilot-Center of Gravity Offset and Velocity Mismatch," *J. Guidance and Control* 2, no. 6 (1979): 538–540.
S-14 Stengel, R. F., *Optimal Control and Estimation*, Dover Publications, New York, 1994 (originally published as *Stochastic Optimal Control: Theory and Application*, J. Wiley & Sons, New York, 1986).
S-15 Stengel, R. F., "Equilibrium Response of Flight Control Systems," *Automatica* 18, no. 3 (1982): 343–348.
S-16 Stengel, R. F., "Toward Intelligent Flight Control," *IEEE Trans. Systems, Man, and Cybernetics* 23, no. 6 (1993): 1699–1717.

S-17 Stevens, B. L., and Lewis, F. L., *Aircraft Control and Simulation*, J. Wiley & Sons, New York, 1992.

S-18 Stinton, D., *Flying Qualities and Flight Testing of the Airplane*, American Institute of Aeronautics and Astronautics, Washington, D.C., 1996.

S-19 Stone, J. R., and Gerken, G. J., "The Prediction of Pilot Acceptance for a Large Aircraft," *Proc. 1973 Joint Automatic Control Conference*, Columbus, Ohio, June 1973, pp. 807–809.

T-1 Taylor, A. S., and Woodcock, D. L., "Mathematical Approaches to the Dynamics of Deformable Aircraft," British Aeronautical Research Council R&M 3776, London, 1976.

T-2 Tischler, M. B., ed., *Advances in Aircraft Flight Control*, Taylor & Francis, London, 1996.

T-3 Tischler, M. B., Lee, J. A., and Colbourne, J. D., "Comparison of Flight Control Design Methods Using the CONDUIT Design Tool," *J. Guidance, Control, and Dynamics* 25, no. 3 (2002): 482–493.

T-4 Trayer, G. W., and March, H. W., "The Torsion of Members Having Sections Common in Aircraft Construction," NACA Report 334, Washington, D.C., Jan. 1930.

T-5 Tustin, A., "The Nature of the Operator's Response in Manual Control and Its Implications for Controller Design," *Proc. IEE* 94, part IIA (1947): 190–202.

T-6 Tyler, J. S., Jr., "The Characteristics of Model-Following Systems as Synthesized by Optimal Control," *IEEE Trans. Automatic Control* 9, no. 5 (1964): 485–498.

V-1 van Paassen, M. M., *Biophysics in Aircraft Control: A model of the neuromuscular system of the pilot's arm*, Doctoral thesis, Delft University of Technology, Delft, The Netherlands, 1994.

V-2 Venneri, S. L., chair, *Transonic Unsteady Aerodynamics and Aeroelasticity*, AGARD-CP-507, Neuilly-sur-Seine, France, Mar. 1992.

V-3 Vincenti, W. G., *What Engineers Know and How They Know It*, Johns Hopkins University Press, Baltimore, 1990.

W-1 Wang, Q., and Stengel, R. F., "Robust Nonlinear Control of a Hypersonic Aircraft," *J. Guidance, Control, and Dynamics* 23, no. 4 (2000): 577–585.

W-2 Waszak, M. R., and Schmidt, D. K., "Flight Dynamics of Aeroelastic Vehicles," *J. Aircraft* 25, no. 6 (1988): 563–571.

W-3 Williams, J., "The USAF Stability and Control Digital DATCOM," AFFDL-TR-76-45, Wright-Patterson AFB, Ohio, Nov. 1976.

W-4 Wykes, J. H., "Flying Qualities with Elastic Airframes," AGARD-CP-106, London, June 1972.

W-5 Wykes, J. H., and Borland, C. J., "B-1 Ride Control," in *Active Controls in Aircraft Design*, P. R. Kourzhals, ed., AGARD-AG-234, Nov. 1978, pp. 11-1–11-15.

Z-1 Zwaan, R. J., chair, *Advanced Aeroservoelastic Testing and Data Analysis*, AGARD-CP-566, Neuilly-sur-Seine, France, Nov. 1995.

Chapter 5

Longitudinal Motions

When a symmetric aircraft is either in steady, level flight or climbing or descending in the vertical plane, longitudinal and lateral-directional motions are uncoupled to first order. The most significant longitudinal dynamic effects can be characterized by only four equations representing perturbations in the translational velocity, flight path angle, pitch angle, and angular rate (Section 5.1). Altitude variation has small effect on dynamic characteristics, and range perturbations have no dynamic effect in the flat-earth model. Thus, significant insights about the nature of motions in the vertical plane can be obtained by studying a fourth-order set of linear, time-invariant ordinary differential equations.

Further simplification is warranted by differences in the time scales of the modes of longitudinal motion. We can, for example, use separate second-order models to examine the phugoid and short-period modes, identifying the primary stability derivatives for each mode and the important effects of the longitudinal controls: elevator, thrust, flaps, and symmetric spoilers. Initial studies of more complex effects take advantage of these simplifications. Long-time-scale phenomena like atmospheric gradient effects or optimal climb and descent profiles are only marginally affected by short-period dynamics. Conversely, interactions arising from the internal dynamics of control systems and from aeroelasticity are largely independent of the phugoid mode. Understanding these reduced-order models aids interpretation of high-fidelity simulations and flight testing results.

This chapter first examines second-order models (Sections 5.2 and 5.3) and then progresses to a fourth-order model that couples short-period and phugoid modes (Section 5.4). Coupling effects are most significant when the frequencies of these two modes are close, as is the case for large transport aircraft. Both reduced-order models contain the effects of aerodynamic lift, and the fourth-order model more accurately portrays lift-induced perturbations than either second-order model alone. Selected longitudinal flying qualities criteria are presented in the section.

Control-system dynamics and structural flexibility may couple with otherwise rigid-body motions, and the final two sections deal with these interactions. Introductory material that applies to lateral-directional as well as longitudinal motions is presented before addressing specifically longitudinal effects. Dynamics of the control system mechanization are important when the control surfaces are not constrained to a fixed position, as when the pilot is not exerting force on the control stick of a mechanical system or a failure has occurred in a powered system (Section 5.5). Design aspects of the control mechanization have strong effect on the aircraft handling qualities perceived by the pilot, particularly in regard to the stick force or displacement required for pitch-axis commands. Aeroelastic distortions of the wing, fuselage, and empennage produced by aerodynamic loads and vibration alter the net forces and moments, which in turn couple into the pitching and heaving of the aircraft (Section 5.6). Because aircraft are designed for light weight and high aerodynamic efficiency, aeroelasticity may have a critical effect on overall stability and performance.

5.1 Longitudinal Equations of Motion

The flat-earth equation summary of Section 3.6 presents translational and rotational velocity equations in a body-axis reference frame and translational and rotational position equations in an earth-fixed (~inertial) frame. In this section, we present the longitudinal equations in an alternative, hybrid reference frame that may be found more intuitive than the body-axis equations.

For flight in the vertical plane, the *longitudinal equations of motion* describe changes in axial and normal velocity u and w, pitch rate and angle q and θ, range x_I, and altitude $-z_I$. The six equations are

$$\dot{u} = X/m - g\sin\theta - qw \tag{5.1-1}$$

$$\dot{w} = Z/m + g\cos\theta + qu \tag{5.1-2}$$

$$\dot{q} = M/I_{yy} \tag{5.1-3}$$

$$\dot{x}_I = u\cos\theta + w\sin\theta \tag{5.1-4}$$

$$\dot{z}_I = -u\sin\theta + w\cos\theta \tag{5.1-5}$$

$$\dot{\theta} = q \tag{5.1-6}$$

The true airspeed components u_A and w_A contain earth-relative and wind velocity components w_x and w_z in the body-axis frame of reference:

$$u_A = u - w_x\cos\theta + w_z\sin\theta \tag{5.1-7}$$

$$w_A = w - w_x \sin\theta - w_z \cos\theta \tag{5.1-8}$$

The true airspeed and air-relative angle of attack are related to these body-axis components as

$$V_A = \sqrt{u_A^2 + w_A^2} \tag{5.1-9}$$

$$\alpha_A = \tan^{-1}(w_A/u_A) \tag{5.1-10}$$

The dynamic pressure is derived from the air density $\rho(z)$ and airspeed,

$$\bar{q} \triangleq \frac{\rho(z)\, V_A^2}{2} \tag{5.1-11}$$

and the Mach number is defined with the sound speed $a(z)$ as

$$\mathcal{M} = \frac{V_A}{a(z)} \tag{5.1-12}$$

The axial force, normal force, and pitching moment contain aerodynamic and thrusting effects:

$$X = C_X\, \bar{q}\, S + T_x \tag{5.1-13}$$

$$Z = C_Z\, \bar{q}\, S + T_z \tag{5.1-14}$$

$$M = C_m\, \bar{q} S\, \bar{c} + M_{T_y} \tag{5.1-15}$$

The corresponding linearized equations for a symmetric aircraft in steady, level flight are given by eqs. 4.1-96–4.1-102.

Two changes to these equations are desirable for the following analyses. u and w are replaced by the inertial speed V and flight path angle γ. The pitch angle θ can be replaced by the inertial (zero-wind) angle of attack α, because, at zero roll angle,

$$\alpha = \tan^{-1}(w/u) = \theta - \gamma \tag{5.1-16}$$

The lift and drag L and D are computed using the true airspeed V_A. They are aligned with and perpendicular to the air-relative velocity vector, which is misaligned from the inertial velocity vector by $\Delta\alpha_w$, where

$$\Delta\alpha_w = \alpha - \alpha_A \tag{5.1-17}$$

Thus, a wind disturbance produces two effects in this coordinate frame: it changes the lift and drag magnitudes, and it rotates their "lines of action."

The resulting *hybrid-axis longitudinal equations of motion* are summarized as

$$\dot{V} = \frac{(-D\cos\Delta\alpha_w - L\sin\Delta\alpha_w + T_x\cos\alpha + T_z\sin\alpha)}{m} - g\sin\gamma \tag{5.1-18}$$

$$\dot{\gamma} = \frac{(-D\sin\Delta\alpha_w + L\cos\Delta\alpha_w + T_x\sin\alpha - T_z\cos\alpha)}{mV} - \frac{g\cos\gamma}{V} \tag{5.1-19}$$

$$\dot{q} = M/I_{yy} \tag{5.1-3}$$

$$\dot{\alpha} = \dot{\theta} - \dot{\gamma} = q - \frac{(-D\sin\Delta\alpha_w + L\cos\Delta\alpha_w + T_x\sin\alpha - T_z\cos\alpha)}{mV} + \frac{g\cos\gamma}{V} \tag{5.1-20}$$

with the range and altitude equations expressed in terms of V and γ:

$$\dot{x}_I = V\cos\gamma \tag{5.1-21}$$

$$\dot{z}_I = -V\sin\gamma \tag{5.1-22}$$

This new equation set allows long and short-period motions to be largely isolated, as reflected in the linearized equations that accompany it. We expand these equations to first order for the state, control, and disturbance perturbation vectors:

$$\Delta\mathbf{x}_{\mathrm{Lo}}^{\mathrm{T}} = [\Delta V\ \Delta\gamma\ \Delta q\ \Delta\alpha\ \Delta x_I\ \Delta z_I]^T \tag{5.1-23}$$

$$\Delta\mathbf{u}_{\mathrm{Lo}}^{\mathrm{T}} = [\Delta\delta E\ \Delta\delta T\ \Delta\delta F]^T \tag{5.1-24}$$

$$\Delta\mathbf{w}_{\mathrm{Lo}}^{\mathrm{T}} = [\Delta V_w\ \Delta\alpha_w]^T \tag{5.1-25}$$

The control vector $\Delta\mathbf{u}_{\mathrm{Lo}}$ contains elevator, thrust, and flap controls. The linearized state equation takes the form

$$\Delta\dot{\mathbf{x}}_{\mathrm{Lo}} = \mathbf{F}_{\mathrm{Lo}}\Delta\mathbf{x}_{\mathrm{Lo}} + \mathbf{G}_{\mathrm{Lo}}\Delta\mathbf{u}_{\mathrm{Lo}} + \mathbf{L}_{\mathrm{Lo}}\Delta\mathbf{w}_{\mathrm{Lo}} \tag{5.1-26}$$

Neglecting altitude gradients in air density and sound speed, the stability and control derivative matrices are

$$
\mathbf{F}_{Lo} = \begin{bmatrix}
-D_V & -g\cos\gamma_o & -D_q & -D_\alpha & 0 & 0 \\
\dfrac{L_V}{(V_o + L_{\dot\alpha})} & \dfrac{g\sin\gamma_o}{V_o} & \dfrac{L_q}{(V_o + L_{\dot\alpha})} & \dfrac{L_\alpha}{(V_o + L_{\dot\alpha})} & 0 & 0 \\
[M_V - a\,L_V] & 0 & [M_q - a(L_q - V_o)] & [M_\alpha - a\,L_\alpha] & 0 & 0 \\
\dfrac{-L_V}{(V_o + L_{\dot\alpha})} & \dfrac{-g\sin\gamma_o}{V_o} & \dfrac{(V_o - L_q)}{(V_o + L_{\dot\alpha})} & \dfrac{-L_\alpha}{(V_o + L_{\dot\alpha})} & 0 & 0 \\
\cos\gamma_o & -V_o\sin\gamma_o & 0 & 0 & 0 & 0 \\
-\sin\gamma_o & -V_o\cos\gamma_o & 0 & 0 & 0 & 0
\end{bmatrix}
\tag{5.1-27}
$$

$$
a \triangleq \frac{M_{\dot\alpha}}{(V_o + L_{\dot\alpha})}
$$

$$
\mathbf{G}_{Lo} = \begin{bmatrix}
-D_{\delta E} & T_{\delta T} & -D_{\delta F} \\
\dfrac{L_{\delta E}}{(V_o + L_{\dot\alpha})} & \dfrac{L_{\delta T}}{(V_o + L_{\dot\alpha})} & \dfrac{L_{\delta F}}{(V_o + L_{\dot\alpha})} \\
[M_{\delta E} - a\,L_{\delta E}] & [M_{\delta T} - a\,L_{\delta T}] & [M_{\delta F} - a\,L_{\delta F}] \\
\dfrac{-L_{\delta E}}{(V_o + L_{\dot\alpha})} & \dfrac{-L_{\delta T}}{(V_o + L_{\dot\alpha})} & \dfrac{-L_{\delta F}}{(V_o + L_{\dot\alpha})} \\
0 & 0 & 0 \\
0 & 0 & 0
\end{bmatrix}
\tag{5.1-28}
$$

$$
\mathbf{L}_{Lo} = -\begin{bmatrix}
-D_V & -D_\alpha \\
\dfrac{L_V}{(V_o + L_{\dot\alpha})} & \dfrac{L_\alpha}{(V_o + L_{\dot\alpha})} \\
[M_V - a\,L_V] & [M_\alpha - a\,L_\alpha] \\
\dfrac{-L_V}{(V_o + L_{\dot\alpha})} & \dfrac{-L_\alpha}{(V_o + L_{\dot\alpha})} \\
0 & 0 \\
0 & 0
\end{bmatrix}
\tag{5.1-29}
$$

where the dimensional stability and control derivatives are defined in Section 4.1. These sensitivity matrices are evaluated at a nominal flight condition defined by

$$
\mathbf{x}_{Lo_o}^{T} = [V_o\ \gamma_o\ q_o\ \alpha_o\ x_{I_o}\ z_{I_o}]^T \tag{5.1-30}
$$

$$
\mathbf{u}_{Lo_o}^{T} = [\delta E_o\ \delta T_o\ \delta F_o]^T \tag{5.1-31}
$$

$$
\mathbf{w}_{Lo_o}^{T} = [V_{w_o}\ \alpha_{w_o}]^T \tag{5.1-32}
$$

which is distinct from the perturbation definitions of eqs. 5.1-23–5.1-25. The first-order effects of wind disturbances are contained in $\mathbf{L}_{\mathrm{Lo}}$ (eq. 5.1-29).

For the hybrid-axis system, the stability matrix $\mathbf{F}_{\mathrm{Lo}}$ is organized in (2×2) blocks of the general form shown in Table 5.1-1. The principal phugoid variables are ΔV and $\Delta\gamma$, the principal short-period variables are Δq and $\Delta\alpha$, and the position variables are Δx_I and Δz_I, as illustrated by the eigenvectors in the following example. From eq. 5.1-27, it is apparent that the position variables are driven by the phugoid variables, but the position is otherwise uncoupled from the longitudinal dynamic model.

Example 5.1-1. Hybrid-Axis Eigenvectors of a Business Jet Aircraft

Eigenvectors of the fourth-order hybrid-axis model are readily compared to those of the body-axis model presented in Example 4.3-2 using SURVEY (Appendix C). The eigenvalues for both fourth-order models are

$$\lambda_{1-4} = -0.007 \pm 0.125j, \ -1.28 \pm 2.81j$$

The magnitudes of phugoid and short-period eigenvectors, ordered as [ΔV (m/s), $\Delta\gamma$ (rad), Δq (rad/s), $\Delta\alpha$ (rad)] are

$$\mathbf{a}_{\mathrm{Ph}} = \begin{bmatrix} |\Delta V_{Ph}| \\ |\Delta\gamma_{Ph}| \\ |\Delta q_{Ph}| \\ |\Delta\alpha_{Ph}| \end{bmatrix} = \begin{bmatrix} 0.999 \\ 0.013 \\ 0.002 \\ 0.0003 \end{bmatrix}$$

$$\mathbf{a}_{\mathrm{SP}} = \begin{bmatrix} |\Delta V_{SP}| \\ |\Delta\gamma_{SP}| \\ |\Delta q_{SP}| \\ |\Delta\alpha_{SP}| \end{bmatrix} = \begin{bmatrix} 0.624 \\ 0.106 \\ 0.73 \\ 0.257 \end{bmatrix}$$

Normalizing the velocity perturbation by V_o produces the following:

$$\mathbf{a}'_{\mathrm{Ph}} = \begin{bmatrix} |\Delta V'_{Ph}| \\ |\Delta\gamma'_{Ph}| \\ |\Delta q'_{Ph}| \\ |\Delta\alpha'_{Ph}| \end{bmatrix} = \begin{bmatrix} 0.605 \\ 0.79 \\ 0.1 \\ 0.016 \end{bmatrix}$$

$$\mathbf{a}'_{\mathrm{SP}} = \begin{bmatrix} |\Delta V'_{SP}| \\ |\Delta\gamma'_{SP}| \\ |\Delta q'_{SP}| \\ |\Delta\alpha'_{SP}| \end{bmatrix} = \begin{bmatrix} 0.008 \\ 0.136 \\ 0.935 \\ 0.329 \end{bmatrix}$$

TABLE 5.1-1 STRUCTURE OF THE LONGITUDINAL DYNAMIC MODEL

Phugoid Parameters	Short-Period-to-Phugoid Coupling	Position-to-Phugoid Coupling
Phugoid-to-Short-Period Coupling	**Short-Period Parameters**	Position-to-Short-Period Coupling
Phugoid-to-Position Coupling	Short-Period-to-Position Coupling	**Position Parameters**

Hybrid coordinates produce better decoupling of the two modes than does the previous model. The phugoid mode is largely represented by ΔV and $\Delta\gamma$ components, while the short-period mode is represented by Δq and $\Delta\alpha$. Still, the two modes are not perfectly decoupled, and accurate representation of longitudinal motions calls for the fourth-order model.

5.2 Reduced-Order Models of Long-Period Modes

We begin our study of reduced-order dynamic models with long-period longitudinal motions, which involve inertial velocity, flight path angle, and altitude excursions from the nominal path. The *phugoid mode* is driven by the interchange of kinetic and potential energy that results when the aircraft's center of mass is perturbed from an equilibrium cruising condition [L-1]. Away from the ground, the *height mode* is produced by altitude gradients in the air density, speed of sound, and gravitational acceleration [S-3, S-10]. Near the ground, a similar effect is produced by the changing boundary condition on the flow around the aircraft (the "ground effect") [P-2]. The principal factors in these modes are the drag, lift, and thrust variations that are induced by airspeed, climb angle, and height perturbations.

SECOND-ORDER PHUGOID-MODE APPROXIMATION

Drawing from the upper left corner of the longitudinal model (eq. 5.1-26), the truncated second-order approximation is

$$\Delta \dot{\mathbf{x}}_P = \mathbf{F}_P \, \Delta \mathbf{x}_P + \mathbf{G}_P \, \Delta \mathbf{u}_P + \mathbf{L}_P \, \Delta \mathbf{w}_P \qquad (5.2\text{-}1)$$

where the state contains the inertial velocity magnitude and the flight path angle:

$$\Delta \mathbf{x}_P = \begin{bmatrix} \Delta V \\ \Delta \gamma \end{bmatrix} \qquad (5.2\text{-}2)$$

The altitude varies considerably in this oscillation; however, if ground effect and altitude gradients in air density, speed of sound, and gravity are neglected, the altitude variation has no effect on the dynamic characteristics of the phugoid mode. The inertial frame may be fixed with respect to the earth or moving at constant horizontal velocity with the mean wind. We consider thrust control and wind disturbance along the velocity vector:

$$\Delta \mathbf{u}_P = \Delta\delta T \tag{5.2-3}$$

$$\Delta \mathbf{w}_P = \Delta V_W \tag{5.2-4}$$

The stability-and-control-derivative matrices are

$$\mathbf{F}_P = \begin{bmatrix} -D_V & -g\cos\gamma_o \\ \dfrac{L_V}{V_o} & \dfrac{g\sin\gamma_o}{V_o} \end{bmatrix} \tag{5.2-5}$$

$$\mathbf{G}_P = \begin{bmatrix} T_{\delta T} \\ \dfrac{L_{\delta T}}{V_o} \end{bmatrix} \tag{5.2-6}$$

$$\mathbf{L}_P = \begin{bmatrix} D_V \\ \dfrac{-L_V}{V_o} \end{bmatrix} \tag{5.2-7}$$

The effects of pitch rate and angle of attack, including unsteady aerodynamics, are neglected. We use this model to demonstrate the linear analytical methods introduced in Chapter 4.

Equilibrium Response to Control and Disturbance

From eq. 4.2-12, the equilibrium velocity and flight path angle perturbations produced by step changes in the control and disturbance are

$$\Delta \mathbf{x}_P{}^* = -\mathbf{F}_P^{-1}(\mathbf{G}_P\,\Delta \mathbf{u}_P{}^* + \mathbf{L}_P\,\Delta \mathbf{w}_P{}^*) \tag{5.2-8}$$

In steady, level flight, $\gamma_o = 0$, and this becomes

$$\begin{bmatrix} \Delta V^* \\ \Delta\gamma^* \end{bmatrix} = -\begin{bmatrix} 0 & V_o/L_V \\ \dfrac{-1}{g} & -\dfrac{V_o D_V}{g\,L_V} \end{bmatrix} \left\{ \begin{bmatrix} T_{\delta T} \\ \dfrac{L_{\delta T}}{V_o} \end{bmatrix} \Delta\delta T^* + \begin{bmatrix} D_V \\ \dfrac{-L_V}{V_o} \end{bmatrix} \Delta V_W{}^* \right\} \tag{5.2-9}$$

leading to the scalar equations

$$\Delta V^* = -\frac{L_{\delta T}}{L_V}\Delta\delta T^* + \Delta V_W{}^* \tag{5.2-10}$$

$$\Delta\gamma^* = \frac{1}{g}\left(T_{\delta T} + L_{\delta T}\frac{D_V}{L_V}\right)\Delta\delta T^* \tag{5.2-11}$$

The model indicates that an incremental thrust change produces a steady-state velocity change only if there is a direct lifting effect $L_{\delta T}$ of the thrust. However, $L_{\delta T}$ is small and often negligible for a conventional aircraft configuration. The dominant thrust effect through $T_{\delta T}$ produces a steady flight path angle change but no change in equilibrium velocity. The horizontal wind perturbation shifts the steady inertial velocity directly, and it has no effect on equilibrium flight path angle.

Controllability and Observability

With just one control and two elements of the state, we may wonder if the model is controllable in the sense of Section 4.2. The controllability matrix (eq. 4.2-48) is

$$\mathscr{C} \triangleq [\mathbf{G_P}\ \mathbf{F_P G_P}] = \begin{bmatrix} T_{\delta T} & -\left(D_V\, T_{\delta T} + \dfrac{g}{V_o}L_{\delta T}\right) \\ \dfrac{L_{\delta T}}{V_o} & \dfrac{L_V\, T_{\delta T}}{V_o} \end{bmatrix} \tag{5.2-12}$$

Its determinant

$$|\mathscr{C}| = \frac{L_V T_{\delta T}^2}{V_o} + \frac{L_{\delta T}}{V_o}\left(D_V T_{\delta T} + \frac{g}{V_o}L_{\delta T}\right) \tag{5.2-13}$$

is nonzero as long as either $T_{\delta T}$ or $L_{\delta T}$ is nonzero and the additive terms do not cancel. Hence, the rank of $\mathscr{C}$ is 2, and the model is completely controllable. This means that both velocity and flight path angle perturbations could be brought from arbitrary initial conditions to zero in finite time by the use of thrusting control.

The model is clearly observable if both velocity and flight path angle are measured, but it remains to be seen if observability is retained with just one measured variable. If velocity alone is measured

$$\mathbf{H_P} = [1\ 0] \tag{5.2-14}$$

the observability matrix (eq. 4.2-49) is

$$\mathbb{O} = \begin{bmatrix} \mathbf{H}_P \\ \mathbf{H}_P\mathbf{F}_P \end{bmatrix} = \begin{bmatrix} 1 & 0 \\ -D_V & -g \end{bmatrix} \tag{5.2-15}$$

the determinant is $-g$, the rank is 2, and the system is completely observable. Measuring flight path angle alone,

$$\mathbf{H}_P = [0\ 1] \tag{5.2-16}$$

$$\mathbb{O} = \begin{bmatrix} 0 & 1 \\ \dfrac{L_V}{V_o} & 0 \end{bmatrix} \tag{5.2-17}$$

the determinant is L_V/V_o, the rank is 2, and the system is completely observable. The coupling of velocity and flight path angle perturbations by the phugoid mode induces complete observability from a single output.

Eigenvalues, Natural Frequency, and Damping Ratio

The characteristic matrix is

$$(s\mathbf{I}_2 - \mathbf{F}_P) = \begin{bmatrix} (s + D_V) & g\cos\gamma_o \\ \dfrac{-L_V}{V_o} & \left(s - \dfrac{g\sin\gamma_o}{V_o}\right) \end{bmatrix} \tag{5.2-18}$$

and the characteristic equation is

$$\Delta(s) = |s\mathbf{I}_2 - \mathbf{F}_P| = s^2 + \left(D_V - \frac{g}{V_o}\sin\gamma_o\right)s + \frac{g}{V_o}(L_V\cos\gamma_o - D_V\sin\gamma_o) = 0 \tag{5.2-19}$$

For level flight ($\gamma_o = 0$), this equation reduces to

$$\Delta(s) = s^2 + D_V s + \frac{g}{V_o}L_V = 0 \tag{5.2-20}$$

where D_V and L_V are typically positive. The eigenvalues of the matrix are given by

$$\lambda_{1,2} = -\frac{D_V}{2} \pm \sqrt{\left(\frac{D_V}{2}\right)^2 - \frac{g}{V_o}L_V} \tag{5.2-21a}$$

As the argument of the square root is normally negative, the roots form a complex pair

$$\lambda_{1,2} = -\frac{D_V}{2} \pm j\sqrt{\frac{g}{V_o}L_V - \left(\frac{D_V}{2}\right)^2} \tag{5.2-21b}$$

with natural frequency and damping ratio

$$\omega_{n_p} = \sqrt{g\frac{L_V}{V_o}} \tag{5.2-22}$$

$$\zeta_P = \frac{D_V}{2\omega_{n_P}} = \frac{D_V}{2\sqrt{(g/V_o)L_V}} \tag{5.2-23}$$

In steady, level flight, the lift equals the weight, leading to some interesting conclusions about the phugoid natural frequency and period. From eq. 5.1-19, with negligible thrust effects, $\Delta\alpha = 0$, and zero mean wind, the flight path angle sensitivity to speed variations is

$$\begin{aligned}
\frac{\partial\Delta\dot{\gamma}}{\partial V} \triangleq \frac{L_V}{V_o} &= \frac{\partial}{\partial V}\left(\frac{L - mg}{mV}\right) = \frac{\partial}{\partial V}\left(\frac{C_L\frac{1}{2}\rho V^2 S - mg}{mV}\right) \\
&= \frac{C_L\frac{1}{2}\rho S}{m} + \frac{g}{V_o^2} + \frac{C_{L_V}\frac{1}{2}\rho V_o S}{m} \\
&= \frac{C_L\frac{1}{2}\rho V_o^2 S}{mV_o^2} + \frac{g}{V_o^2} + \frac{C_{L_V}}{C_L}\frac{C_L\frac{1}{2}\rho V_o^2 S}{mV_o} \\
&= \frac{2g}{V_o^2} + \frac{C_{L_V}}{C_L}\frac{g}{V_o}
\end{aligned} \tag{5.2-24}$$

C_L is the lift coefficient required for steady, level flight, and $C_{L_V} = \partial C_L/\partial V$. From eq. 5.2-22, the phugoid natural frequency ω_{nP} (rad/s) is

$$\omega_{nP} = \sqrt{g\left(\frac{2g}{V_o^2} + \frac{C_{L_V}}{C_L}\frac{g}{V_o}\right)} \tag{5.2-25}$$

and the corresponding undamped period of oscillation T_P (s) is

$$T_P = \frac{2\pi}{\omega_{np}} = \frac{2\pi}{\sqrt{g\left(\frac{2g}{V_o^2} + \frac{C_{L_V}}{C_L}\frac{g}{V_o}\right)}} \tag{5.2-26}$$

In subsonic flight, $C_{L_V} \approx 0$; hence, the natural frequency, period, and wavelength W_P in this approximation depend only on the gravitational constant and equilibrium velocity:

$$\omega_{np}(\text{rad/s}) = \sqrt{2}\frac{g}{V_o} = \frac{13.9}{V_o(\text{m/s})} \tag{5.2-27}$$

$$T_P(\text{s}) = \frac{\sqrt{2}\pi V_o}{g} = 0.453V_o(\text{m/s}) \tag{5.2-28}$$

$$W_P(\text{m}) = 0.453V_o^2(\text{m/s}) \tag{5.2-29}$$

The phugoid wavelength is an important consideration in the response to atmospheric disturbances: an atmospheric wave of length W_P would tend to resonate with the phugoid mode, causing large excursions in velocity, flight path angle, and altitude. Table 5.2-1a summarizes the velocity effects on these quantities.

We can make similar estimates of variations in damping for steady, level flight. The sensitivity of velocity change to velocity perturbations is conveyed by

$$\frac{\partial \Delta \dot{V}}{\partial V} \triangleq -D_V = \frac{\partial}{\partial V}\frac{(T-D)}{m} = \frac{\partial}{\partial V}\left[\frac{(C_T - C_D)\frac{1}{2}\rho V^2 S}{m}\right]$$

$$= \frac{1}{m}\left[(C_T - C_D)\rho V_o S + (C_{T_V} - C_{D_V})\frac{1}{2}\rho V_o^2 S\right] \tag{5.2-30}$$

The thrust, modeled as $C_T \rho V^2 S/2$ (Section 2.5), is assumed to be aligned with the velocity vector. If lift equals weight, the equation can be manipulated into the following form:

$$D_V = -\frac{g}{V_o}\left[\frac{2(C_T - C_D)}{C_L} + V_o\frac{(C_{T_V} - C_{D_V})}{C_L}\right] \tag{5.2-31}$$

TABLE 5.2-1 CHARACTERISTICS OF THE SECOND-ORDER PHUGOID APPROXIMATION, NEGLECTING AIR-DENSITY COMPRESSIBILITY

(a) Natural Frequency, Period, and Wavelength

V_o, m/s	ω_{np}, rad/s	T_P, s	W_P, m
50	0.28	23	1,150
100	0.14	45	4,500
150	0.09	68	10,200
200	0.07	91	18,200
250	0.06	113	20,250

(b) Damping Ratio

L/D	ζ_P (unpowered flight)	ζ_P(propeller; $\eta = -3$)	ζ_P (jet $\eta = -2$)
5	0.14	0.21	0.14
10	0.07	0.11	0.07
20	0.04	0.05	0.04
40	0.0	0.03	0.02

In subsonic, unpowered (gliding) flight, $C_T = C_{T_V} = C_{D_V} = 0$, and

$$D_V = \frac{2g}{V_o}\left(\frac{C_D}{C_L}\right) \tag{5.2-32}$$

Neglecting the nonzero flight path angle of gliding flight (eq. 5.2-9) and using eqs. 5.2-27 and 5.2-32 in eq. 5.2-23, the damping ratio is an inverse function of the lift-to-drag ratio (Table 5.2-1b):

$$\zeta_P = \frac{1}{\sqrt{2}}\left(\frac{C_D}{C_L}\right) = \frac{1}{\sqrt{2}(L/D)} \tag{5.2-33}$$

In cruising flight, thrust equals drag; the first term of eq. 5.2-31 vanishes, and damping of the phugoid mode approximation is due entirely to the difference between the velocity sensitivities of the thrust and drag coefficients. From Section 2.5, C_{T_V} takes the general form $\eta C_T/V_o$, where η is the power of the thrust-coefficient sensitivity to velocity. With $C_{D_V} = 0$,

$$D_V = \frac{-g\,C_{T_V}}{C_L} \tag{5.2-34}$$

and

$$\zeta_P = \frac{-g\, C_{T_V}/C_L}{2\sqrt{2}g/V_o} = \frac{-V_o\, C_{T_V}/C_L}{2\sqrt{2}} = \frac{-\eta\, C_T/C_L}{2\sqrt{2}}$$
$$= \frac{-\eta\, C_D/C_L}{2\sqrt{2}} = \frac{-\sqrt{2}\,\eta}{4\,(L/D)} \tag{5.2-35}$$

Aerodynamically efficient aircraft typically have low phugoid damping. For a propeller-driven aircraft with constant power ($\eta = -3$), the phugoid approximation has greater damping than either the gliding aircraft or one that is powered by a turbojet at constant thrust ($\eta = -2$) (Table 5.2-1b). When $\eta = 0$, there is no damping. The mode becomes a divergent oscillation with positive values of η.

Eigenvectors

Literal formulas for the phugoid eigenvectors can be found by substituting eq. 5.2-21 in eq. 4.3-61. In level flight, the adjoint of the characteristic matrix is

$$\mathrm{Adj}(s\,\mathbf{I}_2 - \mathbf{F}_P) = \begin{bmatrix} s & -g \\ \dfrac{L_V}{V_o} & (s + D_V) \end{bmatrix} \tag{5.2-36}$$

Choosing λ_1 as the root with positive imaginary part,

$$\mathrm{Adj}(\lambda_1\,\mathbf{I}_2 - \mathbf{F}_P) = [a_1\mathbf{e}_1 \;\; a_2\mathbf{e}_1]$$
$$= \begin{bmatrix} -\dfrac{D_V}{2} + j\sqrt{\dfrac{g}{V_o}L_V - \left(\dfrac{D_V}{2}\right)^2} & -g \\ \dfrac{L_V}{V_o} & \dfrac{D_V}{2} + j\sqrt{\dfrac{g}{V_o}L_V - \left(\dfrac{D_V}{2}\right)^2} \end{bmatrix} \tag{5.2-37}$$

The columns of eq. 5.2-37 differ by a complex constant, and either one can serve as the first eigenvector.

$\mathrm{Adj}(\lambda_2\mathbf{I}_2 - \mathbf{F}_P)$ is the complex conjugate of $\mathrm{Adj}(\lambda_1\mathbf{I}_2 - \mathbf{F}_P)$. A modal matrix can be formed using the first columns of the two adjoints:

$$\mathbf{E} = [\mathbf{e}_1\ \mathbf{e}_2]$$

$$= \begin{bmatrix} -\frac{D_V}{2} + j\sqrt{\frac{g}{V_o}L_V - \left(\frac{D_V}{2}\right)^2} & -\frac{D_V}{2} - j\sqrt{\frac{g}{V_o}L_V - \left(\frac{D_V}{2}\right)^2} \\ \frac{L_V}{V_o} & \frac{L_V}{V_o} \end{bmatrix} \tag{5.2-38}$$

Recall that the eigenvectors can be expressed in polar form (eq. 4.3-70):

$$\mathbf{E} = \begin{bmatrix} A_V\, e^{j\phi_V} & A_V\, e^{-j\phi_V} \\ A_\gamma\, e^{j\phi_\gamma} & A_\gamma\, e^{-j\phi_\gamma} \end{bmatrix} \tag{5.2-39}$$

The relative involvement of velocity and flight path angle in the phugoid mode is represented by the ratio of A_V to A_γ, while the phasing of initial condition response is represented by the difference in ϕ_V and ϕ_γ. From eq. 5.2-38, the relative magnitudes and phase angles of the eigenvectors are

$$\begin{bmatrix} A_V \\ A_\gamma \end{bmatrix} = \begin{bmatrix} \sqrt{g\frac{L_V}{V_o}} \\ \frac{L_V}{V_o} \end{bmatrix} \tag{5.2-40}$$

$$\begin{bmatrix} \phi_V \\ \phi_\gamma \end{bmatrix} = \begin{bmatrix} \tan^{-1}\sqrt{\frac{g\,L_V/V_o}{(D_V/2)^2} - 1} \\ 0 \end{bmatrix} \tag{5.2-41}$$

The ratio of A_V to A_γ can be simplified to $g/\sqrt{L_V/V_o}$.

Root Locus Analysis of Parameter Variations

We use the approximate phugoid characteristic equation (eq. 5.2-20) to demonstrate the application of the root locus technique to variations in L_V/V_o and D_V. To explore the effect of L_V/V_o, the equation can be put in the form

$$\Delta(s) = d(s) + k\, n(s) = 0 \tag{5.2-42}$$

with $k = L_V/V_o$, $n(s) = g$ (there is no dependence on s), and

$$d(s) = s^2 + D_V s = s(s + D_V) \tag{5.2-43}$$

The poles of $d(s)$ are

$$\lambda_1 = 0 \tag{5.2-44}$$

$$\lambda_2 = -D_V \tag{5.2-45}$$

and the root locus plot for $\pm$ variations in L_V/V_o is shown in Fig. 5.2-1a. For $L_V/V_o = 0$, the system contains a pure integration (λ_1) and a damped first-order mode (λ_2). The calculations underlying the figures of this chapter are roughly based on the stability and control derivatives for the business jet model at $\mathcal{M} = 0.3$ and an altitude of 3,050 m, with variations made to illustrate parametric effects (Table 5.2-2).

From eq. 4.4-31, the root center of gravity is at $-D_V/2$, there are no zeros, and the two roots must go to infinity as $|L_V/V_o|$ becomes large. The asymptotes are ± 90 deg for positive L_V/V_o and (0 deg, 180 deg) for negative L_V/V_o (not a likely situation). The plot shows that increasing L_V/V_o increases both the damped and undamped natural frequencies. Negative L_V/V_o produces two real modes, one of which is unstable.

To examine the effects of D_V variation, we set $k = D_V$, $n(s) = s$, and

$$d(s) = s^2 + g\frac{L_V}{V_o} \tag{5.2-46}$$

There is a zero at the origin, and, for positive L_V, $d(s)$ has imaginary poles at $\pm j\sqrt{g(L_V/V_o)}$. The corresponding root locus (Fig. 5.2-1b) is a circle in the complex plane, with stable values (to the left) for positive D_V and unstable values for negative D_V. In other words, the natural frequency is constant, and D_V affects only the damping ratio. For very high values of $|D_V|$, the roots become real, with one approaching the zero at the origin, and the other going to $\pm\infty$, depending on the sign of D_V.

TABLE 5.2-2 BASELINE PHUGOID APPROXIMATION FOR FIGURES OF SECTION 5.2

$$\mathbf{F} = \begin{bmatrix} -0.0185 & -9.8067 \\ 0.0019 & 0 \end{bmatrix} \quad \mathbf{G} = \begin{bmatrix} 4.6645 \\ 0 \end{bmatrix}$$

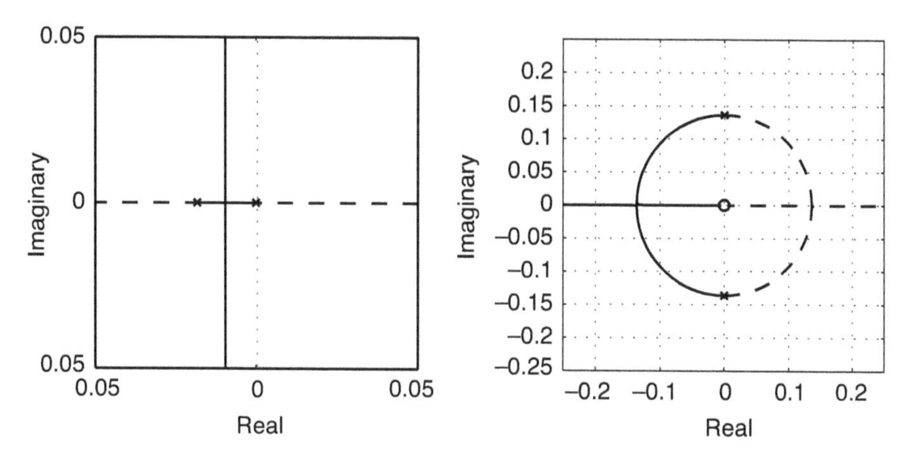

Fig. 5.2-1. Effects of variation in L_v/V_o (left) and D_v (right) on the roots of the second-order phugoid mode approximation. Positive variations shown by solid line, negative variations by dashed line.

Frequency Response

The control and disturbance transfer function matrices $\mathbf{U}_P(s)$ and $\mathbf{W}_P(s)$ relate the output Laplace transform $\Delta\mathbf{y}_P(s)$ to the inputs $\Delta\mathbf{u}_P(s)$ and $\Delta\mathbf{w}_P(s)$; following eq. 4.4-1,

$$\begin{aligned}\Delta\mathbf{y}_P(s) &= \mathbf{H}_P\,(s\mathbf{I}_2 - \mathbf{F}_P)^{-1}[\mathbf{G}_P\,\Delta\mathbf{u}_P(s) + \mathbf{L}_P\,\Delta\mathbf{w}_P(s)] \\ &\triangleq \mathbf{U}_P(s)\,\Delta\mathbf{u}_P(s) + \mathbf{W}_P(s)\,\Delta\mathbf{w}_P(s)\end{aligned} \tag{5.2-47}$$

The characteristic matrix inverse is

$$(s\mathbf{I}_2 - \mathbf{F}_P)^{-1} = \frac{\mathrm{Adj}(s\mathbf{I}_2 - \mathbf{F}_P)}{|s\mathbf{I}_2 - \mathbf{F}_P|} = \frac{\begin{bmatrix} s & -g \\ \dfrac{L_V}{V_o} & (s + D_V) \end{bmatrix}}{s^2 + D_V\,s + g\,(L_V/V_o)} \tag{5.2-48}$$

The individual scalar transfer functions are linear combinations of the elements of this matrix weighted by the elements of $\mathbf{H}_P$, $\mathbf{G}_P$, and $\mathbf{L}_P$. With $\mathbf{H}_P = \mathbf{I}_2$, and $\mathbf{G}_P$ and $\mathbf{L}_P$ defined by eq. 5.2-6 and 5.2-7, the (2×1) transfer function matrices are as follows:

$$\mathbf{U}_P(s) = \begin{bmatrix} \dfrac{\Delta V(s)}{\Delta\delta T(s)} \\ \dfrac{\Delta\gamma(s)}{\Delta\delta T(s)} \end{bmatrix} = \frac{\begin{bmatrix} s\,T_{\delta T} - g\,(L_{\delta T}/V_o) \\ T_{\delta T}\dfrac{L_V}{V_o} + \dfrac{L_{\delta T}}{V_o}(s + D_V) \end{bmatrix}}{s^2 + D_V s + g\,(L_V/V_o)}$$

$$= \frac{T_{\delta T}\begin{bmatrix} s \\ \dfrac{L_V}{V_o} \end{bmatrix}}{s^2 + 2\,\zeta_P\,\omega_{np} s + \omega_{np}^2} \quad (\text{with } L_{\delta T} = 0) \tag{5.2-49}$$

$$\mathbf{W}_P(s) = \begin{bmatrix} \dfrac{\Delta V(s)}{\Delta V_W(s)} \\ \dfrac{\Delta\gamma(s)}{\Delta V_W(s)} \end{bmatrix} = \frac{\begin{bmatrix} s\,D_V + g\,(L_V/V_o) \\ D_V\dfrac{L_V}{V_o} - \dfrac{L_V}{V_o}(s + D_V) \end{bmatrix}}{s^2 + D_V s + g\,(L_V/V_o)}$$

$$= \frac{\begin{bmatrix} D_V\left(s + \dfrac{g}{D_V}\dfrac{L_V}{V_o}\right) \\ -s\dfrac{L_V}{V_o} \end{bmatrix}}{s^2 + 2\zeta_P\omega_{np} s + \omega_{np}^2} \tag{5.2-50}$$

The velocity response to thrust is proportional to the derivative of the input when $L_{\delta T} = 0$, as there is a zero at the origin. Thus, the velocity response leads the flight path angle response by 90 deg. The steady-state response to step inputs can be found from the transfer functions by setting $s = 0$, then multiplying by the input magnitude; compare the results of eqs. 5.2-10 and 5.2-11 with the results using eqs. 5.2-49 and 5.2-50. The corresponding Bode plots could be sketched using the rules of Section 4.4; MATLAB is used to show the plots in Fig. 5.2-2. The light damping produces a sharp peak in the amplitude ratio near the phugoid's natural frequency.

Resonant peaks also occur in the response to sinusoidal horizontal wind input (Fig. 5.2-3), amplifying the hazard of microburst wind shear encounter. While the Bode plot strictly portrays steady-state frequency response, it nevertheless reveals sensitivity to quasi-periodic inputs. The microburst is a large downdraft produced by local cooling at high altitude, as occurs in the formation of rain. When the downdraft impinges on the ground, an outflow is produced. If the microburst is situated on

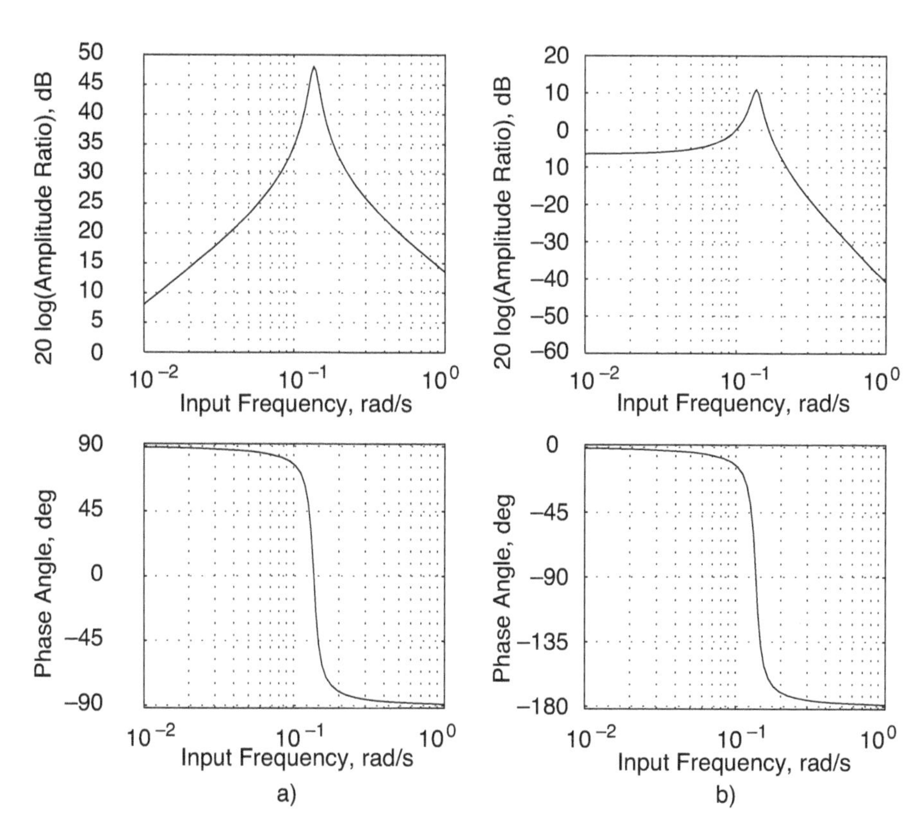

Fig. 5.2-2. Velocity and flight path angle frequency response of the second-order phugoid mode approximation to thrust input. (a) $\Delta V(\omega)/\Delta\delta T(\omega)$. (b) $\Delta\gamma(\omega)/\Delta\delta T(\omega)$.

the approach or departure path, an aircraft first experiences a headwind, then a downdraft, and finally a tailwind, possibly accompanied by secondary vorticity. The horizontal wind component appears as a single cycle of a periodic wave. For a microburst core that is a few kilometers in diameter, the period of the disturbance may be close to the period of a jet transport's phugoid mode. Consequently, maintaining the takeoff climb or the landing glide slope may be difficult using conventional piloting techniques. Optimal flight paths and guidance strategies for minimizing the microburst hazard are presented in [M-5–M-7, P-6–P-8].

Root Locus Analysis of Feedback Control

The stability effects of feeding velocity and flight path angle perturbations back to thrust can be examined using the root locus. With $L_{\delta T} = 0$,

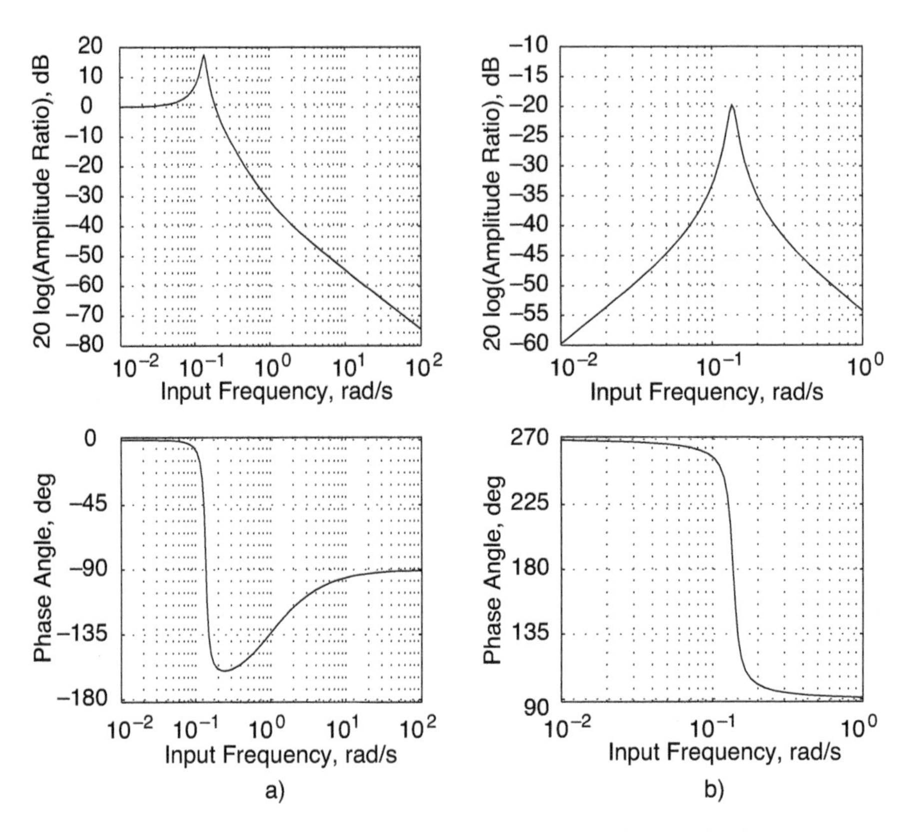

Fig. 5.2-3. Velocity and flight path angle frequency response of the second-order phugoid mode approximation to horizontal wind input. (a) $\Delta V(\omega)/\Delta V_W(\omega)$. (b) $\Delta\gamma(\omega)/\Delta V_W(\omega)$.

the root locus accounts for the dynamic effect of feedback on the closed-loop characteristic equation:

Velocity Feedback to Thrust

$$\Delta(s) = d(s) + k\, n(s) = \left(s^2 + D_V s + g\frac{L_V}{V_o}\right) + k\, T_{\delta T} s = 0 \tag{5.2-51}$$

Flight-Path-Angle Feedback to Thrust

$$\Delta(s) = d(s) + k\, n(s) = \left(s^2 + D_V s + g\frac{L_V}{V_o}\right) + k\, T_{\delta T}\frac{L_V}{V_o} = 0 \tag{5.2-52}$$

By inspection, velocity feedback is analogous to the D_V effect, while flight-path-angle feedback is like the L_V effect, with the root loci beginning at the *open-loop roots* of the model.

Time Response

The time evolution of the state is found by integrating eq. 5.2-1. Because the system is linear, solutions can be expressed as functions of the phugoid mode's state transition matrix $\mathbf{\Phi}_P(\Delta t)$ (eqs. 4.2-3–4.2-10). From eq. 4.3-43, $\mathbf{\Phi}_P(\Delta t)$ is calculated as the inverse Laplace transform of the characteristic matrix inverse:

$$\mathbf{\Phi}_P(\Delta t) = \mathscr{L}^{-1}(s\mathbf{I}_2 - \mathbf{F}_P)^{-1} \triangleq \begin{bmatrix} \phi_{11} & \phi_{12} \\ \phi_{21} & \phi_{22} \end{bmatrix}$$

$$= \frac{e^{-\zeta\omega_n \Delta t}}{\omega_d} \begin{bmatrix} \omega_n \cos(\omega_d \Delta t + \psi) & -g \sin \omega_d \Delta t \\ \dfrac{L_V}{V_o} \sin \omega_d \Delta t & \omega_n \cos(\omega_d \Delta t + \psi) + D_V \sin \omega_d \Delta t \end{bmatrix} \tag{5.2-53}$$

where the *damped natural frequency* and *phase angle* are

$$\omega_d = \omega_n \sqrt{1 - \zeta^2} \tag{5.2-54}$$

$$\psi = \tan^{-1}\left(\frac{\zeta}{\sqrt{1 - \zeta^2}} \right) \tag{5.2-55}$$

and ω_n is found from eq. 5.2-27. The response is oscillatory and is bounded by the decaying envelope $e^{-\zeta\omega_n \Delta t}$.

Initial-condition and step responses characterize the system in the time domain in much the same way that Bode plots do in the frequency domain. The response to an initial condition $\Delta\mathbf{x}(t_0)$ can be portrayed by eq. 4.2-3, with increasing values of Δt, or it can be propagated recursively with a constant value of Δt using eq. 4.2-6:

$$\Delta\mathbf{x}_P(t_k) = \mathbf{\Phi}_P(\Delta t)\, \Delta\mathbf{x}_P(t_{k-1}), \quad \Delta\mathbf{x}_P(t_o) \text{ given} \tag{5.2-56}$$

The response to step control and disturbance inputs $\Delta\delta T^*$ and $\Delta V_W{}^*$ (eq. 4.2-8) is calculated at time steps t_k for zero initial conditions as

$$\begin{bmatrix} \Delta V(t_k) \\ \Delta\gamma(t_k) \end{bmatrix} = \mathbf{\Phi}_P \begin{bmatrix} \Delta V(t_{k-1}) \\ \Delta\gamma(t_{k-1}) \end{bmatrix} + \mathbf{\Gamma}_P\, \Delta\delta T^* + \mathbf{\Lambda}_P\, \Delta V_W{}^*,$$

$$\begin{bmatrix} \Delta V(t_0) \\ \Delta\gamma(t_0) \end{bmatrix} = \begin{bmatrix} 0 \\ 0 \end{bmatrix} \tag{5.2-57}$$

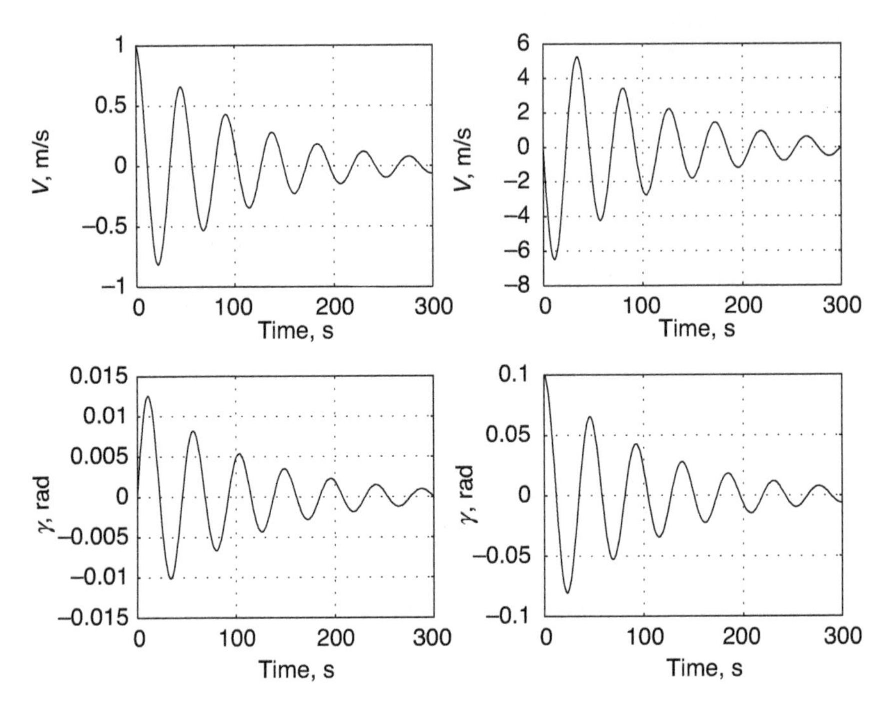

Fig. 5.2-4. Time response to initial velocity and flight path angle perturbations of the second-order phugoid mode approximation. (Left) Initial $\Delta V_o = 1$ m/s. (Right) Initial $\Delta\gamma_o = 0.1$ rad/s.

where

$$\mathbf{\Gamma}_\mathrm{P} = \begin{bmatrix} (\phi_{11}-1) & \phi_{12} \\ \phi_{21} & (\phi_{22}-1) \end{bmatrix} \begin{bmatrix} 0 & V_o/L_V \\ \dfrac{-1}{g} & -\dfrac{V_o D_V}{g L_V} \end{bmatrix} \begin{bmatrix} T_{\delta T} \\ \dfrac{L_{\delta T}}{V_o} \end{bmatrix} \tag{5.2-58}$$

$$\mathbf{\Lambda}_\mathrm{P} = \begin{bmatrix} (\phi_{11}-1) & \phi_{12} \\ \phi_{21} & (\phi_{22}-1) \end{bmatrix} \begin{bmatrix} 0 & V_o/L_V \\ \dfrac{-1}{g} & -\dfrac{V_o D_V}{g L_V} \end{bmatrix} \begin{bmatrix} D_V \\ \dfrac{-L_V}{V_o} \end{bmatrix} \tag{5.2-59}$$

State histories with separate initial perturbations in velocity and flight path angle reveal that lightly damped oscillations continue for several hundred seconds (Fig. 5.2-4). The motions converge toward a stable focus, as shown in Fig. 5.2-5. A 10 percent increase in thrust produces the transients shown in Fig. 5.2-6, with the steady-state behavior predicted by eqs. 5.2-10 and 5.2-11 (with $L_{\delta T} = 0$).

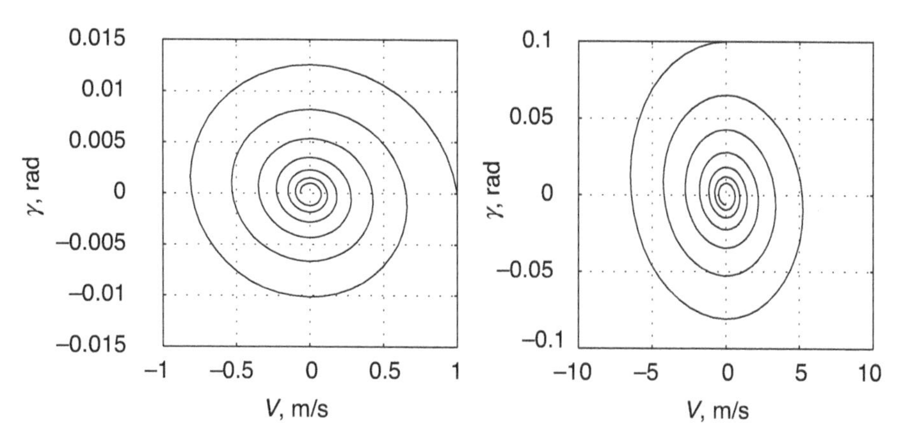

Fig. 5.2-5. Cross-plots of flight path angle versus velocity for the second-order phugoid mode approximation. (Left) Initial $\Delta V_o = 1$ m/s. (Right) Initial $\Delta \gamma_o = 0.1$ rad/s.

Effects of Compressibility

Changes in airspeed affect the aircraft's lift and drag not only by varying the dynamic pressure but by altering the lift and drag coefficients (Section 2.4). Reynolds number effects may be significant for low-speed, lightweight aircraft, and they may lead to either increasing or decreasing forces with airspeed. For typical jet transport and fighter aircraft, Reynolds number effects are incidental, but Mach number variation has a significant effect as the aircraft's true airspeed approaches the speed of sound. These variations are configuration dependent and are documented in the aerodynamic model produced by wind tunnel tests and fluid-dynamic computations.

The qualitative effects of aerodynamic compressibility on the phugoid mode can be estimated assuming that lift and drag coefficients vary according to the Prandtl factor $1/\sqrt{|1 - \mathcal{M}^2|}$ (Section 2.4). The Prandtl factor can be applied for both subsonic and supersonic flight conditions, but it is singular at $\mathcal{M} = 1$. It is not an accurate representation in the range $0.9 < \mathcal{M} < 1.2$, as it neglects important local details of flow over the aircraft, and the forces do not become infinite at $\mathcal{M} = 1$. Assuming that the subsonic lift coefficient C_L takes the form

$$C_L = \frac{k_L}{\sqrt{1 - \mathcal{M}^2}} \tag{5.2-60}$$

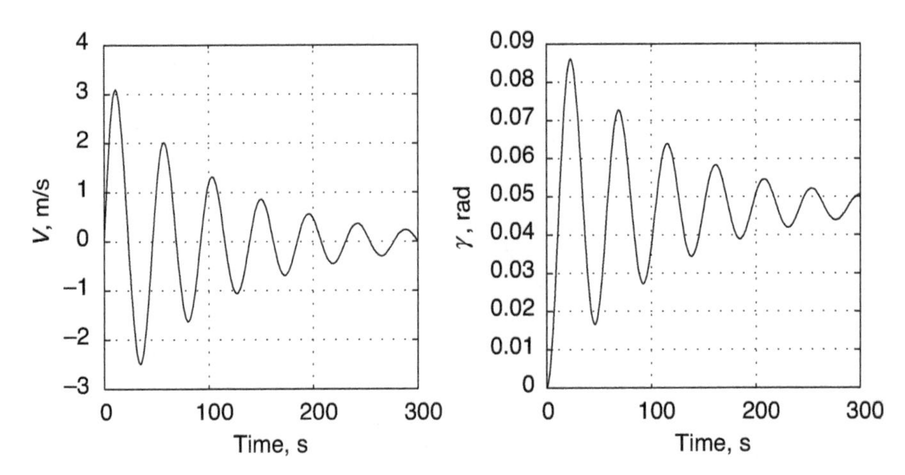

Fig. 5.2-6. Velocity and flight path angle response to a 10 percent thrust increase for the second-order phugoid mode approximation. (Left) Velocity (Right) Flight path angle.

its derivative with respect to Mach number is

$$\frac{\partial C_L}{\partial \mathcal{M}} = C_{L_{\mathcal{M}}} = \frac{k_L \mathcal{M}}{(1 - \mathcal{M}^2)^{3/2}} = \frac{C_L \mathcal{M}}{(1 - \mathcal{M}^2)} \tag{5.2-61}$$

where C_L is the trimmed value. For supersonic flight,

$$C_L = \frac{k_L}{\sqrt{\mathcal{M}^2 - 1}} \tag{5.2-62}$$

The supersonic derivative is identical to eq. 5.2-61; however, the derivative is positive for subsonic airspeed and negative for supersonic airspeed.

The lift coefficient derivative with respect to airspeed is

$$\frac{\partial C_L}{\partial V} = C_{L_V} = C_{L_{\mathcal{M}}} \frac{d\mathcal{M}}{dV} = \frac{C_L \mathcal{M}}{a(1 - \mathcal{M}^2)} \tag{5.2-63}$$

as $\mathcal{M} = V/a$ and $d\mathcal{M}/dV = 1/a$. The corresponding stability derivative (eq. 5.2-24) is

$$\frac{L_V}{V_o} = \frac{2g}{V_o^2} + \frac{C_{L_V}}{C_L} \frac{g}{V_o} = \frac{g}{V_o} \left(2 + \frac{\mathcal{M}^2}{1 - \mathcal{M}^2} \right) \tag{5.2-64}$$

which leads to the following expression for phugoid natural frequency:

$$\omega_{np} = \frac{g}{V_o}\sqrt{2 + \frac{\mathcal{M}^2}{1 - \mathcal{M}^2}} \tag{5.2-65}$$

There is no solution for $1 \le \mathcal{M} \le \sqrt{2}$, because eq. 5.2-64 is singular at $\mathcal{M} = 1$, and the phugoid roots are real in the rest of the region.

The derivation of compressibility effects on D_V and the phugoid damping ratio is similar. Assuming that the drag coefficient variation with Mach number is of the form

$$C_D = \frac{k_D}{\sqrt{1 - \mathcal{M}^2}} \tag{5.2-66}$$

the derivative with respect to velocity is

$$C_{L_V} = C_{D_{\mathcal{M}}}\frac{d\mathcal{M}}{dV} = \frac{C_D \mathcal{M}}{a(1 - \mathcal{M}^2)} \tag{5.2-67}$$

where C_D is the trimmed value. With thrust nominally equal to drag (eq. 5.2-31) and C_{T_V} modeled as before,

$$D_V = -g\left(\frac{C_{T_V} - C_{D_V}}{C_L}\right) = -\frac{g}{V_o}\frac{C_D}{C_L}\left(\eta - \frac{\mathcal{M}}{1 - \mathcal{M}^2}\right) \tag{5.2-68}$$

The damping ratio (eq. 5.2-23) is

$$\zeta_P = -\frac{1}{2}\frac{C_D}{C_L}\frac{\left(\eta \frac{\mathcal{M}}{1 - \mathcal{M}^2}\right)}{\sqrt{2 + \frac{\mathcal{M}}{1 - \mathcal{M}^2}}} \tag{5.2-69}$$

Variations in natural frequency and damping ratio with Mach number and lift-to-drag ratio are shown in Table 5.2-3 for the constant-thrust case. As expected, there is little difference from Table 5.2-1 in the low subsonic region. The natural frequency reaches a minimum in the high subsonic region. Beyond $\mathcal{M} = \sqrt{2}$, the natural frequency passes through a local maximum and then decreases with increasing Mach number. For a given *L/D*, the damping ratio increases with Mach number in the subsonic region, and

TABLE 5.2-3 CHARACTERISTICS OF THE SECOND-ORDER PHUGOID APPROXIMATION, INCLUDING AIR-DENSITY COMPRESSIBILITY (CONSTANT THRUST, SPEED OF SOUND, $a = 300$ M/S)

				L/D		
				5	10	20
V_o ($\mathcal{M}$), m/s (—)	ω_{np}, rad/s	T_P, s	W_P, m	ζ_P, —	ζ_P, —	ζ_P, —
50 (*0.17*)	0.28	22	1,125	0.14	0.07	0.04
100 (*0.33*)	0.14	44	4,400	0.15	0.07	0.04
150 (*0.5*)	0.10	63	9,425	0.15	0.07	0.04
200 (*0.67*)	0.08	77	15,325	0.17	0.08	0.04
250 (*0.83*)	0.08	77	19,375	0.21	0.1	0.05
285 (*0.95*)	0.12	54	15,500	0.34	0.17	0.08
450 (*1.5*)	0.01	645	290,000	0.05	0.02	0.01
525 (*1.75*)	0.013	465	246,000	0.07	0.04	0.02
600 (*2*)	0.013	471	282,500	0.08	0.04	0.02
1,200 (*4*)	0.0079	795	955,000	0.1	0.05	0.02

it is sharply lower at supersonic speed. As $\mathcal{M}$ becomes large, the damping ratio approaches

$$\zeta_P = \frac{1}{2}\frac{C_D}{C_L}(1 + \eta) \tag{5.2-70}$$

The phugoid period does not grow without bound, as implied by this model. At higher velocity, round-earth effects become significant, and the phugoid period becomes the orbital period of the aircraft [E-1].

EFFECTS OF ALTITUDE VARIATION

To this point, the *air density*, *sound speed*, and *gravity variations* that occur as the aircraft climbs and descends in the phugoid oscillation have been neglected because the resulting forces are quite small. In *ground effect*, lift and drag may change with altitude. We can add these subtle effects to the reduced-order model:

$$\Delta\dot{\mathbf{x}}_H = \mathbf{F}_H\Delta\mathbf{x}_H + \mathbf{G}_H\Delta\mathbf{u}_P + \mathbf{L}_H\Delta\mathbf{w}_P \tag{5.2-71}$$

The order is increased by one, as altitude variation is added to the state:

$$\Delta \mathbf{x}_{\mathrm{H}} = \begin{bmatrix} \Delta V \\ \Delta \gamma \\ \Delta z \end{bmatrix} \tag{5.2-72}$$

The inputs are unchanged, and, for $\gamma_o = 0$, the system matrices are

$$\mathbf{F}_{\mathrm{H}} = \begin{bmatrix} -D_V & -g & -D_z \\ \dfrac{L_V}{V_o} & 0 & \dfrac{L_z}{V_o} \\ 0 & -V_o & 0 \end{bmatrix} \tag{5.2-73}$$

$$\mathbf{G}_{\mathrm{H}} = \begin{bmatrix} T_{\delta T} \\ \dfrac{L_{\delta T}}{V_o} \\ 0 \end{bmatrix} \tag{5.2-74}$$

$$\mathbf{L}_{\mathrm{H}} = \begin{bmatrix} D_V \\ \dfrac{-L_V}{V_o} \\ 0 \end{bmatrix} \tag{5.2-75}$$

Air Density, Sound Speed, and Gravity Variations

The altitude-dependent stability derivatives L_z/V_o and D_z are derived in the same manner as L_V/V_o and D_V [S-4]:

$$\frac{L_z}{V_o} = \frac{\partial}{\partial z}\left(\frac{L - mg}{mV}\right) = \frac{g}{V_o}\left[\beta - \frac{\mathcal{M}^2}{1 - \mathcal{M}^2}\left(\frac{a_z}{a}\right) - \frac{2}{R - z_o}\right] \tag{5.2-76}$$

$$D_z = \frac{\partial}{\partial z}\left(\frac{T - D}{m}\right) = g\frac{C_D}{C_L}\left(\frac{a_z}{a}\right)\frac{\mathcal{M}^2}{1 - \mathcal{M}^2} \tag{5.2-77}$$

where $g = g_R[R/(R - z_o)]^2$ follows an inverse square law, g_R is the gravitational acceleration at the earth's surface, R is the earth's radius, $-z_o$ is the aircraft's altitude, a_z is the sound-speed gradient with altitude, and β is the inverse atmospheric scale height (eq. 2.1-5). β is large compared to the other terms in eq. 5.2-76 (except in the near-sonic regime, where the Mach effect is overpredicted), and the drag effect is small compared to the lift effect.

The steady-state response to a constant thrust input is

$$\Delta \mathbf{x}_H{}^* = -\mathbf{F}_H^{-1}\,\mathbf{G}_H\,\Delta \mathbf{u}_P{}^* \tag{5.2-78}$$

With negligible lift due to thrust ($L_{\delta T} = 0$), the solution reveals a very different result from that of the second-order model (eqs. 5.2-9–5.2-11):

$$\begin{bmatrix} \Delta V^* \\ \Delta \gamma^* \\ \Delta z^* \end{bmatrix} = -\frac{T_{\delta T}\begin{bmatrix} L_z/V_o \\ 0 \\ -L_V/V_o \end{bmatrix}}{D_z(L_V/V_o) - D_V(L_z/V_o)}\Delta\delta T^*$$

$$= -\frac{T_{\delta T}}{D_V}\begin{bmatrix} -1 \\ 0 \\ \dfrac{L_V/V_o}{L_z/V_o} \end{bmatrix}\Delta\delta T^* \quad (D_z = 0) \tag{5.2-79}$$

The velocity perturbation reaches a nonzero equilibrium at a new altitude, and the equilibrium flight path angle is zero. The principal reason for this result is that air density decreases with altitude, and the aircraft cannot climb forever in response to a thrust increase.

The third-degree characteristic polynomial of $\mathbf{F}_H$,

$$\Delta(s) = s^3 + D_V s^2 + \left(g\frac{L_V}{V_o} + V_o\frac{L_z}{V_o}\right)s + V_o\left(D_V\frac{L_z}{V_o} - D_z\frac{L_V}{V_o}\right) \tag{5.2-80}$$

factors into a complex pair representing the *phugoid mode* and a real root for the *height mode*. Because the height mode is typically slower than the phugoid mode, the roots of the equation can be approximated by factoring two partitions of the equation:

$$s\left[s^2 + D_V\,s + \left(g\frac{L_V}{V_o} + V_o\frac{L_z}{V_o}\right)\right] \approx 0 \tag{5.2-81}$$

$$\left(g\frac{L_V}{V_o} + V_o\frac{L_z}{V_o}\right)s + V_o\left(D_V\frac{L_z}{V_o} - D_z\frac{L_V}{V_o}\right) \approx 0 \tag{5.2-82}$$

Equation 5.2-81 yields a zero root (representing the height mode as a pure integration) and a complex pair of roots with the following natural frequency and damping ratio:

$$\omega_{np} \approx \sqrt{g\frac{L_V}{V_o} + V_o\frac{L_z}{V_o}} \tag{5.2-83}$$

$$\zeta_P = \frac{D_V}{2\sqrt{g\frac{L_V}{V_o} + V_o\frac{L_z}{V_o}}} \tag{5.2-84}$$

Equation 5.2-82 neglects the phugoid mode and approximates the height mode root as

$$\lambda_H \approx -\frac{V_o\left(D_V\frac{L_z}{V_o} - D_z\frac{L_V}{V_o}\right)}{\left(g\frac{L_V}{V_o} + V_o\frac{L_z}{V_o}\right)} = -\frac{1}{\tau_H} \tag{5.2-85}$$

Positive L_z/V_o increases the phugoid mode's natural frequency and decreases its damping ratio, producing a stable height mode as well. The total damping of the second- and third-order models, which is represented by the coefficient of the second highest power of s in the characteristic equation (D_V), is the same (eqs. 5.2-20 and 5.2-80); therefore, if the height mode is stable, the phugoid mode becomes less stable in the third-order factorization. The effect of air-density altitude gradient on phugoid mode period, damping ratio, and height mode time constant τ_H, derived from eqs. 5.2-83–5.2-85 with $D_z = a_z = 0$ and $\eta = -2$, are presented in Table 5.2-4. The principal change in the phugoid mode is that the supersonic period is shorter and the damping ratio is lower due to the density gradient effect.

It is suspected that regions of inhomogeneous air density at high altitude may disturb the flight of supersonic cruising aircraft. Such effects can be examined by defining the air-density variation as an equivalent change in altitude Δz_{ad} and the disturbance effect matrix as

$$\mathbf{L}_H = \begin{bmatrix} 0 \\ \frac{-L_z}{V_o} \\ 0 \end{bmatrix} \tag{5.2-86}$$

where, from eq. 5.2-76, L_z/V_o is $g\beta/V_o$.

TABLE 5.2-4 ESTIMATES OF PHUGOID AND HEIGHT MODE CHARACTERISTICS, INCLUDING AIR-DENSITY COMPRESSIBILITY AND ALTITUDE VARIATION (CONSTANT THRUST, SPEED OF SOUND, $a = 300$ M/S, SCALE HEIGHT $1/\beta = 9{,}042$ M)

		$L/D = 5$		$L/D = 10$	
V_o ($\mathcal{M}$), m/s (—)	T_P, s	ζ_P, —	τ_H, s	ζ_P, —	τ_H, s
50 (*0.17*)	22	0.14	915	0.07	1835
100 (*0.33*)	43	0.14	475	0.07	950
150 (*0.5*)	60	0.15	335	0.07	670
200 (*0.67*)	71	0.16	260	0.08	525
250 (*0.83*)	72	0.19	210	0.10	420
285 (*0.95*)	52	0.32	170	0.16	345
450 (*1.5*)	183	0.01	1,250	0.01	2,500
525 (*1.75*)	177	0.03	605	0.01	1,210
600 (*2*)	177	0.03	535	0.02	1,070
1,200 (*4*)	186	0.02	695	0.01	1,380

Ground Effect

When aircraft altitude is on the order of the wingspan or less, the *ground effect* introduces a height mode and an altitude dependency in the phugoid characteristics. From eq. 2.4-58, the lift- and drag-coefficient sensitivities to altitude can be estimated as

$$C_{L_z} = \frac{0.126\, b^{1.4}}{\pi\, AR(-z_o)^{2.4}} C_L C_{N_\alpha} \tag{5.2-87}$$

$$C_{D_z} = \frac{0.126\, b^{1.4}}{\pi\, AR(-z_o)^{2.4}} C_D C_{N_\alpha} \tag{5.2-88}$$

and the stability derivatives can be calculated as before. The effects on phugoid and height modes could be estimated from eqs. 5.2-83–5.2-85, but strong pitching sensitivity to height is also likely. Furthermore, C_{L_z} and C_{D_z} change rapidly during landing and takeoff, limiting the value of linearized estimates.

EFFECTS OF WIND SHEAR

A vertical gradient in the horizontal wind gives rise to an altitude-variation effect that is analogous to those considered above. Such gradients commonly occur in the earth's boundary layer (typically the first 1,000 m above ground level), at the vertical boundaries of the jet stream, across weather fronts, downwind of mountain ranges, and, of course, in the vicinity of microbursts. We examine the wind-shear effect without regard to atmospheric density gradients and ignoring control effects:

$$\Delta\dot{\mathbf{x}}_{ws} = \mathbf{F}_P' \Delta\mathbf{x}_{WS} + \mathbf{L}_P'\Delta V_W \tag{5.2-89}$$

where $\Delta\mathbf{x}_{WS} = \Delta\mathbf{x}_H$, and

$$\mathbf{F}_P' = \begin{bmatrix} -D_V & -g & 0 \\ \dfrac{L_V}{V_o} & 0 & 0 \\ 0 & -V_o & 0 \end{bmatrix} \tag{5.2-90}$$

$$\mathbf{L}_P' = \begin{bmatrix} D_L \\ \dfrac{-L_V}{V_o} \\ 0 \end{bmatrix} \tag{5.2-91}$$

With a constant vertical wind shear $\partial V_W/\partial z$,

$$\Delta V_W = \frac{\partial V_W}{\partial z}\Delta z \tag{5.2-92}$$

there is an altitude feedback effect introduced by the gradient. Expressing the wind as

$$\Delta V_W = \begin{bmatrix} 0 & 0 & \dfrac{\partial V_W}{\partial z} \end{bmatrix} \Delta\mathbf{x}_{WS} \triangleq \mathbf{C}\Delta\mathbf{x}_{WS} \tag{5.2-93}$$

eq. 5.2-89 becomes the autonomous differential equation

$$\Delta\dot{\mathbf{x}}_{WS} = (\mathbf{F}_P' + \mathbf{L}_P'\mathbf{C})\Delta\mathbf{x}_{WS} \triangleq \mathbf{F}_{WS}\Delta\mathbf{x}_{WS} \tag{5.2-94}$$

with

$$\mathbf{F}_{WS} \triangleq \begin{bmatrix} -D_V & -g & D_V\dfrac{\partial V_W}{\partial z} \\ \dfrac{L_V}{V_o} & 0 & \dfrac{-L_V}{V_o}\dfrac{\partial V_W}{\partial z} \\ 0 & -V_o & 0 \end{bmatrix} \tag{5.2-95}$$

The characteristic equation takes the same form as eq. 5.2-80:

$$\Delta(s) = |s\mathbf{I}_3 - \mathbf{F}_{WS}| = s^3 + D_V s^2 + \left(g\frac{L_V}{V_o} - V_o\frac{L_V}{V_o}\frac{\partial V_W}{\partial z} \right) s$$
$$= s\left[s^2 + D_V s + g\frac{L_V}{V_o}\left(1 - \frac{V_o}{g}\frac{\partial V_W}{\partial z} \right) \right] \quad (5.2\text{-}96)$$

However, the zeroth-degree term is zero, and the equation factors into the phugoid mode plus a pure integration for altitude variation. The wind shear is seen to affect the natural frequency, leading to a change in damping ratio:

$$\omega_{n_{WS}} = \frac{g}{V_o}\sqrt{2\left(1 - \frac{V_o}{g}\frac{\partial V_W}{\partial z} \right)} \triangleq \sqrt{g\left(\frac{L_V}{V_o} \right)_{eff}} \quad (5.2\text{-}97)$$

and

$$\zeta_{WS} = \frac{D_V}{2\,\omega_{n_{WS}}} \quad (5.2\text{-}98)$$

As defined, a negative shear increases natural frequency and decreases damping ratio, but it never leads to instability. A positive shear increases the damping ratio, which equals 1 when $D_V = 2\omega_{n_{WS}}$. Further increase converts the phugoid oscillation to two real roots. When

$$\frac{V_o}{g}\frac{\partial V_W}{\partial z} = 1 \quad (5.2\text{-}99)$$

$(L_V / V_o)_{eff} = 0$, and one root is at the origin. Higher wind shear produces a positive root and divergent response.

5.3 Reduced-Order Model of the Short-Period Mode

The aircraft's *short-period mode* is principally a longitudinal rotational motion about the center of mass accompanied by a small amount of vertical heaving (or plunging). The dominant effects of both motions are captured in a hybrid second-order model of pitch-rate and angle-of-attack perturbations Δq and $\Delta\alpha$. The model includes the aerodynamic effects of vertical velocity, as $\Delta\alpha \approx \Delta w/V_o$, but it ignores the gravitational effect associated with changing flight path angle. Consequently, both pitching moment and lift are involved in the motion. The force and moment sensitivities are functions of the flight condition—most notably dynamic pressure and Mach number—as well as the aircraft's configuration and balance (i.e., the relative location of the center of mass and the longitudinal aerodynamic center). Unsteady aerodynamic terms also appear in the equations; their effects are small for conventional, subsonic aircraft, but they may be significant for "close-coupled" configurations (e.g., flying wings), ultralightweight aircraft, and all aircraft in the transonic regime.

Second-Order Approximation

The system model is formed by truncating eqs. 5.1-26–5.1-29, retaining elements from the third and fourth rows of $\mathbf{x}_{Lo}$, $\mathbf{F}_{Lo}$, $\mathbf{G}_{Lo}$, and $\mathbf{L}_{Lo}$. The state equation takes the form

$$\Delta\dot{\mathbf{x}}_{SP} = \mathbf{F}_{SP}\,\Delta\mathbf{x}_{SP} + \mathbf{G}_{SP}\,\Delta\mathbf{u}_{SP} + \mathbf{L}_{SP}\,\Delta\mathbf{w}_{SP} \tag{5.3-1}$$

where the state vector contains the pitch rate and the angle of attack:

$$\Delta\mathbf{x}_{SP} = \begin{bmatrix} \Delta q \\ \Delta\alpha \end{bmatrix} \tag{5.3-2}$$

The elevator is the principal control variable, and the angle-of-attack perturbation due to vertical-wind variation provides the most significant disturbance:

$$\Delta\mathbf{u}_{SP} = \Delta\delta E \tag{5.3-3}$$

$$\Delta\mathbf{w}_{SP} = \Delta\alpha_W \tag{5.3-4}$$

The short-period stability-and-control-derivative matrices are

$$\mathbf{F}_{SP} = \begin{bmatrix} \left[M_q - \dfrac{M_{\dot\alpha}}{(V_o + L_{\dot\alpha})}(L_q - V_o)\right] & \left[M_\alpha - \dfrac{M_{\dot\alpha}}{(V_o + L_{\dot\alpha})}L_\alpha\right] \\ \dfrac{(V_o - L_q)}{(V_o + L_{\dot\alpha})} & \dfrac{-L_\alpha}{(V_o + L_{\dot\alpha})} \end{bmatrix} \tag{5.3-5}$$

$$\mathbf{G}_{SP} = \begin{bmatrix} \left[M_{\delta E} - \dfrac{M_{\dot\alpha}}{(V_o + L_{\dot\alpha})}L_{\delta E}\right] \\ \dfrac{-L_{\delta E}}{(V_o + L_{\dot\alpha})} \end{bmatrix} \tag{5.3-6}$$

$$\mathbf{L}_{SP} = -\begin{bmatrix} \left[M_\alpha - \dfrac{M_{\dot{\alpha}}}{(V_o + L_{\dot{\alpha}})} L_\alpha \right] \\ \dfrac{-L_\alpha}{(V_o + L_{\dot{\alpha}})} \end{bmatrix} \tag{5.3-7}$$

The unsteady aerodynamic terms $M_{\dot{\alpha}}$ and $L_{\dot{\alpha}}$ enter not only the stability matrix but the control- and disturbance-effect matrices as well. $L_{\dot{\alpha}}$ and L_q are generally small and, except as noted, are neglected in the analyses that follow.

Equilibrium Response to Control and Disturbance

The equilibrium pitch-rate and angle-of-attack perturbations produced by step changes in the control and disturbance are given by

$$\Delta\mathbf{x}_{SP}^* = -\mathbf{F}_{SP}^{-1}(\mathbf{G}_{SP}\,\Delta\mathbf{u}_{SP}^* + \mathbf{L}_{SP}\,\Delta\mathbf{w}_{SP}^*) \tag{5.3-8}$$

With $L_{\dot{\alpha}} = L_q = 0$, this becomes

$$\begin{bmatrix} \Delta q^* \\ \Delta\alpha^* \end{bmatrix} = -\frac{\begin{bmatrix} \dfrac{L_\alpha}{V_o} & \left(M_\alpha - M_{\dot{\alpha}} \dfrac{L_\alpha}{V_o} \right) \\ 1 & -(M_q + M_{\dot{\alpha}}) \end{bmatrix}}{\left(\dfrac{L_\alpha}{V_o} M_q + M_\alpha \right)} \times \left\{ \begin{bmatrix} M_{\delta E} - M_{\dot{\alpha}} \dfrac{L_{\delta E}}{V_o} \\ -\dfrac{L_{\delta E}}{V_o} \end{bmatrix} \Delta\delta E^* - \begin{bmatrix} \left(M_\alpha - M_{\dot{\alpha}} \dfrac{L_\alpha}{V_o} \right) \\ \dfrac{-L_\alpha}{V_o} \end{bmatrix} \Delta\alpha_W^* \right\} \tag{5.3-9}$$

leading to the scalar equations

$$\Delta q^* = -\frac{\dfrac{L_\alpha}{V_o} M_{\delta E} - \dfrac{L_{\delta E}}{V_o} M_\alpha}{\left(\dfrac{L_\alpha}{V_o} M_q + M_\alpha \right)} \Delta\delta E^* \tag{5.3-10}$$

$$\Delta\alpha^* = -\frac{M_{\delta E} + \dfrac{L_{\delta E}}{V_o} M_q}{\left(\dfrac{L_\alpha}{V_o} M_q + M_\alpha\right)} \Delta\delta E^* + \Delta\alpha_W^* \tag{5.3-11}$$

With an aft control surface and positive static margin, all of the moment derivatives are negative; with negligible lift due to the elevator, positive elevator deflection (trailing edge down) produces negative equilibrium pitch rate and angle of attack. A forward (canard) control surface has positive $M_{\delta E}$ and the opposite effect; hence, the sign of the control-stick-to-control-surface gain is reversed to maintain conventional response to pilot commands. A nonzero $L_{\delta E}$ (positive for both aft elevator and canard) makes Δq^* and $\Delta\alpha^*$ more negative. Should $M_\alpha = -M_q L_\alpha / V_o$, as could occur for a statically unstable vehicle (i.e., with center of mass aft of the aircraft's aerodynamic center), the equilibrium values are infinite in this simplified model. Although its magnitude may change, M_q is normally negative for a rigid airframe. $M_{\dot\alpha}$ is usually negative and about one-third of the magnitude of M_q; however, it may be positive in the transonic regime. A wind-induced angle-of-attack perturbation has direct effect on the aircraft's equilibrium angle of attack but no effect on the equilibrium pitch rate.

Controllability and Observability

As an exercise, follow the phugoid-mode example (Section 5.2) to show the conditions for which the second-order model is controllable by elevator alone and is observable with a single measurement of either Δq or $\Delta\alpha$.

Eigenvalues, Natural Frequency, and Damping Ratio

The characteristic matrix is

$$(s\mathbf{I}_2 - \mathbf{F}_{SP}) = \begin{bmatrix} [s - (M_q + M_{\dot\alpha})] & -\left(M_\alpha - M_{\dot\alpha}\dfrac{L_\alpha}{V_o}\right) \\ -1 & \left(s + \dfrac{L_\alpha}{V_o}\right) \end{bmatrix} \tag{5.3-12}$$

and the characteristic equation is

$$\Delta(s) = |s\mathbf{I}_2 - \mathbf{F}_{SP}| = s^2 - \left(M_q + M_{\dot\alpha} - \frac{L_\alpha}{V_o}\right)s - \left(M_\alpha + M_q\frac{L_\alpha}{V_o}\right) = 0 \tag{5.3-13}$$

The short-period eigenvalues are estimated as

$$\lambda_{1,2} = \left(\frac{M_q + M_{\dot{\alpha}} - \dfrac{L_\alpha}{V_o}}{2}\right) \pm \sqrt{\left(\frac{M_q + M_{\dot{\alpha}} - \dfrac{L_\alpha}{V_o}}{2}\right)^2 + \left(M_\alpha + M_q \frac{L_\alpha}{V_o}\right)} \tag{5.3-14a}$$

For a statically unstable configuration, the roots are real, and one of them is positive. With sufficiently positive static margin (i.e., negative M_α), the square root is negative, and the roots form a complex pair:

$$\lambda_{1,2} = \left(\frac{M_q + M_{\dot{\alpha}} - \dfrac{L_\alpha}{V_o}}{2}\right) \pm j\sqrt{-\left(\frac{M_q + M_{\dot{\alpha}} - \dfrac{L_\alpha}{V_o}}{2}\right)^2 - \left(M_\alpha + M_q \frac{L_\alpha}{V_o}\right)} \tag{5.3-14b}$$

The natural frequency and damping ratio are

$$\omega_{n_{SP}} = \sqrt{-\left(M_\alpha + M_q \frac{L_\alpha}{V_o}\right)} \tag{5.3-15}$$

$$\zeta_{SP} = -\frac{\left(M_q + M_{\dot{\alpha}} - \dfrac{L_\alpha}{V_o}\right)}{2\omega_{n_{SP}}} = -\frac{\left(M_q + M_{\dot{\alpha}} - \dfrac{L_\alpha}{V_o}\right)}{2\sqrt{-\left(M_\alpha + M_q \dfrac{L_\alpha}{V_o}\right)}} \tag{5.3-16}$$

M_α is typically dominant in determining the short-period natural frequency, although $M_q L_\alpha / V_o$ may have important effects for delta-wing aircraft and for configurations with low static margin. M_q and $M_{\dot{\alpha}}$ are normally negative, and L_α / V_o is positive below the stall; the mode is positively damped in these conditions.

Eigenvectors

Literal formulas for the short-period eigenvectors can be found by substituting eq. 5.3-14 in eq. 4.3-61. The adjoint of the characteristic matrix is

$$\mathrm{Adj}(s\mathbf{I}_2 - \mathbf{F}_{SP}) = \begin{bmatrix} \left(s + \frac{L_\alpha}{V_o}\right) & \left(M_\alpha - M_{\dot{\alpha}}\frac{L_\alpha}{V_o}\right) \\ 1 & [s - (M_q + M_{\dot{\alpha}})] \end{bmatrix} \tag{5.3-17}$$

Choosing λ_1 as the root with positive imaginary part,

$$\mathrm{Adj}(\lambda_1\mathbf{I}_2 - \mathbf{F}_{SP}) = [a_1\mathbf{e}_1 \quad a_2\mathbf{e}_1]$$

$$= \begin{bmatrix} \frac{\left(M_q + M_{\dot{\alpha}} + \frac{L_\alpha}{V_o}\right)}{2} + j\sqrt{\cdot} & \left(M_\alpha - M_{\dot{\alpha}}\frac{L_\alpha}{V_o}\right) \\ 1 & \frac{-\left(M_q + M_{\dot{\alpha}} + \frac{L_\alpha}{V_o}\right)}{2} + j\sqrt{\cdot} \end{bmatrix} \tag{5.3-18}$$

where $\sqrt{\cdot}$ represents the radical in eq. 5.3-14b.

$\mathrm{Adj}(\lambda_2\,\mathbf{I}_2 - \mathbf{F}_{SP})$ is the complex conjugate of $\mathrm{Adj}(\lambda_1\mathbf{I}_2 - \mathbf{F}_{SP})$. A modal matrix can be formed using the first columns of the two adjoint matrices:

$$\mathbf{E} = [\mathbf{e}_1 \quad \mathbf{e}_2] = \begin{bmatrix} \frac{\left(M_q + M_{\dot{\alpha}} + \frac{L_\alpha}{V_o}\right)}{2} + j\sqrt{\cdot} & \frac{\left(M_q + M_{\dot{\alpha}} + \frac{L_\alpha}{V_o}\right)}{2} - j\sqrt{\cdot} \\ 1 & 1 \end{bmatrix} \tag{5.3-19}$$

As in Section 5.2, the eigenvectors can be expressed in polar form:

$$\mathbf{E} = \begin{bmatrix} A_q e^{j\phi_q} & A_q e^{-j\phi_q} \\ A_\alpha e^{j\phi_\alpha} & A_\alpha e^{-j\phi_\alpha} \end{bmatrix} \tag{5.3-20}$$

The relative magnitudes and phase angles of the eigenvectors are

$$\begin{bmatrix} A_q \\ A_\alpha \end{bmatrix} = \begin{bmatrix} \sqrt{M_{\dot{\alpha}} L_\alpha / V_o - M_\alpha} \\ 1 \end{bmatrix} \tag{5.3-21}$$

$$\begin{bmatrix} \phi_q \\ \phi_\alpha \end{bmatrix} = \begin{bmatrix} \tan^{-1}\left[\frac{2\sqrt{\cdot}}{(M_q + M_{\dot{\alpha}} - L_\alpha/V_o)}\right] \\ 0 \end{bmatrix} \tag{5.3-22}$$

The pitch rate leads the angle of attack, reflecting the relationship between pitch angle and angle of attack perturbations. (How is this lead related to the damping ratio?)

Root Locus Analysis of Parameter Variations

The short-period characteristic equation

$$\Delta(s) = s^2 - \left(M_q + M_{\dot{\alpha}} - \frac{L_\alpha}{V_o}\right)s - \left(M_\alpha + M_q\frac{L_\alpha}{V_o}\right) = 0 \qquad (5.3\text{-}23)$$

is put in the form

$$\Delta(s) = [d(s)] + k\,n(s) = 0 \qquad (5.3\text{-}24)$$

alternatively using M_α, $M_{\dot{\alpha}}$, M_q, and L_α/V_o as the root locus gain k. To examine the M_α variation, we rewrite eq. 5.3-23 as

$$\Delta(s) = \left[s^2 - \left(M_q + M_{\dot{\alpha}} - \frac{L_\alpha}{V_o}\right)s - M_q\frac{L_\alpha}{V_o}\right] + M_\alpha = 0 \qquad (5.3\text{-}25)$$

With $M_{\dot{\alpha}} = 0$, $d(s)$ factors as

$$d(s) = (s - M_q)\,(s + L_\alpha/V_o) \qquad (5.3\text{-}26)$$

and negative $M_{\dot{\alpha}}$ moves these two negative, real roots farther into the left half of the s plane. Thus, with zero M_α, the short-period roots are real. Figure 5.3-1a is a typical root locus for positive and negative values of M_α, derived from the business jet model with increased M_q and decreased L_α/V_o, increasing the separation of the roots while holding total damping constant (Table 5.3-1). Negative values drive the roots together, causing them to split into a complex pair with increasing damped and undamped natural frequency, as well as decreased damping ratio. Positive values drive the smaller of the two roots into the right half plane, representing divergence.

For $M_{\dot{\alpha}}$ variation, the characteristic equation is written as

$$\Delta(s) = \left[s^2 - \left(M_q - \frac{L_\alpha}{V_o}\right)s - \left(M_\alpha + M_q\frac{L_\alpha}{V_o}\right)\right] + M_{\dot{\alpha}}s = 0 \qquad (5.3\text{-}27)$$

With typical values of the stability derivatives, the roots of $d(s)$ factor as a damped complex pair, and there is a zero at the origin (Fig. 5.3-1b). The root locus is a circle centered on the origin; hence, negative values of $M_{\dot{\alpha}}$ increase the damping ratio, and positive values decrease it.

TABLE 5.3-1 BASELINE SHORT-PERIOD APPROXIMATION FOR FIGURES OF SECTION 5.3

$$\mathbf{F} = \begin{bmatrix} -1.4294 & -7.9856 \\ 0.9901 & -1.1209 \end{bmatrix} \quad \mathbf{G} = \begin{bmatrix} -9.069 \\ 0 \end{bmatrix}$$

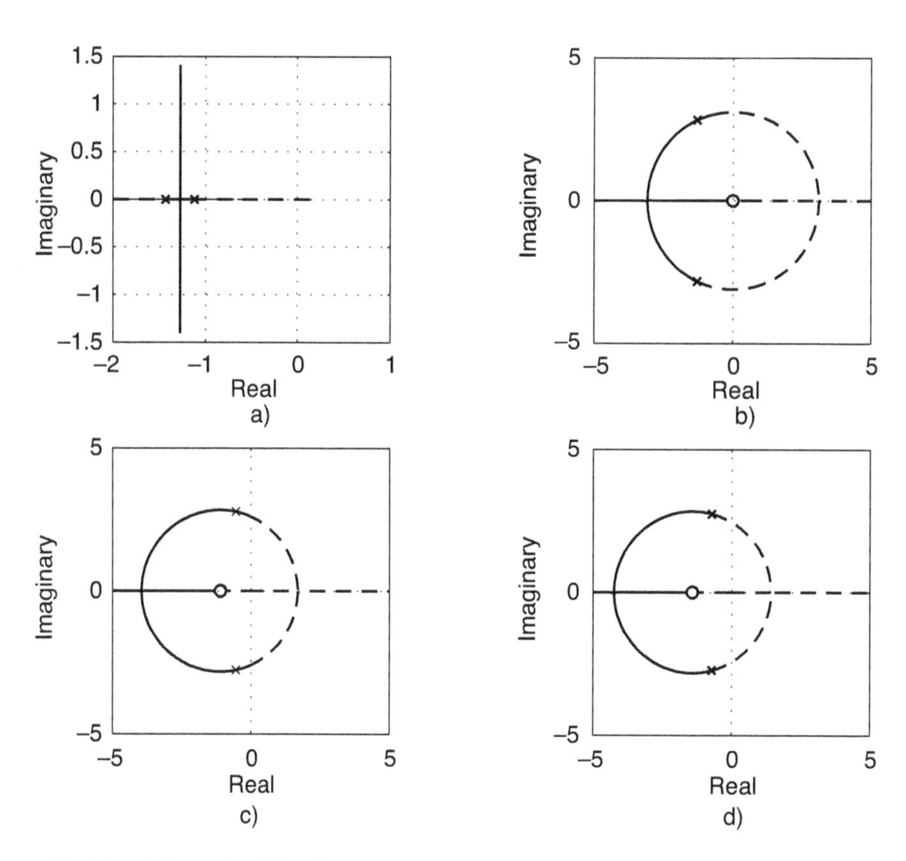

Fig. 5.3-1. Effects of stability-derivative variations on the roots of the second-order short-period approximation. Negative M_α for (b)–(d). Solid lines represent negative moment and positive lift derivatives; dashed lines represent positive moment and negative lift derivatives. (a) M_α variation. (b) $M_{\dot{\alpha}}$ variation. (c) M_q variation. (d) L_α/V_o variation.

M_q has a similar but more complex effect. The characteristic equation is partitioned as

$$\Delta(s) = \left[s^2 - \left(M_{\dot{\alpha}} - \frac{L_\alpha}{V_o} \right) s - M_\alpha \right] - M_q \left(s + \frac{L_\alpha}{V_o} \right) = 0 \tag{5.3-28}$$

$d(s)$ factors into a lightly damped complex pair, and the root locus traces a circle about the point $s = -L_\alpha / V_o$ (Fig. 5.3-1c). Negative values of M_q increase net damping, producing larger values of both $\omega_{n_{SP}}$ and ζ_{SP}.

The root locus for L_α / V_o variation is like the M_q root locus, as the characteristic equation takes the form

$$\Delta(s) = [s^2 - (M_q + M_{\dot{\alpha}})s - M_\alpha] + \frac{L_\alpha}{V_o}(s - M_q) = 0 \tag{5.3-29}$$

However, the circular locus is centered at $s = M_q$ (Fig. 5.3-1d), and positive L_α / V_o is stabilizing.

The root loci for statically unstable aircraft ($M_\alpha > 0$) are qualitatively different. The roots remain real with stabilizing values of the derivatives (M_q, $M_{\dot{\alpha}} < 0$, $L_\alpha / V_o > 0$), and both roots are drawn to the left. The more stable root goes toward $-\infty$, and the unstable root approaches the zero of $n(s)$. For $M_{\dot{\alpha}}$ variation, the unstable root approaches the origin, representing slower divergence but never becoming fully stable (Fig. 5.3-2a). For M_q and L_α / V_o variations, the unstable root is drawn into the left half plane (Figs. 5.3-2b, c).

Frequency Response

The control and disturbance transfer function matrices $\mathbf{U}_{SP}(s)$ and $\mathbf{W}_{SP}(s)$ relate the output Laplace transform $\Delta\mathbf{y}_{SP}(s)$ to the inputs $\Delta\mathbf{u}_{SP}(s)$ and $\Delta\mathbf{w}_{SP}(s)$; following eq. 4.4-1,

$$\begin{aligned}\Delta\mathbf{y}_{SP}(s) &= \mathbf{H}_{SP}(s\mathbf{I}_2 - \mathbf{F}_{SP})^{-1}[\mathbf{G}_{SP}\Delta\mathbf{u}_{SP}(s) + \mathbf{L}_{SP}\Delta\mathbf{w}_{SP}(s)] \\ &\triangleq \mathbf{U}_{SP}(s)\Delta\mathbf{u}_{SP}(s) + \mathbf{W}_{SP}(s)\Delta\mathbf{w}_{SP}(s)\end{aligned} \tag{5.3-30}$$

The characteristic matrix inverse is

$$(s\mathbf{I}_2 - \mathbf{F}_{SP})^{-1} = \frac{\text{Adj}(s\mathbf{I}_2 - \mathbf{F}_{SP})}{|s\mathbf{I}_2 - \mathbf{F}_{SP}|} = \frac{\begin{bmatrix} \left(s + \frac{L_\alpha}{V_o}\right) & \left(M_\alpha - M_{\dot{\alpha}}\frac{L_\alpha}{V_o}\right) \\ 1 & [s - (M_q + M_{\dot{\alpha}})] \end{bmatrix}}{s^2 - \left(M_q + M_{\dot{\alpha}} - \frac{L_\alpha}{V_o}\right)s - \left(M_\alpha + M_q\frac{L_\alpha}{V_o}\right)} \tag{5.3-31}$$

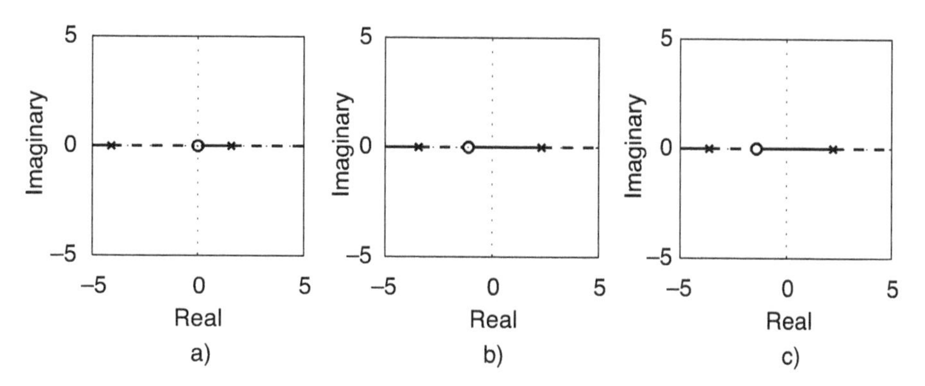

Fig. 5.3-2. Effects of stability-derivative variations on the roots of the second-order short-period approximation, $M_\alpha > 0$. Solid lines represent negative moment and positive lift derivatives; dashed lines represent positive moment and negative lift derivatives. (a) M_α variation. (b) M_q variation. (c) L_α/V_o variation.

The individual scalar transfer functions are linear combinations of the elements of this matrix weighted by the elements of $\mathbf{H}_{SP}$, $\mathbf{G}_{SP}$, and $\mathbf{L}_{SP}$. With $\mathbf{H}_{SP} = \mathbf{I}_2$, $\mathbf{G}_{SP}$ and $\mathbf{L}_{SP}$ defined by eqs. 5.3-6 and 5.3-7, and $L_{\dot{\alpha}} = L_q = L_{\delta E} = 0$, the (2×1) transfer function matrices are as follows:

$$\mathbf{U}_{SP}(s) = \begin{bmatrix} \dfrac{\Delta q(s)}{\Delta \delta E(s)} \\ \dfrac{\Delta \alpha(s)}{\Delta \delta E(s)} \end{bmatrix} = \frac{M_{\delta E}\begin{bmatrix} \left(s + \dfrac{L_\alpha}{V_o}\right) \\ 1 \end{bmatrix}}{s^2 - \left(M_q + M_{\dot{\alpha}} - \dfrac{L_\alpha}{V_o}\right)s - \left(M_\alpha + M_q \dfrac{L_\alpha}{V_o}\right)} \tag{5.3-32}$$

$$\mathbf{W}_{SP}(s) = \begin{bmatrix} \dfrac{\Delta q(s)}{\Delta \alpha_W(s)} \\ \dfrac{\Delta \alpha(s)}{\Delta \alpha_W(s)} \end{bmatrix} = \frac{\begin{bmatrix} -s\left(M_\alpha - M_{\dot{\alpha}} \dfrac{L_\alpha}{V_o}\right) \\ \dfrac{L_\alpha}{V_o}\left[s - \left(M_q + \dfrac{M_\alpha}{L_\alpha/V_o}\right)\right] \end{bmatrix}}{s^2 - \left(M_q + M_{\dot{\alpha}} - \dfrac{L_\alpha}{V_o}\right)s - \left(M_\alpha + M_q \dfrac{L_\alpha}{V_o}\right)} \tag{5.3-33}$$

By inspection of eq. 5.3-32, we see that the amplitude-ratio (AR) frequency response of pitch rate to sinusoidal elevator input, $\Delta q(j\omega)/\Delta \delta E(j\omega)$, has a low-frequency asymptote with 0 dB/dec slope. The level of the asymptote, which has the same value as the equilibrium step response, is found by setting $s = 0$ in the equation and expressing the result in dB (i.e., as 20 $\log_{10}$ AR). The slope increases to +20 dB/dec on reaching the frequency of the zero, L_α/V_o, and it decreases by −40 dB/dec on reaching the natural frequency $\sqrt{-(M_\alpha + M_q L_\alpha/V_o)}$, producing a net roll-off of −20 dB/dec at high frequency.

The actual amplitude response (Fig. 5.3-3a) evidences a resonant peak that is the product of the moderately damped short-period mode and the transfer function zero that occurs at lower frequency (i.e., at the reduced value of L_α/V_o used for the example). The phase angle is −180 deg at zero frequency due to the positive "D.C. response"[1] of positive pitch rate to negative elevator deflection. The phase angle increases as it approaches the minimum-phase zero, and decreases as it approaches the natural frequency, going toward −270 deg as the input frequency becomes large.

The $\Delta\alpha(j\omega)/\Delta\delta E(j\omega)$ amplitude-ratio response also begins with 0 dB/dec slope; however, the transfer function has no zero, so the asymptote shifts to −40 dB/dec on reaching the natural frequency (Fig. 5.3-3b), and the resonant peak is a lower percentage of the steady-state response. MATLAB plots the beginning phase angle at 180 deg arising from the negative static gain from elevator deflection to angle of attack; the angle drops to 90 deg at the natural frequency, and approaches 0 deg at high frequency.

From eq. 5.3-33, the pitch-rate response to sinusoidal wind input $\Delta q(j\omega)/\Delta\alpha_W(j\omega)$ is zero at zero frequency, and the initial asymptotic slope is +20 dB/dec (Fig. 5.3-4a). The slope drops by −40 dB/dec at the natural frequency, producing a high-frequency roll-off of −20 dB/dec. The phase angle begins at +90 deg, due to the differentiation provided by the zero at the origin, and it ends at −90 deg.

The angle-of-attack response (Fig. 5.3-4b) to sinusoidal wind input $\Delta\alpha(j\omega)/\Delta\alpha_W(j\omega)$ tracks the input at very low frequency. The AR plot begins with a 0 dB/dec slope asymptote and shows a typical second-order response to frequencies beyond the natural frequency. As a consequence of the higher-frequency zero, the roll-off slope increases to −20 dB/dec. The phase angle begins at 0 deg, reaches a minimum in the vicinity of the natural frequency, and climbs toward −90 deg at increasing frequency.

When the lift due to elevator deflection is considered, the numerators of $U_{SP}(s)$ become more complicated:

1. "Direct current response" is jargon for zero-frequency or steady-state response, derived from an electric circuit analogy.

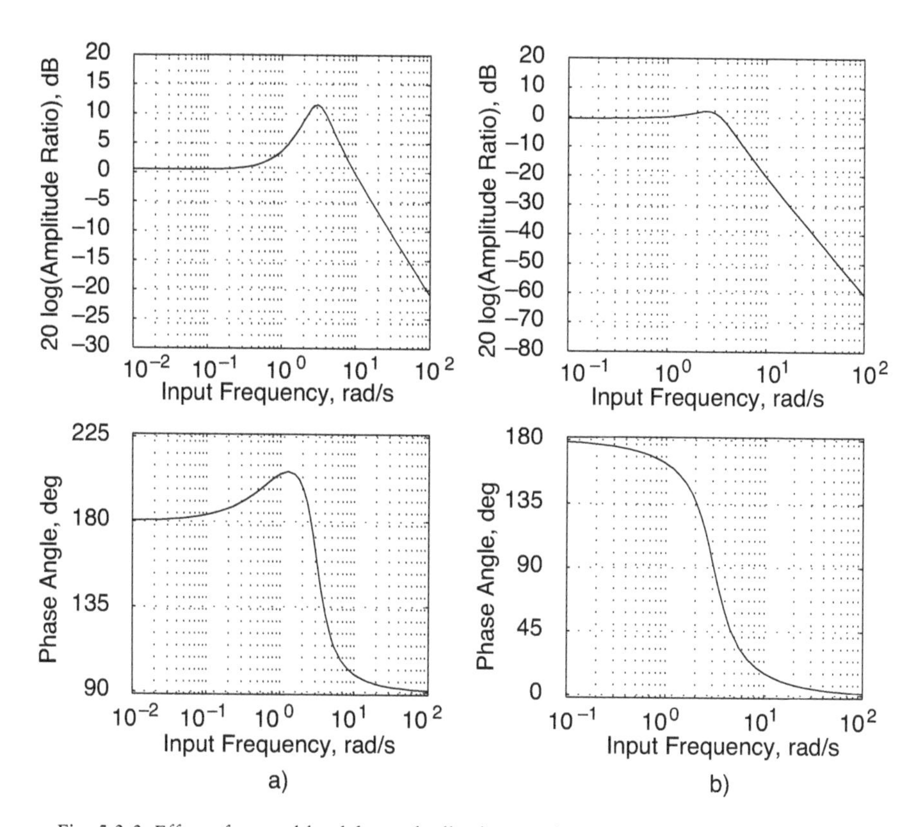

Fig. 5.3-3. Effect of normal-load-factor feedback to pitch control for the second-order short-period approximation. (a) Aft pitch control. (b) Forward pitch control.

$$\begin{bmatrix} n_{\delta E}^{q}(s) \\ n_{\delta E}^{\alpha}(s) \end{bmatrix} = \begin{bmatrix} M_{\delta E}\left(s + \dfrac{L_\alpha}{V_o}\right) - \dfrac{L_{\delta E}}{V_o}(M_{\dot\alpha}s + M_\alpha) \\ M_{\delta E} - \dfrac{L_{\delta E}}{V_o}(s - M_q) \end{bmatrix}$$

$$= \begin{bmatrix} \left(M_{\delta E} - \dfrac{L_{\delta E}}{V_o}M_{\dot\alpha}\right)\left[s + \dfrac{\left(M_{\delta E}\dfrac{L_\alpha}{V_o} - \dfrac{L_{\delta E}}{V_o}M_\alpha\right)}{\left(M_{\delta E} - \dfrac{L_{\delta E}}{V_o}M_{\dot\alpha}\right)}\right] \\ -\dfrac{L_{\delta E}}{V_o}\left[s - \left(M_q + \dfrac{M_{\delta E}}{L_{\delta E}/V_o}\right)\right] \end{bmatrix} \quad (5.3\text{-}34)$$

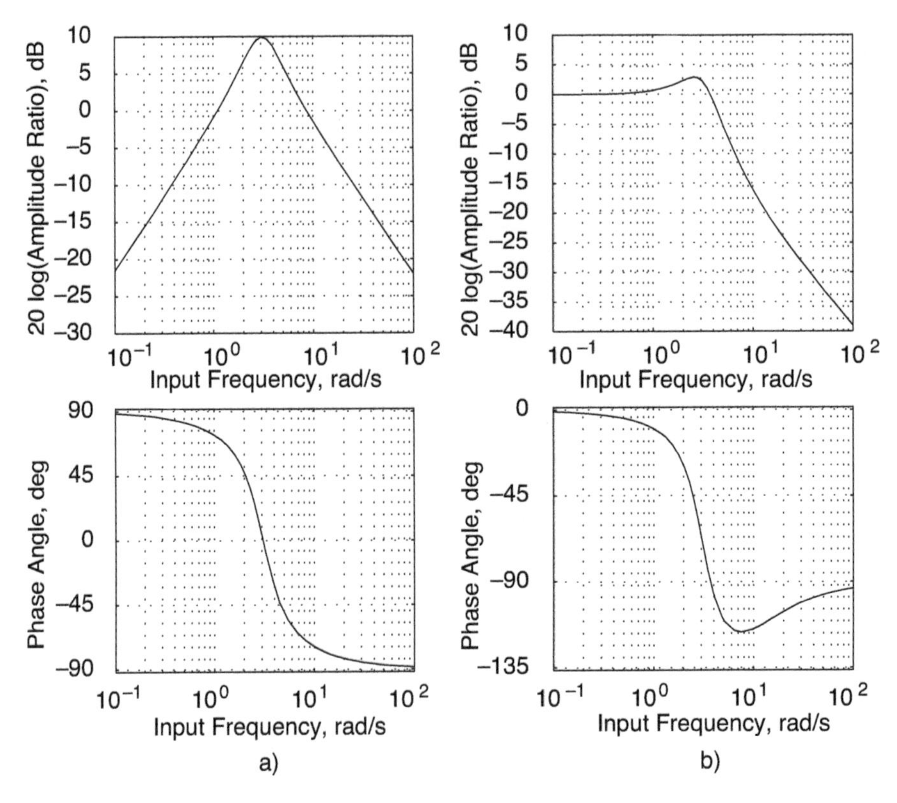

Fig. 5.3-4. Pitch-rate and angle-of-attack frequency response to vertical wind input for the second-order short-period approximation. (a) $\Delta q(\omega)/\Delta \alpha_W(\omega)$. (b) $\Delta \alpha(\omega)/\Delta \alpha_W(\omega)$.

Particular attention is drawn to the angle-of-attack numerator $n^{\alpha}_{\delta E}(s)$. $L_{\delta E}/V_o$ is positive for all configurations, while the sign of $M_{\delta E}$ depends on the location of the pitch control surface relative to the aircraft's center of mass. When the transfer-function zero $[M_q + M_{\delta E}/(L_{\delta E}/V_o)]$ is negative [i.e., when $M_{\delta E} < - M_q/(L_{\delta E}/V_o)$], the initial slope of the response and the steady-state response have the same sign, and the zero is said to have *minimum-phase* effect on the response. When $M_{\delta E} > - M_q/(L_{\delta E}/V_o)$, the zero is positive, producing a *non-minimum-phase* response (i.e., in response to a positive aft-elevator step, $\Delta \dot{\alpha}$ is initially positive, but the final $\Delta \alpha$ is negative). Non-minimum-phase zeros are not in themselves unstable because instability is associated only with positive transfer-function poles (or roots of the characteristic polynomial). However, they can induce instability when a control loop is closed around the transfer function, either manually or automatically, by drawing closed-loop roots into the right half of the s plane as the feedback gain becomes large.

A more significant non-minimum-phase zero is associated with the elevator-to-normal-load-factor transfer function. For this reduced-order model, perturbations in the normal load factor Δn_z measured at the center of mass and expressed in multiples of g (positive down), are given by

$$\Delta n_z = \frac{V_o}{g}(\Delta\dot{\alpha} - \Delta q) = -\frac{V_o}{g}\left(\frac{L_\alpha}{V_o}\Delta\alpha + \frac{L_{\delta E}}{V_o}\Delta\delta E\right) \tag{5.3-35}$$

The normal load factor Δn is measured positive up; hence, the corresponding elevator transfer function is

$$\frac{\Delta n(s)}{\Delta\delta E(s)} = \frac{V_o}{g}\left(\frac{L_\alpha}{V_o}\frac{\Delta\alpha(s)}{\Delta\delta E(s)} + \frac{L_{\delta E}}{V_o}\right) = \left(\frac{V_o}{g}\right)\frac{\frac{L_\alpha}{V_o}n^{\alpha}_{\delta E}(s) + \frac{L_{\delta E}}{V_o}\Delta(s)}{\Delta(s)}$$
$$= \left(\frac{V_o}{g}\right)\frac{\frac{L_{\delta E}}{V_o}\left\{s^2 - M_q s - \left[M_\alpha - \frac{L_\alpha}{V_o}\frac{M_{\delta E}}{L_{\delta E}/V_o}\right]\right\}}{s^2 - \left(M_q + M_{\dot{\alpha}} - \frac{L_\alpha}{V_o}\right)s - \left(M_\alpha + M_q\frac{L_\alpha}{V_o}\right)} \tag{5.3-36}$$

As a consequence of $L_{\delta E}/V_o$, the elevator has a direct effect on Δn, causing the numerator and denominator degrees to be equal. As a consequence, the frequency response does not roll off to zero as frequency approaches infinity. (What is the numerator if $L_{\delta E}/V_o = 0$?)

The zeros of the normal-load-factor transfer function are found by factoring the numerator polynomial:

$$z_{1,2} = \frac{M_q}{2} \pm \sqrt{\left(\frac{M_q}{2}\right)^2 + \left(M_\alpha - \frac{L_\alpha}{V_o}\frac{M_{\delta E}}{L_{\delta E}/V_o}\right)} \tag{5.3-37}$$

When

$$M_{\delta E} < \frac{L_{\delta E}/V_o}{L_\alpha/V_o}\left[\left(\frac{M_q}{2}\right)^2 + M_\alpha\right] \tag{5.3-38}$$

the argument of the square root is positive and greater than $|M_q/2|$. Assuming that M_q is negative, one zero is negative and the other must be positive, or non-minimum phase. Hence, a positive normal-load-factor step command would first produce a negative load-factor rate, as shown in a later figure. $L_{\delta E}/V_o$ is taken to be the nominal value of the aft-tailed business jet model in Fig. 5.3-5a, and the sign of the term is reversed for Fig. 5.3-5b. The two real zeros are positive and negative in the former case, with nearly identical magnitudes; hence, the amplitude-ratio asymptote is raised to 0 dB/dec, but the phase-angle effects cancel. For the latter case, the two zeros form a stable pair in the left half plane, producing an amplitude-ratio "notch" at high frequency and restoring the phase angle.

Root Locus Analysis of Feedback Control

The stability effects of feeding pitch rate and angle-of-attack perturbations back to the elevator can be examined using the root locus. With $L_{\dot{\alpha}} = L_q = L_{\delta E} = 0$, the closed-loop characteristic equations are as follows:

Pitch-Rate Feedback to Elevator

$$\Delta(s) = d(s) + kn(s) = s^2 - \left(M_q + M_{\dot{\alpha}} - \frac{L_\alpha}{V_o}\right)s - \left(M_\alpha + M_q\frac{L_\alpha}{V_o}\right) + kM_{\delta E}\left(s + \frac{L_\alpha}{V_o}\right) = 0 \quad (5.3\text{-}39)$$

Angle-of-Attack Feedback to Elevator

$$\Delta(s) = d(s) + kn(s) = s^2 - \left(M_q + M_{\dot{\alpha}} - \frac{L_\alpha}{V_o}\right)s - \left(M_\alpha + M_q\frac{L_\alpha}{V_o}\right) + kM_{\delta E} = 0 \quad (5.3\text{-}40)$$

Feeding back these two quantities is dynamically equivalent to the effects of M_q and M_α variation (eqs. 5.3-28 and 5.3-25).

Feeding back the normal load factor to pitch control with $L_{\delta E} \neq 0$, the closed-loop characteristic equation is

Normal-Load-Factor Feedback to Elevator

$$\Delta(s) = d(s) + kn(s) = \left\{s^2 - \left(M_q + M_{\dot{\alpha}} - \frac{L_\alpha}{V_o}\right)s - \left(M_\alpha + M_q\frac{L_\alpha}{V_o}\right)\right\} - k\left(\frac{V_o}{g}\right)\frac{L_{\delta E}}{V_o}\left[s^2 - M_q s - \left(M_\alpha - \frac{L_\alpha}{V_o}\frac{M_{\delta E}}{L_{\delta E}/V_o}\right)\right] = 0 \quad (5.3\text{-}41)$$

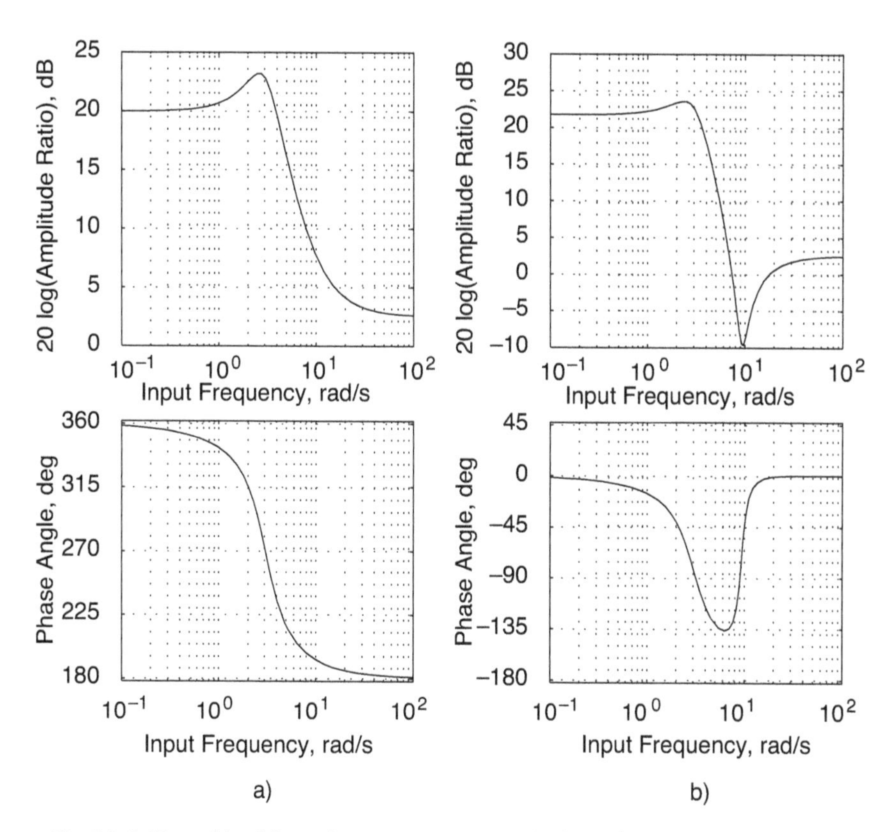

Fig. 5.3-5. Normal-load-factor frequency response with aft and forward pitch control for the second-order short-period approximation. (a) Aft pitch control. (b) Forward pitch control.

Root loci for minimum- and non-minimum-phase zeros are shown in Fig. 5.3-6. With low feedback gains, the feedback effect is analogous to that of angle-of-attack feedback, increasing the natural frequency; however, there is no surplus of poles over zeros, and both closed-loop roots approach the zeros as the gain becomes large. With aft pitch control, high loop gain produces instability, as one root is drawn to the right half plane.

Time Response

The state history is propagated by

$$\Delta \mathbf{x}_{SP}(t_k) = \mathbf{\Phi}_{SP}\, \Delta \mathbf{x}_{SP}(t_{k-1}) + \mathbf{\Gamma}_{SP}\, \Delta \mathbf{u}_{SP}(t_{k-1}) + \mathbf{\Lambda}_{SP}\, \Delta \mathbf{w}_{SP}(t_{k-1}) \tag{5.3-42}$$

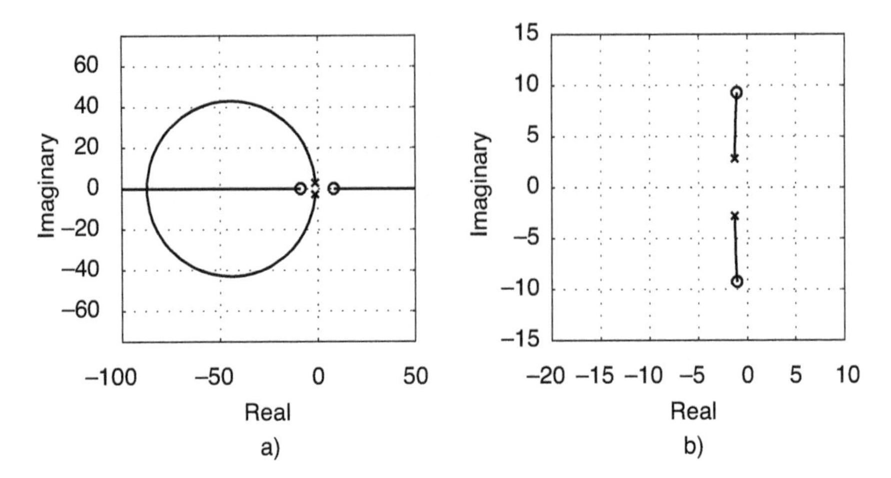

Fig. 5.3-6. Effect of normal-load-factor feedback to pitch control for the second-order short-period approximation. (a) Aft pitch control. (b) Forward pitch control.

The state transition matrix $\Phi_{SP}(\Delta t)$ can be calculated as the inverse Laplace transform of the characteristic matrix inverse (eq. 5.3-31):

$$\Phi_{SP}(\Delta t) = \mathcal{L}^{-1}\,(s\mathbf{I}_2 - \mathbf{F}_{SP})^{-1} \triangleq \begin{bmatrix} \phi_{11} & \phi_{12} \\ \phi_{21} & \phi_{22} \end{bmatrix}$$

$$= \frac{e^{-\zeta\omega_n \Delta t}}{\omega_d}\begin{bmatrix} \omega_n \cos(\omega_d \Delta t + \psi) + \dfrac{L_\alpha}{V_o}\sin\omega_d \Delta t & \left(M_\alpha - M_{\dot\alpha}\dfrac{L_\alpha}{V_o}\right)\sin\omega_d \Delta t \\ \sin\omega_d \Delta t & \omega_n \cos(\omega_d \Delta t + \psi) - (M_q + M_{\dot\alpha})\sin\omega_d \Delta t \end{bmatrix} \quad (5.3\text{-}43)$$

where ω_n and ζ are found from eqs. 5.3-15 and 5.3-16, and ω_d and ψ are calculated by eqs. 5.2-54 and 5.2-55. The state transition relationship for initial-condition and input response is then

$$\begin{bmatrix} \Delta q(t_k) \\ \Delta\alpha(t_k) \end{bmatrix} = \Phi_{SP}\begin{bmatrix} \Delta q(t_{k-1}) \\ \Delta\alpha(t_{k-1}) \end{bmatrix} + \Gamma_{SP}\Delta\delta E(t_{k-1}) + \Lambda_{SP}\Delta\alpha_W(t_{k-1}), \quad \begin{bmatrix} \Delta q(t_0) \\ \Delta\alpha(t_0) \end{bmatrix} \text{ given} \quad (5.3\text{-}44)$$

The discrete-time control- and disturbance-effect matrices Γ_{SP} and Λ_{SP} are calculated by eqs. 4.2-9 and 4.2-10:

$$\Gamma_{SP} = [\Phi_{SP} - \mathbf{I}_2]\,\mathbf{F}_{SP}^{-1}\,\mathbf{G}_{SP} \quad (5.3\text{-}45)$$

$$\Lambda_{SP} = [\Phi_{SP} - \mathbf{I}_2]\,\mathbf{F}_{SP}^{-1}\,\mathbf{L}_{SP} \quad (5.3\text{-}46)$$

or

$$\mathbf{\Gamma}_{SP} = \begin{bmatrix} (\phi_{11}-1) & \phi_{12} \\ \phi_{21} & (\phi_{22}-1) \end{bmatrix} \frac{\begin{bmatrix} \dfrac{L_\alpha}{V_o} & \left(M_\alpha - M_{\dot{\alpha}}\dfrac{L_\alpha}{V_o}\right) \\ 1 & -(M_q + M_{\dot{\alpha}}) \end{bmatrix}}{\left[\dfrac{L_\alpha}{V_o}M_q + M_\alpha\right]} \begin{bmatrix} \left[M_{\delta E} - \dfrac{M_{\dot{\alpha}}}{V_o}L_{\delta E}\right] \\ \dfrac{-L_{\delta E}}{V_o} \end{bmatrix} \quad (5.3\text{-}47)$$

$$\mathbf{\Lambda}_{SP} = \begin{bmatrix} (\phi_{11}-1) & \phi_{12} \\ \phi_{21} & (\phi_{22}-1) \end{bmatrix} \frac{\begin{bmatrix} \dfrac{L_\alpha}{V_o} & \left(M_\alpha - M_{\dot{\alpha}}\dfrac{L_\alpha}{V_o}\right) \\ 1 & -(M_q + M_{\dot{\alpha}}) \end{bmatrix}}{\left[\dfrac{L_\alpha}{V_o}M_q + M_\alpha\right]} \begin{bmatrix} \left[M_\alpha - \dfrac{M_{\dot{\alpha}}}{V_o}L_\alpha\right] \\ \dfrac{-L_\alpha}{V_o} \end{bmatrix} \quad (5.3\text{-}48)$$

Figure 5.3-7 illustrates the effects of pitch-rate and angle-of-attack initial conditions. Oscillations are moderately damped and disappear in a few seconds.

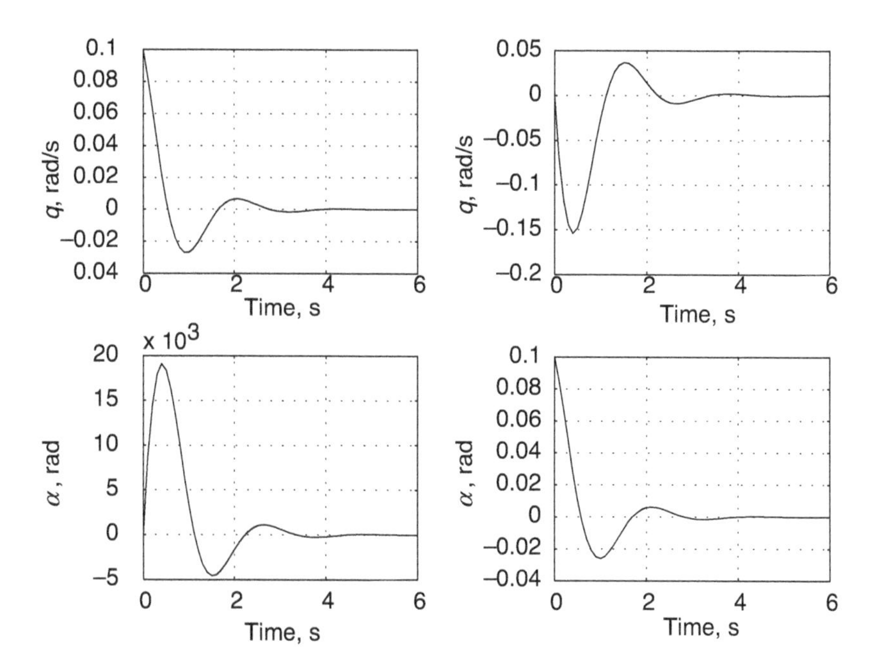

Fig. 5.3-7. Time response to initial pitch-rate and angle-of-attack perturbations of the second-order short-period approximation. (Left) $\Delta q(0) = 0.1$ rad/s. (Right) $\Delta\alpha(0) = 0.1$ rad.

With the reduced value of L_α/V_o used in the example, the pitch-rate response to a 0.1 rad pitch-control step shows more overshoot than that of the angle of attack (Fig. 5.3-8). $L_{\delta E}/V_o = 0$ in this case.

With $L_{\delta E}/V_o \neq 0$, the effects of fore and aft pitch control can be demonstrated. $\Delta n_z(t_k)$ can be found by eq. 5.3-35. A 0.1 rad pitch-control step produces an instantaneous negative load factor with aft control (Fig. 5.3-9a), although the eventual steady-state response is positive. The forward pitch control produces an initial increment that is positive (Fig. 5.3-9b), and the final value is slightly higher than for the previous case.

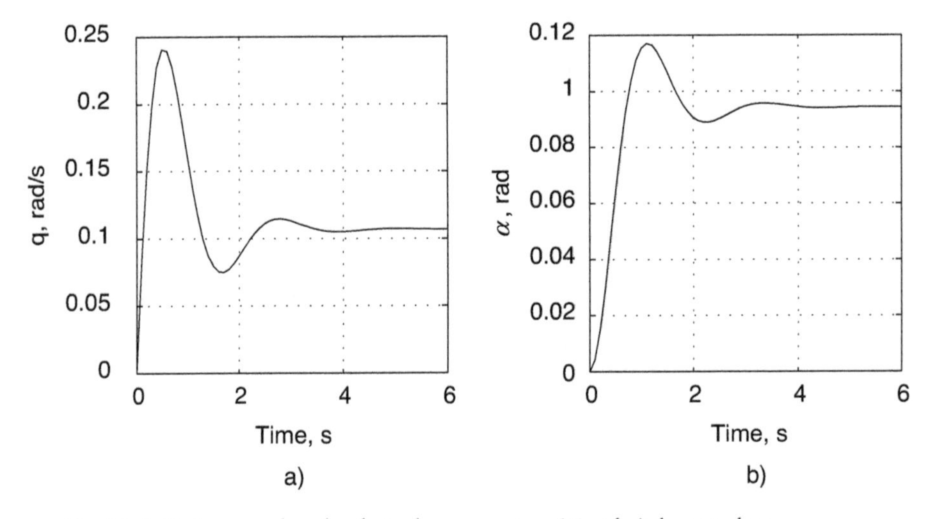

Fig. 5.3-8. Pitch-rate and angle-of-attack response to a 0.1 rad pitch-control step input for the second-order short-period approximation.

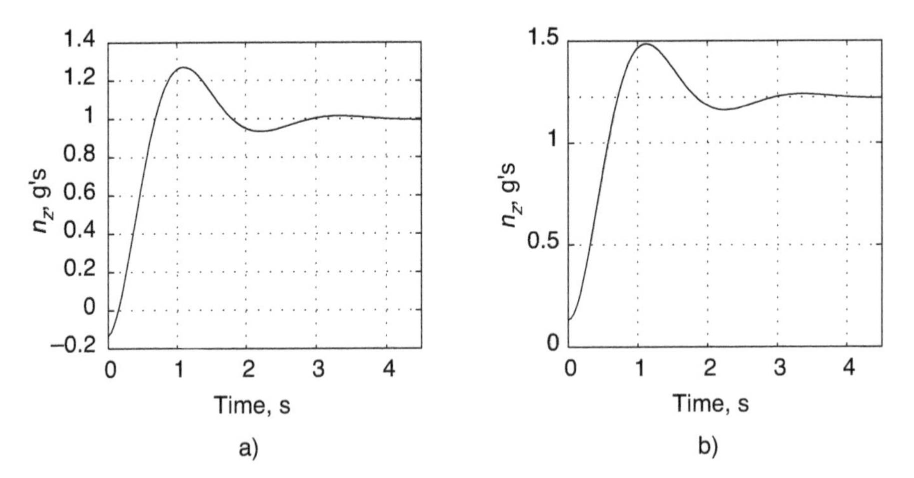

Fig. 5.3-9. Normal-load-factor response to a 0.1 rad pitch-control step input for the second-order short-period approximation. (a) Aft pitch control. (b) Forward pitch control.

Example 5.3-1. Effects of Parameter Variations on Short-Period Natural Frequency and Damping Ratio of a Fighter Aircraft

The effects of parameter variations are demonstrated in the following tables for a statically stable lightweight fighter aircraft configuration [G-4]. The nominal airspeed and altitude are 152 m/s and 6,095 m, with a static margin of 24 percent m.a.c.

VELOCITY VARIATION, NEGLECTING COMPRESSIBILITY EFFECTS

V_o, m/s	$\bar{q}$, P	α_o, deg	$\omega_{n_{SP}}$, s^{-1}	T_{SP}, s	ζ_{SP}
91	2,540	14.6	1.34	4.7	0.3
152	7,040	5.8	2.3	2.74	0.31
213	13,790	3.2	3.21	1.96	0.3
274	22,790	2.2	3.84	1.64	0.3

ALTITUDE VARIATION, CONSTANT DYNAMIC PRESSURE

V_o, m/s	h, m	$\omega_{n_{SP}}$, s^{-1}	T_{SP}, s	ζ_{SP}
122	2,235	2.36	2.67	0.39
152	6,095	2.3	2.74	0.31
213	11,915	2.24	2.8	0.23
274	16,260	2.18	2.88	0.18

MASS VARIATION, CONSTANT MOMENT OF INERTIA

Δm, percent	$\omega_{n_{SP}}$, s^{-1}	T_{SP}, s	ζ_{SP}
−50	2.4	2.62	0.44
0	2.3	2.74	0.31
+50	2.26	2.78	0.26

MOMENT-OF-INERTIA VARIATION, CONSTANT MASS

ΔI_{yy}, percent	$\omega_{n_{SP}}$, s^{-1}	T_{SP}, s	ζ_{SP}
−50	3.25	1.94	0.33
0	2.3	2.74	0.31
+50	1.87	3.35	0.31

Center-of-Mass Variation, Constant Mass and Moment of Inertia

Static margin, percent m.a.c.	ω_{nSP}, s^{-1}	T_{SP}, s	ζ_{SP}
18	2.87	2.19	0.25
24	2.3	2.74	0.31
30	1.51	4.15	0.46

Effects of Compressibility and High Angle of Attack

Some of the problems associated with compressibility and high angle of attack are hinted at in Section 2.4, and while the actual variations in short-period response are configuration dependent, there are general traits to be noted. With increased angle of attack, local areas of flow separation on the wing and horizontal tail can dramatically alter the distribution of lift, and, therefore, the pitching moment. The downwash at the tail due to wing lift and the corresponding sensitivity to angle of attack, $\partial\varepsilon/\partial\alpha$, are strongly affected by the position of the horizontal tail relative to the wing's wake. The tail effectiveness may increase or decrease with $\Delta\alpha$ depending on the vertical positions of the wing and the tail.

A "pitch-up" characteristic, in which M_α changes sign and the aircraft becomes statically unstable, can occur for several reasons [A-1]. High-aspect-ratio, swept wings may experience *tip stall*, moving the wing's aerodynamic center forward. The severity of this effect can be reduced by installing vortex generators or wing fences, changing the tip chord section, adding an outboard leading-edge extension with a mid-span "notch" discontinuity, or by twisting the chord line to lower the outboard angle of attack (called "washout"). The stabilizing influence of the tail may be adversely affected by changes in $\partial\varepsilon/\partial\alpha$. This is especially true for "T" tail configurations, but it can also occur with low-mounted horizontal tails. Hence, there may be regions of static instability as the angle of attack increases. At very high angles of attack, it is likely that static stability is recovered; however, a stable *deep stall* (Section 7.4), in which the aircraft trims at an angle of attack that is too high to sustain level, 1g flight, may result. This is associated not only with the second stable equilibrium point but with elevator effectiveness $M_{\delta E}$ that is too low to pitch the aircraft down through the unstable α region.

With increasing angle of attack, the net lift slope (and, therefore, L_α) decreases, and it changes sign beyond the stall, reducing the dynamic stability of the mode. *Aerodynamic hysteresis* can occur at stall: the angles of attack at which the flow separates and reattaches are different, lowering the effective M_q or M_α.

Compressibility can exacerbate short-period stability problems in the transonic region because local shock waves may induce flow separation. We have already noted that aerodynamic forces tend to change dramatically in the vicinity of Mach 1, and the effect is heightened when a transonic Mach number is combined with a high angle of attack.

There is a general increase of lift and drag in the transonic region, increasing the magnitudes of all the stability and control derivatives; however, straight, thick wings may experience decreased L_α in the high subsonic region. Aft motion of wing and tail aerodynamic centers contributes to a stiffening of the mode (via the more negative M_α) with increasing Mach number, although the straight wing's center of pressure may move forward before moving aft. In this case, the wing's $\partial\varepsilon/\partial\alpha$ may decrease, improving tail effectiveness, and increasing net static stability.

As a rule, static stability for all configurations increases in going from subsonic to supersonic flight, creating a need for retrimming. At higher Mach numbers, intersections of Mach cones from leading discontinuities (e.g., nose, wing tips, and canopy) alter the flows over trailing surfaces in ways that are not readily generalized. The beneficial subsonic carryover effect for aft-flap elevators on fixed horizontal tails is eliminated in supersonic flow, and supersonic configurations often employ all-moving (stabilator or canard) pitch control surfaces.

5.4 Coupled Phugoid/Short-Period Dynamics

The reduced-order models for phugoid and short-period modes provide considerable insights into the nature of translational and rotational motions, but they neglect important coupling and control relationships, such as moments due to velocity variation and wing-flap settings. In certain flight conditions, such as low-speed flight, the natural frequencies may become close enough to produce significant phugoid-mode response in short-period variables (and vice versa). Furthermore, the limiting response behavior at steady equilibrium and at very high frequency may be different for the reduced- and higher-order models. From a control-design perspective, feedback controls generated for reduced-order models may produce different results with the higher-order model, especially if the gains are high. There is, therefore, a need to investigate higher-order models.

A fourth-order model whose state, control, and disturbance vectors are

$$\Delta \mathbf{x}_L^T = [\Delta V\ \Delta\gamma\ \Delta q\ \Delta\alpha]^T \tag{5.4-1}$$

$$\Delta \mathbf{u}_L^T = [\Delta\delta E\ \Delta\delta T\ \Delta\delta F]^T \tag{5.4-2}$$

$$\Delta \mathbf{w}_L^T = [\Delta V_w\ \Delta\alpha_w]^T \tag{5.4-3}$$

can capture all the significant longitudinal dynamic interactions for rigid-body motion except those due to altitude gradients. Altitude effects must be considered in strong wind shear and in ground effect, but air-density and sound-speed variations have little effect on short-period motions. A fifth-order model (including Δz in the state) is required to include the altitude-limiting effects discussed in Section 5.2.

The linearized state equation takes the general form

$$\Delta \dot{\mathbf{x}}_L = \mathbf{F}_L \Delta \mathbf{x}_L + \mathbf{G}_L \Delta \mathbf{u}_L + \mathbf{L}_L \Delta \mathbf{w}_L \tag{5.4-4}$$

Neglecting altitude-gradient effects, the stability-and-control-derivative matrices are

$$\mathbf{F}_L = \begin{bmatrix} -D_V & -g\cos\gamma_o & -D_q & -D_\alpha \\ \dfrac{L_V}{(V_o + L_{\dot\alpha})} & \dfrac{g\sin\gamma_o}{V_o} & \dfrac{L_q}{(V_o + L_{\dot\alpha})} & \dfrac{L_\alpha}{(V_o + L_{\dot\alpha})} \\ [M_V - aL_V] & 0 & [M_q - a(L_q - V_o)] & [M_\alpha - aL_\alpha] \\ \dfrac{-L_V}{(V_o + L_{\dot\alpha})} & \dfrac{-g\sin\gamma_o}{V_o} & \dfrac{(V_o - L_q)}{(V_o + L_{\dot\alpha})} & \dfrac{-L_\alpha}{(V_o + L_{\dot\alpha})} \end{bmatrix}$$

$$a \triangleq \frac{M_{\dot\alpha}}{(V_o + L_{\dot\alpha})} \tag{5.4-5}$$

$$\mathbf{G}_L = \begin{bmatrix} -D_{\delta E} & T_{\delta T} & -D_{\delta F} \\ \dfrac{L_{\delta E}}{(V_o + L_{\dot\alpha})} & \dfrac{L_{\delta T}}{(V_o + L_{\dot\alpha})} & \dfrac{L_{\delta F}}{(V_o + L_{\dot\alpha})} \\ [M_{\delta E} - aL_{\delta E}] & [M_{\delta T} - aL_{\delta T}] & [M_{\delta F} - aL_{\delta F}] \\ \dfrac{-L_{\delta E}}{(V_o + L_{\dot\alpha})} & \dfrac{-L_{\delta T}}{(V_o + L_{\dot\alpha})} & \dfrac{-L_{\delta F}}{(V_o + L_{\dot\alpha})} \end{bmatrix} \tag{5.4-6}$$

$$\mathbf{L}_L = \begin{bmatrix} \dfrac{D_V - L_V}{(V_o + L_{\dot{\alpha}})} & \dfrac{D_\alpha - L_\alpha}{(V_o + L_{\dot{\alpha}})} \\ -[M_V - aL_V] & [M_\alpha - aL_\alpha] \\ \dfrac{L_V}{(V_o + L_{\dot{\alpha}})} & \dfrac{L_\alpha}{(V_o + L_{\dot{\alpha}})} \end{bmatrix} \tag{5.4-7}$$

For demonstration, we assume that the flight path angle is zero in the remainder of the section; however, nonzero γ_o should be taken into account on steep flight paths or where additional precision is called for, as during final landing approach. Unsteady aerodynamic effects ($M_{\dot{\alpha}}$ and $L_{\dot{\alpha}}$), D_q, and L_q are also neglected, as they are generally small values. Some effects of these terms are presented in [S-4].

Residualized Phugoid Mode

From Table 5.1-1, we see that the fourth-order longitudinal model (eqs. 5.4-4–5.4-7) can be partitioned into phugoid and short-period components; neglecting disturbance inputs, the dynamic equation is

$$\begin{bmatrix} \Delta\dot{\mathbf{x}}_P \\ \Delta\dot{\mathbf{x}}_{SP} \end{bmatrix} = \begin{bmatrix} \mathbf{F}_P & \mathbf{F}_{SP}^P \\ \mathbf{F}_P^{SP} & \mathbf{F}_{SP} \end{bmatrix} \begin{bmatrix} \Delta\mathbf{x}_P \\ \Delta\mathbf{x}_{SP} \end{bmatrix} + \begin{bmatrix} \mathbf{G}_P & \mathbf{G}_{SP}^P \\ \mathbf{G}_P^{SP} & \mathbf{G}_{SP} \end{bmatrix} \begin{bmatrix} \Delta\mathbf{u}_P \\ \Delta\mathbf{u}_{SP} \end{bmatrix} \tag{5.4-8}$$

where $\mathbf{F}_{SP}^P$ couples short-period motions into the phugoid variables, and $\mathbf{F}_P^{SP}$ does the reverse. In many circumstances, the phugoid and short-period time scales are widely separated; assuming short-period stability, the fast-time-scale solution appears algebraic on the phugoid time scale:

$$\begin{bmatrix} \Delta\dot{\mathbf{x}}_P \\ 0 \end{bmatrix} = \begin{bmatrix} \mathbf{F}_P & \mathbf{F}_{SP}^P \\ \mathbf{F}_P^{SP} & \mathbf{F}_{SP} \end{bmatrix} \begin{bmatrix} \Delta\mathbf{x}_P \\ \Delta\mathbf{x}_{SP}^* \end{bmatrix} + \begin{bmatrix} \mathbf{G}_P & \mathbf{G}_{SP}^P \\ \mathbf{G}_P^{SP} & \mathbf{G}_{SP} \end{bmatrix} \begin{bmatrix} \Delta\mathbf{u}_P \\ \Delta\mathbf{u}_{SP} \end{bmatrix} \tag{5.4-9}$$

Following Section 4.2, the equation can be residualized, leaving a revised, second-order differential equation for the phugoid mode:

$$\begin{aligned} \Delta\dot{\mathbf{x}}_P &= (\mathbf{F}_P - \mathbf{F}_{SP}^P\mathbf{F}_{SP}^{-1}\mathbf{F}_P^{SP})\Delta\mathbf{x}_P + (\mathbf{G}_P - \mathbf{F}_{SP}^P\mathbf{F}_{SP}^{-1}\mathbf{G}_P^{SP})\Delta\mathbf{u}_P \\ &\quad + (\mathbf{G}_{SP}^P - \mathbf{F}_{SP}^P\mathbf{F}_{SP}^{-1}\mathbf{G}_{SP})\Delta\mathbf{u}_{SP} \\ &\triangleq \mathbf{F}_P'\Delta\mathbf{x}_P + \mathbf{G}_P'\begin{pmatrix} \Delta\mathbf{u}_P \\ \Delta\mathbf{u}_{SP} \end{pmatrix} \end{aligned} \tag{5.4-10}$$

Classifying the flaps as a phugoid control, the matrices of eq. 5.4-10 are

$$\mathbf{F}_{\mathrm{P}} = \begin{bmatrix} -D_V & -g \\ \dfrac{L_V}{V_o} & 0 \end{bmatrix} \tag{5.4-11}$$

$$\mathbf{G}_{\mathrm{P}} = \begin{bmatrix} T_{\delta T} & -D_{\delta F} \\ \dfrac{L_{\delta T}}{V_o} & \dfrac{L_{\delta F}}{V_o} \end{bmatrix} \tag{5.4-12}$$

$$\mathbf{F}_{\mathrm{SP}} = \begin{bmatrix} M_q & M_\alpha \\ 1 & \dfrac{-L_\alpha}{V_o} \end{bmatrix} \tag{5.4-13}$$

$$\mathbf{G}_{\mathrm{SP}} = \begin{bmatrix} M_{\delta E} \\ \dfrac{-L_{\delta E}}{V_o} \end{bmatrix} \tag{5.4-14}$$

$$\mathbf{F}_{\mathrm{SP}}^{\mathrm{P}} = \begin{bmatrix} 0 & -D_\alpha \\ 0 & \dfrac{L_\alpha}{V_o} \end{bmatrix} \tag{5.4-15}$$

$$\mathbf{F}_{\mathrm{P}}^{\mathrm{SP}} = \begin{bmatrix} M_V & 0 \\ \dfrac{-L_V}{V_o} & 0 \end{bmatrix} \tag{5.4-16}$$

$$\mathbf{G}_{\mathrm{SP}}^{\mathrm{P}} = \begin{bmatrix} -D_{\delta E} \\ \dfrac{L_{\delta E}}{V_o} \end{bmatrix} \tag{5.4-17}$$

$$\mathbf{G}_{\mathrm{P}}^{\mathrm{SP}} = \begin{bmatrix} M_{\delta T} & M_{\delta F} \\ \dfrac{-L_{\delta T}}{V_o} & \dfrac{-L_{\delta F}}{V_o} \end{bmatrix} \tag{5.4-18}$$

We limit our examination of this model to the effect on natural frequency and damping ratio. The characteristic equation is

$$\begin{aligned}\Delta(s) &= |s\mathbf{I}_2 - \mathbf{F}_{\mathrm{P}}'| \\ &= s^2 + (D_V + aD_\alpha)s + g\left(\frac{L_V}{V_o} - a\frac{L_\alpha}{V_o}\right)\end{aligned} \tag{5.4-19}$$

where

$$a = \frac{M_V + M_q L_V/V_o}{M_\alpha + M_q L_\alpha/V_o}$$

and the natural frequency and damping ratio are

$$\omega'_{np} = \sqrt{g\left[\frac{L_V}{V_o} - \frac{L_\alpha}{V_o}\frac{M_V + M_q L_V/V_o}{M_\alpha + M_q L_\alpha/V_o}\right]} \tag{5.4-20}$$

$$\zeta'_P = \frac{D_V - D_\alpha(M_V + M_q L_V/V_o)/(M_\alpha + M_q L_\alpha/V_o)}{2\omega'_{np}} \tag{5.4-21}$$

The significant coupling derivatives are L_α, D_α, and M_V. D_α is generally small, except at high angles of attack. M_V can take either sign, depending on the vertical placement and thrust angle of the powerplants. A positive thrust sensitivity to velocity produces a negative M_V for powerplants whose thrust lines are above the center of mass.

Fourth-Order Model

Formulas for equilibrium response, eigenvalues, transfer functions, and root loci reveal the major effects of phugoid/short-period coupling. The fourth-order model also allows the pitch-angle response to be evaluated.

Equilibrium Response to Control

The equilibrium response to step change in the control is given by

$$\Delta \mathbf{x}_L{}^* = -\mathbf{F}_L^{-1}\,\mathbf{G}_L\,\Delta \mathbf{u}_L{}^* \tag{5.4-22}$$

To focus on the separate effects of moment, thrust, and lift control, the control-effect matrix is idealized as

$$\mathbf{G}_L = \begin{bmatrix} 0 & T_{\delta T} & 0 \\ 0 & 0 & \dfrac{L_{\delta F}}{V_o} \\ M_{\delta E} & 0 & 0 \\ 0 & 0 & \dfrac{-L_{\delta F}}{V_o} \end{bmatrix} \tag{5.4-23}$$

With this and the simplifications noted earlier, the steady-state relationship is

$$\begin{bmatrix} \Delta V^* \\ \Delta\gamma^* \\ \Delta q^* \\ \Delta\alpha^* \end{bmatrix} = \frac{\begin{bmatrix} -gM_{\delta E}\dfrac{L_\alpha}{V_o} & 0 & gM_\alpha \dfrac{L_{\delta F}}{V_o} \\ \left(D_V\dfrac{L_\alpha}{V_o} - D_\alpha \dfrac{L_V}{V_o}\right)M_{\delta E} & \left(M_V \dfrac{L_\alpha}{V_o} - M_\alpha \dfrac{L_V}{V_o}\right)T_{\delta T} & (D_\alpha M_V - D_V M_\alpha)\dfrac{L_{\delta F}}{V_o} \\ 0 & 0 & 0 \\ gM_{\delta E}\dfrac{L_V}{V_o} & 0 & -gM_V \dfrac{L_{\delta F}}{V_o} \end{bmatrix}}{g\left(M_V \dfrac{L_\alpha}{V_o} - M_\alpha \dfrac{L_V}{V_o}\right)} \begin{bmatrix} \Delta\delta E^* \\ \Delta\delta T^* \\ \Delta\delta F^* \end{bmatrix} \quad (5.4\text{-}24)$$

ΔV^* and $\Delta\gamma^*$ behave as in the second-order model of Section 5.2, although there is an effect of elevator input, as well as the lifting effect of flaps. The elevator has a strong effect on the initial pitch rate (eqs. 4.2-39 and 5.4-23), but the equilibrium pitch rate is zero, and the steady angle of attack is affected by both elevator and flaps. The flaps also affect ΔV^* and $\Delta\gamma^*$, revealing lift as a coupling control.

The equilibrium response to control surfaces that individually produce pitching moment, lift, and drag (or thrust) can be calculated by superposing the responses of eq. 5.4-24. For example, the elevator on a delta-wing aircraft produces significant lift as well as pitching moment, and its drag effect could be important at large deflection angle. The steady-state response to $\Delta\delta E^*$ could be expressed by setting $T_{\delta T}$ to $-D_{\delta E}$ and $L_{\delta F}/V_o$ to $L_{\delta E}/V_o$ and then summing the numerator columns.

The pitch angle is an important longitudinal variable, although it has less direct effect on the phugoid and short-period modes than either flight path angle or angle of attack. Because

$$\Delta\theta = \Delta\gamma + \Delta\alpha \quad (5.4\text{-}25)$$

in symmetric flight, the equilibrium pitch-angle perturbation is the difference between the corresponding terms in eq. 5.4-24:

$$\Delta\theta^* = \Delta\gamma^* + \Delta\alpha^* = \left[(g - D_\alpha)\frac{L_V}{V_o} - D_V\frac{L_\alpha}{V_o}\right]M_{\delta E}\Delta\delta E^*$$
$$- \left(M_\alpha\frac{L_V}{V_o} - M_V\frac{L_\alpha}{V_o}\right)T_{\delta T}\Delta\delta T^*$$
$$+ \left[(D_\alpha - g)M_V - D_V M_\alpha\right]\frac{L_{\delta F}}{V_o}\Delta\delta F^* \quad (5.4\text{-}26)$$

All three controls affect the equilibrium pitch angle, except when additive terms in their multipliers cancel.

Eigenvalues and Root Locus Analysis of Parameter Variations

The characteristic polynomial for this model is

$$\Delta(s) = |s\mathbf{I}_4 - \mathbf{F}_L|$$
$$= s^4 + \left(D_V + \frac{L_\alpha}{V_o} - M_q\right)s^3$$
$$+ \left[(g - D_\alpha)\frac{L_V}{V_o} + D_V\left(\frac{L_\alpha}{V_o} - M_q\right) - M_q\frac{L_\alpha}{V_o} - M_\alpha\right]s^2$$
$$+ \left\{M_q\left[(D_\alpha - g)\frac{L_V}{V_o} - D_V\frac{L_\alpha}{V_o}\right] + D_\alpha M_V - D_V M_\alpha\right\}s$$
$$+ g\left(M_V\frac{L_\alpha}{V_o} - M_\alpha\frac{L_V}{V_o}\right) \quad (5.4\text{-}27)$$

A literal factorization of Δ(s) is possible, but it is too cumbersome to be of interest, particularly when numerical factorization by computer is so accessible. Nevertheless, we note that the first three (higher-degree) terms of eq. 5.4-27 can be viewed as an approximation for the faster short-period-mode polynomial (after factoring out s^2), and the last three terms approximate the slower phugoid mode.

Factoring eq. 5.4-27 typically produces a pair of oscillatory modes:

$$\Delta(s) = (s^2 + 2\zeta\omega_n s + \omega_n^2)_P(s^2 + 2\zeta\omega_n s + \omega_n^2)_{SP} \quad (5.4\text{-}28)$$

Normally, the phugoid mode is slow and lightly damped, while the short-period mode is quick and moderately damped. The short-period roots are real for statically unstable aircraft, and the qualitative description of modes as traditional phugoid or short-period modes is strained when the periods of motion are similar.

The value of root locus analysis of parameter variations is apparent, as the characteristic polynomial is readily restated in the form

$$\Delta(s) = d(s) + k\, n(s) = 0 \tag{5.4-29}$$

alternatively defining k as each of the stability derivatives. We begin by examining the two short-period stability derivatives M_α and M_q. For the former, the root locus form of the characteristic polynomial is

$$\begin{aligned}
\Delta(s) = {} & s^4 + \left(D_V + \frac{L_\alpha}{V_o} - M_q\right)s^3 \\
& + \left[(g - D_\alpha)\frac{L_V}{V_o} + D_V\left(\frac{L_\alpha}{V_o} - M_q\right) - M_q\frac{L_\alpha}{V_o}\right]s^2 \\
& + \left\{M_q\left[\frac{L_V}{V_o}(D_\alpha - g) - D_V\frac{L_\alpha}{V_o}\right] + D_\alpha M_V\right\}s + g\left(\frac{L_\alpha}{V_o}M_V\right) \\
& - M_\alpha\left(s^2 + D_V\, s + g\frac{L_V}{V_o}\right)
\end{aligned} \tag{5.4-30}$$

The numerator polynomial is the characteristic polynomial for the second-order phugoid approximation (eq. 5.2-20), and a pair of roots approaches the approximate phugoid roots as $|M_\alpha|$ becomes large. Without the M_α terms, $d(s)$ represents the characteristic polynomial of an aircraft with neutral static stability. The surplus of poles over zeros is 2; two roots approach the zeros, and the remaining two roots go toward infinity along asymptotes of ± 90 deg for $M_\alpha < 0$ and (0, 180) deg for $M_\alpha > 0$. A negative pitching moment has the same effect on the short-period mode as predicted by the reduced-order model. For some range of positive M_α, the phugoid and short-period roots are close, and the eigenvectors of both modes are likely to contain significant components of all four state elements.

The pitch-damping derivative affects the short-period mode in the same way for both second- and fourth-order models, principally increasing the mode's damping ratio. The root-locus form of the characteristic polynomial is

$$\Delta(s) = s^4 + \left(D_V + \frac{L_\alpha}{V_o}\right)s^3 + \left[(g - D_\alpha)\frac{L_V}{V_o} + D_V\frac{L_\alpha}{V_o} - M_\alpha\right]s^2$$
$$+ (D_\alpha M_V - D_V M_\alpha)s + g\left(\frac{L_\alpha}{V_o}M_V - \frac{L_V}{V_o}M_\alpha\right)$$
$$- M_q s\left\{s^2 + \left(D_V + \frac{L_\alpha}{V_o}\right)s - \left[\frac{L_V}{V_o}(D_\alpha - g) - D_V\frac{L_\alpha}{V_o}\right]\right\} \quad (5.4\text{-}31)$$

The three numerator zeros are real in the example. The zero at the origin and the smaller of the remaining two are in the neighborhood of the phugoid mode, attracting its roots as $|M_q|$ becomes large.

The phugoid stability derivatives L_V/V_o and D_V are investigated in the same way. Beginning with the lift effect, the characteristic polynomial is written as

$$\Delta(s) = s^4 + \left(D_V + \frac{L_\alpha}{V_o} - M_q\right)s^3 + \left[D_V\left(\frac{L_\alpha}{V_o} - M_q\right) - M_q\frac{L_\alpha}{V_o} - M_\alpha\right]s^2$$
$$+ \left[-M_q\left(D_V\frac{L_\alpha}{V_o}\right) + D_\alpha M_V - D_V M_\alpha\right]s + g\left(\frac{L_\alpha}{V_o}M_V\right)$$
$$+ \frac{L_V}{V_o}(D_\alpha - g)\left[s^2 - M_q s - \frac{gM_\alpha}{(g - D_\alpha)}\right] \quad (5.4\text{-}32)$$

Here, the numerator polynomial is very nearly the reduced-order short-period polynomial. The short-period pair is drawn toward the numerator pair for large $|L_V/V_o|$, and the remaining pair goes toward infinity as in the reduced-order phugoid model (Fig. 5.2-1a).

The effect of drag sensitivity to velocity is similarly defined:

$$\Delta(s) = s^4 + \left(\frac{L_\alpha}{V_o} - M_q\right)s^3 + \left[(g - D_\alpha)\frac{L_V}{V_o} - M_q\frac{L_\alpha}{V_o} - M_\alpha\right]s^2$$
$$+ \left[M_q(D_\alpha - g)\frac{L_V}{V_o} + D_\alpha M_V\right]s + g\left(\frac{L_\alpha}{V_o}M_V - \frac{L_V}{V_o}M_\alpha\right)$$
$$+ D_V s\left[s^2 + \left(\frac{L_\alpha}{V_o} - M_q\right)s - \left(M_q\frac{L_\alpha}{V_o} + M_\alpha\right)\right] \quad (5.4\text{-}33)$$

D_V variation has small effect on the short period, and the phugoid roots follow the reduced-order model (Fig. 5.2-1b).

The significant coupling derivatives L_α/V_o and M_V have more complex effects on the longitudinal roots. The root locus for variations in lift due to the angle of attack is derived from

$$\Delta(s) = s^4 + (D_V - M_q)s^3 - \left[(g - D_\alpha)\frac{L_V}{V_o} - D_V M_q - M_\alpha\right]s^2 + \left[M_q(D_\alpha - g)\frac{L_V}{V_o} + D_\alpha M_V - D_V M_\alpha\right]s - gM_\alpha\frac{L_V}{V_o} + \frac{L_\alpha}{V_o}[s^3 + (D_V - M_q)s^2 - M_q D_V s + gM_V] \quad (5.4\text{-}34)$$

With $M_V = 0$, the equation becomes

$$\Delta(s) = \{\cdot\} + \frac{L_\alpha}{V_o}s(s + D_V)(s - M_q) \quad (5.4\text{-}35)$$

producing three negative real zeros, one of which is at the origin, and a pole-zero surplus of 1. For $L_\alpha/V_o = 0$, the short-period roots have less damping, and the phugoid roots are near the origin. For the baseline model (Table 5.4-1, generic business jet model, with $D_\alpha = D_{\delta E} = L_{\delta E} = L_{\delta T} = M_{\delta T} = 0$), increasing L_α/V_o raises short-period damping and draws the phugoid roots toward the real axis (Fig. 5.4-1). Negative L_α/V_o does not occur in the unstalled regime.

The M_V root locus reveals potential stability problems when the pitching moment sensitivity to velocity variation is large:

$$\Delta(s) = s^4 + \left(D_V + \frac{L_\alpha}{V_o} - M_q\right)s^3 + \left[(gD_\alpha)\frac{L_V}{V_o} + D_V\frac{L_\alpha}{V_o}M_q\right]M_q\frac{L_\alpha}{V_o}M_\alpha s^2 + \left\{M_q\left[\frac{L_V}{V_o}(D_\alpha - g) - D_V\frac{L_\alpha}{V_o}\right] - D_V M_\alpha\right\}s - g\,M_\alpha\frac{L_V}{V_o} + M_V\left(D_\alpha\, s + g\frac{L_\alpha}{V_o}\right) \quad (5.4\text{-}36)$$

With $D_\alpha = 0$, the characteristic polynomial has four poles and no zeros,

$$\Delta(s) \approx \{\cdot\} + M_V g\frac{L_\alpha}{V_o} \quad (5.4\text{-}37)$$

TABLE 5.4-1 BASELINE LONGITUDINAL MODEL FOR FIGURES OF SECTION 5.4

$$\mathbf{F} = \begin{bmatrix} -0.0185 & -9.8067 & 0 & 0 \\ 0.0019 & 0 & 0 & 1.2709 \\ 0 & 0 & -1.2794 & -7.9856 \\ -0.0019 & 0 & 1 & -1.2709 \end{bmatrix} \quad \mathbf{G} = \begin{bmatrix} 0 & 4.6645 \\ 0 & 0 \\ -9.069 & 0 \\ 0 & 0 \end{bmatrix}$$

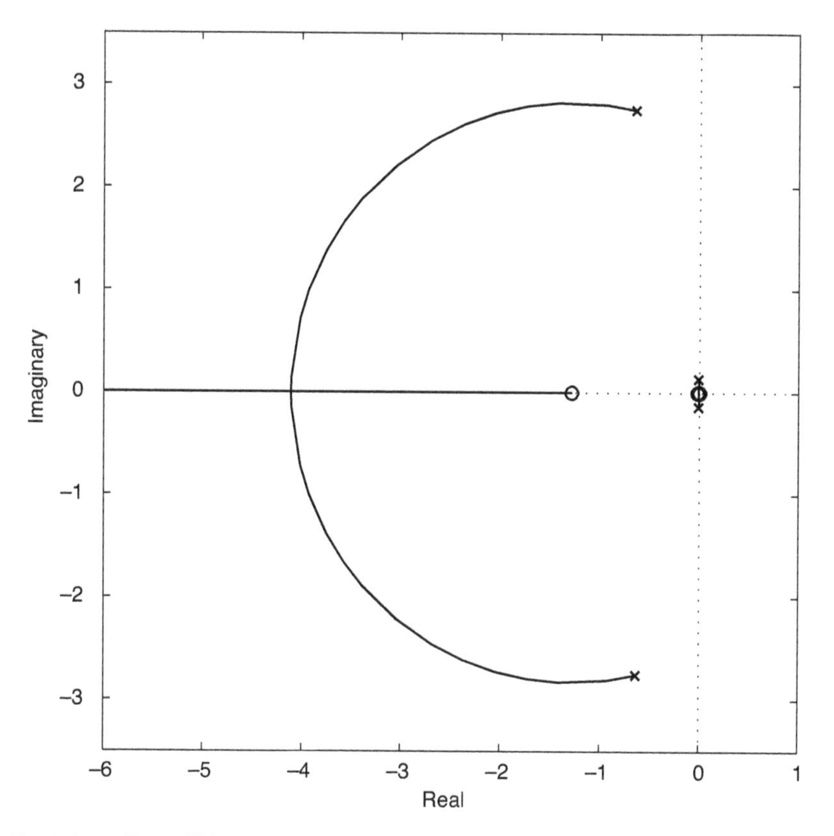

Fig. 5.4-1. Effects of lift sensitivity to angle of attack (L_α/V_o) on the fourth-order longitudinal roots.

For positive M_V, the root asymptotes are at ± 45 and ± 135 deg, while for negative M_V, they are at 0, 90, 180, and 270 deg (Fig. 5.4-2). The phugoid mode is destabilized by either large positive or large negative M_V. In the first case, the phugoid natural frequency and damping initially increase; then the roots migrate to the right half plane. In the second, the phugoid's complex roots become real, with one going to the right, one to the left. The same trends occur when $D_\alpha \neq 0$.

Transfer Functions and Frequency Response

The control and disturbance transfer function matrices $\mathbf{U}_L(s)$ and $\mathbf{W}_L(s)$ relate the output Laplace transform $\Delta\mathbf{y}_L(s)$ to the inputs $\Delta\mathbf{u}_L(s)$ and $\Delta\mathbf{w}_L(s)$; following eq. 4.3-39,

$$\begin{aligned}\Delta \mathbf{y}_L(s) &= \mathbf{H}_L(s\mathbf{I}_4 - \mathbf{F}_L)^{-1}[\mathbf{G}_L\Delta\mathbf{u}_L(s) + \mathbf{L}_L\Delta\mathbf{w}_L(s)] \\ &\quad + \mathbf{J}_L\Delta\mathbf{u}_L(s) + \mathbf{K}_L\Delta\mathbf{w}_L(s) \\ &\triangleq \mathbf{U}_L(s)\Delta\mathbf{u}_L(s) + \mathbf{W}_L(s)\Delta\mathbf{w}_L(s)\end{aligned} \tag{5.4-38}$$

where $\mathbf{J}_L$ and $\mathbf{K}_L$ reflect direct paths from the input to the output, and

$$\begin{aligned}\mathbf{U}_L(s) &= \mathbf{H}_L(s\mathbf{I}_4 - \mathbf{F}_L)^{-1}\mathbf{G}_L + \mathbf{J}_L \\ &= \frac{\mathbf{H}_L \operatorname{Adj}(s\mathbf{I}_4 - \mathbf{F}_L)\mathbf{G}_L + \mathbf{J}_L\,|s\mathbf{I}_4 - \mathbf{F}_L|}{|s\mathbf{I}_4 - \mathbf{F}_L|}\end{aligned} \tag{5.4-39}$$

$$\begin{aligned}\mathbf{W}_L(s) &= \mathbf{H}_L(s\mathbf{I}_4 - \mathbf{F}_L)^{-1}\mathbf{L}_L + \mathbf{K}_L \\ &= \frac{\mathbf{H}_L \operatorname{Adj}(s\mathbf{I}_4 - \mathbf{F}_L)\mathbf{L}_L + \mathbf{K}_L\,|s\mathbf{I}_4 - \mathbf{F}_L|}{|s\mathbf{I}_4 - \mathbf{F}_L|}\end{aligned} \tag{5.4-40}$$

Every scalar transfer function [i.e., each element of $\mathbf{U}_L(s)$ and $\mathbf{W}_L(s)$] has the same denominator and takes the form $n(s)/\Delta(s)$, where $\Delta(s)$ is defined by eq. 5.4-27. We distinguish between the various transfer functions by their numerator polynomials $n_\delta^x(s)$, where δ is the scalar control or disturbance input and x is the scalar output. Thus the control transfer function $u_\delta^x(s)$ is given by $n_\delta^x(s)/\Delta(s)$.

If the corresponding element of $\mathbf{J}_L$ or $\mathbf{K}_L$ is zero, the transfer function is *strictly proper*, that is, the degree of the numerator polynomial is at least one less than the degree of the denominator polynomial. The amplitude frequency response of a strictly proper transfer function rolls off with a slope of -20 dB/dec or more at high frequency. If the direct element is not

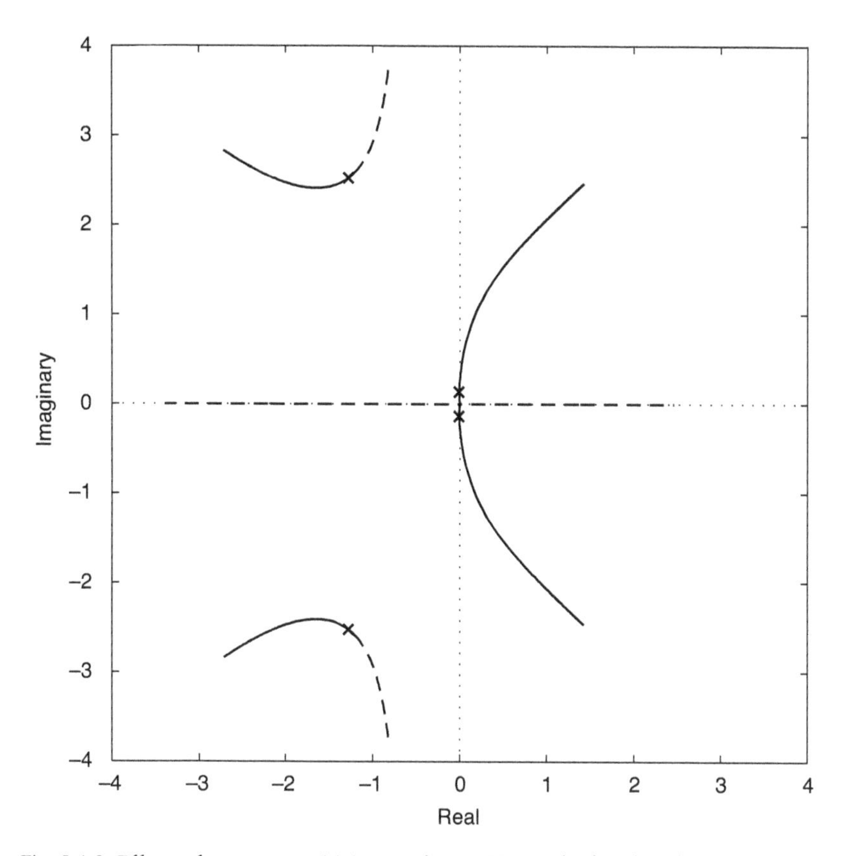

Fig. 5.4-2. Effects of moment sensitivity to velocity (M_V) on the fourth-order longitudinal roots. Positive variations shown by solid line, negative variations by dashed line.

zero, the transfer function is at least *proper*, as $|s\mathbf{I}_4 - \mathbf{F}_L|$ is $\Delta(s)$ and of fourth degree; consequently, the numerator polynomial is never of higher degree than the denominator polynomial.

Here, we define the output as the state (eq. 5.4-1) plus perturbations in pitch angle ($\Delta\theta$) and vertical load factor (Δn_z, measured in units of g at the center of mass, positive down):

$$\Delta\mathbf{y}_L = [\Delta V\ \Delta\gamma\ \Delta q\ \Delta\alpha\ \Delta\theta\ \Delta n_z]^T \tag{5.4-41}$$

$\mathbf{H}_L$ is the (4×4) identity matrix plus two additional rows defining pitch-angle and normal-load-factor output,

$$\mathbf{H}_L = \begin{bmatrix} 1 & 0 & 0 & 0 \\ 0 & 1 & 0 & 0 \\ 0 & 0 & 1 & 0 \\ 0 & 0 & 0 & 1 \\ 0 & 1 & 0 & 1 \\ \frac{-L_V}{g} & 0 & 0 & \frac{-L_\alpha}{g} \end{bmatrix} \tag{5.4-42}$$

where L_V/g is $(V_o/g)(L_V/V_o)$, and so on. The direct-effect matrices are

$$\mathbf{J}_L = \begin{bmatrix} 0 & 0 & 0 \\ 0 & 0 & 0 \\ 0 & 0 & 0 \\ 0 & 0 & 0 \\ 0 & 0 & 0 \\ \frac{-L_{\delta E}}{g} & \frac{-L_{\delta T}}{g} & \frac{-L_{\delta F}}{g} \end{bmatrix} \tag{5.4-43}$$

$$\mathbf{K}_L = \begin{bmatrix} 0 & 0 \\ 0 & 0 \\ 0 & 0 \\ 0 & 0 \\ 0 & 0 \\ \frac{L_V}{g} & \frac{L_\alpha}{g} \end{bmatrix} \tag{5.4-44}$$

With the idealized control effects defined by eq. 5.4-23, the control-transfer-function numerators are as follows:

Elevator (Pitch-Control) Transfer-Function Numerators

$$n_{\delta E}^{V}(s) = -M_{\delta E}\left(D_\alpha s + g\frac{L_\alpha}{V_o}\right) \tag{5.4-45}$$

$$n_{\delta E}^{\gamma}(s) = M_{\delta E}\left[\frac{L_\alpha}{V_o}s + \left(D_V\frac{L_\alpha}{V_o} - D_\alpha\frac{L_V}{V_o}\right)\right] \tag{5.4-46}$$

$$n_{\delta E}^{q}(s) = M_{\delta E}s\left\{s^2 + \left(D_V + \frac{L_\alpha}{V_o}\right)s + \left[(g - D_\alpha)\frac{L_V}{V_o} + D_V\frac{L_\alpha}{V_o}\right]\right\} \tag{5.4-47a}$$

$$n_{\delta E}^{\alpha}(s) = M_{\delta E}\left(s^2 + D_V s + g\frac{L_V}{V_o}\right) \tag{5.4-48}$$

$$n_{\delta E}^{\theta}(s) = M_{\delta E}\left\{s^2 + \left(D_V + \frac{L_\alpha}{V_o}\right)s + \left[(g - D_\alpha)\frac{L_V}{V_o} + D_V\frac{L_\alpha}{V_o}\right]\right\} \tag{5.4-49a}$$

$$n_{\delta E}^{n_z}(s) = -M_{\delta E}s\left[\frac{L_\alpha}{g}s + \left(D_V\frac{L_\alpha}{g} - D_\alpha\frac{L_V}{g}\right)\right] \tag{5.4-50}$$

Thrust Transfer-Function Numerators

$$n_{\delta T}^{V}(s) = T_{\delta T}s\left[s^2 + \left(\frac{L_\alpha}{V_o} - M_q\right)s - \left(M_q\frac{L_\alpha}{V_o} + M_\alpha\right)\right] \tag{5.4-51}$$

$$n_{\delta T}^{\gamma}(s) = T_{\delta T}\left[\frac{L_V}{V_o}s^2 - M_q\frac{L_V}{V_o}s - \left(M_V\frac{L_\alpha}{V_o} - M_\alpha\frac{L_V}{V_o}\right)\right] \tag{5.4-52}$$

$$n_{\delta T}^{q}(s) = T_{\delta T}s\left[M_V s + \left(M_V\frac{L_\alpha}{V_o} - M_\alpha\frac{L_V}{V_o}\right)\right] \tag{5.4-53}$$

$$n_{\delta T}^{\alpha}(s) = -T_{\delta T}s\left[\frac{L_V}{V_o}s - \left(M_V + M_q\frac{L_V}{V_o}\right)\right] \tag{5.4-54}$$

$$n_{\delta T}^{\theta}(s) = T_{\delta T}\left[M_V s + \left(M_V\frac{L_\alpha}{V_o} - M_\alpha\frac{L_V}{V_o}\right)\right] \tag{5.4-55}$$

$$n_{\delta T}^{n_z}(s) = -T_{\delta T}s\left[\frac{L_V}{g}s^2 - M_q\frac{L_V}{g}s - \left(M_\alpha\frac{L_V}{g} - M_V\frac{L_\alpha}{g}\right)\right] \tag{5.4-56}$$

Flap (Lift-Control) Transfer-Function Numerators

$$n_{\delta F}^{V}(s) = -\frac{L_{\delta F}}{V_o}[(g - D_\alpha)s^2 - M_q(g - D_\alpha)s - gM_\alpha] \tag{5.4-57}$$

$$n_{\delta F}^{\gamma}(s) = \frac{L_{\delta F}}{V_o}[s^3 + (D_V - M_q)s^2 - (D_V M_q + M_\alpha)s - (M_V D_\alpha - D_V M_\alpha)] \tag{5.4-58}$$

$$n_{\delta F}^{q}(s) = -\frac{L_{\delta F}}{V_o} s\{M_\alpha s - [M_V(g - D_\alpha) + D_V M_\alpha]\} \tag{5.4-59}$$

$$n_{\delta F}^{\alpha}(s) = -\frac{L_{\delta F}}{V_o}[s^3 + (D_V - M_q)s^2 - D_V M_q s - g M_V] \tag{5.4-60}$$

$$n_{\delta F}^{\theta}(s) = -\frac{L_{\delta F}}{V_o}\{M_\alpha s - [M_v(g - D_\alpha) + D_v M_\alpha]\} \tag{5.4-61}$$

$$n_{\delta F}^{n_z}(s) = \frac{L_{\delta F}}{V_o}\left\{\frac{L_\alpha}{g}s^3 + \left[\frac{L_\alpha}{g}(D_V - M_q)\right]s^2 - M_q\left[\frac{L_V}{g}(g - D_\alpha) + \frac{L_\alpha}{g}D_V\right]s - g\left(\frac{L_\alpha}{g}M_V - \frac{L_V}{g}M_\alpha\right) + \Delta(s)\right\} \tag{5.4-62}$$

These numerators can be factored and put in the general form $k\,(s - z_1)(s - z_2)\cdots$, with complex roots combined in second-degree terms. For example, using the notation of [M-3], the elevator-to-pitch-rate and -pitch-angle numerators typically factor as

$$n_{\delta E}^{q}(s) = M_{\delta E}s(s + 1/T_{\theta_1})(s + 1/T_{\theta_2}) \tag{5.4-47b}$$

$$n_{\delta E}^{\theta}(s) = M_{\delta E}(s + 1/T_{\theta_1})(s + 1/T_{\theta_2}) \tag{5.4-49b}$$

The numerators for controls with compound effects (e.g., $M_{\delta E}$ and $L_{\delta E}$) are readily expressed by summing the appropriate numerators: the elevator-to-pitch-rate numerator would be formed by adding eq. 5.4-47 to eq. 5.4-59, replacing $L_{\delta F}$ by $L_{\delta E}$ in the latter.

For the most part, the zeros have negative real parts and minimum-phase effect. Large D_α or M_V could lead to non-minimum-phase zeros (i.e., with positive real parts) in several instances. More elaborate numerators result when control surfaces have compound effect (e.g., the lifting and pitching effect of elevator deflection discussed in Section 5.3) and when previously neglected stability derivatives are considered. The flap-to-normal-load-factor numerator is fourth degree as a consequence of the flap's direct effect on lift, so the transfer function is not strictly proper.

The pitch-rate numerators are the pitch-angle numerators multiplied by s, as the rate is the derivative of the angle; hence, a zero is added at the origin. Altitude numerators are flight-path-angle numerators multiplied by $-V_o/s$, adding a pole at the origin (see below).

The elevator (pitch-control) numerators contain stability derivatives associated with the phugoid mode, and the thrust numerators contain stability derivatives associated with the short-period mode. Numerator terms approximately cancel denominator terms, so the elevator largely controls the short-period mode and the thrust controls the phugoid mode. The flap (lift-control) effect is less easily categorized, as it is a coupling control.

Figure 5.4-3 compares typical amplitude asymptotes for flight-path-angle, angle-of-attack, and pitch-angle response to elevator input. The pitch-angle frequency response approaches the sum of the flight-path-angle and angle-of-attack asymptotes as the input frequency goes toward zero, following eq. 5.4-26. The angle-of-attack amplitude is flat from the steady state to short-period frequency, with a small resonance and notch near the phugoid natural frequency. Angle-of-attack and pitch-angle frequency responses approach the same asymptote as the input frequency goes toward infinity, indicating that the high-frequency flight-path-angle response to pitch control approaches zero.

A constant elevator angle produces a steady change in velocity, but constant thrust perturbation does not (Fig. 5.4-4). However, the elevator effect rolls off sharply beyond the phugoid resonance. The thrust effect peaks at the phugoid natural frequency and then rolls off at -20 dB/dec.

Flight-path-angle and pitch-angle response to sinusoidal thrust input are virtually identical up to the vicinity of the short-period natural frequency (Fig. 5.4-5). Beyond that frequency, the $\Delta\gamma$ and $\Delta\alpha$ amplitude ratios become similar although -180 deg out of phase, leading to vanishing pitch-angle response. The steady-state angle-of-attack response to thrust variation is zero in this model.

Recent flight research has demonstrated the ability to land aircraft using thrust alone following complete failure of conventional flight controls [B-7]. The thrust-to-angle transfer functions have a critical effect on the ability to perform this maneuver. In the present example, $M_{\delta T}$ is zero, and the thrust effect on angles is small. However, with engines mounted away from the vehicle centerline, the moment due to thrust can be a useful adjunct to conventional pitch control.

In horizontal, wings-level flight, the altitude rate of change (positive down) is

$$\Delta\dot{z} = -V_o\Delta\gamma \tag{5.4-63}$$

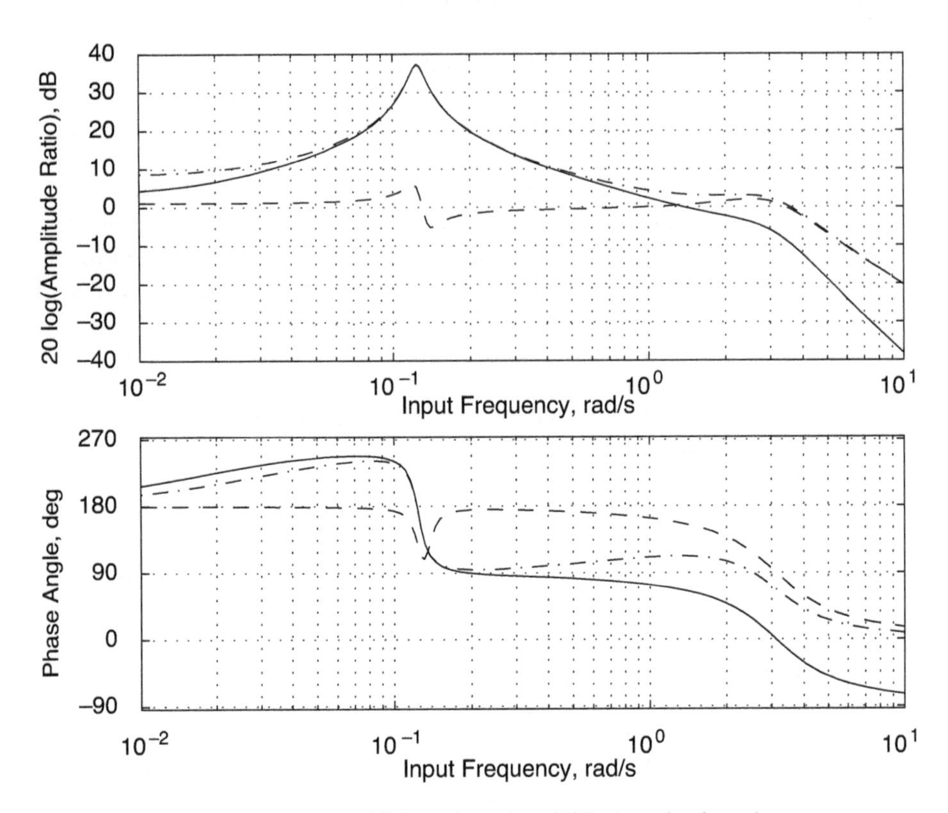

Fig. 5.4-3. Frequency response of flight path angle (solid line), angle of attack (dashed line), and pitch angle (dashed-dotted line) to sinusoidal elevator input.

with the corresponding Laplace transform (neglecting initial conditions)

$$\Delta z(s) = -\frac{V_o}{s}\Delta\gamma(s) \tag{5.4-64}$$

Consequently, the altitude transfer functions are the flight-path-angle transfer functions multiplied by $-V_o/s$:

$$u^{z}_{\delta E}(s) = \frac{-V_o M_{\delta E}\left[\left(\dfrac{L_\alpha}{V_o}\right)s + \left(D_V\dfrac{L_\alpha}{V_o} - D_\alpha\dfrac{L_V}{V_o}\right)\right]}{s\Delta(s)} \tag{5.4-65}$$

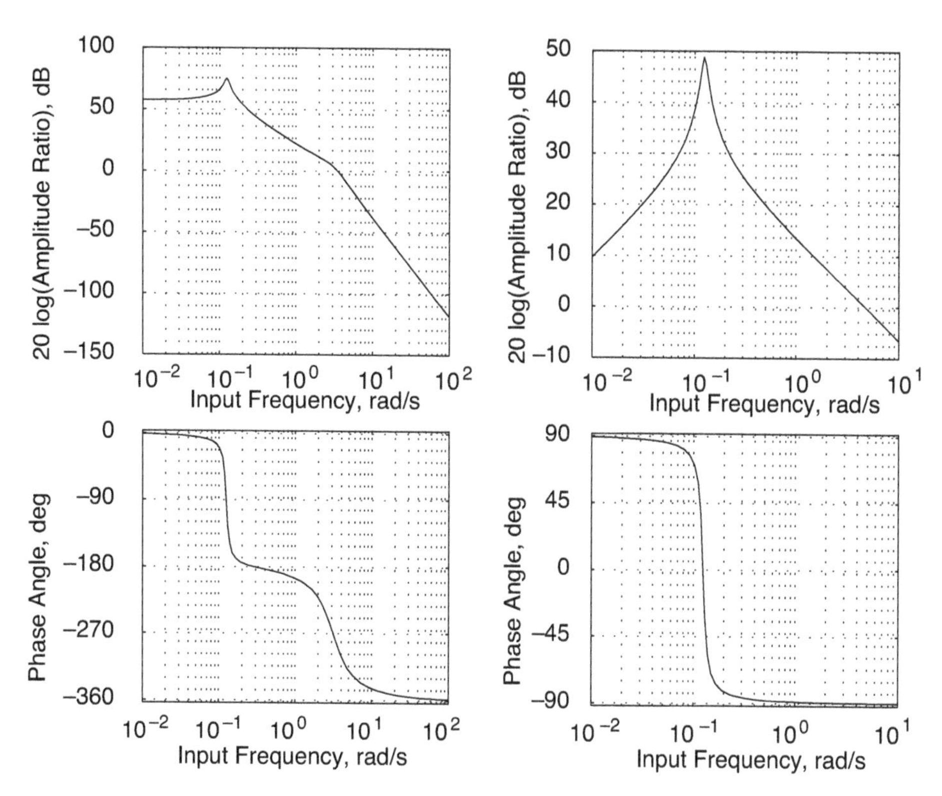

Fig. 5.4-4. Velocity frequency response to elevator and thrust sinusoidal input. (Left) Elevator input. (Right) Thrust input.

$$u^z_{\delta T}(s) = \frac{-V_o T_{\delta T}\left[\left(\frac{L_V}{V_o}\right)s^2 - M_q\left(\frac{L_V}{V_o}\right)s - \left(M_V\frac{L_\alpha}{V_o} - M_\alpha\frac{L_V}{V_o}\right)\right]}{s\Delta(s)} \tag{5.4-66}$$

$$u^z_{\delta F}(s) = \frac{-V_o\left(\frac{L_{\delta F}}{V_o}\right)[s^3 + (D_V - M_q)s^2 - (D_V M_q + M_\alpha)s - (M_V D_\alpha - D_V M_\alpha)]}{s\Delta(s)} \tag{5.4-67}$$

where $\Delta(s)$ is defined by eq. 5.4-27. As a consequence of the integrator, both controls have a large effect on altitude at low frequency (Fig. 5.4-6), as well as resonant response at the phugoid frequency.

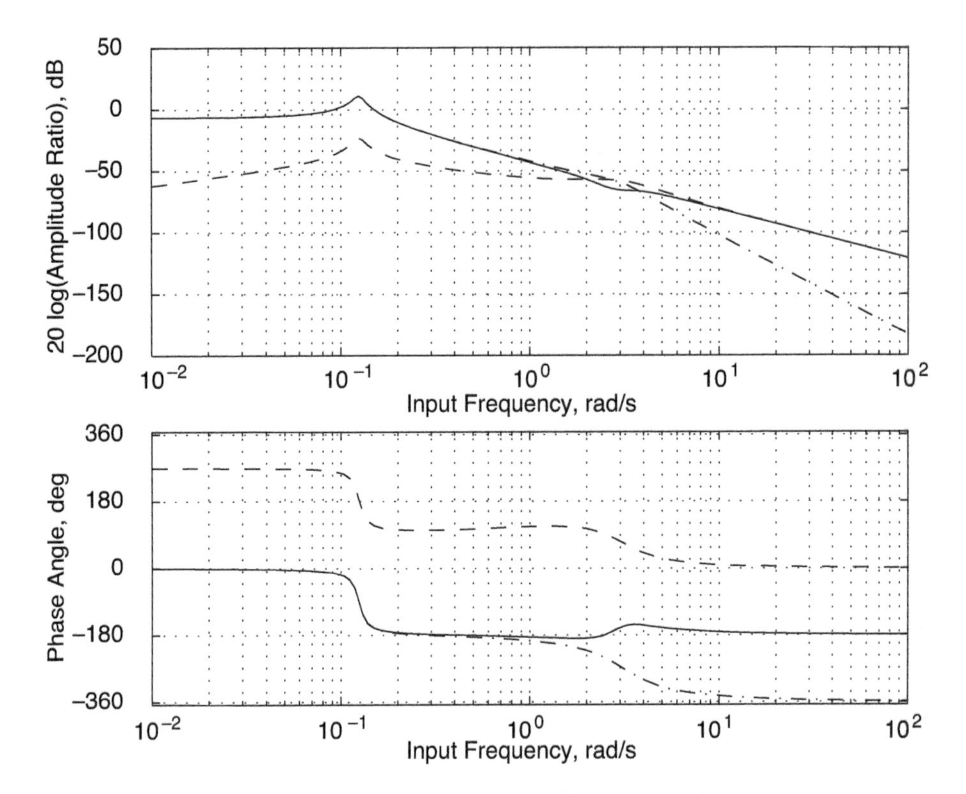

Fig. 5.4-5. Velocity frequency response to elevator and thrust sinusoidal input.

Response to Disturbances

The response to longitudinal wind disturbances of varying frequency is described in the same way as the frequency response to controls, replacing $\mathbf{G}_L$ and $\mathbf{J}_L$ by $\mathbf{L}_L$ and $\mathbf{K}_L$ (eq. 5.4-40). Wind perturbations along the velocity vector ΔV_W and the angle-of-attack perturbations arising from vertical winds $\Delta\alpha_W$ produce pitching moments and lift and drag forces that excite the longitudinal modes of motion. Their effects can be represented by scalar transfer functions of the form $n_\delta^x(s)/\Delta(s)$. The numerators can be constructed from the control numerators, replacing $M_{\delta E}$ by M_V or M_α, $L_{\delta F}/V_o$ by L_V/V_o or L_α/V_o, and $T_{\delta T}$ by D_V or D_α.

Root Locus Analysis of Feedback Control

Feeding back elements of the state to the controls produces effects analogous to the stability-derivative variations considered earlier. For the

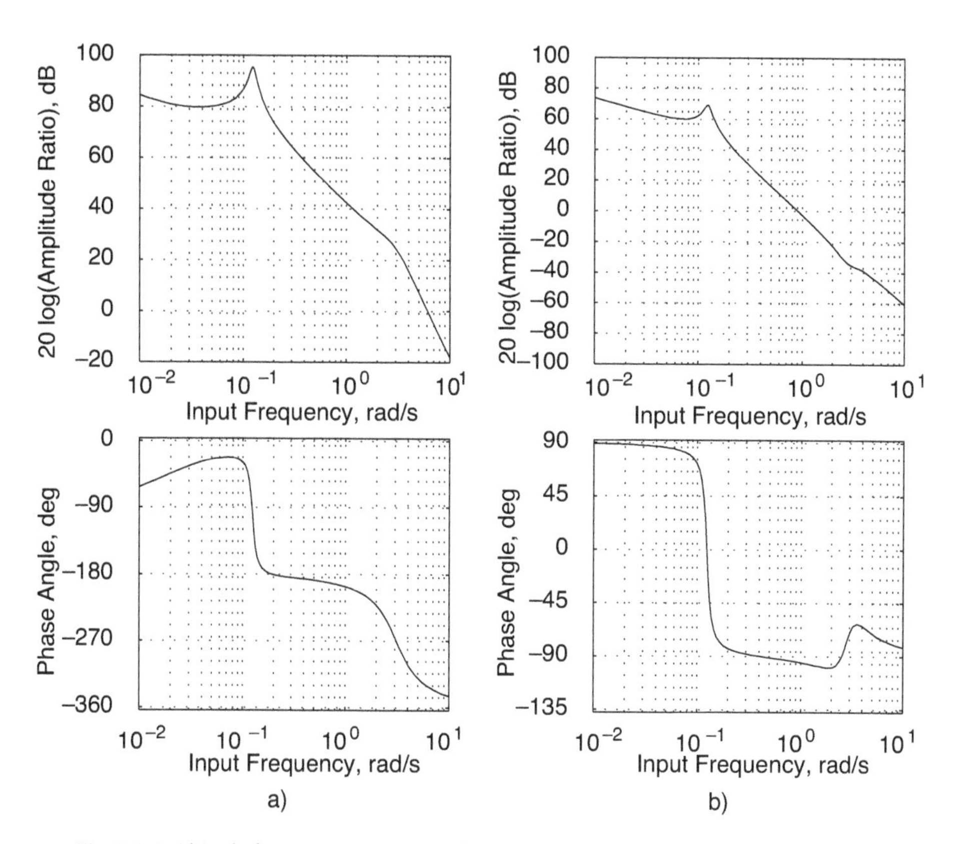

Fig. 5.4-6. Altitude frequency response to elevator and thrust sinusoidal input. (a) Elevator input. (b) Thrust input.

examples, we assume that the elevator and throttle produce pure pitching moment or thrust force. It is assumed that control actuation is instantaneous; the effects of control mechanism dynamics are considered in Section 5.5. Typically, the elevator mechanism is fast in comparison to the rigid-body modes. Thrust actuation is fast in comparison to the phugoid mode but slow compared to the short period. The elements of the output vector are fed back individually and without dynamic compensation. For control system design, multiple outputs and controls can be employed, and signals can be subjected to *dynamic compensation* (i.e., integration and differentiation as well as proportional feedback).

The effects of four scalar pitch-control feedback loops are demonstrated. When the flight path angle $\Delta\gamma$ is the feedback variable (Fig. 5.4-7a), eq. 5.-46 is the transfer function numerator. Feedback with positive loop gain increases the phugoid damping ratio, which is represented by two real

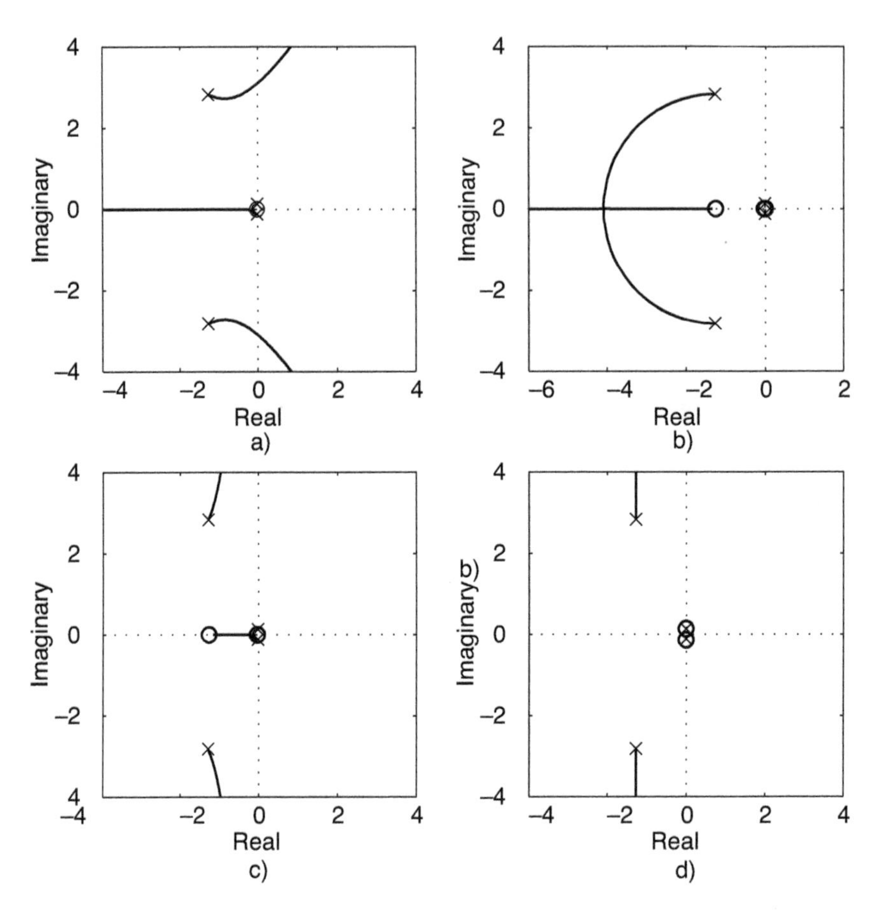

Fig. 5.4-7. Root loci for angular feedback to elevator. (a) Flight path angle. (b) Pitch rate. (c) Pitch angle. (d) Angle of attack.

roots at high gain, and decreases short-period damping. For some range of gains, both modes are stable. The short-period damping ratio is readily increased by the pitch-rate feedback Δq to $\Delta\delta E$ (Fig. 5.4-7b); the phugoid damping ratio is also increased by a lesser amount. Pitch-angle feedback to elevator, $\Delta\theta$ to $\Delta\delta E$, has the same effect on the phugoid mode as flight-path-angle feedback (Fig. 5.4-7c), with reduced destabilizing effect on the short-period mode. Both pitch-angle and angle-of-attack feedback increase the short-period natural frequency; $\Delta\alpha$ feedback has a negligible effect on the phugoid roots and the real part of the short-period roots (Fig. 5.4-7d).

Angle feedback to thrust requires high gain for root modification and is of academic interest only. Flight-path-angle feedback to thrust increases the phugoid natural frequency, but there is a region of conditional instability as the roots migrate into the right half plane before returning to the left half plane zeros (Fig. 5.4-8a). The pitch-rate-to-thrust feedback that stabilizes one mode destabilizes the other (Fig. 5.4-8b). Pitch-angle feedback is similar to flight-path-angle feedback in the low range of gains considered here (Fig. 5.4-8c), although the phugoid mode becomes more unstable at higher gains. Angle-of-attack feedback to thrust increases phugoid damping (Fig. 5.4-8d).

The root locus for velocity feedback to pitch control (Fig. 5.4-9a) is quite similar to the $\Delta\theta$-to-$\Delta\delta T$ locus, with phugoid roots destabilized at low gain. This is effectively the same as the M_V root locus (Fig. 5.4-2). Velocity feedback to thrust produces the same effect as the stability derivative D_V, increasing the phugoid damping ratio; there is negligible effect on the short-period mode (not shown).

Altitude feedback to the elevator is based on the transfer function of eq. 5.4-65, which introduces a pole and a zero that effectively cancel each other. Consequently, the root locus (Fig. 5.4-10a) is like that of the velocity-elevator feedback. Altitude feedback to the thrust produces instability with feedback of either sign. As shown, the phugoid and short-period modes are well behaved; however, the pole at the origin is immediately drawn to the right by the loop closure. With opposite sign, the phugoid roots become unstable. Altitude regulation requires, therefore, a more complex compensation scheme than either of these scalar, proportional loop closures.

Time Response

The state history is propagated by

$$\Delta \mathbf{x}_{\mathrm{L}}(t_k) = \mathbf{\Phi}_{\mathrm{L}}\, \Delta \mathbf{x}_{\mathrm{L}}(t_{k-1}) + \mathbf{\Gamma}_{\mathrm{L}}\, \Delta \mathbf{u}_{\mathrm{L}}(t_{k-1}) + \mathbf{\Lambda}_{\mathrm{L}}\, \Delta \mathbf{w}_{\mathrm{L}}(t_{k-1}) \tag{5.4-68}$$

The state transition matrix $\mathbf{\Phi}_{\mathrm{L}}(\Delta t)$ could be calculated as the inverse Laplace transform of the characteristic matrix inverse; however, the complexity of the literal expression limits its utility, and the ready availability of the numerical version computed by MATLAB provides a practical alternative. Example 4.2-1 illustrates a typical time-response calculation.

Longitudinal Flying Qualities

Steady-state response of the aircraft to pitch control (i.e., the "gearing" or proportionality between steady inputs and outputs) is important to the pilot, although the bulk of flying qualities research has focused on dynamic behavior. The U.S. military flying qualities specification MIL-F-8785C [A-3] sets limits on flight path stability, speed stability, and stick-force

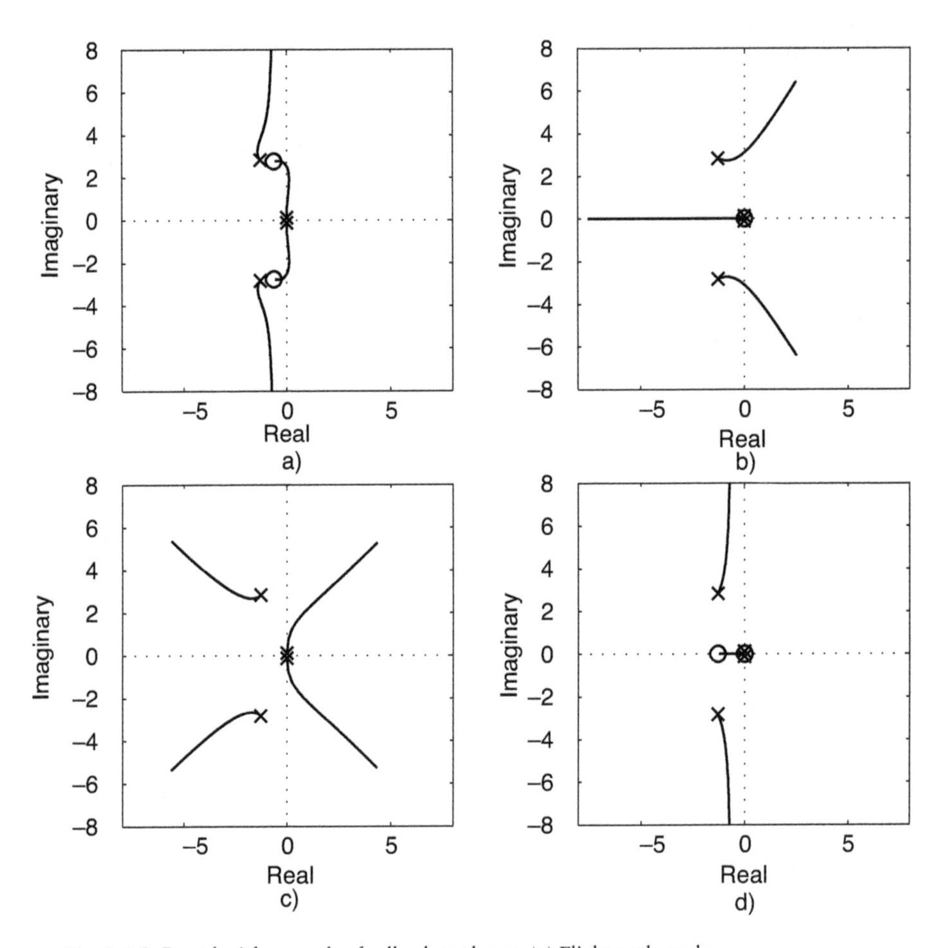

Fig. 5.4-8. Root loci for angular feedback to thrust. (a) Flight path angle. (b) Pitch rate. (c) Pitch angle. (d) Angle of attack.

gradients. For example, the Level 1 flight-path-angle change with airspeed due to pitch control should be less than 0.06 deg/kt. Airspeed should not diverge aperiodically if the aircraft's trim condition is disturbed, and pulling back on the control stick should cause the aircraft to trim at slower speed. The Level 1 control-column force (lb) per "g" of commanded acceleration should not be greater than $240/(n/\alpha)$ for a center stick and $500/(n/\alpha)$ for a control wheel, where (n/α) is the gradient of commanded acceleration with angle of attack ("g"s per rad). Additional restrictions and detailed limits can be found in [A-3].

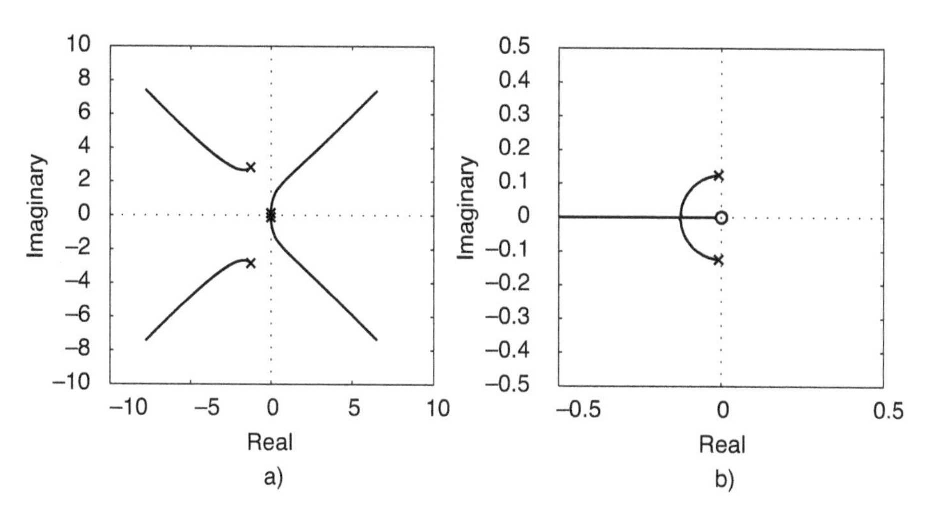

Fig. 5.4-9. Root loci for velocity feedback to elevator and thrust. (a) Elevator. (b) Thrust.

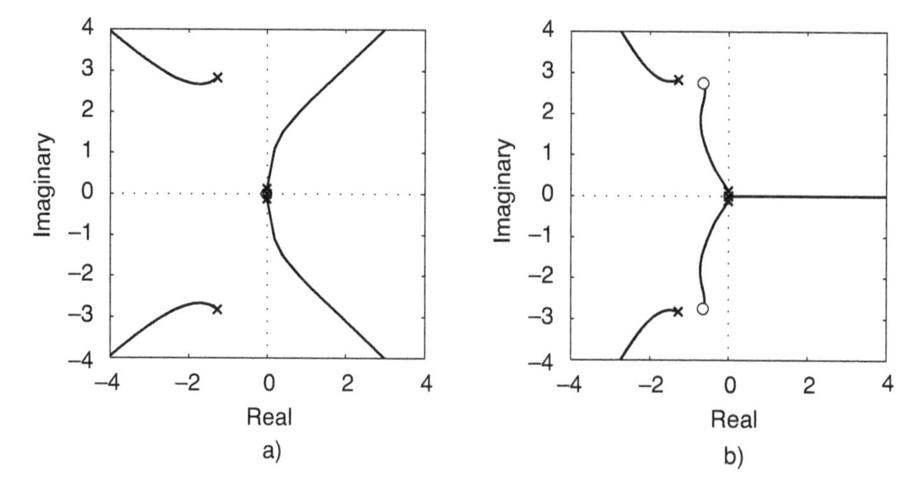

Fig. 5.4-10. Root loci for altitude feedback to elevator and thrust. (a) Elevator. (b) Thrust.

Early aircraft had well-damped short-period modes that did not appear to degrade flying qualities; consequently, flight research was directed at pilot observations of long-period motions. Still, Hartley Soulé found little correlation between the subjective opinions of two pilots and phugoid mode characteristics in an analysis of eight aircraft [S-8]. Later flight tests suggested that the phugoid period was so long that pilots were not sensitive to its actual value but that the damping ratio ζ had an effect [E-1, after O-1]. Negative damping ratio produced divergent flight path response and poor to unsatisfactory pilot ratings, while values greater than 0.15 were judged to have acceptable to good qualities. MIL-F-8785C sets no limitation on the phugoid period; it requires damping ratios greater than 0.04 and 0 for Level 1 and 2 conditions (which are defined in Section 4.7), while a *time to double*, T_2,[2] of up to 55 s is allowed at Level 3. Subsequent flight testing of supersonically cruising aircraft, such as the SR-71 and Concorde, revealed that lightly damped long-period motions can be controlled by the pilot. However, holding airspeed and altitude for long periods of time creates an unacceptable pilot workload and a need for stability augmentation [S-10].

The design evolution of World War II aircraft brought increased concern for short-period flying qualities. Fighter aircraft demanded quicker, more precise maneuvering over expanded flight envelopes, and bombers and transports became heavier, lowering both damping and natural frequency ω_n of the short-period response. Initial research was directed at quantifying the effects of ζ and ω_n on response to pitch inputs. Regions of satisfactory, acceptable, poor, and unacceptable response were identified by *thumbprints* (i.e., nested *iso-opinion contours*) on plots of ζ versus ω_n. Satisfactory ratings were assigned to natural frequencies of 3 to 7 rad/s with damping ratios of 0.4–1.25 in [M-1], while [C-1] presented a narrower range of $\omega_n = 2.5$–3.5 rad/s and $\zeta = 0.4$–1.1. Desired periods are thus 1–2 s. At lower natural frequency, the initial response is perceived to be slow; with light damping, there is a tendency to overshoot the desired response. Higher natural frequency produces oversensitivity to command and a tendency for pilot-induced oscillation. A high damping ratio at any natural frequency produces a sluggish response.

The short-period damping ratio specified in [A-3] is not dependent on the natural frequency; together with minimum and maximum frequencies, the thumbprint is replaced by a rectangular box. For Category A and C flight phases, the Level 1 value is between 0.35 and 1.3, while for Category B, the range is 0.3–2. There is a broad range of satisfactory damping ratios, and the short period may be overdamped ($\zeta > 1$).

Damping and natural frequency alone are incomplete measures of longitudinal flying qualities because they do not identify output variables, response history, or frequency response. For low-speed maneuvering, attitude control and pitch-rate response are important, while at high speed, the normal load factor is a more significant factor. Viewed in the

2. For a second-order system, the time to double, T_2, is defined to be $-0.693/\zeta\omega_n$, while for a first-order system, it is $-0.693/\tau$, where ζ and τ have negative (unstable) values.

frequency domain, control response depends on a transfer function's numerator as well as its denominator. From a time-domain perspective, the initial response to a step command, overshoot and undershoot of the final value, and the settling time are important metrics.

We examine three approaches to portraying short-period response: the control anticipation parameter (CAP), Gibson's dropback criterion, and the C^* criterion. These criteria are based on the second-order short-period model (Section 5.2); their representation of the short period's steady-state behavior neglects the long-period effect of phugoid dynamics.

While changing the flight path angle, the pilot should be able to anticipate the commanded load factor by kinesthetic inputs from the inner ear [B-3]. The ear's semicircular canals sense angular acceleration [Y-1], so the ratio of initial angular acceleration $\Delta\dot{q}(0)$ to steady-state normal load factor Δn_{ss} (positive up), is a good measure of this effect. From eqs. 5.3-6, 5.3-10, and 5.3-35,

$$\Delta\dot{q}(0) = \left(M_{\delta E} - \frac{M_{\dot{\alpha}}}{V_o + L_{\dot{\alpha}}} L_{\delta E} \right) \Delta\delta E^* \tag{5.4-69}$$

$$\Delta n_{ss} = \frac{V_o}{g} \Delta q^* = -\frac{V_o}{g} \frac{\left(\frac{L_\alpha}{V_o} M_{\delta E} - \frac{L_{\delta E}}{V_o} M_\alpha \right)}{\left(\frac{L_\alpha}{V_o} M_q + M_\alpha \right)} \Delta\delta E^* \tag{5.4-70}$$

The *control anticipation parameter* is then

$$\text{CAP} = \frac{-\left(M_{\delta E} - \frac{M_{\dot{\alpha}}}{V_o + L_{\dot{\alpha}}} L_{\delta E} \right)\left(\frac{L_\alpha}{V_o} M_q + M_\alpha \right)}{(L_\alpha M_{\delta E} - L_{\delta E} M_\alpha)/g} \tag{5.4-71}$$

In this approximation, there is a direct proportionality between Δq^* and Δn_{ss}. If $L_{\delta E}$ is negligible, this relationship reduces to the expression

$$\text{CAP} = \frac{-\left(\frac{L_\alpha}{V_o} M_q + M_\alpha \right) M_{\delta E}}{L_\alpha M_{\delta E}/g} = \frac{-\left(\frac{L_\alpha}{V_o} M_q + M_\alpha \right)}{L_\alpha/g} = \frac{\omega_n^2}{n_z/\alpha} \tag{5.4-72}$$

where the short-period natural frequency is given by eq. 5.3-15 and $n_z/\alpha = L_\alpha/g$. (Following convention, n_z is positive up in this usage.) The desired natural frequency is related to the aircraft's normal acceleration sensitivity to angle of attack as

$$\omega_n = \sqrt{\text{CAP}\frac{n_z}{\alpha}} \tag{5.4-73}$$

The upper limit of CAP for Level 1 flying qualities is 3.6 for all flight phases [A-3], while the lower limit is 0.28, 0.085, and 0.16 for Categories A (rapid maneuvering), B (gradual maneuvering), and C (takeoff, approach, and landing) flight phases, with limits on n_z/α and ω_n that depend primarily on aircraft class. The Level 1 CAP limits penalize pitching response that is too abrupt or too sluggish, and they are presented for n_z/α of 1–100 "g"s/rad in [A-3]; corresponding short-period natural frequencies range from 1 to 20 rad/s.

The CAP parameters play a central—but not totally defining—role in the second-order approximation for the elevator-to-pitch-rate transfer function (eq. 5.3-32). With $L_{\delta E} = 0$ and denoting the short-period zero as $1/T_{\theta_2}$ [M-3], the transfer function is

$$\frac{\Delta q(s)}{\Delta \delta E(s)} = \frac{M_{\delta E}\left[s + \left(\frac{g}{V_o}\right)\frac{n_z}{\alpha}\right]}{s^2 + 2\zeta\omega_n s + \omega_n^2} = \frac{M_{\delta E}[s + 1/T_{\theta_2}]}{s^2 + 2\zeta\omega_n s + \omega_n^2} \tag{5.4-74}$$

For a fixed CAP, the pitch-rate response varies with g/V_o, and this, in turn, affects the transient response in attitude-changing maneuvers. Suppose that the flight path angle is to be ramped up to a new value. With a step elevator input that is removed once the target pitch angle is reached, the pitch angle should overshoot slightly before settling at its target value. This observation leads to *Gibson's dropback criterion*: T_{θ_2} should be chosen as the delay (seconds) in flight-path-angle response associated with the simplified model [G-3] τ_γ to achieve this response:

$$\tau_\gamma = 2\zeta/\omega_n \tag{5.4-75}$$

When the dropback criterion is satisfied, the elevator-to-pitch-rate transfer function is a function of ζ and ω_n alone:

$$\frac{\Delta q(s)}{\Delta \delta E(s)} = \frac{M_{\delta E}[s + \omega_n/2\zeta]}{s^2 + 2\zeta\omega_n s + \omega_n^2} \tag{5.4-76}$$

If the aircraft's pitching response is not close enough to its desired value, a stability augmentation system might correct the deficiency. For illustration, the second-order short-period model (eq. 5.3-1) is simplified to

$$\begin{bmatrix} \Delta \dot{q} \\ \Delta \dot{\alpha} \end{bmatrix} = \begin{bmatrix} M_q & M_\alpha \\ 1 & -L_\alpha / V_o \end{bmatrix} \begin{bmatrix} \Delta q \\ \Delta \alpha \end{bmatrix} + \begin{bmatrix} M_{\delta E} \\ -L_{\delta E} / V_o \end{bmatrix} \Delta \delta E \tag{5.4-77}$$

and the feedback control law, with pilot command $\Delta\delta E_c$, is specified as

$$\Delta \delta E = \Delta \delta E_c - \begin{bmatrix} c_1 & c_2 \end{bmatrix} \begin{bmatrix} \Delta q \\ \Delta \alpha \end{bmatrix} \tag{5.4-78}$$

With closed-loop control, the short-period dynamics are described by the differential equation

$$\begin{bmatrix} \Delta \dot{q} \\ \Delta \dot{\alpha} \end{bmatrix} = \begin{bmatrix} (M_q - c_1 M_{\delta E}) & (M_\alpha - c_2 M_{\delta E}) \\ (1 + c_1 L_{\delta E}/V_o) & (-L_\alpha/V_o + c_2 L_{\delta E}/V_o) \end{bmatrix} \begin{bmatrix} \Delta q \\ \Delta \alpha \end{bmatrix} + \begin{bmatrix} M_{\delta E} \\ -L_{\delta E}/V_o \end{bmatrix} \Delta \delta E_c \tag{5.4-79}$$

The corresponding elevator-to-pitch-rate transfer function is

$$\frac{\Delta q(s)}{\Delta \delta E(s)} = \frac{M_{\delta E}\left\{ s + \left[\left(\frac{L_\alpha}{V_o} + c_2 \frac{L_{\delta E}}{V_o} \right) - \frac{(M_\alpha - c_2 M_{\delta E})}{M_{\delta E}} \frac{L_{\delta E}}{V_o} \right] \right\}}{s^2 - \left[(M_q - c_1 M_{\delta E}) - \left(\frac{L_\alpha}{V_o} + c_2 \frac{L_{\delta E}}{V_o} \right) \right] s - \left[(M_\alpha - c_2 M_{\delta E}) + (M_q - c_1 M_{\delta E}) \left(\frac{L_\alpha}{V_o} + c_2 \frac{L_{\delta E}}{V_o} \right) \right]} \tag{5.4-80}$$

The control gains could be chosen to provide the desired transfer function (e.g., eq. 5.4-76).

With the usual dominant effect of $M_{\delta E}$, the desired natural frequency and damping ratio are readily furnished by choosing proper values of c_1 and c_2. However, the transfer function's zero can be altered only by controlling lift. If $L_{\delta E}/V_o$ is not large enough to have significant effect, high-bandwidth wing flaps or spoilers must be driven by Δq and $\Delta\alpha$ to provide the needed response. $\Delta\alpha$ may be difficult to measure with sufficient accuracy for feedback control, in which case, an angle-of-attack estimate could be derived using normal load factor.

The C* *criterion* blends the normal load factor felt by the pilot with pitch rate; pitch rate dominates the term when the airspeed V_o is below a crossover frequency V_c, and the normal load factor has a more significant effect when it is above:

$$C^* = \Delta n_{pilot} + \frac{V_c}{g}\Delta q = l_{pilot}\Delta\dot{q} + \Delta n_{\text{cm}} + \frac{V_c}{g}\Delta q$$

$$= l_{pilot}\Delta\dot{q} + \frac{V_o}{g}(\Delta q - \Delta\dot{\alpha}) + \frac{V_c}{g}\Delta q \quad (5.4\text{-}81)$$

The pilot's location l_{pilot} with respect to the aircraft's center of mass multiplies the pitch acceleration, incorporating the most significant off-c.m. effect presented in eq. 3.2-128. The crossover velocity V_c is typically about 125 m/s. A frequency-domain interpretation of C^* could be developed using eqs. 5.3-32 and 5.3-36; however, the criterion was proposed as a step response envelope, with upper and lower limits on $C^*(t)/C_{ss}^*$ for a step elevator input $\Delta\delta E^*$ [T-1]. The criterion has been embraced by designers of modern flight control systems, but flying qualities specialists have been less enthusiastic, noting that good correlation of time-domain criteria with pilot ratings remains to be established [H-3, M-2].

Nevertheless, the time response is used to make an important point about the differences in response brought about by the characteristics of command-augmentation flight control systems [H-2, H-3]. Assume that the pilot commands a step stick input, then releases it a short time later to increase the aircraft's flight path angle, as in takeoff rotation or the landing flare. Without command augmentation, the pitch angle is dominated by the short-period and phugoid transients (Fig. 4.2-4), increasing until the control is released, then decreasing to its original value. With pitch-rate-command/attitude-hold (RCAH) control, the phugoid is suppressed, and the short period may be faster and more highly damped. When the stick is released, the pitch angle remains at a new value, held precisely with integral compensation. This fundamentally different response to the same pilot input can be confusing; while RCAH command augmentation may improve flying qualities during most flight phases, it may degrade pilot opinion in terminal flight. As noted in [C-2], "Experience with in-flight simulations has shown that pitch rate command/attitude hold flight control systems exhibit mediocre to poor flying qualities for landing. Pilots report poor control of the flight path and tendencies to balloon and float and to exhibit pilot-induced oscillations." This suggests that the pilot needs to develop a different technique to use the RCAH control mode satisfactorily. Pitch-angle and flight-path-angle command system, although "artificial" in response type, can provide precise landing control with good pilot ratings, in part because the pitch-angle response shape is more natural.

Fly-by-wire (FBW) control systems can improve the stability and flying qualities of unaugmented aircraft, but they can also contribute to flying qualities degradation. They may introduce computational time delays between pilot inputs and control actions, artificial feel systems that provide useful force cues to the pilot, low-pass anti-aliasing filters in digital implementation, rate and displacement limiting of control surface deflections, and notch filters to reduce forcing of structural modes [H-2]. The potential for adverse coupling between aircraft and pilot dynamics—including pilot-induced oscillation (PIO) or aircraft-pilot coupling (APC)—is increased. PIOs occur when a pilot's control actions inadvertently destabilize an otherwise stable aircraft. They are categorized as linear, quasilinear, or nonlinear, and criteria for predicting PIO susceptibility have been based on either the aircraft's dynamic modes or more general performance criteria that may include models of pilot control-loop closure. As noted in [H-2], modal criteria are based on limits on pitch transfer function gain, response time constants, and stick force per g (Section 5.5), while nonmodal criteria apply to transfer function bandwidth (with and without modeling of piloting actions), pitch-rate overshoot, phase margin, and variations of these parameters with input frequency.

Thus, it appears that flying qualities criteria depend as much on the flight control system structure and the pilot's training and experience as on the natural dynamics of the aircraft itself. Basic performance measures like response bandwidth, pure time delay, and phase lag remain central, but defining key parameters of reduced-order dynamic models that achieve this performance is less relevant than before.

5.5 Control Mechanisms, Stick-Free Stability, and Trim

To this point, we have assumed that control surfaces could be positioned precisely, without regard to the mechanical characteristics of the control system. If the surfaces were uncommanded by the cockpit control "stick" (or column), they remained at their set points. The aircraft possessed *stick-fixed properties*, that is, the aerodynamic characteristics associated with undeflected control surfaces.

In actuality, the dynamic properties of control systems may couple with aircraft motions. Large modern aircraft have redundant, hydraulically or electrically powered control systems that generate enough force to provide "irreversible" surface positioning and stick-fixed dynamic response when control surfaces are not commanded. If, however, a fully powered system fails, control surfaces may be free to float from commanded positions, producing *stick-free response* to disturbances.

Small aircraft and many medium-sized transports have unpowered or partially powered mechanisms for one or more axes of control, relying on the pilot's strength and skill to position control surfaces. In trimmed flight, control force applied by the pilot is light, and control surface positions may fluctuate. For both powered and unpowered systems, the inherent stability of the control system itself and its coupling to the aircraft's modes of motion are vital concerns for aircraft safety and performance.

We distinguish two subsystems for flight control: the *control system logic* and the *control system mechanism*. The former refers to the human or electronic signaling and computing structure that commands the control system. In this section, we focus on a simple model for the latter, which contains the mechanical components for creating and transmitting the forces or hinge moments required to deflect control surfaces, as well as on the surfaces themselves. Depending on the parameter values chosen, our simple model may represent a manually powered system or the servoactuator of a single fully powered system. The control mechanism possesses inertial properties, and it is affected by hinge moments due to external aerodynamics, mechanical springs and dampers, and command forces that are driven by the control system logic. Having established a generic second-order model of an elevator control mechanism, we examine its coupled interaction with the aircraft's rigid-body motion.

Elevator Control Mechanism

As sketched in Fig. 5.5-1a, the generic control mechanism contains a movable aerodynamic control surface (the elevator) that rotates by the angle δE about its hinge axis. A horn (or bellcrank) is linked to a pushrod that connects to the cockpit control column. The control column is geared so that a forward force Δf_C on the handle (or control wheel or yoke) produces a positive (trailing-edge-down) moment on the elevator. Elevator hinge moments may also be produced by a mechanical spring, mechanical damping, power boost, aerodynamic effects on the elevator, and, if the elevator is not inertially balanced or a bobweight is added, by gravity and maneuvering load factor.

This simple level of abstraction is introduced for illustration, and the details of actual control mechanisms may differ. For example, there may be not one but several linkages between the control column and the elevator, as required for routing through the fuselage. Cables and pulleys may replace the pushrods and bellcranks. At high angle of attack and dynamic pressure, air loads may exceed the force provided by hydraulic actuation, producing a *blowdown effect* that limits displacement below the mechanical stops.

Cockpit control displacement and force characteristics are proportional to each other in an unpowered system. In a *partial-boost system*, force or displacement transducers on the control column electrically command the power-boost unit, but the mechanical linkage is retained, providing a natural pathway by which the pilot can sense elevator hinge moment and position. The position servo of a fully powered system would

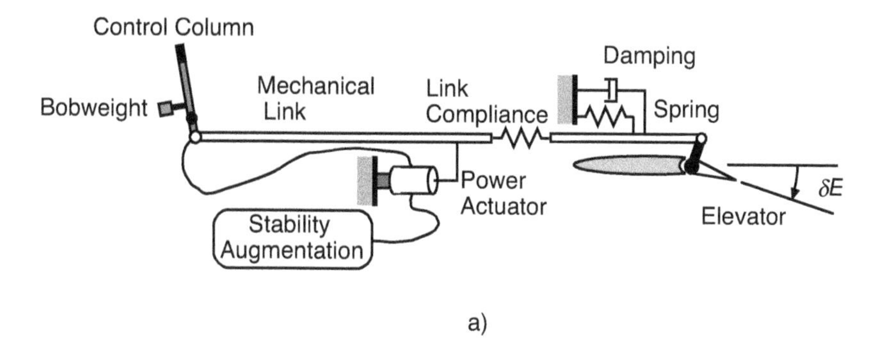

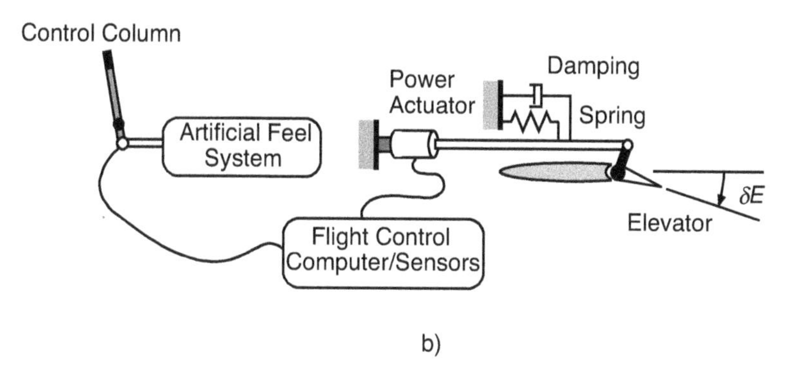

Fig. 5.5-1. Generic mechanical and fly-by-wire systems for longitudinal flight control. (a) Mechanical/power-boosted system with stability augmentation. (b) Fly-by-wire system with artificial feel system.

likely add integral compensation to provide zero steady-state error of the control position with varying air loads on the control surfaces. *Stability augmentation* may be provided by commands to the power actuator from aircraft motion sensors. In a *fly-by-wire control system* (Fig. 5.5-1b), there is no direct mechanical linkage between the elevator and the control column, and the force and displacement characteristics of the stick are not inherently coupled. An *artificial feel system* produces column forces that are proportional to the elevator's displacement or its desired effect, consistent with flying qualities requirements.

There also may be varying levels of mechanical and hydraulic redundancy for protection against system failures. Dual load paths from the cockpit to control surfaces can be implemented in powered and unpowered systems. Powered systems usually have multiple primary actuators driven by independent hydraulic systems, as well as standby (or backup) actuators for each control surface.

The moment balance at the elevator hinge axis is

$$\ddot{\delta E} = \frac{1}{I_\delta} \sum (\text{external moments}) \triangleq \sum_{i=1}^{I} H(i) \tag{5.5-1}$$

where I_δ is an effective moment of inertia that accounts for the mass and inertia of the entire control mechanism, and $H(i)$ is the effective specific moment (i.e., moment per unit of moment of inertia) of the ith contributor. For small perturbations from trimmed equilibrium, control-surface dynamics are described by the second-order linear differential equation

$$\Delta \ddot{\delta E} = H_{\dot{\delta E}} \Delta \dot{\delta E} + H_{\delta E} \Delta \delta E + H_q \Delta q + H_\alpha \Delta \alpha + H_{n_z} \Delta n_z + H_V \Delta V + H_C \Delta f_C \tag{5.5-2}$$

where the H_i are partial derivatives of the specific moment expressing the sensitivity to perturbations in the variables. We assume that the hinge moment is sensitive to changes in elevator displacement and rate ($\Delta\delta E$, $\Delta\dot{\delta E}$), longitudinal motion of the aircraft (ΔV, Δq, $\Delta\alpha$, Δn_z), and the control force command Δf_C.

The sources of the moments may be both mechanical and aerodynamic. For example, mechanical damping and aerodynamic resistance may be sensitive to elevator rate, while mechanical springs and aerodynamic forces can produce restoring moments proportional to elevator deflection. Therefore, we define the partial derivatives as follows:

$$H_{\dot{\delta E}} = \frac{1}{I_\delta}\left(C_{H\dot{\delta E}} \frac{1}{2} \rho V^2 S_E \bar{c}_E + k_{\dot{\delta E}}\right) \tag{5.5-3}$$

$$H_{\delta E} = \frac{1}{I_\delta}\left(C_{H\delta E} \frac{1}{2} \rho V^2 S_E \bar{c}_E + k_{\delta E}\right) \tag{5.5-4}$$

$$H_q = \frac{1}{I_\delta}\left(C_{H_q} \frac{1}{2} \rho V^2 S_E \bar{c}_E + k_q\right) \tag{5.5-5}$$

$$H_\alpha = \frac{1}{I_\delta}\left(C_{H\alpha} \frac{1}{2} \rho V^2 S_E \bar{c}_E + k_\alpha\right) \tag{5.5-6}$$

$$H_{nz} = \frac{1}{I_\delta}\left(C_{H_{nz}}\frac{1}{2}\rho V^2 S_E \bar{c}_E + k_{nz}\right) \tag{5.5-7}$$

$$H_V = \frac{1}{I_\delta}\left(C_{HV}\frac{1}{2}\rho V^2 S_E \bar{c}_E + k_V\right) \tag{5.5-8}$$

$$H_C = k_C / I_\delta \tag{5.5-9}$$

These expressions contain derivatives of the dimensionless hinge-moment coefficient C_H, as defined in [S-4]. S_E and $\bar{c}_E$ are the reference area and length for the elevator, and the k_i are mechanical or kinematic constants.

Defining the control mechanism state vector as

$$\Delta\mathbf{x}_\delta \triangleq \begin{bmatrix} \Delta\delta_1 \\ \Delta\delta_2 \end{bmatrix} = \begin{bmatrix} \Delta\delta E \\ \Delta\delta\dot{E} \end{bmatrix} \tag{5.5-10}$$

the second-order model is represented by a pair of first-order equations

$$\Delta\dot{\delta}_1 = \Delta\delta_2 \tag{5.5-11}$$

$$\begin{aligned}
\Delta\dot{\delta}_2 &= H_{\dot{\delta}E}\Delta\delta_2 + H_{\delta E}\Delta\delta_1 + H_q\Delta q + H_\alpha\Delta\alpha + H_{n_z}\Delta n_z \\
&\quad + H_V\Delta V + H_C\Delta f_C \\
&= H_{\dot{\delta}E}\Delta\delta_2 + \left(H_{\delta E} - H_{n_z}\frac{V_o}{g}\frac{L_{\delta E}}{V_o}\right)\Delta\delta_1 + H_q\Delta q \\
&\quad + \left(H_\alpha - H_{nz}\frac{V_o}{g}\frac{L_\alpha}{V_o}\right)\Delta\alpha + \left(H_V - H_{nz}\frac{V_o}{g}\frac{L_V}{V_o}\right)\Delta V + H_C\Delta f_C \\
&\triangleq H_{\dot{\delta}E}\Delta\delta_2 + H'_{\delta E}\Delta\delta_1 + H_q\Delta q + H'_\alpha\Delta\alpha + H'_V\Delta V + H_C\Delta f_C
\end{aligned} \tag{5.5-12}$$

where the primed derivatives $(\cdot)'$ include the normal-load-factor effect. In standard form,

$$\Delta\dot{\mathbf{x}}_\delta = \mathbf{F}_\delta\Delta\mathbf{x}_\delta + \mathbf{G}_\delta\Delta\mathbf{u}_\delta + \mathbf{L}_\delta\Delta\mathbf{w}_\delta \tag{5.5-13}$$

with

$$\Delta\mathbf{u}_\delta = \Delta f_C \tag{5.5-14}$$

$$\Delta \mathbf{w}_\delta = \begin{bmatrix} \Delta V \\ \Delta \gamma \\ \Delta q \\ \Delta \alpha \end{bmatrix} \tag{5.5-15}$$

$$\mathbf{F}_\delta = \begin{bmatrix} 0 & 1 \\ H'_{\delta E} & H_{\dot{\delta} E} \end{bmatrix} \tag{5.5-16}$$

$$\mathbf{G}_\delta = \begin{bmatrix} 0 \\ H_C \end{bmatrix} \tag{5.5-17}$$

$$\mathbf{L}_\delta = \begin{bmatrix} 0 & 0 & 0 & 0 \\ H'_V & 0 & H_q & H'_\alpha \end{bmatrix} \tag{5.5-18}$$

The control system should possess internal stability, which requires that $H'_{\delta E}$ and $H_{\dot{\delta} E}$ be less than zero. The disturbance vector $\Delta \mathbf{w}_\delta$ is the longitudinal state vector, a critical factor in the dynamic coupling of rigid-body and control-mechanism motions. Neglecting $\Delta \mathbf{w}_\delta$, the steady-state response to control force is

$$\Delta \mathbf{x}_\delta{}^* = -\mathbf{F}_\delta^{-1} \mathbf{G}_\delta \Delta \mathbf{u}_\delta{}^* \tag{5.5-19a}$$

or

$$\begin{bmatrix} \Delta \delta E^* \\ \Delta \delta \dot{E}^* \end{bmatrix} = \begin{bmatrix} -H_C / H'_{\delta E} \\ 0 \end{bmatrix} \Delta f_C{}^* \tag{5.5-19b}$$

Recalling the definition of $H'_{\delta E}$ (eqs. 5.5-4 and 5.5-12), we see that the equilibrium control position depends on the flight conditions, although the effect is reduced by increasing the mechanical spring constant $k_{\delta E}$. An irreversible control system uses a servoactuator to produce a very large $k_{\delta E}$. A corresponding increase in the mechanical leverage or gearing produced by H_C is needed to maintain control force at its original level with high $k_{\delta E}$.

Short-Period/Control-Mechanism Coupling

The structure of dynamic interactions between rigid-body longitudinal motions and the control mechanism is outlined by Table 5.5-1. The dynamic model for the elevator could be appended to the fourth-order longitudinal model (eq. 5.4-4) to create a sixth-order set of equations. However, a fourth-order model combining short-period and control variables provides a simpler introduction to rigid-body/control interactions, neglecting airspeed and altitude variations.

TABLE 5.5-1 STRUCTURE OF THE LONGITUDINAL/CONTROL DYNAMICS MODEL

Longitudinal Dynamics	Control-to-Longitudinal Coupling
Longitudinal-to-Control Coupling	Control Mechanism Dynamics

In the following, we drop the primes from the hinge-moment derivatives and redefine $\mathbf{L}_\delta$ to include just the third and fourth columns of eq. 5.5-18. Combining eqs. 5.3-1 and 5.5-13, while neglecting atmospheric disturbances,

$$\begin{bmatrix} \Delta\dot{\mathbf{x}}_{SP} \\ \Delta\dot{\mathbf{x}}_{\delta} \end{bmatrix} = \begin{bmatrix} \mathbf{F}_{SP} & \mathbf{G}_{SP} \\ \mathbf{L}_{\delta} & \mathbf{F}_{\delta} \end{bmatrix} \begin{bmatrix} \Delta\mathbf{x}_{SP} \\ \Delta\mathbf{x}_{\delta} \end{bmatrix} + \begin{bmatrix} 0 \\ \mathbf{G}_{\delta} \end{bmatrix} \Delta\mathbf{u}_{\delta} \tag{5.5-20}$$

Expanded to conform to the control-rate and displacement vector $\Delta\mathbf{x}_\delta$, the short-period control-effect matrix is

$$\mathbf{G}_{SP} = \begin{bmatrix} \left[M_{\delta E} - \dfrac{M_{\dot{\alpha}}}{(V_o + L_{\dot{\alpha}})} L_{\delta E} \right] & 0 \\ \dfrac{-L_{\delta E}}{(V_o + L_{\dot{\alpha}})} & 0 \end{bmatrix} \tag{5.5-21}$$

If the rate of change of elevator deflection $\Delta\dot{\delta E}$ has significant moment and lifting effect, the second column of eq. 5.5-21 contains the associated control derivatives.

Because $\mathbf{F}_{SP}$ and $\mathbf{F}_\delta$ are both invertible, the *steady-state response* can be expressed as

$$\Delta\mathbf{x}_\delta{}^* = -(\mathbf{F}_\delta - \mathbf{L}_\delta \mathbf{F}_{SP}^{-1}\mathbf{G}_{SP})^{-1}\mathbf{G}_\delta \Delta\mathbf{u}_\delta{}^* \tag{5.5-22}$$

$$\begin{aligned} \Delta\mathbf{x}_{SP}{}^* &= -\mathbf{F}_{SP}^{-1}\mathbf{G}_{SP}\Delta\mathbf{x}_\delta{}^* \\ &= \mathbf{F}_{SP}^{-1}\mathbf{G}_{SP}(\mathbf{F}_\delta - \mathbf{L}_\delta\mathbf{F}_{SP}^{-1}\mathbf{G}_{SP})^{-1}\mathbf{G}_\delta\Delta\mathbf{u}_\delta{}^* \end{aligned} \tag{5.5-23}$$

With $L_q = L_{\dot{\alpha}} = 0$, $\Delta\dot{\delta E}^*$ is zero, the equilibrium control deflection is

$$\Delta\delta E^* = \frac{-H_C \Delta f_C{}^*}{H_{\delta E} - \dfrac{H_q\left[\dfrac{L_{\delta E}}{V_o} M_{\delta E} + \left(M_\alpha - M_{\dot{\alpha}}\dfrac{L_\alpha}{V_o}\right)\dfrac{L_{\delta E}}{V_o}\right] + H_\alpha\left[M_{\delta E} - (M_q + M_{\dot{\alpha}})\dfrac{L_{\delta E}}{V_o}\right]}{(M_q + M_\alpha)}} \tag{5.5-24}$$

and Δq^* and $\Delta\alpha^*$ can be found from eq. 5.3-9. For a given control force, the elevator deflection depends upon flight condition, as reflected by changes in the stability, control, and hinge-moment derivatives.

A *residualized short-period model* accounts for control-mechanism dynamics when the latter is fast and internally stable. Following earlier development, the new second-order model is

$$\Delta\dot{\mathbf{x}}_{SP} = (\mathbf{F}_{SP} - \mathbf{G}_{SP}\mathbf{F}_{\delta}^{-1}\mathbf{L}_{\delta})\Delta\mathbf{x}_{SP} - \mathbf{G}_{SP}\mathbf{F}_{\delta}^{-1}\mathbf{G}_{\delta}\Delta\mathbf{u}_{\delta} \triangleq \mathbf{F}'_{SP}\Delta\mathbf{x}_{SP} + \mathbf{G}'_{SP}\Delta\mathbf{u}_{\delta} \tag{5.5-25}$$

where

$$\mathbf{F}'_{SP} = \begin{bmatrix} \left(M_q - M_{\delta E}\dfrac{H_q}{H_{\delta E}}\right) & \left(M_\alpha - M_{\delta E}\dfrac{H_\alpha}{H_{\delta E}}\right) \\ \left(1 + \dfrac{L_{\delta E}}{V_o}\dfrac{H_q}{H_{\delta E}}\right) & \left(\dfrac{-L_\alpha}{V_o} + \dfrac{L_{\delta E}}{V_o}\dfrac{H_\alpha}{H_{\delta E}}\right) \end{bmatrix} \tag{5.5-26}$$

$$\mathbf{G}'_{SP} = \begin{bmatrix} M_{\delta E}\dfrac{H_C}{H_{\delta E}} \\ -\dfrac{L_{\delta E}}{V_o}\dfrac{H_C}{H_{\delta E}} \end{bmatrix} \tag{5.5-27}$$

Because the control mechanism is assumed to be in equilibrium, the control damping effect $H_{\dot{\delta}E}$ does not appear in the equations.

The corresponding natural frequency and damping ratio (given sufficient positive static margin) can be found from the coefficients of the characteristic polynomial:

$$\begin{aligned} \Delta(s) &= |s\mathbf{I}_2 - \mathbf{F}'_{SP}| \\ &= s^2 - \left[\left(M_q - \frac{L_\alpha}{V_o}\right) - M_{\delta E}\frac{H_q}{H_{\delta E}} + \frac{L_{\delta E}}{V_o}\frac{H_\alpha}{H_{\delta E}}\right]s \\ &\quad - \left[\left(M_\alpha + M_q\frac{L_\alpha}{V_o}\right) - \frac{H_q}{H_{\delta E}}\left(M_\alpha\frac{L_{\delta E}}{V_o} - M_{\delta E}\frac{L_\alpha}{V_o}\right)\right. \\ &\quad \left. - \frac{H_\alpha}{H_{\delta E}}\left(M_q\frac{L_{\delta E}}{V_o} + M_{\delta E}\right)\right] \\ &\triangleq s^2 + 2\zeta\omega_n s + \omega_n^2 \end{aligned} \tag{5.5-28}$$

With negligible control response to pitch rate H_q and lift effect of control $L_{\delta E}/V_o$, the polynomial reduces to

$$\Delta(s) = s^2 - \left(M_q - \frac{L_\alpha}{V_o}\right)s - \left[\left(M_\alpha + M_q \frac{L_\alpha}{V_o}\right) - \frac{H_\alpha}{H_{\delta E}} M_{\delta E}\right] \quad (5.5\text{-}29)$$

This equation reveals the principal effect of the surface's response to angle of attack H_α. A negative value (i.e., corresponding to an underbalanced elevator) causes the surface to float toward the freestream direction, decreasing the natural frequency of the stick-free short-period mode. Both H_α and $H_{\delta E}$ are made less negative by adding an *overhang* or *horn surface* ahead of the hinge line [A-1].

Examination of dynamic interactions between the rigid-body and control-surface oscillations requires use of the higher-order model (eq. 5.5-20). We can anticipate stability problems when either mode is lightly damped or when the uncoupled natural frequencies of the two modes are similar. Although not contained in this model, there is also the potential for a significant nonlinear effect due to localized separation and elevator stall at high combined angles of attack and surface deflection. Reference [A-1] presents an extensive description of the aerodynamics and mechanics of control surface design.

Control Force for Trimmed Flight

The earliest U.S. flight tests to determine stability and control characteristics focused on the longitudinal behavior of aircraft [A-1, after V-1]. In 1919, using the NACA's first experimental aircraft, a Curtiss JN4-H Jenny, Warner, Norton, and Allen assessed the elevator angle and stick force required for equilibrium flight as functions of airspeed, correlated the gradient of elevator angle versus airspeed with longitudinal pitch stability, and correlated the gradient of elevator angle versus airspeed with wind tunnel tests of pitch moment versus angle of attack.

The airspeed sensitivity of the control force required for steady, level flight can be analyzed using a simple aerodynamic model. We assume that the commanded force Δf_C is proportional to the net elevator hinge moment

$$\Delta f_C = g_S\left(C_H \frac{1}{2}\rho V_o^2 S_E \bar{c}_E\right) + \Delta f_D + \Delta f_B \quad (5.5\text{-}30)$$

where g_S is the effective *control stick gearing* (including the effect of power boost), plus the incremental forces due to downsprings and bobweights,

Δf_D and Δf_B (discussed below). The hinge-moment coefficient C_H is described as a linear function of angle-of-attack and elevator-angle perturbations plus a basic hinge moment due to the elevator trim tabs, C_{H_o}:

$$C_H = C_{H_\alpha}\Delta\alpha + C_{H_{\delta E}}\Delta\delta E + C_{H_o} \tag{5.5-31}$$

Following Sections 2.2 and 2.3, the lift and pitching moment coefficients in steady, level flight, can be expressed as

$$C_L{}^* = C_{L_o} + C_{L_\alpha}\Delta\alpha^* + C_{L_{\delta E}}\Delta\delta E^* = \frac{W}{\frac{1}{2}\rho V_o^2 S} \tag{5.5-32}$$

$$C_m{}^* = C_{m_o} + C_{m_\alpha}\Delta\alpha^* + C_{m_{\delta E}}\Delta\delta E^* = 0 \tag{5.5-33}$$

where W is the aircraft's weight. The angle of attack and elevator deflection required for trim are found by inverting eqs. 5.5-32 and 5.5-33:

$$\begin{bmatrix}\Delta\alpha^* \\ \Delta\delta E^*\end{bmatrix} = \begin{bmatrix}C_{L_\alpha} & C_{L_{\delta E}} \\ C_{m_\alpha} & C_{m_{\delta E}}\end{bmatrix}^{-1} \begin{bmatrix}C_L{}^* - C_{L_o} \\ -C_{m_o}\end{bmatrix} \tag{5.5-34}$$

The corresponding control force is found by substitution in eqs. 5.5-30 and 5.5-31 (with $C_{L_o} = 0$):

$$\begin{aligned}\Delta f_c &= g_s(C_{H_\alpha}\Delta\alpha^* + C_{H_{\delta E}}\Delta\delta E^* + C_{H_o})\frac{1}{2}\rho V_o^2 S_E \bar{c}_E + \Delta f_D + \Delta f_B \\ &= g_s\left(\frac{C_{H_\alpha}C_{m_{\delta E}} - C_{H_{\delta E}}C_{m_\alpha}}{C_{L_\alpha}C_{m_{\delta E}} - C_{L_{\delta E}}C_{m_\alpha}}C_L{}^* - \frac{C_{H_\alpha}C_{L_{\delta E}} - C_{H_{\delta E}}C_{L_\alpha}}{C_{L_\alpha}C_{m_{\delta E}} - C_{L_{\delta E}}C_{m_\alpha}}C_{m_o} + C_{H_o}\right) \\ &\quad \times \frac{1}{2}\rho V_o^2 S_E \bar{c}_E + \Delta f_D + \Delta f_B\end{aligned} \tag{5.5-35}$$

Etkin notes that $C_L{}^* \frac{1}{2}\rho V_o^2$ can be replaced by the *wing loading* W/S [E-1], which is invariant with airspeed, in steady, level flight; hence, the equation takes the quadratic form

$$\Delta f_C = a + b\frac{1}{2}\rho V_o^2 \tag{5.5-36}$$

where

$$a = g_S S_E \bar{c}_E \frac{C_{H_\alpha} C_{m_{\delta E}} - C_{H_{\delta E}} C_{m_\alpha}}{C_{L_\alpha} C_{m_{\delta E}} - C_{L_{\delta E}} C_{m_\alpha}} \frac{W}{S} + \Delta f_D + \Delta f_B \tag{5.5-37}$$

$$b = g_S S_E \bar{c}_E \left(C_{H_o} - \frac{C_{H_\alpha} C_{L_{\delta E}} + C_{H_{\delta E}} C_{L_\alpha}}{C_{L_\alpha} C_{m_{\delta E}} - C_{L_{\delta E}} C_{m_\alpha}} C_{m_o} \right) \tag{5.5-38}$$

Figure 5.5-2a illustrates the *trim-tab effect* on the control force and resulting trimmed airspeed. At zero airspeed, Δf_C equals a; the control force decreases with increasing airspeed, and the point at which Δf_C equals zero is the trimmed airspeed. As indicated by eq. 5.5-35, the zero-airspeed control force is actually just the sum of the downspring and bobweight effects. $C_L{}^* \rho V_o^2/2$ is replaced by W/S under the assumption that the aircraft is in 1g flight. Changing C_{H_o} affects b but not a, shifting the intercept to higher or lower speed. Of course, a change in trimmed airspeed also requires rebalancing of the thrust/drag equilibrium.

The coefficient a is proportional to wing-loading, static-margin, downspring, and bobweight effects. To see the static-margin effect, we assume that $C_{L_{\delta E}}$ is negligible and note that

$$C_{m_\alpha} \approx -C_{L_\alpha} \left(\frac{x_{cm}}{\bar{c}} - \frac{x_n}{\bar{c}} \right) = C_{L_\alpha} (h_{cm} - h_n) \tag{5.5-39}$$

h_{cm} is the center-of-mass location, measured positive *aft* by convention, normalized by the mean aerodynamic chord, and h_n is the normalized location of the neutral point. Then

$$a = g_S S_E \bar{c}_E \left[\frac{C_{H_\alpha}}{C_{L_\alpha}} - \frac{C_{H_{\delta E}}}{C_{m_{\delta E}}} (h_{cm} - h_n) \right] \frac{W}{S} + \Delta f_D + \Delta f_B \tag{5.5-40}$$

The effect of varying wing loading, static margin, downspring force, or bobweight force is illustrated in Fig. 5.5-2b. Change in a simply shifts the curve up or down. If, however, the trim tab is concurrently adjusted to maintain a given airspeed, the curve is reshaped (Fig. 5.5-2c). The control-force gradient with respect to trim speed [E-1], where $\Delta f_C = 0$, is

$$\left. \frac{\partial \Delta f_C}{\partial V} \right|_{V = V_{o_{trim}}} = b \rho V_{o_{trim}} = -\frac{2a}{V_{o_{trim}}} \tag{5.5-41}$$

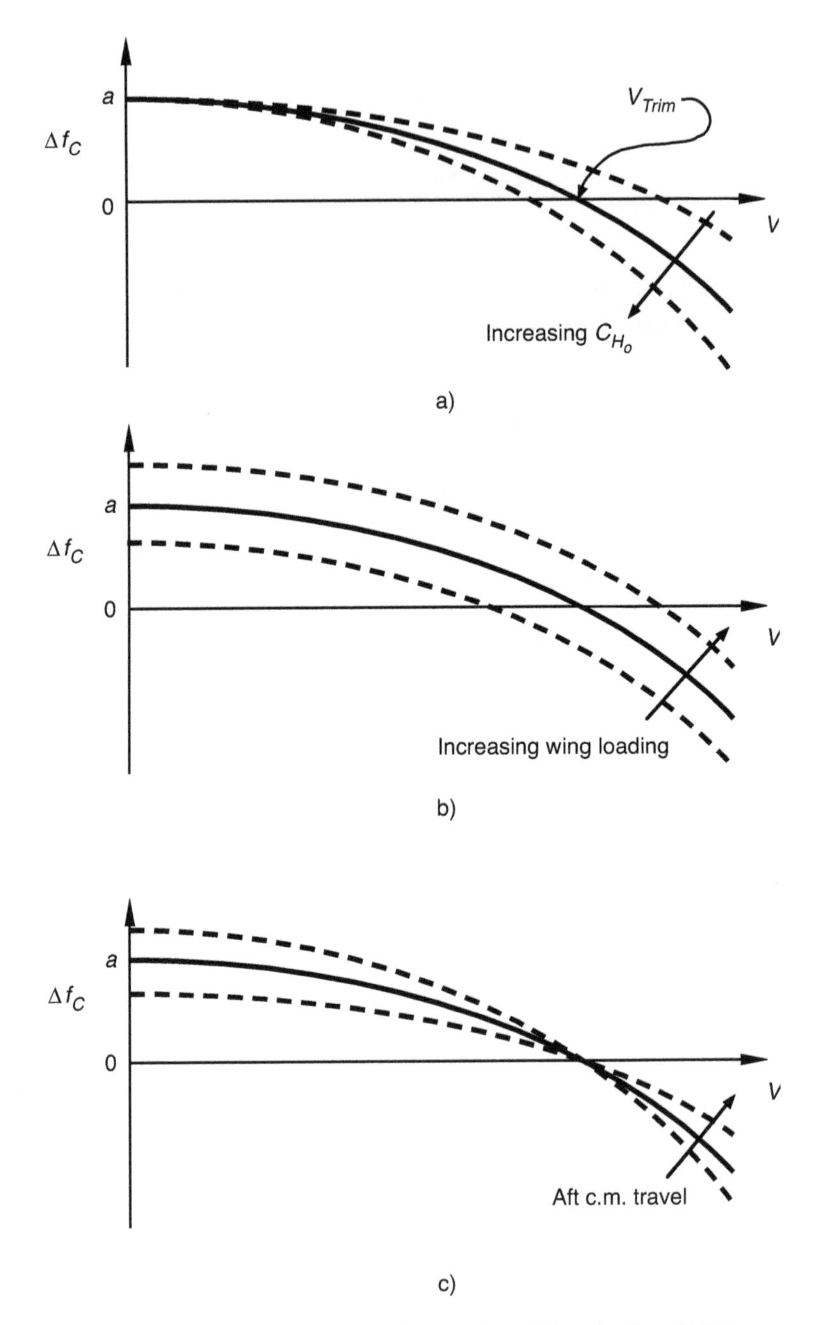

Fig. 5.5-2. Control force Δf_c as a function of airspeed. (a) Trim-tab effect. (b) Wing loading/static margin/downspring/bobweight effect, without retrimming.
(c) Center-of-gravity effect, with retrimming to constant airspeed. (after [E-1])

Downsprings and bobweights have similar effects on the trimmed (i.e., static) control force, although their dynamic effects are different. Conceptually, a *downspring* is a long, linear spring that is stretched to produce a forward (nose-down) preload on the control column. The spring constant k_D is very weak, so that

$$\Delta f_D = \Delta f_{D_o} + k_D \Delta\delta E \approx \Delta f_{D_o} \quad (5.5\text{-}42)$$

In practice, the downspring is likely to be a small nonlinear spring that is rigged for near-zero sensitivity to elevator deflection over its operating range. In a powered system, the downspring effect could be produced by a servoactuator commanded to produce a bias torque.

The *bobweight* is a proof mass mounted on a moment arm that pulls the control column forward under the influence of gravity. Its effect on trim is the same as the downspring's; however, the bobweight also reacts to maneuvering load factor and pitch rate. It is a mechanical accelerometer whose output is a control torque. Because it is offset from the aircraft center of mass, the bobweight also responds to pitch rate. The resulting hinge moments, expressed as H_q and H_{n_z}, couple into short-period dynamics, as described above. The bobweight effect could be produced by accelerometer and pitch-rate sensor feedback in a powered control system.

Elevator Angle and Stick Force per g

The elevator angle and control-column force required to produce an incremental load factor are important elements of maneuvering flight. The solution is similar to that for trimmed flight, with consideration of steady pitch rate and load factors other than one. The lift and pitching moment balances are

$$C_{L_o} + C_{L_\alpha}\Delta\alpha^* + C_{L_{\hat{q}}}\Delta\hat{q}^* + C_{L_{\delta E}}\Delta\delta E^* = \frac{\Delta n_z^* W}{\frac{1}{2}\rho V_o^2 S} \quad (5.5\text{-}43)$$

$$C_{m_o} + C_{m_\alpha}\Delta\alpha^* + C_{m_{\hat{q}}}\Delta\hat{q}^* + C_{m_{\delta E}}\Delta\delta E^* = 0 \quad (5.5\text{-}44)$$

where $C_{L_{\hat{q}}}$ and $C_{m_{\hat{q}}}$ are derivatives with respect to $\hat{q} \triangleq q\bar{c}/2V_o$. The normalized pitch rate in a *steady pullup* is related to the normal load factor by

$$\Delta\hat{q}^* = \left(\frac{\bar{c}}{2V_o}\right)\frac{g}{V_o}\Delta n_z^* \quad (5.5\text{-}45)$$

hence, the sensitivity of elevator angle to load factor can be found by substituting eq. 5.5-45 in eqs. 5.5-43 and 5.5-44 and eliminating $\Delta\alpha^*$. Neglecting C_{L_o} and C_{m_o}, which do not contribute to the sensitivity, the solution is

$$(C_{L_\alpha} C_{m_{\delta E}} - C_{m_\alpha} C_{L_{\delta E}})\Delta\delta E^* = \frac{\Delta n_z^* W}{\frac{1}{2}\rho V_o^2 S}\left[C_{m_\alpha} - (C_{m_\alpha} C_{L_{\hat{q}}} - C_{L_\alpha} C_{m_{\hat{q}}})\frac{g\rho S\bar{c}}{4W}\right] \tag{5.5-46}$$

This equation can be manipulated to express the *elevator angle per g* as

$$\frac{\Delta\delta E^*}{\Delta n_z^*} = \frac{-\left(\frac{W}{S}\right) C_{L_\alpha}(2\mu - C_{L_{\hat{q}}})}{\rho V_o^2 \mu (C_{L_\alpha} C_{m_{\delta E}} - C_{m_\alpha} C_{L_{\delta E}})}\left[(h_{cm} - h_n) + \frac{C_{m_{\hat{q}}}}{(2\mu - C_{L_{\hat{q}}})}\right] \tag{5.5-47}$$

where the *aircraft density factor* or *relative density* μ is defined as

$$\mu = \frac{2m}{\rho S\bar{c}} \tag{5.5-48}$$

It represents the ratio of the aircraft mass to the air mass contained in a volume that is one-half the wing area times the mean aerodynamic chord. The factor is sometimes defined as $m/\rho S\bar{c}$.

The square-bracketed term of eq. 5.5-47 is called the *stick-fixed maneuver margin* of the aircraft ("stick-fixed" because the elevator deflection is specified precisely, without reference to control system dynamics). The center-of-mass location $h_{mp_{fixed}}$ at which the maneuver margin is zero,

$$h_{mp_{fixed}} = h_n - \frac{C_{L_\alpha} C_{m_{\hat{q}}}}{(2\mu - C_{L_{\hat{q}}})} \tag{5.5-49}$$

is called the *stick-fixed maneuver point*, and it is typically behind the neutral point h_n. The lift produced by an aft elevator tends to decrease $\Delta\delta E^*/\Delta n_z^*$, while canard pitch control tends to increase it. However, the pitch-control location has no impact on the maneuver point. We note that $\Delta\delta E^*/\Delta n_z^*$ is an output-to-input relationship, whereas input-to-output pairs are more commonly used in dynamic analysis.

The *stick force per g* is determined in a similar way, adding hinge-moment characteristics and cockpit-control gearing to the equations.

The cockpit-control force is given by eq. 5.5-30, and the hinge-moment coefficient includes the effect of steady pitch rate:

$$C_H = C_{H_\alpha}\Delta\alpha + C_{H_{\hat{q}}}\Delta\hat{q} + C_{H_{\delta E}}\Delta\delta E + C_{H_o} \tag{5.5-50}$$

C_{H_o} does not contribute to the stick force per g, and the effects of downspring and bobweight are neglected below. The steady-state cockpit-control force is

$$\Delta f_C{}^* = g_S\frac{1}{2}\rho V_o^2 S_E\bar{c}_E\left(C_{H_\alpha}\Delta\alpha^* + C_{L_{\hat{q}}}\Delta\hat{q}^* + C_{H_{\delta E}}\Delta\delta E^*\right) \tag{5.5-51}$$

$\Delta\hat{q}^*$ and $\Delta\delta E^*$ are expressed as functions of $\Delta n_z{}^*$ from eqs. 5.5-45 and 5.5-47, while the calculation for $\Delta\alpha^*$ follows that for $\Delta\delta E^*$. Eliminating $\Delta\delta E^*$ in eqs. 5.5-43 and 5.5-44 leads to the following:

$$\Delta\alpha^* = \frac{\left(\dfrac{W}{S}\right)(2\mu - C_{L_{\hat{q}}})}{\rho V_o^2\mu\left(C_{L_\alpha}C_{m_{\delta E}} - C_{m_\alpha}C_{L_{\delta E}}\right)}\left[C_{m_{\delta E}} + \frac{C_{L_{\delta E}}C_{m_{\hat{q}}}}{(2\mu - C_{L_{\hat{q}}})}\right]\Delta n_z{}^* \tag{5.5-52}$$

Substituting in eq. 5.5-51, the *stick-force gradient* is

$$\frac{\Delta f_C{}^*}{\Delta n_z{}^*} = \frac{-g_S\dfrac{1}{2}S_E\bar{c}_E\dfrac{W}{S}(2\mu - C_{L_{\hat{q}}})C_{L_\alpha}C_{H_{\delta E}}}{\left(C_{L_\alpha}C_{m_{\delta E}} - C_{m_\alpha}C_{L_{\delta E}}\right)}\left[(h_{cm} - h_{mp_{fixed}}) - \Delta h_{free}\right] \tag{5.5-53}$$

where Δh_{free} is the increment due to the stick-free condition:

$$\Delta h_{free} = \frac{1}{C_{H_{\delta E}}}\left\{\frac{C_{H_\alpha}C_{m_{\delta E}}}{C_{L_\alpha}} + \frac{1}{2\mu - C_{L_{\hat{q}}}}\left[\frac{C_{L_{\delta E}}}{C_{L_\alpha}}\left(C_{H_\alpha}C_{m_{\hat{q}}} - C_{H_{\hat{q}}}C_{m_\alpha}\right) + C_{H_{\hat{q}}}C_{m_{\delta E}}\right]\right\} \tag{5.5-54}$$

The *stick-free maneuver margin* is defined by $[(h_{cm} - h_{mp}) - \Delta h_{free}]$, and the *stick-free maneuver point* is

$$h_{mp_{free}} = h_{mp_{fixed}} + \Delta h_{free} \tag{5.5-55}$$

Seckel notes [S-4], "The stick force gradient is perhaps the most important single handling characteristic of the airplane. It must be high enough that the pilot will not inadvertently overstress the structure . . . yet not (too high for intended maneuvering)." He also points out that the gradient is independent of dynamic pressure and, therefore, of indicated airspeed. This is a desirable trait, although it changes as the stability and control derivatives change with flight condition. The aft travel of the neutral point as Mach number approaches one increases the stick-force gradient and the force required for trim, a problem first noticed in high-speed dives of World-War-II-era fighter aircraft [A-1]. The unopposed nose-down pitching moment produced the *Mach tuck* or *tuck-under* phenomenon, revealing the need for Mach trim compensation, either by feedback control or, for some supersonic aircraft, by pumping fuel aft to move the center of mass backward.

"Tail-Wags-Dog" Effect

The pitch-control surfaces of aerodynamic reentry vehicles may have high moments of inertia as a consequence of their large size and added weight for heat shielding; hence, the inertial effects of their rotation must be considered. While a detailed investigation of this "tail-wags-dog" effect is beyond the present intent, the avenue for analyzing the resulting coupling between control surface and pitching motions is readily prescribed.

Angular acceleration of the control surface produces a reaction force and a moment on the vehicle that is proportional to the surface moment of inertia and the distance between the surface's hinge point and the vehicle's center of mass. These reaction effects were neglected in eq. 5.5-20. The short-period/control dynamic model is modified to reflect the effect of $\Delta\dot{\mathbf{x}}_\delta$,

$$\begin{bmatrix} \Delta\dot{\mathbf{x}}_{SP} \\ \Delta\dot{\mathbf{x}}_\delta \end{bmatrix} = \begin{bmatrix} \mathbf{F}_{SP} & \mathbf{G}_{SP} \\ \mathbf{L}_\delta & \mathbf{F}_\delta \end{bmatrix} \begin{bmatrix} \Delta\mathbf{x}_{SP} \\ \Delta\mathbf{x}_\delta \end{bmatrix} + \begin{bmatrix} \mathbf{K}_\delta \\ \mathbf{0} \end{bmatrix} \Delta\dot{\mathbf{x}}_\delta + \begin{bmatrix} \mathbf{0} \\ \mathbf{G}_\delta \end{bmatrix} \Delta\mathbf{u}_\delta \tag{5.5-56}$$

where $\mathbf{K}_\delta$ contains the sensitivities of pitch-rate and angle-of-attack rates of change to control-surface acceleration:

$$\mathbf{K}_\delta = \begin{bmatrix} 0 & \left[M_{\delta \ddot{E}} - \dfrac{M_{\dot{\alpha}}}{(V_o + L_{\dot{\alpha}})} L_{\delta \ddot{E}} \right] \\ 0 & \dfrac{-L_{\delta \ddot{E}}}{(V_o + L_{\dot{\alpha}})} \end{bmatrix} \tag{5.5-57}$$

Equation 5.5-56 can be rearranged as

$$\begin{bmatrix} \mathbf{I}_2 & -\mathbf{K}_\delta \\ \mathbf{0} & \mathbf{I}_2 \end{bmatrix} \begin{bmatrix} \Delta\dot{\mathbf{x}}_{SP} \\ \Delta\dot{\mathbf{x}}_\delta \end{bmatrix} = \begin{bmatrix} \mathbf{F}_{SP} & \mathbf{G}_{SP} \\ \mathbf{L}_\delta & \mathbf{F}_\delta \end{bmatrix} \begin{bmatrix} \Delta\mathbf{x}_{SP} \\ \Delta\mathbf{x}_\delta \end{bmatrix} + \begin{bmatrix} \mathbf{0} \\ \mathbf{G}_\delta \end{bmatrix} \Delta\mathbf{u}_\delta \tag{5.5-58}$$

Multiplying both sides of the equation by the inverse of $\begin{bmatrix} \mathbf{I}_2 & -\mathbf{K}_\delta \\ \mathbf{0} & \mathbf{I}_2 \end{bmatrix}$,

$$\begin{aligned}
\begin{bmatrix} \Delta\dot{\mathbf{x}}_{SP} \\ \Delta\dot{\mathbf{x}}_\delta \end{bmatrix} &= \begin{bmatrix} \mathbf{I}_2 & -\mathbf{K}_\delta \\ \mathbf{0} & \mathbf{I}_2 \end{bmatrix}^{-1} \left\{ \begin{bmatrix} \mathbf{F}_{SP} & \mathbf{G}_{SP} \\ \mathbf{L}_\delta & \mathbf{F}_\delta \end{bmatrix} \begin{bmatrix} \Delta\mathbf{x}_{SP} \\ \Delta\mathbf{x}_\delta \end{bmatrix} + \begin{bmatrix} \mathbf{0} \\ \mathbf{G}_\delta \end{bmatrix} \Delta\mathbf{u}_\delta \right\} \\
&= \begin{bmatrix} \mathbf{I}_2 & \mathbf{K}_\delta \\ \mathbf{0} & \mathbf{I}_2 \end{bmatrix} \left\{ \begin{bmatrix} \mathbf{F}_{SP} & \mathbf{G}_{SP} \\ \mathbf{L}_\delta & \mathbf{F}_\delta \end{bmatrix} \begin{bmatrix} \Delta\mathbf{x}_{SP} \\ \Delta\mathbf{x}_\delta \end{bmatrix} + \begin{bmatrix} \mathbf{0} \\ \mathbf{G}_\delta \end{bmatrix} \Delta\mathbf{u}_\delta \right\} \\
&= \begin{bmatrix} (\mathbf{F}_{SP} + \mathbf{K}_\delta \mathbf{L}_\delta) & (\mathbf{G}_{SP} + \mathbf{K}_\delta \mathbf{F}_\delta) \\ \mathbf{L}_\delta & \mathbf{F}_\delta \end{bmatrix} \begin{bmatrix} \Delta\mathbf{x}_{SP} \\ \Delta\mathbf{x}_\delta \end{bmatrix} + \begin{bmatrix} \mathbf{K}_\delta \mathbf{G}_\delta \\ \mathbf{G}_\delta \end{bmatrix} \Delta\mathbf{u}_\delta \\
&\triangleq \begin{bmatrix} \mathbf{F}'_{SP} & \mathbf{G}'_{SP} \\ \mathbf{L}_\delta & \mathbf{F}_\delta \end{bmatrix} \begin{bmatrix} \Delta\mathbf{x}_{SP} \\ \Delta\mathbf{x}_\delta \end{bmatrix} + \begin{bmatrix} \mathbf{K}_\delta \mathbf{G}_\delta \\ \mathbf{G}_\delta \end{bmatrix} \Delta\mathbf{u}_\delta
\end{aligned} \tag{5.5-59}$$

Thus, the modified equation is similar to eq. 5.5-20. The nature of the response to control command is unchanged, but the details of the response are different. The short-period and control-effect matrices $\mathbf{F}'_{SP}$ and $\mathbf{G}'_{SP}$ are changed by terms that are proportional to the control acceleration sensitivity $\mathbf{K}_\delta$. The command $\Delta\mathbf{u}_\delta$ has an immediate effect on the short-period mode as a consequence of $\mathbf{K}_\delta\mathbf{G}_\delta$; without the inertial effect, $\mathbf{K}_\delta$ is zero, and the command effect is filtered through the control mechanism's dynamics.

5.6 Longitudinal Aeroelastic Effects

Structural flexibility allows the wing, fuselage, and empennage to deform in flight, and this deformation can have important effect on flight motions. The *rigid-body model* of an aircraft establishes a body frame of reference (Chapter 2). It assumes that the vehicle's three-dimensional shape and inertial properties are fixed with respect to this frame. Because the airframe is elastic, deformations from the reference shape are caused by the distribution of aerodynamic pressures across the airframe, which depend on the rigid-body state as well as the deformation itself (Section 4.6). The deformations, in turn, contribute to the gross motion by changing forces and moments that drive the rigid-body modes; hence, there is coupling of rigid-body and elastic modes of motion. An *elastic-body model* of the aircraft consists of the rigid-body model *plus* the elastic deformation model *plus* the coupling terms between them.

Coupled motions that are symmetric about the aircraft's vertical plane (i.e., within the plane formed by its x and z axes) are described by the *longitudinal elastic-body model*, an application of the four-block structure presented in Table 4.6-1. We have seen that the longitudinal rigid-body state has six elements (V, γ, q, α, x, z); however, the dynamics associated with V, γ, x, and z are slow compared to the short-period mode, and their interaction with elastic motion is likely to be negligible. Although the linear elastic model has an infinite number of uncoupled second-order modes, there is a finite limit to the number of modes that can be computed (e.g., the I modes associated with the I elements of the lumped-parameter model presented in Section 4.6). All but a few can be neglected because the relative amplitudes and accelerations of their deflections are too small to be of consequence. The most significant modes tend to have the lowest frequencies, some of which may be near the short-period natural frequency; therefore, they may couple with rigid-body motions. Given I computed modes and N significant elastic modes ($N \leq I$), the longitudinal elastic-body model takes the form shown in Table 5.6-1. Thus, the mode shapes and natural frequencies normally associated with rigid-body and vibrational motions are coupled by the off-diagonal blocks.

The general form of the elastic-body model is

$$\Delta\dot{\mathbf{x}} = \mathbf{F}\Delta\mathbf{x} + \mathbf{G}\Delta\mathbf{u} + \mathbf{L}\Delta\mathbf{w} \tag{5.6-1a}$$

or

$$\begin{bmatrix} \Delta\dot{\mathbf{x}}_{SP} \\ \Delta\dot{\mathbf{x}}_{E} \end{bmatrix} = \begin{bmatrix} \mathbf{F}_{SP} & \mathbf{C}_{E}^{SP} \\ \mathbf{C}_{SP}^{E} & \mathbf{F}_{E} \end{bmatrix} \begin{bmatrix} \Delta\mathbf{x}_{SP} \\ \Delta\mathbf{x}_{E} \end{bmatrix} + \begin{bmatrix} \mathbf{G}_{SP} \\ \mathbf{G}_{E} \end{bmatrix} \Delta\mathbf{u} + \begin{bmatrix} \mathbf{L}_{SP} \\ \mathbf{L}_{E} \end{bmatrix} \Delta\mathbf{w} \tag{5.6-1b}$$

where $\Delta\mathbf{x}_{SP}$ is the (2×1) short-period state, $\Delta\mathbf{x}_{E}$ is the $(2I \times 1)$ elastic-mode state, $\Delta\mathbf{u}$ is the control, and $\Delta\mathbf{w}$ is an external disturbance. The uncoupled stability matrices for the short-period and elastic modes are $\mathbf{F}_{SP}$ and $\mathbf{F}_{E}$, the coupling matrices are $\mathbf{C}_{E}^{SP}$ and $\mathbf{C}_{SP}^{E}$, and the direct effects of control and disturbance on rigid-body motions and elastic deformations are $\mathbf{G}_{SP}$, $\mathbf{G}_{E}$, $\mathbf{L}_{SP}$, and $\mathbf{L}_{E}$.

Truncated and Residualized Elastic-Body Models

As detailed in Section 4.2, truncation and residualization are straightforward means for approximating a high-order system by one of lower order. With the system broadly partitioned into two subsystems, the system is *truncated* if the coupling between the two is ignored. This occurs when rigid-body motions are analyzed without regard to elastic deformation or the converse. Similarly, if the $(I - N)$ fastest modes have negligible amplitude, they can be neglected altogether, redefining the dimension of the elastic-mode state $\Delta\mathbf{x}_{E}$ as $(2N \times 1)$ and adjusting the related matrices accordingly.

If the faster subsystem is stable, it may reach steady state on a time scale that is short compared to the time scale of the slower subsystem. In such case, each subsystem sees the other in quasisteady state: transients in the fast subsystem have little impact on the slow subsystem, whose inputs to the fast system are, by definition, slow. If the elastic-body model

Table 5.6-1 Structure of the Longitudinal Elastic-Body Model Stability Matrix

Short-period mode	No. 1 to SP	...	No. N to SP	...	No. I to SP
SP to No. 1	Elastic Mode No. 1	...	0	...	0
...	...	...	...	...	...
SP to No. N	0	...	Elastic Mode No. N	...	0
...	...	...	...	...	...
SP to No. I	0	...	0	...	Elastic Mode No. I

has this fast-slow characteristic, a *residualized* equation for the short-period mode (eq. 4.2-57) can be written as

$$\begin{aligned}\Delta\dot{\mathbf{x}}_{SP} &= (\mathbf{F}_{SP} - \mathbf{C}_{E}^{SP}\,\mathbf{F}_{E}^{-1}\,\mathbf{C}_{SP}^{E})\,\Delta\mathbf{x}_{SP} + (\mathbf{G}_{SP} - \mathbf{C}_{E}^{SP}\,\mathbf{F}_{E}^{-1}\mathbf{G}_{E})\,\Delta\mathbf{u}\\ &\quad + (\mathbf{L}_{SP} - \mathbf{C}_{E}^{SP}\,\mathbf{F}_{E}^{-1}\mathbf{L}_{E})\,\Delta\mathbf{w}\\ &\triangleq \mathbf{F}_{SPE}\Delta\mathbf{x}_{SP} + \mathbf{G}_{SPE}\Delta\mathbf{u} + \mathbf{L}_{SPE}\Delta\mathbf{w} \qquad (5.6\text{-}2)\end{aligned}$$

The new equation for short-period dynamics is second order, and it accounts for the static effects of aeroelastic deformation.

Residualization *within* the elastic model has no impact because the modes of $\mathbf{F}_E$ are already uncoupled (as in Section 4.6), so any low-order approximation of a high-order block-diagonal model is necessarily truncated. However, the static-deformation effects of the higher-order modes on the short-period mode can be considered by residualizing an elastic-body model that is partitioned in a different way. Define the two parts of the state vector as

$$\Delta\mathbf{x} \triangleq \begin{bmatrix}\Delta\mathbf{x}_{S+N}\\ \Delta\mathbf{x}_{I-N}\end{bmatrix} \triangleq \begin{Bmatrix}\text{short period} + 1\text{st } N \text{ elastic modes}\\ \text{remaining } (I-N) \text{ elastic modes}\end{Bmatrix} \qquad (5.6\text{-}3)$$

and partition the system matrices accordingly. Residualization produces a dynamic equation of order $2(N+1)$,

$$\begin{aligned}\Delta\dot{\mathbf{x}}_{SN} &= (\mathbf{F}_{S+N} - \mathbf{C}_{I-N}^{S+N}\mathbf{F}_{I-N}^{-1}\mathbf{C}_{S+N}^{I-N})\,\Delta\mathbf{x}_{SN} + (\mathbf{G}_{S+N} - \mathbf{C}_{I-N}^{S+N}\mathbf{F}_{I-N}^{-1}\mathbf{G}_{I-N})\Delta\mathbf{u}\\ &\quad + (\mathbf{L}_{SN} - \mathbf{C}_{I-N}^{S+N}\mathbf{F}_{I-N}^{-1}\mathbf{L}_{I-N})\,\Delta\mathbf{w}\\ &\triangleq \mathbf{F}_{SN}\,\Delta\mathbf{x}_{SN} + \mathbf{G}_{SN}\,\Delta\mathbf{u} + \mathbf{L}_{SN}\,\Delta\mathbf{w} \qquad (5.6\text{-}4)\end{aligned}$$

where the component matrices are evident from the partitioning of $\Delta\mathbf{x}$.

In the remainder, we examine the effects of adding single, idealized elastic modes to the short-period model. Multiple modes would have additive effects, with each mode having components of both twisting and bending.

Coupling of the Short-Period with a Single Elastic Mode

Because the dynamic model for each elastic stability matrix is the same (eq. 4.6-97b), we begin by examining primary and coupling terms with $N = 1$ and a fourth-order elastic-body model. The state and control vectors are defined as

$$\Delta \mathbf{x}_{SP} = \begin{bmatrix} \Delta q \\ \Delta \alpha \end{bmatrix} \quad \begin{Bmatrix} \text{pitch rate} \\ \text{angle of attack} \end{Bmatrix} \tag{5.6-5}$$

$$\Delta \mathbf{x}_{E} = \begin{bmatrix} \Delta \eta_1 \\ \Delta \eta_2 \end{bmatrix} \quad \begin{Bmatrix} \text{elastic displacement} \\ \text{elastic rate} \end{Bmatrix} \tag{5.6-6}$$

$$\Delta \mathbf{u} = \Delta \delta E \ \{\text{elevator deflection}\} \tag{5.6-7}$$

$$\Delta \mathbf{w} = \Delta \alpha_w \ \{\text{angle of attack due to vertical wind}\} \tag{5.6-8}$$

From eqs. 5.3-5 and 4.6-95b, the primary stability matrices are

$$\mathbf{F}_{SP} = \begin{bmatrix} \left[M_q - \dfrac{M_{\dot\alpha}}{(V_o + L_{\dot\alpha})}(L_q - V_o) \right] & \left[M_\alpha - \dfrac{M_{\dot\alpha}}{(V_o + L_{\dot\alpha})} L_\alpha \right] \\ \dfrac{(V_o - L_q)}{(V_o + L_{\dot\alpha})} & \dfrac{-L_\alpha}{(V_o + L_{\dot\alpha})} \end{bmatrix} \tag{5.6-9}$$

$$\mathbf{F}_{E} = \begin{bmatrix} 0 & 1 \\ -\omega_{n_E}^2 & -2\zeta\omega_{n_E} \end{bmatrix} \tag{5.6-10}$$

while the coupling matrices take the form

$$\mathbf{C}_{E}^{SP} = \begin{bmatrix} \left[M_{\eta 1} - \dfrac{M_{\dot\alpha}}{(V_o + L_{\dot\alpha})} L_{\eta 1} \right] & \left[M_{\eta 2} - \dfrac{M_{\dot\alpha}}{(V_o + L_{\dot\alpha})} L_{\eta 2} \right] \\ \dfrac{-L_{\eta 1}}{(V_o + L_{\dot\alpha})} & \dfrac{-L_{\eta 2}}{(V_o + L_{\dot\alpha})} \end{bmatrix} \tag{5.6-11}$$

$$\mathbf{C}_{SP}^{E} = \begin{bmatrix} 0 & 0 \\ A_q & A_\alpha \end{bmatrix} \tag{5.6-12}$$

The control-effect matrices are

$$\mathbf{G}_{\mathrm{SP}} = \begin{bmatrix} \left[M_{\delta E} - \dfrac{M_{\dot{\alpha}}}{(V_o + L_{\dot{\alpha}})} L_{\delta E}\right] \\ \dfrac{-L_{\delta E}}{(V_o + L_{\dot{\alpha}})} \end{bmatrix} \tag{5.6-13}$$

$$\mathbf{G}_{\mathrm{E}} = \begin{bmatrix} 0 \\ B \end{bmatrix} \tag{5.6-14}$$

and the disturbance-effect matrices are

$$\mathbf{L}_{\mathrm{SP}} = \begin{bmatrix} -\left[M_{\alpha} - \dfrac{M_{\dot{\alpha}}}{(V_o + L_{\dot{\alpha}})} L_{\alpha}\right] \\ \dfrac{L_{\alpha}}{(V_o + L_{\dot{\alpha}})} \end{bmatrix} \tag{5.6-15}$$

$$\mathbf{L}_{\mathrm{E}} = \begin{bmatrix} 0 \\ -A_{\alpha} \end{bmatrix} \tag{5.6-16}$$

Here, as elsewhere in the book, we ignore the separate effect of the rate of vertical wind onset in disturbing the aircraft; that effect is discussed in [S-9].

The resulting equation bears strong resemblance to the short-period/control-mechanism equation developed in Section 5.5 (eq. 5.5-20). The coefficients contained in the primary blocks and the short-period control-effect matrix have been defined in Sections 4.1, 4.6, and 5.1. The sensitivities of the aircraft's lift and pitching accelerations to elastic displacement and rate can be expressed as

$$L_{\eta 1} = C_{L_{\eta 1}} \overline{q}\, S/m \tag{5.6-17}$$

$$L_{\eta 2} = C_{L_{\eta 2}} \overline{q}\, S/m \tag{5.6-18}$$

$$M_{\eta 1} = C_{m_{\eta 1}} \overline{q}\, S\overline{c}/I_{yy} \tag{5.6-19}$$

$$M_{\eta 2} = C_{m_{\eta 2}} \overline{q}\, S\overline{c}/I_{yy} \tag{5.6-20}$$

where $\overline{q}$ is the dynamic pressure, S is the aircraft reference area, $\overline{c}$ is the mean aerodynamic chord, and m and I_{yy} are the aircraft mass and pitching moment of inertia. The coefficients such as

$$C_{L_{\eta 1}} \triangleq \frac{\partial C_L}{\partial \eta_1} \tag{5.6-21}$$

are partial derivatives (with respect to structural displacement and rate) of the lift coefficient (eq. 4.6-106, resolved in the lift direction) and pitching moment (eq. 4.6-107) produced by structural motion. The elastic mode is forced by rigid-body motions and control deflection through A_q, A_α, and B (elements of **A**, **B**, and **L** in eq. 4.6-84), which are proportional to the dynamic pressure.

In the remaining discussion, we assume that angle-of-attack-rate has negligible effect on rigid-body pitching moment and lift and that lift due to pitch rate also can be ignored, allowing the matrices to be simplified. The notation is also condensed, with $(\cdot)'$ representing explicit division by V_o:

$$\mathbf{F}_{\mathrm{SP}} = \begin{bmatrix} M_q & M_\alpha \\ 1 & \dfrac{-L_\alpha}{V_o} \end{bmatrix} \triangleq \begin{bmatrix} M_q & M_\alpha \\ 1 & -L'_\alpha \end{bmatrix} \tag{5.6-22}$$

$$\mathbf{C}_{\mathrm{E}}^{\mathrm{SP}} = \begin{bmatrix} M_{\eta 1} & M_{\eta 2} \\ \dfrac{-L_{\eta 1}}{V_o} & \dfrac{-L_{\eta 2}}{V_o} \end{bmatrix} \triangleq \begin{bmatrix} M_{\eta 1} & M_{\eta 2} \\ -L'_{\eta 1} & -L'_{\eta 2} \end{bmatrix} \tag{5.6-23}$$

$$\mathbf{G}_{\mathrm{SP}} = \begin{bmatrix} M_{\delta E} \\ \dfrac{-L_{\delta E}}{V_o} \end{bmatrix} \triangleq \begin{bmatrix} M_{\delta E} \\ -L'_{\delta E} \end{bmatrix} \tag{5.6-24}$$

$$\mathbf{L}_{\mathrm{SP}} = \begin{bmatrix} -M_\alpha \\ \dfrac{L_\alpha}{V_o} \end{bmatrix} \triangleq \begin{bmatrix} -M_\alpha \\ L'_\alpha \end{bmatrix} \tag{5.6-25}$$

Reactive inertial effects could be modeled along the lines of the "tail-wags-dog" discussion of Section 5.5 (see eq. 5.5-59); the values of $\mathbf{F}_{\mathrm{SP}}$, $\mathbf{G}_{\mathrm{SP}}$, and $\mathbf{C}_{\mathrm{E}}^{\mathrm{SP}}$ would be altered, but the structure of eq. 5.6-1 would remain the same.

We distinguish four idealized elastic modes corresponding to the individual elements of the coupling matrix $\mathbf{C}_{\mathrm{E}}^{\mathrm{SP}}$, with reference to a conventional straight (i.e., unswept) winged aircraft. A *fuselage bending mode*, with vertical deformation of the aircraft's centerline from the reference x axis, produces a pitching acceleration that is principally proportional to the elastic mode's displacement through incremental moments on the nose, wing, and horizontal tail, and its effect on rigid-body motion is characterized by $M_{\eta 1}$. A *wing torsional mode* modifies local angles of attack along the span; hence, it produces lift that is proportional to the elastic displacement through $L'_{\eta 1}$. If the elastic axis is offset from the aircraft's rotational center, the wing torsional mode produces pitching acceleration as well through $M_{\eta 1}$. A *wing bending mode* produces no lift that is proportional to elastic displacement; however, the elastic rate produces a heaving or plunging distribution along the wing that modifies the local angle of attack and thus causes a lift increment, which is reflected by $L'_{\eta 2}$. There is no physical model for a fourth mode that would produce only an elastic-rate-sensitive pitching moment $M_{\eta 2}$, although that term could result from fuselage bending and offset wing torsional modes.

Of course, the actual coupling associated with an elastic mode could involve all four types of coupling. For example, bending of a swept, high-aspect-ratio wing produces changes in local angle of attack that are distributed not only along the y direction but along the x direction as well. Low-aspect-ratio, blended configurations like the cartoon of Fig. 4.6-9 do not have distinct fuselage and wing bending and torsional modes. From Figs. 4.6-5 and 4.6-6, we surmise that the magnitude of $\mathbf{C}_{\mathrm{E}}^{\mathrm{SP}}$ generally becomes smaller with increasing mode number, as distributed positive and negative deformations cancel in generating net lift and moment.

Equilibrium Response to Control

Setting the state rate to zero, the equilibrium response of equation 5.6-1 to constant control is

$$\begin{bmatrix} \Delta \mathbf{x}_{\mathrm{SP}} \\ \Delta \mathbf{x}_{\mathrm{E}} \end{bmatrix}^* = -\begin{bmatrix} \mathbf{F}_{\mathrm{SP}} & \mathbf{C}_{\mathrm{E}}^{\mathrm{SP}} \\ \mathbf{C}_{\mathrm{SP}}^{\mathrm{E}} & \mathbf{F}_{\mathrm{E}} \end{bmatrix}^{-1} \begin{bmatrix} \mathbf{G}_{\mathrm{SP}} \\ \mathbf{G}_{\mathrm{E}} \end{bmatrix} \Delta \mathbf{u}^* \tag{5.6-26}$$

Substituting from eqs. 5.6-9–5.6-22, the steady-state response of the fourth-order model is

$$\begin{bmatrix} \Delta q \\ \Delta \alpha \\ \Delta \eta_1 \\ \Delta \eta_2 \end{bmatrix}^* = \frac{\Delta \delta E^*}{-M_\alpha \omega_{n_E}^2 - L'_\alpha M_q \omega_{n_E}^2 + A_q(L'_{\eta 1} M_\alpha - L'_\alpha M_{\eta 1}) - A_\alpha(M_{\eta 1} + L'_{\eta 1} M_q)}$$
$$\cdot \begin{bmatrix} M_{\delta \mathrm{E}}(A_\alpha L'_{\eta 1} + L'_\alpha \omega_{n_E}^2) - L'_{\delta \mathrm{E}}(A_\alpha M_{\eta 1} + M_\alpha \omega_{n_E}^2) - B(L'_{\eta 1} M_\alpha - L'_\alpha M_{\eta 1}) \\ -M_{\delta \mathrm{E}}(A_q L'_{\eta 1} - \omega_{n_E}^2) + L'_{\delta \mathrm{E}}(A_q M_{\eta 1} + M_q \omega_{n_E}^2) + B(L'_{\eta 1} M_q + M_{\eta 1}) \\ -M_{\delta \mathrm{E}}(A_\alpha - A_q L'_\alpha) + L'_{\delta \mathrm{E}}(A_\alpha M_q - A_q M_\alpha) - B(M_\alpha + L'_\alpha M_q) \\ 0 \end{bmatrix} \tag{5.6-27}$$

Note particularly that the elastic rate $\Delta\eta_2$ is zero at equilibrium and that neither $L'_{\eta 2}$ nor $M_{\eta 2}$ appears in the equation. Consequently, a pure wing bending mode has no effect on the steady-state response to control. Typically, fuselage bending can be expected to reduce the control effect of elevator deflection; however, the actual effect is dependent on a number of factors. *Reversal of the elevator effect* is indicated by the combination of coefficients that render $\Delta\alpha^* = 0$, which occurs when the second row of the numerator in eq. 5.6-27 is zero. The nonzero value of Δq^* predicted by the equation is consistent with the second-order short-period model, although Δq^* would always go to zero if the phugoid mode were included in the model. Dropping all elastic coupling terms, the steady-state values Δq^* and $\Delta\alpha^*$ are the same as those predicted by eqs. 5.3-10 and 5.3-11.

Eigenvalues and Root Locus Analysis of Parameter Variations

The characteristic polynomial for this elastic-body model is

$$\begin{aligned}\Delta(s) = |s\mathbf{I}_4 - \mathbf{F}| = s^4 &+ (L'_\alpha - M_q + 2\zeta\omega_{n_E})s^3 + (A_\alpha L'_{\eta_2} - A_q M_{\eta_2} - M_\alpha \\ &- L'_\alpha M_q + \omega^2_{n_E} + 2\zeta\omega_{n_E} L'_\alpha - 2\zeta\omega_{n_E} M_q)s^2 \\ &+ [A_\alpha(L'_{\eta_1} - L'_{\eta_2}M_q - M_{\eta_2}) + A_q(L'_{\eta_2}M_\alpha - L'_\alpha M_{\eta_2} - M_{\eta_1}) \\ &+ \omega^2_{n_E}(L'_\alpha - M_q) - 2\zeta\omega_{n_E}(M_\alpha + L'_\alpha M_q)]s + [A_q(L'_{\eta_1}M_\alpha - L'_\alpha M_{\eta_1}) \\ &- A_\alpha(M_{\eta_1} + L'_{\eta_1}M_q) - \omega^2_{n_E}(M_\alpha + L'_\alpha M_q)] = 0 \end{aligned} \quad (5.6\text{-}28)$$

Given separately stable short-period and elastic modes, the polynomial factors into two complex pairs of roots that represent the uncoupled short period and elastic oscillations when all the elements of $\mathbf{C}_E^{SP}$ or $\mathbf{C}_{SP}^E$ are zero, as every element of $\mathbf{C}_E^{SP}$ is multiplied by an element of $\mathbf{C}_{SP}^E$:

$$\begin{aligned}\Delta(s)_{uncoupled} &= s^4 + (L'_\alpha - M_q + 2\zeta\omega_{n_E})\, s^3 \\ &+ (-M_\alpha - L'_\alpha M_q + \omega^2_{n_E} + 2\zeta\omega_{n_E} L'_\alpha - 2\zeta\omega_{n_E} M_q)\, s^2 \\ &+ [\omega^2_{n_E}(L'_\alpha - M_q) - 2\zeta\omega_{n_E}(M_\alpha + L'_\alpha M_q)]\, s \\ &+ [-\omega^2_{n_E}(M_\alpha + L'_\alpha M_q)] \\ &= (s^2 + 2\zeta\omega_n s + \omega_n^2)_{SP_{unc}}(s^2 + 2\zeta\omega_n s + \omega_n^2)_{E_{unc}} \end{aligned} \quad (5.6\text{-}29)$$

Because each element of $\mathbf{C}_E^{SP}$ or $\mathbf{C}_{SP}^E$ appears linearly in eq. 5.6-28, the effect of individual coupling terms on system stability can be evaluated using Evans's rules of root locus construction (Section 4.4). Following eq. 4.4-30, equation 5.6-28 can be rearranged as

$$\Delta(s) = d(s) + k\, n(s) = 0 \qquad (5.6\text{-}30a)$$

This is equivalent to the numerator-denominator form

$$\frac{k\, n(s)}{d(s)} = -1 \qquad (5.6.30b)$$

where k is the parameter whose effect is being evaluated. If we examine just one parameter at a time, setting all the others in the same block to zero, then the denominator $d(s) = \Delta(s)_{uncoupled}$, and $n(s)$ takes a simple zeroth- to second-degree form.

We use this approach to examine four cases:

- Fuselage bending mode (variation in M_{η_1}; Fig. 5.6-1)
- Wing torsional mode (variation in $L'_{\eta 1}$; Fig. 5.6-2)
- Wing bending mode (variation in $L'_{\eta 2}$; Fig. 5.6-3)
- Angle-of-attack forcing of the elastic mode (variation in A_α)

The first three cases relate to the idealized modes introduced earlier; the fourth case considers the degree to which a generalized elastic mode is excited by angle-of-attack perturbation, which is likely to be a more significant factor than forcing of the elastic mode by pitch rate through A_q.

It is shown below that elastic/rigid-body coupling is scaled by a product of the elastic input (A_q, A_α) and output (M_{η_1}, M_{η_2}, L'_{η_1}, L'_{η_2}) terms. Because both are proportional to airspeed or airspeed squared, the coupling gain increases as the second to fourth power of the airspeed, all else being fixed. In actuality, the *in vacuo* elastic parameters do remain fixed, but the short-period characteristics also vary with flight condition. Nevertheless, for illustrative purposes, we allow the four parameters (M_{η_1}, L'_{η_1}, L'_{η_2}, A_α) to vary independently.

The four cases are examined using the nominal parameter values contained in Table 5.6-2, which represents the stable short-period mode of Table 5.3-1, and an *in vacuo* elastic mode with 2 Hz natural frequency and damping ratio of 0.05. The short-period mode is loosely representative of a business jet flying at a Mach number of 0.3 and an altitude of 5,000 m. The elastic-mode natural frequency is atypically low to exaggerate the rigid-elastic coupling for the examples. Forcing of the elastic mode is calculated with $A_q = 0$, and A_α is arbitrarily fixed at ω^2_{nE}, except where A_α is allowed to vary. The coupling terms may take large positive or negative values to reveal the underlying shapes of root loci.

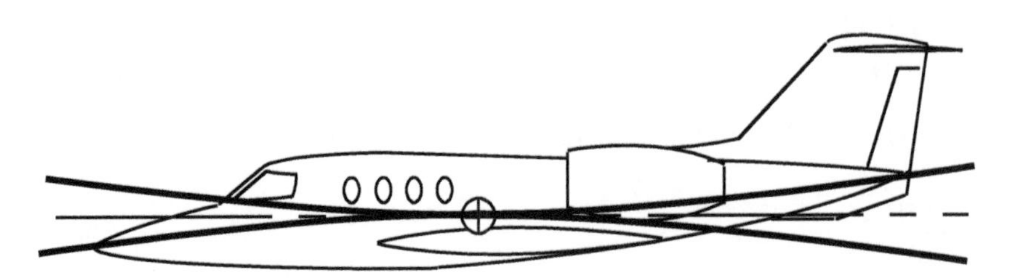

Fig. 5.6-1. Sketch of elastic displacement of aircraft centerline in first symmetric fuselage bending mode.

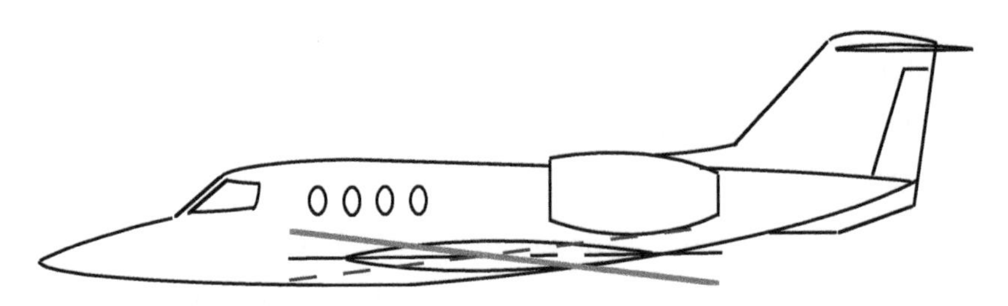

Fig. 5.6-2. Sketch of elastic displacement of wing tip in first symmetric wing torsional mode.

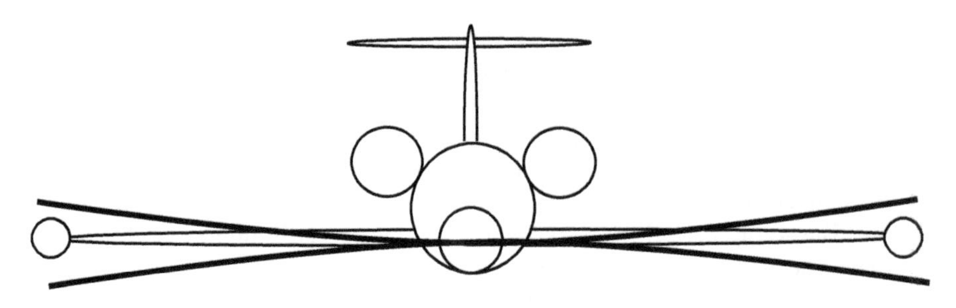

Fig. 5.6-3. Sketch of elastic displacement of wing in first symmetric wing bending mode.

Let us examine the first symmetric fuselage bending mode, which is sketched in Fig. 5.6-1 (after Fig. 4.6-5). The numerator for the fuselage bending mode is

$$kn(s) = M_{\eta_1}(-A_\alpha - A_q L'_\alpha - A_q s) = -M_{\eta_1} A_q \left[s + \left(L'_\alpha + \frac{A_\alpha}{A_q} \right) \right] \tag{5.6-31}$$

TABLE 5.6-2 BASELINE ELASTIC-BODY MODEL WITH SHORT-PERIOD MODE AND SINGLE ELASTIC MODE

$$\mathbf{F} = \begin{bmatrix} -1.4294 & -7.9856 & M_{\eta 1} & M_{\eta 2} \\ 0.9901 & -1.1209 & -L_{\eta 1} & -L_{\eta 2} \\ 0 & 0 & 0 & 1 \\ A_q & A_\alpha & -157.91 & -1.257 \end{bmatrix} \qquad \mathbf{G} = \begin{bmatrix} -9.069 \\ 0 \\ 0 \\ B \end{bmatrix}$$

which equals $-M_{\eta 1} A_\alpha$, if $A_q = 0$. In this case, the root locus asymptotes are either (± 45, ± 135 deg), following the 180 deg phase-angle criterion for negative $M_{\eta 1} A_\alpha$, or (0, ± 90, 180 deg), for the 0 deg negative phase-angle criterion with positive $M_{\eta 1} A_\alpha$. In the first case, the initial effect of coupling is to reduce the damped natural frequency of the elastic mode and to increase frequency of the short-period mode. With high coupling, the ultimate elastic instability is oscillatory (Fig. 5.6-4a). In the second case, the initial trends are reversed, and there is a real divergence of the coupled short-period mode at large $M_{\eta 1} A_\alpha$.

The numerator for the wing torsional mode (Fig. 5.6-2), assuming zero offset of the elastic axis from the aircraft's rotational center, is

$$\begin{aligned} kn(s) &= L'_{\eta 1}(-A_\alpha M_q + A_q M_\alpha + A_\alpha s) \\ &= L'_{\eta 1} A_\alpha \left[s + \left(\frac{A_q}{A_\alpha} M_\alpha - M_q \right) \right] \end{aligned} \tag{5.6-32}$$

and with $A_q = 0$ this becomes

$$kn(s) = L'_{\eta 1} A_\alpha (s - M_q) \tag{5.6-33}$$

The asymptotes for the root locus with varying $L'_{\eta 1}$ (Fig. 5.6-4b) are either (± 60, -180 deg) or (0, ± 120 deg), depending on the sign of $L'_{\eta 1} A_\alpha$. The fourth root ultimately approaches the real zero at M_q. For net positive-gain coupling, damping initially is added to the short-period mode and subtracted from the elastic mode, with the potential for oscillatory instability of the latter. With negative coupling, the converse is true.

For flapping of the wing (Fig. 5.6-3), there is an additional s in the wing bending mode numerator,

$$\begin{aligned} kn(s) &= L'_{\eta 2}[(A_q M_\alpha - A_\alpha M_q)s + A_\alpha s^2] \\ &= L'_{\eta 2} A_\alpha s \left[s + \left(\frac{A_q}{A_\alpha} M_\alpha - M_q \right) \right] \end{aligned} \tag{5.6-34}$$

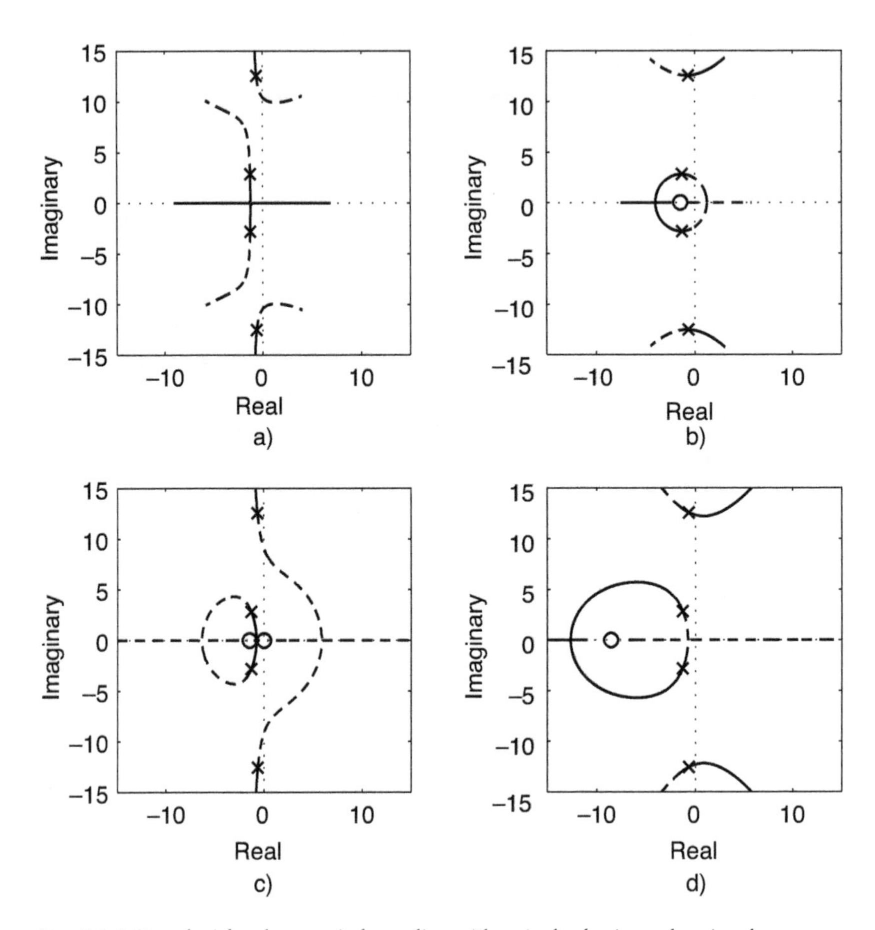

Fig. 5.6-4. Root loci for short-period coupling with a single elastic mode using the parameters of Table 5.6-3. (a) Fuselage bending mode (variation in M_{η_1}). (b) Wing torsional mode (variation in L'_{η_1}). (c) Wing bending mode (variation in L'_{η_2}). (d) Angle-of-attack forcing of the elastic mode (variation in A_α).

which becomes

$$kn(s) = L'_{\eta_2} A_\alpha s(s - M_q) \tag{5.6-35}$$

when $A_q = 0$. One of the zeros is at the origin, and the other occupies the same location as the wing torsional zero M_q. Consequently, the root locus asymptotes are ±90 deg or (0, 180 deg), and the remaining two branches of the root locus are driven to the two zeros at high gain (Fig. 5.6-4c).

Instability can occur only with high negative loop gain, when the oscillatory elastic-mode roots may be driven into the right half plane.

The degree to which the elastic mode is excited by angle-of-attack perturbation is reflected by A_α, whose coupling numerator is

$$\begin{aligned} kn(s) &= A_\alpha[-M_{\eta_1} - L'_{\eta_1}M_q + (L'_{\eta_1} - M_{\eta_2} - L'_{\eta_2}M_q)s + L'_{\eta_2}s^2] \\ &= A_\alpha L'_{\eta_2}\left[s^2 + \left(\frac{L'_{\eta_1} - M_{\eta_2}}{L'_{\eta_2}} - M_q\right)s - \left(\frac{M_{\eta_1} + L'_{\eta_1}M_q}{L'_{\eta_2}}\right)\right] \end{aligned} \tag{5.6-36}$$

If elastic rate effects are negligible, the η_2 derivatives vanish, and the numerator reduces to

$$kn(s) = A_\alpha L'_{\eta_1}\left[s - \left(\frac{M_{\eta_1}}{L'_{\eta_1}} + M_q\right)\right] \tag{5.6-37}$$

For the example (Fig. 5.6-d), $M_{\eta_1} = M_\alpha$ and $L'_{\eta_1} = L'_\alpha/V_o$. The root locus is similar to Fig. 5.6-4b, although there is no possibility of coupled short-period instability with negative loop gain because the zero is farther to the left, exerting a stabilizing effect on the locus.

Control and Disturbance Transfer Functions

Following Section 4.4, transfer function matrices for the elastic-body model are derived from

$$\begin{aligned} \Delta\mathbf{y}(s) &= \mathbf{H}\,\Delta\mathbf{x}(s) + \mathbf{J}\,\Delta\mathbf{u}(s) + \mathbf{K}\,\Delta\mathbf{w}(s) \\ &= \mathbf{H}(s\mathbf{I}_4 - \mathbf{F})^{-1}[\mathbf{G}\,\Delta\mathbf{u}(s) + \mathbf{L}\,\Delta\mathbf{w}(s)] + \mathbf{J}\,\Delta\mathbf{u}(s) + \mathbf{K}\,\Delta\mathbf{w}(s) \\ &\triangleq \mathbf{U}(s)\,\Delta\mathbf{u}(s) + \mathbf{W}(s)\,\Delta\mathbf{w}(s) \end{aligned} \tag{5.6-38}$$

where $\mathbf{J}$ and $\mathbf{K}$ reflect direct paths from the input to the output, and

$$\begin{aligned} \mathbf{U}(s) &= \mathbf{H}(s\mathbf{I}_4 - \mathbf{F})^{-1}\mathbf{G} + \mathbf{J} \\ &= \frac{\mathbf{H}\,\mathrm{Adj}(s\mathbf{I}_4 - \mathbf{F})\mathbf{G} + \mathbf{J}\,|s\mathbf{I}_4 - \mathbf{F}|}{|s\mathbf{I}_4 - \mathbf{F}|} \end{aligned} \tag{5.6-39}$$

$$\begin{aligned} \mathbf{W}(s) &= \mathbf{H}(s\mathbf{I}_4 - \mathbf{F})^{-1}\mathbf{L} + \mathbf{K} \\ &= \frac{\mathbf{H}\,\mathrm{Adj}(s\mathbf{I}_4 - \mathbf{F})\mathbf{L} + \mathbf{K}\,|s\mathbf{I}_4 - \mathbf{F}|}{|s\mathbf{I}_4 - \mathbf{F}|} \end{aligned} \tag{5.6-40}$$

Defining **H** as the identity matrix and **J** and **K** as zero, $\Delta\mathbf{y}(s) = \Delta\mathbf{x}(s)$, and the control transfer function matrix can be written as

$$\mathrm{U}(s) = \frac{[n_{\delta E}^{q}(s)\ n_{\delta E}^{\alpha}(s)\ n_{\delta E}^{\eta_1}(s)\ n_{\delta E}^{\eta_2}(s)]^{\mathrm{T}}}{\Delta(s)} \tag{5.6-41}$$

Literal expressions for the four scalar transfer function numerators are

$$\begin{aligned} n_{\delta E}^{q}(s) = {} & M_{\delta E}[s^3 + (L'_\alpha + 2\zeta\omega_n)s^2 + (A_\alpha L'_{\eta_2} + \omega_n^2 + 2\zeta\omega_n L'_\alpha)s + A_\alpha L'_{\eta_1}] \\ & - L'_{\delta E}[M_\alpha s^2 + (A_\alpha M_{\eta_2} + 2\zeta\omega_n M_\alpha)s + A_\alpha M_{\eta_1}] \\ & + B[M_{\eta_2}s^2 + (M_{\eta_1} + L'_\alpha M_{\eta_2} - L'_{\eta_2}M_\alpha)s + (L'_\alpha M_{\eta_1} - L'_{\eta_1}M_\alpha)] \end{aligned} \tag{5.6-42}$$

$$\begin{aligned} n_{\delta E}^{\alpha}(s) = {} & M_{\delta E}[s^2 - A_q L'_{\eta_2}s + (2\zeta\omega_n - A_q L'_{\eta_1})] - L'_{\delta E}[s^3 - (M_q - 2\zeta\omega_n)s^2 \\ & - (A_q M_{\eta_2} - \omega_n^2 + 2\zeta\omega_n M_q)s - (M_q\omega_n^2 + A_q M_{\eta_1})] \\ & + B[-L'_{\eta_2}s^2 + (L'_{\eta_2}M_q + M_{\eta_2} - L'_{\eta_1})s + (M_{\eta_1} + L'_{\eta_1}M_q)] \end{aligned} \tag{5.6-43}$$

$$\begin{aligned} n_{\delta E}^{\eta_1}(s) = {} & M_{\delta E}[A_q s + (A_\alpha + A_q L'_\alpha)] \\ & + L'_{\delta E}[-A_\alpha s + (A_\alpha M_q - A_q M_\alpha)] \\ & + B[s^2 + (L'_\alpha - M_q)s - (M_\alpha + L'_\alpha M_q)] \end{aligned} \tag{5.6-44}$$

$$n_{\delta E}^{\eta_2}(s) = s n_{\delta E}^{\eta_1}(s) \tag{5.6-45}$$

The numerator expressions are collected in terms of the principal control effects $L'_{\delta E}$, $M_{\delta E}$, and B to illustrate the separate contributions of lift, pitching moment, and elastic forcing. To identify the corresponding transfer function zeros, the terms are assembled by powers of s, and each second- or third-degree polynomial is factored. Little generalization can be accomplished because so many parameters are contained in each numerator. However, Bode and root locus plots reveal a common characteristic: zeros near the open-loop elastic poles in the pitch-rate and angle-of-attack numerators, as shown below. Conversely, for control surfaces with large B and small $M_{\delta E}$ and $L'_{\delta E}$ (exemplified by the ride-control vanes on the forward fuselage of the North American–Rockwell B-1 bomber), the elastic-mode zeros (eqs. 5.6-44 and 5.6-45) cancel the short-period poles, indicating that the control does not excite that mode.

The numerator terms for the disturbance transfer function matrix

$$\mathbf{W}(s) = \frac{[n_w^q(s)\ n_w^\alpha(s)\ n_w^{\eta_1}(s)\ n_w^{\eta_2}(s)]^T}{\Delta(s)} \tag{5.6-46}$$

can be derived from the control numerators by replacing $(L'_{\delta E}, M_{\delta E}, B)$ with $(-L'_\alpha, -M_\alpha, -A_\alpha)$. These transfer functions can be used to determine the power spectra of response to turbulence inputs, as described in Section 4.5 (eq. 4.5-17). Consider the angle-of-attack response to the turbulence spectrum $\Phi_{\alpha_w\alpha_w}(\omega)$ as an example. Substituting $s = j\omega$, the power spectral density is

$$\Phi_{\alpha\alpha}(\omega) = \frac{n_w^\alpha(j\omega)\ n_w^\alpha(-j\omega)}{\Delta(j\omega)\ \Delta(-j\omega)} \Phi_{\alpha_w\alpha_w}(\omega) \tag{5.6-47}$$

If pitch rate and angle of attack are measured at some point other than the center of mass, the observations may include components that are due to the elastic oscillation as well. The (2×4) output matrix $\mathbf{H}$ contains a (2×2) identity matrix and coupling terms that depend upon the shape of the elastic mode and the location of the measurement point with respect to that shape. As an example, assume that the measurements are made at the cockpit, located ahead of the center of mass in the aircraft's plane of symmetry. The idealized wing bending and torsional mode shapes are perpendicular to the aircraft's centerline, and these elastic vibrations are not sensed at the cockpit; hence, in either case,

$$\Delta\mathbf{y} = \mathbf{H}\,\Delta\mathbf{x} \tag{5.6.-48}$$

and

$$\begin{bmatrix} \Delta q \\ \Delta\alpha \end{bmatrix}_{cockpit} = \begin{bmatrix} 1 & 0 & 0 & 0 \\ 0 & 1 & 0 & 0 \end{bmatrix} \begin{bmatrix} \Delta q \\ \Delta\alpha \\ \Delta\eta_1 \\ \Delta\eta_2 \end{bmatrix} \tag{5.6-49}$$

However, the z deflection due to idealized fuselage bending is distributed along the x direction, and it causes elastic displacement and rate at the cockpit. Consequently, the angular rate and angle sensed at the cockpit are given by

$$\begin{bmatrix} \Delta q \\ \Delta\alpha \end{bmatrix}_{cockpit} = \begin{bmatrix} 1 & 0 & 0 & a \\ 0 & 1 & a & 0 \end{bmatrix} \begin{bmatrix} \Delta q \\ \Delta\alpha \\ \Delta\eta_1 \\ \Delta\eta_2 \end{bmatrix} \tag{5.6-50}$$

where a is a constant that converts from linear to angular measure, given the cockpit location and the bending mode shape.

Frequency Response and the Effects of Feedback

Here, we examine typical Bode and root locus plots for the model described in Table 5.6-2 with fuselage bending, wing twisting, and wing bending effects taken one at a time. The Bode plots illustrate the effect of sinusoidal forcing by the elevator on the pitch rate and angle of attack, while the root locus plots show the effect on closed-loop system roots of feeding these variables back to the elevator. In each case, the *in vacuo* elastic natural frequency and damping ratio are 12.54 rad/s and 0.05, while the uncoupled short-period natural frequency and damping ratio are 3.08 rad/s and 0.41. The principal differences between these plots and those shown in Section 5.3 are due to the addition of elastic pole-zero pairs to the pitch-rate and angle-of-attack transfer functions.

With fuselage bending ($M_{\eta_1} = M_\alpha$), the elastic zero is seen to have slightly higher frequency than the elastic pole (Fig. 5.6-5). For the figure, it is assumed that elevator deflection does not force fuselage bending directly. As the forcing frequency increases beyond the short-period natural frequency, there is a resonant peak at the elastic natural frequency. The complex zeros then produce a notch for both pitch-rate (solid line) and angle-of-attack (dashed line) response (Fig. 5.6-5a). If the pitch rate is fed back to the elevator in a simple control system, the elastic pole and zero have negligible effect on the root locus (Fig. 5.6-5b); with increasing loop gain, the short-period roots behave as in the rigid-body case. However, angle-of-attack feedback causes the short-period roots to come near to the elastic roots (Fig. 5.6-5c) for some range of loop gains, and there is a region of *conditional elastic instability* when the roots are in proximity.

The effects of wing twisting are much the same as for wing bending; however, with $L_{\eta_1} = L_\alpha$, the elastic zero is closer to the elastic zero, and the Bode plot resonance notch is slightly altered (Fig. 5.6-6a). Pitch-rate feedback effect is similar to the previous case (Fig. 5.6-6b). For angle-of-attack feedback, the short-period root is captured by the elastic zero at high loop gain, and the elastic root proceeds to higher frequency, with a small range of conditional instability at low gain (Fig. 5.6-6c).

With wing bending ($L_{\eta_2} = 0.1$), the elastic zero frequency is slightly less than the elastic pole frequency (Fig. 5.6-7). Thus, the Bode plot notch occurs at a lower frequency than the resonance. The two root locus plots are similar to those for wing twisting; however, there is no region of conditional instability.

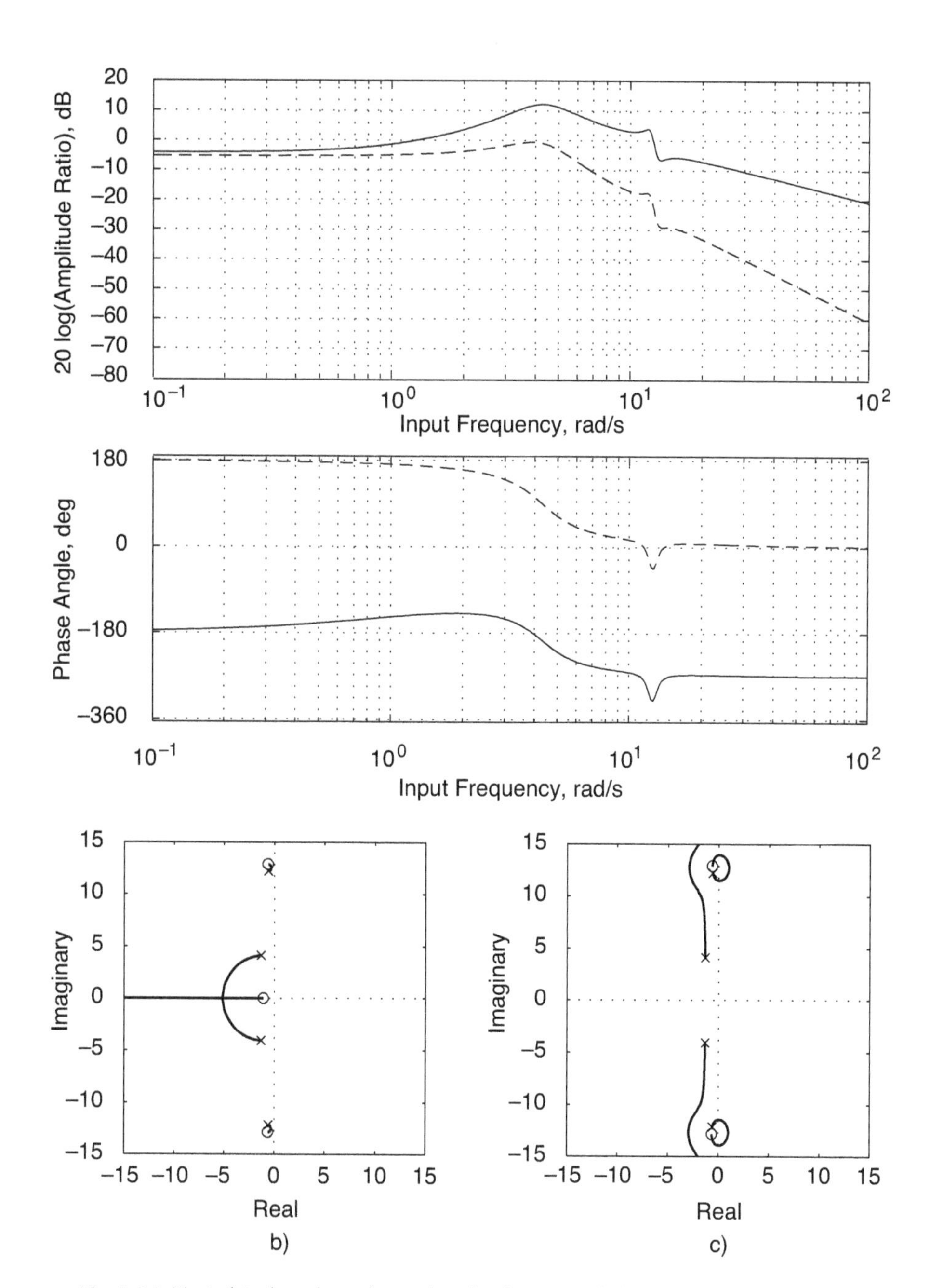

Fig. 5.6-5. Typical Bode and root locus plots for short-period plus fuselage bending mode. (Top) Bode plots. (b) Pitch-rate feedback. (c) Angle-of-attack feedback.

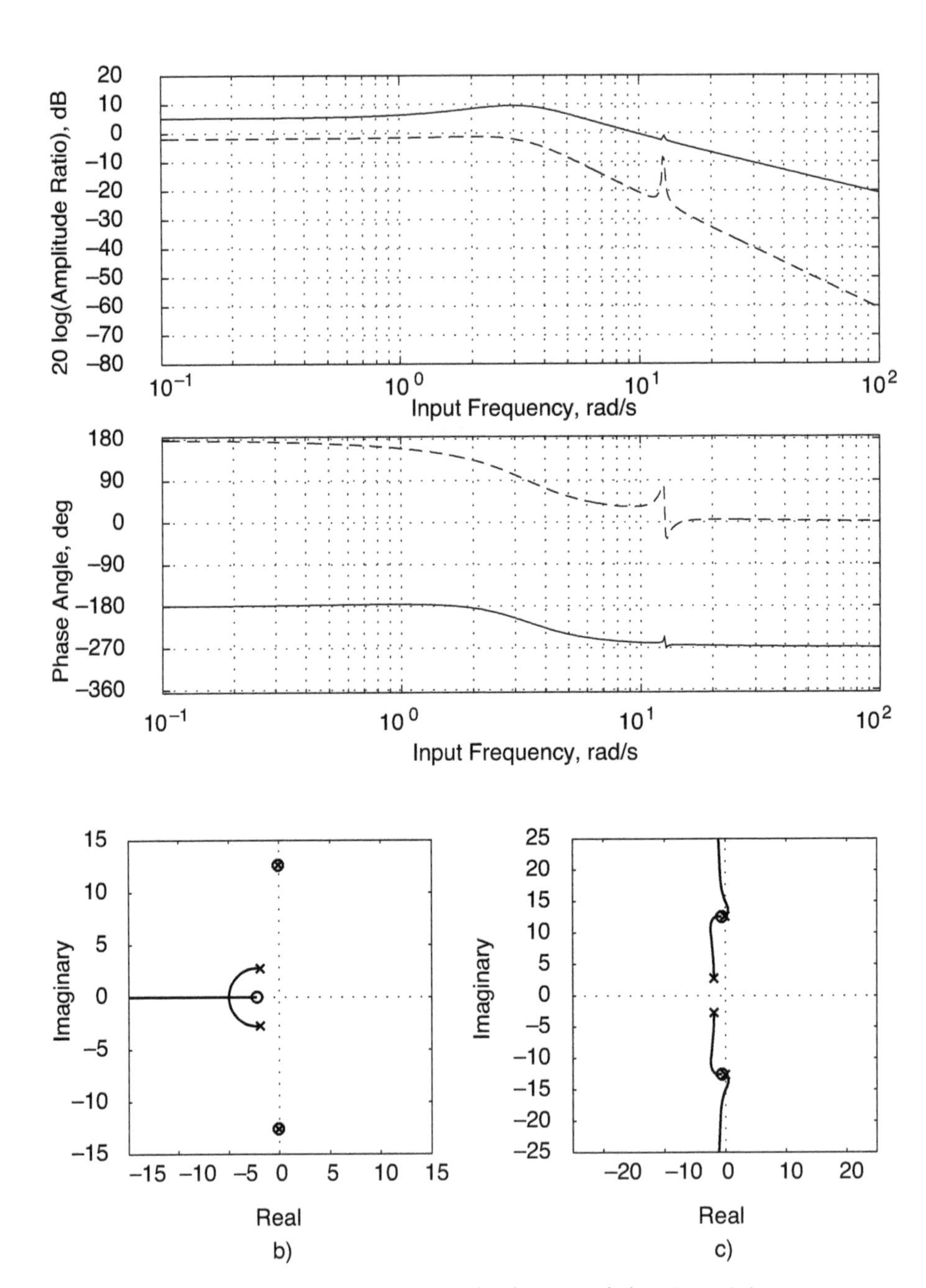

Fig. 5.6-6. Typical Bode and root locus plots for short-period plus wing twisting mode. (Top) Bode plots. (b) Pitch-rate feedback. (c) Angle-of-attack feedback.

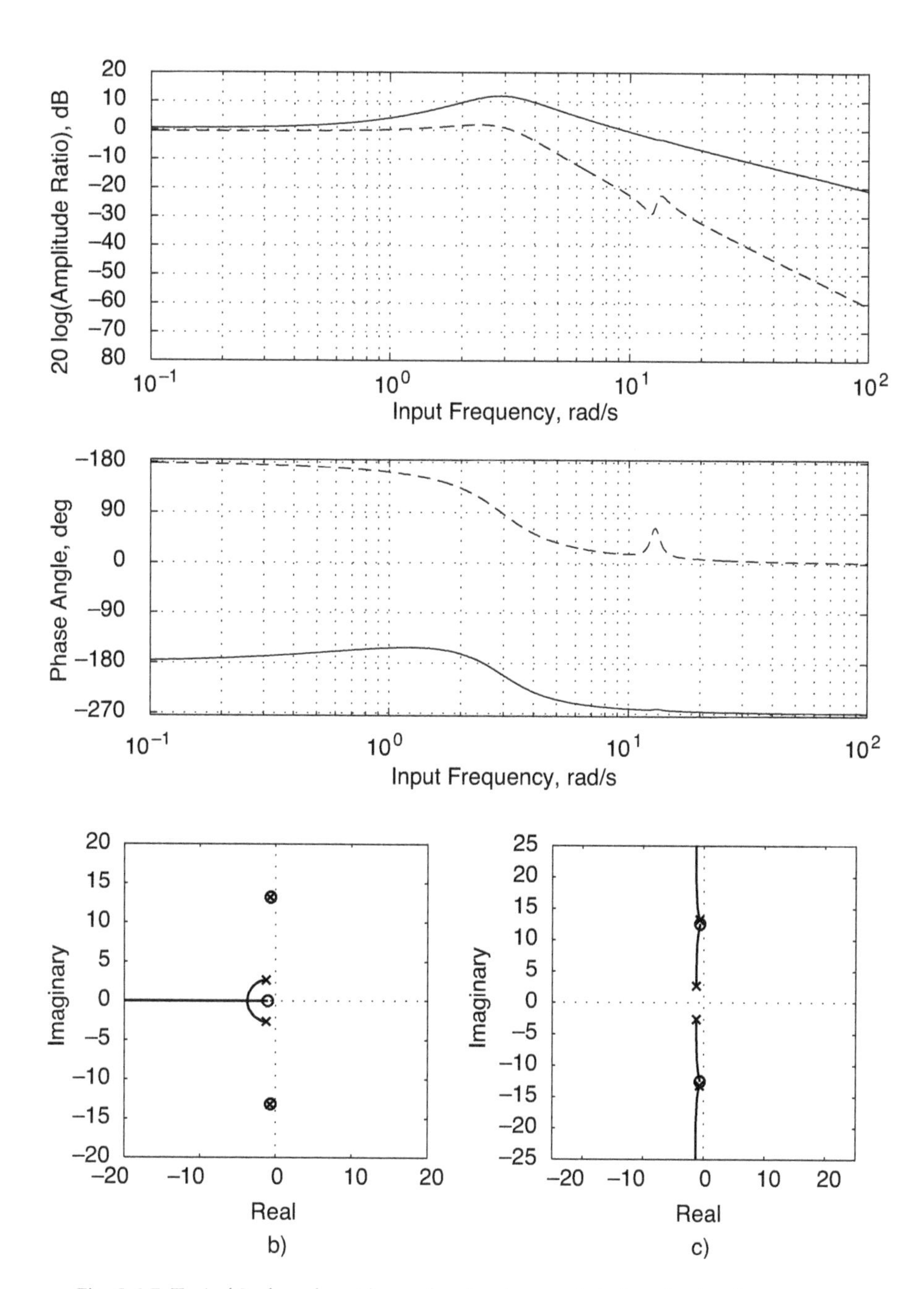

Fig. 5.6-7. Typical Bode and root locus plots for short-period plus wing bending mode. (Top) Bode plots. (b) Pitch-rate feedback. (c) Angle-of-attack feedback.

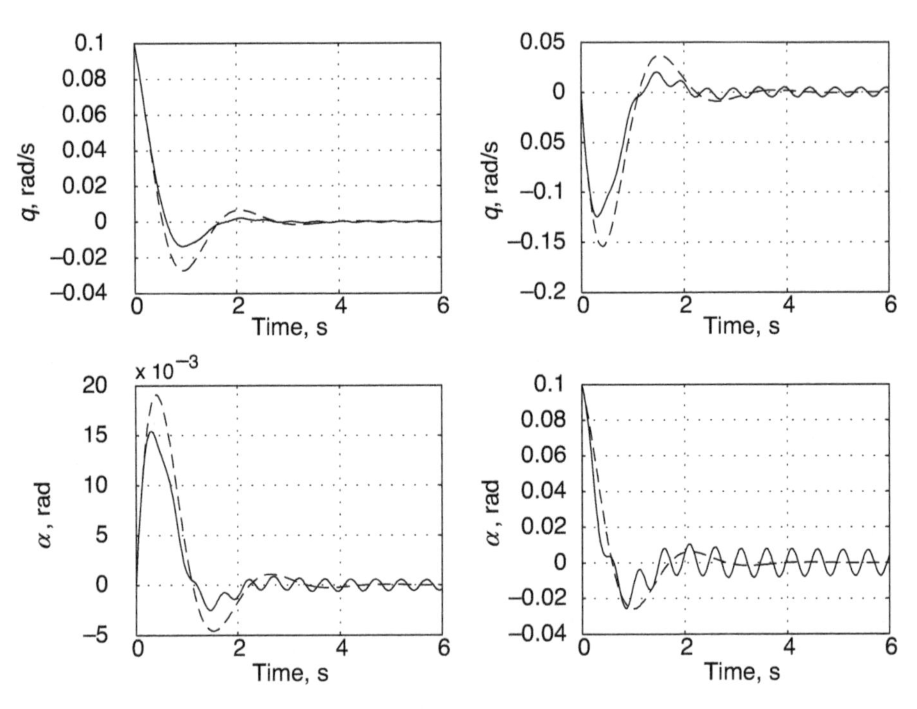

Fig. 5.6-8. Comparison of short-period initial-condition response with and without wing twisting elasticity. Initial pitch rate (left). Initial angle of attack (right).

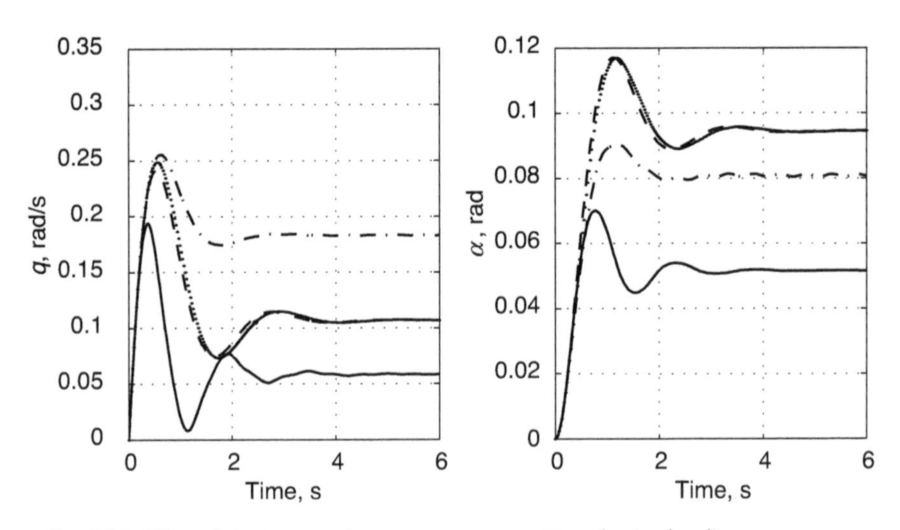

Fig. 5.6-9. Effect of elasticity on elevator step response. Dotted: wing bending; dashed: rigid; solid: fuselage bending; dash-dotted: wing twisting.

Effects of Elasticity on Time Response

Initial-condition and elevator step responses are shown in the next two figures. Figure 5.6-8 presents the responses of the two short-period variables with separate initial conditions on each, given the wing twisting model used above. The elastic oscillation (solid line) is most prominent in the angle-of-attack response and with the angle-of-attack initial condition.

Elasticity is seen to affect both the steady-state and the transient response to an elevator step (Fig. 5.6-9). Wing bending has minimal effect on the responses. Wing twisting causes steady-state pitch rate to increase and angle of attack to decrease. Fuselage bending reduces the steady-state response of both variables and increases the short period oscillation as well. Elastic oscillation is apparent only in the wing twisting case.

References

A-1 Abzug, M. J., and Larrabee, E. E., *Airplane Stability and Control*, Cambridge University Press, Cambridge, 2002.

A-2 Anderson, S. B., and Bray, R. S., "A Flight Evaluation of the Longitudinal Stability Characteristics Associated with the Pitch-Up of a Swept-Wing Airplane in Maneuvering Flight at Transonic Speeds," NACA Report 1237, Washington, D.C., 1955.

A-3 Anon., "Military Specification, Flying Qualities of Piloted Airplanes," MIL-F-8785C, Wright-Patterson Air Force Base, Ohio, Nov. 1980.

B-1 Bar-Gill, A., and Stengel, R. F., "Longitudinal Flying Qualities Criteria for Single-Pilot Instrument Flight Operations," *J. Aircraft* 23, no. 2 (1986): 111–117.

B-2 Beaufrere, H. L., Stratton, D. A., and Shoeder, S., "Control Power Requirements for Statically Unstable Aircraft," AFWAL-TR-87-3018, Wright-Patterson AFB, Ohio, June 1987.

B-3 Bihrle, W., "A Handling Qualities Theory for Precise Flight Path Control," AFFDL-TR-65-198, Wright-Patterson AFB, Ohio, 1966.

B-4 Blakelock, J. H., *Automatic Control of Aircraft and Missiles*, J. Wiley & Sons, New York, 1991.

B-5 Britt, R. T., Jacobson, S. B., and Arthurs, T. D., "Aeroservoelastic Analysis of the B-2 Bomber," *J. Aircraft* 37, no. 5 (2000): 745–753.

B-6 Bryan, G. H., and Williams, W. E., "The Longitudinal Stability of Aerial Gliders," *Proc. Royal Society, Ser. A.* 73, no. 489 (1903).

B-7 Burcham, F. W., Jr., Maine, T. A., Burken, J. J., and Bull, J., "Using Engine Thrust for Emergency Flight Control: MD-11 and B-747 Results," NASA/TM-1998-206552, Washington, D.C., May 1998.

C-1 Chalk, C. R., "Additional Flight Evaluations of Various Longitudinal Handling Qualities in a Variable-Stability Jet Fighter, WADC-TR-57-519, Wright-Patterson AFB, Ohio, 1958.

C-2 Chalk, C. R., "Flying Qualities of Pitch Rate Command/Attitude Hold Control Systems for Landing," *J. Guidance, Control, and Dynamics* 9, no. 5 (1986): 541–545.

C-3 Crimaldi, J. P., Britt, R. T., and Rodden, W. P., "Response of B-2 Aircraft to Nonuniform Spanwise Turbulence," *J. Aircraft* 30, no. 5 (1993): 652–659.

D-1 DiFranco, D. A., "Higher-Order Control System Dynamics and Longitudinal Handling Qualities," *J. Aircraft* 7, no. 5 (1970): 457–464.

E-1 Etkin, B., *Dynamics of Atmospheric Flight*, J. Wiley & Sons, New York, 1972.

G-1 Gates, S. B., "A Survey of Longitudinal Stability Below the Stall," Aeronautical Research Council R & M 1118, London, 1927.

G-2 Gates, S. B., "Proposal for an Elevator Maneouverability Criterion," ARC R & M 2677, London, 1942.

G-3 Gibson, J. C., *The Definition, Understanding and Design of Aircraft Handling Qualities*, Delft University Press, Delft, The Netherlands, 1997.

G-4 Gilbert, W. P., Nguyen, L. T., and Van Gunst, R. W., "Simulator Study of the Effectiveness of an Automatic Control System Designed to Improve the High-Angle-of-Attack Characteristics of a Fighter Airplane," NASA TN-D8176, Washington, D.C., May 1976.

G-5 Gilruth, R. R., "Analysis and Prediction of Longitudinal Stability of Airplanes," NACA Report 711, Washington, D. C., Dec. 1940.

H-1 Heeg, J., Gilbert, M. G., and Pototzky, A. S., "Active Control of Aerothermoelastic Effects for a Conceptual Hypersonic Aircraft," *J. Aircraft* 30, no. 4 (1993): 453–458.

H-2 Hodkginson, J., and Mitchell, D., "Handling Qualities," in *Flight Control Systems*, R. W. Pratt, ed., American Institute of Aeronautics and Astronautics, Reston, Va. 2000, pp. 119–169.

H-3 Hoh, R. H., and Mitchell, D. G., "Handling-Qualities Specification - A Functional Requirement for the Flight Control System," in *Advances in Aircraft Flight Control*, M. B. Tischler, ed., Taylor & Francis, London, 1996, pp. 3–34.

H-4 Hood, M. J., and Allen, H. J., "The Problem of Longitudinal Stability and Control at High Speeds," NACA Report, Washington, D.C., 1943.

K-1 Kemp, W. B., Jr., "Definition and Application of Longitudinal Stability Derivatives for Elastic Airplanes," NASA TN D-6629, Washington, D.C., Mar. 1972.

K-2 Kendall, E. R., "The Design and Development of Flying Qualities for the C-17 Military Transport Airplane," in *Advances in Aircraft Flight Control*, M. B. Tischler, ed., Taylor & Francis, London, 1996, pp. 189–210.

L-1 Lanchester, F. W., *Aerodynamics*, Arnold Constable, London, 1907.

M-1 Mazza, C. J., et al., "Proposal for a Revised Military Specification, 'Flying Qualities of Piloted Airplanes' (MIL F-8785ASG), with Substantiating Text," NADC Report ED-6282, Warminster, Pa., Jan. 1963.

M-2 McLean, D., *Automatic Flight Control Systems*, Prentice-Hall, Englewood Cliffs, N.J., 1990.

M-3 McRuer, D., Ashkenas, I., and Graham, D., *Aircraft Dynamics and Automatic Control*, Princeton University Press, Princeton, N.J., 1973.

M-4 Miller, G. D., Wykes, J. H., and Brosnan, M. J., "Rigid-Body-Structural Mode Coupling on a Forward Swept Wing Aircraft," AIAA Paper No. 82-0683, May 1982.

M-5 Mulgund, S., and Stengel, R. F., "Target Pitch Angle for the Microburst Escape Maneuver," *J. Aircraft* 30, no. 6 (1993): 826–832.

M-6 Mulgund, S., and Stengel, R. F., "Optimal Recovery from Microburst Wind Shear," *J. Guidance, Control, and Dynamics* 16, no. 6 (1993): 1010–1017.

M-7 Mulgund, S., and Stengel, R. F., "Aircraft Flight Control in Wind Shear Using Sequential Dynamic Inversion," *J. Guidance, Control, and Dynamics* 18, no. 5 (1995): 1084–1091.

M-8 Murphy, M., and Stengel, R. F., "Flight Evaluation of Longitudinal Flying Qualities Parameters," AIAA-85-1789CP, New York, Aug. 1985.

N-1 Neal, T. P., and Smith, R. E., "A Flying Qualities Criterion for the Design of Fighter Flight-Control Systems," *J. Aircraft* 8, no. 10 (1971): 803–809.

N-2 Neumark, S., "Longitudinal Stability, Speed and Height," *Aircraft Engineering* 22, Nov. (1950): 323–334.

N-3 Newell, F., and Campbell, G., "Flight Evaluation of Variable Short Period and Phugoid Characteristics in a B-26," WADC-TR-54-594, Wright-Patterson AFB, Ohio, 1954.

O-1 O'Hara, F., "Handling Criteria," *J. Royal Aeronautical Society* 71, no. 676 (1967): 271–291.

O-2 Olsen, J. J., "Coupled Flight Mechanics and Aeroelasticity—Some Effects of Aircraft Maneuvers on Aeroelastic Divergence and Flutter, ICAS-94-9.4.3, Sept. 1994.

P-1 Patil, M. J., Hodges, D. H., and Cesnik, C. E. S., "Nonlinear Aeroelasticity and Flight Dynamics of High-Altitude Long-Endurance Aircraft," *J. Aircraft* 38, no. 1 (2001): 88–94.

P-2 Paulson, J. W., Jr., Kemmerly, G. T., and Gilbert, W. P., "Dynamic Ground Effects," *Aerodynamics of Combat Aircraft Controls and of Ground Effects*, AGARD-CP-465, Neuilly-sur-Seine, April 1990, pp. 21-1–21-12.

P-3 Perkins, C. D., "Development of Airplane Stability and Control Technology," *J. Aircraft* 7, no. 4 (1970): 290–301.

P-4 Powers, B. G., "Space Shuttle Longitudinal Flying Qualities," *J. Guidance, Control, and Dynamics* 9, no. 5 (1986): 566–572.

P-5 Pratt, R., ed., *Flight Control Systems*, American Institute of Aeronautics and Astronautics, Reston, Va., 2000.

P-6 Psiaki, M. L., and Stengel, R. F., "Analysis of Aircraft Control Strategies for Microburst Encounters," *J. Guidance, Control, and Dynamics* 8, no. 5 (1985): 553–559.

P-7 Psiaki, M., and Stengel, R. F., "Optimal Flight Paths Through Microburst Wind Profiles," *J. Aircraft* 23, no. 8 (1986): 629–635.

P-8 Psiaki, M., and Stengel, R. F., "Optimal Aircraft Performance During Microburst Encounter," *J. Guidance, Control, and Dynamics* 14, no. 2 (1991): 440–446.

R-1 Raney, D. L., McMinn, J. D., and Pototzky, A. S., "Impact of Aeroelastic-Propulsive Interactions on Flight Dynamics of a Hypersonic Vehicle," *J. Aircraft* 32, no. 2 (1995): 355–362.

S-1 Sachs, G., "Effects of Thrust/Speed Dependence on Long-Period Dynamics in Supersonic Flight," *J. Guidance, Control, and Dynamics* 13, no. 6 (1990): 1163–1186.

S-2 Scanlan, R. H., and Rosenbaum, R., *Introduction to the Study of Aircraft Vibration and Flutter*, Dover Publications, New York, 1968.

S-3 Scheubel, F. N., "The Effect of Density Gradient on the Longitudinal Motion of an Aircraft," *Luftfahrtforschung* 19, no. 4 (1942): 132–136, R.T.P. Translation 1739.

S-4 Seckel, E., *Stability and Control of Airplanes and Helicopters*, Academic Press, New York, 1964.

S-5 Smith, R. H., "A Theory for Longitudinal Short-Period Pilot Induced Oscillations," AFFDL-TR-77-57, Wright-Patterson AFB, Ohio, June 1977.

S-6 Smith, R. H., and Geddes, N. D., "Handling Qualities Requirements for Advanced Aircraft Design: Longitudinal Mode," AFFDL-TR-78-154, Wright-Patterson AFB, Ohio, 1979.

S-7 Soulé, H. A., and Wheatley, J. B., "A Comparison between the Theoretical and Measured Longitudinal Stability Characteristics of an Airplane," NACA Report 442, Washington, D.C., 1932.

S-8 Soulé, H. A., "Flight Measurements of the Dynamic Longitudinal Stability of Several Airplanes, and a Correlation of the Measurements with Pilots' Observations of Handling Characteristics," NACA Report 578, Washington, D.C., 1936.

S-9 Stengel, R. F., "Wind Profile Measurement Using Lifting Sensors," *J. Spacecraft and Rockets* 3, no. 3 (1966): 365–373.

S-10 Stengel, R. F., "Altitude Stability in Supersonic Cruising Flight," *J. Aircraft* 7, no. 5 (1970): 464–473.

S-11 Stengel, R. F., "Optimal Transition from Entry to Cruising Flight," *J. Spacecraft and Rockets* 8, no. 11 (1971): 1126–1132.

S-12 Stengel, R. F., "Strategies for Control of the Space Shuttle Transition," *J. Spacecraft and Rockets* 10, no. 1 (1973): 77–84.

S-13 Stengel, R. F., "Digital Flight Control Research Using Microprocessor Technology," *IEEE Trans. Aerospace and Electronic Systems* AES-15, no. 3 (1979): 397–404.

S-14 Stengel, R. F., and Nixon, W. B., "Stalling Characteristics of a General Aviation Aircraft," *J. Aircraft* 19, no. 6 (1982): 425–434.

S-15 Stengel, R. F., "A Unifying Framework for Longitudinal Flying Qualities Criteria," *J. Guidance, Control, and Dynamics* 6, no. 2 (1983): 84–90.

S-16 Stengel, R. F., *Optimal Control and Estimation*, Dover Publications, New York, 1994 (originally published as *Stochastic Optimal Control: Theory and Application*, J. Wiley & Sons, New York, 1986).

T-1 Tobie, H. N., Elliott, E. M., and Malcolm, L. G., "A New Handling Qualities Criterion," *Proc. Nat'l. Aerospace Electronics Conf.*, Dayton, Ohio, May 1966.

V-1 Vincenti, W. G., *What Engineers Know and How They Know It*, Johns Hopkins University Press, Baltimore, 1990.

W-1 Weisshaar, T. A., and Zeiler, T. A., "Dynamic Stability of Flexible Forward Swept Wing Aircraft," *J. Aircraft* 20, no. 12 (1983): 1014–1020.

W-2 Williams, J., "The USAF Stability and Control Digital DATCOM," AFFDL-TR-76-45, Wright-Patterson AFB, Ohio, Nov 1976.

Y-1 Young, L. R., "The Current Status of Vestibular System Models," *Automatica* 5, no. 3 (1969): 369–383.

Z-1 Zimmerman, C. H., "An Analysis of Longitudinal Stability in Power-Off Flight with Charts for Use in Design," NACA Report 521, Washington, D.C., 1935.

Chapter 6

Lateral-Directional Motions

For a symmetric flight condition, the lateral-directional and longitudinal motions are uncoupled to first order. Small yawing and rolling motions that occur in steady, level flight can be approximated using linear models that account for lateral-directional dynamics while neglecting longitudinal motions. Whereas large-amplitude longitudinal motions can occur in the vertical plane without disturbing the lateral-directional state, large-amplitude lateral-directional motions perturb the longitudinal axes through nonlinear effects. Although not perturbing lateral-directional motion, a steady vertical pullup does modify the lateral-directional eigenvalues. The lateral-directional stability and control derivatives depend on the longitudinal trimmed state through airspeed, altitude, and angle of attack. For example, a large bank angle reorients the lift vector, disturbing the lift-weight equilibrium and requiring either that the longitudinal control settings change or that the aircraft lose altitude. Therefore, the range over which linear, time-invariant, lateral-directional models are applicable is limited.

Typically, there are three interesting lateral-directional modes of motion (the spiral, roll, and Dutch roll modes) plus two integrations to produce yaw angle and lateral position. While the longitudinal modes often are well separated by time scales, the roll and Dutch roll modes usually occur on similar time scales. Furthermore, yawing motion induces differential lift on the wing, leading to rolling motion. The stability and control derivatives are sensitive to angle of attack, producing significant variations of lateral-directional characteristics with flight condition. In high-angle-of-attack flight at low airspeed, the roll and spiral modes may coalesce to produce a slow, oscillatory ("lateral phugoid") mode with decidedly nonstandard response characteristics. Consequently, most of the lateral-directional state elements are engaged in each of the three important modes.

With these caveats in mind, we begin our study of lateral-directional motions using simplified models. After presenting the complete linearized equations in Section 6.1, we examine reduced-order models of yawing motion and of rolling motion in Sections 6.2 and 6.3. Additional reduced-order and coupled fourth-order models are examined in Section 6.4. Lateral-directional control mechanisms and aerodynamics not only drive yawing and rolling motions, but in the controls-free case, they can couple into the rigid-body motions to produce limit-cycle oscillations. These control effects provide a good opportunity to compare nonlinear, time-delayed, and higher-order linear dynamic models (Section 6.5). The effects of asymmetric aeroelastic vibrations on lateral-directional motions are presented in Section 6.6.

6.1 Lateral-Directional Equations of Motion

The *lateral-directional equations of motion* are drawn from the flat-earth equation summary of Section 3.2. They describe changes in lateral velocity v and roll and yaw rates p and r in the body-axis frame of reference. The roll and yaw angles ϕ and ψ orient the body axes with respect to the inertial frame, and the translational position is expressed by the cross-range y_I. The six nonlinear ordinary differential equations are

$$\dot{v} = Y/m + g \sin\phi \cos\theta - ru + pw \tag{6.1-1}$$

$$\dot{p} = (I_{zz}L + I_{xz}N - [I_{xz}(I_{yy} - I_{xx} - I_{zz})p + [I_{xz}^2 + I_{zz}(I_{zz} - I_{yy})]r\}q) \div (I_{xx}I_{zz} - I_{xz}^2) \tag{6.1-2}$$

$$\dot{r} = (I_{xz}L + I_{xx}N + \{I_{xz}(I_{yy} - I_{xx} - I_{zz})r + [I_{xz}^2 + I_{xx}(I_{xx} - I_{yy})]\,p\}q) \div (I_{xx}I_{zz} - I_{xz}^2) \tag{6.1-3}$$

$$\dot{\phi} = p + (q \sin\phi + r \cos\phi)\tan\theta \tag{6.1-4}$$

$$\dot{\psi} = (q \sin\phi + r \cos\phi)\sec\theta \tag{6.1-5}$$

$$\dot{y}_I = (\cos\theta \sin\psi)u + (\cos\phi \cos\psi + \sin\phi \sin\theta \sin\psi)v + (-\sin\phi \cos\psi + \cos\phi \sin\theta \sin\psi)w \tag{6.1-6}$$

The body-axis force and moments are expressed as sums of aerodynamic and thrusting effects:

$$Y = C_Y \bar{q} S + T_y \tag{6.1-7}$$

$$L = C_l \bar{q} S b + M_{T_x} \tag{6.1-8}$$

$$N = C_n \bar{q} S b + M_{T_z} \tag{6.1-9}$$

As a consequence of coupling induced by the product of inertia (I_{xz}), the roll acceleration ($\dot{p}$) is affected by the yawing moment (N) and the yaw acceleration ($\dot{r}$) is affected by the rolling moment (L).

Following Section 4.1, local linearization of the nonlinear equations produces a vector equation of the form

$$\Delta\dot{\mathbf{x}}_{\mathrm{LD}} = \mathbf{F}_{\mathrm{LD}}\Delta\mathbf{x}_{\mathrm{LD}} + \mathbf{G}_{\mathrm{LD}}\Delta\mathbf{u}_{\mathrm{LD}} + \mathbf{L}_{\mathrm{LD}}\Delta\mathbf{w}_{\mathrm{LD}} \tag{6.1-10}$$

where the sensitivity matrices are given by eqs. 4.1-107–4.1-109. For the present analysis, it is convenient to make three modifications. The yaw angle and cross-range are neglected, as they play no part in defining lateral-directional dynamics. The lateral velocity is replaced by the no-wind sideslip angle

$$\beta = \sin^{-1}(v/V) \tag{6.1-11}$$

so that all the elements of the (4×1) vector $\Delta\mathbf{x}_{\mathrm{LD}}$ are angles or angular rates. The body-axis equations are transformed through the nominal angle of attack α_o to stability axes. The stability x axis is aligned with the velocity vector, while the y axis is unchanged. Hence, the roll rate and angle are measured about the velocity vector, and the axis of yaw rotation is realigned by α_o. As a result, the stability-axis roll angle and bank angle are the same.

At a high angle of attack, there is a significant difference between the body-axis and stability-axis definitions of roll and yaw angles and rates. The linearized state, control, and disturbance vectors are

$$\Delta\mathbf{x}_{\mathrm{LD}}^{\mathrm{T}} = [\Delta r \; \Delta\beta \; \Delta p \; \Delta\phi]^{\mathrm{T}} \tag{6.1-12}$$

$$\Delta\mathbf{u}_{\mathrm{LD}}^{\mathrm{T}} = [\Delta\delta A \; \Delta\delta R \; \Delta\delta SF]^{\mathrm{T}} \tag{6.1-13}$$

$$\Delta\mathbf{w}_{\mathrm{LD}}^{\mathrm{T}} = [\Delta\beta_w \; \Delta p_w]^{\mathrm{T}} \tag{6.1-14}$$

The state elements are reordered, the control vector contains aileron, rudder, and side-force effector, and the disturbance vector includes both side and vortical wind inputs. From eqs. 4.1-125–4.1-127, the corresponding sensitivity matrices are

$$\mathbf{F}_{\mathrm{LD}} = \begin{bmatrix} \left[N_r + \dfrac{(Y_r - V_o)N_{\dot{\beta}}}{V_o}\right] & \left[N_\beta + \dfrac{Y_\beta N_{\dot{\beta}}}{V_o}\right] & \left[N_p + \dfrac{Y_p N_{\dot{\beta}}}{V_o}\right] & N_\phi \\ \dfrac{(Y_r - V_o)}{V_o} & \dfrac{Y_\beta}{V_o} & \dfrac{Y_p}{V_o} & \dfrac{g_o \cos\gamma_o}{V_o} \\ L_r & L_\beta & L_p & L_\phi \\ -\sin\gamma_o & 0 & 1 & 0 \end{bmatrix} \tag{6.1-15}$$

$$\mathbf{G}_{\mathrm{LD}} = \begin{bmatrix} \left[N_{\delta A} + \dfrac{Y_{\delta A} N_{\dot{\beta}}}{V_o}\right] & \left[N_{\delta R} + \dfrac{Y_{\delta R} N_{\dot{\beta}}}{V_o}\right] & \left[N_{\delta SF} + \dfrac{Y_{\delta SF} N_{\dot{\beta}}}{V_o}\right] \\ \dfrac{Y_{\delta A}}{V_o} & \dfrac{Y_{\delta R}}{V_o} & \dfrac{Y_{\delta SF}}{V_o} \\ L_{\delta A} & L_{\delta R} & L_{\delta SF} \\ 0 & 0 & 0 \end{bmatrix} \tag{6.1-16}$$

$$\mathbf{L}_{\mathrm{LD}} = \begin{bmatrix} -\left[N_{\beta} + \dfrac{Y_{\beta} N_{\dot{\beta}}}{V_o}\right] & -\left[N_{p} + \dfrac{Y_{p} N_{\dot{\beta}}}{V_o}\right] \\ \dfrac{-Y_{\beta}}{V_o} & \dfrac{-Y_{p}}{V_o} \\ -L_{\beta} & -L_{p} \\ 0 & 0 \end{bmatrix} \tag{6.1-17}$$

These sensitivity matrices are evaluated at a symmetric nominal flight condition defined by the longitudinal trim state. N_{ϕ} and L_{ϕ} are zero, except when the aircraft is in ground effect.

The stability and control derivatives of eqs. 6.1-15–6.1-17 are expressed in the stability-axis frame of reference; however, the aerodynamic data are likely to be presented in the body-axis frame. A similarity transformation (Section 3.2) through the nominal angle of attack relates the two reference frames. Denoting the stability-axis and body-axis derivatives by subscripts S and B, the roll-rate, sideslip-angle, and yaw-rate effects are transformed by applying eq. 3.2-26 to $\mathbf{F}_{\mathrm{LD}}$ (eq. 6.1-15) and $\mathbf{H}_{\mathrm{B}}^{\mathrm{S}}$ (eq. 3.2-85):

$$N_{r_S} = \cos^2\alpha_o N_{r_B} - \sin\alpha_o \cos\alpha_o (L_{r_B} + N_{p_B}) + \sin^2\alpha_o L_{p_B} \tag{6.1-18}$$

$$N_{\beta_S} = \cos\alpha_o N_{\beta_B} - \sin\alpha_o L_{\beta_B} \tag{6.1-19}$$

$$N_{p_S} = \cos^2\alpha_o N_{p_B} + \sin\alpha_o \cos\alpha_o (N_{r_B} - L_{p_B}) - \sin^2\alpha_o L_{r_B} \tag{6.1-20}$$

$$Y_{r_S} = \cos\alpha_o Y_{r_B} - \sin\alpha_o Y_{p_B} \tag{6.1-21}$$

$$Y_{\beta_S} = Y_{\beta_B} \tag{6.1-22}$$

$$Y_{p_S} = \sin\alpha_o Y_{r_B} + \cos\alpha_o Y_{p_B} \tag{6.1-23}$$

$$L_{r_S} = \cos^2\alpha_o L_{r_B} + \sin\alpha_o \cos\alpha_o (N_{r_B} - L_{p_B}) - \sin^2\alpha_o N_{p_B} \tag{6.1-24}$$

$$L_{\beta_S} = \cos\alpha_o L_{\beta_B} + \sin\alpha_o N_{\beta_B} \tag{6.1-25}$$

$$L_{p_S} = \cos^2 \alpha_o L_{p_B} + \sin \alpha_o \cos \alpha_o (L_{r_B} + N_{p_B}) + \sin^2 \alpha_o N_{r_B} \quad (6.1\text{-}26)$$

$$N_{\delta A_S} = \cos \alpha_o N_{\delta A_B} - \sin \alpha_o L_{\delta A_B} \quad (6.1\text{-}27)$$

$$Y_{\delta A_S} = Y_{\delta A_B} \quad (6.1\text{-}28)$$

$$L_{\delta A_S} = \cos \alpha_o L_{\delta A_B} + \sin \alpha_o N_{\delta A_B} \quad (6.1\text{-}29)$$

The $\dot{\beta}$ derivatives transform in the same way as the β derivatives, and the rudder and side-force effects transform in the same way as the aileron effects. It is clear that there is a major difference between body-axis and stability-axis stability and control derivatives in high angle-of-attack flight. The transformation gives rise to an important simplified indicator of Dutch roll static stability called "$C_{n_{\beta}}$ dynamic" [M-3]. From eq. 6.1-19,

$$C_{n_{\beta_{dyn}}} \triangleq C_{n_\beta} - \frac{I_{zz}}{I_{xx}} \alpha_o C_{l_\beta} \quad (6.1\text{-}30)$$

For the hybrid-axis system, the stability matrix $\mathbf{F}_{LD}$ is organized in (2×2) blocks of the general form shown in Table 6.1-1. The yaw parameters form the basis for the Dutch roll mode, while the roll parameters underlie the roll and spiral modes.

6.2 Reduced-Order Model of the Dutch Roll Mode

The yawing motions of the aircraft are analogous to the pitching motions, in that both are angular rotations about axes perpendicular to the forward axis. However, the lift due to the wing produces fundamentally different effects in the two cases. The lift is the largest contact force on the aircraft, equal to the weight in steady cruising flight; therefore, small perturbations of the lifting force and its distribution have large effects on the aircraft's motion. Symmetric effects produce heaving or plunging motion that couples with the pitching motion to shape the short-period mode. Differential effects produce asymmetric lift, inducing rolling moment (and, to a lesser extent, yawing moment), thereby coupling the yawing motion with rolling motion. Lateral heaving is small because the side

Table 6.1-1 Four-Block Structure of the Fourth-Order Lateral-Directional Dynamic Model

Yaw Parameters	Roll-to-Yaw Coupling
Yaw-to-Roll Coupling	**Roll Parameters**

force is small. Thus, the oscillatory mode that results—called the *Dutch roll mode* from the analogy to a skater smoothly swaying to and fro along a frozen canal in the Netherlands—remains largely rotational and occurs about both yawing and rolling axes. We could, of course, diagonalize the system matrix to find a single axis for the rotation. The axis would lie between the rolling and yawing axes in the aircraft's xz plane.

In spite of the strong yaw-roll coupling, we first look at a weathervane-like model of the Dutch roll mode that ignores rolling motion. This approach introduces the principal parameters of the mode and the main effects of its controls, rudder, and side-force panels, before engaging in coupling issues. To date, only experimental aircraft like Princeton University's Variable-Response Research Aircraft (VRA) the Veridian Total In-Flight Simulator (TIFS), and the USAF AFTI F-16, have been equipped with side-force effectors. Both the VRA and the TIFS have large, movable vertical panels installed at the mid-span of left and right wings. The AFTI F-16 had vertical panels mounted below its air intake.

The system model includes elements from the first and second rows of $\Delta\mathbf{x}_{LD}$, $\mathbf{F}_{LD}$, $\mathbf{G}_{LD}$, and $\mathbf{L}_{LD}$. The state equation takes the form

$$\Delta\dot{\mathbf{x}}_{DR} = \mathbf{F}_{DR}\Delta\mathbf{x}_{DR} + \mathbf{G}_{DR}\Delta\mathbf{u}_{DR} + \mathbf{L}_{DR}\Delta\mathbf{w}_{DR} \tag{6.2-1}$$

where the state vector contains the yaw rate and no-wind sideslip angle:

$$\Delta\mathbf{x}_{DR} = \begin{bmatrix} \Delta r \\ \Delta\beta \end{bmatrix} \tag{6.2-2}$$

The rudder is the principal control variable, and a notional side-force effector, equivalent to a lateral flap, is included. A horizontal-wind perturbation expressed as equivalent sideslip angle provides the most significant disturbance:

$$\Delta\mathbf{u}_{DR} = \begin{bmatrix} \Delta\delta R \\ \Delta\delta SF \end{bmatrix} \tag{6.2-3}$$

$$\Delta\mathbf{w}_{DR} = \Delta\beta_W \tag{6.2-4}$$

The Dutch roll stability-and-control-derivative matrices are

$$\mathbf{F}_{DR} = \begin{bmatrix} \left[N_r + \dfrac{(Y_r - V_o)N_{\dot\beta}}{V_o} \right] & \left(N_\beta + \dfrac{Y_\beta N_{\dot\beta}}{V_o} \right) \\ \dfrac{(Y_r - V_o)}{V_o} & \dfrac{Y_\beta}{V_o} \end{bmatrix} \tag{6.2-5}$$

$$\mathbf{G}_{\mathrm{DR}} = \begin{bmatrix} \left(N_{\delta R} + \dfrac{Y_{\delta R} N_{\dot{\beta}}}{V_o}\right) & 0 \\ 0 & \dfrac{Y_{\delta SF}}{V_o} \end{bmatrix} \tag{6.2-6}$$

$$\mathbf{L}_{\mathrm{DR}} = \begin{bmatrix} -\left(N_{\beta} + \dfrac{Y_{\beta} N_{\dot{\beta}}}{V_o}\right) \\ \dfrac{-Y_{\beta}}{V_o} \end{bmatrix} \tag{6.2-7}$$

Here, we idealize the control effects: the rudder is assumed to produce only yawing moment, and the side-force effector generates only side force. $N_{\dot{\beta}}$ and Y_r are generally small and, except as noted, are neglected in the analyses that follow.

Equilibrium Response to Control and Disturbance

The equilibrium yaw-rate and sideslip-angle perturbations produced by step changes in the control and disturbance are given by

$$\Delta\mathbf{x}^*_{\mathrm{DR}} = -\mathbf{F}^{-1}_{\mathrm{DR}}(\mathbf{G}_{\mathrm{DR}}\Delta\mathbf{u}^*_{\mathrm{DR}} + \mathbf{L}_{\mathrm{DR}}\Delta\mathbf{w}^*_{\mathrm{DR}}) \tag{6.2-8}$$

With $N_{\dot{\beta}} = Y_r = 0$, this becomes

$$\begin{bmatrix} \Delta r^* \\ \Delta\beta^* \end{bmatrix} = -\frac{\begin{bmatrix} \dfrac{Y_{\beta}}{V_o} & -N_{\beta} \\ 1 & N_r \end{bmatrix}}{\left(N_{\beta} + N_r \dfrac{Y_{\beta}}{V_o}\right)} \left\{ \begin{bmatrix} N_{\delta R} & 0 \\ 0 & \dfrac{Y_{\delta SF}}{V_o} \end{bmatrix} \begin{bmatrix} \Delta\delta R^* \\ \Delta\delta SF^* \end{bmatrix} - \begin{bmatrix} N_{\beta} \\ \dfrac{Y_{\beta}}{V_o} \end{bmatrix} \Delta\beta^*_W \right\} \tag{6.2-9}$$

leading to the scalar equations

$$\Delta r^* = \frac{-1}{\left(N_{\beta} + N_r \dfrac{Y_{\beta}}{V_o}\right)} \left(N_{\delta R} \frac{Y_{\beta}}{V_o} \Delta\delta R^* - N_{\beta} \frac{Y_{\delta SF}}{V_o} \Delta\delta SF^* \right) \tag{6.2-10}$$

$$\Delta\beta^* = \frac{-1}{\left(N_{\beta} + N_r \dfrac{Y_{\beta}}{V_o}\right)} \left[N_{\delta R}\Delta\delta R^* + N_r \frac{Y_{\delta SF}}{V_o} \Delta\delta SF^* - \left(N_{\beta} + N_r \frac{Y_{\beta}}{V_o}\right) \Delta\beta^*_W \right] \tag{6.2-11}$$

We see that a *skid turn* (i.e., a turn without rolling) can be commanded by either the rudder or side-force panels, as a steady yaw rate accompanied by sideslip results from a step control. The relative magnitude of the two responses depends on the stability and control derivatives. The side-force panels on Princeton's Variable-Response Research Aircraft (Section 1.2) can produce a lateral load factor of 0.5 g at 105 kt airspeed, producing a much larger turn rate than that produced by the rudder alone.

Rudder and side-force commands can be combined to produce *control-configured-vehicle* (CCV) response characteristics. The proper blend of steady rudder and side-force deflections produces equilibrium yaw rate without sideslip or the converse. (What control ratios produce these results with idealized rudder and side-force effects?)

A side wind induces a no-wind sideslip angle that opposes the wind-induced sideslip. The net air-relative sideslip angle (the sum of the two) is zero; hence, the aircraft stabilizes with the nose into the relative wind, and (earth-relative) side and wind velocities are the same.

Controllability and Observability

It is left as an exercise to show that the reduced-order mode is controllable with either control and observable with measurement of either Δr or $\Delta\beta$.

Eigenvalues, Natural Frequency, and Damping Ratio

The characteristic matrix is

$$(s\mathbf{I}_2 - \mathbf{F}_{DR}) = \begin{bmatrix} s - \left[N_r + \dfrac{(Y_r - V_o)N_{\dot\beta}}{V_o}\right] & -\left(N_\beta + \dfrac{Y_\beta N_{\dot\beta}}{V_o}\right) \\ 1 - \dfrac{Y_r}{V_o} & s - \dfrac{Y_\beta}{V_o} \end{bmatrix} \tag{6.2-12}$$

and, with $N_{\dot\beta} = Y_r = 0$, the characteristic equation is

$$\Delta(s) = |s\mathbf{I}_2 - \mathbf{F}_{DR}| = s^2 - s + \left(N_\beta + N_r\frac{Y_\beta}{V_o}\right) = 0 \tag{6.2-13}$$

The similarity to the short-period characteristic equation is apparent (eq. 5.3-13), although oscillatory roots are produced by negative M_α and by positive N_β. The Dutch roll eigenvalues are estimated as

$$\lambda_{1,2} = \left(\frac{N_r + \dfrac{Y_\beta}{V_o}}{2}\right) \pm \sqrt{\left(\frac{N_r + \dfrac{Y_\beta}{V_o}}{2}\right)^2 - \left(N_\beta + N_r\frac{Y_\beta}{V_o}\right)} \tag{6.2-14a}$$

In most instances, the Dutch roll mode is made statically stable by putting a large enough vertical tail on the aircraft. N_β is positive, the square root is negative, and the roots form a complex pair:

$$\lambda_{1,2} = \left(\frac{N_r + \frac{Y_\beta}{V_o}}{2} \right) \pm j \sqrt{ - \left(\frac{N_r + \frac{Y_\beta}{V_o}}{2} \right)^2 + \left(N_\beta + N_r \frac{Y_\beta}{V_o} \right) } \qquad (6.2\text{-}14b)$$

Aircraft designed for low radar observability (i.e., stealth aircraft) may have little or no vertical tail, leading to low, neutral, or negative yaw stability. N_β may become more negative with increasing supersonic speed, as the vertical tail's center of pressure moves aft; however, unless the fin's chord section is a wedge, as for the X-15 (Section 2.4), the tail's side-force sensitivity to β decreases as the flow becomes more Newtonian, and the magnitude of N_β may decrease.

Given static directional stability, the natural frequency and damping ratio are

$$\omega_{n_{DR}} = \sqrt{N_\beta + N_r \frac{Y_\beta}{V_o}} \qquad (6.2\text{-}15)$$

$$\zeta_{DR} = \frac{-\left(N_r + \frac{Y_\beta}{V_o} \right)}{2\omega_{n_{DR}}} = \frac{-\left(N_r + \frac{Y_\beta}{V_o} \right)}{2\sqrt{N_\beta + N_r \frac{Y_\beta}{V_o}}} \qquad (6.2\text{-}16)$$

N_β is typically dominant in determining the Dutch roll natural frequency. N_r and Y_β/V_o are normally positive, and the mode is positively damped in these conditions.

Eigenvectors

Literal formulas for the Dutch roll eigenvectors can be found by substituting eq. 6.2-14 in eq. 4.3-61. The adjoint of the characteristic matrix is

$$\text{Adj}(s\mathbf{I}_2 - \mathbf{F}_{DR}) = \begin{bmatrix} \left(s - \frac{Y_\beta}{V_o} \right) & N_\beta \\ -1 & (s - N_r) \end{bmatrix} \qquad (6.2\text{-}17)$$

Choosing λ_1 as the root with positive imaginary part,

$$\mathrm{Adj}(\lambda_1 \mathbf{I}_2 - \mathbf{F}_{\mathrm{DR}}) = [a_1\mathbf{e}_1 \quad a_2\mathbf{e}_1] = \begin{bmatrix} \dfrac{\left(N_r - \dfrac{Y_\beta}{V_o}\right)}{2} + j\sqrt{\cdot} & N_\beta \\ -1 & \dfrac{-\left(N_r - \dfrac{Y_\beta}{V_o}\right)}{2} + j\sqrt{\cdot} \end{bmatrix} \tag{6.2-18}$$

where $\sqrt{\cdot}$ represents the radical in eq. 6.2-14b.

$\mathrm{Adj}(\lambda_2 \mathbf{I}_2 - \mathbf{F}_{\mathrm{DR}})$ is the complex conjugate of $\mathrm{Adj}(\lambda_1 \mathbf{I}_2 - \mathbf{F}_{\mathrm{DR}})$. A modal matrix can be formed using the first columns of the two adjoint matrices:

$$\mathbf{E} = \left[\mathbf{e}_1\ \mathbf{e}_2\right] = \begin{bmatrix} \dfrac{\left(N_r - \dfrac{Y_\beta}{V_o}\right)}{2} + j\sqrt{\cdot} & \dfrac{\left(N_r - \dfrac{Y_\beta}{V_o}\right)}{2} - j\sqrt{\cdot} \\ -1 & -1 \end{bmatrix} \tag{6.2-19}$$

Expressing the eigenvectors in polar form:

$$\mathbf{E} = \begin{bmatrix} A_r e^{j\phi_r} & A_r e^{-j\phi_r} \\ A_\beta e^{j\phi_\beta} & A_\beta e^{-j\phi_\beta} \end{bmatrix} \tag{6.2-20}$$

The relative magnitudes and phase angles of the eigenvectors are

$$\begin{bmatrix} A_r \\ A_\beta \end{bmatrix} = \begin{bmatrix} \sqrt{N_\beta} \\ 1 \end{bmatrix} \tag{6.2-21}$$

$$\begin{bmatrix} \phi_r \\ \phi_\beta \end{bmatrix} = \begin{bmatrix} \tan^{-1}\left[\dfrac{2\sqrt{\cdot}}{(N_r - Y_\beta/V_o)}\right] \\ 0 \end{bmatrix} \tag{6.2-22}$$

Root Locus Analysis of Parameter Variations

The Dutch roll characteristic equation

$$\Delta(s) = s^2 - \left(N_r + \frac{Y_\beta}{V_o}\right)s + \left(N_\beta + N_r \frac{Y_\beta}{V_o}\right) = 0 \tag{6.2-23}$$

is put in the form

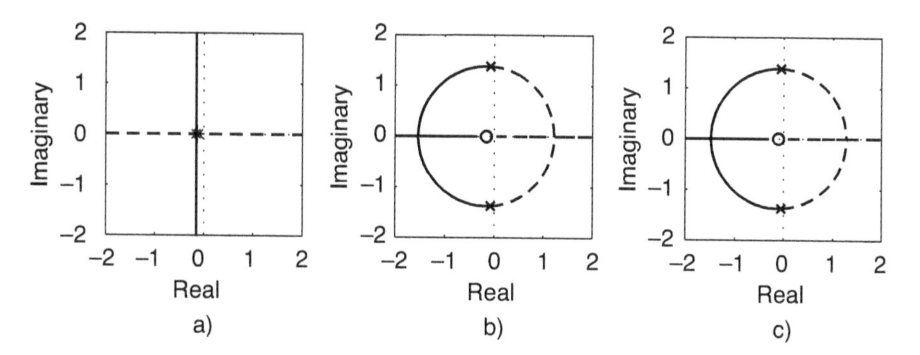

Fig. 6.2-1. Effects of variation in N_β (a), N_r (b), and Y_β/V_o (c) on the roots of the second-order Dutch roll approximation. Positive variations of N_β and negative variations of N_r and Y_β/V_o shown by solid line; opposite variations by dashed line.

$$\Delta(s) = d(s) = kn(s) = 0 \tag{6.2-24}$$

alternatively using N_β, N_r, and Y_β/V_o as the root locus gain k. To examine the N_β variation, we rewrite eq. 6.2-23 as

$$\Delta(s) = \left[s^2 - \left(N_r + \frac{Y_\beta}{V_o}\right)s + N_r\frac{Y_\beta}{V_o}\right] + N_\beta = 0 \tag{6.2-25}$$

$d(s)$ factors as

$$d(s) = (s - N_r)(s - Y_\beta/V_o) \tag{6.2-26}$$

With zero N_β, the Dutch roll roots are real. Figure 6.2-1a is a typical root locus for positive and negative values of N_β, based on the approximation of Table 6.2-1. Positive values drive the roots together, causing them to split into a complex pair with increasing damped and undamped natural frequency, as well as decreased damping ratio. Negative values drive the smaller of the two roots into the right half plane, indicating instability.

The effect of N_r is similar to the effect of M_q on the short-period mode. The characteristic equation is partitioned as

$$\Delta(s) = \left(s^2 - \frac{Y_\beta}{V_o}s + N_\beta\right) - N_r\left(s - \frac{Y_\beta}{V_o}\right) = 0 \tag{6.2-27}$$

Table 6.2-1 Approximate Dutch Roll Stability Matrix for Figure 6.2-1

$$\mathbf{F} = \begin{bmatrix} -0.1079 & 1.9011 \\ -1 & -0.1567 \end{bmatrix}$$

$d(s)$ factors into a lightly damped complex pair, and the root locus traces a circle about the point $s = Y_\beta/V_o$ (Fig. 6.2-1b). Negative values of N_r increase the damping ratio ζ_{DR} and produce slightly larger values of $\omega_{n_{DR}}$ before driving the roots to real values. Positive values introduce oscillatory instability.

The root locus for Y_β/V_o variation is like the N_r root locus, as the characteristic equation takes the form

$$\Delta(s) = (s^2 - N_r s + N_\beta) - \frac{Y_\beta}{V_o}(s - N_r) = 0 \tag{6.2-28}$$

However, the circular locus is centered at $s = N_r$ (Fig. 6.2-1c).

Frequency Response

The control and disturbance transfer function matrices $\mathbf{U}_{\mathrm{DR}}(s)$ and $\mathbf{W}_{\mathrm{DR}}(s)$ relate the output Laplace transform $\Delta\mathbf{y}_{\mathrm{DR}}(s)$ to the inputs $\Delta\mathbf{u}_{\mathrm{DR}}(s)$ and $\Delta\mathbf{w}_{\mathrm{DR}}(s)$:

$$\begin{aligned}\Delta\mathbf{y}_{\mathrm{DR}}(s) &= \mathbf{H}_{\mathrm{DR}}(s\mathbf{I}_2 - \mathbf{F}_{\mathrm{DR}})^{-1}[\mathbf{G}_{\mathrm{DR}}\Delta\mathbf{u}_{\mathrm{DR}}(s) + \mathbf{L}_{\mathrm{DR}}\Delta\mathbf{w}_{\mathrm{DR}}(s)] \\ &\triangleq \mathbf{U}_{\mathrm{DR}}(s)\Delta\mathbf{u}_{\mathrm{DR}}(s) + \mathbf{W}_{\mathrm{DR}}(s)\Delta\mathbf{w}_{\mathrm{DR}}(s)\end{aligned} \tag{6.2-29}$$

The characteristic matrix inverse is

$$(s\mathbf{I}_2 - \mathbf{F}_{\mathrm{DR}})^{-1} = \frac{\mathrm{Adj}(s\mathbf{I}_2 - \mathbf{F}_{\mathrm{DR}})}{|s\mathbf{I}_2 - \mathbf{F}_{\mathrm{DR}}|} = \frac{\begin{bmatrix} \left(s - \dfrac{Y_\beta}{V_o}\right) & N_\beta \\ -1 & (s - N_r) \end{bmatrix}}{s^2 - \left(N_r + \dfrac{Y_\beta}{V_o}\right)s + \left(N_\beta + N_r\dfrac{Y_\beta}{V_o}\right)} \tag{6.2-30}$$

The individual scalar transfer functions are linear combinations of the elements of this matrix weighted by the elements of $\mathbf{H}_{\mathrm{DR}}$, $\mathbf{G}_{\mathrm{DR}}$, and $\mathbf{L}_{\mathrm{DR}}$.

With $\mathbf{H}_{DR} = \mathbf{I}_2$, $\mathbf{G}_{DR}$ and $\mathbf{L}_{DR}$ defined by eqs. 6.2-6 and 6.2-7, and $N_{\dot{\beta}} = 0$, the transfer function matrices are as follows:

$$\mathbf{U}_{DR}(s) = \begin{bmatrix} \dfrac{\Delta r(s)}{\Delta\delta R(s)} & \dfrac{\Delta r(s)}{\Delta\delta SF(s)} \\ \dfrac{\Delta\beta(s)}{\Delta\delta R(s)} & \dfrac{\Delta\beta(s)}{\Delta\delta SF(s)} \end{bmatrix} = \frac{\begin{bmatrix} N_{\delta R}\left(s - \dfrac{Y_\beta}{V_o}\right) & \dfrac{Y_{\delta SF}}{V_o} N_\beta \\ -N_{\delta R} & \dfrac{Y_{\delta SF}}{V_o}(s - N_r) \end{bmatrix}}{s^2 - \left(N_r + \dfrac{Y_\beta}{V_o}\right)s + \left(N_\beta + N_r \dfrac{Y_\beta}{V_o}\right)} \tag{6.2-31}$$

$$\mathbf{W}_{DR}(s) = \begin{bmatrix} \dfrac{\Delta r(s)}{\Delta\beta_W(s)} \\ \dfrac{\Delta\beta(s)}{\Delta\beta_W(s)} \end{bmatrix} = \frac{\begin{bmatrix} \left(s - \dfrac{Y_\beta}{V_o}\right) & N_\beta \\ -1 & (s - N_r) \end{bmatrix} \begin{bmatrix} -N_\beta \\ \dfrac{-Y_\beta}{V_o} \end{bmatrix}}{s^2 - \left(N_r + \dfrac{Y_\beta}{V_o}\right)s + \left(N_\beta + N_r \dfrac{Y_\beta}{V_o}\right)}$$

$$= \frac{\begin{bmatrix} -N_\beta s \\ \dfrac{-Y_\beta}{V_o}\left[s - \left(N_r + \dfrac{N_\beta}{Y_\beta / V_o}\right)\right] \end{bmatrix}}{s^2 - \left(N_r + \dfrac{Y_\beta}{V_o}\right)s + \left(N_\beta + N_r \dfrac{Y_\beta}{V_o}\right)} \tag{6.2-32}$$

The reduced-order Dutch roll transfer functions are analogous to the short-period transfer functions, and their Bode plots follow the same general trends. Reference can be made to Section 5.3 for the shapes of the plots. The yaw rate and sideslip angle respond to rudder and side wind in much the same way that pitch rate and angle of attack respond to elevator and vertical wind. The rudder-to-yaw-rate transfer function contains a zero, while the side-force-to-yaw-rate transfer function does not. Just the opposite is true for the rudder-to-sideslip and side-force-to-sideslip transfer functions. Side wind is differentiated in the yaw-rate response, with zero effect at zero frequency, while there is finite zero-frequency response in sideslip angle.

Root Locus Analysis of Feedback Control

The stability effects of feeding yaw-rate and sideslip-angle perturbations back to the rudder and side-force effector can be examined using the root locus. With $N_{\dot{\beta}} = Y_r = 0$, the root locus solves the following characteristic equations:

Yaw-Rate Feedback to Rudder

$$\Delta(s) = d(s) + kn(s) = s^2 - \left(N_r + \frac{Y_\beta}{V_o}\right)s + \left(N_\beta + N_r\frac{Y_\beta}{V_o}\right) + kN_{\delta R}\left(s - \frac{Y_\beta}{V_o}\right) = 0 \tag{6.2-33}$$

Sideslip-Angle Feedback to Rudder

$$\Delta(s) = d(s) + kn(s) = s^2 - \left(N_r + \frac{Y_\beta}{V_o}\right)s + \left(N_\beta + N_r\frac{Y_\beta}{V_o}\right) - kN_{\delta R} = 0 \tag{6.2-34}$$

Yaw-Rate Feedback to Side-Force Panels

$$\Delta(s) = d(s) + kn(s) = s^2 - \left(N_r + \frac{Y_\beta}{V_o}\right)s + \left(N_\beta + N_r\frac{Y_\beta}{V_o}\right) + k\frac{Y_{\delta SF}}{V_o}N_\beta = 0 \tag{6.2-35}$$

Sideslip-Angle Feedback to Side-Force Panels

$$\Delta(s) = d(s) + kn(s) = s^2 - \left(N_r + \frac{Y_\beta}{V_o}\right)s + \left(N_\beta + N_r\frac{Y_\beta}{V_o}\right) + k\frac{Y_{\delta SF}}{V_o}(s - N_r) = 0 \tag{6.2-36}$$

The root loci take the classical forms. With no zero (eqs. 6.2-34 and 6.2-35), the loop closure increases or decreases the damped natural frequency, while the total damping ($\zeta\omega_n$) remains constant. For sufficiently negative loop gain, the roots become real, and one eventually migrates to the right half plane. With a zero (eqs. 6.2-33 and 6.2-36), the root locus is a circle centered on the zero until the roots converge on the real axis. Then, one root approaches the zero and the other goes to ± infinity, depending on the sign of the loop gain.

Time Response

The state history is propagated by

$$\Delta\mathbf{x}_{DR}(t_k) = \mathbf{\Phi}_{DR}\Delta\mathbf{x}_{DR}(t_{k-1}) + \mathbf{\Gamma}_{DR}\Delta\mathbf{u}_{DR}(t_{k-1}) + \mathbf{\Lambda}_{DR}\Delta\mathbf{w}_{DR}(t_{k-1}) \tag{6.2-37}$$

The state transition matrix $\mathbf{\Phi}_{DR}(\Delta t)$ can be calculated as the inverse Laplace transform of the characteristic matrix inverse (eq. 6.2-30):

$$\begin{aligned}\mathbf{\Phi}_{DR}(\Delta t) &= \mathcal{L}^{-1}(s\mathbf{I}_2 - \mathbf{F}_{DR})^{-1} = \begin{bmatrix}\phi_{11} & \phi_{12}\\ \phi_{21} & \phi_{22}\end{bmatrix}\\ &= \frac{e^{-\zeta\omega_n\Delta t}}{\omega_d}\begin{bmatrix}\omega_n\cos(\omega_d\Delta t + \psi) - \dfrac{Y_\beta}{V_o}\sin\omega_d\Delta t & N_\beta\sin\omega_d\Delta t\\ -\sin\omega_d\Delta t & \omega_n\cos(\omega_d\Delta t + \psi) - N_r\sin\omega_d\Delta t\end{bmatrix}\end{aligned} \tag{6.2-38}$$

where ω_n and ζ are found from eqs. 6.2-15 and 6.2-16, and ω_d and ψ are calculated as before. The state transition relationship for initial-condition and input response is then

$$\begin{bmatrix}\Delta r(t_k)\\ \Delta\beta(t_k)\end{bmatrix} = \mathbf{\Phi}_{DR}\begin{bmatrix}\Delta r(t_{k-1})\\ \Delta\beta(t_{k-1})\end{bmatrix} + \mathbf{\Gamma}_{DR}\begin{bmatrix}\Delta\delta R(t_{k-1})\\ \Delta\delta SF(t_{k-1})\end{bmatrix} + \mathbf{\Lambda}_{DR}\Delta\beta_W^*, \quad \begin{bmatrix}\Delta r(t_0)\\ \Delta\beta(t_0)\end{bmatrix}\ \text{given} \tag{6.2-39}$$

The discrete-time control- and disturbance-effect matrices $\mathbf{\Gamma}_{DR}$ and $\mathbf{\Lambda}_{DR}$ are calculated by eqs. 4.2-9 and 4.2-10:

$$\mathbf{\Gamma}_{DR} = [\mathbf{\Phi}_{DR} - \mathbf{I}_2]\,\mathbf{F}_{DR}^{-1}\,\mathbf{G}_{DR} \tag{6.2-40}$$

$$\mathbf{\Lambda}_{DR} = [\mathbf{\Phi}_{DR} - \mathbf{I}_2]\,\mathbf{F}_{DR}^{-1}\,\mathbf{L}_{DR} \tag{6.2-41}$$

with $\mathbf{G}_{DR}$ and $\mathbf{L}_{DR}$ defined by eqs. 6.2-6 and 6.2-7.

6.3 Reduced-Order Model of Roll and Spiral Modes

The reduced-order rolling model of the aircraft produces the simplest of motions, involving only damping of the roll rate and integration to produce the corresponding bank angle (i.e., roll angle about the velocity vector). The model characterizes the *roll mode* as a first-order convergence and the *spiral mode* as a neutrally stable propagation of the roll rate.

The system model includes elements from the third and fourth rows of $\Delta\mathbf{x}_{LD}$, $\mathbf{F}_{LD}$, $\mathbf{G}_{LD}$, and $\mathbf{L}_{LD}$. The state equation takes the form

$$\Delta\dot{\mathbf{x}}_{RS} = \mathbf{F}_{RS}\Delta\mathbf{x}_{RS} + \mathbf{G}_{RS}\Delta\mathbf{u}_{RS} + \mathbf{L}_{RS}\Delta\mathbf{w}_{RS} \tag{6.3-1}$$

where the state vector contains the roll rate and roll angle:

$$\Delta\mathbf{x}_{RS} = \begin{bmatrix} \Delta p \\ \Delta\phi \end{bmatrix} \tag{6.3-2}$$

The aileron angle is the principal control variable. The angle represents a composite of up and down deflections of opposite control surfaces plus equivalent asymmetric spoiler deflections. A vortical-wind perturbation expressed as equivalent roll rate provides the most significant disturbance:

$$\Delta\mathbf{u}_{RS} = \Delta\delta A \tag{6.3-3}$$

$$\Delta\mathbf{w}_{RS} = \Delta p_W \tag{6.3-4}$$

Out of ground effect, the roll/spiral stability-and-control-derivative matrices are

$$\mathbf{F}_{RS} = \begin{bmatrix} L_p & 0 \\ 1 & 0 \end{bmatrix} \tag{6.3-5}$$

$$\mathbf{G}_{RS} = \begin{bmatrix} L_{\delta A} \\ 0 \end{bmatrix} \tag{6.3-6}$$

$$\mathbf{L}_{RS} = \begin{bmatrix} -L_p \\ 0 \end{bmatrix} \tag{6.3-7}$$

In ground effect (i.e., when the aircraft's altitude is within a span length above the surface), $\mathbf{F}_{RS}$ contains L_ϕ in the (1, 2) element.

Equilibrium Response to Control and Disturbance

We would like to express the equilibrium roll-rate and roll-angle perturbations produced by step changes in the control and disturbance as

$$\Delta\mathbf{x}^*_{RS} = -\mathbf{F}_{RS}^{-1}(\mathbf{G}_{RS}\Delta\mathbf{u}^*_{RS} + \mathbf{L}_{RS}\Delta\mathbf{w}^*_{RS}) \tag{6.3-8}$$

however, $\mathbf{F}_{RS}$ is not invertible out of ground effect, as its determinant is zero. A steady aileron input produces an equilibrium roll rate of

$$\Delta p^* = \frac{-L_{\delta A}}{L_p} \Delta\delta A^* \tag{6.3-9}$$

and the corresponding roll angle is the integral of Δp^*:

$$\Delta\phi^*(t) = -\int_0^t \frac{L_{\delta A}}{L_p} \Delta\delta A \, dt \tag{6.3-10}$$

Consequently, there is no constant-angle response to constant control. This result confirms the identification of the ailerons as a *rate* control, unlike the elevator and rudder, which produce equilibrium aerodynamic *angles* in their reduced-order models. A constant vortical wind would produce a constant roll rate of equal magnitude to the input.

In ground effect, the stability matrix becomes

$$\mathbf{F}_{RS_{GE}} = \begin{bmatrix} L_p & L_\phi \\ 1 & 0 \end{bmatrix} \tag{6.3-11}$$

and the equilibrium solution is

$$\begin{aligned} \begin{bmatrix} \Delta p^* \\ \Delta\phi^* \end{bmatrix} &= -\frac{\begin{bmatrix} 0 & -L_\phi \\ -1 & L_p \end{bmatrix}}{-L_\phi} \left\{ \begin{bmatrix} L_{\delta A} \\ 0 \end{bmatrix} \Delta\delta A^* + \begin{bmatrix} -L_p \\ 0 \end{bmatrix} \Delta p_W^* \right\} \\ &= \begin{bmatrix} 0 \\ \dfrac{-L_{\delta A}}{L_\phi} \end{bmatrix} \Delta\delta A^* + \begin{bmatrix} 0 \\ \dfrac{L_p}{L_\phi} \end{bmatrix} \Delta p_W^* \end{aligned} \tag{6.3-12}$$

Thus, for small inputs, a steady bank angle with zero roll rate results from either step aileron command or constant vortical wind when the aircraft is flying in ground effect.

Controllability and Observability

Controllability of the model using ailerons is shown by substitution of $\mathbf{G}_{RS}$ and $\mathbf{F}_{RS}$ in eq. 4.2-48. The controllability matrix is

$$\mathscr{C} = \begin{bmatrix} L_{\delta A} & L_p L_{\delta A} \\ 0 & L_{\delta A} \end{bmatrix} \tag{6.3-13}$$

The rank of $\mathscr{C}$ is seen to be 2, and the system is controllable. Observability with a roll-rate measurement alone is tested by substituting $\mathbf{H}_{RS} = [1 \; 0]$ and $\mathbf{G}_{RS}$ in eq. 4.2-49:

$$\mathcal{O} = \begin{bmatrix} 1 & 0 \\ L_p & 0 \end{bmatrix} \tag{6.3-14}$$

The observability matrix $\mathcal{O}$ does not have full rank, indicating that the roll rate measurement is not sufficient to observe the state. However, with roll angle measurement, $\mathbf{H}_{RS} = [0 \; 1]$,

$$\mathcal{O} = \begin{bmatrix} 0 & 1 \\ 1 & 0 \end{bmatrix} \tag{6.3-15}$$

the rank is 2, and the system is observable. The difference is that the roll rate can be derived from the roll angle by differentiation, whereas the roll rate must be integrated to obtain the roll angle. The latter requires a constant of integration $\Delta\phi_o$, that is not provided by the rate measurement.

Eigenvalues, Natural Frequency, and Damping Ratio

Out of ground effect, the characteristic matrix is

$$(s\mathbf{I}_2 - \mathbf{F}_{RS}) = \begin{bmatrix} (s - L_p) & 0 \\ -1 & s \end{bmatrix} \tag{6.3-16}$$

The corresponding characteristic equation is

$$\Delta(s) = |s\mathbf{I}_2 - \mathbf{F}_{RS}| = s(s - L_p) = s^2 - L_p s = 0 \tag{6.3-17}$$

and the eigenvalues are seen to be

$$\lambda_{1,2} = 0, L_p \tag{6.3-18}$$

Under normal circumstances, the spiral mode (λ_1) is neutral, and L_p is negative, producing a stable roll mode (λ_2). At high angles of attack, in the vicinity of stall, the differential angle of attack produced by the roll rate could induce the downward-moving wing to stall, reversing the sign of L_p and causing roll-mode instability.

In ground effect, the characteristic equation is

$$\Delta(s) = |s\mathbf{I}_2 - \mathbf{F}_{RS_{GE}}| = s^2 - L_p s + L_\phi = 0 \tag{6.3-19}$$

where L_ϕ is less than zero. The result is a stable *roll-spiral oscillation*, with natural frequency and damping ratio, $\omega_n = \sqrt{-L_\phi}$ and $\zeta = -L_p/2\sqrt{-L_\phi}$.

Recall that L_ϕ is extremely sensitive to altitude; hence, the eigenvalues change rapidly during takeoff or approach to touchdown.

Eigenvectors

Literal formulas for the roll-spiral eigenvectors can be found by substituting eq. 6.3-18 in eq. 4.3-61. The adjoint of the out-of-ground-effect characteristic matrix is

$$\mathrm{Adj}(s\mathbf{I}_2 - \mathbf{F}_{RS}) = \begin{bmatrix} s & 0 \\ 1 & (s - L_p) \end{bmatrix} \tag{6.3-20}$$

For λ_1,

$$\mathrm{Adj}(\lambda_1\mathbf{I}_2 - \mathbf{F}_{RS}) = \begin{bmatrix} a_1\mathbf{e}_1 & a_2\mathbf{e}_1 \end{bmatrix} = \begin{bmatrix} 0 & 0 \\ 1 & -L_p \end{bmatrix} \tag{6.3-21}$$

and the eigenvector of the integral is seen to involve the roll angle alone. The second eigenvector is found from

$$\mathrm{Adj}(\lambda_2\mathbf{I}_2 - \mathbf{F}_{RS}) = \begin{bmatrix} a_1\mathbf{e}_2 & a_2\mathbf{e}_2 \end{bmatrix} = \begin{bmatrix} L_p & 0 \\ 1 & 0 \end{bmatrix} \tag{6.3-22}$$

where the second column is an indeterminate image of the first ($a_2 = 0$). Choosing the first column of each matrix to represent an eigenvector,

$$\mathbf{E} = \begin{bmatrix} \mathbf{e}_1 & \mathbf{e}_2 \end{bmatrix} = \begin{bmatrix} 0 & L_p \\ 1 & 1 \end{bmatrix} \tag{6.3-23}$$

Both eigenvectors are real, and the phase angle for $\mathbf{e}_2$ is zero.

Root Locus Analysis of Parameter Variations

The out-of-ground-effect roll-spiral characteristic equation

$$\Delta(s) = s^2 - L_p s = 0 \tag{6.3-24}$$

takes the form

$$\Delta(s) = d(s) + kn(s) = 0 \tag{6.3-25}$$

using L_p as the root locus gain k. There is a pair of poles at the origin and a zero at L_p. The zero cancels a pole, and the remaining root is moved to the right or left by positive or negative values of L_p.

Frequency Response

The control and disturbance transfer function matrices $\mathbf{U}_{RS}(s)$ and $\mathbf{W}_{RS}(s)$ relate the output Laplace transform $\Delta\mathbf{y}_{RS}(s)$ to the inputs $\Delta\mathbf{u}_{RS}(s)$ and $\Delta\mathbf{w}_{RS}(s)$:

$$\Delta\mathbf{y}_{RS}(s) = \mathbf{H}_{RS}(s\mathbf{I}_2 - \mathbf{F}_{RS})^{-1}[\mathbf{G}_{RS}\Delta\mathbf{u}_{RS}(s) + \mathbf{L}_{RS}\Delta\mathbf{w}_{RS}(s)]$$

$$\triangleq \mathbf{U}_{RS}(s)\Delta\mathbf{u}_{RS}(s) + \mathbf{W}_{RS}(s)\Delta\mathbf{w}_{RS}(s) \quad (6.3\text{-}26)$$

The out-of-ground-effect characteristic matrix inverse is

$$(s\mathbf{I}_2 - \mathbf{F}_{RS})^{-1} = \frac{\mathrm{Adj}(s\mathbf{I}_2 - \mathbf{F}_{RS})}{|s\mathbf{I}_2 - \mathbf{F}_{RS}|} = \frac{\begin{bmatrix} s & 0 \\ 1 & (s - L_p) \end{bmatrix}}{s(s - L_p)}$$

$$= \begin{bmatrix} \dfrac{1}{(s - L_p)} & 0 \\ \dfrac{1}{s(s - L_p)} & \dfrac{1}{s} \end{bmatrix} \quad (6.3\text{-}27)$$

The individual scalar transfer functions are linear combinations of the elements of this matrix weighted by the elements of $\mathbf{H}_{RS}$, $\mathbf{G}_{RS}$, and $\mathbf{L}_{RS}$. With $\mathbf{H}_{RS} = \mathbf{I}_2$, and $\mathbf{G}_{RS}$ and $\mathbf{L}_{RS}$ defined by eqs. 6.3-6 and 6.3-7, the transfer function matrices are as follows:

$$\mathbf{U}_{RS}(s) = \begin{bmatrix} \dfrac{\Delta p(s)}{\Delta\delta A(s)} \\ \dfrac{\Delta\phi(s)}{\Delta\delta A(s)} \end{bmatrix} = \begin{bmatrix} \dfrac{L_{\delta A}}{(s - L_p)} \\ \dfrac{L_{\delta A}}{s(s - L_p)} \end{bmatrix} \quad (6.3\text{-}28)$$

$$\mathbf{W}_{RS}(s) = \begin{bmatrix} \dfrac{\Delta p(s)}{\Delta p_W(s)} \\ \dfrac{\Delta\phi(s)}{\Delta p_W(s)} \end{bmatrix} = \begin{bmatrix} \dfrac{-L_p}{(s - L_p)} \\ \dfrac{-L_p}{s(s - L_p)} \end{bmatrix} \quad (6.3\text{-}29)$$

The aileron-to-roll-rate and aileron-to-roll-angle frequency responses, based on the model of Table 6.3-1, are plotted in Fig. 6.3-1. The roll-rate plots have typical first-order shape. The low-frequency amplitude-ratio asymptote has zero slope, and the high-frequency asymptote is -20 dB/dec; the actual curve has -3 dB attenuation for input frequency equal to $|L_p|$.

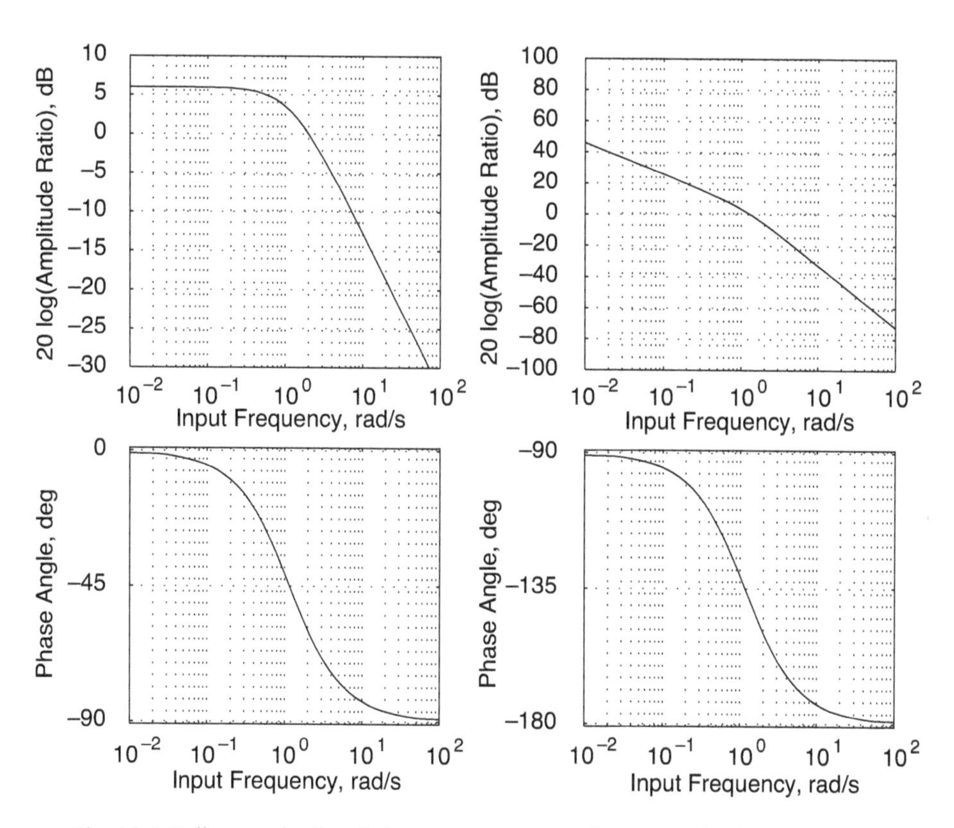

Fig. 6.3-1. Roll-rate and roll-angle frequency response to aileron input for the simplified model. (Left) $\Delta p(\omega)/\Delta\delta A(\omega)$. (Right) $\Delta\phi(\omega)/\Delta\delta A(\omega)$.

TABLE 6.3-1 APPROXIMATE ROLL-SPIRAL MODEL FOR FIGURE 6.3-1

$$\mathbf{F} = \begin{bmatrix} -1.1616 & 0 \\ 1 & 0 \end{bmatrix} \qquad \mathbf{G} = \begin{bmatrix} 2.3106 \\ 0 \end{bmatrix}$$

The corresponding phase angle begins at 0 deg, ends at -90 deg, and it is -45 deg when the input frequency is $|L_p|$. The roll-angle plots shift the amplitude-ratio slopes by -20 dB/dec and the phase-angle plot by -90 deg. The vortical-wind-response plots have the same shapes as the control plots, accounting for the difference in magnitude (and possibly sign) of $L_{\delta A}$ and L_p.

Root Locus Analysis of Feedback Control

The stability effects of feeding roll-rate and roll-angle perturbations back to the ailerons can be examined using the root locus. The root locus solves the following characteristic equations:

Roll-Rate Feedback to Ailerons

$$\Delta(s) = d(s) + kn(s) = s(s - L_p) + kL_{\delta A}s = 0 \tag{6.3-30}$$

Roll-Angle Feedback to Ailerons

$$\Delta(s) = d(s) + kn(s) = s(s - L_p) + k\,L_{\delta A} = 0 \tag{6.3-31}$$

Roll-rate feedback has no effect on the pole at the origin, moving the roll-mode root to the left or right just as L_p variations do. Roll-angle feedback with positive loop gain draws the roots together; with sufficient gain, the roots break away at $s = L_p/2$, creating a roll-spiral oscillation (or *lateral phugoid mode*). The damped natural frequency increases as the gain increases; the damping ratio decreases, but the oscillatory mode never becomes unstable. Feedback with the opposite sign immediately destabilizes the model, pulling the root at the origin into the right half plane, while the root-mode root is pushed to more stable values.

The roll-spiral model provides a good introductory example for *pilot-in-the-loop analysis*. As an exercise, apply the transfer-function pilot models of Section 4.7 (particularly eq. 4.7-1 and 4.7-2) to eq. 6.3-31, replacing k by the pilot transfer function, to show the destabilizing effect that pilot lag can have in simple compensatory roll-angle control. The pilot must limit loop again or introduce discretionary lead compensation to avoid instability.

Time Response

The state history is propagated by

$$\begin{aligned}\Delta\mathbf{x}_{RS}(t_k) = \mathbf{\Phi}_{RS}\Delta\mathbf{x}_{RS}(t_{k-1}) + \mathbf{\Gamma}_{RS}\Delta\mathbf{u}_{RS}(t_{k-1})\\ + \mathbf{\Lambda}_{RS}\Delta\mathbf{w}_{RS}(t_{k-1})\end{aligned} \tag{6.3-32}$$

The state transition matrix $\mathbf{\Phi}_{RS}(\Delta t)$ can be calculated as the inverse Laplace transform of the characteristic matrix inverse (eq. 6.3-27):

$$\mathbf{\Phi}_{RS}(\Delta t) = \mathcal{L}^{-1}(s\mathbf{I}_2 - \mathbf{F}_{RS})^{-1} \triangleq \begin{bmatrix}\phi_{11} & \phi_{12}\\ \phi_{21} & \phi_{22}\end{bmatrix} = \begin{bmatrix} e^{L_p\Delta t} & 0\\ \dfrac{e^{L_p\Delta t} - 1}{L_p} & 1\end{bmatrix} \tag{6.3-33}$$

The state transition relationship for the initial-condition and input response is then

$$\begin{bmatrix} \Delta p(t_k) \\ \Delta \phi(t_k) \end{bmatrix} = \mathbf{\Phi}_{\mathrm{RS}} \begin{bmatrix} \Delta p(t_{k-1}) \\ \Delta \phi(t_{k-1}) \end{bmatrix} + \mathbf{\Gamma}_{\mathrm{RS}} \Delta \delta A(t_{k-1}) + \mathbf{\Lambda}_{\mathrm{RS}} \Delta p_{\mathrm{W}}^{*}, \quad \begin{bmatrix} \Delta p(t_0) \\ \Delta \phi(t_0) \end{bmatrix} \text{ given} \tag{6.3-34}$$

The discrete-time control- and disturbance-effect matrices $\mathbf{\Gamma}_{\mathrm{RS}}$ and $\mathbf{\Lambda}_{\mathrm{RS}}$ cannot be calculated by eqs. 4.2-9a and 4.2-10a because $\mathbf{F}_{\mathrm{RS}}$ is singular. Nevertheless, the two matrices do exist, as they can be calculated from eqs. 4.2-9b and 4.2-10b:

$$\mathbf{\Gamma}_{\mathrm{RS}}(\Delta t) = \left[\mathbf{I}_2 + \frac{1}{2} \mathbf{F}_{\mathrm{RS}} \Delta t + \frac{1}{6} \mathbf{F}_{\mathrm{RS}}^2 \Delta t^2 + \frac{1}{24} \mathbf{F}_{\mathrm{RS}}^3 \Delta t^3 + \cdots \right] \mathbf{G}_{\mathrm{RS}} \Delta t \tag{6.3-35}$$

$$\mathbf{\Lambda}_{\mathrm{RS}}(\Delta t) = \left[\mathbf{I}_2 + \frac{1}{2} \mathbf{F}_{\mathrm{RS}} \Delta t + \frac{1}{6} \mathbf{F}_{\mathrm{RS}}^2 \Delta t^2 + \frac{1}{24} \mathbf{F}_{\mathrm{RS}}^3 \Delta t^3 + \cdots \right] \mathbf{L}_{\mathrm{RS}} \Delta t \tag{6.3-36}$$

with $\mathbf{G}_{\mathrm{RS}}$ and $\mathbf{L}_{\mathrm{RS}}$ defined by eqs. 6.3-6 and 6.3-7.

6.4 Coupled Lateral-Directional Dynamics

Accurate representation of the rolling and yawing motions of an aircraft requires more complicated models than the truncated (2×2) dynamic models used in Sections 6.2 and 6.3. The Dutch roll mode often involves significant rolling, rolling motions induce yaw, and roll angle introduces perturbations due to gravity. The time scales of rolling and yawing motion may be similar, and parameter variations may have complex effect on lateral-directional modes. A linear, fourth-order lateral-directional model portrays these effects when the motions are restricted to small perturbations.

The fourth-order, stability-axis model whose state, control, and disturbance vectors are

$$\Delta \mathbf{x}_{\mathrm{LD}}^T = [\Delta r \ \Delta \beta \ \Delta p \ \Delta \phi]^T \tag{6.4-1}$$

$$\Delta \mathbf{u}_{\mathrm{LD}}^T = [\Delta \delta A \ \Delta \delta R \ \Delta \delta SF]^T \tag{6.4-2}$$

$$\Delta \mathbf{w}_{\mathrm{LD}}^T = [\Delta \beta_w \ \Delta p_w]^T \tag{6.4-3}$$

captures the important lateral-directional dynamic interactions for rigid-body motion. The linearized state equation takes the general form

$$\Delta\dot{\mathbf{x}}_{LD} = \mathbf{F}_{LD}\Delta\mathbf{x}_{LD} + \mathbf{G}_{LD}\Delta\mathbf{u}_{LD} + \mathbf{L}_{LD}\Delta\mathbf{w}_{LD} \tag{6.4-4}$$

With $N_{\dot{\beta}} = N_{\phi} = Y_r = Y_p = L_{\phi} = \gamma_o = 0$, the stability-derivative matrix (eq. 6.1-15) is

$$\mathbf{F}_{LD} = \begin{bmatrix} N_r & N_\beta & N_p & 0 \\ -1 & \dfrac{Y_\beta}{V_o} & 0 & \dfrac{g_o}{V_o} \\ L_r & L_\beta & L_p & 0 \\ 0 & 0 & 1 & 0 \end{bmatrix} \tag{6.4-5}$$

while control and disturbance matrices are

$$\mathbf{G}_{LD} = \begin{bmatrix} N_{\delta A} & N_{\delta R} & N_{\delta SF} \\ \dfrac{Y_{\delta A}}{V_o} & \dfrac{Y_{\delta R}}{V_o} & \dfrac{Y_{\delta SF}}{V_o} \\ L_{\delta A} & L_{\delta R} & L_{\delta SF} \\ 0 & 0 & 0 \end{bmatrix} \tag{6.4-6}$$

$$\mathbf{L}_{LD} = \begin{bmatrix} -N_\beta & -N_p \\ \dfrac{-Y_\beta}{V_o} & \dfrac{-Y_p}{V_o} \\ -L_\beta & -L_p \\ 0 & 0 \end{bmatrix} \tag{6.4-7}$$

Further simplifications are examined before describing the stability and response characteristics of the full equations. The figures of this section derive from this model, with simplified control effects, as summarized in Table 6.4-1. The nominal roots for this system are

$$\lambda_{spiral} = +0.0089 \text{ rad/s}$$

$$\lambda_{roll} = -1.203 \text{ rad/s}$$

$$\lambda_{Dutch\ roll} = -0.116 \pm 1.387j \text{ rad/s}$$

$$\omega_{nDutch\ roll} = 1.392 \text{ rad/s}$$

$$\zeta_{Dutch\ roll} = +0.0832$$

$$\tau_{spiral} = -112 \text{ s}$$

$$\tau_{roll} = +0.83 \text{ s}$$

TABLE 6.4-1 BASELINE LATERAL-DIRECTIONAL MODEL FOR FIGURES OF SECTION 5.4

$$F = \begin{bmatrix} -0.1079 & 1.9011 & 0.0566 & 0 \\ -1 & -0.1567 & 0 & 0.0958 \\ 0.2501 & -2.408 & -1.1616 & 0 \\ 0 & 0 & 1 & 0 \end{bmatrix} \quad G = \begin{bmatrix} 0 & -1.1196 \\ 0 & 0 \\ 2.3106 & 0 \\ 0 & 0 \end{bmatrix}$$

The spiral mode is slightly unstable, but the time constant is nearly two minutes, and the mode is easily controlled. Selected stability derivatives are varied to illustrate parametric effects.

A TRUNCATED DUTCH ROLL/ROLL MODEL

A third-order model contains all of the aerodynamic coupling between rolling and yawing motions, while neglecting the effects of gravity. Dropping the fourth column and row of $\mathbf{F}_{LD}$, the stability matrix is

$$\mathbf{F}_{DRR} = \begin{bmatrix} N_r & N_\beta & N_p \\ -1 & \dfrac{Y_\beta}{V_o} & 0 \\ L_r & L_\beta & L_p \end{bmatrix} \tag{6.4-8}$$

and the third-degree characteristic equation is

$$\begin{aligned} \Delta(s) &= |s\mathbf{I}_3 - \mathbf{F}_{DRR}| \\ &= (s - L_p)\left[s^2 - \left(N_r + \frac{Y_\beta}{V_o}\right)s + \left(N_\beta + N_r\frac{Y_\beta}{V_o}\right)\right] \\ &\quad + \left[L_\beta N_p - L_r N_p\left(s - \frac{Y_\beta}{V_o}\right)\right] \end{aligned} \tag{6.4-9}$$

The characteristic polynomial is the product of the reduced-order roll-mode polynomial and the reduced-order Dutch roll polynomial plus coupling terms. The coupling terms can be rearranged as

$$-L_r N_p\left(s - \frac{Y_\beta}{V_o}\right) + L_\beta N_p = -N_p L_r\left[s - \left(\frac{Y_\beta}{V_o} + \frac{L_\beta}{L_r}\right)\right] \tag{6.4-10}$$

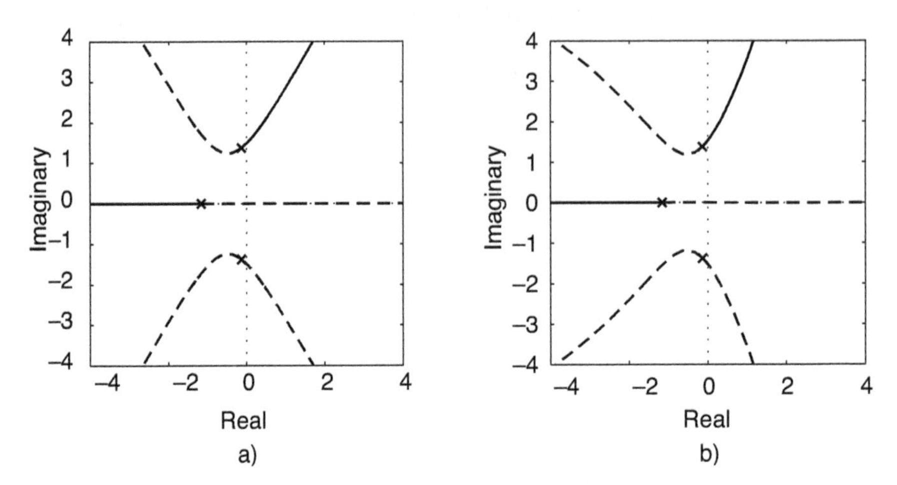

Fig. 6.4-1. Effects of yawing-moment sensitivity to roll rate (N_p) on the roots of the truncated third-order Dutch roll/roll model. Solid line for $N_p < 0$; dashed line for $N_p > 0$. (a) Negligible Y_β/V_o and L_r. (b) With Y_β/V_o and L_r.

This is a suitable form for root locus analysis using N_p as the gain and assuming that all other derivatives (including L_r) are fixed.

If Y_β/V_o and L_r are negligible, the numerator term that produces coupling is simply $N_p L_\beta$. In this case, with $L_\beta < 0$, the Dutch roll roots are stabilized by positive N_p, and the rightward progression of the roll-mode root is unrestrained (Fig. 6.4-1a). For negative N_p, the Dutch roll mode is destabilized, and it remains oscillatory. With the zero located at $(Y_\beta/V_o + L_\beta/L_r) = -9.785$, increasing N_p draws the Dutch roll pair toward the zero (Fig. 6.4-1b, zero not shown), and the high-gain asymptotes become 0 and -180 deg. Otherwise, the root locus is similar to the previous case over the region shown.

Residualized Lateral-Directional Models

Three residualized models may be useful for analysis, depending on the relative time scales of the various modes. If the roll mode is stable and much faster than the Dutch roll mode, the third-order model developed in the last section can be reduced to a second-order model for the Dutch roll mode. Under the same conditions, the fourth-order model can be reduced to a third-order model containing Dutch roll and spiral modes. If, on the other hand, the Dutch roll is stable and fast compared to the roll mode, a second-order roll-spiral model can be formulated.

In all three cases, the residualization structure of Section 4.2 is followed, with the time derivatives of the fast variables replaced by zero, and the corresponding differential equations replaced by algebraic ones. Here, we restrict our attention to the effects on stability; however, the analysis can be extended to forcing effects as well.

Residualized Dutch Roll Mode

For the first case, eq. 6.4-8 is partitioned into (2×2) and (1×1) diagonal blocks, with (2×1) and (1×2) coupling blocks:

$$\mathbf{F}_{\mathrm{DR}} = \begin{bmatrix} N_r & N_\beta \\ -1 & \dfrac{Y_\beta}{V_o} \end{bmatrix} \tag{6.4-11}$$

$$\mathbf{F}_{\mathrm{R}} = [L_p] \tag{6.4-12}$$

$$\mathbf{F}_{\mathrm{R}}^{\mathrm{DR}} = \begin{bmatrix} N_p \\ 0 \end{bmatrix} \tag{6.4-13}$$

$$\mathbf{F}_{\mathrm{DR}}^{\mathrm{R}} = [L_r \ L_\beta] \tag{6.4-14}$$

From eq. 4.2-57, the residualized stability matrix is

$$\mathbf{F}_{\mathrm{DR}_{res}} = \begin{bmatrix} \left(N_r - \dfrac{L_r N_p}{L_p}\right) & \left(N_\beta - \dfrac{L_\beta N_p}{L_p}\right) \\ -1 & \dfrac{Y_\beta}{V_o} \end{bmatrix} \tag{6.4-15}$$

The expressions for natural frequency and damping ratio each contain a term from the second-order model plus a correction for roll-mode effects:

$$\omega_{n_{DR_{res}}} = \sqrt{\left(N_\beta + N_r \dfrac{Y_\beta}{V_o}\right) - \dfrac{N_p\left(L_\beta + L_r \dfrac{Y_\beta}{V_o}\right)}{L_p}} \tag{6.4-16}$$

$$\zeta_{DR_{res}} = \frac{-\left(N_r + \dfrac{Y_\beta}{V_o} - \dfrac{L_r N_p}{L_p}\right)}{2\omega_{n_{DR}}}$$

$$= \frac{-\left(N_r + \dfrac{Y_\beta}{V_o} - \dfrac{L_r N_p}{L_p}\right)}{2\sqrt{\left(N_\beta + N_r \dfrac{Y_\beta}{V_o}\right) - \dfrac{N_p\left(L_\beta + L_r \dfrac{Y_\beta}{V_o}\right)}{L_p}}} \tag{6.4-17}$$

The approximation becomes more accurate as the magnitude of L_p increases, though the correction itself becomes less significant.

Residualized Dutch Roll and Spiral Modes

The second residualization is based on the fourth-order model. The order of Δp and $\Delta\phi$ is reversed in the state vector, and the rows and columns of $\mathbf{F}_{LD}$ are interchanged accordingly:

$$\mathbf{F}'_{LD} = \begin{bmatrix} N_r & N_\beta & 0 & N_p \\ -1 & \dfrac{Y_\beta}{V_o} & \dfrac{g_o}{V_o} & 0 \\ 0 & 0 & 0 & 1 \\ L_r & L_\beta & 0 & L_p \end{bmatrix} \tag{6.4-18}$$

$\mathbf{F}'_{LD}$ is partitioned into (3×3) and (1×1) diagonal blocks and (3×1) and (1×3) off-diagonal blocks. Substituting in eq. 4.2-57, the residualized (3×3) matrix describing Dutch roll and spiral stability is

$$\mathbf{F}_{DRS_{res}} = \begin{bmatrix} \left(N_r - L_r \dfrac{N_p}{L_p}\right) & \left(N_\beta - L_\beta \dfrac{N_p}{L_p}\right) & 0 \\ -1 & \dfrac{Y_\beta}{V_o} & \dfrac{g_o}{V_o} \\ -\dfrac{L_r}{L_p} & -\dfrac{L_\beta}{L_p} & 0 \end{bmatrix} \tag{6.4-19}$$

The characteristic polynomial of the third-order model is readily found from $|s\mathbf{I}_3 - \mathbf{F}_{\mathrm{DRS}_{res}}|$. If L_r and N_p are negligibly small, it is

$$\Delta(s) = s^3 - \left(N_r + \frac{Y_\beta}{V_o}\right)s^2 + \left(N_\beta + N_r\frac{Y_\beta}{V_o} + \frac{g_o}{V_o}\frac{L_\beta}{L_p}\right)s - N_r\frac{g_o}{V_o}\frac{L_\beta}{L_p} \tag{6.4-20}$$

The Dutch roll mode normally is much faster than the spiral mode; hence, eq. 6.4-20 can be approximately factored as

$$\Delta_{\mathrm{DR}_{res}}(s) \approx s^2 - \left(N_r + \frac{Y_\beta}{V_o}\right)s + \left(N_\beta + N_r\frac{Y_\beta}{V_o} + \frac{g_o}{V_o}\frac{L_\beta}{L_p}\right) \tag{6.4-21}$$

$$\Delta_{\mathrm{S}_{res}}(s) \approx s - \frac{N_r\dfrac{g_o}{V_o}\dfrac{L_\beta}{L_p}}{\left(N_\beta + N_r\dfrac{Y_\beta}{V_o} + \dfrac{g_o}{V_o}\dfrac{L_\beta}{L_p}\right)} \tag{6.4-22}$$

The dihedral effect L_β is of particular interest. Within the approximation, the Dutch roll's natural frequency is increased by negative L_β, and the damping ratio is decreased by L_p. The spiral mode root is no longer at the origin, as in the second-order rolling model:

$$\lambda_{\mathrm{S}_{res}}(s) \approx \frac{N_r\dfrac{g_o}{V_o}\dfrac{L_\beta}{L_p}}{\left(N_\beta + N_r\dfrac{Y_\beta}{V_o} + \dfrac{g_o}{V_o}\dfrac{L_\beta}{L_p}\right)} \tag{6.4-23}$$

The denominator is positive for stable Dutch roll, so the sign of the root is seen to depend on the sign of the numerator. L_p and N_r are normally negative, and g_o/V_o is positive; hence, negative L_β stabilizes the spiral mode, and positive L_β destabilizes it.

Residualized Roll-Spiral Modes

The third residualization is predicated on a fast and stable Dutch roll mode. The goal is to derive a corrected second-order model for roll and spiral motions. Equation 6.4-4 is rearranged and partitioned as

$$\begin{bmatrix} \Delta\dot{\mathbf{x}}_{\mathrm{RS}} \\ \Delta\dot{\mathbf{x}}_{\mathrm{DR}} \end{bmatrix} = \begin{bmatrix} \mathbf{F}_{\mathrm{RS}} & \mathbf{F}_{\mathrm{DR}}^{\mathrm{RS}} \\ \mathbf{F}_{\mathrm{RS}}^{\mathrm{DR}} & \mathbf{F}_{\mathrm{DR}} \end{bmatrix}\begin{bmatrix} \Delta\mathbf{x}_{\mathrm{RS}} \\ \Delta\mathbf{x}_{\mathrm{DR}} \end{bmatrix} + \begin{bmatrix} \mathbf{G}_{\mathrm{RS}} & \mathbf{G}_{\mathrm{DR}}^{\mathrm{RS}} \\ \mathbf{G}_{\mathrm{RS}}^{\mathrm{DR}} & \mathbf{G}_{\mathrm{DR}} \end{bmatrix}\begin{bmatrix} \Delta\mathbf{u}_{\mathrm{RS}} \\ \Delta\mathbf{u}_{\mathrm{DR}} \end{bmatrix} \tag{6.4-24}$$

where $\mathbf{F}_{DR}^{RS}$ couples Dutch roll motions into the roll-spiral variables, and $\mathbf{F}_{RS}^{DR}$ does the reverse. Following Section 4.2, we obtain a second-order differential equation for the roll and spiral modes:

$$\begin{aligned}\Delta\dot{\mathbf{x}}_{RS} &= (\mathbf{F}_{RS} - \mathbf{F}_{DR}^{RS}\mathbf{F}_{DR}^{-1}\mathbf{F}_{RS}^{DR})\Delta\mathbf{x}_{RS} + (\mathbf{G}_{RS} - \mathbf{F}_{DR}^{RS}\mathbf{F}_{DR}^{-1}\mathbf{G}_{RS}^{DR})\Delta\mathbf{u}_{RS} \\ &\quad + (\mathbf{G}_{DR}^{RS} - \mathbf{F}_{DR}^{RS}\mathbf{F}_{DR}^{-1}\mathbf{G}_{DR})\Delta\mathbf{u}_{DR} \\ &\triangleq \mathbf{F}'_{RS}\Delta\mathbf{x}_{RS} + \mathbf{G}'_{RS}\begin{bmatrix}\Delta\mathbf{u}_{RS}\\ \Delta\mathbf{u}_{DR}\end{bmatrix}\end{aligned} \tag{6.4-25}$$

where the matrices of eq. 6.4-25 are

$$\mathbf{F}_{RS} = \begin{bmatrix} L_p & 0 \\ 1 & 0 \end{bmatrix} \tag{6.4-26}$$

$$\mathbf{G}_{RS} = \begin{bmatrix} L_{\delta A} \\ 0 \end{bmatrix} \tag{6.4-27}$$

$$\mathbf{F}_{DR} = \begin{bmatrix} N_r & N_\beta \\ -1 & \dfrac{Y_\beta}{V_o} \end{bmatrix} \tag{6.4-28}$$

$$\mathbf{G}_{DR} = \begin{bmatrix} N_{\delta R} & N_{\delta SF} \\ \dfrac{Y_{\delta R}}{V_o} & \dfrac{Y_{\delta SF}}{V_o} \end{bmatrix} \tag{6.4-29}$$

$$\mathbf{F}_{DR}^{RS} = \begin{bmatrix} L_r & L_\beta \\ 0 & 0 \end{bmatrix} \tag{6.4-30}$$

$$\mathbf{F}_{RS}^{DR} = \begin{bmatrix} N_p & 0 \\ 0 & \dfrac{g_o}{V_o} \end{bmatrix} \tag{6.4-31}$$

$$\mathbf{G}_{DR}^{RS} = \begin{bmatrix} L_{\delta R} & L_{\delta SF} \\ 0 & 0 \end{bmatrix} \tag{6.4-32}$$

$$\mathbf{G}_{RS}^{DR} = \begin{bmatrix} N_{\delta A} \\ \dfrac{Y_{\delta A}}{V_o} \end{bmatrix} \tag{6.4-33}$$

We limit our examination of this model to the effect on natural frequency and damping ratio. The characteristic equation is

$$\begin{aligned}\Delta_{\mathrm{RS}_{res}}(s) &= |s\mathbf{I}_2 - (\mathbf{F}_{\mathrm{RS}} - \mathbf{F}_{\mathrm{DR}}^{\mathrm{RS}}\mathbf{F}_{\mathrm{DR}}^{-1}\mathbf{F}_{\mathrm{RS}}^{\mathrm{DR}})| \\ &= s^2 - as + b \end{aligned} \tag{6.4-34}$$

where

$$a = \left[L_p - N_p\left(\frac{L_\beta + L_r Y_\beta/V_o}{N_\beta + N_r Y_\beta/V_o}\right)\right] \tag{6.4-35}$$

$$b = \frac{g_o}{V_o}\left(\frac{L_\beta N_r - L_r N_\beta}{N_\beta + N_r Y_\beta/V_o}\right) \tag{6.4-36}$$

When $(a/2)^2 > b$, the roots are real. Normally, the more negative root represents the roll mode, and the more positive root the spiral mode.

$$\lambda_{\mathrm{R}_{res}} = \frac{\left[L_p - N_p\left(\frac{L_\beta + L_r Y_\beta/V_o}{N_\beta + N_r Y_\beta/V_o}\right)\right]}{2} - \sqrt{\frac{\left[L_p - N_p\left(\frac{L_\beta + L_r Y_\beta/V_o}{N_\beta + N_r Y_\beta/V_o}\right)\right]^2}{4} - \frac{g_o}{V_o}\left(\frac{L_\beta N_r - L_r N_\beta}{N_\beta + N_r Y_\beta/V_o}\right)} \tag{6.4-37}$$

$$\lambda_{\mathrm{S}_{res}} = +\sqrt{\frac{\left[L_p - N_p\left(\frac{L_\beta + L_r Y_\beta/V_o}{N_\beta + N_r Y_\beta/V_o}\right)\right]^2}{4} - \frac{g_o}{V_o}\left(\frac{L_\beta N_r - L_r N_\beta}{N_\beta + N_r Y_\beta/V_o}\right)} \tag{6.4-38}$$

If $(a/2)^2 < b$, the roots are complex, producing a coupled *roll-spiral oscillation* with the following natural frequency and damping ratio:

$$\omega_{n_{\mathrm{RS}_{res}}} = \sqrt{\frac{g_o}{V_o}\left(\frac{L_\beta N_r - L_r N_\beta}{N_\beta + N_r Y_\beta/V_o}\right)} \tag{6.4-39}$$

$$\zeta_{RS_{res}} = -\frac{\left[L_p - N_p\left(\frac{L_\beta + L_r Y_\beta / V_o}{N_\beta + N_r Y_\beta / V_o}\right)\right]}{2\omega_{n_{RS_{res}}}} \tag{6.4-40}$$

For conventional configurations, the roll-spiral oscillation is most likely to occur when the rolling response to yaw rate L_r (normally positive in low-angle-of-attack subsonic flight) becomes negative and N_β is positive and large, making the numerator of eq. 6.4-36 more positive. The roll-spiral mode is sometimes called a *lateral phugoid mode* because it produces a slow, lightly damped oscillation; however, the mechanism for this behavior is unrelated to that of the longitudinal mode.

Fourth-Order Model

The fourth-order model of lateral-directional dynamics is readily analyzed, although generalizing about results is complicated by the fact that several coupling derivatives may take both positive and negative values. This may occur not only for different aircraft configurations but for different flight conditions of the same aircraft (most notably due to variations in Mach number and angle of attack).

Equilibrium Response to Control

The linearized equilibrium response to step change in the control is given by

$$\Delta \mathbf{x}^*_{\mathrm{LD}} = -\mathbf{F}_{\mathrm{LD}}^{-1}\mathbf{G}_{\mathrm{LD}}\Delta \mathbf{u}^*_{\mathrm{LD}} \tag{6.4-41}$$

To focus on the separate effects of roll, yaw, and side-force control, the control-effect matrix is idealized as

$$\mathbf{G}_{\mathrm{LD}} = \begin{bmatrix} 0 & N_{\delta R} & 0 \\ 0 & 0 & \dfrac{Y_{\delta SF}}{V_o} \\ L_{\delta A} & 0 & 0 \\ 0 & 0 & 0 \end{bmatrix} \tag{6.4-42}$$

With this and the simplifications noted earlier, the steady-state relationship between control and state, assuming stability of all modes, is

$$\begin{bmatrix} \Delta r^* \\ \Delta \beta^* \\ \Delta p^* \\ \Delta \phi^* \end{bmatrix} = \frac{\begin{bmatrix} \frac{g_o}{V_o} N_\beta L_{\delta A} & -\frac{g_o}{V_o} L_\beta N_{\delta R} & 0 \\ \frac{g_o}{V_o} N_r L_{\delta A} & \frac{g_o}{V_o} L_r N_{\delta R} & 0 \\ 0 & 0 & 0 \\ \left(N_\beta + N_r \frac{Y_\beta}{V_o} \right) L_{\delta A} & -\left(L_\beta + L_r \frac{Y_\beta}{V_o} \right) N_{\delta R} & (N_\beta L_r - N_r L_\beta) \frac{Y_{\delta SF}}{V_o} \end{bmatrix}}{\frac{g_o}{V_o} (L_\beta N_r - L_r N_\beta)} \begin{bmatrix} \Delta \delta A^* \\ \Delta \delta R^* \\ \Delta \delta SF^* \end{bmatrix}$$

(6.4-43)

The steady-state roll rate is zero for all controls, a consequence of the constant equilibrium roll angle. As in the reduced-order model, the aileron still commands a significant transient roll rate that is roughly proportional to $-L_{\delta A} \, \Delta\delta A / L_p$ during its early evolution. However, the long-term roll-rate response of the fourth-order linear model goes to zero, as long as all modes are stable. With an unstable spiral mode, the initial roll-rate response is well behaved, but it eventually diverges. (The equilibrium response exists for both stable and unstable systems; however, there is no steady-state condition for an unstable system.)

The equilibrium yaw rate and sideslip angle are determined by aileron and rudder settings, and equilibrium roll angle is affected by all three controls. Steady side force has no other effect than to change the equilibrium roll angle. (Of course, a real side-force effector normally would produce some degree of yawing and rolling moment as well.) Given the steady-state yaw rate indicated by eq. 6.4-43, the yaw *angle* increases at a constant rate, producing a *steady turn*. From eq. 6.4-43, the relationship between the constant yaw rate and roll angle produced by aileron alone is independent of $L_{\delta A}$:

$$\frac{\Delta r^*}{\Delta \phi^*} = \frac{\frac{g_o}{V_o} N_\beta}{\left(N_\beta + N_r \frac{Y_\beta}{V_o} \right)} \tag{6.4-44}$$

Following Section 4.2, it is possible to command any three desired combinations of equilibrium yaw rate, sideslip angle, and roll angle with steady aileron, rudder, and side-force-panel settings, or any two combinations using aileron and rudder alone. In the first case, we define the desired three-element output $\Delta\mathbf{y}^*$ as

$$\Delta\mathbf{y}^* = \mathbf{H}_{LD}\Delta\mathbf{x}^* = \begin{bmatrix} 1 & 0 & 0 & 0 \\ 0 & 1 & 0 & 0 \\ 0 & 0 & 0 & 1 \end{bmatrix} \begin{bmatrix} \Delta r^* \\ \Delta\beta^* \\ \Delta p^* \\ \Delta\phi^* \end{bmatrix} \tag{6.4-45}$$

From eq. 4.2-25, the corresponding control settings, taking into account a steady disturbance input $\Delta\mathbf{w}^*$, are given by

$$\Delta\mathbf{u}^* = (-\mathbf{H}_{LD}\mathbf{F}_{LD}^{-1}\mathbf{G}_{LD})^{-1}[\mathbf{H}_{LD}\mathbf{F}_{LD}^{-1}(\mathbf{L}_{LD}\Delta\mathbf{w}^* + \Delta\mathbf{y}^*)] \tag{6.4-46}$$

Given any two desired outputs with aileron and rudder control, eq. 6.4-45 and $\mathbf{G}_{LD}$ are redefined accordingly, and the revised definitions of $\Delta\mathbf{y}^*$ and $\mathbf{H}_{LD}$ are used in eq. 6.4-46. For example, the control settings for a *coordinated turn* are obtained by defining $\Delta\mathbf{y}^* = [\Delta\beta^* \ \Delta\phi^*]^T$, while commanding the desired value of $\Delta\phi^*$ and $\Delta\beta^* = 0$. The corresponding Δr^* for aileron-rudder control of sideslip and roll angle would be generated by eq. 6.4-43. Conversely, Δr^* could be specified, and $\Delta\phi^*$ would be generated by the equation. In all cases, the steady-state stability-axis roll rate is zero.

$(-\mathbf{H}_{LD}\,\mathbf{F}_{LD}^{-1}\,\mathbf{G}_{LD})^{-1}$ specifies the coupling gain for an open-loop *aileron-rudder interconnect* (*ARI*) that could provide steady-state turn coordination. This (2×2) matrix also specifies the aileron and rudder settings that would be required to produce sideslip without roll angle, a useful relationship for crosswind landing. In a prototype manual control system, the pilot's control wheel or lateral stick would command roll angle, and the foot pedals would control sideslip angle. From a piloting perspective, it is generally preferable to command roll rate rather than roll angle, avoiding continued control force during a steady turn. For short periods of flight, such as the final stage of approach and landing, roll angle may be a desirable alternative, especially for less skilled pilots.

Actual control effects are coupled, and eq. 6.4-42 is an oversimplification in most cases. Aileron deflection produces adverse or favorable (sometimes called "proverse") yawing moment as well as rolling moment, and rudder deflection produces rolling as well as yawing. Combining terms from eq. 6.4-43, we see, for example, that steady-state roll-angle response to aileron becomes

$$\frac{\Delta\phi^*}{\Delta\delta A^*} = \frac{\left(N_\beta + N_r \dfrac{Y_\beta}{V_o}\right)L_{\delta A} - \left(L_\beta + L_r \dfrac{Y_\beta}{V_o}\right)N_{\delta A}}{\dfrac{g_o}{V_o}(L_\beta N_r - L_r N_\beta)} \tag{6.4-47}$$

A simplification of this relationship is the basis for the *lateral control divergence parameter* (LCDP), defined as [M-3]

$$\text{LCDP} \triangleq C_{n_\beta} - \frac{C_{n_{\delta A}}}{C_{l_{\delta A}}} C_{l_\beta} \tag{6.4-48}$$

The LCDP should be greater than zero to avoid reversed response due to adverse aileron yaw. Adding an ARI, which commands the rudder in proportion to the aileron deflection with gain, k, the criterion becomes

$$\text{LCDP} \triangleq C_{n_\beta} - \frac{C_{n_{\delta A}} + kC_{n_{\delta R}}}{C_{l_{\delta A}} + kC_{l_{\delta R}}} C_{l_\beta} > 0 \tag{6.4-49}$$

Section 7.4 contains additional discussion of the LCDP.

Eigenvalues and Root Locus Analysis of Parameter Variations

The characteristic polynomial for this model is

$$\begin{aligned}\Delta(s) = |s\mathbf{I}_4 - \mathbf{F}_{\text{LD}}| &= s^4 - \left(L_p + N_r + \frac{Y_\beta}{V_o}\right)s^3 \\ &+ \left[N_\beta - L_r N_p + \frac{Y_\beta}{V_o} L_p + N_r\left(\frac{Y_\beta}{V_o} + L_p\right)\right]s^2 \\ &+ \left[\frac{Y_\beta}{V_o}(L_r N_p - L_p N_r) + L_\beta\left(N_p - \frac{g_o}{V_o}\right) - N_\beta L_p\right]s + \frac{g_o}{V_o}(L_\beta N_r - L_r N_\beta)\end{aligned} \tag{6.4-50}$$

The total damping of the model is portrayed by the coefficient of the cubic term, and the polynomial typically factors into a complex pair plus two real roots:

$$\Delta(s) = (s - \lambda_S)(s^2 + 2\zeta_{DR}\omega_{n_{DR}} s + \omega_{n_{DR}}^2)(s - \lambda_R) \tag{6.4-51}$$

For root locus analysis of parameter variations, $\Delta(s)$ is restated in the form

$$\Delta(s) = d(s) + kn(s) = 0 \tag{6.4-52}$$

alternatively defining k as each of the stability derivatives.

We begin by examining the two Dutch roll stability derivatives N_β and N_r. For N_β, the root locus form of the characteristic polynomial is

$$\Delta(s) = s^4 - \left(L_p + N_r + \frac{Y_\beta}{V_o}\right)s^3 + \left[-L_rN_p + \frac{Y_\beta}{V_o}L_p + N_r\left(\frac{Y_\beta}{V_o} + L_p\right)\right]s^2$$
$$+ \left[\frac{Y_\beta}{V_o}(L_rN_p - L_pN_r) + L_\beta\left(N_p - \frac{g_o}{V_o}\right)\right]s$$
$$+ \frac{g_o}{V_o}L_\beta N_r + N_\beta\left(s^2 - L_ps - \frac{g_o}{V_o}L_r\right) \qquad (6.4\text{-}53)$$

With zero L_r, the numerator polynomial is the characteristic polynomial of the second-order roll-spiral approximation (eq. 6.3-17); hence, a pair of roots approaches the approximate roll and spiral roots as $|N_\beta|$ becomes large. Without the N_β terms, $d(s)$ represents the characteristic polynomial of an aircraft with neutral directional stability. The surplus of poles over zeros is 2; the remaining two roots go toward infinity along asymptotes of ± 90 deg for $N_\beta > 0$ and (0, 180 deg) for $N_\beta < 0$ (Fig. 6.4-2a). Positive N_β introduces the oscillatory Dutch roll and drives the roll and spiral roots to their reduced-order values. The small positive L_r of the model (Table 6.4-1) moves the zero at the origin slightly to the right, and the root locus is similar to the previous case (Fig. 6.4-2b)

Variation in the yaw-damping derivative N_r has the same effect on Dutch roll roots as predicted by the second-order model (Section 6.2). The poles of $d(s)$ reflect negligible Dutch roll damping, with the characteristic polynomial partitioned as follows:

$$\Delta(s) = s^4 - \left(L_p + \frac{Y_\beta}{V_o}\right)s^3 + \left[N_\beta - L_rN_p + \frac{Y_\beta}{V_o}L_p\right]s^2$$
$$+ \left[\frac{Y_\beta}{V_o}L_rN_p + L_\beta\left(N_p - \frac{g_o}{V_o}\right) - N_\beta L_p\right]s - \frac{g_o}{V_o}L_rN_\beta$$
$$- N_r\left[s^3 - \left(\frac{Y_\beta}{V_o} + L_p\right)s^2 + L_p\frac{Y_\beta}{V_o}s - L_\beta\frac{g_o}{V_o}\right] \qquad (6.4\text{-}54)$$

Negative N_r increases Dutch roll damping, and the two complex zeros draw the roll and spiral roots toward roll-spiral oscillation (Fig. 6.4-3). Positive N_r is destabilizing for both the Dutch roll and spiral modes.

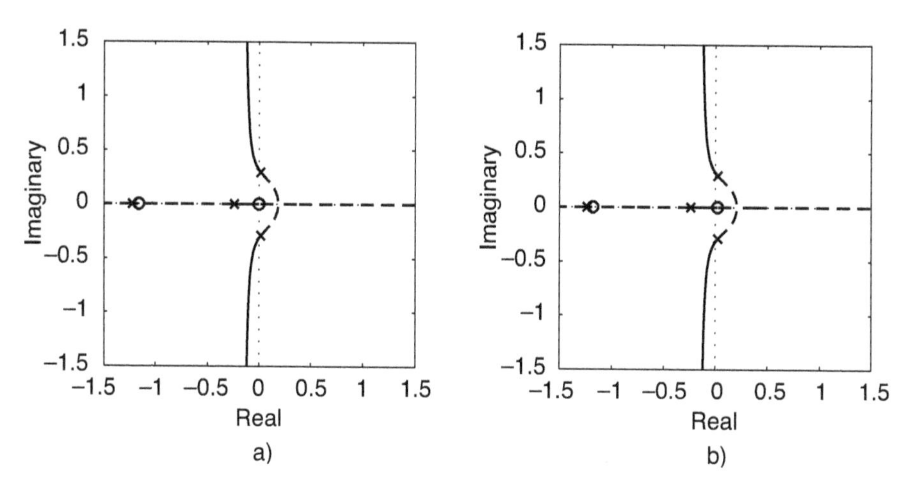

Fig. 6.4-2. Effects of yawing-moment sensitivity to sideslip angle (N_β) on the roots of the fourth-order model. Solid line for $N_\beta > 0$; dashed line for $N_\beta < 0$. (a) $L_r = 0$. (b) $L_r > 0$.

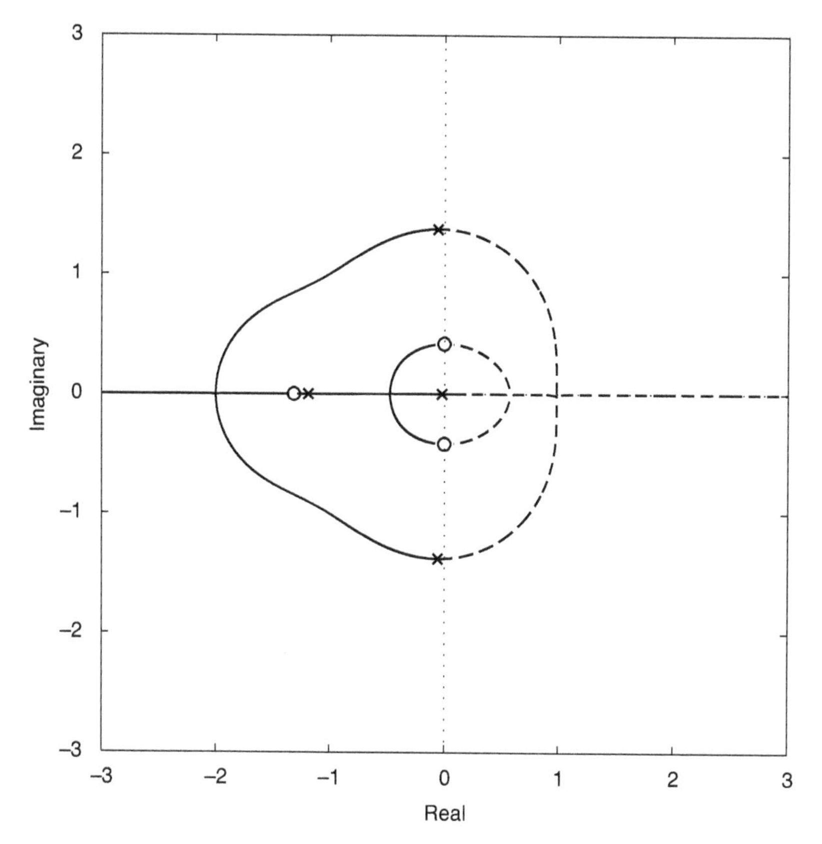

Fig. 6.4-3. Effects of yawing-moment sensitivity to yaw rate (Nr) on the roots of the fourth-order model. Solid line for $N_r < 0$; dashed line for $N_r > 0$.

To examine variations in the roll stability derivative L_p, the characteristic polynomial is written as

$$\Delta(s) = s^4 - \left(N_r + \frac{Y_\beta}{V_o}\right)s^3 + \left[N_\beta - L_r N_p + N_r \frac{Y_\beta}{V_o}\right]s^2 + \left[\frac{Y_\beta}{V_o} L_r N_p + L_\beta\left(N_p - \frac{g_o}{V_o}\right)\right]s + \frac{g_o}{V_o}(L_\beta N_r - L_r N_\beta) - L_p s\left[s^2 - \left(N_r + \frac{Y_\beta}{V_o}\right)s + \left(N_\beta + N_r \frac{Y_\beta}{V_o}\right)\right] \quad (6.4\text{-}55)$$

The numerator polynomial contains the reduced-order Dutch roll polynomial, which attracts the Dutch roll roots of the fourth-order model for both positive and negative values of L_p, and a free s at the origin. The zero at the origin attracts the slightly unstable spiral root for negative L_p, and the roll root is stabilized or destabilized by negative or positive values of L_p (Fig. 6.4-4).

As a coupling parameter, the *dihedral effect* L_β serves to redistribute the total damping of the model rather than to change its magnitude. Thus, negative L_β tends to stabilize the spiral and roll modes and destabilize the Dutch roll mode, while positive L_β does just the opposite. The characteristic equation is arranged as

$$\begin{aligned}\Delta(s) &= s^4 - \left(L_p + N_r + \frac{Y_\beta}{V_o}\right)s^3 + \left[N_\beta - L_r N_p + \frac{Y_\beta}{V_o} L_p + N_r\left(\frac{Y_\beta}{V_o} + L_p\right)\right]s^2 \\ &\quad + \left[\frac{Y_\beta}{V_o}(L_r N_p - L_p N_r) - N_\beta L_p\right]s - \frac{g_o}{V_o} L_r N_\beta + L_\beta\left[\left(N_p - \frac{g_o}{V_o}\right)s + \frac{g_o}{V_o} N_r\right] \\ &= s^4 - \left(L_p + N_r + \frac{Y_\beta}{V_o}\right)s^3 + \left[N_\beta - L_r N_p + \frac{Y_\beta}{V_o} L_p + N_r\left(\frac{Y_\beta}{V_o} + L_p\right)\right]s^2 \\ &\quad + \left[\frac{Y_\beta}{V_o}(L_r N_p - L_p N_r) - N_\beta L_p\right]s - \frac{g_o}{V_o} L_r N_\beta \\ &\quad - L_\beta\left(\frac{g_o}{V_o} - N_p\right)\left[s - \frac{\frac{g_o}{V_o} N_r}{\left(\frac{g_o}{V_o} - N_p\right)}\right]\end{aligned} \quad (6.4\text{-}56)$$

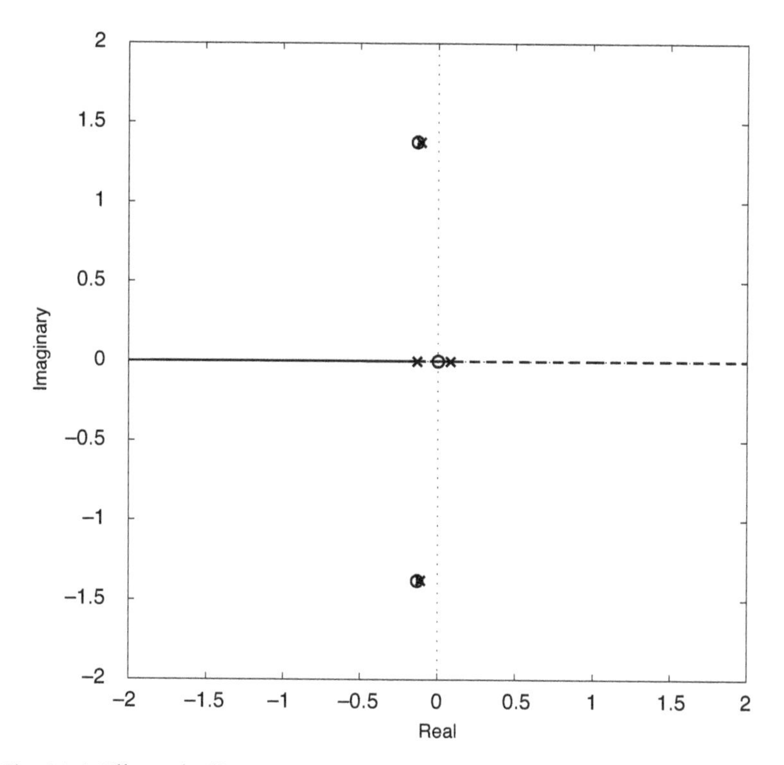

Fig. 6.4-4. Effects of rolling-moment sensitivity to roll rate (L_p) on the roots of the fourth-order model. Solid line for $L_p < 0$; dashed line for $L_p > 0$.

For small N_p, the zero is near N_r. With zero dihedral effect, the poles of $d(s)$ exhibit conventional Dutch roll and roll modes and an unstable spiral mode (Fig. 6.4-5). Thus, a little negative dihedral effect is desirable to stabilize the spiral mode, but too much destabilizes directional motion.

The characteristic equation is arranged to examine L_r variation as follows:

$$\Delta(s) = s^4 - \left(L_p + N_r + \frac{Y_\beta}{V_o}\right)s^3 + \left[N_\beta + \frac{Y_\beta}{V_o}L_p + N_r\left(\frac{Y_\beta}{V_o} + L_p\right)\right]s^2$$
$$+\left[L_\beta\left(N_p - \frac{g_o}{V_o}\right) - \frac{Y_\beta}{V_o}L_pN_r - N_\beta L_p\right]s + \frac{g_o}{V_o}L_\beta N_r$$
$$- L_rN_p\left[s^2 - \frac{Y_\beta}{V_o}s - \frac{g_o}{V_o}\frac{N_\beta}{N_p}\right] \tag{6.4-57}$$

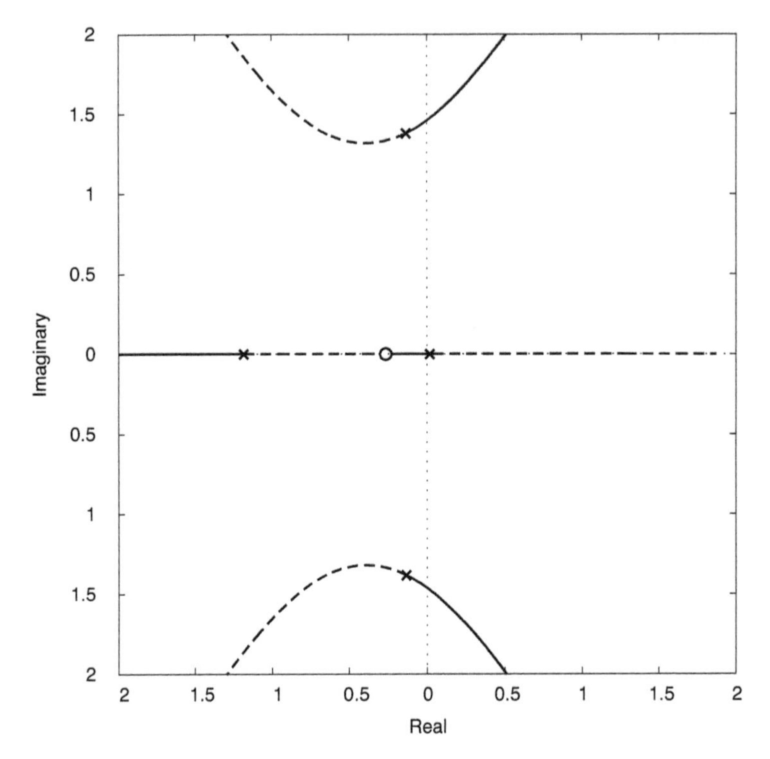

Fig. 6.4-5. Effects of rolling moment sensitivity to sideslip angle (L_β) on the roots of the fourth-order model. Solid line for $L_\beta < 0$; dashed line for $L_\beta > 0$.

This coupling term is generally small and positive, and its effect on the root locus is dependent on the sign of N_p. In the present case, negative values stabilize the Dutch roll and roll modes while destabilizing the spiral mode (Fig. 6.4-6).

Yawing sensitivity to the roll rate N_p is investigated with the characteristic equation expressed as

$$\Delta(s) = s^4 - \left(L_p + N_r + \frac{Y_\beta}{V_o}\right)s^3 + \left[N_\beta + \frac{Y_\beta}{V_o}L_p + N_r\left(\frac{Y_\beta}{V_o} + L_p\right)\right]s^2$$
$$+ \left[-\frac{Y_\beta}{V_o}L_pN_r - L_\beta\frac{g_o}{V_o} - N_\beta L_p\right]s + \frac{g_o}{V_o}(L_\beta N_r - L_r N_\beta)$$
$$+ N_p s\left[-L_r s + \left(\frac{Y_\beta}{V_o}L_r + L_\beta\right)\right] \qquad (6.4\text{-}58)$$

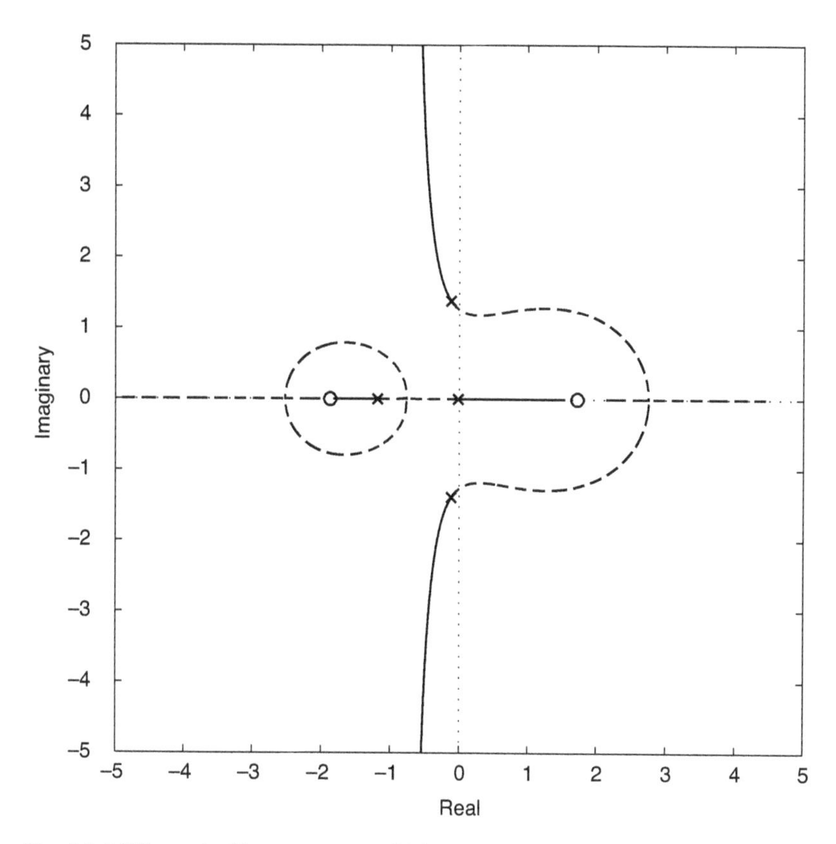

Fig. 6.4-6. Effects of rolling-moment sensitivity to yaw rate (L_r) on the roots of the fourth-order model. Solid line for $L_r < 0$; dashed line for $L_r > 0$.

The root locus is shown in Fig. 6.4-7, and the similarity to the reduced-order result (Fig. 6.4-1b) is apparent. Effects of the free s in the numerator and the nearby spiral mode on the root locus effectively cancel, and the progression of the Dutch roll and roll roots is the same for the two models.

Eigenvectors

The relative magnitude and phase of the state components in each dynamic mode's eigenvector are especially significant for highly coupled lateral-directional motions. From eq. 4.3-61, the eigenvector for a mode with eigenvalue λ_i is any determinate column of the $(n \times n)$ matrix $\mathrm{Adj}(\lambda_i \mathbf{I}_n - \mathbf{F})$, which is the transpose of the matrix of cofactors of

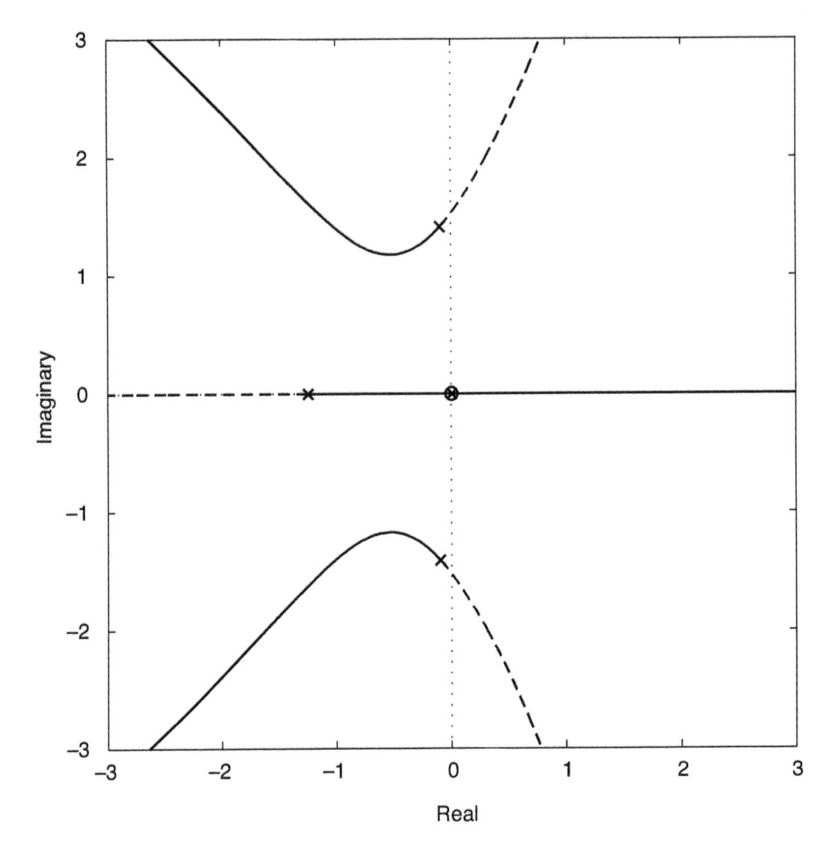

Fig. 6.4-7. Effects of yawing moment sensitivity to roll rate (N_p) on the roots of the fourth-order model. Solid line for $N_p > 0$; dashed line for $N_p < 0$.

$(\lambda_i \mathbf{I}_n - \mathbf{F})$ (eq. 2.2-28). In the present case, symbolic representations of the eigenvectors are easily derived, defining $\mathbf{F}$ by eq. 6.4-5.

Two ratios of eigenvector components are of particular interest: $\Delta\phi/\Delta\beta$ (usually called "ϕ to β" for the Dutch roll mode, and $\Delta r/\Delta p$ for the roll mode. Dividing the fourth component of the first column of $\mathrm{Adj}(\lambda_i \mathbf{I}_n - \mathbf{F})$ by the second,

$$\frac{\Delta\phi}{\Delta\beta} = \left(\frac{V_o}{g}\right)\left[\lambda_i - \left(\frac{Y_\beta}{V_o} + \frac{L_\beta}{L_r}\right)\right] \tag{6.4-59}$$

Identifying λ_i as one of the Dutch roll complex-conjugate values,

$$\lambda_i = -\zeta_{DR}\omega_{n_{DR}} + j\omega_{n_{DR}}\sqrt{1-\zeta_{DR}^2} \tag{6.4-60}$$

the ratio can be written as

$$\frac{\Delta\phi}{\Delta\beta} = \left(\frac{V_o}{g}\right)\left[-\left(\zeta_{DR}\omega_{n_{DR}} + \frac{Y_\beta}{V_o} + \frac{L_\beta}{L_r}\right) + j\omega_{n_{DR}}\sqrt{1-\zeta_{DR}^2}\right] \tag{6.4-61}$$

The magnitude and phase angle of this complex-valued ratio are

$$\left|\frac{\Delta\phi}{\Delta\beta}\right| = \left(\frac{V_o}{g}\right)\left[\left(\zeta_{DR}\omega_{n_{DR}} + \frac{Y_\beta}{V_o} + \frac{L_\beta}{L_r}\right)^2 + \left(\omega_{n_{DR}}\sqrt{1-\zeta_{DR}^2}\right)^2\right]^{1/2} \tag{6.4-62}$$

$$\angle\frac{\Delta\phi}{\Delta\beta} = \tan^{-1}\left[\frac{-\omega_{n_{DR}}\sqrt{1-\zeta_{DR}^2}}{\left(\zeta_{DR}\omega_{n_{DR}} + \frac{Y_\beta}{V_o} + \frac{L_\beta}{L_r}\right)}\right] \tag{6.4-63}$$

In the event that a roll-spiral oscillation occurs, its $\Delta\phi/\Delta\beta$ could be derived from the same equations, using the appropriate values of damping ratio and natural frequency. High $\Delta\phi/\Delta\beta$ which is typical of high-speed flight and configurations with $I_{zz} \gg I_{xx}$ [S-1], indicates that the Dutch roll is largely a rolling motion, while a small value indicates that it is primarily a yawing motion.

The ratio of angular rates in either roll or spiral mode response is derived in similar fashion. Using the fourth column of $\mathrm{Adj}(\lambda_i\mathbf{I}_n - \mathbf{F})$, $\Delta r/\Delta p$ takes the real value

$$\frac{\Delta r}{\Delta p} = \frac{L_\beta N_p + N_\beta(\lambda_i - L_p)}{L_r N_\beta + L_\beta(\lambda_i - N_r)} \tag{6.4-64}$$

where λ_i is the real root for either mode λ_R or λ_S. The ratio defines the angle between the mode's rotational axis and the reference frame's x axis.

Transfer Functions and Frequency Response

The control and disturbance transfer function matrices $\mathbf{U}_{\mathrm{LD}}(s)$ and $\mathbf{W}_{\mathrm{LD}}(s)$ relate the output Laplace transform $\Delta\mathbf{y}_{\mathrm{LD}}(s)$ to the inputs $\Delta\mathbf{u}_{\mathrm{LD}}(s)$

and $\Delta\mathbf{w}_{LD}(s)$:

$$\Delta\mathbf{y}_{LD}(s) = \mathbf{H}_{LD}(s\mathbf{I}_4 - \mathbf{F}_{LD})^{-1}[\mathbf{G}_{LD}\Delta\mathbf{u}_{LD}(s) + \mathbf{L}_{LD}\Delta\mathbf{w}_{LD}(s)] + \mathbf{J}_{LD}\Delta\mathbf{u}_{LD}(s) + \mathbf{K}_{LD}\Delta\mathbf{w}_{LD}(s)$$

$$\triangleq \mathbf{U}_{LD}(s)\Delta\mathbf{u}_{LD}(s) + \mathbf{W}_{LD}(s)\Delta\mathbf{w}_{LD}(s) \tag{6.4-65}$$

where $\mathbf{J}_{LD}$ and $\mathbf{K}_{LD}$ reflect direct paths from the input to the output, and

$$\mathbf{U}_{LD}(s) = \mathbf{H}_{LD}(s\mathbf{I}_4 - \mathbf{F}_{LD})^{-1}\mathbf{G}_{LD} + \mathbf{J}_{LD} = \frac{\mathbf{H}_{LD}\,\mathrm{Adj}(s\mathbf{I}_4 - \mathbf{F}_{LD})\mathbf{G}_{LD} + \mathbf{J}_{LD}\,|s\mathbf{I}_4 - \mathbf{F}_{LD}|}{|s\mathbf{I}_4 - \mathbf{F}_{LD}|} \tag{6.4-66}$$

$$\mathbf{W}_{LD}(s) = \mathbf{H}_{LD}(s\mathbf{I}_4 - \mathbf{F}_{LD})^{-1}\mathbf{L}_{LD} + \mathbf{K}_{LD} = \frac{\mathbf{H}_{LD}\,\mathrm{Adj}(s\mathbf{I}_4 - \mathbf{F}_{LD})\mathbf{L}_{LD} + \mathbf{K}_{LD}\,|s\mathbf{I}_4 - \mathbf{F}_{LD}|}{|s\mathbf{I}_4 - \mathbf{F}_{LD}|} \tag{6.4-67}$$

As before, the scalar transfer functions take the form $n(s)/\Delta(s)$, where $\Delta(s)$ is the lateral-directional characteristic polynomial. Individual transfer functions have numerator polynomials $n_\delta^x(s)$, where δ is the scalar control or disturbance input and x is the scalar output.

We define the output as the state (eq. 6.4-1)

$$\Delta\mathbf{y}_{LD}^T = [\Delta r\ \Delta\beta\ \Delta p\ \Delta\phi\ \Delta\psi]^T \tag{6.4-68}$$

Therefore, $\mathbf{H}_{LD}$ is the (5×5) identity matrix. The direct-effect matrices $\mathbf{J}_{LD}$ and $\mathbf{K}_{LD}$ are zero.

The control-transfer-function numerators are presented for idealized control effectors (eq. 6.4-42). The yaw-rate and roll-rate transfer functions multiplied by 1/s to produce yaw-angle and roll-angle transfer functions.

Aileron (Roll-Control) Transfer-Function Numerators

$$n_{\delta A}^r(s) = L_{\delta A}\left(N_p s^2 - N_p \frac{Y_\beta}{V_o} s + N_\beta \frac{g_o}{V_o}\right) \tag{6.4-69}$$

$$n_{\delta A}^\beta(s) = L_{\delta A}\left[\left(\frac{g_o}{V_o} - N_p\right)s - N_r \frac{g_o}{V_o}\right] \tag{6.4-70}$$

$$n_{\delta A}^p(s) = L_{\delta A} s\left[s^2 - \left(N_r + \frac{Y_\beta}{V_o}\right)s + \left(N_r \frac{Y_\beta}{V_o} + N_\beta\right)\right] \tag{6.4-71}$$

$$n_{\delta A}^{\phi}(s) = L_{\delta A}\left[s^2 - \left(N_r + \frac{Y_\beta}{V_o}\right)s + \left(N_r\frac{Y_\beta}{V_o} + N_\beta\right)\right] \tag{6.4-72}$$

$$n_{\delta A}^{\psi}(s) = \frac{L_{\delta A}}{s}\left(N_p s^2 - N_p\frac{Y_\beta}{V_o}s + N_\beta\frac{g_o}{V_o}\right) \tag{6.4-73}$$

Rudder (Yaw-Control) Transfer-Function Numerators

$$n_{\delta R}^{r}(s) = N_{\delta R}\left[s^3 - \left(\frac{Y_\beta}{V_o} + L_p\right)s^2 + L_p\frac{Y_\beta}{V_o}s - L_\beta\frac{g_o}{V_o}\right] \tag{6.4-74}$$

$$n_{\delta R}^{\beta}(s) = N_{\delta R}\left(s^2 - L_p s - L_r\frac{g_o}{V_o}\right) \tag{6.4-75}$$

$$n_{\delta R}^{p}(s) = N_{\delta R}s\left[L_r s - \left(L_r\frac{Y_\beta}{V_o} + L_\beta\right)\right] \tag{6.4-76}$$

$$n_{\delta R}^{\phi}(s) = N_{\delta R}\left[L_r s - \left(L_r\frac{Y_\beta}{V_o} + L_\beta\right)\right] \tag{6.4-77}$$

$$n_{\delta R}^{\psi}(s) = \frac{N_{\delta R}}{s}\left[s^3 - \left(\frac{Y_\beta}{V_o} + L_p\right)s^2 + \left(N_r\frac{Y_\beta}{V_o} - L_\beta\frac{g_o}{V_o}\right)\right] \tag{6.4-78}$$

Side-Force-Effector Transfer-Function Numerators

$$n_{\delta SF}^{r}(s) = \frac{Y_{\delta SF}}{V_o}s[N_\beta s + (N_p L_\beta - N_\beta L_p)] \tag{6.4-79}$$

$$n_{\delta SF}^{\beta}(s) = \frac{Y_{\delta SF}}{V_o}s[s^2 - (N_r + L_p)s + (N_r L_p - N_p L_r)] \tag{6.4-80}$$

$$n_{\delta SF}^{p}(s) = \frac{Y_{\delta SF}}{V_o}s[L_\beta s - (N_\beta L_r - N_r L_\beta)] \tag{6.4-81}$$

$$n_{\delta SF}^{\phi}(s) = \frac{Y_{\delta SF}}{V_o}[L_\beta s - (N_\beta L_r - N_r L_\beta)] \tag{6.4-82}$$

$$n_{\delta SF}^{\psi}(s) = \frac{Y_{\delta SF}}{V_o}[N_\beta s + (N_p L_\beta - N_\beta L_p)] \tag{6.4-83}$$

Numerators for controls that have compound effects can be assembled from the idealized transfer functions. For example, the yaw-rate response to a rudder that produces rolling moment and side force as well as yawing moment ($L_{\delta R}$, $Y_{\delta R}/V_o$, and $N_{\delta R}$) is, from eqs. 6.4-68, 6.4-73, and 6.4-78,

$$n^r_{\delta R}(s) = N_{\delta R}\left[s^3 - \left(\frac{Y_\beta}{V_o} + L_p\right)s^2 + L_p\frac{Y_\beta}{V_o}s - L_\beta\frac{g_o}{V_o}\right]$$

$$+ L_{\delta R}\left(N_p s^2 - N_p\frac{Y_\beta}{V_o}s + N_\beta\frac{g_o}{V_o}\right) + \frac{Y_{\delta R}}{V_o}s[N_\beta s + (N_p L_\beta - N_\beta L_p)]$$

$$= N_{\delta R}s^3 - \left[N_{\delta R}\left(\frac{Y_\beta}{V_o} + L_p\right) - L_{\delta R}N_p - \frac{Y_{\delta R}}{V_o}N_\beta\right]s^2$$

$$+ \left[\frac{Y_\beta}{V_o}(N_{\delta R}L_p - L_{\delta R}N_p) + \frac{Y_{\delta R}}{V_o}(N_p L_\beta - N_\beta L_p)\right]s$$

$$- \frac{g_o}{V_o}(N_{\delta R}L_\beta - L_{\delta R}N_\beta) \tag{6.4-84}$$

The numerators can be factored and put in the general form $k(s - z_1)(s - z_2)\cdots$, with complex roots combined in second-degree terms.

The lateral-directional controls are less easily categorized than the longitudinal controls (Section 5.4). The roll-rate numerator for idealized aileron (roll-control) contains the reduced-order model of the Dutch roll mode. Approximate cancellation of similar terms in the numerator and denominator suggests that aileron control of the roll-spiral modes is largely unaffected by the Dutch roll mode; however, that is not the case when the Dutch roll mode is lightly damped, and particularly when the aileron produces significant yawing effect through $N_{\delta A}$, as noted in the following section on root locus analysis. Neglecting N_p, the yaw-rate and sideslip-angle response to aileron are seen to be proportional to N_β and N_r, respectively, with only partial cancellation of Dutch roll effects.

For small L_r, the idealized rudder controls sideslip with negligible roll-spiral effect. Roll-spiral effects are attenuated but not entirely eliminated in the roll- and yaw-rate response to rudder. The side-force numerators contain both rolling and yawing derivatives.

Frequency responses for yaw rate, sideslip angle, roll rate, and roll angle are shown in Fig. 6.4-8 and 6.4-9, using the stability and control derivatives presented in Table 6.4-1. All of the transfer functions contain an unstable spiral root at very low frequency. With the sign convention used here, the root makes an initial phase-angle contribution of -180 deg to the aileron Bode plots, and the net change in phase angle due to

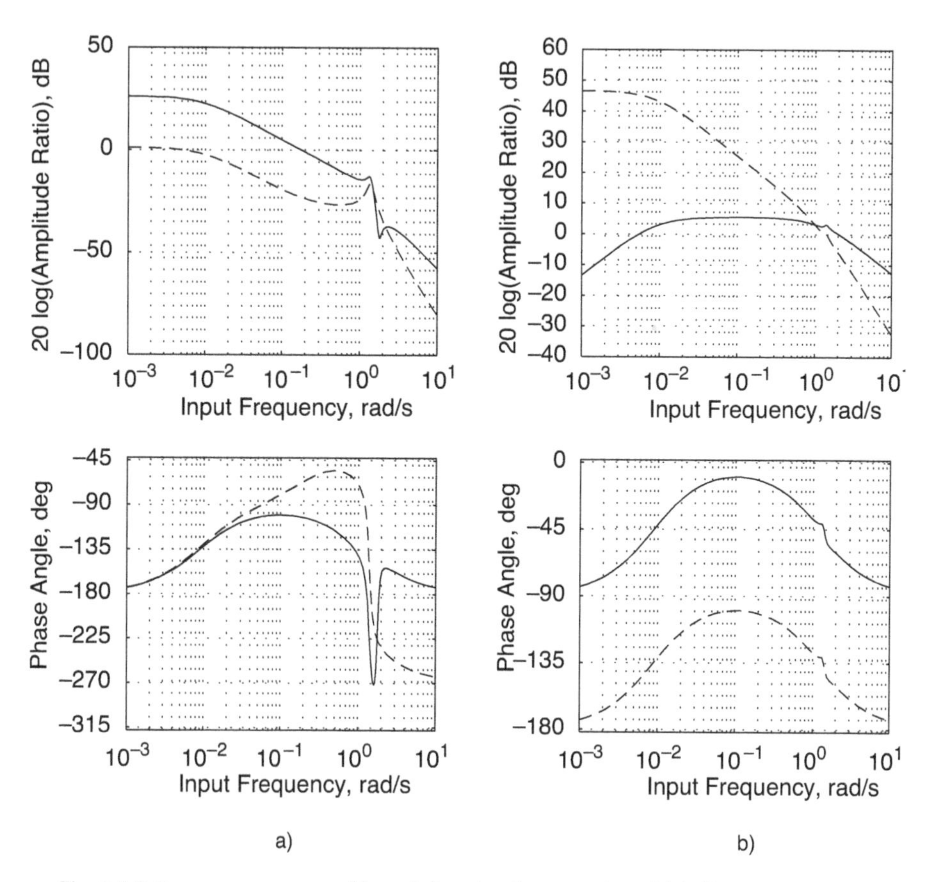

Fig. 6.4-8. Frequency response of lateral-directional state to sinusoidal aileron input. (a) Yaw rate (solid line) and sideslip angle (dashed line). (b) Roll rate (solid) and roll angle (dashed).

the root is +90 deg as input frequency increases (Fig. 6.4-8). The roll and Dutch roll modes are closely spaced; hence, there is a relatively abrupt change in amplitude ratio and phase angle for both yaw rate and sideslip angle at input frequencies between 1 and 1.5 rad/s, with resonance due to the lightly damped Dutch roll mode. Yaw-rate response is notched by complex zeros just above the Dutch roll natural frequency (Fig. 6.4-8a), while a zero below the Dutch roll frequency raises the amplitude and phase angle of the sideslip-angle response (Fig. 6.4-8b). Complex zeros in the roll-rate and roll-angle numerators effectively cancel the Dutch roll component of response to the aileron (Fig. 6.4-8c, d).

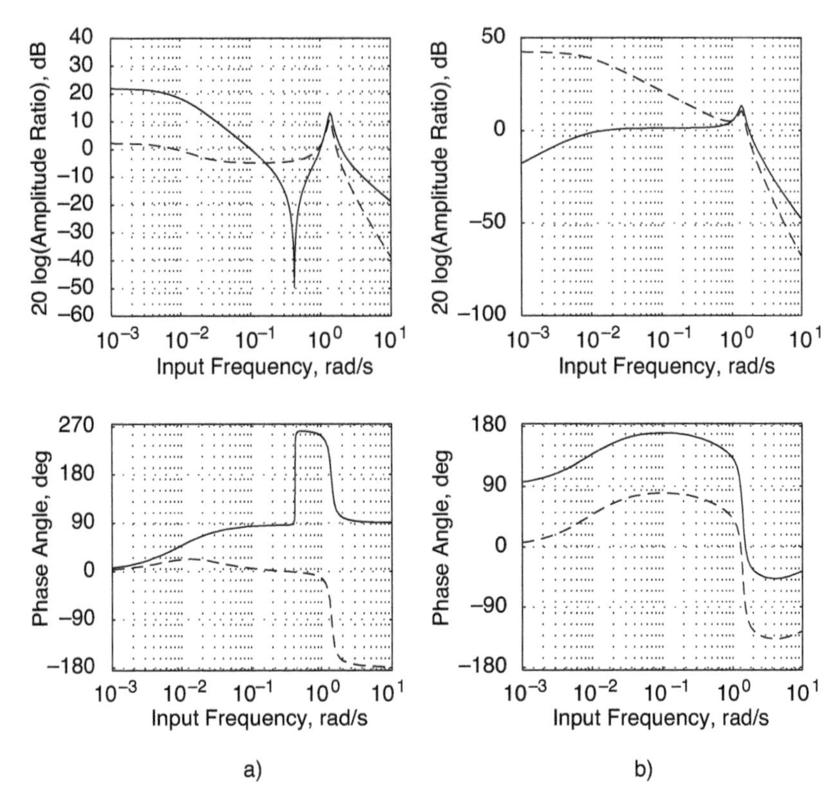

Fig. 6.4-9. Frequency response of lateral-directional state to sinusoidal rudder input. (a) Yaw rate (solid line) and sideslip angle (dashed line). (b) Roll rate (solid) and roll angle (dashed).

The yaw-rate response to sinusoidal rudder input experiences a deep attenuation at $\omega = 0.42$ rad/s, a result of combined rolling and lateral heaving (Fig. 6.4-9a). Zeros in the sideslip-angle transfer function negate the effects of spiral and roll modes (Fig. 6.4-9b), although they do eliminate the unstable long-period mode, and both responses resonate at the Dutch roll natural frequency. The roll-rate response to the rudder is flat between the spiral and Dutch roll frequencies (Fig. 6.4-9c), and roll angle is its integral (Fig. 6.4-9d).

Response to Disturbances

The response to lateral and rolling wind disturbances of varying frequency is described in the same way as the frequency response to controls, replacing $\mathbf{G}_{LD}$ and $\mathbf{J}_{LD}$ by $\mathbf{L}_{LD}$ and $\mathbf{K}_{LD}$. Horizontal wind perturbations induce sideslip, $\Delta\beta_W$, and the roll-rate perturbations expressed as vortical winds, Δp_W, excite the lateral-directional modes of motion. The numerators $n_\delta^x(s)/\Delta(s)$ can be constructed from the control numerators, replacing $L_{\delta A}$ by L_{β_W} and so on.

Root Locus Analysis of Feedback Control

Feeding back elements of the state to the controls produces effects analogous to stability-derivative variations. It is difficult to generalize about the lateral-directional control root loci because of the many alternatives that exist for dynamic coupling. Nevertheless, examples can be based on the model of Table 6.4-1.

Aileron feedback control is principally of interest to augment roll damping or to stabilize the spiral mode. Scalar feedback of yaw rate or sideslip angle destabilizes either the Dutch roll mode (shown in Figs. 6.4-10a, b) or the spiral mode (feeding back with opposite sign). Neither roll-rate nor roll-angle feedback affects the Dutch roll mode, as complex zeros cancel those roots (Figs. 6.4-10c, d). The roll-rate transfer function also contains a zero at the origin, weakly attracting the spiral mode.

Without the extra zero in its transfer function, roll-angle feedback draws the roll and spiral roots together to form a stable roll-spiral oscillation, the underlying mode of "wing leveling" autopilots. From eqs. 6.2-13, 6.4-51, 6.4-72, and 6.4-77, the aileron-to-roll-angle transfer function can be written as

$$\frac{\Delta\phi(s)}{\Delta\delta A(s)} = \frac{L_{\delta A}(s^2 + 2\zeta\omega_n s + \omega_n^2)_{DR_2} + N_{\delta A}\left[L_r s - \left(L_r \dfrac{Y_\beta}{V_o} + L_\beta\right)\right]}{(s - \lambda_s)(s^2 + 2\zeta\omega_n s + \omega_n^2)_{DR_4}(s - \lambda_R)}$$

$$= \frac{k_\phi(s^2 + 2\zeta_\phi\omega_\phi s + \omega_\phi^2)}{(s - \lambda_s)(s^2 + 2\zeta_d\omega_d s + \omega_d^2)(s - \lambda_R)} \qquad (6.4\text{-}85)$$

where the second- and fourth-order models of the Dutch roll polynomial appear in the numerator and denominator. The ratio of natural frequencies ω_ϕ/ω_d is of particular importance in closed-loop control of the roll angle, by either stability augmentation or the pilot. With light damping, ω_ϕ and ω_d are located near the $j\omega$ axis. If $\omega_\phi < \omega_d$, which is equivalent to an otherwise adverse yaw effect, the closed-loop root progresses from ω_d to ω_ϕ with increasing loop gain on an arc to the left of the two singularities; thus the system is more stable at intermediate gain [S-1]. If $\omega_\phi > \omega_d$

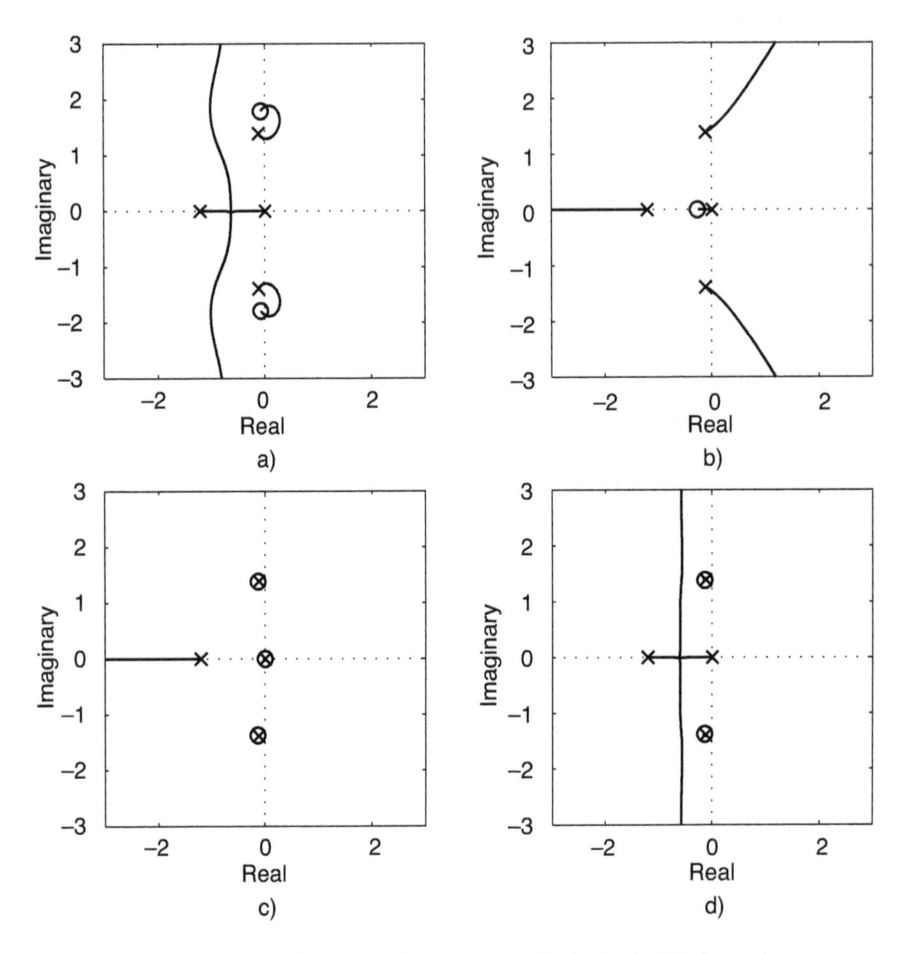

Fig. 6.4-10. Root loci for aileron control. (a) Yaw-rate feedback. (b) Sideslip-angle feedback. (c) Roll-rate feedback. (d) Roll-angle feedback.

(otherswise favorable yaw), the curved root locus between the two goes to the right, decreasing stability, possibly inducing instability if the root crosses the $j\omega$ axis at an intermediate value.

The direct, low-gain effects of yaw-rate and sideslip feedback to the rudder are well predicted by the reduced-order model, with Dutch roll damping being augmented by the former and Dutch roll natural frequency increased by the latter (Figs. 6.4-11a, b). The spiral mode, as well, is stabilized by yaw-rate feedback, although this effect is often purposely

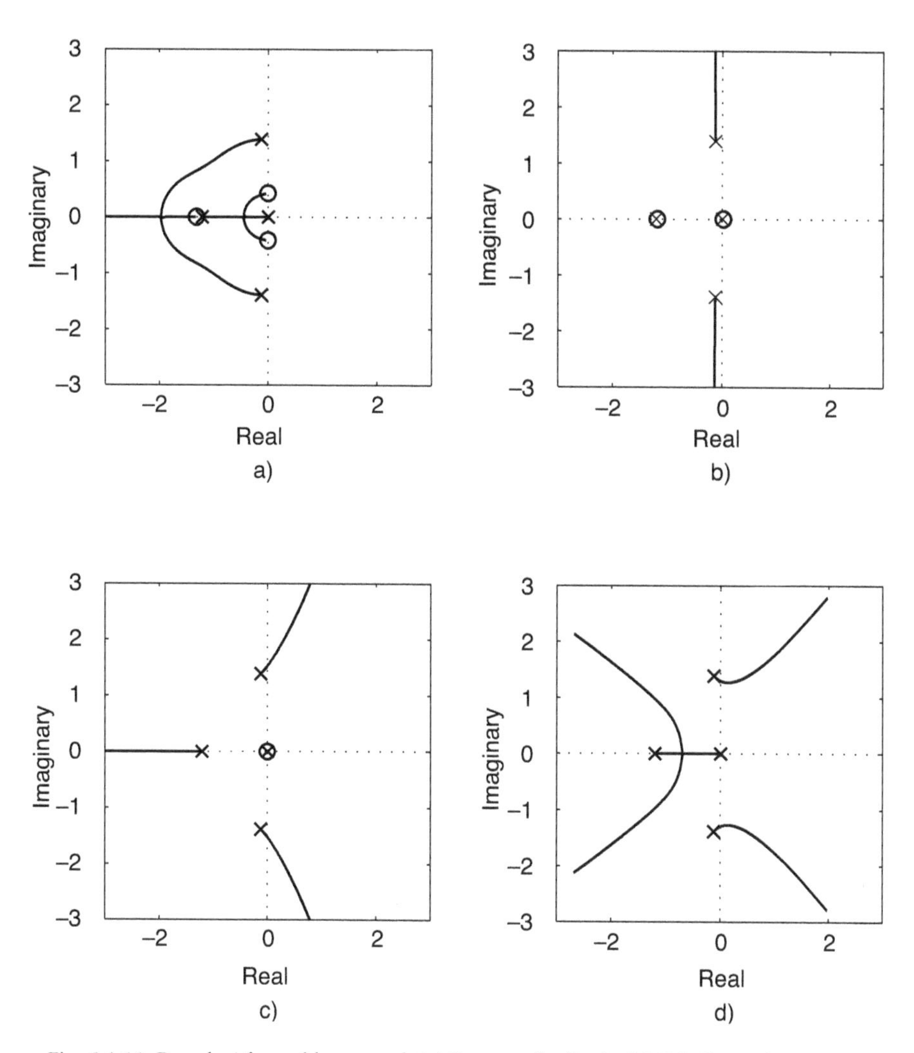

Fig. 6.4-11. Root loci for rudder control. (a) Yaw-rate feedback. (b) Sideslip-angle feedback. (c) Roll-rate feedback. (d) Roll-angle feedback.

"washed out," as described below. Roll-rate and roll-angle feedback to rudder destabilizes either the Dutch roll or spiral mode (Figs. 6.4-11c, d).

Yaw damping has an adverse effect on steady turns. The steady yaw rate that is developed in the turn causes the yaw-rate-to-rudder feedback loop to deflect the rudder in a direction opposing the turn, which is consistent with stabilizing the spiral mode. The problem is solved by adding a *washout* (high-pass) *filter* in the feedback loop. An idealized, first-order high-pass filter has the transfer function

$$H(s) = \frac{s}{s - 1/T_{HP}} \tag{6.4-86}$$

where T_{HP} is the time constant of the filter, typically on the order of one-half s. The filter passes high-frequency signals with negligible distortion while attenuating low-frequency signals (as its steady-state gain is zero). Therefore, the filter passes yaw-rate oscillations on the order of the Dutch roll frequency while holding back the effect of the steady turn.

The closed-loop transfer function of the original yaw damper is

$$\frac{\Delta r(s)}{\Delta r_d(s)} = \frac{\dfrac{\Delta r(s)}{\Delta \delta R(s)}}{1 + k\dfrac{\Delta r(s)}{\Delta \delta R(s)}} \tag{6.4-87}$$

while the closed-loop transfer function with the washout filter becomes

$$\frac{\Delta r(s)}{\Delta r_d(s)} = \frac{\dfrac{\Delta r(s)}{\Delta \delta R(s)}}{1 + k\dfrac{\Delta r(s)}{\Delta \delta R(s)}H(s)} \tag{6.4-88}$$

where k is the feedback gain in both cases. The root locus for the yaw damper with washout solves the equation

$$1 + k\frac{\Delta r(s)}{\Delta \delta R(s)}H(s) = 0 \tag{6.4-89}$$

or

$$\Delta(s) = (s - 1/T_{HP})d(s) + ksn^{r}_{\delta R}(s) = 0 \tag{6.4-90}$$

The washout filter has little effect on Dutch roll damping, requiring at most a slight increase in k to obtain the desired damping ratio.

Lateral-Directional Flying Qualities

The primary lateral-directional piloting action is to control the aircraft's roll angle, and the secondary action is to control sideslip angle. Roll control is implemented by lateral deflection of the control stick or turning of the control wheel, which commands the ailerons, possibly with an interconnect to the rudder to coordinate yaw and to account for high-angle-of-attack effects (Section 7.4). In certain flight regimes, such as slow-speed flight, with large flap deflection, or at high angle of attack, the rudder, commanded by foot pedals in an unaugmented aircraft, produces a rolling moment that is as strong or stronger than the aileron's effect. Rudder deflection is the main method for controlling sideslip angle in a conventional aircraft, either to coordinate a steady turn, to produce a slip during crosswind landing, or to counter disturbances.

The earliest quantitative lateral flying qualities criterion dealt with the minimum acceptable value of an aircraft's roll rate, p, with maximum aileron deflection and rudder fixed. By observing flight test data for a variety of aircraft, Gilruth deduced that the minimum acceptable value of $p_{max}b/2V_o$ was 0.07 [G-3]. (An estimate of maximum steady-state roll rate can be made using eq. 6.3-9.) While this report examined rolling response only, a second report presented judgments about additional lateral-directional (as well as longitudinal) characteristics [G-4], concluding that Dutch roll oscillations should dampen to half amplitude within two cycles and that spiral stability was not a problem for aircraft of the era. Considerable attention was devoted to control force requirements; as controls were unpowered, the maximum force that could be exerted by the pilot was a limiting factor on high-speed maneuverability [T-1].

Transfer Functions and Their Ratios

Neglecting side force due to control, the transfer functions of particular interest are the aileron-to-roll angle (eq. 6.4-85), rudder-to-roll angle

$$\frac{\Delta\phi(s)}{\Delta\delta R(s)} = \frac{L_{\delta R}(s^2 + 2\zeta\omega_n s + \omega_n^2)_{DR_2} + N_{\delta R}\left[L_r s - \left(L_r \dfrac{Y_\beta}{V_o} + L_\beta\right)\right]}{(s - \lambda_s)(s^2 + 2\zeta\omega_n s + \omega_n^2)_{DR_4}(s - \lambda_R)}$$
$$= \frac{k_{\delta R}^{\phi}(s^2 + 2\zeta\omega_n s + \omega_n^2)_{\delta R}^{\phi}}{(s - \lambda_s)(s^2 + 2\zeta\omega_n s + \omega_n^2)_{DR_4}(s - \lambda_R)} \qquad (6.4\text{-}91)$$

and the corresponding roll-rate transfer functions (i.e., the roll-angle transfer functions multiplied by s), aileron-to-sideslip angle

$$\frac{\Delta\beta(s)}{\Delta\delta A(s)} = \frac{L_{\delta A}\left[\left(\dfrac{g}{V_o} - N_p\right)s - N_r \dfrac{g}{V_o}\right] + N_{\delta A}\left(s^2 - L_p s - L_r \dfrac{g}{V_o}\right)}{(s - \lambda_s)(s^2 + 2\zeta\omega_n s + \omega_n^2)_{DR_4}(s - \lambda_R)}$$
$$= \frac{k_{\delta A}^{\beta}(s^2 + 2\zeta\omega_n s + \omega_n^2)_{\delta A}^{\beta}}{(s - \lambda_s)(s^2 + 2\zeta\omega_n s + \omega_n^2)_{DR_4}(s - \lambda_R)} \qquad (6.4\text{-}92)$$

and rudder-to-sideslip angle

$$\frac{\Delta\beta(s)}{\Delta\delta R(s)} = \frac{L_{\delta R}\left[\left(\frac{g}{V_o} - N_p\right)s - N_r\frac{g}{V_o}\right] + N_{\delta R}\left(s^2 - L_p s - L_r\frac{g}{V_o}\right)}{(s-\lambda_s)(s^2+2\zeta\omega_n s+\omega_n^2)_{DR_4}(s-\lambda_R)}$$

$$= \frac{k_{\delta R}^{\beta}(s^2+2\zeta\omega_n s+\omega_n^2)_{\delta R}^{\beta}}{(s-\lambda_s)(s^2+2\zeta\omega_n s+\omega_n^2)_{DR_4}(s-\lambda_R)} \qquad (6.4\text{-}93)$$

The rolling response to the aileron is similar to the second-order model response (Section 6.3), with the added effect of favorable or adverse yaw and superimposed oscillation due to the Dutch roll mode. Except as it may aid turn coordination, the sideslip response to the aileron falls into the "nuisance" category, especially if the Dutch roll is lightly damped. The term $N_{\delta R}L_{\beta}$ normally would dominate in the rudder-to-roll-angle numerator (eq. 6.4-91), and the response is thus proportional to the dihedral effect L_{β}.

Of course, rudder deflection produces both roll and sideslip response. The ratio of their transfer functions is the ratio of their numerators:

$$\left[\frac{\Delta\varphi(s)}{\Delta\beta(s)}\right]_{\delta R} = \frac{L_{\delta R}(s^2+2\zeta\omega_n s+\omega_n^2)_{DR_2} + N_{\delta R}\left[L_r s - \left(L_r\frac{Y_\beta}{V_o} + L_\beta\right)\right]}{L_{\delta R}\left[\left(\frac{g}{V_o} - N_p\right)s - N_r\frac{g}{V_o}\right] + N_{\delta R}\left(s^2 - L_p s - L_r\frac{g}{V_o}\right)}$$

$$= \frac{k_{\delta R}^{\varphi}(s^2+2\zeta\omega_n s+\omega_n^2)_{\delta R}^{\varphi}}{k_{\delta R}^{\beta}(s^2+2\zeta\omega_n s+\omega_n^2)_{\delta R}^{\beta}} \qquad (6.4\text{-}94)$$

Thus, the association between the response to rudder deflection of $\Delta\phi$ and $\Delta\beta$ is frequency dependent. Under the assumption that the polynomials factor as complex pairs, the ratios are constant at very low and high frequencies, exhibiting a resonance at $(\omega_n)_{\delta R}^{\beta}$ and a notch at $(\omega_n)_{\delta R}^{\varphi}$.

The roll and yaw rate responses to aileron deflection are

$$\frac{\Delta p(s)}{\Delta\delta A(s)} = \frac{L_{\delta A}s(s^2+2\zeta\omega_n s+\omega_n^2)_{DR_2} + N_{\delta A}s\left[L_r s - \left(L_r\frac{Y_\beta}{V_o} + L_\beta\right)\right]}{(s-\lambda_s)(s^2+2\zeta\omega_n s+\omega_n^2)_{DR_4}(s-\lambda_R)}$$

$$= \frac{k_\phi s(s^2+2\zeta_\phi\omega_\phi s+\omega_\phi^2)}{(s-\lambda_s)(s^2+2\zeta\omega_n s+\omega_n^2)_{DR_4}(s-\lambda_R)} \qquad (6.4\text{-}95)$$

$$\frac{\Delta r(s)}{\Delta\delta A(s)} = \frac{L_{\delta A}\left(N_p s^2 - N_p \frac{Y_\beta}{V_o} s + N_\beta \frac{g}{V_o}\right) + N_{\delta A}\left[s^3 - \left(\frac{Y_\beta}{V_o} + L_p\right)s^2 + L_p \frac{Y_\beta}{V_o} s - L_\beta \frac{g}{V_o}\right]}{(s-\lambda_s)(s^2 + 2\zeta\omega_n s + \omega_n^2)_{DR_4}(s-\lambda_R)}$$

$$= \frac{k_{\delta A}^r (s - \lambda_{\delta A}^r)(s^2 + 2\zeta\omega_n s + \omega_n^2)_{\delta A}^r}{(s-\lambda_s)(s^2 + 2\zeta\omega_n s + \omega_n^2)_{DR_4}(s-\lambda_R)} \tag{6.4-96}$$

The yaw-rate response can have either favorable or adverse effects on turn coordination, initially swinging the nose into or out of the turn. With negligible yaw due to aileron, $N_{\delta A}$, the ratio of these responses is

$$\frac{\Delta r(s)}{\Delta p(s)} = \frac{L_{\delta A}\left(N_p s^2 - N_p \frac{Y_\beta}{V_o} s + N_\beta \frac{g}{V_o}\right)}{L_{\delta A} s(s^2 + 2\zeta\omega_n s + \omega_n^2)_{DR_2}} = \frac{k_{\delta A}^r (s^2 + 2\zeta\omega_n s + \omega_n^2)_{\delta A}^r}{k_\phi s(s^2 + 2\zeta_\phi\omega_\phi s + \omega_\phi^2)} \tag{6.4-97}$$

The s in the denominator indicates that roll rate leads yaw rate by 90 deg at very low and high input frequencies. As a consequence, the yaw-rate response to the aileron is approximately in phase with the roll angle response, except near the Dutch roll natural frequency.

Flying Qualities Criteria

Modal properties are important for evaluating both longitudinal and lateral-directional flying qualities, but the latter places greater emphasis on time-domain rather than frequency-domain criteria. A possible reason is that longitudinal maneuvers (e.g., load factor command, speed adjustment, and altitude trim) are perceived as more continuous and wavelike, while lateral-directional maneuvers (e.g., wing leveling, coordinated turn, and control against crosswind) are viewed as more discrete. Dutch roll motion is deliberately excited only when commanding sideslip angle [A-4]; otherwise, it is a corrupting factor that is readily described in the time domain. Rolling response is well specified by the time required to achieve a desired state.

Post-World-War-II research concluded that unstable spiral-mode time constants of 20 s or more and stable roll-mode time constants of 1 s or less were satisfactory, with $L_{\delta A}$ in the range of 2–10 rad/s^2 ([E-1], after [O-1]). Values of 2 to 3 s or less were found to be satisfactory for the Dutch roll's *time to half amplitude* $T_{1/2}$, defined as $0.693/\zeta\omega_n$, where $\zeta > 0$. Smaller values of $T_{1/2}$ corresponded to higher values of $|\Delta\phi\,\Delta\beta|/(V_o\sqrt{\rho/\rho_{SL}}$ in the range from 1.6 to 4.9 deg/m/s. The modified $\Delta\phi/\Delta\beta$

criterion provided improved correlation with pilot opinions when velocity and altitude were taken into account, requiring higher damping with higher $\Delta\phi/\Delta\beta$, slower airspeed, or increased altitude. In addition, satisfactory pilot opinions were obtained when ω_ϕ/ω_d was between 1 and 1.2, with Dutch roll damping ratio of 0.1 or higher [A-4]. This result suggests that the pilot prefers favorable aileron yaw if the damping reduction due to roll control-loop closure described after equation 6.4-85 cannot induce instability.

Some of the more salient features of the military specification [A-3] are reviewed briefly, focusing on Level 1 requirements. MIL-F-8785C tightens the roll-mode time constant for all classes of aircraft but relaxes the spiral-mode requirement. The time constant must be 1.4 s or less for all aircraft types in Flight Phase B (gradual maneuvering; Section 4.7) and for medium (land-based) to large aircraft in Flight Phases A and C (rapid maneuvering and terminal flight). Small aircraft, highly maneuverable aircraft, and carrier-based medium aircraft must have roll-mode time constant of 1 s or less in Flight Phases A and C. Minimum times to double amplitude of 12 (A and C) and 20 (B) s or greater satisfy the spiral mode criterion for all aircraft classes, beginning with a bank angle of 20 deg and with cockpit controls free. A roll-spiral oscillation is not permitted in Flight Phase A; if it occurs in B or C, $\zeta\omega_n$ must be greater than 0.5 rad/s, which is consistent with a roll-mode time constant of 2 s or less.

Rolling performance is further specified by the time to achieve a given bank-angle change in [A-3]. Distinctions are made not only for airplane class and flight phase but for airspeed, and the bank-angle change used for evaluation ranges from 30 to 360 deg. Small aircraft (Class I) are allowed 1.3, 1.7, and 1.3 s for 60 deg change in Flight Phases A, B, and C. Medium aircraft (Class II) are evaluated for 45 deg change; increments of 1.4, 1.9, and 1.8 s are allowed for land-based aircraft and 1.4, 1.9, and 1.0 s for carrier-based aircraft in the three flight phases. Large aircraft (Class III) are allowed from 1.5 to 2.5 s to make to 30 deg changes at various speeds and flight phases. Highly maneuverable aircraft (Class IV) have the most detailed breakdown of rolling requirements, ranging from 1.1 s to reach 30 deg at low speed (A and C) to 1.7 s to reach 90 deg at high speed (B) and 2 s to reach 90 deg at very low speed (B). Separate specifications are made for air combat (360 deg in 2.8–4.1 s, depending on speed) and ground attack (180 deg in 3 s). Maximum roll control force to achieve this performance ranges from 90 to 110 N (20 to 25 lb) for lateral stick and 90 to 220 N (20 to 50 lb) for a control wheel.

The aircraft must furnish minimum values of Dutch roll ζ, ω_n, and $\zeta\omega_n$ for both fixed and free controls [A-3]. The most stringent damping-ratio requirement is placed on Class IV in air combat or ground attack ($\zeta \geq 0.4$). Otherwise, the requirement for all classes is $\zeta \geq 0.19$ (A) or 0.08 (B and C). The natural frequency must be greater than 1 rad/s for Classes I and IV (A and C) and for Class II (carrier) aircraft (C), and it must be greater than 0.4 rad/s in the remaining cases. The mode's minimum total damping $\zeta\omega_n$ is 0.35 (A) and 0.15 (B) rad/s for all classes, 0.15 rad/s for I,II (carrier), and IV, and 0.1 rad/s for II (land) and III.

Residual oscillations arising from the Dutch roll mode are treated in a variety of ways. For a step input, the first minimum in roll-rate response should not be less than 60 percent (A and C) or 25 percent (B) of the first overshoot amplitude. Sideslip excursions should be limited to 6 (A) or 10 (B and C) deg for adverse aileron yaw and 2 (A) or 3 (B and C) deg for favorable yaw. Limits on the oscillatory amplitude of roll rate, bank angle, and sideslip angle in response to a roll step command are computed from the first three peaks in step response and are functions of the phase angle of the $\Delta p/\Delta\beta$ eigenvector ratio [A-3].

Like the longitudinal criteria (Section 5.4), lateral-directional flying qualities criteria must take flight control system implementation into account [A-2]. Rate-command/attitude-hold is particularly appropriate for roll control, as the airplane's natural response to aileron is roll rate, a frequent goal is to hold bank angle, and feedback logic can suppress undesirable Dutch roll oscillations. It also becomes possible to command roll about the velocity vector rather than the vehicle's x axis, an important alternative for flight at high angles of attack (Section 7.4). Nevertheless, there is the possibility of pilot-induced oscillation (e.g., roll *ratcheting* [H-3]) when the pilot is subject to significant side force, as in a forward cockpit during velocity-axis roll, and when control gains and time delays are high [G-2].

6.5 Control Mechanisms, Nonlinearity, and Time Delay

Complexities arising from lateral-directional control mechanisms increased as aircraft became larger and more flexible, as they flew faster, and as they required greater maneuverability. Hinge moments grew, and mechanical elasticity allowed control surface motions that were not commanded by the pilot. While many of these problems are now solved by the use of fully powered controls, that option was not available in the early years. The natural response of designers was to reduce hinge moments by reshaping the leading and trailing edges of the movable surfaces, providing overhang at the ends of surfaces, sealing the gaps between moving and fixed surfaces, and locating hinge axes for reduced loads. However, what worked at one flight condition was not always suitable at another, and nonlinear dynamic problems like "aileron buzz" or "control buffeting" and "rudder snaking" often arose at the edges of the flight envelope. Even in an era of fully powered controls, free oscillations of control surfaces should be well behaved; in the event of a system failure that allows a control surface to float, its natural behavior should not be hazardous to the aircraft.

The generic control mechanism of Section 5.5 can be applied to understanding these problems, and modifications are introduced to portray nonlinear effects. Lateral-directional rigid-body/control mechanism coupling is idealized by Table 6.5-1, which contains the primary blocks for independent (or truncated) models of rigid-body aircraft motion and control mechanism dynamics, along with the secondary blocks that couple the dynamic models. The combined model is further truncated in this section to show the coupled yawing/rudder and rolling/aileron motions separately.

Here, we focus on the analysis of *rudder snaking*, which has been attributed to interactions between rigid-body and control-system dynamics [A-1, G-3, G-4, S-1], *aileron buzz*, and the *actuation-delay effect*. For snaking to occur, the Dutch roll mode must be stable but lightly damped, the rudder must be aerodynamically overbalanced to lighten hinge moment, and it must be free to float. With these characteristics, the yaw motions of the aircraft are well behaved until the aircraft encounters a disturbance, causing the free rudder to wiggle. The wiggle turns into a limit-cycle oscillation, with the aircraft and rudder moving synchronously. We use a nonlinear model of rudder dynamics to define the problem, then describe details of the motion using state-space plots. Quasilinear models capture significant elements of the nonlinearities, revealing conditions that are likely to produce a limit cycle [G-3]. The aileron buzz limit cycle can arise when the aileron is aerodynamically overbalanced at low deflection, destabilizing the aileron-actuator mode, and underbalanced at higher deflection (a nonlinear effect), limiting the extent of an otherwise divergent oscillation. Forces produced by spoilers are variously described by nonlinear or time-delayed models, and the effect of time delay on roll-angle regulation via lateral control is demonstrated.

TABLE 6.5-1 FOUR-BLOCK STRUCTURE OF THE LATERAL-DIRECTIONAL/CONTROL DYNAMICS MODEL

Lateral-Directional Dynamics	Control-to-Lateral-Directional Coupling
Lateral-Directional-to-Control Coupling	Control Mechanism Dynamics

Rudder Control Mechanism

As in Section 5.5, a second-order model of rudder dynamics is formed. We admit the possibility of nonlinear effects of rudder damping, restoring moment, and floating tendency, $H(\Delta\dot{\delta}R, \delta R, \beta)$ added to linear effects of yaw rate and control

$$\Delta\ddot{\delta E} = H(\Delta\dot{\delta}R, \delta R, \beta) + H_r\Delta r + H_C\Delta f_C \tag{6.5-1}$$

Nonlinearities in $H(\delta R)$ and $H(\beta)$ contribute to the *rudder lock-in* phenomenon described in [A-1] for unpowered systems, and they establish the *blowdown limit* of deflection in fully powered systems. In the remainder, the yaw-rate (Δr) effect on rudder hinge moment is neglected; because control-free motions are examined, the command input is ignored. For nonlinear analysis of snaking, rudder damping is expressed as *Coulomb friction*, rudder restoring moment and floating tendency are expressed as *threshold functions*, and a mechanical limit on control surface deflection is approximated by a *hardening spring*.

The equivalent linearization of rudder dynamics is expressed as

$$\Delta\ddot{\delta E} = H_{\dot{\delta}R}\Delta\dot{\delta}R + H_{\delta R}\Delta\delta R + H_\beta\Delta\beta \tag{6.5-2}$$

where

$$H_{\dot{\delta}R} = \frac{1}{I_\delta}\left(C_{H\dot{\delta}R}\frac{1}{2}\rho V^2 S_R b_R + k_{\delta\dot{R}}\right) \tag{6.5-3}$$

$$H_{\delta R} = \frac{1}{I_\delta}\left(C_{H\delta R}\frac{1}{2}\rho V^2 S_R b_R + k_{\delta R}\right) \tag{6.5-4}$$

$$H_\beta = \frac{1}{I_\delta}\left(C_{H\beta}\frac{1}{2}\rho V^2 S_R b_R + k_\beta\right) \tag{6.5-5}$$

These expressions contain derivatives of the dimensionless hinge-moment coefficient C_H, as defined in [S-1], and I_δ is the effective inertia of the rudder control system. S_R and b_R are the reference area and length for the rudder, and the k_i are mechanical or kinematic constants.

The nonlinear analyses consider one nonlinearity at a time; therefore, we use eq. 6.5-2 with a single term replaced by its nonlinear counterpart. The control mechanism state vector is defined by deflection and rate:

$$\Delta \mathbf{x}_{\delta R} \triangleq \begin{bmatrix} \Delta\delta_1 \\ \Delta\delta_2 \end{bmatrix} = \begin{bmatrix} \Delta\delta R \\ \Delta\delta\dot{R} \end{bmatrix} \tag{6.5-6}$$

Considering the rudder damping nonlinearity alone, the second-order model is represented by a pair of first-order equations:

$$\Delta\dot{\delta}_1 = \Delta\delta_2 \tag{6.5-7}$$

$$\Delta\dot{\delta}_2 = H(\Delta\delta_2) + H_{\delta R}\Delta\delta_1 + H_\beta \Delta\beta \tag{6.5-8}$$

The *Coulomb friction* function (Fig. 6.5-1a) is simple in form, but it is a "hard" nonlinearity because of its step discontinuity. The model provides a constant moment that opposes the direction of motion; it cannot push or pull, it merely retards other moments. Denoting the corresponding hinge-moment magnitude by h_{cf}, a simple description of the effect is

$$H(\Delta\delta_2) = H_{cf} = -h_{cf}\frac{\Delta\delta_2}{|\Delta\delta_2|} \tag{6.5-9}$$

However, an additional effect is not portrayed by this equation: if

$$|H_{\delta R}\Delta\delta_1 + H_\beta \Delta\beta| < h_{cf} \tag{6.5-10}$$

then $\Delta\dot{\delta}_2 = 0$. There is a *dead zone* in which no motion occurs unless the remaining forces overcome the friction; hence, a better description of the friction effect in this example is

$$H_{cf} = \begin{cases} -h_{cf}\dfrac{\Delta\delta_2}{|\Delta\delta_2|}, & |H_{\delta R}\Delta\delta_1 + H_\beta\Delta\beta| \ge |h_{cf}| \\ -(H_{\delta R}\Delta\delta_1 + H_\beta\Delta\beta), & |H_{\delta R}\Delta\delta_1 + H_\beta\Delta\beta| < |h_{cf}| \end{cases} \tag{6.5-11}$$

The second part of this definition allows the friction model to cancel the remaining moments in the dead zone. This effect is distinct from *stiction*, which describes a "breakout" force that exceeds the constant Coulomb value.

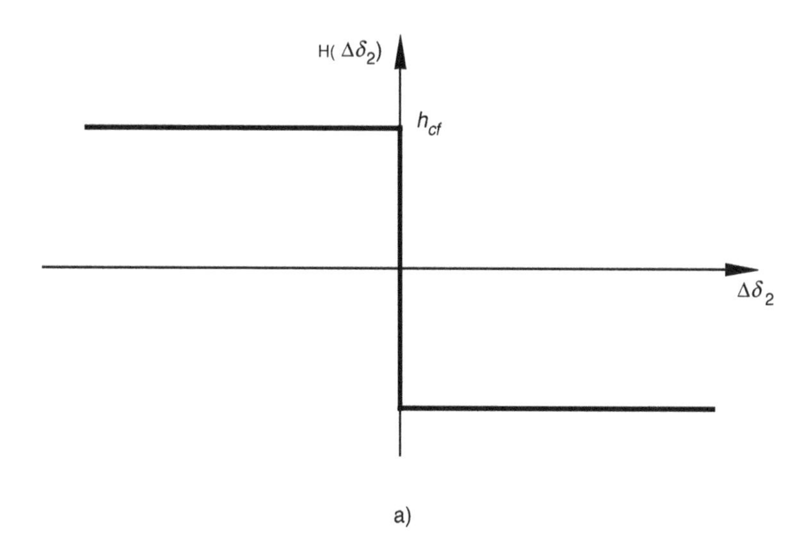

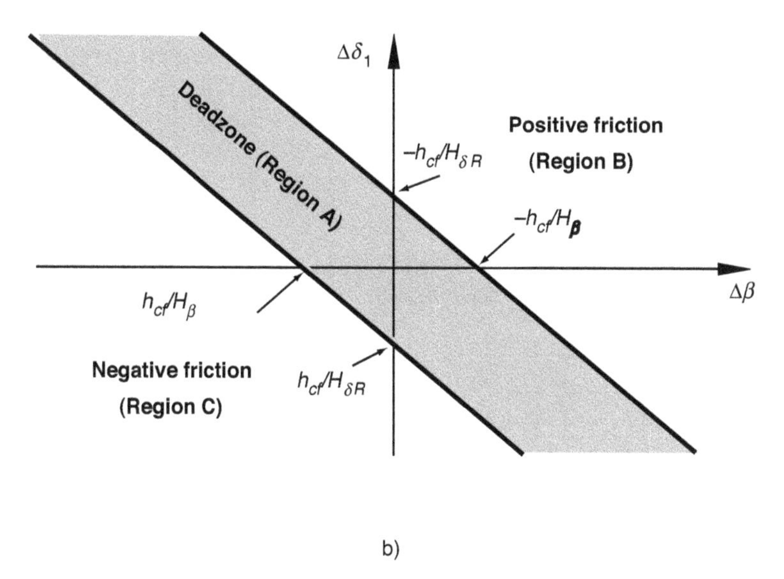

Fig. 6.5-1. Rudder hinge friction model. (a) Simple Coulomb friction function. (b) Dead zone within which no motion occurs.

There are three regions to the rudder friction model: the dead zone (A) and two surrounding regions (B and C) where rudder motion produces positive or negative friction (Fig. 6.5-1b). This is a *piecewise-linear* model due to the nature of the nonlinearity (eq. 6.5-11). Within the central region, $\Delta\delta R$ can be fixed at any value in the range $(\pm h_{cf} - H_\beta\,\Delta\beta_o)/H_{\delta R}$ as long as the sum of the rudder and sideslip torques is below the friction torque h_{cf}. $\Delta\beta_o$ is the sideslip angle that exists when the external torque first falls below the friction torque. The slope of the boundaries between Regions A, B, and C is fixed by $-H_\beta/H_{\delta R}$. For negative $H_{\delta R}$, negative slope is associated with overbalanced floating tendency, and positive slope is associated with underbalance.

Dutch Roll/Rudder Coupling

In Region A, the rudder position is fixed; we assume that the reduced-order Dutch roll model of Section 6.2 applies and is stable. The Dutch roll state $\Delta\mathbf{x}_D$ contains yaw rate Δr and sideslip angle $\Delta\beta$. With the control vector defined by eq. 6.5-6, $\mathbf{F}_{DR}$ and $\mathbf{G}_{DR}$ are given by

$$\mathbf{F}_{DR} = \begin{bmatrix} N_r & N_\beta \\ -1 & \dfrac{Y_\beta}{V_o} \end{bmatrix} \tag{6.5-12}$$

$$\mathbf{G}_{DR} = \begin{bmatrix} N_{\delta R} & 0 \\ 0 & 0 \end{bmatrix} \tag{6.5-13}$$

The characteristic equation governing yaw-rate and sideslip-angle motions in Region A is

$$\Delta(s) = s^2 - \left(N_r + \frac{Y_\beta}{V_o}\right)s + \left(N_\beta + N_r\frac{Y_\beta}{V_o}\right) = 0 \tag{6.5-14}$$

Neglecting steady disturbances, the equilibrium point is defined by

$$\Delta r^* = \frac{-N_{\delta R}\dfrac{Y_\beta}{V_o}}{\left(N_\beta + \dfrac{Y_\beta}{V_o}\right)}\Delta\delta R^* \tag{6.5-15}$$

$$\Delta\beta^* = \frac{-N_{\delta R}}{\left(N_\beta + \dfrac{Y_\beta}{V_o}\right)} \Delta\delta R^* \tag{6.5-16}$$

where $\Delta\delta R^*$ is the rudder position fixed at the point of entering the region. Trajectories within Region A propagate according to eq. 6.2-39, with initial conditions on Δr and $\Delta\beta$ set by the entry point. As long as the system is stable and the trajectory remains in Region A, the state terminates at the equilibrium point.

The coupled aircraft/rudder equation for Regions B and C is

$$\begin{bmatrix} \Delta\dot{\mathbf{x}}_{DR} \\ \Delta\dot{\mathbf{x}}_{\delta R} \end{bmatrix} = \begin{bmatrix} \mathbf{F}_{DR} & \mathbf{G}_{DR} \\ \mathbf{L}_{\delta R} & \mathbf{F}_{\delta R} \end{bmatrix} \begin{bmatrix} \Delta\mathbf{x}_{DR} \\ \Delta\mathbf{x}_{\delta R} \end{bmatrix} + \begin{bmatrix} 0 \\ \mathbf{g}_{\delta R} \end{bmatrix} \tag{6.5-17}$$

where the rudder model is specified by

$$\mathbf{L}_{\delta R} = \begin{bmatrix} 0 & 0 \\ 0 & H_\beta \end{bmatrix} \tag{6.5-18}$$

$$\mathbf{F}_{\delta R} = \begin{bmatrix} 0 & 1 \\ H_\delta & H_{\dot\delta} \end{bmatrix} \tag{6.5-19}$$

$$\mathbf{g}_{\delta R} = \begin{bmatrix} 0 \\ H_{cf} \end{bmatrix} \tag{6.5-20}$$

H_{cf} appears as a forcing bias for a linear system whose characteristic equation is

$$\Delta(s) = s^4 - \left(N_r + \frac{Y_\beta}{V_o}\right)s^3 + \left(N_\beta + N_r\frac{Y_\beta}{V_o} - H_\delta\right)s^2 + H_\delta\left(N_r + \frac{Y_\beta}{V_o}\right)s$$
$$+ N_{\delta R}H_\beta - H_\delta\left(N_\beta + N_r\frac{Y_\beta}{V_o}\right) \tag{6.5-21}$$

with $H_{\dot\delta} = 0$. An otherwise stable system could be driven to instability with large enough adverse values of H_β (<0 for overbalancing) or H_δ (>0 for static instability).

Setting the time derivatives to zero, the system's equilibrium response can be calculated as

$$\begin{bmatrix} \Delta r^* \\ \Delta\beta^* \\ \Delta\delta R^* \\ \Delta\dot{\delta R} \end{bmatrix} = \frac{\pm h_{cf}}{N_{\delta R}H_\beta - H_\delta\left(N_\beta + N_r \dfrac{Y_\beta}{V_o}\right)} \begin{bmatrix} -N_{\delta R}\dfrac{Y_\beta}{V_o} \\ -N_{\delta R} \\ N_\beta + N_r\dfrac{Y_\beta}{V_o} \\ 0 \end{bmatrix} \qquad (6.5\text{-}22)$$

with $+h_{cf}$ corresponding to Region B and $-h_{cf}$ pertaining to Region C. As expected, the equilibrium rudder rate is zero; however, this leads to a contradiction because the friction bias $\pm h_{cf}$ does not apply if $\Delta\delta R$ is fixed. Furthermore, the two points lie outside the regions in which their effects are felt; thus, they are inaccessible to trajectories within those regions. We conclude that solutions to eq. 6.5-22 are not equilibrium points but *singular points*, as defined in Section 4.3. These points affect the trajectories in each region without representing possible equilibria.

Trajectories of the aircraft-rudder system (eq. 6.5-17) are readily calculated using MATLAB. The *separatrices* between regions are four-dimensional *hyperplanes* (flat surfaces of more than three dimensions) that satisfy

$$|H_{\delta R}\Delta\delta R + \mathbf{H}_\beta \Delta\beta| = \pm h_{cf} \qquad (6.5\text{-}23)$$

and whose *intersections* with the $\Delta\delta R$-$\Delta\beta$ plane are shown in Fig. 6.5-1b. The state history is continuous across the separatrices, but the time derivatives of $\Delta\mathbf{x}(t)$ may change discontinuously. For underbalanced or slightly overbalanced rudder ($H_\beta > -\varepsilon$, where ε is a small, positive, problem-dependent constant) and initial condition in Region B or C, switching between regions that is caused by friction aids convergence of the trajectory to the equilibrium point of Region A.

For moderate-to-large rudder overbalance ($H_\beta < -\varepsilon$), the response is typically as follows:

- Small $\Delta\beta_o$ (within Region A): - Convergence to origin with no rudder motion
- Medium $\Delta\beta_o$ (outside Region A): - Convergence to Region A equilibrium with oscillatory rudder motion

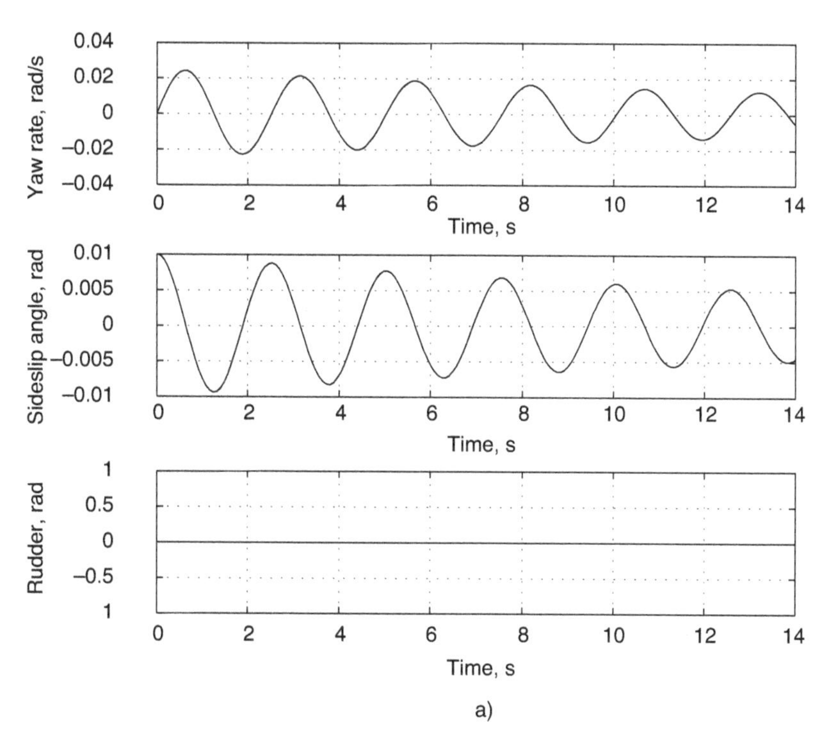

Fig. 6.5-2. Response of the Dutch roll/rudder system to a small initial sideslip angle ($\Delta\beta_o = 0.01$ rad). (a) History of yaw rate, sideslip angle, and rudder deflection.

- Large $\Delta\beta_o$ (outside Region A):
  - Convergence to limit-cycle with periodic rudder motion
  - Chaotic motion (modified parameters)

Friction has an effect that is similar to a child pumping a swing to large amplitude. The combination of friction and overbalance provides a rudder moment that adds energy to the Dutch roll oscillation. The negative separatrix angle (Figure 6.5-1b) allows the amplitude to grow each time a switch out of Region A occurs, until a balance of net effects in the three regions produces a periodic oscillation.

Examples of the yawing motion of a hypothetical aircraft with an overbalanced rudder that is free to float are shown in Figs. 6.5-2–6.5-5. Each figure presents a history of yaw rate, sideslip angle, and rudder

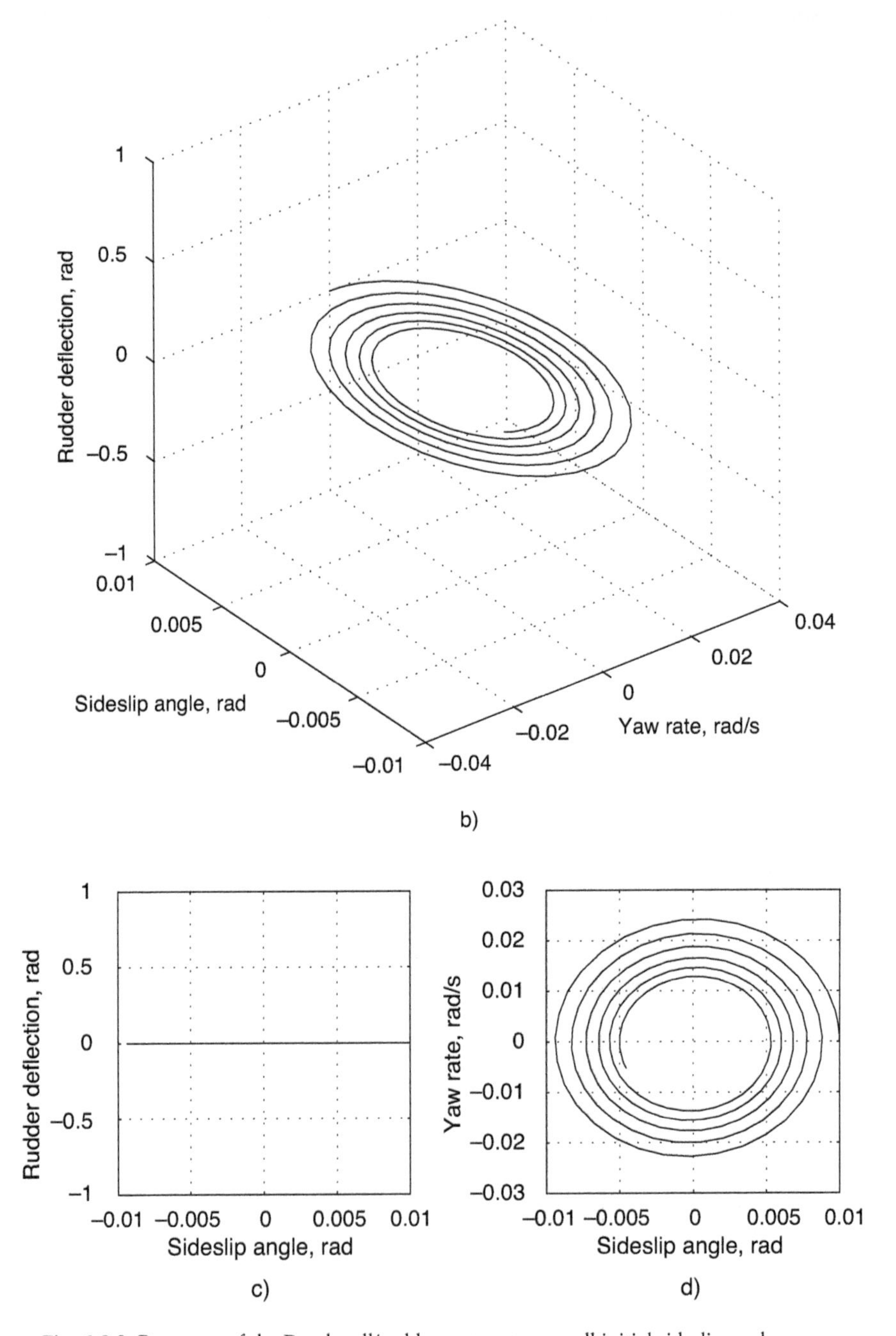

Fig. 6.5-2. Response of the Dutch roll/rudder system to a small initial sideslip angle ($\Delta\beta_o = 0.01$ rad). (b) Yaw rate (x, to right) versus sideslip angle (y, to left) versus rudder deflection (z, vertical), $\Delta\beta_o = 0.01$ rad. (c) Rudder deflection (vertical) versus sideslip angle (horizontal). (d) Yaw rate versus sideslip angle.

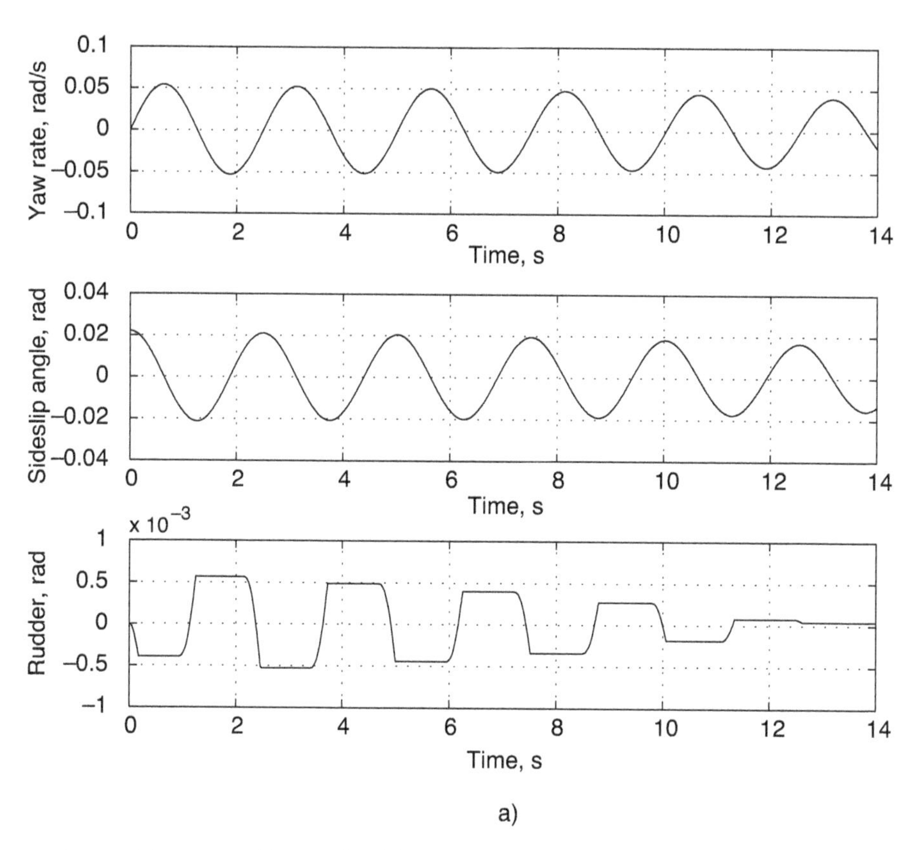

Fig. 6.5-3. Response of the Dutch roll/rudder system to a moderate initial sideslip angle ($\Delta\beta_o = 0.022$ rad). (a) History of yaw rate, sideslip angle, and rudder deflection.

angle, a state-space plot of these three variables, and projections of the trajectory on the $\Delta\delta R$-$\Delta\beta$ and Δr-$\Delta\beta$ planes. The model parameters are

$N_r = -0.1$

$N_\beta = 6.25$

$N_{\delta R} = -6.25$

$Y_\beta / V_o = 0$

$H_\beta = -6$

$H_\delta = -50$

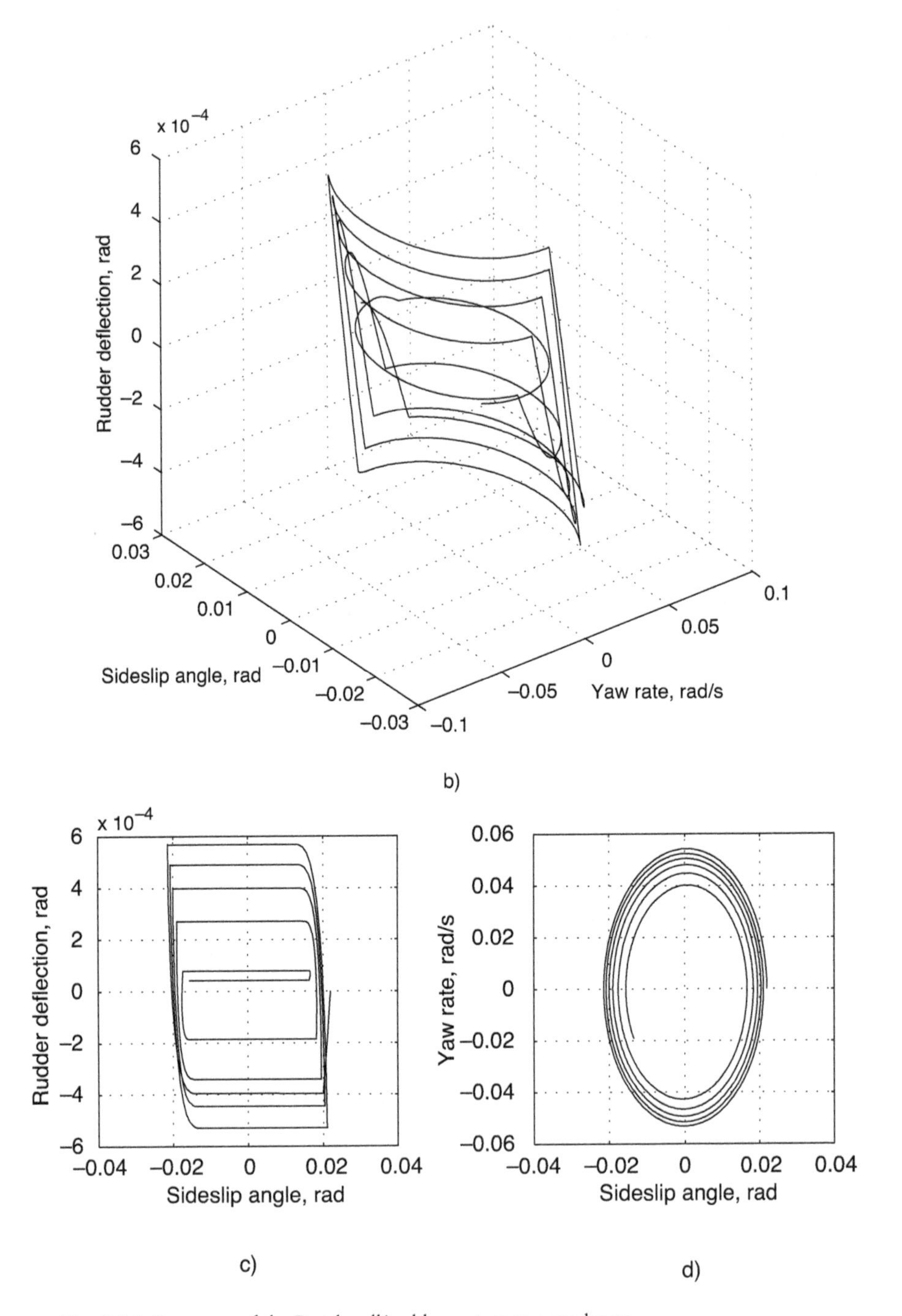

Fig. 6.5-3. Response of the Dutch roll/rudder system to a moderate initial sideslip angle ($\Delta\beta_o = 0.022$ rad). (b) Yaw rate (x, to right) versus sideslip angle (y, to left) versus rudder deflection (z, vertical). (c) Rudder deflection (vertical) versus sideslip angle (horizontal). (d) Yaw rate versus sideslip angle.

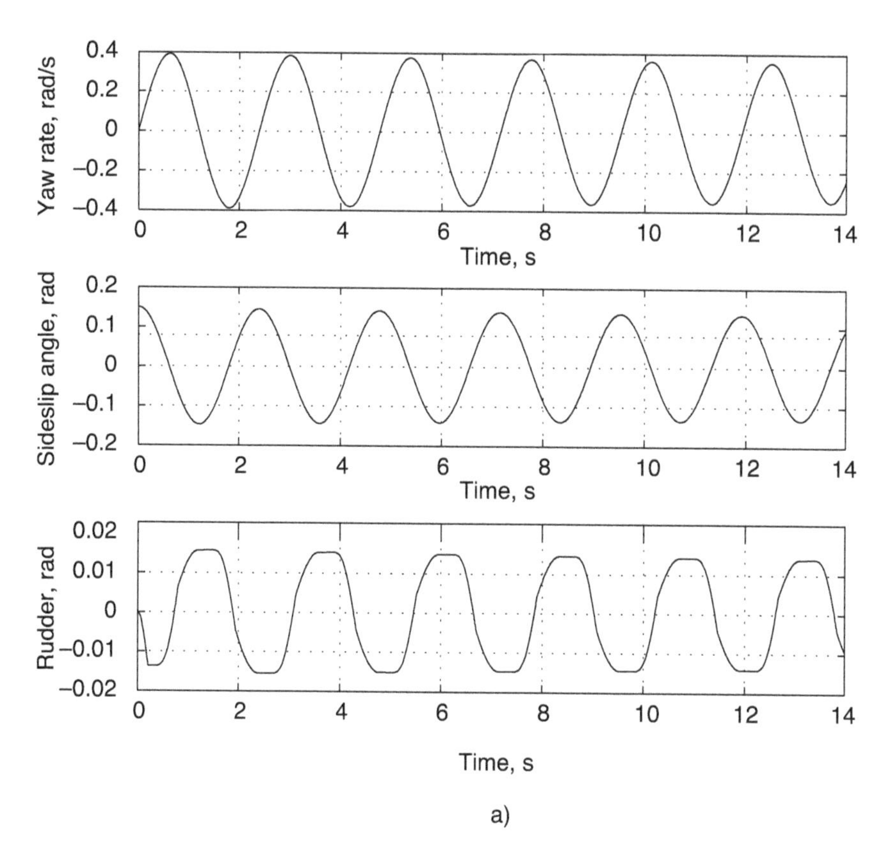

Fig. 6.5-4. Response of the Dutch roll/rudder system to a large initial sideslip angle ($\Delta\beta_o = 0.15$ rad). (a) History of yaw rate, sideslip angle, and rudder deflection.

$$H_{\dot{\delta}} = 0$$

$$b_{cf} = 0.1$$

producing a control-fixed Dutch roll natural frequency of 2.5 rad/s and a rudder natural frequency of about 7 rad/s. The Dutch roll has light linear damping; damping of the rudder mechanism is due to friction alone.

The figures illustrate the progression of system response from linear convergence to limit cycle; with parameter changes, a large initial condition leads to *chaotic motion*, deterministic motion that appears random over time [G-6]. Small changes in initial condition lead to large differences in trajectories over time. Chaos is a phenomenon of nonlinear systems that is

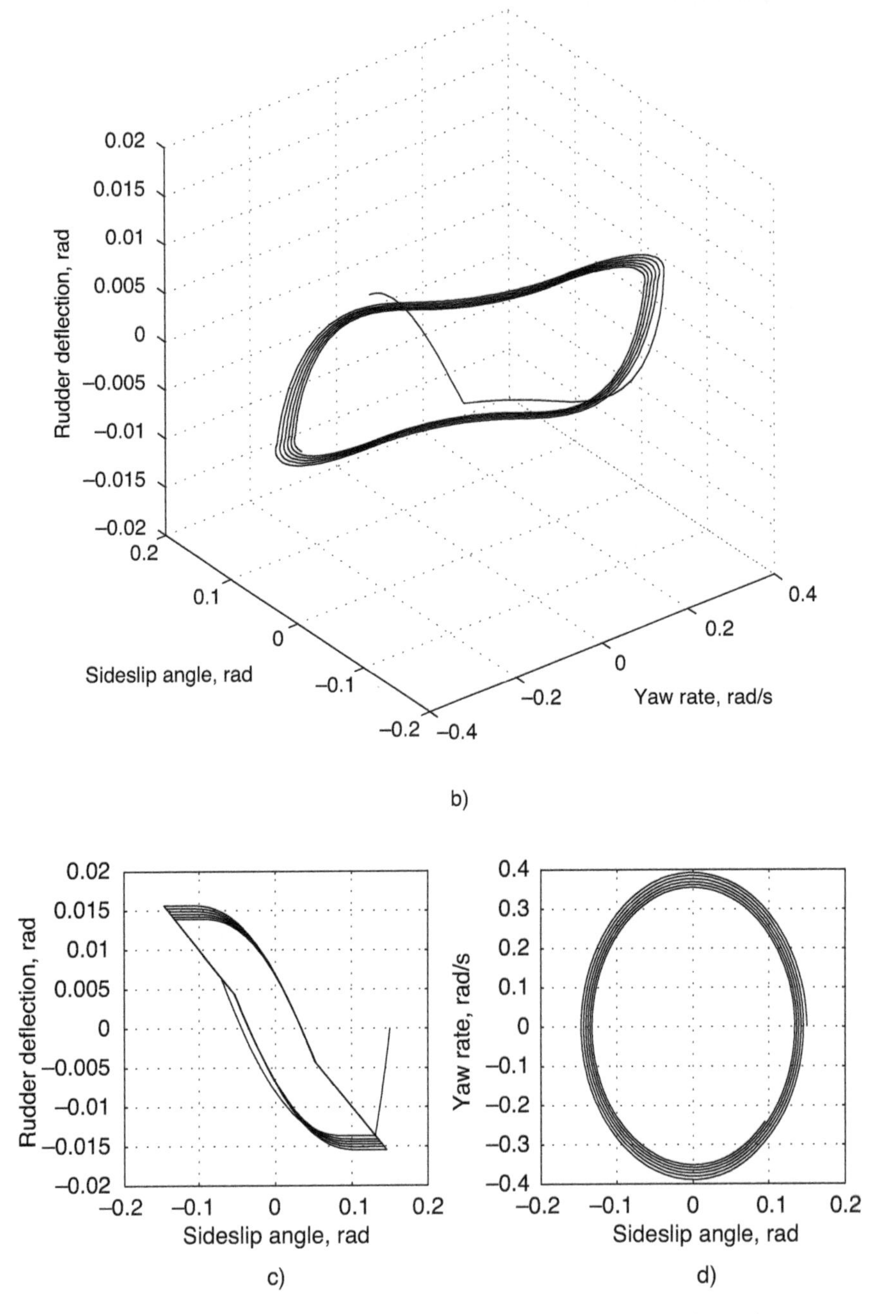

Fig. 6.5-4. Response of the Dutch roll/rudder system to a large initial sideslip angle ($\Delta\beta_o = 0.15$ rad). (b) Yaw rate (x, to right) versus sideslip angle (y, to left) versus rudder deflection (z, vertical). (c) Rudder deflection (vertical) versus sideslip angle (horizontal). (d) Yaw rate versus sideslip angle.

linked to low effective damping and singularity in the parameters of the system model.

For $\Delta\beta_o = 0.01$ rad, the trajectory stays within Region A, and no rudder motion is excited (Fig. 6.5-2). The Δr-$\Delta\beta$ trace progresses in a counterclockwise direction due to sign conventions for yaw rate and sideslip angle. For $\Delta\beta_o = 0.022$ rad, the rudder is nudged away from zero, but the yaw-sideslip oscillation is hardly affected, converging within Region A (Fig. 6.5-3). Nevertheless, state-space and phase-plane plots (Fig. 6.5-3b, c, d) illustrate the rudder switching that results from friction and the free rudder. For $\Delta\beta_o = 0.15$ rad, the trajectory quickly converges to a limit cycle (Fig. 6.5-4). Rudder switching is apparent, and portions of the trajectory follow the separatrices (Fig. 6.5-4c). Increasing overbalance H_β to -25 and control effect $N_{\delta R}$ to -12.25 while introducing a small amount of unstable

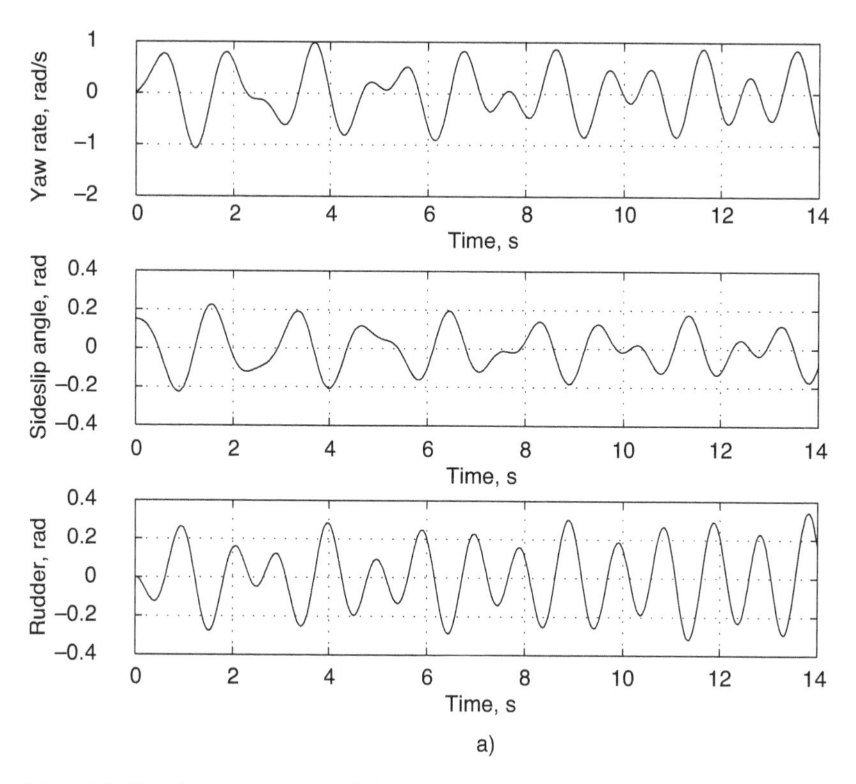

Fig. 6.5-5. Chaotic response of a modified Dutch roll/rudder system to a large initial sideslip angle ($\Delta\beta_o = 0.15$ rad). (a) History of yaw rate, sideslip angle, and rudder deflection.

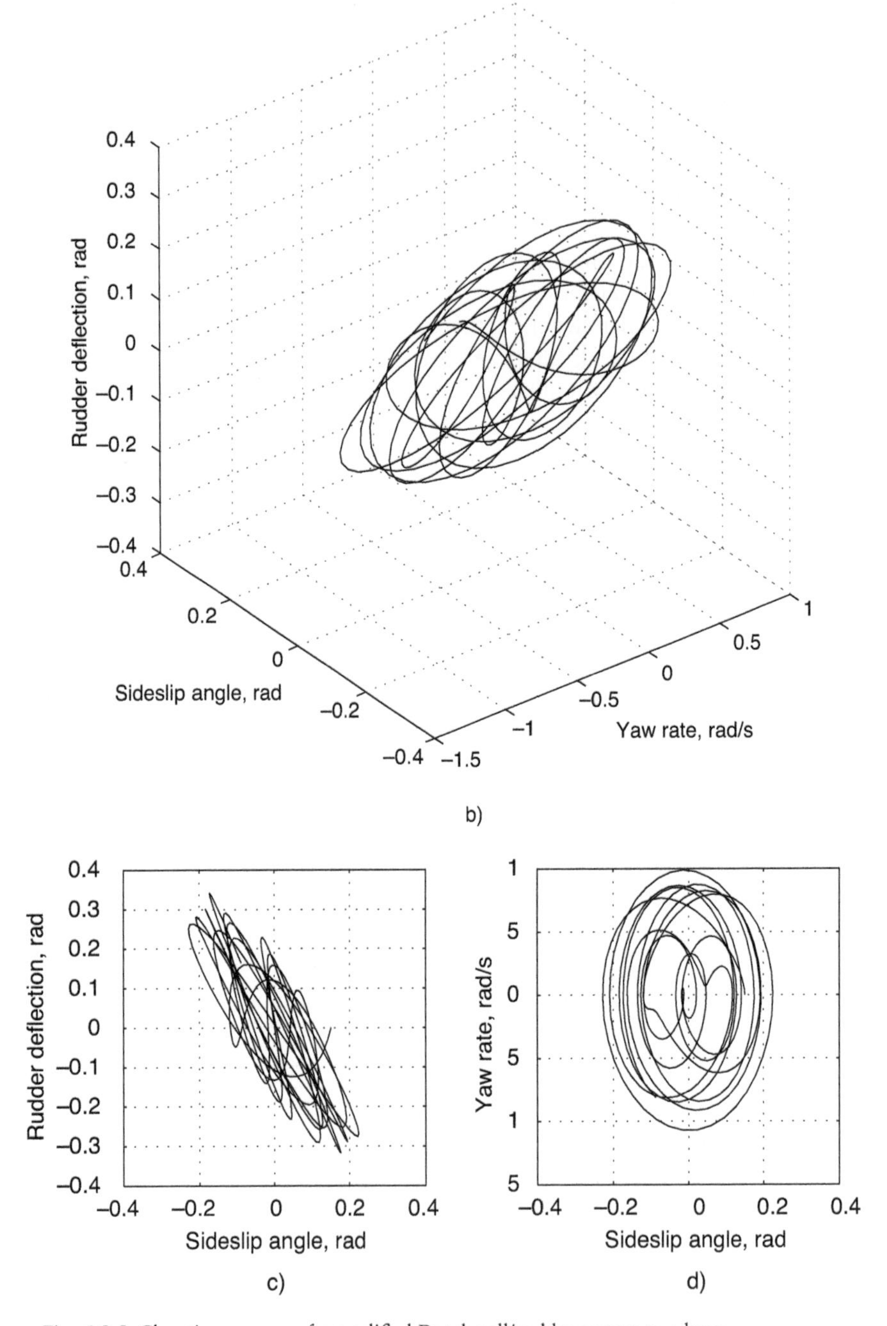

Fig. 6.5-5. Chaotic response of a modified Dutch roll/rudder system to a large initial sideslip angle ($\Delta\beta_o = 0.15$ rad). (b) Yaw rate (x, to right) versus sideslip angle (y, to left) versus rudder deflection (z, vertical). (c) Rudder deflection (vertical) versus sideslip angle (horizontal). (d) Yaw rate versus sideslip angle.

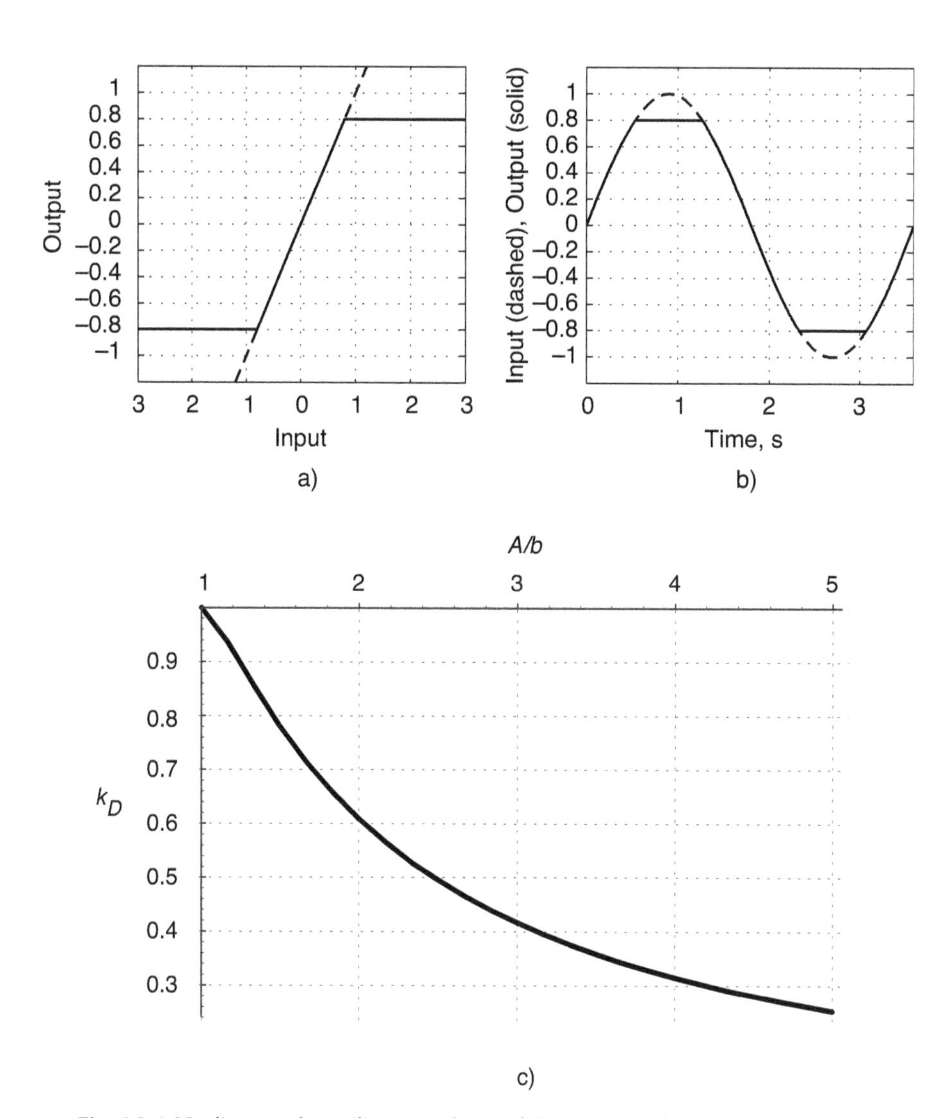

Fig. 6.5-6. Nonlinear and quasilinear attributes of the saturation function. (a) Saturation function (input, dashed; output solid). (b) Clipping due to saturation (input, dashed; output solid). (c) Describing function multiplier k_D versus ratio of sinusoidal amplitude to saturation level A/b.

mechanical damping ($H_{\dot{\delta}} = 0.05$) produces chaotic behavior in the model (Fig. 6.5-5). The motions are quasiperiodic but not strictly repeated.

Quasilinear Representation of Nonlinearity

Even with the "hard" friction nonlinearity, the yaw-rate and sideslip responses shown above have waveforms that are nearly sinusoidal. We can take advantage of this characteristic to build an equivalent linear model of the system. A *quasilinear model* does not capture fine details of the motion: however, it can identify configurations that are likely to experience limit cycles, and it allows the use of linear analysis tools. A principal distinction between "true" linearization and quasilinearization is that the former applies only at a point and ignores the amplitudes of the state and control, while the latter approximates average effects over a region and is amplitude dependent.

The symmetric *saturation function* (Fig. 6.5-6a) provides a good example of the difference between linearization and quasilinearization:

$$f_{sat}(x) = \begin{cases} -b, & x \leq -c \\ kx, & -c < x < c \\ b, & x \geq c \end{cases} \tag{6.5-24}$$

Here, $b = kc$. The saturation nonlinearity portrays an input-output effect that has limits, such as the control moment that a control surface can exert on the aircraft. The *linear model* centered on $x = 0$,

$$\Delta f_L(\Delta x) = k\Delta x \tag{6.5-25}$$

is exact as long as the amplitude of Δx stays within $\pm c$. Let Δx be a sine wave of amplitude A:

$$\Delta x = A \sin \omega t \triangleq A \sin \phi \tag{6.5-26}$$

Then $\Delta f(\Delta x)$ is a perfect sine wave for $A \leq c$ and is a "clipped" sine wave for $A > c$ (Fig. 6.5-6b).

A quasilinear approximation to $\Delta f(\Delta x)$, which is called a *sinusoidal-input describing function* [G-1, G-2], produces a pure sinusoid that is a *least-square-error approximation* to the clipped sine wave. The *quasilinear model* is

$$\Delta f_Q(\Delta x) = k_D(A)\,\Delta x \tag{6.5-27}$$

where k_D is the amplitude-dependent *multiplier* of the describing function. For symmetric, single-valued nonlinearities, k_D is a real number. If the nonlinearity is not symmetric with respect to the input, the describing function takes the form $\Delta f_Q(\Delta x) = k_D(A)\Delta x + b_D$, where b_D is a bias term. If the nonlinearity is multivalued (e.g., hysteresis), k_D is a complex number. Describing functions depend on the input waveform as well as the input magnitudes. *Random-input describing functions* are based on the expected values of inputs and outputs, as described in [G-1, G-2].

The quadratic (or square) error J between the symmetric nonlinear function and its quasilinear approximation over one cycle of the input is

$$J = \int_{-\pi}^{\pi} [f(A \sin \phi) - k_D A \sin \phi]^2 \, d\phi$$
$$= 2 \int_{-\pi/2}^{\pi/2} [f(A \sin \phi) - k_D A \sin \phi]^2 \, d\phi \tag{6.5-28}$$

The change in integration limits is appropriate because the integrated error in $(-\pi/2, \pi/2)$ equals the integrated error that occurs as the oscillation progresses from $\pi/2$ to $-\pi/2$. If the limits were not changed, the two effects would cancel rather than add.

The error must be minimized with respect to the describing-function multiplier. At the minimum value $\partial J/\partial k_D = 0$:

$$\frac{\partial J}{\partial k_D} = 0 = 4 \int_{-\pi/2}^{\pi/2} [f(A \sin \phi) - k_D A \sin \phi] A \sin \phi \, d\phi \tag{6.5-29}$$

leading to

$$\int_{-\pi/2}^{\pi/2} k_D A \sin^2 \phi \, d\phi = \int_{-\pi/2}^{\pi/2} f(A \sin \phi) \sin \phi \, d\phi \tag{6.5-30}$$

The left side of the equation is evaluated as

$$\int_{-\pi/2}^{\pi/2} k_D A \sin^2 \phi \, d\phi = \frac{k_D A \pi}{2} \tag{6.5-31}$$

Hence, the sinusoidal-input describing function multiplier for a symmetric nonlinear function is calculated as

$$k_D = \frac{2}{\pi A} \int_{-\pi/2}^{\pi/2} f(A \sin \phi) \sin \phi \, d\phi \tag{6.5-32}$$

This is the coefficient of the first sine term in a Fourier series representation of $f(x)$.

For the *saturation function*, the integral of eq. 6.5-32 is broken into three parts. Representing the oscillation as rotation about a circle of radius kA, the angle at which saturation occurs is

$$\phi_s = \pm \sin^{-1} \frac{b}{kA} \tag{6.5-33}$$

With eq. 6.5-24, the integral becomes

$$\begin{aligned} k_{D_{sat}} &= \frac{2}{\pi A} \int_{-\pi/2}^{-\phi_s} (-b) \sin\phi \, d\phi + \frac{2}{\pi A} \int_{-\phi_s}^{\phi_s} (kA \sin\phi) \sin\phi \, d\phi \\ &\quad + \frac{2}{\pi A} \int_{-\phi_s}^{\pi/2} (b) \sin\phi \, d\phi \\ &= \frac{1}{\pi} \left[\frac{4b}{A} \sqrt{1 - \frac{b^2}{k^2 A^2}} + 2k \sin^{-1} \frac{b}{kA} + k \sin\left(2 \sin^{-1} \frac{b}{kA} \right) \right] \end{aligned} \tag{6.5-34}$$

for $A > b/k$. For $A \leq b/k$, there is no saturation, and $k_D = k$. A plot of k_D with $k = 1$ (Fig. 6.5-6c) illustrates the decreasing effective gain that comes with increasing input amplitude.

Mechanical friction can occur wherever two surfaces have sliding contact. The describing function for symmetric *Coulomb friction* is derived by substituting eq. 6.5-9 in eq. 6.5-32, with $b = h_{cf}$,

$$k_{D_{cf}} = \frac{2}{\pi A} \left[\int_{-\pi/2}^{0} (b) \sin\phi \, d\phi + \int_{0}^{\pi/2} (-b) \sin\phi \, d\phi \right] = \frac{-4b}{\pi A} \tag{6.5-35}$$

As applied to friction, the signal $A\sin\phi$ represents a rate rather than a displacement. The effective gain goes toward $-\infty$ as the input amplitude goes to zero (Fig. 6.5-7).

A control surface may have negligible effect for small deflections, for example, when the trailing edge moves back and forth within a thick boundary layer without changing the hinge or control moment. The symmetric *threshold function* has zero output when the input amplitude A is less than c, with a linear increase in output amplitude above c (Fig. 6.5-8a):

$$f_{thresh}(x) = \begin{cases} k(x + c), & x \leq -c \\ 0, & -c < x < c \\ k(x - c), & x \geq c \end{cases} \tag{6.5-36}$$

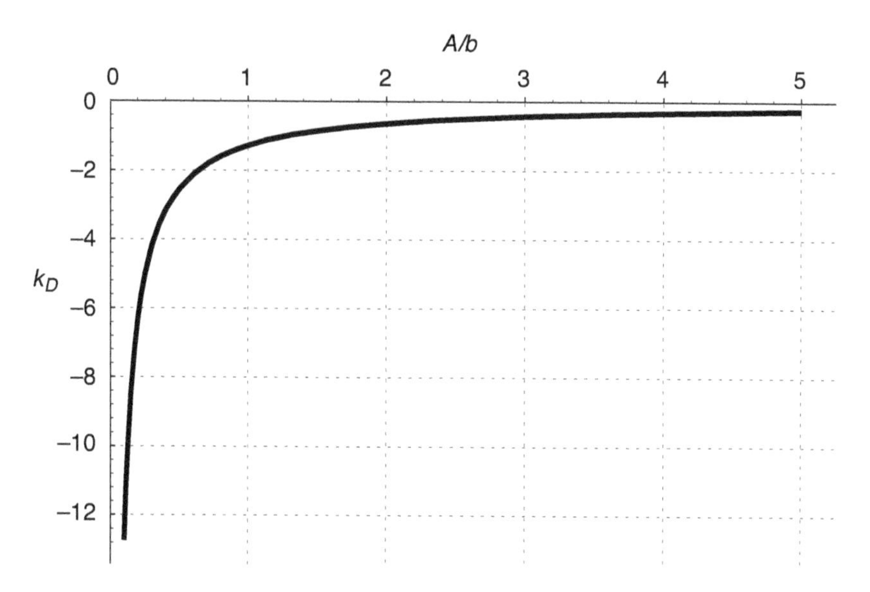

Fig. 6.5-7. Describing function multiplier for the Coulomb friction function.

The phase angle at which the dead zone is exceeded is

$$\phi_t = \pm \sin^{-1} \frac{c}{A} \tag{6.5-37}$$

The describing function multiplier (Fig. 6.5-8b) is

$$\begin{aligned} k_{D_{thresh}} &= \frac{2}{\pi A} \int_{-\pi/2}^{-\phi_t} k(A \sin \phi + c) \sin \phi \, d\phi + \frac{2}{\pi A} \int_{-\phi_t}^{\phi_t} (0) \sin \phi \, d\phi \\ &\quad + \frac{2}{\pi A} \int_{-\phi_t}^{\pi/2} k(A \sin \phi - c) \sin \phi \, d\phi \\ &= \frac{k}{\pi A} \left[\pi A - 4c \sqrt{1 - \frac{c^2}{A^2}} - 2A \sin^{-1} \frac{c}{A} + A \sin\left(2 \sin^{-1} \frac{c}{A} \right) \right] \end{aligned} \tag{6.5-38}$$

for $A > c$, and the multiplier equals zero for $A \le c$.

Control deflections and rates are subject to hard limits on their magnitudes, as a consequence of mechanical stops and flow limits in hydraulic circuits. We cannot numerically model the limit as an infinite hinge moment or flow restriction, but we can impose a large increase in the spring constant at (or just before) the limit. A suitable *hardening-spring*

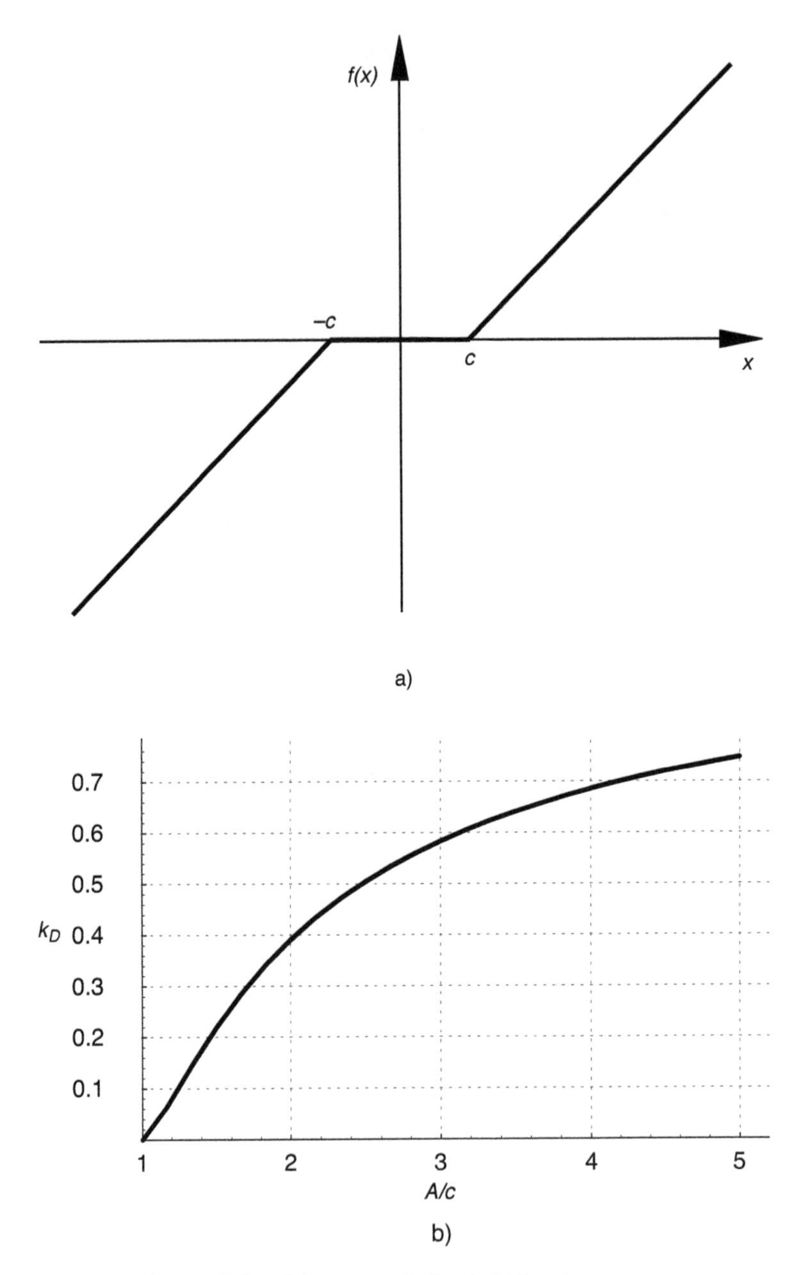

Fig. 6.5-8. Characteristics of the symmetric threshold function. (a) Threshold function. (b) Describing function multiplier.

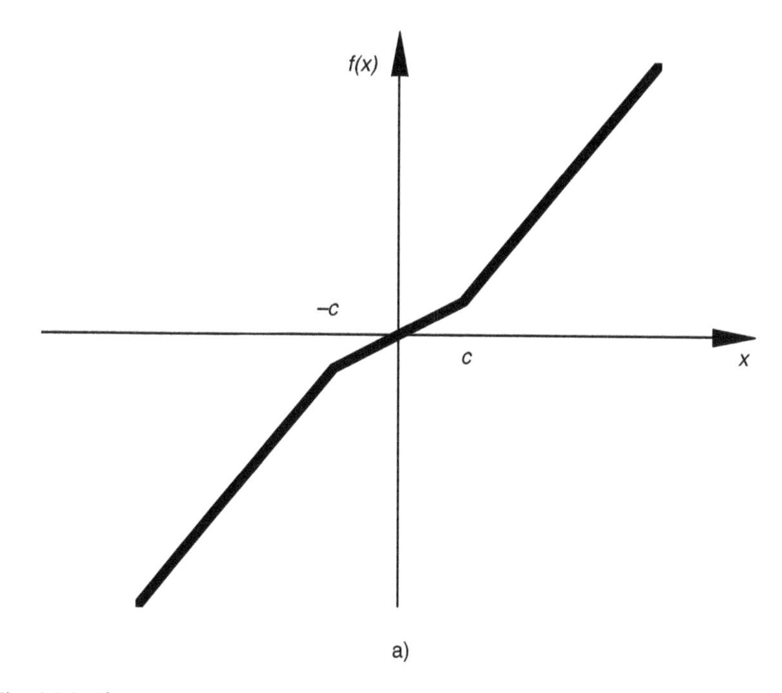

Fig. 6.5-9. Characteristics of the hardening-spring function. (a) Hardening-spring function.

nonlinearity takes the form (Fig. 6.5-9a)

$$f_{hs}(x) = \begin{cases} k_1 c + k_2(x + c), & x \leq -c \\ k_1 c, & -c < x < c \\ -k_1 c + k_2(x - c), & x \geq c \end{cases} \quad (6.5\text{-}39)$$

where k_1 is the unrestrained spring constant and k_2 is a very large constant (both less than zero for restoring effect). The describing function multiplier is

$$k_{D_{hs}} = \frac{1}{\pi A}\left\{\frac{2c}{A}(k_1 - 2k_2)\sqrt{1 - \frac{c^2}{A^2}} + \frac{1}{\pi}(ck_1 + Ak_2) + 2[k_1(A - c) - Ak_2]\sin^{-1}\frac{c}{A} + A(k_2 - k_1)\sin\left(2\sin^{-1}\frac{c}{A}\right)\right\} \quad (6.5\text{-}40)$$

as shown in Fig. 6.5-9b.

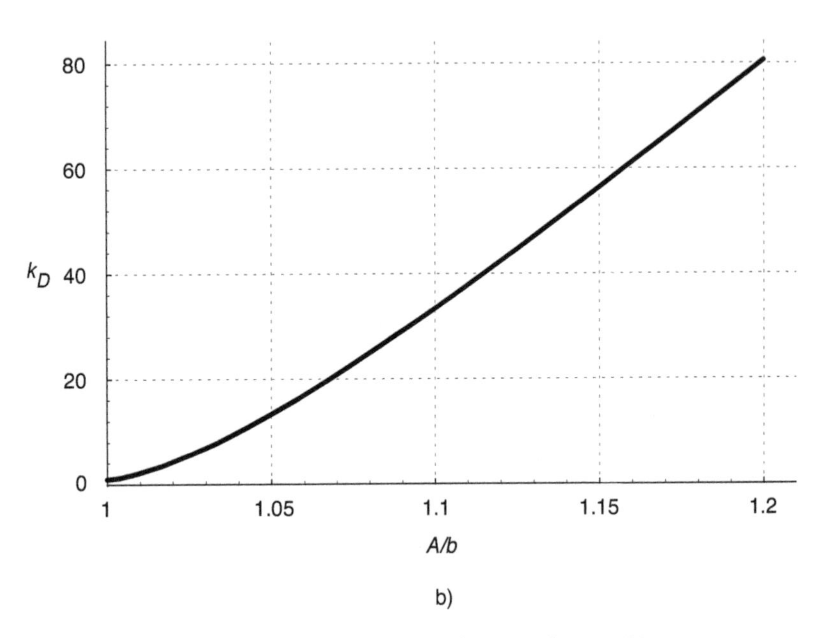

Fig. 6.5-9. Characteristics of the hardening-spring function. (b) Describing function multiplier. $k_1 = 1$, $k_2 = 1{,}000$.

Whereas the hardening-spring nonlinearity introduces a discrete change in the proportionality constant, a more gradual change is provided by the *linear-cubic function*

$$f_{lc}(x) = k_1 x + k_2 x^3 \tag{6.5-41}$$

whose describing function multiplier is

$$k_{D_{lc}} = k_1 + \frac{3}{4} A^2 k_2 \tag{6.5-42}$$

The function represents a stiffening proportionality when both constants have the same sign (Fig. 6.5-10a), or a change in sign as well (Fig. 6.5-10b).

When more than one nonlinear effect occurs concurrently, it may be tempting to multiply describing functions as a representation of the compound nonlinearity. For example, a control surface might be subject to both threshold and saturation nonlinearities. Multiplying the describing functions $k_{D_{sat}}$ and $k_{D_{thresh}}$ does not provide the correct approximation; however, the compound describing function can be calculated using

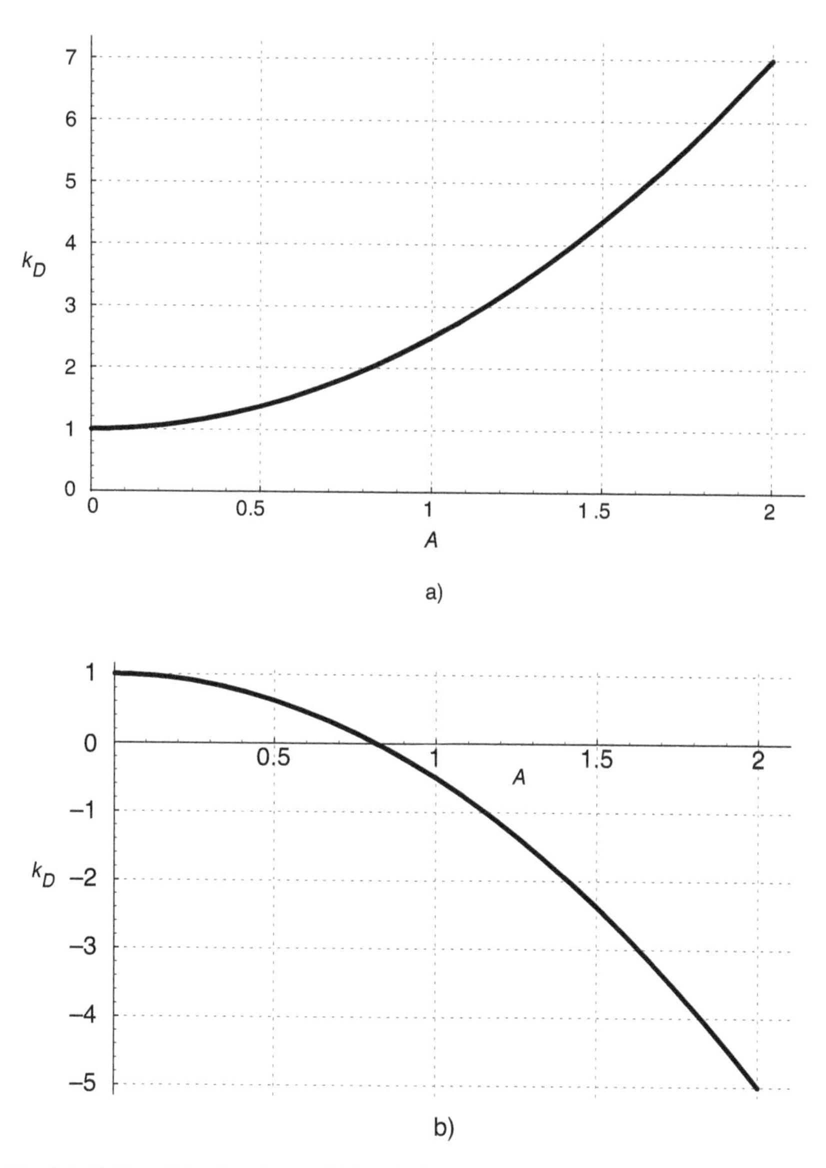

Fig. 6.5-10. Describing function multiplier for linear-cubic spring. $k_1 = 1$, $k_2 = \pm 2$. (a) Stiffening cubic spring, $k_2 = 2$. (b) Softening cubic spring, $k_2 = -2$.

eq. 6.5-32, defining $f(x)$ by the combined characteristics of eq. 6.5-24 and 6.5-36:

$$f(x) = \begin{cases} -b, & x \le -c_{sat} \\ k(x+c), & -c_{sat} < x < -c_{thresh} \\ 0, & -c_{thresh} \le x \le c_{thresh} \\ k(x-c), & c_{thresh} < x \le c_{sat} \\ b, & x \ge c_{sat} \end{cases} \tag{6.5-43}$$

Calculation of the describing function multiplier is left as an exercise.

Quasilinear Root Locus Analysis

A principal application of describing functions is to identify limit cycles via linear methods, under the assumption that the oscillation is nearly sinusoidal. The nonlinearity is represented by a variable parameter (i.e., the describing function multiplier) which, in turn, becomes the root locus gain. In the usual way, each branch of the root locus originates at a pole of the system (gain = zero) and terminates at a transfer function zero or at infinity (gain = $\pm\infty$). Possible limit cycles are indicated by complex-root crossings of the $j\omega$ axis. With roots in the left half of the s plane, oscillations of the mode are stable, and the amplitude of oscillation decays. If the effect of decaying amplitude on the describing function forces the roots to the $j\omega$ axis, a limit cycle is indicated. For roots in the right half plane, the oscillation amplitude grows. If growing amplitude forces the roots to the $j\omega$ axis, a limit cycle is indicated. The *frequency of the limit cycle* is estimated by the frequency at the $j\omega$-axis crossing. The *amplitude of the limit cycle* is estimated by the input-sinusoid amplitude that corresponds to the describing function magnitude at the point of crossing.

Quasilinear root locus analysis is applied to the rudder snaking problem by first writing the characteristic equation of the coupled Dutch roll-rudder system (eq. 6.5-17) in root locus form, with the rudder damping $H_{\dot{\delta}R}$ identified as the root locus gain:

$$\begin{aligned} \Delta(s) = s^4 - \left(N_r + \frac{Y_\beta}{V_o}\right)s^3 + \left(N_\beta + N_r\frac{Y_\beta}{V_o} - H_\delta\right)s^2 + H_\delta\left(N_r + \frac{Y_\beta}{V_o}\right)s \\ + N_{\delta R}H_\beta - H_\delta\left(N_\beta + N_r\frac{Y_\beta}{V_o}\right) \\ - H_{\dot{\delta}R}s\left[s^2 - \left(N_r + \frac{Y_\beta}{V_o}\right)s + \left(N_\beta + N_r\frac{Y_\beta}{V_o}\right)\right] \end{aligned} \tag{6.5-44}$$

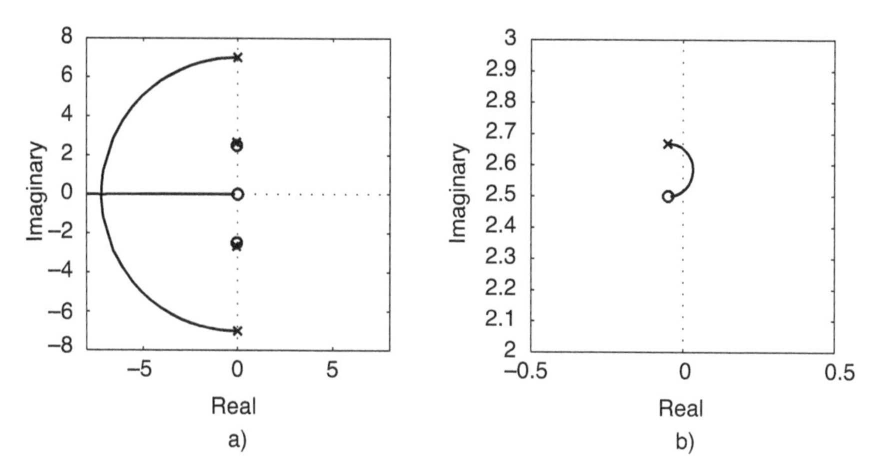

Fig. 6.5-11. Effect of rudder overbalance and mechanical friction on stick-free stability. $H_{\delta R}$ taken as root-locus gain. (a) Dutch roll and rudder mechanism roots. (b) Expanded view to show conditional instability of Dutch roll mode.

Interpreting the characteristic equation as $d(s) + kn(s)$, the denominator $d(s)$ of the root locus contains two pairs of complex poles representing the uncoupled Dutch roll and rudder modes, and the numerator $n(s)$ contains a complex pair of zeros near the Dutch roll poles.

The locus for the overbalanced-rudder case discussed earlier is shown in Fig. 6.5-11. Negative $H_{\delta R}$ increases the rudder-mode damping ratio and draws the rudder-mode roots toward the real axis. The Dutch roll poles and complex roots are near the $j\omega$ axis. As the roots migrate from one pair to the other with changing $H_{\dot{\delta}R}$, they enter the right half plane, and there is a region of *conditional instability*. The Dutch roll mode is stable for both high and low $H_{\dot{\delta}A}$ but is unstable within an intermediate range.

Increasingly negative rudder damping brought about by friction corresponds to decreasing input amplitude (eq. 6.5-35); hence, the roots of the system approach the zeros of $n(s)$ or infinity for low input amplitude and are near the poles of $d(s)$ for high input amplitude. With small input amplitude, the roots remain stable; oscillations decay, and the Dutch roll roots return to the zeros (Fig. 6.5-11b). With larger input amplitude, the roots are drawn to the right half plane and instability, causing the amplitude to grow even further. The roots progress toward the poles until encountering the $j\omega$ axis. Beyond this point, the mode is stable, the amplitude decays, and the roots return to the $j\omega$ axis.

We conclude that the second intercept constitutes a stable state of the system, a continuing oscillation with rudder-rate amplitude determined by eq. 6.5-35. The first $j\omega$-axis crossing predicts an *unstable* limit cycle, a closed contour that does not sustain itself. The second crossing predicts a *stable limit cycle*. Earlier describing-function analyses of rudder snaking [G-3, S-1] neglected the inertial properties of the control mechanism and modeled the coupled system as third order. Nevertheless, they reach similar conclusions about conditional instability and the potential for limit cycles.

The present quasilinear analysis predicts a limit-cycle period of 2.4 s, which is very close to the result shown in Fig. 6.5-4, but the limit-cycle amplitude estimate is off the mark. At the second $j\omega$ crossing, $H_{\dot{\delta}R}$ is -5.3, corresponding to a rudder-rate magnitude of 0.024 rad/s. Given a sinusoidal rudder displacement $A \sin \omega t$, the corresponding rudder rate is $A\omega \cos \omega t$; hence, a rate of 0.024 rad/s corresponds to a displacement of 0.01 rad. The nonlinear simulation gives a rudder displacement of 0.03 rad. The discrepancy can be traced to the rudder waveform, which is not actually sinusoidal, and to the difference between the hinge moments produced by linear and nonlinear friction models. Furthermore, the simple friction model (eq. 6.5-9) was used to calculate the describing function, and the more complex model (eq. 6.5-11) was used for nonlinear simulation.

Other snaking mechanisms can be predicted using quasilinear analysis. If the rudder overbalance is much larger than that used in the example, say -12 rather than -6, the coupled system is unstable. The H_δ root locus (not shown) reveals that increasingly negative values of H_δ draw the destabilized Dutch roll roots into the left half plane. With the slightest initial condition, the rudder oscillation would grow until the surface banged back and forth between its stops. Representing H_δ as a hardening-spring nonlinearity (eq. 6.5-40), progression of the Dutch roll roots from the unstable poles stalls at the $j\omega$ axis, indicating a limit cycle whose period is near that of the Dutch roll mode.

Roll-Spiral/Aileron Coupling

The limit cycle induced by rudder friction depends on the existence of a lightly damped Dutch roll mode, and the limit-cycle period is close to the control-fixed Dutch roll period. Unlike the Dutch roll mode, the roll and spiral motions rarely combine in a lightly damped oscillation. If an aileron limit cycle exists, the mechanism must be qualitatively different, and the period of motion must be related to the only other oscillatory mode of the coupled system, the aileron actuation mode. (If the wing's torsional rigidity is low, then aeroelastic coupling is possible, as discussed in the next section.)

Flight tests of a North American P-51 Mustang revealed an aileron modification that improved roll-rate performance but led to a limit cycle at high Mach number. As reported in [T-1] and summarized in [A-1], a large trailing-edge bevel angle produced aerodynamic overbalance that led to sustained high-amplitude rolling oscillation at a frequency of about 2 Hz (~12.5 rad/s). Unlike rudder snaking, the aileron limit cycle was not associated with a minimum level of excitation or initial condition. An unstable oscillation diverged until a continuing roll-rate oscillation of 30–35 deg/s (~0.5 rad/s) occurred. Quasilinear root locus analysis suggests a limit-cycle mechanism that is compatible with the results of a nonlinear simulation.

An autonomous linear model of the coupled system can be written as

$$\begin{bmatrix} \Delta\dot{\mathbf{x}}_{RS} \\ \Delta\dot{\mathbf{x}}_{\delta A} \end{bmatrix} = \begin{bmatrix} \mathbf{F}_{RS} & \mathbf{G}_{RS} \\ \mathbf{L}_{\delta A} & \mathbf{F}_{\delta A} \end{bmatrix} \begin{bmatrix} \Delta\mathbf{x}_{RS} \\ \Delta\mathbf{x}_{\delta A} \end{bmatrix} \tag{6.5-45}$$

where $\Delta\mathbf{x}_{RS} = [\Delta p \; \Delta\phi]^T$, $\Delta\mathbf{x}_{\delta A} = [\delta A \; \dot{\delta} A]^T$, and the component matrices are

$$\mathbf{F}_{RS} = \begin{bmatrix} L_p & 0 \\ 1 & 0 \end{bmatrix} \tag{6.5-46}$$

$$\mathbf{G}_{RS} = \begin{bmatrix} L_{\delta A} & 0 \\ 0 & 0 \end{bmatrix} \tag{6.5-47}$$

$$\mathbf{L}_{\delta A} = \begin{bmatrix} 0 & 0 \\ H_p & 0 \end{bmatrix} \tag{6.5-48}$$

$$\mathbf{F}_{\delta A} = \begin{bmatrix} 0 & 1 \\ H_{\delta A} & H_{\dot{\delta} A} \end{bmatrix} \tag{6.5-49}$$

The roll rate causes a differential angle of attack at the aileron locations, and H_p expresses the over/underbalance of the aileron control. The characteristic equation is written in root locus form, with H_p taken as the gain:

$$\Delta(s) = s[s^3 - (L_p + H_{\dot{\delta} A})s^2 + (L_p H_{\dot{\delta} A} - H_\delta)s + L_p H_\delta] - H_p L_{\delta A} s \tag{6.5-50}$$

The characteristic equation is fourth order, but the pole at the origin is canceled by the numerator zero and is unaffected by H_p; hence, the remaining roots follow a conventional third-order pattern.

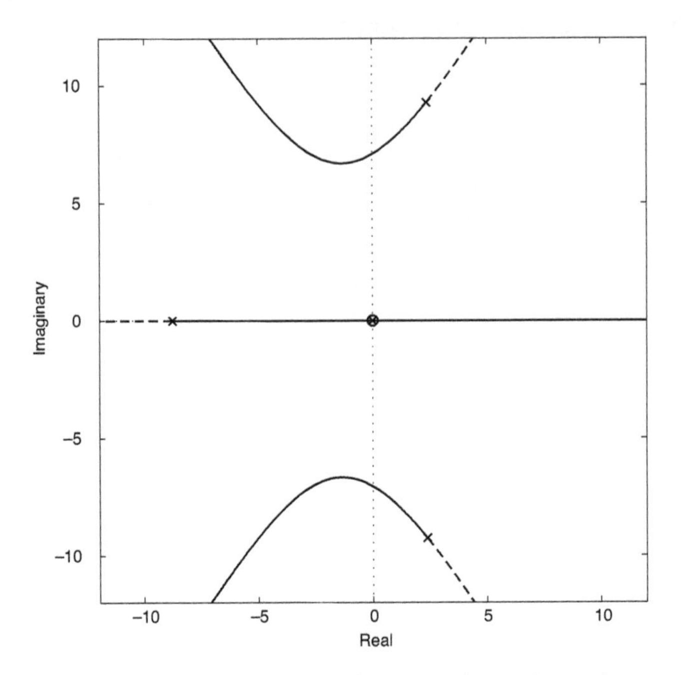

Fig. 6.5-12. Aileron overbalance coupling root locus. H_p taken as the root locus gain; positive values represented by solid line; negative values by dashed line.

The following nominal stability derivatives, representative of a World War II fighter configuration with exaggerated overbalance effect, provide the example:

$L_p = -4$

$L_{\delta A} = 20$

$H_p = -30$

$H_{\delta A} = -144$

$H_{\dot{\delta} A} = 0$

Were it not for the overbalance, the roll, spiral, and aileron modes would be uncoupled, and the aileron mode would be neutrally stable. The negative value of H_p destabilizes the aileron mode, and the linear model has eigenvalues at $(0, -7.8, 1.5 \pm 12.8j)$. More positive values of H_p would restore stability, as shown by Fig. 6.5-12. This suggests that an overbalance nonlinearity whose describing function magnitude decreases and

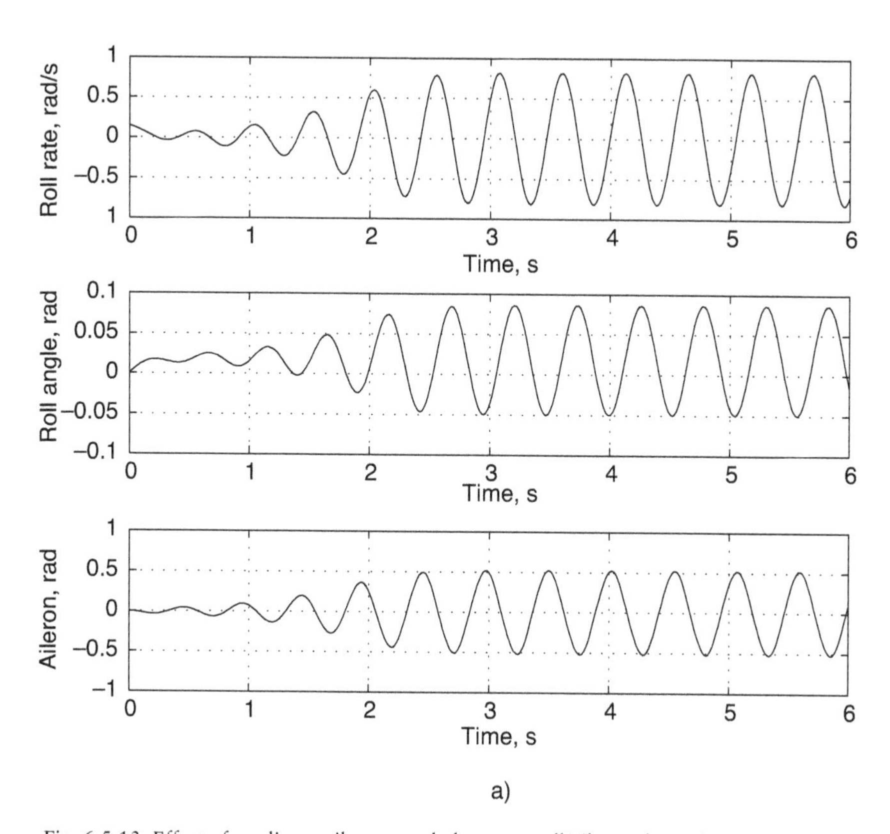

Fig. 6.5-13. Effect of nonlinear aileron overbalance on roll/aileron dynamic system. (a) History of roll rate, roll angle, and aileron deflection.

changes sign with increasing input amplitude, such as the linear-cubic function (eq. 6.5-41), would force the root locus to a stable intercept with the $j\omega$ axis, predicting a limit cycle.

A simulation where $H_p \Delta p$ is replaced by a softening linear-cubic nonlinearity is shown in Fig. 6.5-13. The nonlinear function is

$$H(\Delta p) = H_p \Delta p + H_{p^3} \Delta p^3 = -30\Delta p + 60\Delta p^3 \qquad (6.5\text{-}51)$$

reflecting overbalance at low amplitude, where the bevel effect is most significant, and underbalance at high amplitude, where the hinge moment is less affected by the bevel. An initial roll rate rapidly builds to a limit cycle like that experienced by the P-51 (Fig. 6.5-13a), with the roll rate and angle forced by the aileron oscillation. While the quasilinear model suggested the existence of a limit cycle, it predicted greater cou-

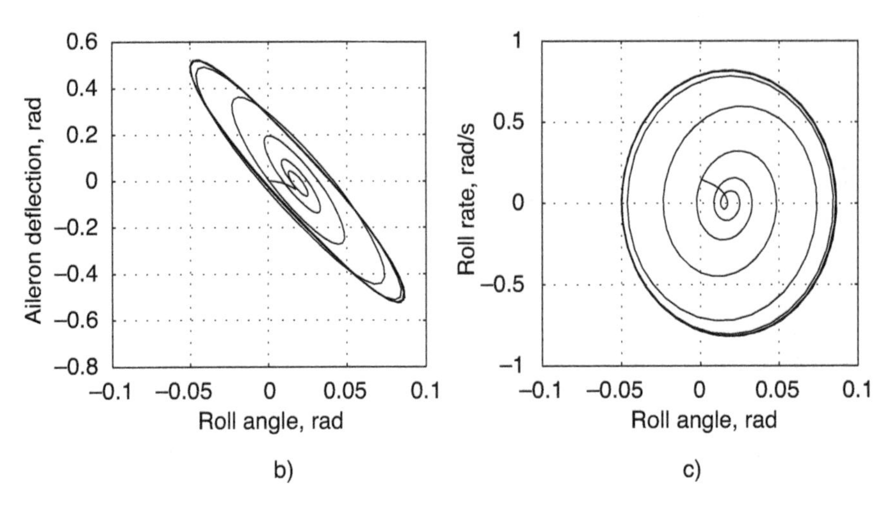

Fig. 6.5-13. Effect of nonlinear aileron overbalance on roll/aileron dynamic system. (b) Aileron deflection versus roll angle. (c) Roll rate versus roll angle.

pling of the modes than occurs in the nonlinear model. Consequently, the oscillation frequencies are quite different, 7 versus 12 rad/s.

Spoiler Nonlinearity and Time Delay

Spoiler flaps mounted on the wing's upper surface produce rolling and yawing moments when activated differentially. Unlike typical aileron deflections, spoilers produce favorable yawing moments; positive rolling moment is accompanied by positive yawing moment, favoring the intended turn.

However, spoilers have two adverse effects. At conventional flight speed and angle of attack, there is a boundary layer above the spoiler, in which the flow velocity is lower than the freestream velocity. The spoiler's initial effect may be small until the trailing-edge deflection exceeds the boundary-layer thickness. The resulting control moment is a *nonlinear* function of the deflection that can be represented as a *threshold* or *hardening spring*. The spoiler acts by interrupting the upper-surface flow over the wing, and there is a *time delay* in the resulting lift and drag increments as the flow readjusts. Rapid deployment of the spoiler produces a starting vortex that may increase lift before decreasing it [B-1]. The transient force may be oscillatory as additional smaller vortices are shed, with the net force approaching the steady value after a short delay. The delay is longer for spoilers mounted forward on the wing than for aft-mounted spoilers. Nonlinearity and time delay may appear similar for a given control input, but they can be distinguished by the response to control inputs that differ in deflection rate, amplitude, and waveform.

A *pure time delay* (also called a *transport lag* or *delay*) has the following input-output characteristic:

$$y(t) = x(t - \tau) \tag{6.5-52}$$

The output lags the input by τ s, but the waveform is unchanged; hence, *time delay is a linear effect. Digital computation delay* has this characteristic, although τ may vary from one cycle to the next. Computing this effect in a digital simulation requires holding the input for τ s before applying it, as illustrated by a delayed step input (Fig. 6.5-14a).

If τ is unchanging, pure time delay is a linear, time-invariant effect that possesses a well-defined Laplace transform. From eqs. 4.3-26 and 6.5-52,

$$\begin{aligned}\mathcal{L}[y(t)] \triangleq y(s) &= \int_0^\infty y(t)e^{-st}dt = \int_0^\infty x(t-\tau)e^{-st}dt \\ &= \int_0^\infty x(u)e^{-s(u+\tau)}du = e^{-\tau s}\int_0^\infty x(u)e^{-su}du \\ &= e^{-\tau s}x(s)\end{aligned} \tag{6.5-53}$$

where u is a variable of integration that equals $(t - \tau)$. Thus, the Laplace transform of the pure time delay is $e^{-\tau s}$, and it is a linear operator.

The frequency response of a pure time delay is evaluated by setting $s = j\omega$, leading to the following amplitude ratio A and phase angle ϕ:

$$A = |e^{-\tau j\omega}| = 1 \; (= 0 \text{ dB}) \tag{6.5-54}$$

$$\phi = \angle e^{-\tau j\omega} = \tan^{-1}\left[\frac{\text{Im}(e^{-\tau j\omega})}{\text{Re}(e^{-\tau j\omega})}\right] = -\tau\omega \,(\text{rad}) \tag{6.5-55}$$

The time delay has no effect on the amplitude of the signal, but it introduces a phase lag that is linearly proportional to the input frequency. The exact effect of time delay is readily portrayed on a Bode plot (Fig. 6.5-14b), Nyquist plot, or Nichols chart using eq. 6.5-55. Time delay hastens the approach to a −180 deg phase angle, and it assures that the phase angle will exceed −180 deg as input frequency increases.

A rational approximation to $e^{-\tau s}$ is required for root locus analysis, allowing poles, zeros, and the resulting closed-loop roots to be plotted. The *Padé approximation* provides a rational function approximation to

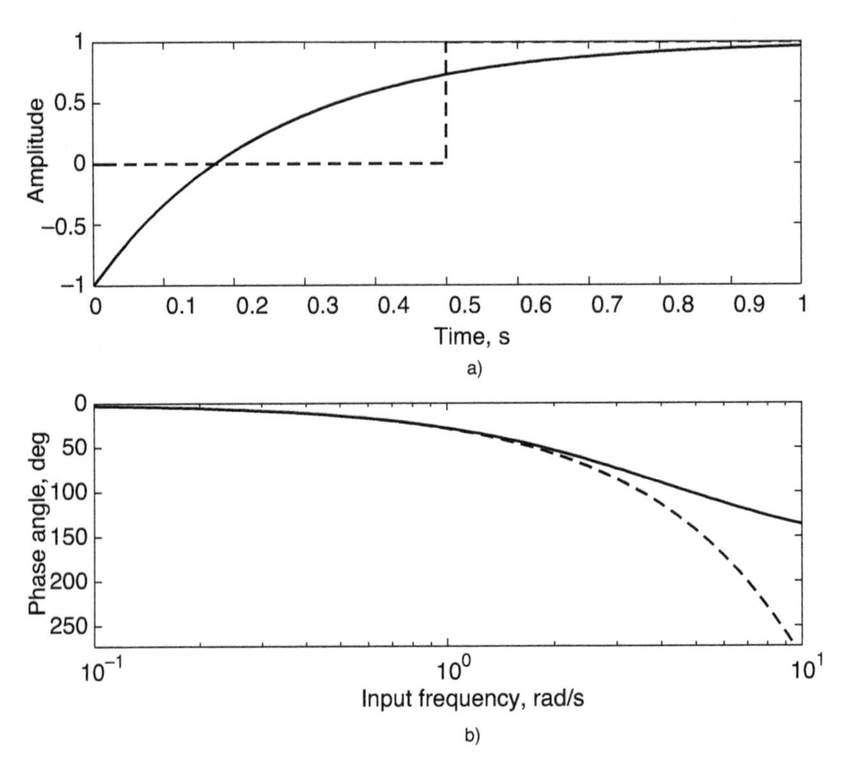

Fig. 6.5-14. Pure time delay and first-degree Padé approximation. (a) 0.5 s pure delay in step output (dashed line) and corresponding first-degree Padé approximation (solid line). (b) Phase angle of pure time delay (dashed line) and of first-degree Padé approximation (solid line).

a power series [R-1], and the linear operator $e^{-\tau s}$ can be expressed as the series

$$e^{-\tau s} = 1 - \tau s + (\tau s)^2/2! - (\tau s)^3/3! + \cdots \tag{6.5-56}$$

The first-degree Padé approximation for this series can be written as

$$e^{-\tau s} \approx \frac{-(s - 2/\tau)}{(s + 2/\tau)} \tag{6.5-57}$$

This transfer function has unit amplitude ratio at all frequencies, 90 deg phase lag when $\omega = 2/\tau$, and -180 deg asymptote at high frequency (Fig. 6.5-14b). It approximates eq. 6.5-55 reasonably well for frequencies

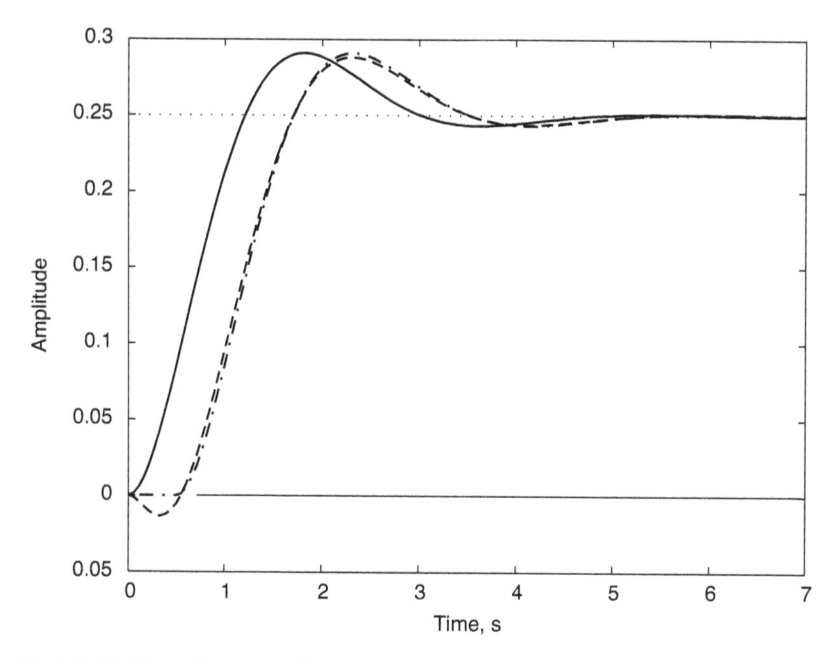

Fig. 6.5-15. Effect of pure time delay and first-degree Padé approximation on the step response of a second-order dynamic system. No delay (solid line), pure delay (dash-dotted line), and Padé approximation (dashed line).

below $2/\tau$, but the error is large at higher frequency. If higher accuracy is required, a higher-degree Padé approximation introduces additional poles and zeros that fit the function more closely.

The first-degree approximation contains a *non-minimum-phase zero* that reverses the initial response of a dynamic system to a step input (Fig. 6.5-15). Once the time exceeds τ, the actual and approximated step responses are quite close. A control loop closure around the pure delay transfer function drives one root to instability at high gain. As an example, consider a simple control law in which roll angle is fed back to the spoilers. From Section 6.3, the reduced-order spoiler-to-roll angle transfer function is

$$\frac{\Delta\phi(s)}{\Delta\delta S(s)} = \frac{L_{\delta S}}{s(s - L_p)} \tag{6.5-58}$$

without pure time delay, where the rolling moment due to the spoiler is denoted by $L_{\delta S}$. Adding the rational approximation for pure time delay,

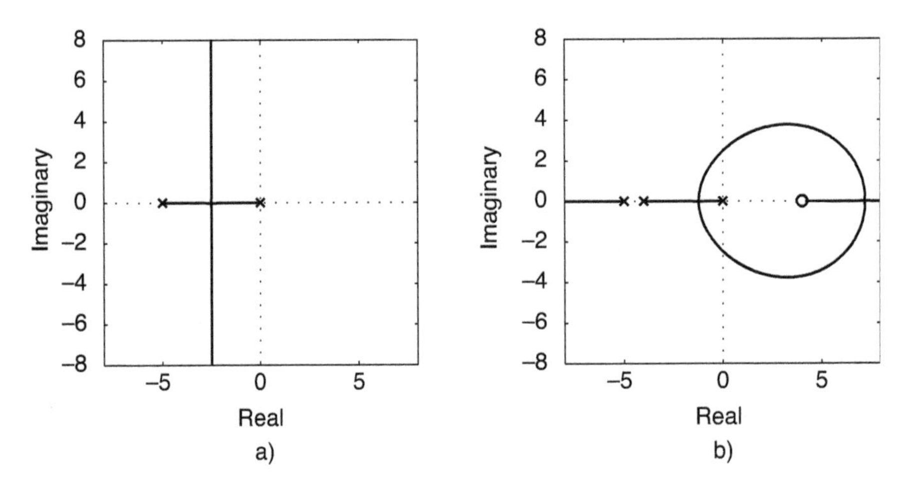

Fig. 6.5-16. Root locus for roll-angle feedback to asymmetric spoiler, including effect of 0.5 s time delay. (a) Without time delay. (b) With time delay.

the transfer function is

$$\frac{\Delta\phi(s)}{\Delta\delta S(s)} = \frac{-L_{\delta S}(s - 2/\tau)}{s(s - L_p)(s + 2/\tau)} \tag{6.5-59}$$

Root loci for the two cases are shown in Fig. 6.5-16. Without time delay, roll-angle feedback causes the real roots to combine in a complex pair, and the damped natural frequency of the closed-loop roll-spiral oscillation increases with increasing gain. With time delay in the feedback, system damping is reduced. At low gain, the two root loci are similar, but as gain increases, the roll-spiral roots are drawn into the right half plane by the numerator zero.

The adverse transient lift reported in [B-1] can be approximated as rolling moment sensitivity to spoiler rate, $H_{\dot{\delta}R}$ With this control-rate effect and no additional time delay, the reduced-order spoiler-to-roll-angle transfer function is

$$\frac{\Delta\phi(s)}{\Delta\delta S(s)} = \frac{L_{\dot{\delta}S}s + L_{\delta S}}{s(s - L_p)} = \frac{L_{\dot{\delta}S}(s + L_{\delta S}/L_{\dot{\delta}S})}{s(s - L_p)} \tag{6.5-60}$$

If the sign of $H_{\dot{\delta}R}$ is opposite to the sign of $L_{\delta S}$, the numerator zero is in the right half plane, and the closed-loop roll/spiral oscillation is destabilized in the same way as by pure time delay. The lack of an additional pole distinguishes the control-rate effect from time delay, allowing spoiler-induced lift to increase before decreasing in the steady state. Through describing function analysis, a gain-increasing nonlinearity like the threshold or hardening spring is seen to have no destabilizing effect on roll-angle feedback unless it is also is accompanied by a time delay or adverse derivative effect.

Lateral-directional control mechanisms have allowed us to compare several approaches for dealing with control complexity: nonlinearity, describing function, time delay, Padé approximation, and derivative effect. For some range of flight conditions and command inputs, these effects are quite similar, allowing flexibility in system descriptions to be used for design and analysis.

6.6 Lateral-Directional Aeroelastic Effects

Methods for analyzing the effects of aeroelasticity on lateral-directional motions follow the approaches taken in Sections 4.6 and 5.6. While the symmetric modes of elasticity interact with longitudinal motions, it is the asymmetric flexible modes that couple with lateral-directional motions. For the idealized case, the elastic motions of particular interest are the first asymmetric modes due to fuselage bending, wing twisting, and wing bending.

Because the Dutch roll, roll, and spiral modes are significant components of all of the lateral-directional motions, we proceed directly to a model that integrates the fourth-order rigid-body model with a single elastic deformation mode. Transfer functions for the sixth-order model are too complex to portray in literal, symbolic fashion, so we restrict much of the discussion to the matrix model and its computational characteristics. As before, the general form of the elastic-body model is

$$\Delta\dot{\mathbf{x}} = \mathbf{F}\,\Delta\mathbf{x} + \mathbf{G}\,\Delta\mathbf{u} + \mathbf{L}\,\Delta\mathbf{w} \tag{6.6-1a}$$

and the lateral-directional version is

$$\begin{bmatrix}\Delta\dot{\mathbf{x}}_{\mathrm{LD}}\\ \Delta\dot{\mathbf{x}}_{\mathrm{E}}\end{bmatrix} = \begin{bmatrix}\mathbf{F}_{\mathrm{LD}} & \mathbf{C}_{\mathrm{E}}^{\mathrm{LD}}\\ \mathbf{C}_{\mathrm{LD}}^{\mathrm{E}} & \mathbf{F}_{\mathrm{E}}\end{bmatrix}\begin{bmatrix}\Delta\mathbf{x}_{\mathrm{LD}}\\ \Delta\mathbf{x}_{\mathrm{E}}\end{bmatrix} + \begin{bmatrix}\mathbf{G}_{\mathrm{LD}}\\ \mathbf{G}_{\mathrm{E}}\end{bmatrix}\Delta\mathbf{u} + \begin{bmatrix}\mathbf{L}_{\mathrm{LD}}\\ \mathbf{L}_{\mathrm{E}}\end{bmatrix}\Delta\mathbf{w} \tag{6.6-1b}$$

$\Delta\mathbf{x}_{\mathrm{LD}}$ is the lateral-directional rigid-body state,

$$\Delta\mathbf{x}_{\mathrm{LD}} = \begin{bmatrix}\Delta r\\ \Delta\beta\\ \Delta p\\ \Delta\phi\end{bmatrix}\begin{Bmatrix}\text{yaw rate}\\ \text{sideslip angle}\\ \text{roll rate}\\ \text{roll angle}\end{Bmatrix} \tag{6.6-2}$$

$\Delta \mathbf{x}_E$ is the elastic-mode state,

$$\Delta \mathbf{x}_E = \begin{bmatrix} \Delta\eta_1 \\ \Delta\eta_2 \end{bmatrix} \begin{Bmatrix} \text{elastic displacement} \\ \text{elastic rate} \end{Bmatrix} \quad (6.6\text{-}3)$$

and $\Delta \mathbf{u}_{LD}$ and $\Delta \mathbf{w}_{LD}$ are the control and external disturbance,

$$\Delta \mathbf{u}_{LD} = \begin{bmatrix} \Delta\delta A \\ \Delta\delta R \end{bmatrix} \begin{Bmatrix} \text{aileron deflection} \\ \text{rudder deflection} \end{Bmatrix} \quad (6.6\text{-}4)$$

$$\Delta \mathbf{w}_{LD} = \begin{bmatrix} \Delta\beta_w \\ \Delta p_w \end{bmatrix} \begin{Bmatrix} \text{wind–induced sideslip} \\ \text{vortical wind} \end{Bmatrix} \quad (6.6\text{-}5)$$

We refer to the "rigid-body" and "elastic" state components to identify their origins, recognizing that they are coupled in the aeroelastic model.

The uncoupled stability matrices for the lateral-directional and elastic modes $\mathbf{F}_{LD}$ and $\mathbf{F}_E$ are

$$\mathbf{F}_{LD} = \begin{bmatrix} N_r & N_\beta & N_p & 0 \\ -1 & \dfrac{Y_\beta}{V_o} & 0 & \dfrac{g_o}{V_o} \\ L_r & L_\beta & L_p & 0 \\ 0 & 0 & 1 & 0 \end{bmatrix} \quad (6.6\text{-}6)$$

$$\mathbf{F}_E = \begin{bmatrix} 0 & 1 \\ -\omega_{n_E}^2 & -2\zeta\omega_{n_E} \end{bmatrix} \quad (6.6\text{-}7)$$

while the direct effects of control and disturbance on rigid-body motions and elastic deformations $\mathbf{G}_{LD}$, $\mathbf{G}_{E,}$ $\mathbf{L}_{LD}$, and $\mathbf{L}_E$ are

$$\mathbf{G}_{LD} = \begin{bmatrix} N_{\delta A} & N_{\delta R} \\ \dfrac{Y_{\delta A}}{V_o} & \dfrac{Y_{\delta R}}{V_o} \\ L_{\delta A} & L_{\delta R} \\ 0 & 0 \end{bmatrix} \quad (6.6\text{-}8)$$

$$\mathbf{G}_E = \begin{bmatrix} 0 & 0 \\ B_{\delta A} & B_{\delta R} \end{bmatrix} \quad (6.6\text{-}9)$$

$$\mathbf{L}_{LD} = \begin{bmatrix} -N_\beta & -N_p \\ \dfrac{-Y_\beta}{V_o} & \dfrac{-Y_p}{V_o} \\ -L_\beta & -L_p \\ 0 & 0 \end{bmatrix} \tag{6.6-10}$$

$$\mathbf{L}_E = \begin{bmatrix} 0 & 0 \\ -A_\beta & -A_p \end{bmatrix} \tag{6.6-11}$$

The coupling matrices take the form

$$\mathbf{C}_E^{LD} = \begin{bmatrix} N_{\eta 1} & N_{\eta 2} \\ \dfrac{Y_{\eta 1}}{V_o} & \dfrac{Y_{\eta 2}}{V_o} \\ L_{\eta 1} & L_{\eta 2} \\ 0 & 0 \end{bmatrix} \tag{6.6-12}$$

$$\mathbf{C}_{LD}^E = \begin{bmatrix} 0 & 0 & 0 & 0 \\ A_r & A_\beta & A_p & A_\phi \end{bmatrix} \tag{6.6-13}$$

EQUILIBRIUM RESPONSE TO CONTROL

Setting the state rate to zero, the equilibrium response of eq. 6.6-1 to constant control is

$$\begin{bmatrix} \Delta\mathbf{x}_{LD} \\ \Delta\mathbf{x}_E \end{bmatrix}^* = -\begin{bmatrix} \mathbf{F}_{LD} & \mathbf{C}_E^{LD} \\ \mathbf{C}_{LD}^E & \mathbf{F}_E \end{bmatrix}^{-1} \begin{bmatrix} \mathbf{G}_{LD} \\ \mathbf{G}_E \end{bmatrix} \Delta\mathbf{u}^* \tag{6.6-14}$$

Using a common matrix identity [S-3], the inverse can be expressed in terms of its components, and the equilibrium relationship is

$$\begin{bmatrix} \Delta\mathbf{x}_{LD} \\ \Delta\mathbf{x}_E \end{bmatrix}^* = -\begin{bmatrix} (\mathbf{F}_{LD} - \mathbf{C}_E^{LD}\mathbf{F}_E^{-1}\mathbf{C}_{LD}^E)^{-1} & -\mathbf{F}_{LD}^{-1}\mathbf{C}_E^{LD}(\mathbf{F}_E - \mathbf{C}_{LD}^E\mathbf{F}_{LD}^{-1}\mathbf{C}_E^{LD})^{-1} \\ -\mathbf{F}_E^{-1}\mathbf{C}_{LD}^E(\mathbf{F}_{LD} - \mathbf{C}_E^{LD}\mathbf{F}_E^{-1}\mathbf{C}_{LD}^E)^{-1} & (\mathbf{F}_E - \mathbf{C}_{LD}^E\mathbf{F}_{LD}^{-1}\mathbf{C}_E^{LD})^{-1} \end{bmatrix} \begin{bmatrix} \mathbf{G}_{LD} \\ \mathbf{G}_E \end{bmatrix} \Delta\mathbf{u}^* \tag{6.6-15}$$

This equilibrium can be expected to exist for typical values of the system parameters, with steady-state rates equal to zero.

Eigenvalues and Root Locus Analysis of Parameter Variations

The effects of variations of system parameters on the coupled eigenvalues can be evaluated using root locus analysis. Here, we focus on three coupling parameters, holding other components of the system model constant. Lateral-directional parameters are taken from Table 6.4-1 and an *in vacuo* elastic mode with 2 Hz natural frequency and damping ratio of 0.05 is proscribed, producing the aeroelastic model described by Table 6.6-1. The aileron and rudder are assumed to force rolling and yawing acceleration, respectively. Elastic deflections create yawing and rolling accelerations and are forced by rigid-body motions as well as control inputs.

The characteristic polynomial for this elastic-body model is of sixth degree, and its roots are the solutions to the determinant equation

$$|s\mathbf{I}_6 - \mathbf{F}| = \Delta(s) = s^6 + a_5 s^5 + a_4 s^4 + a_3 s^3 + a_2 s^2 + a_1 s + a_0 = 0 \tag{6.6-16}$$

The polynomial typically factors into roots that represent the uncoupled Dutch roll, roll, spiral, and elastic modes when all the elements of $\mathbf{C}_{\mathrm{E}}^{\mathrm{LD}}$ or $\mathbf{C}_{\mathrm{LD}}^{\mathrm{E}}$ are zero. With no coupling, the characteristic polynomial and the roots of the model are

$$\begin{aligned}\Delta(s) &= s^6 + 2.6828s^5 + 161.9s^4 + 230.3s^3 + 352.1s^2 \\ &\quad + 366.4s - 3.262 = 0\end{aligned} \tag{6.6-17}$$

$$\lambda_{spiral} = +0.0089\,\text{rad/s}$$

$$\lambda_{roll} = -1.203\,\text{rad/s}$$

$$\lambda_{Dutch\ roll} = -0.116 \pm 1.387j\,\text{rad/s}$$

$$\lambda_{elastic} = -0.628 \pm 12.55j\,\text{rad/s}$$

Table 6.6-1 Baseline Lateral-Directional Elastic-Body Model for Figures of Section 6.6

$$\mathbf{F} = \begin{bmatrix} -0.1079 & 1.9011 & 0.0566 & 0 & N_{\eta_1} & N_{\eta_2} \\ -1 & -0.1567 & 0 & 0.0958 & 0 & 0 \\ 0.2501 & -2.408 & -1.1616 & 0 & L_{\eta_1} & L_{\eta_2} \\ 0 & 0 & 1 & 0 & 0 & 0 \\ 0 & 0 & 0 & 0 & 0 & 1 \\ A_r & A_\beta & A_p & A_\phi & -157.91 & -1.257 \end{bmatrix} \quad \mathbf{G} = \begin{bmatrix} 0 & 0.1196 \\ 0 & 0 \\ 2.3106 & 0 \\ 0 & 0 \\ 0 & 0 \\ B_{\delta A} & B_{\delta R} \end{bmatrix}$$

While the coupling between rigid-body and individual elastic modes may well involve all the elements of $\mathbf{C}_{\mathrm{E}}^{\mathrm{LD}}$ or $\mathbf{C}_{\mathrm{LD}}^{\mathrm{E}}$, we consider simpler coupling through the most important parameters. The effects of fuselage bending and wing twisting are highlighted in three examples:

- Directional fuselage bending mode forced by sideslip perturbation (variation in N_{η_1} with $A_\beta = \omega_{n_\mathrm{E}}^2$)
- Asymmetric wing torsional mode forced by sideslip perturbation (variation in L_{η_1} with $A_\beta = \omega_{n_\mathrm{E}}^2$)
- Asymmetric wing torsional mode forced by roll-rate perturbation (variation in L_{η_1} with $A_p = \omega_{n_\mathrm{E}}^2$)

The aerodynamic effect of asymmetric wing bending is not explored; it is likely to be small, and its symbolic coupling is equivalent to the wing twisting case (i.e., it would principally produce L_{η_1} and be forced by A_β or A_p). All other coupling parameters are assumed to be zero. As in previous illustrations (Section 5.6), the elastic-mode natural frequency is atypically low to exaggerate the rigid-elastic coupling for the examples, and the coupling terms used as root locus gains may take large positive or negative values to portray root locus shapes.

The characteristic polynomials for these cases form the basis for root locus analysis. For fuselage bending, the polynomial is

$$\begin{aligned}\Delta(s) = {} & s^6 + 2.6828s^5 + 161.9s^4 + 230.3s^3 + 352.1s^2 \\ & + 366.4s - 3.262 + N_{\eta_1}(157.9s^2 + 183.4s - 3.783)\end{aligned} \tag{6.6-18}$$

For wing torsion forced by sideslip angle, it is

$$\begin{aligned}\Delta(s) = {} & s^6 + 2.6828s^5 + 161.9s^4 + 230.3s^3 + 352.1s^2 \\ & + 366.4s - 3.262 + L_{\eta_1}(6.19s - 1.632)\end{aligned} \tag{6.6-19}$$

while for wing torsion forced by roll rate, it is

$$\begin{aligned}\Delta(s) = {} & s^6 + 2.6828s^5 + 161.9s^4 + 230.3s^3 + 352.1s^2 + 366.4s \\ & - 3.262 - L_{\eta_1}s(157.9s^2 + 41.78s + 302.9)\end{aligned} \tag{6.6-20}$$

Thus, the denominator of the equivalent transfer function $kn(s)/d(s)$ is the uncoupled polynomial, and there are one to three zeros in the coupling numerators. We note that the total damping for all cases, represented by the coefficient of s^{n-1}, $n = 6$, is fixed in each case; therefore, additions to damping in one mode must necessarily be accompanied by decreased damping in others.

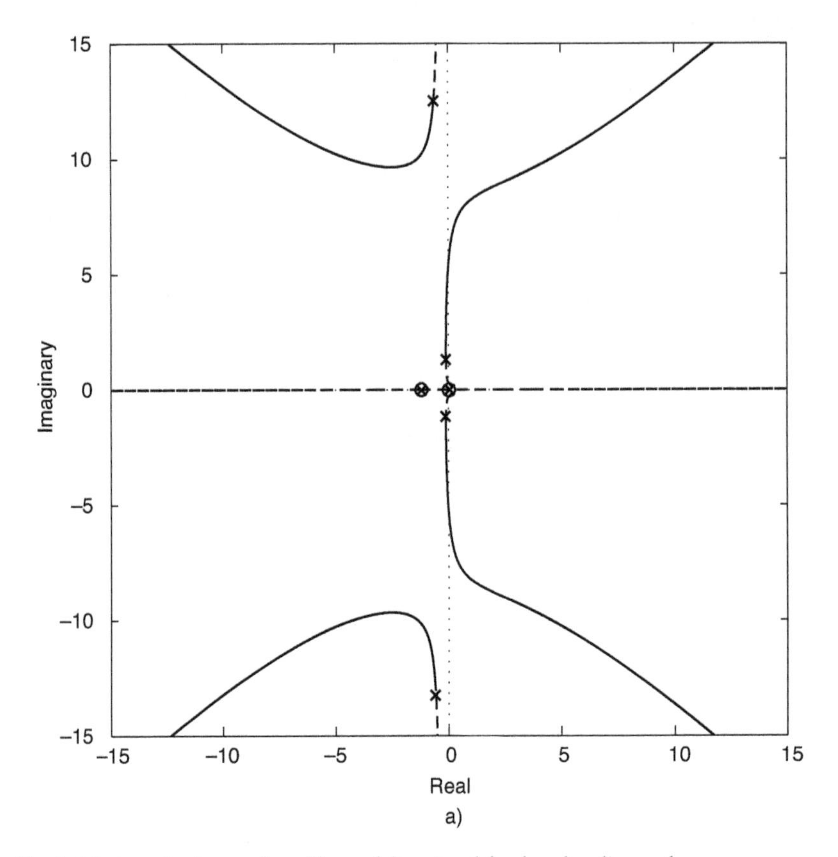

Fig. 6.6-1. Aeroelastic coupling effects of directional fuselage bending and asymmetric wing twisting. (a) Fuselage bending due to sideslip.

Fuselage bending forced by sideslip has negligible impact on the roll and spiral modes (Fig. 6.6-1a), as the zeros of eq. 6.6-18 cancel the corresponding poles. Thus, the principal coupling effect is between the Dutch roll and elastic modes; with a pole-zero surplus of 4, the asymptotic values of the roots are (± 45, ± 135 deg) with positive root locus gain (N_{η_1}) and (0, ± 90, 180 deg) with negative gain. Positive coupling stiffens directional response, initially increasing the Dutch roll natural frequency and decreasing that of the elastic mode; however, the more likely effect is negative coupling, which softens directional stability. The Dutch roll experiences a high-frequency oscillatory instability with high positive coupling and a low-frequency instability (first oscillatory, then

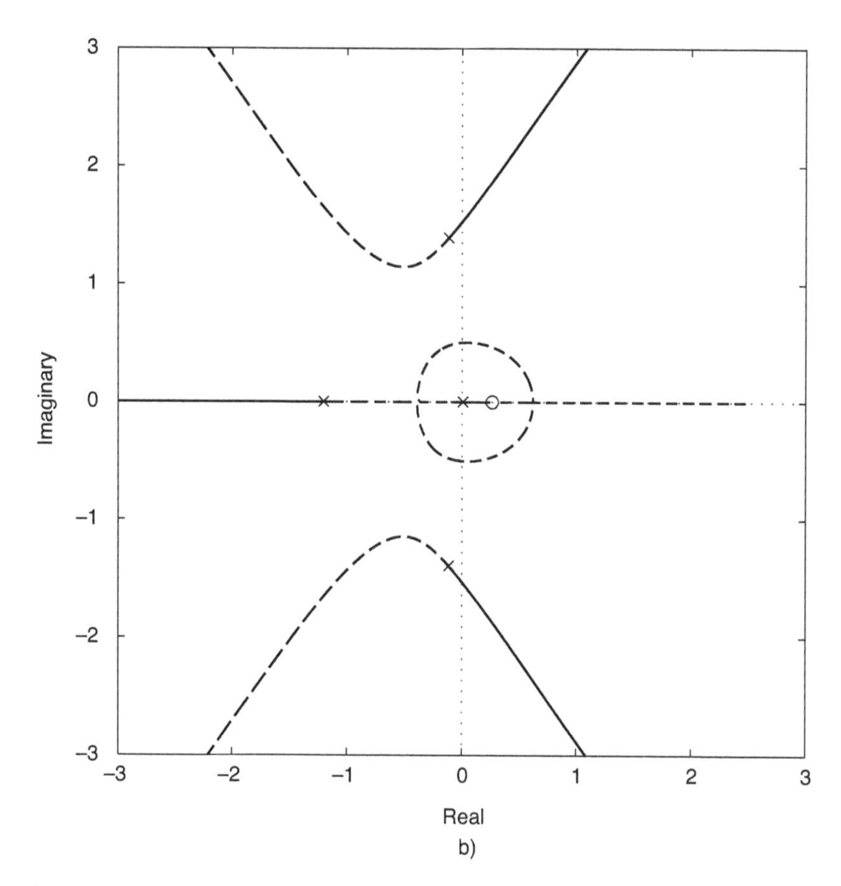

Fig. 6.6-1. Aeroelastic coupling effects of directional fuselage bending and asymmetric wing twisting. (b) Wing twisting due to sideslip.

real) with strong negative coupling. The elastic mode is never destabilized in this case.

For wing twisting forced by sideslip perturbation, there is little change in the elastic-mode roots, and only the otherwise rigid-body modes are shown in Fig. 6.6-1b. With a pole-zero surplus of 5 (eq. 6.6-19), the asymptotes are separated by (360/5) deg = 72 deg. Coupling with a positive root locus gain destabilizes the Dutch roll and spiral modes while stabilizing the roll and elastic roots. With negative coupling, the roll and spiral roots ultimately coalesce in an oscillatory complex pair that becomes unstable at some high value of L_{η_1}, while the Dutch roll is stabilized and the elastic mode is mildly destabilized.

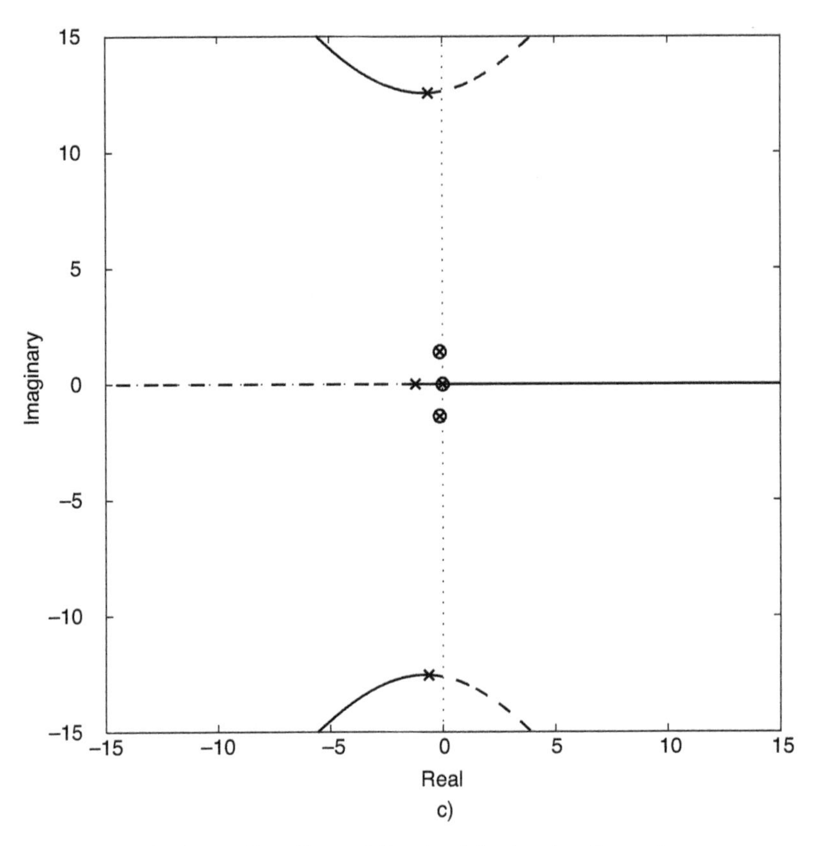

Fig. 6.6-1. Aeroelastic coupling effects of directional fuselage bending and asymmetric wing twisting. (c) Wing twisting due to roll rate.

When wing twisting is forced by roll rate, there are three zeros and a pole-zero surplus of 3 (eq. 6.6-20), producing asymptotes separated by 120 deg. The zeros effectively overlie the Dutch roll and spiral poles; hence these roots are largely unaffected by the coupling. Positive root locus gain stabilizes the twisting mode and forces a spiral-mode divergence. With negative root locus gain, the twisting oscillation becomes unstable because rolling motion tends to amplify wing twisting, and the roll-mode response is quickened.

Response to Initial Conditions and Step Control Inputs

Time responses for the three idealized cases corroborate the underlying trends predicted by root locus analysis and provide additional insight

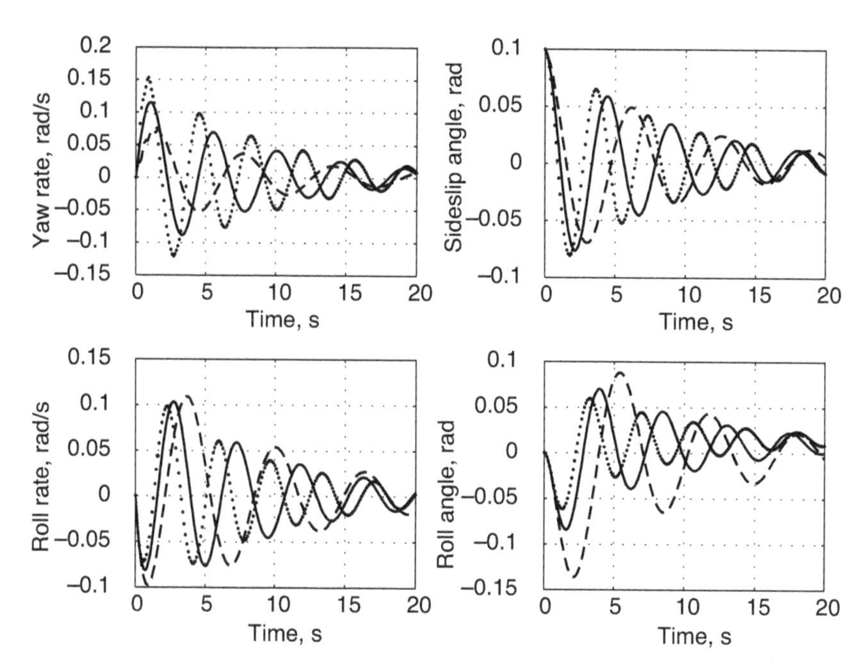

Fig. 6.6-2. Effect of directional fuselage bending on response to initial sideslip angle. No bending (solid), stiffening bending (dotted), and softening bending (double dashed).

into aeroelastic coupling. For the examples considered, elastic vibration is never strongly apparent in the rigid-body variables; however, there are significant effects on Dutch roll, roll, and spiral response. Residualization of aeroelastic effects would be an effective approximation method here.

Directional Fuselage Bending Forced by Sideslip Perturbations

The effects of directional fuselage bending forced by sideslip perturbations reflect changes in Dutch roll natural frequency and modest change in the spiral mode. The numerical value of N_{η_1} is taken to be $\pm N_\beta/2$, with the positive effect increasing $\omega_{n_{DR}}$ and negative decreasing it (Fig. 6.6-2a). There is little change in the Dutch roll damping ratio, and the remaining modes are largely unforced by the initial condition. A high degree of rolling response to sideslip angle is apparent in all cases, a consequence of the rigid-body dynamics. Conversely, initial roll rate induces relatively little directional response, and the bending effects on rolling response are negligible (Fig. 6.6-3). Nevertheless, the stabilizing effect of

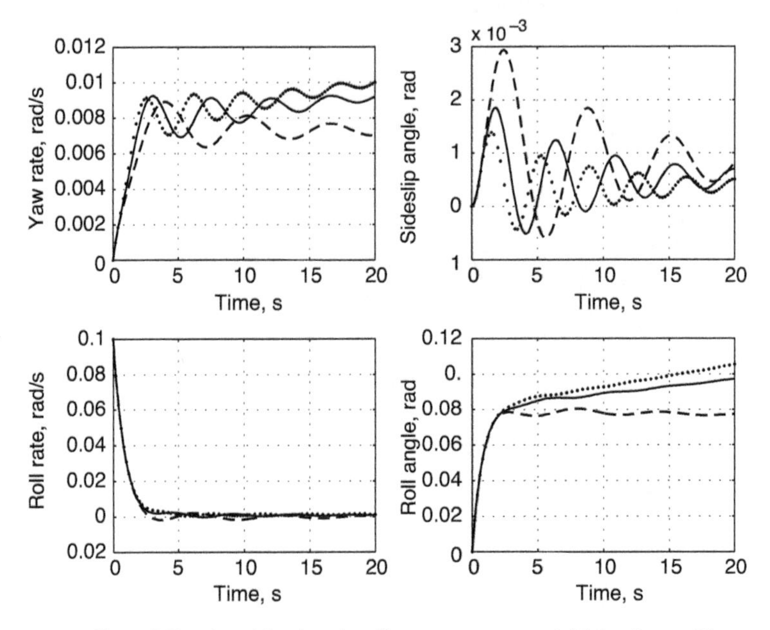

Fig. 6.6-3. Effect of directional fuselage bending on response to initial roll rate. No bending (solid), stiffening bending (dotted), and softening bending (double dashed).

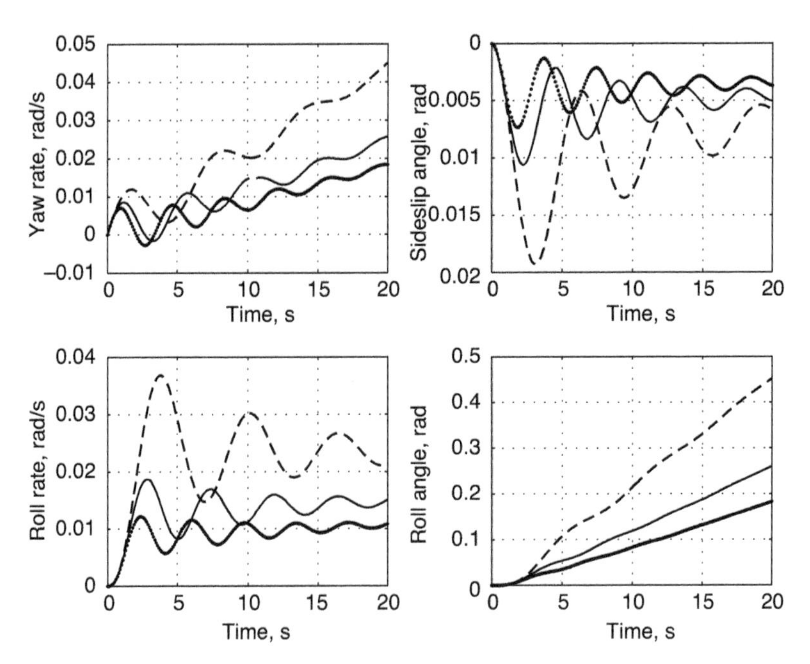

Fig. 6.6-4. Effect of directional fuselage bending on response to rudder step. No bending (solid), stiffening bending (dotted), and softening bending (double dashed).

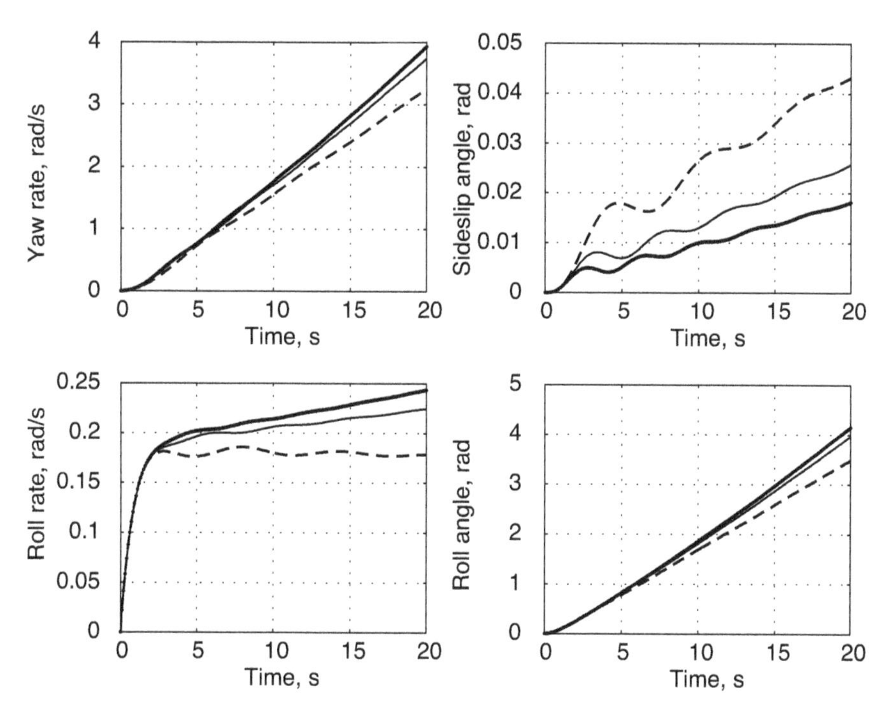

Fig. 6.6-5. Effect of directional fuselage bending on response to aileron step. No bending (solid), stiffening bending (dotted), and softening bending (double dashed).

softening fuselage bending on the spiral mode is evident in the roll-angle response. Softening fuselage bending increases the sideslip-angle and roll-rate overshoots in response to a 0.1 rad rudder step (Fig. 6.6-4), increasing the magnitude of the steady-state response as well. The response to a 0.1 rad aileron step (Fig. 6.6-5) is only slightly affected by the modeled degree of fuselage bending.

Asymmetric Wing Twisting Forced by Sideslip Perturbations

L_{η_1} is taken to be plus or minus four times $|L_p|$ for the example, with the positive value corresponding to positive deflection of the advancing wing's leading edge in response to sideslip perturbation (Fig. 6.6-2b). The Dutch roll natural frequency is hardly affected by wing twisting in this case, although there is a modest effect on its damping ratio, as seen in the response to initial sideslip (Fig. 6.6-6). The roll perturbations due to positive and negative coupling are out of phase with each other, and as before, the roll and spiral modes are little forced by this initial condition.

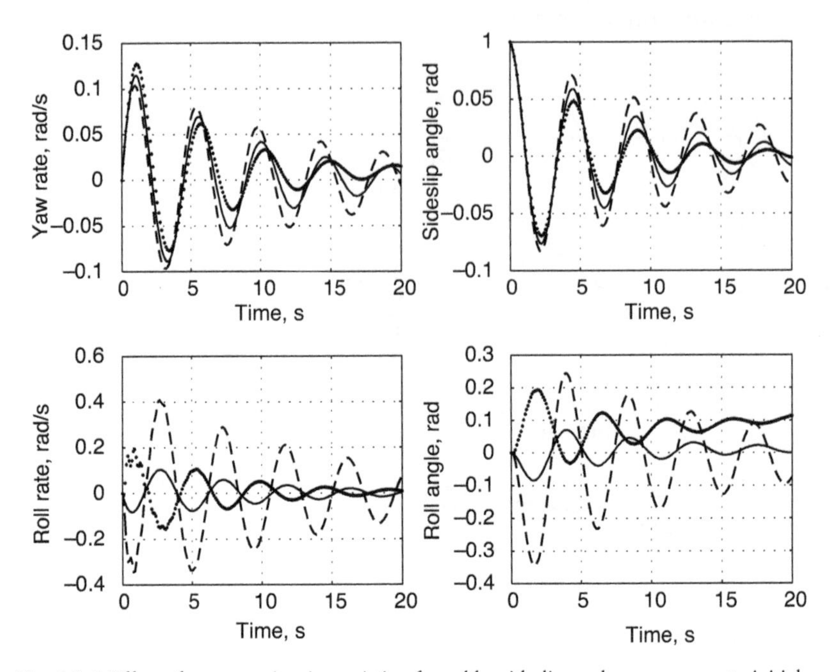

Fig. 6.6-6. Effect of asymmetric wing twisting forced by sideslip angle on response to initial sideslip angle. No bending (solid), positive twist (dotted), and negative twist (double dashed).

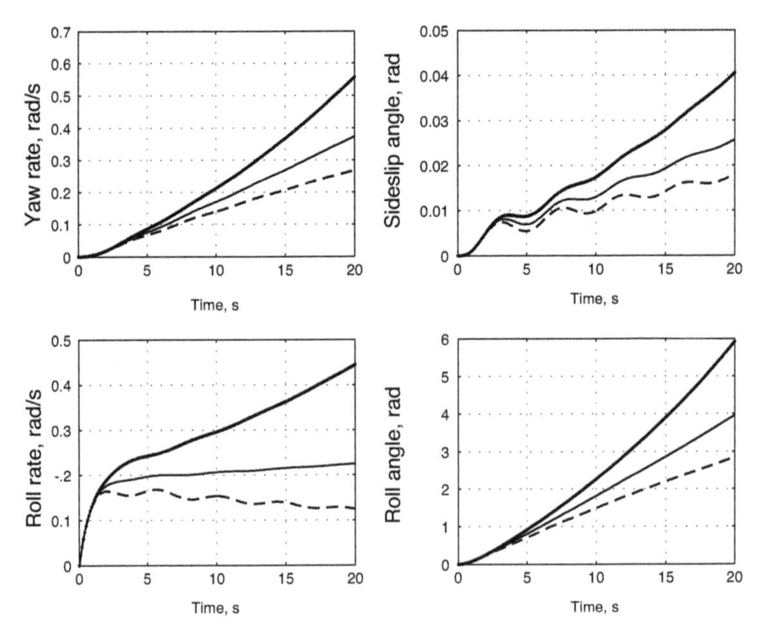

Fig. 6.6-7. Effect of asymmetric wing twisting forced by sideslip angle on response to aileron step. No bending (solid), positive twist (dotted), and negative twist (double dashed).

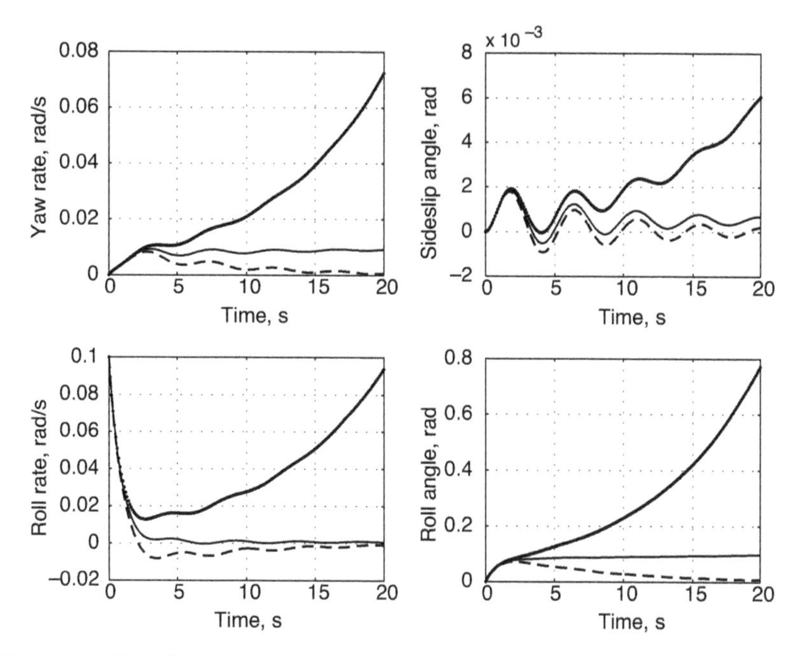

Fig. 6.6-8. Effect of asymmetric wing twisting forced by roll rate on response to initial roll rate. No bending (solid), positive twist (dotted), and negative twist (double dashed).

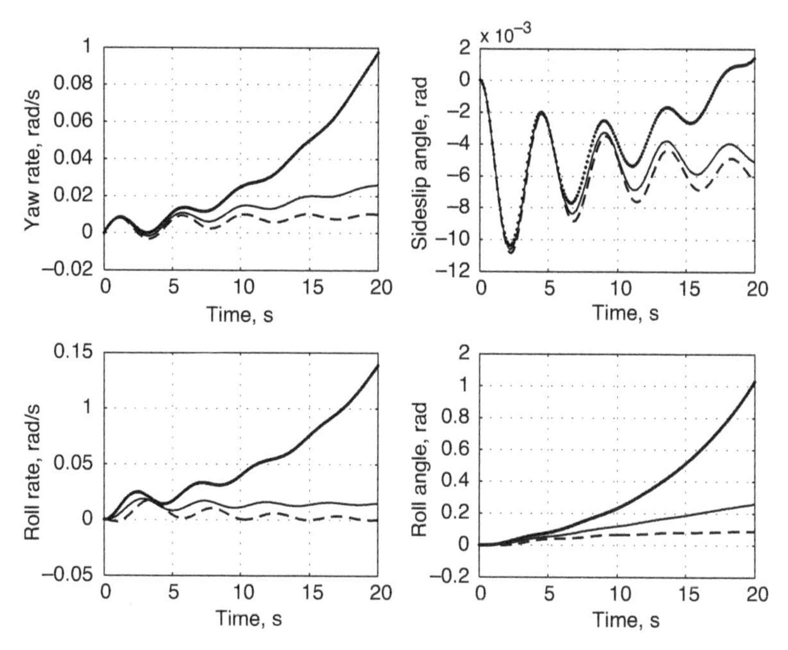

Fig. 6.6-9. Effect of asymmetric wing twisting forced by roll rate on response to rudder step. No bending (solid), positive twist (dotted), and negative twist (double dashed).

The significant effect on the spiral mode is seen in the response to aileron step (Fig. 6.6-7), with positive coupling increasing the rate of divergence and negative coupling introducing stability.

Asymmetric Wing Twisting Forced by Roll-Rate Perturbations

Aeroelastic coupling has relatively little impact on the Dutch roll mode, if wing twisting is induced by roll rate (Fig. 6.6-2c). The most notable effect is on the spiral mode. With $L_{\eta_1} = |L_p|$, the spiral mode becomes divergent, as indicated in both Figs. 6.6-8 and 6.6-9.

References

A-1 Abzug, M. J., and Larrabee, E. E., *Airplane Stability and Control*, Cambridge University Press, Cambridge, 2002.

A-2 A'Harrah, R. C., dir., *Flying Qualities*, AGARD-CP-508, Neuilly-sur-Seine, France, Feb. 1991.

A-3 Anon., "Military Specification, Flying Qualities of Piloted Airplanes," MIL-F-8785C, Wright-Patterson Air Force Base, Ohio, Nov. 1980.

A-4 Ashkenas, I. L., "Some Open-and Closed-Loop Aspects of Airplane Lateral-Directional Handling Qualities," AGARD Report No. 533, Paris, May 1966.

A-5 Atzhorn, D., and Stengel, R. F., "Design and Flight Test of a Lateral-Directional Command Augmentation System," *J. Guidance, Control, and Dynamics* 7, no. 3 (1984): 361–368.

B-1 Bearman, P. W., Graham, J. M. R., and Kalkanis, P., "The Effect of Rapid Spoiler Deployment on the Transient Forces on an Aerofoil," *Aerodynamics of Combat Aircraft Controls and of Ground Effects*, AGARD-CP-465, Neuilly-sur-Seine, 1989.

B-2 Binnie, W. B., and Stengel, R. F., "Flight Investigation and Theory of Direct Side-Force Control," *J. Guidance and Control* 2, no. 6 (1979): 471–478.

B-3 Blakelock, J. H., *Automatic Control of Aircraft and Missiles*, J. Wiley & Sons, New York, 1991.

B-4 Bryant, L. W., and Gandy, R. W. G., "An Investigation of the Lateral Stability of Aeroplanes with Rudder Free," NPL Report S&C 1097, London, 1939.

B-5 Bunton, R. W., and Denegri, C. M., Jr., "Limit Cycle Oscillation Characteristics of Fighter Aircraft," *J. Aircraft* 37, no. 5 (2000): 916–918.

C-1 Coe, P. L., Jr., Graham, A. B., and Chambers, J. R., "Summary of Information on Low-Speed Lateral-Directional Derivatives due to Rate of Change of Sideslip," NASA TN D-7972, Washington, D.C., Sept. 1975.

E-1 Etkin, B., *Dynamics of Atmospheric Flight*, J. Wiley & Sons, New York, 1972.

G-1 Gelb, A., and VanderVelde, W. E., *Multiple-Input Describing Functions and Nonlinear System Design*, M.I.T. Press, Cambridge, Mass., 1968.

G-2 Gibson, J. C., "The Development of Alternate Criteria for FBW Handling Qualities," *Flying Qualities*, AGARD-CP-508, Neuilly-sur-Seine, France, Feb. 1991, pp. 9-1–9-13.

G-3 Gilruth, R. R., and Turner, W. N., "Lateral Control Required for Satisfactory Flying Qualities Based on Flight Tests of Numerous Airplanes," NASA Report 715, Washington, D.C., Apr. 1941.

G-4 Gilruth, R. R., "Requirements for Satisfactory Flying Qualities of Airplanes," NACA Report 755, Washington, D.C., 1943.

G-5 Graham, D., and McRuer, D., *Analysis of Nonlinear Control Systems*, J. Wiley & Sons, New York, 1964.

G-6 Greenberg, H., and Sternfield, L., "Theoretical Investigation of the Lateral Oscillations of an Airplane with Free Rudder with Special Reference to the Effect of Friction," NACA Report 762, Washington, D.C., 1943.

G-7 Greenberg, H., and Sternfield, L., "A Theoretical Investigation of Longitudinal Stability of Airplanes with Free Controls Including the Effect of Friction in the Control System," NACA Report 791, Washington, D.C., 1944.

G-8 Greer, H. D., "Summary of Directional Divergence Characteristics of Several High-Performance Aircraft Configurations," NASA TN D-6993, Washington, D.C., Nov. 1972.

G-9 Grunwald, S., and Stengel, R. F., "Design and Flight Testing of Digital Direct Side-Force Control Laws," *J. Guidance, Control, and Dynamics* 8, no. 2 (1985): 188–193.

G-10 Guckenheimer, J., and Holmes, P., *Nonlinear Oscillations, Dynamical Systems, and Bifurcations of Vector Fields*, Springer-Verlag, New York, 1983.

H-1 Hodgkinson, J., *Aircraft Handling Qualities*, American Institute of Aeronautics and Astronautics, Reston, Va., 1999.

H-2 Hodkginson, J., and Mitchell, D., "Handling Qualities," in *Flight Control Systems*, R. W. Pratt, ed., American Institute of Aeronautics and Astronautics, Reston, Va., 2000, pp. 119–169.

H-3 Höhne, G., "Roll Ratcheting: Cause and Analysis," Doctoral Thesis, Technical University of Braunschweig, DLR-Forschungsbericht 2001-15, Cologne, Germany, 2001.

J-1 Javed, M. A., and Hancock, G. J., "Application of Vortex Lattice Methods to calculate Lv (Rolling Moment due to Sideslip), *Aeronautical Journal* 85, no. 842 (1981): 113–117.

J-2 Jones, R. T., and Cohen, D., "An Analysis of the Stability of an Airplane with Free Controls," NACA Report 709, Washington, D.C., 1941.

K-1 Kempel, R. W., "Analysis of a Coupled Roll-Spiral Mode, Pilot-Induced Oscillation Experienced with the M2-F2 Lifting Body," NASA TN D-6496, Washington, D.C., Sept. 1971.

L-1 Lan, C. E., Chen, Y., and Lin, K.-J., "Experimental and Analytical Investigation of Transonic Limit-Cycle Oscillations of a Flaperon," *J. Aircraft* 32, no. 5 (1995): 905–910.

M-1 McLean, D., *Automatic Flight Control Systems*, Prentice-Hall, Englewood Cliffs, N.J., 1990.

M-2 McRuer, D., Ashkenas, I., and Graham, D., *Aircraft Dynamics and Automatic Control*, Princeton University Press, Princeton, N.J., 1973.

M-3 Moul, M. T., and Paulson, J. W., "Dynamic Lateral Behavior of High-Performance Aircraft," NACA RM L58E16, Washington, D.C., Aug. 6, 1958.

N-1 Nguyen, L. T., "Evaluation of Lateral Acceleration Derivatives in Extraction of Lateral-Directional Derivatives at High Angles of Attack," NASA TN D-7739, Washington, D.C., Oct. 1974.

O-1 O'Hara, F., "Handling Criteria," *J. Royal Aeronautical Society* 71, no. 676 (1967): 271–291.

P-1 Pratt, R., ed., *Flight Control Systems*, American Institute of Aeronautics and Astronautics, Reston, Va., 2000.

R-1 Ralston, A., and Wilf, H. S., *Mathematical Methods for Digital Computers*, J. Wiley & Sons, New York, 1965.

R-2 Rodden, W. P., "Dihedral Effect of a Flexible Wing," *J. Aircraft* 2, no. 5 (1965): 368–374.

S-1 Seckel, E., *Stability and Control of Airplanes and Helicopters*, Academic Press, New York, 1964.

S-2 Shrivastava, P. C., and Stengel, R. F., "Stability Boundaries for Aircraft with Unstable Lateral-Directional Dynamics and Control Saturation," *J. Guidance, Control, and Dynamics* 12, no. 1 (1989): 62–70.

S-3 Stengel, R. F., *Optimal Control and Estimation*, Dover Publications, New York, 1994 (originally published as *Stochastic Optimal Control: Theory and Application*, J. Wiley & Sons, New York, 1986).

T-1 Toll, T. A., "A Summary of Lateral Control Research," NACA Report 868, Washington, D.C., 1947.

W-1 Williams, J., "The USAF Stability and Control Digital DATCOM," AFFDL-TR-76-45, Wright-Patterson AFB, Ohio, Nov. 1976.

Chapter 7

Coupled Longitudinal and Lateral-Directional Motions

Insights gained from the separate descriptions of longitudinal and lateral-directional motions provide excellent guideposts for understanding much of aircraft flight, but these uncoupled models do not accurately portray fully three-dimensional motions. A number of cross-product terms in the equations of motion (Chapters 3 and 4) are neglected, as are aerodynamic asymmetries and the effects of rotating powerplant components. When angles and angular rates are low, these missing effects may not be of concern, but for large-amplitude or high-rate maneuvers, the coupling terms can dominate the motion.

While details of fully coupled motion can be captured only by solving the nonlinear equations, linear models still may describe the qualitative nature of motions within limited regions. Neglecting altitude-gradient effects and the pure integrations of range, crossrange, and yaw angle, the rigid-body equations are eighth order. For small angle and rate perturbations from trim, the eigenvalues describe phugoid, short-period, Dutch roll, roll, and spiral modes, but the eigenvectors of each mode may contain all eight elements of the state. For larger perturbations from steady or quasisteady trim, the mode shapes may become markedly different from the classical forms. For example, short-period and Dutch roll modes are coupled by a steady roll rate.

At this point, we abandon the hybrid- and stability-axis equations of motion used in Chapters 5 and 6, and return to the body-axis equations of Chapter 4. The rationale for hybrid and stability axes is that they decouple the individual longitudinal and lateral-directional linear models into approximately normal modes, and they simplify the models by eliminating explicit dependence on α_o and β_o (or w_o and v_o) in the linear equations of motion. However, stability axes complicate matters for the nonlinear equations, when α_o and β_o may be changing significantly over short periods of time, and when the full transformation between axis systems, $\mathbf{H}_B^S$, (eq. 3.2-83) must be considered. For example, the bookkeeping associated with the angular momentum equation (eq. 3.2-32),

$$\frac{d\boldsymbol{\omega}_B}{dt} = \mathcal{I}_B^{-1}[\mathbf{m}_B - \tilde{\boldsymbol{\omega}}_B \mathcal{I}_B \boldsymbol{\omega}_B] \tag{7.0-1}$$

becomes cumbersome when the transformation

$$\mathbf{H}_B^S = \begin{bmatrix} \cos\alpha_o \cos\beta_o & \sin\beta_o & \sin\alpha_o \cos\beta_o \\ -\cos\alpha_o \sin\beta_o & \cos\beta_o & -\sin\alpha_o \sin\beta_o \\ -\sin\alpha_o & 0 & \cos\alpha_o \end{bmatrix} \tag{7.0-2}$$

must be taken into account. Because the present goal is to explore fully coupled longitudinal/lateral-directional effects, coupling within the longitudinal and lateral-directional states provides no additional burden.

Nonlinearity introduces phenomena that are not readily described by linear models, such as the period doubling demonstrated in Chapter 3, chaotic behavior, and multiple equilibria associated with a single vector of control settings. Plots of possible trim states may bifurcate as the control settings are adjusted. Unrecoverable "deep stall" provides an entirely longitudinal example, while the "flat spin" is a condition involving all axes. Flight at high angles of attack presents a particular challenge for combat aircraft that use this flight regime to gain tactical advantage. Here, the changing lateral-directional characteristics may be more difficult to handle than the longitudinal control itself. Varying transformations between axis systems complicate the description of motions, and nonlinear variations in aerodynamics may be large, sensitive to small changes in the state and control, and not readily unpredictable.

In this chapter, we explore the nature of fully coupled motions. Linear models of coupling effects are presented in Section 7.1. Interactions between pitching and yawing motions brought about by steady roll rate are described in Section 7.2. The circumstances that lead to more than one equilibrium point for a single control setting are discussed in Section 7.3. The special effects of flying at a high angle of attack, which may lead to fully evolved spins, are shown in Section 7.4.

7.1 Small-Amplitude Motions

Coupled motions that appear as small perturbations from a steady flight condition can be investigated using the linear models derived in Section 4.1. While all of the coupling effects could be considered at once, the individual causes of coupling would be hidden. In this section, we focus on the separate dynamic effects of rotating machinery (e.g., powerplant components), inertial and aerodynamic properties, asymmetric flight condition, and coupling control, assuming that trimmed control settings produce coupled equilibrium. The linearized state vector is truncated to the six rates and two angles that have most significant effect on the dynamic modes of motion:

$$\Delta\mathbf{x} = [\Delta u\ \Delta w\ \Delta q\ \Delta\theta\ \Delta v\ \Delta p\ \Delta r\ \Delta\phi]^T \tag{7.1-1}$$

$(\Delta u, \Delta v, \Delta w)$ are the linear velocity components along the body (x, y, z) axes, $(\Delta p, \Delta q, \Delta r)$ are the angular rates about these axes, and $\Delta\theta$ and $\Delta\phi$ are pitch- and roll-angle perturbations. The four principal controls are elevator, thrust, aileron, and rudder:

$$\Delta\mathbf{u} = [\Delta\delta E\ \Delta\delta T\ \Delta\delta A\ \Delta\delta R]^T \tag{7.1-2}$$

Neglecting disturbance effects, the linearized dynamic equation

$$\Delta\dot{\mathbf{x}}(t) = \mathbf{F}\,\Delta\mathbf{x}(t) + \mathbf{G}\,\Delta\mathbf{u}(t) \tag{7.1-3}$$

can be partitioned into longitudinal and lateral-directional components.

There are four state elements and two control elements for each partition; hence, the stability matrix contains four (4×4) primary and coupling (secondary) blocks, and the control matrix contains four (4×2) blocks:

$$\begin{bmatrix} \Delta\dot{\mathbf{x}}_{\mathrm{Lo}} \\ \Delta\dot{\mathbf{x}}_{\mathrm{LD}} \end{bmatrix} = \begin{bmatrix} \mathbf{F}_{\mathrm{Lo}} & \mathbf{F}_{\mathrm{LD}}^{\mathrm{Lo}} \\ \mathbf{F}_{\mathrm{Lo}}^{\mathrm{LD}} & \mathbf{F}_{\mathrm{LD}} \end{bmatrix} \begin{bmatrix} \Delta\mathbf{x}_{\mathrm{Lo}} \\ \Delta\mathbf{x}_{\mathrm{LD}} \end{bmatrix} + \begin{bmatrix} \mathbf{G}_{\mathrm{Lo}} & \mathbf{G}_{\mathrm{LD}}^{\mathrm{Lo}} \\ \mathbf{G}_{\mathrm{Lo}}^{\mathrm{LD}} & \mathbf{G}_{\mathrm{LD}} \end{bmatrix} \begin{bmatrix} \Delta\mathbf{u}_{\mathrm{Lo}} \\ \Delta\mathbf{u}_{\mathrm{LD}} \end{bmatrix} \tag{7.1-4}$$

For a symmetric aircraft in symmetric flight, the secondary matrix blocks are zero, and the eigenvalues and eigenvectors of $\mathbf{F}_{\mathrm{Lo}}$ and $\mathbf{F}_{\mathrm{LD}}$ are uncoupled from each other. Neglecting unsteady aerodynamics, altitude gradients, and the effects of aerodynamic-angle rates on forces, the symmetric-flight blocks of the stability matrix are

$$\mathbf{F}_{\mathrm{Lo}} = \begin{bmatrix} X_u & X_w & -w_o & -g_o\cos\theta_o \\ Z_u & Z_w & u_o & -g_o\sin\theta_o \\ M_u & M_w & M_q & 0 \\ 0 & 0 & 1 & 0 \end{bmatrix} \tag{7.1-5}$$

$$\mathbf{F}_{\mathrm{LD}} = \begin{bmatrix} Y_v & w_o & -u_o & g_o\cos\theta_o \\ L_v & L_p & L_r & 0 \\ N_v & N_p & N_r & 0 \\ 0 & 1 & -\sin\theta_o & 0 \end{bmatrix} \tag{7.1-6}$$

The primary control blocks are

$$\mathbf{G}_{\mathrm{Lo}} = \begin{bmatrix} X_{\delta E} & X_{\delta T} \\ Z_{\delta E} & Z_{\delta T} \\ M_{\delta E} & M_{\delta T} \\ 0 & 0 \end{bmatrix} \tag{7.1-7}$$

$$\mathbf{G}_{\mathrm{LD}} = \begin{bmatrix} Y_{\delta A} & Y_{\delta R} \\ L_{\delta A} & L_{\delta R} \\ N_{\delta A} & N_{\delta R} \\ 0 & 0 \end{bmatrix} \tag{7.1-8}$$

In the remainder of the section, we see how asymmetry leads to nonzero secondary blocks that fully couple the aircraft's motions. The baseline stability and control derivatives, eigenvalues, moments and products of inertia, and linear velocity components used for figures in the next two sections are drawn from the business jet model and are contained in Table 7.1-1. Parameters may be varied and equation order may be reduced to illustrate certain effects. The uncoupled short-period mode is faster and the roll mode is slower than the Dutch roll mode.

TABLE 7.1-1 BASELINE UNCOUPLED LONGITUDINAL AND LATERAL-DIRECTIONAL MODELS FOR FIGURES OF SECTIONS 7.1 AND 7.2

$$\mathbf{F}_{\mathrm{Lo}} = \begin{bmatrix} -0.0121 & 0.096 & -6.45 & -9.787 \\ -0.116 & -1.2773 & -101 & 0 \\ 0.005 & -0.0781 & -1.2794 & 0 \\ 0 & 0 & 1 & 1 \end{bmatrix} \qquad \mathbf{G}_{\mathrm{Lo}} = \begin{bmatrix} 0.0065 & 4.6739 \\ -13.1653 & 0 \\ -9.069 & 0 \\ 0 & 0 \end{bmatrix}$$

$$\mathbf{F}_{\mathrm{LD}} = \begin{bmatrix} -0.1567 & 6.45 & -101 & 9.787 \\ -0.0247 & -1.1768 & 0.1823 & 0 \\ 0.0171 & -0.0112 & -0.0927 & 0 \\ 0 & 1 & 0 & 0 \end{bmatrix} \qquad \mathbf{G}_{\mathrm{LD}} = \begin{bmatrix} -0.1591 & 3.512 \\ 2.3078 & 0.2487 \\ 0.1176 & -1.106 \\ 0 & 0 \end{bmatrix}$$

$\lambda_{spiral} = +0.0035$ rad/s
$\lambda_{phugoid} = -0.0099 \pm 0.1256\,j$ rad/s
$\lambda_{roll} = -1.2035$ rad/s
$\lambda_{Dutch\ roll} = -0.1131 \pm 1.3844\,j$ rad/s
$\lambda_{short\ period} = -1.2745 \pm 2.8117\,j$ rad/s
$I_{xx} = 35{,}927$ kg-m^2
$I_{yy} = 33{,}941$ kg-m^2
$I_{zz} = 67{,}086$ kg-m^2
$I_{xz} = 3{,}418$ kg-m^2
$u_o = 101$ m/s
$w_o = 6.45$ m/s

For the example, the rolling moment of inertia I_{xx} is the middle value, due to both high wing aspect ratio and wing-tip fuel tanks. The yawing moment of inertia I_{zz} is nearly twice as large, while the pitching moment of inertia I_{yy} is only slightly smaller than I_{xx}. For a conventional configuration, I_{zz} is always the largest of the three values because the moment arms about the z axis of both wing and fuselage are large. For a high-aspect-ratio flying wing, I_{zz} is on the order of I_{xx}, and I_{yy} is small. I_{xx} is smaller than I_{yy} for low-aspect-ratio configurations, including most supersonic aircraft.

Effects of Rotating Machinery

Rotating antennas or parts of powerplants, including propellers, compressors, and turbines create angular momentum $\mathbf{h}_R$. This effect adds to the angular momentum of the airframe, coupling longitudinal and lateral-directional motions through gyroscopic moments. The rate of change of the aircraft's total angular momentum $\mathbf{h}_B$ can be expressed in body axes as (eq. 3.2-32)

$$\frac{d\mathbf{h}_B}{dt} = \mathbf{m}_B - \tilde{\boldsymbol{\omega}}_B \mathbf{h}_B \tag{7.1-9}$$

where $\mathbf{m}_B$ is the external moment on the aircraft, and $\boldsymbol{\omega}_B$ is the airframe's angular rate:

$$\boldsymbol{\omega}_B = [p\ q\ r]_B^T \tag{7.1-10}$$

The corresponding cross-product-equivalent matrix of angular rate is

$$\tilde{\boldsymbol{\omega}}_B \triangleq \begin{bmatrix} 0 & -r & q \\ r & 0 & -p \\ -q & p & 0 \end{bmatrix}_B \tag{7.1-11}$$

We assume that flight motions do not change the spin rates of rotating parts and that their axes of rotation are fixed in the body frame of reference:

$$\mathbf{h}_R = [h_x\ h_y\ h_z]_B^T \tag{7.1-12}$$

Thus, $|d\mathbf{h}_R/dt|$ is zero, and $\mathbf{h}_R$ rotates with the aircraft. From Section 3.2, the angular momentum of nonrotating components of the airframe is the product of the inertia matrix $\mathcal{I}_B$ and the angular rate of the airframe $\boldsymbol{\omega}_B$; hence, the dynamic equation becomes

$$\mathcal{I}_B \frac{d\boldsymbol{\omega}_B}{dt} = \mathbf{m}_B - \tilde{\boldsymbol{\omega}}_B(\mathcal{I}_B \boldsymbol{\omega}_B + \mathbf{h}_R) \tag{7.1-13}$$

or

$$\frac{d\boldsymbol{\omega}_B}{dt} = \mathcal{I}_B^{-1}[\mathbf{m}_B - \tilde{\boldsymbol{\omega}}_B(\mathcal{I}_B\boldsymbol{\omega}_B + \mathbf{h}_R)] \tag{7.1-14}$$

The gyroscopic moment produced by the rotating machinery is

$$\tilde{\boldsymbol{\omega}}_B\mathbf{h}_R = \begin{bmatrix} -rh_y + qh_z \\ rh_x - ph_z \\ -qh_x + ph_y \end{bmatrix} \tag{7.1-15}$$

This equation is differentiated with respect to angular rate to reveal the gyroscopic sensitivity to angular-rate perturbations:

$$\frac{\partial(\tilde{\boldsymbol{\omega}}_B\mathbf{h}_R)}{\partial\boldsymbol{\omega}_B} = \begin{bmatrix} 0 & h_z & -h_y \\ -h_z & 0 & h_x \\ h_y & -h_x & 0 \end{bmatrix} \tag{7.1-16}$$

For typical aircraft configurations, the spanwise component of angular momentum h_y is negligible. h_z is the most significant component for helicopters, and h_x is the most significant component for conventional aircraft. A surveillance aircraft with a large antenna rotating about the z axis could have a large h_z as well.

The inertia matrix inverse for an aircraft that is symmetric about its vertical plane is defined by eq. 3.2-43:

$$\mathcal{I}_B^{-1} = \begin{bmatrix} \dfrac{1}{I_{xx} - I_{xz}^2/I_{zz}} & 0 & \dfrac{I_{xz}}{I_{xx}I_{zz} - I_{xz}^2} \\ 0 & \dfrac{1}{I_{yy}} & 0 \\ \dfrac{I_{xz}}{I_{xx}I_{zz} - I_{xz}^2} & 0 & \dfrac{1}{I_{zz} - I_{xz}^2/I_{xx}} \end{bmatrix} \tag{7.1-17}$$

Consequently, the gyroscopic contribution to the body-axis linear model is

$$-\mathcal{I}_B^{-1}\frac{\partial(\tilde{\boldsymbol{\omega}}_B\mathbf{h}_R)}{\partial\boldsymbol{\omega}_B} = -\begin{bmatrix} 0 & \dfrac{I_{zz}h_z - I_{xz}h_x}{I_{xx}I_{zz} - I_{xz}^2} & 0 \\ -\dfrac{h_z}{I_{yy}} & 0 & \dfrac{h_x}{I_{yy}} \\ 0 & \dfrac{I_{xz}h_z - I_{xx}h_x}{I_{xx}I_{zz} - I_{xz}^2} & 0 \end{bmatrix} \tag{7.1-18}$$

This result is transformed to the state-vector ordering of eq. 7.1-1 to produce the coupling blocks of the stability matrix:

$$\mathbf{F}_{\mathrm{LD}}^{\mathrm{Lo}} = \begin{bmatrix} 0 & 0 & 0 & 0 \\ 0 & 0 & 0 & 0 \\ 0 & \dfrac{h_z}{I_{yy}} & -\dfrac{h_x}{I_{yy}} & 0 \\ 0 & 0 & 0 & 0 \end{bmatrix} \tag{7.1-19}$$

$$\mathbf{F}_{\mathrm{Lo}}^{\mathrm{LD}} = \begin{bmatrix} 0 & 0 & 0 & 0 \\ 0 & 0 & -\dfrac{I_{zz}h_z - I_{xz}h_x}{I_{xx}I_{zz} - I_{xz}^2} & 0 \\ 0 & 0 & -\dfrac{I_{xz}h_z - I_{xx}h_x}{I_{xx}I_{zz} - I_{xz}^2} & 0 \\ 0 & 0 & 0 & 0 \end{bmatrix} \tag{7.1-20}$$

Combining eqs. 7.1-19 and 7.1-20 with eqs. 7.1-5 and 7.1-6 to specify **F**, the eighth-order characteristic polynomial of **F** can be written in root locus form. Neglecting h_z, the square of the x-axis angular momentum, h_x^2, appears as the gain, and the polynomial factors as

$$\begin{aligned} \Delta(s) = {} & [(s - \lambda_S)(s^2 + 2\zeta\omega_n s + \omega_n^2)_P(s - \lambda_R) \\ & \cdot (s^2 + 2\zeta\omega_n s + \omega_n^2)_{\mathrm{DR}}(s^2 + 2\zeta\omega_n s + \omega_n^2)_{\mathrm{SP}}] \\ & + h_x^2 s[(s - z_1)(s^2 + 2\zeta\omega_n s + \omega_n^2)_{h_x}(s - z_2)(s - z_3)] \end{aligned} \tag{7.1-21}$$

With $h_x = 0$, the factors represent the uncoupled longitudinal and lateral-directional modes. As $|h_x|$ increases, the short-period roots go toward infinity, and the Dutch roll roots go toward a lower-frequency complex pair (Fig. 7.1-1). The phugoid roots are increasingly damped and eventually become real, the spiral root approaches the zero at the origin, and the roll-mode root quickens slightly. The sign of h_x has no impact on root variation because the gain is h_x^2; however, it does affect the sense of coupled response to initial conditions.

Response to the initial pitch rate for angular momenta that are (numerically) about one-third and three times the rolling moment of inertia is compared to the uncoupled case in Fig. 7.1-2. The smaller angular momentum has negligible effect on longitudinal response, but the larger angular momentum produces the predicted stiffening. Lateral-directional

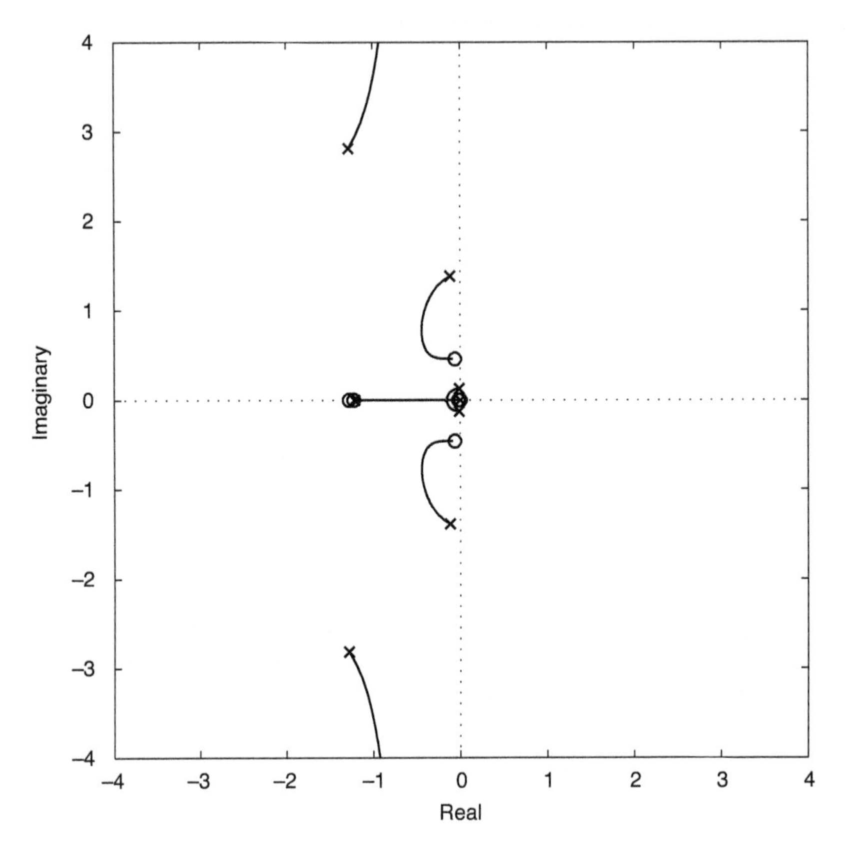

Fig. 7.1-1. Effect of the square of axial angular momentum h_x^2 on longitudinal and lateral-directional roots.

response is zero with no coupling. For the higher axial momentum, the yaw rate and sideslip angle respond at the short-period wavelength, while the roll variables go through a single oscillation at the Dutch roll wavelength. Thus, pitching motions are accompanied by yawing and rolling transients.

The coupling is felt in response to control as well, even when there is no coupling in the control effect matrix **G**. This is most easily seen from the transfer-function relationship between $\Delta\mathbf{u}(s)$ and $\Delta\mathbf{x}(s)$:

$$\Delta\mathbf{x}(s) = (s\mathbf{I} - \mathbf{F})^{-1}\,\mathbf{G}\,\Delta\mathbf{u}(s) \tag{7.1-22}$$

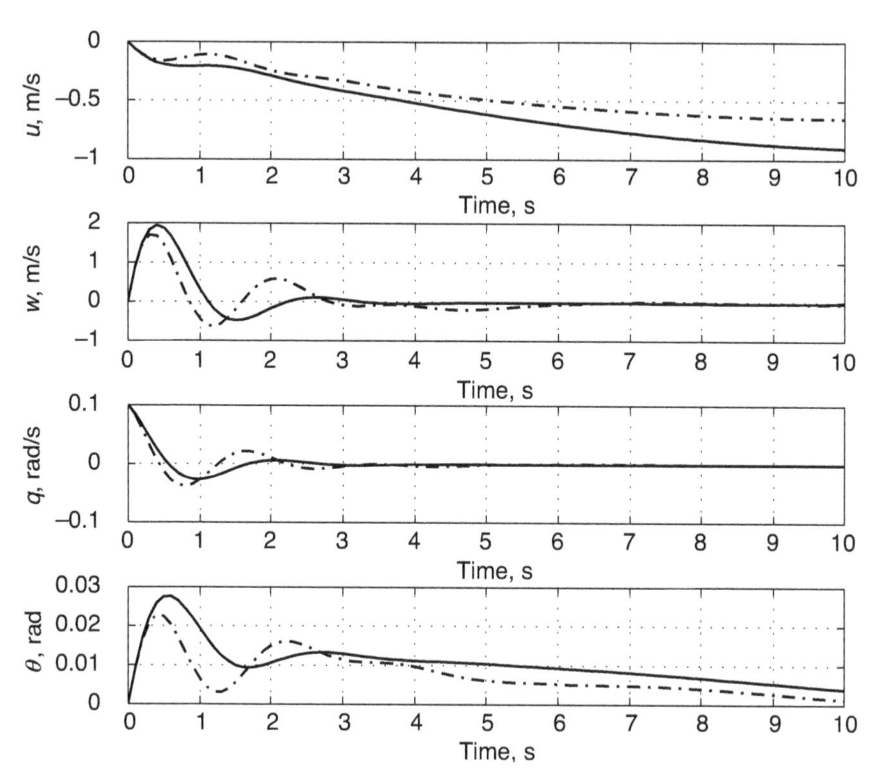

Fig. 7.1-2a–d. Effect of axial angular momentum on the response to an initial pitch rate. $h_x = 0$ (solid line), $h_x = 10{,}000$ (dashed line), $h_x = 100{,}000$ (dash-dotted line): axial velocity, normal velocity, pitch rate, pitch angle.

The coupled (8×8) characteristic matrix $(s\mathbf{I} - \mathbf{F})^{-1}$ typically has few zero elements. It transforms the longitudinal specific forces and moments arising from $\mathbf{G}\Delta\mathbf{u}(s)$ into lateral-directional rates of change and vice versa. Figure 7.1-3 provides an example of the effect, comparing the uncoupled system with systems containing large angular momenta of opposite signs. A half-second aileron pulse is commanded, producing only yawing and rolling in the uncoupled case. The angular momentum effects on yawing and rolling are the same for either sign, but the effects on longitudinal motions are mirror images of each other: with positive h_x, there is a net pitch down, with negative h_x, a net pitch up. Conversely, with fixed angular momentum, rolling to the left and

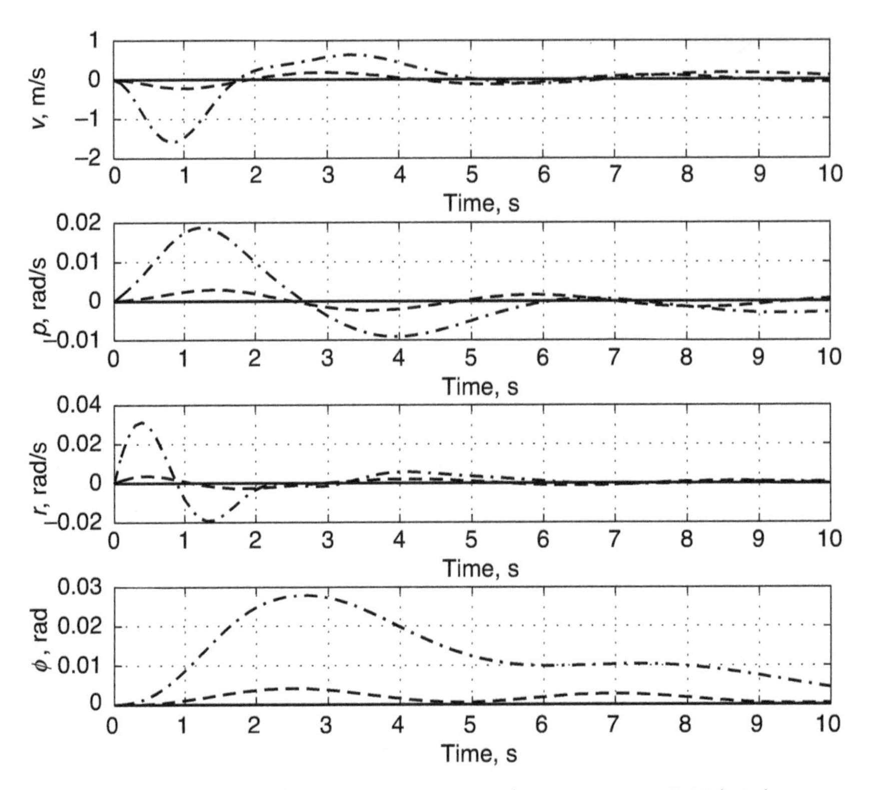

Fig. 7.1-2e–h. Effect of axial angular momentum on the response to an initial pitch rate. $h_x = 0$ (solid line), $h_x = 10{,}000$ (dashed line), $h_x = 100{,}000$ (dash-dot line): side velocity, roll rate, yaw rate, roll angle.

right would produce pitching of opposite sign. World-War-I-era fighters often had rotary engines, whose rotating cylinders produced high angular momentum. Their rolling response was typified by the traces of Fig. 7-1-3 (although their stability and control derivatives were markedly different).

Asymmetric Inertial and Aerodynamic Properties

Neglecting rotating machinery, the aircraft's angular rate is governed by

$$\frac{d\boldsymbol{\omega}_B}{dt} = \boldsymbol{\mathcal{I}}_B^{-1}[\mathbf{m}_B - \tilde{\boldsymbol{\omega}}_B \boldsymbol{\mathcal{I}}_B \boldsymbol{\omega}_B] \tag{7.1-23}$$

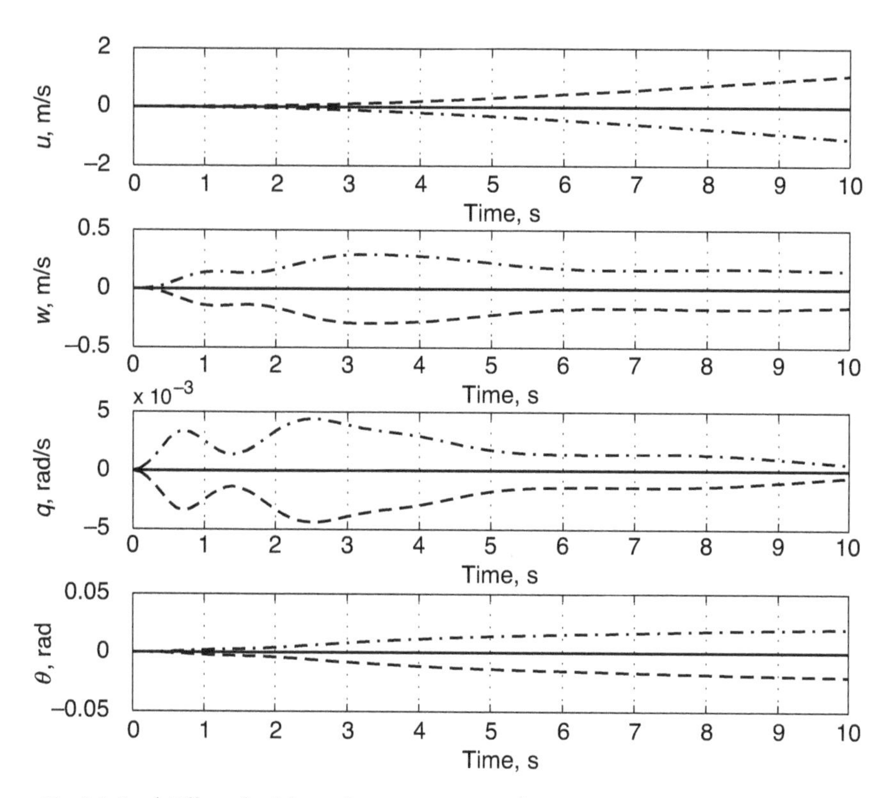

Fig. 7.1-3a–d. Effect of axial angular momentum on the response to a 0.5 s, 0.1 rad aileron pulse. $h_x = 0$ (solid line), $h_x = 100{,}000$ (dashed line), $h_x = -100{,}000$ (dash-dotted line): axial velocity, normal velocity, pitch rate, pitch angle.

In steady, level flight, the cross-product term vanishes because the nominal angular rate is zero, and the linear sensitivity of angular rate to state and control perturbations is

$$\begin{aligned}\frac{d\Delta\boldsymbol{\omega}_B}{dt} &= \mathcal{I}_B^{-1}\left[\frac{\partial \mathbf{m}_B}{\partial \mathbf{x}}\Delta\mathbf{x} + \frac{\partial \mathbf{m}_B}{\partial \mathbf{u}}\Delta\mathbf{u}\right] \\ &\triangleq \mathcal{I}_B^{-1}\begin{bmatrix} L \\ M \\ N \end{bmatrix}_B \end{aligned} \tag{7.1-24}$$

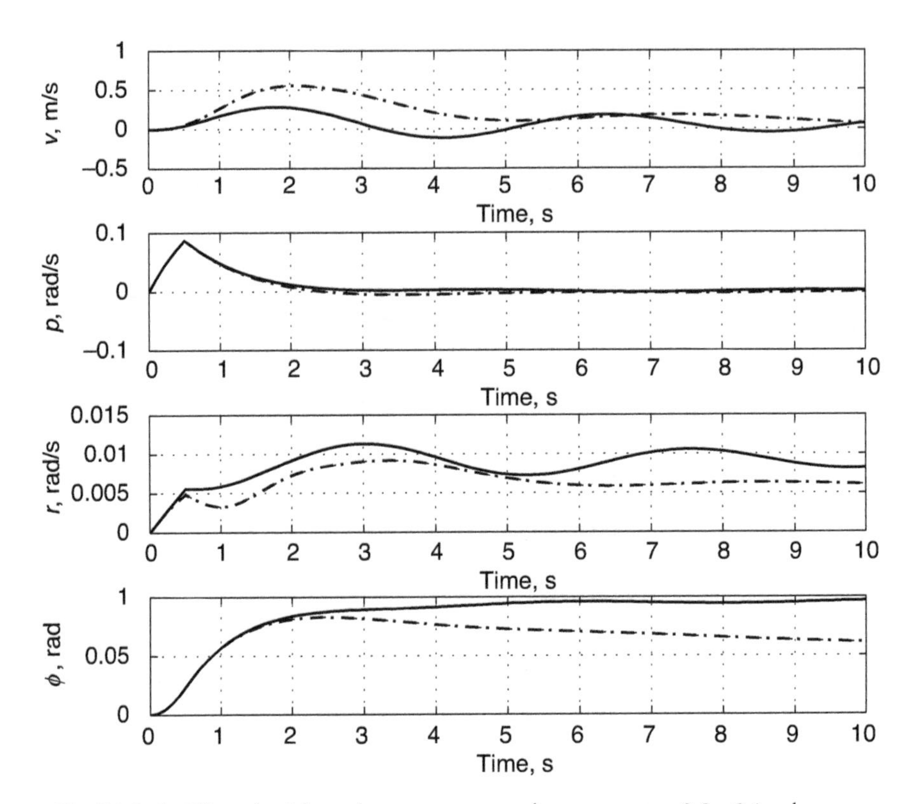

Fig. 7.1-3e–h. Effect of axial angular momentum on the response to a 0.5 s, 0.1 rad aileron pulse. $h_x = 0$ (solid line), $h_x = 100{,}000$ (dashed line), $h_x = -100{,}000$ (dash-dotted line): side velocity, roll rate, yaw rate, roll angle.

If the body axes are principal axes, $\mathcal{I}_B$ is diagonal, and external moments produce accelerations about the input axes only. Inertial asymmetry, expressed by the products of inertia I_{xy}, I_{xz}, and I_{yz} (eq. 3.2-23), causes moments about one axis to produce accelerations about another. Both **F** and **G** are affected by inertial asymmetry. Thus, an elevator deflection may directly produce a rolling or yawing acceleration, or aileron and rudder deflections may cause pitching acceleration.

The effects of each product of inertia are analyzed separately to avoid the complexity associated with symbolic inversion of the full (3×3) inertia matrix. The *nose-high (low) effect* I_{xz} results when the longitudinal principal axis is at a positive (or negative) angle to the aircraft centerline.

The inertia matrix inverse is given by eq. 7.1-17, and the angular acceleration is

$$\frac{d\Delta\boldsymbol{\omega}_B}{dt} = \begin{bmatrix} \dfrac{I_{zz}L + I_{xz}N}{I_{xx}I_{zz} - I_{xz}^2} \\ \dfrac{M}{I_{yy}} \\ \dfrac{I_{xz}L + I_{xx}N}{I_{xx}I_{zz} - I_{xz}^2} \end{bmatrix} \tag{7.1-25}$$

The I_{xz} product of inertia preserves symmetry about the body-axis vertical plane, producing no effect on pitching motion. It does, however, couple yawing moment into rolling acceleration and rolling moment into yawing acceleration. The rolling and yawing effects are absorbed in the conventional definitions of lateral-directional stability derivatives.

I_{xy} is an *oblique-wing inertial effect*, resulting from a longitudinal principal axis that is at an angle to the right or left of the body-axis centerline. From eq. 7.1-24, the angular acceleration is

$$\frac{d\Delta\boldsymbol{\omega}_B}{dt} = \begin{bmatrix} \dfrac{I_{yy}L + I_{xy}M}{I_{xx}I_{yy} - I_{xy}^2} \\ \dfrac{I_{xy}L + I_{xx}M}{I_{xx}I_{yy} - I_{xy}^2} \\ \dfrac{N}{I_{zz}} \end{bmatrix} \tag{7.1-26}$$

Rolling and pitching motions are coupled by I_{xy}, while yawing motions are unaffected by the asymmetry.

I_{yz} is a *wing-high effect*, reflecting a roll orientation of the lateral principal axis with respect to the body y axis. The resulting angular acceleration is

$$\frac{d\Delta\boldsymbol{\omega}_B}{dt} = \begin{bmatrix} \dfrac{L}{I_{xx}} \\ \dfrac{I_{zz}M + I_{yz}N}{I_{yy}I_{zz} - I_{yz}^2} \\ \dfrac{I_{yz}M + I_{yy}N}{I_{yy}I_{zz} - I_{yz}^2} \end{bmatrix} \tag{7.1-27}$$

The rolling acceleration is unaffected, while pitching and yawing motions are coupled by the asymmetry.

From the form of eq. 7.1-24, it can be appreciated that aerodynamic asymmetry has similar effect in introducing longitudinal/lateral-directional coupling. Nevertheless, the aerodynamic and inertial properties are distinct. For example, the oblique wing [H-9], which is swept forward on one side of the centerline and backward on the other, has two effects. The mass distribution couples moments according to eq. 7.1-26, and the flow field introduces coupling in L, M, and N. Mass asymmetry has the same effect on sensitivity to both state and control perturbations (through $\mathcal{I}_B^{-1}$), while the asymmetric aerodynamic state and control sensitivity may be different.

Both the primary and the coupling blocks of **F** and **G** are affected by inertial asymmetry. In the discussion, we assume that the uncoupled longitudinal and lateral-directional stability derivatives are defined for principal axes. I_{xz} has no effect on $\mathbf{F}_{\mathrm{LD}}^{\mathrm{Lo}}$ and $\mathbf{F}_{\mathrm{Lo}}^{\mathrm{LD}}$, as the coupling that it introduces is contained in $\mathbf{F}_{\mathrm{LD}}$.

I_{xy} and I_{yz} do affect $\mathbf{F}_{\mathrm{LD}}^{\mathrm{Lo}}$ and $\mathbf{F}_{\mathrm{Lo}}^{\mathrm{LD}}$, and their effects are similar; therefore, just the former is discussed. The I_{xy} coupling matrices are

$$\mathbf{F}_{\mathrm{LD}}^{\mathrm{Lo}} = \begin{bmatrix} 0 & 0 & 0 & 0 \\ 0 & 0 & 0 & 0 \\ \dfrac{I_{xy} I_{xx} L_v}{I_{xx} I_{yy} - I_{xy}^2} & \dfrac{I_{xy} I_{xx} L_p}{I_{xx} I_{yy} - I_{xy}^2} & \dfrac{I_{xy} I_{xx} L_r}{I_{xx} I_{yy} - I_{xy}^2} & 0 \\ 0 & 0 & 0 & 0 \end{bmatrix} \tag{7.1-28}$$

$$\mathbf{F}_{\mathrm{Lo}}^{\mathrm{LD}} = \begin{bmatrix} 0 & 0 & 0 & 0 \\ \dfrac{I_{xy} I_{yy} M_u}{I_{xx} I_{yy} - I_{xy}^2} & \dfrac{I_{xy} I_{yy} M_w}{I_{xx} I_{yy} - I_{xy}^2} & \dfrac{I_{xy} I_{yy} M_q}{I_{xx} I_{yy} - I_{xy}^2} & 0 \\ 0 & 0 & 0 & 0 \\ 0 & 0 & 0 & 0 \end{bmatrix} \tag{7.1-29}$$

Construction of the stability matrix **F** is completed by adjusting the pitching moments in $\mathbf{F}_{\mathrm{Lo}}$ and the rolling moments in $\mathbf{F}_{\mathrm{LD}}$; in both matrices, the third-row elements are divided by $(1 - I_{xy}^2/I_{xx}I_{yy})$. The primary and secondary control matrices are derived accordingly.

Some effects of I_{xy} can be seen in the response to initial pitch rate (Fig. 7.1-4), which illustrates the effects of asymmetries that are numerically 5 and 10 percent of the rolling moment of inertia. Longitudinal transients are unaffected by the asymmetry. Following and initial roll-rate transients at the short-period wavelength, yawing and rolling motions at the Dutch roll wavelength are apparent. There is a net offset in roll angle during the period shown, with a long-term divergence due to the unstable spiral mode.

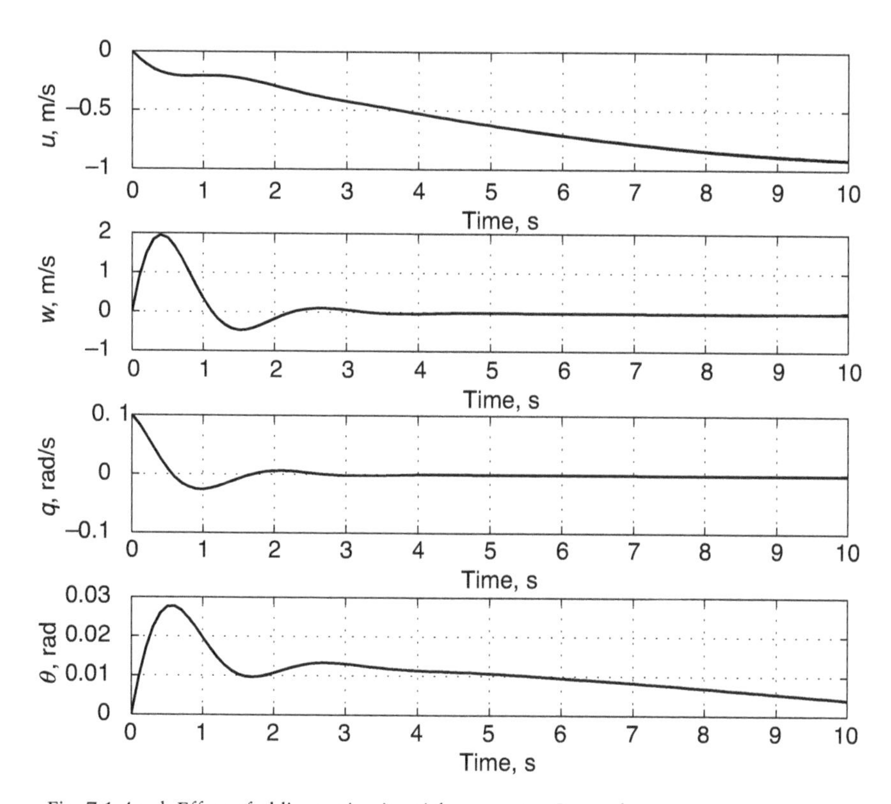

Fig. 7.1-4a–d. Effect of oblique-wing inertial asymmetry I_{xy} on the response to an initial pitch rate. $I_{xy} = 0$ (solid line), $I_{xy} = 1{,}800$ (dashed line), $I_{xy} = 3{,}600$ (dash-dotted line): axial velocity, normal velocity, pitch rate, pitch angle.

Asymmetric Flight Condition and Constant Angular Rate

The longitudinal and lateral-directional motions of a symmetric aircraft are coupled by flying at nonzero sideslip angle, roll angle, or angular rate. Both aerodynamic and inertial effects are involved when β_o is present, while ϕ_o effects are kinematic and gravitational. Lateral-directional rates p_o and r_o have aerodynamic, inertial, and kinematic coupling effects. Although q_o alone does not introduce coupling, it affects the lateral-directional modes. Combined with nonzero roll angle, q_o also contributes to three-axis coupling.

Nonzero Sideslip Angle

We can get an idea of the sources of coupling by examining typical moment coefficients, whose functional relationships to aerodynamic angles are summarized in Section 2.4. For the most part, the wing is the principal

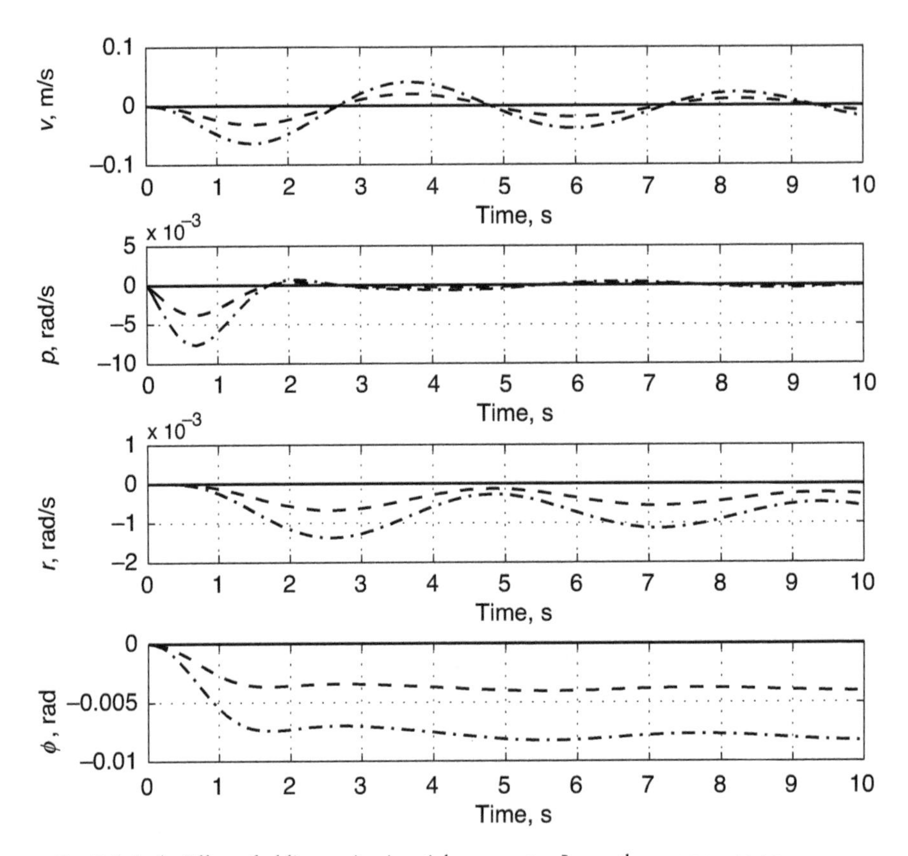

Fig. 7.1-4e–h. Effect of oblique-wing inertial asymmetry I_{xy} on the response to an initial pitch rate. $I_{xy} = 0$ (solid line), $I_{xy} = 1{,}800$ (dashed line), $I_{xy} = 3{,}600$ (dash-dot line): side velocity, roll rate, yaw rate, roll angle.

source of coupling aerodynamics because it produces strong forces, and small asymmetries can lead to large moments. The wing's yawing-moment sensitivity to sideslip angle (eq. 2.4-51) is seen to be a function of the angle of attack:

$$\begin{aligned} C_{n\beta_{\text{wing}}} &= k_0 C_L \Gamma + k_1 C_L^2 \\ &= k_0 C_{L_\alpha} \alpha \Gamma + k_1 (C_{L_\alpha} \alpha)^2 \end{aligned} \tag{7.1-30}$$

For small sideslip angle, the increment in yawing-moment coefficient is

$$\Delta C_n = C_{n\beta_{\text{wing}}} \beta = [k_0 C_{L_\alpha} \alpha \Gamma + k_1 (C_{L_\alpha} \alpha)^2] \beta \tag{7.1-31}$$

With nonzero α_o and β_o, the sensitivity to $\Delta\alpha$ is

$$C_{n_{\alpha_{\text{wing}}}} = (k_0 C_{L_\alpha} \Gamma + 2k_1 C_{L_\alpha} \alpha_o)\beta_o \tag{7.1-32}$$

and as $\Delta\alpha \approx \Delta w/V_o$,

$$C_{n_{w_{\text{wing}}}} = (k_0 C_{L_\alpha} \Gamma + 2k_1 C_{L_\alpha} \alpha_o)\beta_o/V_o \tag{7.1-33}$$

Equation 7.1-32 shows that angle-of-attack perturbations produce yawing moments when the aircraft is flying with a nonzero mean sideslip angle. From eq. 2.4-54, we see that there is a similar effect on the wing's rolling moment, leading to angle-of-attack sensitivity $C_{l_{\alpha_{wing}}}$.

Wind tunnel tests reveal that the normal force and pitching moment of a typical symmetric aircraft are *even functions* of sideslip angle [i.e., $Z(\beta) = Z(-\beta)$ and $M(\beta) = M(-\beta)$], with zero slope for $\beta = 0$ (e.g., [N-1]). For small sideslip perturbation and low angle of attack, the pitching moment coefficient could be modeled as

$$C_m = C_{m_o} + \left(C_{m_\alpha}\big|_{\beta=0} + k\beta^2\right)\alpha \tag{7.1-34}$$

hence, the pitching-moment-coefficient sensitivity to sideslip angle is

$$C_{m_\beta} = 2k\alpha_o\beta_o \tag{7.1-35}$$

With $\Delta\beta \approx \Delta v/V_o$, the sensitivity to side-velocity perturbation is

$$C_{m_v} = 2k\alpha_o\beta_o/V_o = 2kw_o v_o/V_o^3 \tag{7.1-36}$$

Consequently, β_o gives rise to the aerodynamic coupling stability derivatives X_v, Z_v, M_v, Y_w, L_w, and N_w.

Nonzero sideslip angle also affects primary derivatives such as C_{m_α}. From eq. 7.1-34,

$$C_{m_\alpha}(\beta_o) = C_{m_\alpha}\big|_{\beta=0} + k\beta_o^2 \tag{7.1-37}$$

Therefore, X_w, Z_w, M_w, Y_v, L_v, and N_v may be modified by β_o as well.

Side velocity produces a Coriolis effect on translational velocity. From eq. 4.1-33, the sensitivity of translational acceleration $[\Delta\dot{u}\ \Delta\dot{v}\ \Delta\dot{w}]^T$ to angular rate perturbation $[\Delta p\ \Delta q\ \Delta r]^T$ is

$$\mathbf{F}_{12} = \frac{1}{m}\frac{\partial \mathbf{f}'_B}{\partial \boldsymbol{\omega}_B} + \tilde{\mathbf{v}}_{B_o} \tag{7.1-38}$$

where

$$\tilde{\mathbf{v}}_{B_o} = \begin{bmatrix} 0 & -w_o & v_o \\ w_o & 0 & -u_o \\ -v_o & u_o & 0 \end{bmatrix}_B \tag{7.1-39}$$

u_o and w_o have already been incorporated in $\mathbf{F}_{Lo}$ and $\mathbf{F}_{LD}$ (eqs. 7.1-5 and 7.1-6); v_o adds terms to $\mathbf{F}_{LD}^{Lo}$. Together with the aerodynamic effects of β_o, the coupling matrices due to nonzero sideslip are

$$\mathbf{F}_{LD}^{Lo} = \begin{bmatrix} X_v & 0 & v_o & 0 \\ Z_v & -v_o & 0 & 0 \\ M_v & 0 & 0 & 0 \\ 0 & 0 & 0 & 0 \end{bmatrix} \tag{7.1-40}$$

$$\mathbf{F}_{Lo}^{LD} = \begin{bmatrix} 0 & Y_w & 0 & 0 \\ 0 & L_w & 0 & 0 \\ 0 & N_w & 0 & 0 \\ 0 & 0 & 0 & 0 \end{bmatrix} \tag{7.1-41}$$

Nonzero Roll Angle

The roll angle orients gravity's effect on body-axis components of acceleration, and it alters the propagation of Euler-angle perturbations. Aerodynamic forces and moments are not sensitive to roll angle when the aircraft is flying out of ground effect.

The sensitivity of body-axis translational acceleration $[\Delta\dot{u}\ \Delta\dot{v}\ \Delta\dot{w}]^T$ to Euler-angle perturbations $[\Delta\phi\ \Delta\theta\ \Delta\psi]^T$ is (eq. 4.1-35)

$$\mathbf{F}_{14} = g \begin{bmatrix} 0 & -\cos\theta_o & 0 \\ \cos\phi_o \cos\theta_o & -\sin\phi_o \sin\theta_o & 0 \\ -\sin\phi_o \cos\theta_o & -\cos\phi_o \sin\theta_o & 0 \end{bmatrix} \tag{7.1-42}$$

At low roll and pitch angles, gravitational acceleration perturbs $\Delta\dot{u}$ and $\Delta\dot{w}$ through $\Delta\theta$ and $\Delta\dot{v}$ through $\Delta\phi$. At high roll angle, $\Delta\dot{w}$ is more significantly affected by roll-angle variation. At high roll and pitch angle, $\Delta\dot{v}$ responds to $\Delta\theta$. In accessing these effects, recall that pitch angle is always measured in a vertical earth-relative plane, and the roll angle is measured about the aircraft's centerline, which is oriented by the pitch angle.

The translation of body-axis angular rates $[\Delta p\ \Delta q\ \Delta r]^T$ and Euler angles $[\Delta\phi\ \Delta\theta\ \Delta\psi]^T$ to Euler-angle rates is given by $\mathbf{F}_{42}$ and $\mathbf{F}_{44}$

(eqs. 4.1-42 and 4.1-43). Alternatively, the differentials for the nonlinear $\dot{\phi}$ and $\dot{\theta}$ equations (eqs. 3.2-55 and 3.2-56) can be evaluated directly:

$$\Delta\dot{\phi} = \Delta p + [\Delta q \sin\phi_o + \Delta r \cos\phi_o + (q_o \cos\phi_o - r_o \sin\phi_o)\Delta\phi] \tan\theta_o + (q_o \sin\phi_o + r_o \cos\phi_o) \sec^2\theta_o \Delta\theta \tag{7.1-43}$$

$$\Delta\dot{\theta} = \Delta q \cos\phi_o - \Delta r \sin\phi_o - (q_o \sin\phi_o + r_o \cos\phi_o)\Delta\phi \tag{7.1-44}$$

Combining the gravitational and kinematic effects, the coupling matrices are

$$\mathbf{F}_{\mathrm{LD}}^{\mathrm{Lo}} = \begin{bmatrix} 0 & 0 & 0 & 0 \\ 0 & 0 & 0 & -\sin\phi_o \cos\theta_o \\ 0 & 0 & 0 & 0 \\ 0 & 0 & -\sin\phi_o & (q_o \sin\phi_o + r_o \cos\phi_o) \end{bmatrix} \tag{7.1-45}$$

$$\mathbf{F}_{\mathrm{Lo}}^{\mathrm{LD}} = \begin{bmatrix} 0 & 0 & 0 & -\sin\phi_o \sin\theta_o \\ 0 & 0 & 0 & 0 \\ 0 & 0 & 0 & 0 \\ 0 & 0 & -\sin\phi_o \tan\theta_o & (q_o \sin\phi_o + r_o \cos\phi_o)\sec^2\theta_o \end{bmatrix} \tag{7.1-46}$$

Note that sensitivity to Euler-angle perturbation is zero unless there are steady values of pitch and yaw rate. The aircraft is normally turning when rolled, and steady angular rates typically occur. Although not shown, there are corresponding changes in $\mathbf{F}_{\mathrm{Lo}}$ and $\mathbf{F}_{\mathrm{LD}}$ following eqs. 7.1-42 and 7.1-43.

Nonzero Angular Rate

In addition to the effect on Euler angles noted above, a constant angular rate has a significant coupling effect on the dynamic equations. Coriolis and gyroscopic terms are found in both the translational and angular rate equations. There is the possibility of higher-order aerodynamic effects, although such effects are not widely documented.

Constant angular rates occur during steady turns, while quasisteady rates occur during pullups, rolling pullups, windup turns, rolls, and aerobatic maneuvers. In a *steady horizontal turn*, the earth-relative yaw rate $\dot{\psi}$ is constant, and it is equal to the turn rate of the velocity vector ($= W \tan\mu/mV$ from eq. 2.6-86; the bank angle μ is approximately equal to the roll angle ϕ as defined by eq. 2.2-36). $\dot{\psi}$ produces constant angular rates about all three body axes through the Euler-to-body-rate transformation (eq. 2.2-24)

$$\begin{bmatrix} p_o \\ q_o \\ r_o \end{bmatrix} = \mathbf{L}_{E_o}^{B} \begin{bmatrix} 0 \\ 0 \\ \dot{\psi}_o \end{bmatrix} = \begin{bmatrix} 1 & 0 & -\sin\theta_o \\ 0 & \cos\phi_o & \sin\phi_o\cos\theta_o \\ 0 & -\sin\phi_o & \cos\phi_o\cos\theta_o \end{bmatrix} \begin{bmatrix} 0 \\ 0 \\ \dot{\psi}_o \end{bmatrix}$$

$$= \dot{\psi}_o \begin{bmatrix} -\sin\theta_o \\ \sin\phi_o\cos\theta_o \\ \cos\phi_o\cos\theta_o \end{bmatrix} \tag{7.1-47}$$

In a *symmetric pullup*, q_o equals $\dot{\theta}_o$. Both roll and pitch rate occur in the *rolling pullup*. The *windup turn* is a high-μ, high-g maneuver, in which the angle of attack increases to maintain load factor as the airspeed decreases. *Rolling maneuvers* may occur at nearly constant roll rate; however, the roll-angle perturbation is large and not accurately portrayed by a linear model. Further consideration of such maneuvers is given in Sections 7.2 and 7.3.

From eq. 4.1-32, the Coriolis coupling of linear velocity $[\Delta u\ \Delta v\ \Delta w]^T$ to linear acceleration $[\Delta\dot{u}\ \Delta\dot{v}\ \Delta\dot{w}]^T$ is given by

$$-\tilde{\boldsymbol{\omega}}_{B_o} = -\begin{bmatrix} 0 & -r_o & q_o \\ r_o & 0 & -p_o \\ -q_o & p_o & 0 \end{bmatrix}_B \tag{7.1-48}$$

Rotational-rate perturbation $[\Delta p\ \Delta q\ \Delta r]^T$ is coupled to rotational acceleration $[\Delta\dot{p}\ \Delta\dot{q}\ \Delta\dot{r}]^T$ as in eq. 4.1-37. Alternatively, the angular rate coupling can be evaluated from the differentials of eq. 3.2-49–3.2-51 using, for example, Mathematica™ or the MATLAB Symbolic Toolbox. With inertial symmetry about the aircraft's vertical plane, eq. 7.1-24 becomes

$$\frac{d\Delta\omega_B}{dt} = \mathcal{I}_B^{-1} \begin{bmatrix} L \\ M \\ N \end{bmatrix}_B + \mathbf{C} \begin{bmatrix} \Delta p \\ \Delta q \\ \Delta r \end{bmatrix}_B \tag{7.1-49}$$

The elements of the gyroscopic coupling matrix $\mathbf{C}$ are

$$C_p^{\dot{p}} = \left(\frac{\partial\Delta\dot{p}}{\partial\Delta p}\right)_{gyro} = \frac{I_{xz}(I_{xx} - I_{yy} + I_{zz})q_o}{I_{xx}I_{zz} - I_{xz}^2} \tag{7.1-50}$$

$$C_q^{\dot{p}} = \left(\frac{\partial\Delta\dot{p}}{\partial\Delta q}\right)_{gyro} = \frac{[I_{xz}(I_{xx} - I_{yy} + I_{zz})]p_o + [I_{zz}(I_{yy} - I_{zz}) - I_{xz}^2]r_o}{I_{xx}I_{zz} - I_{xz}^2} \tag{7.1-51}$$

$$C_r^{\dot{p}} = \left(\frac{\partial \Delta \dot{p}}{\partial \Delta r}\right)_{gyro} = \frac{[I_{zz}(I_{yy} - I_{zz}) - I_{xz}^2]q_o}{I_{xx}I_{zz} - I_{xz}^2} \tag{7.1-52}$$

$$C_p^{\dot{q}} = \left(\frac{\partial \Delta \dot{q}}{\partial \Delta p}\right)_{gyro} = \frac{-2I_{xz}p_o - (I_{xx} - I_{zz})r_o}{I_{yy}} \tag{7.1-53}$$

$$C_q^{\dot{q}} = \left(\frac{\partial \Delta \dot{q}}{\partial \Delta q}\right)_{gyro} = 0 \tag{7.1-54}$$

$$C_r^{\dot{q}} = \left(\frac{\partial \Delta \dot{q}}{\partial \Delta r}\right)_{gyro} = \frac{(I_{zz} - I_{xx})p_o + 2I_{xz}r_o}{I_{yy}} \tag{7.1-55}$$

$$C_p^{\dot{r}} = \left(\frac{\partial \Delta \dot{r}}{\partial \Delta p}\right)_{gyro} = \frac{[I_{xx}(I_{xx} - I_{yy}) + I_{xz}^2]q_o}{I_{xx}I_{zz} - I_{xz}^2} \tag{7.1-56}$$

$$C_q^{\dot{r}} = \left(\frac{\partial \Delta \dot{r}}{\partial \Delta q}\right)_{gyro} = \frac{[I_{xx}(I_{xx} - I_{yy}) + I_{xz}^2]p_o + [I_{xz}(I_{yy} - I_{xx} - I_{zz})]r_o}{I_{xx}I_{zz} - I_{xz}^2} \tag{7.1-57}$$

$$C_r^{\dot{r}} = \left(\frac{\partial \Delta \dot{r}}{\partial \Delta r}\right)_{gyro} = \frac{I_{xz}(I_{yy} - I_{xx} - I_{zz})q_o}{I_{xx}I_{zz} - I_{xz}^2} \tag{7.1-58}$$

Steady roll and yaw rates are seen to couple longitudinal and lateral-directional motions. Steady pitch rate modifies roll-yaw dynamics but does not couple longitudinal and lateral-directional motions. Equations 7.1-48 and 7.1-50–7.1-58 provide elements of $\mathbf{F}_{\mathrm{LD}}^{\mathrm{Lo}}$ and $\mathbf{F}_{\mathrm{Lo}}^{\mathrm{LD}}$ and modifications to $\mathbf{F}_{\mathrm{Lo}}$ and $\mathbf{F}_{\mathrm{LD}}$. The coupling matrices are

$$\mathbf{F}_{\mathrm{LD}}^{\mathrm{Lo}} = \begin{bmatrix} r_o & 0 & 0 & 0 \\ -p_o & 0 & 0 & 0 \\ 0 & C_p^{\dot{q}} & C_r^{\dot{q}} & 0 \\ 0 & 0 & 0 & 0 \end{bmatrix} \tag{7.1-59}$$

$$\mathbf{F}_{\mathrm{Lo}}^{\mathrm{LD}} = \begin{bmatrix} -r_o & p_o & 0 & 0 \\ 0 & 0 & C_q^{\dot{p}} & 0 \\ 0 & 0 & C_q^{\dot{r}} & 0 \\ 0 & 0 & 0 & 0 \end{bmatrix} \tag{7.1-60}$$

Example 7.1-1. Stability of a Maneuvering Aircraft

The effects of nonzero aerodynamic angles and angular rates are examined using a coupled linear model of a lightweight fighter aircraft [S-16, S-17]. The aircraft is trimmed for horizontal, 1 *g* flight at 6,096 m altitude, 94 m/s airspeed, and 15 deg angle of attack. Eigenvalues and eigenvectors for the twelfth-order equation set are computed over a range of nominal values without retrimming. Stability boundaries for combinations of angle of attack and sideslip angle are presented in Fig. 7.1-5. The phugoid mode is unstable at low α_o, and the Dutch roll mode is unstable at high α_o. The transition from the phugoid mode's complex roots to real roots is indicated by the dashed line. The Dutch roll mode is stabilized by moderate sideslip angle, and damping of the short-period mode decreases.

The eigenvectors indicate the degree of longitudinal/lateral-directional coupling in these asymmetric flight conditions. For variations in the nominal sideslip angle (Fig. 7.1-6), roll-angle response is found in the phugoid mode, and pitch angle is a component of the spiral mode. Coupling of Dutch roll and short-period modes is evident, with $\Delta\alpha$ appearing in the former and $\Delta\beta$ appearing in the latter.

Combined variations in β_o and p_o produce the results shown in Fig. 7.1-7. The stability boundaries depend on the relative signs of β_o and p_o.

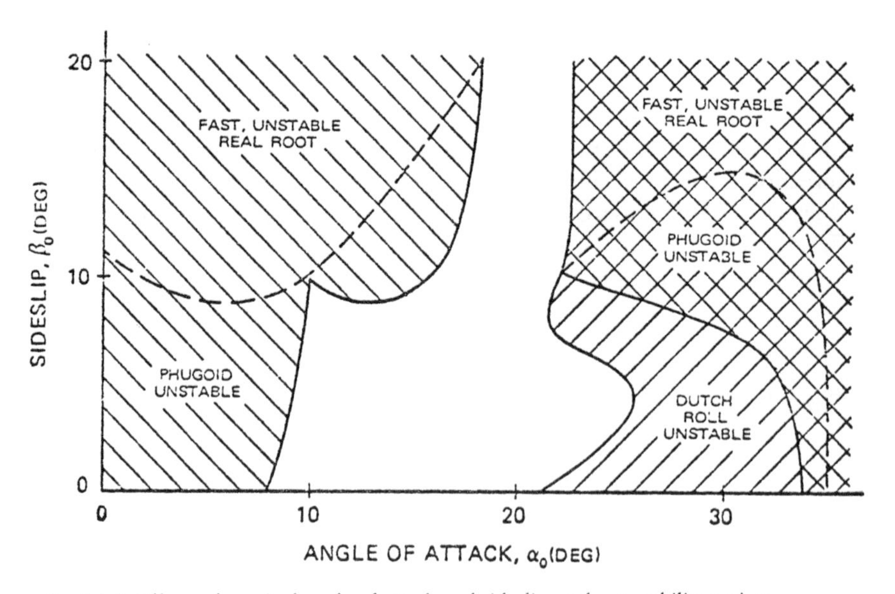

Fig. 7.1-5. Effects of nominal angle of attack and sideslip angle on stability regions. (from [S-16], courtesy of NASA)

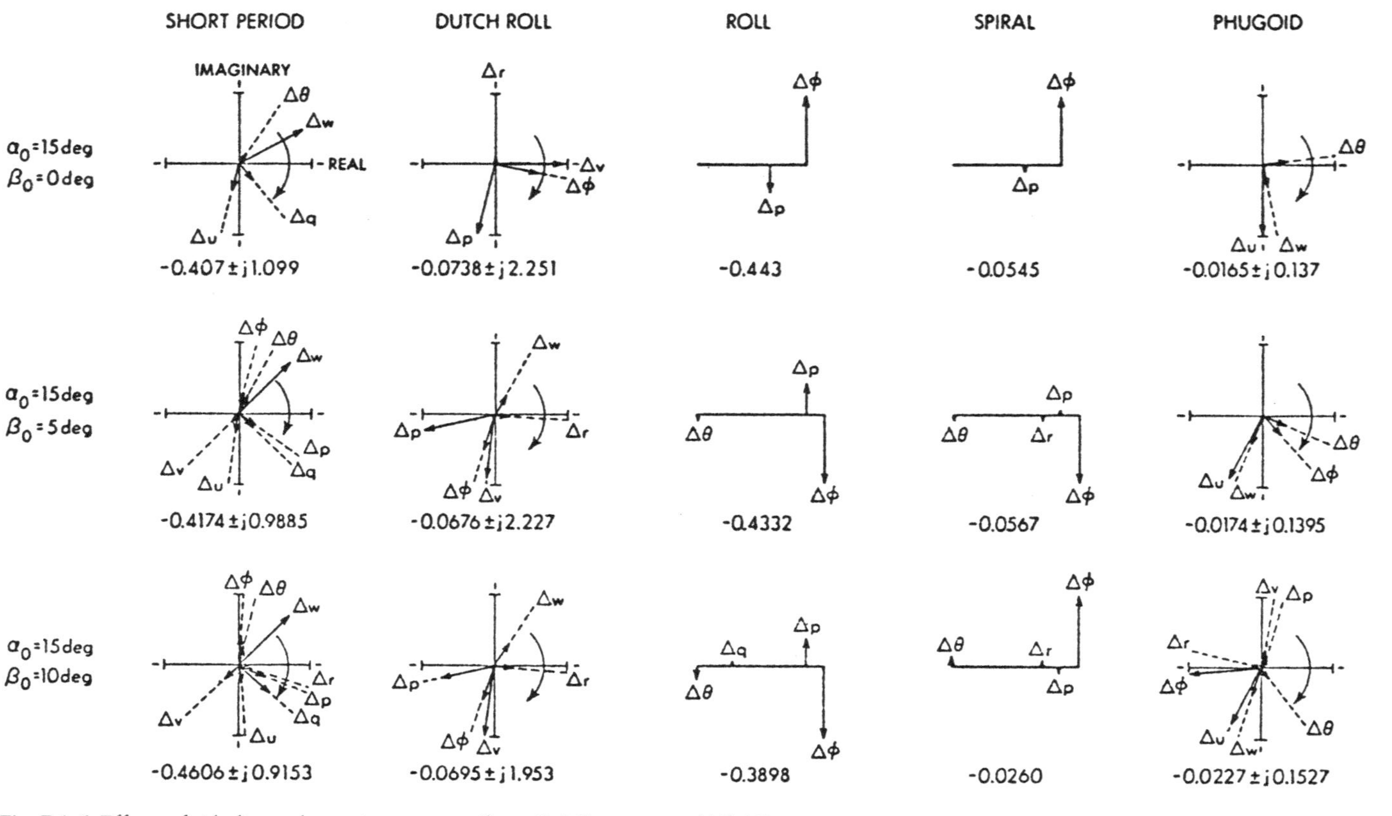

Fig. 7.1-6. Effects of sideslip angle on eigenvectors. (from [S-16], courtesy of NASA)

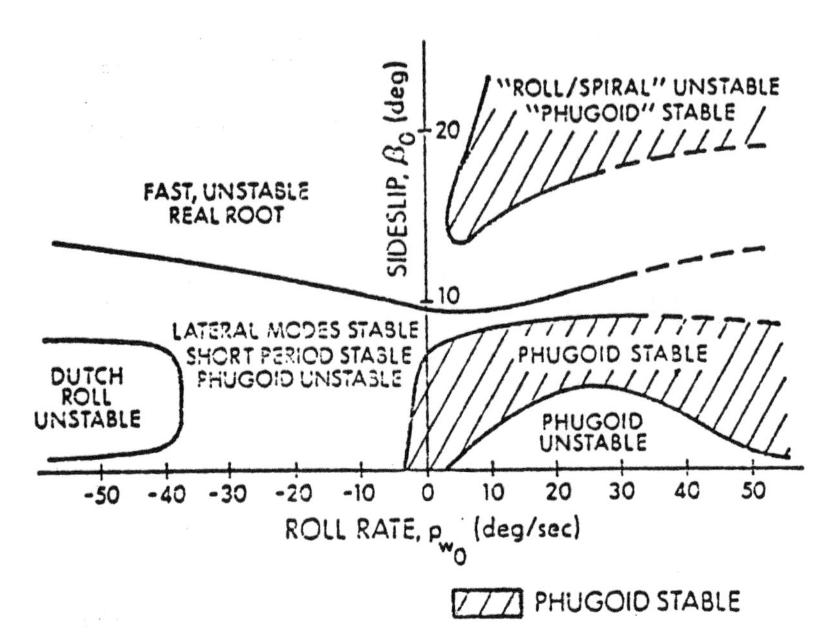

Fig. 7.1-7. Effects of nominal sideslip angle and roll rate on stability regions. (from [S-16], courtesy of NASA)

Positive sideslip and negative roll rate destabilize the Dutch roll mode, while the Dutch roll remains stable when both are positive. Large sideslip angle produces a fast, coupled divergence for all values of roll rate tested.

Coupling Controls

Aircraft control surfaces are generally designed to produce little cross-axis coupling; however, there are counterexamples. V-tail configurations like the early Beechcraft Bonanza and the Fouga Magister replaced the conventional vertical and horizontal surfaces with two diagonal surfaces whose control flap deflections were summed and differenced to provide pitching and yawing moments [A-2]. Future aircraft control systems may be designed with increased redundancy and failure tolerance, and control surfaces may be designed with coupling in mind. The modes of motion are coupled when such control surfaces are used for closed-loop control, either automatically or by the pilot; then, motions about one axis drive the surface to produce forces and moments about others.

Individual spoilers decrease lift when they create rolling moment. Wing-tip petals that control the yaw of tailless aircraft also produce drag. The swept rudder on a typical jet transport produces a download whenever it is deflected. All of these are one-sided (or rectified) effects; therefore, they are nonlinear. The rudder's pitching moment is only positive, drag is only increased by the petals, and spoiler control for positive or negative rolling moment only decreases the lift. Such effects are easily simulated; for example, the pitching moment from the rudder is mode-led as $C_{m_{|\delta R|}}|\Delta\delta R|$. Nevertheless, linear analysis of these effects must use approximations, such as the describing functions introduced in Section 6.5.

7.2 Inertial Coupling of Pitch and Yaw Motions

Combat aircraft of World War II maneuvered at high speeds and roll rates, yet it was not until the postwar years that inertial coupling induced by high roll rate was recognized as a major challenge for aircraft designers [A-2]. The quest for supersonic flight led to new aircraft with thin, low-aspect-ratio wings and low rolling moment of inertia. The early supersonic configurations developed large, uncontrollable pitching and yawing oscillations in high-roll-rate maneuvers, and a number of aircraft were lost in flight. An analysis of the problem was presented by White *et al.* in early 1948 [W-6], and Phillips's 1948 NACA Technical Note [P-4] described a mechanism for these oscillations. Numerous analyses of the problem followed, including [A-1, H-1, P-6, R-2, S-2, Y-1, and Y-2].

The present analysis expands on earlier results. A number of additional factors are considered, and the steady-state response to controls is shown. Root variations are shown in Figs. 7.2-1 to 7.2-8.

Fifth-Order Model of Coupled Dynamics

Inertial coupling produced by high roll rate is best understood through low-order dynamic models. Perturbations in axial velocity have little effect on the evolving motion, and the effects of gravity are small in comparison to the aerodynamic and inertial forces. The phugoid and spiral modes are slow compared to the fast dynamics of the short-period, Dutch roll, and roll modes. Altogether, we can eliminate Δu, $\Delta\theta$, and $\Delta\phi$—and the corresponding phugoid and spiral modes—from the model. The state is truncated to five elements,

$$\Delta\mathbf{x} = [\Delta w\ \Delta q\ \Delta v\ \Delta p\ \Delta r]^T \tag{7.2-1}$$

and we restrict attention to the principal angular controls: elevator, aileron, and rudder:

$$\Delta\mathbf{u} = [\Delta\delta E\ \Delta\delta A\ \Delta\delta R]^T \tag{7.2-2}$$

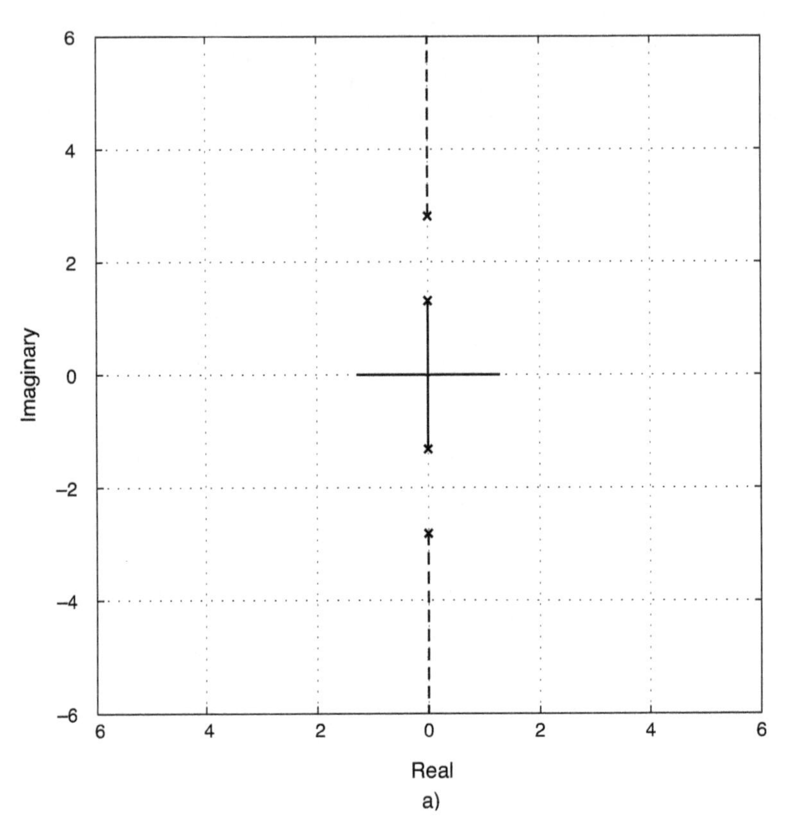

Fig. 7.2-1. Effect of roll rate on fourth-order roots for $I_{xx} = 35{,}927$ kg-m². Dutch roll (solid line); short period (dashed line). (a) Root locus.

Neglecting disturbance effects, the linearized dynamic equation takes its familiar form

$$\Delta\dot{\mathbf{x}}(t) = \mathbf{F}\,\Delta\mathbf{x}(t) + \mathbf{G}\,\Delta\mathbf{u}(t) \tag{7.2-3}$$

The equation can be partitioned into longitudinal, lateral-directional, and coupling components:

$$\begin{bmatrix} \Delta\dot{\mathbf{x}}_{\mathrm{Lo}} \\ \Delta\dot{\mathbf{x}}_{\mathrm{LD}} \end{bmatrix} = \begin{bmatrix} \mathbf{F}_{\mathrm{Lo}} & \mathbf{F}_{\mathrm{LD}}^{\mathrm{Lo}} \\ \mathbf{F}_{\mathrm{Lo}}^{\mathrm{LD}} & \mathbf{F}_{\mathrm{LD}} \end{bmatrix} \begin{bmatrix} \Delta\mathbf{x}_{\mathrm{Lo}} \\ \Delta\mathbf{x}_{\mathrm{LD}} \end{bmatrix} + \begin{bmatrix} \mathbf{G}_{\mathrm{Lo}} & \mathbf{G}_{\mathrm{LD}}^{\mathrm{Lo}} \\ \mathbf{G}_{\mathrm{Lo}}^{\mathrm{LD}} & \mathbf{G}_{\mathrm{LD}} \end{bmatrix} \begin{bmatrix} \Delta\mathbf{u}_{\mathrm{Lo}} \\ \Delta\mathbf{u}_{\mathrm{LD}} \end{bmatrix} \tag{7.2-4}$$

In the absence of cross-axis coupling, the primary blocks of $\mathbf{F}$ are

$$\mathbf{F}_{\mathrm{Lo}} = \begin{bmatrix} Z_w & u_o \\ M_w & M_q \end{bmatrix} \tag{7.2-5}$$

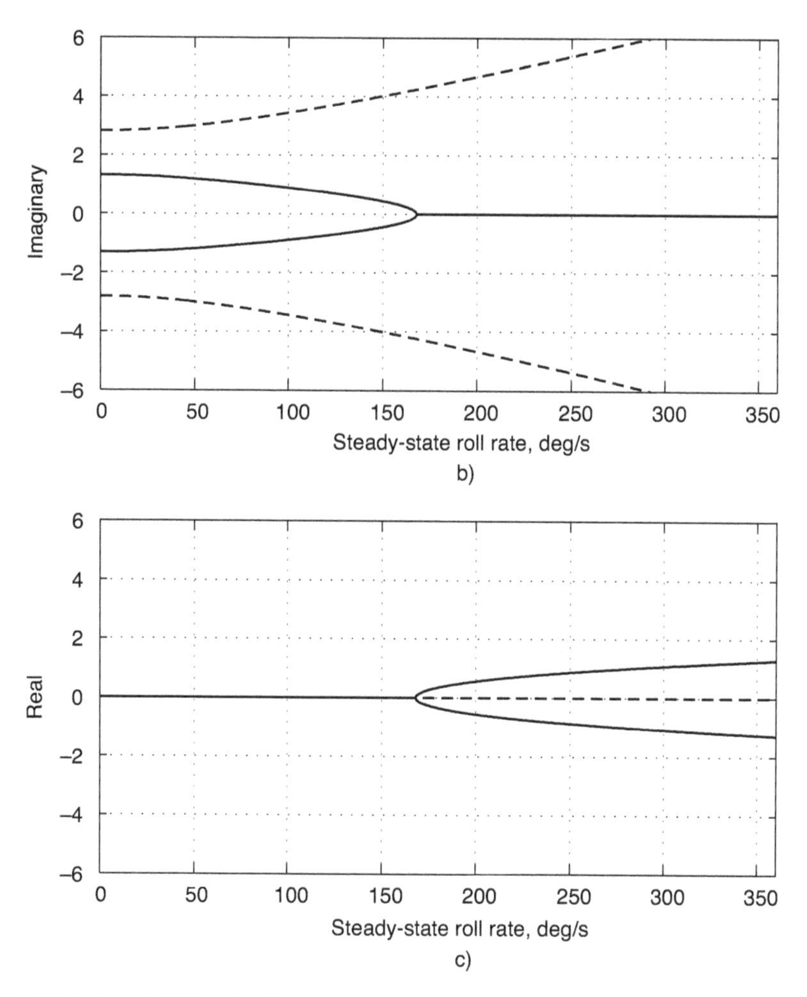

Fig. 7.2-1. (*cont.*). Effect of roll rate on fourth-order roots for $I_{xx} = 35{,}927\,\text{kg-m}^2$. Dutch roll (solid line); short period (dashed line). (b—*top*) Imaginary parts of roots vs. roll rate. (c—*bottom*) Real parts of roots vs. roll rate.

$$\mathbf{F}_{\text{LD}} = \begin{bmatrix} Y_v & w_o & -u_o \\ L_v & L_p & L_r \\ N_v & N_p & N_r \end{bmatrix} \tag{7.2-6}$$

and the secondary blocks are zero. We assume that each control has a direct influence on a single moment. $\mathbf{G}_{\text{LD}}^{\text{Lo}}$ and $\mathbf{G}_{\text{Lo}}^{\text{LD}}$ are zero, and the primary control blocks are

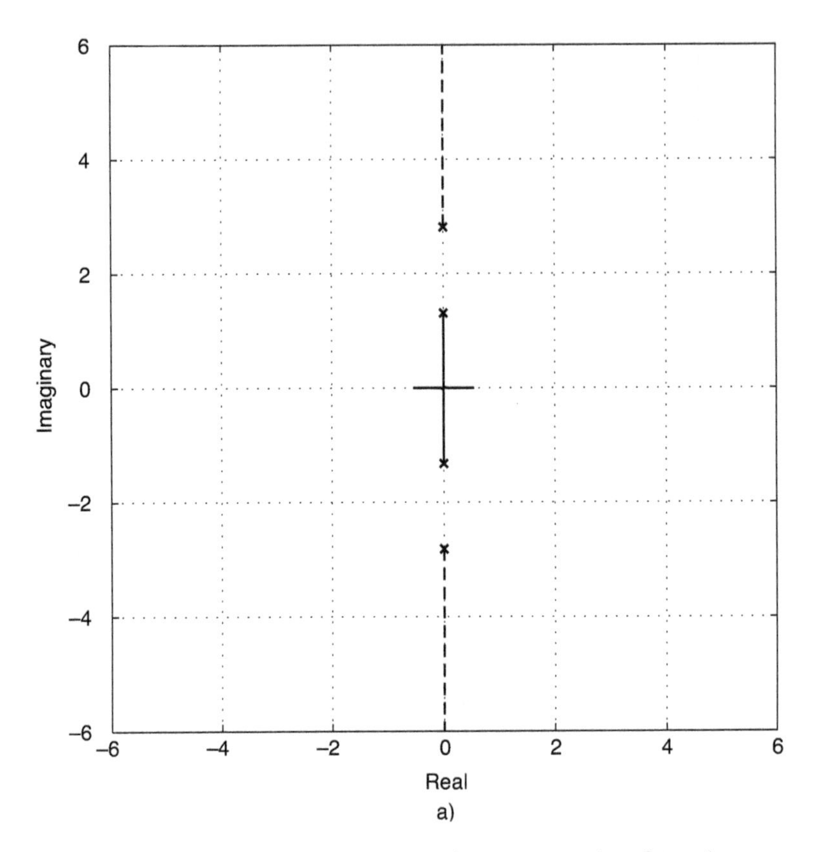

Fig. 7.2-2. Effect of roll rate on fourth-order roots for $I_{xx} = 30{,}000\,\text{kg-m}^2$. Dutch roll (solid line); short period (dashed line). (a) Root locus.

$$\mathbf{G}_{\text{Lo}} = \begin{bmatrix} 0 \\ M_{\delta E} \end{bmatrix} \tag{7.2-7}$$

$$\mathbf{G}_{\text{LD}} = \begin{bmatrix} 0 & 0 \\ L_{\delta A} & 0 \\ 0 & N_{\delta R} \end{bmatrix} \tag{7.2-8}$$

Next, constant angular rates (p_o, q_o, r_o), nonzero sideslip angle (β_o), powerplant angular momentum (h_x), and nose-high product of inertia (I_{xz}) are added to the model. In the coupled case, $\mathbf{F}_{\text{Lo}}$ is unchanged, but $\mathbf{F}_{\text{LD}}$ must be modified to account for steady angular rate:

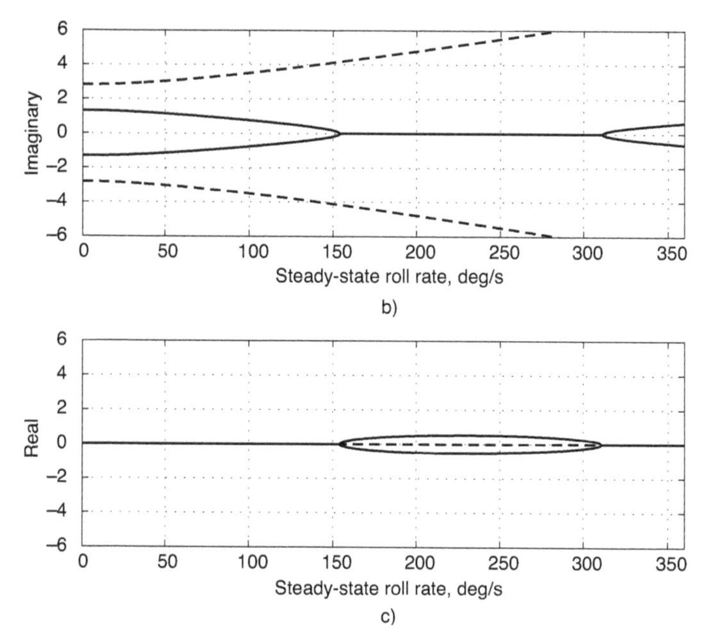

Fig. 7.2-2. (*cont.*). Effect of roll rate on fourth-order roots for $I_{xx} = 30{,}000$ kg-m². Dutch roll (solid line); short period (dashed line). (b—*top*) Imaginary parts of roots vs. roll rate. (c—*bottom*) Real parts of roots vs. roll rate.

$$
\mathbf{F}_{\mathrm{LD}} = \begin{bmatrix} Y_v & w_o & -u_o \\ L_v & (L_p + C_p^{\dot p}) & (L_r + C_r^{\dot p}) \\ N_v & (N_p + C_p^{\dot r}) & (N_r + C_r^{\dot r}) \end{bmatrix} \triangleq \begin{bmatrix} Y_v & w_o & -u_o \\ L_v & L'_p & L'_r \\ N_v & N'_p & N'_r \end{bmatrix} \tag{7.2-9}
$$

Rolling and yawing stability derivatives account for I_{xz} as in eq. 7.1-25, and the gyroscopic coupling elements are defined by eqs. 7.1-50–7.1-58. The corresponding secondary blocks are

$$
\mathbf{F}_{\mathrm{LD}}^{\mathrm{Lo}} = \begin{bmatrix} (Z_v - p_o) & v_o & 0 \\ M_v & C_p^{\dot q} & \left(C_r^{\dot q} - \dfrac{h_x}{I_{yy}}\right) \end{bmatrix} \triangleq \begin{bmatrix} (Z_v - p_o) & v_o & 0 \\ M_v & M'_p & M'_r \end{bmatrix} \tag{7.2-10}
$$

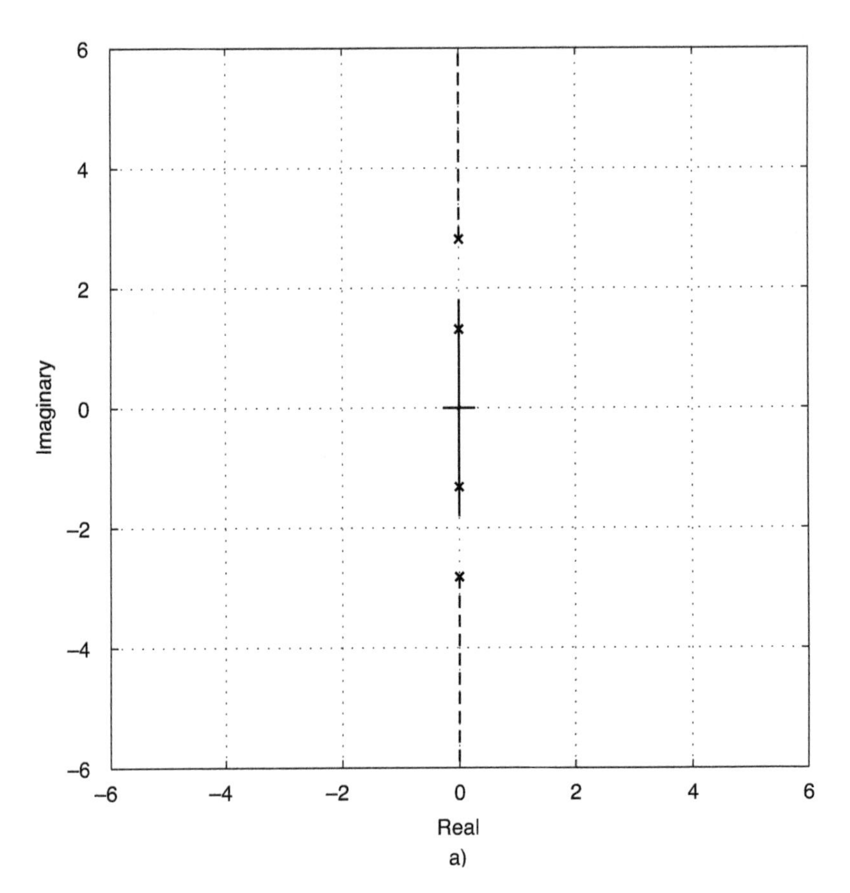

Fig. 7.2-3. Effect of roll rate on fourth-order roots for $I_{xx} = 24{,}000\,\text{kg-m}^2$. Dutch roll (solid line); short period (dashed line). (a) Root locus.

$$\mathbf{F}_{\text{Lo}}^{\text{LD}} = \begin{bmatrix} (Y_w + p_o) & 0 \\ L_w & \left(C_q^{\dot{p}} + \dfrac{I_{xz} h_x}{I_{xx} I_{zz} - I_{xz}^2} \right) \\ N_w & \left(C_q^{\dot{r}} + \dfrac{I_{xx} h_x}{I_{xx} I_{zz} - I_{xz}^2} \right) \end{bmatrix}$$

$$\triangleq \begin{bmatrix} (Y_w + p_o) & 0 \\ L_w & L_q' \\ N_w & N_q' \end{bmatrix} \tag{7.2-11}$$

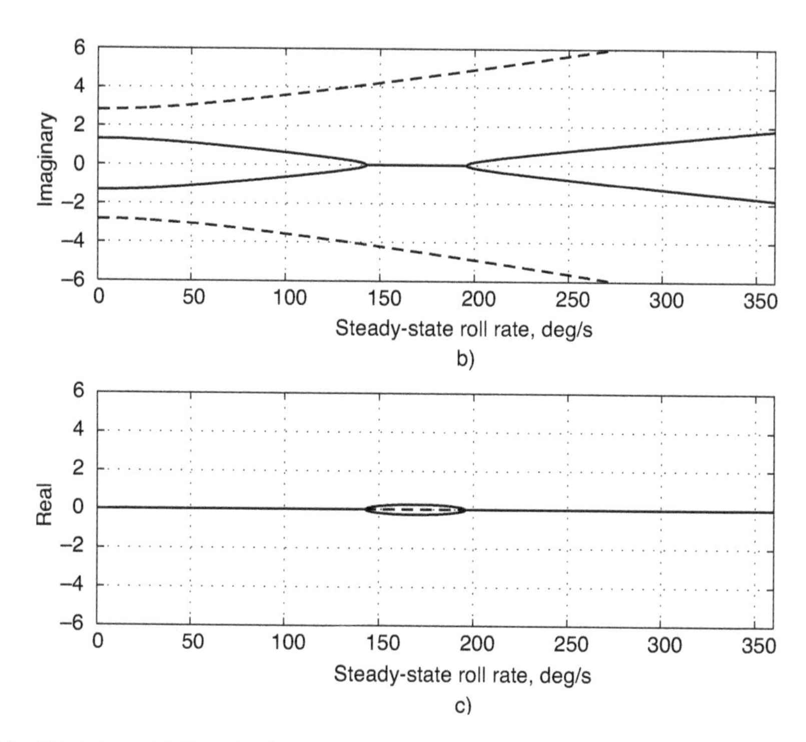

Fig. 7.2-3. (*cont.*). Effect of roll rate on fourth-order roots for $I_{xx} = 24{,}000$ kg-m². Dutch roll (solid line); short period (dashed line). (b—*top*) Imaginary parts of roots vs. roll rate. (c—*bottom*) Real parts of roots vs. roll rate.

where the primed terms $(\cdot)'$ simplify the notation for coupling effects. The body-axis rates corresponding to a constant roll rate about the velocity vector p_{o_s} (measured in stability axes) are, from eq. 3.2-83,

$$\begin{bmatrix} p_o \\ q_o \\ r_o \end{bmatrix}_B = p_{o_s} \begin{bmatrix} \cos\alpha_o \cos\beta_o \\ \sin\beta_o \\ \sin\alpha_o \cos\beta_o \end{bmatrix} \tag{7.2-12}$$

where

$$\alpha_o = \tan^{-1}(w_o/u_o) \tag{7.2-13}$$

$$\beta_o = \sin^{-1}(v_o/V_o) \tag{7.2-14}$$

$$V_o = \sqrt{u_o^2 + v_o^2 + w_o^2} \tag{7.2-15}$$

The angle of attack and sideslip angle can remain constant while rolling about the velocity vector, whereas rolling about the vehicle centerline

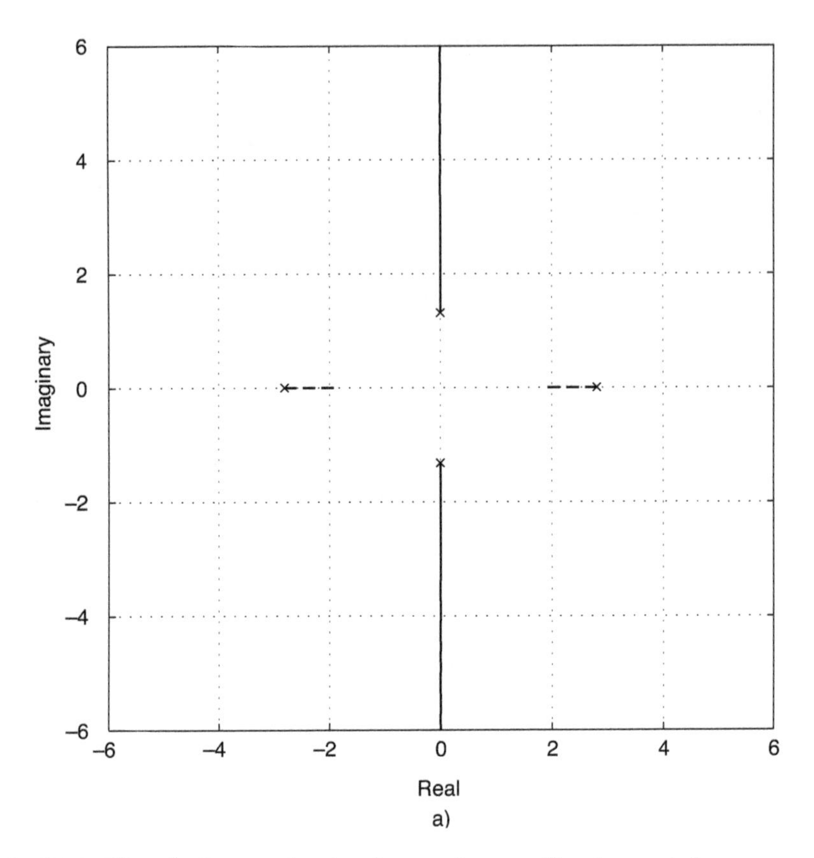

Fig. 7.2-4. Effect of roll rate on fourth-order roots for nonrolling short-period mode unstable. $I_{xx} = 35{,}297\,\text{kg-m}^2$. Dutch roll (solid line); short period (dashed line). (a) Root locus.

implies continually changing aerodynamic angles unless $\alpha_o = \beta_o = 0$, when the two axes are the same.

Truncated and Residualized Fourth-Order Models

If the roll mode is well damped and perturbations in roll rate are not of interest, the fifth-order model can be reduced to fourth order, by either truncation or residualization (Section 4.2). The ordering of Δp and Δr in the state vector is interchanged, and the stability matrix is partitioned into slow (short-period and Dutch roll) and fast (roll) components:

$$\mathbf{F} = \begin{bmatrix} \mathbf{F}_{11} & \mathbf{F}_{12} \\ \mathbf{F}_{21} & \mathbf{F}_{22} \end{bmatrix} \tag{7.2-16}$$

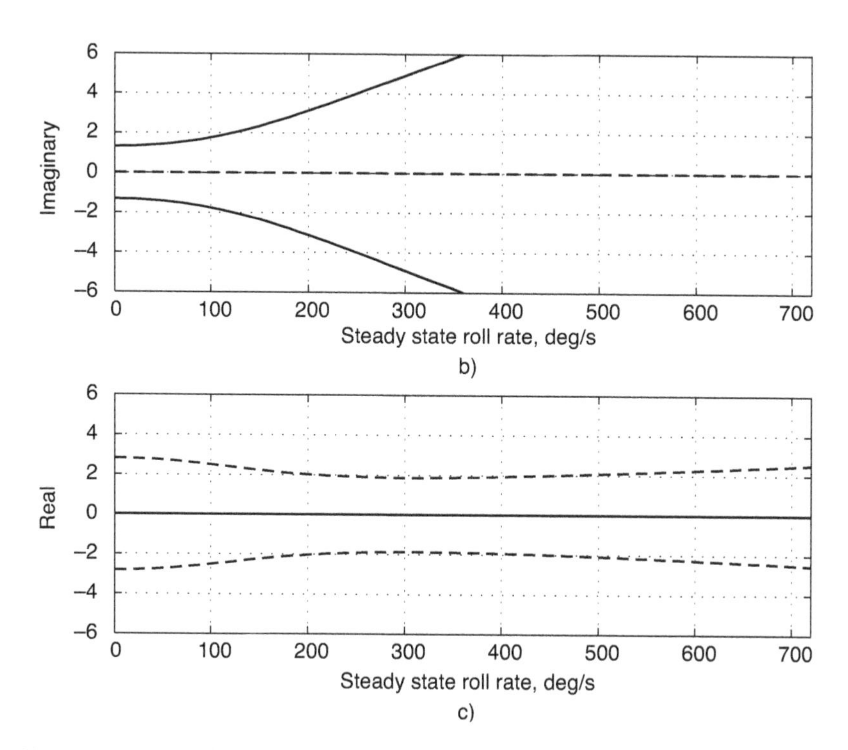

Fig. 7.2-4. (*cont.*). Effect of roll rate on fourth-order roots for nonrolling short period mode unstable. $I_{xx} = 35{,}297\,\text{kg-m}^2$. Dutch roll (solid line); short period (dashed line). (b—*top*) Imaginary parts of roots vs. roll rate. (c—*bottom*) Real parts of roots vs. roll rate.

Simply neglecting the roll-mode effects, the truncated fourth-order stability matrix is

$$\mathbf{F}_{11} = \begin{bmatrix} Z_w & u_o & (Z_v - p_o) & v_o \\ M_w & M_q & M_v & M'_r \\ (Y_w + p_o) & 0 & Y_v & -u_o \\ N_w & N'_q & N_v & N'_r \end{bmatrix} \tag{7.2-17}$$

This model is similar to that used in Phillips's analysis [P-4]. The remaining blocks of the fifth-order model are

$$\mathbf{F}_{12} = \begin{bmatrix} 0 \\ M'_p \\ w_o \\ N'_p \end{bmatrix} \tag{7.2-18}$$

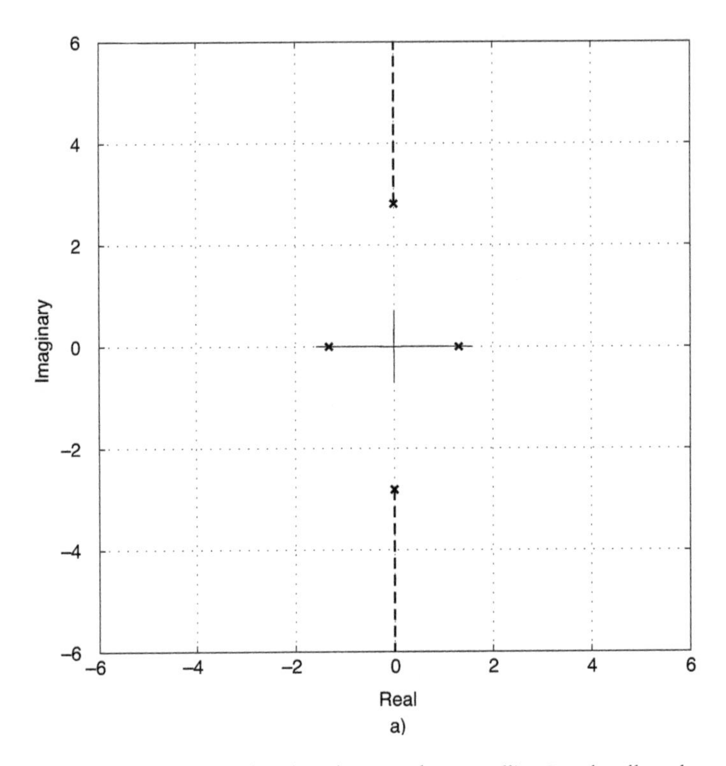

Fig. 7.2-5. Effect of roll rate on fourth-order roots for nonrolling Dutch roll mode unstable. $I_{xx} = 35{,}297\,\text{kg-m}^2$. Dutch roll (solid line); short period (dashed line). (a) Root locus.

$$\mathbf{F}_{21} = [L_w \; L'_q \; L_v \; L'_r] \tag{7.2-19}$$

$$\mathbf{F}_{22} = L'_p \tag{7.2-20}$$

The residualized stability matrix (eq. 4.2-57) becomes

$$\mathbf{F}_{res} = (\mathbf{F}_{11} - \mathbf{F}_{12}\mathbf{F}_{22}^{-1}\mathbf{F}_{21})$$
$$= \begin{bmatrix} Z_w & u_o & (Z_v - p_o) & v_o \\ M_w - \dfrac{M'_p L_w}{L'_p} & M_q - \dfrac{M'_p L'_q}{L'_p} & M_v - \dfrac{M'_p L_v}{L'_p} & M'_r - \dfrac{M'_p L'_r}{L'_p} \\ (Y_w + p_o) - \dfrac{w_o L_w}{L'_p} & -\dfrac{w_o L'_q}{L'_p} & Y_v - \dfrac{w_o L_v}{L'_p} & -u_o - \dfrac{w_o L'_r}{L'_p} \\ N_w - \dfrac{N'_p L_w}{L'_p} & N'_q - \dfrac{N'_p L'_q}{L'_p} & N_v - \dfrac{N'_p L_v}{L'_p} & N_r - \dfrac{N'_p L'_r}{L'_p} \end{bmatrix}$$

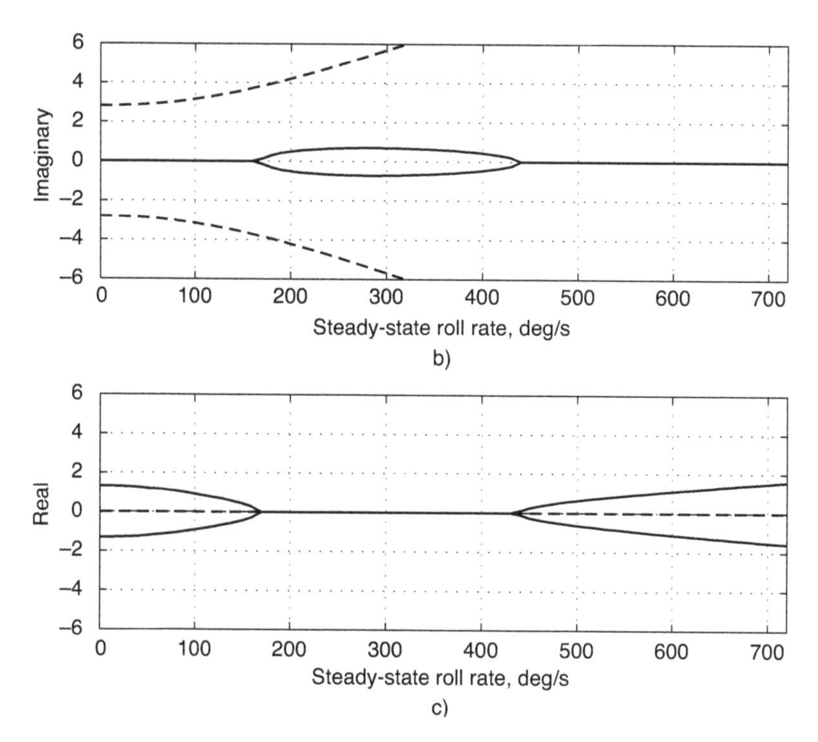

Fig. 7.2-5. (*cont.*). Effect of roll rate on fourth-order roots for nonrolling Dutch roll mode unstable. $I_{xx} = 35{,}297\,\text{kg-m}^2$. Dutch roll (solid line); short period (dashed line). (b—*top*) Imaginary parts of roots vs. roll rate. (c—*bottom*) Real parts of roots vs. roll rate.

$$\triangleq \begin{bmatrix} Z_w & u_o & (Z_v - p_o) & v_o \\ M''_w & M''_q & M''_v & M''_r \\ (p_o + Y''_w) & Y''_q & Y''_v & (-u_o + Y''_r) \\ N''_w & N''_q & N''_v & N''_r \end{bmatrix} \tag{7.2-21}$$

where $(\cdot)''$ denotes a residualized coefficient. Thus, we have alternative fourth-order models for investigating inertial coupling.

We return to the simpler truncated model to illustrate the basic roll coupling phenomenon. At low angle of attack, body-axis and stability-axis roll rates are virtually identical. Assuming that only the body-axis roll rate (p_o) introduces pitch-yaw coupling and that there

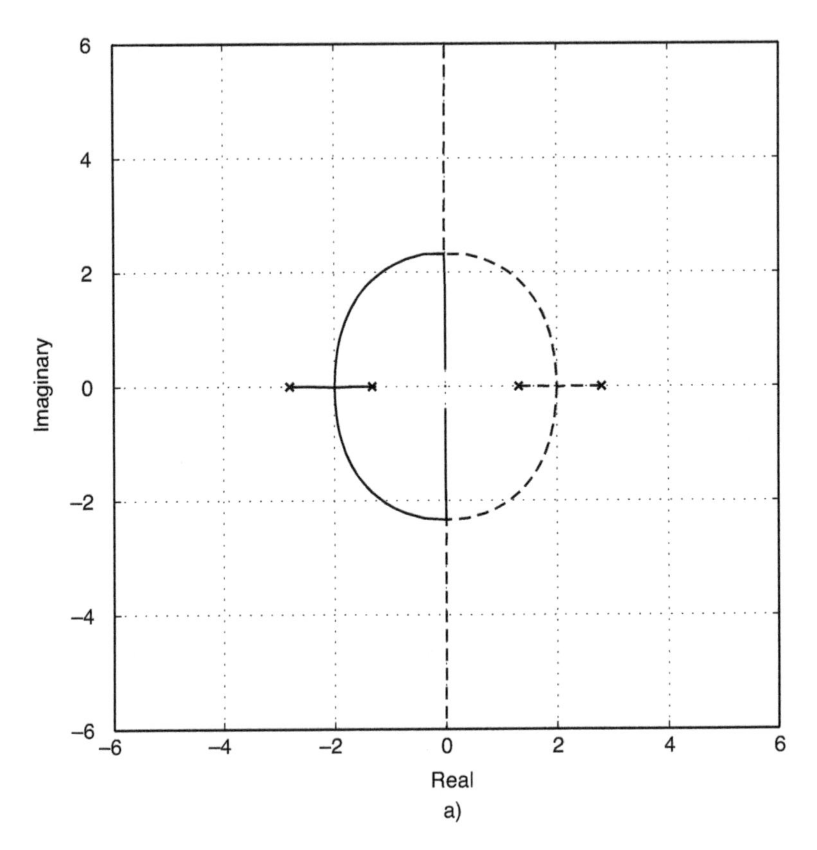

Fig. 7.2-6. Effect of roll rate on fourth-order roots for both rolling modes unstable. $I_{xx} = 35{,}297\,\text{kg-m}^2$. Dutch roll (solid line); short period (dashed line).
(a) Root locus.

are no roll-rate-perturbation (Δp) effects, the fourth-order stability matrix is

$$\mathbf{F}_{tr} = \begin{bmatrix} Z_w & u_o & -p_o & 0 \\ M_w & M_q & 0 & \dfrac{(I_{zz} - I_{xx})p_o}{I_{yy}} \\ p_o & 0 & Y_v & -u_o \\ 0 & \dfrac{(I_{xx} - I_{yy})p_o}{I_{zz}} & N_v & N_r \end{bmatrix} \tag{7.2-22}$$

The characteristic polynomial of the truncated model can be written as

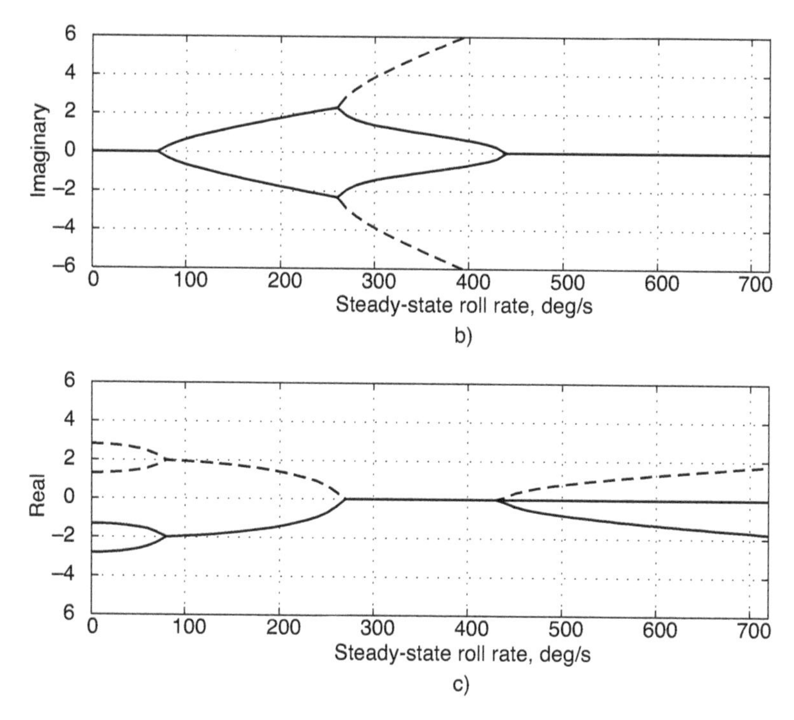

Fig. 7.2-6. (*cont.*). Effect of roll rate on fourth-order roots for both rolling modes unstable. $I_{xx} = 35{,}297\,\text{kg-m}^2$. Dutch roll (solid line); short period (dashed line). (b—*top*) Imaginary parts of roots vs. roll rate. (c—*bottom*) Real parts of roots vs. roll rate.

$$
\begin{aligned}
\Delta(s) = {} & \{[(s - Z_w)(s - M_q) - u_o M_w][(s - Y_v)(s - N_r) + u_o N_v]\} \\
& + p_o^2 \left\{ (s - M_q)(s - N_r) - (s - Z_w)(s - Y_v) \frac{(I_{zz} - I_{xx})(I_{xx} - I_{yy})}{I_{yy} I_{zz}} \right. \\
& \left. - u_o M_w \frac{I_{xx} - I_{yy}}{I_{zz}} - u_o N_v \frac{I_{zz} - I_{xx}}{I_{yy}} - p_o^2 \frac{(I_{zz} - I_{xx})(I_{xx} - I_{yy})}{I_{yy} I_{zz}} \right\} \\
= {} & \{\text{nonrolling characteristic polynomial}\} + \{\text{rolling effects}\}
\end{aligned}
\tag{7.2-23}
$$

The polynomial can be separated into a nonrolling part, which factors as uncoupled short-period and Dutch roll modes, plus a rolling part, which is proportional to both p_o^2 and p_o^4. The Evans rules cannot be used to plot the root locus using roll rate as gain because both second and fourth powers of p_o appear; however, the roots can be calculated and plotted directly. The sign of p_o does not affect the roots, but it does affect the sense

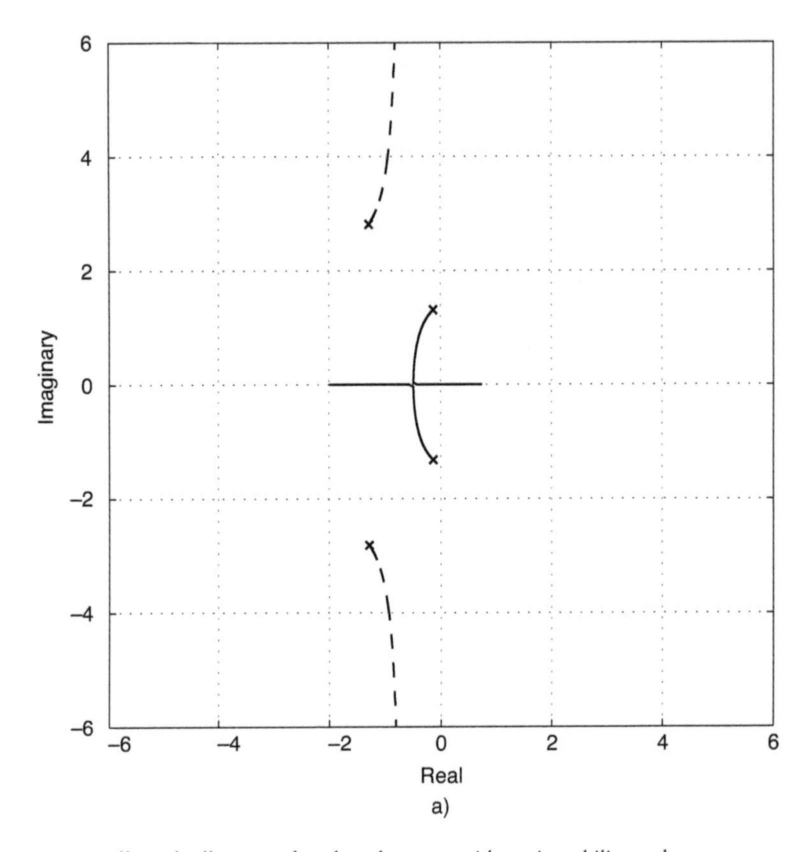

Fig. 7.2-7. Effect of roll rate on fourth-order roots with static stability and aerodynamic damping. Dutch roll (solid line); short period (dashed line): $I_{xx} = 35{,}927\,\text{kg-m}^2$. (a) Root locus.

of coupled response. p_o has no significant effect on the roll mode of the corresponding fifth-order model; hence, its remaining four roots are well described by the roots of eq. 7.2-23.

Let us further suppose that there is no damping in the system, that is, that $Z_w = M_q = Y_v = N_r = 0$. The characteristic polynomial simplifies to

$$\begin{aligned}\Delta(s) = {} & [(s^2 - u_o M_w)(s^2 + u_o N_v)] \\ & + p_o^2 \left\{ s^2 \left[1 - \frac{(I_{zz} - I_{xx})(I_{xx} - I_{yy})}{I_{yy} I_{zz}} \right] \right. \\ & \left. - u_o M_w \frac{I_{xx} - I_{yy}}{I_{zz}} - u_o N_v \frac{I_{zz} - I_{xx}}{I_{yy}} - p_o^2 \frac{(I_{zz} - I_{xx})(I_{xx} - I_{yy})}{I_{yy} I_{zz}} \right\}\end{aligned} \quad (7.2\text{-}24)$$

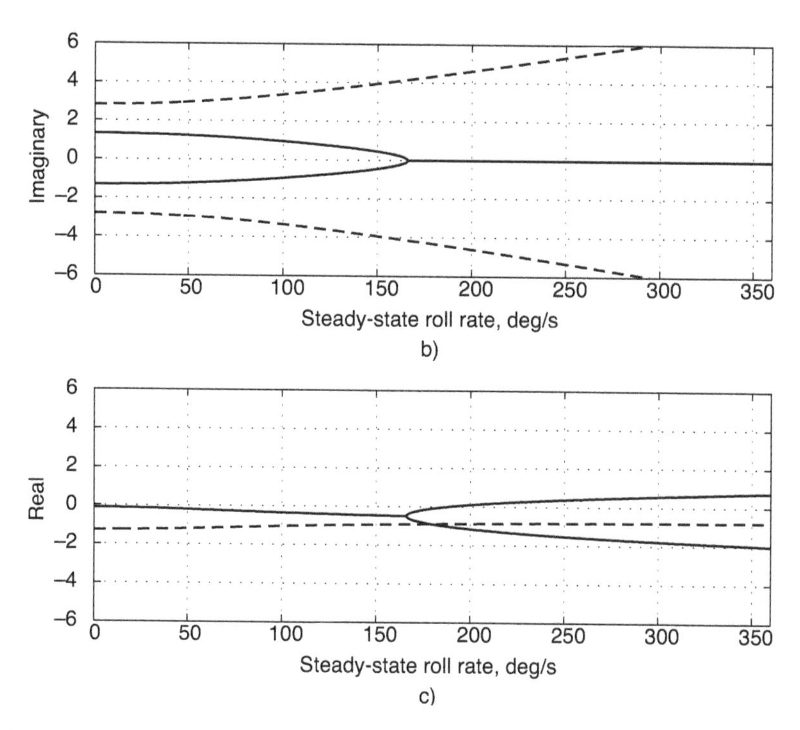

Fig. 7.2-7. (*cont.*). Effect of roll rate on fourth-order roots with static stability and aerodynamic damping. Dutch roll (solid line); short period (dashed line): $I_{xx} = 35{,}927\,\text{kg-m}^2$. (b—*top*) Imaginary parts of roots vs. roll rate. (c—*bottom*) Real parts of roots vs. roll rate.

and the odd powers of s disappear from the characteristic equation. Assuming static stability ($M_w < 0$ and $N_v > 0$), the nonrolling roots for the short-period and Dutch roll approximations lie on the $j\omega$ axis, with

$$\lambda_{1,2} = \pm j\sqrt{-u_o M_w} \tag{7.2-25}$$

$$\lambda_{3,4} = \pm j\sqrt{u_o N_v} \tag{7.2-26}$$

The roots for the nominal undamped case (derived from Table 7.1-1) plus two examples with reduced rolling inertia are presented in Fig. 7.2-1 to 7.2-3. I_{xz} is neglected, and $w_o = 0$. For each case, the root locus is accompanied by separate plots of the real and imaginary components of the roots as functions of roll rate over the range $0 < |p_o| < 360$ deg/s. As $|p_o|$ increases from zero, the faster pair of roots associated with the nonrolling short-period mode remains on the $j\omega$ axis and becomes even faster, while the slower roots move toward the origin. In the nominal case, the complex Dutch roll roots coalesce and bifurcate at the origin to form a pair of

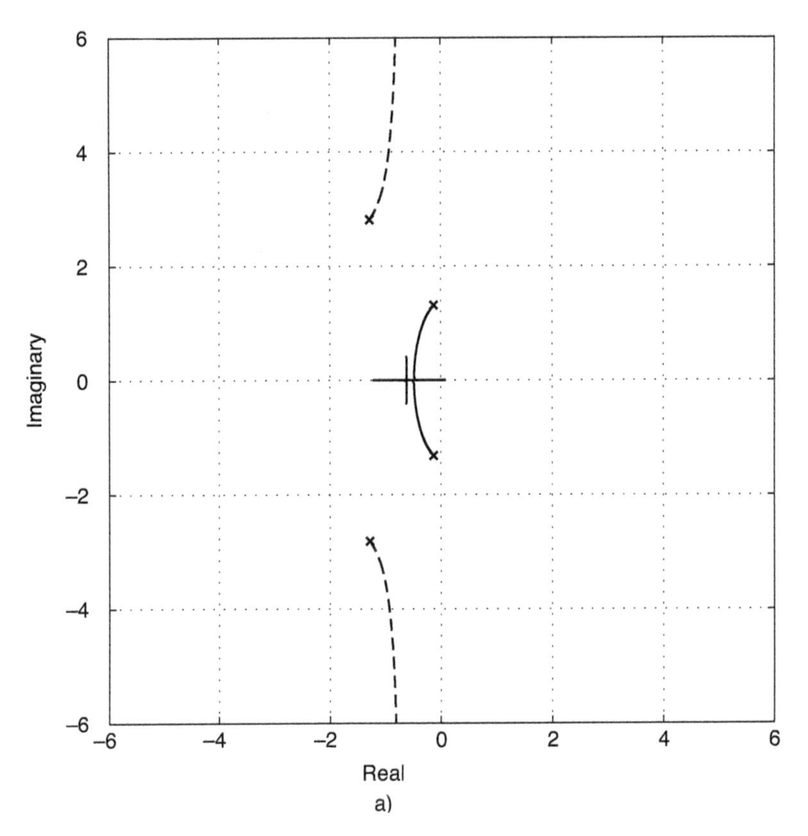

Fig. 7.2-8. Effect of roll rate on fourth-order roots with static stability and aerodynamic damping. Dutch roll (solid line); short period (dashed line): $I_{xx} = 30{,}000\,\text{kg-m}^2$. (a) Root locus.

real roots that move farther into the left and right half planes without limit as the roll rate increases. The *critical roll rate* at which instability occurs is ± 167 deg/s (Fig. 7.2-1). For this undamped model, the critical roll rate and the *bifurcation roll rate*—the point at which two real roots become complex, or the reverse—are the same.

Reducing I_{xx} to $30{,}000\,\text{kg-m}^2$ causes the roll axis to have the smallest moment of inertia. While the root locus appears similar to the previous one over the range of roll rates considered (Fig. 7.2-2), the real roots stop their progression away from the origin, reverse course, and bifurcate at a second critical roll rate to reform a complex pair. The two critical roll rates are 154 and 311 deg/sec. At intermediate roll rates, there is a real instability; however, at higher roll rate, the system again becomes neutrally stable.

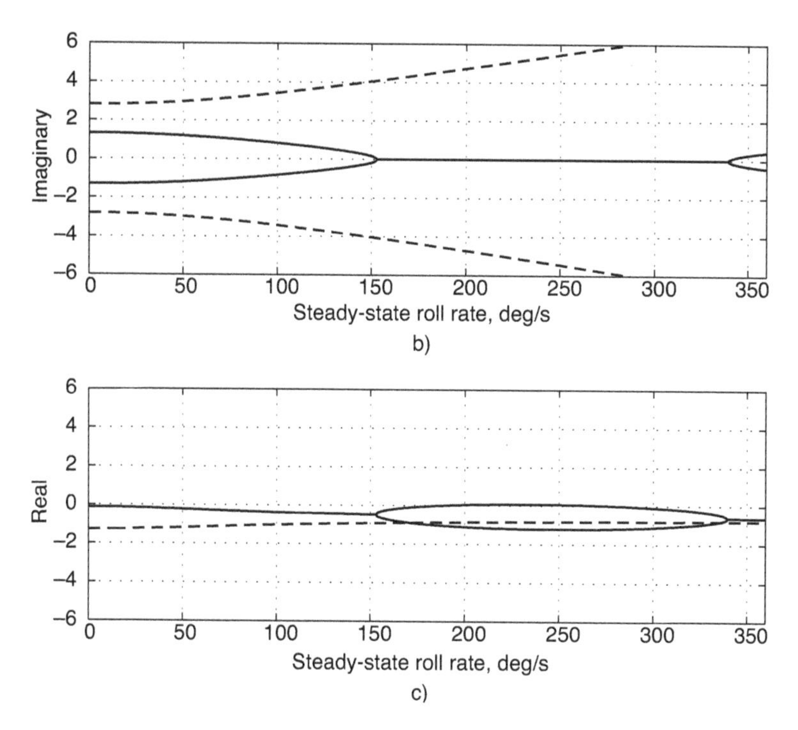

Fig. 7.2-8. (*cont.*). Effect of roll rate on fourth-order roots with static stability and aerodynamic damping. Dutch roll (solid line); short period (dashed line): $I_{xx} = 30{,}000$ kg-m². (b—*top*) Imaginary parts of roots vs. roll rate. (c—*bottom*) Real parts of roots vs. roll rate.

Further reduction of I_{xx} narrows the band of instability; at 24,000 kg-m², the critical roll rates are 143 and 196 deg/s (Fig. 7.2-3).

The characteristic polynomial for the undamped case can be put in the form

$$\Delta(s) = s^4 + \left\{\left[1 - \left(\frac{(I_{xx} - I_{yy})(I_{zz} - I_{xx})}{I_{yy}I_{zz}}\right)p_o^2 + u_o(N_v - M_w)\right]\right\}s^2$$
$$+ \left\{u_o p_o^2\left[M_w \frac{I_{xx} - I_{yy}}{I_{zz}} + N_v \frac{I_{zz} - I_{xx}}{I_{yy}}\right] - p_o^4 \frac{(I_{zz} - I_{xx})(I_{xx} - I_{yy})}{I_{yy}I_{zz}}\right\}$$
$$\triangleq s^4 + c_2 s^2 + c_o \qquad (7.2\text{-}27)$$

hence, the four roots occur in complex or real pairs. Critical roll rates occur when $c_o = 0$ because bifurcation occurs at the origin, causing at

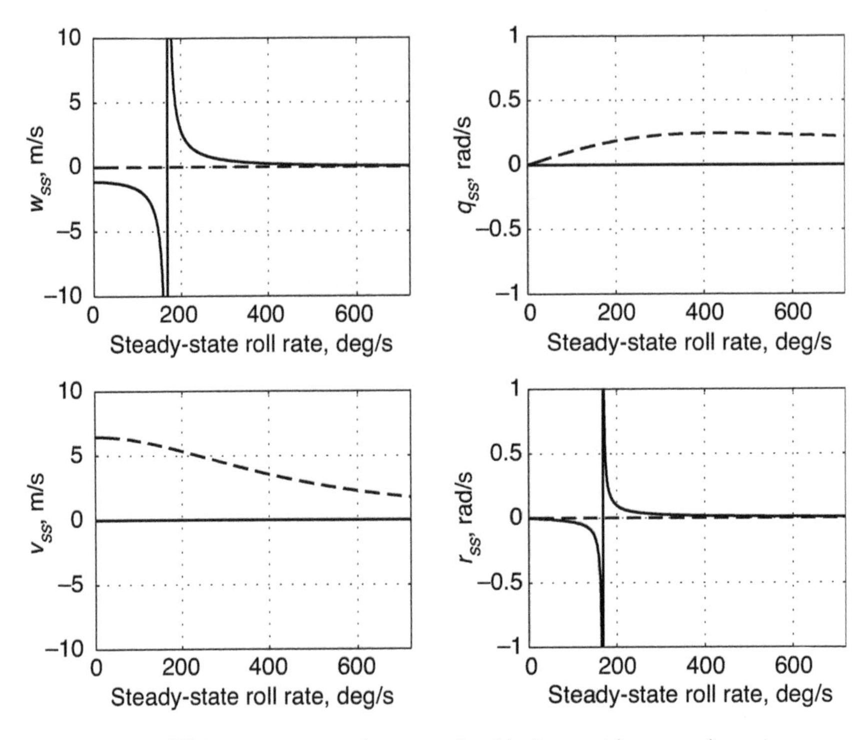

Fig. 7.2-9. Equilibrium response to elevator and rudder input with no aerodynamic damping. Elevator (solid line); 10 × rudder (dashed line): $I_{xx} = 35{,}927\,\text{kg-m}^2$.

least one pair of roots to be zero [S-3]. In this undamped case, the roots are zero when the roll rate is,

$$\begin{aligned} p_o &= \pm\sqrt{\frac{-I_{yy}u_o M_w}{I_{zz} - I_{xx}}} \quad \text{or} \pm\sqrt{\frac{I_{zz}u_o N_v}{I_{yy} - I_{xx}}} \\ &= \pm\, p_{cSP} \quad \text{or} \pm\, p_{cDR} \end{aligned} \tag{7.2-28}$$

requiring the argument of the radical to be positive. The critical roll rates can be either positive or negative. One critical roll rate can be associated with short-period static stability p_{cSP} and the other with Dutch roll static stability p_{cDR}. When both nonrolling modes are stable, there is always a critical roll rate associated with the pitching motion p_{cSP} because I_{zz} must be greater than I_{xx}. However, the critical roll rate associated with yawing motion p_{cDR} occurs only if I_{yy} is greater than I_{xx}, as the second solution of eq. 7.2-28 is imaginary otherwise.

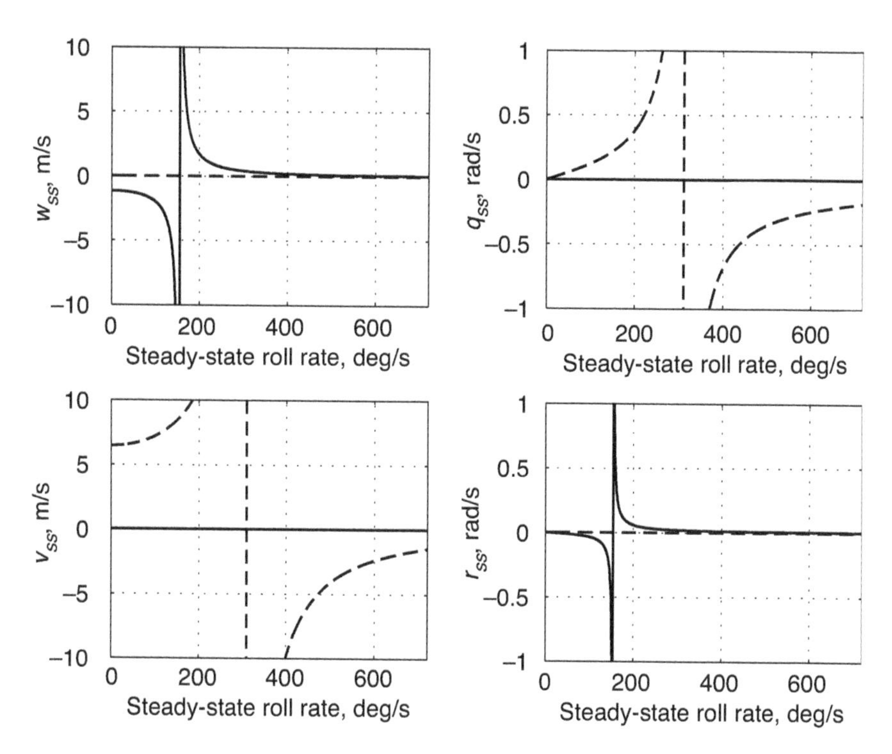

Fig. 7.2-10. Equilibrium response to elevator and rudder input with no aerodynamic damping. Elevator (solid line); 10 × rudder (dashed line): $I_{xx} = 30{,}000$ kg-m^2.

When a nonrolling mode is unstable, the sign of the radical's argument is reversed. With $M_w = 1.2773$ (unstable) and $N_v = 0.0171$ (stable), there are no critical points. One of the roots associated with the nonrolling short-period mode remains in the right half plane for roll rates up to two revolutions per second (Fig. 7.2-4). If the short period is stable ($M_w = -1.2773$) but the Dutch roll is unstable ($N_v = -0.0171$), there are two critical roll rates because $I_{xx} > I_{yy}$, and the radical argument is positive (Fig. 7.2-5). Unlike the previous examples with two critical roll rates, the rolling dynamics are neutrally stable at the intermediate roll rates and unstable at higher and lower rates.

With both nonrolling modes unstable, the system develops three bifurcation roll rates, and the pattern of bifurcation becomes intricate (Fig. 7.2-6). At zero roll rate, there are four real roots, two of which are in the right half plane. The root locus evolves symmetrically about the $j\omega$ axis. With increasing roll rate, each pair of roots in a half plane bifurcates to form complex roots. The pairs of complex roots approach

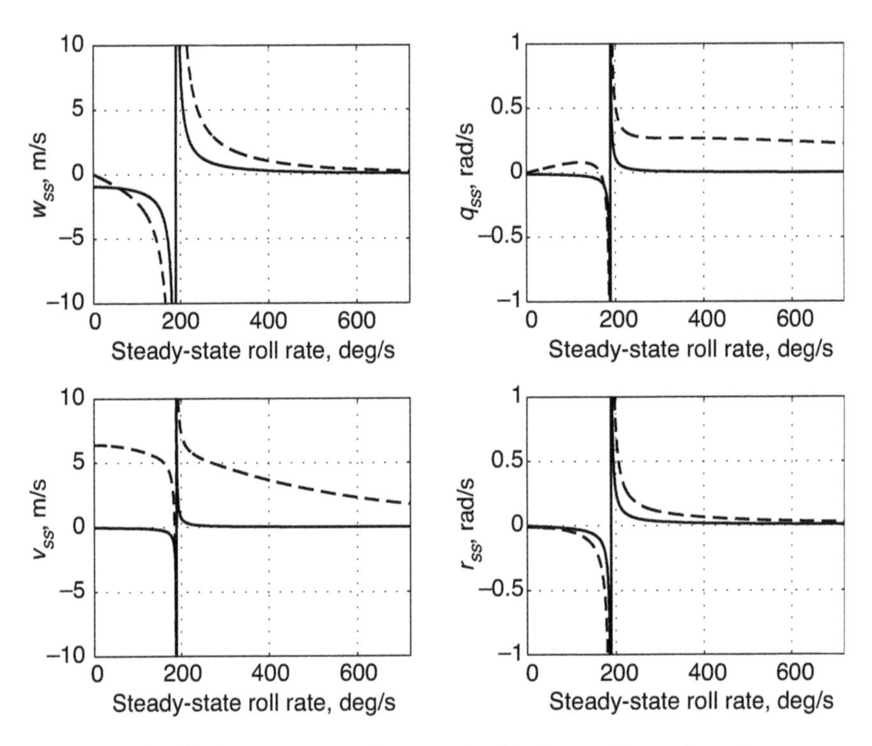

Fig. 7.2-11. Equilibrium response to elevator and rudder input with aerodynamic damping. Elevator (solid line); 10 × rudder (dashed line): $I_{xx} = 35{,}927\,\text{kg-m}^2$.

and bifurcate on the $j\omega$ axis at the first critical roll rate, forming four imaginary roots. One pair of roots increases frequency as roll rate increases further, while the other pair decreases frequency, bifurcating into two real roots at the second critical rolling rate. The system is neutrally stable between the critical roll rates, and only the third is predicted by eq. 7.2-28, which defines roll rates that produce roots at the origin.

With positive static stability and aerodynamic damping, the root locus is displaced to the left in the s plane. I_{xz} and w_o remain equal to zero for the example. Bifurcation trends are similar to those of the undamped cases, but roots are no longer confined to the real and imaginary axes. Consequently, bifurcation and critical roll rates are not the same. Bifurcations occur where the number of branches changes on the Im(λ) versus p_o or Re(λ) versus p_o plot, while critical roll rates occur where Re(λ) changes sign. There is one bifurcation and one critical roll rate when $I_{xx} > I_{yy}$, (Fig. 7.2-7) and there are two of each when $I_{xx} < I_{yy}$ (Fig. 7.2-8).

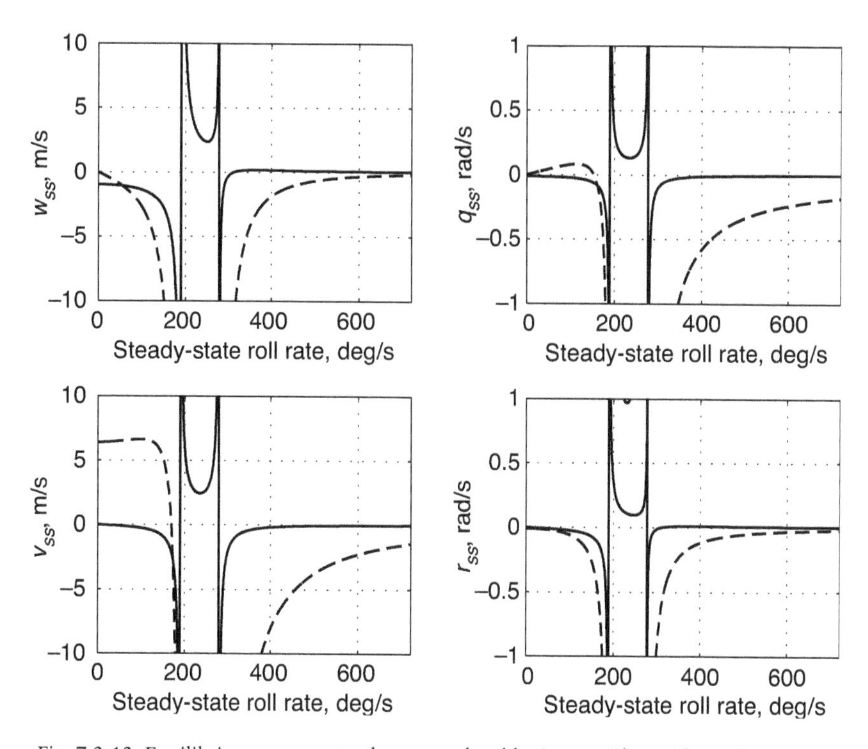

Fig. 7.2-12. Equilibrium response to elevator and rudder input with aerodynamic damping. Elevator (solid line); 10 × rudder (dashed line): $I_{xx} = 30{,}000\,\text{kg-m}^2$.

The combined effects of roll-dependent aerodynamic and inertial stability raise the possibility of *autorotation* or *roll lock-in* in the vicinity of the critical roll rates, as examined in Section 7.3. Regions of conditional instability have been demonstrated in the last several figures. In the same way that control-system nonlinearity can produce limit cycles at a stability boundary (Section 6.5), dynamic coupling that causes roll rate to increase or decay can produce similar effect.

Response to Controls during Steady Rolling

As noted in Section 7.1, asymmetry introduces transient and steady cross-axis control effects even when the direct effects of the primary controls are uncoupled. The equilibrium response to controls in the linear model is given by

$$\Delta \mathbf{x}^* = -\mathbf{F}^{-1}\,\mathbf{G}\,\Delta \mathbf{u}^* \tag{7.2-29}$$

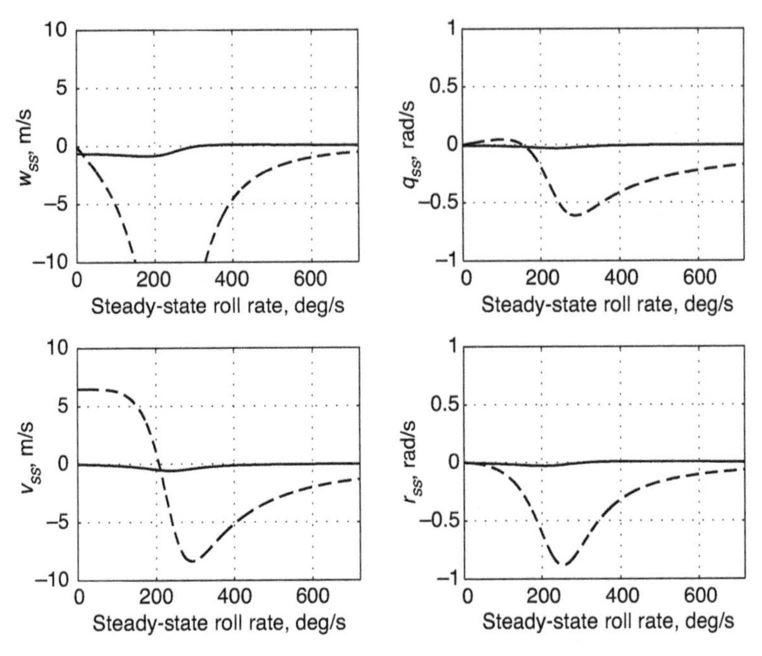

Fig. 7.2-13. Equilibrium response to elevator and rudder input with aerodynamic damping. Elevator (solid line); 10 × rudder (dashed line): $I_{xx} = 30{,}000\,\text{kg-m}^2$, $M_q = -5$.

and the coupling blocks of $\mathbf{F}^{-1}$ contain both even and odd powers of p_o. There are too many terms in eq. 7.2-29 to show the literal expression, except in the case of zero damping. For the truncated fourth-order model (eq. 7.2-22) with elevator and rudder inputs, the equation is simply

$$\begin{bmatrix} \Delta w^* \\ \Delta q^* \\ \Delta v^* \\ \Delta r^* \end{bmatrix} = \begin{bmatrix} \dfrac{I_{yy}u_o M_{\delta E}}{(I_{xx} - I_{zz})p_o^2 - I_{yy}u_o M_w} & 0 \\ 0 & \dfrac{I_{zz}p_o N_{\delta R}}{(I_{yy} - I_{xx})p_o^2 - I_{zz}u_o N_v} \\ 0 & \dfrac{I_{zz}u_o N_{\delta R}}{(I_{yy} - I_{xx})p_o^2 - I_{zz}u_o N_v} \\ \dfrac{I_{yy}p_o M_{\delta E}}{(I_{xx} - I_{zz})p_o^2 - I_{yy}u_o M_w} & 0 \end{bmatrix} \begin{bmatrix} \Delta\delta E^* \\ \Delta\delta R^* \end{bmatrix} \tag{7.2-30}$$

A clear demarcation of coupled and uncoupled effects is revealed. A constant elevator deflection changes the equilibrium normal-velocity (or

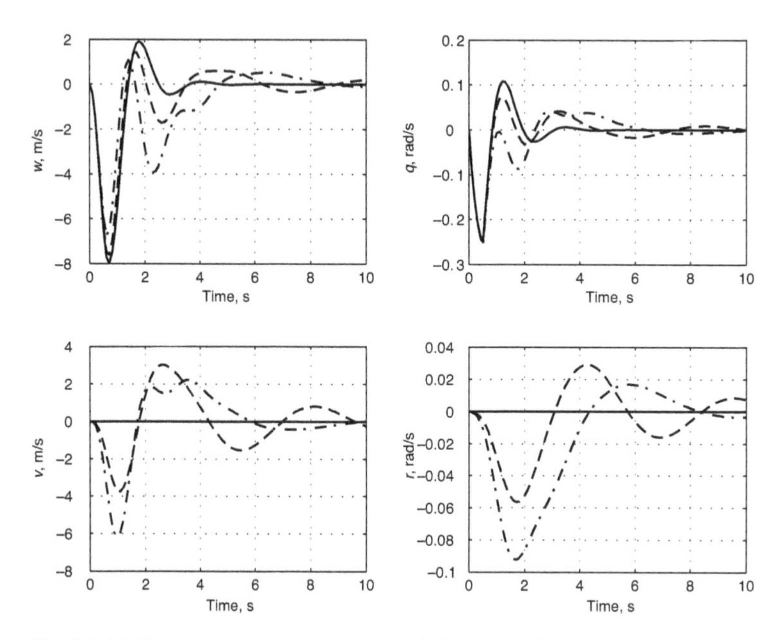

Fig. 7.2-14. State response to a 0.5 s, 0.1 rad elevator pulse, with nominal aerodynamics and inertia. $p_o = 0$ deg/s (solid line), $p_o = 60$ deg/s (dashed line), $p_o = 120$ deg/s (dash-dotted line). (a) Normal velocity. (b) Pitch rate. (c) Side velocity. (d) Yaw rate.

angle-of-attack) perturbation, but it has no effect on the equilibrium pitch rate. It has the cross-axis effect of introducing an equilibrium yaw rate in direct proportion to the equilibrium roll rate. Both elevator effects are attenuated by high roll rate, as gyroscopic stability stiffens the response. The converse occurs with steady rudder deflection: side velocity (or sideslip angle) and pitch rate are induced, but equilibrium yaw-rate perturbation is zero.

On closer examination of eq. 7.2-30, we see that the denominator for elevator effects goes to zero when p_o equals $p_{c_{SP}}$, while the rudder-effect denominator goes to zero when p_o equals $p_{c_{DR}}$. The equilibrium response for the undamped model becomes infinite at the critical roll rates (Fig. 7.2-9), and it changes sign as the roll rate increases through the singularity. Of course, the equilibrium response is observable as a steady-state response only when the system is stable. For nominal I_{xx}, the elevator response is singular at $p_{c_{SP}}$ and the rudder response is never singular. For reduced I_{xx}, the elevator response is singular at $p_{c_{SP}}$, and the rudder response is singular at $p_{c_{DR}}$ (Fig. 7.2-10).

With nominal aerodynamic damping, both elevator and rudder response become singular at the single critical roll rate that occurs when $I_{xx} = 35{,}927\,\text{kg-m}^2$ (Fig. 7.2-11). Reducing the rolling moment of inertia to 30,000 kg-m^2 causes both responses to be singular at both critical roll rates (Fig. 7.2-12). Further increases in damping do not prevent roll-induced instability at the higher I_{xx}; however, for $I_{xx} < I_{yy}$, increased damping can eliminate the critical points. For example, raising pitch damping to $M_q = -5$ eliminates instability and singular equilibrium response to control (Fig. 7.2-13). The lower critical roll rate sets an upper limit on the operational maneuvering roll rate for the aircraft. The final example suggests that the operational roll rate limit can be increased by feedback stability augmentation when $I_{xx} < I_{yy}$.

The transient response to control for the nominal case is illustrated in Fig. 7.2-14. As expected, there is no directional response to elevator when the steady roll rate is zero. Increasing p_o to 60 deg/s or more produces directional response of the same order as the longitudinal response.

7.3 Multiple Equilibrium Points

Unlike their linear counterparts, nonlinear models of aircraft motion may have more than one equilibrium point corresponding to a single control setting. This comes about because forces and moments can be multivalued functions of the state and control and because products of state and control elements occur. The stability of small perturbations from these equilibrium points determines whether the equilibria are "observable," that is, whether they represent quasi-steady-states of the system. Unstable equilibrium points can be enclosed by stable limit cycles, which themselves represent a condition of dynamic equilibrium.

In some instances, small changes in control settings can lead to large jumps in the equilibrium point and associated transient response. Such behavior is not merely theoretical; it has been observed in flight [R-2, B-12]. Jumps limit the controllable response envelope of the aircraft, and they can lead to hazardous flight conditions from which recovery may be impossible. The underlying causes of jumps to new equilibrium points and limit cycles can be described using bifurcation analysis and catastrophe theory.

Second-Order Examples of Multiple Equilibria

The linear, time-invariant system

$$\dot{\mathbf{x}} = \mathbf{f}(\mathbf{x}, \mathbf{u}) \triangleq \mathbf{F}\mathbf{x} + \mathbf{G}\mathbf{u} \tag{7.3-1}$$

possesses the unique equilibrium point

$$\mathbf{x}^* = -\mathbf{F}^{-1}\mathbf{G}\mathbf{u}^* \tag{7.3-2}$$

provided that $\mathbf{F}$ is nonsingular. Because $\mathbf{F}^{-1}\mathbf{G}$ is a constant matrix, smooth variations in $\mathbf{u}^*$ produce smooth variations in $\mathbf{x}^*$. As an example, the second-order system

$$\begin{bmatrix} \dot{x}_1 \\ \dot{x}_2 \end{bmatrix} = \begin{bmatrix} 0 & 1 \\ a & b \end{bmatrix} \begin{bmatrix} x_1 \\ x_2 \end{bmatrix} + \begin{bmatrix} 0 \\ 1 \end{bmatrix} u \tag{7.3-3}$$

possesses the single equilibrium point

$$\begin{bmatrix} x_1{}^* \\ x_2{}^* \end{bmatrix} = \begin{bmatrix} -u^*/a \\ 0 \end{bmatrix} \tag{7.3-4}$$

The stability of perturbations from the equilibrium is governed by the eigenvalues of **F**, which are solutions to the characteristic equation,

$$\Delta(\lambda) = \lambda^2 - b\lambda - a = 0 \tag{7.3-5}$$

Stability depends only on the coefficients a and b; neither the control input nor the equilibrium state has any effect on this result. For $b < 0$, the system is stable if $a < 0$.

Cubic-Spring Effect

Adding a "cubic-spring" term to the model

$$\begin{bmatrix} \dot{x}_1 \\ \dot{x}_2 \end{bmatrix} = \begin{bmatrix} 0 & 1 \\ a & b \end{bmatrix} \begin{bmatrix} x_1 \\ x_2 \end{bmatrix} + \begin{bmatrix} 0 & 0 \\ c & 0 \end{bmatrix} \begin{bmatrix} x_1^3 \\ x_2^3 \end{bmatrix} + \begin{bmatrix} 0 \\ 1 \end{bmatrix} u \tag{7.3-6}$$

does not change the equilibrium solution for $x_2{}^*$, but the solution for $x_1{}^*$ becomes

$$a x_1{}^* + c x_1{}^{*3} + u^* = 0 \tag{7.3-7a}$$

or

$$x_1{}^{*3} + \frac{a}{c} x_1{}^* + \frac{u^*}{c} \triangleq x_1{}^{*3} + p_1 x_1{}^* + p_2 = 0 \tag{7.3-7b}$$

This equation has three roots; either one or three are real, depending on the values of the parameters p_1 and p_2, where $p_1 = a/c$ and $p_2 = u^*/c$. For zero control input, the linear system's $x_1{}^*$ is zero (eq. 7.3-4). The nonlinear equation also has a zero equilibrium for zero control (i.e., $p_2 = 0$), but there are two other possibilities when $p_1 < 0$:

$$x_1{}^* = \pm\sqrt{-p_1} = \pm\sqrt{-a/c} \tag{7.3-8}$$

The forces produced by linear and cubic terms of opposite sign cancel each other at this point. When $p_1 > 0$, the linear and cubic effects have the same sign and never cancel, so there is just one equilibrium point.

The second-order system could represent a flight mode, such as the short period or Dutch roll, whose coefficients change with flight conditions and control settings. In these applications, the cubic term represents a nonlinear sensitivity of pitching or yawing moment to angle of attack or sideslip angle.

If a parameter variation causes a *change in the number of equilibrium points*, a *static bifurcation* is said to occur. For zero control input, the bifurcation occurs when $p_1 = 0$, providing an example of the *pitchfork bifurcation* [G-8], so called because the three equilibrium branches resemble a pitchfork (Fig. 7.3-1). The plot of $x_1{}^*(p_1, p_2)$ versus p_1 and p_2 describes an *equilibrium surface* (or *manifold*, as in Fig. 7.3-2); hence, Fig. 7.3-1 is a slice through Fig. 7.3-2 in the $x_1{}^*$-p_1 plane. The present example produces a surface that has a fold in the region of three equilibrium values of $x_1{}^*$ and is single valued elsewhere. Control effects on $x_1{}^*$ are illustrated in Fig. 7.3-3 for positive and negative values of p_1.

The $x_1{}^*$ equilibrium surface is vertical along the bifurcation boundary, that is, for the *bifurcation set* (p_{1_b}, p_{2_b}). Small variations in p_1 or p_2 in the vicinity of (p_{1_b}, p_{2_b}) can lead to a dramatic jump (or *catastrophe*) in the equilibrium condition. The static bifurcation set is readily portrayed on the control-parameter surface (p_1 versus p_2) by projecting the vertical folds of the surface onto the plane (Fig. 7.3-4). The bifurcation set has a *cusp* at $a = 0$. Circled numbers identify the number of equilibria within bifurcation bounds. Calculations and plotting of equilibrium surfaces and bifurcation sets are aided by a reparameterization of the equilibrium solution [K-11].

The stability of small perturbations determines whether these equilibria can be sustained (or be *observable*) and whether catastrophes occur at the boundaries. Following Lyapunov's first theorem (Section 4.3), we characterize system stability on the equilibrium manifold by the stability of the corresponding locally linearized model. Linear model stability is determined by the eigenvalues of $\partial \mathbf{f}/\partial \mathbf{x}$ at the equilibrium point. For the linear system (eq. 7.3-3), $\partial \mathbf{f}/\partial \mathbf{x}$ is $\mathbf{F}$, and stability is determined by eq. 7.3-4. For the nonlinear system (eq. 7.3-6), $\partial \mathbf{f}/\partial \mathbf{x}$ is

$$\left.\frac{\partial \mathbf{f}}{\partial \mathbf{x}}\right|_{\mathbf{x}=\mathbf{x}^*} = \begin{bmatrix} 0 & 1 \\ (a + 3cx_1{}^{*2} & b \end{bmatrix} \tag{7.3-9}$$

where $x_1{}^*$ is a real solution of eq. 7.3-7. Stability at the equilibrium point is determined by the roots of

$$\Delta(\lambda) = \lambda^2 - b\lambda - (a + 3cx_1{}^{*2}) = 0 \tag{7.3-10}$$

With $b < 0$, the system is stable when $(a + 3cx_1{}^{*2}) < 0$.

For zero control input, $x_1{}^*$ is either zero or $\pm\sqrt{-a/c}$ (eq. 7.3-8); in the latter case,

$$\left.\frac{\partial \mathbf{f}}{\partial \mathbf{x}}\right|_{\mathbf{x}=\mathbf{x}^*} = \begin{bmatrix} 0 & 1 \\ -2a & b \end{bmatrix} \tag{7.3-11}$$

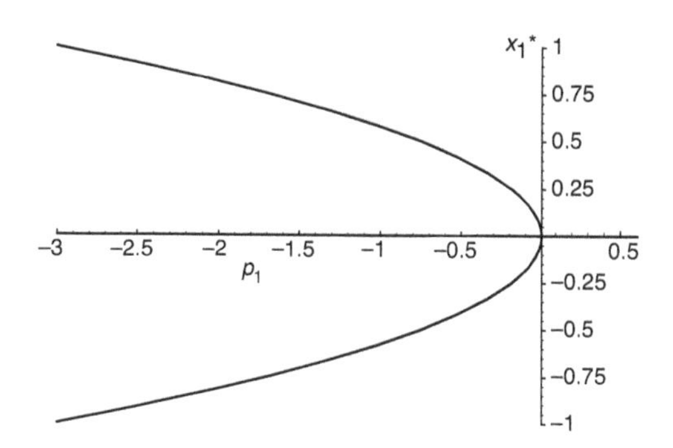

Fig. 7.3-1. The pitchfork bifurcation.

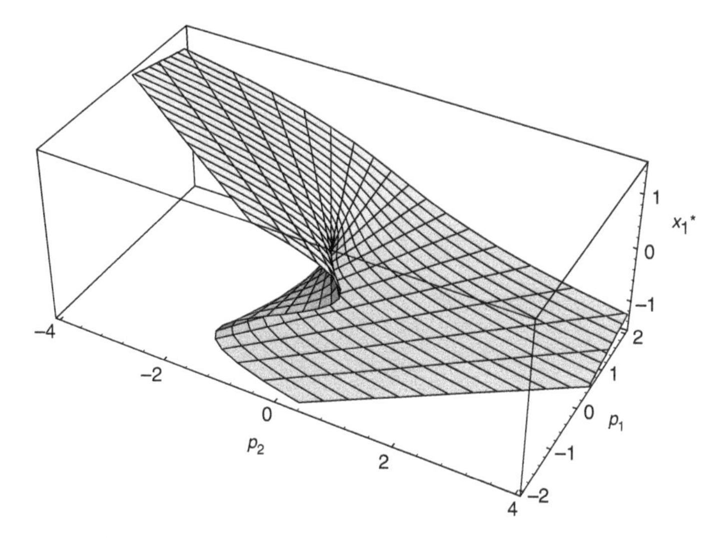

Fig. 7.3-2. Equilibrium surface for a cubic-spring system.

and

$$\Delta(\lambda) = \lambda^2 - b\lambda + 2a = 0 \tag{7.3-12}$$

Given $b < 0$, this equation is stable for $2a > 0$, while the zero equilibrium point is stable for $a < 0$; hence, stability of the zero equilibrium solution is accompanied by instability at the nonzero equilibria (or the converse).

An alternative explanation for the appearance of multiple observable equilibria is based on the stabilizing characteristic of a hardening cubic spring. As a progresses from negative to positive values, the response of the linear system (eq. 7.3-3) becomes divergent. The negative cubic term (eq. 7.3-6) increases system stiffness as the amplitude grows, eventually limiting the divergence.

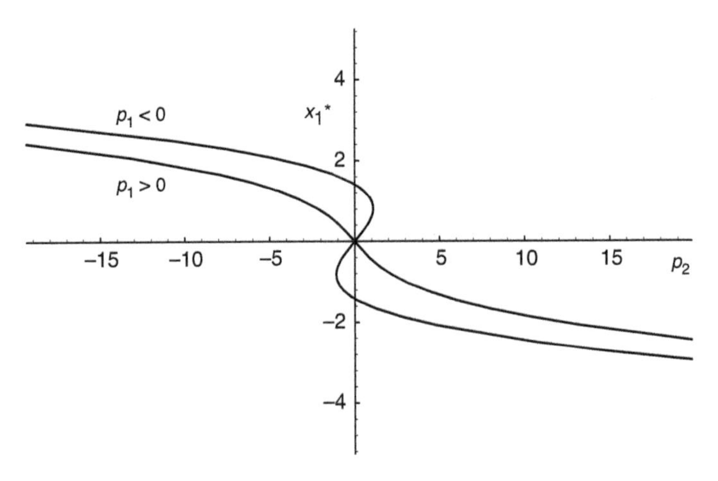

Fig. 7.3-3. Effects of control for positive and negative values of p_1.

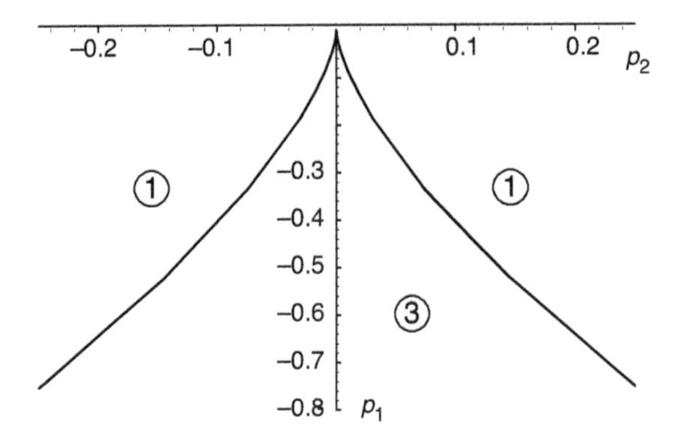

Fig. 7.3-4. Static bifurcation set for cubic-spring example.

The possibilities for catastrophic response to control input are best seen in Fig. 7.3-3, labeling the regions of the equilibrium surface as A through D. We consider two cases: $c > 0$ and $c < 0$. In both cases, a varies from positive to negative values. In the first case, p_1 has the same sign as a and p_2 the same sign as u^*. With $p_2 = 0$, the system is unstable for $p_1 > 0$ (Region A) and stable for $p_1 < 0$ (Region B). Consequently, the system is unstable in Regions C and D. Only equilibria in Region B are stable. Divergent response occurs in Regions A, C, and D, but there is no possibility of jumping from one x_1^* to another at a constant control setting.

In the second case, p_1 and a have opposite signs, as do p_2 and u^*. The system is thus stable in Regions A, C, and D and unstable in Region B. For positive p_1 (Region A), the origin of Fig. 7.3-3 is stable; as the control progresses from large positive to negative values, corresponding to p_2 progression of opposite sign, the equilibrium state also progresses smoothly. With negative p_1, the origin is unstable. $x_1{}^*$ moves along the Region C branch until Bifurcation Point (BP) 1 is reached. Further increase in p_2 causes the equilibrium to jump to Region D. A catastrophe has occurred.

If the control direction is reversed, the equilibrium remains in Region D until BP 2 is reached. Continued control motion forces the equilibrium to switch to Region C. Thus, within the bifurcation values of p_2, which equilibrium solution occurs depends upon the prior state of the system. Furthermore, bifurcation effects are one sided: an equilibrium point on Region C in the vicinity of the negative bifurcation value of p_2 is unaffected by BP 2. With large swings in control output, the nonlinearity produces a *hysteresis effect* in the equilibrium state.

The actual trajectory exhibits transient response, and the state history depends upon the rate of change and waveform of the control input. The *catastrophe* described above assumes a negligibly slow change in control setting. Nevertheless, a jump in x_1 produces a large rate (x_2); this introduces a damping effect in eq. 7.3-6 (Fig. 7.3-5), and the jump is not instantaneous. For the example, $a = 1$, $b = -1$, $c = -1$, $x_1(0) = x_2(0) = 0$, and the control is changing slowly ($u = -2 + 0.2t$). The trajectory begins at the unstable equilibrium point, is attracted to one stable region of the manifold, skips across the unstable fold in the $x_1{}^*$ manifold, and approaches the second stable manifold.

The locations of equilibrium points are unaffected by negative damping, but no values of a, c, and u^* produce stable equilibrium points. Feedback control can stabilize the system and eliminate multiple equilibria within the limits of control displacement and rate. Suppose that $a > 0$ and $c < 0$, which admits the possibility of multiple equilibria, and $b > 0$, which destabilizes all equilibria. Assume that the control law takes the form

$$u = u^* - c_1 x_1 - c_2 x_2 \tag{7.3-13}$$

Choosing $c_2 < -b$ stabilizes Regions A, C, and D, while choosing $c_1 < -a$ assures that the equilibrium is in Region A, where just one equilibrium point can occur.

Cubic-Damper Effect

Adding a "cubic-damper" term to the linear model,

$$\begin{bmatrix} \dot{x}_1 \\ \dot{x}_2 \end{bmatrix} = \begin{bmatrix} 0 & 1 \\ a & b \end{bmatrix} \begin{bmatrix} x_1 \\ x_2 \end{bmatrix} + \begin{bmatrix} 0 & 0 \\ 0 & c \end{bmatrix} \begin{bmatrix} x_1^3 \\ x_2^3 \end{bmatrix} + \begin{bmatrix} 0 \\ 1 \end{bmatrix} u \tag{7.3-14}$$

does not change the equilibrium solution (eq. 7.3-4). However, it does affect the stability of the system, and it introduces the possibility of a *dynamic*

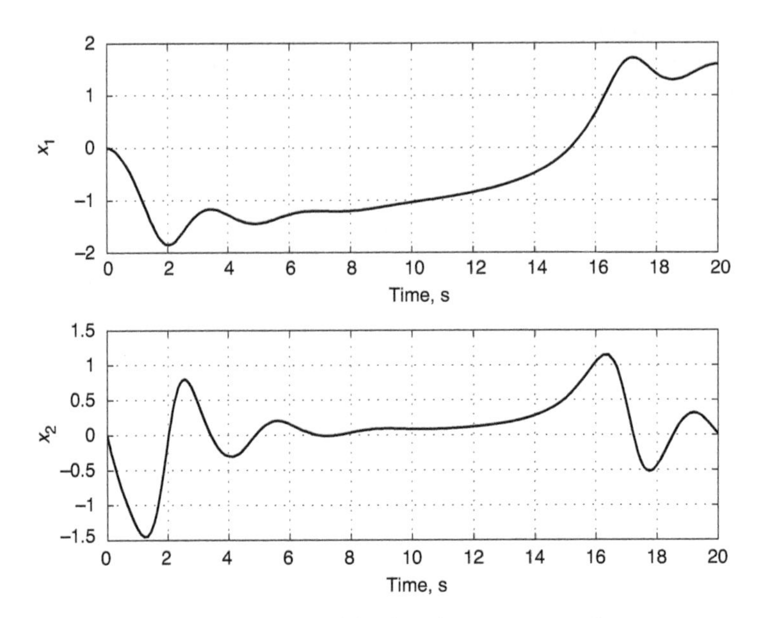

Fig. 7.3-5a. Transient response to control for the cubic-spring example ($u = -2 + 0.2t$): state history.

bifurcation from a stationary point to a stationary oscillation, that is, to a limit cycle. The resulting system is called a *van der Pol oscillator* [G-3]. We assume that $a < 0$; for $b < 0$, the single equilibrium point at $\mathbf{x}^* = [0\ 0]^T$ is stable, while for $b > 0$, it is unstable. Similarly, $c < 0$ always has a stabilizing effect as the magnitude of x_2 increases, while $c > 0$ is destabilizing.

Given a sinusoidal oscillation with maximum rate $x_{2_{max}}$, the cubic damping effect can be approximated by the describing function of eq. 6.5-42:

$$k_{D_c} = \frac{3}{4} c x_{2_{\max}}^2 \tag{7.3-15}$$

The corresponding characteristic equation becomes

$$\begin{aligned}\Delta(\lambda) &= \lambda^2 - (b + k_{D_c})\lambda - a \\ &= \lambda^2 - \left(b + \frac{3}{4} c x_{2_{max}}^2\right)\lambda - a = 0\end{aligned} \tag{7.3-16}$$

Zero damping occurs when

$$x_{2_{max}} = \sqrt{\frac{-4b}{3c}} \tag{7.3-17}$$

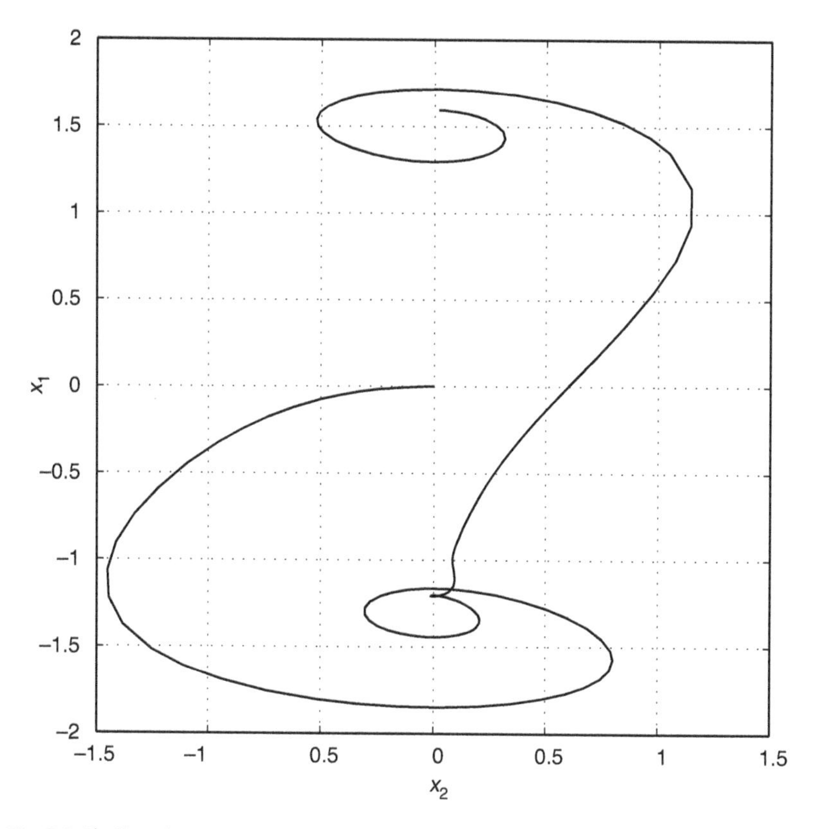

Fig. 7.3-5b. Transient response to control for the cubic-spring example. ($u = -2 + 0.2t$): $x_1(t)$ versus $x_2(t)$.

providing an estimate of the maximum magnitude of the limit-cycle rate. Equation 7.3-17 has a real solution only when b and c have opposite signs. Thus, when b is positive, a negative c introduces a stable limit cycle; when $b < 0$, a positive c produces an unstable limit cycle. The progression from a zero equilibrium point to a stable limit cycle, sketched using eq. 7.3-17 in Fig. 7.3-6a, produces a *Hopf bifurcation* [G-3].

An example of the limit cycle with $a = -1, b = 1$, and $c = -1$ is shown in Fig. 7.3-6b. The plot of x_2 versus x_1 is periodic but not sinusoidal; hence the magnitude estimate of eq. 7.3-17 is only an approximation. The combination of unstable linear damping and stable cubic damping is used to model *wing rock* in [S-3].

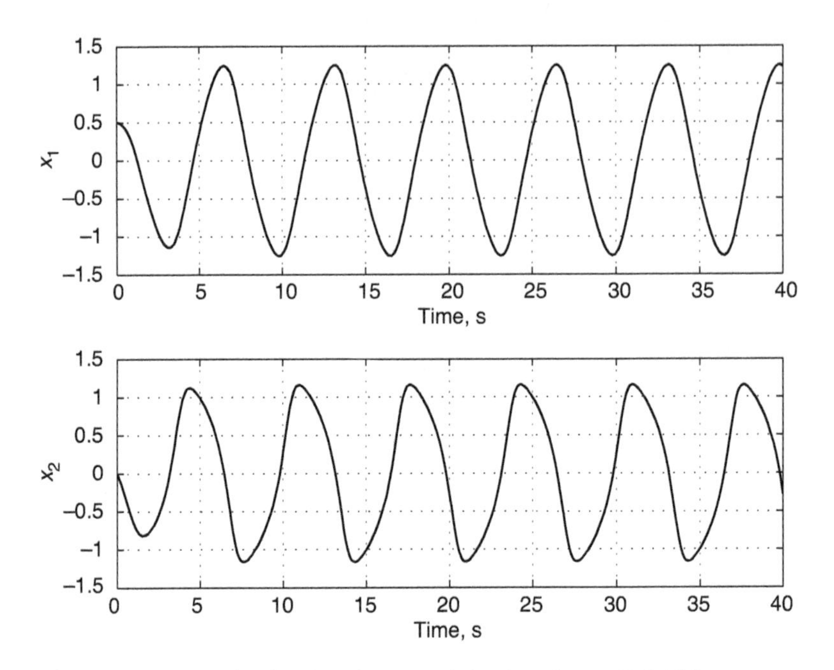

Fig. 7.3-6a. A system with cubic damping ($a = -1$, $b = 1$, and $c = -1$): initial-condition response.

Effects of Cross-Coupling and Control on Rolling Equilibrium

Rhoads and Schuler were among the first to apply extensive digital computation to flight dynamics analysis, identifying the potential for multiple equilibrium states associated with a single control setting. Their paper [R-2] compares computed results with the flight test maneuvers of a Lockheed F-80A aircraft, an early jet fighter with a straight wing, whose inertial properties were similar to those of the propeller-driven fighters of World War II. The paper also presents a computational analysis of the supersonic North American F-100 prototype, whose slender, swept wing gave it a low rolling moment of inertia. The prototype aircraft had an undersized vertical tail that allowed large sideslip excursions to occur in rolling pullups [A-2]. The production aircraft was fitted with a larger tail to reduce the problem.

Description of the Phenomenon

The calculations of [R-2] were based on a fifth-order model that could be expressed as $\mathbf{f}(\mathbf{x}^*, \mathbf{u}^*) = 0$, with symmetric aerodynamics and nonlinear inertial effects. Figure 7.3-7 (from [R-2]) illustrates the existence of nine equilibrium roll rates (four not shown) for the swept-wing aircraft, with

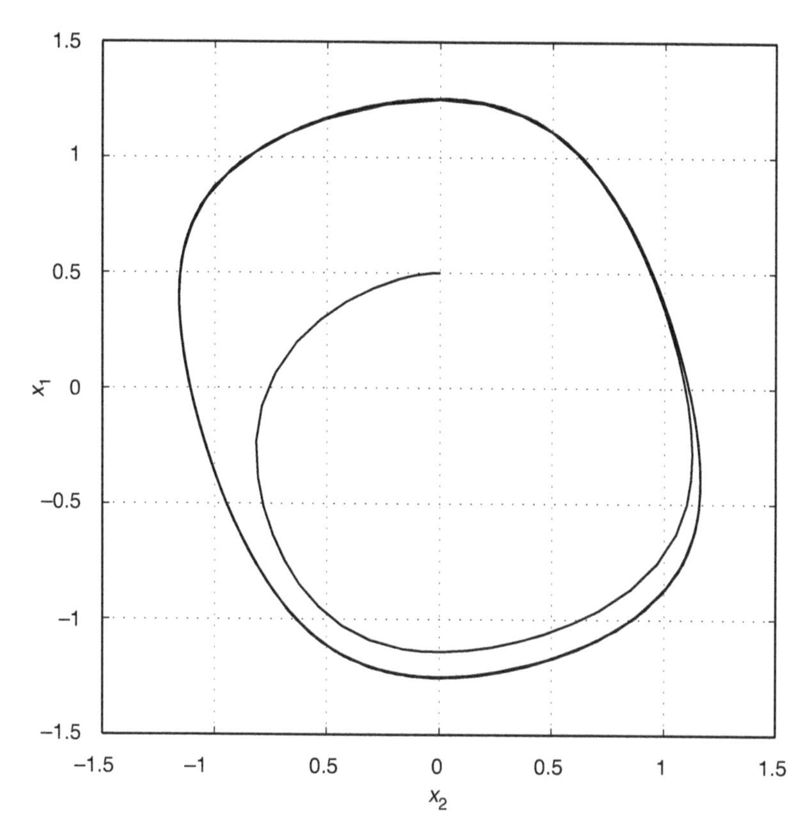

Fig. 7.3-6b. A system with cubic-damping. ($a = -1$, $b = 1$, and $c = -1$): development of limit cycle.

aileron angles in the range of ±7 deg and an elevator angle of −1 deg. Eight of the equilibria are near the positive and negative critical roll rates (Section 7.2). The ninth equilibrium represents the conventional response to the aileron, where the roll rate is below the lower critical roll rate. The corresponding equilibrium values of angle of attack and sideslip angle go to $\pm\infty$ at the critical roll rates.

There are seven possible equilibria for $|\delta A|$ from 7 to 8 deg, and there are five equilibria for higher aileron settings. Beginning with zero roll rate, increasing aileron setting would command conventional rolling response up to $|\delta A| = 8$ deg. At that point, further aileron increase would cause a jump in roll-rate equilibrium from 150 to 285 deg/s. Assuming a stable equilibrium at the higher rate, decreasing δA would have little effect on roll rate: the aircraft would "autorotate," continuing to roll even with zero or negative aileron setting. Without additional dynamic or

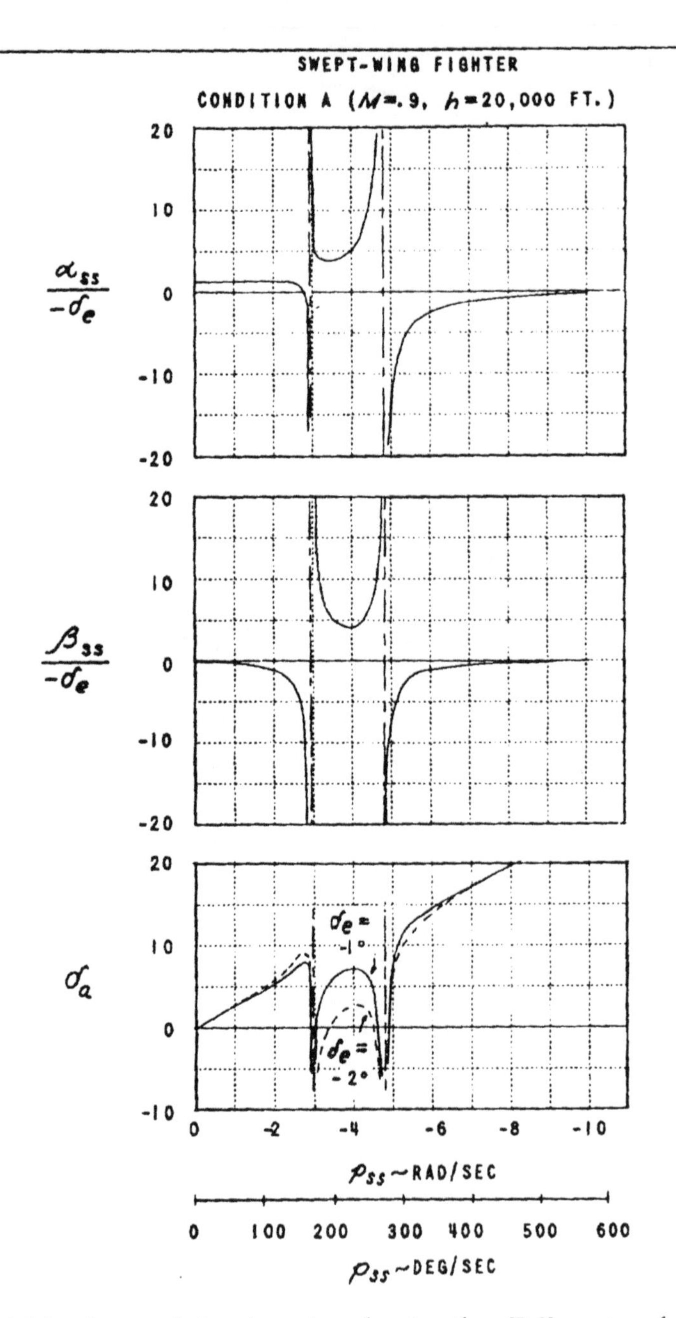

Fig. 7.3-7. Steady-state solution of equations of motion. (from [R-2], courtesy of AIAA)

control effects, this is an unrecoverable situation. An aircraft with this characteristic would have to be restricted from exceeding the lower critical roll rate.

Decreasing δE to -2 deg decreases the aileron range within which nine equilibria occurs and increases the region of seven equilibria (Fig. 7.3-7). This amount of elevator change has little effect on the *p-jump* phenomenon described above, but it suggests that coordinated control settings can influence both the location and the stability of achievable equilibrium points. Furthermore, the evolution of motions over time differs from the simplified notion of jumping from one equilibrium to the next, as demonstrated by Fig. 7.3-5. These dynamic and control effects offer a possibility for identifying control strategies for satisfactory recovery.

Characterization of Solutions

Schy and Hannah developed a simplified method for calculating the coupled equilibrium conditions [S-2], which is adapted to the notation of this book. We assume that aerodynamic effects are linear, that the body axes are principal axes, and that the effects of gravity and changing airspeed can be neglected. (Schy and his co-authors included nonlinear angle-of-attack effects in later solutions to the problem [Y-1, Y-2].) Following Section 3.6, the differential equations $\dot{\mathbf{x}} = \mathbf{f}(\mathbf{x}, \mathbf{u})$ for side and normal velocity and for roll, pitch, and yaw rate are as follows:

$$\dot{v} = Y_v v + Y_{\delta A}\delta A + Y_{\delta R}\delta R - u_o r + pw \tag{7.3-18}$$

$$\dot{w} = Y_w w + Z_{\delta E}\delta E + u_o q - pv \tag{7.3-19}$$

$$\dot{p} = L_v v + L_p p + L_r r + L_{\delta A}\delta A + L_{\delta R}\delta R - \frac{I_{xx} - I_{yy}}{I_{xx}} qr \tag{7.3-20}$$

$$\dot{q} = M_q q + M_w w + M_{\delta E}\delta E - \frac{I_{xx} - I_{zz}}{I_{yy}} pr \tag{7.3-21}$$

$$\dot{r} = N_v v + N_p p + N_r r + N_{\delta A}\delta A + N_{\delta R}\delta R + \frac{I_{xx} - I_{yy}}{I_{zz}} pq \tag{7.3-22}$$

At equilibrium, the accelerations are zero. Given an equilibrium roll rate p^*, equilibrium of the four remaining variables is specified by

$$Y_v v^* - u_o r^* + p^* w^* = -Y_{\delta A}\delta A^* - Y_{\delta R}\delta R^* \tag{7.3-23}$$

$$Y_w w^* + u_o q^* - p^* v^* = -Z_{\delta E}\delta E^* \tag{7.3-24}$$

$$\frac{I_{yy}}{I_{xx}-I_{zz}}(M_q q^* + M_w w^*) - p^* r^* = -\frac{I_{yy}}{I_{xx}-I_{zz}} M_{\delta E}\delta E^* \tag{7.3-25}$$

$$\frac{I_{zz}}{I_{xx}-I_{yy}}(N_v v^* + N_r r^*) + p^* q^* = -\frac{I_{zz}}{I_{xx}-I_{yy}}(N_{\delta A}\delta A^* - N_{\delta R}\delta R^* - N_p p^*) \tag{7.3-26}$$

These four equations can be written in the form

$$(p^*\mathbf{I}_4 - \mathbf{A})\,\mathbf{x}_4{}^* = \mathbf{a} + p^*\mathbf{b} \tag{7.3-27}$$

where

$$\mathbf{x}_4{}^* = [v^*\ w^*\ q^*\ r^*]^T \tag{7.3-28}$$

$$\mathbf{A} = \begin{bmatrix} 0 & Z_w & u_o & 0 \\ -Y_v & 0 & 0 & u_o \\ -N'_v & 0 & 0 & -N'_r \\ 0 & -M'_w & -M'_q & 0 \end{bmatrix} \tag{7.3-29}$$

$$\mathbf{a} = \begin{bmatrix} Z_{\delta E}\delta E^* \\ -Y_{\delta A}\delta A^* - Y_{\delta R}\delta R^* \\ -N'_{\delta A}\delta A^* - N'_{\delta R}\delta R^* \\ M'_{\delta E}\delta E^* \end{bmatrix} \tag{7.3-30}$$

$$\mathbf{b} = \begin{bmatrix} 0 \\ 0 \\ N'_p \\ 0 \end{bmatrix} \tag{7.3-31}$$

The primed stability and control derivatives $(\cdot)'$ are the original derivatives multiplied by the appropriate ratio of moments of inertia [e.g., $N_v' = N_v' I_{zz}/(I_{xx} - I_{yy})$]. If coupling aerodynamic effects are known, they can be included in $\mathbf{A}$. Equation 7.3-27 has the solution

$$\mathbf{x}_4{}^* = (p^*\mathbf{I}_4 - \mathbf{A})^{-1}(\mathbf{a} + p^*\mathbf{b}) \tag{7.3-32}$$

which can be written as a (4×1) ratio of polynomials in p^*:

$$\mathbf{x}_4{}^* = \frac{[n_v{}^*(p^*)n_w{}^*(p^*)n_q{}^*(p^*)n_r{}^*(p^*)]^T}{d(p^*)} \tag{7.3-33}$$

The denominator is a fourth-degree polynomial in p^*:

$$d(p^*) = p^{*4} + [u_o(M'_w + N'_v) + Y_v Z_w - M'_q N'_r]p^{*2} + [u_o^2 M'_w N'_v + u_o(M'_w N'_r Y_v - M'_q N'_v Z_w) - M'_q N'_r Y_v Z_w] \quad (7.3\text{-}34)$$

$d(p^*)$ is the zeroth-degree term of the characteristic polynomial (in s) for the fourth-order pitch-yaw model of Section 7.2 (eq. 7.2-27). The four numerator elements $[n_v^*(p^*)\ n_w^*(p^*)\ n_q^*(p^*)\ n_r^*(p^*)]^T$ are third- or fourth-degree polynomials in p^*.

The equilibrium solution for roll rate depends on the control settings and on the equilibrium values of the remaining variables:

$$L_{\delta A}\delta A^* + L_{\delta R}\delta R^* + L_p p^* + L_v v^* + L_r r^* - \frac{I_{xx} - I_{yy}}{I_{xx}} q^* r^* = 0 \quad (7.3\text{-}35)$$

Collecting the rolling effects of aileron and rudder in L_C and using eq. 7.3-33, this equation can be written as

$$L_C + L_p p^* + L_v \frac{n_v^*(p^*)}{d(p^*)} + L_r \frac{n_r^*(p^*)}{d(p^*)} - \left[\frac{I_{xx} - I_{yy}}{I_{xx}}\right]\left[\frac{n_q^*(p^*) n_r^*(p^*)}{d^2(p^*)}\right] = 0 \quad (7.3\text{-}36)$$

Multiplying by $d^2(p^*)$ produces a ninth-degree polynomial that defines the equilibrium roll rates p^*:

$$(L_C + L_p p^*)\, d^2(p^*) + [L_v n_v^*(p^*) + L_r n_r^*(p^*)]\, d(p^*) - \frac{I_{xx} - I_{yy}}{I_{xx}} n_q^*(p^*) n_r^*(p^*) = 0 \quad (7.3\text{-}37)$$

This equation always has at least one real solution, corresponding to the conventional rolling response to controls, plus as many as eight other real solutions, the equilibria described by Rhoads and Schuler. Once the real solutions for p^* are computed, the corresponding values of v^*, w^*, q^*, and r^* can be found from eq. 7.3-33. If the $n_q^*(p^*) n_r^*(p^*)$ product is negligibly small or zero, eq. 7.3-33 reduces to a fifth-degree polynomial in p^*.

The stability of the equilibrium solutions is determined by the five eigenvalues of $\partial \mathbf{f}/\partial \mathbf{x}$, derived from eqs. 7.3-18–7.3-22 with $\mathbf{x} = \mathbf{x}^*$ and $\mathbf{u} = \mathbf{u}^*$:

$$\frac{\partial \mathbf{f}}{\partial \mathbf{x}}(\mathbf{x}^*, \mathbf{u}^*) = \begin{bmatrix} Y_v & p^* & w^* & 0 & u_o \\ -p^* & Z_w & -v^* & u_o & 0 \\ L_v & 0 & L_p & -r^* \dfrac{(I_{zz} - I_{yy})}{I_{xx}} & \left[L_r - q^* \dfrac{(I_{zz} - I_{yy})}{I_{xx}}\right] \\ 0 & M_w & -r^* \dfrac{(I_{xx} - I_{zz})}{I_{yy}} & M_q & -p^* \dfrac{(I_{xx} - I_{zz})}{I_{yy}} \\ N_v & 0 & \left[N_p + q^* \dfrac{(I_{xx} - I_{yy})}{I_{zz}}\right] & p^* \dfrac{(I_{xx} - I_{yy})}{I_{zz}} & N_r \end{bmatrix} \tag{7.3-38}$$

Although the constant control settings do not appear explicitly in eq. 7.3-38, they effect the equilibrium values of the state and, therefore, the stability of the system.

Bifurcation Analysis

Mehra, Carroll, and others [M-4 to M-6, C-2, G-9, J-1, L-5] recognized that the explanation of aircraft equilibrium states could take advantage of concepts from functional analysis and theoretical applied mechanics [e.g., G-3, G-4, G-8, I-1, P-10]. Thom's characterization of the *elementary catastrophes* of systems is a motivating factor [T-6]. Equilibrium surfaces and static bifurcation sets take canonical forms for systems whose force is the gradient of a potential field. Linearized models of such systems have real eigenvalues and are thus subject to static bifurcation. The cubic-spring second-order model considered above provides an example of the elementary *cusp catastrophe*. Although all forces and moments on aircraft are not the result of potential fields, static equilibrium points of the fifth-order dynamic model follow the elementary bifurcation classifications. Catastrophe theory does not address dynamic bifurcations.

The *implicit function theorem*, as described in [I-1], establishes the circumstances under which a dynamic system possesses a unique relationship between constant control and the equilibrium state. For the n-dimensional system

$$\dot{\mathbf{x}} = \mathbf{f}(\mathbf{x}, \mathbf{u}) \tag{7.3-39}$$

the equilibrium point defined by

$$\mathbf{f}(\mathbf{x}^*, \mathbf{u}^*) = 0 \tag{7.3-40}$$

is unique if

$$\left|\frac{\partial \mathbf{f}}{\partial \mathbf{x}}(\mathbf{x}^*, \mathbf{u}^*)\right| \triangleq |\mathbf{F}(\mathbf{x}^*, \mathbf{u}^*)| \neq 0 \tag{7.3-41}$$

The existence of the determinant guarantees that smooth changes in the constant control produce smooth variations in the equilibrium point in the vicinity of $(\mathbf{x}^*, \mathbf{u}^*)$. Expanding eq. 7.3-40 to first degree, the perturbed solution is unique and smoothly varying because $\mathbf{F}^{-1}(\mathbf{x}^*, \mathbf{u}^*)$ exists:

$$\begin{aligned}\Delta\mathbf{x} \triangleq (\mathbf{x} - \mathbf{x}^*) &= -\left(\frac{\partial \mathbf{f}}{\partial \mathbf{x}}\right)^{-1}_{\substack{\mathbf{x}=\mathbf{x}^*\\ \mathbf{u}=\mathbf{u}^*}} \left(\frac{\partial \mathbf{f}}{\partial \mathbf{u}}\right)_{\substack{\mathbf{x}=\mathbf{x}^*\\ \mathbf{u}=\mathbf{u}^*}} (\mathbf{u} - \mathbf{u}^*)\\ &\triangleq -\mathbf{F}^{-1}(\mathbf{x}^*, \mathbf{u}^*)\,\mathbf{G}(\mathbf{x}^*, \mathbf{u}^*)\,\Delta\mathbf{u}\\ &\rightarrow 0 \text{ as } \Delta\mathbf{u} \rightarrow 0\end{aligned} \tag{7.3-42}$$

Static bifurcation occurs only at points where the implicit function theorem is violated, that is, where $|\mathbf{F}(\mathbf{x}^*, \mathbf{u}^*)| = 0$. Whether or not a static bifurcation actually does occur depends upon the nature of the nonlinearities. The twelfth-order equations of motion or the uncoupled sixth-order longitudinal and lateral-directional equations never satisfy eq. 7.3-42 in forward flight; hence, they do not formally exhibit bifurcations. Nevertheless, static bifurcations may occur in uncoupled nonlinear models up to fourth order and coupled nonlinear models up to eighth order. The implicit function theorem gives no information about the possibility of dynamic bifurcations because $|\mathbf{F}(\mathbf{x}^*, \mathbf{u}^*)|$ need not be zero at the bifurcation (see eq. 7.3-14 and discussion).

The *center manifold theorem* [K-4] suggests that bifurcations are low-order phenomena, even when they occur in high-order systems. The eigenvalues of $\mathbf{F}(\mathbf{x}^*, \mathbf{u}^*)$ fall into three categories according to the sign of their real parts: stable (negative), neutrally stable (zero), and unstable (positive). The nonlinear dynamic equation can be partitioned in three parts as

$$\dot{\mathbf{x}}_S = \boldsymbol{\Lambda}_S\mathbf{x}_S + \Delta\mathbf{f}_S(\mathbf{x}_S, \mathbf{x}_U, \mathbf{x}_N, \mathbf{u}) \tag{7.3-43}$$

$$\dot{\mathbf{x}}_N = \boldsymbol{\Lambda}_N\mathbf{x}_N + \Delta\mathbf{f}_S(\mathbf{x}_S, \mathbf{x}_U, \mathbf{x}_N, \mathbf{u}) \tag{7.3-44}$$

$$\dot{\mathbf{x}}_U = \boldsymbol{\Lambda}_U\mathbf{x}_U + \Delta\mathbf{f}_S(\mathbf{x}_S, \mathbf{x}_U, \mathbf{x}_N, \mathbf{u}) \tag{7.3-45}$$

Λ_S, Λ_N, and Λ_U are block-diagonal matrices containing the stable, neutrally stable, and unstable linear models, and $\mathbf{x}_S$, $\mathbf{x}_N$, and $\mathbf{x}_U$ are the corresponding modal coordinates with $\dim(\mathbf{x}_S) = n_S \times 1$, $\dim(\mathbf{x}_N) = n_N \times 1$, $\dim(\mathbf{x}_U) = n_U \times 1$, and $n_S + n_N + n_U = n$. The nonlinear residuals are assumed to vanish smoothly at the equilibrium point:

$$\Delta \mathbf{f}_S(\mathbf{x}_S^*, \mathbf{x}_U^*, \mathbf{x}_N^*, \mathbf{u}^*) \to \mathbf{0} \quad \text{as } \|\mathbf{x}_S^*, \mathbf{x}_U^*, \mathbf{x}_N^*, \mathbf{u}^*\| \to 0 \qquad (7.3\text{-}46)$$

$$\Delta \mathbf{f}_N(\mathbf{x}_S^*, \mathbf{x}_U^*, \mathbf{x}_N^*, \mathbf{u}^*) \to \mathbf{0} \quad \text{as } \|\mathbf{x}_S^*, \mathbf{x}_U^*, \mathbf{x}_N^*, \mathbf{u}^*\| \to 0 \qquad (7.3\text{-}47)$$

$$\Delta \mathbf{f}_U(\mathbf{x}_S^*, \mathbf{x}_U^*, \mathbf{x}_N^*, \mathbf{u}^*) \to \mathbf{0} \quad \text{as } \|\mathbf{x}_S^*, \mathbf{x}_U^*, \mathbf{x}_N^*, \mathbf{u}^*\| \to 0 \qquad (7.3\text{-}48)$$

Hence, the nonlinear response manifolds and their linear counterparts are tangent at the equilibrium point.

Bifurcations occur where stable modes become unstable or vice versa, that is, only on the neutrally stable *center manifold*. Thus, bifurcations of the n-dimensional system are described by bifurcations of a reduced-order central subsystem (eq. 7.3-44). Because the linear component of response is neither stable nor unstable, stability on the center manifold is determined by the nonlinear term alone.

For distinct roots, variations of system parameters typically produce no more than one real root or two complex roots at a time; hence, the central subsystem usually is first or second order. The static bifurcation associated with a real root is between single and multiple equilibrium points; the dynamic bifurcation associated with a complex pair of roots is between the equilibrium point of a converging (diverging) oscillation and a stable (unstable) limit cycle. Any of the reduced-order models presented in Chapters 4–6 could become the linear counterpart of a center-manifold subsystem when parameter variation causes their eigenvalues to become neutrally stable.

Finding a family of static equilibrium solutions can be an arduous process. Solving eq. 7.3-40 for a completely nonlinear model requires a general numerical solution technique, such as the steepest-descent, Newton-Raphson, or gradient-free algorithm. The process is iterative and must be repeated for each branch of the equilibrium surface. Because a family of neighboring solutions is sought, *continuation methods* are of particular interest. Once an equilibrium solution is found, it is used as the starting point to search for additional neighboring solutions.

The implicit function theorem provides the foundation for continuing the search up to points of bifurcation. The Schy-Hannah solution could provide starting solutions, as it allows all the linear-aerodynamic solutions corresponding to a given control setting to be calculated at once. The Jacobian, expressed by eq. 7.3-38 with stability and control derivatives evaluated at ($\mathbf{x}^*$, $\mathbf{u}^*$), provides a first estimate of the new state through eq. 7.3-42, and the search algorithm converges to an improved estimate. Previous researchers have used the approach suggested in [K-3], with modifications to bridge bifurcation points described by [K-9].

Having established the surface of stable and unstable equilibrium points, it remains to be seen whether limit cycles occur. Stable limit cycles are associated with the unstable manifold of the system (eq. 7.3-45); hence, the order of the phenomenon generally is less than the order of the system. Solving the differential equations of motion (either eq. 7.3-39 or 7.3-45) for initial conditions in the neighborhood of the unstable equilibrium point is one approach to finding a limit cycle. It also is possible to make preliminary predictions based on describing-function analysis [H-5, S-18, T-4]. Unstable limit cycles may define the regions of attraction of equilibrium points on the stable manifold. The boundaries can be estimated by solving the equations of motion in reverse time, with initial conditions in the neighborhood of the stable equilibrium [S-7], or by analytical methods [P-11].

An example based on [M-5] illustrates the combined effects of aileron and elevator setting on equilibrium roll rate. For negative elevator, aileron control produces a single "conventional" p^* response (Fig. 7.3-8a). For less negative elevator, a single aileron setting can produce either one or three equilibria (Fig. 7.3-8b). At low and high δA^*, there is a single equilibrium; within an intermediate range, there are three. The middle equilibrium is unstable, as the eigenvalues of the associated Jacobian matrix reveal. p^* would jump from the first to the third branch with increasing aileron angle. Having reached the third branch, if the aileron deflection were decreased, the roll rate would remain on the third branch until that solution no longer existed, at which point, the roll rate returns to the first branch, where the response is conventional.

With further increase in the elevator setting (trailing edge down), aileron deflection produces one, three, or five equilibria, and a "no-return" situation may arise (Fig. 7.3-8c). The equilibrium roll rate could have an opposite sign from that associated with the conventional response, and the magnitude of this reversed response could be nearly as large as the response in the usual sense. For the condition shown, increasing aileron deflection would produce a "p-jump" to an extreme branch, returning the aileron to zero would have little effect, and reversing the aileron initially would not arrest the roll rate but then would force a jump to the opposite extreme branch. There is no path back to the origin at this elevator setting; however, pulling back on the stick would reestablish a recoverable condition. With further elevator increase, there are one or three equilibria, and the origin is unstable; hence, autorotation is the only possibility (Fig. 7.3-8d).

These four graphs are cuts through the equilibrium surface, and can be inferred from the three-dimensional view shown in Fig. 7.3-9, which

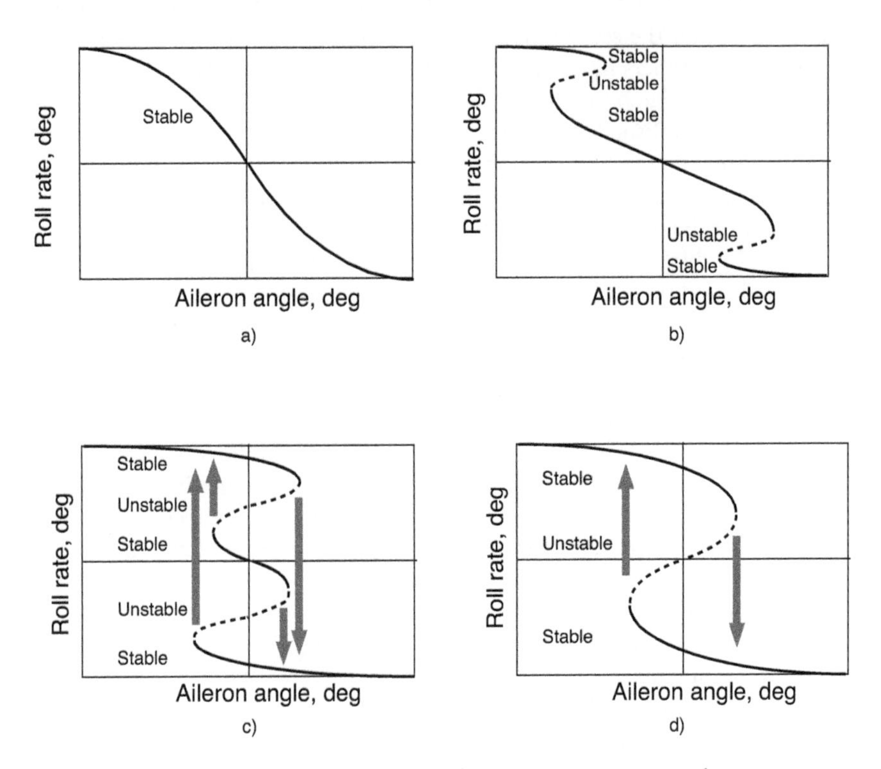

Fig. 7.3-8. Example of "p-jump" response to aileron setting. Progression of equilibrium state with increasing elevator angle (trailing edge down). (a) Conventional aileron response, negative elevator setting. (b) One or three equilibrium responses, less negative elevator setting. (c) One, three, or five equilibrium responses, positive elevator setting. (d) One or three equilibrium responses, more positive elevator setting. (after [M-5])

plots the equilibrium roll rate against constant aileron and elevator settings. The control-parameter space for the bifurcation set associated with this problem is the base plane of the figure. The bifurcation set is seen to be the *butterfly catastrophe* of [T-6], which is replotted in Fig. 7.3-10. Although not shown, multiple equilibria could also evolve for large negative values of δE^* and the associated high angles of attack.

Figure 7.3-11 illustrates the potential for entering autorotation from a symmetric flight condition. The plot of p^* versus δE^* is a cut through a simpler equilibrium surface (e.g., one characterized by Figs. 7.3-8a and d only or by Fig. 7.3-2) with $\delta A^* = 0$. The equilibrium roll rate remains zero with increasing δE^* up the point of bifurcation (BP 1). If δE^* is increased further, the zero solution is unstable, the roll rate must jump to

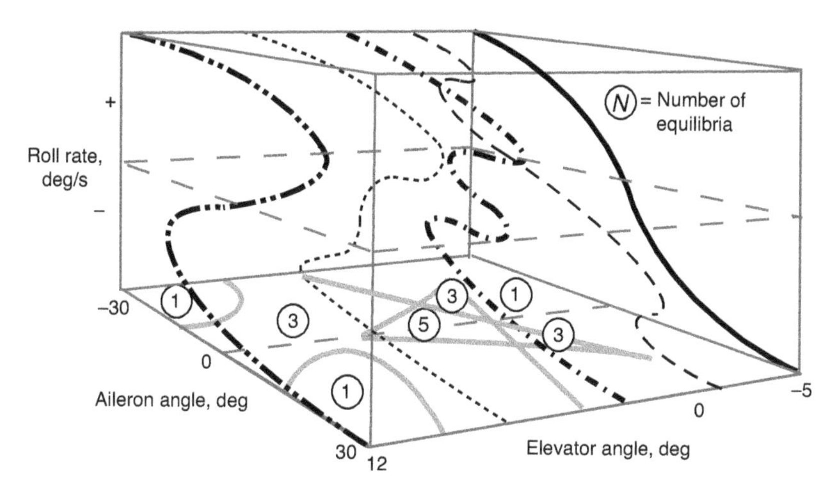

Fig. 7.3-9. Cuts through the roll-rate equilibrium surface plotted above the bifurcation surface. (after [M-5])

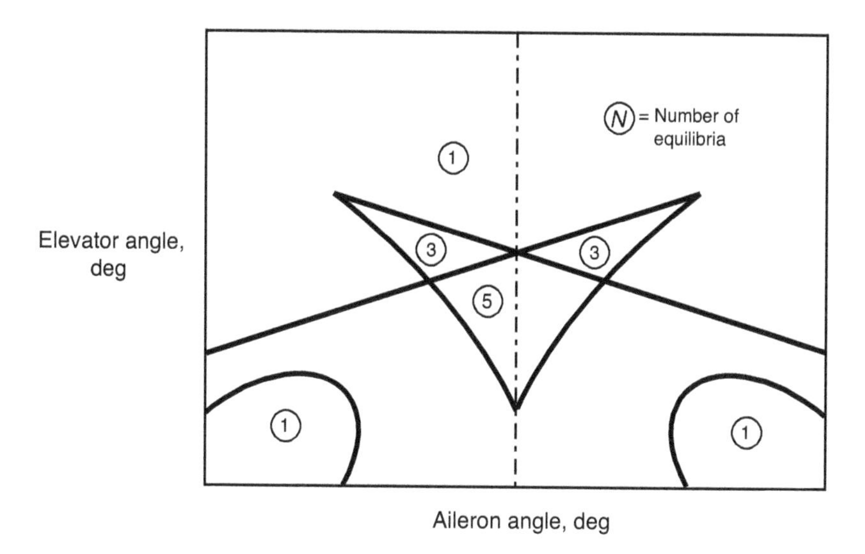

Fig. 7.3-10. Bifurcation surface corresponding to the roll-rate equilibrium surface, illustrating the "butterfly catastrophe." (after [M-5])

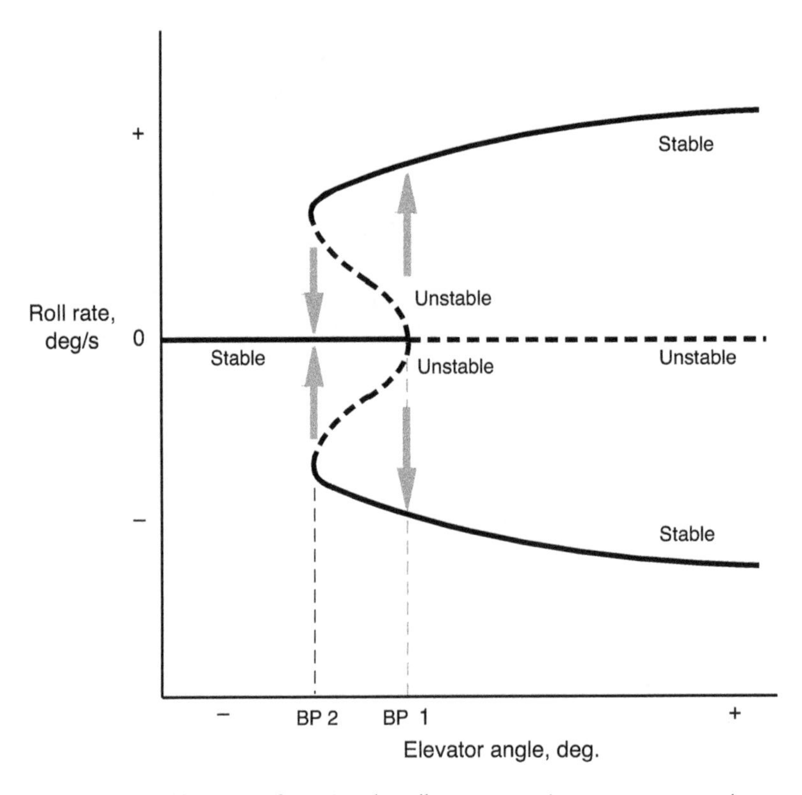

Fig. 7.3-11. Equilibrium configuration that allows symmetric entry to autorotation. (after [M-5])

a positive or negative value, whichever is triggered by a random disturbance. The autorotation would end when δE^* was reduced below BP 2.

7.4 Flight at High Angle of Attack

Aircraft intended for aerobatics, air combat, and reentry from orbit are designed for high-angle-of-attack flight, but all others enter this region only during testing and training flights or under abnormal conditions. We distinguish between the two regions by the nature of the airflow over the aircraft. At low angle of attack, the air passes smoothly over the skin, and the streamlines are said to be attached to its surfaces. At high angle of attack, the flow is attached on the lower (or "windward") side of the airplane but is detached from a significant portion of the upper (or "leeward") side; hence, the patterns of surface pressures that create gross forces and moments are markedly different. Conventional aircraft spend most of their time flying at low angles of attack, where cruising efficiency and speed can be high, dynamics and response to controls are well behaved, and safety margins are good. Whether flight at high angle is intended or not, an ultimate goal is to return safely to low angle of attack for flight to the destination and landing.

Flight at high angles of attack and angular rates has been widely studied, as attested to by the list of references for this chapter. It is differentiated from low-angle flight by large, nonlinear variations of aerodynamic forces and moments [O-3], increased inertial coupling of motions, large dissimilarity between body- and wind-axis frames of reference, and noticeable changes in the aircraft's response to control [N-2]. Thrust and control settings may be near their limits, reducing the options for changing flight condition or stabilizing the path. Flight is more likely to be "accelerated" than to be trimmed for $1g$ load factor; hence, airspeed may be changing rapidly as well. If flown to their maneuvering limits, many aircraft types may depart from controlled flight, requiring special recovery procedures to restore safe conditions. The *departure* can be symmetric, as in deep stalling, or it may be asymmetric, producing wing drop, wing rock, nose slice, or spin. There is usually a loss of altitude during departure and recovery; if a departure occurs too near the ground, a crash is inevitable.

Important aerodynamic effects and the governing equations of motion have been presented earlier. Here, we focus on the effects of high-angle-of-attack aerodynamics and inertial coupling on aircraft flight, examining both planned and unintended maneuvers through computation and reference to the literature. We see that control actions play a critical role in safe and effective flight at high angle of attack. The treatment is necessarily introductory and incomplete, as the scope is broad and many characteristics of flight are aircraft specific. Underlying principles are reviewed, and the means for predicting flight motions are illustrated.

High-Angle-of-Attack Aerodynamics and Control Effects

As the angle of attack α varies from 0 to 90 deg, the flight regime passes through several regions that are categorized by the wing's lifting characteristic (Table 7.4-1). Where these regions begin and end is heavily dependent on the aircraft configuration, particularly the nose shape, wing aspect ratio, leading-edge sweep angle and contour, and Mach number. The angle-of-attack ranges listed are loosely representative of the aircraft types shown, but they are approximate. While wings with low sweep angle and high aspect ratio normally have a well-defined maximum lift

TABLE 7.4-1 TYPICAL FLIGHT REGIMES FOR GENERAL AVIATION (GA), JET TRANSPORT (JT), AND FIGHTER (F) AIRCRAFT

Aerodynamic Region	Angle-of-Attack Range (deg)	Possible Flight Attributes
Low Angle of Attack	0–15 (GA, F) 0–10 (JT)	Conventional flight
Prestall	15–20 (GA) 10–15 (JT) 15–25 (F)	Unsteady effects (buffet, wing drop, wing rock)
Stall, Stall Break	20–30 (GA) 15–25 (JT) 25–35 (F)	First lift peak, loss of lift, porpoising, loss of longitudinal and directional stability, adverse yaw
Poststall	30–40 (GA) 25–40 (JT) 35–50 (F)	Departure, post-stall gyrations, incipient spin
Superstall	40–90 (GA, JT) 50–90 (F)	Second lift peak, deep stall, spin, supermaneuverability

coefficient at the point of stall, many fighter aircraft have no traditional stall break, as vortex lift from the wing's swept leading edge induces a more gradual transition to separated flow.

At *low angle of attack*, the airflow is attached over most of the surface, and small changes in aerodynamic angles or rates produce proportional changes in the integrated forces and moments. With increasing α, local regions of pressure- or shock-induced flow separation may form on the wing, producing unsteady *buffeting* forces. Buffet in the *prestall region* can be heavy enough to make reading panel instruments or performing precision tracking difficult, even if average stability and control effects are acceptable. Differential effects, including stalling of one wing before the other, hysteresis in the regions of separated flow, asymmetric bursting of vortices shed by strakes and the nose, and rapidly changing sidewash and downwash on downstream surfaces can lead to *wing drop* or *wing rock*. In the first instance, construction asymmetries, rolling motion, or the effect of a swirling propeller slipstream causes the angle of attack of one wing to be higher than the other's, stalling that wing and producing a roll-off that complicates stall recovery. In the second, periodically changing rolling moments are phased to produce an α-dependent limit cycle that degrades air-combat tracking or stall recovery.

In the *stall/stall break* region, the upper-surface flow separates, creating a number of effects, including loss of lift and change in the sensitivity

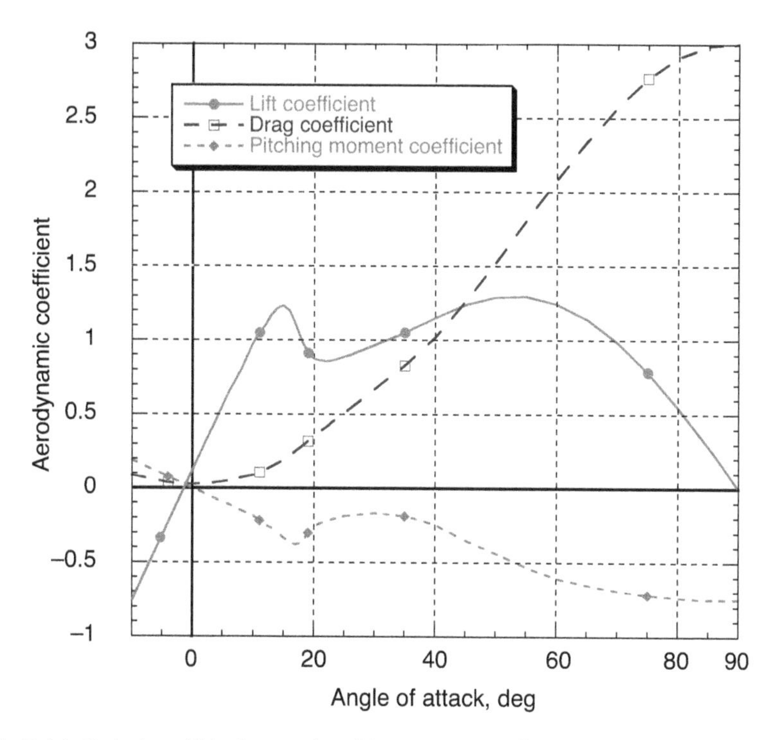

Fig. 7.4-1. Variation of lift, drag, and pitching-moment coefficients with angle of attack for a generic business jet aircraft.

of downwash angle to angle of attack. The break in lift may be particularly sharp for configurations with straight, high-aspect-ratio wings, while the change in downwash sensitivity that occurs as the wing wake sweeps across conventional tail surfaces affects their stabilizing influence, causing a pitch-up moment over some range of α (Fig. 7.4-1). Unless the aircraft is close to the ground, the changes in moments are of greater concern than the loss of lift, as angular stability and control are critical. The smallest asymmetry may lead one wing to drop before the other, producing a rolling event that must be controlled quickly. There may be buffet and hysteresis in flow separation and reattachment that leads to loss of effective damping of the short-period or Dutch roll modes, allowing *porpoising* or wing rock to occur as the trim equilibrium bifurcates to a limit cycle.

Aircraft with highly swept wings, strakes, or leading-edge extensions may not evidence a sharp stall break in lift coefficient (Fig. 7.4-2) [N-2]; however, the lateral-directional variations may still be large (Fig. 7.4-3)

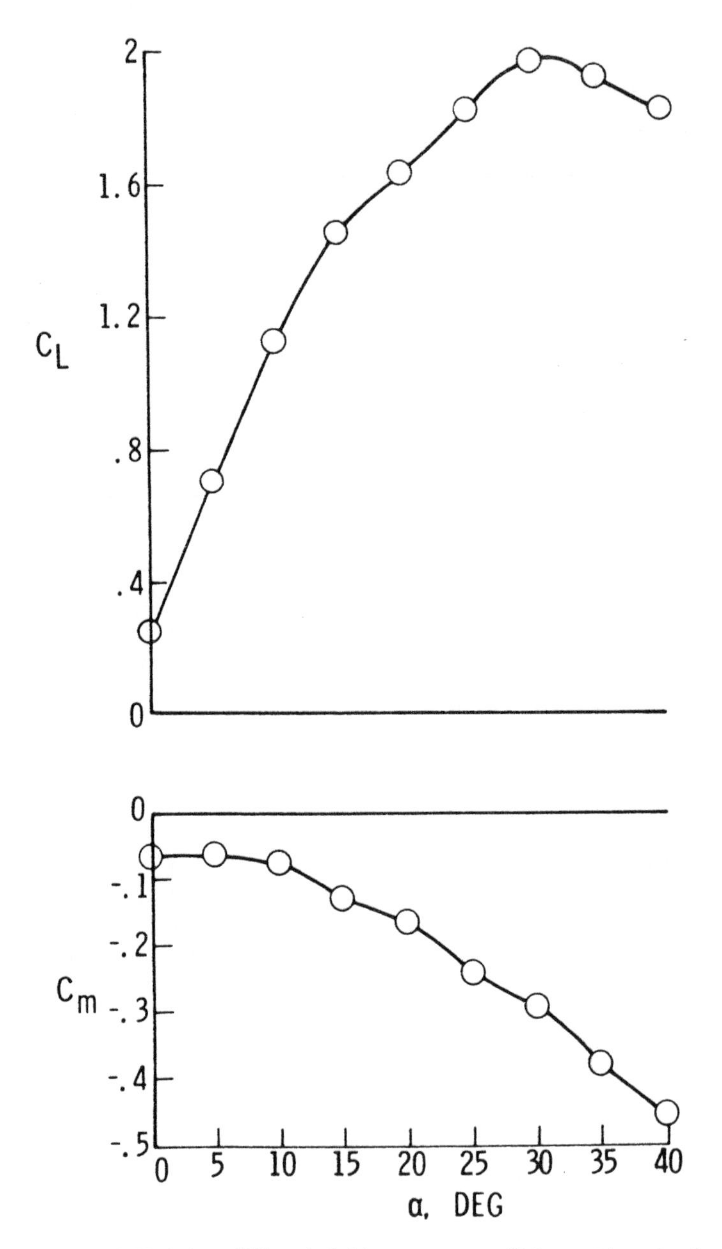

Fig. 7.4-2. Variation of lift and pitching-moment coefficients with angle of attack for a typical fighter aircraft configuration. (from [N-2], courtesy of NASA)

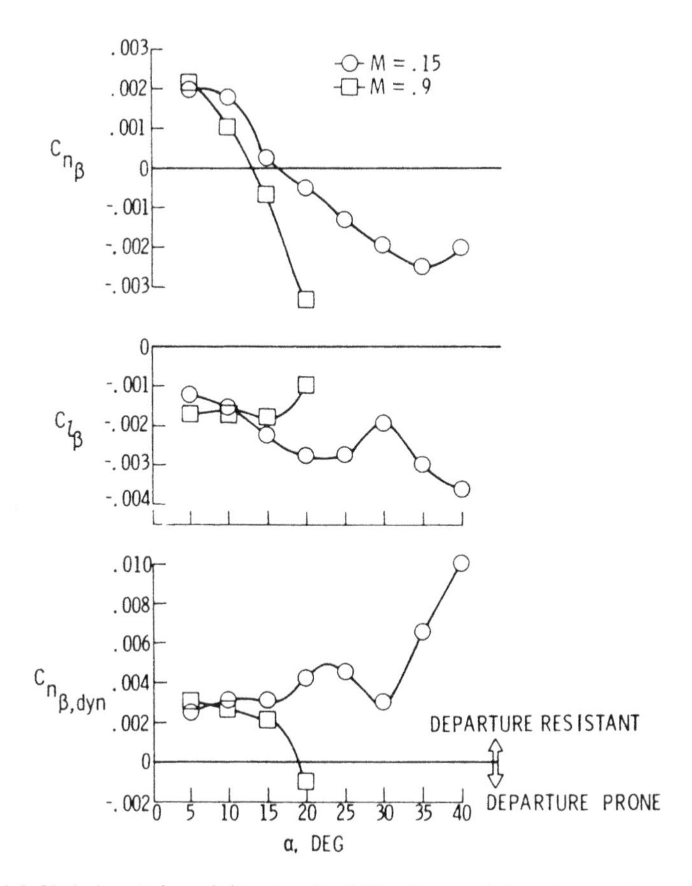

Fig. 7.4-3. Variations in lateral-directional stability characteristics of a typical fighter aircraft configuration, including $C_{n_{\alpha\text{dyn}}}$. (from [N-2], courtesy of NASA)

[N-2]. With increasing angle of attack and Mach number, this configuration experiences a loss of directional stability, as C_{n_β} changes sign, and yaw response to lateral (differential stabilator) control becomes adverse (Fig. 7.4-4) [N-2]. Over the same α range, the yaw moment due to rudder decreases with both angle of attack and Mach number, although it does not change sign (Fig. 7.4-5) [N-2].

Linear models reveal the potential for stability and control difficulty, as reflected by the "C_{n_β}-dynamic" and the lateral control divergence parameter (LCDP). Both are normally expressed in terms of body-axis

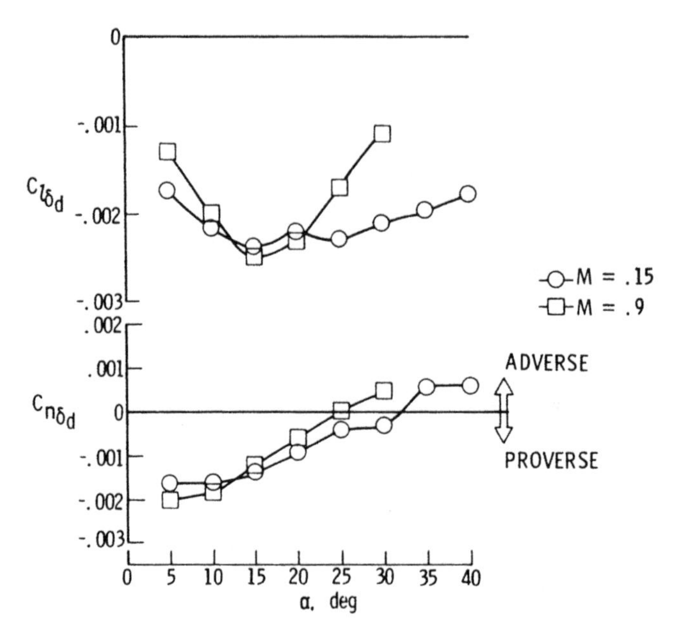

Fig. 7.4-4. Variations in effect of lateral control (differential stabilator δ_d) for a typical fighter aircraft configuration. (from [N-2], courtesy of NASA)

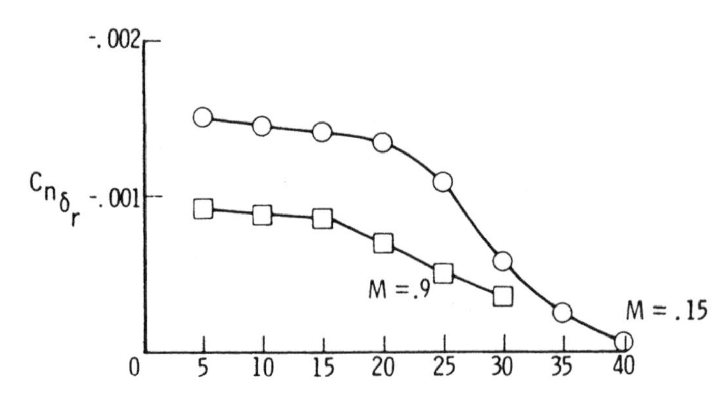

Fig. 7.4-5. Variations in effect of directional control (rudder δ_r) for a typical fighter aircraft configuration. (from [N-2], courtesy of NASA)

stability and control derivatives that vary with the trim angle of attack. The $C_{n_{\beta}}$ dynamic is

$$\begin{aligned} C_{n_{\beta_{dyn}}} &\triangleq C_{n_{\beta}} \cos\alpha_o - \frac{I_{zz}}{I_{xx}} \sin\alpha_o C_{l_{\beta}} \\ &\approx C_{n_{\beta}} - \frac{I_{zz}}{I_{xx}} \alpha_o C_{l_{\beta}} \end{aligned} \tag{7.4-1a}$$

and it can be written in terms of the dimensional stability-axis derivative N_{β_s}(eq. 6.1-19) as

$$C_{n_{\beta_{dyn}}} \triangleq N_{\beta_s} / S\bar{q}b \tag{7.4-1b}$$

Consequently, $C_{n_{\beta_{dyn}}}$ portrays the static stability of the Dutch roll mode. If $C_{n_{\beta_{dyn}}}$ is negative, as is the case for $\alpha > 18$ deg in Fig. 7.4-4, there is a good possibility of *nose slice* departure from controlled flight. One solution to the problem is to augment Dutch roll static stability using closed-loop control, with gains scheduled on the angle of attack.

A negative value of the LCDP criterion

$$\text{LCDP} \triangleq C_{n_{\beta}} - \frac{C_{n_{\delta A}}}{C_{l_{\delta A}}} C_{l_{\beta}} \tag{7.4-2}$$

predicts reversal of the rolling response to lateral control, such as aileron, δA, or the differential stabilator deflection shown in Fig. 7.4-4. This aircraft's LCDP becomes negative for $\alpha > 23$ deg at $\mathcal{M} = 0.15$ deg and for $\alpha > 15$ deg at $\mathcal{M} = 0.9$ (Fig. 7.4-6) [N-2]. With conventional controls, the pilot would be forced either to reverse lateral control inputs in these regimes or to roll the aircraft through the combined effect of rudder deflection, sideslip angle, dihedral effect $C_{l_{\beta}}$, and the rolling moment produced by rudder deflection directly.

The rationale for LCDP is most easily understood by referring to the truncated Dutch roll/roll model of Section 6.4 (eq. 6.4-8). Defining the corresponding control-effect matrix as

$$\mathbf{G}_{\text{DRR}} = \begin{bmatrix} N_{\delta A} & N_{\delta R} \\ 0 & 0 \\ L_{\delta A} & L_{\delta R} \end{bmatrix} \tag{7.4-3}$$

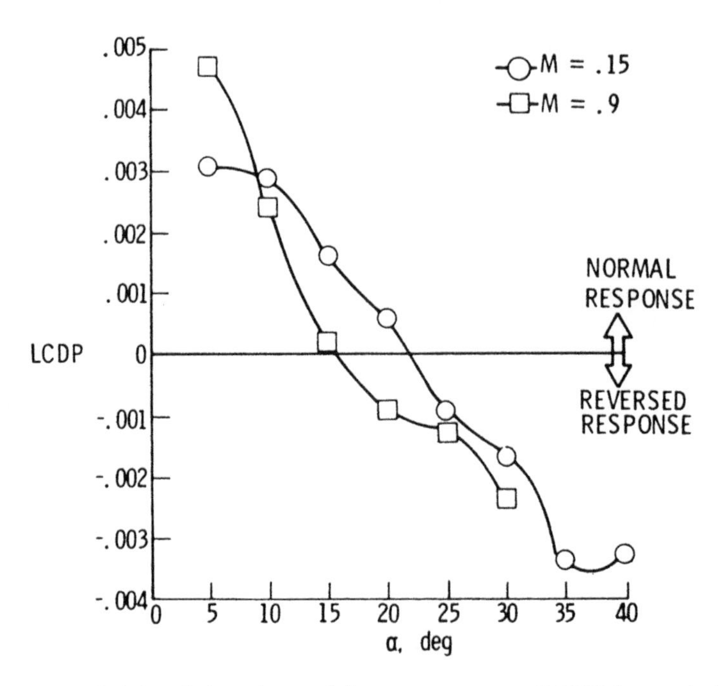

Fig. 7.4-6. Variations in lateral control divergence parameter (LCDP) for a typical fighter aircraft configuration. (from [N-2], courtesy of NASA)

the steady-state roll-rate response Δp^* to constant aileron command $\Delta\delta A^*$ is

$$\frac{\Delta p^*}{\Delta\delta A^*} = \frac{N_{\delta A}(L_\beta - L_r Y_\beta/V_o) - L_{\delta A}(N_\beta + N_r Y_\beta/V_o)}{L_p N_\beta - L_\beta N_p - L_r N_p Y_\beta/V_o + L_p N_r Y_\beta/V_o} \tag{7.4-4}$$

The equation numerator contains the dimensional equivalents of eq. 7.4-2 when Y_β/V_o is ignored. A change in sign of this ratio predicts a reversal of rolling response to lateral control. As presented in Section 6.4, eq. 7.4-4 is a stability-axis equation; however, the same form applies when the derivatives and the roll rate are expressed in body axes.

Aileron-rudder interconnects (ARI) are implemented on most fighter aircraft to prevent roll reversal at high angles of attack. With the simple expedient of deflecting the rudder in proportion to aileron deflection,

$$\Delta\delta R = k\,\Delta\delta A \tag{7.4-5}$$

the steady-state roll response becomes

$$\Delta p^* = \frac{(N_{\delta A} + kN_{\delta R})(L_\beta - L_r Y_\beta / V_o) - (L_{\delta A} + kL_{\delta R})(N_\beta + N_r Y_\beta / V_o)}{L_p N_b - L_\beta N_p - L_r N_p Y_\beta / V_o + L_p N_r Y_\beta / V_o} \Delta \delta A^* \tag{7.4-6}$$

As long as the aileron and rudder moments are linearly independent, k can be scheduled with angle of attack to reduce the likelihood of control-induced departure.

The distinction between body and stability axes is especially important for flight at high aerodynamic angles. If an aircraft that is flying at high angle of attack and zero sideslip angle is rolled about its body x axis, the angle of attack is converted geometrically to sideslip angle in what is called *kinematic coupling*. With instantaneous 90 deg roll about the body axis, α and β are interchanged completely, and large transient motions about pitching and yawing axes would be induced. Conversely, if the aircraft was being held at nonzero sideslip angle, as in an air-combat pointing task, β would be converted to α, and track might be lost.

If the aircraft is rolled about its velocity vector—a so-called *stability-axis roll*, though "wind-axis roll" is more accurate—perturbations to the initial α and β are reduced, and the lift vector is efficiently reoriented to change the aircraft's flight path with smaller transients. This is what a well-designed ARI, aided by suitable closed-loop control, is intended to do. From the pilot's perspective, the advantages of the stability-axis roll may not be obvious, as the rolling motion is not rolling in his or her cockpit-oriented reference frame. The roll appears to have a large yawing component, and the rotational axis may not be in the line of sight. Furthermore, considerable side force can be generated if the cockpit is located far ahead of the center of mass, as in large fighter aircraft, degrading the sense of coordination normally associated with efficient turning flight.

An additional phenomenon associated with stability-axis roll is the *dumbbell effect*, an inertially induced pitch-up moment that can increase the angle of attack. The name derives from the behavior of a dumbbell (two masses on the opposite ends of a shaft) that is rotated about the mid-point of the shaft. In the absence of external moments, steady rotation can occur either along the shaft axis (the axis of minimum moment of inertia) or perpendicular to it. However, only the latter equilibrium is stable, so a rotating dumbbell gravitates toward a "flat spin" about a transverse axis. Most fighter aircraft would be characterized as "fuselage heavy," with higher moment of inertia about the z axis than about the x axis; therefore, the shaft of an equivalent dumbbell would be aligned with the x axis. Assuming negligible product of inertia I_{xz}, the inertial contribution to pitching acceleration (eq. 3.2-50) is

$$\dot{q} = \frac{(I_{zz} - I_{xx})pr}{I_{yy}} \tag{7.4-7}$$

where p and r are measured in body axes. With a stability-axis roll rate of p_S, the body-axis roll and yaw rates are $p \approx p_S \cos\alpha$ and $r \approx p_S \sin\alpha$; hence, the pitching acceleration due to roll rate is

$$\dot{q} = \frac{(I_{zz} - I_{xx})}{2I_{yy}} p_s^2 \cos 2\alpha \tag{7.4-8}$$

Given $I_{zz} > I_{xx}$, the acceleration is always positive because the effect depends on the square of the roll rate. The inertial effect must be countered by aerodynamic control to prevent angle-of-attack increase. A similar directional effect can occur during a *rolling pullup* maneuver, where the pq product in eq. 3.2-51 induces a yawing acceleration $\dot{r}$.

At intermediate angles of attack, control effects may not be easily predictable because the effects of small pressure variations or gradients and flow interference from adjacent surfaces are not merely additive. Examples are given in Fig. 7.4-7 for an unspecified aircraft type [N-1], where three results of control surface deflection are shown. The first graph illustrates the effect of sideslip angle on the incremental pitching moment coefficient ΔC_m, due to horizontal tail deflection δ_h. While the control effect is not linear, the nonlinear variation with δ_h is reminiscent of the isolated control surface effects predicted by Newtonian theory in Section 2.4. The second graph displays the yawing moment increment ΔC_n, predicted by simply adding the separate effects of two control surfaces with the effect measured in a wind tunnel. There is good correspondence at low and high α, but not at intermediate angles. The third graph shows that deflecting the horizontal tail for pitch control has a large destabilizing effect on directional stability $C_{n\beta}$. In the vicinity of a 30 deg angle of attack, a pitch-up command could contribute to a nose slice.

Combinations of high α and β lead to large variations in aerodynamics as well (Fig. 7.4-8) [N-1]. At $\alpha = 40$ deg, the C_m variation due to sideslip angle is roughly symmetric and destabilizing. At $\alpha = 80$ deg, $C_{n\beta}$ is negative (unstable), and there is a large bias in C_n; otherwise, the sideslip effect is approximately antisymmetric and stabilizing (i.e., $C_{n\beta}$ becomes less negative). Airplanes of all types rarely fly with such large sideslip angles, and aerodynamic data are often not available for β beyond a few degrees, complicating the prediction of *poststall gyrations* (irregular, nonrepeatable motions, like those of a falling leaf) or *fully developed spin.*

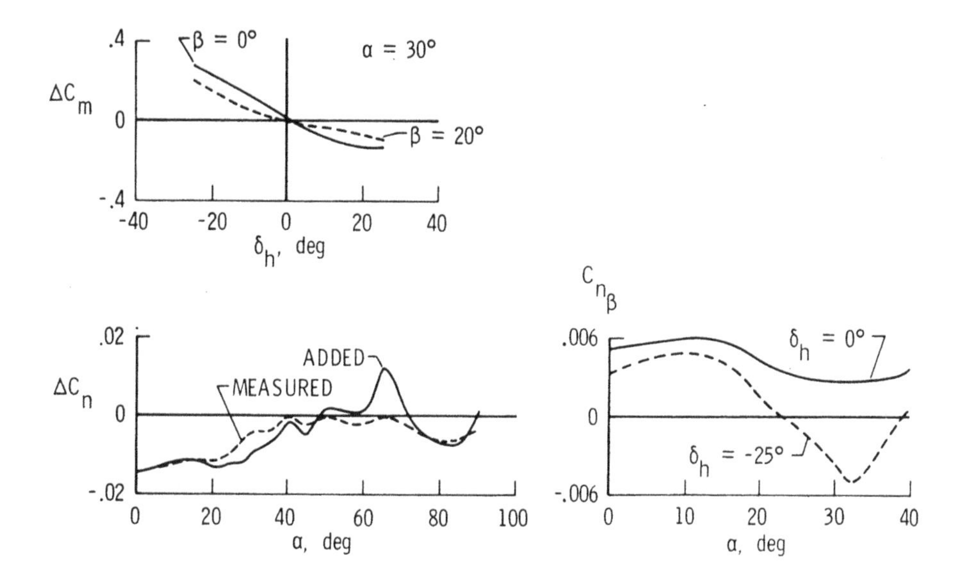

Fig. 7.4-7. Some nonlinear control effects on yawing- and pitching-moment coefficients. (from [N-1], courtesy of NASA)

An even more perplexing problem is that static aerodynamic coefficients may not be reliably predicted at high angles of attack; hence, computed flight paths may not correlate well with flight test results. Data collected using different wind tunnel models of the same aircraft may differ, and the same model may produce different results from one run to the next. The problem is especially acute for aircraft with long, slender noses, which generate bound vortices emanating from the tip (Fig. 7.4-9) [C-8]. At low α and zero β, the vortices are symmetric, and they produce no net yawing moment. At intermediate angles, one vortex is closer to the nose than the other, and different pressures on the left and right sides of the nose result; a significant yawing moment is produced, even when the sideslip angle is zero. Once the vortex pattern is established, it remains until the angle of attack is changed; however, on repeating the condition, the opposite pattern may occur, producing the opposite yawing moment. The pattern that actually occurs may be a function of small differences in initial condition or imperfections in the nose surface. Figure 7.4-10 illustrates the variability in zero-β yawing effect that is obtained for four different models of a single variable-sweep configuration (with sweep angle Λ of 50 deg) [A-6]. While the C_n excursions all lie within the same angle-of-attack region, their magnitudes and details are markedly different.

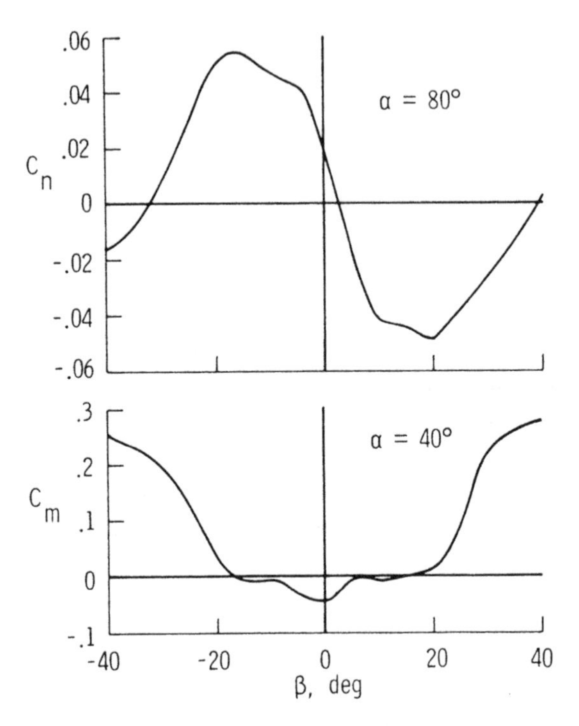

Fig. 7.4-8. Variation of yawing- and pitching-moment coefficients with sideslip angle. (from [N-1], courtesy of NASA)

Examples of zero-sideslip yawing moment coefficient for three aircraft types are shown in Fig. 7.4-11 [N-1]. Note that the two configurations with the most slender forebodies have the greatest extremes in C_n. These extremes exceed the ΔC_n that can be generated by maximum deflection of the rudder, so they are not controllable. The zero-sideslip yawing moment is strongly affected by the cross-sectional shape of the nose, as well as its fineness ratio; an elliptical cross section with horizontal major axis has smaller C_n excursions than one with vertical major axis. Alternatively, horizontal nose strakes have similar beneficial effect, as does the air-data nose boom frequently used during flight testing.

Aerodynamic damping is strongly affected by changes in angle of attack, as demonstrated in the next three figures. Figure 7.4-12 presents the combined effects of body angular rate and aerodynamic rate on moment derivatives from forced-oscillation wind tunnel tests of a wing-body configuration at $\mathcal{M} = 0.6$ [O-3]. In this example, pitch and yaw damping effects become stronger with increasing angle of attack. Although not

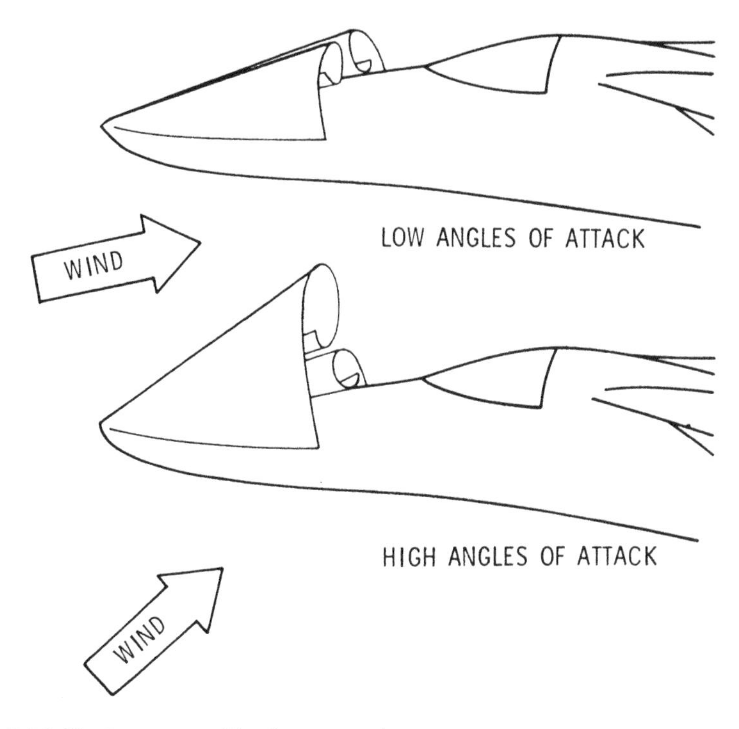

Fig. 7.4-9. Vortices generated by sharp nose. (from [C-7], courtesy of NASA)

differentiated from q and r effects here, $\dot{\alpha}$ and $\dot{\beta}$ effects can be significant in the stall region. Experimental damping results may well depend on the amplitude of oscillation: Fig. 7.4-13 illustrates that the roll damping of a fighter configuration may even change sign in the stall region as oscillation magnitude increases, bringing opposing wings in and out of stall [C-7]. Aerodynamic hysteresis in the stall region, as suggested by Fig. 7.4-14 [T-11], may also affect directional damping. C_n switches from one branch to the other during oscillation, creating a phase lag in the moment.

At low angle of attack, the slope of the pitching-moment coefficient C_m is negative for conventionally balanced aircraft; however, it may become positive at intermediate angle of attack, then negative again after passing through a second peak (Fig. 7.4-1). With no control input, there is a single equilibrium where C_m crosses the horizontal axis (in this case, at $\alpha = 0$ deg), and it is stable because the slope is negative. Negative deflection of an aft elevator translates the C_m curve up, or aft movement of the center of mass rotates the curve up, and at some point there may be

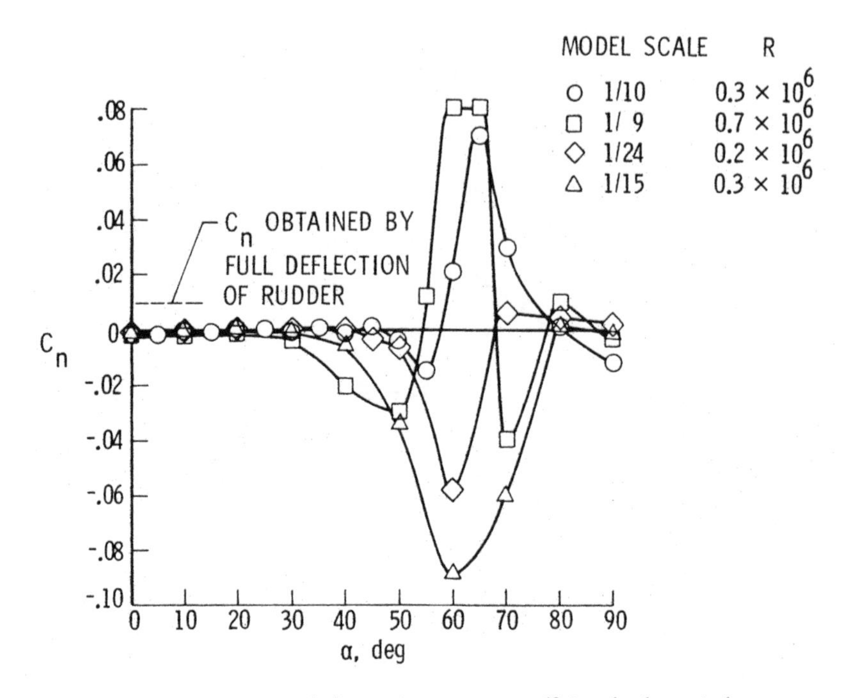

Fig. 7.4-10. Variation in zero-sideslip yawing-moment coefficient for four wind-tunnel models of the same configuration. (from [A-6], courtesy of NASA)

not one but three equilibrium angles of attack. The outer values are stable, and the inner one is unstable because its slope is positive. For any initial angle of attack greater than the middle value, C_m is positive, and the aircraft will rotate up to the high-α equilibrium, entering a *deep stall*. Without a very high thrust-to-weight ratio, the aircraft would develop a high rate of descent. If the $-\Delta C_m$ due to elevator cannot exceed the peak positive C_m, the aircraft cannot be returned to low-α flight by simply repositioning the pitch control. Pumping the elevator in phase with a short-period oscillation could rock the aircraft back to an angle of attack below the middle α, allowing capture by the low-α equilibrium; however, additional time and altitude are lost in this maneuver, and recovery is not assured.

Aircraft that are statically unstable at low angle of attack may be statically stable at high α, as the aerodynamic center shifts aft (Fig. 7.4-15) [N-2]. Such aircraft rely on closed-loop control for stable flight at low angle of attack and have adequate control effect for maneuvering; however, the full limits of pitch control may not be sufficient to break

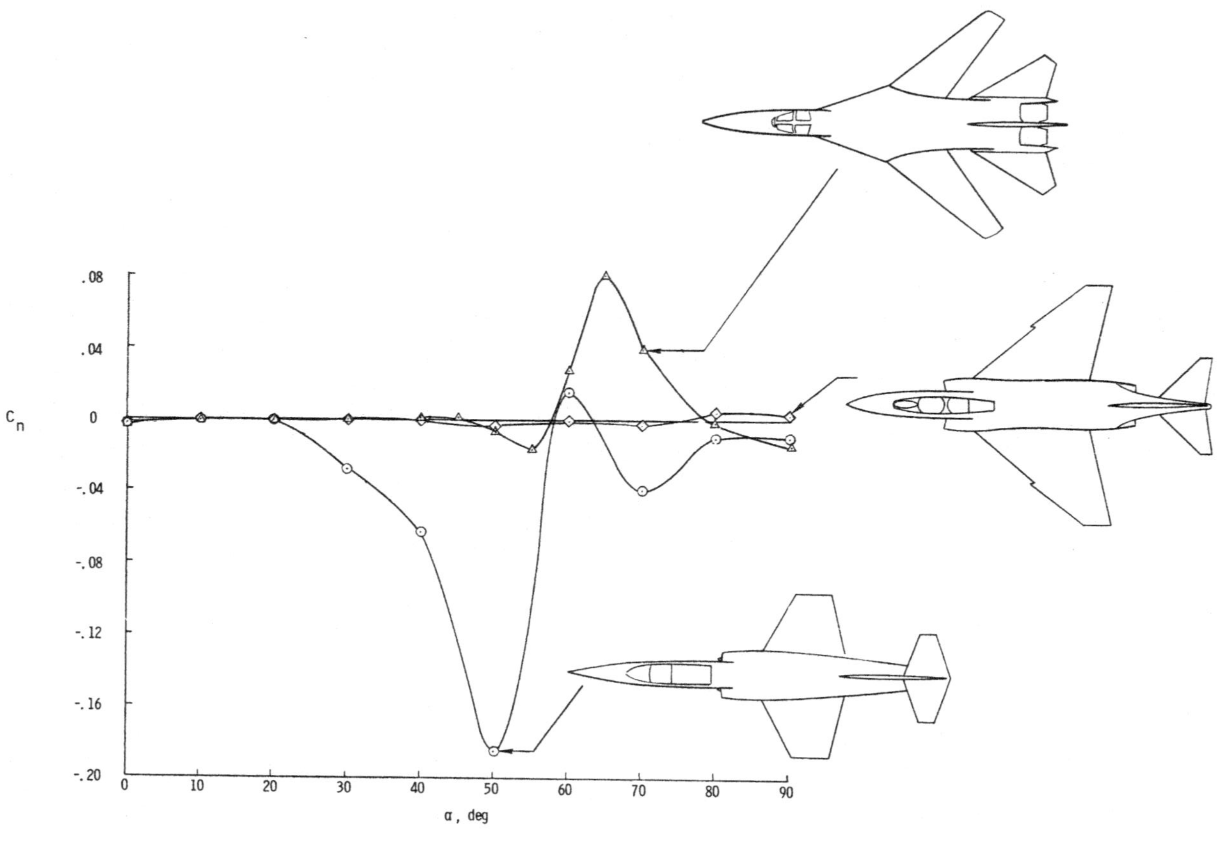

Fig. 7.4-11. Yawing-moment coefficients for three aircraft types at zero sideslip angle. ([N-1], courtesy of NASA)

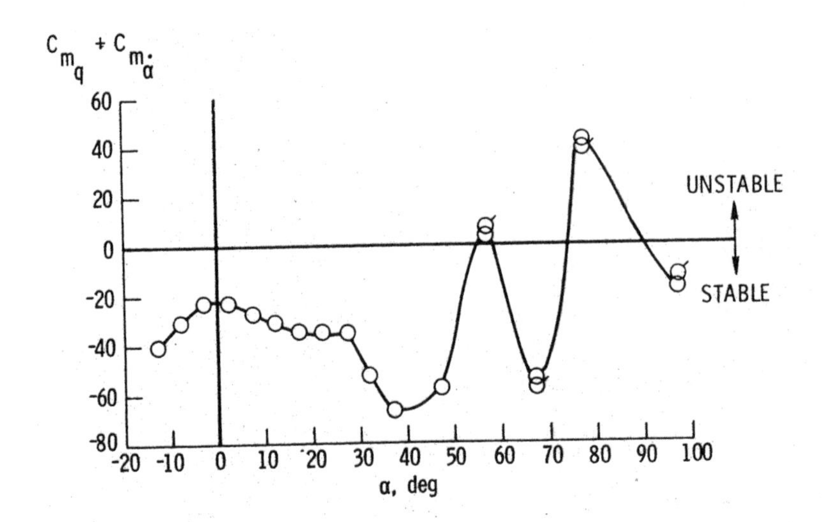

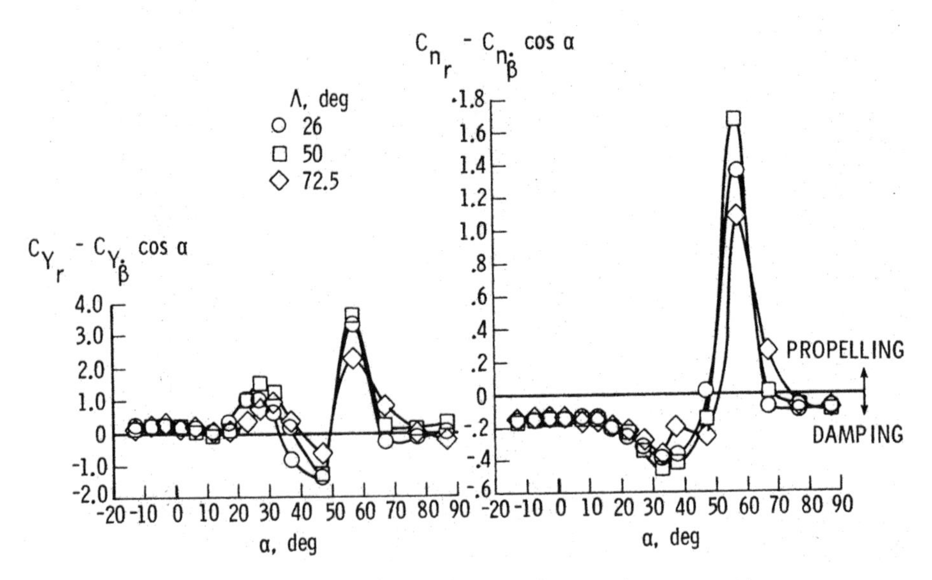

Fig. 7.4-12. Typical variation in pitching and yawing damping derivatives with angle of attack. ([O-1], courtesy of AIAA)

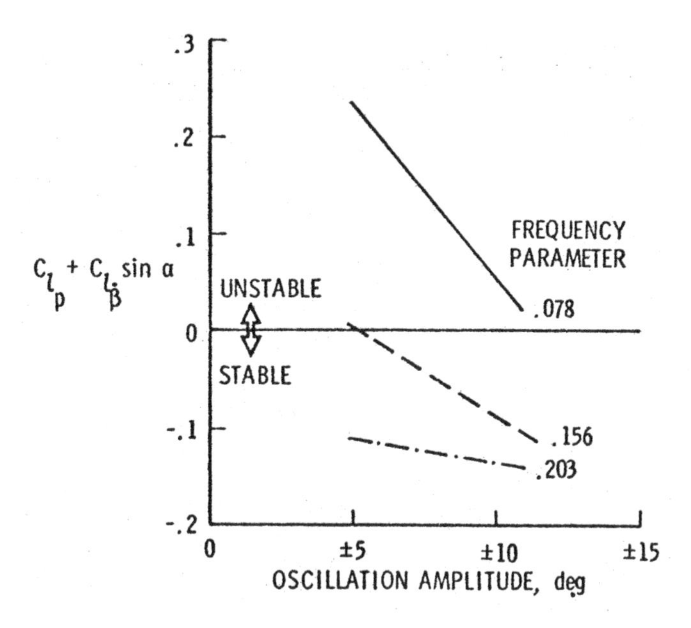

Fig. 7.4-13. Effect of oscillation amplitude on damping in roll with angle of attack. (from [O-3], courtesy of AIAA)

through the peak C_m if the flight condition enters the deep stall region. For this reason, the aircraft is limited to maneuver below some critical angle of attack α_{crit} that is defined by the maximum pitch-down controllability of the aircraft.

Most fighters are equipped with α-limiting control systems to reduce the likelihood of departure, and high-α air-combat maneuvering occurs below the lift peak. Examples of such maneuvers include the wind-up turn, a high-g, maximum-thrust turn with increasing α, and the rolling reversal, a high-g, rolling, looping climb to reverse the direction of flight. Some high-performance aircraft can execute poststall maneuvers using conventional controls, such as the impressive Pugachev Cobra maneuver, where the aircraft is pitched up to a maximum α of more than 90 deg in symmetric flight. However, controlled air-combat maneuvering up to angles of attack of 90 deg almost certainly requires thrust vectoring [H-7]. With fully separated flow on the leeward side of the wing, lift is low but drag is quite high, and the time spent at high α is short. Similar to the rolling reversal, the *Herbst maneuver* allows an aircraft to reverse its flight direction in minimum time by pulling up to high α and increased altitude, rolling about the reduced-magnitude velocity vector (requiring a

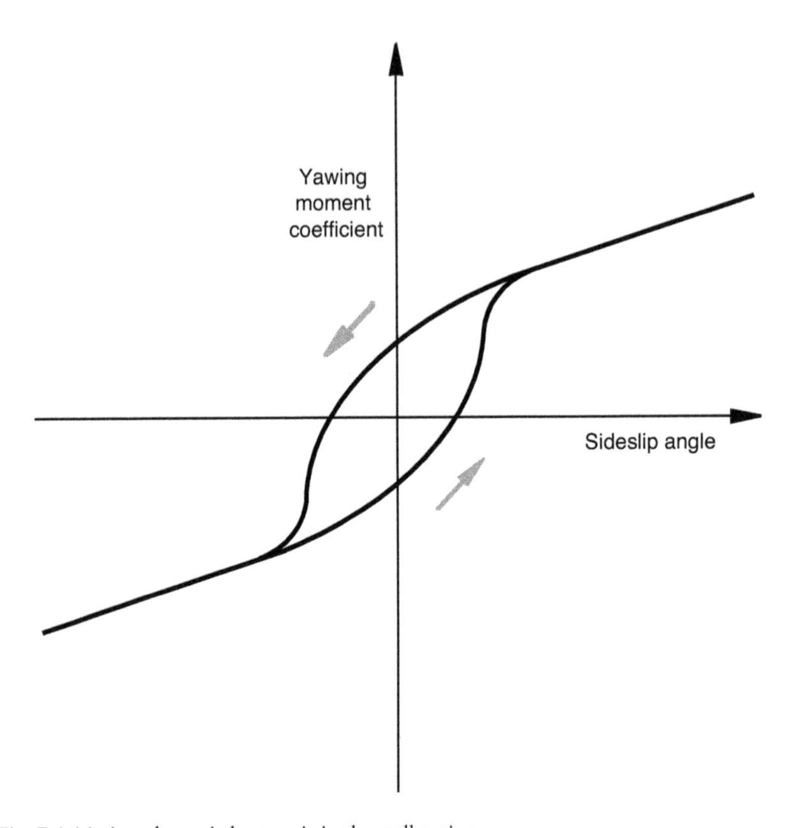

Fig. 7.4-14. Aerodynamic hysteresis in the stall region.

high degree of body-axis yaw), and swooping down to the original altitude and airspeed in opposite direction.

The space shuttle reenters the atmosphere from orbit with an α of 40 deg, allowing flight energy to be dissipated as heat without exceeding the temperature limits of leeward surfaces. The vehicle performs stability-axis rolls for range and cross-range control, then pitches down to arrive at a runway for an unpowered landing at a low angle of attack.

The aerodynamics of fully developed spin are complicated by the rotary flow field that the aircraft encounters. In a steady spin, the aircraft's center of mass follows a helical path about a vertical axis, and the Euler-angle yaw rate is constant. As a consequence, the angle of attack, sideslip angle, and airspeed vary from left to right wing tip and from nose to tail. In the spin of a typical general aviation airplane, the geometric nose-to-tail variation in airspeed and sideslip angle are 37 to 41 m/s and -5

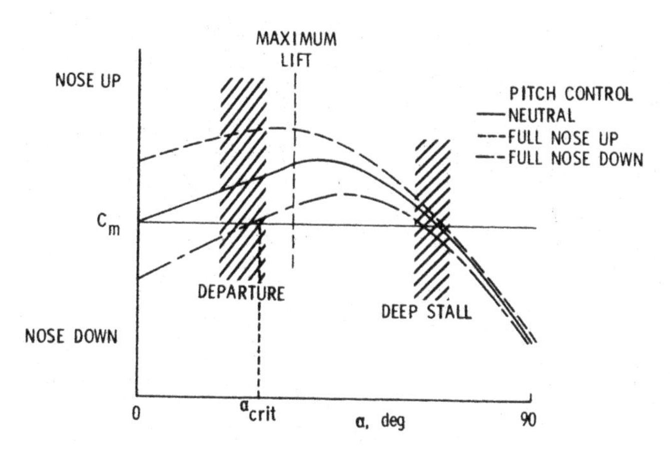

Fig. 7.4-15. Pitching moment variation with angle of attack and control setting for a statically unstable aircraft. (from [N-2], courtesy of NASA)

to −20 deg, while the left-to-right wing tip angle-of-attack variation is 35 to 65 deg [I-2]. Conventional wind-tunnel measurements with a model supported by a nonmoving sting are made in a geometrically linear flow field and cannot be transformed to this rotating system. Wind-tunnel measurements can be obtained using a rotary balance, which mimics the spin by rotating the aircraft model in a conventional wind tunnel [B-8]. Flight test data from spinning tests of actual or scale-model aircraft can be processed to identify aerodynamic coefficients [T-5]. The complex flow field and changing regions of separation present large impediments to coefficient estimation by computational fluid dynamics; strip theory has been applied to the problem, and it is likely that solutions via the Navier-Stokes equations will be developed.

Fully Developed Spins

Spins normally are classified as flat or steep, upright or inverted, steady or oscillatory. Steep spins typically occur at mean angles of attack between the first and second lift peaks, while flat-spin angle of attack is beyond the second peak. Given satisfactory data, steady spinning characteristics can be predicted by finding equilibrium solutions to eqs. 3.2-45–3.2-47 and 3.2-49–3.2-51. The first three equations describe the force balance that determines the helix angle, spin radius, and vertical velocity; however, it is more useful to resolve them into earth-fixed coordinates.

Assuming that the side force is zero, the total (or resultant) aerodynamic force coefficient is

$$C_R = \sqrt{C_L^2 + C_D^2} \tag{7.4-9}$$

The helix angle γ is expressed in terms of the spin radius R, spin rate Ω, and vertical velocity V [T-8]:

$$\gamma = \tan^{-1} \frac{\Omega R}{V} \tag{7.4-10}$$

The force equations can be written as

$$C_R \bar{q} S \sin \alpha = mg \tag{7.4-11}$$

$$C_R \bar{q} S \cos \alpha = m\Omega^2 R \tag{7.4-12}$$

leading to a relationship between spin radius and rate:

$$R = \frac{g \cot \alpha}{\Omega^2} \tag{7.4-13}$$

Subject to these conditions and to the constraint that

$$\Omega^2 = p^2 + q^2 + r^2 \tag{7.4-14}$$

the moment balance of the second three equations determines the aerodynamic and Euler angles of the spin. From [B-9], the angular rates can be expressed as

$$p = \Omega \cos \alpha \cos (\beta + \gamma) \tag{7.4-15}$$

$$q = \Omega \sin (\beta + \gamma) \tag{7.4-16}$$

$$r = \Omega \sin \alpha \cos (\beta + \gamma) \tag{7.4-17}$$

With zero product of inertia, the moment equations are

$$0 = L + (I_{zz} - I_{yy}) rq \tag{7.4-18}$$

$$0 = M + (I_{zz} - I_{xx}) pr \tag{7.4-19}$$

$$0 = N + (I_{xx} - I_{yy}) pq \tag{7.4-20}$$

where the aerodynamic moments (L, M, N) are derived from rotary balance data.

These equations can be solved in a number of ways, including the functional minimization described in Section 3.6. The stability of the equilibrium and the likelihood of significant oscillatory behavior can be evaluated by computing the eigenvalues of the locally linearized model for the equilibrium spin condition (Section 4.1). Appendix E identifies sources of rotary balance data for typical configurations.

While the interplay between aerodynamic and inertial moments in producing the spin is clear, the relative importance of the two once the spin is fully developed is dependent on the mass-density ratio of the aircraft, $\mu = m/\rho S\bar{c}$. The spin characteristics of an aircraft with low mass density and wing loading are dominated by aerodynamics, while those of an aircraft with high wing loading are dominated by inertial effects. In [L-1], it is shown that the angular rates of a spinning fighter aircraft maintain essentially constant kinetic energy $\mathcal{T}$, where

$$\mathcal{T} = \frac{1}{2}(I_{xx}p^2 + I_{yy}q^2 + I_{zz}r^2) \tag{7.4-21}$$

In this case, the magnitude of the angular rate vector is constrained to lie on the surface of a *Poinsot ellipsoid*, precessing along surface curves called *Polhodes* that are functions of the spin rate. Thus, the oscillations in p, q, and r are coupled, and their magnitudes are closely predicted by the applicable curve. The difference in the actual oscillatory spin traces for the aircraft and those generated by the corresponding Polhode are small perturbations driven by aerodynamics, particularly C_{l_β}, C_{n_β}, and C_{m_q}.

Spin recovery techniques are aircraft specific, but they are strongly affected by the aircraft's mass distribution (Fig. 7.4-16) [C-8]. A common goal should be to remove rotational kinetic energy from the spinning aircraft. For conventional, straight-wing configurations, the rudder is deflected to slow the spin rate, and the aircraft is pitched down to regain low α. Having established a symmetric descent, the aircraft is then brought back to level flight. Slender, fuselage-heavy aircraft may be recovered using an aileron yawing moment as well to break the spin. It is desirable for the rudder and vertical tail surface to be fully exposed to the airflow to generate strong yawing moments against the spin. Blocking of the high-α flow by the horizontal tail surface can reduce the speed of recovery. Nevertheless, wing design features, such as drooped outer leading edges for general aviation aircraft and leading-edge flaps for fighter aircraft, may be even more important in determining spin modes and recovery procedures.

Because spinning can be very disorienting, automatic systems for spin recovery have been proposed and tested. With the world seeming to turn at high rate and little time to act, pilots may have difficulty identifying the spin type and the proper recovery procedures. Proper control inputs may be hampered by a control system designed for departure prevention; for

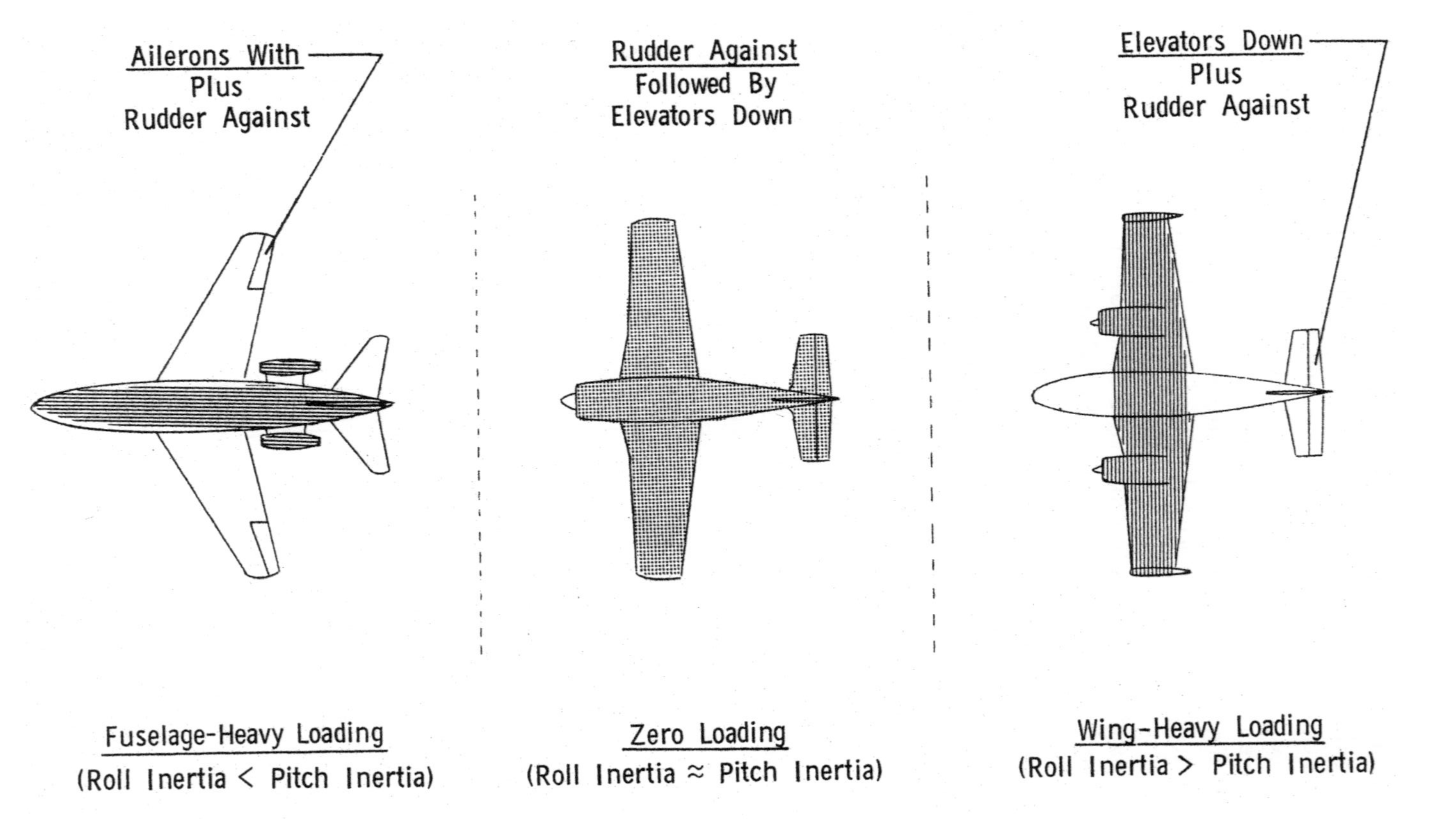

Fig. 7.4-16. Effect of mass distribution on primary spin recovery controls. (from [C-7], courtesy of NASA)

example, crossed controls may be called for, but the control system may not allow it. Nevertheless, most attention has been devoted to stall and departure prevention, as it is best not to enter the spin in the first place.

Simulated Motions of a Business Jet Aircraft

The generic business jet model presented in Appendix B provides a basis for examining some of the effects described in this section. Its static longitudinal characteristics for -10 deg $< \alpha <$ 90 deg are shown in Fig. 7.4-1, with the first lift peak at $\alpha = 15$ deg, the stall break minimum at $\alpha = 22$ deg, and the second lift peak at $\alpha = 55$ deg. The pitching-moment coefficient for center of mass at the quarter mean aerodynamic chord has negative slope until changing sign at $\alpha = 17$ deg, changing sign again at $\alpha = 30$ deg. The model's control surface effects are linear in their respective deflections, and static lateral-directional effects are linear in sideslip angle. The derivatives for both vary with angle of attack to the limits of the data presented in [S-12] and are constant at higher angle. Angular rate derivatives are constant throughout the angle-of-attack range. Numerical values for -10 deg $< \alpha <$ 0 deg are extrapolated from the positive-α data. The FLIGHT program presented in Appendix B is used for these simulations.

Beginning at two different trimmed airspeeds, the elevator is gradually deflected -12 deg (trailing edge up) from zero over five seconds, producing the pitchups shown in Fig. 7.4-17. With an initial airspeed of 102 m/s, the stabilator setting for trim is -1.96 deg, and the angle of attack is 4.12 deg. The maneuver begins at $t = 5$ s, with the pitch angle following a near-sinusoidal curve to a maximum value of more than 50 deg. Airspeed decreases and altitude increases during the next 15 s. The angle of attack climbs, levels off, then shoots up to a 45 deg peak at 23 s, a kinematic consequence of the aircraft falling through its maximum altitude with high pitch angle. Pitching stability is strong at the maximum angle of attack, and α quickly returns to 15 deg, avoiding deep-stall equilibrium. There is a short-period transient as the new equilibrium for $\delta E = -12$ deg is reached.

In the second case, the initial airspeed is 75 m/s, the stabilator setting for trim is -4.25 deg, and the angle of attack is 8.94 deg (Fig. 7.4-17). The α excursion is earlier and a little higher than before, but the pitch-angle excursion is less. In this case, the aircraft is captured in deep stall, producing an altitude sink rate of about 100 m/s. Because the deep stall is induced by up elevator, α can be returned to a low value by trailing-edge down elevator. For the figure, δE is returned to zero over five seconds beginning at $t = 35$ s, and the initial trim condition would eventually be

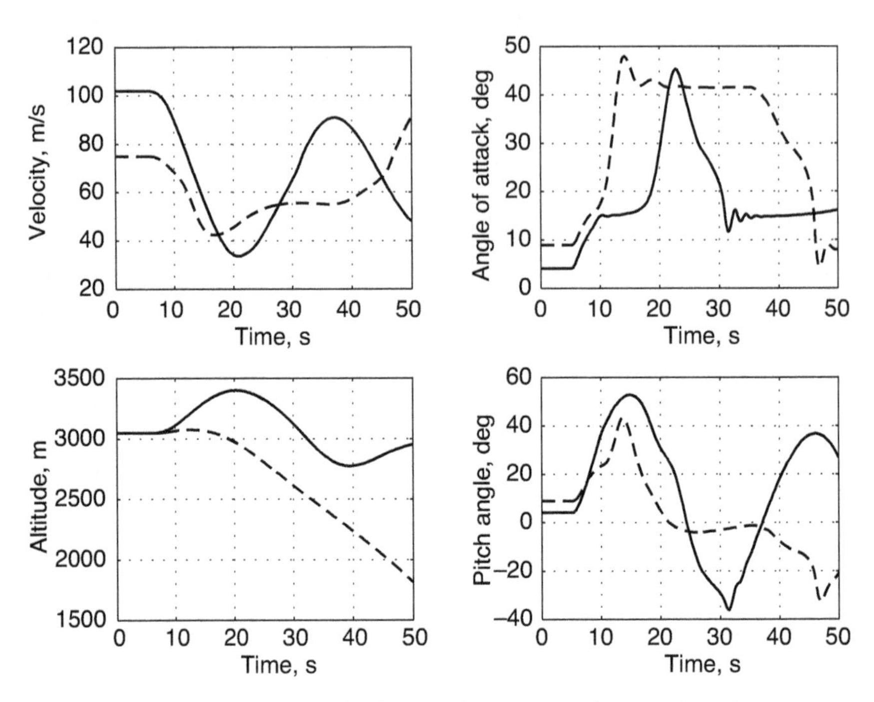

Fig. 7.4-17. Comparison of − 12 deg elevator change over 5 s for two trimmed flight conditions: $V = 102$ m/s and $h = 3{,}050$ m (solid), $V = 75$ m/s and $h = 3{,}050$ m (dashed).

restored. While the aircraft is in the deep stall, the aircraft attitude is nearly horizontal.

The effect of reduced static margin is presented in Fig. 7.4-18, with two examples at the 102 m/s flight condition. The first of these is the same as before, while the static margin is moved aft 20 percent for the comparison, with no change in the modeled elevator moment arm. The second case is retrimmed with $\delta S = 0.48$ deg and $\alpha = 3.6$ deg. This configuration is captured at the deep-stall equilibrium, not returning to low α on its own. At $t = 40$ s, the elevator is returned to zero over five seconds, but the control increment is too small to break the deep stall, and α settles at 36 deg. Although not shown, further pitch-down control (e.g., $\delta E = +12$ deg) is sufficient to return to low-α flight.

Small rudder deflections that are unopposed over time can lead to large roll-offs because this configuration has a slightly unstable spiral mode ($\tau = -112$ s, as in Section 6.4). The sideslip and roll-angle responses to $\delta R = 1$ and 2 deg are relatively linear (Fig. 7.4-19) in control setting.

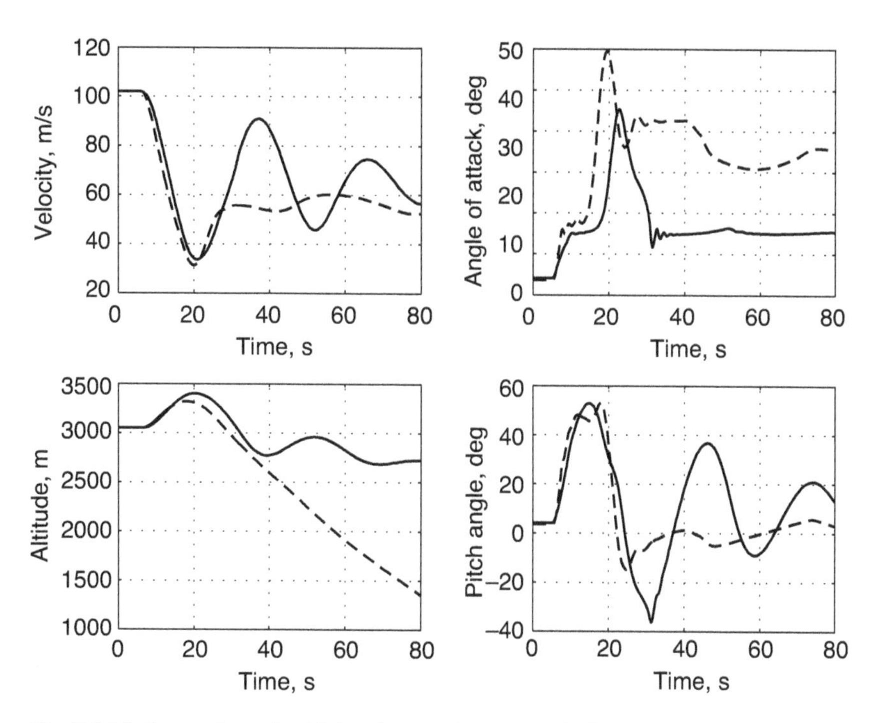

Fig. 7.4-18. Comparison of − 12 deg elevator change over 5 s for two static margins at $V = 102$ m/s and $h = 3{,}050$ m. Static margin = 25% mac (solid) or 5% mac (dashed).

Significant longitudinal response evolves as the roll angle becomes large, and, in the second case, the aircraft enters a steep dive; pitch angle reaches −37 deg and roll angle approaches −100 deg after 40 s. The velocity, altitude, angle-of-attack, pitch-angle, and roll-angle excursions due to $\delta R = 2$ deg are more than double those shown for $\delta R = 1$ deg.

The last two figures illustrate the effects of stability- and body-axis roll-rate initial conditions at two angles of attack. These are transient responses, not responses to step control inputs (i.e., δA and δR are zero). The first flight condition in Fig. 7.4-20 is the $V = 102$ m/s case considered before, while the second is trimmed at $V = 50$ m/s, with $\delta S = -5.94$ deg and $\alpha = 39.28$ deg. An initial condition of $p_S = 40$ deg/s is added to both. The throttle setting required for steady level flight at the second condition exceeds the maximum thrust of the model ($\delta T = 146.5$ percent) and is used only to illustrate angular motions at high α. The low-α responses for stability- and body-axis conditions (Fig. 7.4-21) are nearly the same, with little perturbation of pitching response over 10 s. For the high-α

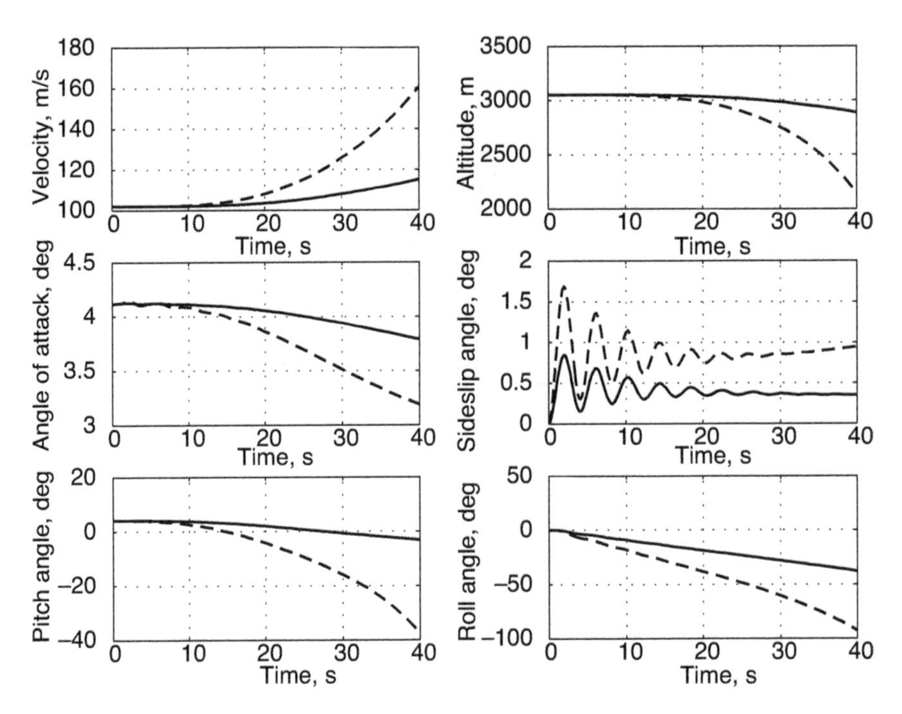

Fig. 7.4-19. Effect of 1 and 2 deg rudder deflections at $V = 102$ m/s, $h = 3{,}050$ m.

case, the roll-induced pitch-up is seen in angle-of-attack and pitch-rate traces, and the roll rate persists rather than going to zero (Fig. 7.4-20). In addition to the roll angle increasing, the pitch angle decreases and the yaw angle increases as the aircraft cones about the velocity vector. For the body-axis roll-rate initial condition (Fig. 7.4-21), the primary response is a large-amplitude rolling and yawing Dutch roll oscillation, with somewhat lower pitching response. Whereas the first high-α response could form the basis of a useful turning maneuver, the second would be far more difficult to control without rudder input.

Stability of High-Angle-of-Attack Maneuvers

While aerobatic aircraft can perform extreme maneuvers with conventional controls, effective maneuvering of modern fighter aircraft at high angle of attack almost certainly requires control augmentation, and proper piloting actions are critical for both types of aircraft. The manual maneuvering task is compounded beyond the conventional compensatory

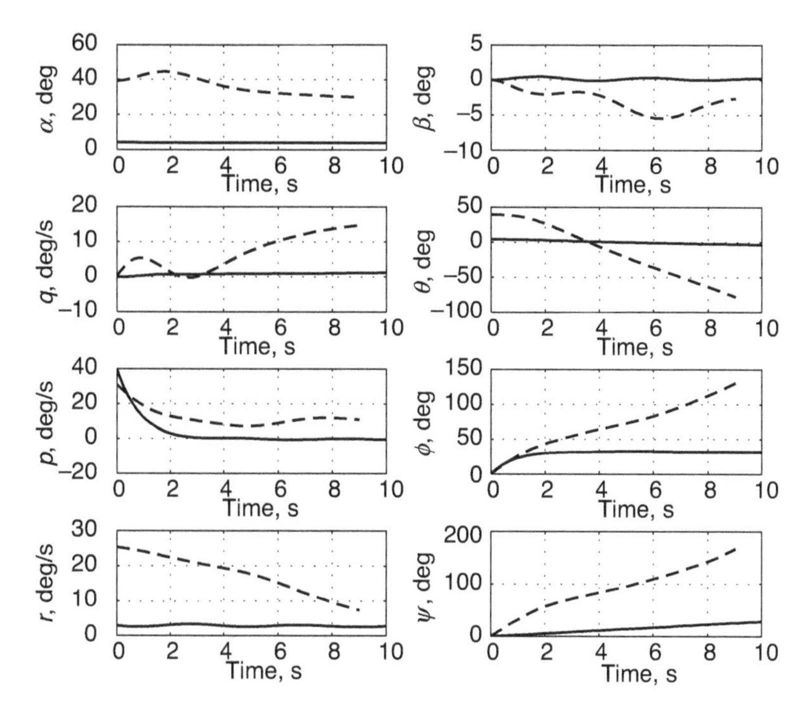

Fig. 7.4-20. Stability-axis roll-rate initial condition = 40 deg/s. Solid line represents aircraft trimmed at $V = 102$ m/s, $h = 3{,}050$ m. Dashed line represents aircraft trimmed at deep stall condition, $V = 50$ m/s.

tracking task because a complicated flight path must be followed. The pilot either is following a three-dimensional figure as a set maneuver or is tracking another aircraft, whose motion is unpredictable. He or she must plan a nominal strategy, command baseline control inputs, adjust to unanticipated motions of a target, and account for (possibly unstable) transient motions of his or her aircraft.

In order to be successful, the pilot must prevent more than momentary departure from controlled flight. We distinguish between *unforced* and *forced departures*. In the first instance, the pilot fails to stabilize an unstable aircraft, while in the second, the pilot destabilizes an otherwise stable aircraft. Because flight characteristics may be changing quickly and maneuver times are short, "stability" in the strict sense of approaching some limit set or point as time goes to infinity is not a particular concern. However, conventional stability metrics (e.g., eigenvalues of locally linearized models) that vary over the course of a maneuver are indicators

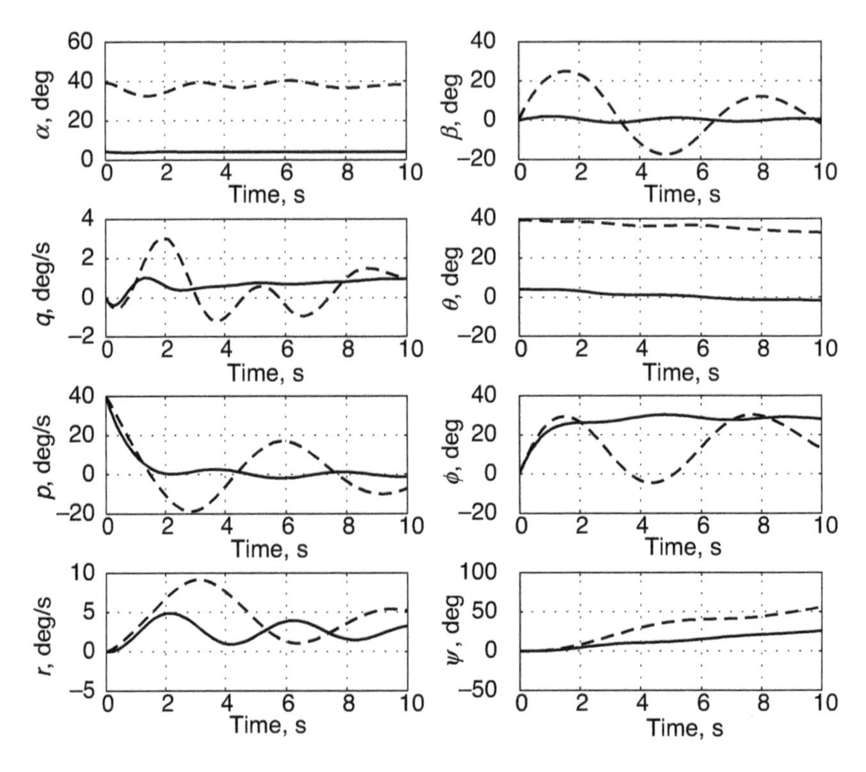

Fig. 7.4-21. Body-axis roll-rate initial condition = 40 deg/s. Solid line represents aircraft trimmed at $V = 102$ m/s, $h = 3{,}050$ m. Dashed line represents aircraft trimmed at deep stall condition.

of the changing sensitivities to small perturbations in state and control inputs that occur.

Two examples illustrate the aircraft state to be followed and the variation in stability characteristics that may occur during high-α maneuvers. Working points on a *windup-turn* flight path for a lightweight fighter (Aircraft A) are shown in Table 7.4-2 [S-16], while five points on a *rolling reversal* history for a twin-jet, variable-sweep fighter (Aircraft B) are given in Table 7.4-3 [S-18]. In both cases, there are periods of high-α flight with rapidly changing velocity and significant angular rates and attitudes. Angle of attack is continually increasing in the windup turn, and even with maximum thrust, Aircraft A must reduce airspeed and lose altitude up to the point that the maneuver is terminated. Angular rates and angle of attack are higher in the rolling reversal of Aircraft B, with extreme pitch angles and a 360 deg roll being encountered over a shorter period of time.

TABLE 7.4-2 WINDUP TURN HISTORY FOR AIRCRAFT A [S-16]

Time, s	V, m/s	α, deg	ϕ, deg	θ, deg	p, deg/s	q, deg/s	r, deg/s
0	217	5	45	5	5	ng	5
13	217	11	85	−15	ng	10	5
30	177	15	70	−20	ng	10	5
52	134	22	70	−20	10	15	10
75	116	27	60	−25	12.5	12.5	ng

ng indicates not given.

TABLE 7.4-3 ROLLING REVERSAL HISTORY FOR AIRCRAFT B [S-18]

Time, s	V, m/s	α, deg	ϕ, deg	θ, deg	p, deg/s	q, deg/s	r, deg/s
0	217	25	0	30	ng	15	ng
4	168	30	−90	50	−10	15	ng
10	107	30	−180	−30	−15	15	−5
15	117	23	−315	−55	ng	15	5
22	152	20	−360	−10	0	10	ng

ng indicates not given.

Thus, these maneuvers expose the aircraft to many of the potential aerodynamic problems discussed earlier.

The system eigenvalues computed at these working points give some indication of the time scale and stability of short-term response. Identifying modes by their most significant eigenvector components, the short period is well behaved, and, though they are briefly unstable, the spiral and phugoid modes are inconsequential due to their long time scales (Tables 7.4-4 and 7.4-5). The Dutch roll mode becomes dynamically unstable toward the end of both maneuvers, suggesting the possibility of wing rock, and the roll mode is briefly unstable during the rolling reversal maneuver of Aircraft B, suggesting possible uncommanded wing drop. The latter maneuver provides three instances of nonconventional modes, as well as two examples of real phugoid response.

PILOT-AIRCRAFT INTERACTIONS

The ability of pilots to cope with these maneuvers is investigated in [S-18, S-19], where an optimal-control pilot model is applied to the analysis. Following [K-5–K-7], the pilot model takes the form of a modified

TABLE 7.4-4 EIGENVALUES FOR WINDUP TURN OF AIRCRAFT A [S-16]

Time, s	Short Period	Dutch Roll	Roll	Spiral	Phugoid
0	$-0.9 \pm 1.8j$	$-0.4 \pm 3.4j$	-1.6	$+0.06$	$-0.5 \pm 0.1j$
13	$-1.0 \pm 2.2j$	$-0.3 \pm 4.4j$	-1.0	-0.04	$+0.0 \pm 0.2j$
30	$-0.8 \pm 1.7j$	$-0.2 \pm 4.0j$	-0.6	-0.07	$-0.0 \pm 0.2j$
52	$-0.5 \pm 2.7j$	$+0.08 \pm 3.3j$	-0.8	-0.1	$-0.0 \pm 0.3j$
75	$-0.4 \pm 2.2j$	$+0.2 \pm 2.6j$	-0.5	-0.2	$-0.1 \pm 0.2j$

TABLE 7.4-5 EIGENVALUES FOR ROLLING REVERSAL OF AIRCRAFT B [S-18]

Time, s	Short Period	Dutch Roll	Roll	Spiral	Phugoid
0	$-0.7 \pm 3.2j$	$-0.6 \pm 3.3j$	-1.3	$+0.1$	$(-0.05, -0.2)$
4	$-0.5 \pm 3.0j$	$-1.5 \pm 0.5j$	-0.5	$+0.2$	$+0.1 \pm 0.5j$
10	$-0.3 \pm 1.8j$	$-0.4 \pm 0.5j$	$+0.4$	-1.1^*	$(-0.4, 0.2)$
15	$-0.4 \pm 1.6j$	$-0.0 \pm 2.0j$	-0.9	$-0.2 \pm 0.4j^\dagger$	$(-0.1)^\ddagger$
22	$-0.7 \pm 1.4j$	$+0.1 \pm 2.8j$	-1.2	-0.07	$-0.1 \pm 0.1j$

* Sideslip-yaw mode.
† Pitch-yaw mode.
‡ Speed mode.

linear-quadratic-Gaussian (LQG) regulator (Section 4.7), as shown in Fig. 7.4-22 [S-19]. The principal modifications to the LQG controller are the addition of the pilot's observation delay and neuromuscular dynamics, which have limiting effects on the bandwidth and stability of control. While it is understood that a human operator does not carry out a mathematical optimization to learn how to fly, there is good correspondence of the input-output characteristics of such a model with the actual performance of a skilled human operator. Furthermore, the functional blocks of the model have analogues in the tasks that a pilot must perform, like estimating the dynamic state of the system and generating control actions to minimize a tracking error. Thus, the modular nature of the model allows system solutions to be isolated, such as improvements in displays or cues to improve estimation or modifications in system dynamics and effectors to improve control. The pilot model closes the control loop for an unaugmented aircraft in Fig. 7.4-22; however, it can easily be applied to an aircraft model that contains stability and command augmentation.

The pilot model is used to generate a turn entry using lateral stick alone with and without an aileron-rudder interconnect and with no stability

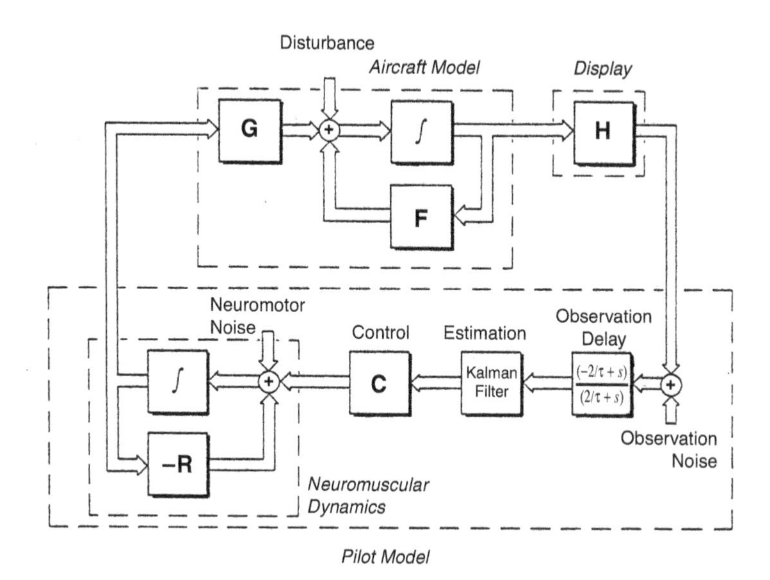

Fig. 7.4-22. Dynamic model of pilot-aircraft interactions. (from [S-19], courtesy of ONR)

augmentation system (SAS) (Fig. 7.4-23) [S-18]. The model is adapted to each of three dynamic conditions, that is, the modified LQG regulator is designed for the three conditions. At 30 deg angle of attack without the ARI, Aircraft B is turned primarily by the differential stabilator's adverse yaw, and the required pilot input is in the opposite direction from the command at $\alpha = 10$ deg. With the ARI on, differential stabilator and rudder deflections are blended to produce the same result at $\alpha = 30$ deg using a pilot input of the normal sense.

Alternative adaptation strategies can also be studied using the pilot model. Assuming that the pilot is instantaneously and correctly aware of changing characteristics and modifies piloting actions optimally, the closed-loop system remains stable at all flight conditions. This is portrayed by a 45 deg line in a plot of stability regions with ARI and SAS off (Fig. 7.4-24) [S-19]. The vertical axis represents the aircraft's actual angle of attack, and the horizontal axis represents the angle of attack assumed by the pilot model. Using lateral stick alone, the region of closed-loop stability narrows in the vicinity of $\alpha = 15$–18 deg, where the pilot is forced to reverse lateral-stick strategy for roll control. Using lateral stick and rudder pedals, the stable region is expanded greatly, reducing the pilot's need to adapt to retain stability (Fig. 7.4-24), and with rudder pedals alone, there is no region of instability (not shown). With the ARI

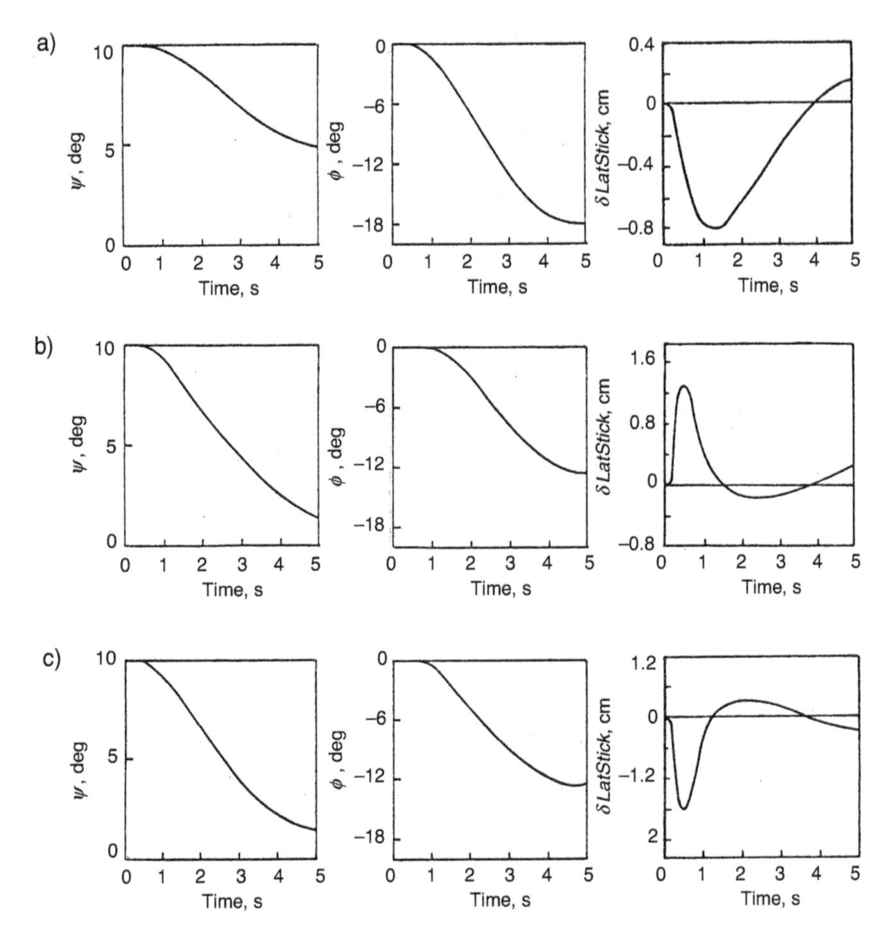

Fig. 7.4-23. Turn entry of Aircraft B commanded by the adapted pilot model with and without aileron-rudder interconnect at two angles of attack. (from [S-18], courtesy of ONR)

engaged, the unstable regions shrink, allowing stability to be retained with greater pilot-aircraft mismatch.

The pilot model provides a basis for predicting the tracking accuracy and level of control activity required as functions of pilot-aircraft mismatch. Applying the covariance analysis of Section 4.5 with nominal values of disturbance input, observation error, and neuromuscular noise, contours of constant standard deviation can be generated (Fig. 7.4-25) [S-19]. The figure shows the variation in lateral-stick and foot-pedal standard deviations with angle of attack and model mismatch. The standard

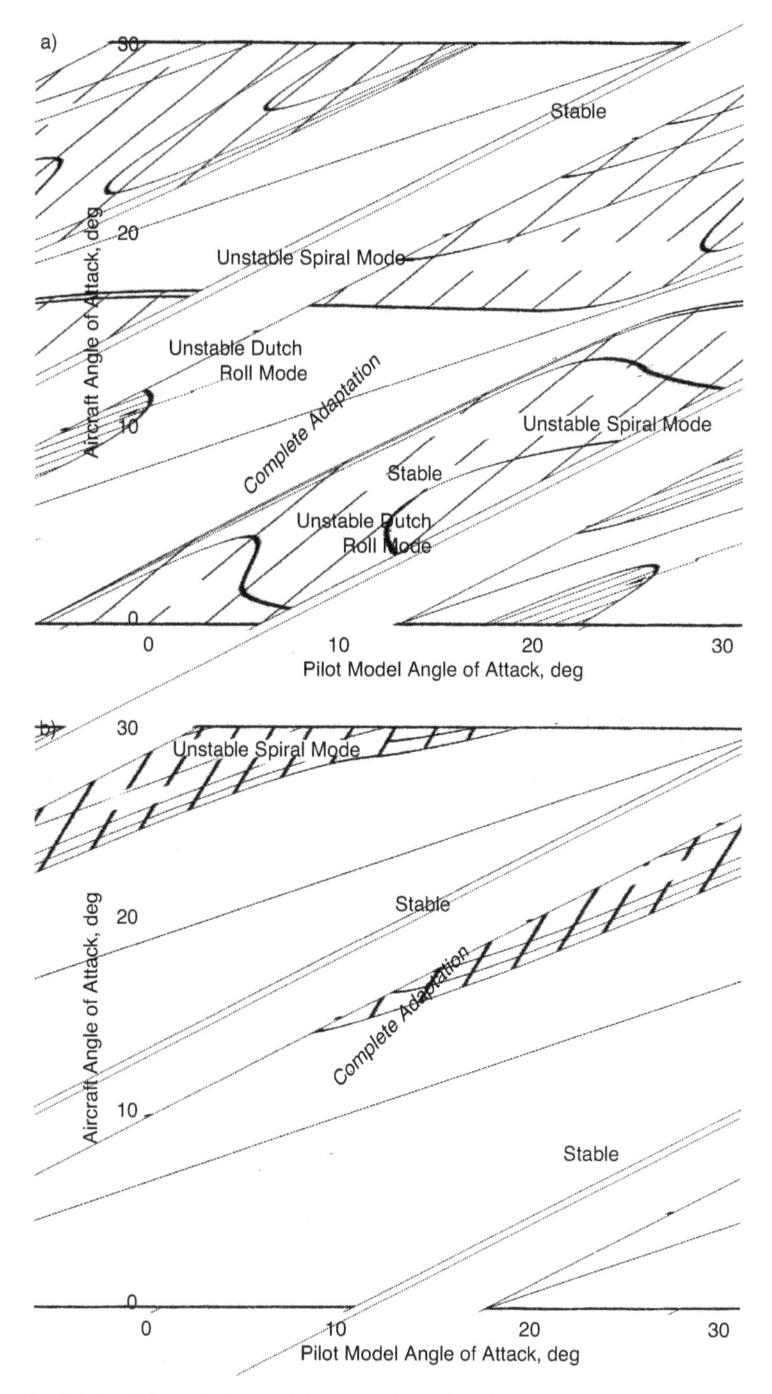

Fig. 7.4-24. Effect of pilot model mismatch on closed-loop stability of Aircraft B with increasing angle of attack, ARI and SAS off. (from [S-19], courtesy of ONR)

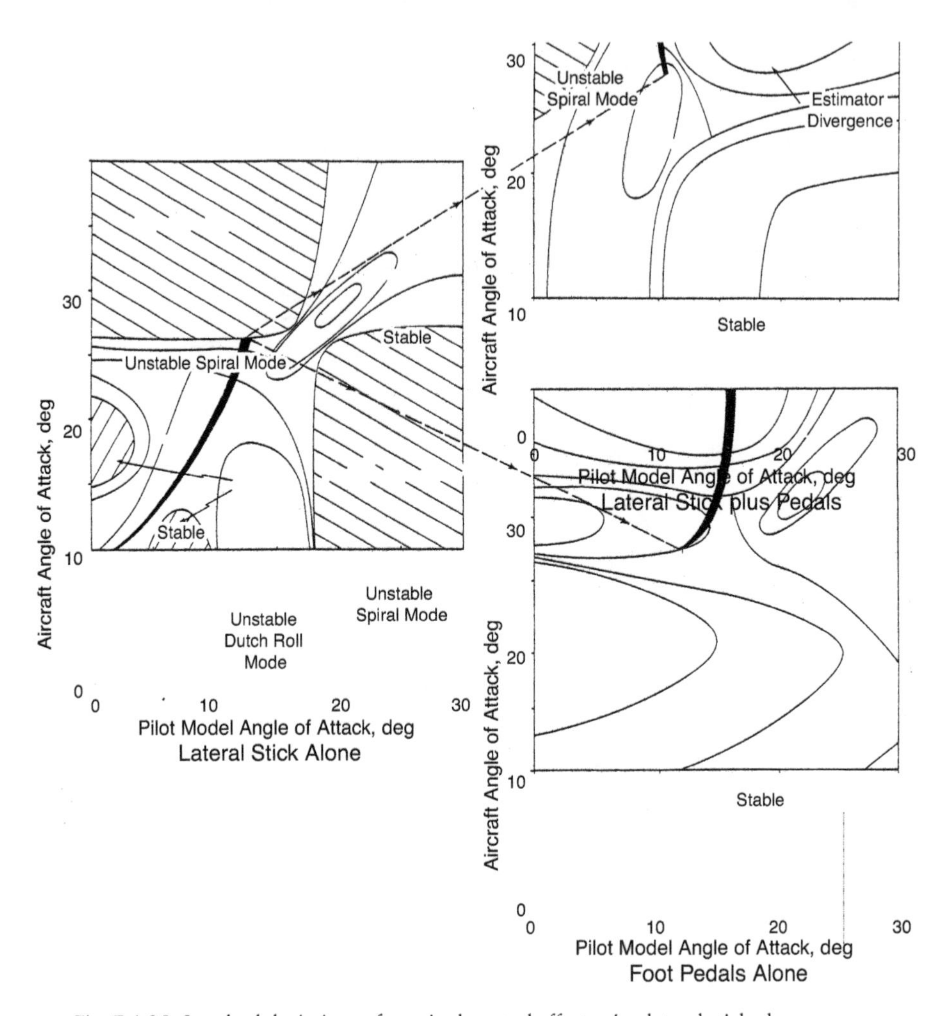

Fig. 7.4-25. Standard deviations of required control effort using lateral stick alone, lateral stick plus foot pedals, and foot pedals alone, with contours of minimum control effort (MCE). (from [S-19], courtesy of ONR)

deviations of roll rate and tracking error are loosely similar to the control standard deviations.

Figure 7.4-25 gives rise to an interesting adaptation strategy based on minimizing the control effort required to perform a particular maneuver, such as the windup turn. Following the contours, the minimum-control-effort (MCE) strategy predicts that the pilot will use lateral stick alone until the unstable spiral mode boundary is encountered. At this point, the pilot can control either using pedals alone or using stick and pedals.

The latter requires less pedal input, so that option is chosen. As angle of attack continues to increase, pedal effort increases and instability is encountered, requiring the pilot to switch to pedals alone.

The MCE analytical results help interpret tests made in a cockpit simulator. Results of a human-piloted simulation of Aircraft B in a windup turn with ARI and SAS off are presented in Fig. 7.4-26 [S-19]. The pilot primarily uses lateral stick alone up to the region of 15–18 deg, where there are increasing sideslip excursions. Foot pedal action begins, and sideslip perturbations are controlled. As the simulation continues, lateral stick motion diminishes, and the control is effectively exerted by foot pedals alone, with good sideslip control to termination of the maneuver.

Gain-Scheduled Stability and Command Augmentation

Stability and command response at high angles of attack can be markedly improved by feedback control. The optimal linear control design techniques presented in Section 4.7 can be applied at any operating point in the flight envelope, improving stability and furnishing desirable flying qualities. However, there are several issues in applying linear design methods to a nonlinear system that must be addressed to achieve satisfactory control across the flight envelope. The most significant one is that the gain matrices for optimal control and estimation vary with flight conditions. The feedback control required to follow a single ideal model at high and low dynamic pressure, angle of attack, and roll rate would clearly change as stability and control derivatives change. Thus, there must be a mechanism for adapting the linear controller to changes in flight condition.

The most common approach to adaptation is *gain scheduling*. Constant-gain controllers and estimators are designed at a number of operating points throughout the flight envelope, producing separate gain matrices at each point. The aircraft's stability and control matrices $\mathbf{F}$ and $\mathbf{G}$ are assumed to be constant at each trimmed point. Using the LQG design approach, the corresponding control and estimation gain matrices $\mathbf{C}$ and $\mathbf{K}$ are constant as well. The operating points are characterized by *scheduling variables*, such as dynamic pressure, angle of attack, and roll rate, and the individual elements of $\mathbf{C}$ and $\mathbf{K}$ are scheduled by curve fitting. For example, each element could be characterized by a least-squares fit of a polynomial in the scheduling variables. The scheduled gains should have correlations with the training gains of 80 percent or higher in order to represent the original linear designs well.

This approach can be applied with or without implicit model following, and in the latter case, there is no need to be fixed on a single ideal model. In fact, the model probably should change either continuously or through pilot-selectable mode control to reflect changes in task requirements, flight conditions, and controllability. The rationale for gain scheduling is that

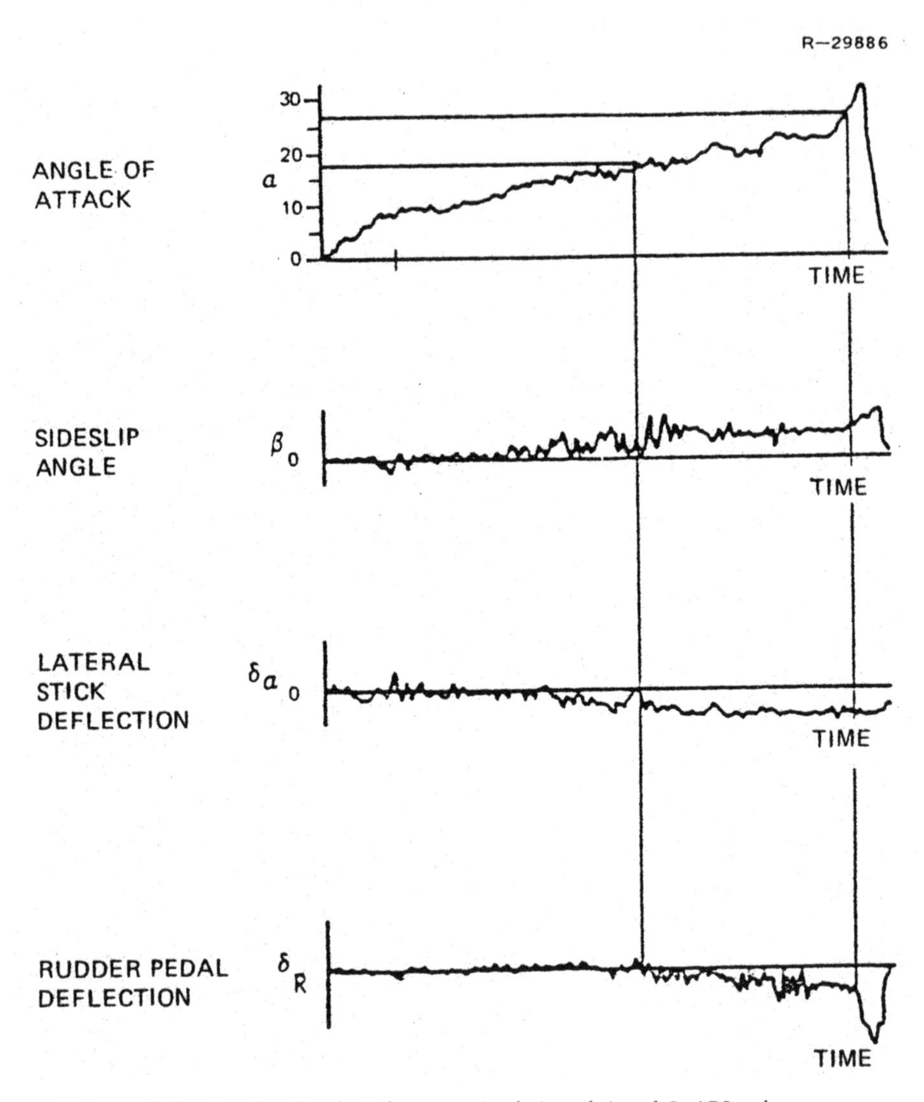

Fig. 7.4-26. Results of a piloted wind-up turn simulation of aircraft B, ARI and SAS off. (from [S-19], courtesy of ONR)

the nonlinear model can be adequately represented by changing trim points, which are inherently nonlinear but static, plus a linear, parameter-varying model whose parameter variations are slow on the time scale of important transient motions.

Examples of the effect of LQ closed-loop control at an operating point and of gain variations with flight condition are given. Figure 7.4-27 [S-18] illustrates the change in directional response for Aircraft B flying at $\alpha = 30$ deg. The slow, rolling response to sideslip initial condition is replaced by a quick response with little rolling motion. Figure 7.4-28 shows the effects of wind-axis roll rate and sideslip angle on four differential stabilator gains. Two of the gains are purely lateral-directional, and two are crossfeeds of longitudinal motions. The crossfeed gains are zero when the flight condition is symmetric.

Control system examples with gain variation and gain scheduling can be found in a number of papers and reports. Effects of optimal LQ SAS gain variation for Aircraft A and B are shown in [S-16, S-18]; optimal LQ SCAS gain variations for Aircraft B are demonstrated in [S-19], and a scheduled, digital LQ SCAS for Airplane B is presented and evaluated in [B-6]. A scheduled, digital LQG controller for a tandem-rotor helicopter, which has an equally challenging range of flight characteristics, is reported in [S-20, S-21]. This controller was successfully flight tested at NASA Langley Research Center. A scheduled, digital LQ controller is shown to prevent wing rock in the stalling flight tests of a general aviation aircraft in [E-1].

Adaptive Neural Network Control

High-α maneuvering could be improved by more flexible control logic that preserves the positive features of gain-scheduled linear controllers while explicitly accounting for nonlinear effects. Replacing the gain matrices of a linear controller with neural networks (Section 3.6) produces a nonlinear controller that adapts to changing flight conditions. For example, a neural-network controller motivated by proportional-integral control (Fig. 4.7-3) is shown in Fig. 7.4-29. Each gain matrix has been replaced by a neural network, $\mathbf{NN}_B$ for $\mathbf{C}_B$, $\mathbf{NN}_F$ for $\mathbf{C}_F$, and $\mathbf{NN}_I$ for $\mathbf{C}_I$. The input-output structure is unchanged, and the command error is integrated, as in the linear system, to produce satisfactory steady-state response. The *scheduling variable generator* (**SVG**) produces auxiliary inputs $\mathbf{a}(t)$ to the neural networks based on command and exogenous variables $\mathbf{e}(t)$ such as air data. The *command state generator* (**CSG**) produces a system state vector $\mathbf{x}_c(t)$ that is compatible with the command vector $\mathbf{y}_c(t)$.

The network of networks $\{\mathbf{NN}_B, \mathbf{NN}_F, \mathbf{NN}_I\}$ produces a nonlinear control command of the form

$$\mathbf{u}(t) = \mathbf{c}[\mathbf{y}_c(t), \Delta\mathbf{x}(t), \boldsymbol{\xi}(t), \mathbf{a}(t)] \tag{7.4-22}$$

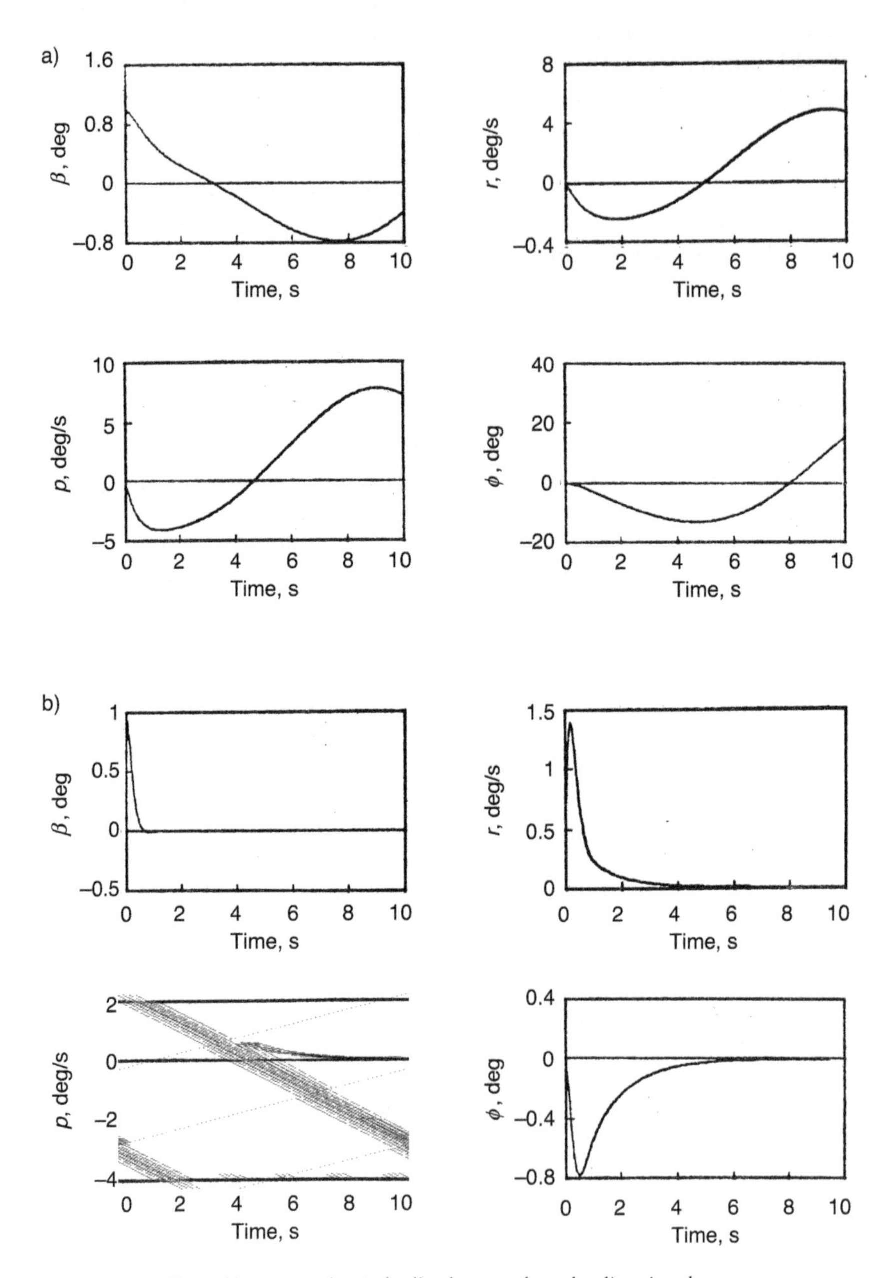

Fig. 7.4-27. Effect of linear-quadratic feedback control on the directional response of aircraft B at $\alpha = 30$ deg. (from [S-18], courtesy of ONR)

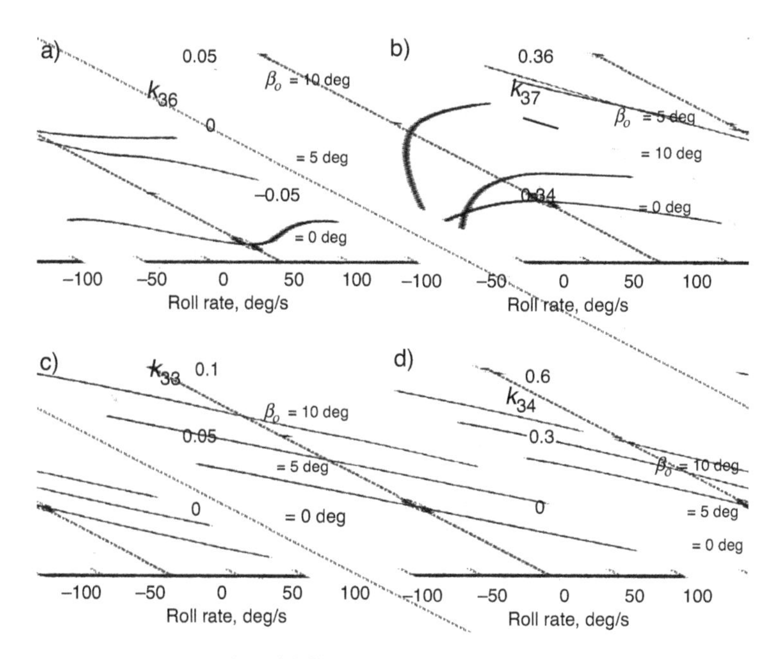

Fig. 7.4-28. Variation in selected differential stabilator gains for linear-quadratic control of airplane B with wind-axis roll rate and sideslip angle. (from [S-18], courtesy of ONR)

At any operating point specified by $\mathbf{a}(t)$, perturbations in the control signal can be expressed as

$$\begin{aligned}\Delta\mathbf{u}(t) &= \frac{\partial\mathbf{c}}{\partial\mathbf{y}_c}\Delta\mathbf{y}_c + \frac{\partial\mathbf{c}}{\partial\Delta\mathbf{x}}\Delta\mathbf{x} + \frac{\partial\mathbf{c}}{\partial\boldsymbol{\xi}}\Delta\boldsymbol{\xi} \\ &\triangleq \mathbf{C}_F\Delta\mathbf{y}_c + \mathbf{C}_B\Delta\mathbf{x} + \mathbf{C}_I\Delta\boldsymbol{\xi}\end{aligned} \tag{7.4-23}$$

In other words, the Jacobian matrices of $\mathbf{c}$, which represent the slopes of the function with respect to its arguments, are the linear gain matrices at the operating point.

In general, $\mathbf{c}[\mathbf{y}_c(t), \Delta\mathbf{x}(t), \boldsymbol{\xi}(t), \mathbf{a}(t)]$ is not known, but its Jacobians at all operating points can be specified by the linear point designs discussed in the previous section. Once the control gain matrices have been defined, they can be used to train the corresponding neural networks algebraically, a fast alternative to back propagation [F-2]. With the number of sigmoid nodes equal to the number of operating conditions used for training, the neural networks produce an output (control) function whose

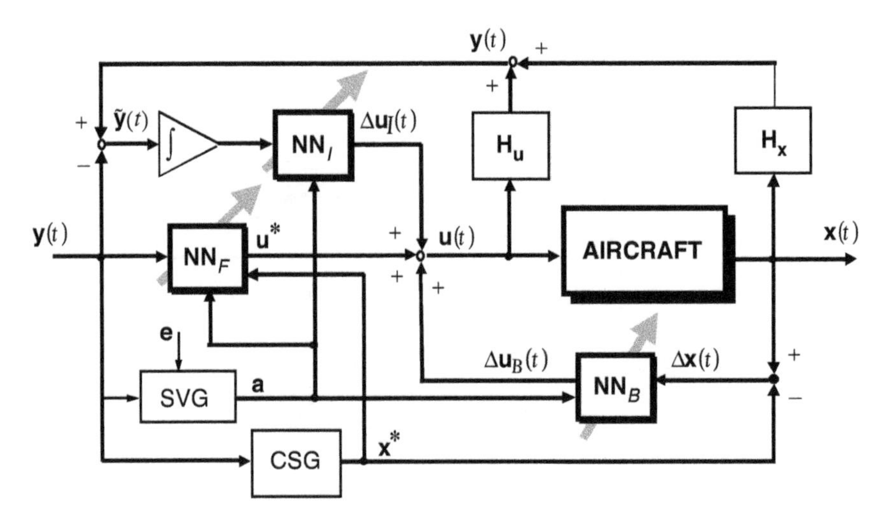

Fig. 7.4-29. Proportional-integral neural-network control law.

gradients match the linear gains precisely, that is, with 100 percent correlation. The networks can also be structured with fewer nodes than training points, in which case the fit is not exact but can have least-square error. Consequently, the neural network flight controller is a superior alternative to ad hoc gain scheduling. Algebraic neural network training is not restricted to use with a LQ training set; it could be used with any linear control system specified at a number of operating points.

If the training set is a controller that optimizes a cost function, on-line training by an *adaptive critic* may further improve performance, during either extended simulation or actual flight [F-2]. Adaptive-critic designs model the optimal cost and the control law using neural networks and apply the principle of dynamic programming to optimization [B-5]. A *critic network* provides an approximate solution for the "cost to go" or its gradient with respect to the state using the Hamilton-Jacobi-Bellman equation, and an *action network* implements the optimal control law. Each network is used to improve the parameters of the other as the aircraft performs maneuvers, adjusting for errors in the specification of aircraft properties, and accounting for limits on control effects. For example, the controller could be pretrained for symmetric operating points, establishing a good topology for the networks, and then be trained with the adaptive critic in simulated asymmetric flight conditions to learn the coupling effects. Control limiting effects could also be learned during simulation. If implemented for adaptation during actual flight, the adaptive-critic neural-network controller may implicitly detect some system failures, and reconfigure control actions to make the most of remaining controllability [F-3].

Robust Nonlinear-Inverse-Dynamic Control

An alternative approach to high-α flight control is to compute commands that are inverse to the natural motions of the aircraft, and then to drive the combined system with the desired response. If the first step is done precisely, the sequence of the inverting control followed by the aircraft's actual dynamics would provide an identity relationship between the input and output [S-8, S-9]. The desired dynamics could then be specified in the input to the inverting control law. Of course, there must be no singularities or instabilities in the inversion, and, as a practical matter, the dimensions of the control $\mathbf{u}$ and the chosen output $\mathbf{y}$ must be the same so that the identity relationship is square. Thus the inversion takes place not for the entire state $\mathbf{x}$, but for an output vector $\mathbf{y}$.

The aircraft's nonlinear equations of motion are separated into a nonlinear homogeneous term $\mathbf{f}(\mathbf{x})$, plus the effect of control

$$\dot{\mathbf{x}} = \mathbf{f}(\mathbf{x}) + \mathbf{G}(\mathbf{x})\mathbf{u} \tag{7.4-24}$$

with $\dim(\mathbf{x}) = n \times 1$ and $\dim(\mathbf{u}) = m \times 1$. The $n \times m$ matrix of control sensitivity $\mathbf{G}(\mathbf{x})$ may be a nonlinear function of the state, but control deflections enter linearly. For simplicity, we assume that the $m \times 1$ output vector is a linear function of the state,

$$\mathbf{y} = \mathbf{H}\mathbf{x} \tag{7.4-25}$$

The output matrix $\mathbf{H}$ is selected by the designer, as $\mathbf{y}$ represents the desired response to command input.

The immediate goal is to define a nonlinear feedback control law that decouples the response of $\mathbf{y}$ from aircraft dynamics. We cannot specify the desired value of the output, $\mathbf{y}_c(t)$, directly, but we can find a control law that decouples time derivatives of its elements. Let us start with the simplest case first. From eqs. 7.4-24 and 7.4-25, the derivative of $\mathbf{y}$ is

$$\dot{\mathbf{y}} = \mathbf{H}\dot{\mathbf{x}} = \mathbf{H}[\mathbf{f}(\mathbf{x}) + \mathbf{G}(\mathbf{x})\mathbf{u}]$$

$$\triangleq \mathbf{f}^*(\mathbf{x}) + \mathbf{G}^*(\mathbf{x})\mathbf{u} \tag{7.4-26}$$

where $\dim(\mathbf{f}^*) = m \times 1$ and $\dim(\mathbf{G}^*) = m \times m$. Denoting the derivative of $\mathbf{y}_c$ as $\mathbf{v}$, the inverting control law is

$$\mathbf{u} = \mathbf{G}^{*-1}(\mathbf{x})\,[\dot{\mathbf{y}} - \mathbf{f}^*(\mathbf{x})] = \mathbf{G}^{*-1}(\mathbf{x})\,[\mathbf{v} - \mathbf{f}^*(\mathbf{x})] \tag{7.4-27}$$

provided that $\mathbf{G}^{*-1}$ exists. Substituting eq. 7.4-27 in eq. 7.4-26, the net effect is that the output is the integral of the input:

$$\dot{\mathbf{y}} = \mathbf{v} \tag{7.4-28}$$

The control law includes a model of the aircraft's homogeneous dynamics in the feedback loop and the inverse of its control effects in the forward loop (Fig. 7.4-30). The final step is to design a controller, such as a low-order linear prefilter and compensator, or rate-command/attitude-hold controller, for the system described by eq. 7.4-28.

$\mathbf{HG(x)}$ is often not invertible because some elements of $\mathbf{y}$ are not linear functions of $\mathbf{u}$; however, repeated differentiation eventually reveals a linear relationship. Suppose that this is the case for the first element of $\mathbf{y}$, y_1. Denoting the first row of $\mathbf{H}$ by $\mathbf{h}_1$, the first derivative of y_1 is

$$\dot{y}_1 = \mathbf{h}_1\mathbf{f(x)} \tag{7.4-29}$$

and the second derivative is

$$\ddot{y}_1 = \mathbf{h}_1 \frac{\partial \mathbf{f(x)}}{\partial \mathbf{x}}\dot{\mathbf{x}} = \mathbf{h}_1 \frac{\partial \mathbf{f(x)}}{\partial \mathbf{x}}[\mathbf{f(x)} + \mathbf{G(x)u}] \tag{7.4-30}$$

If eq. 7.4-30 establishes the necessary linear relationship between $\mathbf{u}$ and $\ddot{y}_1$, the first rows of $\mathbf{f}^*(\mathbf{x})$ and $\mathbf{G}^*(\mathbf{x})$ are redefined as $\mathbf{h}_1[\partial\mathbf{f(x)}/\partial\mathbf{x}]\mathbf{f(x)}$ and $\mathbf{h}_1[\partial\mathbf{f(x)}/\partial\mathbf{x}]\mathbf{G(x)}$, and the second element of $\mathbf{y}$ is checked. If eq. 7.4-30 does not produce a linear relationship, the next derivative of y_1 is evaluated, and so on. We denote the first satisfactory derivative of $y_1$as $y_1{}^{(d)}$, where d is the order of differentiation required. The vector of suitable derivatives of $\mathbf{y}$ is denoted by $\mathbf{y}^{(d)}$, and $\mathbf{f}^*(\mathbf{x})$ and $\mathbf{G}^*(\mathbf{x})$ are defined row by row, yielding

$$\mathbf{y}^{(d)} = \mathbf{f}^*(\mathbf{x}) + \mathbf{G}^*(\mathbf{x})\,\mathbf{u} \tag{7.4-31}$$

where $\mathbf{G}^*(\mathbf{x})$ is invertible for linearly independent controls. Thus, $\mathbf{v}$ represents desired values of $\mathbf{y}^{(d)}$, and the inverting control law is described by eq. 7.4-27 and Fig. 7.4-30.

While the expression of the inverse control law appears simple, its implementation can be quite complex. Evaluation of $\mathbf{f}^*(\mathbf{x})$ and $\mathbf{G}^*(\mathbf{x})$ requires that a full, d-differentiable model of the aircraft dynamics be included in the control system. Real-time execution of a full inverting control law that includes control limiting effects is demonstrated in [L-3], but it may be desirable to simplify the dynamic model for inversion. This is done by partitioning the system into slow- and fast-time-scale

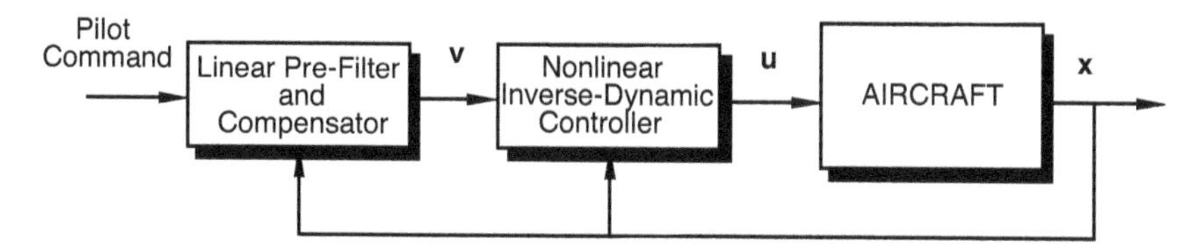

Fig. 7.4-30. Nonlinear-inverse-dynamic control law.

subsystems [M-8, M-11]. The effects of feeding back estimates of the state rather than the state itself are presented in [M-12].

A robust, nonlinear-inverse-dynamic controller for a high-performance aircraft design is developed in [W-2], using this two-time-scale approach. The angular-rate dynamics are assumed to evolve on the fast time scale, producing commands to canard and tail surfaces. The slow-time-scale dynamics for sideslip angle and velocity produce throttle and body-axis roll-rate and yaw-rate commands. The fast dynamics are fed by the pilot's pitch-rate command, which is subject to α-limiting effects, and by the slow-time-scale outputs. The pilot's inputs to the slow dynamics partition are stability-axis roll rate, sideslip angle, and velocity. The pilot's commands are prefiltered by simple models of desirable dynamics, whose natural frequencies, damping ratios, and time constants are elements of a stochastic robustness design parameter vector $\mathbf{d}$ (Section 4.7). The stability and performance metric J that is minimized by a genetic algorithm search of Monte Carlo evaluation results contains 18 components that are motivated by the flying qualities design requirements of [M-1]. The resulting design compares favorably with the benchmark solutions contained in the report.

References

A-1 Abzug, M. J., "Effects of Certain Steady Motions on Small-Disturbance Airplane Motions," *J. Aeronautical Sciences* 21, no. 11 (1954): 749–762.

A-2 Abzug, M. J., and Larrabee, E. E., *Airplane Stability and Control*, Cambridge University Press, Cambridge, 2002.

A-3 Adams, W. M., Jr., "Analytical Prediction of Airplane Equilibrium Spin Characteristics," NASA TN D-6926, Washington, D.C., Nov. 1972.

A-4 Anathkrishnan, N., and Sudhakar, K., "Prevention of Jump in Inertia-Coupled Roll Maneuvers of Aircraft," *J. Aircraft* 31, no. 4 (1994): 981–983.

A-5 Anderson, C. A. "Stall/Post-Stall Characteristics of the F-111 Aircraft," *Fluid Dynamics of Aircraft Stalling*, AGARD-CP-102, Neuilly-sur-Seine, Nov. 1972, pp. 18-1–18-9.

A-6 Anderson, C. A., "The Stall/Spin Problem—American Industry's Approach," *Stall/Spin Problems of Military Aircraft*, AGARD-CP-199, Neuilly-sur-Seine, June 1976, pp. 2-1–2-8.

A-7 Anderson, S. B., "Historical Overview of Stall/Spin Characteristics of General Aviation Aircraft," *J. Aircraft* 16, no. 7 (1979): 455–461.

A-8 Anderson, S. B., Enevoldson, E. K., and Nguyen, L. T., "Pilot Human Factors in Stall/Spin Accidents of Supersonic Fighter Aircraft," *Flight Mechanics and System Design Lessons from Operational Experience*, J. E. Cayot and J. Renaudie, chairs, AGARD CP-347, Neuilly-sur-Seine, Oct. 1983, pp. 27-1– 27-10.

A-9 Anglin, E. L., and Scher, S. H., "Analytical Study of Aircraft-Developed Spins and Determination of Moments Required for Satisfactory Spin Recovery," NASA TN D-2181, Washington, D.C., Feb. 1964.

B-1 Babister, A. W., "Aircraft Longitudinal Motion at High Incidence," *Aeronautical Journal* 83, no. 822 (1979): 230–232.

B-2 Banks, D. W., Gatlin, G. M., and Paulson, J. W., Jr., "Low-Speed Longitudinal and Lateral-Directional Characteristics of the X-31 Configuration," NASA TM-4351, Washington, D.C., Oct. 1992.

B-3 Barham, R. W., "Thrust Vector Aided Maneuvering of the YF-22 Advanced Tactical Fighter Prototype," *Technologies for Highly Manoeuverable Aircraft*, AGARD-CP-548, Neuilly-sur-Seine, Mar. 1994, pp. 5-1–5-14.

B-4 Bates, G. P., Jr., and Czinczenheim, J., eds., *Stall/Spin Problems of Military Aircraft*, AGARD-CP-199, Neuilly-sur-Seine, June 1976.

B-5 Bellman, R. E., *Dynamic Programming*, Princeton University Press, Princeton, N.J., 1957.

B-6 Berry, P. W., Broussard, J. R., and Gully, S., "Validation of High Angle-of-Attack Analysis Methods," ONR-CR-215-237-3F, Arlington, Va., Sept. 1979.

B-7 Beyers, M. E., and Ericsson, L. E., "Implications of Recent Rotary Rig Results for Flight Prediction," *J. Aircraft* 37, no. 4 (2000): 545–553.

B-8 Bihrle, W., Jr., "Influence of the Static and Dynamic Aerodynamic Characteristics on the Spinning Motion of Aircraft," *J. Aircraft* 8, no. 10 (1971): 764–768.

B-9 Bihrle, W., Jr., "Spin Prediction Techniques," AIAA-80-1564, Danvers, Mass., Aug. 1980.

B-10 Bornemann, W. E., and Surber, T. E., "Aerodynamic Design of the Space Shuttle Orbiter," *High Angle of Attack Aerodynamics*, J. L. Jones, chair, AGARD-CP-247, Neuilly-sur-Seine, Jan. 1979, pp. 11-1–11-24.

B-11 Bowman, J. S., Jr., "Summary of Spin Technology as Related to Light General-Aviation Airplanes," NASA TN D-6575, Washington, D.C., Dec. 1971.

B-12 Broussard, J. R., and Stengel, R. F., "Stability of the Pilot-Aircraft System in Maneuvering Flight," *J. Aircraft* 14, no. 10 (1977): 959–965.

B-13 Bridges, R. D., "YA-10 Stall/Post-Stall/Spin Avoidance Flight Test Program," *Technical Review, Society of Experimental Test Pilots* 13, no. 3 (1977): 15–31.
B-14 Bryant, L. W., and Jones, I. M. W., "Recovery from a Spin," Aeronautical Research Committee R & M No. 1426, London, Mar. 1932.
B-15 Burris, W. R., and Lawrence, J. T., "Aerodynamic Design and Flight Test of US Navy Aircraft at High Angles of Attack," *Fluid Dynamics of Aircraft Stalling*, L. G. Napolitano, ed., AGARD-CP-102, Neuilly-sur-Seine, Nov. 1972, pp. 25-1–25-10.
B-16 Byushgens, G. S., and Studney, R. V., "Dynamics of the Spatial Motion of an Aircraft," NASA TT F-555, Washington, D.C., Apr. 1969.
C-1 Calkins, D. E., "Aircraft Accident Flight Path Simulation and Animation," *J. Aircraft* 31, no. 2 (1994): 376–386.
C-2 Carroll, J. V., and Mehra, R. K., "Bifurcation Analysis of Nonlinear Aircraft Dynamics," *J. Guidance and Control* 5, no. 5 (1982): 529–536.
C-3 Chambers, J. R., and Anglin, E. L., "Analysis of Lateral-Directional Stability Characteristics of a Twin-Jet Fighter Aircraft at High Angles of Attack," NASA TN D-5361, Washington, D.C., Aug. 1969.
C-4 Chambers, J. R., Bowman, J. S., Jr., and Anglin, E. L., "Analysis of the Flat-Spin Characteristics of a Twin-Jet Swept-Wing Fighter Airplane," NASA TN D-5409, Washington, D.C., Sept. 1969.
C-5 Chambers, J. R., Anglin, E. L., and Bowman, J. S., Jr., "Effects of a Pointed Nose on Spin Characteristics of a Fighter Airplane Model Including Correlation with Theoretical Calculations," NASA TN D-5921, Washington, D.C., Sept. 1970.
C-6 Chambers, J. R., and Bowman, J. S., Jr., "Recent Experience with Techniques for Prediction of Spin Characteristics of Fighter Aircraft," *J. Aircraft* 8, no. 7 (1971): 548–553.
C-7 Chambers, J. R., and Grafton, S. B., "Aerodynamic Characteristics of Airplanes at High Angles of Attack," NASA TM 74097, Washington, D.C., Dec. 1977.
C-8 Chambers, J. R., "Overview of Stall/Spin Technology," AIAA-80-1580, Danvers, Mass., Aug. 1980.
C-9 Chevalier, H., and Brusse, J. C., "A Stall/Spin Prevention Device for General-Aviation Aircraft," SAE Paper No. 730333, Warrendale, Pa., 1973.
C-10 Chevalier, H. L., "Some Theoretical Considerations of a Stall Proof Airplane," SAE Paper No. 790604, Warrendale, Pa., Apr. 1979.
C-11 Clark, C. K., Walker, J. E., and Buckner, J. K., "The Evolution of the High-Angle-of-Attack Features of the F-16 Flight Control System," *Proc. Ninth Annual Symposium*, Society of Flight Test Engineers, Oct. 1978, pp. 17-1–17-21.
C-12 Coe, P. L., Jr., Chambers, J. R., and Letko, W., "Asymmetric Lateral-Directional Characteristics of Pointed Bodies of Revolution at High Angles of Attack," NASA TN D-7095, Washington, D.C., Nov. 1972.
C-13 Coe, P. L., Jr., and Newsom, W. A., Jr., "Wind-Tunnel Investigation to Determine the Low-Speed Yawing Stability Derivatives of a Twin-Jet Fighter

Model at High Angles of Attack," NASA TN D-7721, Washington, D.C., Aug. 1974.

C-14 Cornish, J. J., III, and Jenkins, M. W. M., "The Application of Spanwise Blowing for High Angle of Attack Spin Recovery," *High Angle of Attack Aerodynamics*, J. L. Jones, chair, AGARD-CP-247, Neuilly-sur-Seine, Jan. 1979, pp. 9-1–9-12.

C-15 Crabbe, E. N., "Simple Theory of Spinning with Particular Reference to the HS Gnat Trainer," Paper No. 317, *Aeronautical Journal* 82, (1978): 223–228.

C-16 Crimi, P., *Dynamic Stall*, AGARD-AG-172, Neuilly-sur-Seine, Nov. 1973.

E-1 Ehrenstrom, W. A., "A Lateral-Directional Controller for High-Angle-of-Attack Flight," M.S.E. thesis, Princeton University, Princeton, N.J., 1983.

E-2 Eichhorn, G., "International Aerobatic Club (IAC) Aerobatics Figures," http://acro.harvard.edu/IAC/acro_figures.html, 1998.

E-3 Ericsson, L. E., and Beyers, M. E., "Conceptual Fluid/Motion Coupling in the Herbst Maneuver," *J. Aircraft* 34, no. 3 (1997): 271–277.

F-1 Ferrari, S., and Stengel, R. F., "Classical/Neural Synthesis of Nonlinear Control Systems," *J. Guidance, Control, and Dynamics* 25, no. 3 (2002): 442–448.

F-2 Ferrari, S., and Stengel, R. F., "An Adaptive Critic Global Controller," *Proc. 2002 American Control Conference*, Anchorage, Ala., May 2002, pp. 2665–2670.

F-3 Ferrari, S., "Algbraic and Adaptive Learning in Neural Control Systems," Ph.D. thesis, Princeton University, MAE-306-T, Princeton, N.J., Nov. 2002.

F-4 Fremaux, C. M., "Estimation of the Moment Coefficients for Dynamically Scaled, Free-Spinning Wind-Tunnel Models," *J. Aircraft* 32, no. 6 (1995): 1407–1409.

G-1 Garner, H., and Wright, K., "The Longitudinal Control of an Aeroplane Beyond the Stall," ARC R & M 1193, London, May 1928.

G-2 Gates, S., "Notes on Longitudinal Stability at Stalling in Gliding Flight," ARC R & M 1189, London, July 1928.

G-3 Gilmore, R., *Catastrophe Theory for Scientists and Engineers*, J. Wiley & Sons, New York, 1981.

G-4 Glendenning, P., *Stability, Instability, and Chaos*, Cambridge University Press, Cambridge, 1995.

G-5 Goman, M., and Khrabrov, A., "State-Space Representation of Aerodynamic Characteristics of an Aircraft at High Angles of Attack," *J. Aircraft* 31, no. 5 (1994): 1109–1115.

G-6 Gousman, K. G., Juang, J. C., Loschke, R. C., and Rooney, R. H., "Aircraft Deep Stall Analysis and Recovery," AIAA-91-2888-CP, Reston, Va., 1991.

G-7 Grafton, S. B., and Anglin, E. L., "Dynamic Stability Derivatives at Angles of Attack from $-5°$ to $90°$ for a Variable-Sweep Fighter Configuration with Twin Vertical Tails," NASA TN D-6909, Washington, D.C., Oct. 1972.

G-8 Guckenheimer, J., and Holmes, P., *Nonlinear Oscillations, Dynamical Systems, and Bifurcations of Vector Fields*, Springer-Verlag, New York, 1983.
G-9 Guicheteau, P., "Stability Analysis through Bifurcation Theory (1) and (2), NonLinear Flight Dynamics," *Non Linear Dynamics and Chaos*, AGARD-LS-191, Neuilly-sur-Seine, June 1993, pp. 3-1–3-10, 4-1–4-11, 5-1–5-13.
H-1 Hacker, T., and Oprisiu, C., "A Discussion of the Roll Coupling Problem," *Progress in Aerospace Sciences* 15 (1974): 151–180.
H-2 Hacker, T., "Attitude Instability in Steady Rolling and Roll Response," *J. Aircraft* 14, no. 1 (1977): 23–31.
H-3 Hacker, T., "Constant-Control Rolling Maneuver," *J. Guidance and Control* 1, no. 5 (1978): 313–318.
H-4 Hancock, G., *Problems of Aircraft Behaviour at High Angles of Attack*, AGARDograph 136, Neuilly-sur-Seine, Apr. 1969.
H-5 Hedrick, J. K., "Nonlinear System Response: Quasi-Linearization Methods," *Nonlinear System Analysis and Synthesis: Volume 1—Fundamental Principles*, American Society of Mechanical Engineers, New York, 1978, pp. 97–124.
H-6 Heffley, R. K., and Jewell, W. F., "Aircraft Handling Qualities Data," NASA CR-2144, Washington, D.C., Dec. 1972.
H-7 Herbst, W. B., "Future Fighter Technologies," *J. Aircraft* 17, no. 8 (1980): 561–566.
H-8 Hill, R., and Stephens, H., "Notes on Stalled Flying," ARC R & M 963, London, Oct. 1922.
H-9 Hopkins, E. J., and Yee, S. C., "A Comparison of the Experimental Aerodynamic Characteristics of an Oblique Wing with Those of a Swept Wing," NASA-TM-X-3547, Washington, D.C., Jun. 1977.
H-10 Hwang, C., and Pi, W. S., "Some Observations on the Mechanism of Aircraft Wing Rock," *J. Aircraft* 16, no. 6 (1979): 366–373.
I-1 Iooss, G., and Joseph, D. D., *Elementary Stability and Bifurcation Theory*, Springer-Verlag, New York, 1990.
I-2 Imbrie, A. P., "A Geometrical Study of the Steady-State Spin for a Typical Low-Wing General Aviation Aircraft," *J. Aircraft* 18, no. 6 (1981): 510–512.
J-1 Jahnke, C. C., and Culick, F. E. C., "Application of Bifurcation Theory to the High-Angle-of-Attack Dynamics of the F-14," *J. Aircraft* 31, no. 1 (1994): 26–34.
J-2 John, H., and Kraus, W., "High Angle of Attack Characteristics of Different Fighter Configurations," *High Angle of Attack Aerodynamics*, AGARD-CP-247, Neuilly-sur-Seine, Jan. 1979, pp. 2-1–2-15.
J-3 Johnston, D. E., and Hogge, J. R., "The Effect of Non-Symmetric Flight on Aircraft High Angle of Attack Handling Qualities and Departure Characteristics," AIAA Paper 74-792, New York, Aug. 1974.
J-4 Jones, B. M., chair, "The Lateral Control of Stalled Aeroplanes. General Report by the Stability and Control Panel," ARC R & M 1000, London, Nov. 1925.

J-5 Jordan, F. L., Jr. and Hahne, D. E., "Wind Tunnel Static and Free-Flight Investigation of High-Angle-of-Attack Stability and Control Characteristics of a Model of the EA-6B Airplane," NASA TP-3194, Washington, D.C., May 1992.

K-1 Kalviste, J., "Use of Rotary Balance and Forced Oscillation Test Data in Six Degrees of Freedom Simulation," AIAA-82-1364, San Diego, Calif., Aug. 1982.

K-2 Kao, H. C., "Side Forces on Unyawed Slender Inclined Aerodynamic Bodies," *J. Aircraft* 12, no. 3 (1975): 142–150.

K-3 Keller, H. B., "Numerical Solution of Bifurcation and Nonlinear Eigenvalue Problems," *Applications of Bifurcation Theory*, P. H. Rabinowitz, ed., Academic Press, New York, 1977, pp. 359–384.

K-4 Kelley, A., "The Stable, Centre-Stable, Centre, Centre-Unstable, Unstable Manifolds," *J. Differential Equations* 3, Oct. (1967): 546–570.

K-5 Kempel, R. W., McNeill, W. E., and Maine, T. A., "Oblique-Wing Research Airplane Motion Simulation with Decoupling Control Laws," AIAA Paper No. 88-0402, Reno, Nev., Jan. 1988.

K-6 Kleinman, D. L., Baron, S., and Levison, W. H., "An Optimal Control Model of Human Response," *Automatica* 6, no. 3 (1970): 357–383.

K-7 Kleinman, D. L., Baron, S., and Levison, W. H., "A Control Theoretic Approach to Manned-Vehicle Systems Analysis," *IEEE Trans. on Automatic Control* AC-16, no. 12 (1971): 824–832.

K-8 Kleinman, D. L., and Baron, S., "A Control Theoretic Model for Piloted Approach to Landing," *Automatica* 9, no. 3 (1973): 339–347.

K-9 Kraus, W., "X-31, Discussion of Steady State and Rotary Derivatives," *Maneouvring Aerodynamics*, AGARD-CP-497, Neuilly-sur-Seine, Nov. 1991, pp. 13-1–13-32.

K-10 Kubicek, M., "Algorithm 502, Dependence of Solution of Nonlinear Systems on a Parameter," *ACM-TOMS*, Vol. 2, 1976, pp. 98–107.

K-11 Kwatny, H. G., Bennett, W. H., and Berg, J., "Regulation of Relaxed Static Stability Aircraft," *IEEE Trans. Automatic Control* 36, no. 11 (1991): 1315–1323.

K-12 Kwatny, H. G., and Chang, B.-C., "Constructing Linear Families from Parameter-Dependent Nonlinear Dynamics," *IEEE Trans. Automatic Control* 43, no. 8 (1998): 1143–1147.

L-1 LaBurthe, C., "Une Nouvelle Analyse de la Vrille Basée sur L'experience Francaise sur les Avions de Combat," *Stall/Spin Problems of Military Aircraft*, AGARD-CP-199, Neuilly-sur-Seine, June 1976, pp. 15A-1–15A-9.

L-2 Lamers, J. P., "Design for Departure Prevention in the YF-16," AIAA-74-794, *Mechanics and Control of Flight Conf.*, Anaheim, Calif., Aug. 1974.

L-3 Lane, S., and Stengel, R. F., "Flight Control Design Using Non-linear Inverse Dynamics," *Automatica* 24, no. 4 (1988): 471–483.

L-4 Lindahl, J. H., "Application of the Automatic Rudder Interconnect (ARI) Function in the F-14 Aircraft," *U.S. ONR Symposium on Optimal Control*, Monterey, Calif., July 1975.

L-5 Littleboy, D. M., and Smith, P. R., "Using Bifurcation Methods to Aid Nonlinear Dynamic Inversion Control Law Design," *J. Guidance, Control, and Dynamics* 21, no. 4 (1998): 632–638.

M-1 Magni, J.-F., Bennani, S., and Terlouw, J., *Robust Flight Control, A Design Challenge*, Lecture Notes in Control and Information Sciences, Springer-Verlag, Berlin, 1997.

M-2 Martin, C., "The Spinning of Aircraft—A Discussion of Spin Prediction Techniques Including a Chronological Bibliography," Aeronautical Research Laboratory ARL-AERO-R-177, Melbourne, Australia, Aug. 1988.

M-3 Martin, C.A., and Thompson, D. H., "Scale Model Measurements of Fin Buffet due to Vortex Bursting on F/A-18," *Maneouvring Aerodynamics*, AGARD-CP-497, Neuilly-sur-Seine, Nov. 1991, pp. 12-1–12-10.

M-4 Mehra, R. K., "Catastrophe Theory, Nonlinear System Identification and Bifurcation Control," *Proc. 1977 Joint Automatic Control Conference*, San Francisco, June 1977, pp. 823–831.

M-5 Mehra, R. K., and Carroll, J. V., "Application of Bifurcation Analysis and Catastrophe Theory Methodology (BACTM) to Aircraft Stability Problems at High Angles-of-Attack," *Proc. 1978 Conference on Decision and Control*, San Diego, Calif., Jan. 1979, pp. 186–192.

M-6 Mehra, R. K., and Carroll, J. V., "Bifurcation Analysis of Aircraft High Angle-of-Attack Flight Dynamics," *Proc. Atmospheric Flight Mechanics Conference*, AIAA Paper No. 80-1599, Aug. 1980, pp. 358–371.

M-7 Mendenhall, M. R., Hegedus, M. C., Budd, G. D., and Frackowiak, A. J., "Aerodynamic Analysis of the Deep-Stall Flight Characteristics of the Pegasus XL Booster," AIAA-97-3583, Reston, Va., 1997.

M-8 Menon, P. K. A., Chatterji, G., and Cheng, V., "A Two-Time-Scale Autopilot for High-Performance Aircraft," AIAA-91-2674, New Orleans, Aug. 1991.

M-9 Moore, W. A., Skow, A. M., and Lorincz, D. J., "Enhanced Departure/ Spin Recovery of Fighter Aircraft through Control of the Forebody Vortex Orientation, AIAA-80-0173, Pasadena, Calif., Jan. 1980.

M-10 Moul, M. T., and Paulson, J. W., "Dynamic Lateral Behavior of High-Performance Aircraft," NACA RM L58E16, Washington, D.C., Aug. 1958.

M-11 Mulgund, S., and Stengel, R. F., "Aircraft Flight Control in Wind Shear Using Sequential Dynamic Inversion," *J. Guidance, Control, and Dynamics* 18, no. 5 (1995): 1084–1091.

M-12 Mulgund, S., and Stengel, R. F., "Optimal Nonlinear Estimation for Aircraft Flight Control in Wind Shear," *Automatica* 32, no. 1 (1996): 3–13.

N-1 Nguyen, L. T., Anglin, E. L., and Gilbert, W. P., "Recent Research Related to Prediction of Stall/Spin Characteristics of Fighter Aircraft," *Proc. of the AIAA Atmospheric Flight Mechanics Conference*, Arlington, Va., June 1976, pp. 79–91.

N-2 Nguyen, L. T., "Control System Techniques for Improved Departure/Spin Resistance for Fighter Aircraft," SAE Paper No. 791083, Warrendale, Pa., Dec. 1979, pp. 1–23.

N-3 Nguyen, L. T., Ogburn, M. E., Gilbert, W. P., Kibler, K. S., Brown, P. W., and Deal, P. L., "Simulator Study of Stall/Post-Stall Characteristics of a Fighter Airplane With Relaxed Longitudinal Static Stability," NASA TP-1538, Washington, D.C., Dec. 1979.

N-4 Nguyen, L. T., Gilbert, W. P., Gera, J., Iliff, K. W., and Enevoldson, E. K., "Application of High-α Control System Concepts to a Variable-Sweep Fighter Airplane," AIAA-80-1582, Danvers, Mass., Aug. 1980.

O-1 Orlik-Rückemann, K. J., "Aerodynamic Coupling Between Lateral and Longitudinal Degrees of Freedom," *AIAA Journal* 15, no. 12 (1977): 1792–1799.

O-2 Orlik-Rückemann, K. J., dir., *Dynamic Stability Parameters*, AGARD-LS-114, Neuilly-sur-Seine, May 1981.

O-3 Orlik-Rückemann, K. J., "Aerodynamic Aspects of Aircraft Dynamics at High Angle of Attack," AIAA-82-1363, New York, Aug. 1982.

O-4 Orlik-Rückemann, K. J., *Aircraft Dynamics at High Angles of Attack: Experiments and Modeling*, AGARD-R-776, Neuilly-sur-Seine, Mar. 1991.

P-1 Pamadi, B. N., and Taylor, L. W., Jr., "An Estimation of Aerodynamic Forces and Moments on an Airplane Model Under Steady State Spin Conditions," AIAA-82-1311, *AIAA Atmospheric Flight Mechanics Conf.*, San Diego, Calif., Aug. 1982.

P-2 Pamadi, B. N., *Performance, Stability, Dynamics, and Control of Airplanes*, American Institute of Aeronautics and Astronautics, Reston, Va., 1998.

P-3 Pedreiro, N., Takahara, Y., and Rock, S. M., "Aileron Effectiveness at High Angles of Attack: Interaction with Forebody Blowing," *J. Aircraft* 36, no. 6 (1999): 981–987.

P-4 Phillips, W. H., "Effects of Steady Rolling on Longitudinal and Directional Stability," NACA TN 1627, Washington, D.C., June 1948.

P-5 Phillips, W. H., "Simulation Study of the Oscillatory Longitudinal Motion of an Airplane at the Stall," NASA TP-1242, Washington, D.C., Aug. 1978.

P-6 Pinsker, W. J. G., "Aileron Control of Small Aspect Ratio Aircraft, in Particular, Delta Aircraft," ARC R&M 3188, London, 1953.

P-7 Pinsker, W. J. G., "Critical Flight Conditions and Loads Resulting from Inertial Cross-Coupling and Aerodynamic Stability Deficiencies," RAE Technical Note Aero 2502, London, 1957.

P-8 Polhamus, E. C., "Effect of Flow Incidence and Reynolds Number on Low-Speed Aerodynamic Characteristics of Several Noncircular Cylinders with Applications to Directional Stability and Spinning," NACA TN-4176, Washington, D.C., Jan. 1958.

P-9 Porter, R. F., and Loomis, J. P., "Examination of an Aerodynamic Coupling Phenomenon," *J. Aircraft* 2, no. 6 (1965): 553–556.

P-10 Postin, T., and Stewart, I., *Catastrophe Theory and Its Applications*, Pittman, London, 1978.

P-11 Psiaki, M. L., and Luh, Y. P., "Nonlinear System Stability Boundary Approximation by Polytopes in State Space," *International Journal of Control* 57, no. 1 (1993): 197–224.

R-1 Ray, E. J., McKinney, L. W., and Carmichael, J. G., "Maneuver and Buffet Characteristics of Fighter Aircraft," *Fluid Dynamics of Aircraft Stalling*, -AGARD-CP-102, Neuilly-sur-Seine, Nov. 1972, pp. 24-1–24-10.
R-2 Rhoads, D. W., and Schuler, J. M., "A Theoretical and Experimental Study of Airplane Dynamics in Large-Disturbance Maneuvers." *J. Aeronautical Sciences* 24, no. 7 (1957): 507–526, 532.
R-3 Ross, A. J., "Flying Aeroplanes in Buffet," *Aeronautical Journal* 81, no. 802 (1977): 427–436.
R-4 Ross, A. J., "Lateral Stability at High Angles of Attack, Particularly Wing Rock," *Stability and Control*, J. Buhrmann and R. Shevell, chairs, AGARD-CP-260, Neuilly-sur-Seine, May 1979, pp. 10-1–10-19.
S-1 Scher, S., "Post-Stall Gyrations and Their Study on a Digital Computer," AGARD Report 359, Neuilly-sur-Seine, Apr. 1961.
S-2 Schy, A. A., and Hannah, M. E., "Prediction of Jump Phenomena in Roll-Coupled Maneuvers of Airplanes," *J. Aircraft* 14, no. 4 (1977): 375–382.
S-3 Schmidt, L. V., *Introduction to Aircraft Flight Dynamics*, American Institute of Aeronautics and Astronautics, Reston, Va., 1998.
S-4 Schoenstadt, A. L., "Nonlinear Relay Model for Post-Stall Oscillations," *J. Aircraft* 12, no. 7 (1975): 572–577.
S-5 Seddon, J, ed., *Fluid Dynamics of Aircraft Stalling*, AGARD-CP-102, Neuilly-sur-Seine, Nov. 1972.
S-6 Seltzer, R. M., and Calvert, J. F., "Application of Current Departure Resistance Criteria to the Post-Stall Maneoeuvering Envelope," *Technologies for Highly Manoeuverable Aircraft*, AGARD-CP-548, Neuilly-sur-Seine, Mar. 1994, pp. 17-1–17-17.
S-7 Shrivastava, P. C., and Stengel, R. F., "Stability Boundaries for Aircraft with Unstable Lateral-Directional Dynamics and Control Saturation," *J. Guidance, Control, and Dynamics* 12, no. 1 (1989): 62–70.
S-8 Singh, S. N., and Rugh, W. J., "Decoupling in a Class of Nonlinear Systems by State Variable Feedback," *Trans. of the ASME, J. Dynamic Systems, Measurement, and Control* 94, no. 4 (1972): 323–329.
S-9 Singh, S. N., and Schy, A., "Output Feedback Nonlinear Decoupled Control Synthesis and Observer Design for Maneuvering Aircraft," *Int'l. J. Control* 31, no. 4 (1980): 781–806.
S-10 Skow, A. M., Titiriga, A., Jr., and Moore, W. A., "Forebody/Wing Vortex Interactions and Their Influence on Departure and Spin Resistance, *High Angle of Attack Aerodynamics*, AGARD-CP-247, Neuilly-sur-Seine, Jan. 1979, pp. 6-1–6-26.
S-11 Skow, A. M., and Erickson, G. E., "Modern Fighter Aircraft Design for High-Angle-of-Attack Maneuvering," *High Angle-of-Attack Aerodynamics*, AGARD-LS-121, Neuilly-sur-Seine, Dec. 1982, pp. 4-1–4-59.
S-12 Soderman, P. T., and Aiken, T. N., "Full-Scale Wind-Tunnel Tests of a Small Unpowered Jet Aircraft with a T-Tail," NASA TN D-6573, Washington, D.C., Nov. 1971.
S-13 Soulé, H., and Gough, M., "Some Aspects of the Stalling of Modern Low-Wing Monoplanes," NACA TN 645, Washington, D.C., Apr. 1938.

S-14 Staudacher, W., Poisson-Quinton, Laschka, B., and Ledy, J. P., "Aerodynamic Characteristics of a Fighter-Type Configuration During and Beyond Stall," *High Angle of Attack Aerodynamics*, AGARD-CP-247, Neuilly-sur-Seine, Jan. 1979, pp. 8-1–8-15.

S-15 Stengel, R. F., "Effect of Combined Roll Rate and Sideslip Angle on Aircraft Flight Stability," *J. Aircraft* 12, no. 8 (1975): 683–685.

S-16 Stengel, R. F., and Berry, P. W., "Stability and Control of Maneuvering High-Performance Aircraft," NASA CR-2788, Washington, D.C., Apr. 1977.

S-17 Stengel, R. F., and Berry, P. W., "Stability and Control of Maneuvering High-Performance Aircraft," *J. Aircraft* 14, no. 8 (1977): 787–794.

S-18 Stengel, R. F., Taylor, J. H., Broussard, J., and Berry, P., "High Angle-of-Attack Stability and Control," ONR-CR215-237-1, Arlington, Va., Apr. 1976.

S-19 Stengel, R. F., Broussard, J., Berry, P., and Taylor, J. H., "Modern Methods of Aircraft Stability and Control Analysis," ONR-CR215-237-2, Arlington, Va., May 1977.

S-20 Stengel, R. F., Broussard, J., and Berry, P., "Digital Controllers for VTOL Aircraft," *IEEE Trans. Aerospace and Electronic Systems* AES-14, no. 1 (1978): 54–63.

S-21 Stengel, R. F., Broussard, J., and Berry, P., Digital Flight Control Design for a Tandem-Rotor Helicopter, *Automatica* 14, no. 4 (1978): 301–311.

S-22 Stengel, R. F., and Nixon, W. B., "Stalling Characteristics of a General Aviation Aircraft," *J. Aircraft* 19, no. 6 (1982): 425–434.

S-23 Stevens, H., "The Behaviour of Certain Aeroplanes when the Controls are Abandoned in Stalled Flight," ARC R & M 1020, London, Nov. 1925.

S-24 Stough, H. P., III, and Patton, J. M., Jr., "The Effects of Configuration Changes on Spin and Recovery Characteristics of a Low-Wing General Aviation Research Airplane, AIAA-79-1786, New York, Aug. 1979.

S-25 Stough, H. P., DiCarlo, D. J., and Stewart, E. C., "Wing Modification for Increased Spin Resistance," SAE Paper No. 83070, Warrendale, Pa., Apr. 1983.

T-1 Takao, K., Baba, Y., Nishiyama, T., and Ogawa, Y., "A Method for Prediction of Spin Characteristics," *Trans. Japan Society for Aeronautical and Space Technology* 19, Dec. (1976): 202–209.

T-2 Tavoularis, S., Marineau-Mes, S., Woronko, A., and Lee, B. H. K., "Tail Buffet of F/A-18 at High Incidence with Sideslip and Roll," *J. Aircraft* 38, no. 1 (2001): Part 1: pp. 10–16, Part 2: pp. 17–21.

T-3 Taylor, C. R., ed., *Aircraft Stalling and Buffeting*, AGARD-LS-74, Neuilly-sur-Seine, Feb. 1975.

T-4 Taylor, J. H., "General Describing Function Method for Systems with Many Nonlinearities, with Application to Aircraft Performance," *Proc. 1980 Joint Automatic Control Conf.*, San Francisco, Aug. 1980, p. FP9-1/1-6.

T-5 Taylor, L. W., Jr., "Applications of Parameter Estimation in the Study of Spinning Airplanes," AIAA-82-1309, San Diego, Calif., Aug. 1982.

T-6 Thom, R., *Structural Stability and Morphogenesis*, Addison-Wesley, Reading, Mass., 1974.
T-7 Thomas, H., and Collingbourne, J., "Longitudinal Motions of Aircraft Involving High Angles of Attack," ARC R & M 3753, London, 1954.
T-8 Tischler, M. B., and Barlow, J. B., "Application of the Equilibrium Spin Technique to a Typical Low-Wing General Aviation Airplane," AIAA-79-1625, Boulder, Colo., Aug. 1977.
T-9 Tischler, M. B., and Barlow, J. B., "Determination of the Spin and Recovery Characteristics of a Typical Low-Wing General Aviation Design," AIAA-80-0169, New York, Jan. 1980.
T-10 Tischler, M. B., and Barlow, J. B., "A Dynamic Analysis of the Motion of a Low-Wing General Aviation Aircraft About Its Calculated Equilibrium Flat Spin Mode," AIAA-80-1565, Danvers, Mass., Aug. 1980.
T-11 Titiriga, A., Jr., Ackerman, J. S., and Skow, A. M., "Design Technology for Departure Resistance of Fighter Aircraft," *Stall/Spin Problems of Military Aircraft*, AGARD-CP-199, Neuilly-sur-Seine, June 1976, pp. 5-1–5-14.
T-12 Tobak, M., and Schiff, L. B., "Generalized Formulation of Nonlinear Pitch-Yaw-Roll Coupling: Part I—Nonaxisymmetric Bodies and Part II—Nonlinear Coning-Rate Dependence," *AIAA Journal* 13, no. 3 (1975): 323–332.
W-1 Walchli, L. A., "X-29: Longitudinal Instability at High Angle of Attack," *Stability in Aerospace Systems*, AGARD-R-789, Neuilly-sur-Seine, Feb. 1993, p. 18-1–18-13.
W-2 Wang, Q., and Stengel, R. F., "Robust Nonlinear Flight Control of a High-Performance Aircraft," AIAA-2001-4101, Montreal, Aug. 2001.
W-3 Wardlaw, A. B., Jr., and Morrison, A. M., "Induced Side Forces at High Angles of Attack," *J. Spacecraft and Rockets* 13, no. 10 (1976): 589–593.
W-4 Weissman, R., "Preliminary Criteria for Predicting Departure Characteristics/Spin Susceptibility of Fighter-Type Aircraft," *J. Aircraft* 10, no. 4 (1973): 214–219.
W-5 Westbrook, C., ed., *Stall/Post-Stall/Spin Symposium*, Wright-Patterson AFB, Ohio, Dec. 1971.
W-6 White, R. J., Uddenberg, R. C., Murray, D., and Graham, F. D., "The Dynamic Stability and Control Equations of a Pivoted-Wing Supersonic Pilotless Aircraft, with Downwash, Wake and Interference Effects Included," Boeing Aircraft Co. Document No. D-8510, Seattle, Wash., Jan. 1948.
W-7 White, D. A., and Sofge, D., *Handbook of Intelligent Control*, Van Nostrand Reinhold, New York, 1992.
W-8 Willen, T. B., and Johnson, K., "Anticipated Spin Susceptibility Characteristics of the A-10 Aircraft," AIAA Paper No. 75-33, New York, Jan. 1975.
W-9 Wimpress, J. K., "Predicting the Low Speed Stall Characteristics of the Boeing 747," *Fluid Dynamics of Aircraft Stalling*, L. G. Napolitano, ed., AGARD-CP-102, Neuilly-sur-Seine, Nov. 1972, pp. 21-1–21-9.
W-10 Winkelman, A. E., Barlow, J. B., Saini, J. K., Anderson, J. D., Jr., and Jones, E., "The Effects of Leading-Edge Modifications on the Post-Stall Characteristics of Wings," AIAA-80-0199, Pasadena, Calif., Jan. 1980.

W-11 Wykes, J. H., "An Analytical Study of the Dynamics of Spinning Aircraft," WADC TR 58-381, Feb. 1960.

Y-1 Young, J. W., Schy, A. A., and Johnson, K. G., "Prediction of Jump Phenomena in Aircraft Maneuvers of Aircraft, Including Nonlinear Aerodynamic Effects," *J. Guidance and Control* 1, no. 1 (1978): 26–31.

Y-2 Young, J. W., Schy, A. A., and Johnson, K. G., "Pseudosteady State Analysis of Nonlinear Aircraft Maneuvers," AIAA Paper No. 80-1600, New York, 1980.

Chapter 8

Epilogue

In the hundred years since the Wright brothers' first flight, the technology of aircraft design and operation has progressed on every front, but the physical principles of flight dynamics have remained the same. What has changed is our knowledge of those principles and our ability to use them in the earliest stages of aircraft development. Today, it is possible to predict performance, stability, and control characteristics long before first flight and to analyze flight test data with remarkable precision.

Throughout the history of aviation, methods of analysis and design have necessarily conformed to the tools at hand, beginning with paper, pencil, T-square, and slide rule. Today's ubiquitous computer can implement all of the earlier manual tasks electronically, more accurately and more quickly, changing the skill sets required of engineers and vastly increasing their power to generate results.

These technological gains come at a price. Computational tools can be employed without requiring the user to develop intuition, the ability to approximate and to visualize, or an awareness of the validity of results. Working without modern tools, engineers had to acquire deep knowledge and to understand minute details of the physics of flight. Getting useful results required a high degree of sophistication and elegance in problem solving because numerical computation was limited. Today, it seems simpler to generate new results than to find and refer to comprehensive reports of completed research. Consequently, corporate memory is lost, the audit trail from theory to accomplishment is obscured, and a great tradition is diminished.

Our goal should not be to lament the passing of an era but to surpass prior capabilities, using modern tools to overcome these deficiencies. To do this, we must develop new paradigms for system analysis and design, new ways of using computation to gain knowledge, and new avenues for educating future generations of engineers in these practices. We should use unifying theory and computational power to develop intuition, approximation, and visualization skills, and the ability to discriminate truth from fiction. We need not become giants to stand on the shoulders of giants, but we must have good footing and balance, so that we do not fall back to a lower level.

Referring to the related field of control system development, we note that the methods of "modern control" were first introduced about 50 years ago! There need be no contradiction between sound classical and modern approaches: the latter should extend the former. In the scheme of things, modern concepts become classical over time, or they vanish.

A major objective of this book is to provide the building blocks for this new synthesis. Modular, linear, time-invariant dynamic models provide a strong foundation for building intuition, but the new synthesis must provide a smooth transition to nonlinearity, stochastic methods, and intelligent/adaptive control. The figures of merit used in analysis and design should aid this transition.

Part of the process is to shift attention from the key parameters of a system to the key features of its response. For example, the damping ratio that produces a satisfactory overshoot in a simple second-order system may be used as a surrogate for the actual overshoot. However, if the transfer function has a zero, or the system is high order or nonlinear, the damping ratio alone is a poor indicator of system characteristics while the overshoot remains a good one. Then, it is a small step to take from a classical step-response overshoot criterion to the probability of satisfactory overshoot, given likely variation in uncertain system parameters and any description of the system. Another example: the likelihood of performing a complex maneuver (e.g., a rolling reversal) satisfactorily and with low pilot workload may be a better metric than a classical measure of stability would be. And a third: emulating the declarative, procedural, and reflexive functions of a skilled pilot, perhaps through the use of expert systems or neural networks, may be a useful adjunct for control system design and evaluation.

Unlimited computation alone is not the answer; however, intelligent use of large-scale computation is a central feature of the new synthesis. Computers support theory and empiricism for high-order systems, as recognized in countless applications, such as computational fluid dynamics and deep-space trajectory generation. Nevertheless, using the computer places great importance on having good data and good analytical tools to begin with; it is far too easy to generate realistic pictures that do not represent reality. To avoid error, the functions of computer code must be transparent, accessible, and expressed in a manner that is not subject to misinterpretation.

Among computational approaches to analysis and design, stochastic methods are the wave of the future. They integrate tried-and-true engineering methods with large-scale computation and with traditional engineering goals, such as maximizing the robustness of system performance over a range of expectable variations. Deterministic criteria are replaced by numerical estimates of the probability of achieving satisfactory values of those criteria. Although numerical evaluation is computationally intensive, the continued growth in computer speed assures that what takes hours to compute today will take minutes tomorrow and seconds the day after. Stochastic methods help identify what must be known about a system in order to make it robust. If there is not sufficient information to do a stochastic simulation, there is not sufficient information to judge the robustness of any deterministic solution.

Presenting the new synthesis in a flight dynamics course fits well into modern engineering curricula because upper-year students have already been exposed to calculus and linear algebra in their mathematics courses. The use of state-space methods and the opportunity to compute state histories may provide a student's first real applications of materials taught earlier. Given the typical engineering student's mathematics background, it may well be easier for students to learn time-domain solution techniques for multivariate, nonlinear equations of motion before they are introduced to linear, frequency-domain scalar equations. Experience suggests that Bode and root locus plots remain mysterious to many students long after their instructor has assumed these concepts are solid, yet the students understand the time evolution of motion variables almost immediately.

The course instructor has a major responsibility to demonstrate the value of earlier work to students. That is one reason this book contains bibliographic references. Still, these books, reports, and papers are just the tip of the iceberg; there are many more to be found in the world's technical libraries, and an increasing number are becoming available over the internet. The instructor should refer to the literature and give case study examples to amplify the importance of critical aspects of performance, stability, and control. Revealing the successes and failures, as well as the people responsible for them, gives texture to the subject, encouraging the student to become a participant in a larger, identifiable endeavor. The pioneers deserve to be remembered, and we benefit from their wisdom.

The next generation of aircraft must be better than the previous generations in every way: They must be developed at reasonable cost. They must be safer, more reliable, more efficient, more cost effective. These goals connote improved failure tolerance, improved performance, and excellent flying qualities in both normal and unexpected conditions. A select cadre of engineers must fully understand flight dynamics and control to assure that these goals are achieved.

Appendix A

Constants, Units, and Conversion Factors

Constants

Air density at sea level, ρ_o	1.225 kg/m^3
Air density scale height in lower atmosphere, $1/\beta$	9,042 m
Air pressure at sea level, p_o	101,325 Pa
Earth flattening, f	1/298.3
Earth gravitational constant, μ_E	3.986×10^{14} m/s^2
Earth radius (equator), R_E	6,378.4 km
Earth radius (pole), R_P	6,356.9 km
Earth rotational rate, Ω	7.2924×10^{-5} rad/s
Gas constant (air), R	287.05 N-m/kg-K
Gravitational acceleration at sea level, g_o, 45° latitude	9.80665 m/s^2
Ratio of specific heats (air), γ	1.4

Quantity	*Multiply*	*by*	*to Obtain*
Angle	deg	0.01745329252	rad
Angle	rad	57.29577951	deg
Area	ft^2	0.092893	m^2
Area	m^2	10.765	ft^2
Density	kg/m^3	515.379	slugs/ft^3
Density	slugs/ft^3	0.00194032	kg/m^3
Energy	Btu	1,055.89	J (N-m)
Energy	Cal	4.1868	J (N-m)
Energy	kWh	3.6×10^6	J (N-m)
Force	lb	4.448	N
Force	N	0.2248	lb
Frequency	Hz	0.15915	rad/s
Frequency	rad/s	6.2832	Hz
Inertia	kg-m^2	0.73763	slug-ft^2

Quantity	*Multiply*	*by*	*to Obtain*
Inertia	slug-ft^2	1.3556	kg-m^2
Length	ft	0.3048	m
Length	m	3.281	ft
Length	nm	6,076.1	ft
Length	nm	1,852	m
Mass	kg	0.068522	slugs
Mass	slugs	14.5939	kg
Mass to weight (SL)	kg	2.2046	lb
Mass to weight (SL)	slugs	32.174	lb
Momentum	kg-m/s	0.2248	slug-ft/s
Momentum	slug-ft/s	4.448	kg-m/s
Power	W	745.7	hp
Power	W	1.356	ft-lb/s
Power	ft-lb/s	0.7376	W
Pressure	lb/ft^2	47.88	P (N/m^2)
Pressure	P (N/m^2)	0.02089	lb/ft^2
Speed	ft/s	0.3048	m/s
Speed	ft/s	0.5925	kt
Speed	ft/s	0.9113	km/hr
Speed	km/hr	1.097	ft/s
Speed	km/hr	0.54	kt
Speed	km/hr	0.2777	m/s
Speed	kt	0.5144	m/s
Speed	kt	1.853	km/hr
Speed	kt	1.6878	ft/s
Speed	m/s	3.6	km/hr
Speed	m/s	1.944	kt
Speed	m/s	3.281	ft/s
Temperature	°C + 273.15	1	K
Temperature	°F − 32	5/9	°C
Temperature	°F + 459.7	1	°R
Temperature	K	9/5	°R
Thrust	lb	0.2248	N
Thrust	N	4.448	lb
Torque	ft-lb	1.356	m-N
Torque	m-N	0.7376	ft-lb
Volume	ft^3	35.32	m^3
Volume	m^3	0.0283	ft^3
Weight (SL) to Mass	lb	0.45359	kg
Weight (SL) to Mass	lb	0.03108	slugs
Wing loading	kg/m^2	0.2048	lb/ft^2
Wing loading	lb/ft^2	4.883	kg/m^2

Reference for Appendix A

Fogiel, M., ed., *Handbook of Mathematical, Scientific, and Engineering Formulas, Tables, Functions, Graphs, And Transforms*, Research and Educational Association, Piscataway, N.J., 1988.

Appendix B

Mathematical Model and Six-Degree-of-Freedom Simulation of a Business Jet Aircraft

A number of the examples in this book are based on the dynamic characteristics of a generic business jet aircraft. The mathematical models given below are motivated by full-scale wind-tunnel tests of an early configuration [B-1]; however, they should not be construed as an accurate representation of any specific aircraft, as considerable liberties have been taken in the derivation of the mathematical models. Specifically, Mach effects are estimated using the methods of Section 2.5, rotary derivatives are calculated with the equations of Section 3.4, power effects are limited to a simple model of thrust (with no modeling of associated aerodynamic effects), and moments and products of inertia are estimated using simplified mass distributions. Furthermore, the aerodynamic data of [B-1] are simplified considerably for inclusion in the model.

The aircraft dynamic models are presented as MATLAB functions that are called by the program *FLIGHT*. FLIGHT illustrates the essential features of flight analysis and simulation using a modular framework. Aircraft characteristics are described in the function, AeroModel and may be augmented by tables stored in the main program.

FLIGHT is a tutorial program, heavily commented to make interpretation easy. It provides a full six-degree-of-freedom simulation of an aircraft, as well as trimming calculations and the generation of a linearized model at any flight condition chosen by the user. Changes to aircraft control histories, initial conditions, flag settings, and other program control actions are made by changing the numbers contained in the code; there is no separate user interface. The code has been designed for simplicity and clarity, not for speed of execution, leaving a challenge to the reader to find ways to make the program run faster. Numerous additions could also be made to the code, including implementation of feedback control logic, simulation of random turbulence or microburst wind shear, and interfaces for real-time execution. No explicit or implicit warranties are made regarding the accuracy or correctness of the computer code, which can be found on-line at http://www.pupress.princeton.edu/stengel.

B.1 Main Program for Analysis and Simulation (FLIGHT)

FLIGHT is the executive file that calls program functions. Initial conditions are defined here, the three primary features (trim, linearization, and simulation) are enabled, and output is generated. Initial perturbations to trim state and control allow transient effects to be simulated. As shown, trimming for steady, level flight is accomplished by first defining a cost function J that contains elements of the state rate, then minimizing the cost using the downhill simplex (Nelder-Mead) algorithm contained in `fmins`. The longitudinal trimming parameters are stabilator angle, throttle setting, and pitch angle. The linear model is generated by `numjac`, a numerical evaluation of the Jacobian matrices associated with the equations of motion. The linear model is saved to disk files in the variables Fmodel and Gmodel. MATLAB's `ode23` integrates the equations of motion to produce the state history. The state history is displayed in time plots, with angles converted from the radians used in calculation to degrees. The reader can readily change the units of plotted quantities or add additional plots through minor modifications to the code. Any result (e.g., numerical values of the state history) can be displayed in the MATLAB command window simply by removing the semicolon at the end of the line.

B.2 Low-Angle-of-Attack, Mach-Dependent Model (LoAeroModel)

This version of AeroModel uses aerodynamic and dimensional data contained in Ref. B-1 with estimates of inertial properties of the generic business jet. Details of the configuration, such as sweep and aspect ratio of the wing and tail, are used in the estimates of Mach effects. Wind tunnel results were reviewed for all test conditions and control settings, then reduced to simple linear and quadratic coefficients that describe significant aerodynamics for low angles of attack and sideslip (within about ± 8 degrees of zero). Increments for landing gear, symmetric spoilers, and flaps are added, as appropriate. Estimates of Mach effects are based on the Prandtl factor (Section 2.5) or the modified Helmbold equation (eq. 2.5-9). Thrust available is assumed to be proportional to the air density ratio raised to a power that fits both static thrust at sea level and the thrust required for flight at maximum Mach number (0.81) at an altitude of 13,720 m (45,000 ft).

B.3 High-Angle-of-Attack, Low-Subsonic Model (HiAeroModel)

This version of AeroModel uses aerodynamic and dimensional data contained in Ref. [B-1] with estimates of inertial properties of the generic business jet. The angle-of-attack range extends from -10 to 90 deg, with tunnel data used where it is available and estimates where it is not. Linear interpolation of tabulated data is used to model nonlinearities in aerodynamic coefficients. Control surface effects are linear in their respective deflections, and static lateral-directional effects are linear in sideslip angle. The derivatives for both vary with angle of attack to the limits of data presented in the reference and are constant at higher angle. Angular rate derivatives are constant throughout the angle of attack range. No Mach, landing gear, or flap effects are modeled. The same thrust model used for the Mach-dependent AeroModel (Section B.2) is used here.

B.4 Supporting Functions

B.4.1 Equations of Motion (EoM)

The equations of motion for function *EoM* are written using the flat-earth assumption, as described in Section 3.2. Linear and translational rates are expressed in body axes, linear position is expressed in earth-relative axes, and the Euler angles describe the orientation of the body frame with respect to the earth frame. The lift and drag coefficients produced by AeroModel are transformed to body-axis coefficients; the remaining coefficients and thrust produced by the function are expressed in body axes. Effects of the wind and control settings are calculated by WindField and ControlSystem prior to calling AeroModel; hence, they are accounted for in the aerodynamic calculations. An ad hoc limit on the cosine of the pitch angle is imposed to prevent singular calculations in near-vertical flight. This expediency introduces a small error when the pitch angle is near $\pm 90°$.

B.4.2 Cost Function for Aerodynamic Trim (TrimCost)

Trim control settings are calculated by minimizing the quadratic function of longitudinal accelerations (i.e., rates of change of axial velocity, normal velocity, and pitch rate) contained in TrimCost. As shown, equal weight is given to each component in the cost function J, because the defining matrix $\mathbf{R}$ is an identity matrix. The parameters that are adjusted to minimize J are the stabilator angle, the throttle setting, and the pitch angle (which equals the angle of attack in steady, level flight). TrimCost calls EoM to generate the needed accelerations. x_1 and x_3 are varied to maintain constant airspeed with varying angle of attack.

B.4.3 Direction Cosine Matrix (DCM)

The direction-cosine matrix given by eq. 2.1-11 is implemented in the function DCM.

B.4.4 Linear System Matrices (LinModel)

The function LinModel calls function EoM to generate $\mathbf{f}(\mathbf{x};\mathbf{u})$ and appends seven dummy elements to account for the controls. Thus, `xdotj` represents $\mathbf{f}(\mathbf{x}')$, where $\mathbf{x}'^T = [\mathbf{x}^T\ \mathbf{u}^T]$, and `numjac` calculates the Jacobian matrix $\partial\mathbf{f}/\partial\mathbf{x}'$, evaluated at the nominal values of state and control. The stability matrix $\mathbf{F} = \partial\mathbf{f}/\partial\mathbf{x}$ is contained in the upper left (12×12) partition, and the control-effect matrix $\mathbf{G} = \partial\mathbf{f}/\partial\mathbf{u}$ is contained in the upper right (12×7) partition of $\partial\mathbf{f}/\partial\mathbf{x}'$.

B.4.5 Wind Field (WindField)

WindField produces a three-component wind vector as a function of altitude, with linear interpolation between tabulated points. The first point is tabulated at negative altitude to assure that no computational problems occur at zero altitude. The wind vector is rotated to body axes for application in FLIGHT and EoM.

B.4.6 Atmospheric State (*Atmos*)

Air density, air pressure, air temperature, and sound speed are generated as functions of altitude by *Atmos*. The first point is tabulated at negative altitude to assure that no computational problems occur at zero altitude. As neither air pressure nor air temperature are used in FLIGHT, the function could be abbreviated for computational efficiency.

Reference for Appendix B

B-1 Soderman, P. T., and Aiken, T. N., "Full-Scale Wind-Tunnel Tests of a Small Unpowered Jet Aircraft with a T-Tail," NASA TN D-6573, Washington, D.C., Nov. 1971.

Appendix C

Linear System Survey

The SURVEY program accepts the linear, time-invariant stability and control-effect matrices, Fmodel and Gmodel, from FLIGHT as inputs and performs a variety of analyses on them. It first provides the option of reducing these matrices into smaller models using the function LonLatDir. The function produces a single model at each call for either longitudinal or lateral-directional motions, in body or stability axes, and with four or six elements in the state vector. The remaining SURVEY functions operate on either the original twelve-dimensional model or on the reduced model produced by LonLatDir. The program calculates transient response, static response, controllability and observability matrices, natural frequencies, damping ratios, and real roots. It forms single-input/single-output transfer functions selected by the user and makes their Bode, Nyquist, and Nichols plots. SURVEY is tutorial, and the user must control program flow by making changes in the code. The interested reader could consider adding root locus plots and Monte Carlo evaluation to the main program. No explicit or implicit warranties are made regarding the accuracy or correctness of the computer code, which can be found on-line at http://www.pupress.princeton.edu/stengel.

C.1 Main Program for Analysis and Simulation (SURVEY)

As shown, SURVEY is set up to make calculations based on the 12×12 Fmodel and 12×7 Gmodel produced by FLIGHT, with the generic business jet model trimmed for flight at $V = 102$ m/s and $h = 3{,}050$ m. The flags are set to generate a fourth-order, body-axis, lateral-directional model from the data set; to calculate eigenvalues, eigenvectors, and the transfer function from aileron to roll rate; and to show the Bode plot, Nyquist plot, and Nichols chart for the transfer function.

C.2 Supporting Functions

C.2.1 Reduced-Order Models (LonLatDir)

LonLatDir accepts stability and control-effect matrices for combined longitudinal and lateral-directional dynamics as input and either passes the matrices unchanged, reorders elements of the matrices, or generates a selected reduced-order model from them. It presents matrices for body, stability, or hybrid axes, for longitudinal or lateral-directional partitions, and for fourth- or sixth-order computations. As shown, there are twelve original state elements, four longitudinal controls, and three lateral-directional controls.

C.2.2 Transient Response (Trans)

Trans calculates and plots transient response to an arbitrary array of initial conditions and step control inputs, calling the MATLAB function `lsim` to perform the simulation.

C.2.3 Static Response (Static)

Static computes static control and command response for fourth-order systems with no disturbances and two control inputs (elevator and throttle in the longitudinal case, aileron and rudder for the lateral-directional model). Static control equilibrium does not exist for the higher-order models presented here because the stability matrix is singular.

C.2.4 Controllability and Observability (ConObs)

The matrix rank tests described in Chapter 4 are performed by calling the MATLAB functions `ctrb` and `obsv` in ConObs.

C.2.5 Natural Frequency (NatFreq)

NatFreq presents eigenvalues sorted in descending order and natural frequencies, damping ratios, and real roots using the MATLAB functions `esort` and `damp`.

C.2.6 Stability and Modes of Motion (StabMode)

StabMode presents eigenvalues and eigenvectors, as well as amplitudes of the eigenvector components for both original and velocity-weighted sets.

Appendix D

Paper Airplane Program

This appendix presents the MATLAB code for Example 1.3-1. The purpose is tutorial, and no explicit or implicit warranties are made regarding the accuracy or correctness of the computer code.

```
%     Example 1.3-1 Paper Airplane Flight Path Calculation

      global CL CD S m g rho
      S          =    0.017;          % Reference Area, m^2
      AR         =    0.86;           % Wing Aspect Ratio
      e          =    0.9;            % Oswald Efficiency Factor;
      m          =    0.003;          % Mass, kg
      g          =    9.8;            % Gravitational acceleration, m/s^2
      rho        =    1.225;          % Air density at Sea Level, kg/m^3
      CLa        =    3.141592 * AR/(1 + sqrt(1 + (3.141592 * AR / 2)^2));
                                      % Lift-Coefficient Slope, per rad
      CDo        =    0.02;           % Zero-Lift Drag Coefficient
      epsilon    =    1 / (3.141592 * e * AR);
                                      % Induced Drag Factor
      CL         =    sqrt(CDo / epsilon);
                                      % CL for Maximum Lift/Drag Ratio
      CD         =    CDo + epsilon * CL^2;
                                      % Corresponding CD
      LDmax      =    CL / CD;        % Maximum Lift/Drag Ratio
      Gam        =    -atan(1 / LDmax);% Corresponding Flight Path Angle, rad
      V          =    sqrt(2 * m * g /(rho * S * (CL * cos(Gam)
                      - CD * sin(Gam))));
                                      % Corresponding Velocity, m/s
      Alpha      =    CL / CLa;       % Corresponding Angle of Attack, rad

%     a) Equilibrium Glide at Maximum Lift/Drag Ratio
      H          =    2;              % Initial Height, m
      R          =    0;              % Initial Range, m
      to         =    0;              % Initial Time, sec
      tf         =    6;              % Final Time, sec
      tspan      =    [to tf];
      xo         =    [V;Gam;H;R];
      [ta,xa]    =    ode23('EqMotion',tspan,xo);

%     b) Oscillating Glide due to Zero Initial Flight Path Angle
      xo         =    [V;0;H;R];
      [tb,xb]    =    ode23('EqMotion',tspan,xo);

%     c) Effect of Increased Initial Velocity
      xo         =    [1.5*V;0;H;R];
      [tc,xc]    =    ode23('EqMotion',tspan,xo);

%     d) Effect of Further Increase in Initial Velocity
      xo         =    [3*V;0;H;R];
      [td,xd]    =    ode23('EqMotion',tspan,xo);

      figure
      plot(xa(:,4),xa(:,3),xb(:,4),xb(:,3),xc(:,4),xc(:,3),xd(:,4),xd(:,3))
      xlabel('Range, m'), ylabel('Height, m'), grid
```

```
	figure
	subplot(2,2,1)
	plot(ta,xa(:,1),tb,xb(:,1),tc,xc(:,1),td,xd(:,1))
	xlabel('Time, s'), ylabel('Velocity, m/s'), grid
	subplot(2,2,2)
	plot(ta,xa(:,2),tb,xb(:,2),tc,xc(:,2),td,xd(:,2))
	xlabel('Time, s'), ylabel('Flight Path Angle, rad'), grid
	subplot(2,2,3)
	plot(ta,xa(:,3),tb,xb(:,3),tc,xc(:,3),td,xd(:,3))
	xlabel('Time, s'), ylabel('Altitude, m'), grid
	subplot(2,2,4)
	plot(ta,xa(:,4),tb,xb(:,4),tc,xc(:,4),td,xd(:,4))
	xlabel('Time, s'), ylabel('Range, m'), grid

	function xdot = EqMotion(t,x)
%	Fourth-Order Equations of Aircraft Motion

	global CL CD S m g rho

	V	=	x(1);
	Gam	=	x(2);
	q	=	0.5 * rho * V^2;% Dynamic Pressure, N/m^2
	xdot	=	[(-CD * q * S - m * g * sin(Gam)) / m
			 (CL * q * S - m * g * cos(Gam)) / (m * V)
			 V * sin(Gam)
			 V * cos(Gam)];
```

Appendix E

Bibliography of NASA Reports Related to Aircraft Configuration Aerodynamics

Representative aerodynamic data is of great value for understanding flight dynamics. The models presented in Appendix B allow calculations to be made for a typical business jet aircraft. While it is not practical to provide many data packages in this book, the references cited in this appendix allow models to be developed for a wide range of aircraft types. These NASA Technical Papers and Memoranda, which date from 1960, are readily available in technical libraries throughout the world, and they can be ordered via the internet at the NASA Center for Aerospace Information Technical Report Server, http://ntrs.nasa.gov/. The list was compiled by searching the NASA literature logged at this web site using a variety of criteria that concentrated on the net aerodynamic characteristics of full configurations. The list is not guaranteed to be a complete rendering of the open NASA literature on the topic, although it is extensive.

At this time, the NASA reports cannot be downloaded directly; however, the reports of its predecessor, the National Advisory Committee for Aeronautics (NACA), have been scanned and are available on-line at the *NACA Digital Library*, accessed by browsing at http://naca.larc.nasa.gov/.

In the following, reports are grouped within categories. The largest section is devoted to configuration aerodynamics, while the other sections contain ancillary data. Many reports identify aircraft types in their titles; the names in square brackets at the end of some citations identify aircraft that appear to be similar to those described in the report.

Aeroservoelasticity

Breitbach, E. J., *Flutter analysis of an airplane with multiple structural nonlinearities in the control system*, NASA-TP-1620, Mar 01, 1980.

Brenner, M. J., *Actuator and aerodynamic modeling for high-angle-of-attack aeroservoelasticity*, NASA-TM-4493, Jun 01, 1993.

Brenner, M. J., *Aeroservoelastic modeling and validation of a thrust-vectoring F/A-18 aircraft*, NASA-TP-3647, Sep 01, 1996.
Housner, J. M., and Stein, M., *Flutter analysis of swept-wing subsonic aircraft with parameter studies of composite wings*, NASA-TN-D-7539, Sep 01, 1974.
Mori, A. S., Nardi, L. U., and Wykes, J. H., *XB-70 structural mode control system design and performance analyses*, NASA-CR-1557, Jul 01, 1970.
Smith, J. P., *X-38 Vehicle 131 flutter assessment*, NASA-TP-3683, May 01, 1997.

Component Aerodynamics

Coe, P. L., Jr., Chambers, J. R., and Letko, W., *Asymmetric lateral-directional characteristics of pointed bodies of revolution at high angles of attack*, NASA-TN-D-7095, Nov 01, 1972.
Gainer, T. G., *Subsonic wind-tunnel investigation of the aerodynamic effects of pivoting a low-aspect-ratio wing to large yaw angles with respect to the fuselage to increase lift-drag ratio*, NASA-TN-D-225, Mar 01, 1960.
Gatlin, G. M., and McGrath, B. E., *Low-speed longitudinal aerodynamic characteristics through poststall for 21 novel planform shapes*, NASA-TP-3503, Aug 01, 1995.
Henderson, W. P., *Studies of various factors affecting drag due to lift at subsonic speeds*, NASA-TN-D-3584. Oct 01, 1966.
Hopkins, E. J., and Yee, S. C., *A comparison of the experimental aerodynamic characteristics of an oblique wing with those of a swept wing*, NASA-TM-X-3547, Jun 01, 1977.
Jorgensen, L. H., *Prediction of static aerodynamic characteristics for slender bodies alone and with lifting surfaces to very high angles of attack*, NASA-TR-R-474, Sep 01, 1977.
Keenan, J. A., and Kuhlman, J. M., *The effects of winglets on low aspect ratio wings at supersonic Mach numbers*, NASA-CR-4407, Nov 01, 1991.
Luckring, J. M., *Theoretical and experimental analysis of longitudinal and lateral aerodynamic characteristics of skewed wings at subsonic speeds to high angles of attack*, NASA-TN-D-8512, Dec 01, 1977.
Polhamus, E. C., and Sleeman, W. C., Jr., *The rolling moment due to sideslip of swept wings at subsonic and transonic speeds*, NASA-TN-D-209, Feb 01, 1960.
Ray, E. J., and Taylor, R. T., *Buffet and static aerodynamic characteristics of a systematic series of wings determined from a subsonic wind-tunnel study*, NASA-TN-D-5805, Jun 01, 1970.

Configuration Aerodynamics

Andrews, W. H., *Summary of preliminary data derived from the XB-70 airplanes*, NASA-TM-X-1240, Jun 01, 1966.
Anglin, E. L., and Chambers, J. R., *Analysis of lateral directional stability characteristics of a twin jet fighter airplane at high angles of attack*, NASA-TN-D-5361, Aug 01, 1969. [F-4]

Anglin, E. L., *Static force tests of a model of a twin-jet fighter airplane for angles of attack from minus 10 deg to 110 deg and sideslip angles from minus 40 deg to 40 deg*, NASA-TN-D-6425, Aug 01, 1971. [F-4]

Anon., *Selected advanced aerodynamics and active controls technology concepts development on a derivative B-747*, NASA-CR-3164, Jul 01, 1980.

Aoyagi, K., Cook, A. M., and Greif, R. K., *Large-scale wind-tunnel investigation of the low-speed aerodynamic characteristics of a supersonic transport model having variable-sweep wings*, NASA-TN-D-2824, May 01, 1965.

Aoyagi, K., and Tolhurst, W. H., Jr., *Large-scale wind tunnel tests of a subsonic transport with AFT engine nacelles and high tail*, NASA-TN-D-3797, Jan 01, 1967.

Arnaiz, H. H., *Flight-measured lift and drag characteristics of a large, flexible, high supersonic cruise airplane*, NASA-TM-X-3532, May 01, 1977. [B-70]

Arnaiz, H. H., Peterson, J. B., Jr., and Daugherty, J. C., *Wind-tunnel/flight correlation study of aerodynamic characteristics of a large flexible supersonic cruise airplane (XB-70-1). 3: A comparison between characteristics predicted from wind-tunnel measurements and those measured in flight*, NASA-TP-1516, Mar 01, 1980.

Averett, B. T., and Wright, B. R., *Transonic aerodynamic damping and oscillatory stability in yaw and pitch for a model of a variable-sweep supersonic transport airplane*, NASA-TM-X-1207, Mar 01, 1966.

Banks, D. W., Fisher, D. F., Hall, R. M., Erickson, G. E., Murri, D. G., Grafton, S. B., and Sewall, W. G., *The F/A-18 High-Angle-of-Attack Ground-to-Flight Correlation; Lessons Learned*, NASA-TM-4783, Jan 01, 1997.

Banks, D. W., Gatlin, G. M., and Paulson, J. W., Jr., *Low-speed longitudinal and lateral-directional aerodynamic characteristics of the X-31 configuration*, NASA-TM-4351, Oct 01, 1992.

Bartlett, D. W., and Re, R. J., *Wind-tunnel investigation of basic aerodynamic characteristics of a supercritical-wing research airplane configuration*, NASA-TM-X-2470, Feb 01, 1972.

Bihrle, W., Jr., Barnhart, B., and Pantason, P., *Static aerodynamic characteristics of a typical single-engine low-wing general aviation design for an angle-of-attack range of −8 deg to 90 deg*, NASA-CR-2971, Jul 01, 1978.

Blackwell, J. A., Jr., Brooks, E. N., Jr., and Decker, J. P., *Static aerodynamic characteristics of a model of a typical subsonic jet-transport airplane at Mach numbers from 0.40 to 1.20*, NASA-TM-X-1345, Feb 01, 1967. [B-707]

Boyden, R. P., *Aerodynamic roll damping of a T-tail transport configuration*, NASA-TM-X-3115, Dec 01, 1974.

Brower, M. L., and Wright, B. R., *Aerodynamic damping and oscillatory stability in pitch for a model of a typical subsonic jet-transport airplane*, NASA-TN-D-3159, Jan 01, 1966.

Byrdsong, T. A., and Hallissy, J. B., *Longitudinal and lateral static stability and control characteristics of a 1/6-scale model of a remotely piloted research vehicle with a supercritical wing*, NASA-TP-1360, May 01, 1979.

Capone, F. J., Bare, E. A., and Arbiter, D., *Aerodynamic characteristics of a supersonic fighter aircraft model at Mach 0.40 to 2.47*, NASA-TP-2580, Apr 01, 1986.

Carr, P. C., and Gilbert, W. P., *Effects of fuselage forebody geometry on low-speed lateral-directional characteristics of twin-tail fighter model at high angles of attack*, NASA-TP-1592, Dec 01, 1979. [F-15]

Curry, R. E., and Sim, A. G., *In-flight total forces, moments and static aeroelastic characteristics of an oblique-wing research airplane*, NASA-TP-2224, Oct 01, 1984.

Damato, R., Freeman, D. C., Jr., and Grafton, S. B., *Static and dynamic stability derivatives of a model of a jet transport equipped with external-flow jet-augmented flaps*, NASA-TN-D-5408, Sep 01, 1969.

Daugherty, J. C., *Wind-tunnel/flight correlation study of aerodynamic characteristics of a large flexible supersonic cruise airplane (XB-70-1). 1: Wind-tunnel tests of a 0.03-scale model at Mach numbers from 0.6 to 2.53*, NASA-TP-1514, Nov 01, 1979.

Dillon, J. L., Creel, T. R., Jr., *Aerodynamic characteristics at Mach number 0.2 of a wing-body concept for a hypersonic research airplane*, NASA-TP-1189, Jun 01, 1978.

Dillon, J. L., Pittman, J. L., *Aerodynamic characteristics at Mach numbers from 0.33 to 1.20 of a wing-body design concept for a hypersonic research airplane*, NASA-TP-1044, Dec 01, 1977.

Dillon, J. L., Pittman, J. L., *Aerodynamic characteristics at Mach 6 of a wing-body concept for a hypersonic research airplane*, NASA-TP-1249, Aug 01, 1978.

Dollyhigh, S. M., *Stability and control characteristics at Mach numbers 1.60 to 2.86 of a variable-sweep fighter configuration with supercritical airfoil sections*, NASA-TM-X-2284, Jun 01, 1971.

Dollyhigh, S. M., *Static longitudinal aerodynamic characteristics of close-coupled wing-canard configurations at Mach numbers from 1.60 to 2.86*, NASA-TN-D-6597, Dec 01, 1971.

Dollyhigh, S. M., *Stability and control characteristics of a fighter configuration with a cranked-leading-edge wing planform at Mach numbers 1.60 to 2.86*, NASA-TM-X-2695, Jan 01, 1973.

Dollyhigh, S. M., *Wing-camber effects on longitudinal aerodynamic characteristics of a variable-sweep fighter configuration at Mach numbers from 1.60 to 2.86*, NASA-TM-X-2826, Nov 01, 1973.

Dollyhigh, S. M., *Subsonic and supersonic longitudinal stability and control characteristics of an aft-tail fighter configuration with cambered and uncambered wings and cambered fuselage*, NASA-TN-D-8472, Sep 01, 1977.

Dollyhigh, S. M., Monta, W. J., and Sangiorgio, G., *Longitudinal aerodynamic characteristics at Mach 0.60 to 2.86 of a fighter configuration with strut braced wing*, NASA-TP-1102, Dec 01, 1977.

Dunham, D. M., Gentry, C. L., Jr., Manuel, G. S., Applin, Z. T., and Quinto, P. F., *Low-speed aerodynamic characteristics of a twin-engine general aviation configuration with aft-fuselage-mounted pusher propellers*, NASA-TP-2763, Oct 01, 1987.

Ferris, A. T., and Kelly, H. N. *Free-flight and wind-tunnel studies of deployment of a dynamically and elastically scaled inflatable parawing model*, NASA-TN-D-4724, Sep 01, 1968.

Ferris, J. C. , *Static aerodynamic characteristics of a model with a 17 percent thick supercritical wing*, NASA-TM-X-2551, May 01, 1972.

Ferris, J. C., *Static longitudinal aerodynamic characteristics of a model with a modified 17 percent thick supercritical wing*, NASA-TM-X-3211, May 01, 1975.

Feryn, M. O., and Fournier, R. H., *Aerodynamic characteristics at Mach numbers from 2.50 to 4.63 of a fighter configuration with various vertical- and horizontal-tail arrangements*, NASA-TM-X-1378, May 01, 1967.

Feryn, M. O., and Fournier, R. H., *Aerodynamic characteristics at Mach numbers from 2.50 to 4.63 of a fighter configuration with various ventral-fin arrangements*, NASA-TM-X-1486, Dec 01, 1967.

Fournier, P. G., *Low speed aerodynamic characteristics of a transport model having 42.33 deg swept low wing with supercritical airfoil, double-slotted flaps, and T-tail or low tail*, NASA-TM-X-3276, Nov 01, 1975.

Fox, M. C., and Forrest, D. K., *Supersonic aerodynamic characteristics of an advanced F-16 derivative aircraft configuration*, NASA-TP-3355, Jul 01, 1993.

Gatlin, G. M., *Ground effects on the low-speed aerodynamics of a powered, generic hypersonic configuration*, NASA-TP-3092, Jan 01, 1990.

Gilyard, G. B., Matheny, N. W., Strutz, L. W., and Wolowicz, C. H., *Preliminary flight evaluation of the stability and control derivatives and dynamic characteristics of the unaugmented XB-70-1 airplane including comparisons with predictions*, NASA-TN-D-4578, May 01, 1968.

Grafton, S. B., and Anglin, E. L., *Dynamic stability derivatives at angles of attack from minus 5 deg to 90 deg for a variable-sweep fighter configuration with twin vertical tails*, NASA-TN-D-6909, Oct 01, 1972. [F-14]

Grafton, S. B., Chambers, J. R., and Coe, P. L., Jr., *Wind-tunnel free-flight investigation of a model of a spin-resistant fighter configuration*, NASA-TN-D-7716, Jun 01, 1974.

Grafton, S. B., and Libbey, C. E., *Dynamic stability derivatives of a twin-jet fighter model for angles of attack from −10 deg to 110 deg*, NASA-TN-D-6091, Jan 01, 1971. [F-4]

Hahne, D. E., *Evaluation of the Low-Speed Stability and Control Characteristics of a Mach 5.5 Waverider Concept*, NASA-TM-4756, May 01, 1997.

Hahne, D. E., *Low-speed static and dynamic force tests of a generic supersonic cruise fighter configuration*, NASA-TM-4138, Oct 01, 1989.

Hahne, D. E., Wendel, T. R., and Boland, J. R., *Wind-tunnel free-flight investigation of a supersonic persistence fighter*, NASA-TP-3258, Feb 01, 1993.

Hassell, J. L., Jr., Newsom, W. A., Jr., and Yip, L. P., *Full-scale wind tunnel-investigation of the Advanced Technology Light Twin-Engine airplane (ATLIT)*, NASA-TP-1591, May 01, 1980.

Heath, A. R., Jr., *Longitudinal aerodynamic characteristics of a high-subsonic-speed transport airplane model with a cambered 40 deg sweptback wing of aspect ratio 8 at Mach numbers to 0.96*, NASA-TN-D-218, Feb 01, 1960.

Henderson, W. P., and Abeyounis, W. K., *Aerodynamic characteristics of a high-wing transport configuration with an over-the-wing nacelle-pylon arrangement*, NASA-TP-2497, Dec 01, 1985.

Hoad, D. R., and Gentry, G. L., Jr., *Longitudinal aerodynamic characteristics of a low-wing lift-fan transport including hover characteristics in and out of ground effect*, NASA-TM-X-3420, Feb 01, 1977. [Do-31]

Iliff, K. W., Maine, R. E., and Shafer, M. F., *Subsonic stability and control derivatives for an unpowered, remotely piloted 3/8-scale F-15 airplane model obtained from flight test*, NASA-TN-D-8136, Jan 01, 1976.

Iliff, Kenneth W., Wang, K.-S., *X-29A Lateral-Directional Stability and Control Derivatives Extracted From High-Angle-of-Attack Flight Data*, NASA-TP-3664, Dec 01, 1996.

Jacobs, P. F., and Flechner, S. G., *The effect of winglets on the static aerodynamic stability characteristics of a representative second generation jet transport model*, NASA-TN-D-8267, Jul 01, 1976.

Jacobs, P. F., Flechner, S. G., and Montoya, L. C., *Effect of winglets on a first-generation jet transport wing. 1: Longitudinal aerodynamic characteristics of a semispan model at subsonic speeds*, NASA-TN-D-8473, Jun 01, 1977.

Jacobs, P. F., and Gloss, B. B., *Longitudinal aerodynamic characteristics of a subsonic, energy-efficient transport configuration in the National Transonic Facility*, NASA-TP-2922, Aug 01, 1989.

Jordan, F. L., Jr., and Hahne, D. E., *Wind-tunnel static and free-flight investigation of high-angle-of-attack stability and control characteristics of a model of the EA-6B airplane*, NASA-TP-3194, May 01, 1992.

Kilgore, R. A., and Adcock, J. B., *Supersonic aerodynamic damping and oscillatory stability in pitch and yaw for a model of a variable-sweep fighter airplane with twin vertical tails*, NASA-TM-X-2555, May 01, 1972.

Kilgore, R. A., *Some transonic and supersonic dynamic stability characteristics of a variable-sweep-wing tactical fighter model*, NASA-TM-X-2163, Feb 01, 1971. [F-111]

Kilgore, R. A., *Aerodynamic damping and oscillatory stability in pitch and yaw of a variable-sweep supersonic transport configuration at Mach numbers from 0.40 to 1.80*, NASA-TM-X-2164, Feb 01, 1971.

Lamar, J. E., and Frink, N. T., *Experimental and analytical study of the longitudinal aerodynamic characteristics of analytically and empirically designed strake-wing configurations at subcritical speeds*, NASA-TP-1803, Jun 01, 1981.

Lamb, M., *Additional study of a fixed-wing twin-vertical-tail fighter model including the effects of wing and vertical tail modifications*, NASA-TM-X-2619, Aug 01, 1972. [Mig-25]

Lamb, M., Sawyer, W. C., and Thomas, J. L.. *Experimental and theoretical supersonic lateral-directional stability characteristics of a simplified wing-body configuration with a series of vertical-tail arrangements*, NASA-TP-1878, Aug 01, 1981.

Langhans, R. A., and Flechner, S. G., *Wind-tunnel investigation at Mach numbers from 0.25 to 1.01 of a transport configuration designed to cruise at near-sonic speeds*, NASA-TM-X-2622, Aug 01, 1972.

Leavitt, L. D., *Longitudinal aerodynamic characteristics of a vectored-engine-over-wing configuration at subsonic speeds*, NASA-TP-1533, Oct 01, 1979.

Luoma, A. A., *Transonic wind-tunnel investigation of the static longitudinal stability and performance characteristics of a supersonic fighter-bomber airplane*, NASA-TM-X-513, Jul 01, 1961.

Maine, R. E., *Aerodynamic derivatives for an oblique wing aircraft estimated from flight data by using a maximum likelihood technique*, NASA-TP-1336, Oct 01, 1978.

Mann, M. J., Huffman, J. K., and Fox, C. H., Jr., *Subsonic longitudinal and lateral-directional characteristics of a forward-swept-wing fighter configuration at angles of attack up to 47 deg*, NASA-TP-2727, Sep 01, 1987.

Mann, M. J., and Langhans, R. A., *Transonic aerodynamic characteristics of a supercritical-wing transport model with trailing-edge controls*, NASA-TM-X-3431, Oct 01, 1977.

Mclemore, H. C., Parlett, L. P., *Low speed wind tunnel tests of 1/10-scale model of a blended-arrow supersonic cruise aircraft*, NASA-TN-D-8410, Jun 01, 1977.

Mclemore, H. C., Parlett, L. P., and Sewall, W. G., *Low-speed wind tunnel tests of 1/9-scale model of a variable-sweep supersonic cruise aircraft*, NASA-TN-D-8380, Jun 01, 1977.

Monaghan, R. C., *Flight-measured buffet characteristics of a supercritical wing and a conventional wing on a variable-sweep airplane*, NASA-TP-1244, May 01, 1978.

Morris, O. A., *Subsonic and supersonic aerodynamic characteristics of a supersonic cruise fighter model with a twisted and cambered wing with 74 deg sweep*, NASA-TM-X-3530, Aug 01, 1977.

Murri, D. G., Nguyen, L. T., and Grafton, S. B., *Wind-tunnel free-flight investigation of a model of a forward-swept-wing fighter configuration*, NASA-TP-2230, Feb 01, 1984.

Newsom, W. A., Jr., Satran, D. R., and Johnson, J. L., Jr., *Effects of wing-leading-edge modifications on a full-scale, low-wing general aviation airplane: Wind-tunnel investigation of high-angle-of-attack aerodynamic characteristics*, NASA-TP-2011, Jun 01, 1982. [Yankee]

Nguyen, L. T., Ogburn, M. E., Gilbert, W. P., Kibler, K. S., Brown, P. W., and Deal, P. L., *Simulator study of stall/post-stall characteristics of a fighter airplane with relaxed longitudinal static stability*, NASA-TP-1538, Dec 01, 1979. [F-16]

Parlett, L. P., *Free-flight wind-tunnel investigation of a four-engine sweptwing upper-surface blown transport configuration*, NASA-TN-D-8479, Aug 01, 1977.

Paulson, J. W., Jr., and Thomas, J. L., *Summary of low-speed longitudinal aerodynamics of two powered close-coupled wing-canard fighter configurations*, NASA-TP-1535, Dec 01, 1979.

Peterson, J. B., Jr., Mann, M. J., Sorrells III, R. B., Sawyer, W. C., and Fuller, D. E., *Wind-tunnel/flight correlation study of aerodynamic characteristics of a large flexible supersonic cruise airplane (XB-70-1) 2: Extrapolation of wind-tunnel data to full-scale conditions*, NASA-TP-1515, Feb 01, 1980.

Petroff, D. N., Scher, S. H., and Cohen, L. E., *Low speed aerodynamic characteristics of an 0.075-scale F-15 airplane model at high angles of attack and sideslip*, NASA-TM-X-62360, Jul 01, 1974.

Petroff, D. N., Scher, S. H., and Sutton, C. E., *Low-speed aerodynamic characteristics of a 0.08-scale YF-17 airplane model at high angles of attack and sideslip*, NASA-TM-78438, Apr 01, 1978.

Pittman, J. L., and Riebe, G. D., *Experimental and theoretical aerodynamic characteristics of two hypersonic cruise aircraft concepts at Mach numbers of 2.96, 3.96, and 4.63*, NASA-TP-1767, Dec 01, 1980.

Powers, B. G., *Phugoid characteristics of a YF-12 airplane with variable-geometry inlets obtained in flight tests at a Mach number of 2.9*, NASA-TP-1107, Dec 01, 1977.

Pyle, J. S., *Lift and drag characteristics of the HL-10 lifting body during subsonic gliding flight*, NASA-TN-D-6263, Mar 01, 1971.

Ray, E. J., and Taylor, R. T., *Effect of configuration variables on the subsonic longitudinal stability characteristics of a high-tail transport configuration*, NASA-TM-X-1165, Oct 01, 1965.

Re, R. J., *Stability and control characteristics: Including aileron hinge moments of a model of a supercritical-wing research airplane*, NASA-TM-X-2929, Apr 01, 1974.

Saltzman, E. J., Hicks, J. W., and Luke, S., *In-flight lift-drag characteristics for a forward-swept wing aircraft and comparisons with contemporary aircraft*, NASA-TP-3414, Dec 01, 1994.

Satran, D. R., Gilbert, W. P., and Anglin, E. L., *Low-speed stability and control wind-tunnel investigations of effects of spanwise blowing on fighter flight characteristics at high angles of attack*, NASA-TP-2431, May 01, 1985.

Satran, D. R., *Wind-tunnel investigation of the flight characteristics of a canard general-aviation airplane configuration*, NASA-TP-2623, Oct 01, 1986. [Vari-Eze]

Shrout, B. L., *Aerodynamic characteristics at Mach numbers from 0.6 to 2.16 of a supersonic cruise fighter configuration with a design Mach number of 1.8*, NASA-TM-X-3559, Sep 01, 1977.

Sim, A. G., and Curry, R. E., *Flight-determined stability and control derivatives for the F-111 TACT research aircraft*, NASA-TP-1350, Oct 01, 1978.

Sim, A. G., *Flight characteristics of a modified Schweizer SGS1-36 sailplane at low and very high angles of attack*, NASA-TP-3022, Jul 01, 1990.

Soderman, P. T., and Aiken, T. N., *Full-scale wind-tunnel tests of a small unpowered jet aircraft with a T-tail*, NASA-TN-D-6573, Nov 01, 1971. [Learjet 23]

Taylor, A. B., *Development of selected advanced aerodynamics and active control concepts for commercial transport aircraft*, NASA-CR-3781, Feb 01, 1984.

White, E. R., *Wind-tunnel investigation of effects of wing-leading-edge modifications on the high angle-of-attack characteristics of a T-tail low-wing general-aviation aircraft*, NASA-CR-3636, Nov 01, 1982.

Yip, L. P., *Wind-tunnel free-flight investigation of a 0.15-scale model of the F-106B airplane with vortex flaps*, NASA-TP-2700, May 01, 1987.

Control Effects

Bartlett, D. W., *Effects of differential and symmetrical aileron deflection on the aerodynamic characteristics of an NASA supercritical-wing research airplane model*, NASA-TM-X-3231, Oct 01, 1975.

Cannaday, R.L., *Effects of control inputs on the estimation of stability and control parameters of a light airplane*, NASA TP-1043, Dec 01, 1977.

Capone, F. J., and Reubush, D. E., *Effect of thrust vectoring and wing maneuver devices on transonic aeropropulsive characteristics of a supersonic fighter*, NASA-TP-2119, Feb 01, 1983.

Erickson, G. E., and Campbell, J. F., *Improvement of maneuver aerodynamics by spanwise blowing*, NASA-TP-1065, Dec 01, 1977.

Powell, R. W., *Aileron roll hysteresis effects on entry of space shuttle orbiter*, NASA-TN-D-8425, Jul 01, 1977.

Reubush, David E., and Berrier, B. L., *Effects of the installation and operation of jet-exhaust yaw vanes on the longitudinal and lateral-directional characteristics of the F-14 airplane*, NASA-TP-2769, Dec 01, 1987.

Yip, L. P., and Paulson, J. W., Jr., *Effects of deflected thrust on the longitudinal aerodynamic characteristics of a close-coupled wing-canard configuration*, NASA-TP-1090, Dec 01, 1977.

Flying Qualities

Berry, D. T., and Powers, B. G., *Handling qualities of the XB-70 airplane in the landing approach*, NASA-TN-D-5676, Feb 01, 1970.

Cox, T. H., Sachs, G., Knoll, A., and Stich, R., *A flying qualities study of longitudinal long-term dynamics of hypersonic planes*, NASA-TM-1999-104308, Apr 01, 1995.

Matheny, N. W., and Gatlin, D. H., *Flight evaluation of the transonic stability and control characteristics of an airplane incorporating a supercritical wing*, NASA TP-1167, Feb. 1978.

Powers, B. G., *Statistical survey of XB-70 airplane responses and control usage with an illustration of the application to handling qualities criteria*, NASA-TN-D-6872, Jul 01, 1972.

Powers, B. G., *Analytical study of ride smoothing benefits of control system configurations optimized for pilot handling qualities*, NASA TP-1148, Feb 01, 1978.

Smetana, F. O., Summery, D. C., and Johnson, W. D., *Riding and handling qualities of light aircraft: A review and analysis*, NASA-CR-1975, Mar 01, 1972.

Numerical Methods

Adcock, J. B., Wahls, R. A., Witkowski, D. P., and Wright, F. L., *A Longitudinal Aerodynamic Data Repeatability Study for a Commercial Transport Model Test in the National Transonic Facility*, NASA-TP-3522, Aug 01, 1995.

McDonnell, J. D., Berg, R. A., Heimbaugh, R. M., and Felton, C. A., *Modeling and parameter uncertainties for aircraft flight control system design*, NASA-CR-2887, Sep 01, 1977.

Steinmetz, G. G., and Wilson, J. W., *Analysis of numerical integration techniques for real-time digital flight simulation*, NASA-TN-D-4900, Nov 01, 1968.

Wolowicz, C. H., Brown, J. S., Jr., and Gilbert, W. P., *Similitude requirements and scaling relationships as applied to model testing*, NASA-TP-1435, Aug 01, 1979.

Reentry Vehicles

Ashby, G. C., Jr., *Longitudinal, directional, and lateral aerodynamic characteristics of an inflatable-type reentry vehicle at a Mach number of 6.0*, NASA-TM-X-972, Jun 01, 1964.

Campbell, J. F., and McShera, J. T., Jr., *Stability and control characteristics of a manned lifting entry vehicle at Mach numbers from 2.29 to 4.63*, NASA-TM-X-1019, Jan 01, 1964.

Campbell, J. F., and Silvers, H. N., *Stability characteristics of a manned lifting entry vehicle with various fins at Mach numbers from 1.50 to 2.86*, NASA-TM-X-1161, Oct 01, 1965.

Chyu, W. J., Cavin, R. K., and Erickson, L. L., *Static and dynamic stability analysis of the space shuttle vehicle orbiter*, NASA TP-1179, Mar 01, 1978.

Freeman, D. C., Jr., Boyden, R. P., and Davenport, E. E., *Supersonic dynamic stability characteristics of a space shuttle orbiter*, NASA-TN-D-8043, Jan 01, 1976.

Freeman, D. C., Jr., *Dynamics stability derivatives of space shuttle orbiter obtained from wind-tunnel and approach and landing flight tests*, NASA-TP-1634, Apr 01, 1980.

Harris, J. E., *Longitudinal aerodynamic characteristics of a manned lifting entry vehicle at a Mach number of 19.7*, NASA-TM-X-1080, Apr 01, 1965.

Harris, C. D., *Transonic aerodynamic characteristics of a manned lifting entry vehicle with modified tip fins*, NASA TM X-1918, Jan 01, 1970.

Henderson, W. P., *Static stability characteristics of a manned lifting entry vehicle at high subsonic speeds*, NASA-TM-X-1349, Feb 01, 1967.

Jackson, B. E., *Preliminary subsonic aerodynamic model for simulation studies of the HL-20 lifting body*, NASA-TM-4302, Aug 01, 1992.

Kilgore, R. A., and Wright, B. R., *Aerodynamic damping and oscillatory stability in pitch and yaw of Gemini configurations at Mach numbers from 0.50 to 4.63*, NASA-TN-D-3334, Mar 01, 1966.

Ladson, C. L., *Aerodynamic characteristics of the HL-10 manned lifting entry vehicle at a Mach number of 10.5*, NASA-TM-X-1504, Jan 01, 1968.

Olstad, W. B., *Static longitudinal aerodynamic characteristics at transonic speeds of a flat delta wing hypersonic glider for angles of attack up to 100 deg*, NASA-TM-X-557, Sep 01, 1961.

Shanks, R. E., *Investigation of the low-subsonic flight characteristics of a model of an all-wing hypersonic boost-glide configuration having very high sweep*, NASA-TN-D-369, Jun 01, 1960.

Spencer, B., Jr., *Effects of elevon planform on low-speed aerodynamic characteristics of the HL-10 manned lifting entry vehicle*, NASA-TM-X-1409, Jul 01, 1967.

Ware, G. M., *Full-scale wind-tunnel investigation of the aerodynamic characteristics of the HL-10 manned lifting entry vehicle*, NASA-TM-X-1160, Oct 01, 1965.

Spin and Stall Aerodynamics

Anglin, E. L., Bowman, J. S., Jr., and Chambers, J. R., *Analysis of the flat-spin characteristics of a twin-jet swept-wing fighter airplane*, NASA-TN-D-5409, Sep 01, 1969. [F-4]

Anglin, E. L., Bowman, J. S., Jr., and Chambers, J. R., *Effects of a pointed nose on spin characteristics of a fighter airplane model including correlation with theoretical calculations*, NASA-TN-D-5921, Sep 01, 1970. [F-111]

Barnhart, B., *Analysis of rotary balance data for the F-15 airplane including the effect of conformal fuel tanks*, NASA-CR-3479, Apr 01, 1982.

Barnhart, B., *Rotary balance data for an F-15 model with conformal fuel tanks for an angle-of-attack range of 8 deg to 90 deg*, NASA-CR-3516, May 01, 1982.

Barnhart, B., *Rotary balance data for a typical single-engine general aviation design for an angle-of-attack range of 8 deg to 90 deg. 2: Influence of horizontal tail location for Model D*, NASA-CR-3247, Nov 01, 1982.

Bihrle, W., Jr., Hultberg, R. S., and Mulcay, W., *Rotary balance data for a typical single-engine low-wing general aviation design for an angle-of-attack range of 30 deg to 90 deg*, NASA-CR-2972, Jul 01, 1978.

Bihrle, W., Jr., and Hultberg, R. S., *Rotary balance data for a typical single-engine general aviation design for an angle-of-attack range of 8 deg to 90 deg. 1: High-wing model B*, NASA-CR-3097, Sep 01, 1979.

Bihrle, W., Jr., and Hultberg, R. S., *Rotary balance data for a typical single-engine general aviation design for an angle-of-attack range of 8 deg to 90 deg. 2: Low-wing model B*, NASA-CR-3098, Sep 01, 1979.

Bihrle, W., Jr., and Mulcay, W., *Rotary balance data for a typical single-engine general aviation design for an angle-of-attack range of 8 degrees to 35 degrees, 3. Effect of wing leading-edge modifications, model A*, NASA-CR-3102, Nov 01, 1979.

Burk, S. M., Jr., Bowman, J. S., Jr., and White, W. L., *Spin-tunnel investigation of the spinning characteristics of typical single-engine general aviation airplane designs. 1. Low-wing model A: Effects of tail configurations*, NASA-TP-1009, Sep 01, 1977. [Yankee]

Grafton, S. B., *A study to determine effects of applying thrust on recovery from incipient and developed spins for four airplane configurations*, NASA-TN-D-3416, Jun 01, 1966.

Grafton, S. B., and Grantham, W. D., *Effects of aircraft relative density on spin and recovery characteristics of some current configurations*, NASA-TN-D-2243, Mar 01, 1965.

Grantham, W. D., *Effects of static pitching moment and of rolling moment due to sideslip in enabling or preventing spin entry*, NASA-TM-X-1088, Apr 01, 1965.

Hultberg, R. S., and Chu, J., *Rotary balance data for a typical single-engine general aviation design for an angle-of-attack range of 8 deg to 90 deg. 2: High-wing model C*, NASA-CR-3201, Oct 01, 1980.

Hultberg, R. S., and Mulcay, W., *Rotary balance data for a typical single-engine general aviation design for an angle-of-attack range of 8 deg to 90 deg. 1: Low-wing model A*, NASA-CR-3100, Feb 01, 1980.

Mulcay, W. J., and Chu, J., *Rotary balance data for a single-engine agricultural airplane configuration for an angle-of-attack range of 8 deg to 90 deg*, NASA-CR-3311, Dec 01, 1980. [Thrush]

Mulcay, W., and Rose, R., *Rotary balance data for a typical single-engine general aviation design for an angle-of-attack range of 8 deg to 90 deg. 2: High-wing model A*, NASA-CR-3101, Sep 01, 1979.

Mulcay, W. J., and Rose, R. A., *Rotary balance data for a typical single-engine general aviation design for an angle of attack range of 8 deg to 90 deg. 1: Low wing model C*, NASA-CR-3200, Oct 01, 1980.

Mulcay, W. J., and Rose, R., *Rotary balance data for a single engine general aviation design having a high aspect-ratio canard for an angle-of-attack range of 30 deg to 90 deg*, NASA-CR-3170, Dec 01, 1980. [Vari-Eze]

Pantason, P., and Dickens, W., *Rotary balance data for a single-engine trainer design for an angle-of-attack range of 8 deg to 90 deg*, NASA-CR-3099, Aug 01, 1979. [T-34C]

Ralston, J., *Rotary balance data for a typical single-engine general aviation design for an angle-of-attack range of 8 deg to 90 deg. 1: Influence of airplane components for model D*, NASA-CR-3246, Mar 01, 1983.

Ralston, J. N., and Barnhart, B. P., *Rotary balance data for a typical single-engine general aviation design for an angle-of-attack range of 20 to 90 deg. 3: Influence of control deflection on predicted model D spin modes*, NASA-CR-3248, Jun 01, 1984.

Turbulence Effects

Incrocci, T. P., and Scoggins, J. R., *An investigation of the relationships between mountain-wave conditions and clear air turbulence encountered by the XB-70 airplane in the stratosphere*, NASA-CR-1878, Jul 01, 1971.

Larson, R. R., Love, B. J., and Wilson, R. J., *Evaluation of effects of high-altitude turbulence encounters on the XB-70 airplane*, NASA-TN-D-6457, Jul 01, 1971.

Index

www.ingramcontent.com/pod-product-compliance
Ingram Content Group UK Ltd.
Pitfield, Milton Keynes, MK11 3LW, UK
UKHW032246140125
453474UK00005B/16/J